中国地质调查局
CHINA GEOLOGICAL SURVEY

国家出版基金项目
NATIONAL PUBLICATION FOUNDATION

中国区域地质志

上海志

上海市地质调查研究院　编著

地质出版社
·北　京·

内 容 简 介

本书是《中国区域地质志》省（市、区）区域地质志之一。本书在1988年版《上海市区域地质志》的基础上，充分收集了不同单位近30年来在区内开展的基础地质、城市地质调查和研究成果，并针对性地补充了适量新的测试资料。按照“新、全、准”的要求，对资料进行了核实。在此基础上运用当代最新地质科学理论进行了基础地质资料系统集成与综合研究，发现了一些新的地质事实，提出了一些新的认识。本志除绪言、结语外，共计8篇38章。系统总结了区内地层及沉积作用、岩浆岩及岩浆作用、变质岩及变质作用、岩石圈结构构造、地质构造、第四纪地质与地貌等内容，并融合了城市地质调查和研究成果。同时编制了上海市第四纪地质及地貌图、上海市基岩地质图、上海市地质构造图、上海市布格重力异常图、上海市航磁（ΔT）异常图、上海市第Ⅳ承压含水层综合水文地质图、上海市工程地质结构综合分区图、上海市累计地面沉降等值线图（1921—2016年）、上海市地质遗迹图等9幅1：25万系列地质图件。

本书既适用于专门的地质工作者，也适用于了解区内地质条件的一般读者，可供地质调查、工程勘察和高校、政府管理等有关部门人员参考使用。

图书在版编目（CIP）数据

中国区域地质志. 上海志 / 上海市地质调查研究院编著. — 北京 ：地质出版社，2023. 12
ISBN 978 - 7 - 116 - 13901 - 5

Ⅰ. ①中… Ⅱ. ①上… Ⅲ. ①区域地质-概况-上海 Ⅳ. ①P562

中国国家版本馆 CIP 数据核字（2023）第228822号

Zhongguo Quyu Dizhizhi · Shanghaizhi

责任编辑：李 莉 田 野 张世刚
责任校对：陈 曦
出版发行：地质出版社
社址邮编：北京市海淀区学院路31号，100083
电　　话：（010）66554528（邮购部）；（010）66554631（编辑室）
网　　址：https://www. gph. clmpg. com
传　　真：（010）66554686
印　　刷：中煤地西安地图制印有限公司
开　　本：889mm×1194mm 1/16
印　　张：52. 25 **附图**：8幅
字　　数：1580千字
版　　次：2023年12月北京第1版
印　　次：2023年12月西安第1次印刷
定　　价：980. 00元
审 图 号：沪S（2021）045号
书　　号：ISBN 978 - 7 - 116 - 13901 - 5

《中国区域地质志》编纂指导委员会
（第二届）

The second advisory committee of *Regional Geology of China*

《中国区域地质志》编纂指导委员会
（第三届）

The third advisory committee of
Regional Geology of China

《中国区域地质志》编纂指导委员会
（第四届）

主　任：李金发

副主任：徐学义　叶建良　吴登定　侯增谦　肖桂义

成　员（按姓氏笔画排序）：

毛晓长	任收麦	许振强	李　军	李建星
杨志明	汪大明	邱海峻	张海啟	张开军
赵洪伟	郭兴华	姚华军	高延光	夏　鹏
贾伟光	顾晓华	熊自力		

The fourth advisory committee of
Regional Geology of China

Director：Li Jinfa

Associate directors：Xu Xueyi　Ye Jianliang　Wu Dengding

Hou Zengqian　Xiao Guiyi

Committee members（in Chinese-character stroke order of surname）：

Mao Xiaochang	Ren Shoumai	Xu Zhenqiang	Li Jun
Li Jianxing	Yang Zhiming	Wang Daming	Qiu Haijun
Zhang Haiqi	Zhang Kaijun	Zhao Hongwei	Guo Xinghua
Yao Huajun	Gao Yanguang	Xia Peng	Jia Weiguang
Gu Xiaohua	Xiong Zili		

《中国区域地质志》编辑委员会
（第一届）

《中国区域地质志》编辑委员会
(第三届)

The third editorial board of *Regional Geology of China*

项目承担单位

中国地质科学院地质研究所

Project organization institution

Institute of Geology, Chinese Academy of Geological Sciences

项目办公室

主　任：庞健峰

（耿元生，2010 年—2013 年 8 月；

耿树方，2013 年 8 月—2015 年 12 月）

成　员：张　聪　闫　斌　张　恒　石成龙　王　军

郑　宁　王振洋　郑洪伟　金　铭　张继彪

Project office

Director: Pang Jianfeng

(Geng Yuansheng from 2010 to August, 2013;

Geng Shufang from August, 2013 to December, 2015)

Staff:

Zhang Cong	Yan Bin	Zhang Heng	Shi Chenglong
Wang Jun	Zheng Ning	Wang Zhenyang	Zheng Hongwei
Jin Ming	Zhang Jibiao		

责任编辑组

郑长胜	祁向雷	孙亚芸	刘亚军	蔡卫东
吴宁魁	柳　青	李　莉	王作勇	袁亚琴
宋永存	肖莹莹	田　野	贺秋梅	张　诚
程　静	关会梅	杨　艺	吴金键	张世刚
孙　琳	顾秋蓓	费详惠		

Responsible editorial group

Zheng Changsheng	Qi Xianglei	Sun Yayun	Liu Yajun
Cai Weidong	Wu Ningkui	Liu Qing	Li Li
Wang Zuoyong	Yuan Yaqin	Song Yongcun	Xiao Yingying
Tian Ye	He Qiumei	Zhang Cheng	Cheng Jing
Guan Huimei	Yang Yi	Wu Jinjian	Zhang Shigang
Sun Lin	Gu Qiubei	Fei Xianghui	

《中国区域地质志·上海志》组织协调委员会

主　任：杨联萍

副主任：方国安　张先林　王寒梅　严学新　季善标

成　员：方　正　李勤奋　杨建刚　何中发　史玉金

谢建磊　杨天亮　杨丽君

Organization and Coordination Committee of *Regional Geology of China*, *Shanghai Municipality*

Director: Yang Lianping

Associate Directors: Fang Guoan　Zhang Xianlin　Wang Hanmei

Yan Xuexin　Ji Shanbiao

Committee Members: Fang Zheng　Li Qinfen　Yang Jiangang

He Zhongfa　Shi Yujin　Xie Jianlei

Yang Tianliang　Yang Lijun

《中国区域地质志·上海志》编辑委员会

主　编： 王寒梅

副主编： 谢建磊　史玉金

编　委： 赵宝成　俞俊英　陈大平　郭兴杰　温晓华
吴建中　何中发　李　晓　王　远　王治华
战　庆　郑　磊　司永峰　黎　兵　刘　婷
张士宽　洪斌兰

Editional Committee of *Regional Geology of China, Shanghai Municipality*

Chief Editor: Wang Hanmei

Associate Editors: Xie Jianlei　Shi Yujin

Committee Members:

Zhao Baocheng	Yu Junying	Chen Daping	Guo Xingjie
Wen Xiaohua	Wu Jianzhong	He Zhongfa	Li Xiao
Wang Yuan	Wang Zhihua	Zhan Qing	Zheng Lei
Si Yongfeng	Li Bing	Liu Ting	Zhang Shikuan
Hong Binlan			

总　序

20世纪80~90年代，我国编纂出版了第一代省级区域地质志，系统总结了20世纪80年代以前各省（区、市）区域地质调查研究成果，为促进我国地质事业和服务经济社会的发展发挥了重要作用。

随着改革开放的不断深入，地质调查和各类科研项目的实施，我国地质事业又获得蓬勃发展，取得了丰硕的调查研究成果。在青藏高原、大兴安岭等地，完成了1：25万区域地质调查和航空磁测，实现了全国陆域中比例尺区域地质调查和航空磁测的全覆盖。开展了各种比例尺的区域地球物理和区域地球化学的调查研究，实施了“深部探测技术与实验研究”专项，拓展了环境地质、农业地质、城市地质等新领域，获得了丰富的基础地质资料，解决了许多基础地质问题，找到了一大批矿产，大幅度地提高了我国的地质研究程度和地质科学技术水平。

《国务院关于加强地质工作的决定》（国发〔2006〕4号）指出，要提高基础地质调查（研究）程度，建立地质图文更新机制。《全国地质勘查规划》（2005—2020）提出，要开展全国基础地质资料综合研究和成果集成工作，编制新一代全国系列基础地质图件，努力创造有宏观影响的大成果。为此，专门组织有关专家，分层次编纂各省（区、市）、跨省（区）成矿区带和重要经济区，以及全国层次的新一代系列地质志书，集中反映我国区域地质调查和地质科学研究工作的重大进展和成果，创造一项有宏观影响的大成果，为地质工作和国家经济社会发展提供一套基础地质资料。

为国家经济建设和社会发展提供矿产资源和地质资料，是地质工作服务经济社会发展和人民生产生活的两项基本任务。通过编纂新一代区域地质志书，把全国各部门多年地质调查研究所积累的丰富地质、地球物理、地球化学等资料进行综合研究和集成，使之系统化、规律化和理论化，不仅为地质工作和经济社会发展提供了一套整装的地质资料，而且为发展地质科学理论奠定了扎实的资料基础。

党的十八届三中全会指出，要紧紧围绕建设美丽中国、深化生态文明体制改革，加快建立生态文明制度，健全国土空间开发、资源节约利用、生态环境保护的体制机制，推动形成人与自然和谐发展的现代化建设新格局。这对地质工作提出了新的更高要求。举凡矿产资源的合理开发利用、国土空间的规划、地质环境的保护、地下水污染的治理、地质公园的建设、地球化学对人体健康的影响，以及土壤性质及农产品的优化等，都需要丰富的基础地质数据和地质信息服务。《中国区域地质志》既可以满足一个省（区、市）、一个成矿区（带）及经济区（带）和全国地质矿产勘查、地质工作规划、部署和管理的需求，又可以在城乡建设、交通、水利、能源和环境保护等重大工程规划、设计、建设和科学管理中发挥重要作用。

新一代《中国区域地质志》是对新时期我国基础地质调查工作的系统总结和高度浓缩，注重在结构和体例上尽量与第一代区域地质志保持有机衔接、汲取其优点和精粹的同时，着力坚持创新，突出反映近30年来取得的新成果和新资料，体现了新的水平和研究高度。

这套志书的编纂和陆续出版是一项浩繁的科学工程，参加单位和人员多、涉及专业领域广、持续时间长，浓缩了几代地质工作者数十年辛勤劳动的成果，是我国地质界的一件大事。参与编纂出版的地质工作者以对历史、对人民负责的高度责任感，孜孜以求，辛勤劳作，力争使志书做到资料翔实、内容精湛、质量上乘、富于创新，打造成高水平的精品著作。在这里，我要向为这项工作付出艰苦努力、做出贡献的地质工作者以及关心、支持这项工作的各有关方面致以崇高敬意和衷心感谢！

中华人民共和国国土资源部副部长

中国地质调查局局长

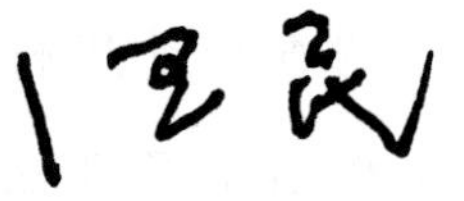

2014年2月

General Preface

Compiled in the 1980s and 1990s, Regional Geology of China of the first generation systematically summarized the research results of regional geology before the 1980s in provinces, autonomous regions or municipalities, and played vital roles in serving social and economic development, and geological advancement in China.

With the deepening of reforms and opening-up, the geological undertakings in China embrace another chance of prosperity. Abundant results of geological survey and research have been achieved along with the implementation of geological survey programs and various scientific research programs. In areas such as the Qinghai-Tibet Plateau and Great Khingan Mountains, regional geological survey and aeromagnetic survey at the scale of 1 : 250000 have been completed, which means that geological and aeromagnetic survey on medium scales have achieved a full coverage of the whole land area in China. Regional geophysical and geochemical surveys on different scales have been conducted. Sino-Probe, a special program for deep exploration research in China, has been carried out as well. New research fields such as environmental geology, agricultural geology and urban geology have been expanded. All these efforts have brought us with substantial fundamental geological materials, and put a series of geological problems into settlement. Besides, a number of mineral deposits have been found. The intensity of geological survey and the level of geological science and technology have been improved significantly.

As is stated in *Decision of the State Council on Strengthening Geological Work*, the intensity of geological survey and research should be enhanced, and an update mechanism concerning geological materials, including geological maps and reports, should be established. *National Geological Exploration Plan* (*Program*) (2005—2020) proposes that comprehensive studies of nationwide geological data and integration of research results should be conducted, national series of fundamental geological maps of a new generation should be compiled, and we should strive to create outstanding achievements with macro impact. Under the above mentioned directions, experts in relative disciplines are organized to compile the new series of Regional Geology that successively cover provincial (municipal or autonomous regional) level, cross-provincial (or cross-regional) metallogenic belts or important economic zones level, and national level. The new series of Regional Geology principally represent the significant advancements and achievements of geological survey and research in China. The compilation of the series of Regional Geology constitutes an achievement with remarkable impact, and provides a set of geological data for economic and social development.

Providing mineral resources and supplying geological data for economic construction and social development of China, are two basic tasks of geological work which serves social and economic development and people's production and life. By means of compiling the new series of Regional Geology, abundant geological, geophysical and geochemical data that have been accumulated by different departments of the entire country for many years are integrated systematically, regularly and theoretically. The comprehensive study not only provides an entire set of geological data for both geology itself and economic and social development, but also lays concrete foundation for developing geological theory.

In the Third Plenary Session of the 18th CPC Central Committee, it is proposed that we must deepen ecological environmental management reform by centering on constructing a beautiful China, and that we should accelerate system building to promote ecological progress, improve institutions and mechanisms for developing land space, conserving resources and protecting the ecological environment, and promoting modernization featuring harmonious development between human and nature. This raises greater demand on geological work. The tasks such as rational development and utilization of mineral resources, land space planning, protection of geological environment, groundwater pollution control, geo-park construction, evaluation of geochemical impact on human health and optimization of soil properties and agricultural products, all depend on rich fundamental geological data and geological information services. The *Regional Geology of China* can not only fulfill the requirement for mineral resources exploration and planning, deployment and management of geological work in provincial (municipal or autonomous regional) level, metallogenic zone (belt) and economic zone level and national level, but also hold an important role in planning, design, construction and scientific management of major projects in fields such as urban and rural construction, transportation, water conservancy, energy development and environmental protection.

The *Regional Geology of China* of the new generation systematically summarizes and highly condenses the fundamental geological survey work of the new era. To consistent with the first generation in structure and style as much as possible, the new series follow its advantages. Moreover, the series stick to innovation, and focus on presenting the new advancement and new materials acquired in past 30 years, which stands for the new height and achievement in research.

The compilation and successive publication of the Series is a complex scientific project, in which tremendous time and labor have been input. Scholars of a wide range of professional fields from different institutions were organized for the project. Results of hard work obtained by geologists of several generations in the past decades are condensed in the Series. It is a significant event for the geological community in China. Geologists participating in the project, with strong sense of responsibility for history and people, devoted themselves earnestly and unreservedly to creating excellent works with a high quality. They strive to make the books faithful, informative, sophisticated and innovative. Here, I would like to express my heartfelt thanks to and sincere respect for all the geologists who worked hard and contributed to the project and all concerned who wholeheartedly support this project.

Wang Min,
Vice Minister of the Ministry of Land and Resources, PRC
Director of China Geological Survey

王民

February 2014

总 前 言

2008年，中国地质调查局下达了重编《中国区域地质志》的任务，计划项目名称为“中国地质构造区划综合研究与区域地质调查综合集成”。其总体目标任务是：通过对区域地质调查资料和地学研究成果的综合研究与集成，全面总结30多年来我国区域地质调查、矿产勘查和专题研究所取得的成果，特别是地质大调查实施以来的新资料、新进展、新成果。以活动论和先进的地质科学理论为指导，应用地质编图制图新技术、新方法，按省（区、市）、构造单元（成矿区带、经济区等）和全国3个层次编纂《中国区域地质志》，编制相关地质图件，建立空间数据库；针对综合研究过程中存在的全国性、区域性重大基础地质问题，有重点地开展野外调查和专题研究；进行区域地质调查和地质科学研究资料的系统总结，提高全国地质研究程度和水平，实现地质调查资源共享，提升地质调查和科学研究成果的应用效果和影响，为地质调查、矿产勘查、有关经济和工程建设、地质环境及地质灾害评估，以及地质科研和地质教育等，提供基础地质资料，使地质调查研究成果更好地服务于地质事业和经济社会的可持续发展。

2006年，《国务院关于加强地质工作的决定》（国发〔2006〕4号）指出，要“提高基础地质调查（研究）程度”“系统开展区域地质、地球物理、地球化学和遥感地质等调查，建立地质图文更新机制，为社会提供有效快捷的地质信息服务”。经国务院同意颁布的《全国地质勘查规划》（2005—2020）指出，要“开展重大基础地质问题研究和重点地区基础地质综合研究，力争实现基础地质研究的重大突破。开展全国基础地质资料综合研究和成果集成工作。编制新一代全国系列基础地质图件，开展区域基础地质、矿产、地质环境等系列图件编制与各时代主要地层以及多学科的综合性立典研究”“努力创造有宏观影响的大成果，不断提高地质勘查的精度、深度和广度”。

为经济建设和社会发展提供矿产资源和基础地质资料，一向是地质工作服务经济和社会发展的两项基本任务。通过编纂区域地质志，把各个部门、各种类型地质调查研究取得的丰富地质、地球物理、地球化学、遥感地质等资料进行综合研究和集成，使之系统化、规律化和理论化，可以为国家有关部门和广大民众提供一套整装的地质资料，并为发展地球科学理论开辟通途。因此，国土资源部、中国地质调查局决定重编《中国区域地质志》是一个具有远见卓识的英明决定，是一项带有里程碑意义的战略性综合研究工作，必将为我国地质事业和国家经济社会发展做出重大贡献，在我国地质事业发展史上谱写出光辉的一页。

如果从1916年国民政府农商部改中央地质调查所为地质调查局算起，中国区域地质调查研究已经走过了100多年的时间。在中国地质调查研究的历史上，中国地质界具有编纂地质矿产志的光荣传统，早在20世纪前半叶，就先后编辑出版了《苏浙皖矿志》《中国矿产志》《福建矿务志略》《江苏地质志》《北京西山地质志》《四川西康地质志》等（陈宝国等，2011）。20世纪80~90年代，地质矿产部主持编纂的《中国区域地质志》系统地总结了20世纪80年代末以前我国区域地质调查及专题的研究成果，为我国地质事业和经济社会发展做出了重要贡献，建树了历史的功绩。

近30多年来，我国各有关部门进行了各种类型的地质矿产调查研究，积累了丰富的实际资料，获得大量地质、地球物理、地球化学、遥感地质，特别是高精度同位素年代及地球化学、岩石地球化学数据，解决了一系列长期悬而未决的地质问题，大幅度地提高了我国的地质研究程度，为《中国区域地质志》的编纂创造了良好条件。

为适应各方面的需求，提供更多的地质信息，项目任务要求开展3方面的调查研究，提交3方面的成果：① 32个省（区、市）地质志（广东省地质志含香港、澳门特别行政区），15个左右重要地区（构造单元、成矿区带、重要经济区、行政大区等）地质志，全国性地质志（地质、地球物理、地球化学系列专著）。②系列基础地质图件及空间数据库。地质图件包括地质图、地质构造图、岩浆岩地质图、第四纪地质与地貌图、航磁异常图、重力异常图（内部使用），各省（区、市）及各地区可根据地区特点与需要编制其他地质图件，如基岩地质图、岩相古地理图、变质地质图、断裂系统图等。③结合志书编写和系列图件编制开展必要的专题研究，解决一些重要的疑难地质问题，旨在统一地质划分对比和深化对这些疑难地质问题的认识，提高志书的质量。

编纂《中国区域地质志》涉及的范围大、学科领域广，参加的单位和人员多。这是一项对百万地质大军几十年乃至上百年地质调查研究成果的一次大综合、大集成。这样大规模的综合地质研究可以而且应该创造出有宏观影响的大成果。因此，我们提出以下5方面的要求，作为提交有宏观影响大成果的主要标志。

第一，在地质成果上要大综合和大集成。在充分搜集、分析各个部门、各种类型地质调查研究成果的基础上，进行地质、地球物理、地球化学相结合的综合研究和成果集成，使志书内容丰富、资料翔实、数据准确、特色鲜明，如实反映当前地质调查研究的程度，反映志书地域内各种地质现象的认知水平。

第二，在地质研究程度上有大幅度的提高。不断提高全国和一个地区的地质研究程度，是地质工作永恒追求的目标，也是地质工作服务于经济社会发展的必然要求。地质研究程度的提高，是靠不断地进行地质调查研究和持续地开展综合研究和成果集成而实现的。因此，通过编纂地质志，从地质、地球物理、地球化学、遥感地质以及地表与深部地质结合上开展精细的综合研究，大幅度地提高编志地区的地质研究程度和水平，为地质工作和经济社会发展提供高质量的地质资料。

第三，在地质工作和经济社会发展中发挥大的作用。随着人类社会的进步，地质工作的社会功能和服务领域日益宽广，地质工作几乎是兴利于各个部门和行业，造福于人类的方方面面。作为一套完整而翔实的基础地质资料，既可以满足一个省（区、市）、一个成矿区（带）及经济区（带）和全国地质矿产勘查、地质工作规划、部署和管理的需求，又可以在城乡建设、交通、水利、资源和环境保护等重大工程规划、设计、建设和科学管理中发挥重要作用；既可以为地质科研和地质教育开辟捷径，又可以为丰富地质信息库和普及地质科学知识提供翔实的地质资料，充分体现地质工作服务于经济社会发展的强度和广度。

第四，在地质规律的概括和地质理论创新上有大的提升。地质工作是一项科学技术密集的调查研究工作，地质科学基本上属于认识性科学。矿产资源勘查能否取得重大突破，地质环境及地质灾害的评估和预测是否准确，对诸多工程建设规划、设计和实施能否发挥更大作用，归根结底取决于对地质规律的认识水平和地质理论及技术方法的创新程度。地质调查研究的目的，在很大程度上是总结地质规律、认识地质规律和利用地质规律。抓住了地质规律，地质理论的创新也就应运而生了。地质志质量的好坏、水平的高低、贡献的大小，在很大程度上取决于对编志地区地质规律和地质理论概括、认识的深度和高度。因此，在地质志编纂过程中，在详细记录地质事实的基础上，通过实际资料的归纳、分析和总结，概括出地质规律，凝练出新发现、新认识，大幅度地提高对编志地区地质规律和地质总体特征的认知水平。

第五，在国内外地质界得到广泛应用和产生大的影响。一项科学成果，其在社会上应用的广度和影响的大小，主要取决于其实用价值和科学价值，其实用价值和科学价值越大，它的影响力就越大。中国地质构造十分复杂而又十分特殊，有许多地质现象为世界罕见，如高大的青藏高原、北方新生代的黄土堆积、南方岩溶（喀斯特）地质及地貌、中部延展数千千米的高压-超高压变质带、某些标准的地层剖面和罕见的古生物群落、某些特殊类型的矿床等，既是中国地质的特色和区位优势，又是诞生地质科学新理论的摇篮，引起国内外地质科学家的关注和竞相探索，以期地质科学理论上的突破。通过编纂区域地质志，认真总结中国地质，特别是上述特色地质的总体特征及其演化规律，给予理论的概括和解释，不但会对地质勘查实践和经济社会发展发挥更大作用，而且会对发展地质科学理论奠定基础和做出重要贡献，也必将引起国内外地质界的广泛兴趣，产生重大影响。

编纂《中国区域地质志》是地质工作中一项最基础的综合研究工作，其成果将在自然资源部门及其他有关部门和外国各有关地质单位长期使用。因此，确保质量是对这套志书的必然要求。为了确保志书质量，要求做到“高起点、高要求、高水平”。

所谓“高起点”，就是在充分掌握编志地区地质研究程度、地质特点和地质规律的高点上着手进行地质志的编写。为此，首先，要广泛收集编志地区地质资料和成果，包括报告、专著、图件和论文，特别是近些年来的地质调查研究的成果。其次，对收集到的资料进行地质、地球物理、地球化学相结合的综合研究，力求洞悉本区地质特点、地质规律和存在的主要地质问题。最后，利用先进的地质理论和技术方法进行志书编写。

所谓“高要求”，就是在编志过程中，对各项工作，包括资料的收集、整理、样品分析测试、志书编写和图件编制等，都要高标准、严要求，严格按“项目总体设计”“工作指南”“项目管理办法”和自然资源部、中国地质调查局的有关规程、规范执行，切实保证成果质量。

所谓“高水平”，就是力争把这套志书打造成高水平的精品著作，在科学内容上精益求精，忠实记录地质事实，科学总结地质特征和地质规律，使志书做到资料翔实，内容精湛，质量上乘，富于创新。在表现形式上也要精益求精，文字描述要言简意赅，文图并茂；附图、附表要主题鲜明，清晰美观；计量单位和名词术语使用等要符合国标、行标和国际标准及惯例。

为了丰富志书内容，提高志书水平，为地矿行业、社会和公众提供更多的地质信息，充分发挥志书的功能，要求：一要实行地质、地球物理、地球化学、遥感等探测研究资料相结合，从宏观和微观尺度鉴别、认识地质现象，解决地质问题；二要实行地表地质与深部地质相结合，从三维角度揭示各种地质现象、成矿作用和地质灾害的深部根源；三要实行大陆地质与海洋地质相结合，把海陆作为一个整体加以综合研究；四要实行志书、图件和数据库建设相结合，更全面、形象地显示编志地区地质、地貌特点，更好地发挥志书功能；五是实行综合研究与专题研究相结合，在开展综合研究的同时，就一些重大的、有争议的地质问题进行专题研究，统一认识，统一划分对比，解决争论问题，提高志书质量。在指导思想上坚持继承和创新相结合，在继承基础上创新，在创新指导下继承，既要继承所有好的成果，又要大胆创新，敢于改变传统的认识和结论。同时，通过编志，以老带新，互教互学，培养一批学科性和综合性地质科技人才。

地质工作和国民经济发展规划、部署，地质资源勘查及地质环境与地质灾害评估、预测，城乡建设及重大工程规划、设计，普及地质科学知识，提高全民族科学文化素质和水平，开展国内外地质科学技术合作与交流，以及发展地球科学理论等，都需要丰富的地质信息。《中国区域地质志》可以为上述诸方面提供厚实的基础地质资料。

这次地质志的编纂规模大、要求高，参加的单位和人员多，持续的时间长，特别需要加强领导和组织协调，加强项目科学管理，全国、各省（区、市）及各地质区地质志的编写，都要求建立领导机构和学术指导组织。

在项目编纂指导委员会和编辑委员会的领导下，设立了《中国区域地质志》综合研究项目（即“中国地质构造区划综合研究与区域地质调查综合集成”“全国陆域及海区地质图件更新与共享”），由中国地质科学院地质研究所承担实施，负责《中国区域地质志》编纂的科学技术指导和组织协调工作，参加人员及分工如下。

一、项目负责人

项目负责人：丁孝忠　李廷栋　刘　勇　庞健峰

二、特聘专家

组长：肖庆辉

成员（按姓氏笔画排序）：邓晋福　陆松年　张克信　潘桂棠

三、项目专业组主要人员

区域地质与洋板块地质组：李廷栋　丁孝忠　刘　勇　石成龙

区域地层组：张克信　杨崇辉　高林志　万渝生　张　恒　董春燕

区域岩浆岩组：王　涛　薛怀民　刘　翠　张建军　童　英　王晓霞　张　磊　寇彩化

区域构造组：游国庆　王　军　张传恒

第四纪地质与地貌组：王　永　迟振卿　董　进　袁路朋　田　飞　叶梦旎　凌　媛

沉积与岩相古地理组：刘燕学　旷红伟　柳永清　郑　宁　彭　阳

变质地质与前寒武纪地质组：任留东　杨崇辉　杜利林　王彦斌　宋会侠

蛇绿岩组：肖序常　翟庆国　胡培远　高　俊　宋述光

同位素年代组：万渝生　董春燕　颉颃强　刘守偈　朱文斌

地球物理及深部地质组：管　烨　郑洪伟　金　胜　张贵宾

地理底图、数据库与图件设计组：韩坤英　庞健峰　剧远景　张庆合　王丽亚　王振洋　姜　兰　金　铭　丁伟翠

六大区与构造单元区域地质综合研究组：计文化　朱小辉　尹福光　任　飞　杨祝良　余明刚　牛志军　谭满堂　赵凤清　李承东

杨晓平　杨雅军　邢德和　李　敏　邢光福

肖国强　刘燕学

海域区域地质综合研究组：温珍河　李学杰　王中波　姚永坚　侯方辉　杨楚鹏

王保军　高红芳　徐子英　祁江豪

《中国区域地质志》的编纂获得了自然资源部（国土资源部）、财政部、科技部及有关司局、中国地质调查局及有关部（室）和中国地质科学院及有关处（室）的指导和大力支持；获得了中国地质调查局6个地质调查中心、中国地质科学院有关研究所和地质研究所及有关处（室），各省（区、市）自然资源部门（国土资源厅）、地质矿产勘查局、地质调查院（所）和有关地质院校的积极参与和大力支持；地矿行业各部门、全国及各省（区、市）地质资料、地质图书保管等单位为志书编写提供了丰硕的地质资料和信息；参与志书编写的专家、学者和管理人员孜孜以求、精益求精，付出了艰辛的劳动和智慧。在这里，向所有关怀、指导志书编写的部门和领导，向所有关心、帮助志书编写的部门和人士，向所有参与志书编写的专家、学者和管理者，致以衷心的感谢和敬意。

地质工作是一项永恒的事业，只要有人类存在，就有地质工作；哪里有人类存在，哪里就有地质工作。这是因为地质工作服务的领域很广，而且越来越广。地质工作为经济社会发展服务的主要途径之一就是不断地为社会提供有效快捷的地质信息，这也是编纂《中国区域地质志》的初衷和主要宗旨之一。

我们希望这套志书能够集我国地质调查研究成果之大成，走向志书的深处、高处和远处，打造出志书的精品，为我国地质事业的发展做出更大贡献！为优化国土生态环境、建设美丽中国做出更大贡献！为我国经济社会的可持续发展和实现中华民族伟大复兴的中国梦做出更大贡献！为世界地球科学的繁荣昌盛做出更大贡献！

2020年5月

General Foreword

In 2008, a project concerning recompilation of the *Regional Geology of China* entitled "Comprehensive Study of Geological and Tectonic Division and Synthesized Integration of Regional Geological Survey in China", was assigned by China Geological Survey (CGS). The overall objective of the project is to fully summarize achievements acquired in regional geological survey, mineral resources exploration and monographic studies, especially the new materials, new advancement and new achievement since the implementation of the Great Geological Survey in China, by means of comprehensive study and integration of regional geological survey data and geoscience research results. Under the guidance of the Mobilism and advanced geoscience theories, the *Regional Geology of China* and relative geological maps are compiled at three levels respectively: provincial (municipal or autonomous regional) level, tectonic unit (metallogenic belts or economic zones) level and national level, while the spatial database are established, with new technologies and methods of geological mapping. Field surveys and monographic studies are conducted with priority to solve the national or regional fundamental geological problems encountered in comprehensive study. Systematical summaries of regional geological survey data and geoscience research materials are done so as to improve intensity and level of geological survey and research, to realize the sharing of geological survey resources and to raise the applied effectiveness and impact of geological survey and research achievements. The *Regional Geology of China* can provide fundamental geological materials for geological survey, mineral exploration, relative economic development and engineering construction, assessment of geological environment and hazard, as well as geological research and education. That enable achievements of geological survey to better serve geological cause and sustainable development of economy and society.

In 2006, *Decision of the State Council on Strengthening Geological Work* put forward that "the intensity of geological survey and research should be enhanced", and that "regional geological, geophysical, geochemical and geological remote sensing surveys should be conducted systematically and an update mechanism concerning geological maps and texts should be established, to provide effective and efficient geological information service". According to *National Geological Exploration Program* (2005—2020) promulgated with the State Council's consent, "researches on major geological problems and comprehensive study of fundamental geology in key areas should be carried out so as to achieve crucial breakthrough in fundamental geological researches. Comprehensive studies of nationwide geological data and integration of research results should be conducted. The new series of fundamental geological maps should be compiled; compilation of map series of regional fundamental geology, mineral resources, geological environment and multi-disciplinary comprehensive typical studies of various major strata should be conducted", and we should "strive to create outstanding achievements with great impact, and to improve accuracy, depth and breadth of geological exploration".

Providing mineral resources and supplying geological data for economic construction and social development of the country, have always been two basic tasks of geological work that serves social and economic development. Through the compilation of the *Regional Geology of China* abundant geological, geophysical, geochemical and remote sensing data collected in all kinds of geological surveys from various institutions are used for comprehensive research and integration. Geological materials that have been systematized, regularized and theorized, can serve relevant government departments and the general public with a whole set of data, and beat a path for developing theories of earth science. Therefore, to recompile the *Regional Geology of China* is a wise decision with visionary transcendence by China Geological Survey, former Ministry of Land and Resources (MLR) of China; it is a strategic synthetic study with monumental significance, which is bound to contribute greatly to geological cause and development of economy and society, and to write a glorious page in the history of geological development in China.

Dating back to 1916 when the Ministry of Commerce and Agriculture renamed the Central Geological Institute as the Geological Survey, the geological survey and research of China have gone through a history of more than 100 years. Compiling chronicles or records of regional geology has always been a glorious tradition upheld by geological community in the history of geological survey of China. Early in the first half of the 20th century, *Minerals Geology of Jiangsu, Zhejiang and Anhui*, *Minerals Geology of China*, *Mining Records of Fujian*, *Geology of Jiangsu*, *Xishan Geology of Beijing*, *Xikang Geology of Sichuan* were published successively (Chen Baoguo et al., 2011). The *Regional Geology of China* compiled in the 1980s and 1990s, hosted by the former Ministry of Geology and Mineral Resources, systematically summarized results of regional geological survey and monographic studies before the late 1980s in China, which made great contributions to geological cause and development of economy and society, and acquired historic achievements in development of China.

In the past three decades, geological mineral survey and research conducted by relevant departments have accumulated a wealth of practical materials, and have acquired generous data of geology, geophysics, geochemistry, remote sensing geology, and in particular data of high-precision isotopic chronology and rock geochemistry. Besides, a series of pending geological problems have been solved and the intensity of geological survey and research have been enhanced tremendously. All these have created favorable conditions for compiling the new *Regional Geology of China*.

For the purpose of providing more geological data and meeting requirements of various aspects, surveys and researches in 3 aspects should be conducted and 3 kinds of achievements should be submitted in the project: ①Regional geology of 32 provinces (municipalities and autonomous regions) (the Regional Geology of Guangdong Province covers the Hong Kong Special Administrative Region and Macao Special Administrative Region), regional geology of 15 major regions (tectonic units, metallogenic belts, key economic zones, and major administrative regions), and national geology (series of geological, geophysical and geochemical monographs). ②The series of fundamental geological maps and spatial database. The geological map series include geological map, tectonic map, magmatic rock map, Quaternary geology and geomorphology map, aeromagnetic anomaly map and gravity anomaly map (restricted for internal use only). Each province (municipality or autonomous region) or region can adjust to compile other types of maps according to its features and demands, such as geological map of bed rocks, lithofacies and paleogeographic map, metamorphic map and map of fault systems. ③To address major problems of geology, necessary monographic studies should be carried out in the course of compiling regional geology and maps. It is intended to unify recognition on geological division and correlation, to deepen recognition on difficult geological problems, and finally to improve the quality of the series of regional geology.

This compilation work involves a wide range of regions and scholars of broad discipline fields from many departments. The project synthetically integrates achievements of geological survey and researches carried out by a million geologists in decades of years or even a hundred years. It is believed that a comprehensive geological study on such a large scale might and should create achievements of significant influence. Therefore, we propose the following 5 requirements, as major indicators of achievements with macro-influence.

Firstly, geological achievements should be broadly synthesized and integrated. Comprehensive study and integration of geological, geophysical and geochemical data should be conducted on the basis of fully collecting and analyzing various results of geological survey and research from different departments, to make the books informative, faithful, accurate and featured, and to present truthfully the level of geological survey as well as research and cognition of all geological phenomena in target areas.

Secondly, the level of geological survey and research should be highly raised. To continuously raise the level of geological survey and research of the whole country and a particular area is the eternal pursuit of geological work and an inevitable requirement of economic and social development. The raise of geological research level lies in continuous geological survey and comprehensive study as well as synthetic integration. By way of compiling the series of regional geology, sophisticated researches are conducted with integration of geological data, geophysical data, geochemical data, remote sensing data and combination of surface and deep geological data, which will significantly improve the intensity and level of geological researches and provide geological work and social and economic development with geological data of high quality.

Thirdly, the *Regional Geology of China* should play a vital role in geological development and economic and social development. With the progress of human society, the social functions and service sectors of geological work have been expanded. Geology facilitates almost all departments and industries and benefits all aspects of humankind. As a complete set of detailed and faithful fundamental geological materials, the series of the *Regional Geology of China* can not only meet the requirements of mineral exploration, and planning, arrangement and management of geological work in provincial (municipal or autonomous regional) level, in level of metallogenic zone (belt) and economic zone (belt), or in national level, but also play important roles in planning, design, construction and scientific management of major projects in such fields as urban and rural construction, transportation, water conservancy, energy development and environmental protection. Meanwhile, it will not only beat a shortcut for geological research and education, but also provide informative data for enrichment of geological repository and popularization of geological knowledge. All these embody the service intensity and breadth of geology to social and economic development.

Fourthly, there should be great improvement in summarization of geological regularity and innovation of geological theory. Geological work is an investigative work with intensive feature of science and technique. Geoscience essentially belongs to cognitive science. Whether a significant breakthrough in mineral resources exploration will be made or not, whether evaluation and prediction of geological environment and hazards will be accurate or not, and whether geology will play a greater role in planning, design and implementation of project construction, all these questions, in the last analysis, depend on the cognitive level of geological regularity and innovative extent of geological theory and technical method. The purpose of geological survey is to summarize, to recognize and to utilize geological regularity to a large degree. Once geological regularity is obtained, innovation in geological theory will come into being naturally. The quality of the series of the *Regional Geology of China* and their achievements and contribution, largely depend on the depth and breadth of summarization and cognition of geological regularity and theory for the study areas. Therefore, in the course of compiling the series of Regional Geology, practical materials are concluded and analyzed on the basis of detailed records of geological facts so as to summarize the geological regularity and to extract new discoveries and new recognition, resulting in significant enhancement of cognition on geological regularity and general geological features in target areas of the regional geology.

Fifthly, the achievements should be widely used by the geological community both in China and abroad, and great impact should be achieved. The applied breadth and impact of a scientific achievement mainly depend on its practical value and scientific value. The higher its practical value and scientific value are, the greater its influence will be. Completely complex and quite special, geological structures in China hold lots of geological phenomena rare in the world, including the lofty Qinghai-Tibet Plateau, Cenozoic loess accumulation in North China, karst geology and landform in South China, high- or ultrahigh-pressure metamorphic belt that extend thousands of kilometers in Central China, standard stratigraphic sections and rare paleo-biocenosis, and deposits of special types. These geological phenomena not only embody features and regional advantages of geology in China, but also serve as a cradle where new theories of geoscience are incubating. They attract attention of geoscientists both of China and abroad who race to the quest for theoretical breakthroughs in geoscience. General features and evolution regularities of geology of China, especially the geological phenomena mentioned above, are concluded and explained theoretically by means of compiling this series. Such practices will not only play a greater role in geological exploration and economic and social development, but also lay concrete theoretical foundation for development of geoscience and have important contribution to it. It is believed that interests of geoscientists both at home and abroad will be attracted and enormous influence will be extended.

Compiling the *Regional Geology of China* is the most fundamental synthetic study in geological work; its achievements will be used for a long time by departments in relation to natural resources as well as other relevant fields, or even geological departments of other countries. So, to ensure the quality of this series is an inevitable requirement. For this purpose, requirements of high starting point, high requirement and high level are imposed.

The so called "high starting point" means that the compilation of the books and maps should be commenced from the starting point that the intensity of geological researches, geological features and regularities of the target areas have been quite grasped. Hence geological materials and achievements of the target areas, including reports, monographs, geological maps, papers and especially achievements of geological survey in recent years, are collected first of all. And then in the compilation of the books and maps, the researchers should conduct synthetic studies integrated with geological, geophysical and geochemical data that have been gathered and seek to thoroughly understand the geological features, regularities and major geological problems that existed in the target areas. Finally, the books are written and the maps are compiled with advanced geological theories and technical methods.

The so called " high requirement" means that all procedures, including data collecting and sorting, sample analyzing and testing, book and map compiling, should be conducted strictly according to the "General Design of the Project", "Work Guide", "Measures for Project Administration" and relevant regulation issued by the Ministry of Natural Resources (MNR) and CGS, to ensure the quality of the results.

The so called "high level" means that we should strive to create an excellent series of books with high level, to make sure that the books are excellent, informative in scientific content and faithful in geological facts, and should scientifically and creatively summarize geological features and regularities. We pursuit excellence in manifestation of the books as well. The descriptive text should be concise and illustrated with proper figures. The figures and tables should be outstanding in theme, and clear and beautiful in form. All the units of measure and terminology used in the books should conform to national, sectoral and international standards and convention.

In order to enrich the contents and improve the quality of the books, and to fully perform their functions so as to provide more geological information for the geological and mineral industry as well as the public, the following demands are proposed: firstly, integrated researches of geological, geophysical, geochemical and remote sensing data should be done so that geological phenomena can be distinguished and recognized on both macroscopic and microscopic scales and geological problems can be solved; secondly, surface geology and deep geology should be combined with each other so that the deep roots of geological phenomena, mineralization and geohazards can be uncovered from three-dimensional perspective; thirdly, continental geology and marine geology should be associated as a whole for comprehensive study; fourthly, book and map compiling should be connected with construction of database, so that geological and geomorphological features of the target areas are manifested completely and vividly and the functions of this series are better performed; fifthly, comprehensive studies and monographic studies should be integrated; while conducting comprehensive studies, we should carry out monographic studies concerning the major controversial geological problems as well, to unify cognition and to solve disputed problems, and to improve the quality of the books. Adhering to the ideology that inheritance combines with innovation, we seek to innovate on the basis of inheriting and to inherit with the guidance of innovation. We will not only inherit all the great achievements, but also innovate boldly to change traditional understanding and conclusions. Meanwhile, through the compilation of the books and maps, a number of comprehensive geoscientists in various disciplines have been cultivated by means that the senior instruct the junior and they learn from each other.

Aspects such as planning and deployment of geological work and national economic development, geological mineral exploration, geological environment evaluation and hazards forecast, planning and design of urban and rural construction and major engineering projects, popularization of geoscience, improvement of scientific and cultural quality and level of the people, international cooperation and communication of geoscience and technology, and development of theories of earth science, all require abundant geological information. The *Regional Geology of China* can provide substantial fundamental geological data for the aspects mentioned above.

The project of compiling the series of the books, because of its large scale and strict requirements, claims tremendous time and labor input. Hence it is necessary to strengthen leadership and coordination, and to reinforce scientific management of the project. It is required that leading organs and academic guiding organization should be established at the national level, provincial (municipal or autonomous regional) level and the level of geological areas.

Under the leadership of Advisory Committee and Editorial Board, a team responsible for Comprehensive Study Project of the *Regional Geology of China* (known as "Comprehensive Study of Tectonic Division and Synthesized Integration of Regional Geological Survey in China" and "Updating and sharing of national lands and seas geological maps" as well) was set up and by the Institute of Geology, Chinese Academy of Geological Sciences (CAGS) to implement. Participants of the project and members of the team who are responsible for academic guiding, coordinating and organizing in the compilation of the *Regional Geology of China* are as follows:

a. Project Leaders

Chief Project Leaders: Ding Xiaozhong, Li Tingdong, Liu Yong and Pang Jianfeng

b. Distinguished Expert Group

Group Leader: Xiao Qinghui

Members (by surname stroke order): Deng Jinfu, Lu Songnian, Zhang Kexin and Pan Guitang

c. Project Professional Groups

Group of Regional Geology and Ocean Plate Geology: Li Tingdong, Ding Xiaozhong, Liu Yong and Shi Chenglong

Group of Regional Stratigraphy: Zhang Kexin, Yang Chonghui, Gao Linzhi, Wan Yusheng and Zhang Heng

Group of Regional Magmatic Rocks: Wang Tao, Xue Huaimin, Liu Cui, Zhang Jianjun, Tong Ying, Wang Xiaoxia, Zhang Lei, Kou Caihua

Group of Regional Tectonics: You Guoqing, Wang Jun and Zhang Chuanheng

Group of Quaternary Geology and Geomorphology: Wang Yong, Chi Zhenqing, Dong Jin, Yuan Lupeng, Tian Fei, Ye Mengni, Ling Yuan

Group of Sedimentology and Lithofacies Palaeogeography: Liu Yanxue, Kuang Hongwei, Liu Yongqing, Zheng Ning and Peng Yang

Group of Metamorphic Geology and Precambrian Geology: Ren Liudong, Yang Chonghui, Du Lilin, Wang Yanbin and Song Huixia

Group of Ophiolite: Xiao Xuchang, Zhai Qingguo, Hu Peiyuan, Gao Jun and Song Shuguang

Group of Isotopic Chronology: Wan Yusheng, Dong Chunyan, Xie Hangqiang, Liu Shoujie and Zhu Wenbin

Group of Geophysics and Deep Geology: Guan Ye, Zheng Hongwei, Jin Sheng and Zhang Guibin

Group of Geographic Basemap, Database and Map Design: Han Kunying, Pang Jianfeng, Ju Yuanjing, Zhang Qinghe, Wang Liya, Wang Zhenyang, Jiang Lan, Jin Ming and Ding Weicui

Comprehensive Study Group of Regional Geology for the Six Administrative Regions and Tectonic Units: Ji Wenhua, Zhu Xiaohui, Yin Fuguang, Ren Fei, Yang Zhuliang, Yu Minggang, Niu Zhijun, Tan Mantang, Zhao Fengqing, Li Chengdong, Yang Xiaoping and Xing Dehe

Comprehensive Study Group of Marine Geology: Wen Zhenhe, Li Xuejie, Wang Zhongbo, Yao Yongjian, Hou Fanghui, Yang Chupeng, Wang Baojun, Xu Ziying, Gao Hongfang, and Qi Jianghao

The compilation work has gained full support and energetic assistance from the MNR, former MLR, MOF, MOST and its relevant departments and bureaus, from the CGS and its relevant departments, and from the CAGS and its relevant departments. Six Centers of the CGS, relevant institutes of the CAGS, relevant departments of the Institute of Geology, CAGS, provincial (municipal or autonomous regional) Departments of Natural and Resources, provincial (municipal or autonomous regional) Bureau of Geology and Mineral Exploration, provincial (municipal or autonomous regional) Bureau (Institute) of Geological Survey and relevant universities or colleges have actively participated in the job. Various departments of the geological and mining industry, National Geological Library of China, provincial (municipal or autonomous regional) Geological Libraries, National Geological Archives of China, and provincial (municipal or autonomous regional) Geological Archives have provided substantial geological materials and information. All experts, scholars and management personnel who participated in the compilation have devoted their wisdom and hard work in pursuit of excellence. Hereby, I would like to express my sincere gratitude to and respect for departments and leaders who exercise care and guide the compilation of the series, to departments and persons who concern and help in the compilation, and to all experts, scholars and management personnel who have been involved in the compilation of the series.

Geological work is an eternal cause, and where there are human beings, there is geological work. The reason of this lies in that the fields which geological work serve are wide and will be wider and wider. One of the main paths by means of which geological work serve social and economic development is to provide effective and efficient geological information. It is one of the major reasons why we strive to compile the series of books.

It is our hope that the series of the *Regional Geology of China* can integrate all great achievements of geological survey in China and mark the culmination of years of hard work. It is our pursuit to create a series of books with excellent quality, to make greater contribution to the development of geological cause in China, to make a great contribution to the optimization of ecological environment and the construction of beautiful China, to make greater contribution to sustainable development of economy and society and realization of the Chinese dream of Chinese Nation's great rejuvenation, and to make greater contribution to the prosperity of Earth Science in the world.

Li Tingdong

李廷栋

May 2020

前　　言

在全国实施新一代《中国区域地质志》编制的背景下，受中国地质科学院地质研究所委托，上海市地质调查研究院承担了“上海区域地质调查片区总结与服务产品开发”项目，编制新一代《中国区域地质志·上海志》。项目起止时间为2016—2018年。总体目标任务为全面收集和整理上海市近30多年的各类基础地质调查成果，按照技术要求，开展重大地质问题综合研究、区域地质成果总结和集成。在系统总结和研究地层、岩浆岩、断裂构造、深部构造、第四纪地质等区域地质特征基础上，编制《上海市区域地质志》和地质系列图件，建立地质图空间数据库，为国土资源规划及管理、城市防灾减灾等提供基础地质资料，为社会公众提供公益性区域地质信息。

第一代《上海市区域地质志》编撰以来，上海的地质工作转向以城市地质调查研究为主，工作区也从陆域逐渐扩展到了近岸海域。在此过程中，积累了一批新的地质、地球物理资料，发现了一些新的地质事实，提出了一些新的认识。与此同时，随着地球科学理论的发展和各种测试技术的进步，上海及其邻区江苏、浙江、安徽等在地层、岩浆岩、构造等方面都取得了一定进展。通过新一代区域地质志的编制，把相关资料进行综合研究和集成，使之系统化、规律化和理论化，具有重要的意义。本志在充分吸收第一代区域地质志成果的基础上，认真按照中国地质调查局、中国地质科学院地质研究所的相关部署和《中国区域地质志工作指南》技术要求，完成了对上海市相关地质、地球物理资料的收集、整理和成果集成。通过补充适量新的测试资料，开展了岩浆作用期次及形成的构造环境等地质问题研究。编撰过程中，坚持在详细记录地质事实基础上，评述不同观点，并进行归纳、总结，力求有新发现、新认识。

在基础地质条件认识上，本志重新厘定了上海基岩的岩石地层系统，将金山群划为古—中元古界，划分出河上镇群，并指出古—中元古界和青白口系为不同阶段的岛弧-弧后盆地建造；将原上侏罗统修正为下白垩统；重新定义了原夏驾桥组含义；总结了不同阶段的沉积作用特征和地质演化史；发现晋宁期、燕山中期花岗岩，修正了侵入作用和火山作用不同步的认识；划分出新村—兴隆断陷盆地，将天马山构造火山盆地修正为天马山构造火山隆起；厘定了第四纪海侵期次和古河道演化特征等。这些新的认识将上海的区域地质研究提升到了一个新的阶段，为城市地质、地震地质背景、地震危险性、中深层地热资源、水文地质和工程地质等方面的调查研究提供了新的资料，夯实了地质基础。

相比第一代《上海市区域地质志》，本志对区内近30多年的城市地质工作概况和成果进行了系统总结，部分资料和成果可以上溯到20世纪60年代。主要包括水文地质条件、工程地质条件、地面沉降、环境地球化学、岸带冲淤、地下水资源、滩涂资源、地下空间资源和旅游地质资源等内容。城市地质调查和研究成果的集成大大丰富了志书内容，增强了新一代志书的实用性，使其成为一部全面了解区内地质条件的工具书，既适用于专门的地质工作者，也适用于了解区内地质条件的一般读者，可供地质勘查单位、工程勘察单位、科研部门、政府管理部门等有关部门和人员参考使用。

本志全面、系统、客观地反映了上海市基础地质和城市地质调查研究的最新成果，资料丰富详实，涉及资料历史周期较长、工作主体较多，是有关调查、实验部门许多地学、科技工作者长期以来共同劳动的结晶。在编撰过程中，得到了中国地质科学院地质研究所、中国地质调查局南京地质调查中心、上海市规划和自然资源局领导和有关研究所及处（室）领导的关心和大力支持。同时，得到了邻近江苏、浙江、安徽等兄弟单位和本市各类地质工作单位的支持。不同阶段成果的汇报、咨询过程中，《中国区域地质志》编委会的各位专家提出了较多的宝贵意见和建议，使得志书内容不断完善、成形。在此谨向曾为本志做出过贡献的所有领导、专家致以谢忱。

谢建磊

2021年3月

Foreword

In answer to the country's call for compiling a new generation of *Regional Geology of China*, at the request of Institute of Geology, Chinese Academy of Geological Sciences, Shanghai Institute of Geological Survey undertook the "Summary on the Regional Geological Survey Area of Shanghai and Development of Service Products" project and the task of compiling a new generation of the *Regional Geology of China*, *Shanghai Municipality*. The project was scheduled to span from 2016 to 2018. The overall objective is to collect and collate all fundamental geological survey results of Shanghai over the past 30 years; conduct comprehensive research on major geological concerns and summarize and integrate regional geological outputs according to the technical requirements; after systematic summarization and research of regional geological characteristics such as stratigraphy, magmatic rock, fault structure, deep structure and Quaternary geology, compile the regional geological annals of Shanghai and the associated geological series maps, and establish a geological map spatial database to provide fundamental geological data for land resources planning and management and urban disaster prevention and mitigation, and to provide nonprofit regional geological information for the social public.

As the compilation of the *Regional Geology of Shanghai* proceeded, our work gradually moved from onshore to near-shore area. During this process, we gained a set of new geological and geophysical data, discovered new geological facts, and acquired new understandings. In the meantime, with new achievements made in geoscientific theories and various testing techniques, the neighboring Jiangsu and Zhejiang provinces have both made progress in stratigraphy, magmatic rock and structure. Hence it is of great significance to make comprehensive research and research of these new data through the compilation of a new generation of regional geology so as to systemize, find the law of, and summarize the theory on these data. In addition to absorbing all meaningful contents of the first generation of the regional geological annals, we also completed the collection and collation of related geological and geophysical data of Shanghai and the integration of related research results as deployed by China Geological Survey and Institute of Geology, Chinese Academy of Geological Sciences and as required by the *Work Guide on the Regional geology of China*. By adding a proper amount of new testing data, we investigated the episodes of magmatism and the tectonic setting of its formation. During the compilation, in addition to recording the geological facts in full detail, we also reviewed and summarized different observations in an endeavor to identify new discoveries and new insights.

Based on the understanding of fundamental geological conditions, we redeterminaed the rock stratigraphic system of the bedrock of Shanghai, redefined the Jinshan Group as a Paleoproterozoic-Mesoproterozoic formation, marked off the Heshangzhen Group, and pointed out that the Paloproterozoic-Mesoproterozoic and Qingbaikou System are islandarc-backarc basin formations of different periods; redefined the original Upper Jurassic as Lower Cretaceous; redefined the meaning of the original Xiajiaqiao Formaiton; summarized the sedimentation characteristics and geological evolution in different periods; discovered Jinningian and middle Yanshanian granites, and corrected the misunderstanding that the intrusion did not occur simultaneously as the volcanism; marked off the Xincun-Xinglong faulted basin, and redefined the Tianmashan tectono-volcanic basin as the Tianmashan tectono-volcanic uplift; and determined the episodes of Quaternary transgressions and the evolution of paleochannels. These new understandings upgrade the regional geological research of Shanghai to a new stage. They provide new information for the survey and investigation of urban geology, seismic geological background, seismic risk, medium and deep geothermal resources, hydrological geology and engineering geology, and consolidating the geological foundation.

Compared with the first generation of the *Regional Geology of Shanghai*, this updated version includes a systematic summarizations of the urban geological activities and outputs within the area over the past three decades. Some of the data and findings can trace back to the 1960s, covering the hydrological geological conditions, engineering geological conditions, surface subsidence, environmental geochemistry, coastal erosion and deposition, groundwater resources, tidal flat resources, underground spatial resources and tourist resources. The integrations of urban geological survey and research results greatly enriched the contents of the annuals, enhances the practicality of the new generation of the annals and makes it a useful tool for getting a comprehensive insight on the geological conditions in the area. This book is applicable for both professional geological workers and general readers who are interested in the geological conditions in the area. It can serve as a reference book geological investigation institutions, engineering exploration institutions, government authorities and individuals working at these institutions.

This book gives a comprehensive, systematic and objective presentation of the latest research results of the fundamental geology and urban geology in Shanghai. With abundant, accurate data, extended coverage of history and multiple working subjects, data contained herein are all the fruit of the concerted work of many geoscience related survey and experimental workers from related survey and experimental departments. During the compilation, we received kind concerns and strong support from leaders of Chinese Academy of Geological Sciences, Nanjing Geological Survey Center, Shanghai Planning and Natural Resource Bureau, as well as related research institutes and sections. We are also grateful to the sister institutions from neighboring Jiangsu, Zhejiang and Anhui provinces and to the various geological workers within the city. During results presentation and consultation in different phases, the compilers of the *Regional Geology of China* offered many valuable ideas and opinions, helping refining and finalizing the book. Hence we want to thank all those leaders and experts who have contributed to this book.

Xie Jianlei

March, 2021

目　　录

第二篇　岩浆岩及岩浆作用

第三篇　变质岩及变质作用

第四篇 地质构造

第五篇 岩石圈结构构造

第六篇　第四纪地质与地貌

第七篇　城市地质

第八篇　区域地质发展史

CONTENTS

Part Ⅱ Magmatic Rock and Magmatism

Part Ⅲ Metamorphic Rock and Metamorphism

Part Ⅳ Geological Structure

Part Ⅴ Texture and Structure of the Lithosphere

Part Ⅵ Quaternary Geology and Geomorphology

Part Ⅶ Urban Geology

Part Ⅷ Regional Geological History

Attached Maps

1. Quaternary Geological and Geomorphological Map of Shanghai Municipality (1 : 250000)
2. Bedrock Geological Map of Shanghai Municipality (1 : 250000)
3. Geological Structure Map of Shanghai Municipality (1 : 250000)
4. Aeromagnetic (ΔT) Anomaly Map of Shanghai Municipality (1 : 250000)
5. Integrated Hydrogeological Map of the IV Confined Aquifer of Shanghai Municipality (1 : 250000)
6. Integrated Engineering Geological Structure Zoning Map of Shanghai Municipality (1 : 250000)
7. Cumulative Land Subsidence Contours Map of Shanghai Municipality (1921–2016) (1 : 250000)
8. Geological Relics Map of Shanghai Municipality (1 : 250000)

绪　　言

上海市（简称“沪”，别称“申”）位于长江口与杭州湾之间。北滨长江，东临东海，南依杭州湾，西接江苏、浙江两省，全市西起东经120°51′，东至122°05′，南自北纬30°42′，北抵31°52′，面积约6833km²，为我国面积最小的一个省级行政单位（图1）。

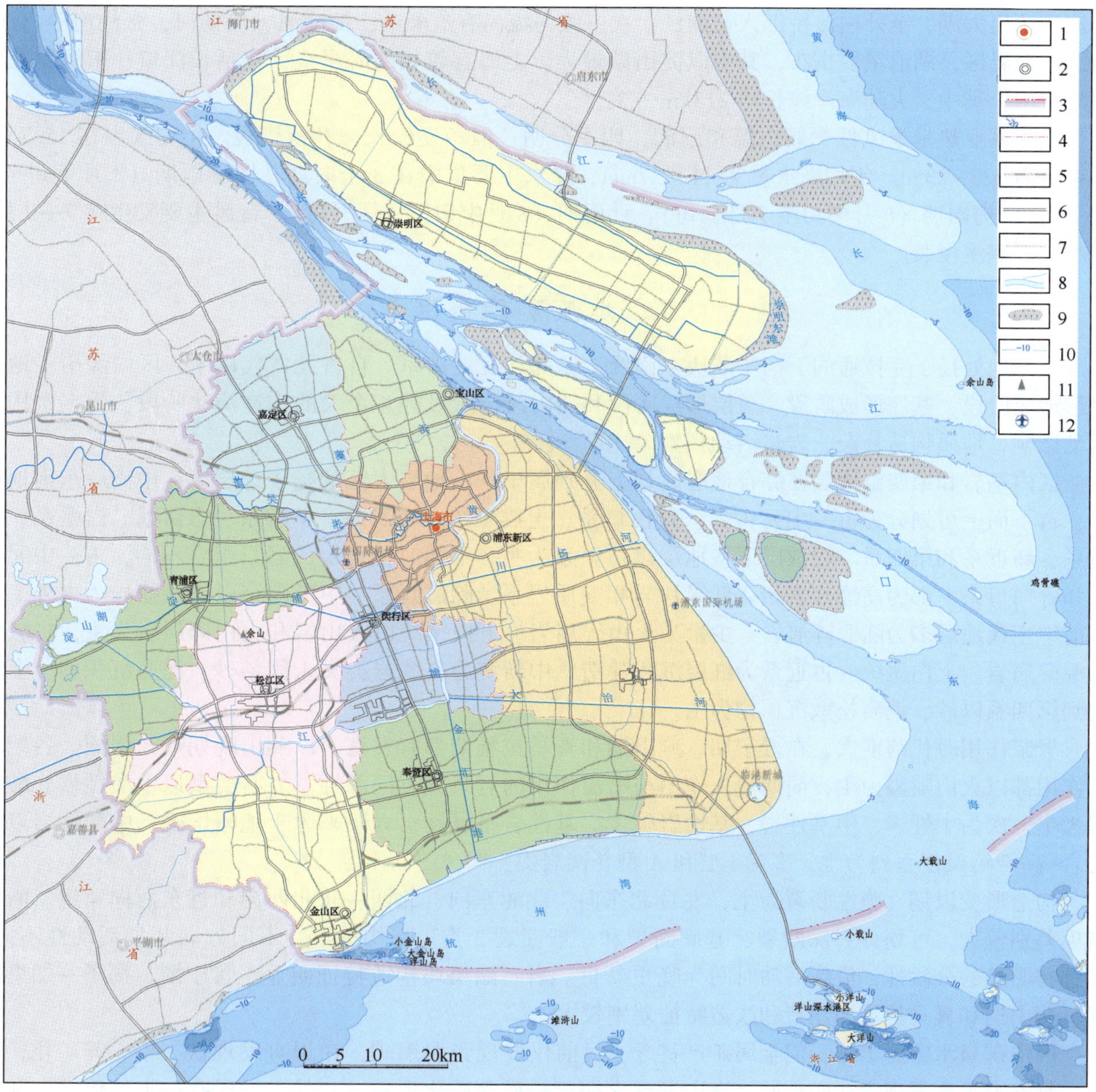

图1　上海市行政区划图

Fig. 1　Administrative division map of Shanghai

1—市政府驻地；2—区、县政府驻地；3—省、直辖市界；4—区、县界；
5—铁路；6—高速公路；7—主要道路；8—河流；9—滩涂；10—等深线（m）；11—山峰；12—机场

一、自然地理概况

上海位于长江三角洲东端，为宽广的冲积平原区，地势平坦，海拔一般在3~5m之间。仅在金山、松江有10余处由火山岩、火山碎屑岩等组成的小山丘，以大金山、天马山、佘山最高，海拔在100m左右。小山丘累计面积约为2.5km^2，占全市总面积的万分之四。长江口有崇明、长兴与横沙等冲积沙岛，杭州湾内又有由大、小金山等火山岩、火山碎屑岩组成的小岛，岛屿总面积约为1400km^2，约占全市面积的20%。

上海河网密布，水量丰富，终年不冻，除长江外，黄浦江为区内最大河流，源于淀山湖，至吴淞口汇入长江，江宽水深，可泊巨轮，为国内著名良港；其次为苏州河（又称吴淞江）是黄浦江的主要支流，源出太湖，至外白渡桥汇入黄浦江。次一级河流尚有蕴藻浜、淀浦河、川杨河、张泾河等。江、浙、沪交界区，湖泊星罗棋布，其中以淀山湖最大。上海有较长的海岸线，包括长江口、东海岸及杭州湾在内，其中，大陆海岸线长达213km左右。

上海属亚热带季风性气候，温和湿润，四季分明，但冬夏较长，春秋较短。年均气温在15.7℃，无霜期为223~235d。年均降水量为1123.7mm，降水主要集中在3个时期，即春雨期（3月下旬至5月下旬）、梅雨期（6月中旬至7月上旬）、秋雨期（8月中旬至9月下旬）。台风主要发生在7—9月。冬季常有寒流侵袭。

二、地质矿产概况

上海地处扬子克拉通的下扬子陆块东南缘，紧邻华夏造山系。自古元古代以来，区内经历了地壳雏形形成阶段、基底形成阶段、盖层形成阶段和陆内改造阶段等4个大的构造演化阶段，受多次构造运动影响，地质构造复杂、岩浆活动频繁。

区内基岩和第四纪地层在综合地层区划上分别属于华南地层大区的扬子地层区、中南—东南地层区，自下而上可划分为古—中元古界、青白口系、上震旦统、寒武系、奥陶系、志留系、白垩系、古近系、新近系和第四系。松散层下伏地层以白垩系为主。不同时代地层的岩石组合不同，古—中元古界和青白口系下部为高绿片岩相—低绿片岩相变质岩系。除震旦系—中奥陶统以碳酸盐岩为主外，其他前中生代地层多为陆源碎屑岩，少量为火山碎屑岩。下白垩统以火山岩和火山碎屑岩为主，少量为陆源碎屑岩。上白垩统—古近系为红层沉积建造。中新统为玄武岩类夹未固结砂、泥质沉积物。上新统到第四系以砂、泥质松散沉积物为主。

岩浆作用时代跨度大，在晋宁期、燕山期和喜马拉雅期都有发育，以燕山晚期最为强烈。各期岩浆作用都以火山喷溢为主，同期侵入作用与喷溢作用的岩浆具有同源性。岩石类型以中酸性岩、酸性岩为主，次为中性岩、中基性岩，基性岩较少。基性岩呈碱性，岩浆来源于地幔熔融。中酸性岩以钾玄岩-高钾钙碱性系列为主，多为I型和A型花岗岩类。

构造形变以韧、脆性断裂为主，发育北东向、北北东向、北西—北北西向和近东西向4组。断裂规模差别较大，可划分为深断裂、基底断裂和一般断裂3类。新构造期，不同方向断裂活动性不同，且具有明显的分段性，最新活动时间为晚更新世。岩石圈层块构造特征明显，晋宁期、扬子—加里东期、燕山期和喜马拉雅期建造和改造特征差别较大。

区内矿泉水资源丰富，但金属矿产贫乏，目前仅发现张堰铜矿，在坦直发现铁、铅、锌矿化，在周浦赵家宅等地发现含钛磁铁矿，查山发现含铜黄铁矿等矿化现象。此外，发现有一定的建筑砂和石材资源，区域上局部具有水热型和干热型地热资源的潜力。

三、地质调查和研究简史

上海的地质工作始于20世纪20年代，据记载最初由刘季辰、赵汝钧等在金山、松江的佘山、天马山及大、小金山等地进行地质调查。其后是英、美学者如克里赛等人利用英商上海自来水公司开凿的一些深井，对上海市区的第四系进行了初步描述和水质分析。

新中国成立以来，地质、石油、冶金部门和有关的高等院校与科研单位，在区内开展了大量的地质调查和找矿工作。1949—1950 年，上海石油公司（中国石油公司前身）翁文波、李德生、王纲道等，在上海及苏南进行了系统的地质调查、重力测量及少量钻探工作后，认为上海基底为变质岩和火成岩。20 世纪 50 年代末，在全市开展了石油、天然气资源调查，初步了解了中、新生代地质情况及石油地质条件等，对石油、天然气资源潜力做出了评价。20 世纪 60 年代，又进行了全市 1∶20 万水文地质调查，基本上查明了地下水资源的数量、质量、分布以及补径排规律，提出了合理开发利用方案。20 世纪 50—60 年代，于本区积极开展了找矿工作，通过对磁异常进行钻探验证，发现中型铜矿一处和众多的矿点、矿化地段。与此同时，地质部华东石油物探大队、航空物探大队九〇九队等先后对本区进行了 1∶100 万航磁及重力概查，1∶20 万和 1∶5 万重力、磁力普查，1∶2.5 万航磁详查，1∶1 万地磁测量等工作，发现异常 141 处。20 世纪 80 年代，又系统进行了全市 1∶10 万工程地质调查和 1∶20 万环境水文地质调查，重点研究了浅层工程地质特征，开展了分区评价。环境水文地质调查摸清了全市环境水文地质背景值及其污染现状、原因与途径，并进行了质量评价，提出了防治措施意见。同时又联合邻省开展了长江三角洲地区的水文地质、工程地质、环境地质综合评价。专门性地质工作始于 20 世纪 60 年代，对地面沉降开展了详细的调查和研究工作，查明了引起地面沉降的主导原因，掌握了沉降机理，提出了相关防治措施。20 世纪 80 年代，在实践基础上，发展了地下含水层储能技术。上述工作中，先后共施工钻孔 6000 多个，基岩累计进尺 9.5 万余米；但空间上分布很不均匀，大多集中在张堰矿区和早期的地铁沿线。此外，上海高等院校和科研单位，在本市也做了大量工作。如华东师范大学陈吉余等对长江三角洲的地貌、第四纪地质和新构造运动进行过全面调查；同济大学建立了以上海面粉厂为代表的第四纪地层层序，并进行了古气候演化研究，划分了海侵层。

第一代《上海市区域地质志》出版以来，区内地质工作转向以城市地质调查研究为主，工作区也从陆域逐渐扩展到了近岸海域。上海市地震局、上海市地质矿产局等部门围绕地震地质背景、区域地壳稳定性、活动性断裂等开展了大量工作，主要有：1988—1989 年，上海市地质矿产局实施了金山卫—常熟大地电磁测深工作，了解了沿线深部地质构造特征和垂向电性结构，验证了张堰—南汇断裂等断裂的存在。1987—1990 年，上海市地震局开展了上海地震危险性分析与基本烈度复核项目，系统论述了上海及邻近地区的地震地质背景、地震活动性规律，开展了地震危险性分析和基本烈度复核。1994—1995 年，上海市地质技术开发总公司开展了张堰铜矿百家村矿段的地质详查，项目实施钻探 12501.86m，计算了矿石和金属储量，开展了开发的技术经济可行性研究。1995 年上海即开始进行活动断层探测，至 2006 年，上海市地震局先后实施了陆域和近岸海域的活动断裂调查，论述了活动断裂的分布情况。2001—2003 年，上海市地质调查研究院和浙江省地质调查院分别承担了 1∶25 万上海幅和杭州幅区域地质调查工作，覆盖了上海西部地区。在基础地质调查基础上，对国土资源、地质灾害等情况进行了调查。2004—2005 年，上海市地震局通过大地电磁测深、深地震反射、高分辨率地震折射等方法开展了泥城—平湖剖面的深部地球物理探测。2004—2008 年，上海市人民政府和国土资源部联合开展的“上海市三维城市地质调查”项目，对历史资料进行了系统收集和整理，开展了全面的三维城市地质结构、环境地质和环境地球化学调查，并开展了工程建设适宜性、地质环境容量等评价工作，建立了城市地质信息平台。2006—2008 年，上海市地质调查研究院开展了“上海市近岸海域多目标区域地球化学调查”，计算了表层、深层沉积物 54 项指标的系列地球化学基准值和背景值，查明了测区元素地球化学分布和分配特征。

1999 年和 2017 年出版的《上海历史图集》《上海城市地图集成》分别汇集了公元前 221 年到 1948 年间 19 个时期上海的政区图、217 种与上海相关的古舆图与近现代地图，集中展现了区内 2000 多年来的地貌变迁特征。此外，21 世纪以来，尤其是“十一五”和“十二五”期间，上海市开展了地面沉降防治、地热勘查等项目，在全市实施了大量钻探工程，为基岩地质构造研究积累了宝贵资料。

四、本次工作概况

本次志书修编的工作周期为2016—2018年。其间，收集整理了区内近30年的各种地质、地球物理资料，包括基岩钻孔、具有综合测试资料的第四纪地质孔、深部地球物理剖面、相关地质调查和专题研究成果及公开发表的期刊论文和出版的专著等。在此基础上，针对岩浆岩和变质岩补充了少量同位素年龄和岩石化学测试数据。资料引用时间截至2017年12月。需要指出的是，近30年，由于基岩地质调查工作开展较少，很多内容新资料和新成果较少，为保持志书的完整性，沿用了第一版志书的相关内容。本次主要根据邻区进展，以板块构造、洋板块地质思想为指导，对部分章节进行了修编。第四纪地质与地貌部分在《上海市第四纪地层与沉积环境》专著基础上补充了最近10多年的相关资料。为提高对水文地质、工程地质的指导作用，松散层沉积物名称保留了原沉降法与筛析法的命名方案，与谢帕德三角图解分类命名法则的详细对应情况见第二十三章第一节。此外，根据全国地质志编撰委员会和首席科学家要求，为突出特色和增加志书的适用性，本志较第一版志书，增加了“城市地质”篇。

通过近3年的工作，在基础地质方面获得了部分实际资料，并取得了一定进展，主要有：重新厘定了岩石地层系统，将金山群划为古—中元古代地层，划分出河上镇群，并指出古—中元古界和青白口系为不同阶段的岛弧-弧后盆地建造；将原上侏罗统修正为下白垩统；重新定义了原夏驾桥组含义；总结了不同阶段的沉积作用特征和地质演化史；发现晋宁期、燕山中期花岗岩，修正了侵入作用和火山作用不同步的认识；划分出新村—兴隆断陷盆地，将天马山构造火山盆地修正为天马山构造火山隆起；厘定了第四纪海侵期次和古河道演化特征等。

本套图、志由上海市地质调查研究院编著，是在中国地质调查局、中国地质科学院地质研究所、南京地质调查中心领导下进行的。上海市地质调查院王寒梅副院长、严学新总工程师、方正副总工程师负责组织领导和技术指导，地质环境研究所史玉金所长负责日常协调工作，项目负责人谢建磊主持编撰日常工作。

本志除绪言、结语外，共计8篇38章。第一篇、第二篇、第三篇、第四篇、第五篇和第八篇由谢建磊执笔。第六篇和第三十五章由赵宝成执笔。第七篇第二十七章和第三十二章由俞俊英执笔；第二十八章和第三十三章由陈大平执笔；第二十九章由吴建中执笔；第三十章由温晓华执笔；第三十一章和第三十四章由郭兴杰执笔。绪言与结语由谢建磊执笔。全志及附图由谢建磊负责统稿，由王寒梅副院长和方正副总工分别进行了审阅。

附图比例尺为1∶25万，共计9幅。《上海市基岩地质图》和《上海市地质构造图》由谢建磊编制，《上海市第四纪地质及地貌图》和《上海市地质遗迹图》由赵宝成编制，《上海市布格重力异常图》和《上海市航磁（ΔT）异常图》由王治华编制，《上海市第Ⅳ承压含水层综合水文地质图》由俞俊英编制，《上海市工程地质结构综合分区图》由陈大平等编制，《上海市累计地面沉降等值线图（1921—2016年）》由吴建中编制。此外，李晓、王远、洪斌兰等同志参加了该项工作。

为提高编写质量，特聘请顾澎涛、许长海和王张华为本志技术顾问。在本志编写过程中，得到了各位顾问的悉心指导，在此谨致谢意。此外，感谢在资料搜集过程中，上海市地矿工程勘察院、上海市地震局等同仁的协助和指导。

第一篇 区域地层

概　　述

上海几乎均被晚新生代晚期疏松沉积物覆盖，基岩露头面积约 2.5km²，主要分布在西南部的佘山、天马山、辰山等地。基岩面埋深起伏明显，整体上自西南向东北方向明显变大（图Ⅰ-1）。松江、金山、青浦等地埋深多在 160~220m；浦东、宝山、嘉定等沿江沿海地区埋深多在 280~360m；河口砂岛地区埋深总体大于 400m，崇明岛局部可达近 500m。

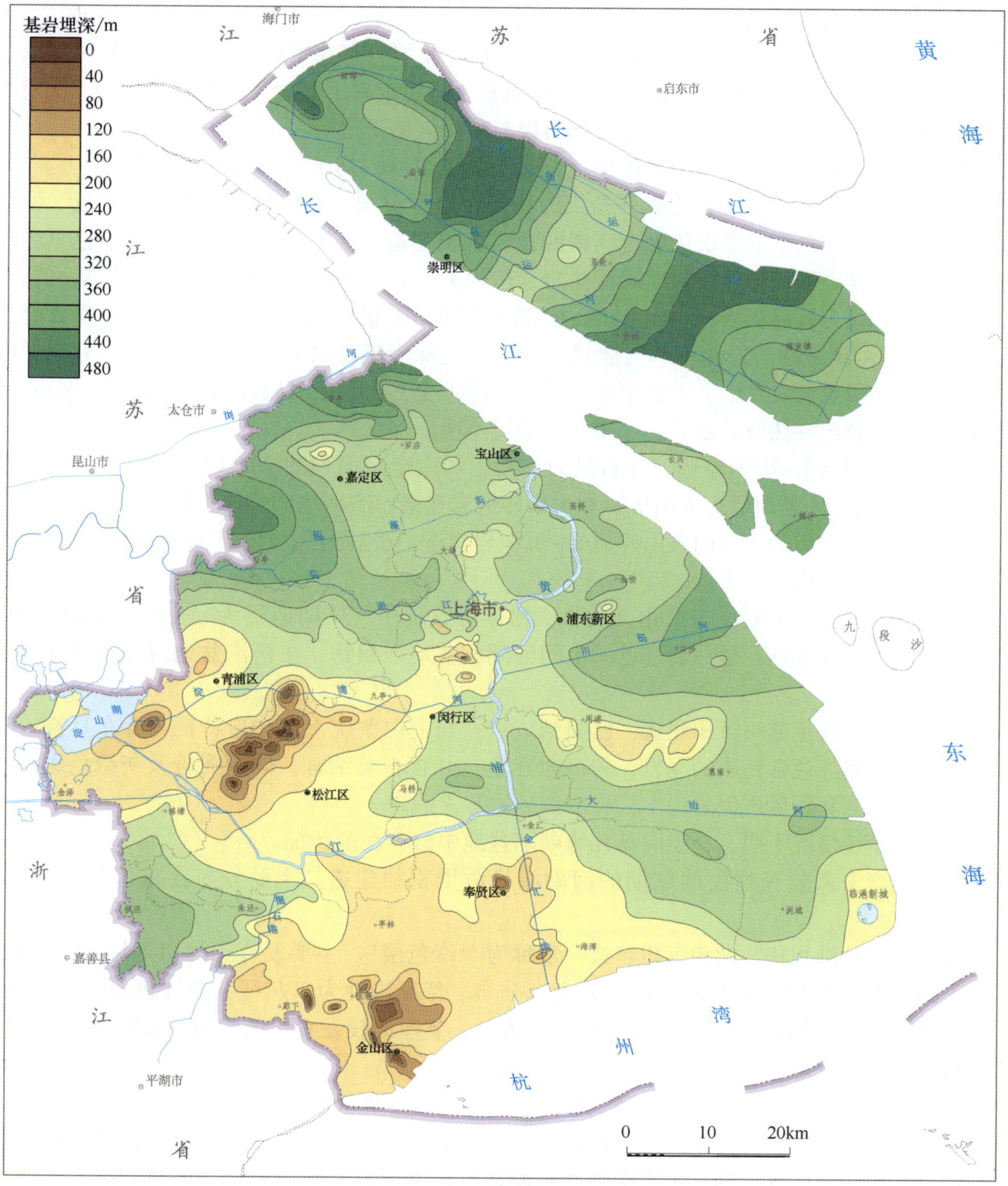

图Ⅰ-1　上海基岩埋深分布图

Fig. Ⅰ-1　Distribution of bedrock depth in Shanghai

上海地层的划分和空间分布认识主要来自各类钻孔资料，主要发育有古—中元古界、青白口系、震旦系、寒武系、奥陶系、志留系、白垩系、古近系、新近系和第四系，暂未发现上古生界和三叠系。自下而上划分有33个岩石地层单位（表Ⅰ-1），其中基岩层有23个，松散层有10个。根据中国岩石地层区划方案，上海基岩地层的主体归属于华南地层大区扬子地层区的下扬子地层分区和江南地层分区之上海—杭州地层小区（图Ⅰ-2）。在张克信（2015）划分的中国构造-地层区划方案中，上海位于下扬子构造-地层区的怀玉山—天目山构造-地层分区和下扬子（宁国—南通）构造-地层分区。根据地层的总体发育特征，自中元古代开始，上海又可以划分为5个基本地层区划阶段。中元古代—青白口纪地层属于华南地层大区之江南地层区，南华纪—志留纪地层属于扬子—华南地层大区之扬子地层区，白垩纪—古近纪地层属于西北—华南地层大区之华南地层区，新近纪地层属于华南地层大区之东南地层区，第四纪地层属于华南地层大区之中南—东南地层区。其中，第四纪地层分区又可划分为长江三角洲和太湖平原两个沉积分区。

古—中元古界和青白口系构成区内结晶或褶皱基底，主要分布在张堰—钱圩、朱泾—吕巷、周浦、朱行等地。古—中元古界为一套变质的火山-沉积岩系，青白口系下部为一套岩性单一的浅变质细粒次深海沉积建造，青白口系上部为一套滨浅海碎屑岩与陆相火山岩建造。

南华系、下震旦统—上震旦统下部层位（相当于陡山沱组）迄未发现，情况不明。

震旦系上部层位—志留系发育比较齐全，为稳定台地型沉积。在钱桥、北新泾、大场、罗店、庙镇、港西等地分布。震旦系上部层位为含镁碳酸盐岩沉积建造。寒武系除下部地层为黑色页岩与碳质页岩外，其余以含镁碳酸盐岩及含泥质碳酸盐岩建造为主。下—中奥陶统下部由岩性较单一的白云岩、白云质灰岩逐渐转化为微晶灰岩、含泥灰岩，上部由含泥质条带状生物骨屑灰岩组成，富含介壳相化石（碎片）。上奥陶统下部为岩性单一的瘤状灰岩层，上部为笔石页岩相沉积建造，具有典型的类复理石沉积特征。志留系仅发育中、下志留统，且顶、底不全，由砂、泥质碎屑岩组成。

晚古生代与三叠纪—侏罗纪地层迄今未见，情况不明。

下白垩统为陆相火山-沉积建造，分布最为广泛，约占全市面积的70%以上。早白垩世早期地层为火山-沉积岩建造，中期地层为火山熔岩和火山碎屑岩建造，晚期地层整体上逐渐由局部夹炭质条带和煤线的砂、砾岩，发展为中部的火山碎屑岩-沉积岩建造，最后为火山熔岩-火山碎屑岩，局部层位为含粉砂灰岩、泥灰岩或白云岩。

上白垩统分布局限，主要见于安角、北桥、松江等盆地中。早期地层为一套紫色类磨拉石建造，晚期地层由下向上粒度变细，是一套红色的山间河湖相盆地堆积物，在松江盆地内经常见薄层状或网脉状石膏。

古新统为一套紫红色陆相沉积建造，已固结成岩，与上白垩统上部建造特征与分布具有一定相似性。下部为棕红、暗红色含钙质的泥岩、粉砂岩建造，并含石英细砾，同生泥砾及灰绿色斑块。上部以棕红、灰褐色泥岩和粉砂质泥岩为主，夹粉细砂岩和玄武岩。

始新统及渐新统情况不明，在崇明岛北缘断陷盆地中似存在渐新统。

中新统为一套玄武岩夹喷溢间歇期的河湖相松散堆积物建造。主要分布在东部和北部的滨江、滨海。

上新统为一套冲洪积相、湖相沉积建造，常称为“绿色层”。主要发育于河口地区，其次在宝山、浦东的滨岸地带及嘉定西北部等地凹陷区内也有分布。特征岩性为以灰绿、黄绿色为主，夹杂紫红、紫褐、褐黄等色的黏土和粉质黏土，黏土呈坚硬状态，节理比较发育。与大都为泥质胶结的含砾砂、砂砾石组成互层，形成多个沉积旋回。

下更新统为一套河流—泛滥湖泊相沉积建造，常称为“杂色层”。主要分布于安亭—崇明凹陷，其次分布在枫泾—三墩凹陷。特征岩性是以褐黄、褐色为主，有灰绿、黄绿、蓝灰色网纹或斑纹穿插的杂色网纹状、斑杂状的黏土和黏土质粉砂，普遍含有多寡不一的钙质结核、网脉和铁锰质结核。与浅灰、灰黄、褐黄色砂、含砾砂、砂砾石互层。

表Ⅰ-1　上海岩石地层单位划分表

Table Ⅰ-1　Division of the lithostratigraphic units in Shanghai

地质年代				岩石地层
新生代	第四纪	Qh	全新世	青浦组 Qh_3q ～ 如东组 Qh_3r 上海组 Qh_2s 娄塘组 Qh_1l
		Qp_3	晚更新世	南汇组 Qp_3n ～ 漏湖组 Qp_3g 川沙组 Qp_3c
		Qp_2	中更新世	嘉定组 Qp_2j
		Qp_1	早更新世	安亭组 Qp_1a
	新近纪	N_2	上新世	崇明组 N_2c
		N_1	中新世	白龙港玄武岩 N_1b
	古近纪	E_3	渐新世	?
		E_2	始新世	
		E_1	古新世	北桥组 E_1b 夏驾桥组 E_1x
中生代	白垩纪	K_2	晚白垩世	赤山组 K_2c 浦口组 K_2p
		K_1	早白垩世	? 寿昌组 K_1s 黄尖组 K_1h 劳村组 K_1l
	侏罗纪	J_3	晚侏罗世	
		J_2	中侏罗世	?
		J_1	早侏罗世	
	三叠纪	T_3	晚三叠世	
		T_2	中三叠世	
		T_1	早三叠世	
古生代	二叠纪	P_3	晚二叠世	
		P_2	中二叠世	?
		P_1	早二叠世	
	石炭纪	C_2	晚石炭世	
		C_1	早石炭世	
	泥盆纪	D_3	晚泥盆世	
		D_2	中泥盆世	
		D_1	早泥盆世	
	志留纪	S_4	末志留世	
		S_3	晚志留世	
		S_2	中志留世	?
		S_1	早志留世	康山组 S_1k 河沥溪组 S_1h
	奥陶纪	O_3	晚奥陶世	长坞组 O_3c 黄泥岗组 O_3h
		O_2	中奥陶世	牯牛潭组 $O_{1\text{-}2}g$
		O_1	早奥陶世	红花园组 O_1h 仑山组 O_1l
	寒武纪	$\in_3$	晚寒武世	超峰组 $\in_3cf$
		$\in_2$	中寒武世	杨柳岗组 $\in_2y$
		$\in_{0\text{-}1}$	始—早寒武世	大陈岭组 $\in_1d$ 超山组 $\in_{0\text{-}1}c$
新元古代	震旦纪	Z_2	晚震旦世	灯影组 Z_2d
		Z_1	早震旦世	
	南华纪	Nh_3	晚南华世	
		Nh_2	中南华世	?
		Nh_1	早南华世	
	青白口纪	Qb_2	晚青白口世	河上镇群 Pt_3H
		Qb_1	早青白口世	惠南板岩 Pt_3h
古—中元古代				金山群 $Pt_{1\text{-}2}J$

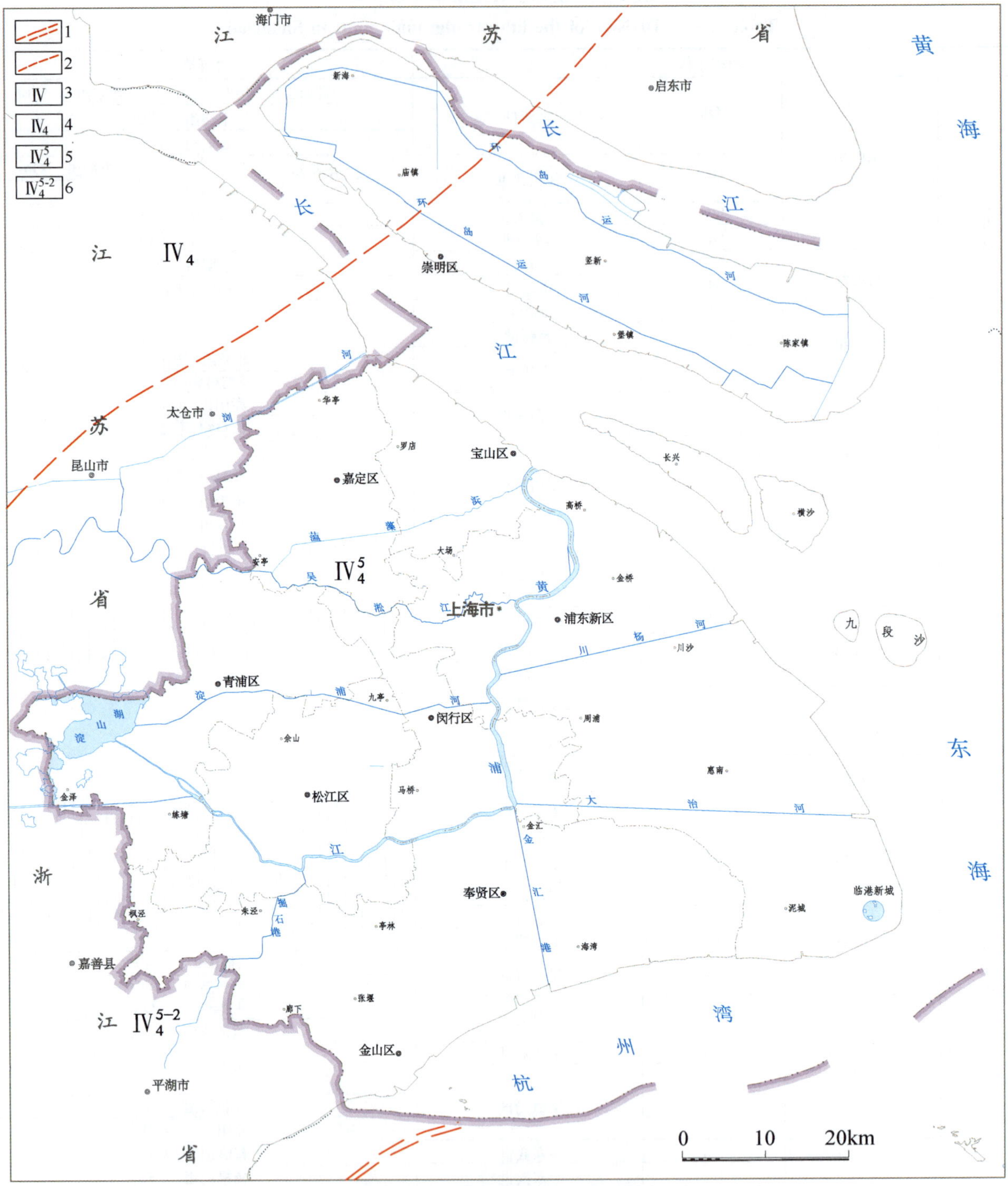

图Ⅱ-2 上海基岩地层综合区划图

Fig. Ⅱ-2 Integrated devision map of the bedrock stratigraphy in Shanghai

1—江绍断裂带；2—湖苏断裂带；3—华南地层大区；4—扬子地区层；5—江南地层分区；6—上海-杭州地层小区（上海部分）

中更新统为一套三角洲河流—泛滥平原、湖泊相沉积建造。在全区分布广泛，但东、西部地区地层发育有所差异。西部存在粗、细相间两个沉积旋回形成的上、下两段地层；东部和河口地区，上段地层因遭受侵蚀而断续出现，大都只保存下段地层。特征岩性为蓝灰、灰绿色黏土，褐黄杂蓝灰、灰绿色细网纹状、条纹状黏土、粉质黏土，也有灰色粉质黏土和粉土。与灰、黄灰色含砾砂、粉细砂互层。

上更新统下部基本为一套河流—三角洲河流相沉积建造。大致沿外冈至燎原农场一线以西地区分布，一般为灰、棕灰色粉细砂和砂质粉土夹含砾砂。在青浦、金山、松江接壤区为河口湾堆积的灰、棕灰色粉质黏土与浅灰色粉砂、砂质粉土互层。河口和东部地区以细砂和含砾砂、砾质砂为主，后者夹粉细砂或呈互层，向上夹粉细砂、砂质粉土、粉质黏土。

上更新统上部为陆相—海相/海、陆交替相—陆相沉积建造，相应分为下、中、上3段。在全区分布广泛，岩性、岩相变化较大。下段，主要为泛滥平原—湖泊相的灰绿、灰黑、深灰色及局部的杂色黏土、粉质黏土和黏质粉土，局部夹砂质粉土或粉砂层，含钙质结核、腐木碎块。中段，在现滨海平原区北部地区主要为浅海—海湾相灰色粉质黏土频夹浅灰色细砂、砂质粉土纹层或呈薄互层；在现滨海平原区南部地区为滨海—浅海相灰、灰黄色粉细砂、砂质粉土互层夹粉质黏土。青浦及与之接壤的松江、金山的部分地区为滨海—潟湖相与河流边滩、泛滥平原—湖泊相交替的暗灰、褐黄和杂色黏土、粉质黏土，时夹黄色粉砂、砂质粉土或呈互层。河口区，下部通常为灰、黄灰色中细砂、粉细砂夹粉土或呈互层，底部或现砂砾层；上部多为青灰、灰色粉细砂夹褐灰色粉质黏土、粉土或呈互层。上段，下部通常为姜黄、褐黄色粉质黏土，局部夹砂质粉土或粉细砂；上部为暗绿、灰绿或黄绿色黏土和粉质黏土。

全新统下部可分为上、下段。下段在三角洲沉积区多为河口相，岩性常为灰色砂质粉土、黏质粉土与灰白色粉砂互层；在太湖沉积区为湖沼相的灰黑、灰褐色粉质黏土，局部形成泥炭或碳质黏土层。上段普遍为滨海堆积。大都为深灰、褐灰色（淤泥质）黏土、粉质黏土，可见大量植物茎叶与黄色泥灰质结核，局部形成泥炭或碳质黏土层。

全新统中部，在长江三角洲沉积区呈滨岸浅海（下部）—河口滨海（上部）堆积，为深灰色淤泥质黏土和淤泥质粉质黏土，灰色砂质粉土、黏土粉土夹粉砂。在太湖平原沉积区呈潟湖—滨海沼泽堆积，通常为较单一的深灰色淤泥质黏土、黏土和淤泥质粉质黏土。

全新统上部，在太湖平原沉积区为夹有泥炭层的深灰、棕灰色淤泥质黏土、粉质黏土的湖沼沉积和褐灰、褐黄色粉质黏土夹粉砂，局部可见泥炭或碳质黏土夹层的湖泊相沉积建造。在长江三角洲沉积区，主要类型为河口湾—滨岸潮滩堆积的灰、灰黄色砂质粉土、黏质粉土夹粉砂，其上部普遍为滨岸潮滩堆积的褐黄、褐灰色粉质黏土；次要类型为吴淞江所在三角洲支流河道堆积的青灰色粉砂和砂质粉土、古吴淞江滨岸堆积的灰褐色砂质粉土。经过几千年的耕作改造，全新统顶部形成了一层薄薄的耕作土，其间保留有宝贵的古文化遗址。

第一章　古元古界—青白口系

古—中元古界和青白口系构成区内基底地层，可进一步划分为金山群、惠南板岩和河上镇群。金山群和惠南板岩为变质岩系建造，河上镇群揭露地层为一套陆源碎屑岩和火山碎屑岩建造。主要分布在中部枫泾—川沙断裂带以南的地区，即有吕巷、钱圩、张堰、朱行、惠南等地（图1-1）。

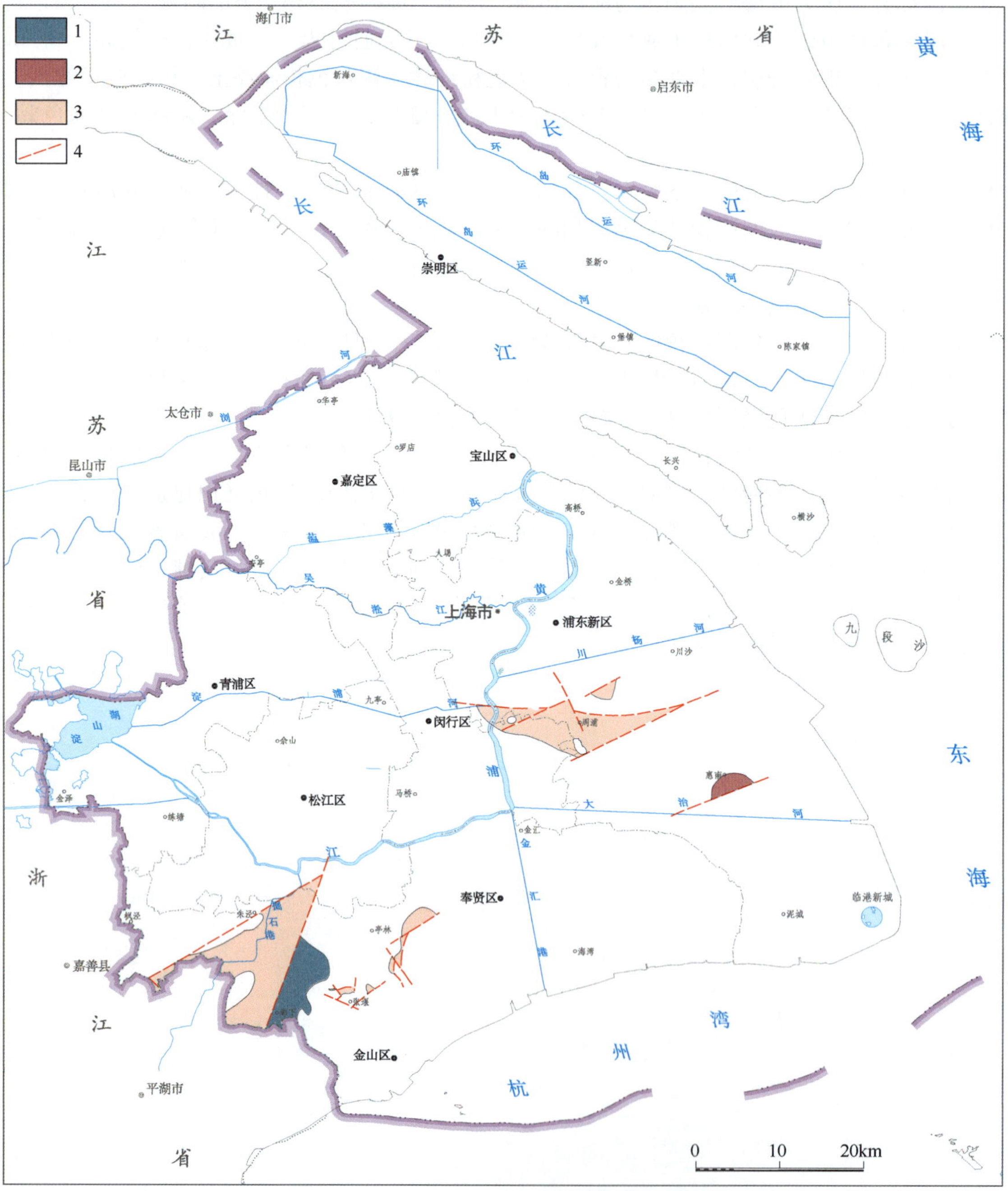

图1-1　上海古元古界—青白口系分布图

Fig. 1-1　Distribution of the Mesoproterozoic-Qingbaikouan System in Shanghai

1—河上镇群；2—惠南板岩；3—金山群；4—断裂

1958—1962 年，于金山朱泾—吕巷等地的部分钻孔中发现一套变质地层，当时建组并命名为“金山组”，推测时代为前震旦纪。

1964 年以后，又陆续在浦东的周浦、三林等地的钻孔中相继发现了这套地层，并在后期开展了大量工作。随着研究程度的提高，发现这一地层比较复杂，用“组”这一级岩石地层单位加以命名并不确切，改称“金山群”。

1988 年，《上海市区域地质志》根据地层岩性、岩相特征以及变质程度，在金山群之上建立了一个新的岩石地层单位，即惠南板岩，系指惠南镇南汇中学钻孔所发现的一套灰色钙质绢云母板岩。推测厚度大于 180m，变质程度低于金山群。

河上镇群为本次新引进的一个岩石地层单位，系指在钱圩 502 孔发现的一套凝灰质砂岩、熔凝灰岩、凝灰岩等岩石组合。长期以来，将该套地层划分为黄尖组下段，本次根据锆石 U-Pb 年龄、岩石类型特征，将其单独划分出来，置于惠南板岩之上（表 1-1）。

表 1-1　上海前震旦系岩石地层单位划分沿革表

Table 1-1　Previous divisions of the Presinian lithostratigraphic units in Shanghai

<table>
<tr><th>地层</th><th>《上海地区矿产地质普查工作总结报告》，1962</th><th>《上海地区石油地质图集》，1966</th><th>《1:20万上海地区基岩地质推断图》，1979</th><th>《上海地区地层概述》，1981</th><th>《上海市区域地质志》，1988</th><th>“上海市三维城市地质调查”，2010</th><th>本志</th></tr>
<tr><td>上覆地层</td><td>下震旦统</td><td>寒武系—震旦系</td><td>上震旦统</td><td>第四系</td><td>?</td><td>第四系</td><td>上震旦统</td></tr>
<tr><td>南华系</td><td rowspan="4"></td><td rowspan="5">金山群</td><td rowspan="6">金山群</td><td rowspan="6">金山群</td><td rowspan="4">?</td><td rowspan="4">?</td><td>?</td></tr>
<tr><td>青白口系</td><td>河上镇群
惠南板岩</td></tr>
<tr><td>中元古界</td><td rowspan="2">金山群</td></tr>
<tr><td>中元古界</td></tr>
<tr><td>古元古界</td><td rowspan="2">金山组</td><td>惠南板岩</td><td>惠南板岩</td><td></td></tr>
<tr><td>太古宇</td><td></td><td>金山群</td><td>金山群</td><td></td></tr>
</table>

第一节　岩石地层划分

一、金山群（$Pt_{1-2}J$）

（一）创名及原始定义

原称“金山组”，由上海市地质勘察局（1961）所建，命名地点在上海金山朱泾、吕巷。原始定义：“金山组”为一套深变质岩系，主要由混合片麻岩、绢云母石英片岩、石英片岩、碳质石英片岩、绿片岩、白粒岩、黑云母角闪石石英片岩等组成。原岩建造大体亦由正常砂、泥岩建造与火山岩建造两部分。

（二）现在定义

金山群为一套变质岩系，由 3 个岩性段组成。上段（c 段）以白色白云母片岩、云母石英片岩及绢云母石英片岩为主；中段（b 段）以灰黑、深灰色斜长角闪岩、绿帘斜长角闪岩、角闪斜长片麻岩及二云斜长片麻岩等组成；下段（a 段）由白色白云母浅粒岩、黑云片岩、阳起片岩及含磁铁绢云母片岩组成。

（三）层型剖面描述

金山群按岩性组合，自下向上可划分为 a、b、c 三个岩性段。其层型剖面分别由沈 zk201 孔、赵

zk1 孔、康 zk1 孔 3 个钻孔剖面代表（图 1-2）。

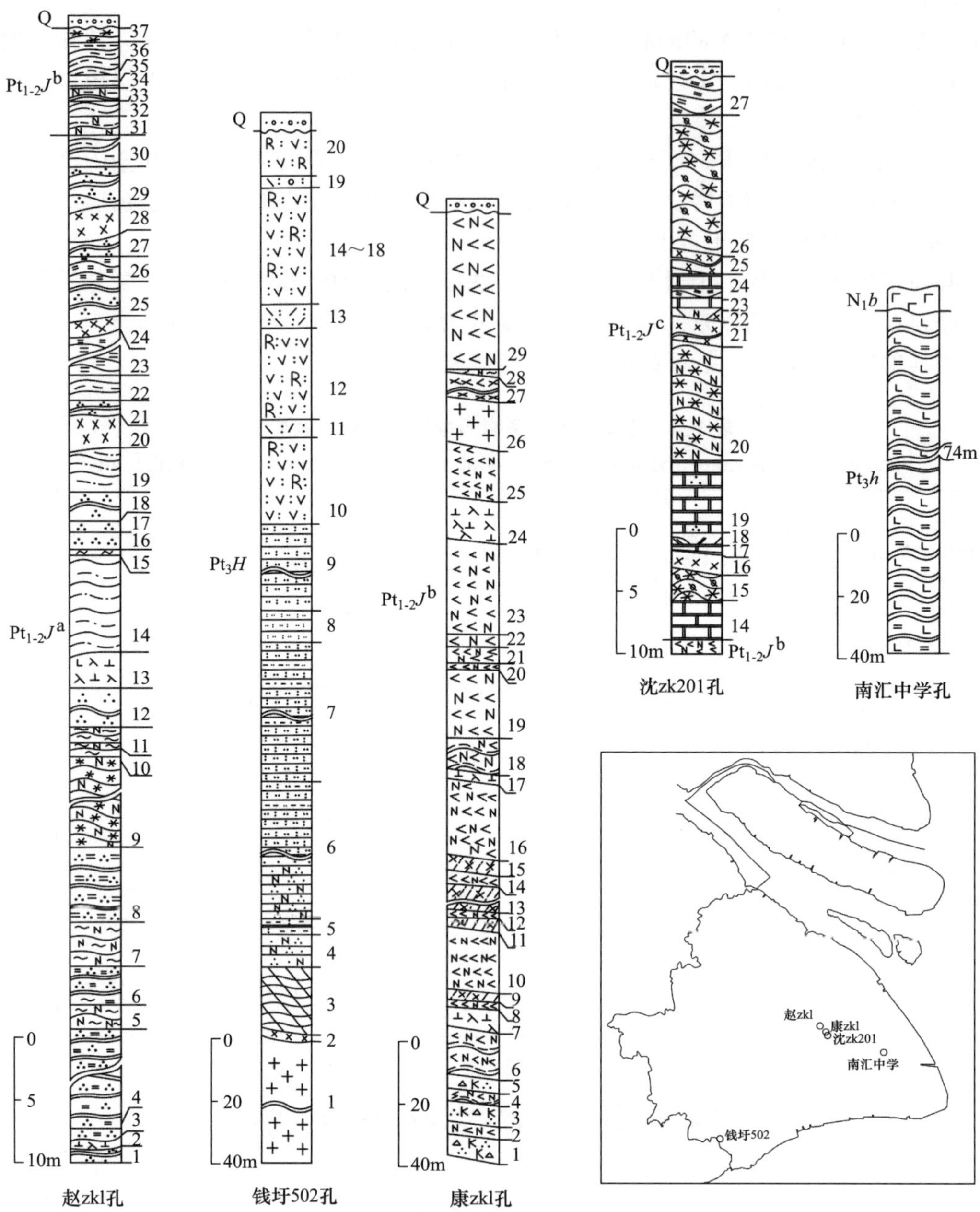

图 1-2　典型钻孔柱状图

Fig. 1-2　Typical borehole histograms

1. 沈 zk201 孔剖面

上覆地层：第四系黏土、砂砾层

~~~~~~~~~~ 不　整　合 ~~~~~~~~~~

金山群 c 岩性段（$Pt_{1-2}J^c$）	总厚度>39.3m
27. 灰白色白云母片岩	>3.1m
26. 灰白、黄褐及深绿色绿帘阳起片岩	11.4m
25. 蚀变辉绿岩脉	(5.4m)
24. 灰白色含石英大理岩与绿泥石化白云母片岩互层	1.8m
23. 灰白色微—细粒大理岩	1.1m
~~~~~~~~~~

22. 绿帘钠长阳起片岩 0. 9m
21. 蚀变辉绿岩脉 (6. 7m)
20. 灰白色黝帘钠长阳起片岩 9. 0m
19. 灰白色微晶含石英大理岩 6. 0m
18. 白云母片岩 0. 8m
16. 蚀变辉绿岩脉 (2. 0m)
15. 绿帘钠长阳起片岩 2. 0m
14. 灰白色大理岩 2. 7m

——————— 整　合 ———————

下伏地层：金山群 b 岩性段绿帘斜长角闪岩、绿帘角闪斜长片麻岩等

2. 康 zk1 孔剖面

上覆地层：第四系

~~~~~~~~ 不　整　合 ~~~~~~~~

金山群 b 岩性段（$Pt_{1-2}J^b$） 总厚度>107. 9m

29. 灰黑、深灰色斜长角闪岩，顶部色较浅 >24. 9m
28. 浅灰色绿泥石化斜长角闪岩 0. 5m
27. 变角闪辉绿岩脉 (20. 8m)
26. 不等粒花岗岩脉 (12. 2m)
25. 灰黑色斜长角闪岩 8. 4m
24. 绢云母化闪长玢岩脉 (7. 5m)
23. 深灰斜长角闪岩 14. 1m
22. 灰色斜长角闪岩，含铜、铅、锌矿脉 2. 1m
21. 深灰色斜长角闪岩 3. 1m
20. 浅灰色蚀变斜长角闪岩 0. 4m
19. 灰色斜长角闪岩，层间有岩脉穿插 11. 7m
18. 深灰色角闪斜长片麻岩 6. 2m
17. 灰略带肉红色花岗闪长玢岩脉 (0. 9m)
16. 深灰—灰黑色斜长角闪岩 12. 7m
15. 灰微带肉红色糜棱岩化花岗岩脉（?） (2. 4m)
14. 深灰色斜长角闪岩 1. 7m
13. 灰微带肉红色糜棱岩化花岗岩脉（?） (21. 4m)
12. 灰、灰黑色斜长角闪岩 0. 5m
11. 灰微带肉红色糜棱岩化花岗岩脉（?） (2. 2m)
10. 灰、灰黑色斜长角闪岩 10. 3m
9. 灰略带肉红色糜棱岩化花岗岩脉（?） (1. 6m)
8. 灰、深灰色斜长角闪岩 0. 4m
7. 灰、浅灰色闪长玢岩脉 (4. 3m)
6. 灰、深灰色角闪斜长片麻岩 7. 4m
5. 长英质碎斑岩脉 (3. 1m)
4. 角闪斜长片麻岩 1. 4m
3. 长英质碎斑岩脉 (3. 8m)
2. 灰黑色斜长角闪岩 2. 1m
1. 长英质碎斑岩脉 (3. 7m)

未　见　底
~~~~~~~~

3. 赵 zk1 孔剖面

上覆地层：金山群 b 岩性段绿帘阳起片岩、斜长片麻岩

———— 整　合 ————

金山群 a 岩性段（$Pt_{1-2}J^a$）		总厚度>119.7m
30~37.	灰白色硅化白云母浅粒岩	8.1m
29.	灰色白云母石英片岩	6.6m
28.	蚀变辉绿岩脉	(3.0m)
27.	石英岩	0.8m
26.	灰色白云母片岩	2.2m
25.	白云母长石石英岩	3.0m
24.	蚀变辉绿岩脉	(1.8m)
23.	绢云母片岩	7.0m
22.	灰白色浅粒岩	2.5m
21.	白云母长石石英岩	0.8m
20.	蚀变辉长辉绿岩脉	(3.1m)
19.	浅粒岩	3.4m
18.	绿泥长石石英岩	2.2m
17.	白云母长石石英岩	0.9m
16.	白云母石英片岩	1.3m
15.	深灰、绿色磁铁矿斜长绿泥片岩	0.1m
14.	变糜棱岩化浅粒岩	7.8m
13.	蚀变闪长玢岩脉	(3.8m)
12.	糜棱岩化石英岩，层间有蚀变辉长辉绿岩脉穿插	3.1m
11.	深绿色磁铁矿阳起片岩	1.8m
10.	斜长绿泥片岩	1.2m
9.	含磁铁长石阳起片岩	12.8m
8.	白云母石英岩	6.4m
7.	磁铁斜长绿泥片岩	3.6m
6.	深灰绿色白云母石英岩	3.3m
5.	深绿色磁铁斜长绿泥片岩	2.1m
4.	白云母石英岩，层内穿插有 2 条蚀变辉绿岩脉	37.9m
3.	碎裂绢云母石英片岩	0.4m
2.	角闪石闪长玢岩脉	(0.3m)
1.	碎裂绢云母石英岩	>0.5m

未　见　底

（四）地质特征及区域变化

金山群主要见于闵行华泾，浦东三林塘、天花庵、赵家宅、康家宅、沈庄以及金山朱泾、吕巷等地钻孔中，在浦东瓦屑和横沔、奉贤烟墩头和肖塘等地钻孔中亦有所见。

金山群为一套变质岩系，主要岩性为云母片岩、绿泥片岩、长石石英岩以及斜长角闪岩、斜长角闪片麻岩、角闪片岩、角闪斜长片麻岩、浅粒岩等。按其岩性可划分成 3 个岩性段（即 a、b、c 岩性段），钻孔揭露厚度一般大于 260m，未见顶、底。

c 岩性段：以灰白色白云母片岩、云母石英片岩以及绢云母石英片岩等为主，夹绿帘钠长阳起片岩、钠长绿泥片岩及厚度不一的含石英大理岩（在三林塘—康家宅地段有增多加厚趋势）。大理岩呈透镜状产出，并普遍具有条带状构造。已见钻孔揭露厚度在百米以上。

b 岩性段：由灰黑、深灰及绿黑色斜长角闪岩、绿帘斜长角闪岩、含磁铁角闪岩以及灰黑、灰绿色的角闪斜长片麻岩、角闪二长片麻岩、二云斜长片麻岩等组成，夹有黑色角闪片岩、长石石英岩与灰白色云母浅粒岩。具有色暗，富含角闪石和斜长石之特征，钻孔揭露厚度亦在百米以上。

a 岩性段；该地层仅在周浦赵家宅一带钻孔中发现，已见钻孔揭露厚度大于 150m（未见底）。上部主要为灰白色白云母浅粒岩、含磁铁绢云母片岩、黑云片岩及白云母石英岩；下部由白色大理岩、结晶灰岩夹阳起片岩与绿泥片岩等组成。

二、惠南板岩（$Pt_3 h$）

（一）创名及原始定义

该组于 1988 年编著《上海市区域地质志》时所创建，命名地点在惠南镇南汇中学。原始定义为一套成分很单一的钙质绢云母板岩，变质程度低于金山群。

（二）现在定义

惠南板岩为一套成分很单一的钙质绢云母板岩，推测角度不整合于金山群之上。

（三）层型剖面描述

正层型剖面为浦东惠南南汇中学地震观察孔剖面（见图 1-2）

上覆地层：中新统白龙港玄武岩

═══════ 断　层 ═══════

惠南板岩（Pt_3h）　　总厚度>194. 39m

浅绿、棕红色等，构造角砾岩。岩芯松散、破碎，方解石脉、石英脉穿插　　24. 14m

灰、深灰色钙质绢云母板岩。产微古植物：*Laminarits* sp. （少量）、*Leiominuscula minuta* Naum、*Margominuscula* sp. 、*Lophominuscula* sp. 、*Asperatophosphaera umishanensis* Sin et Lin、*Margominuscula* aff. *tennela* Naum、*Paleamorpha* sp. 、*Asperatopsophosphaera umishanensis* var. *minar* Sin et Lin、*Leiopsophosphaerasolida* Sin et Lin、*Dictyosphaera sinica* Sin et Lin、*Dictysphaera* sp. 、*Asperatopsophosphaera* sp. 等　　170. 25m

未　见　底

（四）地质特征及区域变化

该套地层仅见于浦东惠南的一个钻孔剖面，这套板岩的主要特征是：①色浅，以灰色为主，岩石成分极为单一，以绢云母为主，含少量的石英粉砂；②变质程度浅，局部可称之“板状泥岩”（或钙质泥岩）。

三、河上镇群（$Pt_3 H$）

（一）创名及原始定义

由浙江省区域地质调查队（1990）所创，命名地点在浙江省萧山市（现称萧山区）西南河上镇。原始定义为一套海相复理石建造和陆相火山岩建造，岩性为岩屑砂岩、粉砂质泥岩、泥岩和中基性熔岩、中酸性火山碎屑岩。自下而上划分为骆家门组、虹赤村组和上墅组。

（二）现在定义

河上镇群为一套海相复理石和陆相火山岩建造，岩性为岩屑砂岩、粉砂质泥岩、泥岩和中基性熔岩、中酸性火山碎屑岩。自下而上划分为骆家门组、虹赤村组和上墅组。

（三）层型剖面描述

本市次层型剖面为金山钱圩 502 孔剖面（见图 1-2）

上覆地层：第四系

~~~~~~~~~~ 不　整　合 ~~~~~~~~~~

河上镇群（Pt_3H）		总厚度>466.83m
20.	暗绿色绿泥石化安山质熔结凝灰岩，局部为含砾凝灰岩，含少量角砾，砾石成分为凝灰质砂岩、次生石英岩，少量杏仁构造	14.1m
19.	土黄色含砾凝灰岩，角砾较多	4.92m
18.	灰微带棕红色安山质熔凝灰岩，含少量角砾	5.73m
17.	暗绿色绿泥石化安山质熔凝灰岩，含角砾，呈半圆状	2.30m
16.	灰微带棕红色安山质熔凝灰岩，含少量角砾，多呈棱角状	22.84m
15.	暗绿色或灰微带青色绿泥石化安山质熔凝灰岩，显绿泥石化，含角砾	6.44m
14.	灰微带棕色安山质熔凝灰岩，含角砾	6.09m
13.	绿微带灰色凝灰岩，碳酸盐化、绿泥石化、黄铁矿化	9.07m
12.	深灰色安山质熔凝灰岩，局部为凝灰岩，显轻微绿泥石化，下部绿泥石化、碳酸盐化强烈，见少量角砾	32.07m
11.	暗绿色凝灰岩，绿泥石化、碳酸盐化和黄铁矿化，岩石破碎	6.97m
10.	深灰色含砾晶屑岩屑凝灰岩，见少量角砾，绿泥石化、碳酸盐化强烈，局部少量黄铁矿化	31.07m
9.	灰、浅灰色凝灰质砂岩，局部为粉砂质泥岩和泥质粉砂岩，下部见极少量黄铁矿化，倾角30°~45°	60.74m
8.	以浅灰色为主，顶部微带棕色，245m左右呈深灰—灰黑色的凝灰质粉砂岩，具少量黄铁矿化，底部见黄铜矿化	2.11m
7.	灰、浅灰—灰白色凝灰质砂岩和含砂泥岩，局部黄铁矿化和黄铜矿化	81.84m
6.	浅灰—灰白色、局部深灰色、中部带微棕色的长石石英砂岩、凝灰质砂岩，局部为粉砂质泥岩，局部黄铁矿化、绿泥石化，底部夹煌斑岩脉	135.67m
5.	灰色，斑点状凝灰质硅质泥岩，裂隙内见黄铁矿化	5.00m
4.	浅灰—灰白色，局部微带棕色的长石石英砂岩，底部轻微绿泥石化、角岩化	10.21m
3.	浅灰色长英质角岩，夹数层厚1m左右的绢云母化花岗斑岩岩脉	21.53m
2.	灰绿色辉绿岩，显绿泥石化	（0.65m）
1.	浅灰微带棕红色花岗岩，底部夹0.23m厚的灰黑色煌斑岩脉	61.47m

未　见　底

（四）地质特征及区域变化

该套地层仅见于金山钱圩的一个钻孔剖面，其主要岩性特征是上部为以灰、绿色为主的火山碎屑岩建造，岩性以熔结凝灰岩为主，下部为灰、浅灰色的陆源碎屑岩建造，以含凝灰质砂岩、长石石英砂岩为主，少量含砂泥岩和泥质粉砂岩等，普遍见黄铁矿化。

第二节　沉积作用和沉积建造

前震旦系主要以变质岩系为主，变质岩岩石类型见变质岩章节描述。变质岩原岩为海洋环境正常沉积的砂、泥质碎屑岩夹碳酸盐岩以及火山-火山碎屑岩等建造。此外，在河上镇群见有凝灰质砂岩、凝灰质粉砂岩、长石石英砂岩、粉砂质泥岩、泥质粉砂岩和含砂泥岩等陆源碎屑岩。

一、沉积类型

古元古代—青白口纪，区内沉积作用以海相沉积为主，主要的沉积类型为滨海沉积、浅海沉积、次深海沉积等。

（一）滨海沉积

滨海沉积主要发育于河上镇群的下部，岩石组合以粗粒的陆源碎屑岩为主，夹少量粉砂质泥岩、
~~~~~~~~~~

泥质粉砂岩等细粒碎屑岩。局部纹层特征清楚，表现出韵律层理。

（二）浅海沉积

浅海沉积主要发育于金山群，岩石组合为砂质、泥质-碳酸盐岩类。

（三）次深海沉积

次深海沉积主要发育于惠南板岩，岩石组合为一套岩性单一的泥岩或粉砂质泥岩。

二、主要岩石类型

（一）长石石英砂岩

浅灰—灰白色，中粗粒砂状结构，块状构造，长石含量为30%~35%，石英约为40%，胶结物为25%~30%。胶结物由长石、石英和黏土矿物组成。

（二）凝灰质砂岩

浅灰、灰白色，块状构造，局部层理构造，长石含量为30%~40%，石英为3%~12%，胶结物由脱玻化的隐晶质长石、石英和次为类云母的黏土矿物组成，占40%~55%。

（三）凝灰质粉砂岩

浅灰、灰白色，块状构造，局部层理构造，长石含量为15%~25%，石英为8%~12%，胶结物由脱玻化的隐晶质长石、石英和次为类云母的黏土矿物组成，占60%~65%。常有不同程度的硅化现象。

（四）含砂泥岩

浅青灰色，纹层发育，韵律层理，具有弱变质，砂级碎屑颗粒成分主要为石英和白云母，石英碎屑颗粒含量约为15%，呈定向排列，基质为泥质，形成纹层。

第三节　古生物群组合和地层时代

截至目前，本区仅在惠南板岩中发现较多的微体古生物化石，共计10个属种。主要有 *Leiominuscula minuta* Naum（小型光面小球藻）、*Margominuscula* sp.（厚缘小球藻未定种）、*Lophominuscula* sp.（瘤面小球藻未定种）、*Leiopsophosphaerasolida* Liu et Sin（坚壁光球藻）、*Asperatopsophosphaera umishanensis* Sin et Liu（雾迷山糙面球形藻）、*Paleamorpha* sp.（膜片藻未定种）、*Margominuscula* aff. *tennela* Naum（薄壁后缘小球藻近似种）、*Leiopsophosphaera* sp.（光球藻未定种）、*A. umishanensis* var. *minor* Sin et Liu（小雾迷山糙面球形藻）、*Dictyosphaera macroreticulata* Sin et Liu（大网球藻）。

早期，根据部分属种与蓟县系层型剖面中所含属种相似的特征，认为惠南板岩形成于长城纪晚期—蓟县纪早期（上海市地质矿产局，1988）。目前看，该组合中当时认为时代较早的大多数分子都可延续到青白口纪早期到中期，甚至更新，且该组合中的很多属种在邻近江西的双桥山群中都有发现（文子才，2001）。近年来，在邻区上述浅变质岩中的一系列高质量锆石U-Pb年龄数据已经证明相关地层为青白口纪早期沉积。

河上镇群在浙西北地区被归入新元古代—青白口纪晚期。本志在区内钱圩502孔160~170m的含砾晶屑岩屑凝灰岩、罗店西南JD7-1孔328~339m的灰色岩屑晶屑凝灰岩中分别获得774±12Ma、801±13Ma的锆石U-Pb年龄。

20世纪80—90年代，针对金山群，相继开展了全岩Rb-Sr、全岩Sm-Nd、角闪石Ar-Ar测年（上海市地质矿产局，1988；周新华，1993；许保桐，1995），Rb-Sr、Sm-Nd等时线年龄为1123±27Ma、1364±113Ma，t_{DM}在1826~1941Ma，角闪石Ar-Ar视年龄为1259±30Ma、585.3±15Ma、496±13Ma、379.9±10Ma、303.5±8.4Ma、238.2±6.7Ma、223.5±6.3Ma、190.3±9.0Ma、140.1±4.1Ma、124.4±3.6Ma、118.2±3.4Ma、112.2±3.3Ma。本次在周浦康zk1孔约480m和康zk4孔约630m的斜长

角闪岩、糜棱岩化花岗斑岩中分别获得 1078±8.9Ma、1781±16Ma 的锆石 U-Pb 年龄，在朱 zk4 孔约 423m 中获得大量古元古代年龄。后者年龄与江西东南部天井坪组斜长角闪岩的锆石 U-Pb 年龄（1766±19Ma）相当（李献华，1998）。另外，根据金山群变质程度明显高于惠南板岩的特征推断，金山群的地层年代应早于惠南板岩。综上所述，将本区金山群地层年代划归为古—中元古代。

第四节　区域地层对比

一、区域对比

区内古—中元古界、青白口系分布局限，且揭露厚度有限。根据前述微体古生物组合和同位素年龄特征，将区内金山群、惠南板岩、河上镇群分别划为古—中元古界、青白口系下部和青白口系上部，与浙西北和赣东北、皖南地区的前南华纪地层具有可对比性（表 1-2）。

表 1-2　区域地层对比表

Table 1-2　Regional stratigraphic correlation

地区		上海	浙江 浙西北	浙江 浙东南	皖南 祁门	皖南 歙县	赣北 万年	赣北 鄣公山	赣北 九岭	赣北 信江	赣北 德兴	湘北—鄂东南
上覆地层		灯影组	蓝田组/陡山沱组	乌灶组	蓝田组	蓝田组	华山岭组	黄龙组	蓝田组	陡山沱组		陡山沱组
新元古代	南华纪	?	雷公坞组 休宁组/志棠组	陈蔡俯冲增生杂岩	雷公坞组 休宁组	雷公坞组 休宁组			南沱组 莲沱组	南沱组 大塘坡组 古城组 莲沱组		南沱组 大塘坡组 古城组
新元古代	青白口纪	河上镇群 推测	河上镇群：上墅组 虹赤村组 骆家门组		沥口群：小安里组 铺岭组 邓家组 葛公镇组	昱岭关群：井潭组 周家村组			落可岽组(汉阳峰组)(马涧桥组)	广丰群：听门组 罗村组；桃源组 翁家岭组	登山群：叶家组 拔竹坑组 高山组	渫水河组 张家湾组；板溪群
新元古代	青白口纪	惠南板岩 推测	双溪坞群：章村组 岩山组 北坞组 平水组		溪口岩群：牛屋岩组 木坑岩组 板桥岩组 樟前岩组(未见底)	歙县岩群：昌前岩组 西村岩组	万年群：程源组 牛头岭组 枫树岭组(未见底)	溪口岩群：牛屋岩组 木坑岩组 板桥岩组(未见底)	双桥山群：修水组 安乐林组 记林组 横涌组(未见底)		F 张村岩片	冷家溪群：大药姑组 小木坪组 黄浒洞组 雷神庙组(未见底)
中元古代		金山群(未见底)								田里岩组(未见底)	铁沙街岩组(未见底)	
古元古代				八都群								

金山群长期以来被认为与浙东南原陈蔡群相当。近年，在原陈蔡群中相继获得比较复杂的锆石 U-Pb 年龄信息，主要有角闪岩 1781±21Ma（Li et al.，2010），黑云斜长片麻岩 848±10Ma～431.4±7.8Ma（高林志，2014），黑云斜长片麻岩 2.4～2.0Ga、1848±23Ma、1072～787Ma、220±3Ma（王存智，2015），斜长角闪岩 420.6±1.8Ma（王存智，2016），认为其为一套早古生代晚期形成的俯冲增生杂岩。金山群的锆石年龄特征与之差别较大，未获得新元古代以来的原岩形成时代信息。从岩石组合上看，金山群的变质程度较陈蔡俯冲增生杂岩低，混合岩化特征不明显，且变粒岩较少（仅占全部揭露岩芯长度的 0.7%左右）。位于萍乡—绍兴断裂带北侧的赣东北上饶信江发育有中元古界田里岩组和铁沙街岩组，田里岩组以青灰、灰绿和灰白色云母片岩、石英片岩、石英岩为主，夹大理岩和变石英砂岩等。锆石年龄有最年轻的碎屑锆石 SHRIMP U-Pb 年龄为 1535Ma（江西省地质矿产勘查开发局，2017）。铁沙街岩组以千枚岩和变流纹岩、变细碧岩和变角斑岩为主，夹大理岩和变粉砂岩等，锆石 SHRIMP U-Pb 年龄为 1172.3±9.7Ma～1132±8Ma（高林志，2013）。所以，从岩石组合和年代特征，推断金山群似可与田里岩组和铁沙街岩组对比，受断裂构造影响，变质程度稍深。

惠南板岩在区域上似可与江南隆起带东部双溪坞群、溪口岩群、双桥山群和冷家溪群等相对比。

主要为片理化中基性—酸性火山岩-火山碎屑岩、变质沉积岩、凝灰质沉积岩夹砂砾岩、砂岩和泥岩等，局部有片岩和变火山岩。扬子地层区南缘广泛分布的这套以复理石沉积为主的浅变质岩系无论从地层层序还是地质年龄上均表现出相同的属性。

河上镇群在区内层位与浙西北地区中上部层位相当，未进行细分。浙西北骆家门组和虹赤村组以岩屑砂岩、粉砂质泥岩、泥岩为主，夹基性火山岩；上墅组为中基性火山岩和火山碎屑岩、酸性火山岩夹酸性火山碎屑岩，在流纹岩和斑脱岩中获得的锆石 U-Pb 年龄为 767±5Ma（高林志，2008）。皖南沥口群和昱岭关群、广丰群等上部都发育一套滨海或陆相沉积-火山岩系，其层位与区内相当。井潭组凝灰岩和英安岩中获得的锆石 U-Pb 年龄分别为 779±7Ma～773±7Ma、835±9Ma～820±16Ma。

二、存在问题

（一）金山张堰和朱行等地钻孔揭露地层的归属

金山张堰和朱行等地钻孔的岩石组合与吕巷、周浦、三林等地差别较大。张堰为接触热变质和热液变质岩石组合，主要岩性为角闪石斜长角岩、黑云母斜长角岩、透辉石斜长角岩、黑云母堇青石角岩以及云母斜长角岩、矽卡岩等，并夹有厚薄不一的大理岩。局部有页岩、片岩，靠上部地层尚有长英质角岩。朱行主要为变流纹岩、变安山岩、变熔结凝灰岩和长英质板岩等，局部为角岩类，部分层位呈现明显的片状构造和糜棱结构。

长期以来，大多数成果都将其划为金山群，也有部分成果划为下古生界。本志在朱 zk4 和戚家浜 11201 孔中发现了大量 2700～1800Ma 的年龄记录，明显与北部金山群年龄特征不同。此外，在界山 1701 孔发现较多 *Dictyolidium* sp.（网面藻未定种）、*Cymatiogalea* sp.（波口藻未定种）、*Rhopaliophara* sp.（腔突藻未定种）和 *Eliasum* sp.（伊里藻未定种）等，似应归属早奥陶世。本志暂将其按照金山群处理，有待进一步研究解决。

（二）河上镇群的划分和对比

长期以来，钱圩 502 孔上部、查山 ck1 孔等地层被划为黄尖组下段。本次基于地质年代和区域对比，将钱圩 502 孔地层厘定为河上镇群，但未进行组级别地层划分。根据区域上的河上镇群层序，也不排除区内惠南板岩为河上镇群下部层位。

第五节　岩相古地理

鉴于上海已发现的金山群、惠南板岩、河上镇群，地层剖面不完整，分析资料有限，不可能对当时、当地的岩相古地理环境做出完整的分析，目前只能根据现有的零星资料和邻区岩相古地理演化认识，对其沉积条件做一概略推测。

中元古代主要为具有类复理石建造特征的浅海相沉积，伴随有间歇性的（海底）火山喷发活动。由早期向晚期，海水似有加深趋势，并且岩浆作用似有加剧特征。

青白口纪早期，海水逐渐加深，属次深海沉积环境；晚期，海水逐渐退缩，逐渐转化为滨浅海—陆地环境，其间火山活动逐渐加剧。

第二章 震旦系

本区震旦系主要分布于浦东坦直、三灶一带（图 2-1），在宝山罗店、大场、江湾等地钻孔中也有发现。根据区域重力场和邻区地层分布特征，推断在崇明港东一带也可能存在。整体上为一套蚀变较强烈的以灰岩、白云质灰岩、白云岩为主的地层。

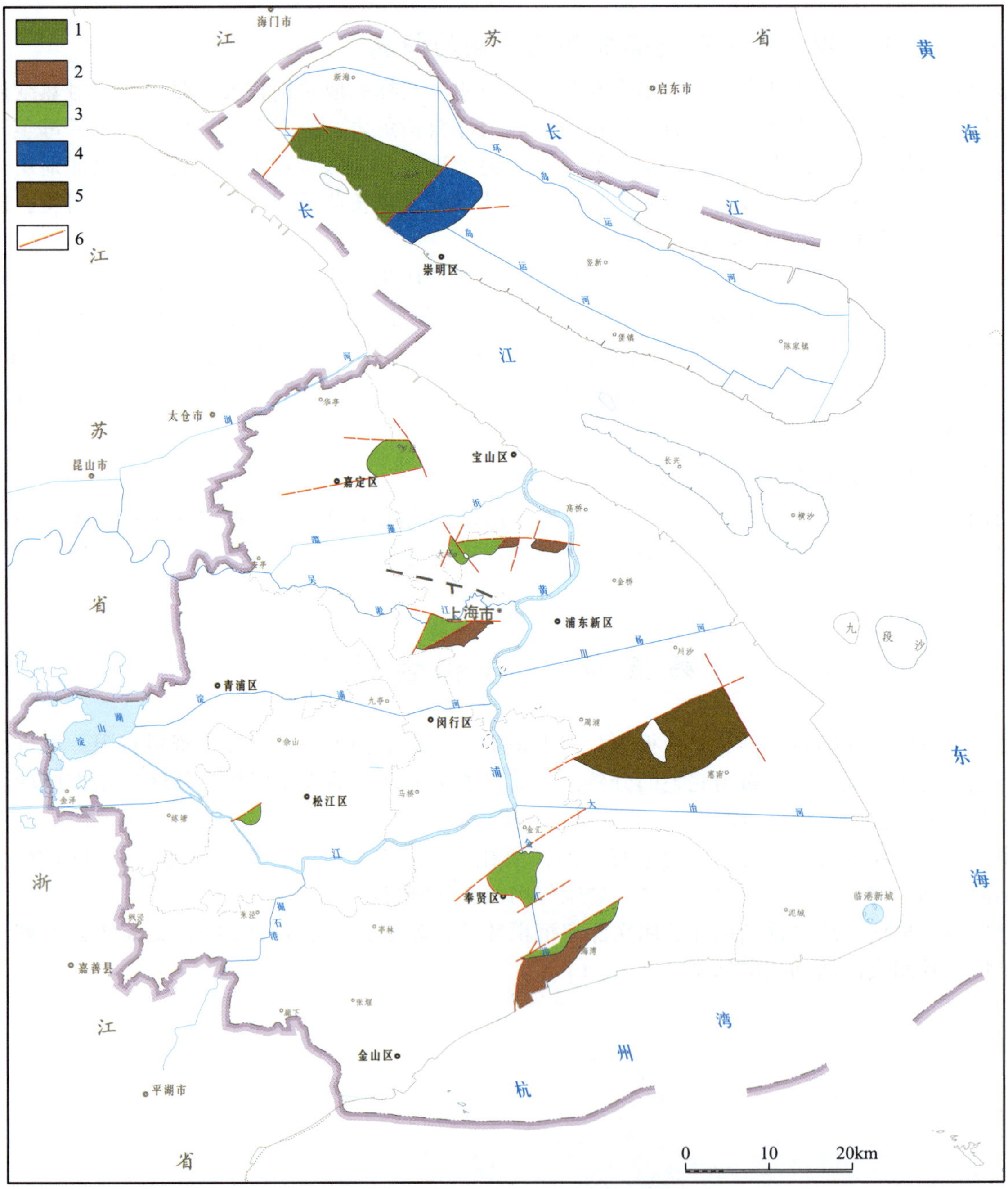

图 2-1　上海震旦系与下古生界分布图

Fig. 2-1　Sketch map of the distribution the Sinian and Lower Paleozoic Systems in Shanghai

1—志留系；2—奥陶系；3—寒武系；4—震旦系—奥陶系；5—震旦系；6—断裂

1962 年《上海地区矿产地质普查工作总结报告》中将本区震旦系划分为上统及下统：上统指见于坦直钻孔，并首次创建“坦直组”，当时曾将其与浙西地区的上震旦统西峰寺组对比；下统见于江湾钻孔，称为“雷公坞组”。

1966 年出版的《上海地区石油地质图集》、1979 年编制的《1∶20 万上海地区基岩地质推断图》中基本上维持以上观点，唯对是否存在下震旦统，即相当于浙西地区雷公坞组、志棠组有疑，而不予采用。又在宝山大场、江湾一带深孔内，于下寒武统之下发现一套含硅、镁质碳酸盐岩地层，其间虽未发现化石，认为其层位与岩性特征，大体可与邻区上震旦统对比。

1988 年编制《上海市区域地质志》时，将上述地层重新命名为灯影组（表 2-1）。此后，该划分方案沿用至今。

表 2-1　上海震旦系岩石地层单位划分沿革表

Table 2-1　Previous divisions of the Sinian lithostratigraphic units in Shanghai

<table>
<tr><th>地层</th><th>《上海地区矿产地质普查工作总结报告》，1962</th><th>《上海地区石油地质图集》，1966</th><th>《1∶20 万上海地区基岩地质推断图》，1979</th><th>《上海地区地层概述》，1981</th><th>《上海市区域地质志》，1988</th><th>“上海市三维城市地质调查”，2010</th><th>本志</th></tr>
<tr><td>上覆地层</td><td>寒武系</td><td>?</td><td>寒武系</td><td>寒武系</td><td>寒武系</td><td>寒武系</td><td>底—下寒武统</td></tr>
<tr><td>上震旦统</td><td>坦直组</td><td rowspan="2">坦直群</td><td rowspan="2">坦直组</td><td>西峰寺组</td><td>灯影组</td><td>灯影组</td><td>灯影组</td></tr>
<tr><td>下震旦统</td><td>雷公坞组/志棠组</td><td>?</td><td>?</td><td>?</td><td>?</td></tr>
</table>

第一节　岩石地层划分

灯影组（Z_2d）

（一）创名及原始定义

灯影组原称之“灯影石灰岩”，是由李四光等（1924）创名，命名地点在湖北宜昌灯影峡。其原义是：“上震旦统灯影石灰岩”，白色块状呈峭壁形的石灰岩，多少带白云质；在风化表面上有坚硬的硅质夹层突出。“在很多地方包含特殊而且成群的层状体，有时呈似圆筒状，但更多的是不对称圆锥状。”是震旦纪浅水中繁殖的钙质藻类遗骸，特地命名为 *Collenia cylindrica*、*Colleniaangulata*。灯影石灰岩，从灯影峡下口的寒武纪页岩下突起。

（二）现在定义

岩性以白云岩或藻白云岩、泥质白云岩为主，夹白云质灰岩、硅质泥岩等。含叠层石和微古植物化石。与下伏陡山沱组紫红色含钾粉砂质泥岩呈整合接触，与上覆寒武系荷塘组石煤层之下的磷块岩或皮园村组硅质岩呈平行不整合或整合接触。

（三）层型剖面描述

次层型剖面为宝山大场少年村路岩溶 8#孔、岩溶 16#孔和坦 308 孔剖面（图 2-2）。剖面地层分述如下：

1. 岩溶 8#孔剖面

上覆地层：下寒武统超山组黑色碳质页岩夹灰岩

———— 整　合 ————

灯影组（Z_2d）　　总厚度>72.9m

7.　深灰色硅质岩，具同生角砾，顶部地层较破碎　　20.9m

6. 灰黑色（主）与灰白色碎裂灰岩。近底部有硅质交代现象似硅质层　　8.3m
5. 浅灰白色细晶白云岩　　6.9m
4. 蚀变岩脉　　（8.3m）
3. 深灰略带灰绿色细晶质灰岩　　6.7m
2. 灰绿色透辉石化硅质灰岩　　7.2m
1. 灰黑—深灰色中粗粒灰岩，夹白云质灰岩薄层　　>22.9m

未　见　底

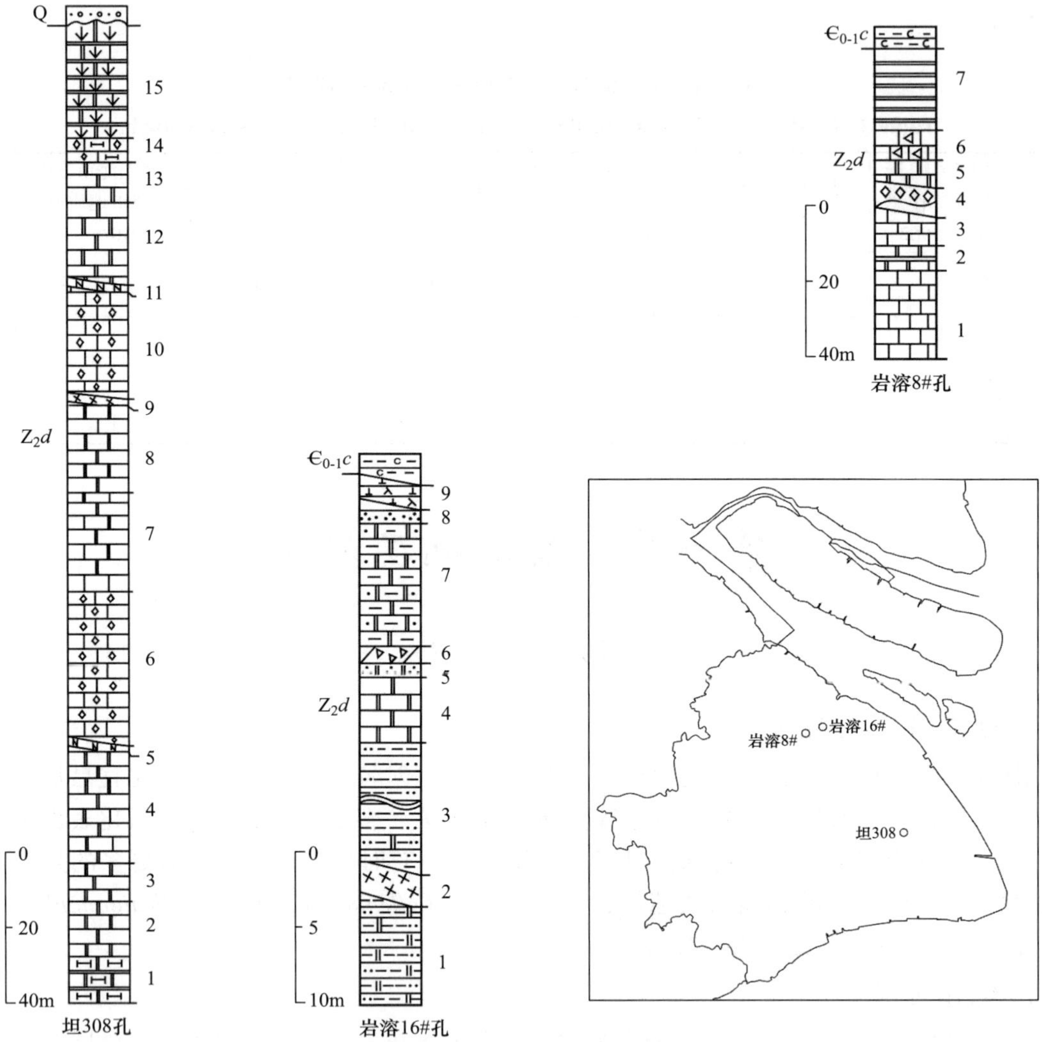

图 2-2　典型钻孔柱状图

Fig. 2-2　Typical borehole histograms

2. 岩溶 16#孔剖面

上覆地层：下寒武统超山组灰黑、黑色碳质粉砂质页岩及碳质页岩

---------------　不　　明　---------------

灯影组（Z_2d）　　总厚度>63.2m

9. 绢云母化闪长玢岩脉　　（1.6m）
8. 灰黑色石英砂岩　　0.6m
7. 灰黑色含粉砂泥质白云岩　　8.1m

6. 蚀变辉绿岩脉，呈角砾岩状 （1.0m）

5. 灰色白云质石英砂岩 0.8m

4. 灰、灰白色碎裂微晶白云岩 4.0m

3. 灰黑色白云质泥岩，含细砂、粉砂，上部夹有数层薄层页岩 43.1m

2. 蚀变辉绿岩脉 （2.3m）

1. 灰黑色白云质泥岩，含细砂、粉砂 >6.6m

未 见 底

3. 坦 308 孔剖面

上覆地层：第四系

~~~~~~~~~ 不 整 合 ~~~~~~~~~

灯影组（Z_2d） 总厚度>260.6m

15. 灰—灰白色蛇纹石化大理岩，含白云石，中、下部夹黑色炭质条带 >28.1m

14. 灰黑—灰白色透辉石化结晶灰岩 5.8m

13. 灰黑—灰白色白云质灰岩，含硅质，并构成黑白相间的条带构造 10.1m

12. 灰白夹灰黑色钙质白云岩 21.1m

11. 蚀变斜长岩脉 （1.3m）

10. 灰白夹灰黑色结晶灰岩，含白云石 28.0m

9. 闪长岩脉 （0.8m）

8. 灰白—灰黑色钙质白云岩 22.4m

7. 灰白色白云质灰岩，局部呈灰色 37.9m

6. 浅灰白—灰白色结晶灰岩 41.9m

5. 绿帘石化斜长岩脉 （0.7m）

4. 浅灰白色、白色钙质白云岩 30.5m

3. 浅灰—灰白色白云岩 11.3m

2. 灰黄、肉红色白云岩，破碎 10.3m

1. 浅灰白—灰白色透辉石大理岩 >13.2m

未 见 底

（四）地质特征及区域变化

岩性以灰白色白云岩、灰色灰岩与灰黑色白云质灰岩为主。宝山大场一带的灯影组顶部还见有硅质层，其间夹有泥质白云岩、白云质泥岩及薄层石英砂岩等。坦直—三灶等地钻孔中的一套浅灰—灰白色厚层状白云岩、钙质白云岩、灰岩与白云质灰岩，局部因脉岩穿插或受岩体的影响，部分已蚀变成结晶灰岩和蛇纹石化大理岩。钻孔揭露厚度在 60m（大场一带）~260m（坦直）。未见底，与下伏地层接触关系不明，与上覆寒武系则呈整合接触。

总体上，具有下列诸特征：①以含硅的镁质碳酸盐岩为主；②普遍具有不明显的纹层理构造；③含磷。据岩石化学分析，在宝山大场岩溶 8#孔中发现的碳酸盐岩地层，其 P_2O_5 含量可达 0.65%~1.13%，而区内其他时代的碳酸盐岩地层中 P_2O_5 含量的平均值仅为 0.01%~0.03%。

第二节 沉积作用和沉积建造

一、沉积类型

震旦系为海相沉积，主要为一套碳酸盐岩建造，顶部或局部夹有少量陆源碎屑岩。主要的沉积类
~~~~~~~~~

型有滨海沉积、浅海沉积等。

（一）滨海沉积

滨海沉积主要为台地边缘浅滩、局限海台地和蒸发岩台地，分布于上海坦直—六灶以南地区，岩石组合以白云岩和钙质白云岩为主，夹少量白云质灰岩和白云质泥岩。厚层状，含有石英等碎屑颗粒。

（二）浅海沉积

浅海沉积主要为开阔海台地相，分布于上海市区和青浦、宝山等地，岩石组合以白云质灰岩、灰岩为主，顶部为硅质岩。局部发育层理构造。

二、主要岩石类型

上述沉积环境中形成的主要岩石类型有白云岩、泥质白云岩、白云质灰岩、灰岩、白云质泥岩、石英砂岩等。主要特征如下：

（一）白云岩

白云岩呈浅灰、白色，不等粒变晶细粒结构，致密块状，局部具有层理构造，大部分由微—细晶白云石组成，占98%，粒径为0.02~0.05mm，自形或半自形微粒，常有菱形断面，有很少量的褐铁矿、高岭石和黄铁矿。

（二）泥质白云岩

泥质白云岩呈灰黑色，粒屑结构，块状构造。黏土矿物含量占35%，主要为水白云母、高岭石，碳酸盐矿物占63%，主要为白云石，石英约2%，少量的磁铁矿、黄铁矿、云母、锆石和磷灰石。白云石按形态分为碎屑状、泥晶集合体，局部重结晶现象明显。

（三）白云质灰岩

白云质灰岩呈灰白色，不等粒结构，块状构造。方解石含量占70%，不规则状或云雾状，质不纯。白云石约占20%，粒径为1~2.5mm，分布不均。绢云母和绿泥石约占7%。见少量不规则状石英、黄体矿、黏土矿物等。

（四）灰岩

灰岩呈黑、灰、灰黑、深灰色，多为粉—细晶结构、中—粗晶结构，局部水平纹层层理构造。主要由方解石组成，含少量石英、绢云母和泥质团块、褐铁矿化的黄铁矿等。粒径在0.02mm左右、0.05mm左右和0.05~0.2mm，见显微缝合线构造。

（五）白云质泥岩

白云质泥岩呈灰黑色，局部含砾，泥状结构，基底式、孔隙式胶结。碎屑含量占30%~50%，局部砾石含量可占5%~20%，泥质含量可高达70%。大于0.1mm的碎屑主要为白云质灰岩、砂岩和泥岩等岩屑，0.03~0.08mm的碎屑以石英为主，另有云母和碳酸盐矿物。以次棱角—半磨圆状为主。碳酸盐矿物部分重结晶，以白云质为主。

（六）石英砂岩

石英砂岩呈灰、灰黑色，砂状结构，以孔隙式胶结为主，局部为接触式胶结。碎屑含量为65%~80%，胶结物为20%~35%。局部胶结物中含碳酸盐矿物，为白云质石英砂岩。碎屑成分以石英为主，其次为黏土化的长石或岩屑、白云母，少量的黄铁矿、锆石、磷灰石、电气石等。部分层位石英大多具有次生加大边结构，长石蚀变严重。粒径多在0.08~0.2mm，少数为0.2~0.5mm或小于0.05~0.1mm。以半磨圆—次棱角状为主，个别磨圆状。胶结物以泥质为主，含少量的硅质、碳质和碳酸盐矿物。

（七）硅质岩

硅质岩呈灰、灰黑色，微晶结构，块状构造，主要由隐—微晶石英组成，粒度一般小于0.01mm，成分和结构比较均匀，有很少量的微体化石骨屑。局部碳酸盐化。石英含量为82%~88%，方解石为5%~15%，黄铁矿为2%~3%，个别层位明显含有机质。

（八）灰质白云岩

灰质白云岩呈灰白、暗灰色，致密块状，嵌晶结构，由白云石及少量方解石、褐铁矿等组成。白云石自形，含量为70%~80%，直径为0.25mm×0.11mm，聚片双晶发育，晶形完整；方解石含量为20%~30%，直径在0.25mm左右，他形。见少量磁铁矿（褐铁矿）和极微量的石英。

第三节　地层时代和区域地层对比

一、地层时代和区域对比

整体上，从地层层序、岩性组合或构造特征来看，区内震旦系均可与江苏、浙江、江西等地的上震旦统对比，目前所见层位可能仅相当于灯影组的中、上部。与皮园村组、板桥山组、灯影组岩性组合大体相同或相似，为同期异相沉积。浙西北地区，以余杭泰山—开化一线为界，板桥山组向西北厚度变薄，砂岩逐渐尖灭，白云岩逐渐为灰岩替代，泥质岩增多；向东南地层厚度也变薄，砂岩和砂质迅速减少，白云岩明显增多，逐渐相变为灯影组（表2-2）。区内灯影组岩石组合中，相对浙江灯影组，灰岩和白云质灰岩明显偏多，且在北部砂岩、泥岩等夹层明显增多，顶部出现硅质岩。这种组合特征更接近于板桥山组，且自南东向北西的演化特征与浙西北地区同期沉积的变化规律较为一致。

表2-2　区域地层对比表

Table 2-2　Regional stratigraphic correlation

<table>
<tr><td colspan="2" rowspan="2">地区</td><td rowspan="2">上海</td><td colspan="3" rowspan="2">浙西北</td><td colspan="2">安徽</td><td colspan="3">江西</td><td colspan="2">江苏</td><td rowspan="2">湘北</td></tr>
<tr><td>滁州（皖中）</td><td>休宁（皖南）</td><td>修水（赣西北）</td><td>上饶（赣东北）</td><td>德兴（赣东北）</td><td>南京</td><td>南通</td></tr>
<tr><td colspan="2">上覆地层</td><td>超山组</td><td colspan="3">超山组/荷塘组</td><td>荷塘组</td><td>荷塘组</td><td>荷塘组</td><td>荷塘组</td><td>荷塘组</td><td>荷塘组</td><td>超山组</td><td>牛蹄塘组</td></tr>
<tr><td rowspan="3">震旦系</td><td rowspan="2">上震旦统</td><td>灯影组</td><td rowspan="2">皮园村组</td><td rowspan="2">板桥山组</td><td rowspan="2">灯影组</td><td>皮园村组</td><td rowspan="2">皮园村组</td><td rowspan="2">灯影组</td><td rowspan="2">灯影组</td><td rowspan="2">皮园村组</td><td rowspan="2">灯影组</td><td rowspan="2">灯影组</td><td rowspan="2">皮园村组</td></tr>
<tr><td rowspan="3">?</td><td>灯影组</td></tr>
<tr><td>下震旦统</td><td>蓝田组</td><td colspan="2">陡山沱组</td><td>黄墟组</td><td>蓝田组</td><td>陡山沱组</td><td>陡山沱组</td><td>陡山沱组</td><td>黄墟组</td><td>黄墟组</td><td>金家洞组</td></tr>
<tr><td colspan="2">下伏地层</td><td colspan="3">雷公坞组</td><td>苏家湾组</td><td>雷公坞组</td><td>南沱组</td><td>南沱组</td><td>南沱组</td><td>苏家湾组</td><td>?</td><td>洪江组</td></tr>
</table>

二、存在问题

（一）坦直—三灶等地的地层归属

坦直—三灶等地的该套地层，本次沿用长期以来划分为灯影组的方案。但因其具条带状构造、纹层理构造以及原生角砾状构造等特征，早期曾提出该区地层似应与浙西地区的寒武系对比。上述两种认识的依据都略显不足，有待进一步深入研究。

（二）南华系—震旦系下部地层是否存在

新元古代中晚期以来，区域上发生被动陆缘裂解，整体发生沉降，开始南华纪—震旦纪的陆缘裂谷和滨浅海建造（徐磊，2012；何卫红，2014）。邻近赣浙皖地区的休宁组、南沱组、苏家湾组、周岗组、雷公坞组、志棠组、陡山沱组、蓝田组等都是南华纪—早震旦世建造的记录。从区域构造演化看，区内似乎应该存在相对应的沉积记录。

（三）崇明港西是否存在震旦系

崇明港西为一布格重力异常高值区，长期以来都被解释为志留系或下古生界，但该异常值与区内坦直基本相当。1995 年，江苏省地质矿产局第四地质大队在进行 1∶5 万区域地质调查时，将该异常的江苏段解释为震旦系—奥陶系。本次在港西推断存在震旦系，但还需要更多资料的验证。

第四节 岩相古地理

灯影组沉积时期，在宝山大场一带沉积了深灰、灰黑色白云岩，夹白云质石英砂岩与白云质泥岩薄层，近顶部处尚有硅质层沉积，总体上表现出海进沉积序列，似属台地边缘浅滩相—开阔台地相沉积；而坦直、三灶一带则以浅色白云岩为主，夹薄层白云质灰岩，可能为局限海台地—蒸发岩台地相沉积。

第三章 寒武系

寒武系为区内下古生界揭露最多的地层，目前已于宝山大场和罗店、长宁北新泾—天山五村、奉贤齐贤桥—钱桥、崇明东滩西侧、青浦重固等地的钻孔中发现（见图 2-1）。被揭露的地层层序也较齐全，自下而上可分为：底—下寒武统超山组、大陈岭组，中寒武统杨柳岗组，上寒武统超峰组。已揭露厚度为 872.5m，未见顶。除超峰组与其上、下层位尚未见有接触关系外，其他各组之间均为整合接触关系。

自 1958 年开始，已确定本区存在早古生代寒武纪地层；但由于缺乏生物化石依据和详细的区域对比资料，长期以来划分为中、上寒武统与下统荷塘组两部分，而且建组含义亦不确切。在编著江苏和上海区域地层表期间，曾认为区内中—上寒武统与浙江杨柳岗组一致。1981 年，许保桐首次系统总结了上海的地层发育概况，将寒武系划分为超山组、杨柳岗组和超峰群。1988 年，在上述方案基础上，《上海市区域地质志》进一步将下寒武统划分为大陈岭组和超山组。1996 年，《浙江省岩石地层》将“超峰群”改为“超峰组”，并将“超山组”定义为仅指下段的灰黑色白云质泥岩。故在此之后，本区亦改称之“超峰组”，一直沿用至今。

在中国地层表（2014）中，寒武系开始采用 4 分方案。根据浙西北地区年代地层与生物地层的对比关系，本志对区内岩石地层的地质年代归属进行了相应调整，将超山组归为底—下寒武统，大陈岭组归为下寒武统（表 3-1）。

表 3-1 上海寒武系岩石地层单位划分沿革表

Table 3-1 Previous divisions of the Cambrian lithostratigraphic units in Shanghai

<table>
<tr><th>地层</th><th>《上海地区矿产地质普查工作总结报告》，1962</th><th>《上海地区石油地质图集》，1966</th><th>《1：20 万上海地区基岩地质推断图》，1979</th><th>《上海地区地层概述》，1981</th><th>《上海市区域地质志》，1988</th><th>“上海三维城市地质调查”，2010</th><th colspan="2">本志</th></tr>
<tr><td>上覆地层</td><td>上侏罗统</td><td>?</td><td>奥陶系</td><td>奥陶系</td><td>奥陶系</td><td>奥陶系</td><td>奥陶系</td><td>上覆地层</td></tr>
<tr><td rowspan="2">上寒武统</td><td rowspan="3">未分</td><td rowspan="5">坦直群</td><td rowspan="3">未分</td><td rowspan="2">超峰群</td><td rowspan="2">超峰群</td><td rowspan="2">超峰组</td><td>超峰组</td><td>上寒武统</td></tr>
<tr><td>杨柳岗组</td><td>中寒武统</td></tr>
<tr><td>中寒武统</td><td>杨柳岗组</td><td>杨柳岗组</td><td>杨柳岗组</td><td>大陈岭组</td><td>下寒武统</td></tr>
<tr><td rowspan="2">下寒武统</td><td rowspan="2">荷塘组</td><td rowspan="2">荷塘组</td><td rowspan="2">超山组</td><td>大陈岭组</td><td>大陈岭组</td><td rowspan="2">超山组</td><td rowspan="2">底寒武统</td></tr>
<tr><td>超山组</td><td>超山组</td></tr>
</table>

第一节 岩石地层划分

一、超山组（$C_{0-1}c$）

（一）创名及原始定义

由鞠天吟（1979）创建，命名地点在浙江余杭超山。原义是据产状认为假整合于上震旦统之上，上部与“大陈岭组”连续沉积。厚 31.0m。分上、下两段：下段为黑色碳质页岩；上段为灰黑色泥灰岩、钙质页岩。本组无硅质层、石煤层和磷块岩，含三叶虫、腕足类及海绵骨针，软舌螺，时代为早寒武世。

（二）现在定义

超山组岩性以黑色碳质页岩为主，其夹有不规则之泥质或白云质灰岩体，与上覆大陈岭组及下伏上震旦统灯影组均呈整合接触。

（三）层型剖面描述

次层型剖面为岩溶 8#、岩溶 16#和罗店 701 孔剖面（图 3-1）。岩溶 8#孔、罗店 701 孔剖面地层分述如下：

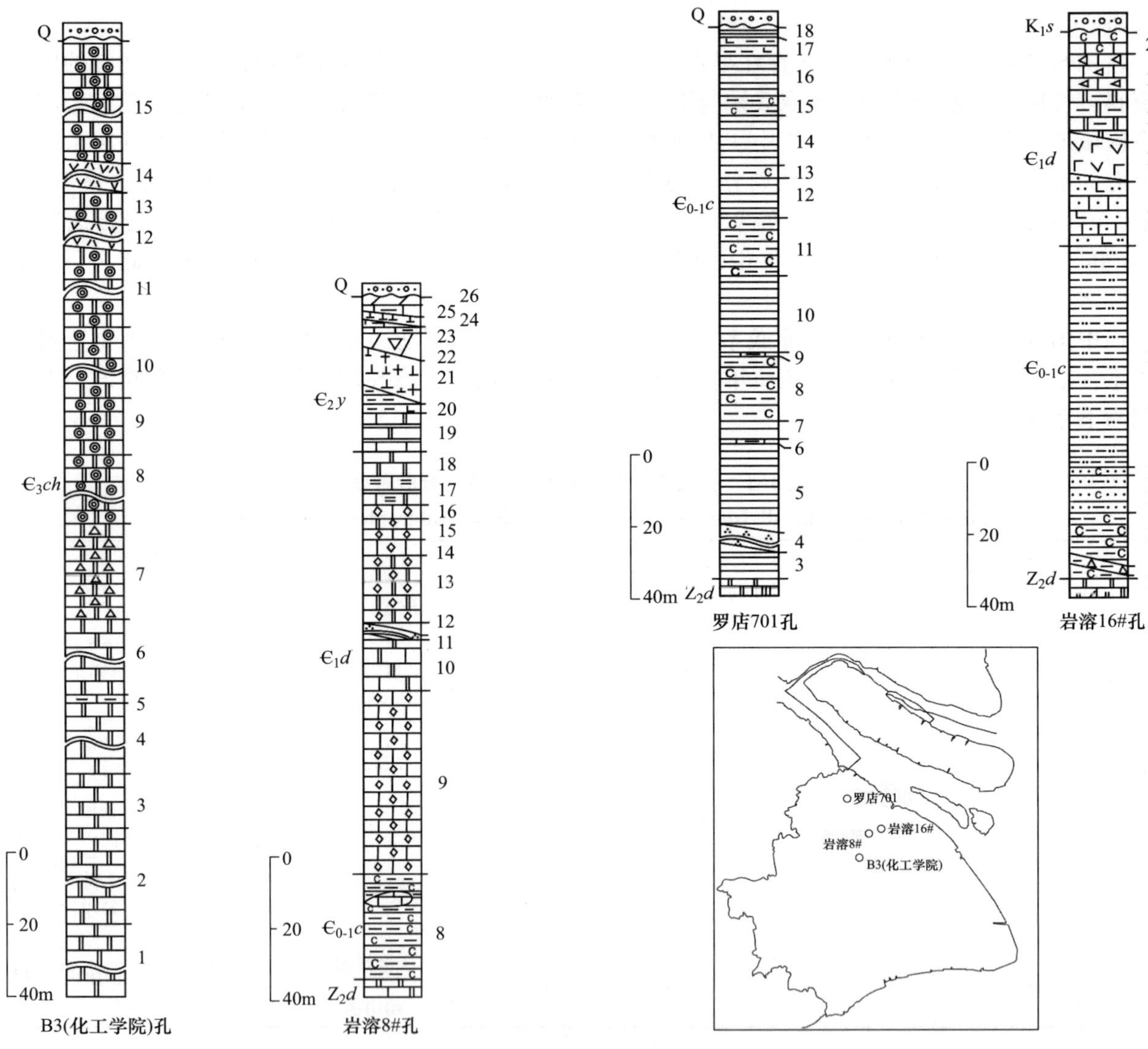

图 3-1　次层型钻孔剖面柱状图

Fig. 3-1　Hypostratotype borehole section histograms

1. 岩溶 8#孔剖面

上覆地层：下寒武统大陈岭组灰白、灰黄及灰黑色灰岩、泥质白云岩等

———— 整　合 ————

超山组（$\in_{0-1}c$）　　　　总厚度 27.6m

8. 灰黑色碳质（或钙质）页岩，中、上部夹有透镜状泥质或硅质灰岩薄层。产三叶虫：*Redlichia*（*Pteroredlichia*）*shanghaiensis* Ju et Xu（sp. nov.），*Shanghaiatypicalis* Ju et Xu，*Shanghaia* sp.，*Shanghaiaspisusualis* Ju et Xu，*Shanghaiaspislatilimbatus* Ju et Xu，*Shanghaiaspisdecipiens* Ju et Xu；含腕足类：*Homotreta*? sp.，此外，尚有一些不明之管状物（痕迹化石）及少量古片藻等

———— 整　合 ————

下伏地层：上震旦统灯影组硅质岩、白云岩及灰岩等

2. 罗店 701 孔剖面

上覆地层：第四系

~~~~~~ 不　整　合 ~~~~~~

超山组（$\epsilon_{0-1}c$）	总厚度> 147.9m
18. 黑色页岩，顶部因风化而呈棕黄色	>1.6m
17. 黑色钙质页岩，含石英砂粒	4.5m
16. 黑色页岩，含星散状黄铁矿	10.7m
15. 黑色碳质页岩，含碳量较高，黄铁矿呈细脉状分布	4.7m
14. 灰黑色页岩，层内有较多石英脉穿插	14.4m
13. 黑色碳质页岩	3.4m
12. 黑色页岩	11.9m
11. 黑—灰黑色碳质页岩	16.8m
10. 黑色页岩	21.4m
9. 灰色泥质灰岩	0.6m
8. 灰黑—黑色碳质页岩	18.2m
7. 黑色页岩	4.9m
6. 灰色泥质灰岩	1.0m
5. 灰黑—黑色页岩	25.3m
4. 浅灰色叶蜡石化次生石英脉岩	（4.8m）
3. 灰黑色页岩	8.5m

———— 整　合 ————

下伏地层：上震旦统灯影组，浅灰、灰白色白云岩及硅化灰岩

（四）地质特征及区域变化

代表钻孔剖面资料显示（见图 3-1），本区超山组以灰黑色碳质页岩为主，局部为泥岩，间夹有钙质页岩及含粉砂、细砂质页岩薄层，亦具有下列 3 个特征：①层间普遍含有星散状或小透镜状黄铁矿；②层间多薄层状或透镜状之泥、硅质灰岩夹层，在这些灰岩夹层中富含有三叶虫、腕足类等化石。

钻孔揭露厚度约 148m，具有南薄北厚之趋势。顶、底地层均已揭露，下伏地层为灯影组，上覆地层为大陈岭组，且均呈整合接触。

二、大陈岭组（$\epsilon_1 d$）

（一）创名及原始定义

由李尉秾（1965）创建，命名地点在浙江江山大陈东南大陈岭。原始定义：江山大陈岭附近原杨柳岗组底部之含三叶虫 *Arthricocephalus* 化石的白云质灰岩，其层位可与皖南黄柏岭组上部及滇东龙王庙阶上部进行对比。
~~~~~~

（二）现在定义

大超岭组系指碳质页岩之上一套含镁碳酸盐岩建造，以灰、灰黄及灰黑色条带状白云质灰岩为主，夹 2 层黑色碳、硅质泥岩。境内条带状构造不明显。与下伏超山组碳质页岩，上覆杨柳岗组泥质、砂质灰岩均为整合接触。

（三）层型剖面描述

次层型剖面为岩溶 8#孔、岩溶 16#孔剖面（见图 3-1）。岩溶 8#孔剖面地层分述如下：

上覆地层：中寒武统杨柳岗组泥质、砂质灰岩

———— 整　合 ————

大陈岭组（$\epsilon_1 d$）	总厚度 113.9m
18. 灰—灰黑色钙质白云岩	6.6m
17. 灰白—微黄色白云岩化大理岩	7.8m
16. 灰—浅灰黑色结晶灰岩，局部夹薄层白云质灰岩	2.8m
15. 灰白—土黄色假砾状白云质灰岩	5.7m
14. 灰黑夹灰白色白云质结晶灰岩	4.4m
13. 浅灰黑、灰白色结晶白云岩，有脉状石膏穿插	18.0m
12. 灰黄、黄褐色（假）砾状白云质灰岩	1.9m
11. 灰黑色硅质岩	1.1m
10. 浅黄、灰黑色白云质灰岩	14.3m
9. 灰黑及浅灰色结晶灰岩，富含黄铁矿，呈斑状、脉状及浸染状，并具有明显的条带状构造	51.3m

———— 整　合 ————

下伏地层：下寒武统超山组碳质页岩

（四）地质特征及区域变化

区内大陈岭组主要由灰、灰黑色白云质灰岩、白云岩及灰岩组成，局部夹薄层硅质岩、碳质页岩和钙质砂岩。归纳该组地层亦具以下特征：①以含镁碳酸盐岩为主；②在灰岩中有泥质条带，局部呈纹层状构造；③常含有斑块状、细脉状或星点状黄铁矿。

钻孔已揭露厚度在 33~114m 不等。顶、底地层均已揭露，下伏地层为超山组，上覆地层为杨柳岗组，且均呈整合接触。

三、杨柳岗组（$\epsilon_2 y$）

（一）创名及原始定义

杨柳岗组原称之“杨柳岗石灰岩”，由卢衍豪、穆恩之等于 1955 年创建，命名地点在浙江江山大陈东北 3.5km 处的杨柳岗。原定义是：杨柳岗石灰岩整合于荷塘组硅质页岩之上，以石灰岩为主，间夹少量之页岩层，厚度为 127.0m，可分为两部分：下部为深灰褐色厚层白云质石灰岩及灰色中至厚层石灰岩；上部由薄层石灰岩组成，夹少量页岩。

（二）现在定义

杨柳岗组由条带状白云质灰岩、饼条状灰岩和含灰岩透镜体泥质灰岩、泥质灰岩及碳质硅质泥岩组成。以泥质灰岩或黑色碳、硅质泥岩与下伏大陈岭组白云质灰岩，以含灰岩透镜体泥质灰岩与上覆华严寺组薄层状灰岩或超峰组厚层状白云岩均为整合接触。

（三）层型剖面描述

次层型剖面为岩溶 8#孔剖面（见图 3-1）。该钻孔剖面地层分述如下：

上覆地层：第四系疏松层

~~~~~~~~~~ 不　整　合 ~~~~~~~~~~

杨柳岗组（$\epsilon_2 y$）	总厚度>21.2m
26. 灰黄、灰褐色含粉砂泥灰岩，夹粉砂岩薄层，顶部风化强烈	>2.1m
25. 浅灰褐、灰绿色含泥质灰岩，局部夹粉砂质薄层	3.1m
24. 花岗闪长岩脉	(1.7m)
23. 浅灰褐色含泥质灰岩	3.0m
22. 构造角砾岩	(6.6m)
======== 断　层 ========	
21. 花岗闪长斑岩脉	(13.0m)
20. 灰—灰绿色钙质泥岩，局部具薄层状构造	2.5m
19. 灰白色大理岩化灰岩	10.5m

———— 整　合 ————

下伏地层：下寒武统大陈岭组灰—灰黑色白云质灰岩、白云岩及灰岩

（四）地质特征及区域变化

区内杨柳岗组岩性以浅灰、深灰色泥灰岩、含泥质灰岩与含粉砂灰岩为主，局部夹钙质页岩或硅质灰岩薄层。泥质灰岩中可见有缝合线构造，含砂灰岩中粉砂常聚成团块状。

钻孔揭露厚度在150m以上。与下伏地层大陈岭组呈整合接触。顶部已被剥蚀，出露不全，为第四系疏松层覆盖。

四、超峰组（$\epsilon_3 cf$）

（一）创名及原始定义

超峰组原称“超峰群”，由鞠天吟于1979年创名，命名地点在浙江余杭超山之顶超峰。原始定义：整合于杨柳岗组之上，为灰、深灰色白云岩，厚度在200m以上，仅获腕足类 Obolus 化石。

（二）现在定义

超峰组由灰白色厚层块状白云岩组成。与下伏杨柳岗组薄—中层灰岩呈整合接触，未见顶。

（三）层型剖面描述

次层型剖面为本市化工学院B3孔剖面（见图3-1）。该钻孔剖面地层分述如下：

上覆地层：第四系疏松层

~~~~~~~~~~ 不　整　合 ~~~~~~~~~~

超峰组（$\epsilon_3 cf$）	总厚度>762.0m
15. 灰、灰黑与黄色硅化白云岩，含燧石结核，溶蚀现象特别发育	60.8m
14. 蚀变英安玢岩脉	(32.0m)
13. 灰色硅化白云岩	8.5m
12. 蚀变英安玢岩脉	(20.3m)
11. 灰白—灰黑色硅化白云岩，局部夹1.0m左右的硅质层，溶蚀现象亦较发育	139.8m
10. 灰黑夹灰黄色硅化白云岩，间夹有轻微硅化白云质灰岩夹层	46.1m
9. 灰、灰黄、灰白色硅化白云岩	25.8m
8. 灰、青灰色硅化白云岩	119.4m
7. 青灰色白云岩，具同生角砾	44.4m
6. 灰色白云岩	92.7m
5. 浅灰绿色含泥质微晶灰岩，底部为白云质灰岩	3.3m
4. 灰、灰白夹青灰色白云岩	53.0m

3. 青灰与灰黑色白云岩 27.7m

2. 灰色白云岩 47.5m

1. 青灰色白云岩 >44.7m

未 见 底

(四) 地质特征及区域变化

区内所见超峰组主要由一套浅灰与青灰色白云岩、同生角砾状白云岩组成，局部夹灰白与灰黑色白云质灰岩、硅质岩。具有下列特征：①中、上部其溶洞发育，相互沟通；②上部普遍硅化，色浅，以灰白、灰褐色为主，下部以青灰色为主，且具同生角砾构造；③呈厚层状，夹硅质层与燧石结核；④岩石中 CaO 含量为 28.64%~31.5%，MgO 为 18.21%~21.1%，SiO_2 约为 4.26%。

钻孔揭露厚度在 248.0~762m，较厚，且尚未见顶、底，与上、下层接触关系不明。

第二节 沉积作用和沉积建造

一、沉积类型

寒武系为海相沉积，底—下寒武统主要为细粒的泥质沉积建造，中—上寒武统主要为一套碳酸盐岩建造。主要的沉积类型有滨海沉积、浅海沉积等。

(一) 滨海沉积

滨海沉积主要为台地边缘斜坡和局限台地潟湖沉积。台地边缘斜坡沉积主要出现在杨柳岗组，沉积了一套以泥质灰岩和泥灰岩为主的碳酸盐岩建造，粉砂、泥质等陆源碎屑含量较多；后者主要出现在超峰组，形成厚层状白云岩，同生角砾状白云岩，局部夹白云质灰岩、硅质岩及遂石结核。

(二) 浅海沉积

浅海沉积主要为滞留平缓盆地—盆地斜坡沉积，出现在超山组和大陈岭组，自下而上由碳质页岩、含碳页岩的细粒碎屑岩建造发育为局部发育（假）砾状的白云质灰岩夹碳质页岩建造，星散状、团块状黄铁矿发育。

二、主要岩石类型

上述沉积环境中形成的主要岩石类型有页岩、碳质页岩、白云质灰岩、泥质灰岩、钙质泥岩、硅质岩、白云岩等。主要特征如下：

(一) 碳质页岩

黑色，泥质结构，层理构造。由泥质集合体组成，内有石英、黄铁矿和绢云母。有机泥质约占93%，呈条带状。石英约占4%，最大粒径 0.25mm，磨圆度好，零星分布。方解石约 1%，极微量绢云母和白云母，呈鳞片状。次生黄铁矿，条状，最大宽度 0.25~0.3mm。

(二) 泥岩

灰黑色，泥状结构，黑白相间的纹层构造。深色纹层中碎屑呈两个粒级，一种粒径 0.05mm 左右，成分以石英为主，其次为长石，多为次棱角状。一种粒径小于 0.02mm，成分为石英、云母、长石，常混于碳泥质胶结物中。浅色层，以长石为主，石英次之，少量云母。长石全部水白云母化，粒径 0.1mm 左右。

(三) 白云质灰岩

深灰色，泥状结构，块状构造。方解石多在 60%左右，白云石约 39%，黄铁矿等约 1%。方解石大多呈泥晶状聚集体，粒径 0.02~0.05mm。白云石自形—半自形，粒径和白云石相似。

（四）泥灰岩

灰绿色，泥状结构，条带状构造和叶状层理。主要由方解石和水白云母组成，方解石占45%，绢云母-水云母约40%，绿泥石5%，金红石5%，黄铁矿等3%，石英2%等。

（五）白云岩

不等粒结构，主要由白云石组成，粒径变化较大，最大3.5mm×0.7mm。部分呈菱形，部分呈他形，极少数有双晶。局部硅化明显，石英呈隐—微晶质，呈脉状和侵染状。

第三节　地层时代和区域地层对比

一、地层时代和区域对比

本区寒武系生物化石，仅在超山组中、上部有所发现。所采获的 *Redlichia shanghaiensia* Ju et Xu 属于 Redlichiidae Poulsen 超科，定名为上海翼形莱德利基虫（新种）。此翼形莱德利基虫除在我国常见外，在朝鲜、澳大利亚等国亦有发现。*Redlichia*（*Pteroredlichia*）*shanghaiensis* Ju et Xu 和产于湖北宜都下寒武统的 *Redlichia*（*R.*）*hupehensis*，*Redlichia chinensis*（幼虫）比较，颇为相似；和产于湖北宜昌下寒武统石牌组中的 *Redlichia*（*R.*）*kobayashii* 亦十分接近。此外，还找到了 *Shanghaia* Ju et Xu（属 Oryctocephalidae Beecher 科），其特征介乎浙江相应层位中的 *Changaspis* 及 *Arthricocephalus* 之间。从化石的发现层位看，推测区内超山组地层时代当应属底—早寒武世。从岩性组合上看，与浙西北、皖南和赣东北、苏南地区的荷塘组为同期异相沉积，未见硅质层与石煤层，且所含化石也不一致。在苏南地区，超山组沉积的中后期与幕府山组下部的灰岩、白云质灰岩夹钙质泥岩形成明显的相变关系。相比杭州—嘉兴超山组，区内灰岩透镜体夹层明显增多，白云质含量较少。相比于安徽芜湖的黄柏岭组和江西九江的观音堂组碳质和有机质含量、灰岩透镜体夹层明显较多（表3-2）。

大陈岭组岩性特征大致可与余杭超山下寒武统大陈岭组相当，与宁镇下寒武统幕府山组上部岩性段对比，为同期异相沉积。从邻区的化石组合看，其地层时代相当于龙王庙期，应属于早寒武世晚期。

杨柳岗组是以泥质灰岩、泥灰岩为主的一套地层，与浙西、皖南的杨柳岗组岩性组合相似。与宁镇以白云岩、泥质白云岩夹白云岩为主的中寒武统炮台山组为同期异相沉积。其时代应为中寒武世。

鉴于区内上寒武统的岩性组合特征（包括层中的常量元素丰度值，表3-3）与杭嘉湖地区可以对比，将其归于上寒武统超峰组。

表3-2　区域地层对比表

Table 3-2　Regional stratigraphic correlation

地区		上海	浙西北		安徽 滁州	安徽 巢湖	安徽 青阳	安徽 宁国	江西 九江(赣北)	江西 上饶(赣东北)	江苏 南京	江苏 南通	湘中
上覆地层		仑山组	留下组	印渚埠组	上欧冲组	仑山组	大坞埠组	印渚埠组	仑山组/印渚埠组	印渚埠组	仑山组	仑山组	白水溪组
寒武系	上寒武统	超峰组	超峰组	西阳山组 / 华严寺组	琅琊山组	观音台组	青坑组 / 团山组	西阳山组 / 华严寺组	西阳山组 / 华严寺组	西阳山组 / 华严寺组	观音台组	超峰组	探溪组
寒武系	中寒武统	杨柳岗组	杨柳岗组	杨柳岗组	杨柳岗组	炮台山组	杨柳岗组	杨柳岗组	杨柳岗组	杨柳岗组	炮台山组	杨柳岗组	探溪组 / 污泥塘组
寒武系	下寒武统	大陈岭组	超山组	大陈岭组 / 荷塘组	大陈岭组 / 黄柏岭组	幕府山组	大陈岭组 / 黄柏岭组	大陈岭组 / 荷塘组	观音堂组	荷塘组	幕府山组	大阵岭组	污泥塘组 / 牛蹄塘组
寒武系	底寒武统	超山组	超山组	荷塘组	荷塘组	幕府山组	荷塘组	荷塘组	荷塘组	荷塘组	荷塘组	超山组	牛蹄塘组
下伏地层		灯影组	灯影组 / 板桥山组	皮园村组	皮园村组	灯影组	皮园村组	皮园村组	皮园村组/灯影组	皮园村组/灯影组	灯影组	灯影组	灯影组/留茶坡组

表 3-3　区内与杭嘉湖地区上寒武统白云岩岩石化学成分对比

Table 3-3　Correlation of the petrochemical compositions of the Upper Cambrian dolomite between Shanghai and the Hangzhou-Jiaxing-Huzhou area

w（B）/%

指标	余杭超山	肖山赭山	海宁东山	平湖瓦山	昆山马鞍山	上海市区化工学院	奉贤钱桥 ck2 孔
CaO	29.41	28.69	27.53	28.22	29.24	30~31.5	28.64
MgO	20.02	18.41	19.60	20.54	19.86	21.1	18.21
Fe_2O_3	0.55	0.35	0.28	0.2	—	—	—
SiO_2	4.76	6.03	9.13	6.20	5.58	—	4.26

二、存在问题

（一）崇明港西是否存在寒武系

崇明港西为一个布格重力异常高值区，长期以来都被解释为志留系或下古生界，但该异常值与区内坦直基本相当。1995 年，江苏省地质矿产局第四地质大队在进行 1∶5 万区域地质调查时，将该异常的江苏段解释为震旦系—奥陶系。本次推断港西存在寒武系，但还需要更多资料的验证。

（二）罗店超山组的划分

该区 701 孔和 702 孔以页岩、泥岩为主，夹少量灰岩，但 704 孔和 705 孔发育大量的砂岩和粉砂岩，且夹有板岩。目前，这些钻孔地层都被划分为超山组，但从岩石组合上看其差别较大，似乎更支持归属不同的地层单位。

第四节　岩相古地理

区内寒武纪从早—晚期，经历了由盆地相—开阔海台地相—局限台地潟湖相的全过程，海水深度发生由深变浅的变化，海水的咸化度相应发生变化。

一、底—早寒武世

底寒武世开始，海侵范围扩大，海水加深，生物也有较大发展。超山组中的碳质页岩、含碳页岩及星散状、团块状黄铁矿的出现，反映当时有一个适合有机体大量繁衍的湿热气候和还原性质的沉积环境，推测当时是一滞留平缓盆地。

至早寒武世晚期，大陈岭组以含镁碳酸盐岩建造为主，与超山组类似，层间也夹有碳质页岩薄层，并见星散状、团块状黄铁矿出现。说明早寒武世晚期的沉积环境与早期有明显继承性。在白云岩中局部具有同生角砾，可能为盆缘斜坡环境。

二、中寒武世

区内的杨柳岗组沉积了一套以泥灰岩及含粉砂质灰岩为主的磷酸盐岩建造，陆源碎屑物质较多，具薄层状构造，推测为开阔海台地沉积环境。

三、晚寒武世

晚寒武世，起峰组为厚层状白云岩、同生角砾状白云岩间夹白云质灰岩，为局限盆地潟湖相沉积。可能因海水中含盐度较高，生物不易繁衍、生存，层中化石稀少。

第四章　奥陶系

奥陶系分布局限，目前已于上海市区北部、市区西南部和奉贤海湾西部等地的钻孔中发现（图 2-1）。被揭露的地层层序也较齐全，中奥陶统发育大量海百合、角石等生物碎屑，由上部碎屑岩与下部碳酸盐岩两套地层组成。自下而上可分为：下奥陶统仑山组、红花园组；下—中奥陶统牯牛潭组和上奥陶统黄泥岗组和长坞组。已揭露钻孔厚度大于 250m，未见顶、底，与上覆、下伏地层接触关系不明。

1975 年，为编著华东地区区域地层表，在复查市区部分钻孔岩芯时，初次发现了奥陶纪地层，将其定为上奥陶统，并划分两个岩性段。1979 年，在编制《1∶20 万上海地区基岩地质图》时，又进行了合并。1981 年，许保桐根据地层、化石资料，首次进行了岩石地层单位划分，自下而上为下—中奥陶统荆山群、上奥陶统长坞组。1988 年，编制《上海市区域地质志》时，将具有泥、硅质瘤状（或团块状）构造的碳酸盐岩单独划出，建立黄泥岗组。并对奥陶系的地层特征、岩相特征以及区域对比情况等，做了系统研究和总结。许保桐等（2003）在奉贤东南部地区 B52（岩 2）孔碳酸盐岩地层中发现了较为丰富的牙形石、腹足类化石，结合岩石地层特征，经过进一步对比将原荆山群解体为仑山组、红花园组和牯牛潭组。

区域上，在晚奥陶世发育一套碳酸盐岩类建造，在浙西北地区为大田坝组和砚瓦山组。根据区内地层发育特点和以往认识，本志将此做缺失处理（表 4-1）。

表 4-1　上海奥陶系岩石地层单位划分沿革

Table 4-1　Previous divisions of the Ordovician lithostratigraphic units in Shanghai

<table>
<tr><th>地层</th><th>《华东地区区域地层表（江苏市及上海市分册）》，1975</th><th>《1∶20 万上海地区基岩地质推断图》，1979</th><th>《上海地区地层概述》，1981</th><th>《上海市区域地质志》，1988</th><th>“上海市三维城市地质调查”，2010</th><th>本志</th></tr>
<tr><td>上覆地层</td><td>上侏罗统</td><td>上侏罗统</td><td>志留系</td><td>志留系</td><td>河沥溪组</td><td>河沥溪组</td></tr>
<tr><td rowspan="3">上奥陶统</td><td rowspan="3">上岩性段
下岩性段</td><td rowspan="7">未分</td><td rowspan="3">长坞组</td><td>长坞组</td><td>长坞组</td><td>长坞组</td></tr>
<tr><td rowspan="2">黄泥岗组</td><td rowspan="2">黄泥岗组</td><td>黄泥岗组</td></tr>
<tr><td></td></tr>
<tr><td>中奥陶统</td><td rowspan="4">?</td><td rowspan="4">荆山群</td><td rowspan="4">荆山群</td><td rowspan="2">牯牛潭组</td><td rowspan="2">牯牛潭组</td></tr>
<tr><td rowspan="3">下奥陶统</td></tr>
<tr><td>红花园组</td><td>红花园组</td></tr>
<tr><td>仑山组</td><td>仑山组</td></tr>
</table>

第一节　岩石地层划分

一、仑山组（$O_1 l$）

（一）创名及原始定义

关于仑山组命名时间和命名人，争论较大。张守信（2005）介绍了不同说法，认为系丁文江（1919）创建。原始定义：江苏省句容县（今句容市）东北仑山灰岩。李毓尧等（1935）将含 *Hopeioceras* 化石之下的部分称为仑山灰岩。

（二）现在定义

仑山组为一套灰色厚层含灰质白云岩、白云质灰岩和灰、暗灰色厚层状灰岩、鲕状灰岩，局部夹燧石条带与燧石结核层。区内指一套灰色含灰质白云岩和白云质灰岩，一般厚度较大。与上覆红花园组呈整合接触。

（三）层型剖面描述

次层型剖面为奉贤奉新星火交管站 B52（岩 2）孔剖面（图 4-1）。剖面地层分述如下：

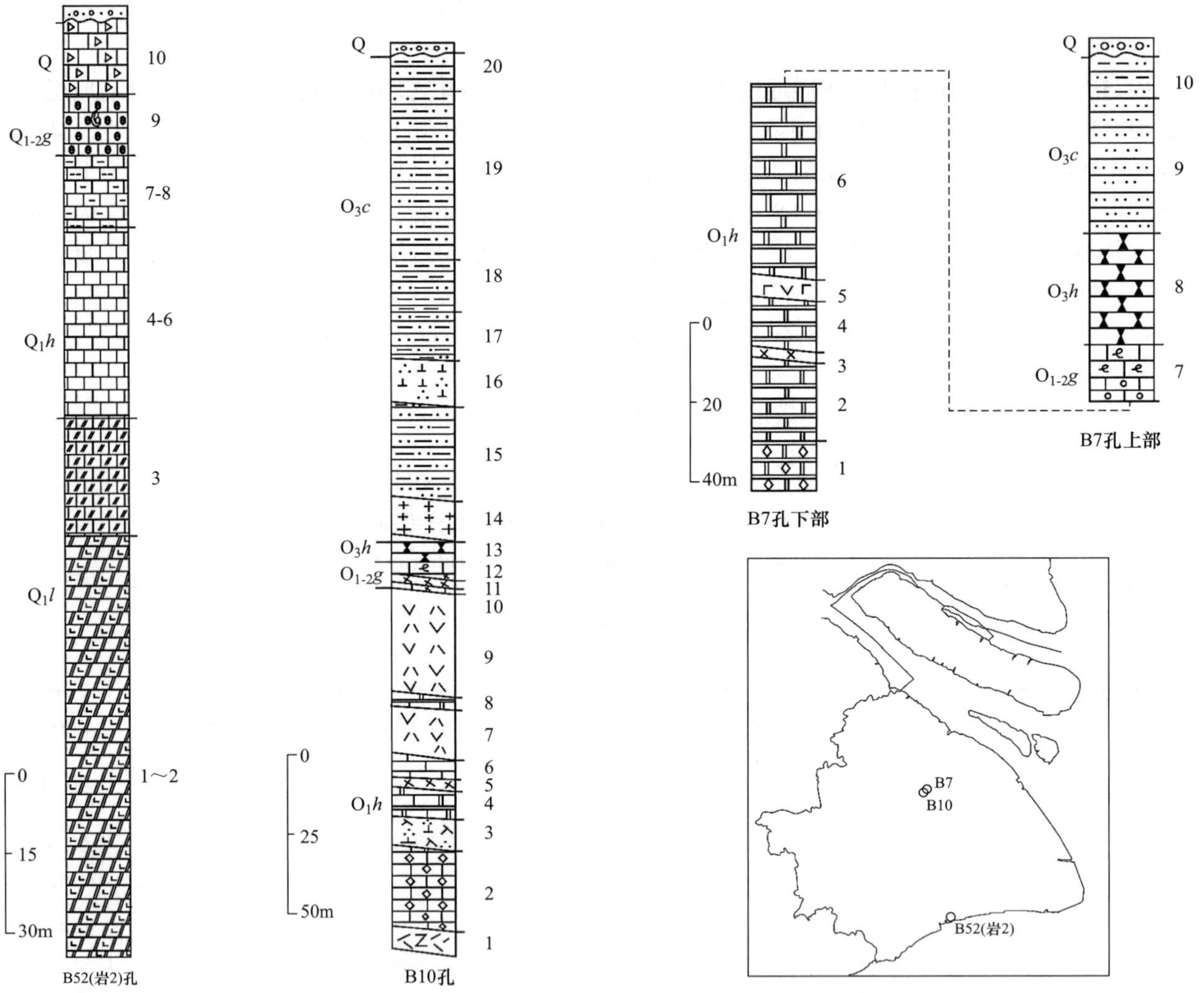

图 4-1　次层型钻孔剖面柱状图

Fig. 4-1　Hypostratotype borehole section histograms

上覆地层：下奥陶统红花园组灰、灰白色微晶灰岩

———— 整　合 ————

仑山组（O_1l）	总厚度>102.8m
3. 灰紫色白云质灰岩（云灰岩）	21.8m
2. 灰色含灰质白云岩	35.0m
1. 灰色含灰质细晶白云岩	>46.0m

未 见 底

（四）地质特征及区域变化

仑山组主要在奉贤南部地区被揭露，从现已揭露岩性来看，主要以灰色白云质灰岩与灰质白云岩为主，比较单一，单层厚度较大。揭露最大钻孔厚度为 198.42m，与上覆下奥陶统红花园组微晶灰岩呈整合接触。

二、红花园组（O_1h）

（一）创名及原始定义

红花园组原称红花园石灰岩，由张鸣韶和盛莘夫于 1940 创建，并于 1958 年正式发表。命名地点在贵州省桐梓县南 7km 处的红花园东山坡。原始定义：红花园灰岩，整合于桐梓层之上，岩性为灰色厚层灰岩，厚 40m，含化石丰富，时代为早奥陶世。

（二）现在定义

区内红花园组为以灰黑、灰色为主的微晶灰岩，上部层内夹有泥灰岩，层内普遍含有生物骨屑和泥质，有时呈页片状夹层。其与上覆牯牛潭组和下伏仑山组均呈整合接触。

（三）层型剖面描述

次层型剖面为奉贤奉新星火交管站 B52（岩 2）孔剖面（见图 4-1）。剖面地层分述如下：

上覆地层：中—下奥陶统牯牛潭组紫红色微晶生物碎屑灰岩

——————整　合——————

红花园组（O_1h）　　总厚度 50.2m

8.	灰及灰黑色粉晶团块灰岩，层内见有少量生物骨屑，已白云石化	8.9m
7.	紫红与蓝灰色含泥灰岩、微晶灰岩互层	4.7m
6.	蓝灰色微晶灰岩，局部具角砾状构造	12.7m
5.	灰与灰白色微晶灰岩，夹有泥质微晶灰岩薄层，其中灰白色为含泥质呈页片状夹层，但不稳定，时而消失或尖灭	4.2m
4.	青灰、灰黑色微晶灰岩，其间夹有含泥碳质云灰岩薄层	19.7m

——————整　合——————

下伏地层：下奥陶统仑山组灰色白云质灰岩、白云岩

（四）地质特征及区域变化

在奉贤南部和市区北部、西南地区都有揭露。本组地层岩性以灰、灰白色微晶灰岩为主，层间夹有蓝灰色含泥灰岩，呈互层状，本层最大特征是层内普遍含有泥质，分布极不规则，时而尖灭或消失。于顶部地层中尚见有少量生物骨屑。已揭露最大钻孔厚度为 176.99m。其与下伏仑山组和上覆牯牛潭组呈整合接触。

三、牯牛潭组（$O_{1-2}g$）

（一）创名及原始定义

牯牛潭组原称牯牛潭石灰岩，由张文堂于 1957 年创建，命名地点在湖北省宜昌市分乡场牯牛潭。原始定义：牯牛潭石灰岩厚约 23.0m，为灰色泥质干裂纹石灰岩，夹灰黄色泥质瘤状石灰岩，与上覆庙坡页岩和下伏扬子贝层页岩夹瘤状灰岩，均呈整合接触。

（二）现在定义

区域上牯牛潭组为一套紫灰、黄绿、灰色薄至中厚层瘤状灰岩，与上覆庙坡组黑色页岩和下伏大

湾组页岩夹薄层瘤状灰岩均呈整合接触。区内该组为紫红、灰和灰黑色微晶生物碎屑灰岩与微晶灰岩互层，见有硅质条带或呈脉状出现，与下伏红花园组呈整合接触。

（三）层型剖面描述

次层型剖面为奉贤奉新星火交管站 B52（岩 2）孔剖面（见图 4-1）。剖面地层分述如下：

上覆地层：第四系疏松层

~~~~~~~~ 不　整　合 ~~~~~~~~

牯牛潭组（$O_{1-2}g$） 总厚度 25.9m

10. 紫红色含碎屑微晶（泥晶）灰岩，角砾状构造 14.3m

9. 紫红色微晶生物碎屑灰岩与微晶灰岩互层，层内见条带状或脉状硅质岩穿插，含腹足类 *Umbonellina* cf. *inflata* Yu、*Maclurites* sp.，*Sinuites* sp.、*Raphispira* sp.、*Liospira* sp. 和含牙形石 *Periodon aculeatus* Hadding，*Protopanderodus rectus*（Lindström），*Walliserodus ethingtom*（Fähraeus）。另外，还含有海百合茎、介形虫等化石碎片 11.6m

———— 整　合 ————

下伏地层：下奥陶统红花园组灰、蓝灰等色微晶灰岩、泥灰岩等

（四）地质特征及区域变化

除市区酒精一、二厂孔外，另在奉贤星火农场亦有分布。其岩性在奉贤南部地区主要为紫红色微晶碎屑生物灰岩夹微晶灰岩，顶部有角砾状结构。在上海市区主要为灰白色含生物碎屑灰岩。以生物碎屑大量发育为特征，局部穿插有硅质条带或脉状体。钻孔揭露厚度多在十数米，个别至三四十米，最厚约 60m。与下伏红花园组整合接触。

四、黄泥岗组（O_3h）

（一）创名及原始定义

由卢衍豪于 1963 年创名，命名地点在浙江省江山市黄泥岗村。原始定义：在江山、常山一带砚瓦山石灰岩之上常有一层紫红色石灰质页岩，其中亦有黄色页岩。厚度达 45m。

（二）现在定义

区域上黄泥岗组为灰绿、紫红色含钙质结核泥岩，局部夹瘤状泥灰岩。其与下伏砚瓦山组瘤状泥灰岩和上覆长坞组泥岩、粉砂质泥岩均呈整合接触。区内黄泥岗组为泥质、硅质条带状（或瘤状）微晶灰岩，与下伏砚瓦山组（含）生物碎屑灰岩整合接触。

（三）层型剖面描述

次层型剖面为上海市延安西路 1047 号酒精一厂（B10）孔剖面（见图 4-1）。剖面地层分述如下：

上覆地层：上奥陶统长坞组青灰、灰及灰褐色粉砂质泥岩、泥岩与泥质粉砂岩

———— 整　合 ————

黄泥岗组（O_3h） 总厚度 5.5m

13. 青灰、灰白色泥质、硅质条带（或瘤状）微晶灰岩，含三叶虫等化石碎片 5.5m

----- 平行不整合 -----

下伏地层：中奥陶统牯牛潭组泥、硅质条带状灰岩与（含）生物骨屑灰岩

（四）地质特征及区域变化

黄泥岗组仅在上海市区西南部被揭露，岩性是一套厚度不大的瘤状灰岩层，岩性较为简单，以具
~~~~~~~~

有泥质或硅质瘤状体为其特征，而瘤状体颜色较深，灰岩“基体”部分颜色一般较浅，两者相间。钻孔揭露厚度达 39.2m，与下伏牯牛潭组平行不整合接触，与上覆长坞组呈整合接触。

五、长坞组（O_3c）

（一）创名及原始定义

长坞组系 1963 年由卢衍豪创名，命名地点在浙江省江山市城北南山——长坞东山。原始定义：长坞组下和黄泥岗组为连续沉积，上与中石炭统藕塘底群底砾岩不整合。全部以黄绿色页岩为主，间夹砂岩薄层。含笔石，厚 280m。

（二）现在定义

区域上长坞组为由黄绿色泥岩、粉细砂岩组成的复理石、类复理石韵律层，其间夹块状砂岩和泥岩。含笔石、腕足类等化石。其与下伏黄泥岗组含钙结核泥岩和上覆文昌组含钙质砂岩均为整合接触。区内长坞组岩石组合与之相似。

（三）层型剖面描述

次层型剖面为上海市延安西路 1047 号酒精一厂（或 B10）孔剖面（见图 4-1）。剖面地层分述如下：

上覆地层： 第四系疏松层

~~~~~~~~~ 不　整　合 ~~~~~~~~~

长坞组（O_3c）	总厚度>123.3m
20. 灰、青灰色含粉砂质泥岩，局部粉砂富集呈条带状，具微细层理构造	>10.3m
19. 深灰、灰黑色粉砂质泥岩，局部粉砂呈小扁豆体产出，页理发育，产笔石 *Climacograptus typicalis* Hall，*Climacograptus* sp.，*C. latus* Elles et Wood 等	52.6m
18. 黄绿色，局部为青灰色的泥岩，夹泥质粉砂薄层	13.4m
17. 灰紫、灰褐色泥质粉砂岩，下部含硅质，产笔石 *Climacograptus* cf. *latus* Elles et Wood	16.5m
16. 石英闪长岩脉	(15.0m)
15. 青灰、灰绿色泥质粉砂岩，泥质富集处呈团块状，下部层位已绿泥石化、碳酸盐化，蚀变较强烈，产笔石 *Orthograptus gracilis* (Roemer)	30.5m
14. 花岗斑岩脉	(13.3m)

———— 整　合 ————

下伏地层： 上奥陶统黄泥岗组泥质、硅质条带状（或瘤状）灰岩

（四）地质特征及区域变化

长坞组仅在上海市区西南部被揭露，岩性由黄绿—灰黑色泥岩、泥质粉砂岩、粉砂质泥岩与粉—细砂岩等组成。地层颜色较单调，沉积韵律非常清晰，具有典型的类复理石沉积特征。层内富含以 *Climacograptus* 为主的笔石化石。已见钻孔地层厚度大于 123.0m，未见顶，上部直接与第四纪疏松层接触，底部与黄泥岗组呈整合接触。

第二节　沉积作用和沉积建造

一、沉积类型

奥陶系为海相沉积，晚奥陶世初期之前沉积了一套以碳酸盐岩为主的台地沉积，之后演化为浅海陆棚类复理石沉积。主要的沉积类型有开阔海台地、台缘斜坡和浅海沉积等。
~~~~~~~~~

（一）滨海沉积

滨海沉积主要为台地边缘浅滩，出现于红花园组到黄泥岗组。自红花园组开始，局部层位开始出现明显的角砾结构，生物化石由缺乏至丰富，碎片较多。主要岩石组合为以灰岩和生物碎屑灰岩为主的碳酸盐岩类，在上海南部主要呈紫与灰、灰黑色，北部主要呈灰、灰白色，普遍含有硅质条带（或瘤状），发育不均，局部有网纹状构造。

（二）浅海沉积

浅海沉积主要为开阔海台地和陆棚浅海，分别出现于仑山组和长坞组。仑山组生物化石比较贫乏，为碳酸盐岩建造；长坞组以细碎屑陆源沉积物为主。

二、主要岩石类型

上述沉积环境中形成的主要岩石类型有灰质白云岩、白云质灰岩、生物碎屑灰岩、含生物碎屑灰岩、灰岩、泥质粉砂岩和粉砂质泥岩、细砂岩。主要特征如下：

（一）灰质白云岩

灰质白云岩主要出现在仑山组。灰色，块状构造，微—细晶结构，主要由方解石和白云石组成，两者比在1/4~1/3，含少量黄铁矿（少量褐铁矿化），平均粒径一般在0.05~0.3mm。白云石粒径较粗，多呈自形，方解石多呈他形。

（二）白云质灰岩

白云质灰岩主要出现在仑山组。灰紫色，晶粒结构，块状构造。方解石含量约占65%，白云石约占30%，铁氧化物含量小于5%，很少量的黏土矿物。白云石，自形晶，具环带构造，平均粒径为0.05~0.15mm，分布不均匀。方解石多呈泥晶，少数为粉晶，个别为内碎屑状，充填于白云石颗粒之间。少量黏土矿物混杂于方解石中。

（三）生物碎屑灰岩

生物碎屑灰岩主要出现于牯牛潭组。暗紫红色，粒屑结构，块状构造。生物碎屑约占70%，骨屑粒度多在0.1~0.8mm，全部由方解石组成，多为碎片，个别完整。主要生物有介形虫、海绵骨针、棘皮动物等，以棘皮动物最多。骨屑以泥晶方解石胶结为主，局部为亮晶方解石。胶结物铁染较明显。

（四）含生物碎屑灰岩

含生物碎屑灰岩主要出现于牯牛潭组。灰黑、灰白色，微晶结构、生物骨屑结构，以块状构造为主，局部网纹状构造。网纹主要为含钙质和生物碎片较多的帘石化泥岩。方解石含量在90%以上，其中生物骨屑含量在20%~40%，多为贝壳碎屑。平均粒径多在0.01~0.03mm，生物骨屑中方解石平均粒径可达0.05~0.1mm。含有少量的赤铁矿、磁铁矿等，含量约2%。

（五）灰岩

灰岩从红花园组到黄泥岗组的局部层位都有发育。肉红、灰白、青灰—灰黑色，多为微晶结构，少量粉晶结构和粒屑结构，块状构造，局部夹白云质灰岩或含泥量较高，表现出层理构造。白云质灰岩中局部含泥炭质。红花园组等见少量化石碎片。微晶结构中，方解石含量在90%~95%以上，无明显晶形，平均粒径多为0.01mm，局部由更小的晶体形成0.1~1mm的集合体。粒屑结构中内碎屑主要由生物碎屑或碳酸盐岩碎屑构成，形成平均粒径在1.5~5.5mm的团块，由泥质、铁质和钙质胶结物胶结。

（六）泥质粉砂岩

泥质粉砂岩主要出现在长坞组中。灰绿、青灰、深灰带褐红、灰黄绿色，粉砂状结构，块状构造。碎屑成分以石英、长石为主，石英含量多在20%~35%，长石含量多在1%~5%，局部层位岩屑含量可达45%，方解石含量可达10%。水白云母和高岭土等黏土矿物含量多在45%~60%。孔隙式、接触式胶结，碎屑呈次棱角—次磨圆状，平均粒径一般在0.05~0.08mm，大者在0.1~0.2mm，分选中等。

（七）粉砂质泥岩

粉砂质泥岩主要出现在长坞组中。深灰、灰略带黄绿色，泥状结构，以块状构造为主，局部有水平层理。水白云母和高岭石等黏土矿物含量多超过80%，局部层位在60%左右。含有少量石英、云母、方解石等碎屑，碎屑呈棱角—次磨圆状，粒径多在0.02~0.05mm，少量为0.1mm，分选中等。

第三节　地层时代和区域地层对比

一、生物群特征与地层时代

本区奥陶系在红花园组到砚瓦山组中均发现生物化石，但多为碎片，有海百合茎、介形虫、三叶虫等。

南部牯牛潭组中发现牙形石化石 *Eoconodontus notchoeakensis*，*Periodon aculeatus* Hadding，*Protopanderodus rectus*（Lindstöm），*Walliserodus ethingtoni*（Fahraeus），*Strophamena* sp.，此动物群常见于牯牛潭期，为其主要组成分子，其时代为早奥陶世晚期—中奥陶世。层内见腹足类化石 *Umbonellina* cf. *inflata* Yu，*Sinuites* sp.，*Maclurites* sp.，*Raphispira* sp.，*Liospira* sp. 为中奥陶世常见动物群。市区可见海百合茎 *Cyclocyclincus* sp.，*Pentagonocyclincus* 等。综合认为，牯牛潭组时代为早奥陶世晚期—中奥陶世。

市区长坞组中发现笔石化石 *Climacograptus typicalis*，*C.* sp.，*C. latas*，*Orthograptus gracilis*（Roemer）等，均为浙西地区 *Pseudoclimacograptus anhuiensis* 笔石带，*Dicellograptus szechuanensis* 为笔石带中主要组成分子，其地层时代为晚奥陶世。

二、区域地层对比

本区奥陶系，无论是岩石组合、沉积特点，或从所含化石（碎片）的组合类型来看，均可与浙西北、江苏和皖、赣地区对比（表4-2）。

仑山组层型剖面在邻区江苏镇江、浙西北杭州—嘉兴化石贫乏，与区内相似。与皖南和浙西北地区西阳山组顶部和印渚埠组下部、皖南上欧冲组、分乡组和大坞圩组构成同期异相，由开阔海台地相变为台地浅滩和浅海盆地沉积。

红花园组层型剖面在贵州，但在邻区浙、赣、苏已广泛应用该地层单位，在浙西北地区相当于留下组的中上部。岩性相似，为一套灰岩类组合，主要为台地边缘斜坡——浅滩环境，在皖南—赣东北—浙西北的江南主体区相变为细粒的浅海陆棚沉积。

牯牛潭组到黄泥岗组在邻近下扬子地层区被广泛共用，同样为一套台地沉积，与皖南、赣北和江苏南部的汤山组、汤头组、大湾组、紫台组和宝塔组相当。呈灰岩类岩石组合，但在晚期，皖南和江西北部、浙西北地区已经逐渐相变为类浅海陆棚沉积。相比红花园组，岩性都以出现明显大量的含瘤状构造为特征。生物相较一致，由三叶虫、头足类和腹足类、海百合等组成混合相。

长坞组为一套类复理石沉积，与五峰组和新开岭组岩石组合相似，并发现有笔石化石 *Climacograptus* 及 *Orthograptus*。上述属种是浙西北、皖南和赣北地区钱塘江阶（原五峰阶）典型的主要组成分子，尤其是 *Climacograptus*，在浙西北和皖南、湘中地区为带化石。

表 4-2　区域地层对比表

Table 4-2　Regional stratigraphic correlation

<table>
<tr><th colspan="2" rowspan="2">地区</th><th rowspan="2">上海</th><th colspan="3" rowspan="2">浙西北</th><th colspan="4">皖南</th><th colspan="3">江西</th><th>江苏</th><th rowspan="2">湘中</th></tr>
<tr><th>滁州</th><th>巢湖</th><th>青阳</th><th>宁国</th><th>九江(赣北)</th><th colspan="2">上饶(赣东北)</th><th>南京</th></tr>
<tr><td colspan="2">上覆地层</td><td>河沥溪组</td><td>西湖组</td><td colspan="2">霞乡组</td><td colspan="3">高家边组</td><td>霞乡组</td><td>梨树窝组</td><td colspan="2">霞乡组</td><td rowspan="2">高家边组</td><td>龙马溪组</td></tr>
<tr><td rowspan="19">奥陶系</td><td rowspan="7">上奥陶统</td><td>?</td><td></td><td colspan="2">文昌组</td><td colspan="3" rowspan="2">五峰组</td><td rowspan="2">长坞组</td><td rowspan="2">新开岭组</td><td rowspan="2">长坞组</td><td rowspan="2">三衢山组</td><td rowspan="2">天马山组</td></tr>
<tr><td>长坞组</td><td>长坞组</td><td>三衢山组</td><td>长坞组</td><td rowspan="5">汤头组</td></tr>
<tr><td>黄泥岗组</td><td colspan="3">黄泥岗组</td><td colspan="3">汤头组</td><td>黄泥岗组</td><td rowspan="3">汤头组</td><td colspan="2">黄泥岗组</td><td rowspan="2">宝塔组</td></tr>
<tr><td rowspan="4"></td><td>大田坝组</td><td colspan="2">砚瓦山组</td><td colspan="3">宝塔组</td><td>砚瓦山组</td><td colspan="2">砚瓦山组</td></tr>
<tr><td rowspan="3">牯牛潭组</td><td colspan="2" rowspan="3">胡乐组</td><td colspan="3" rowspan="3">庙坡组/大田坝组</td><td rowspan="3">胡乐组</td><td colspan="2" rowspan="3">胡乐组</td><td rowspan="4">烟溪组</td></tr>
<tr><td rowspan="5">汤山组</td></tr>
<tr><td rowspan="2">汤山组</td></tr>
<tr><td rowspan="5">中奥陶统</td><td rowspan="7">牯牛潭组</td><td rowspan="5"></td><td colspan="2" rowspan="5">宁国组</td><td colspan="2" rowspan="2">牯牛潭组</td><td rowspan="6">里山圩组</td><td rowspan="5">宁国组</td><td colspan="2" rowspan="8">宁国组</td></tr>
<tr><td rowspan="3">牯牛潭组</td><td rowspan="5">桥亭子组</td></tr>
<tr><td colspan="2" rowspan="4">大湾组/紫台组</td></tr>
<tr><td rowspan="5">红花园组</td></tr>
<tr><td rowspan="2">大湾组</td></tr>
<tr><td rowspan="7">下奥陶统</td><td rowspan="2">荆山组</td><td colspan="2" rowspan="6">印渚埠组</td><td rowspan="6">印渚埠组</td></tr>
<tr><td colspan="2" rowspan="2">红花园组</td><td rowspan="5">大坞圩组</td><td rowspan="2">红花园组</td><td rowspan="6">白水溪组</td></tr>
<tr><td rowspan="2">红花园组</td><td rowspan="5">留下组</td></tr>
<tr><td rowspan="2">分乡组</td><td rowspan="3">仓山组</td><td rowspan="3">仓山组</td><td colspan="2" rowspan="3">印渚埠组</td><td rowspan="3">仓山组</td></tr>
<tr><td rowspan="3">仓山组</td></tr>
<tr><td>上欧冲组</td></tr>
<tr><td colspan="2" rowspan="2">西阳山组</td><td rowspan="2">琅琊山组</td><td colspan="2" rowspan="2">青坑组</td><td rowspan="2">西阳山组</td><td colspan="3" rowspan="2">西阳山组</td><td rowspan="2">观音台组</td></tr>
<tr><td colspan="2">下伏地层</td><td>超峰组</td><td>超峰组</td><td>探溪组</td></tr>
</table>

三、存在问题

（一）奥陶系层序划分的不确定性

区内长坞组时代为晚奥陶世晚期无疑，且根据其他层位岩石组合、牙形石和腹足类化石等特征，应该存在早—中奥陶世地层；但由于钻孔有限，上述对奥陶系内部层序划分、与上覆和下伏地层的接触关系等都认识需要更多资料支撑。

（二）崇明岛西部是否存在奥陶系

崇明港西的布格重力异常高值区，长期以来都被解释为志留系或下古生界。1995 年，江苏省地质矿产局第四地质大队在进行 1：5 万区域地质调查时，将该异常的江苏段划分出奥陶系。2012—2014 年，上海市地质调查研究院在崇明岛西部近新村施工的钻孔中揭露了一套厚近 1050m 的地层，岩性主要为以灰、深灰色为主的泥质粉砂岩、粉细砂岩夹泥岩。其岩屑的孢粉特征显示该套地层的地层年代不会早于中生代；但从岩石组合上看，其与奥陶纪晚期和志留纪早期地层很相似。综上，本次推断港西存在奥陶系，但还需要更多资料的验证。

第四节　岩相古地理

仑山组沉积期，继承了寒武纪晚期的台地环境，但岩石组合显示水体应该变深，更接近开阔海台地。红花园组沉积期，出现灰黑、灰紫色碳质碳酸盐岩，并且局部夹泥质纹层，开始出现角砾状结构，表明其水深进一步加大，可能演化为潮下区的台地边缘斜坡环境。

牯牛潭组沉积期，区内继承早期的台地边缘斜坡环境，头足类、腹足类、三叶虫、海百合、牙形石生物丰富，其主要生物分子介于宁镇与浙西地区混生的生物群特征。形成一套含泥、硅质条带（或瘤状）的微晶灰岩及生物骨屑微晶灰岩，其中生物化石多呈碎片，瘤状构造和局部角砾状结构发育，反映水体较浅，水动力能量较强。

晚奥陶世早期，本区黄泥岗组碳酸盐岩建造和邻省昆山的黄泥岗组及宁镇的汤头组相似，均具有瘤状构造或夹有较多的泥岩，灰岩中泥质增多。在黄泥岗组沉积期，邻区皖南、江西和浙西北地区已结束台地沉积，演化为陆棚沉积；但区内主要仍呈台地边缘斜坡环境，但水深增加。

长坞组以泥砂质碎屑沉积为主，具纹层理，并富含笔石化石。表明随着奥陶纪海侵自西南向东北的逐渐推进，区域上海侵达到高峰，区内已演化为陆棚浅海环境，水体明显变深，形成较厚的类复理石沉积建造。

第五章　志留系

志留系目前仅见于崇明岛西北部庙镇—草棚一带（图 2-1），是一套以砂岩、泥岩和页岩为主的碎屑岩建造，钻孔已揭露厚度约为 656. 7m。迄今只有两个钻孔揭露，分别施工于 20 世纪 50 年代末期、21 世纪初。地层上未见顶，下未穿底。

志留系研究程度很低，对上述揭露的这套地层的时代归属，历年来争论不休。1962 年，在《上海地区石油及天然气普查总结》中将其划定为白垩系。20 世纪 70 年代，又将其划为寒武系—震旦系，并认为区内缺失志留系。1981 年，许保桐在《上海地区地层概述》一文中首次将其划为志留系，但依据不足。在《上海市区域地质志》中延续了这一认识。2003 年，许保桐等根据新施工钻孔中的孢粉化石，将志留系划分为下志留统河沥溪组和康山组。2010 年，在“上海市三维地质调查”项目中，将康山组划为中—下志留统（表 5-1）。

表 5-1　上海志留系岩石地层单位划分沿革表

Table 5-1　Previous divisions of the Silurian lithostratigraphic units in Shanghai

	《上海地区石油及天然气普查总结》，1962	《1：20万上海地区基岩地质推断图》，1979	《上海地区地层概述》，1981	《上海市区域地质志》，1988	“上海市三维城市地质调查”，2010	本志	
上覆地层	上侏罗统	上侏罗统	上侏罗统	?	上侏罗统	下白垩统	上覆地层
上志留统	?		未分	未分			顶志留统
							上志留统
中志留统					?	?	中志留统
					康山组	康山组	
下志留统					河沥溪组	河沥溪组	下志留统

第一节　岩石地层划分

一、河沥溪组（$S_1 h$）

（一）创名及原始定义

由安徽省地质局 317 队（1965）创名，命名地点在安徽宁国河沥溪。原始定义：指整合于下伏霞乡组与太平群之间的一套浅海相具类复理石韵律的碎屑岩沉积。

（二）现在定义

指分别整合于下伏霞乡组与上覆康山组之间的一套灰绿、黄绿夹紫红色中厚层状岩屑石英砂岩或石英细砂岩、粉砂岩或泥质粉砂岩、粉砂质泥页岩组成的类复理石沉积。区内主要为一套以灰、微绿色泥质砂岩、粉砂质泥岩为主的类复理石沉积。

（三）层型剖面描述

次层型剖面为崇明庙镇崇西中学操场内庙镇孔剖面（图 5-1）。剖面地层分述如下：

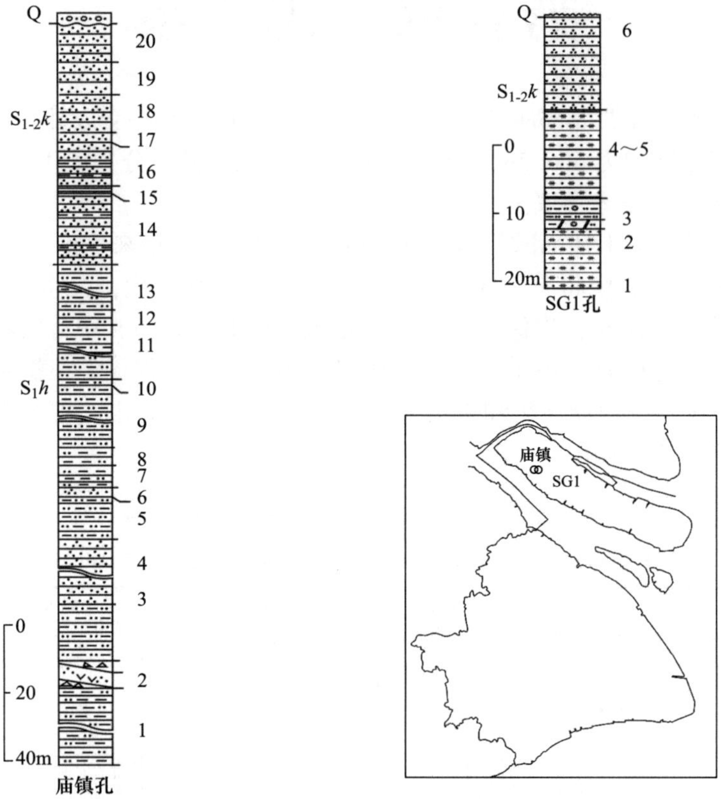

图 5-1　次层型钻孔剖面柱状图

Fig. 5-1　Hypostratotype borehole section histograms

上覆地层：康山组褐灰、灰色岩屑石英杂砂岩，间夹有浅灰色泥质粉砂岩与泥质细晶白云岩等。

———— 整　合 ————

河沥溪组（S_1h）	总厚度>489.3m
13. 灰色泥质砂岩，夹薄层暗灰绿色砂质泥岩，间有岩脉穿插	33.2m
12. 灰色粉砂质泥岩和页岩，微带绿色	6.1m
11. 灰、微绿色泥质砂岩，夹薄层粉砂质泥岩，含泥砾及紫红色斑点	66.1m
10. 灰褐色粉砂质泥岩	2.4m
9. 灰微带绿色泥质砂岩，夹薄层粉砂质泥岩（中部可能为断裂破碎带?）	95.0m
8. 灰黑色粉砂质泥岩，夹灰绿色泥质砂岩	16.7m
7. 灰、微绿色石英砂岩	4.0m
6. 灰、微绿色泥质砂岩与粉砂质泥岩互层	18.7m
5. 暗绿灰色石英砂岩，夹粉砂质泥岩	79.6m
4. 灰、微绿色泥质砂岩，夹粉砂质泥岩，含泥砾	25.3m
3. 粉砂质泥岩夹泥质砂岩，因受挤压而破碎，呈角砾状	5.5m
2. 灰白色石英安山岩脉	(17.2m)
1. 烟灰色粉砂质泥岩，夹灰、微绿色石英砂岩	>136.7m

未　见　底

（四）地质特征及区域变化

区内该组地层一般颜色较深，以灰、微绿、灰绿色为主，局部呈灰黑和灰褐色。岩石组合主要为泥质砂岩、粉砂质泥岩互层，局部夹有页岩或石英砂岩薄层。层内普遍见有泥砾，大小不一。已见该组地层厚度较大，钻孔揭露最大厚度仅500m左右，未见底，与上覆康山组呈整合接触关系。

二、康山组（$S_{1-2}k$）

（一）创名及原始定义

原称之“康山层”，系浙江省煤炭工业厅科学研究所（1961）创名，命名地点在浙江安吉康山—郭笑山一带。原始定义：岩性为黄绿色页岩与浅色页岩、细砂岩互层。自下而上分为第一页岩层、第二砂岩层、第二页岩层和第三砂岩层、第三页岩层。

（二）现在定义

康山组下部以黄绿色厚层-块状长石石英砂岩为主，夹粉砂岩、粉砂质泥岩；中部为黄绿色粉砂质泥岩、粉砂岩与砂岩互层；上部为灰绿、紫红色砂岩、粉砂岩、泥岩互层。以大套砂岩与下伏河沥溪组、以紫红色层与上覆唐家坞组均呈整合接触。区内以灰、灰黑、灰褐、紫红色岩屑石英杂砂岩、石英砂岩为主，夹泥质粉砂岩、泥岩等或成互层状。

（三）层型剖面描述

次层型剖面为崇明庙镇崇西中学操场内庙镇孔和中心幼儿园边SG1孔剖面（见图5-1）。剖面地层分述如下：

1. 新庙镇孔（SG1）剖面

上覆地层：第四系疏松层

~~~~~~~~~~ 不　整　合 ~~~~~~~~~~

康山组（$S_{1-2}k$）		总厚度>40.5m
6.	风化之紫红色石英砂岩	14.1m
5.	浅褐色细粒岩屑石英杂砂岩，受轻微绢云母化，略显纹理，含 *Dictyotidium* sp. 化石	9.8m
4.	褐灰色纹层状岩屑石英杂砂岩，含有褐铁矿而呈褐色	3.0m
3.	浅灰色纹层状泥质粉砂岩，泥质粉砂岩与泥岩、粉砂质泥岩呈纹层状互层，液化流动构造发育，含孢粉 *Leiosphaeridia* sp.，*Synsphaeridium* sp.，*Dactylofusa cabottii*，*Nodospora* sp.，*Tetrahedraletes* sp. 等化石	3.3m
2.	灰白色含粉砂质泥质细晶白云岩，局部夹富细砂的砂层，呈斑块状分布，似生物潜穴充填（?），含孢粉 *Dactylofusa cabottii*，*Nodospora* sp.，*Leiosphaeridia* sp. 化石	1.5m
1.	灰黑、深灰色细粒岩屑石英杂砂岩，轻微变质，含 *Dactylofusa cabottii* 化石	>8.8m

未　见　底

2. 庙镇孔剖面

上覆地层：第四系疏松层

~~~~~~~~~~ 不　整　合 ~~~~~~~~~~

康山组（$S_{1-2}k$）		总厚度120.4m
20.	褐紫红色石英砂岩（顶部因风化而破碎）	15.7m
19.	紫红与灰色石英砂岩互层，含泥砾	13.6m
18.	灰黑、紫红色薄层石英砂岩，夹少量薄层页岩，含泥砾	16.3m
17.	灰黑、紫红色石英砂岩	3.6m
16.	灰黑色石英砂岩，夹薄层砂质页岩，近下部见较多的泥砾，在砂质页岩层面上有黑色炭质薄膜	18.5m
15.	黑色页岩	4.0m

14. 灰黑色石英砂岩，夹薄层粉砂质页岩，近下部有较多的泥砾　　28.3m

（四）地质特征及区域变化

区内岩性上部以紫红、灰黑色岩屑石英杂砂岩、石英砂岩为主，下部以灰褐—灰、灰黑色岩屑石英杂砂岩为主，夹有粉砂质泥岩、泥岩之薄层，普遍受到轻微变质影响。胶结物以硅质和泥质为主。层内见有丰富藻类及孢子等微古植物化石。所见地层未见顶，厚度大于40.0m。

第二节　沉积作用和沉积建造

一、沉积作用

志留系主要为滨海沉积，具体为有障壁的潮坪相，形成一套碎屑物粒度由细变粗的砂岩、泥岩或页岩建造。主要岩性有石英砂岩、泥质砂岩、岩屑石英杂砂岩、粉砂岩、泥（页）岩、泥质粉砂岩、砂质泥岩、粉砂质泥岩。

二、主要岩石类型

上述沉积环境中形成的主要岩石类型特征如下：

（一）石英砂岩

石英砂岩在河沥溪组和康山组都有出现，多呈灰、灰褐、灰白色，块状构造，部分具平行层理构造。多呈细粒砂质结构，粒径一般在0.1mm×0.14mm～0.18mm×0.24mm。部分中粒砂质结构，粒径一般在0.22mm×0.32mm～0.24mm×0.36mm。碎屑主要成分为石英，具有波状消光，分选好，含量多达90%以上，个别层位大幅减少。其次为长石，部分绢云母化。还可见白云母、角闪石、磁铁矿等。碎屑多呈次棱角—次磨圆状，少数棱角状、磨圆状。长石磨圆度一般较石英差。基底式胶结，部分为接触式胶结，胶结物以硅质和泥质胶结为主，局部以泥质为主。

（二）泥质砂岩

泥质砂岩主要出现在河沥溪组，深灰色，块状构造，砂质结构，基底式胶结，局部层位接触式胶结，胶结物以泥质、硅质、绢云母质为主。碎屑主要为石英，以棱角—次棱角状为主，局部具波状消光。次为长石，次棱角状。见有次棱角—次圆状硅质岩岩屑。

（三）岩屑石英杂砂岩

岩屑石英杂砂岩主要出现在康山组，灰色，碎屑结构，平行层理。碎屑颗粒为细砂级，多为石英、岩屑，多数碎屑颗粒直径为0.15～0.25mm，少数碎屑颗粒为极细砂级，碎屑颗粒磨圆程度一般，以次棱角—次圆状为主。碎屑颗粒中石英约占40%，部分石英颗粒有次生加大边，岩屑颗粒占60%。杂基含量较高，达到20%，为泥质黏土矿物，岩石为杂基支撑，基底式胶结。

（四）粉砂岩

粉砂岩主要出现在河沥溪组，灰黑色，粉砂质结构，块状构造，基底式胶结，部分接触式胶结，胶结物主要为硅质和泥质。石英碎屑呈棱角—次圆状，粒径一般为0.06mm×0.08mm，部分具波状消光。长石，半次圆状，具钠长石双晶。还可见白云母、黑云母、磁铁矿等碎屑。

（五）泥（页）岩

泥（页）岩在河沥溪组和康山组都有出现，灰褐色，泥质结构，块状构造或水平层理构造。基质主要由黏土矿物、绢云母组成，粒径小于0.01mm。微含粉砂级石英、长石、磁铁矿等，粒径一般为0.02mm×0.02mm。

（六）泥质粉砂岩

泥质粉砂岩主要出现在河沥溪组，灰褐色，块状构造，泥质粉砂结构，以基底式胶结为主，胶结物主要为泥质，个别为硅质和鳞片状云母。碎屑主要由石英、长石组成，分布不均匀，分选不好。棱角—次圆状，部分具波状消光。粒径一般为 0.06mm×0.1mm。长石个别具聚片双晶。此外，碎屑还见燧石、磁铁矿和云母条带。

（七）粉砂质泥岩

粉砂质泥岩在河沥溪组和康山组都有出现，黑、灰褐色，粉砂泥质结构，块状构造，局部具有微层理构造。基质主要由泥质微粒组成，粒径多小于 0.01mm。主要碎屑矿物为石英，棱角—次棱角状，粒径一般为 0.04mm×0.08mm，部分具波状消光。还见有磁铁矿、黄铁矿和白钛石碎屑。

（八）砂质泥岩

砂质泥岩在河沥溪组和康山组都有出现，灰褐色，砂泥质结构，块状构造。基质主要由泥质微粒组成，粒径多小于 0.01mm。基质中含细砂粒级石英、长石，石英呈棱角—次圆状，长石呈次棱角—次圆状。

第三节　地层时代和区域地层对比

本区于新庙镇孔（SG1）中发现众多微古化石，有节球孢未定种 *Nodospora* sp.、光面球藻未定种（*Leiosphaeridia* sp.）、连球藻？未定种（*Synsphaeridium*? sp.）、四胞体未定种（*Tetrahedraletes* sp.）、卡博特指棱藻（*Dactylofusa cabottii*）以及网面球藻未定种（*Dictyotidium* sp.），上述微古化石均为在国内外已知早志留世常有的典型孢粉型化石。在浙江余杭康山组中，就含有以上诸多化石，特别是卡博特指棱藻、首球孢、四胞体等。

区内河沥溪组仅在旧庙镇孔被揭露，该孔中、下部揭露的一套地层与上部康山组截然不同，因无化石依据，仅以岩性对比，应与浙江余杭下志留统下部河沥溪组相一致，均以灰、灰褐色砂岩与粉砂质泥岩、泥岩互层为主。

早志留世晚期—中志留世，区域上主要为潮坪、过渡带和陆棚浅海沉积。河沥溪组、康山组与坟头组、茅山组和清水组在不同时段构成上述相变沉积（表 5-2）。随着海水变浅，气候变暖，扬子地区的志留系自上而下发育了 3 套浅水红层，分别为中志留统上部（罗德洛统红层）、下志留统的上红层（特列奇阶下部）和下红层（特列奇阶中下部）（戎嘉余，2012）。区域上，河沥溪组、清水组和溶溪组构成下红层，回星哨组和唐家坞组、茅山组构成上红层。

表 5-2　区域地层对比表

Table 5-2　Regional stratigraphic correlation

地区		上海	浙西北	皖南 巢湖	皖南 滁州	皖南 宁国	江西 九江(赣北)、武宁-修水	江西 上饶(赣东北)	江苏 南京	湘西北
上覆地层		寿昌组	西湖组/五通群	五通群		黄龙组	五通群	五通群	五通群	云台观组
志留系	顶志留统									
	上志留统									
	中志留统	?	唐家坞组		茅山组	唐家坞组	茅山组	唐家坞组	茅山组	回星哨组
							坟头组	康山组		吴家院组
	下志留统	康山组	康山组	坟头组		康山组	清水组	河沥溪组	坟头组	辣子壳组
		河沥溪组	河沥溪组			河沥溪组				溶溪组
		?	霞乡组	高家边组		霞乡组	殷背组	霞乡组	高家边组	小河坝组
							梨树窝组			新滩组
下伏地层		长坞组	长坞组/文昌组	五峰组		长坞组	新开岭组	长坞组	汤头组	龙马溪组

奥陶纪晚期，区内已演化为浅海环境，从区域地层的发育特点和构造演化看，区内下—中志留统似基本可与邻区对比。尤其是志留纪初期，区域上整体为陆架、过渡带和临滨环境，霞乡组或高家边组沉积缺失的可能性很小。然而，目前根据岩性对比和有限的化石资料，仅仅初步划分出了河沥溪组和康山组，这一认识还需要更多资料的支撑。

第四节　岩相古地理

本区已发现的志留纪地层，颜色由下部的灰绿、灰黑色往上变成紫红或暗红色，碎屑物粒度则由细变粗，显示出一个较完整的海退序列。其下部由灰绿色砂岩、粉砂岩、泥岩组成一套类复理石建造，局部层位中含有泥砾，可能属潮下高能环境的沉积。上部，以紫红色砂岩为主，具有强氧化作用的潮坪沉积特点，几乎普遍含有泥砾，这显然是由于水体变浅，水动力能量增加造成的。

第六章　白垩系

区内迄今未见晚古生代与三叠纪—侏罗纪地层。根据区域地层发育特点，本志暂将下—中泥盆统、中—上三叠统和上侏罗统做缺失处理，其他地层作未发现处理。

现揭露中生界全部为白垩系。其中，以下白垩统分布最广，约占全市面积的70%以上。上白垩统分布范围局限，主要分布于上海西北部与江苏毗邻的安角盆地、中部的松江盆地、崇明岛西段的北缘（图6-1）。下白垩统为中、酸性火山岩及火山碎屑-沉积岩系，自下而上划分为劳村组、黄尖组和寿昌

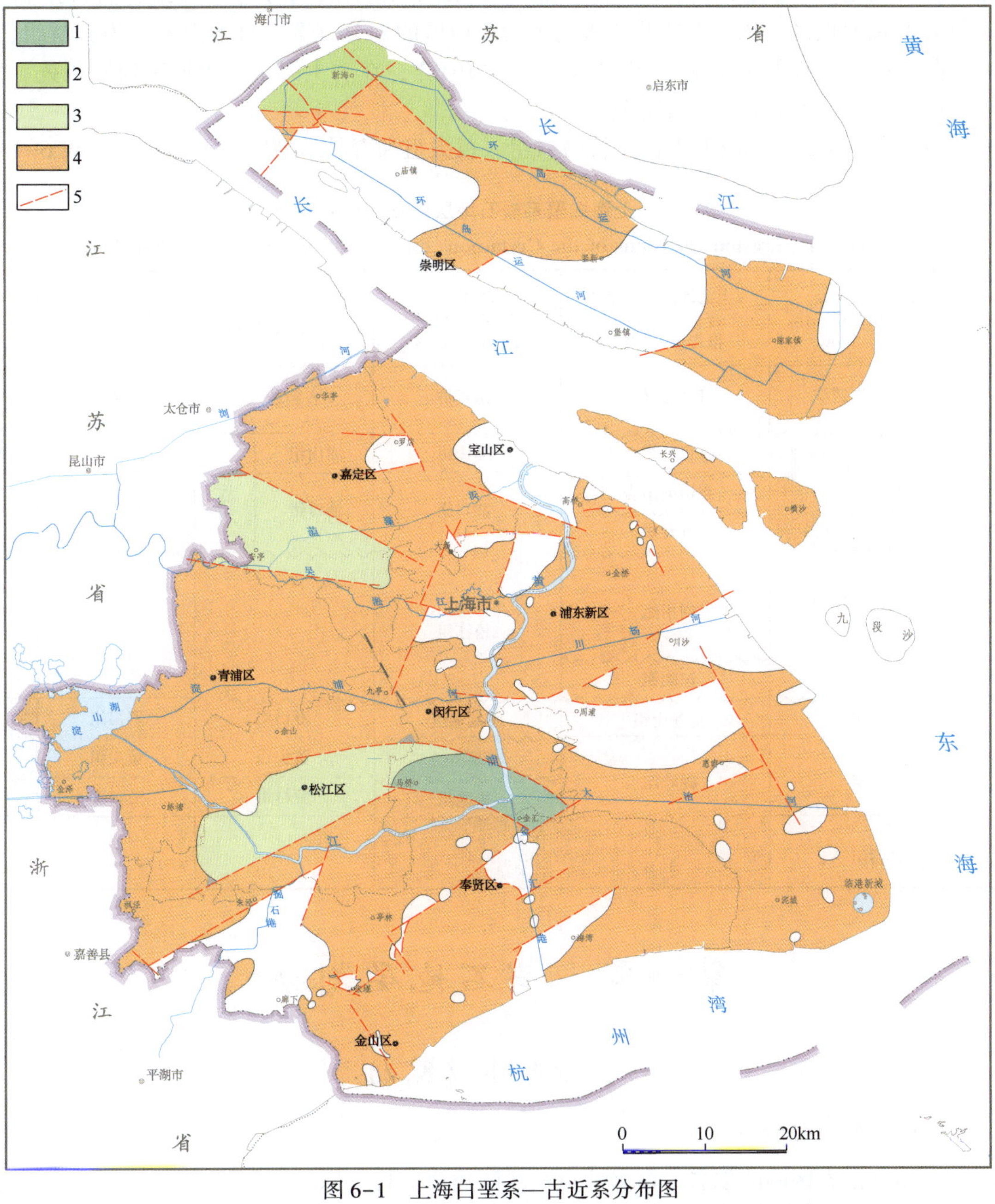

图6-1　上海白垩系—古近系分布图

Fig. 6-1　Distribution of the Cretaceous-Paleogene Systems in Shanghai

1—古近系；2—上白垩统—古近系；3—上白垩统；4—下白垩统；5—断裂

组。其中，劳村组以沉积岩为主；黄尖组为火山岩和火山碎屑岩建造；寿昌组则由下部河湖—沼泽相及上部火山岩-火山碎屑岩组成。上白垩统为一套断陷盆地陆源碎屑岩沉积，自下而上划分为浦口组和赤山组。浦口组以砂砾岩-砾岩为主；赤山组以泥质粉砂岩与粉砂质泥岩为主，在松江等地夹有石膏。

白垩系尽管揭露钻孔较多，但多揭露厚度不大，下白垩统和上白垩统揭露的最大钻孔厚度分别为1850.7m、789.56m（邻区安深1#钻孔厚可达2614.5m）。各组往往顶、底不全。

区内中生代地层研究程度较差，划分方案多种多样。长期以来，将全市大面积分布的地层划分为侏罗系，松江、金山卫和安亭等局部钻孔中划分出白垩系。

1960年，上海市人民委员会地质处，首次将中生界划分为中侏罗统儒岙组、上侏罗统砚岭组和寿昌组，下白垩统砂砾岩和流纹岩两个岩性段。1962年，上海市地质勘察局重新将其划分为上侏罗统括苍山组、砚岭组，下白垩统馆陶组、朝川组和上白垩统方岩组、壳山火山岩和金芒组。20世纪70年代，上侏罗统又被划分为松隐组、天马山组和唐家宅组，白垩系划分为松江组、浦口组和赤山组。1988年，通过与浙西地区对比，上侏罗统采用了浙西地区的划分方案，自下而上分为劳村组、黄尖组和寿昌组；白垩系划分为浦口组/松江组和赤山组。2010年，在“上海市三维城市地质调查”项目中，进一步将松江组归并入赤山组，并沿用至今。

根据最新研究进展，本志将原上侏罗统劳村组、黄尖组和寿昌组划归为下白垩统（表6-1）。

表6-1　上海白垩系岩石地层单位划分沿革表

Table 6-1　Previous divisions of the Cretaceous lithostratigraphic units in Shanghai

地层	《上海市石油及天然气地质普查工作报告》，1960	《上海地区矿产地质普查工作总结报告》，1962	《1∶20万上海地区基岩地质推断图》，1979	《上海市区域地质志》，1988	“上海市三维城市地质调查”，2010	本志
上覆地层	始新统—渐新统	巨江红层	甪直群	古新统	古新统	古新统
上白垩统		金芒组 壳山火山岩 方岩组	赤山组 浦口组	赤山组 浦口组 松江组	赤山组 浦口组	赤山组 浦口组
下白垩统	流纹岩段 砂砾岩段	朝川组 馆陶组	松江组	?	?	? 寿昌组 黄尖组 劳村组
上侏罗统	寿昌组 砚岭组	括苍山组 砚岭组	唐家宅组 天马山组 松隐组	寿昌组 黄尖组 劳村组	寿昌组 黄尖组 劳村组	
中侏罗统	儒岙组			?	?	?

第一节　岩石地层划分

一、劳村组（K_1l）

（一）创名及原始定义

1965年，由浙江省区调队创名。命名地点在浙江建德劳村。原始定义：岩性为一套暗紫色泥质粉砂岩，夹不稳定的流纹质凝灰岩和少量砾岩、黄绿色砂岩、粉砂岩等。局部含钙质结核及透镜状灰岩，底部有不稳定的砾岩。

（二）现在定义

岩性为暗紫色泥质粉砂岩、砂岩夹不稳定的流纹质凝灰岩、流纹岩和少量砾岩、黄绿色砂岩、粉砂岩。在寿昌—临浦一带的东段以火山岩为主，夹沉积岩。区内主要为灰白、灰紫色相间的凝灰质砂岩、凝灰质砂砾岩、凝灰质砾岩、流纹质凝灰熔岩及砂岩。

（三）层型剖面描述

次层型剖面为金山松隐孔剖面（图 6-2）。剖面地层分述如下：

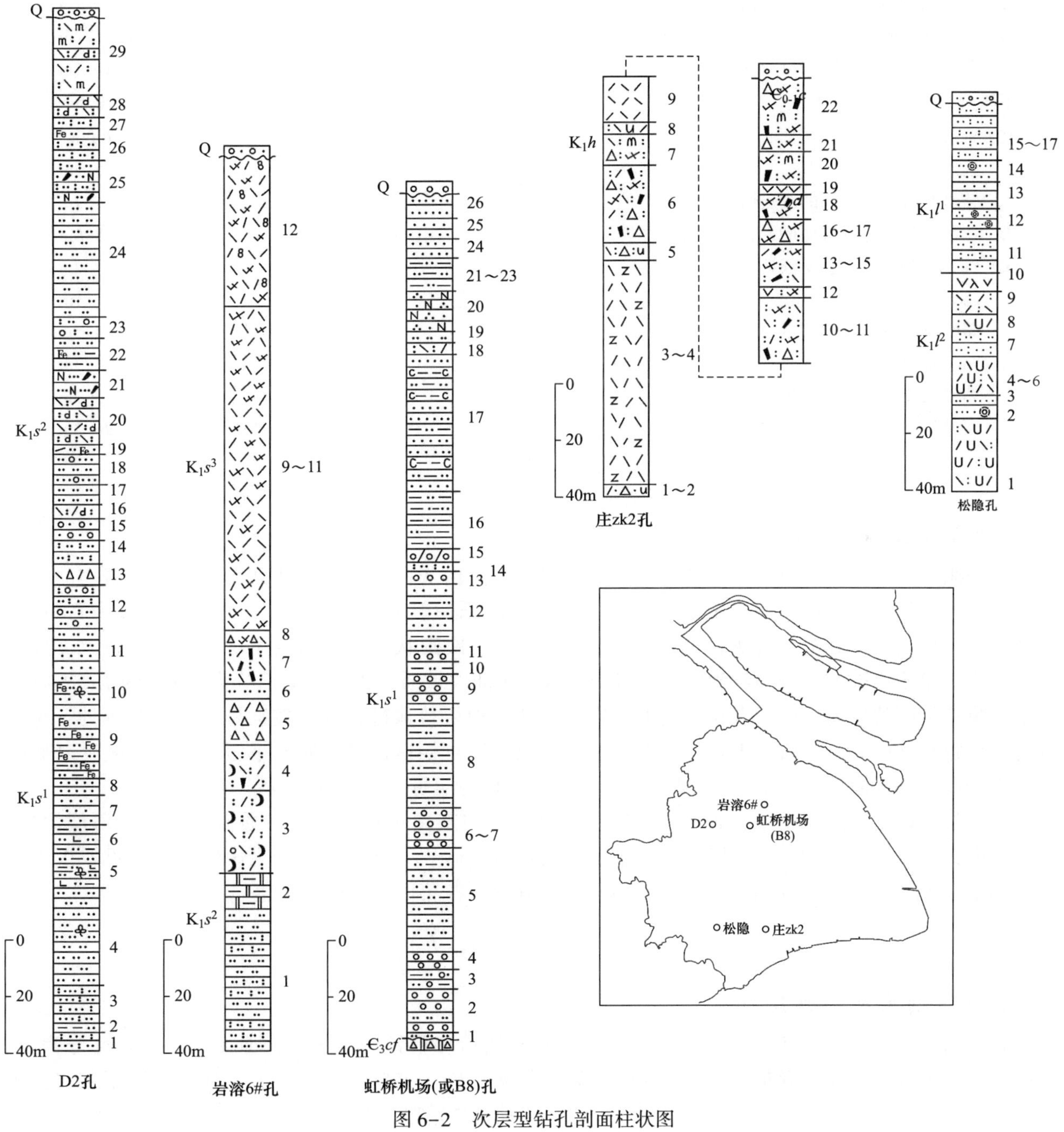

图 6-2 次层型钻孔剖面柱状图

Fig. 6-2 Hypostratotype borehole section histograms

上覆地层：第四系疏松层

~~~~~~~~~~ 不 整 合 ~~~~~~~~~~

劳村组（K_1l）

上段（K_1l^2）	总厚度 61.7m
17. 紫红与灰白色相间的凝灰质细砂岩，薄层状，层理清晰	5.2m
16. 紫红及棕红色凝灰质细砂岩，底部有一层细砾岩	8.0m
15. 棕红色凝灰质细砂岩，底部有一层细砾岩	8.0m
14. 紫红、肉红与灰白色相间薄层状硅质粉砂岩夹泥岩，具交错层理	8.0m
13. 暗紫红色粗砂岩，常具灰白色斑点	10.2m
12. 暗紫红色硅化石英砂岩，具交错层理	6.2m
11. 紫红、青白色相间的凝灰质砂岩	16.1m
下段（K_1l^1）	总厚度>70.5m
10. 青灰色安山玢岩	（7.0m）
9. 紫红色凝灰岩	8.2m
8. 紫红与青白色相间的凝灰熔岩	6.7m
7. 紫红与赭红色凝灰质砂岩，夹紫红色泥岩薄层	8.6m
6. 红、青灰及灰白色凝灰熔岩	7.2m
5. 暗紫红色硅质砂岩	1.1m
4. 紫红与青灰色凝灰熔岩	4.5m
3. 暗紫红色细砂岩	4.3m
2. 紫红与青灰白色硅质砂岩	3.7m
1. 暗紫红夹灰白色凝灰熔岩	>26.2m

未 见 底

（四）地质特征及区域变化

劳村组下段广布于上海东南部奉贤、金山境内，主要由灰白、灰紫色相间的凝灰质砂岩、凝灰质砂砾岩、凝灰质砾岩、流纹质凝灰熔岩及砂岩组成。砂砾一般为长石、石英；砾石成分多为石英岩、花岗岩、安山岩、片麻岩、凝灰岩等。砾石大小不一，呈次圆状，少数为棱角状，分选性较差，局部可见砾石，呈定向排列。胶结物为硅质、凝灰质岩、钙质及铁质。以细碎屑组分为主的凝灰质砂岩中，一般均见有清晰的层理。层间有较多安山玢岩脉穿插，地层普遍硅化，且较强烈。在松隐等处，见紫红与灰白色相间的凝灰熔岩，由石英及火山玻璃组成，流纹构造清晰，倾角平缓。所见厚度受基底隆拗构造控制，如吴泾厚逾 237.6m 尚未见底，而在肖塘附近，厚仅 18.0m 之多。

劳村组上段主要见于松隐、张堰。松隐主要为紫红与灰白色相间的凝灰质砂岩，间夹薄层泥岩，层理清晰，并见波状层理和交错层理；张堰多为长石石英砂岩，基本上不含火山物质。已见厚度大于 167.9m，劳村组与下伏金山群呈不整合接触关系，其上多被第四系疏松层覆盖。

二、黄尖组（K_1h）

（一）创名及原始定义

1964 年，由邹鑫佑等创名。创名地点在浙江省建德市寿昌镇南黄尖山。原始定义：岩性分为上、下两部分，上部为流纹质凝灰熔岩、玻屑凝灰岩、凝灰角砾岩夹紫红色流纹岩，顶部为层凝灰岩夹紫红色粉砂岩；下部为紫、灰色流纹岩、流纹斑岩，流动构造清晰。

（二）现在定义

本组岩性以酸性熔岩或酸性火山碎屑岩夹中性熔岩为主，偶夹沉积岩薄层。区域上，在不同的火山活动区或同一区的不同火山构造部位，岩性和厚度变化较大。区内主要为英安岩、安山岩、英安质
~~~~~~~~~~

凝灰岩等中酸性火山岩和火山碎屑岩。

(三) 层型剖面描述

次层型剖面为奉贤庄行庄 zk2 孔剖面（见图 6-2）。剖面地层分述如下：

上覆地层：第四系疏松层

~~~~~~~~~~ 不　整　合 ~~~~~~~~~~

黄尖组（K_1h）	总厚度>249.7m
22. 英安质含角砾晶屑熔结凝灰岩，具流纹构造，见安山岩角砾	19.9m
21. 英安质含角砾晶屑凝灰岩	5.5m
20. 英安质晶屑熔结凝灰岩	11.4m
19. 安山岩	2.1m
18. 英安质含岩屑晶屑凝灰岩，岩屑成分为安山岩、花岗斑岩、片岩等	10.5m
17. 英安质含角砾晶屑凝灰岩	7.9m
16. 碳酸盐化英安质晶屑凝灰岩	0.6m
15. 流纹英安质晶屑凝灰岩	3.9m
14. 碳酸盐化英安质含角砾凝灰岩	0.8m
13. 流纹英安质晶屑凝灰岩	11.1m
12. 安山英安质角砾凝灰岩	2.6m
11. 流纹英安质含角砾晶屑凝灰岩	23.0m
10. 英安质含角砾凝灰熔岩	0.7m
9. 流纹岩，偶见安山岩尖棱角状角砾，最大砾径 1cm	16.4m
8. 流纹质凝灰熔岩	3.5m
7. 流纹英安质含角砾熔结凝灰岩，含安山岩岩屑 2%~3%	11.5m
6. 流纹英安质含角砾晶屑凝灰岩，角砾为安山岩	27.3m
5. 流纹质含角砾凝灰熔岩	6.3m
4. 霏细斑岩	76.0m
3. 霏细岩	5.2m
2. 碳酸盐化流纹英安质含角砾凝灰熔岩，角砾成分为安山玢岩	1.9m
1. 流纹英安质含角砾晶屑熔结凝灰岩	>1.6m

未　见　底

(四) 地质特征及区域变化

黄尖组主要分布上海中部和南部地区，岩性主要为英安岩、流纹英安岩、英安质含角砾熔结凝灰岩、流纹英安质熔结凝灰岩、熔结凝灰岩等中酸性火山岩和火山碎屑岩。并往往和浅成、超浅成的次火山岩组成火山杂岩。在金桥及天马山（图 6-3）为安山玢岩、安山岩和英安玢岩、流纹岩。单孔所见最大厚度可达 484.0m。与上覆寿昌组接触关系不明。

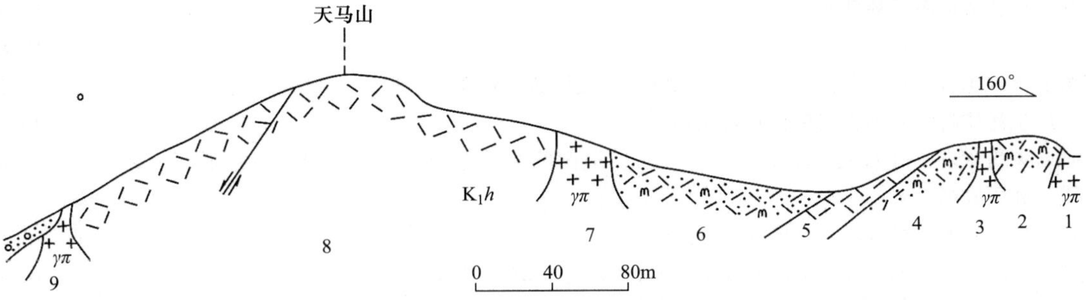

图 6-3　松江天马山下白垩统黄尖组实测剖面图

Fig. 6-3　Measured section of the Lower Cretaceous Huangjian Formation in Tianmashan, Songjiang District

1，3，7，9—花岗斑岩；2，4，6—熔结凝灰岩；5，8—流纹斑岩
~~~~~~~~~~

三、寿昌组（K_1h）

（一）创名及原始定义

1962年，由顾知微等人创名。命名地点为浙江省建德市寿昌镇。原义是指“寿昌城南剖面发育于蔡郎岗组之上的一套以杂色砂、页岩、黏土夹凝灰质沉积为特征，仅顶部50m为红色砂、页岩”的地层。

（二）现在定义

寿昌组岩性为灰绿、黄绿、紫红等杂色砂、页岩，中部和顶部各有一层厚度不稳定的酸性火山岩，以灰色凝灰质砂岩、含砾砂岩与下伏黄尖组熔结凝灰岩，以流纹质晶屑玻屑凝灰岩、熔结凝灰岩与上覆横山组紫红色含砂粉砂岩均呈整合接触。

（三）层型剖面描述

次层型剖面为虹桥机场孔、D2孔、岩溶6#孔剖面（见图6-2）。剖面地层分述如下：

1. 虹桥机场孔剖面

上覆地层：第四系疏松层

～～～～～～ 不　整　合 ～～～～～～

寿昌组下段（K_1h^1）	总厚度300.7m
26. 黄灰色中细砂岩夹灰紫色粉细砂岩，钙质及泥质胶结	9.7m
25. 黄灰色粗砂岩，含少量细砾，微含钙质	6.3m
24. 灰绿夹灰紫色中细砂岩	6.3m
23. 紫红色泥质粉砂岩	2.4m
22. 灰白色含砾砂岩	2.4m
21. 紫红色泥质粉砂岩夹细砂岩，具灰黑色条带和团块	6.8m
20. 青灰色石英长石中细砂岩	14.5m
19. 灰黑色具黑色条带粉砂岩，钙结胶结	4.3m
18. 酸性凝灰岩，夹有黑色条带和暗红色斑点	4.3m
17. 青灰色中细砂岩夹黑色泥质粉砂岩和黑色炭质条带	48.3m
16. 灰紫、紫红色夹青灰色泥质粉砂岩与粉砂质泥岩互层	20.1m
15. 灰白色含细砂泥灰岩	4.5m
14. 凝灰质砂岩夹砾岩，砾石滚圆状，大的砾径5~6cm，成分为灰岩、石英岩、燧石等，钙质胶结	3.1m
13. 灰白色砾岩夹紫红色泥质砂岩，砾石大小不等，大的5~6cm，滚圆状，成分以灰岩为主	4.7m
12. 紫红色粉细砂岩及粉砂质泥岩	23.7m
11. 灰红色砾岩，砾石成分有灰岩、砂岩、燧石等，大的5~6cm，钙质、砂质胶结	3.9m
10. 紫红色泥质粉砂岩夹粉砂质泥岩	4.8m
9. 砾岩，砾石大者7~8cm，滚圆状，成分有灰岩、紫红色泥岩、凝灰质细砂岩、石英岩等，钙质胶结	9.7m
8. 紫红夹青灰色泥质粉砂岩	37.8m
7. 黄灰色砂砾岩、砾岩夹细砂岩	10.6m
6. 黄灰色砂砾岩、砾岩夹粉细砂岩，具细脉状、浸染状黄铁矿	3.9m
5. 暗红色泥质粉砂岩夹青灰色粉砂岩和粉砂质泥岩	36.7m
4. 砾岩夹粉砂岩，砾石大小不等，在6~7cm，滚圆状，成分为灰岩、砂岩及少量燧石，钙质、砂质胶结	6.8m
3. 黄灰色含砾泥质粉砂岩，见星散状黄铁矿	6.8m
2. 砾岩夹粉砂岩，砾石大小不均，棱角状和圆球状，钙质、砂质、铁质胶结	16.4m
1. 青灰、白略带红色粉砂岩	1.9m

～～～～～～ 不　整　合 ～～～～～～

下伏地层：上寒武统超峰组灰黑、灰色角砾状白云岩、灰质白云岩

2. D2 孔剖面

上覆地层：第四系疏松层

~~~~~~~~ 不 整 合 ~~~~~~~~

寿昌组（K_1h）

中段（K_1h^2） 总厚度 214.1m

29. 棕黄色熔结凝灰岩夹灰紫色沉凝灰岩 26.0m
28. 紫红及灰黄色沉凝灰岩，局部见水平层理 8.0m
27. 暗红色凝灰质粉砂岩及泥质粉砂岩 7.2m
26. 灰黄色凝灰质中粗砂岩，顶部为含砾粗砂岩 8.1m
25. 灰紫色凝灰质砂岩，由上至下逐渐变粗，局部夹灰绿色薄层凝灰质结晶灰岩 15.1m
24. 紫红色粉砂岩 38.6m
23. 紫红色凝灰质含砾中粗砂岩，砾石含量为 6%~10%，砾径为 0.2~2cm，次圆状 7.7m
22. 紫红色铁泥质粉砂岩，顶部为细砂岩 11.0m
21. 黏土化岩屑细砂岩 9.6m
20. 绿灰色沉凝灰岩，下部为灰白色含砾凝灰质粗砂岩 16.6m
19. 紫红色泥质粉砂岩 3.3m
18. 灰白色含砾中粗砂岩，层理清晰 10.7m
17. 青灰—灰黑色粉砂岩，含植物化石碎片，微层理清晰 6.9m
16. 紫红色沉凝灰岩 5.3m
15. 青灰—灰红色砂砾岩，砾石含量为 40%~50%，棱角状—半圆状，成分为安山岩、流纹岩、花岗细晶岩等 7.7m
14. 灰白—灰黑色凝灰质砂岩 8.9m
13. 灰白色火山角砾岩，顶部有厚为 0.8m 的泥质粉砂岩 7.3m
12. 灰白色含砾凝灰质粗砂岩 16.2m

下段（K_1h^1） >152.9m

11. 青灰、灰黑色粉细砂岩，含植物化石碎片 16.0m
10. 灰白色中砂岩夹暗红色泥质粉砂岩，含植物化石碎片 15.5m
9. 紫红色泥质粉砂岩，微层理清晰，见植物化石碎片 22.0m
8. 青灰色中砂岩 5.5m
7. 灰红色中粗砂岩，上部为灰红色粉砂岩夹泥岩 11.2m
6. 暗红色泥质粉砂岩夹青灰色钙质粉砂质泥岩 10.1m
5. 青灰色钙泥质粉砂岩，含植物化石碎片 12.8m
4. 灰色粉砂岩，含较多的植物化石碎片 36.3m
3. 浅棕黄色含凝灰质粗砂岩 13.2m
2. 灰色粉砂质泥岩 4.4m
1. 碳酸盐化、绿泥石化凝灰质中砂岩 5.9m

未 见 底

3. 岩溶 6#孔剖面

上覆地层：第四系疏松层

~~~~~~~~ 不 整 合 ~~~~~~~~

寿昌组（K_1h）

上段（K_1h^3） 总厚度 250.7m

12. 球泡流纹英安岩，主要由霏细状长英质及部分球泡组成，球泡有时呈串珠状排列，局部显示流纹构造 51.6m

11.	流纹英安岩，浅灰—紫红色，具斑状结构，流纹构造，其中夹厚为 1.76m 的灰绿色晶屑岩屑凝灰岩	63.0m
10.	含砾流纹英安岩，砾石为同源流纹岩碎块，其中夹一层厚约 1m 的安玄岩	2.5m
9.	流纹英安岩	47.1m
8.	紫红色流纹英安质碎裂岩，顶部为含角砾凝灰岩，角砾主要由流纹岩、流纹英安岩碎块组成	5.7m
7.	浅灰绿色晶屑凝灰岩，顶部为紫红色沉凝灰岩	13.7m
6.	紫红色粉砂岩	5.5m
5.	紫红色流纹质火山角砾岩	15.7m
4.	浅灰绿色岩屑玻屑凝灰岩夹安玄岩	16.0m
3.	浅灰绿色含砾玻屑凝灰岩，顶部为角砾凝灰岩，角砾成分有晶屑凝灰岩、酸性火山碎屑岩、安玄岩、砂岩；中部夹一层紫红色晶屑玻屑凝灰岩；底部为碎裂流纹英安岩	29.9m

中段（K_1h^2） 总厚度>64.2m

2.	灰或灰黑色白云质泥岩、灰岩（?），紫红、灰绿色粉砂岩夹晶屑凝灰岩等	13.2m
1.	紫红夹灰绿色粉砂岩、凝灰质粉砂岩，夹岩屑玻屑凝灰岩，层理清晰，偶见植物化石碎屑	>51.0m

未 见 底

（四）地质特征及区域变化

该组上、中、下 3 个岩性段均为渐变过渡关系。其岩性特征为：

1）寿昌组上段：主要由球泡流纹英安岩、流纹英安岩、含角砾熔结凝灰岩、岩屑晶屑玻屑凝灰岩等中酸性火山岩与火山碎屑岩组成。已见最大厚度为 326.9m。

2）寿昌组中段：主要由紫红、青灰、灰白色凝灰岩、凝灰质砂岩以及泥岩、粉砂岩、中粗砂岩组成，局部为含粉砂灰岩或晶屑玻屑白云岩。含较多植物化石碎片，层理清晰。已见最大厚度为 256.1m。

3）寿昌组下段：为紫红、青灰、灰黑色粉细砂岩、砂岩和杂色砾岩组成的韵律层，夹灰黑色炭质条带和煤线，局部为泥灰岩，基本不含火山物质，层理发育。已见最大厚度为 300.7m。

已揭露最大厚度达 883.7m，顶、底出露不全，与上、下层的地层接触关系尚不清楚。

四、浦口组（K_2p）

（一）创名及原始定义

1924 年，刘季辰和赵汝钧于江苏南京浦口创“浦口层”（指赭色岩层或赭色砂粒层）。其原始义为赭色砂岩、页岩、粗砾岩，下与斑岩层，上与赤山层均呈不整合接触。

（二）现在定义

浦口组系指一套紫棕色（俗称猪肝色）、紫灰色的粗粒陆相沉积，以砾岩、砂砾岩为主，间夹有厚度不等的粉—细砂岩和粉砂质泥岩层。其砾石成分大部分是火山岩砾石。

（三）层型剖面描述

次层型剖面为嘉定黄渡 I 孔剖面（图 6-4）。剖面地层分述如下：

上覆地层：上新统黏土、砂砾岩

~~~~~~~~~~ 不　整　合 ~~~~~~~~~~

浦口组（K_2p） 厚 356.5m

6.	棕红色砂质泥岩及棕紫色砂砾岩互层，泥岩中见蓝色圆形斑点，砾石呈棱状，以火山岩为主，分选差	63.8m
5.	紫灰色砾岩及砂砾岩互层	124.6m
~~~~~~~~~~

4. 紫或微带黄色砂质泥岩，中下部夹砂岩薄层 24.6m

3. 紫红色砂砾岩与砂岩互层，间夹紫红色泥岩，胶结疏松，分选差 91.3m

2. 紫红色含砾砂岩，中部夹紫或棕红色砂质泥岩薄层 40.3m

1. 暗紫红色砂质泥岩 11.6m

~~~~~~ 不 整 合 ~~~~~~

下伏地层：寿昌组紫灰色凝灰岩

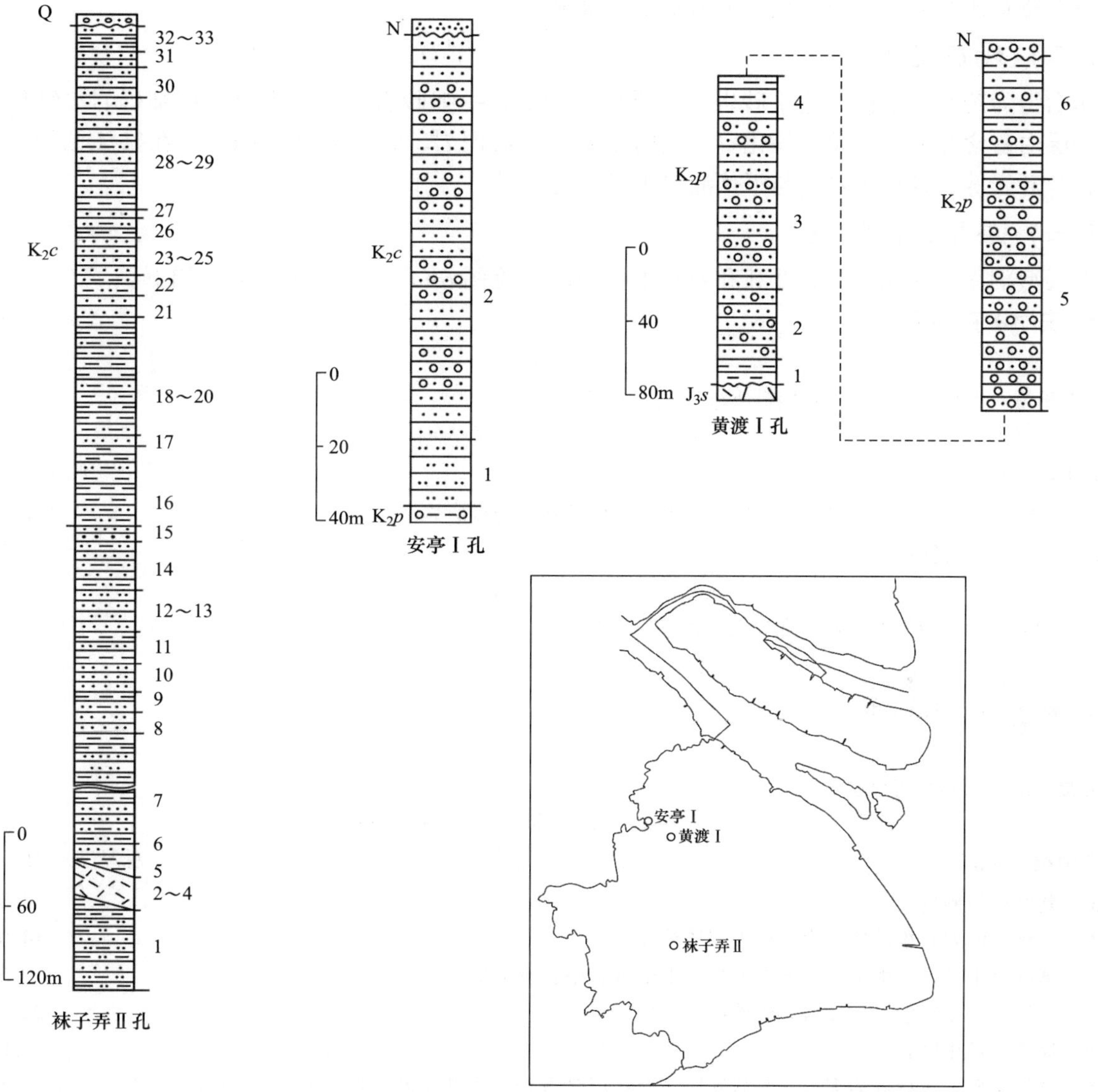

图 6-4 次层型钻孔剖面柱状图

Fig. 6-4 Hypostratotype borehole section histograms

（四）地质特征及区域变化

区内安角断陷盆地南缘分布广泛，为一套较为典型的粗粒陆相沉积建造，其岩性以大套块状紫棕、紫灰、杂色砂砾岩、砾岩为主，夹有厚薄不一的红棕色粉砂岩、含砾粉—细砂岩和泥岩。以发育大套的粗碎屑岩为其显著特征，砾石磨圆度差，分选性差，砾石成分以火山岩成分为主，如安山岩、流纹岩、凝灰质岩类等，为典型的近源山麓堆积。厚度变化极大，靠盆地边部显著变薄，如安亭为548.6m，黄渡为356.3m，至小村仅32.2m。与下伏下白垩统呈不整合接触。
~~~~~~

五、赤山组（K_2c）

（一）创名及原始定义

1924 年，刘季辰和赵汝钧于江苏句容赤山创赤山层。之后，逐渐演变为赤山组。其下部为紫红、砖红色细砂岩、粉砂岩，灰绿色钙质泥岩夹薄层砂砾岩、含砾砂岩；上部为红棕、紫红色，局部为灰白色的钙、泥、铁质细砂岩、粉砂岩、页岩或为互层，并夹有泥岩。在区域上与下伏浦口组呈整合接触，与上覆泰州组或阜宁组、盐城组为不整合接触。

（二）现在定义

赤山组为砖红、暗棕红色粉砂岩、泥质粉砂岩及粉—细砂岩，局部夹有薄层细砾岩。钙质、泥质胶结，层内富含介形虫、轮藻等化石。区内主要为砂质泥岩、粉砂岩夹砂砾岩，在松江见较多的粉砂质泥岩、砂岩等，且夹石膏。与下伏浦口组呈整合接触。

（三）层型剖面描述

次层型剖面为松江袜子弄Ⅱ孔和安亭Ⅰ孔剖面（见图 6-4）。剖面地层分述如下：

1. 安亭 I 孔剖面

上覆地层：上新统浅黄色石英粗砂夹砂质黏土薄层

~~~~~~~~~ 不　整　合 ~~~~~~~~~

赤山组（K_2c）	厚 126. 6m
2. 棕红色砂岩及砂砾岩互层，碎屑呈棱角状，少量砾石为变质岩	111. 1m
1. 棕红、灰白色粉砂岩	15. 5m

———— 整　合 ————

下伏地层：浦口组棕红色砂质泥岩与杂色砂砾岩互层

2. 袜子弄 II 孔剖面

上覆地层：第四系疏松层

~~~~~~~~~ 不　整　合 ~~~~~~~~~

赤山组（K_2c）	总厚度>771. 9m
33. 紫红色粉砂岩	4. 0m
32. 紫红色泥岩和粉砂质泥岩互层夹泥质粉砂岩	14. 0m
31. 紫红色细砂岩，由上至下逐渐变粗，下部砂岩中含有泥砾	10. 5m
30. 紫红、淡紫红色钙质砂岩和砂质泥岩互层	22. 5m
29. 淡紫红色中砂岩	4. 0m
28. 紫红色砂岩、粉砂质泥岩、泥岩互层，交错层理发育，见薄层状和脉状石膏（石膏层厚为 1~2mm）	6. 5m
27. 紫灰带绿色不等粒砂岩，由上至下粒度变粗，下部含泥砾，大者逾 1cm	5. 5m
26. 紫红色泥质砂岩和泥岩互层	14. 0m
25. 紫红及紫灰色细砂岩，含泥砾，裂隙中有石膏细脉	10. 0m
24. 暗紫红—灰黑色泥质砂岩和薄层泥岩互层	6. 0m
23. 紫灰色细砂岩，含黑色炭质碎块	12. 0m
22. 紫红色砂质泥岩和泥质砂岩互层，夹薄层石膏，具交错层理和波状层理	17. 5m
21. 紫红色不等粒砂岩	12. 0m
20. 紫红色泥岩和砂质泥岩互层，富钙，夹薄层状和脉状石膏	21. 5m
19. 紫灰色砂岩，胶结、分选较好，有较多石膏脉穿插	2. 5m
18. 紫红色泥岩和砂质泥岩互层，夹紫红色砂岩，见薄层状与脉状石膏	64. 5m

17.	紫红色砂岩，含泥砾、泥团及薄层石膏	7.0m
16.	紫红色泥质砂岩、泥岩或砂质泥岩互层夹砂岩，顶部含石英质砾石，呈圆形，表面光滑，夹有较多的薄层状、脉状石膏	63.0m
15.	紫灰、灰黑色中、粗粒砂岩，含石膏脉及泥砾	13.0m
14.	紫红色砂岩、泥质砂岩、砂质泥岩互层，夹薄层状、脉状石膏	35.0m
13.	紫红、暗紫红色粉、细砂岩，泥质粉细砂岩，下部含绿色形状不定的砾石	32.0m
12.	紫红、紫灰色细砂岩	3.3m
11.	紫红色泥岩和泥质粉砂岩互层，具有石膏脉和交错层理	25.7m
10.	浅紫红色细砂岩，见方解石脉和石英脉	22.3m
9.	紫红色泥岩和泥质粉砂岩互层	16.6m
8.	紫红色细砂岩、中细砂岩夹泥质粉砂岩，具绿色含钙质小砾石，节理面上有石膏	16.9m
7.	紫红色泥岩与泥质粉砂岩、粉砂岩互层，夹紫灰色细砂岩多层，普遍见交错层理与斜层理，石膏脉多见，底部有灰黑色含钙质的小砾石	141.8m
6.	暗紫红色细砂岩，发育水平层理及斜交层理	10.0m
5.	紫红色泥岩、粉砂岩互层	10.0m
4.	花岗斑岩	(4.2m)
3.	紫红色细、粉砂岩和粉砂质泥岩互层，层理清晰	4.6m
2.	花岗斑岩	(20.7m)
1.	紫红色泥岩、泥质粉砂岩互层，夹紫红、紫灰色细砂岩，具钙质、泥质小砾石和交错层理	>73.7m

未　见　底

（四）地质特征及区域变化

区内揭露赤山组的钻孔较少，北桥盆地主要由棕红色粉砂质泥岩与粉砂岩组成，夹紫红色细砾岩及含砾砂岩。安亭、南翔等地粒度变粗，以砂岩、砂砾岩、砂质泥岩等互层为主。在松江，赤山组是一套紫红色陆相碎屑沉积建造，粒度较细，以紫色粉砂质泥岩、泥质粉砂岩为主，夹中细砂岩及个别粗砂岩层。碎屑组分以石英为主，长石次之，并含云母、赤铁矿碎屑。分选性及磨圆度均较差。层中常可见薄层状或网脉状石膏（厚度一般仅数毫米）。部分层位中可见钙质砾石、泥砾或炭质碎块。已见厚度为771.9m，并分别与下白垩统寿昌组、黄尖组，上寒武统超峰组及前震旦系金山群等不同时代的地层呈不整合（或断层）接触。

第二节　沉积作用和沉积建造

一、沉积作用

白垩系为陆相沉积，以地面流水沉积作用为主，按沉积环境可以分为冲积扇、河流、湖泊、沼泽等4种沉积类型，形成洪积扇、河流、湖泊等沉积体系。

（一）冲积扇沉积

冲积扇沉积主要出现于浦口组，以发育砾岩、砂砾岩为主的类磨拉石建造为特征。

（二）河流沉积

河流沉积主要出现在劳村组，发育滞留河道和边滩亚相。滞留河道沉积以粗砂岩、细砂岩等为主，上段出现较薄的细砾沉积。砾石分选性较差，多呈次圆状，少数为棱角状。局部砾石定向排列。

（三）湖泊、沼泽沉积

湖泊、沼泽沉积主要出现于赤山组和寿昌组中—下段，湖湘发育滨湖和浅湖亚相。见有交错层理，碎屑分选较差，砾石多呈棱角状，部分层位以次圆状为主，不同层位会见植物化石碎片、灰黑色炭质条带和煤线，局部为泥灰岩或砂质灰岩。

二、主要岩石类型

上述沉积环境中形成的主要岩石类型有粉砂质泥岩、粉砂岩、砂岩、砂砾岩、砾岩、泥灰岩、含砾砂岩、凝灰质粉砂岩、凝灰质砂岩。

（一）粉砂质泥岩

粉砂质泥岩主要分布在赤山组中，其他层位中偶见。局部粉砂、砂含量变化较大，呈泥岩或砂质泥岩。以紫红色为主，偶见青灰、砖红色等。泥质结构，块状构造。碎屑分布不均，主要由石英组成，少量长石和磁铁矿，长石风化严重。石英含量为5%～15%，长石为0～2%。呈不规则棱角状，平均粒径为0.05～0.1mm。基底式胶结，胶结物主要为黏土矿物、钙质和铁质。

（二）粉砂岩

粉砂岩主要分布在赤山组中，在寿昌组中—下段中少见，在劳村组中偶见。局部层位泥质含量高，呈泥质粉砂岩。主要呈紫灰、紫红色，下白垩统偶见灰、灰绿色等。粉砂结构，以块状构造为主，局部见层理构造。碎屑矿物以石英和长石为主，少量磁铁矿。石英含量在30%～75%，长石含量多为5%～30%，磁铁矿含量小于5%。石英、长石平均粒径多在0.05～0.1mm，分选性差，棱角状。以接触式胶结为主，胶结物为黏土矿物、钙质和铁质，局部见较多绢云母。

（三）砂岩

砂岩分布于劳村组、寿昌组、浦口组和赤山组，局部层位含砾石，为含砾砂岩。赤山组中局部泥质含量较高，呈泥质砂岩。主要呈紫红色，下白垩统局部为青灰、灰白色。细粒、中粒、粗粒结构，以块状构造为主，局部见层理构造。碎屑主要为石英、长石，次棱角—次圆状。部分见云母碎片、磁铁矿等碎屑，偶见电气石和绿泥石。粒径多在0.05～0.1mm、0.05～0.5mm、0.5～1.5mm，个别层位分选性好。石英含量多在40%～70%，局部可达80%；长石含量多小于15%。以接触式胶结为主，局部孔隙式胶结。胶结物以泥质、铁质和钙质为主，局部为硅质。

（四）砂砾岩

砂砾岩主要分布在浦口组中，在寿昌组中—下段、劳村组和赤山组中也可见。主要呈紫灰、棕红和紫红色。砾状结构，块状构造和斑杂状构造。砾石多占50%～60%，多呈次棱角—次圆状，砾石大小不一，局部可达6～7cm，成分复杂。填隙物以长石、石英碎屑和钙、泥质胶结物为主，少量绢云母和褐铁矿化的磁铁矿。固结程度较差，风化强烈。

（五）砾岩

砾岩主要分布在浦口组中，在劳村组和寿昌组局部可见。以灰紫、灰红和紫红色为主，局部为灰、灰白色。以细砾结构为主，块状和斑杂状构造。局部砾石含量高达70%～80%，多呈次棱角—次圆状，少数圆状。砾石大小多在3～7mm，局部可达6～7cm，分选性差。砾石岩性和颜色很杂，有泥岩、砂岩、灰岩和多种火山岩、变质岩，填隙物为泥质、钙质胶结物，局部含铁质。

（六）泥灰岩

泥灰岩主要分布在寿昌组局部层位。灰黑、灰紫色，泥状结构，局部鲕状结构，块状构造，局部层理构造。鲕粒泥灰岩由鲕状方解石和不规则晶粒状方解石组成，鲕粒呈规则球形，粒径多在0.2～0.3mm，内部有黏土矿物混入。鲕粒间被结晶方解石和非晶质云雾状方解石充填。粉砂质泥灰岩中，粉砂质含量约为35%，泥质含量为65%。粉砂主要由石英组成，副矿物含量较少，不到2%。泥质中含有较多泥晶方解石。

（七）凝灰质砂岩

凝灰质砂岩主要分布在劳村组，少部分分布在寿昌组中—下段。紫红、青灰、灰白色。砂状结构，块状构造，局部层理构造。局部含少量砾石。碎屑物质主要为石英和玻璃质岩屑，局部有长石、磁铁矿、云母、绿泥石。石英多呈棱角状，含量为15%~40%。玻璃质岩屑多脱玻化。碎屑粒径多为0.1~2mm，局部层位长轴见定向排列。胶结物主要为硅质、铁质和凝灰质。

（八）凝灰质粉砂岩

凝灰质粉砂岩主要分布在寿昌组中—下段。以紫红色为主，夹灰绿、灰黄色。粉砂状结构，块状构造、层理构造。碎屑占60%~70%，主要为蚀变严重的岩屑和长石假象、石英，其次为玻屑和云母。粒径一般在0.01~0.05mm，分选不甚好，磨圆度较差，多为次棱角状，少量棱角状和次磨圆状，分布杂乱。胶结物占30%~40%，大多为黏土矿物（高岭石、水白云母），并有铁质散布在其间。呈孔隙式和接触式胶结。常含少量大小2mm左右的砾石。

第三节 古生物群组合和地层时代

一、古生物群组合

仅在凤溪D2孔寿昌组下段发现大量孢粉化石。化石类型较单调，并以裸子植物花粉含量为主，占组合总数的97.94%，蕨类植物孢子仅占组合总数的2.06%。裸子植物花粉中又以*Classopollis*花粉含量最高，可占整个孢粉组合总数的91.76%。与*Classopollis*伴生的有*Psophosphaera*、*Alisponites*、*Monocolpollenites*、*Araucariacites*、*Sphaeripollenites*和*Podocarpidites*花粉。蕨类孢子仅见少量的*Cirratriradites*、*Lygodiumsporites*（海金砂孢属）、*Deltoidospora*，以及*Cyathidites*、*Schizacoisporites*等。均为在干热气候条件之产物。

二、地层时代

区内下白垩统发育了广泛分布的火山岩-火山碎屑岩建造，在不同地区中获得了大量锆石U-Pb年龄。如，奉贤庄行庄zk2孔晶屑凝灰岩136.1±2.4Ma、松江天马山安山岩129.0±0.8Ma、浦东曹路基23孔岩屑凝灰岩135.1±0.9Ma、崇明城桥J60孔晶屑凝灰岩132.6±1.7Ma、宝山大场12#孔岩屑凝灰岩130.5±1.4Ma、东海大桥LQ17孔晶屑凝灰岩130.9±1.1M。这些年龄都显示劳村组、黄尖组为早白垩世。在D2孔寿昌组中获得的孢粉以裸子植物为主，均为早白垩世常见分子。尤其是，具孔类被子植物花粉*Monocolpollenites*在中国白垩系中仅产出于阿尔必期（万传彪，2004）。D2孔顶部凝灰岩中锆石$^{206}Pb/^{238}U$年龄比较分散，多集中在120Ma左右，推断寿昌组应该在早白垩世晚期。

上白垩统浦口组和赤山组均中未见化石，但马桥LQ11孔玄武岩夹层获得K-Ar年龄为78.31±1.77Ma。

第四节 区域地层对比

一、区域对比

区内早白垩世和晚白垩世的建造特征差别较大，与邻近江苏、浙江、安徽和江西的总体特征十分接近（表6-2）。早白垩世，总体上都呈现出以陆源碎屑岩为主、火山-火山碎屑岩、陆源碎屑岩为主的三段式建造特征。晚白垩世总体上都呈现为断陷红盆沉积，除浙东南地区塘上组和两头塘组下部发育较多火山-火山碎屑建造外，多为陆源碎屑岩类，局部夹有一层到数层的火山岩；但由于都为陆相建造，局部岩性组合又存在明显的变化。如皖南白垩系下部的龙门院组和龙王山组、中分村组以火山-火山碎屑岩为主。

浙西北的劳村组以陆相紫红色碎屑沉积岩为主，夹少量火山碎屑岩和中酸性熔岩，底部常见砾岩层。在淳安—临安一带以陆源碎屑岩为主，夹火山岩。向西至皖南的红花桥组、彭家口组、炳丘组和江西的如意厅组，向东至江苏的西横山组等都是以陆源碎屑岩为主，局部含凝灰质较多，与区内劳村组上段相当。浙江的大爽组和寿昌—临浦一带东段的劳村组中火山-火山碎屑岩建造较多，与区内劳村组下段相似。寿昌盆地和浦江盆地中劳村组锆石 U-Pb 年龄为 136±1Ma~131±1Ma（Liu et al.，2014）。在浙西北层型剖面上，劳村组锆石 U-Pb 年龄为 135.1±1.6Ma~129.1±1.4Ma（王加恩，2016）。以火山岩为主的皖南龙门院组、龙王山组、中分村组分别获得锆石 U-Pb 年龄为 134.8±1.8Ma（周涛发，2008）、134.8±1.3Ma（Zhou，2011）、134.4±2.9Ma（袁峰等，2010）。江西打鼓顶组的锆石 U-Pb 年龄多在 136.6±2.7Ma~134±1.9Ma（江西省地质矿产勘查开发局，2017）。

浙西北地区黄尖组与皖南地区黄石坝组、江镇组和汪公庙组下段、双庙组、砖桥组和姑山组、大王山组、蝌蚪山组、赤沙组都为一套火山岩、火山碎屑岩及其过渡类型，均与上海的黄尖组相似。浙江劳村组层型剖面上的黄尖组的锆石 U-Pb 年龄为 128.0±1.9Ma（王加恩，2016）。寿昌盆地和浦江盆地黄尖组的锆石 U-Pb 年龄为 130±1Ma~127±1Ma（Liu et al.，2014）。衢江大坞山黄尖组的锆石 U-Pb 年龄为 135.5±1.2Ma~128.2±2.9Ma（王洪作，2017）。

区内寿昌组与浙西北地区寿昌组的岩性不尽相同，其顶部为火山熔岩与火山碎屑岩段，在流纹英安岩中含有较多球泡，可与浙东地区的九里坪组、馆头组对比；其下的沉凝灰岩段和正常沉积碎屑岩段的岩性及孢粉组合特征，则可与茶湾组对比（见表 6-2）。向西与广德组、岩塘组和石溪组、浮山组等大体相似，以陆源碎屑岩为主，且区内大量出现的 *Classopollis* 化石为江西石溪和冷水坞中的带化石。寿昌组层型剖面中寿昌组的锆石 U-Pb 年龄为 123.8±1.0Ma（李祥辉，2011）。在寿昌盆地和浦江盆地寿昌组的年龄在 123±2Ma（Liu et al.，2014）。

表 6-2　区域地层对比表

Table 6-2　Regional stratigraphic correlation

地层		上海	浙江			皖南							赣中和赣北			江苏
			浙西北	浙东南		滁州黄石坝盆地	怀宁盆地	庐枞盆地	宁芜盆地	繁昌盆地	屯溪盆地	皖浙交界	武宁	德兴—抚州	贵溪	南京
上覆地层		夏驾桥组	长河组	嵊县组		舜山集组/望虎墩组/第四系							武宁群	新余组/第四系		泰州组
白垩系	上白垩统	赤山组	桐乡组	衢县组	赤诚山组	赤山组					小岩组		莲荷组			赤山组
				金华组	两头塘组								塘边组			
		浦口组		中戴组	塘上组	七房村组					齐云山组		河口组			浦口组
													周田组			
													茅店组			
	下白垩统	?	横山组	方岩组		杨湾组					徽州组			冷水坞组		上党组
				朝川组												
		寿昌组	寿昌组	馆头组		黄石坝组	汪公庙组	浮山组	娘娘山组	广德组	岩塘组			石溪组		葛村组
		黄尖组	黄尖组	九里坪组				双庙组	姑山组	蝌蚪山组	石岭组	黄尖组		打鼓顶组	鹅湖岭组	龙王山组
				茶湾组												
				西山头组			江镇组	砖桥组	大王山组	赤沙组						
		劳村组	劳村组	高坞组		红花桥组	彭家口组	龙门院组	龙王山组	中分村组	炳丘组	劳村组		如意亭组		西横山组
				大爽组												
下伏地层		?	渔山尖组	毛弄组		罗岭组/洪琴组/南陵湖组							双桥山群	张村群	万年群	象山群

在上海和浙西北、赣中和赣北地区，浦口组与茅店组、河口组、齐云山组、中戴组等岩性总体特征相似，都为一套以砂砾岩为主的建造类型，但在局部粉砂、泥岩等细粒建造稍厚。茅店组锆石 U-Pb 年龄在 91.03±3.0Ma~99±0.7Ma（江西省地质矿产勘查开发局，2017）。赤山组所见层位与塘边组、莲荷组和小岩组相当，岩性偏细，局部夹有玄武岩。

二、存在问题

（一）庄行和金桥构造火山盆地区地层归属需进一步确定

区内庄行钻孔地层呈火山岩-火山碎屑岩建造，长期作为区内黄尖组的次层型剖面，但本志获得其锆石 U-Pb 年龄为 136.1±2.4Ma，与金桥火山构造盆地区的凝灰岩年龄相当。区域上，大量的锆石 U-Pb 年龄多支持黄尖组形成于约 130Ma 之后，但不同剖面上的劳村组和黄尖组年龄也存在明显的重叠特征。所以，将大于130Ma左右的火山岩-火山碎屑岩建造作为原始定义劳村组的同期异相来处理，更符合劳村组的空间分布特征。但受资料和目前区域认识所限，本志仍继续以岩石组合为依据，将其划归黄尖组。

（二）早白垩世晚期地层划分方案还需进一步对比

区内早白垩世地层划分方案与邻区还存在不一致。如区内寿昌组局部层位发育泥灰岩、砂粒灰岩，而邻区只在劳村组、朝川组、冷水坞组、葛村组中上部、上白垩统内部发育。此外，区域上，寿昌组还上覆一套以砂、泥岩类为主的建造，与横山组、徽州组和冷水坞组层位相当。李祥辉（2011）在寿昌组层型剖面顶部发育的横山组中获得的锆石U-Pb年龄为120.9±1.2Ma。Liu（2014）在寿昌盆地和浦江盆地获得横山组的年龄为121±2Ma~118±1Ma，这些年龄特征与本区D2孔顶部凝灰岩中获得的年龄相当。

第五节　岩相古地理

早白垩世，气候炎热，形成巨厚的陆相火山-陆源碎屑沉积岩。在强烈岩浆喷发活动时期，以火山岩堆积为主，火山喷发间歇期，在火山断陷洼地常常堆积河湖相火山碎屑岩、沉积岩；晚白垩世，在局部断陷盆地中充填巨厚的冲积扇和河湖相沉积物。

劳村组沉积早期，主要为湖泊环境，局部有酸性火山活动。晚期，在松隐、亭林，地层层理清晰，见波状层理及交错层理，显示了较强的水动力环境，主要表现为河流沉积。黄尖组形成时期，区内火山活动急剧加强，形成青浦、天马山、查山等火山构造隆起和庄行、金桥、重固等构造火山盆地等地貌。寿昌组早期沉积时期，本区是温湿气候条件下的滨湖—沼泽相沉积环境，形成一套以砂岩局部含砾的沉积岩建造，局部还可见炭质条带与煤线。寿昌组中晚沉积期，葛陵—南翔—大场以北以浅湖相沉凝灰岩与凝灰质粉砂岩为主，夹薄层泥岩与灰岩，地层层理及微细层理清晰；以南则属于湖盆边缘的陆相堆积，发育砂砾岩和砾岩。

晚白垩世，区内断陷盆地发育，从而堆积了一套巨厚的红色陆相碎屑建造。早期，以炎热气候条件下的冲洪积沉积为主。晚期，安角盆地的沉积中心向北、北西迁移，盆中积水成湖。松江盆地区发育紫红色砂岩、粉砂质泥岩与泥质粉砂岩建造，部分层位含砾石、有交错层理和钙质结核、泥砾，局部间夹薄层石膏。表明局部滨湖地区似曾出现河流冲积扇前缘部位的沉积环境。

第七章　古近系

本区古近纪地层主要分布于断陷盆地中，如安角断陷盆地及北桥断陷盆地。在崇明北缘盆地中推断也有分布。迄今仅发现古新统，自下而上划分为夏驾桥组及北桥组，为一套紫红色陆相沉积建造，已固结成岩。根据区域地层发育特点，将区内始新世、渐新世地层作为暂未发现处理。

1960 年，上海市人民委员会地质处首次将古近系划分为始—渐新统，并认为古新统缺失。1962 年，上海市地质勘察局延续上述方案，并称之为巨江红层。1966 年，地质部石油地质局中心实验室创名“甪直群”，时代定为晚白垩世—渐新世。20 世纪 70 年代，都延续这一划分方案，但时间定为古近纪或始—渐新世。1983 年，刘若愚根据邻区昆山安角 3#建立了夏驾桥组，并下伏赤山组。1983 年，上海市地质处在北桥 D3 孔中，于上新统崇明组之下的棕红、灰褐色砂泥岩中采集到丰富的介形类、轮藻及植物孢粉化石，时代为古新世中—晚期。1988 年，在编制《上海市区域地质志》时，将夏驾桥组引入区内，将古新统划分为夏驾桥组和北桥组，始—渐新统作为未发现处理。之后，这一划分方案一直沿用至今（表 7-1）。

表 7-1　上海古近系岩石地层单位划分沿革表

Table 7-1　Previous divisions of the Paleogene lithostratigraphic units in Shanghai

<table>
<tr><th>地层</th><th>《上海市石油及天然气地质普查工作报告》，1960</th><th>《上海地区矿产地质普查工作总结报告》，1962</th><th>地质部石油地质局中心实验室，1966</th><th>《1：20万上海地区基岩地质推断图》，1979</th><th>《上海市区域地质志》，1988</th><th>本志</th></tr>
<tr><td>上覆地层</td><td>上新统</td><td>中新统</td><td>中新统</td><td>上新统</td><td>中新统</td><td>中新统</td></tr>
<tr><td>始—渐新统</td><td>未分</td><td>巨江红层</td><td rowspan="3">甪直群</td><td>甪直群</td><td>?</td><td>?</td></tr>
<tr><td rowspan="2">古新统</td><td rowspan="2"></td><td rowspan="2"></td><td rowspan="2"></td><td>北桥组</td><td>北桥组</td></tr>
<tr><td>夏驾桥组</td><td>夏驾桥组</td></tr>
</table>

第一节　岩石地层划分

一、夏驾桥组（E_1x）

（一）创名及原始定义

1983 年，由刘若愚和朱荣依据昆山夏驾桥安角 3 孔的资料，首创“夏驾桥组”一名，时代定为古新世。原义是指砖红色砂、泥岩的赤山组上的一套以棕红色泥岩为主的地层。

（二）现在定义

2005 年，上海幅 1∶25 万区域地质调查修测时，将安角 3 孔原赤山组划为泰州组，与夏驾桥组相当；原夏驾桥组划为阜宁组，与北桥组相当。本志沿用该认识，将夏驾桥组现在定义为以棕红色（局部砖红色）粉细砂岩、粉砂岩、泥质粉砂岩为主，局部含砾，夹薄层砂砾岩，少量粉砂质泥岩和泥砾，下部见夹薄层砂质灰岩。

（三）层型剖面描述

主要正层型剖面为昆山的夏驾桥安用3孔剖面（图7-1）。剖面地层分述如下：

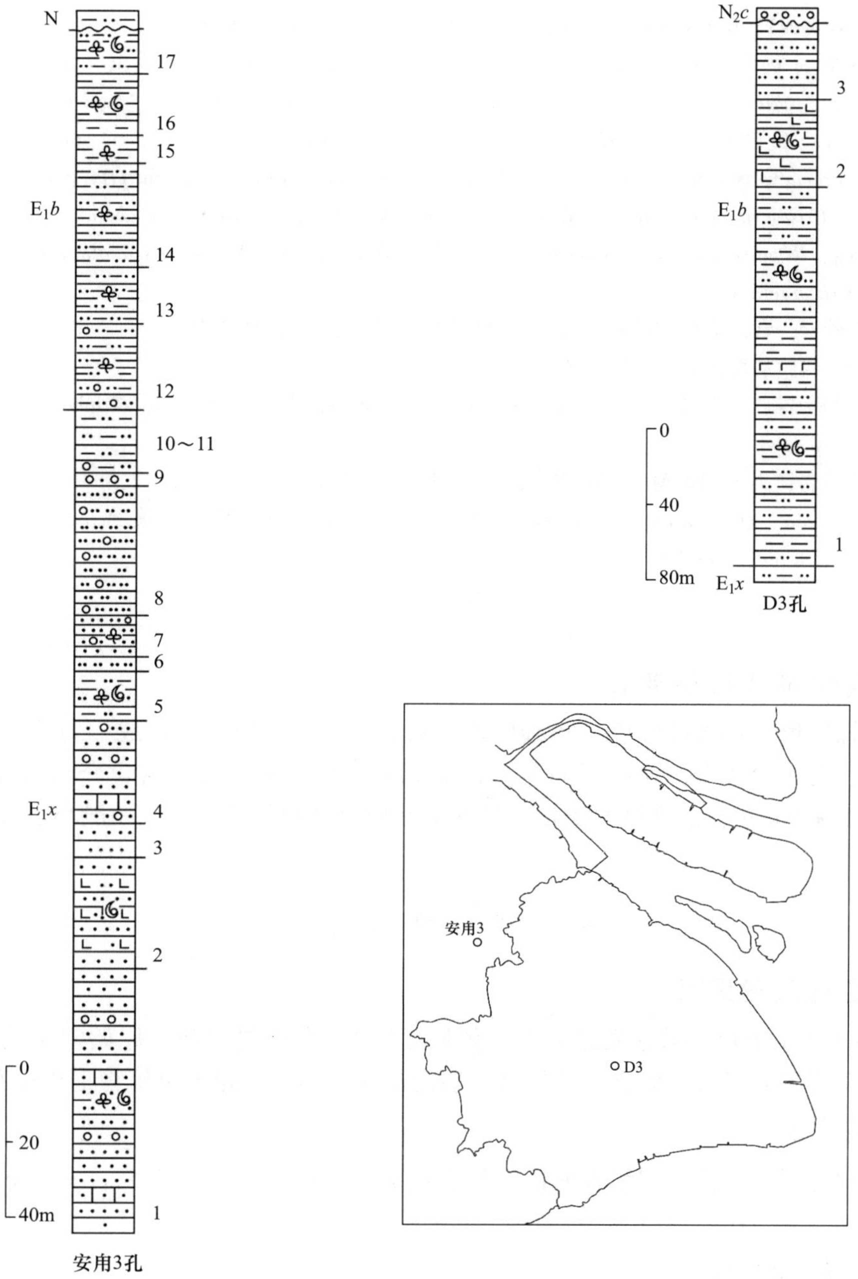

图7-1　正层型钻孔剖面柱状图

Fig. 7-1　Holostratotype borehole section histograms

上覆地层：古近系北桥组棕红、暗紫红色泥质粉砂岩和粉砂质泥岩

——————　整　　合　——————

夏驾桥组（E_1x）　　总厚度>213.7m

11. 砖红色含细砾泥质粉砂岩，具灰绿色斑块　　9.2m
10. 暗棕红色泥质粉砂岩，钙质胶结，含少量中粗砂及细砾石　　2.3m
9. 暗棕红色砂砾岩，夹少量粉砂岩等，胶结疏松，产介形虫 *Ziziphocypris simakovi*　　2.3m
8. 砖红（浅棕）色含砾粉砂岩，砾径在4~5cm，次圆状，成分为岩浆岩和灰岩　　33.7m

7. 砖红—浅棕色含砾粉—细砂岩，夹薄层钙质胶结砂砾岩，产轮藻 *Latochara biacumnata*、*L. specialis*、*L. Corollacea*、*Charites acuminata* Wang、*C.* sp. 11.5m

6. 暗棕红色粉砂岩夹薄层钙质胶结的砂砾岩，产介形虫 *Cristocypridea cylindrica*、*Cyclocypris* sp.、*Cyprois symmetrica*、*C.* sp. 等，轮藻 *Charites stipitata* Wang、*Grambastichara subcylindrica* (Reid et Groves)、*G. longiconica*、*G. parcula*、*Obtusochara elliptica* Wang 3.8m

5. 暗棕红色泥质粉砂岩和粉砂质泥岩，具微细层理，见同生泥砾，产介形虫（上部）*Cristocypridea chinensis*、*C. brevis*、*C. longa*、*C. ovata*、*C. latiovata*、*C. quadrata*、*C. jiangsuensis*、*C. obesa*、*C. kunshanensis*、*C. misera*、*C. prona*、*C. concta*、*C. gracila*、*C. propria*、*Proeucypris inflata*、*P. longa*、*P. extenda*、*P. suvbstriangula*、*Ruficypris sinensis* 和（下部）*Cristocypridea amoena*、*C. longa*、*C. kunshanensis*，*C. prona*、*C. obesa*、*Ruficypris sinensis*、*R. postslina*、*R. subsinensis* 等，轮藻 *Latocharacurata* Wang、*Pseudolatochara* sp.、*Maedlerisphaera* sp. 13.4m

4. 棕红色粉细砂岩，含细砾粉细砂岩，夹薄层砂砾岩和砂质灰岩，层理清晰 27.2m

3. 棕红色粉细砂岩，局部夹薄层砂质灰岩 9.2m

2. 棕灰色粉细砂岩、钙质砂岩，夹薄层砂砾岩和砂质灰岩，产介形虫 Ziziphocypris jiangsuensis (sp. nov.) 等 29.9m

1. 棕红色粉细砂岩，夹薄层砂砾岩和砂质灰岩，产介形虫 *Cristocyprideacomisa*、*C. subcomica*、*Cypridea cavernosa*、*Quadracypris jiangsuensis*、*Candoniella* aff. *suruni* 等，轮藻 *Actistochara micorosphaera*、*A. Subtrianguformis* >71.2m

未　见　底

（四）地质特征及区域变化

本组主要由棕红色（局部砖红色）粉细砂岩、粉砂岩、泥质粉砂岩组成，局部层位夹砂砾岩、粉细砂岩、粉砂岩、泥质粉砂岩，含砾石明显。自下而上沉积变细，层内富含介形类及轮藻微体化石。区内所见厚度在130.0m左右，其厚度变化在靠断陷盆地边缘和西部变薄，一般小于百米。未见与下伏赤山组的接触关系。

二、北桥组（E_1x）

（一）创名及原始定义

1988年，编著《上海市区域地质志》时，依据北桥D3孔资料创建。原始定义是指整合于夏驾桥组之上的棕红、灰褐色砂泥岩，夹有玄武岩层。层内富含介形类、轮藻及植物孢粉化石。

（二）现在定义

由棕红、灰褐色泥岩、粉砂质泥岩夹粉细砂岩组成，夹一层到数层玄武岩，与下伏夏驾桥组为整合接触关系。

（三）层型剖面描述

正层型剖面为北桥部队营房后D3孔剖面（见图7-1）。剖面地层分述如下：

1. D3孔剖面

上覆地层：上新统崇明组黄绿、灰白色砂砾层

～～～～～～ 不　整　合 ～～～～～～

北桥组（E_1x） 总厚度289.4m

3. 棕红及灰色粉砂质泥岩、粉砂岩，局部含砂、砾和火山岩岩屑 38.3m

2. 棕红色钙质泥岩及棕红、灰褐色钙质粉细砂岩，内含少量硅质小砾石与植物化石碎屑，底部夹碳质泥岩，产轮藻 *Rhabodochara* sp.、*Latochara curtula* Wang，*L. antingensis* Hu、*L. yizhenesis* Wang et Zhang、*L. tenus*、*Peckichara subsphaeica* Lin et Wang、*Stephanochara jiangsuensis* Wang et Zhang、*Charitessadleri straub*、*Collichara xiaohekouensis* Wang et Zhang，介形类 *Paracandona caudata* Ho、*Ilyocypris subhuangqiaoensis* Yang、*I. huangqiaoensis* Yang、*Eucypris* sp.、*E. longicallida* Yang、*E.* spp.、*Homoeucypris* sp.、*H. ovata* Ho、*Paraiyocypris multispinosa* Yang、*Quadracypris ovata* Ho、*Q.* sp.，孢粉 *Quercoidites-Ulmipollenites-Triporo pollenites* 组合带，其中木本花粉占67%~75%（阔叶树42%~57.8%，针叶树仅10%左右），草木占15%~20%，蕨类偶见，阔叶花粉以栎粉、多种小亨氏栎粉、亨氏栎粉及粗糙栎粉为主，还有榆粉（如小榆粉、榆粉属）多量，三孔粉型有亚三孔粉及褶皱粉，热—亚热带成分有木兰、樟、梭形栗粉、三角山核桃等 46.4m

1. 棕红色粉砂质泥岩、泥质粉砂岩夹灰褐、灰白色泥岩，中部夹厚4m的玄武岩，偶见含砂的隐晶质灰岩薄层，含灰绿色及蓝灰色斑块，钙质团块和姜结石，局部为粗砂、砾石层，底部砾石增多，砾石呈次棱角状，砾径在4~6cm，成分有熔结凝灰岩、石英岩、花岗斑岩、英安岩等，泥岩中微层理发育，并见交错层理，产轮藻 *Latochara yizhengensis* Wang et Zhang、*L. certula* Wang、*L. antingensis* Hu、*L. tenus*、*Grovesichara changzhouensis* Wang et Huang、*Gyrogona wubaoensis* Wang et Lin、*Sinochara rudongensis* Lin et Wang、*Gobichara deserta* Karczemska et Zimbinska、*Spheqrochara parvula* Horn et Rantzien、*Charites sadleri* var. *convexa* Zhang et Wang，介形类 *Ilyocypris subhuangqiaoensis* Yang、*I. sublevis* Yang、*I.* sp.、*I. spinosa* Yang、*Candona*（*Candona*）*wangdianonsis* Ho、*Candona*（*Candona*）sp.、*Eucypris fuscata*（Jurine）、*Candona*（*Candona*）*bellula* Yang、*Homoeucypris* sp.、*H. privis* Ho et Chen、*Quadracypris longa*、*Q. ovata* Ho、*Cypris extenda* Zhou、*Metacypris elongata* Ho、*M. changzhouensis* Chen，另含少量腹足类、双壳类化石，孢粉 *Ulmipollenites minor-Triporopollenites* 组合带。其中木本花粉占90%（阔叶74%，针叶16%），草本占6%，优势成分是小榆粉、克氏脊榆粉、粒纹亚三孔粉、褶皱粉等，此外，柳粉、拟桦粉、黄杞粉、朴粉出现较多，针叶树花粉以松、落叶松及皱球粉较常见 204.7m

———— 整　合 ————

下伏地层：古新统夏驾桥组砖红色粉砂岩

2. 安用3孔剖面

上覆地层：新近系砾、砂和砂质黏土，底部为卵石层

~~~~~~ 不　整　合 ~~~~~~

北桥组（E_1x） 总厚度 101.9m

17. 暗红色泥质粉砂岩，富含钙质，常具钙质团块和灰绿色斑点，下部泥质增多，渐变为粉砂质泥岩，含小砾石，产介形虫 *Parailyocypris obesa*、*P. lingtangqiaoensis*、*Eucypri sillofrmis*、*E. stagnalis* Mandelstam、*Cypris* sp.、*Cyprinotus* sp. 等，轮藻 *Latochara cylindrica*、*Stephanocharacompta*、*S. forites*、*S. huangjianensis*、*S.* cf. *funingensis*、*S. longiosata*、*S. breviosalis*、*S. stipitata*、*S. cylindrical*、*Sinochara rudongensis*、*Rhabdochara changzhouensis*、*R. kunshanensis*、*R. kisgyouensis*、*Gyrogona wubaoensis*、*G. huajiachangensis*、*Gobicharaluata*、*Harrisichara yunlongensis*、*H. poculiformis*、*Amblyocharacasta*、*Nemegtichara prima*、*N. paraovalis*、*Obtusochara lanpingensis*、*Grambastichara tornata* 等 12.3m

16. 灰褐夹暗红色泥岩，含粉砂及钙质，产介形虫 *Parailyocypris obesa*，*Eucypris micida*、*E. decora*、*Cypris extenda*、*C.* sp. 等，轮藻 *Latochara antingensis*、*L. yizhenensis*、*Stephanochara tenuis*、*S. macroporosa*、*S. cuneiformis* 以及上层所见种属 15.3m

15. 灰、灰带红色泥岩，产轮藻 *Peckichara cronata*、*P. longa*、*P. subsphaera*、*P. varians*、*Neocharahuananensis*、*N. sinuolata*、*Gobicharagloba* 7.7m
~~~~~~

14. 棕红、暗棕红色粉砂质泥岩，含钙质，并常含同生泥砾和石英细砾石，具微层理构造，产轮藻 *Grovesichara changzhouensis*、*Gobichara deserta*、*Gyrogona gianjiangica*、*Croftlella lasis*、*Obtusochara bresicylindrica*、*O. latacylindrica* 27.6m

13. 棕红、暗棕红色泥质粉砂岩，富钙，有钙质结核和团块，普遍含石英细砾石，砾径为2~3mm，产轮藻 *Gobichara alboides*、*Sinochara caopiensis* 15.3m

12. 棕红、暗紫红色含砾泥质粉砂岩和粉砂质泥岩，含少量砾石和钙质团块，砾径一般为3~4mm，大者6~7mm，成分有石英岩、灰岩、钙质页岩等，产轮藻 *Lacochara globa* 23.7m

———— 整　合 ————

下伏地层：古新统夏驾桥组砖红色含砾泥质粉砂岩

（四）地质特征及区域变化

本组以棕红、暗（紫）红、灰褐色泥岩、粉砂质泥岩为主，局部含砾石，夹粉细砂岩和玄武岩为特征。钙质含量明显，局部含钙质团块、泥砾。层内轮藻、介形类、植物孢粉化石丰富。与下伏夏驾桥组为整合渐变接触关系。

第二节　古生物群组合和地层时代

一、介形虫

夏驾桥组：昆山安角3孔发现介形类化石共8属46种，其中有冠女星介（*Cristocypridea*）、原真星介（新属）（*Proeucypris* gen. nov.）、赤星介（新属）（*Ruficypris* gen. nov.）、枣星介（*Ziziphocypris*）、女星介（*Cypridea*）、柔星介（*Cyprois*）以及球星介（*Cyclocypris*）等属。冠女星介属最为典型，共19种，无论种的数量或个体数量均占有绝对优势。仅次于冠女星介属，新属真星介及赤星介两属也居相当重要的地位，以 *Cristocypridea*-*Proeucypris*-*Ruficypris* 为组合特征。

冠女星介，保存完美，形态多样，大致可分为3种类型：

Cristocypridea ovata 型：壳体呈卵形或椭圆形，冠状突起强，分布广泛；

Cristocypridea concta 型：壳体在背部及后部有1~2条与周缘平行的围脊；

Cristocypridea comisa 型：壳体背腹平行，背部有明显或不明显的背槽，冠状突起在前中间。在层位上比上述类型略低。

冠女星介属地质历程较短，目前所知，在晚白垩世之前未曾发现它的代表。

北桥组：介形虫种属比较单调，个体数量较少。昆山安角3孔只发现4属8种，其中真金星介属较多，其次是金星介属及似土星介属，而美星介属仅见个别分子。就现发现的介形类化石的面貌，与下伏夏驾桥组有极明显的差别，以 *Cypridea*-*Cyprois*-*Parailyocypris* 为组合，与邻区古—渐新统特点较一致。上海市内仅在一个钻孔中所见，可划分出 *Candona*-*Eucypris*-*Ilyocypris* 组合，常见的介形虫有 *Candona*（*Candona*）*wangdianensis*、*Ilyocypris subhuangqiaoensis*、*Eucypris* 等。

二、轮藻

夏驾桥组：在昆山安角3孔发现少量，组合中以大量具顶孔的孔轮藻科宽轮藻属分子为主。此外，还可见国内外常见的种属，如 *Charites*，*Obtusochara*，*Actistochara* 等，计7属15种，其中8新种，大多数都为一些小型藏卵器。轮藻植物类化石迥然不同于北桥组中的轮藻化石组合面貌。

北桥组：轮藻化石属种和数量十分丰富，计有14属45种，其中7个新种。以大量的表面具多种类型装饰的大型藏卵器为主，不具表面装饰的大型厚壁藏卵器和小型藏卵器次之，还有少量中生代孑遗分子如孔轮藻科宽轮藻（*Latochara*）。前者以冠轮藻（*Stephanochara*）、培克轮藻（*Peckichara*）、哈氏轮藻（*Harrisichara*）为代表；后者以厚球轮藻（*Grovesichara*）、迟钝轮藻（*Ambylochara*）、钝头轮藻（*Obtusochara*）和莱曼格特轮藻（*Nemegtichara*）为代表。这种化石组合反映了一种中、新生代混合过渡型。

中生代孑遗分子孔轮藻科宽轮藻属已呈特化类型，其个体宽而大，在形态上和夏驾桥组中大量见到的该属分子具有明显的差异。鉴于本组上述轮藻化石类群大量的新生代分子的出现，而又见少量或行将灭绝的中生代孑遗分子，该组地层时代应为古新世早期。

北桥组中的轮藻化石按个体形态与演化特征，具有以下化石组合：*Gobichara deserta-Peckichara-Stephanochara jiangsuensis* 组合。层内主要有 *Gobichara deserta*、*Stephanochara jiangsuensis*、*Peckichara subsphaeica* 等，相当于江苏阜宁组二—四段中该轮藻化石组合。

三、孢粉

孢粉仅在北桥组中发现，划分为以下两个组合带：

（一）*Ulmipollenites minor-Triporopollenites* 组合带

该组合带特征：①以木本花粉占绝对优势，占总量的 90%（以阔叶树及针叶树为主）；②优势成分以小榆粉、克氏脊榆粉等为主；③含少量孑遗分子如陀孔粉、扁三沟粉等。可以认定其时代为古—始新世早期。

（二）*Quercoidites-Ulmipollenites-Triporopollenites* 组合带

本组合特征：①木本花粉占优势（67%~75%），仍以阔叶树为主；②阔叶树花粉为栎粉多种、小亨氏栎粉等，另有一些热—亚热带的成分，如木兰、樟、梭形栗粉等。根据该组合特征显然与始新世早期的相似。

第三节　区域地层对比

一、区域对比

区域上，古近系建造总体上以泥岩、粉砂岩或粉细砂岩等细粒岩为主，局部层位含砾或夹砂砾岩，尤其是底部砂砾岩较多；但在皖南阜阳盆地和赣北武宁—南昌地区，以砂岩、砂砾岩、砾岩等粗碎屑建造为特征。

区内夏驾桥组岩性特征与相邻江苏泰州组极为一致（表 7-2），尤其是本组地层所含的轮藻化石，以新生代分子为主，少量中生代的孑遗分子。其化石组合广泛见于我国古近纪盆地中，与邻近江苏泰州组、阜宁组一段所含轮藻化石组合极为相似。介形虫中，*Eucypris* 为江苏阜宁组和三垛组、安徽瞬山集组、望虎墩组和痘姆组的带化石，并且在江苏泰州组也大量分布。此外，*Eucypris*、*Parailyocypris* 是赣南古新统池江组的主导属种。*Parailyocypris* 广泛见于江苏泰州组、阜宁组底部和下部，是其带化石。

区内北桥组上部以 *Sinocypris* 介形类动物群和具瘤形装饰的 *Grovesichara* 等轮藻为主；下部为 *Quadracypris* 介形类和 *Latochara* 轮藻组合，该组合特征与江苏泰州组上部、阜宁组一段及浙江长河组相似。划分的两个孢粉组合带相当于阜宁组一段、二段。华南东南部的 *Sinocypris-Eucypris-Ilyocypris-Cypris* 介形虫组合，与区内北桥组建立的组合很相似，前者区域上发现的层位有舜山集组上部、阜宁组三段和四段、长河组二段等。江苏阜宁组一段为棕红、灰白色泥质粉砂岩、粉细砂岩互层，开始明显发育玄武岩和泥灰岩夹层，其岩性组合特征与北桥组十分相似。阜宁组玄武岩夹层年龄为 56.3Ma（K-Ar 法）和 58.6±1.1Ma（Ar-Ar 法）。因此，区内将北桥组置于夏驾桥组之上，并与阜宁组可对比，为古新世晚期地层。

表 7-2　区域地层对比表

Table 7-2　Regional stratigraphic correlation

地层		上海	浙江		皖南		江西			江苏
			浙西北	浙东南	来安盆地	阜阳盆地等	武宁—南昌		赣州	南京
上覆地层		白龙港玄武岩/松散层	嵊县组	第四系	下草湾组	洞玄馆组	第四系			洞玄馆组/盐城组
古近系	渐新系	?	长河组	嵊县组 长河组		吴雪岭组	奉新砾岩 郑家渡组	临江组	下虎组	三垛组
	始新系	?	长河组	长河组	张山集组 狗头山组	照明山组 双塔寺组	郑家渡组 磨下组	新余组	下虎组 池江组	戴南组
	古新系	北桥组 夏驾桥组	长河组	长河组	舜山集组	痘姆组 望虎墩组	磨下组	新余组	池江组	阜宁组 泰州组
下伏地层		赤山组	上白垩统		赤山组		莲荷组	未见底	未见底	赤山组

二、存在问题

尽管近 30 年来针对区内古近纪地层的划分方案基本达成共识，但由于揭露钻孔有限，依据并不充分。2006 年，书院镇 SG7 孔揭露了厚约 15.72m 的灰褐、紫红色砂砾岩和泥质粉砂岩。孢粉中以裸子植物花粉占绝对优势，绝大多数为松科，单/双束花粉 *Abietineaepollenites/Pinuspollenites* 大量出现；被子植物花粉为胡桃科和榆科的一些分子；蕨类以新近纪石松孢 *Lycopodiumsporites neogenicus* 为多。主要组分大多为古近纪中晚期—新近纪的常见类型。未见古近纪早期出现的一些古老分子，少见主要繁盛于新近纪的一些草本和灌丛花粉。松科花粉大量出现、典型的热—亚热带分子稀少、草本花粉缺乏等特征都表明，揭露的地层置于早—中渐新世较为合适。2012 年，在新海农场一个深 600.45m 的钻孔中，于基岩面（399.17m）以下揭露了一层厚达 30m 的玄武岩。在玄武岩下部为一套砂泥岩沉积，461.4m 以浅以棕红、灰色、淡紫色为主，下部以紫红色为主，为一套断陷盆地沉积。该玄武岩层顶部 K-Ar 年龄为 31.6±1.31Ma，底部年龄为 38.57±2.05Ma，为始新世末期—渐新世初期，相当于三垛组沉积期；但由于岩芯缺失，对下部红盆沉积的具体层位缺乏进一步资料的支撑。

第四节　岩相古地理

古新世早期基本继承了晚白垩世的沉积环境，在安角盆地和北桥盆地中形成了一套湖相的细碎屑岩。主要岩性为粉砂质泥岩、泥质粉砂岩、含砾泥质粉砂岩、泥岩、粉砂岩、粉细砂岩，含钙质，层理清晰，富含介形类及轮藻化石。

北桥组中除棕色砂泥岩堆积外，灰色（或灰褐色）以及含碳夹层的出现，反映当时出现了周期性的温湿气候和还原性质的沉积环境。

鉴于上海迄今未发现始新世和渐新世地层，当时的岩相古地理环境情况不明。

第八章 新近系

上海新近系是由下部的玄武岩和上部的冲洪积和湖泊堆积两部分地层组成。下部以见于本区西北侧昆山附近的钻孔剖面较为典型，大致相当于中新统；上部地层以崇明、嘉定安亭较为发育，称之“崇明组”，属上新统。中新统主要分布于东南部的长江河口滨岸地带，沿灯塔—赵家宅断裂和罗店—新场断裂两条北西向断裂带呈串珠状分布。上新统主要发育于上海市内晚新生代沉陷最深的崇明等河口地区，其次在宝山、浦东的滨岸地带及嘉定西北部等地的凹陷区内也有分布（图 8-1）。

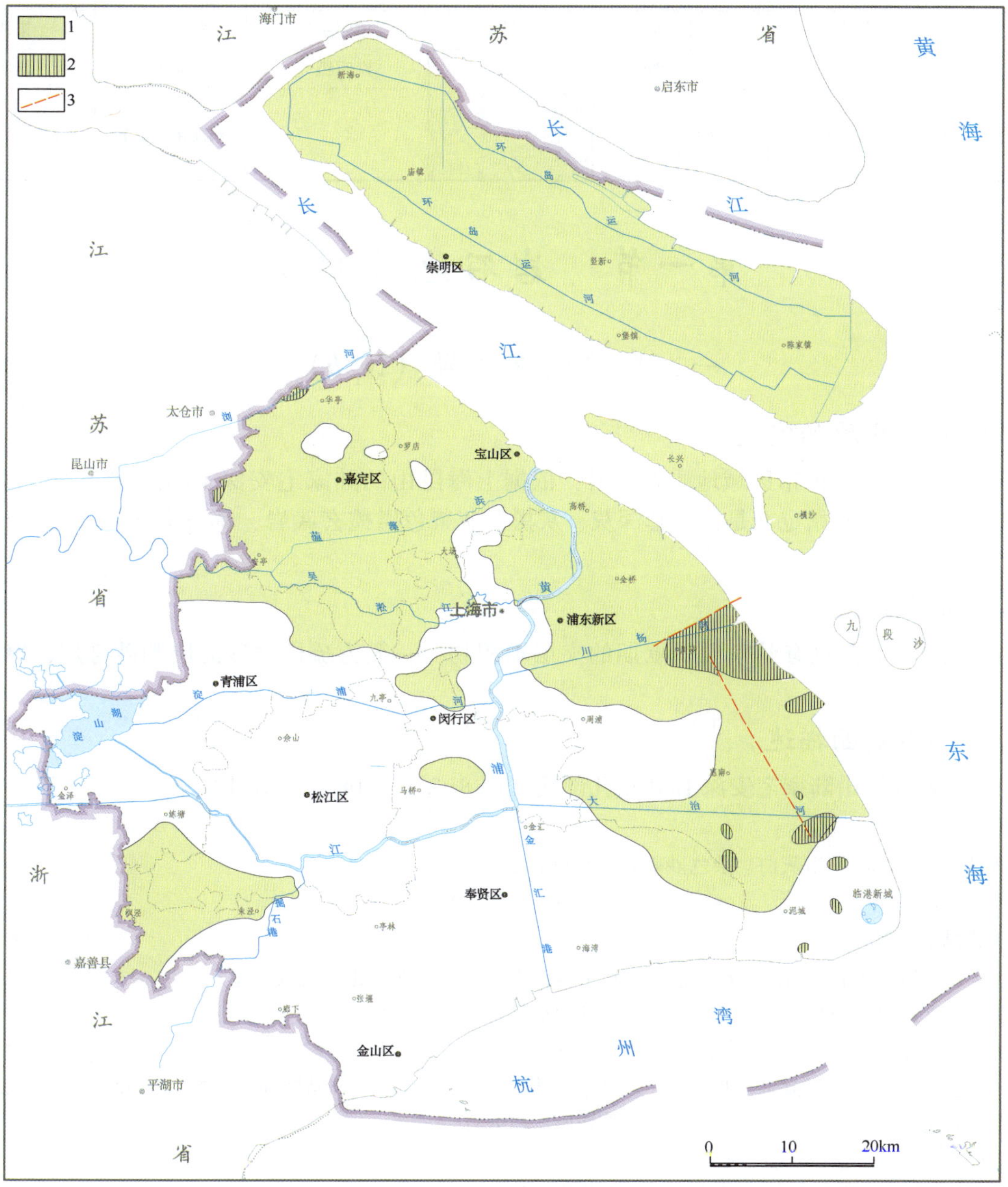

图 8-1 上海中—上新统分布图

Fig. 8-1 Distribution of the Miocene-Pliocene in Shanghai

1—上新统；2—中新统；3—断裂

1960年，上海市人民委员会地质处认为区内发育上新统，中新统缺失。1962年，上海市地质勘察局首次将区内划分出中新统和上新统。20世纪70年代末，为验证川沙磁异常，在川沙白龙港钻遇厚约260m以上仍未见底的玄武岩。1979年，在编制《上海地区基岩地质推断图》时，将其划为上新统。1984年，上海市水文地质队编写《上海第四纪地层研究报告》时将上新统创名为“崇明组”。1988年，编著《上海市区域地质志》时，创名“白龙港玄武岩”，置于中新统，并划分出上新统崇明组。之后，这一划分方案一直沿用至今（表8-1）。

表8-1　上海新近系岩石地层单位划分沿革表

Table 8-1　Previous divisions of the Neogene lithostratigraphic units in Shanghai

地层	《上海市石油及天然气地质普查工作报告》，1960	《上海地区矿产地质普查工作总结报告》，1962	《1：20万上海地区基岩地质推断图》，1979	《上海第四纪地层研究报告》，1984	《上海市区域地质志》，1988	本志
上覆地层	第四系	第四系	第四系	第四系	第四系	第四系
上新统	未分	3个旋回	未分	崇明组	崇明组	崇明组
中新统		嵊县玄武岩		?	白龙港玄武岩	白龙港玄武岩

第一节　岩石地层划分

一、白龙港玄武岩（N_1b）

（一）创名及原始定义

1988年，编著《上海市区域地质志》时，依据上海昆山的陆家宅安深1#钻孔资料而创名，称之“白龙港玄武岩”。其原始定义是：一套深灰、灰黑及灰紫色橄榄玄武岩、辉石玄武岩及气孔状（杏仁状）玄武岩。

（二）现在定义

区内所见玄武岩都见有其喷溢间歇期的夹层堆积物，一般为疏松之砂泥层和砂砾层，厚度不等。多者见有3个夹层。

（三）层型剖面描述

正层型剖面为昆山陆家宅安深1#孔钻孔剖面（图8-2）。剖面地层分述如下：

上覆地层：上新统崇明组灰白、杂色砂砾石夹灰色黏土

-----假　整　合-----

白龙港玄武岩（N_1b）　　总厚度38.5m

8. 黄褐色橄榄玄武岩，局部呈灰黑色，斑状结构，橄榄石斑晶沸石化、蛇纹石化　　3.5m
7. 灰黑色玄武岩，隐晶质，粗玄结构，局部具细小气孔构造　　5.0m
6. 灰黑微带绿色橄榄玄武岩，斑晶为橄榄石、辉石　　4.0m
5. 灰紫色玄武岩，局部棕紫、灰黑色，具杏仁状构造，斑状结构，斑晶橄榄石已蛇纹石化、伊丁石化，中部夹一层浅棕色泥岩，含菱铁矿、白云石　　2.0m
4. 浅棕色泥岩，含显微状菱铁矿、方解石　　1.0m
3. 暗红棕色黏土，软塑性，含较多钙质及少量岩屑粉砂　　4.0m
2. 深灰、灰色黏土、粉砂质黏土，夹棕色泥质条带或团块状灰色含砾粉砂质黏土。见少量介形虫化石碎片、腹足类化石碎片和鱼骨化石碎片　　14.5m

1. 杂色砂砾岩，砾石成分主要为石英岩、石英砂岩，其次为燧石、石灰岩及少量的岩浆岩，多棱角状或次棱角状，砾径一般为 3~5mm，砂粒大部分为岩屑组成的粗砂，胶结较紧　　4.5m

～～～～～～ 不　整　合 ～～～～～～

下伏地层：赤山组浅棕色粉砂岩、泥质粉砂岩与棕色泥岩、含砾粉细砂岩互层

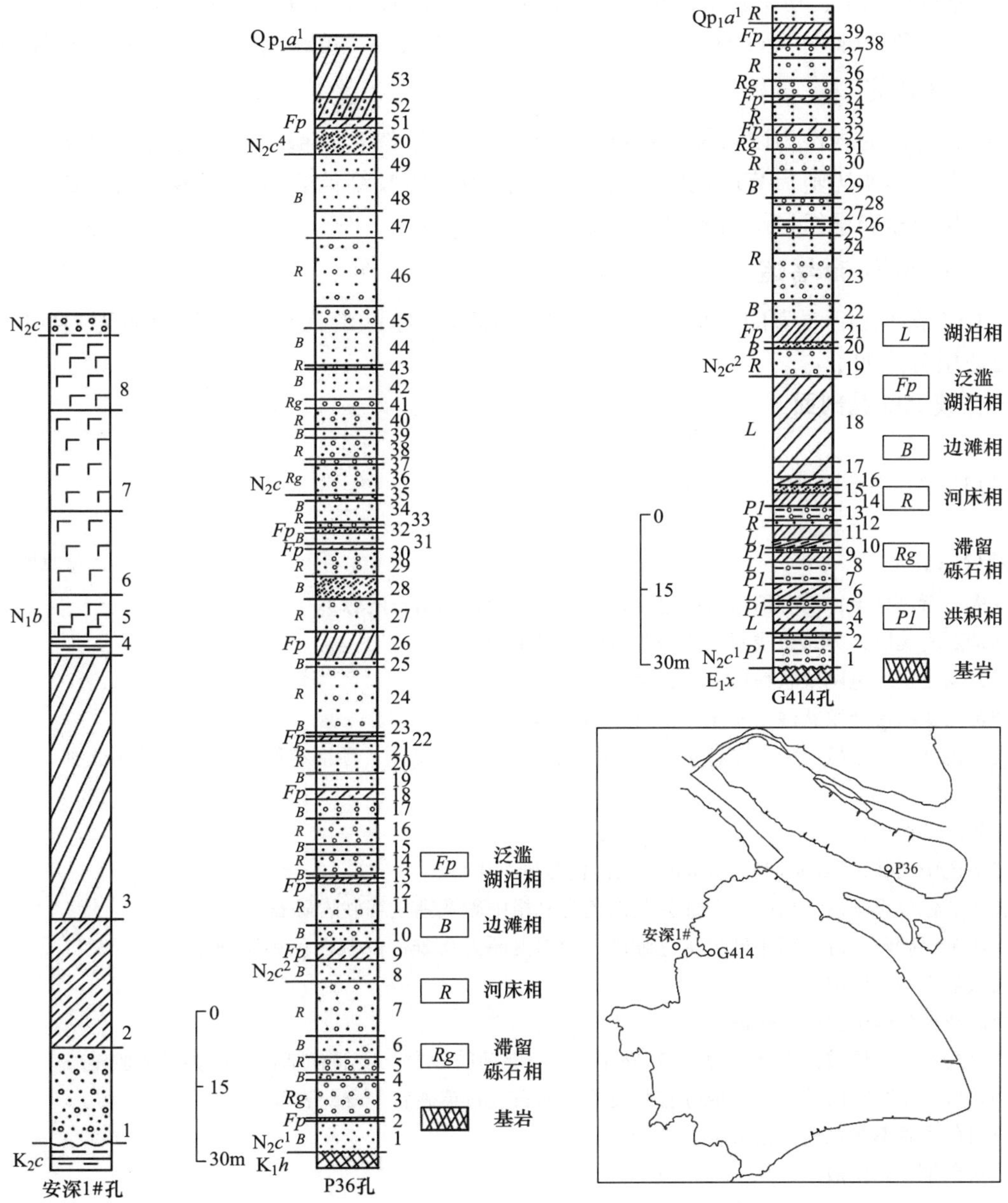

图 8-2　正层型和次层型钻孔剖面柱状图

Fig. 8-2　Holostratotype and hypostratotype boreohle section histograms

（四）地质特征及区域变化

在浦东一些钻孔中曾发现多处玄武岩，厚度一般为 23~69m。在白 zk2 孔见有深灰色辉石玄武岩、橄榄玄武岩和灰绿色玻质玄武岩，已见厚度达 260.0m，但仍未见底，近顶部尚见有厚约 9.0m 之粉砂质黏土和砂层质夹层。在昆山附近的钻孔中尚见有多个喷溢间歇期的堆积层，如昆山 901、902 等孔。

区内玄武岩全岩 K-Ar 法年龄为 26.7~18.7Ma，与其下伏不同时代地层角度不整合接触。

二、崇明组（N_2c）

（一）创名及原始定义

1984年，邱金波等人据崇明向化P36孔创建，并于1988年编著《上海市区域地质志》时首次公布。其原始定义：系指冲洪积相、湖相沉积。岩性为灰白、灰黄、灰绿色含砾砂、泥质砂砾石及浅灰绿、黄绿杂褐黄、紫红色黏土、粉质黏土，粗细相间组成两个旋回，上以黄绿、灰绿色黏土与上覆第四系下更新统安亭组灰白色含砾砂、砂砾石层呈假整合接触；下以灰绿、黄绿色砂砾层与下伏地层接触。

（二）现在定义

崇明组具特征的岩性为以灰绿、黄绿色为主，夹杂紫红、紫褐、褐黄等色的黏土和粉质黏土，黏土呈坚硬状态，不能搓塑，节理比较发育。与大都为泥质胶结的含砾砂、砂砾石组成互层，形成多个沉积旋回。与上覆安亭组呈平行不整合接触。

（三）层型剖面描述

该组正层型钻孔剖面为向化镇P36孔，上海陆域平原区的次层型钻孔剖面为G414孔（见图8-2）。剖面地层分述如下：

1. 向化镇P36孔剖面

上覆地层： 第四系下更新统安亭组黄色细砂层

----- 假 整 合 -----

崇明组（N_2c）	总厚度218.65m
53. 浅褐、紫色黏土，均有稀疏灰绿色网纹，顶、底较多钙结石网脉	9.51m
52. 钙结石富集层，分布于浅紫杂淡绿色网纹状黏土中	4.27m
51. 浅紫褐带灰绿色网纹状粉质黏土，富含钙结石网脉	2.00m
50. 锈黄杂灰绿色网纹状砂质粉土，富含钙结石和网脉	5.00m
49. 浅灰色细砂，松散，夹有厚各4cm两小层钙质细砂岩透镜体，顶部有厚为0.26m的钙质细砂岩	4.31m
48. 深灰间杂淡黄色细砂，松散	6.90m
47. 浅灰色中细砂，顶部有厚为0.6m的灰白色含砾细中砂	5.42m
46. 浅灰黄色含砾中砂，松散，顶部为灰白色含砾粗中砂及钙质细砂岩砾石层	13.50m
45. 浅灰黄色砂砾、砾质粗中砂夹灰色细砂，中部夹厚为0.26m的灰白色钙质细砂岩	4.35m
44. 浅灰色细砂，松散	7.65m
43. 浅肉灰色含砾中粗砂，松散	0.65m
42. 灰色粉细砂，松散，夹有厚为1~2.5cm的3小层钙质砂岩，顶部有厚0.15m为铁质胶结	5.75m
41. 卵石层，卵圆—次棱角状，成分有脉石英、燧石、石英砂岩、凝灰岩等	1.70m
40. 灰白色含砾不等粒砂，松散	4.32m
39. 浅灰色细砂，松散	1.78m
38. 浅黄灰色含砾不等粒砂与浅灰色粉细砂互层	4.00m
37. 卵石层，多卵圆—扁平次圆状，成分有脉石英、燧石、石英砂岩、硅板岩、黑云片麻岩等	1.00m
36. 灰白色含砾不等粒砂，松散，粗、中、粉细砂各占1/3左右，底为厚0.24m的卵石层	6.20m
35. 浅黄绿色含砾细中砂，多为石英质砾，并含黄绿色泥砾	1.10m
34. 浅绿色细砂、灰白色中细砂，为泥质略胶结	4.20m
33. 灰白色含砾粗砂、含砾粗中砂，含少许黄绿色泥砾及半炭化木片	1.25m
32. 浅灰绿色砂质粉土夹粉质黏土	1.00m
31. 灰白色细砂，局部夹灰绿色粉质黏土，顶部为厚0.30m的灰白色含砾细中砂	2.05m
30. 浅灰绿色粉质黏土、黏质粉土，夹有灰白色粉细砂	0.95m
29. 灰绿色中砂夹砾质粗中砂，泥质略胶结	5.59m

28. 浅绿灰色砂质粉土夹粉砂，顶部有钙积层（钙质粉砂岩）薄层 4.31m

27. 灰白色含砾中粗砂，泥质胶结，砾径1.2~2.5cm，次棱角—扁平状，成分有燧石、长石石英砂岩等 6.72m

26. 棕黄、黄绿色黏土，有浅紫色杂斑和网纹，顶部有浅灰色钙积层（钙质细砂岩） 5.53m

25. 浅黄绿色细中砂，上部较松散，中下部为泥质略胶结 1.40m

24. 浅黄绿、黄灰色含砾粗中砂，泥质胶结较密，砾石呈次棱角状，砾径为0.2~2cm，主要有燧石、凝灰岩、石英砂岩，较多绿色泥砾 13.10m

23. 灰白色细砂，泥质弱胶结，底部含砾较多，成分复杂 0.60m

22. 浅灰色粉质黏土夹砾质粗砂，砂中含较多泥砾 1.05m

21. 灰白色细砂、细中砂和黏质粉土互层，含少量半炭化植物碎屑 1.85m

20. 浅灰绿色中粗砂，含大量灰绿色泥砾，少量凝灰岩、石英质砾 4.40m

19. 灰白色细砂，为泥质弱胶结 3.28m

18. 浅灰色粉质黏土，含钙质结核，局部有钙积层（钙质砂岩） 1.82m

17. 灰白色细中砂、灰绿色砂砾石互层，砂砾层中含大量灰绿色泥砾，并有较少的石英、凝灰岩、钙积石、千枚岩等砾石 4.04m

16. 浅灰色含砾中粗砂（上部）、灰绿色砂砾石，含较多灰绿色泥砾 4.96m

15. 灰白色中细砂，泥质胶结，偶见灰绿色泥砾、石英砾。顶部含半炭化植物碎屑 2.35m

14. 浅灰绿色砾质中粗砂夹中细砂，含较多石英砾和灰绿色泥砾 3.50m

13. 灰白色细中砂，含少量石英砾及灰绿色泥砾 0.81m

12. 灰绿色黏土，偶见半炭化植物碎屑 1.04m

11. 灰白色含砾中粗砂、砂砾石，中部夹薄层钙积层 8.45m

10. 浅灰黄、灰白色中细砂，夹有砾质中粗砂，砾石成分为石英、凝灰岩，并含泥砾 3.43m

9. 浅黄绿色粉质黏土、砂质粉土互层 3.47m

8. 浅绿色细砂，局部夹灰绿色砂质粉土 4.15m

7. 灰白色含砾粗中砂、砾质中粗砂夹含砾中细砂，下部为砂砾层 10.92m

6. 浅灰绿、灰白色细砂，偶含砾 4.08m

5. 浅灰黄色砂砾石，砾石多呈次圆状，成分有石英、硅质页岩、钙积石、凝灰岩等 3.32m

4. 灰白色细砂（上部）、浅灰黄色砾质粗中砂（下部） 1.23m

3. 卵石层，多呈次圆—次棱角状，砾径在1.5~8cm，成分有花岗斑岩、石英正长斑岩、燧石、石英、凝灰岩、硅质白云岩等 7.63m

2. 褐黄杂灰绿色黏土，具网纹构造 0.32m

1. 浅绿色细砂，含少量半炭化植物碎屑 6.44m

~~~~~~~~ 不　整　合 ~~~~~~~~

下伏地层：下白垩统黄尖组灰绿色流纹岩

2. 安亭G414孔剖面

上覆地层：下更新统安亭组下段锈黄色中细砂、含砾粗中砂

----- 假　整　合 -----

崇明组（N_2c） 总厚度128.36m

上段（N_2c^2） 厚70.46m

39. 绿灰杂褐色黏土夹纹层状粉砂，具不规则水平层理 3.00m

38. 褐杂灰绿色黏土与灰色细砂互层 1.59m

37. 灰色含砾中粗砂，松散，砾石呈次棱角状，多扁平砾，大小较匀，一般为1~5cm，成分较杂，有安山岩、砂岩、凝灰岩等 2.56m

36. 灰色含砾粗砂，砾石含量、粒度自上向下递增，大小较匀，成分较杂 4.42m

35. 砾石层，砾石大小不一，成分杂，有燧石、石英、砂岩、凝灰岩 3.03m
~~~~~~~~

34. 灰黄色粉质黏土，下部厚 0.4m 变为砂质粉土，含砾 1.00m
33. 锈黄色中粗砂，含少量细砾 4.55m
32. 灰黄色粉质黏土，含钙质结核 2.15m
31. 砾石层，砾石大小较匀，成分有石英、砂岩、燧石等 2.90m
30. 砂砾石，砾石大小较匀，呈次棱角—浑圆状，成分多为石英、燧石，顶部厚 0.3m 为泥质砾石层，其下 3m 为灰黄色泥质砂砾石 4.80m
29. 灰色细中砂，松散 5.00m
28. 砂砾石，砾石大小较匀，多呈次圆状，成分有石英、燧石等 1.00m
27. 灰黄色含砾粗中砂，松散，砾石大都呈次圆状，成分多为石英质 3.58m
26. 灰黑色泥质粗砂，局部含细砾，顶部夹有粉质黏土薄层 1.10m
25. 棕黄色含砾粗砂，砂中含较多已黏土化的长石 1.50m
24. 浅灰黄色中砂，松散，顶部含少量砾石 3.57m
23. 砂砾石，松散，砾石大小较匀，砾径一般为 2~7cm，成分多为石英、燧石，上部夹有厚为 1m 的灰黄色中粗砂，底部砾石较大 9.82m
22. 浅绿灰、黄灰色中砂，上部厚 1.2m 为细中砂 3.83m
21. 蓝灰杂褐黄色黏土，下部 1.7m 为褐黄杂蓝灰色粉质黏土 4.30m
20. 褐黄色砂质粉土 1.24m
19. 灰黄色含砾中粗砂，松散，向下有变粗趋势，砾石成分为石英质 5.52m

下段（N_2c^1） 厚 57.90m

18. 浅黄、绿灰色黏土，上部见少量长石碎屑，下部见少量铁质结核 16.94m
17. 浅绿灰色黏土，含不规则钙质团块，与上覆层黏土均具硬塑性，细腻似蜡状 3.05m
16. 浅黄色粉质黏土，含有较多长石砂粒 1.75m
15. 灰色砂质粉土 1.30m
14. 浅黄、绿灰色黏土，节理发育，具擦痕 2.70m
13. 浅棕色泥质砾石，砾石和黏土混杂，分选很差，砾石大都为次棱角状，成分有凝灰岩、安山岩等 2.70m
12. 灰色中粗砂，顶部有厚为 0.3m 的浅黄、浅绿灰色黏土 1.10m
11. 褐黄色黏土 2.90m
10. 浅黄灰色粉质黏土夹纹层状粉砂 1.60m
9. 灰色泥质砂砾石，夹有浅黄灰色粉质黏土 0.80m
8. 浅黄、浅灰绿色黏土 2.20m
7. 灰棕、褐色泥质砾石层，泥质胶结，分选很差，中部夹厚为 0.9m 的浅灰绿色砂质粉土 4.20m
6. 浅棕色粉质黏土，向下渐变为砂质粉土 3.40m
5. 泥质砾石，砾石呈次棱角状，成分有凝灰岩、斑岩等 1.54m
4. 褐黄杂灰色网纹状粉质黏土，下部夹有厚为 0.2m 的砂质粉土 2.76m
3. 浅黄色粉质黏土，底部有厚为 0.1m 的砂质粉土，顶部有厚为 0.4m 的灰色泥质砾石 2.20m
2. 灰色含砾泥质中粗砂，分选差，松散 0.71m
1. 灰色泥质砂砾石，为中粗砂、砾石和泥质混杂，砾石呈次棱角状，成分以砂岩、凝灰岩为主 6.76m

~~~~~~~~ 不　整　合 ~~~~~~~~

下伏地层：上白垩统赤山组棕红色泥质长石粉细砂岩

该孔下段底部（第 1~4 层）只有少量的裸子植物 *Pinus*、*Abies* 和草本植物 *Gramineae*、水生草本植物 *Cyperaceae* 孢粉。下段下部（第 5~6 层）含落叶阔叶植物花粉占 45.8%，主要有 *Quercus*、五加科（Araliaceae）、*Salix*、*Zelekova* 等；常绿阔叶植物花粉占 6%左右，有 *Q. glauca*、*Eurya*、*Magnolia* 等；还有少量的 *Pinus*、*keteleeria* 等针叶植物花粉；草本植物花粉占 34%，以 *Gramineae*、*Typha* 为主；还有较少的蕨类孢子 *Ceratopteris*、里白属（*Gleichenia*）和槐叶苹属（*Salvinia*）。下段中部（第 7~12
~~~~~~~~

层）草本植物花粉占55%，以 *Gramineae*、毛茛科（Ranunculaceae）、*Artemisia* 为主；阔叶植物花粉占30.6%，有 *Quercus*、*Castanea*、*Ulmus*、朴属（*Celtis*）、山毛榉属（*Fagus*）等；并见少量的裸子植物花粉，有 *Pinus*、*Cupressaceae*，个别的常绿阔叶植物 *Lauraceae*；蕨类孢子较少，有 *Polypodiaceae*、铁线蕨属（*Adiantum*）。下段上部（第13~18层）以木本植物花粉最多，主要有 *Quercus* spp.，其次有 *Liquidambar*、*Castanea*、*Zelekova*、*Salix*、*Alnus* 等，见少量的常绿阔叶成分 *Quercus glauca*、*Lauraceae* 及针叶植物 *Pinus*、*Taxodiaceae* 花粉；草本植物花粉以 *Gramineae*、*Artemisia*、*Polygonum* 较多见，还有一些水生草本植物 *Typha*、水鳖属（*Hydrocharis*）；尚见较少的蕨类孢子 *Pteridium*、鳞盖蕨属（*Microlepia*）、*Polypodiaceae* 和膜蕨属（*Hymenophyllum*）。反映下段由底—下—中—上部，古气候经历了湿冷—热湿—暖干—热湿的变化。

上段中下部（第19~31层）为灰黄或灰色砂砾石与细中砂互层，偶夹杂色黏土。孢粉比较贫乏。上段下部（第19~25层）草本植物占显著优势，出现大量的中旱生草本植物 *Gramineae*、*Artemisia* 花粉；其次为针叶植物 *Cupressaceae*、*Pinus* 花粉，出现暗针叶树 *Abies*、*Picea* 花粉；蕨类孢子只有一些 *Polypodiaceae* 和 *Pteridium*。上段中部（第26~34层）以草本植物花粉为主，主要有 *Gramineae*、*Typha*；木本植物花粉居次，多半为落叶阔叶植物 *Quercus*、*Liquidambar*，也有相当数量的常绿阔叶成分和较少的针叶植物花粉，并广泛出现 *Ceratopteris* 孢子。上段上部（第35~39层）含木本植物花粉达80%，以落叶栎为主，并含较多的常绿阔叶植物 *Quercus glauca*、*Castanopsis* 及针叶植物花粉；而含草本植物花粉较少，蕨类孢子贫乏。反映上段由下—中—上部，古气候经历了冷干—暖湿—热湿的变化。

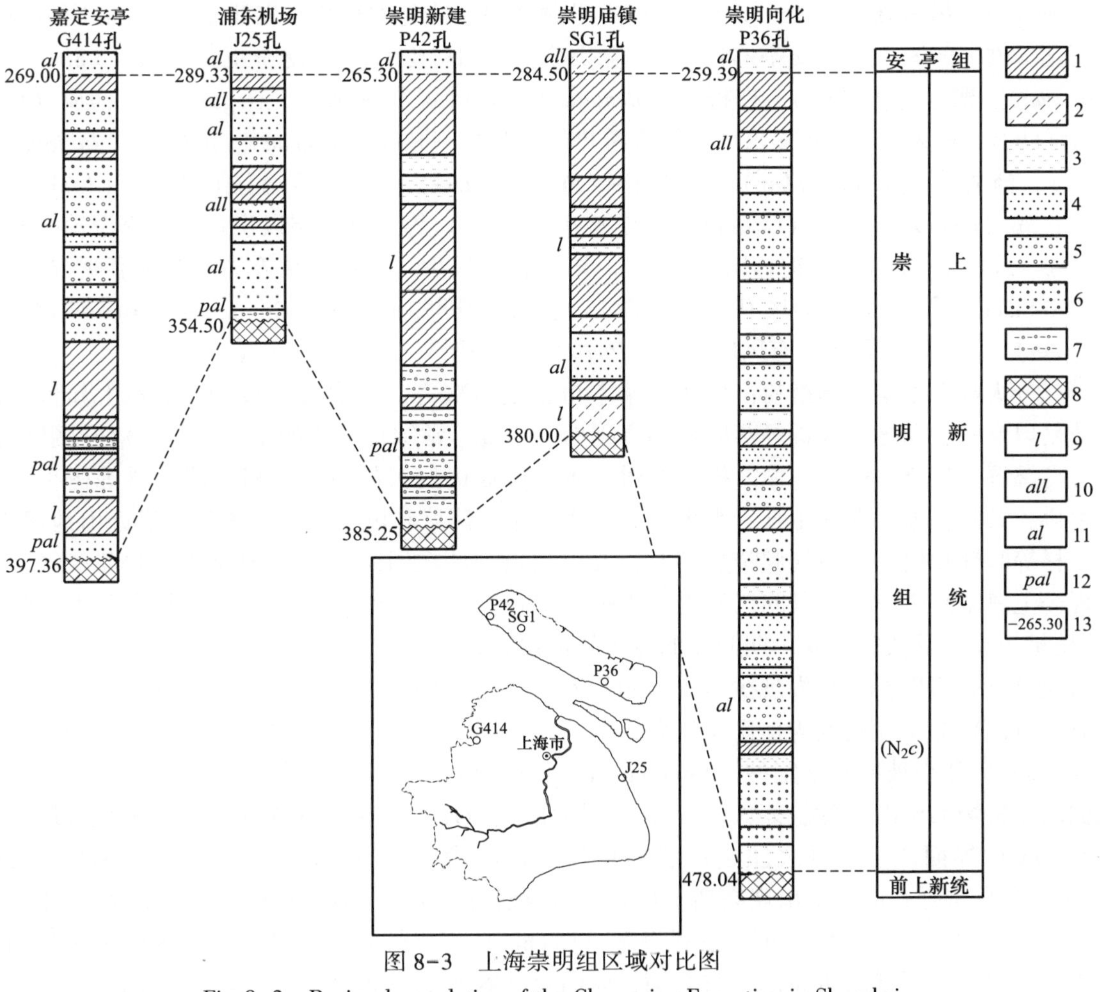

图8-3 上海崇明组区域对比图

Fig. 8-3 Regional correlation of the Chongming Formation in Shanghai

1—黏土、粉质黏土；2—砂质粉土、黏质粉土；3—粉砂、细砂；4—细中砂、中粗砂；5—含砾砂；6—砂砾石；7—泥质砾石；8—基岩；9—湖泊相；10—泛滥湖泊相；11—河流相；12—冲洪积相；13—深度（m）

（四）地质特征及区域变化

在河口沙岛上，崇明岛东、西端地区地层发育较全，可见上、下两套地层：下部为砂砾石、含砾砂和灰绿色黏土频繁重现组成的辫状河流沉积，或以湖积的灰绿等色黏土为主夹洪积含泥砾质砂或砂砾石；上部以含砾砂与细砂互层为主，向上渐变成以黏土为主具有二元结构的曲流河道沉积；或以湖积黄绿、褐黄色黏土为主与洪积泥质砾石互层。两套地层分界深度起伏于350~400m之间，其顶界埋深通常为260~290m，厚度介于60~220m间（图8-3）。

在上海陆域的宝山、浦东滨岸地带和嘉定西北部的安亭凹陷内也有本组地层分布，其岩性、岩相序列与崇明所见相似：通常由下部的洪积灰黄色泥质砂砾石频夹黏土和上部的洪积—湖积灰绿色含砾黏土、黏土层组成。顶界埋深也为260~290m，厚度大都为10~65m，唯在安亭凹陷内厚者可达120m（图8-3）。

据浦东机场J25孔和崇明新建P42孔的黏土矿物分析，本组均出现以蒙脱石占优势、伊利石居次的黏土矿物组合带，与上覆骤变为伊利石占优势、蒙脱石骤减的黏土矿物组合带的分界，恰与第四系与新近系的界线重合。

第二节　古生物群组合和地层时代

本区新生代仅于崇明组中发现孢粉和介形虫化石，其中以孢粉为主。

孢粉仅在少数钻孔中检出，以落叶栎（*Quercusdeciduous*）-常绿栎（*Quercus evergreen*）-莎草科（Cyperaceae）-凤尾蕨属（*Pteris*）组合为主。以木本花粉占绝对优势，在浦东J25孔和庙镇SG1孔木本花粉高达85%~90%；嘉定安亭G414孔、金山朱泾G418孔的木本花粉也达46%~>80%，以麻栎（*Q. acutissima*）、槲栎（*Q. alina*）、柞栎（*Q. dentata*）等落叶栎和青刚栎（*Q. glauca*）、栲属（*Castanopsis*）等常绿阔叶花粉为主，并较多出现一般喜暖热的大戟科（Euphorbiaceae）、胡桃属（*Juglans*）、漆树属（*Rhus*）、枫香属（*Liquidambar*）、木兰科（Magnoliaceae），还有喜温湿的榆属（*Ulmus*）、朴属（*Ciltis*）、桦属（*Betula*）、栗属（*Castanea*）、榉属（*Zelekova*）等；有的层段出现较多喜温凉—寒冷的松属（*Pinus*）、云杉属（*Picea*）、冷杉属（*Abies*）。草本花粉较少，仅占10%左右，多为喜暖湿的莎草科、茜草科（Rubiaceae）；在安亭G414孔、朱泾G418孔本带内则多禾本科（Gramineae）和一些水生草本香蒲属（*Typha*）。蕨类孢子很少，有凤尾蕨属（*Pteris*）、水蕨属（*Ceratopteris*）等。

孢粉成分大都以木本植物孢粉占优势，草本植物孢粉次之，蕨类孢子较少。显示上新世，区内覆盖的植被是以麻栎、槲栎、柞栎等落叶栎和青刚栎、栲属等常绿栎为主的常绿阔叶、落叶阔叶混交林，林中杂有喜温湿的胡桃、枫香、榆、桦、栗树，还有喜暖热的大戟科、漆树、茜草科等树种，林下草被为禾本科，其间湿地、湖泊生长着水生草本植物香蒲、莎草和凤尾蕨、水蕨类，附近山地混生有松、云杉、冷杉等暗针叶树。反映气候暖热潮湿，大致相当于现在浙闽地区的中亚热带北部气候，年平均气温比现今上海略高2~3℃。其间已有以草本植物孢粉占显著优势，与暗针叶植物成分组成的暗针叶林-草原植被出现，反映曾遭受一期明显的冷干气候侵袭。

在新建P42孔崇明组顶部发现少量介形虫化石，以奇异白花介（*Leucocythere mirabilis*）、浪游土星介（*I. errabundis*）为主，其次为柯氏土星介相似种（*I.* cf. *carnae*）、光滑白花介（*L. luculenta*）、白花介未定种（*L.* sp.）和小玻璃介未定种（*Candoniella* sp.）的介形虫化石组合。

根据浦东机场J25孔、嘉定安亭G414孔、崇明庙镇SG1孔和前哨农场P35孔等磁性地层分析，其顶界略在Ma/G界面之下，据之厘定地层的顶界时代约为2.60Ma B.P.。根据安亭G414孔在孔深379m处灰褐色泥质砾石层底部出现高斯（Gauss）正极性时与吉尔伯特负极性时转换界面（年龄约为3.40Ma B.P.），孔深387~390m记录到科奇蒂（Cochiti）正极性亚时（年龄为3.86~3.98Ma B.P.）；孔深397.36m处为与下伏古近系的不整合面；而向化P36孔的其底界更深达478m。由此推算，区内底界年龄应大于4.00Ma B.P.。谢建磊等（2017）根据横沙岛磁性地层推断其底界年龄可达近6.3Ma。于振江（2004）将崇明红星农场P42孔的底界划至中新统。

第三节　区域地层对比

白龙港玄武岩为一套火山-沉积建造，以玄武岩为主，局部夹砂泥层和砂砾、泥岩。该地层在川沙白龙港孔中的 K-Ar 年龄为 18.7Ma、在南汇中学地震观测孔中的 K-Ar 年龄为 26.7Ma（于振江，2004）。区域上，在该时段的玄武岩主要见于江苏盐城组下段和黄岗组、嵊县组（表 8-2）。苏北盆地盐城组一段内玄武岩夹层的 Ar-Ar 年龄为 18.3±1.1Ma（钱勤，1996）。浙江新昌嵊县组玄武岩的成岩年龄在 20.7±1.7Ma~4.9±0.3Ma（李相传，2014）。

表 8-2　区域地层对比表

Table 8-2　Regional stratigraphic correlation

地层		上海	浙江	皖南		赣北	江苏	
				来安盆地	安庆		南京	南通
上覆地层		第四系	第四系	第四系	第四系	第四系	第四系	第四系
新近系	上新统	崇明组	嵊县组	桂五组	安庆组		方山组	盐城组
							雨花台组	
	中新统	白龙港玄武岩		下草湾组	洞玄观组		黄岗组	
							六合组	
							洞玄观组	
下伏地层		前新近系	前新近系	张山集组	吴雪岭组	前新近系	三垛组	三垛组

从崇明组所含介形虫化石来看，与国内上新世地层如华北稻地组 *Candona* sp.、*Candoniella* sp. 组合，苏皖盐城组上部 *Ilyocypris radiata*、*Cyprinotus*（*Hyterocypris*）*chiuhsiensis* 组合，以至南疆的阿图什组 *Ilyocypris manasensis*、*I. errabundis*、*Candona comprissafomis*、*Cypridopsisvidua* 组合等所含介形虫化石相似。与之相当层位，在皖南局部地区为桂五组的玄武岩建造。盐城组上段为黄棕、棕红色黏土、粉砂质黏土与绿灰、灰白色中、细砂岩，含砾砂岩互层，局部夹多层玄武岩。在浙江嵊州、温岭和诸暨等地有相当于该阶段的玄武岩建造记录，成岩时间在 4.77±0.7Ma~2.5±0.1Ma（李相传，2014）。

由于对区内中新统玄武岩的顶界年龄认识不清，且受深部松散层缺乏有效测年手段的制约，如上所述，目前关于崇明组底界和顶界年龄的认识还存在争论，本志暂将其归为上新世。

第四节　岩相古地理

中新世，区内西南部主要被广袤的剥蚀低山所占据，东部形成几个面积不等（从几至 50 余平方千米）的火山熔岩盆地，其中以川沙盆地规模最大，形成玄武岩类建造。在火山喷发间歇，受暖湿气候和盆地沉降影响，原为熔岩堆积的火山盆地一度演化为湖盆，沉积了厚数米至 25m 不等，以深灰、灰绿和棕色黏土为主的湖相地层，其底部往往有杂、灰白色砂砾沉积。

上新世时，本区出现剥蚀低山与堆积盆地相间分布的现象。上海长期处在相当于现浙闽地区的中亚热带气候环境下，地面广为常绿阔叶落叶阔叶混交林植被覆盖，古气候暖热潮湿。3.6~3.2Ma，气候转冷，在上海安亭孔深逾 300m 的上新统崇明组辫状河流沉积物中留下了反映这次寒冷气候的暗针叶林-草原植被的孢粉记录，推算降温的幅度可达 6~8°C，与全球的记录一致。上海中南部陆域的青浦、松江、金山、奉贤、闵行以及浦东与市区的接壤区仍为广袤的低山所占据，山峰比高大都起伏于 300~400m 之间。其间在闵行的北桥和七宝—虹桥有山间湖盆和山上残积洼地嵌布。在现陆域滨岸的嘉定—宝山—市区东北部—浦东—南汇新城一带出现北西-南东向连续分布、波状起伏的沉降带，还有金山枫泾—朱泾与松江新浜、青浦蒸淀的山间沉降盆地，在这些沉降区内形成山前洪积平原和湖泊洼地（图 8-4），堆积了一套常呈胶结状、含较多风化岩屑、长石的冲洪积相灰绿、灰黄色泥质砂砾石和泥质含砾中粗砂，与湖泊相灰绿、黄绿、褐黄色为主间杂赭红、紫褐色的黏土、粉质黏土的互层，沉积物累厚介于 10~130m 不等，其富水性较差。处在沉降幅度较大的若干地段如嘉定安亭、浦东机场、金山枫泾等地的地层剖面揭示，上新世沉积序列具有两个旋回，在多为泥质砂砾石、泥质含砾砂、灰

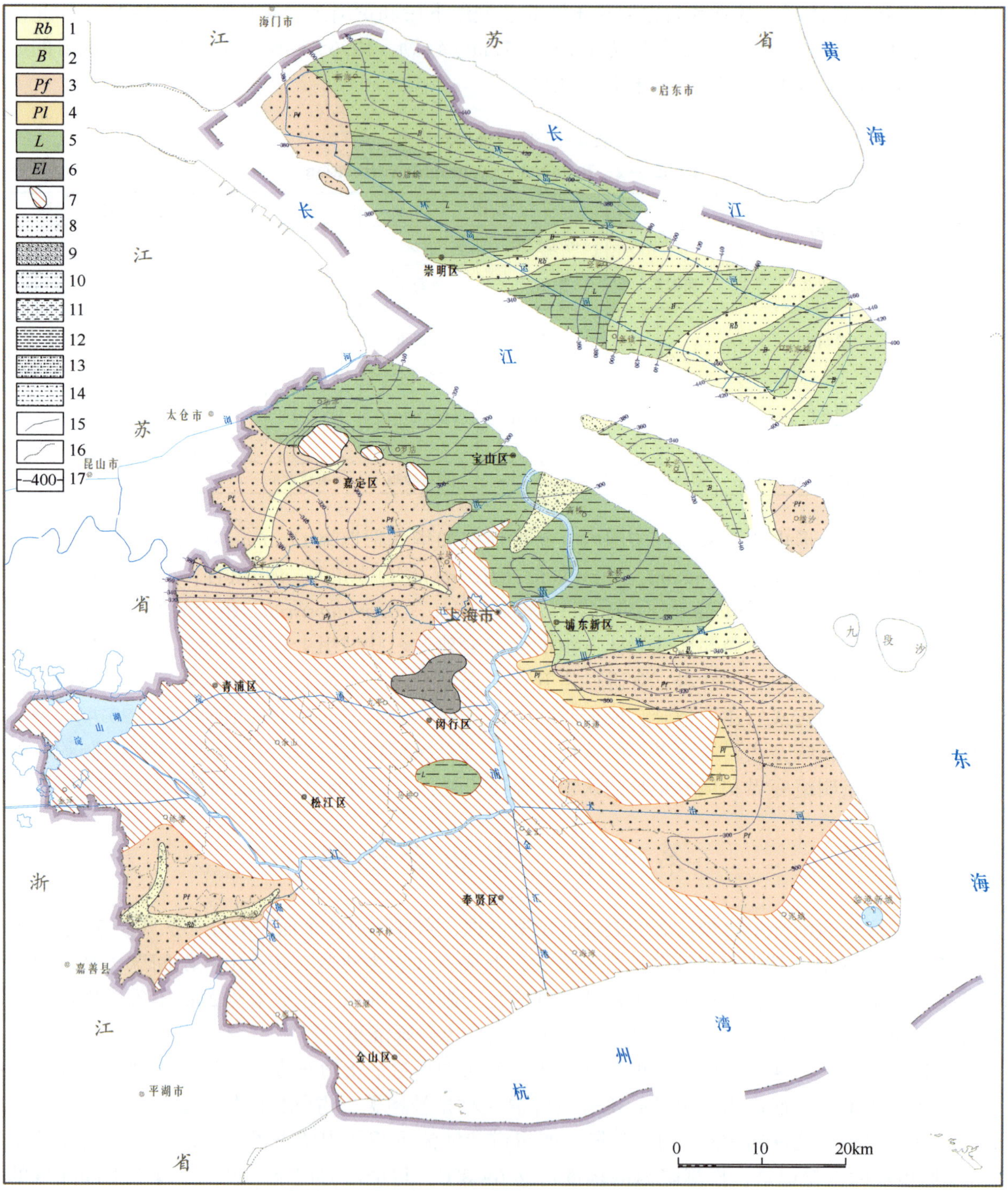

图 8-4　上新世岩相古地理图

Fig. 8-4　Lithofacies paleogeographic map of the Pliocene

1—河流相；2—河流心滩、边滩相；3—洪积扇（裙）相；4—山前、扇缘湖泊洼地相；5—湖泊相；6—山上残积洼地相；7—基岩剥蚀区；8—中细砂与含砾中粗砂互层；9—砂砾石或砾质砂；10—含砾砂；11—黏性土与中细砂互层；12—黏性土；13—泥质砂砾石；14—含岩屑、角砾黏土；15—岩相分区界线；16—岩性分区界线；17—层底等高线（m）

绿色黏土组成的冲洪积—扇上湖泊相的下旋回之上；又堆积了以砂砾石、含砾中粗砂与中细砂互层频夹黏土为主和粉土、黏土层组成的辫状河流—湖泊相的上旋回。前者厚为40~60m，后者厚在20~70m不等。

上新世，上海的沉降盆地中心在崇明岛上。当时在崇明岛东端的向化、前哨及迤北一带、中部的城桥—富民农场以及西端的绿华—红星农场一带，在广阔的河间滩地之上出现数条河流。长江口三岛，尤其是崇明岛东部地区为河床相砾砂层的主要发育场所，其埋深一般在280~350m之间，厚度一般有50~60m。这一砂砾层内富含松散岩类孔隙水，构成上海广义上所称的第Ⅴ含水层。

第九章　第四系

上海第四纪地层十分发育，除少数基岩残丘以外，均有第四纪沉积物覆盖，覆盖厚度主要在200~280m，下伏新近纪松散沉积物。20世纪60年代以来，随着水文和工程地质勘查的开展以及第四纪地质技术日新月异的发展，上海第四系划分也经历了多次变化（表9-1）。与此同时，上海市地质调查研究院（上海市环境地质站）结合城市建设需求，持续地开展不同比例尺的区域第四纪地质调查，及时地将科研理论成果应用到地层划分，实现了对区内第四纪地层更为详细的划分，将100m以浅地层与工程地质分层进行了对应，将100m以深地层与水文地质分层进行对应，有效地支撑了水文工程地质的发展。

1962年，上海市地质勘查局在《上海市石油天然气地质普查总结报告》中将埋深一般至90~110m、崇明埋深至150m以上的海陆交替堆积，与苏北东台群对比，将之归属第四系；认为其下系弱胶结的、基本为陆相的地层，可与苏北盐城组对比，将之划为上第三系。

1964年，地质部石油地质中心实验室在编纂石油地质图集时，将上述分层方案赋予岩石地层单位名称，在"第四系"内建立了"上海组"（属全新统）、"南汇组"（属中、上更新统）和"川沙组"（属下更新统）；将"上第三系"名之为"宝山组"（属上新统）。

20世纪70—80年代初期，在岩石地层划分方面变化不大；但在气候地层和年代地层方面提出了一些新的认识（刘金陵和叶萍宜，1977；闵秋宝和汪品先，1979；郭蓄民等，1979；竹淑贞等，1980，1983，1984；孙永福，1981）。依据当时将国内外第四纪下限放在距今350万左右的倾向，将覆盖基岩之上松散层均划入第四系。竹淑贞等（1979）根据气候旋回将上海第四系内分为6个统，并认为上海第四纪曾出现5次海侵。郭蓄民、竹淑贞、孙永福等（1979，1981）分别提出了上海全新世三分或四分的意见。

1984年，上海市地质处水文地质队在《1∶20万上海幅、启东幅区域水文地质普查报告》中，依据诸多钻孔的微体古生物化石、孢粉分析、古地磁、^{14}C定年等成果，综合岩石地层特征，将上海松散层划分为第四系（包括上、中部的"灰色层"和"杂色层"）和上第三系（绿色层）两部分，其分界时限约为3.00Ma（3.15~2.92Ma），埋深在300m上下；将第四系内分为4个统和10个组。之后，在1988年出版的《上海市区域地质志》中和2000年出版的《中国地层典·第四系》都延续该划分方案和认识。

1986年，江、浙、沪3省、市地质矿产局和上海海洋地质调查局合作完成的《长江三角洲地区水文地质工程地质综合评价报告》，也做出了与相同的晚新生代地层划分。

2007年，在邱金波和李晓编著的《上海市第四纪地层与沉积环境》中和随后完成的"上海市三维城市地质调查"项目成果中，将第四纪下限划在高斯（Gauss）与松山（Matuyama）极性时的分界，埋深在260~300m，将中更新世底界划在松山与布容（Brunhes）极性时的分界，埋深约为150m，在1987年地层单位基础上梳理了中更新统和下更新统岩石单位。之后，这一方案一直沿用至今。

表 9-1　上海第四纪地层划分方案对比表

Table 9-1　Comparison of different stratigraphic division schemes of the Quaternary in Shanghai

<table>
<tr><th>岩层</th><th>层底埋深/m</th><th>含水层</th><th colspan="2">上海市地质勘查局，1962</th><th colspan="2">地质部石油地质中心实验室，1964</th><th colspan="2">刘金陵，1977</th><th colspan="2">上海市地质处，1979</th><th colspan="2">同济大学，1974，1979</th><th colspan="2">竹淑贞等，1979</th><th colspan="2">上海市水文地质队，1984</th><th colspan="3">邱金波和李晓，2007</th><th colspan="3">本志</th></tr>
<tr><td>(1)黄褐色粉质黏土、砂质粉土夹黄灰色粉砂</td><td>6±</td><td rowspan="2">潜水</td><td rowspan="7">第四系</td><td rowspan="3">全新统</td><td rowspan="7">第四系</td><td rowspan="3">上海组</td><td rowspan="3">全新统</td><td rowspan="3">冰后期</td><td rowspan="3">全新统</td><td>Qh^4</td><td rowspan="3">全新世</td><td rowspan="3">上海组</td><td rowspan="3">全新统</td><td rowspan="3">冰后期</td><td rowspan="3">全新统</td><td>青浦组</td><td rowspan="3">全新统</td><td>青浦组</td><td>如东组</td><td rowspan="3">全新统</td><td>青浦组</td><td>如东组</td></tr>
<tr><td>(2)上部黄灰色砂质粉土，下部淤泥质粉质黏土</td><td>20±</td><td>Qh^3</td><td>上海组</td><td colspan="2">上海组</td><td colspan="2">上海组</td></tr>
<tr><td>(3)上部淤泥质黏土，下部褐灰色砂质粉土、粉细砂</td><td>30±</td><td></td><td>Qh^2
Qh^1</td><td>娄塘组</td><td colspan="2">娄塘组</td><td colspan="2">娄塘组</td></tr>
<tr><td>(4)暗绿色和褐黄色粉质黏土、黄色粉细砂</td><td>40±</td><td>微承压水 I</td><td rowspan="4">更新统</td><td>南汇组</td><td rowspan="7">上更新统</td><td rowspan="3">大理冰期</td><td rowspan="7">上更新统</td><td>Qp_3^3</td><td rowspan="6">晚更新世</td><td>南汇组</td><td>新更新统</td><td>苏州河寒冷期</td><td rowspan="4">上更新统</td><td rowspan="3">南汇组</td><td rowspan="4">上更新统</td><td rowspan="3">漏湖组</td><td rowspan="3">南汇组</td><td rowspan="4">上更新统</td><td rowspan="3">漏湖组</td><td rowspan="3">南汇组</td></tr>
<tr><td>(5)灰色粉质黏土、粉细砂薄互层，下部中细砂</td><td>70±</td><td>$\mathrm{II}_{\text{上}}$</td><td rowspan="3">川沙组</td><td rowspan="5">Qp_3^2</td><td rowspan="5">川沙组</td><td rowspan="6">新更新统</td><td rowspan="5">闸北—苏州河温暖期</td></tr>
<tr><td>(6)深灰、灰绿色粉质黏土，灰色粉细砂</td><td>75±</td><td></td></tr>
<tr><td>(7)浅灰色细砂、含砾中细砂</td><td>95±</td><td>$\mathrm{II}_{\text{下}}$</td><td rowspan="3">庐山—大理间冰期</td><td>川沙组</td><td colspan="2">川沙组</td><td colspan="2">川沙组</td></tr>
<tr><td>(8)蓝灰色黏土、粉质黏土</td><td>100±</td><td></td><td rowspan="3">第三系</td><td rowspan="3">上第三系</td><td rowspan="3">上第三系</td><td rowspan="3">宝山组</td><td rowspan="3">中更新统</td><td rowspan="3">嘉定组</td><td rowspan="3">中更新统</td><td colspan="2" rowspan="3">嘉定组</td><td rowspan="3">中更新统</td><td colspan="2" rowspan="3">嘉定组</td></tr>
<tr><td>(9)灰色细砂、含砾中细砂夹粉质黏土</td><td>115±</td><td>$\mathrm{III}_{\text{上}}$</td></tr>
<tr><td>(10)黄绿、黄褐色粉质黏土夹黄褐色粉细砂</td><td>130±</td><td></td><td>庐山冰期</td><td>Qp_3^1</td><td>早—中更新世</td><td>宝山组</td><td>闸北寒冷期</td></tr>
</table>

续表

<table>
<tr><th>岩层</th><th>层底埋深/m</th><th>含水层</th><th colspan="2">上海市地质勘查局，1962</th><th colspan="2">地质部石油地质中心实验室，1964</th><th colspan="2">刘金陵，1977</th><th colspan="2">上海市地质处，1979</th><th colspan="2">同济大学，1974，1979</th><th colspan="2">竹淑贞等，1979</th><th colspan="2">上海市水文地质队，1984</th><th colspan="2">邱金波和李晓，2007</th><th colspan="2">本志</th></tr>
<tr><td>(11)灰色细砂、中粗砂、砂砾石</td><td>145±</td><td>Ⅲ下</td><td rowspan="11">第三系</td><td rowspan="11">上第三系</td><td rowspan="11">上第三系</td><td rowspan="11">宝山组</td><td rowspan="6">中更新统</td><td rowspan="3">大姑—庐山间冰期</td><td></td><td></td><td rowspan="6">早—中更新世</td><td rowspan="6">宝山组</td><td rowspan="2">中更新统</td><td>沪东—闸北温暖期</td><td></td><td>宝山组</td><td></td><td></td><td></td><td></td></tr>
<tr><td>(12)杂色网纹状黏土夹褐黄色细砂</td><td>165±</td><td></td><td rowspan="4">中更新统</td><td>Qp_2^2</td><td>沪东寒冷期</td><td rowspan="6">下更新统</td><td rowspan="2">周浦组</td><td rowspan="5">下更新统</td><td rowspan="5">安亭组</td><td rowspan="5">下更新统</td><td rowspan="5">安亭组</td></tr>
<tr><td>(13)灰、灰黄色细砂、含砾中粗砂夹砂质粉土</td><td>190±</td><td>Ⅳ上</td><td rowspan="3">Qp_2^1</td><td rowspan="4">早更新统</td><td rowspan="3">沪西—沪东温暖期</td></tr>
<tr><td>(14)杂色斑纹状黏土夹褐黄色中细砂</td><td>220±</td><td></td><td rowspan="3">大如冰期</td><td rowspan="2">朱泾组</td></tr>
<tr><td>(15)褐黄、灰白色细砂、含砾中粗砂</td><td>240±</td><td>Ⅳ下</td></tr>
<tr><td>(16)上、中部灰、灰黄色粉质黏土、细中砂互层，下部灰黄、姜黄色含砾中粗砂</td><td>280±</td><td rowspan="6">Ⅴ</td><td rowspan="2">下更新统</td><td>Qp_1^3</td><td>沪西寒冷期</td><td rowspan="2">安亭组</td></tr>
<tr><td>(17)上部灰绿粉质黏土、砂质粉土，下部灰白色含砾泥质中粗砂</td><td>310±</td><td>下更新统(Qp_1)</td><td>鄱阳—大如间冰期</td><td>Qp_1^2
Qp_1^1</td><td></td><td></td><td rowspan="5">古更新统</td><td>崇明—沪西温暖期</td><td rowspan="5">上新统</td><td rowspan="5">崇明组</td><td rowspan="5">上新统</td><td rowspan="5">崇明组</td></tr>
<tr><td>(18)灰绿、局部杂、紫红色黏土夹砂质粉土</td><td>330±</td><td></td><td></td><td></td><td></td><td></td><td></td><td rowspan="4">崇明寒冷期</td><td rowspan="4">上新统</td><td rowspan="4">崇明组</td></tr>
<tr><td>(19)灰白、黄绿色中细砂、含砾中粗砂夹绿色黏土</td><td>400±</td><td></td><td></td><td></td><td></td><td></td><td></td></tr>
<tr><td>(20)褐黄、紫灰、灰绿色黏土、含砾黏土</td><td>440±</td><td></td><td></td><td></td><td></td><td></td><td></td></tr>
<tr><td>(21)灰黄、淡绿色泥质砂砾、卵石和细砂互层</td><td>478</td><td></td><td></td><td></td><td></td><td></td><td></td></tr>
<tr><td>(22)橄斑玄武岩夹红棕、深灰色黏土，底部砂砾石</td><td>456</td><td></td><td></td><td></td><td></td><td></td><td></td><td></td><td></td><td></td><td></td><td></td><td></td><td></td><td>中新统</td><td>白龙港玄武岩</td><td>中新统</td><td>白龙港玄武岩</td><td>中新统</td><td>白龙港玄武岩</td></tr>
</table>

第一节　岩石地层划分

考虑同期异相沉积作用，区内第四纪地层共划分有 9 个岩石地层单位，自下而上为下更新统安亭组，中更新统嘉定组，上更新统南汇组和川沙组/滆湖组，全新统娄塘组、上海组和青浦组/如东组（表 9-2）。

表 9-2　上海第四纪地层划分简表

Table 9-2　Summary of the stratigraphic division of the Quaternary in Shanghai

年代地层	岩石地层		代号	层底埋深/m	层厚/m	岩性特征	沉积相
全新统	青浦组/如东组		Qh_3q / Qh_3r	1~14	1~14	褐黄色粉质黏土、砂质粉土。河口区中下部多粉细砂，太湖区下部夹泥炭	湖沼-河口滨海
	上海组		Qh_2s	10~20	0~15	灰色淤泥质粉质黏土夹粉砂，河口区粉质黏土、粉砂互层。太湖区褐黄色粉质黏土夹粉砂	潟湖-河口滨海
				18~26	0~12	深灰色淤泥质黏土，下部夹含贝壳粉砂纹层。太湖区灰色黏土，多植物化石碎屑，偶夹泥炭，下部粉质黏土夹砂质粉土，含贝壳	潟湖-滨海-浅海
	娄塘组		Qh_1l	23~42	0~20	深灰色淤泥质粉质黏土、黏土，多植物碎片，含泥钙质结核，西南部地区下部夹泥炭质层	河口滨海
				39~61	0~18	灰、黄灰色粉土、粉细砂夹粉质黏土，较多软体动物介壳	支流河道、河口
上更新统	南汇组/滆湖组	上段	Qp_3n^3 / Qp_3g^3	5~43	0~14	暗绿、褐黄色粉质黏土、黏土，含铁锰质结核、斑纹	泛滥平原、湖泊
		中段	Qp_3n^2 / Qp_3g^2	48~85	10~54	上部姜黄色粉质黏土，黄、黄灰色粉细砂，中部灰色粉质黏土频夹粉细砂、砂质粉土纹层，下部多为灰色细砂、含砾中细砂。河口区全组为粉细砂、含砾砂。太湖区为褐黄、暗绿、杂色褶纹状粉质黏土、黏土夹粉土，下部灰色细砂、含砾砂	滨海-浅海
		下段	Qp_3n^1 / Qp_3g^1	54~91	0~16	深灰、灰绿色粉质黏土、砂质粉土，含钙核。河口区与嘉定灰色粉质黏土、粉细砂薄互层	泛滥平原
	川沙组		Qp_3c	85~133	16~75	上部浅灰、黄灰色粉细砂夹粉土，河口区夹含砾砂，中下部以含砾砂为主，太湖区灰色粉质黏土夹粉砂	河流-三角洲 河流-河口
中更新统	嘉定组	上段	Qp_2j^2	95~123	0~22	灰绿、蓝灰色黏土，东南部深灰色细砂夹粉质黏土	泛滥平原-湖泊
				109~134	0~33	灰、黄灰色中细砂、含砾砂、砂砾石。西部夹灰色砂质粉土、粉质黏土，含贝、螺壳碎片	河流-三角洲河流
		下段	Qp_2j^1	111~143	0~20	灰、蓝灰杂色黏土、粉质黏土，具细网纹构造，含钙质、铁锰质小结核及螺、蚌壳碎片，上部夹灰、褐黄色细砂、粉土	泛滥平原-湖泊
				131~165	1~35	灰、黄灰色粉细砂、含砾砂，底部含砾砂、砂砾石，河口区为砾质砂、砂砾石与含砾砂互层。太湖区部分为杂色网纹状黏土、粉砂互层	三角洲河流- 泛滥平原

续表

年代地层	岩石地层		代号	层底埋深/m	层厚/m	岩性特征	沉积相
下更新统	安亭组	上段	Qp_1a^3	159~203	6~62	蓝灰、灰绿、褐黄色黏土、粉质黏土，网纹、斑纹状，中下部夹褐黄色粉细砂，较多钙质结核。中下部见较大淡水螺壳、丽蚌碎片。黏土层内现节理	三角洲河流-泛滥平原
				173~217	0~48	灰、褐黄色细砂夹粉质黏土，杂色网纹状粉质黏土，富钙核；下部灰、灰黄色、含砾砂、砾质砂和砂砾石，含半炭化木块，顶部常见钙质细砂岩	泛滥湖泊
		中段	Qp_1a^2	196~238	0~49	杂色、网纹斑杂状黏土、粉质黏土、粉土，夹中细砂，较多钙质结核、网脉。顶部常为蓝灰、黄绿色黏土，下部夹1~3层钙质砂岩薄层。局部有紫灰色黏土，黏土内见有擦痕的节理	泛滥湖泊
				213~255	0~31	浅灰、灰黄色中细砂，含砾砂，松散或弱胶结。中部部分地段为绿、蓝灰色斑杂状黏土、粉质黏土与砂互层	河流-三角洲河流
		下段	Qp_1a^1	259~296	0~60	中上部褐黄、褐杂灰绿、黄绿、紫灰、褚红等色网纹状、斑杂状黏土、粉质黏土、黏质粉土。含大量钙质结核、网脉，常见具擦痕节理。常夹姜黄、灰色粉细砂、含砾中细砂，含砾石、泥砾，砂中岩屑较多，常弱胶结，夹有细砂岩钙板层。下部棕黄、灰色中细砂、含砾中粗砂夹砂质粉土、粉质黏土	泛滥湖泊-河流

一、安亭组（Qp_1a）

1984年，上海市水文地质队编写《上海第四纪地层研究报告》时创建。2001—2003年上海市地质调查研究院进行1∶25万上海市幅区域地质调查时，将下更新统原来的3个岩石地层单位合并，统称为“安亭组”，分为下段、中段、上段，分别对应于早期的安亭组、朱泾组和周浦组。曾按其中黏性土的颜色特征称之为“杂色层”。

现在定义为一套河流—泛滥湖泊沉积。其特征的岩性是以褐黄、褐色为主，有灰绿、黄绿、蓝灰色网纹或斑纹穿插的杂色网纹状、斑杂状黏土、粉质黏土，中下段黏土一般呈坚硬—硬塑状态，上段则呈硬塑—可塑状，常可见具擦痕的节理，普遍含有多寡不一的钙质结核、网脉和铁锰质结核；与浅灰、灰黄、褐黄色砂、含砾砂、砂砾石间互成层。由下而上大体形成砾砂层与黏土层交替出现的3个沉积旋回，与冷、暖交替周期性变化的气候旋回基本相随，因而可相应划分为下段、中段、上段。主要分布于上海东北部的安亭—崇明凹陷内，其次分布在南部的枫泾—三墩凹陷内。

二、嘉定组（Qp_2j）

1984年，上海市水文地质队编写《上海第四纪地层研究报告》时创建。2001—2003年上海市地质调查研究院进行1∶25万上海市幅区域地质调查时，将中更新统原来的2个岩石地层单位合并，统称为“嘉定组”，分为下段、上段，分别对应于早期的宝山组和嘉定组。

现在定义为一套三角洲河流—泛滥平原、湖泊沉积，其岩性、岩相具有早、晚更新世地层之间的过渡特点。其特征的岩性为蓝灰、灰绿色黏土、褐黄杂蓝灰、灰绿色细网纹状、条纹状黏土、粉质黏土，也有灰色粉质黏土和粉土，大都具可塑状，少部分为硬塑状；与灰、黄灰色含砾砂、粉细砂间互成层。在全区分布广泛，但上海东、西部地区地层发育有所差异。在“冈身带”以西的太湖沉积区，中更新世地层保存齐全，存在粗、细相间两个沉积旋回形成的上、下两段地层，其埋深通常在95~150m之间，全厚为50~60m。而在“冈身带”以东的滨海平原、河口沙岛等地区，由于晚更新世地层底界的下切，中更新统上段地层因遭受侵蚀而断续出现，大都只保存下段地层，其埋深介于120~160m之间，厚度为25~35m。

三、川沙组（$Qp_3\,c$）

1964年，地质部石油地质中心实验室创建。原始定义是指上海陆域中东部地区埋深在40~95m的一套海陆过渡相浅灰色砂、含砾砂和海相粉质黏土、粉细砂薄互层。当时划归为下更新统。1984年，上海市水文地质队编写《上海第四纪地层研究报告》时，将其限定为晚更新世早期，即末次冰期前一个温暖期堆积的一套海陆过渡相砂层。

现在定义基本为一套河流—弱海陆过渡的三角洲河流相细砂和含砾砂，在与浙江毗邻的青浦、金山、松江接壤区，为河口湾堆积的黏土质粉砂与粉砂互层。

四、南汇组（$Qp_3\,n$）

1964年，地质部石油地质中心实验室创建。原始定义是指上海陆域中东部地区埋深在20~40m陆相沉积的暗绿、褐黄色黏土和下伏的黄色粉细砂层，认为系中—晚更新世沉积。1984年，上海市水文地质队编写《上海第四纪地层研究报告》时，将其限定为晚更新世晚期整个末次冰期堆积的海、陆交替地层。

现在定义：区内中—东北部地区末次冰期堆积的以海相和海陆过渡相为主的地层，分为下段、中段和上段，分别以黏土质粉砂和粉砂、（含砾）粉细砂和粉砂、黏土质粉砂为主。

五、滆湖组（$Qp_3\,g$）

1987年，吴标云和李从先创建，命名剖面位于江苏太湖区寨桥D114孔。原始定义为：全组分为3段，下段以灰绿、青灰、灰黄色粉质黏土为主；中段为灰、深灰、灰黄色粉砂、砂质粉土和淤泥质黏土；上段为灰黄、灰绿、青灰色粉质黏土。2001—2003年，上海市地质调查研究院进行1∶25万上海市幅区域地质调查时，取代“南汇组”。2008年，在《上海市第四纪地层与沉积环境》专著中将其限定为仅指青浦在与松江、金山接壤区分布的一套与苏南地区相似的海、陆相交替地层。

现在定义为青浦与松江、金山接壤区分布的海、陆相交替地层，自上而下分为3段。上、下段为寒冷气候条件下的陆相沉积，与南汇组相同；只是中段地层内出现惯称上海西部“第二硬土层”的陆相沉积暗绿色黏土为标志。

六、娄塘组（$Qh_1\,l$）

1984年，上海市水文地质队编写《上海第四纪地层研究报告》时创建。原始定义：分上段、下段：下段在东部和河口砂岛地区多为河口—溺谷堆积，岩性常为灰色砂质粉土、黏质粉土与灰白色粉砂互层。在金山朱泾、青浦赵巷等西部地区为湖沼堆积的灰黑、灰褐色粉质黏土，常含较多的植物碎片，局部形成泥炭或碳质黏土层。上段普遍为滨海堆积，大都为深灰、褐灰色（淤泥质）黏土、粉质黏土，具特征的标志是，可见大量的植物茎叶与黄色泥灰质结核，局部形成泥炭或碳质黏土层，广泛分布于金山、奉贤、浦东南部等地和市区、宝山等地的部分地段，部分为深灰、褐灰色粉质黏土夹灰白色粉砂或成互层。现在定义不变。

七、上海组（Qh_2s）

1964 年，地质部石油地质中心实验室创建。原始定义指区内的全新世地层。1984 年，上海市水文地质队编写《上海第四纪地层研究报告》时，将其限定为中全新统海相层，主要以淤泥质黏性土为特征。

现在定义为东部和河口砂岛区的滨岸浅海—河口滨海堆积，下段为滨岸浅海堆积的深灰色淤泥质黏土和淤泥质粉质黏土；上段为滨海堆积的灰色砂质粉土、黏质粉土夹粉砂，其上部常为灰色砂质粉土与粉质黏土互层或为黏土、粉质黏土夹砂质粉土和粉砂。在西部为潟湖—滨海沼泽堆积，通常为较单一的深灰色淤泥质黏土、黏土和淤泥质粉质黏土。

八、如东组（Qh_3r）

1987 年，由吴标云和李从先以江苏如东孔剖面创建。原始定义：按岩性分为 3 段：中、下段为灰或灰黑色粉砂、细砂、淤泥质亚黏土；上段为灰黄色亚砂土、亚黏土、粉砂。岩相由下而上为河口—浅海—河床、三角洲平原相，反映一个完整的冰后期海侵旋回。实际上该组原始定义划分的 3 个段与上海长江三角洲沉积区全新世的 3 个基本地层单位相当。2007 年，在《上海市第四纪地层与沉积环境》专著将该组引入上海，仅指上海中东部长江三角洲沉积区的晚全新世地层。

现在定义：主要类型为河口湾—滨岸潮滩堆积的灰、灰黄色砂质粉土、黏质粉土夹粉砂，其上部普遍为滨岸潮滩堆积的褐黄、褐灰色粉质黏土。次要类型为古吴淞江所在三角洲支流河道堆积的青灰色粉砂和砂质粉土、古吴淞江滨岸堆积的灰褐色砂质粉土。

九、青浦组（Qh_3q）

1984 年，上海市水文地质队编写《上海第四纪地层研究报告》时创建。原始定义：全区晚全新世的一套沉积建造。2007 年，在《上海市第四纪地层与沉积环境》专著中仅指上海以假整合超覆于上更新统南汇组上段暗绿色黏土层之上的夹有泥炭层的深灰、棕灰色淤泥质黏土、粉质黏土的湖沼沉积。现在定义同原始定义。

第二节　区域地层对比

江苏南部下更新统为海门组，岩性分 3 段，为一套河流相，局部河口相沉积。黏性土中含钙质结核及铁锰质结核，夹灰、青灰色条带，岩性硬塑，具古土壤化特征。与区内安亭组三分和杂色黏土发育的特征十分接近。至浙江称为嘉兴组（表 9-3）。

表 9-3　上海及邻近地区第四纪岩石地层划分方案对比表

Table 9-3　Comparison of the lithostratigraphic division scheme of the Quaternary in Shanghai and its neighboring areas

<table>
<tr><td colspan="2" rowspan="3">地质年代</td><td colspan="11">岩石地层</td></tr>
<tr><td colspan="3">江苏</td><td colspan="4" rowspan="2">上海</td><td colspan="4">浙江</td></tr>
<tr><td>丘岗区</td><td colspan="2">平原区</td><td colspan="2">山麓区</td><td colspan="2">平原区</td></tr>
<tr><td rowspan="3">全新世</td><td>晚</td><td rowspan="3">?</td><td rowspan="3">如东组</td><td>上段 Qhr^3</td><td>青浦组</td><td>如东组</td><td>Qh_3q</td><td>Qh_3r</td><td rowspan="3">鄞江桥组</td><td rowspan="3">Qhy</td><td rowspan="3">镇海组</td><td>上段 $Qhzh^3$</td></tr>
<tr><td>中</td><td>中段 Qhr^2</td><td colspan="2">上海组</td><td colspan="2">Qh_2s</td><td>中段 $Qhzh^2$</td></tr>
<tr><td>早</td><td>下段 Qhr^1</td><td colspan="2">娄塘组</td><td colspan="2">Qh_1l</td><td>下段 $Qhzh^1$</td></tr>
</table>

续表

<table>
<tr><th colspan="2" rowspan="3">地质年代</th><th colspan="11">岩石地层</th></tr>
<tr><th colspan="3">江苏</th><th colspan="4" rowspan="2">上海</th><th colspan="4">浙江</th></tr>
<tr><th>丘岗区</th><th colspan="2">平原区</th><th colspan="2">山麓区</th><th colspan="2">平原区</th></tr>
<tr><td rowspan="10">更新世</td><td rowspan="6">晚</td><td rowspan="7">下蜀组 $Qp_{2-3}x$</td><td rowspan="3">滆湖组</td><td>上段 Qp_3g^3</td><td rowspan="3">滆湖组</td><td rowspan="3">南汇组</td><td>上段 Qp_3g^3</td><td>上段 Qp_3n^3</td><td rowspan="5">莲花组</td><td rowspan="5">Qp_3l</td><td rowspan="3">宁波组</td><td>上段 Qp_3n^3</td></tr>
<tr><td>中段 Qp_3g^2</td><td>中段 Qp_3g^2</td><td>中段 Qp_3n^2</td><td>中段 Qp_3n^2</td></tr>
<tr><td>下段 Qp_3g^1</td><td>下段 Qp_3g^1</td><td>下段 Qp_3n^1</td><td>下段 Qp_3n^1</td></tr>
<tr><td rowspan="2">昆山组</td><td>上段 Qp_3k^2</td><td colspan="2" rowspan="2">川沙组</td><td colspan="2" rowspan="2">Qp_3c</td><td rowspan="2">东浦组</td><td>上段 Qp_3d^2</td></tr>
<tr><td>下段 Qp_3k^1</td><td>下段 Qp_3d^1</td></tr>
<tr><td rowspan="2">启东组</td><td>上段 Qp_2q^2</td><td colspan="2" rowspan="2">嘉定组</td><td colspan="2">上段 Qp_2j^2</td><td rowspan="2">之江组</td><td rowspan="2">Qp_2z</td><td rowspan="2">前港组</td><td>上段 Qp_2q^2</td></tr>
<tr><td rowspan="2">中</td><td>下段 Qp_2q^1</td><td colspan="2">下段 Qp_2j^1</td><td>下段 Qp_2q^1</td></tr>
<tr><td>柏山组 Qp_2b</td><td rowspan="3">海门组</td><td>上段 Qp_1h^3</td><td colspan="2" rowspan="3">安亭组</td><td colspan="2">上段 Qp_1a^3</td><td rowspan="3">汤溪组</td><td rowspan="3">Qp_1t</td><td rowspan="3">嘉兴组</td><td>上段 Qp_1j^3</td></tr>
<tr><td rowspan="2">早</td><td rowspan="2"></td><td>中段 Qp_1h^2</td><td colspan="2">中段 Qp_1a^2</td><td>中段 Qp_1j^2</td></tr>
<tr><td>下段 Qp_1h^1</td><td colspan="2">下段 Qp_1a^1</td><td>下段 Qp_1j^1</td></tr>
</table>

江苏南部中更新统为启东组，一般有 2~3 个旋回，自下而上表现为典型的二元沉积结构，与区内嘉定组以河流沉积体系为主的沉积特征比较接近。自上而下都可以划分为两段。

江苏昆山组为一套灰、灰黄或灰褐色含砾粗砂、细砂、粉细砂、亚砂土，含腐殖质及螺、贝壳。孢粉反映温暖潮湿气候。含海相有孔虫、介形虫，为一海侵层。与区内川沙组十分一致，都为一套末次间冰期三角洲河流沉积体系。

对于全新世寒温、高温、低温 3 个气候分期，辽东半岛地区全新统由下而上划分的普兰店组（现改名为泡子组）、大孤山组和庄河组，成为我国全新世地层对比的标准剖面。北京、河北平原等地也将全新统划分为 3 个基本地层单位。邻区江苏和浙江尽管都分别采用了一个岩石地层单位，但其都根据岩性岩相变化，都进行了三分，对应于区内的娄塘组、上海组和如东组/青浦组。

第二篇

岩浆岩及岩浆作用

第十章　侵入岩和侵入作用

本区已发现各类侵入岩体 33 个（图 10-1；表 10-1），岩石类型主要有基性岩、中性岩和酸性岩。其中，酸性岩类约占侵入岩面积的 94%，岩体规模较大，主要分布在本区的北部；中性岩类和基性岩

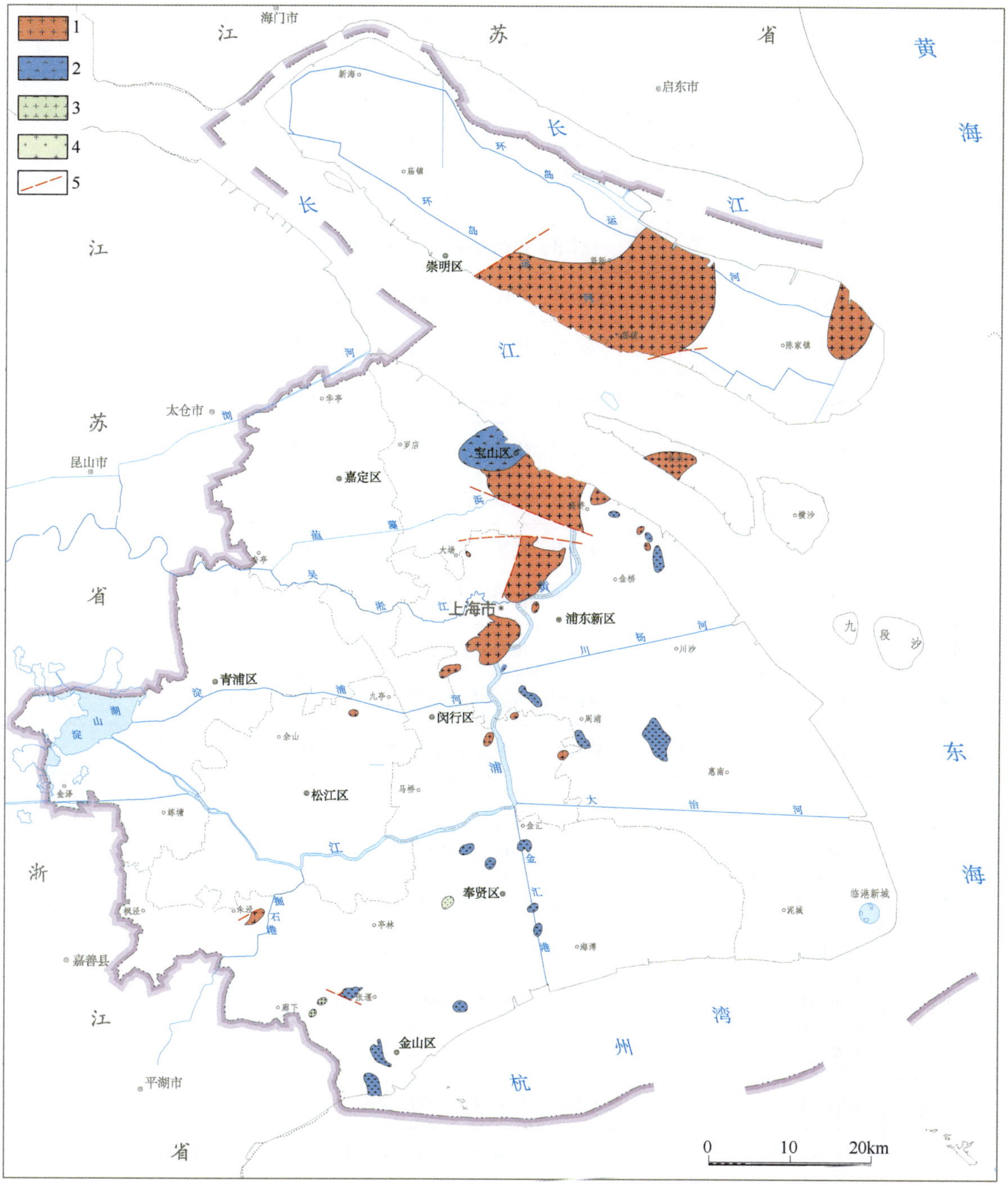

图 10-1　上海主要侵入岩体分布图

Fig. 10-1　Distribution of the main intrusive rocks in Shanghai

1—花岗岩；2—石英闪长岩、辉石闪长岩；3—花岗闪长岩；4—花岗斑岩；5—断裂

类仅占侵入岩面积的6%，岩体多而规模较小，多分布于本区的东部和南部。岩体产状以岩株、岩枝、岩脉为主，大于100km^{2}的岩基只有一个，即崇明花岗岩体。

表10-1　上海主要岩体特征表

Table 10-1　Characteristics of the main rock bodies in Shanghai

编号	岩体名称	主要岩性	面积/km^2	围岩时代及蚀变
1	堡镇	花岗岩	>225	寿昌组、黄尖组，接触关系未见，蚀变特征不明
2	东胜沙	花岗岩	33	寿昌组，接触关系未见，蚀变特征不明
3	吴淞	花岗岩	54.7	寿昌组、黄尖组，接触关系未见，蚀变特征不明
4	市北	花岗岩	30	奥陶系、寿昌组、黄尖组，角岩化
5	市南	花岗岩	25.8	红花园组、仑山组、寿昌组、黄尖组
6	虹桥	花岗岩	3	黄尖组，接触关系未见，蚀变特征不明
7	曹行	花岗岩	1	劳村组，接触关系未见，蚀变特征不明
8	朱泾	花岗岩	4.5	金山群、劳村组，接触关系未见，蚀变特征不明
9	乳山	花岗岩	0.7	黄尖组，接触关系未见，蚀变特征不明
10	黄家湾	花岗岩	0.5	黄尖组，接触关系未见，蚀变特征不明
11	顾路	花岗岩	0.5	黄尖组，接触关系未见，蚀变特征不明
12	泗泾	花岗岩	0.7	黄尖组，接触关系未见，蚀变特征不明
13	杜行	花岗岩	0.5	劳村组，接触关系未见，蚀变特征不明
14	庄行	花岗斑岩	1	劳村组、黄尖组，接触关系未见，蚀变特征不明
15	外高桥	花岗闪长岩	3	黄尖组，接触关系未见，蚀变特征不明
16	钱圩	花岗闪长岩	0.5	劳村组、黄尖组，接触关系未见，蚀变特征不明
17	月浦	闪长岩	26	寿昌组，接触关系未见，蚀变特征不明
18	杨思	闪长岩	0.5	黄尖组，接触关系未见，蚀变特征不明
19	查山	闪长岩	4	黄尖组，接触关系未见，蚀变特征不明
20	漕泾	闪长岩	2	黄尖组，接触关系未见，蚀变特征不明
21	长兴	石英闪长岩	13	寿昌组，接触关系未见，蚀变特征不明
22	曹路	石英闪长岩	4	黄尖组，接触关系未见，蚀变特征不明
23	三林塘	石英闪长岩	3	金山群，角岩化、石英岩化
24	肖塘	石英闪长岩	2	劳村组，绢云母化、碳酸盐化、绿泥石化
25	张堰	石英闪长岩	2.3	劳村组，角岩化、矽卡岩化、角岩化、铜矿化、铜钼矿化
26	石化	石英闪长岩	3	黄尖组，接触关系未见，蚀变特征不明
27	荷花塘	石英闪长岩	1	黄尖组，绢云母化、碳酸盐化、绿泥石化

续表

编号	岩体名称	主要岩性	面积/km^2	围岩时代及蚀变
28	钱桥	石英闪长岩	2	黄尖组、劳村组，绿泥石化、方解石化
29	高东	辉石闪长岩	0.7	黄尖组，接触关系未见，蚀变特征不明
30	坦直	辉石闪长岩	9.8	前寒武系金山群、灯影组，接触关系清晰，绢云母化、高岭土化、绿帘石化
31	沈庄	辉石闪长岩	1.5	前寒武系金山群，接触关系清晰
32	齐贤桥	辉石闪长岩	2	劳村组，高岭土化、绿泥石化、绢云母化
33	杨园	辉石闪长岩	0.5	黄尖组，接触关系未见，蚀变特征不明

燕山期是上海地质发展史上最重要的岩浆活动期，有多次侵入活动的特点。相比之下，目前已发现的晋宁期及喜马拉雅期的侵入活动规模则较小。

在岩石成因上，有直接来自上地幔的玄武岩浆和由不断分熔的安山质岩浆形成的基性、中性岩类；但都不同程度地受到陆壳物质的混染，形成同熔型花岗岩类。

1962 年，上海市地质勘察局编写了《上海地区矿产地质普查工作总结报告》，这是本区最早研究侵入岩的报告。报告把区内侵入岩划分为燕山早期、燕山晚期和喜马拉雅期等 3 期。嗣后，金山张堰矿区成矿地质条件评价、成矿区划研究等项目都有对区内侵入岩和侵入作用的论述。如，1978 年，华东地质研究所编写的《金山县张堰铁铜矿床地质特征的初步研究》；1979 年，上海市地质处编写的《上海成矿区划说明书》；1982 年，上海市地质处编写的《上海市金山县张堰铜矿区普查评价报告》；1995 年，上海市地质技术开发总公司编制的《上海市金山县张堰铜矿床百家村矿段地质详查报告》等。报告都把本区的侵入岩划为燕山晚期，并有喜马拉雅期的辉绿岩脉侵入。尤其是对张堰岩体进行了详细论述。2013 年，谢建磊等对齐贤桥岩体进行了 LA-ICP-MS 锆石 U-Pb 测年，并论述了区内岩浆侵入作用分期。

第一节　侵入期次划分与岩石类型

一、侵入期次划分

(一) 侵入岩体地质年代确定原则

1. 以地质依据为基础

以岩体与岩体、岩体与地层之间的侵入、覆盖、穿插等接触关系作为基础，当年龄数值与地质依据相矛盾时，以后者为准。

2. 以锆石 U-Pb 同位素年龄为基础，综合其他同位素年龄数据

同一岩体内有 U-Pb 同位素年龄和其他年龄数据时，以前者为准；用 K-Ar 法测定的同位素年龄数据，一般选用最老的年龄值代表岩体年龄；期次划分时，用 K-Ar 法测定的年龄数据的准确程度一般按角闪石、黑云母、全岩顺序依次降低。

3. 对比法

对于缺少接触关系和同位素年龄资料的多数岩体，通过与时代已知的岩体进行对比确定其时代归属。

除上述 3 原则外，尚需考虑岩体与区域构造关系，岩浆岩的共生组合特点等。

(二) 侵入期次与年代格架

根据岩体的地质年代，将区内侵入岩和岩浆侵入作用划分为晋宁期、燕山期和喜马拉雅期 3

期（表10-2）。

表10-2 侵入期次划分表

Table 10-2 Division of the intrusion periods

时代			期	次	代表地点	岩性	产状	同位素年龄/Ma	
								主要年龄（方法）	其他年龄（方法）
新生代	古近纪	渐新世	喜马拉雅期		张堰	煌斑岩	岩脉		
						辉绿岩		40.7（全岩K-Ar）	
中生代	白垩纪	晚白垩世	燕山期	4	界山	霏细斑岩	岩脉		
						花岗斑岩		88.1（全岩K-Ar）	
						花岗细晶岩			
					市区	花岗斑岩	岩脉	81.0、84.0（全岩K-Ar）	
		早白垩世		3	界山	石英闪长岩	岩株	95.7（全岩K-Ar）	
					坦直	辉石闪长岩	岩株	97.1（黑云母K-Ar）	
					齐贤桥	辉石闪长岩	岩株	100.6±0.6、99.5±1.2（锆石U-Pb）	
					张堰	花岗闪长岩	岩株	101.6±1.1（锆石U-Pb）	
					界山	辉绿岩	岩脉/岩株	106（全岩K-Ar）	
					沈庄	闪长岩	岩株	104.2±1.0（锆石U-Pb）	86.2（黑云母K-Ar）
						辉石闪长（玢）岩	岩株	106.9±1.2、104.3±1.0（锆石U-Pb）	
					竖河	花岗岩	岩基	107.1（钾长石K-Ar）	
				2	张堰	辉绿岩	岩脉/岩株	112.7（全岩K-Ar）	102（角闪石K-Ar）
						石英闪长岩	岩株	119.6（黑云母K-Ar）	102.2（金云母K-Ar）
					劳动公园	花岗岩	岩株	132.9±1.1（锆石U-Pb）	112.2（黑云母K-Ar）
					肖塘	闪长岩	岩株	142.6±2.5（锆石U-Pb）	
	侏罗纪	中侏罗世		1	港城路	花岗岩	岩株	178.8±2.3（锆石U-Pb）	131.7（全岩K-Ar）
					三林	石英闪长岩	岩株	163.4±1.5（锆石U-Pb）	
中元古代	蓟县纪		晋宁期	2	朱泾	花岗岩	岩株	1082.4±7.9（锆石U-Pb）	
	长城纪			1	康家宅	糜棱岩化花岗斑岩	岩脉	1781±16（锆石U-Pb）	

1. 晋宁期

在朱泾揭露有蓟县纪晚期花岗岩，康家宅金山群中发育大量的糜棱岩化花岗斑岩。此外，在金山群中局部夹已变质的辉绿岩。

2. 燕山期

燕山期为区内最重要的岩浆侵入作用期。共有4次侵入活动，其中1次发生于中侏罗世，2次发生于早白至世，1次发生于晚白垩世。

3. 喜马拉雅期

目前在张堰等地的钻孔有揭露，主要为脉状辉绿岩。

二、岩石分类与定名

依据SiO_2含量的多少，分为基性岩、中性岩和酸性岩。采用国际地科联（IUGS）推荐的QAP分类命名方案。暗色矿物主要为辉石和角闪石时，用图 10-2 分类；超镁铁质岩和镁铁质岩以外的其余岩石，按浅色矿物的含量进一步细分，用 QAP 三角图分类（图 10-3）。暗色矿物含量小于 90%的岩石分类方案中一个明显的不足是，没有考虑暗色矿物的种属、含量和一些特殊的矿物和结构特征。确定岩石的全名还需考虑暗色矿物和特征矿物的含量和种类。岩石中暗色矿物含量很少，可以加“浅色”作为定名前缀。岩石中的暗色矿物亦应作为前缀参加定名。

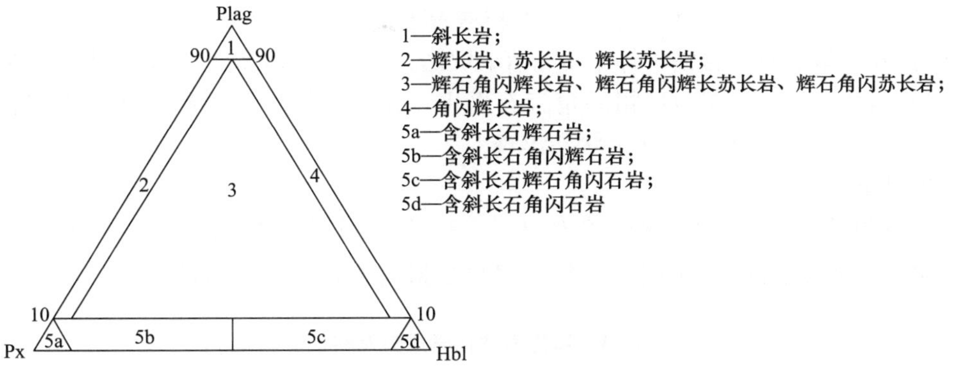

图 10-2 辉长岩矿物分类图

Fig. 10-2 Mineral classification diagram of the gabbro

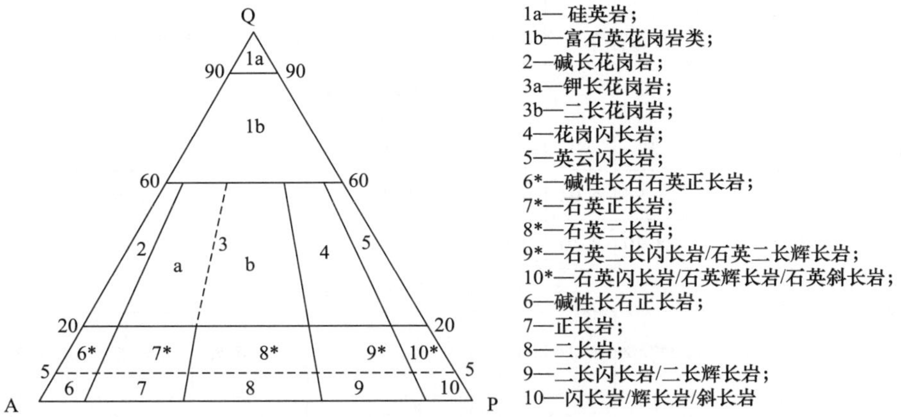

图 10-3 侵入岩的 QAP 分类图

Fig. 10-3 QAP classification diagram of the intrusive rocks

第二节 基性侵入岩

一、晋宁期

在前震旦系金山群中，见部分斜长角闪岩具有变余辉绿结构（如天花庵等地区），说明原岩可能为辉绿岩、角闪辉绿岩。岩石多呈黑绿、深绿和绿灰色，片状构造比较明显，矿物组成主要为中更长石，次为角闪石，具明显的变余结构，有定向性，斜长石和角闪石相对集中。

二、燕山期

常见岩石有角岩化辉绿岩、角闪石辉绿岩及角岩化次闪石化辉绿玢岩、辉长岩等，分布于金山张堰、界山等地。岩体规模较小，多呈岩床、岩脉产出，多产生明显的角岩化。

（一）角闪石辉绿岩

1. 地质特征

见于金山张堰，顺层侵入于劳村组流纹质疑灰岩及前震旦系金山群黑云母斜长角岩中，呈岩床产出。

2. 岩石学特征

岩石经蚀变作用，呈灰黑略带绿色，中—细粒嵌晶含长结构，矿物成分由中长石（50%）、角闪石（30%~45%）及少量辉石、磁铁矿等组成。角闪石的形体极不规则，边缘弯曲，多色性强；斜长石（An=38），呈条柱状，镶嵌于他形粗晶角闪石之中。在粗晶角闪石颗粒之间有许多较细的板条状斜长石和他形晶角闪石、黑云母，呈交生状态出现。

3. 岩石化学特征

岩石中 SiO_2 含量为 48.3%，K_2O/Na_2O 值为 0.33，镁铁比为 1.05（表 10-3）。标准矿物以斜长石、透辉石、磁铁矿含量较高及无石英为特征。里特曼指数为 4.55，属碱性。

表 10-3　基性岩体化学成分及特征值

Table 10-3　Chemical compositions and eigenvalues of the basic rock bodies

岩石名称	氧化物/%												
	SiO_2	TiO_2	Al_2O_3	Fe_2O_3	FeO	MnO	MgO	CaO	Na_2O	K_2O	P_2O_5	烧失量	合计
角闪石辉绿岩	48.3	1.85	15.38	4.12	7.89	0.27	6.85	8.31	3.68	1.23	0.27	1.09	99.24
辉长岩	48.72	1.40	17.39	5.52	5.63	0.14	5.76	8.44	3.25	1.70	0.24	1.49	99.68

岩石名称	CIPW 标准矿物/%										
	Q	Or	Ab	An	Mt	Ap	Il	Di	Hy	C	Ol
角闪石辉绿岩	—	7.24	30.94	21.98	6.02	0.67	3.49	14.18	0.46		13.27
辉长岩	—	10.02	27.27	28.10	8.10	0.67	2.73	9.52	8.31		3.63

岩石名称	其他特征参数								
	a	*b*	(Na_2O+K_2O)/%	K_2O/Na_2O	$Mg^{2+}/(Fe^{3+}+Fe^{2+})$	氧化系数	分异指数	里特曼指数	碱度指数
角闪石辉绿岩	9.96	27.82	4.91	0.33	1.05	0.34	38.18	4.55	1.52
辉长岩	9.99	24.22	4.95	0.52	0.97	0.495	37.29	4.28	1.47

岩体中微量元素以 Ba、Cu、Pb、Sn、Cr、Ga、Be、Yb、Ta、Mo10 种元素含量高于同类岩石维氏值为特征，其中 Ba 高出 26 倍，Sn 高出近 8 倍，Ga 高出 4 倍，Yb 高出 3.5 倍，Mo 高出 8 倍，Ta 高出 83 倍。

岩体局部受接触热变质作用影响，发生角岩化，并出现黑云母、绿泥石、阳起石等新生矿物。

（二）辉长岩

主要在坦直辉石闪长岩体的下部见到，向上与辉石闪长岩、辉长闪长岩过渡。在江湾等地钻孔也可见。

岩石呈灰黑色，细—中粒结构、辉长-辉绿结构与反应边结构，由中—拉长石（55%~60%）、辉石（20%~35%）、黑云母（5%）、橄榄石（少量~5%）、角闪石及副矿物磁铁矿、磷灰石等组成。斜长石呈板柱状，聚片双晶发育，双晶纹宽，An=49~51；橄榄石呈短柱状、粒状，有时呈自形晶，常具反应边结构，中心为橄榄石，边部为辉石，辉石圈外又有黑云母。岩体普遍钾化，出现少量钾长石、石英，呈粒状分布于斜长石与辉石之间。

岩石中 SiO_2 含量为 48.72%，K_2O/Na_2O 值为 0.52，镁铁比为 0.97（见表 10-3）。标准矿物以正长石、斜长石、透辉石、紫苏辉石、磁铁矿含量较高及无石英为特征。里特曼指数为 4.28，属碱性。

三、喜马拉雅期

目前所见都只是一些辉绿岩脉，分布广。并曾见其入侵于燕山晚期石英闪长岩体中。

岩石呈绿灰黑色，少斑—斑状结构，基质为辉绿结构。矿物成分主要为斜长石、普通辉石、角闪石及少量磁铁矿、磷灰石等。斜长石普遍脱钙成更—中长石，部分暗色矿物已绿泥石化及碳酸盐化。

岩石中 SiO_2 含量为 52.48%，全碱含量为 7.16%、K_2O/Na_2O 值为 0.64、镁铁比为 1.18，镁铁组分含量为 13.14%（表 10-4）。标准矿物以正长石、斜长石、透辉石、紫苏辉石、磁铁矿、橄榄石含量较高及无石英为特征。里特曼指数为 5.41，属碱性。

表 10-4 基性岩体化学成分及特征值

Table 10-4 Chemical compositions and eigenvalues of the basic rock bodies

岩石名称	氧化物/%												
	SiO_2	TiO_2	Al_2O_3	Fe_2O_3	FeO	MnO	MgO	CaO	Na_2O	K_2O	P_2O_5	烧失量	合计
辉绿岩	52.48	1.12	16.76	3.02	5.00	0.14	5.12	5.22	4.36	2.80	0.36	3.48	99.86

岩石名称	CIPW 标准矿物/%										
	Q	Or	Ab	An	Mt	Ap	Il	Di	Hy	C	Ol
辉绿岩	—	16.70	35.71	17.80	4.40	1.01	2.12	4.49	7.13		6.19

岩石名称	其他特征参数								
	a	*b*	(Na_2O+K_2O) /%	K_2O/Na_2O	$Mg^{2+}/(Fe^{3+}+Fe^{2+})$	氧化系数	分异指数	里特曼指数	碱度指数
辉绿岩	14.10	18.69	7.16	0.64	1.18	0.38	53.41	5.41	1.97

第三节 中—酸性侵入岩

一、晋宁期

在前震旦系金山群中局部见较多的糜棱岩化花岗斑岩，锆石 U-Pb 年龄为 1781±16Ma。此外，在朱泾发现蓟县纪晚期的花岗岩，锆石 U-Pb 年龄为 1082.4±7.9Ma。

（一）糜棱岩化花岗斑岩

主要见于浦东康家宅、奉贤肖塘等地钻孔中。岩石呈肉红色，间杂灰黑色、暗绿色条痕或残影，具变余斑状结构，基质呈变余细粒花岗结构、糜棱结构、鳞片变晶结构，条带状构造、眼球状构造，条带方向和与其交替产出的角闪质类岩石条带方向基本一致。

斑晶以微斜长石为主，条纹长石次之，两者总含量最低为 3%，最高达 45%，呈肉红色，自形程度差，多为不规则椭圆状、碎斑状、眼球状，呈定向排列，粒径一般为 0.4～3mm，最大粒径达 7mm，部分具格子双晶，此外尚有少量斜长石碎斑，其表面多已泥化、绢云母化或被石英、绿帘石交代。基质成分为钾长石、斜长石、石英等。以钾长石为主，平均含量为 40%，最高达 65%；斜长石含量一般为 15%～20%，最高达 30%；石英含量在 10%～40%不等。基质矿物粒径为 0.02～

0.6mm，常呈条带状集合体绕过长石碎斑作定向排列，其中石英颗粒较长石更细，多以粉状出现，且多经不同程度的重结晶。岩石中常含一定数量的黑云母、绿泥石、白云母、方解石、绿帘石等。副矿物有锆石、磷灰石。

岩石中 SiO_2 含量为 65.56%～69.26%，全碱含量为 7.55%～8.31%、K_2O/Na_2O 值为 1.49～1.74、镁铁比为 0.087～0.45，镁铁组分含量为 5.37%～5.68%（表 10-5）。标准矿物以石英、正长石、斜长石、透辉石为主，含少量的紫苏辉石、磁铁矿。里特曼指数为 2.43～2.61，属钙碱性。

表 10-5　糜棱岩化花岗斑岩化学成分及特征值

Table 10-5　Chemical compositions and eigenvalues of the mylonitized granite porphyry

岩石名称	氧化物/%												
	SiO_2	TiO_2	Al_2O_3	Fe_2O_3	FeO	MnO	MgO	CaO	Na_2O	K_2O	P_2O_5	烧失量	合计
糜棱岩化花岗斑岩	65.56	0.42	10.12	1.84	2.48	0.20	1.05	5.63	3.03	4.52	0.05	4.75	99.65
	69.26	0.56	12.42	2.30	3.13	0.05	0.25	2.17	3.03	5.28	0.09	1.33	99.87

岩石名称	CIPW 标准矿物/%										
	Q	Or	Ab	An	Mt	Ap	Il	Di	Hy	C	Ol
糜棱岩化花岗斑岩	21.81	26.72	25.70	0.56	2.78	0.34	0.76	11.04	5.38		6.19
	26.44	31.17	25.70	4.73	3.24	0.34	1.06	4.59	1.39		

岩石名称	其他特征参数								
	a	*b*	(Na_2O+K_2O)/%	K_2O/Na_2O	$Mg^{2+}/(Fe^{3+}+Fe^{2+})$	氧化系数	分异指数	里特曼指数	碱度指数
糜棱岩化花岗斑岩	13.13	12.46	7.55	1.49	0.45	0.43	78.07	2.43	2.25
	14.10	6.85	8.31	1.74	0.087	0.42	84.55	2.61	2.42

（二）花岗岩

仅见于金山朱泾，推测为一小型岩珠。

岩石呈肉红色，破碎，较新鲜。花岗伟晶结构、嵌晶结构，块状构造。矿物晶体较大，为石英斑晶和长石大斑晶镶嵌产出，粒径为 0.5～1.5cm。石英占 40%～54%，长石占 35%～60%。微斜长石斑晶粗大，可见格子双晶，一级灰干涉色，发育完全解理。条纹长石具文象结构，一级灰干涉色，发育较完全解理。石英可见波状消光，表面干净，半自形晶，呈一级灰白干涉色。较细小的石英晶体颗粒互相镶嵌，分布在长石与石英晶体之间。含有少量暗色矿物，以黑云母为主，占 0～10%，锆石、帘石少量。

岩石中 SiO_2 含量为 80.13%，全碱含量为 7.70%、K_2O/Na_2O 值为 1.94、镁铁比为 0.17，镁铁组分含量为 0.87%（表 10-6）。标准矿物以石英、正长石、斜长石为主，刚玉、紫苏辉石、磁铁矿极少量。里特曼指数为 1.60，属钙碱性。

表 10-6　朱泾花岗岩化学成分及特征值

Table 10-6　Chemical compositions and eigenvalues of the Zhujing granite

岩石名称	氧化物/%												
	SiO_2	TiO_2	Al_2O_3	Fe_2O_3	FeO	MnO	MgO	CaO	Na_2O	K_2O	P_2O_5	烧失量	合计
花岗岩	80.13	0.08	10.51	0.66	0.15	0.009	0.06	0.01	2.62	5.08	0.02	0.85	100.19

续表

岩石名称	CIPW标准矿物/%									
	Q	Or	Ab	An	Mt	Ap	Il	Di	Hy	C	Ol
花岗岩	45.67	30.22	22.32	52.54	0.28	0.05	0.15		0.15	0.71	

岩石名称	其他特征参数									
	a	*b*	(Na_2O+K_2O) /%	K_2O/Na_2O	Mg^{2+}/ ($Fe^{3+}+Fe^{2+}$)	氧化系数	分异指数	里特曼指数	碱度指数
花岗岩	12.49	0.78	7.70	1.94	0.17	0.81	98.21	1.60	2.98

岩石中微量元素Ba、W含量最高，可达668×10^{-6}、351×10^{-6}。Rb/Sr值为3.69，Ba/Sr值为15.6。与酸性维氏值比，Ba、Cu、Pb、Zn、Ni、Cd、Be、V、Cr、Ni等大多数元素含量都偏低，W、Co含量偏大，W含量可达234倍、Co含量达10.3倍（表10-7）。

表10-7　朱泾花岗岩微量（含稀土）元素含量

Table 10-7　Contents of trace elements (incl. REE) in the Zhujing granite　　*w*(B) $/10^{-6}$

Sc	Li	Be	Co	Cu	Zn	Ga	Ge	Rb	Zr	Nb	Mo	Cd	Cs	Hf	Ta
1.64	1.54	3.03	51.3	2.45	23.5	21.2	1.15	158	126	21.1	0.22	0.041	0.26	3.2	2.04
W	Tl	Pb	Th	U	Ba	Cr	Ni	Sr	V	As	Sb	Bi	Sn	Ag	
351	0.52	14.7	12.6	1.43	668	1.75	4.94	42.8	3.65	1.16	0.08	0.043	4.00	0.060	
La	Ce	Pr	Nd	Sm	Eu	Gd	Tb	Dy	Ho	Er	Tm	Yb	Lu	Y	ΣREE
37.3	68.0	7.95	28.8	5.93	1.28	5.66	0.98	5.68	1.20	3.90	0.64	4.54	0.69	33.0	205.55

稀土元素（含Y）总量中等，$\sum REE=205.55\times10^{-6}$，其中$\sum LREE=149.26\times10^{-6}$，$\sum HREE+Y=56.29\times10^{-6}$。$\delta Eu=0.68$，具负Eu异常，表明岩石分异程度较高。$\delta Ce=0.93$，Ce亏损很少。$(La/Yb)_N=5.55$，Sm/Nd=0.21，$\sum LREE/(\sum HREE+Y)=2.65$，稀土元素配分曲线中等倾斜，重稀土部分则近水平，轻、重稀土分馏较强，属轻稀土富集型（图10-4）。

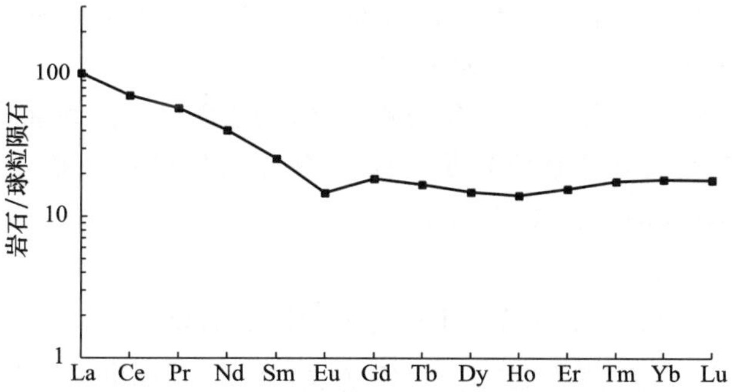

图10-4　朱泾花岗岩稀土元素配分模式图

Fig. 10-4　REE pattern of the Zhujing granite

二、燕山期

燕山期侵入岩主要形成于中侏罗世、早白垩世和晚白垩世，岩性以中—酸性较多。其中，中性岩体规模较小，主要有月浦闪长岩、坦直辉石闪长岩、齐贤桥辉石闪长岩、三林塘石英闪长岩、张堰石英闪长岩等。酸性岩体规模较大，主要有崇明花岗岩、吴淞花岗岩、市区花岗岩等。

（一）坦直辉石闪长岩

1. 地质特征

岩体位于坦直以东，面积约$10km^2$，侵位于上震旦统。围岩已大理岩化与矽卡岩化。岩体垂直分异现象显著，由上至下出现花岗闪长岩—二长闪长岩—石英闪长岩—闪长岩—辉石闪长岩—辉长闪长岩—辉长岩，其间均为渐变过渡关系。

2. 岩石学特征

（1）石英闪长岩-闪长岩

岩石呈灰白色与绿灰色，中—细粒（以细粒为主）半自形粒状结构。主要矿物有更—中长石（55%左右）、钾长石（少量~15%）、角闪石（5%~15%）、石英（3%~15%）、黑云母（3%）、辉石（少量~10%）等，副矿物有磁铁矿、磷灰石、锆石，次生与蚀变矿物有绿泥石、绢云母、高岭石、石英及黄铁矿等。

斜长石呈板柱状，An=28~50，聚片双晶发育，具环带构造，粒径一般为1mm×2mm；钾长石呈粒状，粒径一般为0.5mm×1mm。

（2）辉石闪长岩

岩石呈黑灰色，细—中粒半自形粒状结构。矿物成分由中长石（65%~78%）、辉石（10%~20%）、黑云母（3%~10%）、角闪石（少量~4%，局部达15%）、石英（少量~5%）、钾长石（少量）及副矿物磷灰石、磁铁矿、榍石等组成，次生与蚀变矿物有绢云母、高岭石、绿泥石、绿帘石等。

斜长石呈板柱状，An=40~50，聚片双晶发育，有时具环带构造，粒径大者为0.5~2.5mm，一般为0.5~1mm；钾长石呈他形粒状，沿斜长石颗粒之间及边缘分布，局部交代斜长石，有蠕英石析出；辉石以普通辉石为主，其次为紫苏辉石，呈短柱状、粒状，有些辉石晶体的四周出现普通角闪石反应边，部分辉石蚀变为绿泥石、方解石、纤闪石、黑云母等。

3. 副矿物特征

副矿物有11种，属磷灰石型或磷灰石-锆石型组合，并伴有磁铁矿、钛铁矿、黄铁矿等，其中以磁铁矿含量较高。

锆石以浅红色为主，次为浅灰褐色、浅灰色或无色，透明度差，表面粗糙，具极发育的环带状构造，晶面上生长有三角阶梯状凹形，并充填有长石。锆石晶形发育程度较差，晶形以｛110｝与｛111｝组成的聚形为主，其次由｛111｝组成正方双锥之聚形，晶体长为0.04~0.5mm，宽为0.04~0.3mm，长宽比主要为1∶1.1~1∶2，其次为1∶2.1~1∶3。

磷灰石以深—浅烟灰色为主，次为无色及浅粉红色，玻璃光泽，表面光滑。晶形由柱面 $\{10\bar{1}0\}$ 与锥面 $\{10\bar{1}1\}$ 组成带有双锥六方柱状之聚形。磷灰石大部分呈不完整六方柱状，大者为0.91mm×0.26mm，一般为0.26~0.04mm×0.13~0.02mm。

4. 岩石化学特征

岩石中 SiO_2 含量为53.44%~57.37%，Fe_2O_3+FeO 为7.09%~9.03%，Na_2O+K_2O 为6.15%~7.41%，且 $Na_2O>K_2O$，K_2O/Na_2O 值为0.71~0.81（表10-8）。标准矿物以斜长石、透辉石、紫苏辉石含量高为特征。里特曼指数为3.62~3.83，属碱性岩石。

表10-8　基性岩体化学成分及特征值

Table 10-8　Chemical compositions and eigenvalues of the basic rock bodies

岩石名称	氧化物/%												
	SiO_2	TiO_2	Al_2O_3	Fe_2O_3	FeO	MnO	MgO	CaO	Na_2O	K_2O	P_2O_5	烧失量	合计
辉石闪长岩	53.44	1.36	18.83	3.62	5.41	0.13	4.00	7.03	3.60	2.55	0.02	0.80	100.79
	57.03	1.18	17.76	1.95	5.14	0.11	2.92	5.38	3.95	3.20	0.62	0.31	99.55
	57.37	1.25	17.12	2.65	5.22	0.05	2.24	5.76	4.27	3.14	0.44	0.29	99.80
	57.18	1.25	17.55	2.53	5.25	0.07	2.27	5.83	4.24	3.12	0.44	0.45	100.18
	56.89	1.25	17.49	2.43	5.34	0.06	2.57	5.80	4.26	3.03	0.49	0.29	99.90

续表

岩石名称	CIPW标准矿物/%										
	Q	Or	Ab	An	Mt	Ap	Il	Di	Hy	C	Ol
辉石闪长岩	1. 26	15. 03	30. 42	27. 82	5. 33	0. 34	2. 58	4. 96	12. 42		
	4. 93	18. 93	33. 456	21. 14	2. 78	1. 34	2. 28	1. 83	12. 53		
	5. 17	8. 37	36. 18	18. 36	3. 94	1. 01	2. 43	6. 44	7. 85		
	4. 93	18. 37	35. 66	19. 75	3. 70	1. 01	2. 43	5. 52	8. 44		
	4. 09	17. 81	36. 18	19. 75	3. 47	1. 01	2. 43	5. 27	9. 64		

岩石名称	其他特征参数								
	a	*b*	(Na_2O+K_2O) /%	K_2O/Na_2O	$Mg^{2+}/(Fe^{3+}+Fe^{2+})$	氧化系数	分异指数	里特曼指数	碱度指数
辉石闪长岩	11. 95	17. 30	6. 15	0. 71	0. 83	0. 495	46. 71	3. 62	1. 62
	13. 68	13. 26	7. 15	0. 81	0. 75	0. 40	57. 42	3. 64	1. 89
	14. 22	13. 87	7. 41	0. 74	0. 53	0. 28	59. 72	3. 82	1. 96
	14. 13	13. 57	7. 36	0. 74	0. 53	0. 34	58. 96	3. 82	1. 92
	14. 00	14. 00	7. 29	0. 71	0. 62	0. 33	58. 08	3. 83	1. 91

表 10-9 坦直辉石闪长岩体微量（含稀土）元素含量

Table 10-9 Contents of trace elements (incl. REE) in the Tanzhi pyroxene diorite bodies

$w(B)/10^{-6}$

Ba	Cu	Pb	Zn	Sn	V	Cr	Ni	Sr	Ga	Co	Be	Sc	Ta	Mo	Nb	Zr
1302	146	33	102	10	820	102	63	859	90	26	1	10	40	10	10	74

La	Ce	Pr	Nd	Sm	Eu	Gd	Tb	Dy	Ho	Er	Tm	Yb	Lu	Y	ΣREE
52. 79	103. 22	13. 06	45. 50	8. 69	2. 06	6. 71	1. 08	5. 76	1. 10	3. 11	0. 47	2. 78	0. 44	28. 35	275. 12

与中性岩类维氏值相比，Ba、Cu、Pb、Zn、V、Cr、Ga、Co、Sc、Mo、Ag 等含量较高，其中 Cu、Sc 高出 4 倍，Ga 高出 4. 5 倍，V 高出 8 倍，Mo 高出 11 倍，Sr、Ni 相近（表 10-9）。

稀土元素（含 Y）含量 $\Sigma REE = 275.12\times10^{-6}$，$\Sigma LREE = 225.32\times10^{-6}$，$\Sigma HREE+Y = 49.8\times10^{-6}$，$\Sigma LREE/(\Sigma HREE+Y) = 4.52$，轻稀土富集，其模式曲线为右倾型（图 10-5）。$\delta Eu=0.82$，Eu 亏损较少。$\delta Ce=0.92$，Ce 亏损很少。$(La/Yb)_N = 12.83$，$Sm/Nd=0.19$，$\Sigma LREE/(\Sigma HREE+Y) = 4.52$，稀土元素配分曲线中等倾斜，重稀土部分则近水平，轻、重稀土分馏较强，属轻稀土富集型。

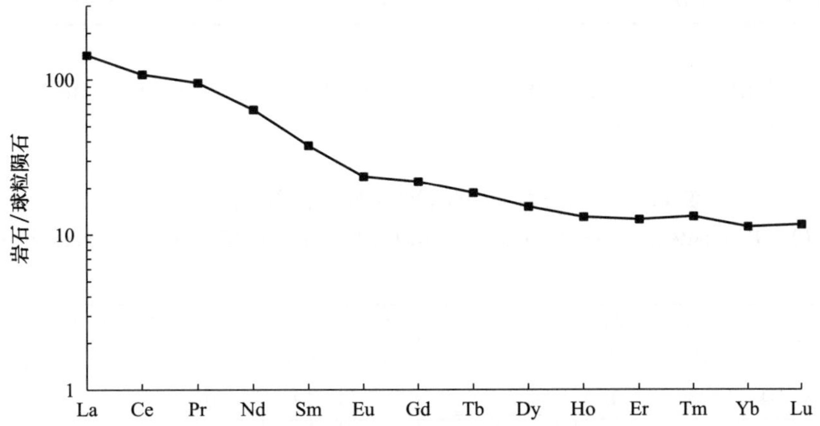

图 10-5 坦直辉石闪长岩稀土元素配分模式图

Fig. 10-5 REE pattern of the Tanzhi pyroxene diorite

（二）沈庄辉石闪长岩

1. 地质特征

见于沈庄，呈东西向延伸的小岩株产出。岩体受罗店—周浦断裂和枫泾—川沙断裂带控制，侵入于前震旦系金山群变粒岩与片岩中，见有金山群片岩捕虏体，使围岩角岩化。锆石 U-Pb 年龄为 106.9~104.2Ma。布格重力等值线显示该岩体由南、北两个封闭的相对负异常构成。岩体碳酸盐化、绿泥石化、黄铁矿化明显。

2. 岩石学特征

岩石呈灰色与灰白色，中细粒（以细粒为主）半自形结构、辉绿结构。主要矿物有中长石（60%~80%）、石英（少量~5%）、辉石（5%~15%）、角闪石（少量~9%）及黑云母（少量~5%）等，副矿物有磁铁矿、磷灰石、锆石、榍石等，蚀变矿物有方解石、黄铁矿等。

斜长石呈板柱状，聚片双晶，环带构造发育，为中长石，粒径 1.3~4mm，以中粒为主，细粒次之；辉石呈淡褐色，为不规则粒状，以普通辉石为主，紫苏辉石次之，粒度以细粒为主，中粒较少，部分辉石已纤闪石化；角闪石部分已蚀变成绿泥石、方解石。

3. 岩石化学特征

岩石中 SiO_2 含量为 53.79%~58.01%，Fe_2O_3+FeO 为 6.49%~7.78%，Na_2O+K_2O 为 5.05%~6.18%，且 Na_2O>K_2O，K_2O/Na_2O 值为 0.51~0.71（表 10-10）。标准矿物以斜长石特别是钙长石及透辉石、紫苏辉石含量较高为特点。里特曼指数为 2.36~2.44，属钙碱性岩石。

表 10-10　沈庄辉石闪长岩体岩石化学成分及特征值

Table 10-10　Chemical compositions and eigenvalues of the Shenzhuang pyroxene diorite bodies

岩石名称	氧化物/%												
	SiO_2	TiO_2	Al_2O_3	Fe_2O_3	FeO	MnO	MgO	CaO	Na_2O	K_2O	P_2O_5	烧失量	合计
辉石闪长岩	53.79	0.84	19.92	3.09	4.69	0.10	5.07	7.88	3.35	1.70	0.18		100.87
	58.01	0.82	16.81	3.42	3.07	0.10	3.19	5.88	3.62	2.56	0.25	3.85	101.59

岩石名称	CIPW 标准矿物/%										
	Q	Or	Ab	An	Mt	Ap	Il	Di	Hy	C	Ol
辉石闪长岩	2.40	10.02	28.32	34.22	4.40	0.34	1.67	3.34	15.79		
	11.2	15.48	31.34	22.57	5.07	0.59	1.59	4.45	7.70		

岩石名称	其他特征参数								
	a	*b*	(Na_2O+K_2O) /%	K_2O/Na_2O	$Mg^{2+}/(Fe^{3+}+Fe^{2+})$	氧化系数	分异指数	里特曼指数	碱度指数
辉石闪长岩	10.13	17.44	5.05	0.51	1.21	0.40	40.74	2.36	1.44
	12.04	11.76	6.18	0.71	0.94	0.53	58.02	2.44	1.75

与中性岩类维氏值相比，Ba、Cu、Pb、Zn、V、Cr、Ga、Ag、Co、Sc、Mo 等含量较高，其中 Cu 高出 7 倍，Pb 高出 5 倍（表 10-11）。

表 10-11　沈庄辉石闪长岩体微量（含稀土）元素含量

Table 10-11　Contents of trace elements (incl. REE) in the Shenzhuang pyroxene diorite

w（B）$/10^{-6}$

Sc	Li	Be	Co	Cu	Zn	Ga	Ge	Rb	Zr	Nb	Mo	Cd	Cs	Hf	Ta
13.8	14.8	1.52	31.2	32.6	82.3	21.6	1.36	71.7	193	9.95	0.81	0.065	1.73	4.7	1.01
5		2	34	236	144	94			84	<10	<5				<40
W	Tl	Pb	Th	U	Ba	Cr	Ni	Sr	V	As	Sb	Bi	Sn	Ag	
101	0.36	11.0	8.03	1.89	660	22.7	13.8	597	131	0.85	0.06	0.016	1.46	0.076	

续表

W	Tl	Pb	Th	U	Ba	Cr	Ni	Sr	V	As	Sb	Bi	Sn	Ag	
		75			2160	190	22	554	339				11		
La	Ce	Pr	Nd	Sm	Eu	Gd	Tb	Dy	Ho	Er	Tm	Yb	Lu	Y	ΣREE
30.1	58.7	7.29	28.0	5.78	1.29	5.29	0.81	4.37	0.87	2.48	0.39	2.57	0.37	23.6	171.92

稀土元素（含 Y）含量中等，$\Sigma REE=171.92\times10^{-6}$，$\Sigma LREE=131.16\times10^{-6}$，$\Sigma HREE+Y=40.75\times10^{-6}$。$\delta Eu=0.71$，具负 Eu 异常。$\delta Ce=0.93$，Ce 亏损很少。$(La/Yb)_N=7.91$，$Sm/Nd=0.21$，$\Sigma LREE/(\Sigma HREE+Y)=3.22$，稀土元素配分曲线中等倾斜，轻、重稀土分馏较强，属轻稀土富集型。与坦直辉石闪长岩相比，轻稀土富集下降且轻、重稀土分馏减弱（图 10-6）。

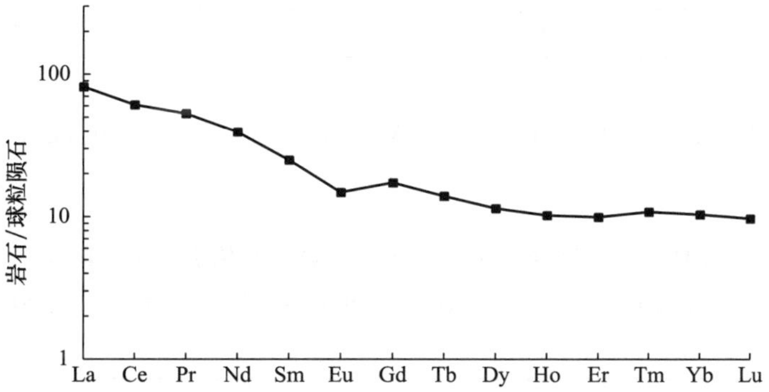

图 10-6　沈庄辉石闪长岩稀土元素配分模式图

Fig. 10-6　REE pattern of the Shenzhuang pyroxene diorite

（三）三林塘石英闪长岩

1. 地质特征

岩体位于浦东三林，梅隆—邓镇断裂带上。长约 3.5km，宽约 1km，面积约 $3km^2$，呈一北西-南东向延伸的小岩株。北部为劳村组角度不整合覆盖；南部侵位于金山群变质岩中，围岩已角岩化、大理岩化。锆石 U-Pb 同位素年龄约为 163.4Ma。

2. 岩石学特征

岩石呈灰自微带浅红色，中细粒（以细粒为主）半自形粒状结构，局部具有似斑状结构。主要矿物有斜长石（50%~70%）、钾长石（5%~15%）、石英（5%~15%）、黑云母及角闪石（<10%），副矿物为磁铁矿、榍石、磷灰石，次生矿物有高岭石、绿泥石、绢云母、方解石等。

斜长石呈板状、柱状，具环带构造，为中—更长石，表面绢云母化；钾长石呈粒状、不规则状，分布于斜长石之间；部分角闪石已绿泥石化，部分黑云母已褪色，析出铁质，生成榍石、绿泥石等矿物。

3. 岩石化学特征

岩石中 SiO_2 含量为 61.76%~65.54%，Fe_2O_3+FeO 为 4.37%~5.3%，Na_2O+K_2O 为 7.29%~8.22%，且 $Na_2O>K_2O$，K_2O/Na_2O 值为 0.51~0.68。标准矿物以斜长石和似长石、石英及紫苏辉石含量较高为特点。里特曼指数为 2.32~3.60，属钙碱性—碱性岩石（表 10-12）。

表 10-12　三林塘石英闪长岩体岩石化学成分及特征值

Table 10-12　Chemical compositions and eigenvalues of the Sanlintang quartz diorite bodies

岩石名称	氧化物/%											
	SiO_2	TiO_2	Al_2O_3	Fe_2O_3	FeO	MnO	MgO	CaO	Na_2O	K_2O	P_2O_5	烧失量	合计
石英闪长岩	61.76	0.75	16.18	1.64	2.73	0.04	2.74	4.52	4.90	3.32	0.23	1.05	99.86
	62.20	0.74	16.19	1.13	4.17	0.06	2.02	3.28	5.16	2.63	0.23	2.20	100.01
	65.54	0.62	15.91	1.79	1.75	0.04	1.65	3.55	4.50	2.79	0.20	2.36	100.70

续表

岩石名称	CIPW 标准矿物/%										
	Q	Or	Ab	An	Mt	Ap	Il	Di	Hy	C	Ol
石英闪长岩	8.17	19.48	41.43	12.52	2.32	0.67	1.37	6.71	6.24		
	10.57	15.59	43.53	13.35	1.62	0.67	1.37	0.92	10.23		
	19.44	16.77	38.72	15.22	2.64	0.47	1.20	1.07	4.47		

岩石名称	其他特征参数								
	a	*b*	(Na_2O+ K_2O) /%	K_2O /Na_2O	Mg^{2+}/ (Fe^{3+}+Fe^{2+})	氧化系数	分异指数	里特曼指数	碱度指数
石英闪长岩	15.47	11.13	8.22	0.68	1.15	0.38	69.08	3.60	2.32
	15.40	9.19	7.79	0.51	0.69	0.21	69.69	3.16	2.33
	12.02	5.16	7.29	0.62	0.89	0.51	74.93	2.32	2.20

与中性岩类维氏值相比，Ba、Cu、Pb、Ag、V、Cr、Ga、Sc 等含量较高，其中 Cu 高出 4.8 倍，Cr 高出 6 倍（表 10-13）。

表 10-13　三林塘石英闪长岩体微量（含稀土）元素含量

Table 10-13　Contents of trace elements (incl. REE) in the Sanlintang quartz diorite

w (B) /10^{-6}

Sc	Li	Be	Co	Cu	Zn	Ga	Ge	Rb	Zr	Nb	Mo	Cd	Cs	Hf	Ta
5.98	12.4	1.91	23.5	52.3	52.0	23.9	1.14	74.3	157	8.16	18.9	0.076	1.47	4.2	0.95
<5		2	11	169	38	59			69	<10	33				<40
W	**Tl**	**Pb**	**Th**	**U**	**Ba**	**Cr**	**Ni**	**Sr**	**V**	**As**	**Sb**	**Bi**	**Sn**	**Ag**	
118	0.26	10.7	5.94	1.91	1065	21.5	14.6	837	73.4	1.01	0.14	0.020	1.34	0.178	118
		<30			1874	302	19	633	115				<10		
La	**Ce**	**Pr**	**Nd**	**Sm**	**Eu**	**Gd**	**Tb**	**Dy**	**Ho**	**Er**	**Tm**	**Yb**	**Lu**	**Y**	**ΣREE**
43.67	86.58	10.65	35.71	6.11	1.49	4.13	0.64	3.33	0.60	1.68	0.28	1.57	0.24	16.21	212.89
28.1	53.5	6.52	24.5	4.54	1.14	3.75	0.49	2.23	0.40	1.08	0.15	0.99	0.14	10.5	138.03

稀土元素（含 Y）含量中等，$\sum REE = 138.03\times10^{-6} \sim 212.89\times10^{-6}$，$\sum LREE = 118.3\times10^{-6} \sim 184.21\times10^{-6}$，$\sum HREE+Y = 19.73\times10^{-6} \sim 28.68\times10^{-6}$。$\delta Eu = 0.84 \sim 0.91$，具负 Eu 异常。$\delta Ce = 0.93 \sim 0.94$，Ce 亏损很少。La/Yb = 18.80～19.18，Sm/Nd = 0.17～0.19，$\sum LREE/(\sum HREE+Y) = 6.00 \sim 6.42$，稀土元素配分曲线中等倾斜，轻、重稀土分馏较强，属轻稀土明显富集型（图 10-7）。

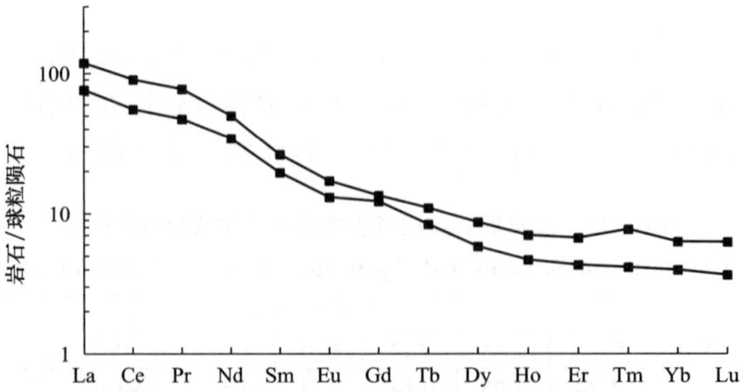

图 10-7　三林塘石英闪长岩稀土元素配分模式图

Fig. 10-7　REE pattern of the Sanlintang quartz diorite

（四）张堰石英闪长岩

1. 地质特征

岩体见于金山张堰，位于张堰—南汇断裂带上。大致呈北东东向，东西长约 3. 05km，南北最宽处宽 1. 25km，上小下大，为一岩株。侵位于金山群等，围岩已角岩化。岩体内 2 组节理发育，在高角度节理面上广泛发育铜矿化或铜钼矿矿化。锆石 U-Pb 年龄 101. 6±1. 1Ma。

2. 岩石学特征

岩石呈灰与灰白色，具不等粒半自形粒状结构、似斑状结构，主要矿物为斜长石（55%）、钾长石（15%）、石英（17%）、角闪石（5%）及黑云母（7%），局部角闪石含量增多可达 18%，为花岗闪长岩。副矿物有 12 种，主要有榍石、磷灰石、锆石等。此外，磁铁矿含量很高。次生与蚀变矿物为绢云母、绿泥石、方解石和石英。粒径 0. 1~6mm，以细粒为主。

斜长石呈板柱状，An=32~34 或 26~28，粒径一般 0. 4~1mm，大者达 5~6mm，以密集型酸性斜长石聚片双晶为特征，部分有环带结构，有不同程度绢云母化、钾长石化；钾长石，呈半自形—他形粒状，无双晶或简单双晶，可见次生蚀变，主要为高岭土化。角闪石，半自形，常含有磁铁矿、磷灰石包裹体，部分已被交代成方解石、绿帘石和绿泥石的集合体；黑云母，片状，多填充在长石斑晶间隙，局部有少量黄铜矿、黄铁矿、辉钼矿等浸染。

锆石为浅红色，透明度良好，表面不平坦—光滑，晶面上有熔融凹陷，并镶嵌有磁铁矿、磷灰石等，且具环带及再生长构造。晶形呈复四方柱聚形，晶体长 0. 04~0. 74mm，宽 0. 02~0. 23mm，长宽比大者为 5. 3~6，一般为 1. 1~3. 0。

3. 岩石化学特征

岩石中 SiO_2 含量为 62. 2%~65. 18%，Fe_2O_3+FeO 为 4. 2%~5. 39%，Na_2O+K_2O 为 6. 52%~7. 35%，且 $Na_2O>K_2O$，K_2O/Na_2O 值为 0. 50~0. 68（表 10-14）。标准矿物中石英、斜长石和钾长石等含量均接近于实测矿物含量。里特曼指数为 2. 18~2. 69，属钙碱性岩石。

表 10-14　张堰岩体化学成分及特征值

Table 10-14　Chemical compositions and eigenvalues in the Zhangyan rock

岩石名称	氧化物/%												
	SiO_2	TiO_2	Al_2O_3	Fe_2O_3	FeO	MnO	MgO	CaO	Na_2O	K_2O	P_2O_5	烧失量	合计
石英闪长岩	62. 24	1. 14	17. 18	2. 05	3. 22	0. 06	2. 06	4. 49	4. 64	2. 56	0. 22	0. 99	100. 85
	63. 68	0. 71	17. 16	1. 80	2. 37	0. 06	1. 89	4. 09	4. 64	2. 60	0. 18	0. 93	100. 61
	62. 51	0. 59	10. 46	2. 19	3. 20	0. 07	4. 44	2. 48	4. 03	2. 49	0. 24	0. 96	99. 68
	65. 18	0. 72	16. 14	1. 88	2. 70	0. 07	2. 30	3. 80	4. 50	2. 58	0. 20	0. 86	99. 75
	62. 20	0. 72	16. 64	1. 38	2. 82	0. 07	1. 69	3. 70	4. 44	2. 88	0. 16	0. 46	100. 14
	62. 20	0. 76	16. 47	2. 47	2. 91	0. 07	2. 34	4. 67	4. 42	2. 21	0. 25	1. 60	100. 37
	63. 65	0. 67	16. 43	1. 53	3. 15	0. 05	2. 12	3. 90	4. 43	2. 92	0. 20	0. 94	99. 99
花岗闪长岩	64. 76	0. 69	16. 21	1. 95	1. 88	0. 04	1. 74	4. 02	4. 23	2. 88	0. 21	1. 84	100. 44

岩石名称	C I P W 标准矿物/%										
	Q	Or	Ab	An	Mt	Ap	Il	Di	Hy	C	Ol
石英闪长岩	12. 86	15. 03	39. 33	18. 36	3. 01	0. 67	2. 12	1. 79	6. 76		
	14. 36	15. 59	39. 33	18. 08	2. 55	0. 34	1. 37	1. 14	6. 93		
	17. 00	14. 47	34. 09	10. 57	3. 24	0. 67	1. 06		14. 34	3. 26	
	16. 04	15. 03	38. 28	16. 14	2. 78	0. 34	1. 21	1. 57	7. 46		
	16. 82	17. 26	37. 76	16. 69	2. 08	0. 34	1. 37	0. 68	6. 79		
	15. 08	12. 80	37. 23	18. 92	3. 47	0. 67					
	14. 48	17. 26	37. 23	16. 41	2. 32	0. 34	1. 21	1. 83	7. 99		
花岗闪长岩	18. 82	17. 26	36. 30	16. 97	2. 87	0. 49	1. 33	1. 47			

续表

岩石名称	其他特征参数								
	a	b	(Na_2O+K_2O) /%	K_2O/Na_2O	$Mg^{2+}/(Fe^{3+}+Fe^{2+})$	氧化系数	分异指数	里特曼指数	碱度指数
石英闪长岩	14.00	9.33	7.20	0.55	0.72	0.39	67.22	2.69	1.995
	14.06	8.09	7.24	0.56	0.75	0.39	69.29	2.53	2.03
	12.13	15.57	6.52	0.62	1.53	0.41	65.56	2.18	2.05
	13.68	8.82	7.08	0.57	0.92	0.41	69.35	2.38	2.10
	13.94	7.11	7.32	0.65	0.75	0.33	71.84	2.42	2.12
	13.11	10.14	6.63	0.50	0.81	0.46	65.11	2.29	1.91
	14.05	8.77	7.35	0.66	0.84	0.33	68.97	2.62	2.13
花岗闪长岩	11.65	5.60	7.11	0.68	0.88	0.69	72.38	2.29	2.08

微量元素含量与中性岩石维氏平均值相比，岩体中部有 Ba、Cu、Pb、Cr、Ga、Mo 等 6 种元素含量偏高，岩体边部有 Ba、Cu、Pb、Cr、Ga、Ag、Mo 等 7 种元素含量偏高（表 10-15），其中 Cu 高出 18 倍，Mo 高出 40 倍，且岩体边部 Cu、Mo 含量明显高于岩体中心。

表 10-15　张堰岩体微量（含稀土）元素含量

Table 10-15　Contents of trace elements (incl. REE) in the Zhangyan rock　$w(B)/10^{-6}$

岩体部位	样品数/件	Ba	Ti	Cu	Pb	Zn	V	Cr	Sr	Ga	Ag	Co	Mo				
中部	65	1365	5200	65	31	34	92	139	632	67	<1	8	5				
边部	76	1970	5580	630	41	53	93	161	437	80	1	9	36				
不明		La	Ce	Pr	Nd	Sm	Eu	Gd	Tb	Dy	Ho	Er	Tm	Yb	Lu	Y	ΣREE
		23.67	46.03	5.87	20.69	4.04	1.09	2.81	0.40	1.97	0.33	0.86	0.12	0.75	0.11	8.92	117.66
3702 孔	1	20.5	38.6	4.75	18.0	3.40	1.02	2.73	0.36	1.63	0.29	0.75	0.11	0.72	0.10	7.59	100.55

稀土元素（含 Y）总量 $\sum REE=100.55\times10^{-6}\sim117.66\times10^{-6}$，其中 $\sum LREE=86.27\times10^{-6}\sim101.39\times10^{-6}$，$\sum HREE+Y=14.28\times10^{-6}\sim16.27\times10^{-6}$。$\delta Eu=0.99\sim1.02$，$\delta Ce=0.96$，亏损很少，属幔源型。$(La/Yb)_N=19.24\sim21.33$，$Sm/Nd=0.19\sim0.20$，$\sum LREE/(\sum HREE+Y)=6.04\sim6.23$，属轻稀土富集型，其分布模式曲线为右倾型（图 10-8）。

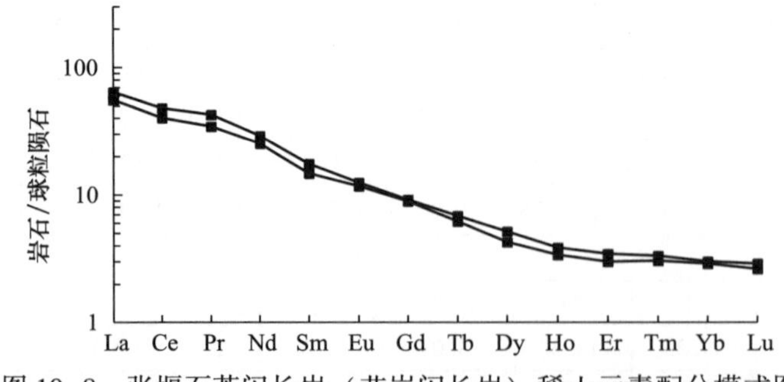

图 10-8　张堰石英闪长岩（花岗闪长岩）稀土元素配分模式图

Fig. 10-8　REE pattern of the Zhangyan quartz diorite (granodiorite)

（五）崇明花岗岩

1. 地质特征

岩体分布在崇明岛中部新开河—堡镇一带，呈岩基状产出。岩体与围岩黄尖组接触关系不明。钾长石 K-Ar 同位素年龄为 107.1Ma。

2. 岩石学特征

岩石呈肉红色，具细—粗粒不等粒结构，以细粒为主，粒径一般为0.9~1.7mm，大者达5~7mm，主要矿物成分为碱性长石（50%）、中—更长石（15%）、石英（30%）、黑云母（2%）、角闪石（2%~3%），副矿物有磁铁矿、磷灰石、榍石及锆石。

钾长石呈粒状与不规则状，属正长石和微斜长石，表面常见高岭石等黏土矿物，粒径一般为1.2~1.7mm，大者达5~7mm；斜长石呈板状、柱状，具钠长石律双晶，An=27，部分斜长石具环带构造，多为正环带，长石牌号An可达到30，表面附有绢云母及高岭石，矿物粒径一般为0.7~1.1mm；石英为他形不规则粒状，粒径一般为0.3~0.35mm；黑云母呈褐色，部分已绿泥石化、碳酸盐化；角闪石显绿色，呈柱状，粒径为0.46~0.7mm。

副矿物多达14种，副矿物组合属榍石-褐帘石-锆石型，此外，尚伴有磁铁矿、钛铁矿、黄铁矿、萤石、磷灰石等，其中以磷灰石含量较高。锆石以浅红色为主，次为浅灰褐色、浅黄色，钻孔中见有不均匀的红色斑点。晶内有包裹体，晶面有熔蚀现象，且具环带构造。晶体长为0.04~0.61mm，宽为0.018~0.312mm，长宽比大者为1∶5.2、1∶4.5，一般为1∶1~1∶3。

3. 岩石化学特征

岩石中SiO_2含量为69.7%，Fe_2O_3+FeO为2.30%，Na_2O+K_2O为9.46%，且$Na_2O<K_2O$，K_2O/Na_2O值为1.38（表10-16）。标准矿物中石英、钾长石等含量低于实测统计值，但石英含量大于20%。里特曼指数为3.35，属碱性岩石。

表10-16　崇明花岗岩化学成分及特征值

Table 10-16　Chemical compositions and eigenvalues of the Chongming granite

岩石名称	氧化物/%												
	SiO_2	TiO_2	Al_2O_3	Fe_2O_3	FeO	MnO	MgO	CaO	Na_2O	K_2O	P_2O_5	烧失量	合计
花岗岩	69.70	0.22	14.92	0.58	1.72	0.075	0.70	1.44	3.98	5.48	0.095	0.56	99.47

岩石名称	CIPW标准矿物/%										
	Q	Or	Ab	An	Mt	Ap	Il	Di	Hy	C	Ol
花岗岩	21	32.29	33.56	6.40	0.93	0.34	0.46		4.08	0.10	

岩石名称	其他特征参数									
	a	*b*	(Na_2O+K_2O) /%	K_2O/Na_2O	$Mg^{2+}/(Fe^{3+}+Fe^{2+})$	氧化系数	分异指数	里特曼指数	碱度指数
花岗岩	16.46	3.37	9.46	1.38	0.55	0.25	86.85	3.35	3.74

微量元素含量与酸性岩石维氏平均值相比，Cr偏高，其他元素相近或偏低（表10-17）。

表10-17　崇明花岗岩微量元素含量

Table 10-17　Contents of trace elements in the Chongming granite　　　$w(B)/10^{-6}$

含量	Cu	Pb	Zn	Ti	V	Cr	Ni	Co	Mn	Cd	Be
平均含量（12件）	7	19	61	1333	16	43	3	5	679	0.45	1.95
变化范围	4~14	15~24	42~80	844~1860	10~40	4~69	1~5	2~8	506~1550	0.1~1.0	1.4~3.4

稀土元素（含Y）含量$\sum REE=302.01\times10^{-6}$，$\sum LREE=262.06\times10^{-6}$，$\sum HREE+Y=39.95\times10^{-6}$（表10-18）。$\delta Eu=0.48$，具负Eu异常，表明岩浆起源于地壳浅部重熔。$\delta Ce=0.87$，Ce亏损很少。$(La/Yb)_N=19.44$，$Sm/Nd=0.17$，$\sum LREE/(\sum HREE+Y)=6.56$，稀土元素配分曲线右倾，轻、重稀土分馏较强，属轻稀土明显富集型。穿插于该岩体中的晚期花岗细晶岩脉，其中的稀土也属轻稀土元素富集型，其分布模式曲线与岩体相似，$\delta Eu=0.34$，Eu亏损也极明显，说明细晶岩脉与崇明花岗岩体出自同一岩浆源（图10-9）。

表 10-18 崇明花岗岩稀土元素含量

Table 10-18 Contents of REE in the Chongming granite w (B) $/10^{-6}$

	La	Ce	Pr	Nd	Sm	Eu	Gd	Tb	Dy	Ho	Er	Tm	Yb	Lu	Y	ΣREE
花岗岩	75.37	122.44	14.47	41.72	7.12	0.94	4.95	0.83	4.39	0.83	2.40	0.41	2.62	0.40	23.12	302.01
花岗细晶岩	25.94	50.52	6.18	19.39	3.68	0.34	2.59	0.48	2.94	0.57	1.86	0.33	2.47	0.39	18.13	135.81

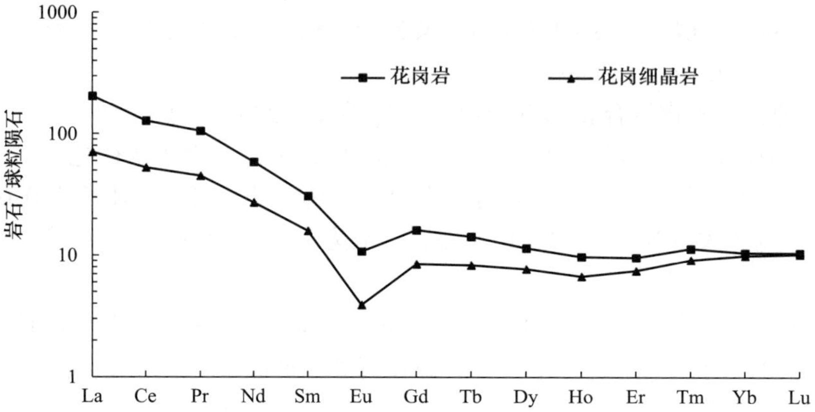

图 10-9 崇明花岗岩稀土元素配分模式图

Fig. 10-9 REE pattern of the Chongming granite

(六) 吴淞花岗岩

1. 地质特征

岩体分布于宝山东部吴淞—凌桥一带，位于金桥火山盆地北侧，其西南边界为淞南断裂所控制，面积约 55km^2，呈岩株状产出。布格重力异常等值线呈一中心位于凌桥的菱形低异常区，向（陆域）外侧缓慢升高。岩体与围岩接触关系不明，推测为黄尖组、寿昌组角度不整合覆盖。锆石 U-Pb 年龄为 178.8±2.3Ma。

2. 岩石学特征

岩石呈肉红色，具细—粗粒不等粒结构，粒径为 0.28~7.0mm。主要矿物成分为钾长石（30%~60%）、斜长石（5%~35%）、石英（25%~30%）、黑云母（2%~4%）、角闪石（2%）等，副矿物有磁铁矿（1%）、磷灰石、榍石、锆石、金红石等。岩体边部为二长花岗岩（斜长石含量高达 35%）。钾长石呈他形，有微纹长石、条纹长石，表面已黏土化，粒径一般为 1.75~2.1mm，大者达 5~7mm；斜长石呈板柱状，具钠长石律双晶，少部分斜长石具环带构造，An=25~30，表面布有绢云母及高岭石，粒径一般为 0.6~1.3mm，大者达 2.5~4.9mm，石英呈不规则粒状，粒径一般为 0.5~0.6mm，大者达 3.5~5.1mm；黑云母呈片状，已被方解石、绿泥石交代，角闪石呈柱状，部分已碳酸盐化，并析出氧化铁。

3. 岩石化学特征

岩石中 SiO_2 含量为 71.18%，Fe_2O_3+FeO 为 2.56%，Na_2O+K_2O 为 6.93%，且 Na_2O<K_2O，K_2O/Na_2O 值为 1.02（表 10-19）。标准矿物中石英含量略大于实测统计值，钾长石低于实测统计值，斜长石与实测统计值相当。里特曼指数为 1.68，属钙碱性岩石。

表 10-19 吴淞花岗岩化学成分及特征值

Table 10-19 Chemical compositions and eigenvalues of the Wusong granite

岩石名称	氧化物/%												
	SiO_2	TiO_2	Al_2O_3	Fe_2O_3	FeO	MnO	MgO	CaO	Na_2O	K_2O	P_2O_5	烧失量	合计
花岗岩	71.18	0.39	14.16	1.19	1.37	0.03	0.99	1.13	3.43	3.50	0.14	4.77	102.28

续表

岩石名称	CIPW标准矿物/%										
	Q	Or	Ab	An	Mt	Ap	Il	Di	Hy	C	Ol
花岗岩	34.77	21.21	29.76	4.81	1.77	0.33	0.76		3.50	3.09	

岩石名称	其他特征参数								
	a	*b*	(Na_2O+K_2O)/%	K_2O/Na_2O	$Mg^{2+}/(Fe^{3+}+Fe^{2+})$	氧化系数	分异指数	里特曼指数	碱度指数
花岗岩	12.71	4.05	6.93	1.02	0.74	0.46	85.74	1.68	2.63

微量元素含量与酸性岩石维氏平均值相比，Cu、Pb、Co偏高，其他元素相近或偏低（表10-20）。

表10-20 吴淞花岗岩微量（含稀土）元素含量

Table 10-20 Contents of trace elements (incl. REE) in the Wusong granite $w(B)/10^{-6}$

Sc	Li	Be	Co	Cu	Zn	Ga	Ge	Rb	Zr	Nb	Mo	Cd	Cs	Hf	Ta
5.05	6.46	1.70	25.0	36.0	33.6	16.9	1.15	105	112	18.9	0.18	0.032	3.65	2.8	1.73
W	Tl	Pb	Th	U	Ba	Cr	Ni	Sr	V	As	Sb	Bi	Sn	Ag	
183	0.54	38.0	16.1	2.73	606	11.4	9.22	187	40.5	0.75	0.07	0.061	1.26	0.051	183
La	Ce	Pr	Nd	Sm	Eu	Gd	Tb	Dy	Ho	Er	Tm	Yb	Lu	Y	ΣREE
31.5	51.3	5.42	18.2	3.17	0.81	2.90	0.42	2.19	0.43	1.28	0.20	1.34	0.21	11.6	131
27.16	45.31	5.28	16.57	3.10	0.62	2.15	0.34	2.12	0.40	1.22	0.21	1.39	0.21	12.15	118.23

稀土元素总量$\Sigma REE=118.23\times10^{-6}\sim131\times10^{-6}$，其中$\Sigma LREE=98.04\times10^{-6}\sim110.4\times10^{-6}$，$\Sigma HREE+Y=20.19\times10^{-6}\sim20.57\times10^{-6}$。$\delta Eu=0.73\sim0.82$，$\delta Ce=0.93\sim0.96$，Eu、Ce亏损都很少。$La/Yb=14.02\sim16.86$，$Sm/Nd=0.17\sim0.19$，$\Sigma LREE/(\Sigma HREE+Y)=4.86\sim5.37$，其分布模式曲线为右倾型，属轻稀土元素富集型（图10-10）。

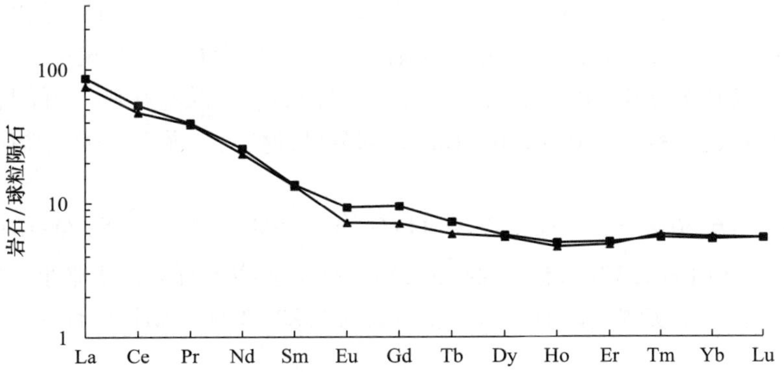

图10-10 吴淞花岗岩稀土元素配分模式图

Fig. 10-10 REE pattern of the Wusong granite

（七）市区花岗岩

1. 地质特征

岩体分布在市区，由南、北两岩体组成，位于近北东—北北东向较宽缓的布格重力异常带上，均呈岩株状产出，处于北新泾—大场下古生界隆起与金桥火山盆地交界处，侵位于下奥陶统和下白垩统寿昌组、黄尖组中（图10-11）。市北花岗岩分布于五角场—外白渡桥，面积约$22km^2$，西部和北部分别为大场—廊下断裂、大场—江湾断裂控制。岩体绢云母化、高岭土化及碳酸盐化较强烈。锆石U-Pb同位素年龄为132.9±1.1Ma。市南花岗岩分布于南市—龙华，面积约$26km^2$。

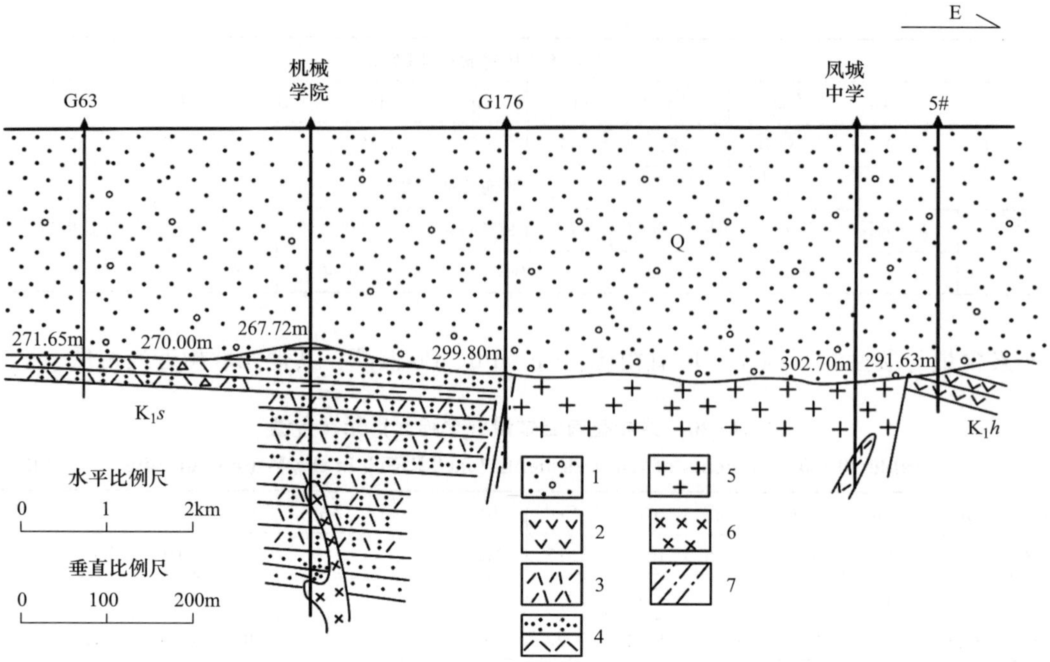

图 10-11　市区北部花岗岩体剖面图

Fig. 10-11　Section of the northern downtown granite bodies

1—第四系；2—安山岩；3—花岗斑岩；4—凝灰质砂岩、凝灰岩；5—花岗岩；6—辉绿岩；7—糜棱岩

2. 岩石学特征

岩石呈肉红色、灰白色，具细—粗粒不等粒结构，以细粒为主，粒径在 0.5mm 左右，大者可达 7mm。由钾长石（45%）、更—中长石（18%）、石英（25%）、角闪石（5%）、黑云母（4%）及磁铁矿（1%）、磷灰石、榍石、锆石等组成。

钾长石呈他形粒状，以正长石为主，高岭土化，粒径一般为 0.53~1.44mm，大者达 7~7.7mm；斜长石呈板柱状，具钠长石律双晶，部分斜长石具环带构造，多为正环带，已绢云母化和高岭土化，粒径为 0.7~1.05mm，大者达 1.23~4.55mm；石英为不规则粒状，粒径为 0.46~0.63mm，大者达 3.5~5.1mm；角闪石显绿色，柱状，粒径为 0.53~0.95mm，部分已被绿泥石、碳酸盐矿物、纤闪石等交代，并析出榍石、磁铁矿；黑云母呈褐色，粒径为 0.53~0.81mm，部分已蚀变为绿泥石、碳酸盐矿物。

3. 岩石化学特征

岩石中 SiO_2 含量为 66.36%~67.47%，Fe_2O_3+FeO 为 2.79%~4.04%，Na_2O+K_2O 为 8.46%~9.04%，且 Na_2O<K_2O，K_2O/Na_2O 值为 1.35~1.51（表 10-21）。标准矿物中石英含量略小于实测统计值，钾长石低于实测统计值，斜长石大于实测统计值。里特曼指数为 2.87~3.50，属钙碱性—碱性岩石。

表 10-21　市北花岗岩化学成分及特征值

Table 10-21　Chemical compositions and eigenvalues of the northern downtown granite

岩石名称	氧化物/%												
	SiO_2	TiO_2	Al_2O_3	Fe_2O_3	FeO	MnO	MgO	CaO	Na_2O	K_2O	P_2O_5	烧失量	合计
花岗岩	66.36	0.54	14.74	1.27	2.77	0.10	1.31	2.76	3.84	5.20	0.13	0.59	99.61
	67.47	0.41	14.88	1.21	1.58	0.05	0.88	2.22	3.37	5.09	0.14	4.76	102.07

岩石名称	CIPW 标准矿物/%										
	Q	Or	Ab	An	Mt	Ap	Il	Di	Hy	C	Ol
花岗岩	16.28	30.62	32.51	7.79	1.85	0.34	1.06	4.15	4.55		
	22.73	30.91	29.31	10.38	1.80	0.33	0.80		3.61	0.13	

续表

岩石名称	其他特征参数								
	a	*b*	(Na_2O+K_2O) /%	K_2O/Na_2O	$Mg^{2+}/(Fe^{3+}+Fe^{2+})$	氧化系数	分异指数	里特曼指数	碱度指数
花岗岩	15.77	7.41	9.04	1.35	0.60	0.31	79.41	3.50	3.14
	15.00	4.15	8.46	1.51	0.60	0.43	82.95	2.87	2.30

微量元素含量与酸性岩石维氏平均值相比，Zn、Co 偏高，其他元素相近或偏低（表 10-22）。

表 10-22 市北花岗岩微量（含稀土）元素含量

Table 10-22 Contents of trace elements (incl. REE) in the northern downtown granite

w (B) $/10^{-6}$

Sc	Li	Be	Co	Cu	Zn	Ga	Ge	Rb	Zr	Nb	Mo	Cd	Cs	Hf	Ta
7.10	13.0	2.89	32.9	5.43	41.1	17.9	1.41	178	197	17.0	1.39	0.046	6.00	5.0	1.27
W	Tl	Pb	Th	U	Ba	Cr	Ni	Sr	V	As	Sb	Bi	Sn	Ag	
230	0.76	20.6	31.6	6.46	637	7.46	5.88	270	44.4	0.36	0.11	0.146	2.27	0.036	230
La	Ce	Pr	Nd	Sm	Eu	Gd	Tb	Dy	Ho	Er	Tm	Yb	Lu	Y	ΣREE
43.9	77.9	8.42	28.5	5.28	0.95	4.92	0.77	4.18	0.90	2.65	0.42	2.86	0.43	24.1	206.18
43.84	75.66	9.16	27.65	5.22	0.83	3.99	0.69	4.12	0.81	2.58	0.44	2.95	0.46	23.49	201.89

稀土元素（含 Y）含量 $\sum REE=201.89\times10^{-6}\sim206.18\times10^{-6}$，$\sum LREE=162.36\times10^{-6}\sim164.95\times10^{-6}$，$\sum HREE+Y=39.53\times10^{-6}\sim41.23\times10^{-6}$。$\delta Eu=0.56\sim057$，Eu 亏损明显。$\delta Ce=0.88\sim0.95$，Ce 亏损很少。$(La/Yb)_N=10.04\sim10.37$，$Sm/Nd=0.19$，$\sum LREE/(\sum HREE+Y)=4.00\sim4.11$，其分布模式曲线为右倾型，属轻稀土元素富集型（图 10-12）。

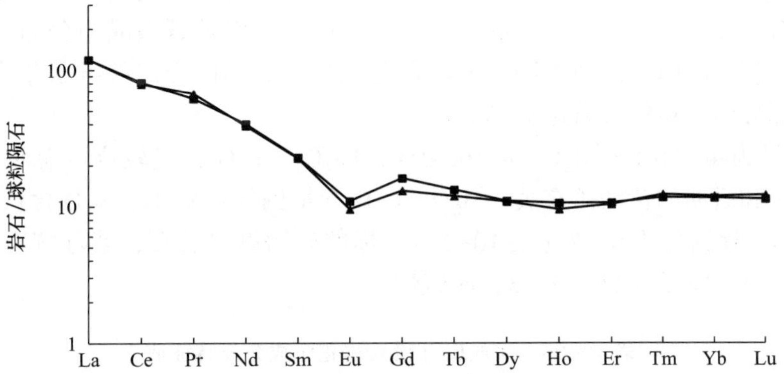

图 10-12 市北花岗岩稀土元素配分模式图

Fig. 10-12 REE pattern of the northern downtown granite

第四节 浅成斑（玢）岩类

区内斑（玢）岩类比较发育，主要有角岩化、次闪石化辉绿玢岩、煌斑岩、辉石闪长玢岩、石英闪长玢岩、闪长玢岩、安山玢岩和花岗斑岩、霏细斑岩、正长斑岩等。主要形成于燕山晚期，多呈岩脉产出，相互间的穿插关系难以见到，但其穿插岩体的现象较多。见有花岗斑岩、安山玄武玢岩等穿插到石英闪长岩体中；霏细斑岩、花岗斑岩、安山玢岩、煌斑岩、辉绿岩、安山玄武玢岩等穿插于花岗岩体中。

一、角岩化、次闪石化辉绿玢岩

岩体主要分布于金山张堰东城惶庙一带，并顺层侵位于劳村组流纹质凝灰岩及前震旦系金山群中，呈岩床产出（图 10-13）。

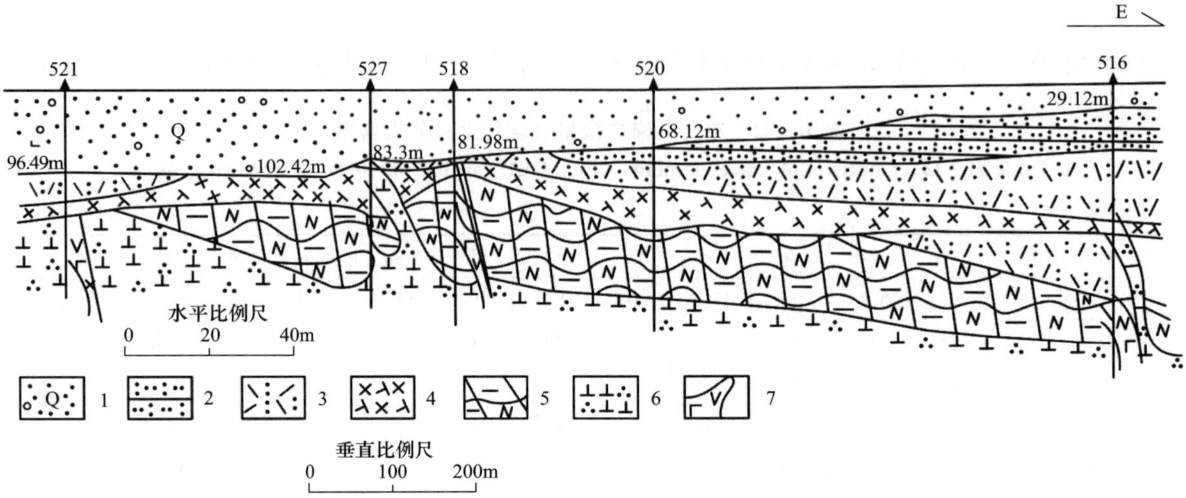

图 10-13 金山张堰东城隍庙次闪石辉绿玢岩与围岩接触关系剖面图

Fig. 10-13 Section showing the contact relation between the Dongchenghuang miao ouralite diabase-porphyrite and the surrounding rock in Zhangyan, Jinshan District

1—第四系；2—凝灰质砂岩；3—流纹质凝灰岩；4—次闪石辉绿玢岩；5—黑云母斜长角岩；6—石英闪长岩；7—安玄玢岩

岩石呈灰黑稍带绿色，具嵌晶含长结构，矿物成分由次闪石（35%），中长石（40%~50%）、黑云母（约 15%）、磁铁矿（约 5%）等组成。在粒径为 3~4mm 的次闪石晶体中包含有呈交织状的条状中长石微晶（0.2mm×0.04mm，An=31~39），一个次闪石晶体中可以包含几十个条状斜长石和后生的黑云母鳞片及阳起石微晶，局部尚可见镶嵌于次闪石晶体中的斜长石微晶与分布在次闪石晶体之间的斜长石微晶呈平行排列，次闪石出现不同程度的褪色现象，有的已变成纤维状集合体。斜长石表面有绿泥石化小斑点，部分尚可见钠长石律双晶。

岩石中 SiO_2 含量为 46.20%，Al_2O_3 为 16.41%，Fe_2O_3、FeO 含量较高，总和为 15.55%，这与岩石中含较多的磁铁矿和后期的黄铁矿有关；MgO、CaO 含量偏低；Na_2O、K_2O 含量与一般基性岩相近，K_2O/Na_2O 值为 0.19；镁铁比为 0.72（表 10-23）。标准矿物以斜长石、紫苏辉石、磁铁矿含量较高，石英含量很低为特征。里特曼指数为 7.35，属碱性。

表 10-23 浅成斑（玢）岩化学成分及特征值

Table 10-23 Chemical compositions and eigenvalues of the hypabyssal porphyry (porphyrite)

岩石名称	氧化物/%											
	SiO_2	TiO_2	Al_2O_3	Fe_2O_3	FeO	MnO	MgO	CaO	Na_2O	K_2O	P_2O_5	烧失量	合计
次闪石化辉绿玢岩（张堰）	46.20	1.70	16.41	8.84	6.71	0.28	5.89	4.73	4.07	0.78	0.23	3.96	99.80
辉石闪长玢岩（沈庄）	53.04	0.92	17.15	3.99	4.60	0.12	4.63	8.30	3.50	1.90	0.10	1.76	100.01
花岗斑岩（界山）	73.66	0.12	12.53	0.14	1.72	0.01	0.70	1.15	4.00	3.64	0.02	2.22	99.91
煌斑岩（张堰）	49.10	1.25	16.05	4.28	4.32	0.13	5.68	5.17	3.38	4.56	0.53	10.52	104.99
辉石闪长玢岩（沈庄）	52.18	1.03	17.55	3.25	4.83	0.14	4.75	8.35	3.19	1.67	0.22	4.77	101.93
花岗斑岩（庄行）	73.93	0.16	12.52	1.08	0.52	0.09	0.29	0.69	4.28	4.37	0.03	3.93	101.9

续表

岩石名称	CIPW 标准矿物/%										
	Q	Or	Ab	An	Mt	Ap	Il	Di	Hy	C（Ne）	Ol（Hm）
次闪石化辉绿玢岩（张堰）	0.12	4.45	34.61	21.70	12.73	0.67	3.19		17.43	C0.92	
辉石闪长玢岩（沈庄）	2.76	11.13	29.37	25.59	5.79	0.34	1.82	11.83	9.70		
花岗斑岩（界山）	31.66	21.71	34.09	5.01	0.23	0.34	0.30		4.61	C0.10	
煌斑岩（张堰）		28.53	26.46	16.04	6.57	1.30	2.51	5.90		Ne 2.07	Ol 10.62
辉石闪长玢岩（沈庄）	3.26	10.16	27.78	29.47	4.85	0.52	2.01	9.4	12.53		
花岗斑岩（庄行）	31.38	26.36	36.97	2.09	1.54	0.07	0.31	0.94	0.3		Hm 0.04

岩石名称	其他特征参数								
	a	*b*	(Na_2O+K_2O) /%	K_2O/Na_2O	$Mg^{2+}/(Fe^{3+}+Fe^{2+})$	氧化系数	分异指数	里特曼指数	碱度指数
次闪石化辉绿玢岩（张堰）	10.71	26.05	4.85	0.19	0.72	0.57	39.18	7.35	1.595
辉石闪长玢岩（沈庄）	10.72	20.11	5.40	0.54	1.01	0.46	43.26	2.90	1.54
花岗斑岩（界山）	13.74	3.00	7.64	0.91	0.65	0.08	87.46	1.90	3.53
煌斑岩（张堰）	14.83	18.50	7.94	1.38	1.25	0.50	57.06	7.87	1.93
辉石闪长玢岩（沈庄）	9.87	16.31	4.86	0.52	1.10	0.40	41.20	2.34	1.46
花岗斑岩（庄行）	15.34	1.93	8.65	1.02	0.33	0.68	94.71	2.40	4.68

次闪石辉绿玢岩在张堰曾见石英闪长岩入侵于其中，明显受到接触热变质作用影响，发生重结晶，部分斜长石由条状变成细粒状，暗色矿物（辉石）次闪石化，并被微鳞片状黑云母、绿泥石取代。因此，推测其为燕山晚期第一阶段的产物。

二、煌斑岩

煌斑岩主要见于张堰，在南丹路和淮海中路个别钻孔中见有煌斑岩穿插于花岗岩体中。

岩石呈灰色、褐色和绿色，具有斑状结构、块状构造。斑晶为半自形单斜辉石和板状斜长石，局部有角闪石。单斜辉石为普通辉石，短柱状，斜消光。斑晶粒径为0.2~0.5mm。基质多为隐晶结构，由斜长石组成，杂乱分布。填充在斑晶周围，少数斜长石呈针毡状排布。局部由0.1mm大小的斜长石及角闪石杂乱组成，构成交织结构。斜长石为板状和长柱状，发生次生变化，主要为碳酸盐化和绿泥石化。辉石含量占45%、斜长石占50%、方解石占5%。

岩石中 SiO_2 含量为49.10%，Fe_2O_3+FeO 为8.60%，Na_2O+K_2O 为7.94%，且 $Na_2O<K_2O$，K_2O/Na_2O 值为1.38（见表10-23）。标准矿物以似长石和斜长石为主，出现少量的磁铁矿、透辉石、霞石和橄榄石。里特曼指数为7.87，属碱性岩石。

与基性岩石维氏平均值相比，Rb、Ba、Sr含量较高，Ba为4.9倍、Sr为1.5倍，Nb、Th、Zr、U等偏低（表10-24）。

表10-24 浅成斑（玢）岩微量（含稀土）元素含量

Table 10-24 Contents of trace elements (incl. REE) in the hypabyssal porphyry (porphyrite)

w(B) /10^{-6}

岩性	Sc	Li	Be	Co	Cu	Zn	Ga	Ge	Rb	Zr	Nb	Mo	Cd	Cs	Hf	Ta
煌斑岩	19.0	47.8	1.35	31.1	34.6	84.1	17.3	1.15	146	162	7.91	0.48	0.068	2.19	3.8	0.71
辉石闪长玢岩	23.3	16.8	1.12	33.2	44.1	108	23.9	1.51	51.3	126	6.49	0.91	0.157	2.89	3.4	0.77
花岗斑岩	4.58	6.05	1.36	86.6	2.24	45.6	16.6	1.03	101	182	16.2	0.27	0.139	1.43	4.7	1.42
岩性	W	Tl	Pb	Th	U	Ba	Cr	Ni	Sr	V	As	Sb	Bi	Sn	Ag	
煌斑岩	14.0	1.06	6.17	3.36	0.60	1478	110	56.3	639	169	2.46	0.20	0.020	0.93	0.066	

续表

岩性	Sc	Li	Be	Co	Cu	Zn	Ga	Ge	Rb	Zr	Nb	Mo	Cd	Cs	Hf	Ta
辉石闪长玢岩	38.9	0.23	30.0	4.08	1.05	546	62.0	20.9	593	218	0.48	0.14	0.021	1.14	0.295	
花岗斑岩	365	0.56	8.07	14.8	2.97	324	2.44	18.9	37.3	6.30	0.42	0.20	0.030	2.12	0.022	
岩性	La	Ce	Pr	Nd	Sm	Eu	Gd	Tb	Dy	Ho	Er	Tm	Yb	Lu	Y	ΣREE
煌斑岩	24.9	55.9	7.42	30.7	5.98	1.74	5.10	0.70	3.60	0.68	1.78	0.26	1.71	0.25	17.0	158
辉石闪长玢岩	21.1	41.7	5.55	22.6	5.06	1.33	4.99	0.76	4.12	0.84	2.43	0.36	2.33	0.35	21.1	135
花岗斑岩	44.8	85.0	9.64	34.2	7.12	0.59	6.71	1.12	6.53	1.35	4.24	0.69	4.61	0.70	38.0	245

稀土元素总量$\sum REE=158\times10^{-6}$，其中$\sum LREE=126.64\times10^{-6}$，$\sum HREE+Y=31.08\times10^{-6}$。$\delta Eu=0.96$，Eu亏损不明显。$\delta Ce=1.01$，Ce无亏损。La/Yb=10.44，Sm/Nd=0.19，$\sum LREE/(\sum HREE+Y)=4.07$，其分布模式曲线为缓右倾型，属轻稀土元素富集型（图10-14）。

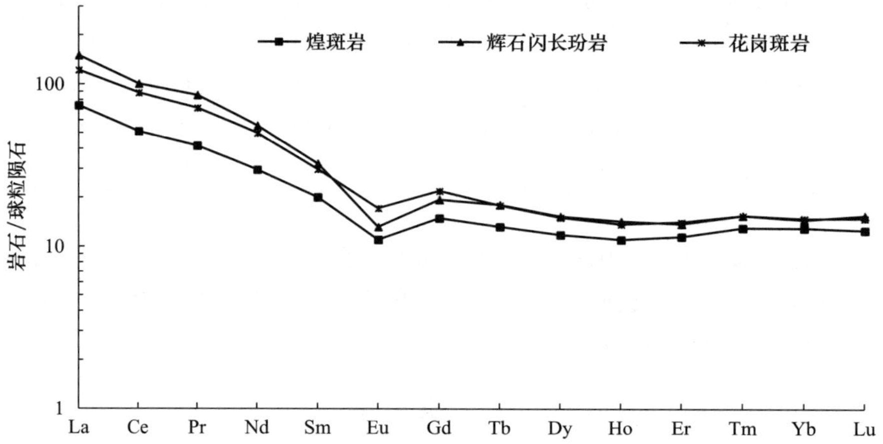

图10-14　浅成斑（玢）岩稀土元素配分模式图

Fig. 10-14　REE pattern of the hypabyssal porphyry (porphyrite)

三、闪长玢岩、石英闪长玢岩、辉石闪长玢岩

岩石分布较广，市区和宝山、川沙、崇明、金山、奉贤等地钻孔中均有分布，其成岩时代可能和石英闪长岩、闪长岩相近。

闪长玢岩和石英闪长玢岩呈灰—浅灰色，斑状结构，基质为半自形细—微粒粒状结构，斑晶约占10%，主要为中长石及少量角闪石，部分岩体中斜长石斑晶环带极发育，具有聚片双晶。基质以更—中长石为主（70%），次为角闪石、黑云母、钾长石及石英等。岩石发生绿帘石化、绿泥石化，绿帘石常见，普通角闪石蚀变成绿泥石。在界山可见辉石斑晶，角闪石含量可达75%。辉石被普通角闪石包裹，普通角闪石呈半自形晶。基质由角闪石和斜长石细晶组成，杂乱分布。

辉石闪长玢岩呈灰色、灰绿色，斑状结构、半自形粒状结构，块状构造，由斜长石、黑云母、角闪石和普通辉石组成。斜长石斑晶较自形，主要为中长石，具有聚片双晶和环带构造，次生蚀变强烈，发生钠黝帘石化，含量约为50%。角闪石呈半自形，含量约为15%。普通辉石呈短柱状，含量约为15%。在长石周围可见黑云母，呈半自形晶。黑云母发生强烈的绿泥石化。局部有微弱的黄铁矿化和磁铁矿化。

辉石闪长玢岩SiO_2含量为52.18%~53.04%，Fe_2O_3+FeO为8.08%~8.59%，Na_2O+K_2O为4.86%~5.4%，且$Na_2O>K_2O$，K_2O/Na_2O值为0.52~0.54（见表10-23）。标准矿物以似长石和斜长石为主，出现少量的磁铁矿、透辉石、紫苏辉石。里特曼指数为2.34~2.90，属钙碱性岩石。

与基性岩维氏平均值相比，Rb、Ba、Sr、Nb、Th、Zr等含量都偏高，尤其是Ba（见表10-24）。

稀土元素总量$\sum REE = 135\times10^{-6}$，其中$\sum LREE = 113.52\times10^{-6}$，$\sum HREE + Y = 37.28\times10^{-6}$。$\delta Eu = 0.81$，亏损不明显。$\delta Ce = 0.94$，亏损亦不明显。La/Yb = 6.50，Sm/Nd = 0.22，$\sum LREE/(\sum HREE + Y) = 3.05$，其分布模式曲线为缓右倾型，属轻稀土元素富集型（见图 10-14）。

四、花岗斑岩

岩石分布也较广泛，成岩时代晚于石英闪长岩及闪长岩，同位素年龄值为 84~88.1Ma。最新入侵围岩为上白垩统赤山组。

岩石呈肉红色，斑状结构，基质为隐—微晶显微粒状嵌晶结构或球粒结构，斑晶有钾长石（9%）、斜长石（更—中长石，3%~6%）、石英（5%~8%）、角闪石及黑云母（1%~2%），基质由隐—微晶状长石及石英组成，次生矿物有少量绢云母。

岩石中 SiO_2 含量为 73.66%~73.93%，Fe_2O_3+FeO 为 1.60%~1.86%，Na_2O+K_2O 为 7.64%~8.65%，且 $Na_2O<K_2O$，K_2O/Na_2O 值为 0.91~1.02（见表 10-23）。标准矿物以似长石和斜长石为主，见刚玉或赤铁矿。里特曼指数为 1.9~2.4，属钙碱性岩石。

与酸性岩维氏平均值相比，Cu、Pb、Zn、V、Cr 等含量都偏低，Ni 和 Co 稍微偏高（见表 10-24）。

稀土元素总量$\sum REE = 245\times10^{-6}$，其中$\sum LREE = 181.35\times10^{-6}$，$\sum HREE+Y = 63.95\times10^{-6}$。$\delta Eu = 0.26$，Eu 亏损显著。$\delta Ce = 1.00$，Ce 无亏损。La/Yb = 6.97，Sm/Nd = 0.21，$\sum LREE/(\sum HREE+Y) = 2.84$，其分布模式曲线为缓右倾型，属轻稀土元素富集型（见图 10-14）。

五、霏细斑岩

见于市区及松江、金山等地，侵入于奥陶系及上白垩统中，同位素年龄值为 81~89Ma。

岩石呈灰、灰白、浅肉红及褐黄色，具流纹构造。基质具球粒结构、显微粒状嵌晶结构、霏细结构。斑晶由钠长石（8%）、钾长石（2%）、黑云母（3%~4%）等组成；基质由钾长石和石英组成。岩石中 SiO_2 含量为 68.56%，Al_2O_3 为 14.33%，全碱为 7.13%。

第十一章　火山岩和火山作用

上海火山岩、火山碎屑岩类分布极为广泛，但露头稀少，仅 2.5km^2，主要分布在区内的西南部，如小昆山、横山、砚山、天马山、钟家山、罗山、辰山、东佘山、西佘山、雪山、凤凰山、北竿山及固山，南部的查山、秦皇山、甸山，西部的淀山等。

20 世纪 20 年代，刘季辰、赵汝钧等在佘山、天马山及大金山、小金山等地进行了地质调查，并将调查资料载于 1924 年的《江苏地质志》中。新中国成立以来，上海市积极开展各类找矿工作，结合磁异常验证，拉开了对区内火山岩整体研究的序幕。

火山作用可以划分为晋宁早—中期、晋宁晚期、燕山晚期和喜马拉雅期。岩石类型复杂，基性、中性、中酸性、酸性岩类均有发现，其中以酸性、中酸性岩类分布面积较广。

晋宁早—中期基性、中酸性及酸性火山岩发育，后经多次变质作用，成为金山群斜长角闪岩、黑云母斜长角岩及长石石英角岩等区域变质岩系或热变质岩系。

晋宁晚期中性、中基性火山岩、火山碎屑岩主要分布在钱圩、查山等上海西南部地区，主要为青白口纪晚期建造。

燕山晚期火山作用始于白垩纪初期，呈现出弱—强—弱的喷发特征，形成一套钙碱系列的安玄岩-安山岩-英安岩-流纹岩组合及相应的火山碎屑岩。

喜马拉雅期仅在局部地区有基性岩浆喷溢，形成了一套碱性系列的玄武岩。

燕山晚期陆相火山岩岩相发育较齐全，以溢流相和空落相为主，还可见沸溢相、火山碎屑流相、火山爆发崩塌相、火山喷发沉积相、火山颈和火山通道相、潜火山岩相等。喜马拉雅旋回火山岩则以溢流相为主。

第一节　火山岩类型及火山作用期次

一、分类命名原则

（一）火山岩

采用化学成分分类即 IUGS 提出的火山岩 TAS 分类方案（图 11-1）。另外，为了强调熔岩的结构特征，在命名时还可以加上前缀，如杏仁状玄武岩、气孔状玄武岩、球泡流纹岩等。中酸性岩中，当次要矿物含量超过 5%时参与定名，如辉石安山岩、石英安山英等。玄武岩结构分为粒玄岩、玄武岩、玻基玄武岩（火山玻璃在 50%~90%），且橄榄石含量在 5%以上及磁铁矿含量在 5%以上时均参加命名。

（二）火山碎屑岩

根据火山碎屑物的类型和含量来定名，并且当火山碎屑岩中的外来物质较少时，参照火山岩的化学分类方法确定其化学成分类型（流纹质、英安质、安山质、玄武质），同时参与命名，如流纹质熔接凝灰岩、安山质凝灰岩等。

根据火山碎屑物大小分为火山集块岩、火山角砾岩和凝灰岩 3 大类。火山集块岩：主要由大的集块（>50mm）构成，集块所占整个岩石的体积大于等于 50%。火山角砾岩：岩石中火山角砾（2~50mm）的含量大于等于 50%。凝灰岩：岩石主要由细小的火山物质（<2mm）组成，其含量大于等于 50%。上述 3 类岩石之间都有过渡类型，如集块角砾岩、凝灰角砾岩、角砾凝灰岩等，前缀代表碎屑物的含量在 25%~50%之间；含集块角砾岩、含角砾凝灰岩等，前缀代表碎屑物的含量小于 25%。

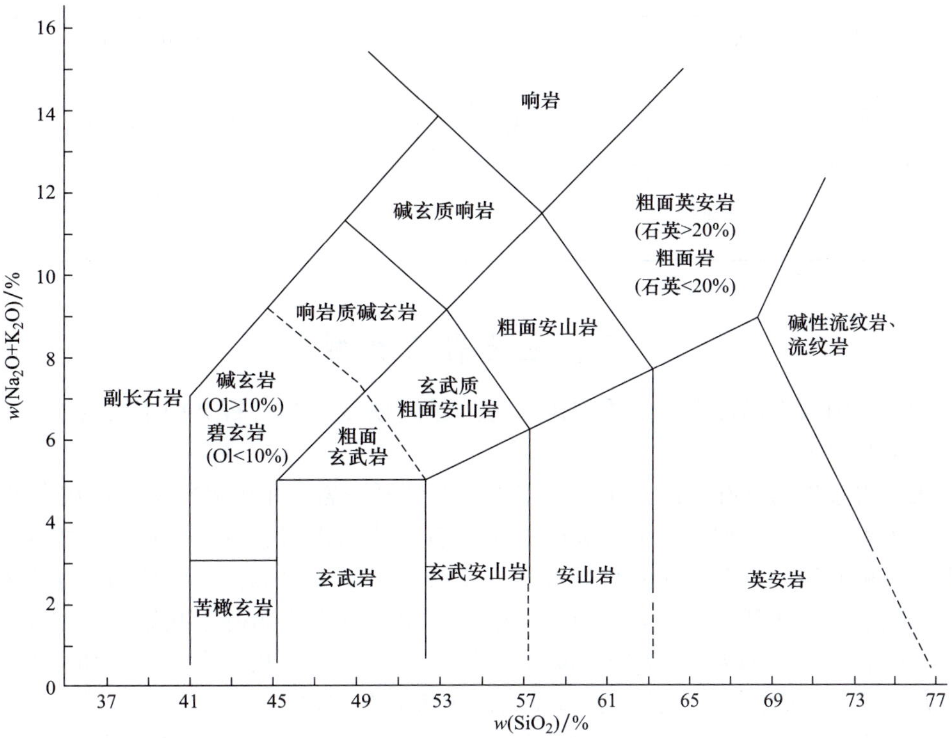

图 11-1 火山岩 TAS 分类

Fig. 11-1 TAS classification of the volcanic rocks

对于粒度类似的火山碎屑物，还可以根据碎屑物种类进一步细分，如凝灰岩可分为晶屑凝灰岩、玻屑凝灰岩、晶玻屑凝灰岩等，含量一般也以 25%、50%、75% 为界。

根据熔结程度分类，分为强熔结、熔结和弱熔结 3 类。如强熔结角砾岩、强熔结凝灰岩、熔结角砾岩、熔结凝灰岩、弱熔结角砾岩、弱熔结凝灰岩等。

此外，还包括向熔岩的过渡类型，有集块熔岩、角砾熔岩、凝灰熔岩等。向沉积岩的过渡类型，有沉火山角砾岩、沉凝灰岩等，以沉凝灰岩最常见。

二、火山作用期次

根据地层接触关系和锆石 U-Pb 年龄、K-Ar 年龄，本志将上海火山作用划分为晋宁早—中期、晋宁晚期、燕山晚期和喜马拉雅期 4 期（表 11-1），暂未发现与中侏罗世和晚白垩世侵入作用相对应的火山作用记录。

表 11-1 上海火山期次划分

Table 11-1 Division of volcanic periods in Shanghai

时代	构造期	地层	主要火山岩类	火山岩岩相		火山岩系列
新近纪	喜马拉雅期	白龙港玄武岩	碱性橄榄玄武岩	陆相	以溢流相为主	碱性系列
古近纪		渐新统	气孔状玄武岩			?
		北桥组	碳酸盐化玄武岩			碱性系列
早白垩世	燕山晚期	寿昌组	流纹英安岩-酸性凝灰岩-酸性火山碎屑岩		以溢流相和空落相为主	钙碱性系列
		黄尖组	玄武安山岩-安山岩-英安岩-流纹岩-酸性火山岩			
		劳村组	流纹质凝灰熔岩-酸性火山碎屑岩			

续表

时代	构造期	地层	主要火山岩类	火山岩岩相		火山岩系列
青白口纪	晋宁晚期	河上镇群	安山质熔结凝灰岩、角砾凝灰岩、凝灰岩	陆相	以溢流相和空落相为主	
中元古代	晋宁早—中期	金山群	中—酸性火山岩、基性火山岩		以溢流相为主	

三、主要岩石类型

晋宁早—中期火山岩系已经发生变质作用，经过原岩恢复，其岩石类型主要为玄武质凝灰岩、流纹岩等，可能有流纹斑岩。

晋宁晚期火山岩系揭露钻孔有限，主要有安山质熔结凝灰岩、角砾凝灰岩、凝灰岩等。

燕山晚期火山岩系岩石类型最复杂，火山岩主要有玄武安山岩、辉石安山岩、安山岩、角闪安山岩、石英安山岩、粗面岩、英安岩和流纹英安岩、流纹岩、霏细岩等，火山碎屑岩主要有向熔岩过渡类型（流纹凝灰熔岩、安山质凝灰熔岩、英安质凝灰熔岩、英安质角砾凝灰熔岩、安山质角砾凝灰熔岩等）、熔接凝灰岩类（晶屑熔结凝灰岩、玻屑熔结凝灰岩等）、凝灰岩类（流纹质凝灰岩、晶屑凝灰岩、含角砾凝灰岩等）、沉凝灰岩。此外，偶见火山角砾岩和集块岩。

喜马拉雅期火山作用主要形成玄武岩类，主要有伊丁石化橄榄玄武岩、气孔状玄武岩等。

第二节　前震旦纪火山岩和火山作用

火山岩主要分布于上海中部北东向构造带中，如周浦、张堰、钱圩等地。

中元古代为基性、中—酸性火山岩建造，变质程度达高绿片岩相。玄武质凝灰岩主要变质为斜长角闪岩、角闪斜长片麻岩、斜长角闪片麻岩、角闪片岩和角闪岩等角闪质岩类。主要为拉斑玄武岩系列，次为钙碱性玄武岩系列。不同类型的岩石地球化学特征不同（表 11-2，表 11-3）。角闪质类岩石 SiO_2 平均含量为 47.73%、Al_2O_3 为 14.15%、Na_2O+K_2O 为 4.35%且 $Na_2O>K_2O$。富 Cr、Ti、Ba、Cu，Ni、Co、Sr、Ba、V、Cu 含量与金山群中的辉绿岩脉相近。中—酸性火山岩主要变质为黑云斜长片麻岩、变粒岩，局部为浅粒岩。绿泥斜长片麻岩中 SiO_2 含量为 61.68%、Na_2O+K_2O 为 5.71%。稀土元素总量 $\sum REE=119\times10^{-6}$，其中 $\sum LREE=103.14\times10^{-6}$，$\sum HREE+Y=15.53\times10^{-6}$。$\delta Eu=0.84$，Eu 轻微亏损。La/Yb=22.46，其分布模式曲线为缓右倾型，属轻稀土元素富集型（图 11-2）。黑云斜长片麻岩中 SiO_2 含量为 71.04%、Al_2O_3 为 13.68%、Na_2O+K_2O 为 8.84%，且 $Na_2O<K_2O$。

表 11-2　中元古代变质基性、中—酸性火山岩主量元素含量

Table 11-2　Major elements in the Mesoproterozoic metamorphic basic and intermediate-acidic volcanic rocks

岩石	氧化物/%												
	SiO_2	TiO_2	Al_2O_3	Fe_2O_3	FeO	MnO	MgO	CaO	Na_2O	K_2O	P_2O_5	烧失量	合计
角闪质类岩石	43.73	1.77	14.15	4.73	8.51	0.21	6.82	7.11	2.53	1.82	0.28	2.84	
绿泥斜长片麻岩	61.68	0.41	12.82	1.06	3.65	0.07	3.80	3.34	3.88	1.83	0.22	13.55	106.31
黑云斜长片麻岩	71.04	0.26	13.68	0.72	1.51	0	1.05	1.45	3.74	5.10	0.288	1.65	94.958

表 11-3　变质基性、中—酸性火山岩微量（含稀土）元素含量

Table 11-3　Contents of trace elements (incl. REE) in the metamorphic basic and intermediate-acidic volcanic rocks

w（B）$/10^{-6}$

岩石	Sc	Li	Be	Co	Cu	Zn	Ga	Ge	Rb	Zr	Nb	Mo	Cd	Cs	Hf	Ta
角闪质类岩石				30	114											
绿泥斜长片麻岩（朱 zk4）	9.31	19.8	1.40	38.5	8.60	65.0	16.8	0.96	55.1	101	4.63	0.10	0.134	2.33	2.3	0.40
岩石	W	Tl	Pb	Th	U	Ba	Cr	Ni	Sr	V	As	Sb	Bi	Sn	Ag	
角闪质类岩石						1169	229	63	247	187						
绿泥斜长片麻岩（朱 zk4）	118	0.23	11.4	4.64	0.96	704	239	98.1	337	73.4	0.40	0.09	0.046	1.14	0.023	118
岩石	La	Ce	Pr	Nd	Sm	Eu	Gd	Tb	Dy	Ho	Er	Tm	Yb	Lu	Y	ΣREE
绿泥斜长片麻岩（朱 zk4）	26.6	46.8	5.51	19.9	3.54	0.88	2.89	0.37	1.74	0.32	0.91	0.13	0.85	0.13	8.19	119

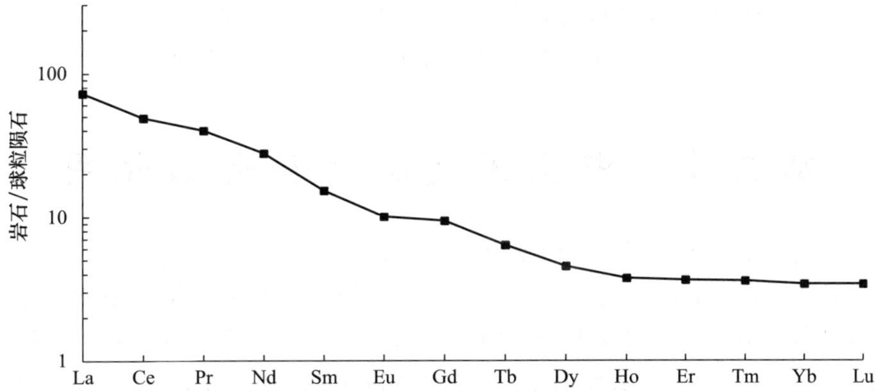

图 11-2　中元古代绿泥斜长片麻岩稀土元素配分模式图

Fig. 11-2　REE pattern of the Mesoproterozoic chlorite plagioclase gneiss

青白口纪火山作用形成中性火山岩建造，以熔结凝灰岩、角砾凝灰岩为主，属于高钾钙碱性系列。岩石 SiO_2 含量为 64.31%、Al_2O_3 为 14.55%、Na_2O+K_2O 为 3.64%，且 $Na_2O<K_2O$（表 11-4）。Ba、Pb、Zn、V、Cr、Ni、Co、Ga、Ag、Zr 含量偏高于地壳克拉克值（表 11-5）。稀土元素总量 $\Sigma REE=245\times10^{-6}$，其中 $\Sigma LREE=214.15\times10^{-6}$，$\Sigma HREE+Y=54.02\times10^{-6}$。$\delta Eu=0.63$，Eu 具有明显亏损。$La/Yb=9.36$，其分布模式曲线为右倾型，属轻稀土元素富集型（图 11-3）。

表 11-4　青白口纪火山碎屑岩主量元素含量

Table 11-4　Major elements in the Qingbaikouan volcanoclastic rocks　　w（B）/%

岩石	SiO_2	TiO_2	Al_2O_3	Fe_2O_3	FeO	MnO	MgO	CaO	Na_2O	K_2O	P_2O_5	烧失量	合计
含砾晶屑岩屑凝灰岩	64.31	0.82	14.55	0.95	5.35	0.20	2.95	5.04	0.44	3.20	0.15	3.15	101.11

表 11-5　青白口纪火山碎屑岩微量（含稀土）元素含量

Table 11-5　Contents of trace elements (incl. REE) in the Qingbaikouan volcanoclastic rocks

w（B）$/10^{-6}$

岩石	Sc	Li	Be	Co	Cu	Zn	Ga	Ge	Rb	Zr	Nb	Mo	Cd	Cs	Hf	Ta
含砾晶屑岩屑凝灰岩	16.2	33.4	2.89	27.4	29.0	193	21.5	2.43	143	211	17.2	0.68	0.110	11.5	5.1	1.20
	W	Tl	Pb	Th	U	Ba	Cr	Ni	Sr	V	As	Sb	Bi	Sn	Ag	
	50.6	0.80	17.0	13.3	2.12	715	198	114	123	118	18.2	0.61	0.342	1.95	0.137	50.6
	La	Ce	Pr	Nd	Sm	Eu	Gd	Tb	Dy	Ho	Er	Tm	Yb	Lu	Y	ΣREE
	46.2	87.7	10.5	38.0	7.37	1.46	6.82	1.05	5.81	1.16	3.48	0.54	3.54	0.52	31.1	245

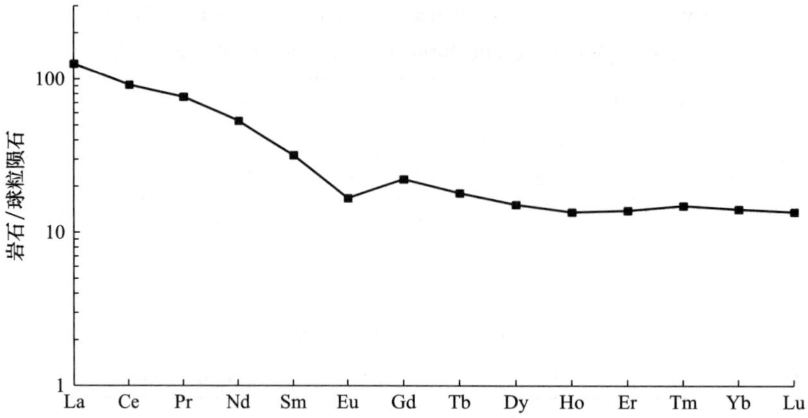

图 11-3 青白口纪含砾晶屑岩屑凝灰岩稀土元素配分模式图

Fig. 11-3 REE pattern of the Qingbaikouan pebbly crystallitic tuff

第三节 中生代火山岩和火山作用

在上海目前仅见白垩纪火山作用，早白垩世火山作用可划分为 3 个亚旋回，第一亚旋回发生于初期，即劳村组下段形成时期，有酸性凝灰质和熔岩喷发；第二亚旋回发生于中期，相当黄尖组形成时期，此时火山活动强烈；第三亚旋回发生于末期，相当于寿昌组上段形成时期，此时火山活动渐趋减弱。上白垩统在赤山组局部层位见夹有玄武岩，马桥 LQ11（311.45~311.60m）玄武岩年龄为 78.31±1.77Ma。

一、岩石类型及岩石学特征

岩石类型比较齐全。在空间分布上，南部以酸性火山碎屑岩分布较广，伴有中性和中酸性火山岩；北部除中性和中酸性火山岩外，尚有较多的酸性火山岩；中部火山岩露头区，则见有偏中性和酸性熔岩及其火山碎屑岩类。

（一）火山岩

1. 中基性、中性火山岩

主要见于黄尖组，分布于查山、佘山、金桥等地。

中基性火山岩常见有玄武安山岩、辉石安山岩等（表 11-6）。岩石呈深灰绿—灰黑色，斑状结构，基质具交织结构和间隐结构。斑晶为斜长石和辉石，基质主要由斜长石、普通辉石、磁铁矿和火山玻璃等组成，含少量磷灰石，常见绿帘石化、绿泥石化和碳酸盐化。斜长石 An=50~55，钠长石聚片双晶带较宽，部分斜长石已脱钙，变为中长石，同时形成绿帘石、绿泥石及碳酸盐矿物，粒径为 0.35~1.23mm，大者达 3.5~5.25mm；辉石为普通辉石，淡绿色，少部分向角闪石变化，粒径多在 0.42mm 左右，大者达 0.7~1.12mm，部分辉石具暗化边。

表 11-6 基性—中性火山岩矿物含量和结构构造特征

Table 11-6 Mineral contents, texture and structure characteristics of the basic-intermediate volcanic rocks

岩石	斑晶/%			基质		
	斜长石	辉石	角闪石	含量/%	结构构造	成分/%
安山岩	5~10			90	玻晶交织结构，部分具气孔构造	斜长石（65）、角闪石或阳起石（10）、磁铁矿（4）、火山玻璃（5）、磷灰石（少量）、绿帘石、绿泥石、碳酸盐矿物等

续表

岩石	斑晶/%			基质		
	斜长石	辉石	角闪石	含量/%	结构构造	成分/%
安山岩	5~10		0~5	85	玻晶交织结构	斜长石（70）、角闪石（5）、磁铁矿（2）、磷灰石（少量）、绿帘石、绿泥石、碳酸盐矿物等
角闪安山岩	15~20		6~10	75	交织结构	斜长石（50~55）、角闪石（10）、榍石（3）、磷灰石（少量）、绿帘石、绿泥石、碳酸盐矿物、石英等
辉石安山岩	15~20	5~10		75	玻晶交织结构、交织结构	斜长石（45~50）、辉石（15）、角闪石（0~3）、磁铁矿（5）、火山玻璃（0~5）、磷灰石（少量）、绿帘石、绿泥石、碳酸盐矿物等
玄武安山岩	10	0~5		85	交织结构、气孔构造	斜长石（72）、磁铁矿+白钛石（5）、阳起石（3）、次生石英（5）
玄武岩	10	8		82	微晶交织结构	斜长石（68）、辉石（2）、钛磁铁矿+白钛石（3）、绿帘石（7）、阳起石（1）、石英（<1）

中性火山岩以安山岩和角闪安山岩为主，并有少量辉石安山岩和石英安山岩，与中基性火山熔岩呈渐变关系。南部火山岩露头区见有粗面岩。

安山岩呈暗灰、暗灰绿和褐灰色，斑状结构，局部具气孔构造，基质具交织结构与玻晶交织结构。斑晶含量多在15%以下，少数超过20%，主要为中长石和更长石，次为辉石、角闪石。斜长石多具环带构造及钠长石聚片双晶，An=29~45，多为正环带，核心部位往往有蚀变矿物，如绢云母、绿泥石、方解石、黏土矿物等，外环呈净边状。角闪石呈绿色，为普通角闪石，往往蚀变为阳起石、绿帘石、绿泥石、碳酸盐矿物、磁铁矿等，部分具暗化边，粒径在0.7~0.88mm，大者达1.93~2.98mm。基质主要为中—更长石微晶，次为普通辉石、磁铁矿。普遍有绿帘石化、碳酸盐化、绿泥石化、阳起石化及少量的硅化、绢云母化现象。

粗面岩呈褐紫色与紫灰色，斑状结构。基质具霏细结构、包含霏细结构、球粒结构及粗面结构。斑晶由钾长石、酸性斜长石及少量黑云母以及微量磁铁矿、磷灰石、锆石等组成，基质由长英质组成，由不同结构与色泽相间成带，构成不明显的流动构造。常遭受不同程度的绢云母化、黏土化及碳酸盐化。

2. 中酸性、酸性火山岩

主要见于黄尖组和寿昌组上段，分布于庄行、天马山、佘山和真如等地。

中酸性火山岩常见英安岩、流纹英安岩等（表11-7）。岩石呈灰紫、暗灰和灰绿等色，少斑或斑状结构，基质具包含霏细结构、包含微晶结构和粗面交织结构等，并偶见流纹构造。斑晶含量一般在20%以下，以酸性斜长石为主，次为碱性长石、角闪石和少量的石英、黑云母。斜长石An=23~28，具钠长石律双晶，双晶纹细密，碳酸盐化、泥化，粒径为0.5~0.7mm。碱性长石呈宽板状与粒状，具格子双晶与卡式双晶，表面黏土化，粒径为0.53~0.7mm，大者达1.47~2.28mm。角闪石呈柱状，往往只保留晶体轮廓，其成分已变为氧化铁、方解石等，粒径为0.21~0.28mm，大者达0.7~0.87mm。石英含量少，往往具熔蚀边。黑云母常具暗化边，多被氧化铁、方解石等交代。基质矿物组分主要为碱性长石和石英，碱性长石呈针状、似毡状或交织状分布，略显定向排列，包含在石英小颗粒中，形成显微包含结构，其间有少量石英、斜长石、碱性长石、角闪石及磁铁矿等分布。岩石不同程度地遭受碳酸盐化、绿泥石化、磁铁矿和褐铁矿化、黏土化、硅化等。

表11-7 中酸性—酸性火山岩矿物含量和结构构造特征

Table 11-7 Mineral contents, texture and structure characteristics of the intermediate-acidic volcanic rocks

岩石	斑晶/%					基质		
	斜长石	钾长石	石英	角闪石	黑云母	含量/%	结构构造	成分/%
英安岩	5~20	0~5	少量或<1	0~5	少量	60~70	部分具流动构造，基质具交织结构、显微包含结构、隐晶结构	主要由长石、石英组成，少量的针状斜长石、角闪石，暗色矿物多蚀变为氧化铁，具暗化边

续表

岩石	斑晶/%					基质		
	斜长石	钾长石	石英	角闪石	黑云母	含量/%	结构构造	成分/%
流纹英安岩	2~3	3~6	少量或<1			90	部分具流动构造，基质具霏细结构、显微花岗结构、显微包含结构	主要由长石、石英组成，少量的针状斜长石、磁铁矿
英安流纹岩	8	3			2	87	具流动构造，霏细结构等	主要由长石、石英组成，少量的磷灰石、锆石、磁铁矿
流纹岩（寿昌组）	2	5	3~5	1~5		85	具流纹构造、球泡构造，基质具球粒结构、显微包含结构、霏细结构	由隐晶质长石、石英及长英质球粒组成
流纹岩（黄尖组）	<10	5	1~10	少量	少量	75~80	基质具球粒结构、显微包含结构、霏细结构、显微文象结构	由隐晶质长石、石英及长英质球粒组成

酸性岩以流纹岩为主，次为霏细岩。岩石呈紫、紫红、紫灰及灰白等色，常具流纹构造，呈斑状或少斑结构，基质具霏细结构、显微包含结构、显微文象结构、微粒结构、球粒结构等。斑晶含量多在20%以下，主要由酸性斜长石、碱性长石和石英组成，有时含少量的黑云母或角闪石。斜长石 An=23，钠长石律双晶发育，粒径为0.35~0.87mm，大者达2.5~5mm。钾长石具卡式双晶，多为正长石，表面多高岭土化，粒径为0.35~0.7mm。黑云母呈片状，具不明显的暗化边。石英多数位于球粒的中心，往往具熔蚀现象，呈港湾状或浑圆状。基质由隐晶质或微晶质碱性长石和石英组成，往往含少量黑云母、磁铁矿、磷灰石等。

酸性火山常以有、无流纹构造和斑状结构以及斑晶含量多少再细分为流纹岩、霏细岩等。根据特征性结构构造，又分球粒流纹岩、球泡流纹岩等。球泡构造见于寿昌组上段的流纹岩中，其球泡大小不一，在1.5mm左右。球泡由放射状长英质纤维组成，球泡中心多被次生石英、方解石、绿泥石等矿物充填。

（二）火山碎屑岩

岩石种类多，熔结疑灰岩分布颇广，且以酸性、中酸性及中性为主（其中以玻屑为主的凝灰岩常呈酸性，而以岩屑为主的火山碎屑岩则多为中性与中酸性）。火山碎屑粒度以凝灰级占绝对优势，角砾级次之，集块级极少。

1. 凝灰熔岩

指火山碎屑含量在10%~90%，以熔浆胶结的火山碎屑岩。按粒级可分为角砾熔岩和凝灰熔岩，按岩性分为酸性、中酸性和中性。本类岩石分布在劳村组下段、黄尖组及寿昌组上段。

（1）酸性凝灰熔岩

见于松隐、张堰劳村组中的酸性凝灰熔岩，呈灰白色、青灰色及紫红色，具晶屑岩屑结构。晶屑为反条纹长石、钠长石及石英碎屑；岩屑为石英岩及变质岩等，杂乱地分布在基质中。碎屑大小与形状不定，最大者达3mm，最小者仅0.1~0.2mm。石英熔蚀现象常见。熔岩物质主要由长英质及少量的黑云母、磁铁矿组成，并见有锆石及金红石。基质具霏细结构、球粒结构及微晶结构。已不同程度地重结晶。

（2）中酸性凝灰熔岩

见于奉庄行等地，岩石呈灰色、暗红色，斑状凝灰质结构。基质具微晶结构及霏细结构，呈斑杂状构造。斑晶和晶屑共占35%~40%。斜长石为更—中长石，含量为5%~25%，钾长石少量，暗色矿物占5%~8%，主要为黑云母及角闪石，多数已绿泥石化，石英占1%~2%，具熔蚀边，另含少量磷灰石及粒状磁铁矿，占3%~5%。岩屑为安山岩和霏细岩碎屑，占1%~2%。胶结物由长英质组成，含量约60%。

（3）中性凝灰熔岩

主要见于查山、庄行等地。岩石呈黑绿色，斑状凝灰质结构，斑杂状构造，斑晶占20%~25%，主要为更—中长石，黑云母、角闪石少量，晶屑含量为25%~30%，成分同上，胶结物为脱玻化隐晶质长石及石英，还有少量的磁铁矿等。

（4）中酸性—中性角砾凝灰熔岩

英安质角砾凝灰熔岩、安山质角砾凝灰熔岩与同类凝灰熔岩不同处在于：①岩屑粒度大于2mm的角砾含量多达10%~40%（若角砾含量为40%，则为角砾熔岩，但极少见到）；②随着角砾含量的增加，角砾成分将影响岩石性质与定名，如安山质角砾凝灰熔岩角砾多为安山岩碎块；而英安质角砾凝灰熔岩，其角砾成分则为安山岩、霏细岩、凝灰岩、英安岩碎块等。

2. 熔结凝灰岩类

在庄行、金桥、真如及火山岩露头区均有发现。火山碎屑含量在90%以上，以熔结方式成岩，岩石以中酸性为主，酸性次之。

岩石呈黑灰色、黑绿色、灰黑带紫红色斑点及紫褐色等，具假流纹构造。火山碎屑有晶屑、岩屑、塑变玻屑、浆屑及尘屑等，根据碎屑成分及含量可再细分为晶屑、玻屑、岩屑、晶屑玻屑、玻屑晶屑及多屑熔结凝灰岩。

常见晶屑成分有斜长石、钾长石、石英、角闪石、黑云母等；岩屑成分有安山玢岩、霏细岩、凝灰岩、凝灰熔岩等；塑变玻屑呈弧面棱角状、撕裂状；浆屑外形呈透镜状、条带状，脱玻后形成长石、石英等，浆屑的边缘结晶程度较低，向中间结晶程度逐渐增高，往往中间出现石英、长石斑晶，尘屑已脱玻化成隐晶质长英质。

如见于庄行的英安质熔结凝灰岩，岩石呈紫褐色，多具屑凝灰质结构，假流纹构造。晶屑含量约为23%，由中酸性斜长石（15%）、钾长石（2%左右）、石英（3%~5%）、角闪石及黑云母（1%~2%）组成。长石多棱角状，岩屑含量在5%~15%，有安山岩、英安岩、霏细岩及凝灰岩，岩屑呈棱角状与近椭圆状，塑变玻屑含量在15%~20%，多为撕裂状，经蚀变已黏土化，浆屑较少，胶结物约占40%，为长英质，已脱玻化。

3. 凝灰岩类

系指火山碎屑含量在90%以上，火山爆发物落地后以压结方式形成的正常火山碎屑岩类。按岩性可分为酸性、中酸性和中性3类。根据碎屑性质、粒级及含量划分，本区以玻屑凝灰岩、晶屑玻屑凝灰岩、玻屑晶屑凝灰岩及含角砾凝灰岩较多见。其中晶屑以石英及长石晶屑为主。

本区凝灰级火山碎屑岩分布较广。劳村组下段以酸性凝灰岩为主，黄尖组上段以中酸性凝灰岩为主，寿昌组中、上段则以酸性凝灰岩为主，次为中酸性凝灰岩。粗粒级的火山碎屑岩发现不多，以含角砾凝灰岩较多见，火山角砾岩偶见，火山集块岩则仅见于个别钻孔中。

凝灰岩呈灰黑色、墨绿色、灰黑带草绿黄色或呈紫褐色斑点，具火山碎屑结构，碎屑呈棱角状、次棱角状与贝壳状，分选性差。酸性凝灰岩晶屑以石英、碱性长石和酸性斜长石为主。中性凝灰岩石英明显减少，斜长石明显增多，斜长石牌号也有增加。酸性凝灰岩岩屑成分以霏细岩、石英岩及凝灰岩碎屑为主，中性凝灰岩则以安山质为主。酸性凝灰岩玻屑以弧面棱角状玻屑、尘屑为主，中酸性凝灰岩以撕裂状玻屑和尘屑为主，中性凝灰岩则以同成分的细小火山碎屑和铁质为主。

中酸性—中性凝灰岩已普遍绿泥石化、碳酸盐化、绿帘石化、黄铁矿化和褐铁矿化。

4. 沉凝灰岩类

为凝灰岩类向正常沉积岩类过渡的产物，火山碎屑物含量在50%~90%。岩石色杂，呈紫红、紫灰、灰褐、浅灰绿与黄绿灰色，具凝灰结构，略显层状构造，极少数呈沉火山角砾结构。晶屑有长石、石英及少量黑云母碎屑，岩屑有安山岩、英安岩、流纹岩及凝灰岩等，胶结物由细小的火山碎屑物和正常沉积物组成，其成分为长英质，有绢云母、铁质、泥质、硅质、碳酸盐矿物、高岭石、水云母等。

庄行见有含火山泥球的沉凝灰岩，火山泥球含量为5%~10%，呈椭球体与球体，直径不等，最大者约为1.5mm，主要由火山灰及其他细火山碎屑（晶屑和弧面棱角状玻屑）组成。

二、岩石化学特征

（一）劳村组

流纹质凝灰熔岩SiO_2平均含量为76.93%，Fe_2O_3+FeO为2.97%，Na_2O+K_2O为5.85%，且$Na_2O>K_2O$，K_2O/Na_2O值为0.59（表11-8）。里特曼指数平均值为1.01，呈钙碱性。碱度率为2.81。凝灰岩SiO_2含量为66.07%~71.45%，Fe_2O_3+FeO为1.03%~3.75%，Na_2O+K_2O为3.21%~4.16%，且$Na_2O<K_2O$，K_2O/Na_2O值为19.1~58.4。里特曼指数为0.36~0.75，呈钙碱性。碱度率为1.46~1.65。

凝灰岩微量元素中Rb、Ba、Th、U相对富集，尤其是Rb、Ba（表11-9）。稀土元素总量$\sum REE=198\times10^{-6}\sim217\times10^{-6}$，其中$\sum LREE=160.09\times10^{-6}\sim179.80\times10^{-6}$，$\sum HREE+Y=37.15\times10^{-6}\sim38.03\times10^{-6}$。$\delta Eu=0.59\sim0.69$，Eu具有明显亏损。La/Yb=11.63~13.52，其分布模式曲线为右倾型，属轻稀土元素富集型（表11-10；图11-4）。

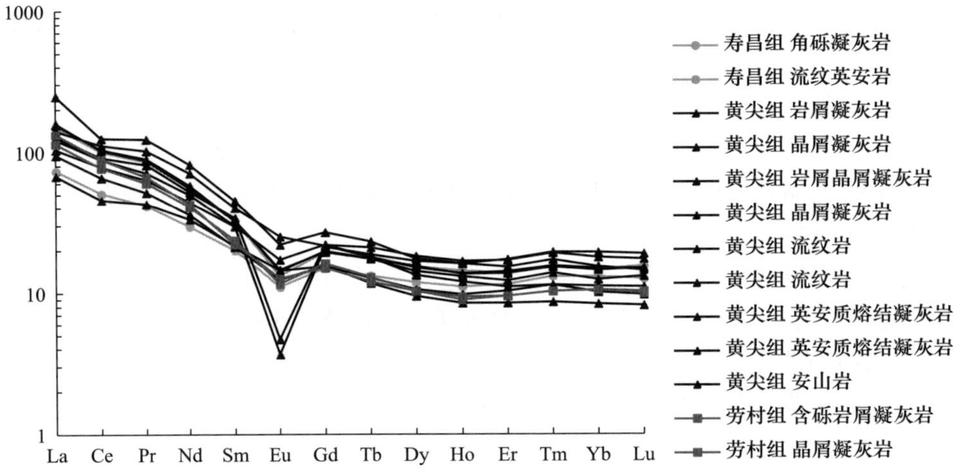

图11-4 中生代火山岩-火山碎屑岩稀土配分模式图

Fig. 11-4 REE pattern of the Mesozoic volcanic-volcanoclastic rocks

（二）黄尖组

1. 玄武岩

SiO_2平均含量为51.07%，Fe_2O_3+FeO为8.60%，Na_2O+K_2O为5.20%，且$Na_2O>K_2O$，K_2O/Na_2O值为0.45。里特曼指数平均值为3.22~3.48，属钙碱性—碱性。碱度率为1.51。

2. 玄武安山岩

SiO_2平均含量为54.57%，Fe_2O_3+FeO为7.82%，Na_2O+K_2O为5.90%，且$Na_2O>K_2O$，K_2O/Na_2O值为0.34。里特曼指数平均值为3.01，属钙碱性。碱度率为1.60。

3. 安山岩

SiO_2平均含量为60.82%，Fe_2O_3+FeO为5.65%，Na_2O+K_2O为6.76%，且$Na_2O>K_2O$，K_2O/Na_2O值为0.56。里特曼指数平均值为2.56，属钙碱性。碱度率为1.87。

4. 英安岩

偏酸性，呈流纹英安岩，以高硅、低铝、全碱偏高为特征。SiO_2平均含量为67.12%，Fe_2O_3+FeO为3.50%，Na_2O+K_2O为8.23%，且$Na_2O<K_2O$，K_2O/Na_2O值为1.51。里特曼指数平均值为2.81，属钙碱性。碱度率为3.22。

表 11-8 中生代火山岩-火山碎屑岩主量元素含量

Table 11-8 Major elements in the Mesozoic volcanic-volcanoclastic rocks

地层	位置	岩性	含量/%													里特曼指数	碱度率
			SiO_2	TiO_2	Al_2O_3	Fe_2O_3	FeO	MnO	MgO	CaO	Na_2O	K_2O	P_2O_5	烧失量	合计		
寿昌组	真如	流纹英安岩	68.56	0.24	14.83	1.25	1.57	0.04	0.53	1.37	0.20	6.93	0.06	3.97	99.55	1.99	2.57
		流纹英安岩	67.19	0.23	14.67	1.00	1.93	0.07	0.58	2.34	0.51	6.81	0.07	3.73	99.13	2.22	2.51
	12#	角砾凝灰岩	66.35	0.47	14.93	0.51	2.46	0.11	1.22	2.31	0.93	4.26	0.18	5.38	99.11	1.15	1.86
黄尖组	庄行	流纹岩	74.29	0.10	12.42	0.86	1.73	0.04	0.11	0.56	4.16	4.43	0.03	0.37	99.10	2.36	4.91
		流纹岩	73.83	0.07	12.44	0.94	2.00	0.04	0.16	1.00	4.93	3.43	0.04	0.80	99.68	2.27	4.29
		流纹岩	72.36	0.27	13.10	1.32	1.37	0.07	0.34	1.64	6.18	1.54	0.04	1.66	99.89	2.03	3.20
	松江钟家山	流纹岩	72.65	0.22	13.60	1.83	0.95	0.01	0.06	0.60	3.64	5.15	0.01	1.10	99.82	2.61	4.25
	松江罗山	碱流岩	71.60	0.19	14.38	0.57	2.34	0.06	0.38	0.53	6.81	3.57	0.02	1.00	101.39	3.77	5.58
	松江辰山	碱流岩	69.49	0.40	14.79	1.96	0.90	0.07	0.12	0.83	4.33	5.78	0.10	0.76	99.53	3.86	4.67
	松江凤凰山	碱流岩	68.16	0.45	14.06	1.80	2.51	0.03	0.50	1.29	3.76	6.53	0.07	0.81	99.97	4.21	5.07
	松江东佘山	碱流岩	68.02	0.44	14.58	0.82	2.30	0.05	0.06	2.03	3.73	6.20	0.09	1.34	99.66	3.94	3.97
	松江肖山	碱流岩	67.52	0.45	15.05	0.70	2.98	0.03	0.38	1.58	4.44	5.94	0.07	1.43	100.57	4.39	4.32
	松江秦皇山	碱流岩	67.50	0.48	15.21	0.47	2.50	0.07	0.53	1.68	4.66	5.46	0.08	1.21	99.85	4.18	3.99
	松江天马山	碱流岩	67.37	0.26	14.91	0.56	2.93	0.08	0.48	1.49	4.01	5.28	0.08	1.07	98.52	3.54	3.61
	松江天马山	碱流岩	67.20	0.38	15.48	1.27	1.11	0.06	1.62	1.13	4.65	5.10	0.16	3.47	101.62	3.93	3.84
	松江独山	碱流岩	67.11	0.42	15.13	1.14	2.15	0.09	0.53	2.12	4.91	4.70	0.11	0.84	99.25	3.83	3.52
	松江钟家山	碱流岩	66.42	0.50	15.24	1.30	1.58	0.05	0.59	2.20	3.56	5.73	0.05	2.52	99.74	3.69	3.28
	松江横山	碱流岩	66.24	0.47	14.68	0.38	3.83	0.08	0.33	1.83	3.85	5.75	0.08	2.14	99.66	3.97	3.78
	松江凤凰山	英安流纹岩	71.99	0.20	13.53	1.79	1.50	0.06	0.08	0.60	4.17	5.14	0.03	0.65	99.74	2.99	4.86
	松江罗山	英安流纹岩	70.28	0.19	13.06	0.52	2.77	0.13	0.30	1.35	2.65	6.70	0.05	2.00	100.00	3.20	4.70
	松江西佘山	流纹英安岩	67.89	0.40	14.48	1.73	1.40	0.04	0.10	2.34	3.22	5.17	0.09	2.56	99.42	2.83	2.99
	松江东佘山	流纹英安岩	67.03	0.32	14.31	1.30	2.60	0.08	0.06	2.34	3.51	5.14	0.11	1.89	98.69	3.11	3.16
	松江天马山	流纹英安岩	66.44	0.55	15.23	2.51	0.94	0.07	1.53	2.13	3.10	4.55	0.21	5.10	102.36	2.50	2.58

续表

地层	位置	岩性	含量/%													里特曼指数	碱度率
			SiO_2	TiO_2	Al_2O_3	Fe_2O_3	FeO	MnO	MgO	CaO	Na_2O	K_2O	P_2O_5	烧失量	合计		
黄尖组	松江淀山	粗面岩	61.55	0.56	14.58	1.73	6.05	0.09	1.17	2.31	3.89	4.95	0.15	1.49	98.52	4.21	3.20
	松江北竿山	粗面岩	65.22	0.60	15.06	1.42	1.63	0.07	1.11	2.23	3.67	5.76	0.13	2.57	99.47	4.00	3.40
	松江西佘山	粗面岩	64.11	0.25	14.08	1.18	2.55	0.08	0.79	2.73	3.93	6.09	0.07	2.50	98.36	4.76	3.95
	松江罗山	粗面岩	65.74	0.49	15.44	1.49	1.96	0.05	0.38	2.29	4.41	5.14	0.10	2.81	100.30	4.01	3.33
	金山查山	石英安山岩	63.25	0.76	16.62	3.62	1.58	0.02	1.56	3.34	4.84	2.64	0.02	1.05	99.30	2.76	2.20
	金山查山	安山岩	60.82	0.91	17.09	4.17	1.48	0.16	1.73	5.18	4.34	2.42	0.11	2.16	100.57	2.56	1.87
	金山查山	玄武安山岩	54.68	1.00	18.44	3.49	3.44	0.18	3.32	7.27	3.86	2.60	0.18	2.59	101.05	3.57	1.67
	金山查山	玄武安山岩	54.45	0.69	18.19	7.33	1.38	0.16	2.33	7.40	4.93	0.40	0.65	0.67	98.58	2.48	1.53
	金山查山	碱性玄武岩	51.08	1.16	17.03	5.61	2.89	0.17	4.91	8.97	3.54	1.76	0.15	3.40	100.67	3.48	1.51
	金山查山	碱性玄武岩	51.05	0.58	16.78	4.19	4.49	0.12	5.82	8.37	3.64	1.45	0.34	0.75	97.58	3.22	1.51
	莘庄323孔	粗安岩	52.40	1.03	14.49	6.79	2.38	0.12	6.53	6.73	3.10	2.35	0.32	4.16	100.40	3.16	1.69
	莘庄323孔	石英安粗岩	58.51	0.81	13.66	5.16	2.01	0.13	4.31	4.38	3.57	3.55	0.28	4.06	100.43	3.27	2.30
	莘庄323孔	石英安粗岩	61.98	0.57	15.24	2.04	3.15	0.10	2.93	2.85	3.40	4.27	0.22	3.51	100.26	3.10	2.47
	西佘山	粗面英安岩	67.17	0.56	15.13	1.85	1.29	0.06	0.93	2.14	4.22	5.50	0.13	0.99	99.97	3.91	3.57
	基23	岩屑凝灰岩	62.99	0.85	15.91	2.44	2.32	0.12	1.65	3.61	3.79	4.01	0.26	3.27	101.22	3.04	2.33
	J60	晶屑凝灰岩	62.82	0.60	16.02	4.13	0.58	0.06	1.76	2.48	3.48	3.77	0.20	9.01	104.91	2.65	2.29
	基33	岩屑晶屑凝灰岩	51.42	0.94	15.98	7.31	0.20	0.13	2.33	8.86	2.78	1.43	0.36	18.92	110.67	2.11	1.41
	庄zk2	晶屑凝灰岩	77.55	0.08	11.23	0.59	0.18	0.02	0.05	0.37	2.86	5.42	0.02	2.86	101.22	1.98	5.99
劳村组	G404	含砾岩屑凝灰岩	71.45	0.16	13.46	0.88	0.15	0.06	0.79	3.66	0.16	3.05	0.08	12.60	106.49	0.36	1.46
	LQ17	晶屑凝灰岩	66.07	0.57	14.62	2.92	0.82	0.04	1.59	2.43	0.07	4.09	0.16	13.24	106.62	0.75	1.65
	不明	流纹质凝灰熔岩（4件平均）	76.93	0.28	11.72	1.78	1.19	0.05	0.30	0.59	3.68	2.17	0.04	1.18	99.91	1.01	2.81

表 11-9　中生代火山岩-火山碎屑岩微量元素含量

Table 11-9　Contents of trace elements in the Mesozoic volcanic-volcanoclastic rocks

w (B) $/10^{-6}$

位置	岩性	Sc	Li	Be	Co	Cu	Zn	Ga	Ge	Rb	Zr	Nb	Mo	Cd	Cs	Hf	Ta	W	Tl	Pb	Th	U	Ba	Cr	Ni	Sr	V	As	Sb	Bi	Sn	Ag
12#	角砾凝灰岩	6.12	4.61	3.05	11.9	8.95	61.8	20.7	1.97	179	127	18.2	0.23	0.074	10.2	3.0	1.79	25.9	0.71	10.0	13.0	3.26	375	9.82	5.41	93.2	44.9	0.26	0.14	0.192	1.88	0.069
基 23	岩屑凝灰岩	12.1	16.1	2.22	21.4	12.3	85.3	20.0	1.56	119	269	15.5	1.83	0.137	3.64	6.6	1.16	109	0.56	23.7	13.7	3.36	1025	14.7	6.26	459	76.0	2.06	0.36	0.090	2.15	0.123
J60	晶屑凝灰岩	8.55	13.7	1.87	20.0	9.78	63.5	18.0	1.27	106	179	10.6	0.84	0.061	3.70	4.4	0.88	40.8	0.52	17.5	11.7	2.61	908	7.77	9.58	281	75.3	10.1	0.15	0.141	1.78	0.078
基 33	岩屑晶屑凝灰岩	16.3	21.8	1.06	33.8	37.0	47.3	18.6	1.08	36.5	156	6.73	0.18	0.098	0.95	4.0	0.50	35.0	0.11	5.40	4.99	1.28	356	48.1	17.1	367	110	2.86	0.14	0.018	1.92	0.101
庄 zk2	晶屑凝灰岩	3.04	7.76	1.29	64.5	11.2	67.3	14.6	1.09	125	117	15.1	0.22	0.277	1.26	2.9	1.30	369	0.55	92.2	15.1	4.61	233	2.16	2.91	40.7	3.66	0.39	0.13	0.547	2.69	0.357
G404	含砾岩屑凝灰岩	3.31	79.0	1.53	12.0	1.69	36.1	14.6	0.93	80.1	120	10.3	0.32	0.068	6.09	3.3	0.57	85.7	0.42	17.5	9.49	2.21	544	2.78	6.72	139	4.65	7.08	0.26	0.118	1.32	0.047
LQ17	晶屑凝灰岩	8.37	10.4	1.66	24.3	9.24	33.2	17.8	1.48	127	210	9.15	1.94	0.086	13.8	5.3	0.79	101	0.73	16.4	17.6	6.72	262	9.81	8.83	46.3	66.6	6.20	0.42	1.555	1.74	0.063
庄行	流纹岩（4件）				3	113	63	75			145		5							100			300	195	10	143	16					1
	流纹岩（25件）				1	53	32	44			143		5							44			944	222	10	120	13					1
	安山质凝灰熔岩（8件）				7	55	49	43			34		5							52			1850	80	10	400	49					1
	安山英安质凝灰熔岩（16件）				13	297	64	81			28		5							135			1431	89	13	644	138					1
	英安质凝灰熔岩（12件）				20	66	51	61			72		5							38			2590	123	12	683	153					1
	英安质凝灰熔岩（4件）				6	38	38	53			93		14							30			875	110	11	168	143					1
	英安质凝灰熔岩（18件）				9	73	69	54			38		5							35			1272	80	10	511	75					1

续表

位置	岩性	Sc	Li	Be	Co	Cu	Zn	Ga	Ge	Rb	Zr	Nb	Mo	Cd	Cs	Hf	Ta	W	Tl	Pb	Th	U	Ba	Cr	Ni	Sr	V	As	Sb	Bi	Sn	Ag
庄行	英安质含角砾凝灰熔岩（28件）				13	136	94	71			56		5							158			1850	172	18	542	200					1
	英安质含角砾凝灰熔岩（32件）				6	125	38	54			100		5							30			794	160	13	200	125					1
	英安质熔结凝灰熔岩（3件）				10	57	43	47			80		5							30			700	130	10	140	163					1
	安山英安质含角砾熔结凝灰岩（3件）				3	53	37	43			80		5							30			1670	367	12	157	47					1
	英安质熔结凝灰熔岩（8件）				2	49	34	51			243		5							40			2400	115	10	174	44					1
	流纹英安质含角砾熔结凝灰岩（4件）				3	55	30	48			160		5							33			2700	103	11	188	36					1
	英安质含角砾凝灰岩（5件）				5	62	44	52			158		5							42			840	274	10	164	106					1
	英安质含角砾凝灰岩（7件）				2	37	37	44			309		5							30			1814	171	10	156	41					1
	英安质含角砾凝灰岩（10件）				6	64	84	45			100		5							31			860	140	11	208	122					1
	英安质含角砾凝灰岩（26件）				11	90	52	60			57		5							30			1742	201	13	619	120					1

续表

位置	岩性	Sc	Li	Be	Co	Cu	Zn	Ga	Ge	Rb	Zr	Nb	Mo	Cd	Cs	Hf	Ta	W	Tl	Pb	Th	U	Ba	Cr	Ni	Sr	V	As	Sb	Bi	Sn	Ag
朱行	流纹质熔结凝灰岩（42件）				10	283	48	31			300		6							39			1433	390	12	185	19					1
	流纹质凝灰岩（11件）				10	185	50	32			327		5							32			1127	144	10	183	22					1

表 11-10　中生代火山碎屑岩稀土元素含量

Table 11-10　Contents of REE in the Mesozoic volcanoclastic rocks

w（B）/10^{-6}

地层	位置	岩性	La	Ce	Pr	Nd	Sm	Eu	Gd	Tb	Dy	Ho	Er	Tm	Yb	Lu	Y	ΣREE
寿昌组	12#	角砾凝灰岩	27.1	48.9	5.71	21.1	4.64	0.96	4.59	0.77	4.52	0.94	2.87	0.47	3.23	0.48	26.3	152
	真如	流纹英安岩	55.19	96.26	11.70	39.52	7.47	1.15	5.95	1.05	5.86	1.22	3.45	0.56	3.63	0.59	31.31	264.91
黄尖组	基 23	岩屑凝灰岩	44.8	84.7	9.77	35.5	6.92	1.50	6.74	1.04	5.76	1.17	3.54	0.56	3.70	0.57	31.7	238
	J60	晶屑凝灰岩	34.9	63.0	7.09	25.6	4.91	1.08	4.73	0.74	4.04	0.83	2.57	0.41	2.77	0.42	22.4	175
	基 33	岩屑晶屑凝灰岩	24.9	43.8	5.88	23.7	5.00	1.25	4.74	0.68	3.62	0.72	2.11	0.31	2.07	0.31	18.3	137
	庄 zk2	晶屑凝灰岩	38.3	74.9	8.69	31.4	6.86	0.32	6.50	1.09	6.41	1.37	4.30	0.70	4.81	0.72	37.7	224
	庄行	流纹岩	57.28	99.79	12.25	41.24	7.53	1.24	6.06	1.01	5.44	1.11	3.04	0.50	3.08	0.50	27.99	268.06
	庄行	流纹岩	46.90	87.41	11.15	37.37	7.88	0.41	6.72	1.22	6.97	1.43	4.24	0.69	4.42	0.66	39.78	257.25
	东佘山	英安质熔结凝灰岩	58.37	98.43	12.01	39.44	7.54	1.27	5.95	1.01	5.89	1.16	3.43	0.55	3.61	0.57	32.18	271.41
	天马山	英安质熔结凝灰岩	91.01	119.41	16.86	58.05	10.40	1.92	8.24	1.34	6.68	1.37	3.81	0.62	3.80	0.55	38.02	362.08
	查山	安山岩	52.01	106.65	14.03	50.52	9.31	2.19	6.61	1.04	5.10	1.02	2.75	0.41	2.51	0.37	25.85	280.37
劳村组	G404	含砾岩屑凝灰岩	49.2	84.8	9.17	30.4	5.13	1.10	4.62	0.70	3.93	0.76	2.37	0.37	2.61	0.39	21.4	217
	LQ17	晶屑凝灰岩	42.0	74.2	8.30	29.1	5.48	1.01	4.95	0.75	3.93	0.80	2.36	0.37	2.59	0.38	21.9	198

5. 粗面岩

SiO_2 平均含量为 64.16%，Fe_2O_3+FeO 为 4.51%，Na_2O+K_2O 为 9.47%，且 Na_2O<K_2O，K_2O/Na_2O 值为 1.38。里特曼指数平均值为 4.24，属碱性。碱度率为 3.46。

6. 碱流岩

碱流岩以低硅、低钠、高钾为特征。SiO_2 平均含量为 67.88%，Fe_2O_3+FeO 为 3.28%，Na_2O+K_2O 为 9.89%，且 Na_2O<K_2O，K_2O/Na_2O 值为 1.23。里特曼指数平均值为 3.93，属碱性。碱度率为 4.05。

7. 流纹岩

流纹岩以高钠、低钾为特征。SiO_2 平均含量为 73.28%，Fe_2O_3+FeO 为 2.75%，Na_2O+K_2O 为 8.37%，且 Na_2O>K_2O，K_2O/Na_2O 值为 0.77。里特曼指数平均值为 2.31，属碱性。碱度率为 4.06。

Ag、Mo、Ga、Cr、Pb、Cu、Ba 等元素含量普遍高于克拉克值，其中 Cu、Pb、Ba 含量高和火山岩中钾含量较高有关；Ti、Cr 为铁族元素，这和火山岩中铁的含量高有关，且常和安山质角砾及岩屑含量多少相关。

稀土元素总量 $\sum REE=137\times10^{-6}\sim362.08\times10^{-6}$，其中 $\sum LREE=104.53\times10^{-6}\sim297.65\times10^{-6}$，$\sum HREE+Y=32.86\times10^{-6}\sim66.13\times10^{-6}$。$\delta Eu=0.15\sim0.85$，Eu 具有明显亏损。La/Yb=5.71~17.18，其分布模式曲线为右倾型，属轻稀土元素富集型。

（三）寿昌组

流纹英安岩 SiO_2 平均含量为 67.88%，Fe_2O_3+FeO 为 2.88%，Na_2O+K_2O 为 7.23%，且 Na_2O<K_2O，K_2O/Na_2O 值为 19.08。里特曼指数平均值为 2.10，属钙碱性。碱度率为 2.54。角砾凝灰岩 SiO_2 含量为 66.35%，Fe_2O_3+FeO 为 3.14%，Na_2O+K_2O 为 5.44%，且 Na_2O<K_2O，K_2O/Na_2O 值为 5.89。里特曼指数平均值为 1.15，属钙碱性。碱度率为 1.86。

凝灰岩微量元素中 Zn、Ba、V、Zr 相对富集。稀土元素总量 $\sum REE=264.91\times10^{-6}$，其中 $\sum LREE=108\times10^{-6}$，$\sum HREE+Y=44.2\times10^{-6}$。$\delta Eu=0.64$，Eu 具有明显亏损。La/Yb=6.01，其分布模式曲线为右倾型，属轻稀土元素富集型（见图 11-4）。

三、岩石系列

在硅钾图（图 11-5）上，火山岩和凝灰岩主要为钾玄岩系列，少量高钾钙碱性系列。从劳组到寿昌组钾含量似有增高的趋势。

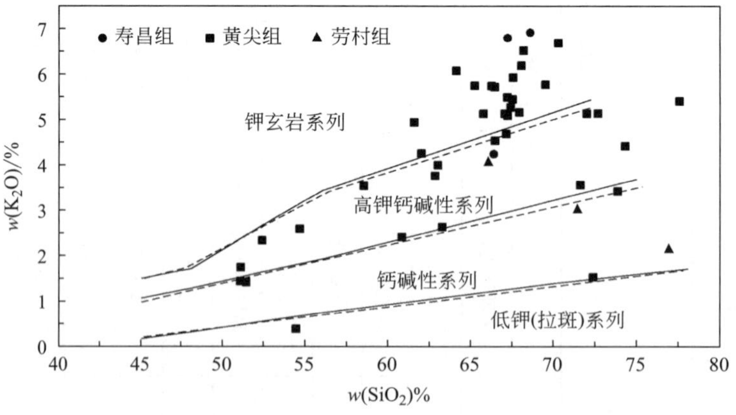

图 11-5 中生代火山岩的 SiO_2-K_2O 图解

Fig. 11-5 SiO_2-K_2O diagram of the Mesozoic volcanic rocks

在硅碱图（图 11-6）上，上述岩石既有落入钙碱性岩区内也有落入碱性岩区，其中寿昌组和劳村组主要呈钙碱性系列，黄尖组大部分落在碱性岩区，由于黄尖组 SiO_2 含量差异较大，使得硅碱图解应用有一定误差。在 AFM 图中，主要为钙碱性系列。综合判断中生代火山岩建造为高钾钙碱性-钾玄岩系列。

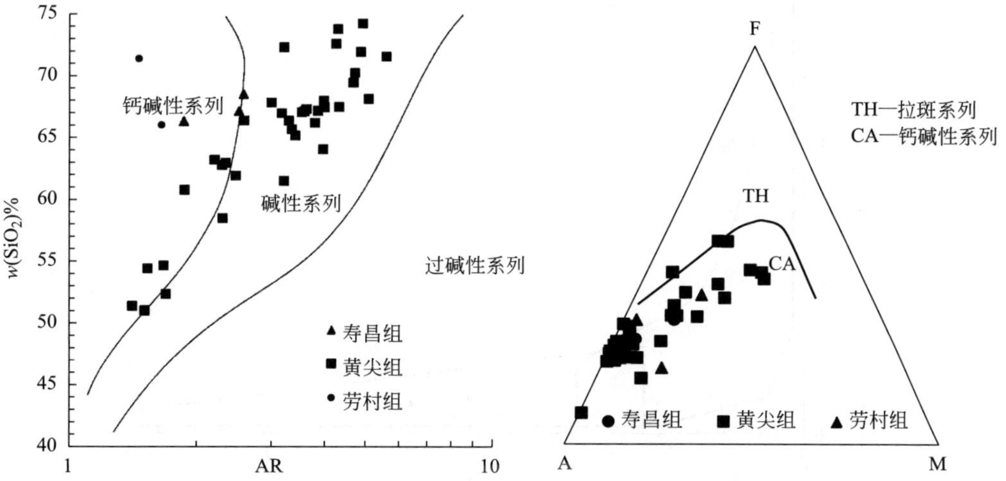

图 11-6　中生代火山岩的 SiO_2-AR（左）和 AFM（右）图解

Fig. 11-6　SiO_2-AR (left) and AFM (right) diagrams of the Mesozoic volcanic rocks

四、构造环境

安山岩、粗面岩、流纹岩和安山岩及个别凝灰熔岩的多个构造环境判别图解都显示，该时期火山作用建造都表现为造山带及岛弧地区火山岩特征（图 11-7，图 11-8），但偏离现代日本岛弧型火山岩。在戈蒂尼-里特曼图解中样品投点大多处于 B 区和 C 区交界处，个别落入 C 区。C 区岩石主要为 B 区火山作用派生成因，为碱性岩。从碱度率（见表 11-8）看，其变化区间在 1.41~5.99 间，绝大多数小于 4.0，与一般岛弧和活动大陆边缘火山岩的碱度率偏低、变化范围较小的特点相吻合。综合判断，该时期的构造环境是活动大陆边缘区，为后造山期挤压向伸展转换的阶段。

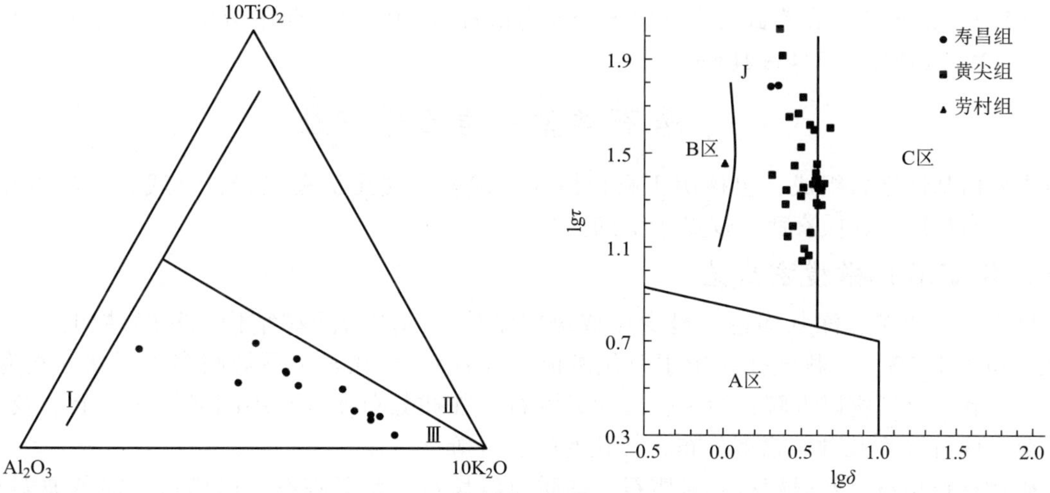

图 11-7　中生代火山岩构造环境判别图解

（左图为中基性火山岩；右图为全部火山岩）

Fig. 11-7　Tectonic environment discrimination diagrams of the Mesozoic volcanic rocks

(left: intermediate-basic volcanic rocks; right: all volcanic rocks)

Ⅰ—大洋玄武岩；Ⅱ—大陆玄武岩、安山岩；Ⅲ—岛弧、造山带玄武岩、安山岩。

A 区—非造山带地区火山岩；B 区—造山带及岛弧地区火山岩；C 区—A 区、B 区派生的碱性、富碱性岩；J—日本火山岩

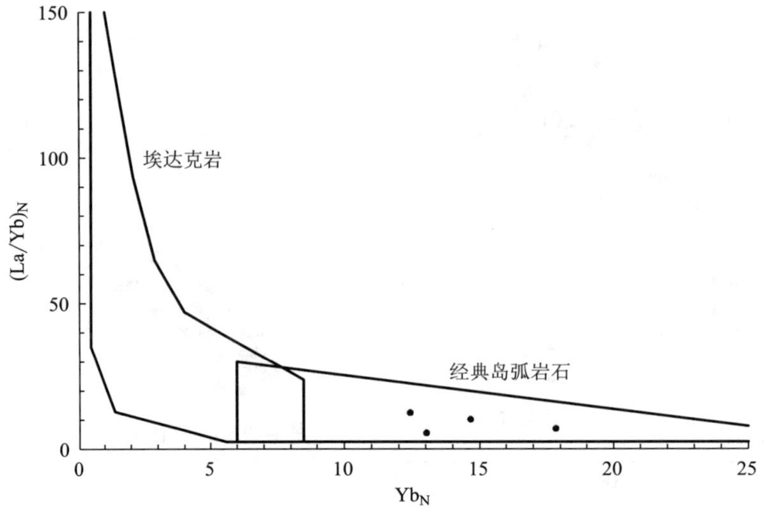

图 11-8　中生代火山岩构造环境 Yb-La/Yb 判别图解

Fig. 11-8　Yb-La/Yb discrimination diagram of the tectonic environment of the Mesozoic volcanic rocks

第四节　新生代火山岩和火山作用

新生代火山岩在上海东部比较发育，特别是沿海一带，从金桥、白龙港到芦潮港，似可连成一片。本区中部（马桥、北桥）和北部（崇明岛）也有零星分布。

从活动时间上可划分为两个阶段：古新世的碳酸盐化玄武岩主要分布在北桥、马桥等地，夹于北桥组中；始新世末期—中新世的玄武岩成片分布，在航磁图上显示强正异常。全岩 K-Ar 年龄有 38.57±2.05Ma~31.6±1.31Ma（崇明新海农场）、26.7Ma（南汇中学）、18.7Ma（白龙港）。从区域上嵊县组和盐城组中玄武岩的年龄看，不排除区内喜马拉雅期的活动时间可至中新世晚期（钱勤，1996；李相传，2014）。2008 年，在玄武岩中获得 5 个捕虏锆石的 $^{206}Pb/^{238}U$ 年龄分别为 132±2Ma、838±9Ma、1701±17Ma、1837±18Ma、2243±21Ma。

一、岩石类型及岩石学特征

岩石主要由基性熔岩组成，包括伊丁石化橄榄玄武岩、气孔状玄武岩、玄武岩、铁质玄武岩等。其中玄武岩分布最广，铁质玄武岩仅见于个别钻孔中。

（一）伊丁石化橄榄玄武岩

岩石呈深灰、灰黑、深灰褐色，呈少斑或斑状结构，基质为间粒结构、间隐结构，具气孔状构造（气孔含量大于 5%者，称气孔状伊丁石化橄榄玄武岩），气孔多呈不规则多边形状，次为椭圆形，大小不一。气孔中充填物以方解石为主，次为绿泥石（往往分布在气孔中间或气孔壁上）及沸石、透闪石、石英、燧石（后几种矿物多分布在气孔的中心）等。

斑晶矿物有斜长石、普通辉石、橄榄石；基质由斜长石、普通辉石、橄榄石、褐色玻璃及副矿物等组成。斑晶中斜长石含量为 0~5%，为中—拉长石，呈宽板状，往往具环带构造，多为正环带；辉石含量为 1%~10%，一般为 4%，为普通辉石及含钛普通辉石，呈柱状、粒状；橄榄石含量为 4%~10%，多为柱状，两端呈锐角或呈近似菱形，裂隙粗而发育，多已蚀变，大部分只保留橄榄石的晶形，蚀变往往沿边缘开始，蚀变矿物为伊丁石、绿高岭石、赤铁矿、绿泥石、蛇纹石及碳酸盐矿物等。基质总含量为 80%~90%，其中斜长石为 55%左右，为中长石，呈柱状；普通辉石为 15%~18%，呈微粒状、短柱状，橄榄石为少量~6%，呈微粒状；褐色玻璃为 0~15%，往往和铁质混杂出现。副矿物有磁铁矿及赤铁矿。

（二）气孔状玄武岩

岩石呈深灰、灰黑及黑褐色，少斑或斑状结构，基质具间隐-间粒结构，普遍发育气孔构造，最多可达30%，气孔大者可达1.2～3.7mm，形状有不规则的多边形和规则的椭圆形，气孔中的充填物以方解石为主，次为水白云母、硅质及褐铁矿。气孔状玄武岩与伊丁石化橄榄玄武岩的矿物成分相似，其中橄榄石含量为0～5%，气孔含量大于5%。

（三）玄武岩

橄榄石含量为0～4%，气孔含量小于5%，斑晶含量少，多数在2%～4%，斜长石和辉石含量增加。

（四）铁质玄武岩

铁质玄武岩的主要特征为含铁量高，磁铁矿与赤铁矿含量可达15%，呈少斑结构，基质具间隐结构。斑晶含量约6%，为橄榄石、普通辉石及斜长石；基质由55%的条状中长石，10%的微粒辉石、15%的磁铁矿、赤铁矿以及10%的褐色玻璃组成。

二、岩石化学特征

岩石具有高硅、低铁和高碱特征，SiO_2 含量为44.98%～54.74%，Fe_2O_3+FeO为6.36%～12.42%，Na_2O+K_2O 为3.45%～6.09%，且 $Na_2O>K_2O$，K_2O/Na_2O 值为0.24～0.76（表11-11）。里特曼指数平均值为1.53～11.19，岩石多属钙碱性，少量过碱性。碱度率为1.36～1.93。

玄武岩中铁族元素含量普遍低于维氏基性岩的平均值；Zn、Nb、La及Ce略高；Ba、Sr的丰度值分别为维氏平均值的1倍或1/2（表11-12）。

古新统玄武岩稀土总量 $\sum REE=95.92\times10^{-6}$，其中 $\sum LREE=68.99\times10^{-6}$，$\sum HREE+Y=26.93\times10^{-6}$。渐新世和中新世之交的玄武岩稀土总量 $\sum REE=134.05\times10^{-6}$，其中 $\sum LREE=97.04\times10^{-6}$，$\sum HREE+Y=37.01\times10^{-6}$。中新世玄武岩稀土总量 $\sum REE=274.52\times10^{-6}\sim291\times10^{-6}$，其中 $\sum LREE=225.49\times10^{-6}\sim234.15\times10^{-6}$，$\sum HREE+Y=56.65\times10^{-6}$。轻稀土总量及重稀土总量均随地层自老至新增高（表11-13；图11-9）。δEu值均大于0.92～1.25，Eu呈轻微正异常。古新统最大为1.25，指示玄武岩岩浆主要为幔源型。稀土元素配分曲线自老至新都表现为轻稀土富集型，互相平行，位置比较接近，其分布模式呈缓右倾。

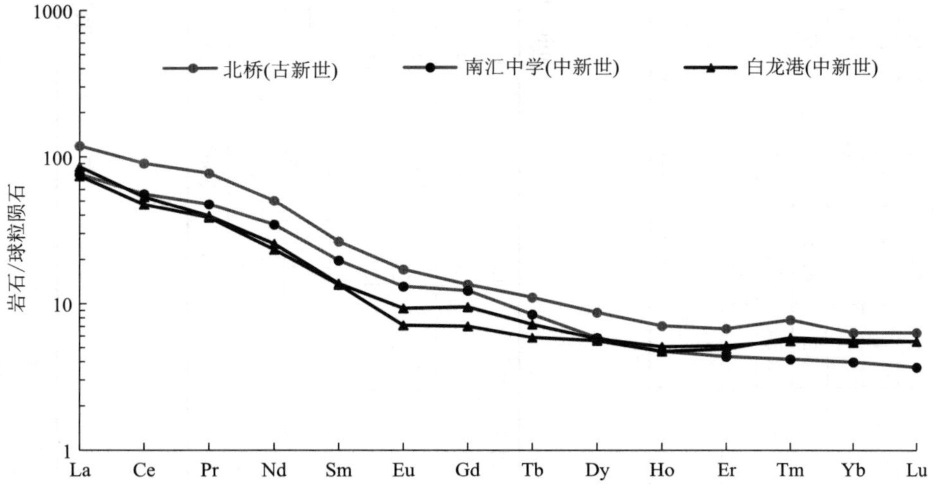

图11-9 新生代火山岩稀土元素配分模式图

Fig. 11-9 REE pattern of the Cenozoic volcanic rocks

表 11-11　中新世火山岩主量元素含量及特征值

Table 11-11　Major elements contents and eigenvalues of Miocene vocanic rocks

岩石	样品数/件	氧化物/%													σ	AR
		SiO_2	TiO_2	Al_2O_3	Fe_2O_3	FeO	MnO	MgO	CaO	Na_2O	K_2O	P_2O_5	烧失量	合计		
玄武岩	2	54.74	1.81	15.38	2.67	4.67	0.18	3.90	7.59	3.79	1.68	0.65	2.58	99.64	2.55	1.63
气孔状伊丁石化橄榄玄武岩	1	53.91	1.16	16.32	6.61	1.45	0.08	3.88	6.29	3.49	2.60	0.40	2.88	99.07	3.40	1.74
玄武岩	2	51.49	0.95	16.47	6.42	1.85	0.10	4.99	7.40	2.92	2.13	0.62	3.71	99.05	3.00	1.54
玄武岩	2	45.51	1.81	12.67	3.54	8.88	0.23	8.90	9.08	2.97	1.95	0.40	3.06	99.00	9.64	1.58
玄武岩	2	53.36	0.65	16.55	2.88	3.48	0.12	6.24	6.94	3.23	2.47	0.62	3.01	99.55	3.14	1.64
气孔状玄武岩	1	50.76	2.20	13.90	3.52	7.18	0.38	6.00	8.90	2.78	0.67	0.06	2.35	98.70	1.53	1.36
玄武岩	2	50.38	2.03	12.98	3.48	7.51	0.19	7.62	8.72	3.00	1.14	0.57	2.00	99.62	2.32	1.47
玄武岩	1	44.98	2.83	12.30	4.37	7.35	0.15	9.27	9.13	3.08	1.63	1.01	7.09	103.2	11.19	1.93

表 11-12　中新世火山岩微量元素含量

Table 11-12　Trace elements contents of Miocene vocanic rocks

$w(B)/10^{-6}$

孔号	样品数/件	V	Cr	Ni	Y	Zn	Nb	La	Ce	Ba	Rb	Sr	Cu	Pb	Mo	Co
白龙港	2	147	80	51	16	151	14	48	68	900	40	871	150	25	3	15
某孔	2	152	129	153	18	223	50	37	85	816	34	851	100	25	3	25
42 号钻孔	1	148	182	218	14	117	30	25	30	321	14	474	200	40	3	50
白 zk2	1	Sc	Li	Be	Co	Cu	Zn	Ga	Ge	Rb	Zr	Nb	Mo	Cd	Cs	Hf
		17.1	8.27	2.24	53.9	61.1	149	24.0	1.35	41.9	307	79.1	1.29	0.109	0.90	7.6
		W	Tl	Pb	Th	U	Ba	Cr	Ni	Sr	V	As	Sb	Bi	Sn	Ag
		24.0	0.02	3.46	7.13	1.59	579	268	211	921	175	0.45	0.07	0.012	2.24	0.052

表 11-13　新生代玄武岩稀土元素含量

Table 11-13　REE contents of Cenozoic vocanic rocks

$w(B)/10^{-6}$

时代	岩性(位置)	La	Ce	Pr	Nd	Sm	Eu	Gd	Tb	Dy	Ho	Er	Tm	Yb	Lu	Y	ΣREE
古新世	玄武岩(北桥)	14.86	28.70	3.92	16.22	3.88	1.41	3.53	0.57	3.42	0.59	1.55	0.21	1.20	0.16	15.70	95.92
中新世	玄武岩(南汇中学)	20.42	40.29	5.47	23.19	5.77	1.90	5.67	0.94	4.75	0.88	2.12	0.33	1.65	0.25	20.42	134.05
	玄武岩(白龙港)	55.71	94.68	12.34	48.31	11.03	3.42	8.72	1.29	6.49	1.07	2.52	0.32	1.66	0.21	26.75	274.52
		55.6	99.5	12.6	51.5	11.6	3.35	10.6	1.50	7.22	1.23	3.05	0.39	2.35	0.31	30.0	291

三、岩石系列

在硅钾图（图 11-10）上，岩石主要呈钙碱性-钾玄岩系列。硅碱图（图 11-11 左）上，上述岩石主要属钙碱性系列，个别至过碱性系列。在 AFM 图（图 11-11 右）中，主要为钙碱性系列。综合判断新生代火山岩建造为钙碱性-钾玄岩系列。

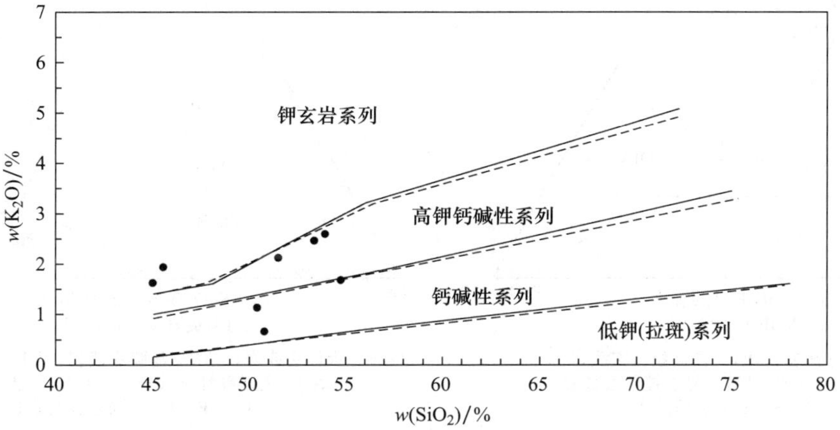

图 11-10　新生代火山岩 SiO_2-K_2O 图解

Fig. 11-10　SiO_2-K_2O diagram of the Cenozoic volcanic rocks

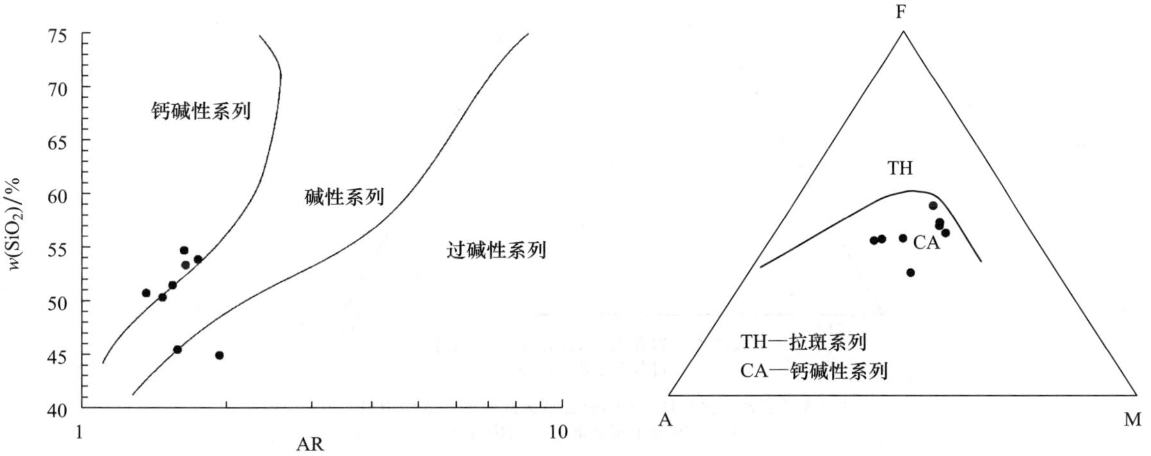

图 11-11　新生代火山岩的 SiO_2-AR（左）和 AFM（右）图解

Fig. 11-11　SiO_2-AR (left) and AFM (right) diagrams of the Cenozoic volcanic rocks

四、构造环境

新生代以来，受喜马拉雅运动影响，区内断陷盆地进一步发展，幔源型岩浆活动加剧。对该期岩浆活动环境的分析表明，该区总体上处于一种大陆裂谷型的构造环境（图 11-12）。沉积特征表明该时期为陆相环境，但在火山岩的判别图解上还显示了岛弧-洋岛火山岩组合，这体现出了一种大陆边缘和陆壳强烈拉张减薄的背景。

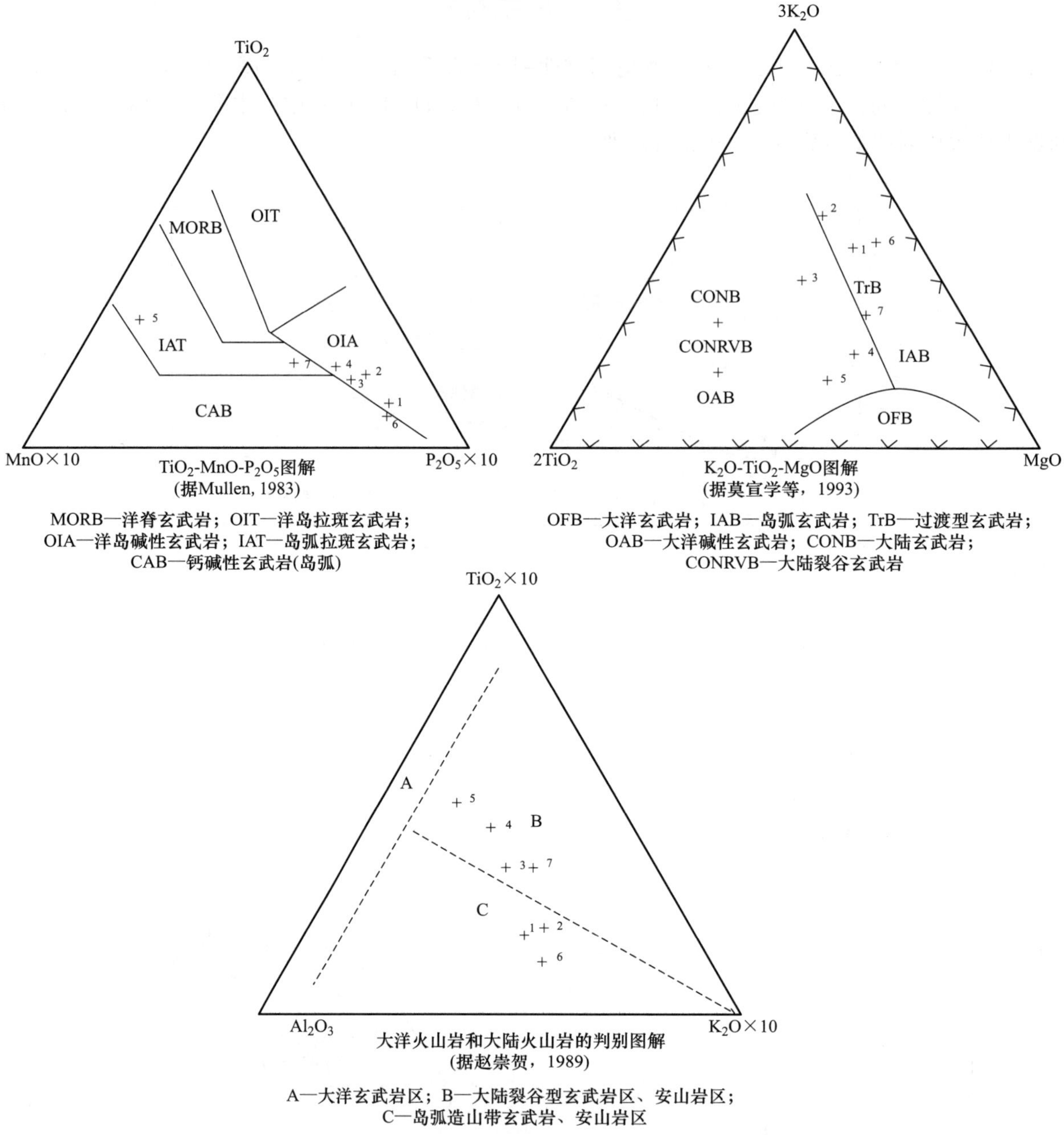

图 11-12　上海白龙港玄武岩构造环境判别

Fig. 11-12　Tectonic environment discrimination of the Bailonggang basalt in Shanghai

第五节　火山构造

早白垩世火山作用喷发类型兼有裂隙式及中心式两类，主要受北东向断裂的控制，此外，还受北西向断裂的影响。根据火山活动及其在空间上的分布与组合特征，上海火山构造可划分为下列 4 级：火山岩区、火山喷发带、火山盆地和隆起、火山机构。

火山岩区和火山喷发带划分与浙西北地区相似。以枫径—川沙北东向断裂为界，又可划分为两个亚带，断裂带南部为金山—惠南火山喷发亚带，断裂带北部为青浦—宝山火山喷发亚带。两个火山喷发亚带均沿北东向呈带状分布，两带的差异有：南部火山喷发亚带，早期为早白垩世劳村期酸性火山碎屑岩夹熔岩，嗣后有黄尖期中基性—中性—酸性火山岩浆的强烈喷发，以黄尖组的喷溢相为主，爆发相次之。北部火山喷发亚带，黄尖期中性—酸性火山岩浆强烈喷发，尔后为寿昌期中酸性—酸性火山熔岩及相应火山碎屑岩的堆积，喷溢相和爆发相几乎相当，也有喷发-沉积相。各亚带内主要的三级和四级火山构造特征如下：

一、松隐火山盆地

（一）分布形态及重磁场特征

在布格重力异常图上，松隐、亭林、叶榭等地为相对重力低缓地带，周边地区为相对重力高带，如北西及北侧有吕巷重力高、朱家庄重力高和车墩重力高。其外圈则为松江—新侨—北桥—闸港弧形重力低所组成的环带。盆地南缘因受张堰—南汇断裂带破坏和叠置，重力场的完整性已被破坏。盆地南北最长处长约15km，东西最宽处宽约10km，面积约为117.8km^2，整个形态呈北西-南东向展布。

盆地所在地是较为稳定的航磁负异常区，而盆地边缘地区，则在负异常背景上出现局部的正异常。经张堰、肖塘、齐贤等地钻探证实，均为中性—酸性岩株所引起。

（二）物质组成

盆地主要由劳村组组成。肖塘、叶榭、吴泾等地是以紫红、灰白色凝灰质砂岩、砂砾岩为主的沉积岩系。它们的共同特点是碎屑组分较杂，由硅质、凝灰质、铁质胶结。在以细碎屑组分为主的凝灰质砂岩、凝灰岩中，均见有清晰的层理，地层倾角比盆地中心大，一般为30°~40°，个别大于40°。局部钻孔下部层位见紫红与青灰色相间的流纹质凝灰熔岩，单层视厚度从数米至20余米不等，并与紫红及灰白色凝灰岩、凝灰质砂岩等相间出现。

除在肖塘附近见劳村组下段超履于前震旦系金山群之上以外，其他诸孔均未见底。劳村组下段在横向上的厚度变化尚不清楚。

晚劳村期火山活动渐趋减弱，以至衰亡。相应沉积物的分布范围大大减小。物质组成可以该地区常见的劳村组上段为代表，为紫红、灰白色相间的凝灰质砂岩（局部为细砂岩和泥岩）。张堰以灰白色长石石英砂岩和疑灰质砂岩为主，层理清晰，并见有波状层理和交错层理。和早劳村期的沉积物比较，颗粒较细，火山物质也趋减少。由松隐、亭林往张堰方向，沉积物的厚度增大，其中松隐仅60余米，张堰一带可大于900m；沉积物的颜色由深变浅，如松隐、亭林一带主要是紫红与灰白色相间，张堰一带则以灰白或浅灰、深灰色为主；分选性变好，如松隐、亭林等地以复陆屑、火山碎屑为主，张堰以较纯的长石石英砂岩为主，并渐变成正常碎屑沉积。钻孔所见地层倾角则由小变大，其中松隐一般5°~12°，而张堰一般均大于45°，甚至达60°~65°。此外，新场也见有劳村组上段，说明晚劳村期松隐火山盆地已转化成积水的湖盆，且由北而南水体渐次加深。上述岩性、岩相及厚度变化与张堰—南汇断裂带的重新活动有关，是由于断裂北西盘下陷引起沉积中心往南迁移的结果。

二、查山锥状火山

查山锥状火山位于钱圩—查山断裂南段的东北侧，地貌上为一火山岩残丘，北陡南缓，平面上呈等轴状，直径约为1km。航磁图上为一强正异常区，地面异常最大峰值为1700nT，呈北西-南东向展布。异常已经钻探验证，中心为一套偏基性的、富含角砾的安山岩类，主要有辉石安山岩、含角砾安山质熔岩、粗斑角闪安山岩与安山岩，偶见安山质凝灰岩夹层。

综合有关地质资料，有以下几个特点。

1）岩性较单一，同时具备浅成相和喷出相的特点：①呈少斑—斑状结构。上部基质很细，具玻晶质—微晶针状交织结构，下部基质粗，具半自形晶粒状-间粒间片结构。上述结构的变化呈渐变关系，无明显的界线可分。②斑晶由中—拉长石、角闪石、普通辉石组成。基质中暗色矿物较多，辉石含量为15%~20%，角闪石为5%~10%，黑云母为6%~15%。③角闪石具暗化边，基质中长石微晶较斑晶略呈酸性，其粒间常被玻璃质和脱玻后的显微弱光性物质所充填。④岩石常含杏仁体和不等量角砾。角砾大小不一，最大砾径大于10cm，小者仅0.1cm×0.2cm，成分为辉石安山岩或安山岩，与原岩（胶结物）成分一致，角砾无分选、无定向，呈棱角状至浑圆状。

2）在平面图上，外围出现安山集块岩，再向外变为安山岩和安山玢岩。经钻孔揭露，于中心部位见辉石安山岩、含角砾安山质熔岩、粗斑角闪安山岩，厚953.8m，尚未见底，为火山通道相。推测在垂直剖面上呈筒状。

3）具一般岩颈常见蚀变现象，如青磐岩化、绿帘石化、钠长石化、硅化、绢云母化等。

4）在火山通道附近施工的钻孔中，常见浅成—超浅成脉岩，如闪长玢岩、流纹斑岩等，为火山活动晚期略偏酸性的同源次火山岩或浅成侵入体。

5）地貌上为一锥状山丘，据钻孔揭露，最浅处基岩埋深仅32m（附近锥顶已裸露地表），而平距150m以外其埋深即变为163.50m，高差达131.50m。

三、金桥构造火山盆地

（一）分布范围

金桥构造火山盆地分布于上海市区以东的川沙境内，盆地东缘淹没于长江口水域之下。盆地中心位于川沙金桥附近，周边为断裂所围限，外围有中酸性岩株侵入和古老褶皱岩层出露，组成复杂的环带状火山构造盆地。

（二）地质背景及构造控制条件

金桥构造火山盆地受区域性大断裂枫径—川沙断裂带的控制。盆地北西侧上叠于由早古生代地层组成的印支期褶皱之上，南半部为前震旦系基底褶皱岩系分布区。其外围北西侧和南侧后期岩浆活动的特征也有所差异。

（三）重磁场特征

在布格重力异常图上，盆地表现为一个完整的、较为规则的圆形至椭圆形相对重力低。梯度变化平缓而稳定，由重力等值线所圈闭的盆地中心位于金桥附近。盆内局部地区存在重力正异常，盆地外围为重力高所包围。盆地内部有紊乱的航磁负异常，盆地西部及北部多为航磁负异常区，东缘白龙港为强正异常区。单个异常方向为北东向，整体呈北西向排列。

（四）物质组成

经钻孔揭露，盆地的内部主要由黄尖组中性火山岩组成，岩性为安山岩、安山玢岩，其次为安山质凝灰熔岩，部分地区钻孔中见英安岩和英安玢岩。局部岩性组合较为复杂，由辉长闪长岩、安山玢岩、安山质熔岩等组成。推测英安玢岩、石英闪长岩是构造火山盆地发育较后期的岩浆活动产物。

（五）断裂构造系统

金桥构造火山盆地断裂系统较为复杂，具有环状和放射状断裂。盆地边缘，由于差异升降运动，导致断凸和断凹的形成。

1. 高东断裂

高东断裂分布于高桥高东一带，略呈向北突出的东西向弧形展布，长约8km。沿断裂出现重力梯度密集带，断裂附近中性侵入岩发育，岩石破碎，蚀变强烈。

2. 梅陇—邓镇断裂

梅陇—邓镇断裂位于莘庄北梅陇、周浦和瓦屑、邓镇一带，呈近东西向略显向南突出的弧形延伸，据物探、地质资料推断为向北倾斜的正断层。断层线全长约为36km，向东北延入海域。

断裂东段北侧为重力低，南侧为重力高。沿断裂有多个呈串珠状排列的航磁异常，由中性—中偏基性侵入岩及前震旦系角闪质岩类所引起。断裂西段北侧为紊乱的航磁负异常区，其下为下白垩统沉积火山岩系，南侧为航磁正异常区，其下为前震旦系变质岩系及上震旦统白云岩、白云质灰岩。中、酸性脉岩穿插频繁，构造破碎，断层角砾岩、糜棱岩化岩石十分普遍，热液蚀变也颇强烈。

3. 淞南断裂

淞南断裂位于金桥构造火山盆地西北部的淞南—殷行一线，走向292°，长约12km，沿断裂有较明显的重力异常梯度带，且见有较多的闪长岩及浅成辉绿岩脉，它控制了吴淞花岗岩体南部边界。

4. 大场—江湾断裂

大场—江湾断裂位于大场—江湾，走向呈东西向，长约13km，推测为断面北倾的正断层。

大场—江湾重力高呈东西向，断裂位于重力异常梯度密集带略偏北的位置上，附近钻孔中均有构造破碎现象，北侧是寿昌组分布区，南侧为由下古生界组成的断凸。

5. 静安寺断裂

静安寺断裂西起北新泾，沿长风公园、静安寺和浦东塘桥一线分布，走向北西西。推测断面北倾，长约19km。北盘为厚度很大的下白垩统火山岩系，南盘为下古生界组成的断凸，东段切割市南花岗岩体。

断裂于北新泾重力高北缘通过，东段重力等值线出现向南东东方向的扭曲。附近常有构造破碎带以及后期侵入的脉岩和热液蚀变现象。在钻孔钻进过程中失水现象普遍，据史料记载，静安寺原有一只热水井，被称之“泛泡泡井”，可能与该断裂有关。

（六）与火山构造发育有关的环状构造迹象

1. 呈半环状分布的侵入岩

岩体围绕盆地外围西半部和南部呈环带状分布，大致可以枫径—川沙断裂为界，西半部以酸性侵入岩为主，南部以中性侵入岩为主。西半部的酸性侵入体有吴淞花岗岩体、市北花岗岩体和市南花岗岩体。岩石普遍具不等粒结构，矿物组成及岩石化学成分也基本一致。此外，靠近盆地一侧，脉岩十分发育，辉绿岩、花岗斑岩等沿环状断裂侵入，构成一个呈断续分布的环状脉岩带。

侵入岩环带的南半部为一些大小不等的岩株，如三林塘石英闪长岩体、赵家宅闪长岩体、坦直辉石闪长岩体和沈庄辉石闪长岩体等。此外，闪长玢岩、辉绿岩、正长斑岩、二长斑岩、细晶岩等浅成脉岩也颇多见。

2. 呈环带状分布的基底地层

在构造火山盆地西部，于侵入岩环带外围分布着下古生界岩层。北新泾—大场重力高大体反映了西部褶皱岩层环带的展布形态。由于静安寺断裂、大场—江湾断裂的差异升降运动，致使北新泾和大场—江湾呈近东西向的断凸隆起。褶皱岩环带的南半部主要由前震旦系变质岩系及上震旦统组成，与侵入岩环带重叠在一起构成断凸上隆带。断凸的展布方向与区域构造线的方向明显斜交，说明在成因上与构造火山盆地的发育有着密切的关系。

3. 当今基岩地貌的环带状特征

盆地外侧为断续分布的环状高地所围绕。以南半部为例，地貌上显示出3个从低到高的呈环带分布的平台：埋深大于320m为盆地内部平台，向外320~280m为二级平台高地，再向外240~200m为三级高地。说明燕山期火山作用及其后的块断差异升降运动是塑造本区当今基岩地貌十分重要的地质作用。

四、庄行构造火山盆地

（一）分布范围

位于上海南部金山朱行、漕泾及奉贤庄行、南桥一带。北缘叠覆于劳村期松隐火山盆地之上，西部是黄尖组安山岩分布区，南濒杭州湾，平面上呈椭圆形至卵形，略向东北方向拉长，长约18km，宽约13km，面积为191.2km^2。南部在漕泾-柘林之间，向海域开口。钻孔揭示盆地内堆积了厚247m以上的黄尖组中酸性火山熔岩和火山碎屑岩。根据地质、物探资料推断其周边有一环状断裂，大致可以圈定火山构造的范围。浅成脉岩沿环状断裂侵入，并在盆地的一侧分布。盆地外围发育有较为完好的古老褶皱岩层上隆带，呈半环状围绕盆地分布。

（二）重磁场特征

在布格重力异常图上，盆地显示一个十分规则的椭圆形相对重力低，它的东侧、北—北西侧和西侧都是相对重力高。相对重力高与重力低之间有一明显的重力梯度密集带，西北部及东南侧的最大梯度变化可达4mGal/km。盆地内部梯度变化平缓而稳定，中心部位是由重力极低值等值线所圈闭的重力低，呈南北方向拉长的椭圆状。

盆地内部属平静航磁正异常区，局部航磁正异常围绕盆地中心重力低值圈外围分布，它们与庄行、烟墩头等地的正磁异常的长轴延伸都显示极为规则的北东向。盆地东缘光明以北为航磁负异常区，以南基本为航磁正异常区，张堰—南汇断裂带南支即位于其交变带上。盆地西侧磁异常跳动比较紊乱。

（三）物质组成

盆地内部主要由黄尖组中酸性火山熔岩和火山碎屑岩组成，所有钻孔均未见底，最大厚度大于247.6m。岩石组合主要有含砾流纹英安岩、含角砾晶屑熔结凝灰岩、英安质晶屑熔结凝灰岩、英安质含角砾晶屑凝灰岩、熔结凝灰岩和凝灰熔岩等，具明显的假流纹构造。角砾大小不一，无分选，无定向，杂乱无章，多为棱角状。其组分复杂，大部分是周围火山岩的破碎产物，部分来自早期岩层。个别钻孔中见沉凝灰岩及凝灰质粉砂岩薄层，是火山喷发间隙的正常沉积物。

（四）外围环带构造

1. 浅成脉岩沿环状断裂贯入

经少量钻孔了解，沿环状断裂常有浅成脉岩贯入，主要有石英闪长玢岩、花岗闪长斑岩、花岗斑岩、闪长玢岩等。

2. 呈环状分布的基底地层

于盆地北侧、东侧有呈环带状分布的古老地层，东侧由上寒武统超峰群白云岩、白云质灰岩、角砾状白云岩组成，与区域构造线近乎正交，推测西陡东缓。钻孔揭示，环状断裂钱桥附近的基岩，在地貌上是一个陡峻的斜坡。

盆地北部的叶榭—肖塘一带，为一个北东向的重力高，其下见前震旦系金山群变质岩系，上覆劳村组厚度仅18.1m。烟墩头—朱行一带，金山群也沿环状断裂外侧分布，推测因火山盆地的塌陷，外围地区成为一个相对上隆且遭受侵蚀的环带。

五、天马山构造火山隆起

（一）分布范围

天马山构造火山隆起位于上海中部的青浦，松江、上海等境内，北以青浦—龙华断裂为界，南与松江—北桥断陷盆地相邻，东北终止于浦东杨思一带，西南延入浙江。区内呈北东60°方向狭长形延伸。它与金桥构造火山盆地组成了区内中部十分醒目的中酸性火山岩带。

（二）重磁场特征

天马山构造火山隆起主体为小昆山—佘山—泗泾北东向狭长的异常带所占有，北侧为重固重力高，南侧为松江重力低，航磁上为低缓的正异常区。

（三）物质组成

隆起北带主要由安山岩和安山玢岩组成。中心主体由中酸性火山岩及火山碎屑岩组成，在天马山、佘山、凤凰山等火山锥上均已露出地表，主要岩石有熔结凝灰岩、英安岩、流纹英安岩、流纹岩、含砾玻屑凝灰岩等，见塑变玻屑、浆屑和假流纹构造。各种浅成脉岩也很多见，主要有花岗斑岩、正长斑岩、闪长岩和安玄岩。天马山地表显示为一倾向北西的单斜构造，地层倾向310°~330°，倾角为25°~30°。

六、重固—顾村构造火山盆地

重固—顾村构造火山盆地位于昆山—嘉定断裂以南，青浦—龙华断裂以北的广大地区，是在黄尖期火山构造背景上继承发育起来的火山-沉积盆地，盆地内寿昌组分布广泛。寿昌期由于昆山—嘉定断裂的活动，使寿昌组由南至北大致有由老变新的趋势。寿昌组下段沿盆地的南缘分布，为内陆河湖—沼泽相沉积，以富含钙质、白云母片、煤线、炭质条带及基本不含火山物质为特征，是继黄尖期大规模火山喷发活动后的一个间隙期，寿昌组中段为含火山物质的内陆湖泊相沉积，分布最广。寿昌组上段分布局限，为酸性火山岩及火山碎屑岩。

此外，区内喜马拉雅期火山活动似受北东向枫泾—川沙断裂带、张堰—南汇断裂带和北西向灯塔—赵家宅断裂控制。在两组断裂交汇处，常形成火山通道相，如川沙白龙港三甲村，玄武岩厚逾252m，未见底。据钻孔揭示，本区沿海地带可能存在一玄武岩“岩被”。

第十二章　岩浆活动总体特征及演化规律

岩浆岩的形成、演化与大地构造环境和地壳演化有密切关系。不同时代岩浆岩具有明显的旋回性特点。中元古代末期发育中性岩和酸性火山岩、侵入岩。金山群斜长角闪岩接近零的 Nd 同位素初始比值表明其原岩源自弱亏损地幔或未分异地幔（周新华，1993）。朱泾花岗岩稀土元素含量超过 200×10^{-6}，铕明显负异常，轻、重稀土元素分馏作用明显，表明岩浆经历了明显分异作用或为古老地壳物质的部分熔融。河上镇群在区内仅揭露有中性火山碎屑岩，但在邻近浙西北地区同样发育基性、酸性火山岩和中酸性侵入岩。稀土元素含量较高，且具有轻微铕异常，可能为岩浆分异或幔源岩浆受浅部地壳混染所致。早白垩世，岩浆作用最强烈，形成大量基性、中性和酸性火山岩、侵入岩。晚白垩世，主要形成了基性、酸性侵入岩类。新生代以来，形成少量的基性岩脉和局部呈面状分布的基性火山岩，未见中酸性岩浆作用。

第一节　中生代岩浆岩成因和演化探讨

燕山中期发育有石英闪长岩和花岗岩，总体上都呈钙碱性系列。稀土元素配分曲线特征相似，都表现为轻微的铕负异常且轻、重稀土元素分馏明显（图 12-1）。轻稀土元素富集特征相似，重稀土元素富集特征差别明显，表明其岩浆源区具有一定的相似性。La/Yb 值在 19.5～28.4 之间，推测岩浆来源深度较深，且岩浆在结晶演化过程中发生了较低程度的地壳混染。

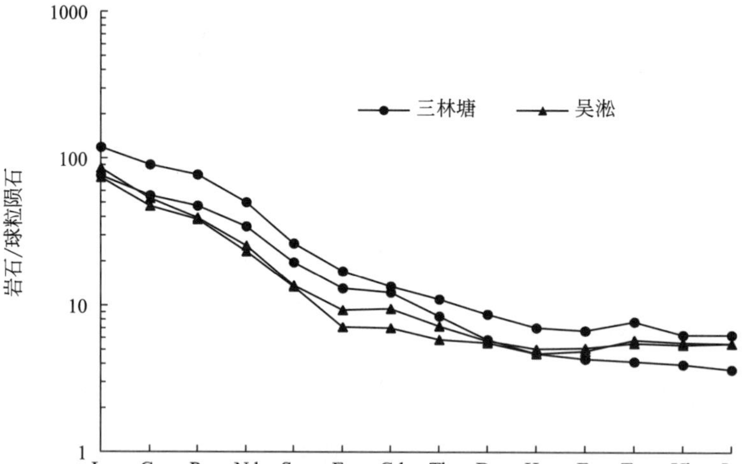

图 12-1　燕山中期岩体稀土元素配分模式图

Fig. 12-1　REE pattern of the Middle Yanshanian intrusive rocks

燕山晚期侵入岩以中酸性岩为主，少量基性岩，属钙碱性—碱性系列。稀土元素特征与燕山中期类似（图 12-2），个别岩体铕负异常明显。K_2O-Na_2O 等图解（图 12-3）显示，石英闪长岩和花岗岩成因有一定差异。三林塘和张堰岩体呈 I 型，吴淞岩体逐渐接近 S 型和 A 型，市北和崇明岩体则呈 A 型。微量元素 Ti、Zn、Y、Cr、Ca、Ni、V 及 Zr 含量由高逐渐降低，而 Cu、Pb（除酸性岩）、Ag、Sn、Be、Sc、Nb 和 Ga 相对稳定，与同类岩石维氏值比较，中基性岩石中的 Ba、Cu、Pb、Zn、Ag、Sn、V、Sr、Ga、Ta、Mo 含量偏高，中性岩中 Ba、Cu、Pb、Ag、Cr、Ga、Ta、Mo 等元素含量较高。而酸性岩则以 Cr 含量偏高为特征。

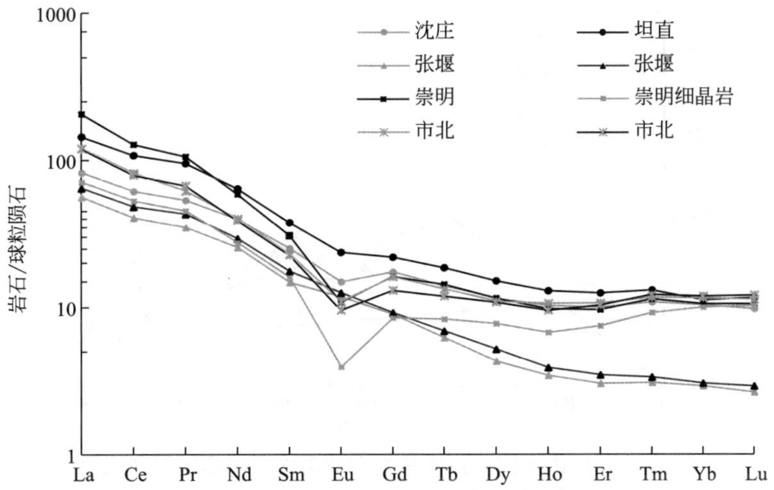

图 12-2　燕山晚期岩体稀土元素配分模式图

Fig. 12-2　REE pattern of the Late Yanshanian intrusive rocks

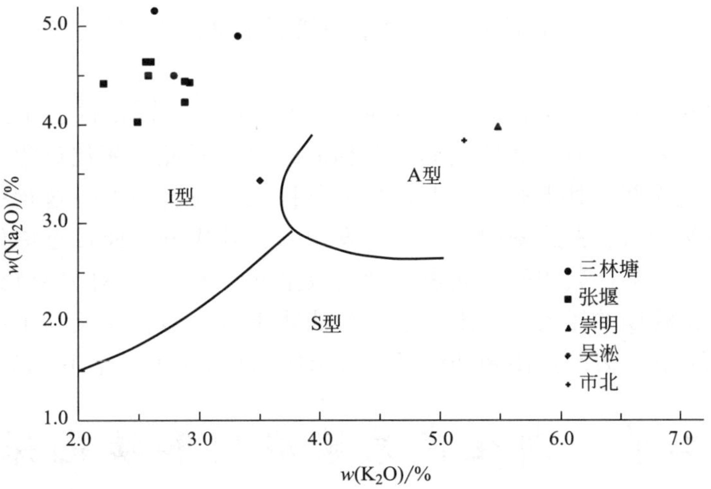

图 12-3　燕山中期岩体岩石成因类型 K_2O-Na_2O 图解

Fig. 12-3　K_2O-Na_2O diagram of the petrogenetic type of the Middle Yanshanian rock bodies

中酸性侵入岩主要形成于同造山—后造山阶段（图 12-4），稀土元素分布模式有 4 类，且均为轻稀土元素富集型，模式曲线为右倾型。第一类如坦直和沈庄辉石闪长岩，为一些中基性岩体，Eu 呈弱亏损，具幔源型稀土元素配分曲线特征。第二类如市北花岗岩、崇明花岗岩等，为一些酸性岩体。Eu 亏损中等至明显，属上地壳同熔型曲线特征。第三类如崇明花岗细晶岩，Eu 亏损强烈，表明岩浆演化已进入晚期残余岩浆的富碱阶段。第四类如张堰石英闪长岩体，具有高 Sr、低 Y 和 Yb 特征，铕无亏损，显示岩浆未经历明显的斜长石分异作用。

张堰的 Cu、Ag（Fe、Mo、Zn）矽卡岩型矿床，产生于岩体与围岩（主要为大理岩夹层）的接触带中，以外带为主。根据岩体所含微量元素统计，石英闪长岩体中的 Cu、Pb、Zn、Ag、Mo 含量高于同类岩石维氏值的 2~40 倍。铁铜矿矿体中的铜平均品位约为 1%，全铁平均品位在 40%以上，伴生组分银含量达 21×10^{-6}，钼为 0.066%，锌为 1.36%。

燕山晚期火山岩-火山碎屑岩的酸度整体上有高—低—高的波动，其中劳村组 SiO_2 平均含量达 76.93%，黄尖组平均为 63.70%，寿昌组火山岩酸度回升，达 67.88%。SiO_2 含量变化于 51.05%~77.55%之间，Na_2O+K_2O 为 5.20%~9.89%，Fe_2O_3+FeO 为 2.75%~8.59%，燕山旋回的早期和晚期均

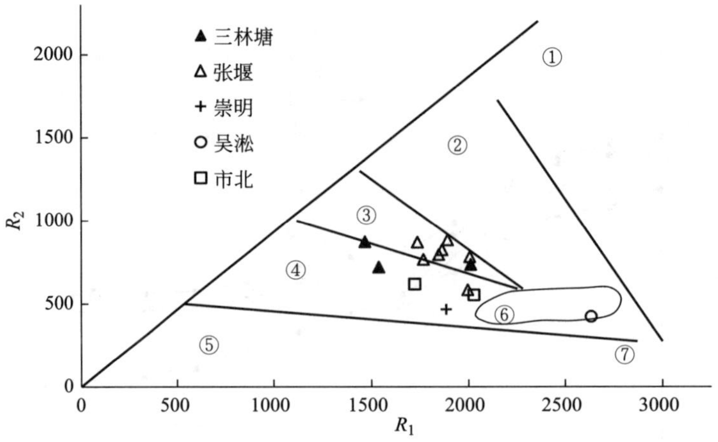

图 12-4 区内不同阶段闪长岩和花岗岩构造环境判别图解

Fig. 21-4 Tectonic environment discrimination diagrams of different periods of diorite and granite in Shanghai

①幔斜长花岗岩；②破坏性活动板块边缘（板块碰撞前）花岗岩；③板块碰撞后隆起期花岗岩；④晚造山期花岗岩；⑤非造山区 A 型花岗岩；⑥同碰撞（S 型）花岗岩；⑦造山期后 A 型花岗岩

有流纹岩发育。火山岩中 Ni、Co、V 铁族元素含量一般低于维氏平均值，Cu、Pb、Zn、Mo 含量分别较维氏值高 2~3 倍。

稀土元素的配分模式均为富集型，模式曲线为右倾型。而且不同地区、不同岩类（除流纹岩有明显铕亏损外）的曲线形态基本一致，表明它们具有相似的地质环境和演化过程。推测燕山旋回火山岩的原始岩浆可能来自地壳深部，其中有部分上地幔成分掺入，在岩浆上升过程中，有壳源成分的不断加入。早期旋回出现的玄武岩、玄武安山岩、辉石安山岩，其成分可能接近原始岩浆的成分。稀土元素配分曲线呈相互平行的特征，表明稀土元素含量大致呈同步变化，不同层位火山岩是同源岩浆不同期次喷发的产物。与燕山晚期的侵入岩稀土元素配分曲线十分类似，尤其是市区花岗岩、崇明花岗岩和崇明花岗岩中的细晶岩脉，表明燕山晚期广泛的火山作用与上述侵入作用岩浆源区的相似性。

第二节 新生代岩浆成因和演化探讨

喜马拉雅旋回火山活动从古近纪初—新近纪初由弱增强。其基本特征为：

1）火山岩以钙碱性系列为主，岩石组合为伊丁石化橄榄玄武岩-玄武岩。

2）酸度偏高，SiO_2 含量平均达 51.45%，全碱为 4.97%，而 FeO、Fe_2O_3、CaO 相对偏低。

3）中新世和古新世玄武岩其稀土元素的配分模式均为轻稀土富集型，模式曲线表现为向右倾斜，Eu 正异常。不同地区、不同时代，其配分曲线形态基本一致，表明它们具有相似的地质环境和演化过程，可能是同源岩浆分异作用的产物。

4）地层自老至新，其稀土元素总量、轻稀土元素总量和重稀土元素总量也随之增高。

5）微量元素含量也有某些差异，如 42 号钻孔中的 Cr、Ni 含量比白龙港某孔偏高，而 Zn、Ce、Ba、Rb、Sr 明显偏低。

第三节 构造-岩浆演化特点和规律

纵观整个地质历史，岩浆演化总的特点是都以钙碱性和碱性系列为主，且同时期侵入作用和火山作用岩浆源区具有一致性。

中生代岩浆作用最为强烈，区域分布规律明显受区域构造的制约，演化规律最为明显。侵入作用建造表现为中酸性—中基性—中酸性的演化特征。火山作用建造由安山岩-英安岩-流纹岩及其相应的火山碎屑岩组合而成。自下而上共有3个亚旋回。第一亚旋回主要由劳村组酸性火山碎屑岩及其熔岩组成，主要分布在本区南部；第二亚旋回由黄尖组玄武安山岩-安山岩-英安岩-流纹岩及粗面岩-碱流岩等组成，分布在查山、金桥、庄行、朱行、天马山等地；第三亚旋回由寿昌组英安岩-流纹岩组成，见于真如一带。火山活动强度，有弱—强—弱的发展趋势。总体演化特点和规律如下。

一、大规模活动时向接近

区内中生代岩浆作用强度有由弱—强—弱—强的演化趋势，侵入岩和火山岩主要发育于130Ma左右，两者规模相当，波动趋势合拍。

喜马拉雅旋回的岩浆活动较弱，主要分布在本区东部，火山岩与侵入岩在岩石系列、矿物组分、化学成分以及稀土元素含量特征等方面均颇相似。

二、同受基底构造控制、空间分布一致

燕山旋回的岩浆活动主要受北东向基底断裂的控制。其中，火山岩带及亚带受二、三级断裂构造控制呈北东向展布，火山构造盆地等次级火山单元，受更次一级构造控制，常发育于几组断裂的交汇部位。

侵入岩也受北东向、北北东向及北西向几组断裂控制，部分分布在火山盆地的边缘。

喜马拉雅旋回岩浆活动受北东向和北西向两组断裂控制。火山岩多分布在这些断裂的交汇处，而脉岩多分布在环状断裂及放射状断裂中。

三、岩浆成分及演化特征相似

火山岩和侵入岩具有相似的岩浆演化特征，可能为同源岩浆的两种形成方式。

（一）岩石种类

燕山晚期旋回火山岩分3个亚旋回，早期火山活动较微弱，堆积了一些酸性火山碎屑岩，为火山-沉积相产物，尚未发现相应的侵入活动，火山岩似属陆壳改造型；中期的火山活动最强，形成一套基性—中基性—中性—中酸性、酸性火山岩，岩性为玄武岩-玄武安山岩-安山岩-英安岩、流纹岩，与其相对应的侵入岩为角闪石辉长辉绿岩-辉长岩-辉石闪长岩-闪长岩-石英闪长岩-花岗闪长岩；晚期的火山活动已趋向尾声，岩性为中酸、酸性火山岩-英安岩、流纹岩，其对应的侵入岩为一系列中酸性脉岩和浅成岩体。

喜马拉雅旋回火山岩为玄武岩，其对应的侵入脉岩为辉绿岩。

（二）化学成分

燕山旋回火山岩和侵入岩的同类岩石，其化学成分十分相近，主要为钾玄岩系列。

喜马拉雅旋回火山岩和侵入岩（脉岩），岩石化学成分也相近，属钙碱性-钾玄岩系列。

（三）稀土元素特征

燕山期第二次侵入岩和黄尖组火山岩相比，其稀土元素的配分模式十分相似，其中查山玄武安山岩与坦直辉石闪长岩的稀土元素配分曲线位置相近，几乎重叠。

上述情况说明侵入岩和火山岩形成于相似的地质条件中，有相似的演化趋势。

第三篇
变质岩及变质作用

第十三章　区域变质岩系和变质作用

区内变质岩形成于晋宁中期、晋宁晚期和燕山晚期 3 个主要变质作用时期，前两期为区域变质作用，燕山期为接触变质作用。

晋宁中期区域变质岩组成前震旦系金山群。其分布明显受断裂构造带控制，多产于断块隆起部位。

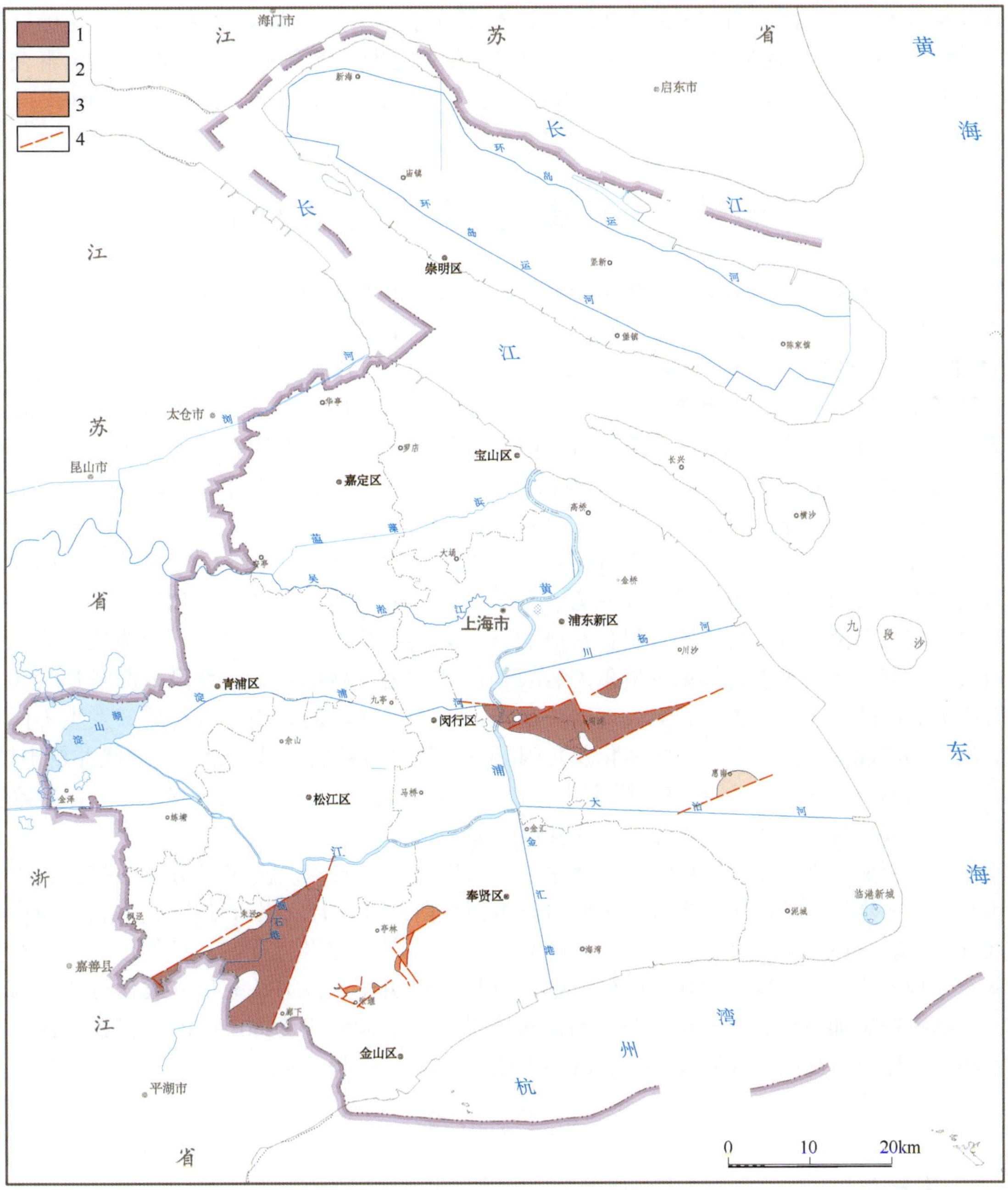

图 13-1　上海变质岩分布图

Fig. 13-1　Distribution of metamorphic rocks in Shanghai

1—晋宁中期；2—晋宁晚期；3—燕山晚期；4—断裂

已揭示的变质岩断续分布于枫泾—川沙断裂带与张堰—南汇断裂带之间（图 13-1），即华泾、三林塘、天花庵、赵家宅、康家宅、沈庄、肖塘、烟墩头、车墩、朱泾、吕巷、朱行等地，其中以三林塘—康家宅一带揭露厚度最大。变质作用类型为低压区域变质作用，形成低绿片岩相和高绿片岩相递增变质岩系。局部叠加后期的动力变质作用，形成构造角砾岩及糜棱岩化岩石。

晋宁晚期区域变质岩由前震旦系惠南板岩组成，仅见于浦东惠南，为新近纪玄武岩覆盖。已揭示部分的变质岩由单一的低绿片岩相板岩组成。

1959 年首次于天花庵钻遇变质岩，命名为“石英角闪石片岩”“角闪石石英片岩”“石英黑云母角闪石片岩”“黑云母石英片麻岩”“花岗片麻岩”等。嗣后，又相继在奉贤肖塘、松江车墩、金山吕巷等地钻孔中发现“片麻岩”“绿泥石云母石英片岩”“石英片岩”“碳酸盐化绢云母片岩”等变质岩。迄今，并未开展过系统的、专题性的变质岩研究工作，仅在矿产普查和磁异常验证工作中，从地层学工作角度对区内已发现的变质岩进行了定名，并对其原岩及隶属时代进行过一些推断。1986 年在编制第一轮地质志时，首次对金山群变质岩进行了系统研究，首次在金山群变质岩中获得 1123±27Ma 的变质年龄数据，并在岩石类型、岩石化学特征、变质类型等方面获得了系统认识。

第一节　主要岩石类型及特征

变质岩分类命名采用结构构造和矿物组合特征相结合的原则，当两者有矛盾时，以矿物组合特征为主要依据。根据变质岩自然共生组合，自下而上可以分为 a、b、c 3 个岩性段。

a 岩性段见有浅粒岩、长石石英岩、石英岩、云母片岩、绿片岩组合及大理岩、绿片岩、石英岩组合等两种变质岩岩石自然共生组合。主要岩石类型有浅粒岩、白云母长石石英岩、绿泥石长石石英岩、石英岩、白云母石英岩、白云母石英片岩、白云母片岩、斜长绿泥片岩、钠长阳起片岩、阳起片岩、绿帘阳起片岩、（含）磁铁阳起片岩、含蛇纹石大理岩、石榴子石大理岩、硅灰石大理岩等。岩石普遍糜棱岩化。见于赵家宅。厚度大于 117. 60m。

b 岩性段以角闪质岩石夹浅粒岩、长石石英岩、石英岩及少量黑云母片岩、变粒岩组合为特征。主要岩石类型有斜长角闪岩、角闪斜长片麻岩、斜长角闪片麻岩、角闪片岩、角闪石英片岩、含绿帘角闪片岩等，其次有黑云母片岩、黑云斜长片麻岩、二云斜长片麻岩、浅粒岩、含榴浅粒岩、长石石英岩、白云母石英岩及石英岩，偶见少量角闪岩及角闪变粒岩。顶部与 c 岩性段呈渐变过渡关系，角闪质岩石中央有厚薄不等的白色含石英大理岩。受燕山期岩浆活动的影响，岩石局部发生不同程度地热接触变质及热液蚀变，如岩石发生角岩化及角闪石不同程度地被黑云母、绿泥石取代并析出铁质等。见于三林塘、天花庵、厍里、赵家宅、康家宅、肖塘、朱行等地。厚度大于 107. 9m。在康家宅、肖塘、朱行等地的 b 岩性段中见有大量糜棱岩化花岗斑岩。

c 岩性段以云母片岩、云母石英片岩、绿片岩、大理岩、石英岩、浅粒岩组合为特征，朱泾、吕巷为云母片岩、大理岩夹碳质页岩组合。主要岩石类型有白云母片岩、绢云母片岩、白云母石英片岩、绿帘阳起片岩、绿帘钠长阳起片岩、白云母石英岩、含蛇纹石大理岩、石榴子石大理岩、含石英大理岩、硅灰石大理岩等。岩石成层性好，层理清晰，沉积韵律明显，特别是绿片岩或云母片岩与大理岩互层时，层状构造更加清晰。岩石片理与岩石层理产状一致，倾角多为 60°～70°，岩石已强烈糜棱岩化。见于华泾、厍里、沈庄、横沔及朱泾、吕巷等地。厚度大于 40. 20m。

一、金山群

（一）片岩类

片岩类岩石由云母、绿泥石、阳起石、角闪石和长石、石英等组成，含帘石类矿物。有明显的片状构造。片柱状矿物含量大于 25%，粒状矿物以石英为主，长石含量小于 25%。

1. 云母片岩及石英片岩

（1）白云母片岩

白云母片岩呈灰色、灰白色，鳞片变晶结构，片状构造。主要由白云母、绢云母和石英组成。白云母含量在25%~70%不等，绢云母为15%~20%，石英小于50%，另含3%~5%的绿泥石及少量微粒状磁铁矿。石英呈粒状、拉长状，长宽比为2∶1~3∶1。见于a、c岩性段中。

（2）绢云母片岩

绢云母片岩呈灰—灰白色、浅棕黄色，鳞片变晶粒状结构，片状构造。绢云母含量为60%，呈鳞片状，粒径为1.0~0.5mm，平行排列；石英含量为10%，呈粒状变晶，粒径为0.1~0.5mm，波状消光，拉长状，与绢云母相间排列；含10%的棕红色铁质。20%的细晶方解石呈微层状夹于绢云母片岩之间，构成绢云母片岩夹细晶大理岩。见于吕巷。

（3）白云母石英片岩

白云母石英片岩呈灰色、灰白色，呈鳞片微—细粒变晶结构、多边镶嵌结构，片状构造，主要由白云母和石英组成，前者含量为25%~30%，后者为65%~75%，另含少量绢云母及微量长石（长石已被绢云母集合体交代）。其中，白云母无色，呈板片状，绢云母为细小鳞片状，两者常一起形成细带状集合体定向分布。石英常呈近等轴多边形粒状体，颗粒直径为0.02~0.5mm，以0.05~0.2mm为主，呈透镜—条带状集合体定向排列。岩石局部受挤压，破碎强烈。所见副矿物以锆石为主，磷灰石次之，电气石仅几小颗。锆石多为棕、浅棕色，少数无色透明，具有金刚光泽，多为长柱状，短柱状少见，“磨圆度”较好，呈浑圆形，仅少数晶体棱角比较清楚，长为0.1~0.23mm，宽为0.03~0.1mm，长∶宽为3.3。磷灰石为六方柱状，无色透明，玻璃光泽，颗粒较小，一般小于0.17mm。电气石为褐绿色小柱状体，玻璃光泽，晶棱比较清楚，晶粒长为0.07mm，宽为0.05mm。见于a、c岩性段中。

（4）黑云母片岩

黑云母片岩呈深绿色，微晶鳞片变晶结构，片状构造。矿物成分以黑云母为主，含量在50%~85%不等，具有多色性，多被挤压成破碎细小的鳞片，定向排列明显，有的已被绿泥石交代并析出铁质，少数未破碎或破碎较轻者则发生弯曲、变形。次要矿物成分有钠长石、石英、绿帘石等，钠长石含量10%左右，石英含量为10%~15%，常集中呈显微条带状或透镜状，其分布方向与片理方向一致，绿帘石最高含量达20%。副矿物有少量榍石、锆石、金红石及磷灰石。岩石已不同理度地糜棱岩化，其中石英、长石被挤压破碎，具波状消光。由于后来燕山期岩浆活动的叠加，形成角岩化黑云母片岩或黑云母斜长角岩。

根据矿物成分及岩石结构的差异，尚可细分为含绿帘黑云片岩、绿帘黑云片岩、绿帘钠长黑云片岩、含磁铁黑云片岩、微晶黑云片岩等。

黑云母片岩与黑云石英片岩、角闪片岩、斜长角闪岩、角闪斜长片麻岩等共生。分布于b岩性段中。

（5）黑云石英片岩

岩石特征及分布情况同黑云母片岩，不同之处是石英含量大于等于50%，最高达65%。岩石微角岩化，石英的重结晶现象比较明显，多呈近等轴粒状，少数呈不规则粒状，粒径在0.02~0.25mm，以0.1~0.2mm为主。黑云母含量为25%，棕褐色，有的已褪色，呈板片状，常弯曲，粒径为0.1~0.35mm。另含8%~10%的长石。岩石经受糜棱岩化后又发生过重结晶作用。

（6）二云母片岩

二云母片岩呈花岗鳞片变晶结构，片状构造。主要由黑云母、白云母和石英组成。以黑云母为主，白云母次之，两者含量共70%~75%，石英含量为25%~30%，有微量金属矿物及少量次生绿泥石、碳酸盐矿物。岩石糜棱岩化明显，石英大部分被拉长、破碎，具波状消光，其集合体呈条带状和透镜状分布，云母亦有破碎成鳞片状者，多数发生弯曲、变形，黑云母有被绿泥石交代现象。偶见于b岩性段中。

2. 绿片岩

（1）阳起片岩

阳起片岩具鳞片变晶结构、纤状变晶结构、柱状和粒状变晶结构，片状构造。主要组成矿物为阳起石，一般含量为30%~45%，偶尔高达90%，浅绿色，微晶呈柱状、纤维状，集合体有时呈束状或放射状；次要矿物有绿帘石、长石（主要为钠长石）、绿泥石、石英；副矿物有榍石、磷灰石、磁铁矿及白钛矿。岩石糜棱岩化，矿物大部分破碎或弯曲、变形。可分为绿帘阳起片岩、磁铁阳起片岩、绿帘磁铁阳起片岩，含磁铁阳起片岩、钠长阳起片岩、含磁铁长石阳起片岩、绿帘钠长阳起片岩等。见于a、c岩性段中。

（2）斜长绿泥片岩

斜长绿泥片岩呈淡绿、深灰绿—深绿色，鳞片粒状变晶结构，片状构造。主要由绿泥石和斜长石组成，其次有绢云母、石英、磁铁矿及少量绿帘石。绿泥石含量为30%~95%，呈定向排列，少量绿泥石有受力扭曲现象，斜长石含量为10%~25%，绢云母含量为15%~20%，多与绿泥石相伴出现，石英最高含量达35%，呈细—微晶粒状、拉长状，长宽比为2∶1~3∶1，波状消光，定向排列，磁铁矿含量最高达15%（磁铁斜长绿泥片岩），绿帘石含量为3%~5%。见于a、c岩性段中。

（二）片麻岩类

该类岩石由长石、石英、黑云母、白云母组成，其中长英质矿物含量大于50%（长石含量大于等于25%），具片麻状构造。

1. 黑云斜长片麻岩

黑云斜长片麻岩具鳞片花岗变晶结构，略显片麻状构造。长石含量为58%~70%，以更钠长石为主，钾长石次之，石英含量为20%~30%，黑云母为10%~12%，副矿物有微量磷灰石及榍石。矿物粒径为0.05~1.5mm，以0.25~0.3mm居多。长石为不规则粒状变晶，局部见钾长石交代斜长石及石英交代长石的现象；黑云母为片状，定向排列。岩石经受糜棱岩化，发生不同程度的破碎，有时具糜棱结构。偶见于b岩性段中。

2. 二云斜长片麻岩

二云斜长片麻岩具鳞片花岗变晶结构，片麻状构造。由长石、石英、黑云母、白云母及微量磷灰石、锆石、榍石等组成。长石含量为25%~58%，以更钠长石为主，钾长石次之；石英含量为27%~55%，黑云母+白云母含量为13%~20%。矿物粒径为0.01~1.6mm，以0.3mm者居多。由于糜棱岩化作用，上述矿物均不同程度地发生破碎。石英破碎最为强烈，破碎的石英又经重结晶，呈糖粒状。矿物基本上已定向排列。见于b岩性段中。

（三）长英质粒岩类

该类岩石主要由长石、石英等粒状矿物组成，两者占矿物总含量的70%以上。片柱状矿物含量小于30%，组成近等粒的粒状变晶结构。

1. 变粒岩

变粒岩呈灰色、淡绿色，具花岗变晶结构、糜棱结构，片状、微片状或条带状构造。长英质粒状矿物含量大于等于70%，其中长石含量大于等于25%，片柱状矿物含量为10%~30%。矿物粒径为0.02~1.5mm，大多在0.3mm左右。片柱状矿物主要为角闪石、黑云母。副矿物有少量榍石、磷灰石。有些角闪石、黑云母被绿泥石交代，并析出铁质。有角闪变粒岩和黑云变粒岩之分。岩石已糜棱岩化，强烈破碎的石英又发生重结晶现象。见于b岩性段中。

2. 浅粒岩

浅粒岩呈浅灰色、灰白色、微带肉红色，具等轴粒状多边镶嵌变晶结构、细粒花岗变晶结构、残斑微晶粒状变晶结构，不明显的片状构造及条带状构造。暗色矿物含量小于10%，有透辉石、角闪石、黑云母、绿帘石、绿泥石等。长石含量大于等于25%，最高可达60%，为钠长石、微斜长石和条纹长石。石英含量在20%~62%不等。有时含白云母及1%~2%的铁铝榴石。铁铝榴石为粒状或近等轴粒状，浅黄色，裂隙发育，无异常干涉色，晶体中见有石英、长石、黑云母等包裹物。长石、石英均已不同程度重结晶。粒径为0.02~1.5mm，以0.1~0.5mm为多。长石粒径较石英稍粗，呈近等轴的多边形粒状或不规则粒状。偶见2%~5%的板状钾长石残斑，这种钾长石多已泥化，部分被绿泥石交代。

副矿物有微量磷灰石、锆石。分为浅粒岩、钾长浅粒岩、含榴浅粒岩、白云母浅粒岩等。岩石已不同程度糜棱岩化，组成岩石的矿物被挤压破碎并具定向排列现象，破碎后的矿物往往发生重结晶，长英质矿物重结晶后常具等轴粒状多边镶嵌结构。多见于 a、c 岩性段中，b 岩性段中偶有所见。

3. 长石石英岩

长石石英岩呈灰色、灰白色、灰绿色，具微晶粒状花岗变晶结构，有时具不明显的片状构造。主要矿物长石、石英总含量大于等于 70%，其中长石含量小于 25%，一般为 13%~14%，石英含量最高达 86%。次要矿物有阳起石、绿泥石、白云母，故有阳起石长石石英岩、绿泥石长石石英岩及白云母长石石英岩之分。副矿物有磷灰石及锆石。偶见长石残斑。岩石发生轻微糜棱岩化，石英、长石常被压扁、拉长，并具定向排列。石英波状消光及再度重结晶现象明显。见于 a、b 岩性段中。

4. 石英岩

石英岩呈灰白色，有时微带淡绿色，具鳞片粒状变晶结构与等轴粒状多边镶嵌结构，微具片状构造。主要矿物石英含量大于等于 70%，最高达 90%，次要矿物有白云母、绢云母、黑云母、绿泥石、长石等。片状矿物总量小于 25%，其中白云母含量较高，有时可达 20%（称为白云母石英岩）；长石含量小于 10%。矿物粒径在 0.03~1.14mm，一般多为 0.3mm。岩石已糜棱岩化，石英被挤压、拉长以致破碎，长宽比为 2∶1~3∶1，个别达 5∶1。见于 b 岩性段中。

（四）角闪质岩类

该类岩石颜色较深，具片状、片麻状、条带状、眼球状等构造。主要由角闪石、斜长石、石英组成，次要矿物有帘石类、辉石、石榴子石等。主要副矿物为黄铁矿、磷灰石，有时可见极少量的锆石、榍石及磁铁矿。锆石呈破碎状，其表面多为麻点状，显浅褐色，具玻璃光泽—金刚光泽。磷灰石为六方柱状，在肖塘见双锥状的六方柱状磷灰石，锥体比较发育，晶棱、晶面比较清楚，浑圆度差，最大粒径为 0.25mm×0.60mm，最小为 0.05mm×0.10mm，长宽比多为 2∶1，个别为 3∶1 或 1∶1，无色透明，具玻璃光泽。所见副矿物大部分已破碎，反映岩石经受过强烈的动力作用，已难以保留原岩的副矿物特征。

1. 角闪片岩

角闪片岩呈深灰色、灰黑色，具粒状、柱状变晶结构，片状构造。主要矿物成分为角闪石、石英；次要矿物成分有绿帘石、辉石、斜长石；副矿物有 2%~3%的磁铁矿，小于 1%的磷灰石及微—少量榍石；蚀变矿物有黑云母、绿泥石、方解石等。角闪石含量一般大于 50%，最高达 85%，有时为 30%~50%，绿色，纤维状、柱状，横切面呈菱形，解理角为 124°，呈定向排列，石英含量一般小于 30%，局部在 50%以上，绿帘石含量一般小于 5%，个别达 7%，辉石含量一般小于 5%，偶尔高达 20%，呈短柱状、粒状。磁铁矿多沿角闪石边缘分布，这时角闪石多被黑云母或绿泥石取代。根据岩石中石英含量的多少、是否含次要矿物及其种类等，又可分为角闪片岩、角闪石英片岩、含绿帘角闪片岩及辉石角闪片岩等。见于 b 岩性段中。

2. 斜长角闪岩

斜长角闪岩呈深灰色、灰黑色，具花岗柱状变晶结构、粒状和柱状变晶结构，片状构造、条带状构造，为角闪质岩类的主要岩石类型之一。主要由角闪石和斜长石组成，角闪石含量大于 50%，最高达 80%，斜长石含量为 10%~25%，角闪石呈柱状、粒状，定向排列，绿色，多色性强，矿物粒径在 0.03~5.5mm，一般为 0.4~1.0mm，大颗粒角闪石中见有石英及斜长石包裹体，具包含变晶结构，斜长石为粒状，粒径为 0.05~0.4mm，粗大者表面多被泥化而显混浊，并有石英和绿帘石包裹体，微细者表面光洁明亮。次要矿物有石英、帘石类，石英含量一般小于 10%，有时大于 10%，帘石类呈柱状、粒状，以绿帘石为主，黝帘石少量，含量一般小于 5%，偶尔高达 25%。岩石偶含辉石，但个别含量可达 12%。副矿物有榍石、磷灰石、磁铁矿、锆石、白钛矿等。榍石呈信封状，含量一般为 2%~5%，最高达 10%，磷灰石含量在微量~2%，磁铁矿含量一般小于 5%，偶尔达 6%~10%。岩石后期蚀变较强，黑云母化、绿泥岩化普遍，角闪石常被黑云母、绿泥石、方解石取代并析出铁质，斜长石的绢云母化也较强。根据次要矿物的含量及后期蚀变可分为斜长角闪岩、含绿帘斜长角闪岩、绿帘斜长角闪岩、辉石斜长角闪岩、黑云母化斜长角闪岩及绿泥石化斜长角闪岩等。见于b岩性段中。

3. 角闪斜长片麻岩

角闪斜长片麻岩呈深灰色、灰绿色，具鳞片花岗变晶结构与柱状、粒状、纤状变晶结构，条带状、片麻状及似眼球状构造。主要矿物成分为角闪石、斜长石和石英。角闪石含量在10%~50%不等，以20%~50%居多，常蚀变为黑云母或绿泥石并析出铁质，粗粒角闪石中有时含斜长石包裹体，具包含变晶结构。斜长石含量大于25%，以35%~50%居多，最高达55%，呈眼球状、扁豆状，不连续定向排列。石英含量为15%~24%，颗粒细小。次要矿物有绿帘石、石榴子石，绿帘石含量一般小于5%，个别为5%~8%，石榴子石仅在个别情况下见到，但其含量可在10%~15%。副矿物有榍石、磷灰石及极少量锆石，榍石最高含量达6%，磷灰石最高含量亦达4%~5%。蚀变矿物有黑云母、绿泥石及方解石。根据次要矿物成分尚可细分为角闪斜长片麻岩、含绿帘角闪斜长片麻岩及石榴角闪斜长片麻岩。见于b岩性段中。

4. 斜长角闪片麻岩

斜长角闪片麻岩岩性基本同角闪斜长片麻岩，所不同的是斜长角闪片麻岩中的角闪石含量大于50%，最高达66%，斜长石含量则相应减少，在25%~38%之间。角闪石最大粒径可达3mm，一般为0.2~0.8mm。见于b岩性段中。

5. 角闪岩

角闪岩呈黑色，变晶结构，片状构造。主要由角闪石组成，含量大于等于95%，另含少量石英、磁铁矿、绿泥石及方解石。仅见于天花庵b岩性段中。

本类岩石普遍糜棱岩化，岩石中的长石、石英均已破碎，石英破碎最为强烈，片柱状矿物则以弯曲、变形为主，有些也具有破碎现象。岩石经糜棱岩化后又曾发生重结晶作用。

（五）大理岩类

1. 含蛇纹石大理岩

含蛇纹石大理岩呈白色、灰色、灰白色，具微晶均粒粒状变晶结构，块状构造。方解石含量为70%~95%，蛇纹石含量为2%~15%，局部含少量透辉石。见于a岩性段中。

2. 石榴子石大理岩

石榴子石大理岩呈白色、灰白色、淡绿色，具粒状变晶结构。方解石含量为50%~53%，石榴子石含量为35%~40%，局部含10%透辉石或10%更钠长石，另含少量硅灰石、绿帘石及白云母。方解石为细—中粒状，石榴子石显浅绿色，粒状，主要为钙铝榴石，常呈集合体产出，硅灰石以小柱状分散分布。见于a、c岩性段中。

3. 含石英大理岩

含石英大理岩见于沈庄、吕巷c岩性段及沈庄b岩性段顶部。

沈庄所见含石英大理岩为灰白色、灰—深灰色，粒状变晶结构，层状构造。方解石含量为50%~95%，石英含量一般为5%~10%，最高达25%，绿帘石含量偶达30%，有时含2%的透闪石。方解石呈不规则的多边形或圆形，粒径为0.04~0.1mm。吕巷所见含石英大理岩呈灰—深灰色，粒状变晶结构，层状构造。主要矿物方解石含量为85%，大多呈0.1mm×0.2mm的他形粒状变晶，少数为0.02mm×0.05mm左右的微晶，长轴大多呈定向或半定向分布，次要矿物有石英（7%）、绢云母（3%）等。石英呈他形粒状变晶，粒径在0.1mm左右，或呈分散状分布在岩石中，或集中分布并与绢云母构成绢云母片岩夹层，绢云母多为0.01~0.02mm×0.1~0.15mm的微鳞片，常与石英等一起集中分布于含石英大理岩的夹层——绢云母片岩中。

4. 硅灰石大理岩

硅灰石大理岩呈灰白—灰黑色及淡绿色，粒状变晶结构、柱状和放射状变晶结构，层状构造。方解石含量为70%~80%，硅灰石含量为15%，有时含3%~5%的石榴子石，3%~8%的透闪石及少量磷灰石。方解石稍有拉长并显定向排列，粒径为0.15~0.45mm，硅灰石为纤维状，多呈放射状集合体分布在岩石中。见于c岩性段。

（六）糜棱岩化花岗斑岩

岩石呈肉红色，间杂灰黑色、暗绿色条痕或残影，具变余斑状结构，基质呈变余细粒花岗结构、糜棱结构、鳞片变晶结构，条带状构造、眼球状构造，条带方向和与其交替产出的角闪质类岩石条带方向基本一致。斑晶以微斜长石为主，条纹长石次之，两者总含量最低为3%，最高达45%，呈肉红色，自形程度差，多为不规则椭圆状、碎斑状、眼球状，定向排列，粒径一般为0.4～3mm，最大粒径达7mm，部分具格子双晶，此外尚有少量斜长石碎斑，其表面多已泥化、绢云母化或被石英、绿帘石交代。基质成分为钾长石、斜长石、石英等。以钾长石为主，平均含量为40%，最高达65%，斜长石含量一般为15%～20%，最高达30%，石英含量在10%～40%不等。基质矿物粒径为0.02～0.6mm，常呈条带状集合体绕过长石碎斑作定向排列，其中石英颗粒较长石更细，多以粉状出现，且多发生不同程度的重结晶。岩石中常含一定数量的黑云母、绿泥石、白云母、方解石、绿帘石等。副矿物有钻石、磷灰石。

该类岩石具有下列特征：①均见于金山群b岩性段中，与一套变质程度较深的角闪质岩石以渐变和突变两种接触关系频繁交替产出，单层厚从数厘米到数十厘米、数米或数十米不等；②岩石具有一种隐约可见的阴影构造和条带构造，这种阴影和条带由暗色矿物组成；③微斜长石与条纹长石呈粗细不等的团粒状充填岩石；④具交代蠕英结构、交代条纹结构及交代净边结构、包含变晶结构（即粗粒微斜长石及条纹长石中包含具有定向排列的角闪石及具聚片双晶的斜长石）。

二、惠南板岩

岩石组合简单，由钙质绢云母板岩、粉砂质绢云母板岩及粉砂质钙质绢云母板岩组成。板理发育，板理面平滑，沿板理面可击成极薄的瓦板状薄片。

1. 钙质绢云母板岩

钙质绢云母板岩呈灰色，具变余泥质结构，板状构造。由52%～55%的泥质、绿泥石、绢云母，40%的方解石和6%的粉砂及极少量隐晶状硅质组成。绢云母呈很淡的绿色，鳞片状，大致呈定向排列，粉砂质成分为长石、石英等，颗粒细小，粒径常在0.05mm以下，一般呈滚圆—次圆状。

2. 粉砂质绢云母板岩

粉砂质绢云母板岩呈灰色，具变余粉砂状结构，板状构造。主要矿物成分为泥质、绿泥石及绢云母，三者总含量约为85%，其次为粉砂，含量约为15%，另有少量方解石及磁铁矿微粒。粉砂呈棱角状、半棱角状。有后期方解石脉斜交板理穿插，脉中有粒状、立方体状黄铁矿。

3. 粉砂质钙质绢云母板岩

粉砂质钙质绢云母板岩呈灰色，具板状构造，由40%～45%的泥质和绢云母、30%的方解石、20%的粉砂及4%的绿泥石组成。绢云母大致呈平行排列，方解石为隐晶—微粒状，粉砂成分以石英为主，长石及绿泥石为次。长英质粉砂呈棱角—次棱角状，粒径在0.02mm左右。

第二节　原岩恢复和大地构造环境分析

一、岩石化学特征

（一）金山群

1. 斜长角闪岩

岩石中SiO_2含量介于42.29%～53.04%之间，平均含量为47.93%；K_2O平均含量为1.79%，Na_2O平均含量为2.02%，$K_2O<Na_2O$，K_2O/Na_2O平均值为0.38；CaO平均含量为7.14%，MgO平均含量为8.24%，CaO<MgO（表13-1），al-alk值介于7.87～16.83，平均值为11.88（<16）；c值介于14.56～20.28，平均值为18.24（<26）。岩石兼具沉积岩和玄武质或安山玄武质岩浆岩岩石化学特征。

表 13-1 金山群主要变质岩氧化物含量、尼格里特征值及分子数对比表

Table 13-1 Correlation of oxide content, Niggli number and molecule number among the main metamorphic rocks in the Jinshan Group

序号	岩石	地点	氧化物含量/%														尼格里特征值										分子数	
			SiO_2	TiO_2	Al_2O_3	Fe_2O_3	FeO	MnO	MgO	CaO	Na_2O	K_2O	P_2O_5	烧失量	H_2O^-	总计	si	al	fm	c	alk	mg	k	al-alk	al+fm-c-alk	Al_2O_3	Na_2O+ K_2O+ CaO
1	黑云母化斜长角闪岩	三林塘	53.04	1.36	17.03	4	6.36	0.12	6.32	5.04	1.72	3.32	0.26	1.01	0.22	99.58	142.88	27.02	48.22	14.56	10.19	0.53	0.56	16.83	50.49	167	153
2	黑云母化斜长角闪岩	三林塘	51.57	1.34	15.01	4.75	7.39	0.11	6.79	6.51	1.37	2.84	0.26	0.75		98.91	132.61	22.72	51.31	17.93	8.04	0.51	0.58	14.68	48.06	147	168
3	斜长角闪岩	赵家宅	50.38	1.18	14.74	6.22	6.61	0.22	6.69	7.06	3.02	1.4	0.08	1.56		99.16	124.33	21.51	50.30	18.69	9.50	0.49	0.23	12.01	43.62	145	190
4	斜长角闪岩	沈庄	46.96	0.79	14.44	4.48	7.9	0.10	8.06	8.19	2.34	2.52	0.05	3.27		99.10	108.47	19.72	50.97	20.28	9.03	0.54	0.42	10.69	41.38	142	211
5	含绿帘斜长角闪岩	肖塘	42.29	1.3	9.48	3.61	9.40	0.19	13.83	8.69	0.74	0.44	0.24	5.92	0.21	96.34	89.45	111.82	66.33	19.69	2.16	0.66	0.29	9.66	56.30	93	172
6	斜长角闪岩	肖塘	43.87	1.24	9.19	3.12	9.6	0.18	14.63	8.88	0.64	1.51	0.20	4.75	0.24	98.05	89.79	11.07	66.3	19.43	3.2	0.67	0.62	7.87	54.74	90	184
7	绿帘斜长角闪岩	肖塘	49.26	1.65	13.46	5.41	9.38	0.19	5.83	6.6	2.94	1.33	0.39	1.42	0.05	97.91	124.62	20.06	52.74	17.93	9.27	0.42	0.23	10.79	45.6	132	179
8	绿帘斜长角闪岩	肖塘	48.67	1.74	13.68	4.91	8.7	0.22	5.75	6.72	3.25	1.1	0.32	1.37	0.11	96.54	125.39	20.74	50.77	18.58	9.91	0.44	0.19	10.83	43.02	134	184
9	含绿帘斜长角闪岩	肖塘	45.32	1.92	14.78	7.59	8.44	0.2	6.27	6.57	2.19	1.63	0.33	2.37	0.14	97.75	110.07	21.17	54.16	17.08	7.59	0.42	0.33	13.58	50.66	145	169
10	斜长角闪片麻岩	康家宅	46.34	2.20	15.53	3.51	9.48	0.19	7.55	7.68	2.16	2.60	0.39	1.89		99.52	107.38	21.17	50.98	19.08	8.77	0.51	0.44	12.4	44.3	151	200
11	绿泥石化斜长角闪片麻岩	康家宅	43.32	2.81	15.03	4.58	10.38	0.22	5.37	7.27	2.63	2.31	0.39	3.84	0.09	98.24	105.87	21.58	49.49	19.09	9.84	0.39	0.37	1.74	42.14	147	197
12	绿泥石化斜长角闪片麻岩	康家宅	43.62	2.34	15.24	4.41	10.03	0.14	6.41	6.02	2.92	1.69	0.35	4.52	0.13	97.82	107.24	22.01	52.58	15.58	9.6	0.45	0.28	12.41	49.18	149	172
13	绿泥石化角闪斜长片麻岩	康家宅	39.70	2.12	14.59	5.64	7.04	0.80	6.49	6.9	2.16	2.65	0.41	8.18		96.68	98.66	2.34	50.9	18.36	9.4	0.47	0.44	11.94	44.48	143	186
14	角闪斜长片麻岩	康家宅	48.42	2.54	16.39	3.60	9.92	0.20	4.17	7.68	3.58	2	0.34	1.16		100.0	121.02	24.18	43.39	20.57	11.86	0.36	0.27	12.32	35.14	161	216

续表

序号	岩石	地点	氧化物含量/%														尼格里特征值										分子数	
			SiO_2	TiO_2	Al_2O_3	Fe_2O_3	FeO	MnO	MgO	CaO	Na_2O	K_2O	P_2O_5	烧失量	H_2O^-	总计	si	al	fm	c	alk	mg	k	al−alk	al+fm−c−alk	Al_2O_3	Na_2O+K_2O+CaO
15	绿泥石化角闪斜长片麻岩	三林塘	51.62	1.55	15.28	4.7	6.76	0.13	5.44	6.45	3.84	1.74	0.27	1.67		99.45	135.28	23.62	45.67	18.11	12.6	0.47	0.23	11.02	38.58	150	195
16	角闪斜长片麻岩	三林塘	49.88	1.17	16.41	5.13	6.36	0.14	6.58	7.33	3.66	1.56	0.17	1		99.39	120.99	23.47	46.36	19.09	11.08	0.51	0.22	12.39	39.66	161	207
17	绿泥石化角闪斜长片麻岩	康家宅	44.89	2.26	11.5	3.7	10.05	0.43	5.16	9.56	2.25	1.19	0.25	5.53	0.07	96.84	114.57	17.33	49.08	26.07	7.52	0.4	0.27	9.81	32.82	113	219
18	黑云母化角闪斜长片麻岩	康家宅	53.62	2.44	13.33	4.29	8.74	0.17	3.41	5.97	4.22	1.24	0.49	1.15	0.16	99.23	153.53	22.55	45.27	18.24	13.94	0.32	0.16	8.61	35.64	131	187
19	角闪片岩	天花庵	50.96	1.84	13.96	6.21	8.36	0.16	6.92	5.42	2.18	1.59	0.288		0.64	98.53	129.66	20.95	56.27	14.83	7.95	0.47	0.33	13	54.44	137	149
20	角闪片岩	天花庵	50.85	1.7	13.85	4.66	9.25	0.16	4.71	7.72	2.72	1.79	0.03	0.93	0.14	98.51	131.57	21.15	47.59	21.42	9.8	0.38	0.3	11.35	37.48	136	201
21	黑云斜长片麻岩	沈庄	71.04	0.26	13.68	0.72	1.51	0	1.05	1.45	3.74	5.1	0.04	1.65		100.24	358.18	40.61	16.97	7.88	34.54	0.46	0.47	0.46	15.16	134	140
22	绿帘阳起片岩	沈庄	64.58	0.84	10.78	3.1	5.05	0.12	1.79	5.31	0.42	2.63	0.02	4.18	0.43	99.3	274.94	27.11	39.64	24.3	8.95	0.28	0.80	18.16	33.5	106	130
23	浅粒岩	康家宅	75.3	0.15	12.27	0.8	0.93	0.04	0	1	4.46	4.46	0.02			99.43	445.91	42.7	8.54	6.41	42.35	0	0.39	0.35	2.48	120	137
24	浅粒岩	赵家宅	70.8	0.54	10.87	2.48	4.67	0.04	0.85	2.52	2.70	2.36	0.06	1.48		99.37	347.49	31.56	34.81	13.28	20.35	0.18	0.36	11.21	32.74	107	114
25	白云母石英岩	赵家宅	88.16	0.18	5.42	0.19	1.29	0.01	0.81	0.67	0.22	1.40	0.03	1.35		99.73	1183.1	42.74	32.26	9.68	15.32	0.5	0.79	27.2	50.00	53	31
26	白云母石英片岩	华泾	81.98	0.31	7.29	0.24	2.18	0.01	1.29	0.66	0.12	2.47	0.09	1.26	0.29	99.19	721.69	37.57	41.27	6.35	44.81	0.41	0.93	22.76	57.68	71	40
27	白云母石英片岩	华泾	89.31	0.16	4.73	0.62	1.48	0.02	0.91	0.31	0.06	1.54	0.06	0.71	0.30	100.11	1227.3	38.02	42.97	4.96	14.05	0.44	0.94	23.97	61.98	46	23
28	白云母石英片岩	华经	84.77	0.22	6.62	1.01	1.29	0.01	1.32	0.48	0.08	2.10	0.10	1.26	0.11	99.37	876.4	40.37	39.75	5.59	14.29	0.52	0.96	26.08	60.24	65	32
29	白云母石英片岩	华泾	84.16	0.19	6.60	1.02	2.1	0.02	1.48	0.48	0.08	2.14	0.13	1.07	0.24	99.71	791.5	36.72	44.63	5.09	13.56	0.47	0.96	23.16	62.7	65	33
30	含石英大理岩	吕巷	40.46	0.41	4.33	0.67	1.98	0.06	1.48	26.32	0.52	0.96	0.13	21.65	0.16	99.13	111.61	6.97	12.27	77.78	2.98	0.5	0.56	3.99	−61.52		
31	含石英大理岩	吕巷	30.69	0.43	5.07	0.88	2.38	0.12	2.12	29.63	0.38	1.2	0.10	24.43	0.16	97.59	73.42	7.18	14.23	75.86	2.73	0.54	0.68	4.45	−57.18		
32	含石英大理岩	吕巷	7.99	0.12	1.85	0.55	0.78	0.04	0.84	44.88	0.35	0.47	0.12	38.49	0.31	96.69	15.3	2.07	4.6	92.06	1.27	0.53	0.45	0.8	−80.66		
33	板岩	惠南	57.24	0.74	16.05	1.06	5.35	0.11	3.18	3.16	0.92	3.90	0.16	7.00	0.65	107.50	217.3	28.7	38.7	12.76	12.76	0.47	0.74	15.95	41.91	157	113

2. 斜长角闪片麻岩

岩石中 SiO_2 含量介于43.32%~46.34%，平均含量为44.43%；K_2O 平均含量为2.20%，Na_2O 平均含量为2.57%，$K_2O<Na_2O$，K_2O/Na_2O 平均值为0.36；CaO 平均含量为6.99%，MgO 平均含量为6.44%，CaO>MgO，al-alk 值介于11.74~12.41，平均值为12.18（<l6）；c 值介于15.81~19.09，平均值为17.99（<26）。岩石具玄武质或安山玄武质岩浆岩岩石化学特征。

3. 角闪斜长片麻岩

岩石中 SiO_2 含量介于39.70%~53.62%，平均含量为48.02%；K_2O 平均含量为1.73%，Na_2O 平均含量为3.29%，$K_2O<Na_2O$，K_2O/Na_2O 平均值为0.27；CaO 平均含量为7.32%，MgO 平均含量为5.21%，CaO>MgO，al-alk 值介于8.61~12.39，平均值为11.02（<l6）；c 值介于18.11~26.07，平均值为20.07（<26）。具玄武质或安山玄武质岩浆岩岩石化学特征。

4. 角闪片岩

岩石中 SiO_2 含量介于50.85%~50.96%，平均含量为50.91%；K_2O 平均含量为1.69%，Na_2O 平均含量为2.45%，$K_2O<Na_2O$，k 平均值为0.32；CaO 平均含量为6.57%，MgO 平均含量为5.82%，CaO>MgO，al-alk 平均值为12.18<l6；c 平均值为18.15<26。具玄武质或安山玄武质岩浆岩岩石化学特征。

5. 黑云斜长片麻岩

SiO_2 含量为71.04%，K_2O 含量为5.10%，Na_2O 含量为3.74%，$K_2O>Na_2O$，k 值为0.47，CaO 含量为1.45%，MgO 含量为1.05%，CaO>MgO。兼具沉积岩和岩浆岩岩石化学特征。

6. 绿帘阳起片岩

SiO_2 含量为64.58%，K_2O 含量为22.68%，Na_2O 含量为0.42%，$K_2O>Na_2O$，k 值为0.80，CaO 含量为5.31%，MgO 含量为1.79%，CaO>MgO，兼具沉积岩和岩浆岩岩石化学特征。

7. 浅粒岩

岩石中 SiO_2 含量介于70.8%~75.3%，平均含量为73.05%，K_2O 平均含量为3.41%，Na_2O 平均含量为3.58%，$K_2O>Na_2O$，k 平均值为0.38，CaO 平均含量为1.76%，MgO 平均含量为0.43%，CaO>MgO。具岩浆岩岩石化学特征。

8. 白云母石英岩

SiO_2 含量为88.16%，K_2O 含量为1.40%，Na_2O 含量为0.22%，$K_2O>Na_2O$，k 值为0.79，CaO 含量为0.67%，MgO 含量为0.81，CaO<MgO。具正常沉积成因岩石化学特征。

9. 白云母石英片岩

SiO_2 含量为81.98%~89.31%，平均含量为85.03%，K_2O 平均含量为2.06%，Na_2O 平均含量为0.09%，$K_2O>Na_2O$，k 平均值为0.95，CaO 平均含量为0.48%，MgO 平均含量为1.25%，CaO<MgO。具正常沉积成因岩石化学特征。

10. 含石英大理岩

SiO_2 含量为7.99%~40.46%，平均含量为26.38%，CaO 含量为26.32%~44.88%，平均含量为3.61%，K_2O 平均含量为0.88%，Na_2O 平均含量为0.42%，$K_2O>Na_2O$。

总体上具有如下特征：①斜长角闪岩、斜长角闪片麻岩、角闪斜长片麻岩、角闪片岩等角闪质类岩石低硅、富镁铁、富钙，相对贫碱，且 $Na_2O>K_2O$、CaO>MgO。除一个样品外，其余19个样品氧化铝的分子数比碱加钙的量小。②黑云斜长片麻岩富硅、铝及碱，相对贫钙、贫镁铁，氧化铝的分子数比碱加钙的量小。③绿帘阳起片岩富镁铁、富钙，低碱，氧化铝的分子数比碱加钙的量小。④白云母石英岩与白云母石英片岩相似，富硅、贫碱，贫钙、贫镁铁，且 $K_2O \gg Na_2O$、Mg0≫CaO，SiO_2 含量大于78%，氧化铝分子数比碱加钙的分子数大是这类岩石的另一特征。⑤同类岩石的化学成分是相近的。鉴于所有变质岩都是一定岩石在相对封闭条件下经变质作用的产物，其成分变化基本上是等化学的，在同一变质条件下，具相似岩石化学特征的几种变质岩应视为同一类型原岩的变质产物。⑥角闪质类岩石以富 Cr、Ti、Ba、Cu 为特征，Ni、Co、Sr、Ba、V、Cu 的含量与金山群中的辉绿岩脉相近，各类岩石中 Ba 的平均含量均高于维诺格拉多夫值，同时高于同类岩石 Sr 的平均含量，即 Sr/Ba 值均小于 1，其中大理岩、灰岩中的 Ba 含量高达 2781×10^{-6}，为维氏值的 4.28 倍，各类岩石的 Cr 平均含量均高于同类岩石的 Ni 平均含量，即 Cr/Ni 值均大于 1（表 13-2）。

表 13-2 金山群不同岩石微量元素含量

Table 13-2 Contents of trace elements among different rocks in the Jinshan Group

序号	岩石	微量元素/10^{-6}								元素比值	
		Cr	Ni	Co	Ti	Sr	Ba	V	Cu	Cr/Ni	Sr/Ba
1	角闪质类岩石	229	63	30	9105	247	1169	187	144	3.63	0.21
2	浅粒岩	281	25	12	4483	141	1083	77	57	11.24	0.13
3	白云母石英岩	277	29	12	3533	129	736	105	50	9.55	0.18
4	绿片岩	210	57	30	5369	214	903	218	64	3.68	0.24
5	二云片岩	290	113	25	3333	360	1200	150	67	2.57	0.30
6	大理岩、灰岩	129	39	12	1538	58	2781	48	82	3.31	0.02
7	辉绿岩(侵入金山群中的)	366	62	31	5804	239	1011	262	116	5.9	0.24
8	克拉克值	83	58	18	4500	340	650	90	47		
9	玄武岩平均含量	200	160	45	9000	440	300	200	100		

注：克拉克值和玄武岩平均含量据 A. Л. 维诺格拉多夫，1962。

（二）惠南板岩

岩石中 SiO_2 含量为 57.24%，K_2O 含量为 3.90%，Na_2O 含量为 0.92%，k 值为 0.74，CaO 含量为 3.16%，MgO 含量为 3.18%。al-alk 值为 15.95，c 值为=12.76（见表 13-1）。

岩石相对富集 Ba、Sr、V 和 Zn，Sr/Ba 值为 0.25，比金山群稍高（表 13-3）。

稀土元素总量 $\sum REE=220\times10^{-6}$，其中，$\sum LREE=172.55\times10^{-6}$，$\sum HREE+Y=47.14\times10^{-6}$，$\delta Eu=0.92$，Eu 亏损不明显。$\delta Ce=0.94$，Ce 亏损不明显。$(La/Yb)_{北美页岩}=1.30$，分布模式曲线近水平，轻、重稀土元素分馏不明显。

表 13-3 惠南板岩微量（含稀土）元素含量表

Table 13-3 Contents of trace elements (incl. REE) in the Huinan slate $w(B)/10^{-6}$

岩石	Sc	Li	Be	Co	Cu	Zn	Ga	Ge	Rb	Zr	Nb	Mo	Cd	Cs	Hf	Ta
板岩	15.8	37.9	2.75	20.5	37.1	109	23.2	1.95	160	151	13.5	0.35	0.085	8.22	3.6	1.02
	W	Tl	Pb	Th	U	Ba	Cr	Ni	Sr	V	As	Sb	Bi	Sn	Ag	
	29.8	0.74	22.2	13.8	2.81	411	87.4	47.8	101	124	5.03	0.29	0.366	2.68	0.041	29.8
	La	Ce	Pr	Nd	Sm	Eu	Gd	Tb	Dy	Ho	Er	Tm	Yb	Lu	Y	∑REE
	43.4	79.3	9.14	33.0	6.49	1.22	5.99	0.91	4.98	1.03	3.05	0.47	3.14	0.47	27.1	220

二、主要变质矿物特征

（一）斜长石

斜长石是金山群的主要变质矿物之一，在低绿片岩相岩石中主要为钠长石，在高绿片岩相岩石中为更长石及中长石，受后期蚀变影响，斜长石去钙化现象明显，导致斜长石牌号普遍偏低。经测定，共获得 5 组角闪质类岩石斜长石的牌号及光轴角：①An=28，$2V=-86°$；②An=15~24，$2V=+85°\sim87°$；③An=15~23，$2V=+85°\sim84°$；④An=16~20；⑤An=18~32，$2V=+90°\sim85°$。用 X 光粉晶法，求得第 3 组斜长石的有序度$S=0.7$，An=38，较显微镜下测得的 An=15~23 高得多。

（二）角闪石

角闪石为金山群中角闪质类岩石的主要变质矿物之一，柱状、粒状，多色性强，N_g 为淡绿—深绿色或带有蓝色色调的绿色，N_p 为淡黄色、淡黄绿色，N_m 为中等程度绿色或微弱的淡黄色。光轴角 $2V$ 为-75°~80°，消光角 $C \wedge N_g$ 为 15°~25°。

根据角闪石的化学成分（表 13-4），角闪石标准晶体化学式为：

表 13-4 角闪石化学成分

Table 13-4 Chemical composition in the hornblende

样号	地点	氧化物含量/%												
		SiO_2	TiO_2	Al_2O_3	Fe_2O_3	FeO	MnO	MgO	CaO	Na_2O	K_2O	P_2O_5	H_2O^+	合计
Df_4	肖塘	50.00	1.13	8.05	2.83	9.15	0.22	14.10	9.76	0.87	1.11	0.35	2.52	100.09
Df_5	肖塘	49.20	0.78	8.80	3.38	9.88	0.17	13.38	8.99	0.94	0.58	0.23	2.55	98.88
Df_6	康家宅	46.60	2.82	14.76	3.94	10.86	0.28	5.86	5.92	2.52	1.74	0.49	3.34	99.13
Df_7	康家宅	43.80	2.88	11.85	4.92	14.20	0.34	6.39	10.15	1.84	0.79	0.31	2.84	100.31

样号	地点	按 23 个氧原子法计算的阳离子系数										分类
		Si	Ti	Al	Fe^{3+}	Fe^{2+}	Mn	Mg	Ca	Na	K	
Df_4	肖塘	7.186	0.122	1.364	0.306	1.099	0.027	3.022	1.503	0.242	0.204	钙质角闪石类
Df_5	肖塘	7.151	0.089	1.507	0.370	1.201	0.021	2.900	1.400	0.266	0.108	钙质角闪石类
Df_6	康家宅	6.875	0.313	2.567	0.438	1.340	0.035	1.289	0.936	0.722	0.328	钠-钙角闪石类
Df_7	康家宅	6.573	0.325	2.096	0.555	1.782	0.043	1.430	1.632	0.536	0.151	钙质角闪石类

Df_4：$(K_{0.204})_{0.204}(Fe_{0.099}^{2+}, Mn_{0.027}, Ca_{1.503}, Na_{0.242})_{1.871}(Al_{0.550}, Fe_{0.306}^{3+}, Ti_{0.122}, Mg_{3.022}, Fe_{1.000}^{2+})_{5.00}[(Si_{7.186}, Al_{0.184})_{8.00}O_{22}](OH)_{2.00}$；

Df_5：$(K_{0.108})_{0.108}(Fe_{0.215}^{2+}, Mn_{0.021}, Ca_{1.400}, Na_{0.266})_{1.902}(Al_{0.658}, Fe_{0.370}^{3+}, Ti_{0.086}, Mg_{2.900}, Fe_{0.986}^{2+})_{5.00}[(Si_{7.151}, Al_{0.849})_{8.00}O_{22}](OH)_{2.00}$；

Df_6：$(K_{0.328})_{0.328}(Ca_{0.936}, Na_{0.722})_{1.658}(Al_{1.442}, Fe_{0.438}^{3+}, Ti_{0.313}, Mg_{1.289}, Fe_{1.340}^{2+}, Mn_{0.035})_{4.857}[(Si_{6.875}, Al_{1.125})_{8.00}O_{22}](OH)_{2.00}$；

Df_7：$(Na_{0.168}, K_{0.151})_{0.319}(Ca_{1.632}, Na_{0.358})_{2.00}(Al_{0.669}, Fe_{0.555}^{3+}, Ti_{0.325}, Mg_{1.430}, Fe_{1.782}^{2+}, Mn_{0.043})_{4.804}[(Si_{6.573}, Al_{1.427})_{8.00}O_{22}](OH)_{2.00}$；

根据标准晶体化学式计算，Df_4 的 $(Ca+Na)_B=1.745>1.34$、$Na_B=0.242<0.67$，Df_5 的 $(Ca+Na)_B=1.666>1.34$、$Na_B=0.266<0.67$，均属于钙质角闪石类；Df_4 的 $(Na+K)_A=0.204<0.50$、$Ti=0.122<0.50$、$Mg/(Mg+Fe)=0.73$、$Si=7.186$，Df_5 的 $(Na+K)_A=0.108<0.50$、$Ti=0.086<0.50$、$Mg/(Mg+Fe)=0.71$、$Si=7.151$，两个样品均属钙质角闪石类阳起角闪石亚类；Df_6 的 $(Ca+Na)_B=1.658>1.34$、$0.67<Na_B=0.722<1.34$，属钠-钙角闪石类，其 $(Na+K)_A=0.328<0.50$、$Mg/(Mg+Fe)=0.49$、$Si=6.875$，属钠-钙角闪石类的亚铁冻蓝闪石；Df_7 的 $(Ca+Na)_B=2.00>1.34$、$Na_B=0.368<0.67$、$(Na+K)_A=0.319<0.50$、$Ti=0.325<0.50$、$Mg/(Mg+Fe)=0.45$、$Si=6.573$，属钙质角闪石类的亚铁角闪石。

钙质闪石可分为 2 组：阳起石组和普通角闪石组。阳起石组包括透闪石、阳起石、铁透闪石（铁阳起石）。其余的钙质闪石统称为普通角闪石。都城秋穗把 7.2 的 Si 离子数看作阳起石组和普通角闪石组之间的边界，上述角闪石中以 23（O）为基础计算的 Si 离子数均低于 7.2，据此认为金山群所见的角闪石主要为普通角闪石。

三、原岩恢复

（一）金山群

金山群 a、c 岩性段层状构造清晰，具有较明显的沉积岩特征。原岩由泥质、砂泥质、砂质及碳酸盐岩岩石组成。b 岩性段层状构造不明显，但条带构造、片状构造、片麻状构造较清晰，且条带和片理倾角与 a、c 岩性段中的片岩、大理岩的片理或层理倾角一致。兼具岩浆岩类及沉积岩类的构造特征。

1. 尼格里四面体图解

在尼格里四面体图解中（图 13-2），角闪质类岩石、黑云斜长片麻岩、绿帘阳起片岩及浅粒岩样品投点落在火成岩区域，3 个白云母石英片岩及 1 个白云母石英岩样品投点落在残余沉积、黏土质沉积物区域，1 个白云母石英片岩样品投点落在火成岩区域与残余沉积、黏土质沉积物区域的边界线靠近火成岩一侧；3 个含石英大理岩样品投点落在化学沉积物区石灰岩小区。

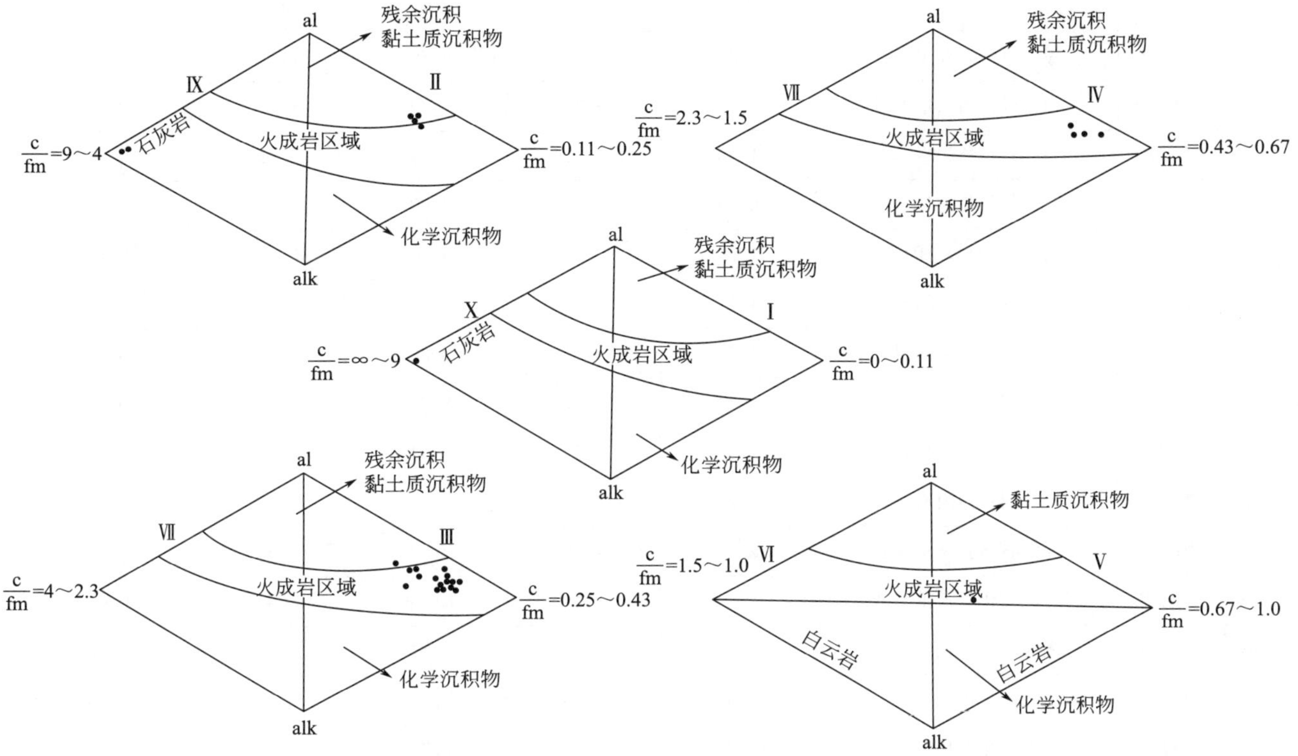

图 13-2　尼格里四面体图解

Fig. 13-2　Niggli's tetrahedron classification diagram

2. 西蒙南［（al+fm）-（c+alk）］-Si 关系图解

10 个角闪质类岩石及黑云斜长片麻岩样品投点落在该图的火山岩区域（图 13-3）；10 个角闪质类岩石、1 个绿帘阳起片岩、2 个浅粒岩及 1 个白云母石英岩样品投点落在厚层泥岩区；4 个白云母石英片岩样品投点落在砂岩区。

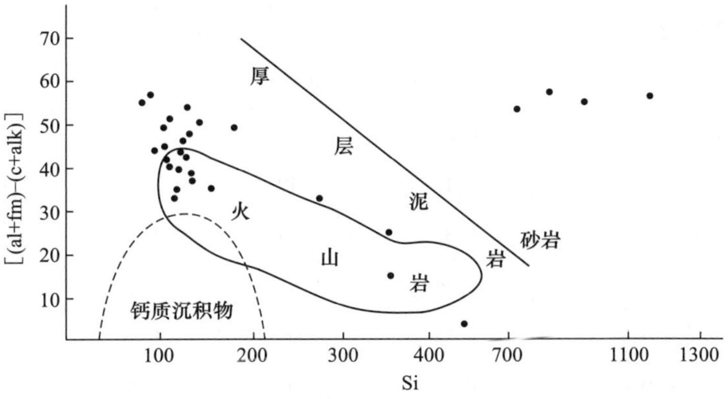

图 13-3　西蒙南［（al+fm）-（c+alk）］-Si 关系图解

Fig. 13-3　Simonen［（al+fm）-（c+alk）］-Si relation diagram

3. S. N. 米斯拉 TiO_2-F 图解

9 个角闪质类岩石样品投点落在火成岩区（表 13-5；图 13-4），11 个角闪质类岩石及黑云斜长片麻岩、绿帘阳起片岩、浅粒岩、白云母石英岩、白云母石英片岩等 20 个样品投点均落在沉积岩区。

表 13-5　TiO_2-F 表

Table 13-5　TiO_2-F table

序号	地点	岩石	氧化物含量/%				F 值
			TiO_2	Fe_2O_3	FeO	MgO	
1	三林塘	黑云母化斜长角闪岩	1.36	4.00	6.36	6.32	0.62
2	三林塘	黑云母化斜长角闪岩	1.34	4.75	7.39	6.79	0.64
3	赵家宅	斜长角闪岩	1.18	6.22	6.61	6.69	0.66
4	沈庄	斜长角闪岩	0.79	4.48	7.90	8.06	0.61
5	肖塘	含绿帘斜长角闪岩	1.30	3.61	9.40	13.83	0.48
6	肖塘	蚀变斜长角闪岩	1.24	3.12	9.60	14.63	0.47
7	肖塘	绿帘斜长角闪岩	1.65	5.41	9.38	5.83	0.72
8	肖塘	绿帘斜长角闪岩	1.74	4.91	8.70	5.75	0.70
9	肖塘	含绿帘斜长角闪岩	1.92	7.59	8.44	6.27	0.72
10	康家宅	斜长角闪岩	2.20	3.51	9.48	7.55	0.63
11	康家宅	绿泥石化斜长角闪岩	2.81	4.58	10.38	5.37	0.74
12	康家宅	绿泥石化斜长角闪岩	2.34	4.41	10.03	6.41	0.69
13	康家宅	绿泥石化斜长角闪岩	2.12	5.64	7.04	6.49	0.66
14	康家宅	斜长角闪岩	2.54	3.60	9.92	4.17	0.76
15	三林塘	绿泥石化斜长角闪岩	1.55	4.70	6.76	5.44	0.68
16	三林塘	角闪斜长片麻岩	1.17	5.13	6.36	6.58	0.64
17	康家宅	绿泥石化斜长角闪岩	2.26	3.70	10.05	5.16	0.73
18	康家宅	黑云母化斜长角闪岩	2.44	4.29	8.74	3.41	0.79
19	天花庵	角闪片岩	1.84	6.21	8.36	6.92	0.68
20	天花庵	角闪片岩	1.70	4.66	9.25	4.71	0.75
21	沈庄	黑云斜长片麻岩	0.26	0.72	1.51	1.05	0.68
22	沈庄	绿帘阳起片岩	0.84	3.10	5.05	1.79	0.82
23	康家宅	浅粒岩	0.15	0.80	0.93	0	1.0
24	赵家宅	浅粒岩	0.54	2.48	4.67	0.85	0.89
25	赵家宅	白云母石英岩	0.18	0.19	1.29	0.81	0.65
26	华泾	白云母石英片岩	0.31	1.24	2.18	1.29	0.73
27	华泾	白云母石英片岩	0.16	0.62	1.48	0.91	0.70
28	华泾	白云母石英片岩	0.22	1.01	1.29	1.32	0.64
29	华泾	白云母石英片岩	0.19	1.02	2.10	1.48	0.68

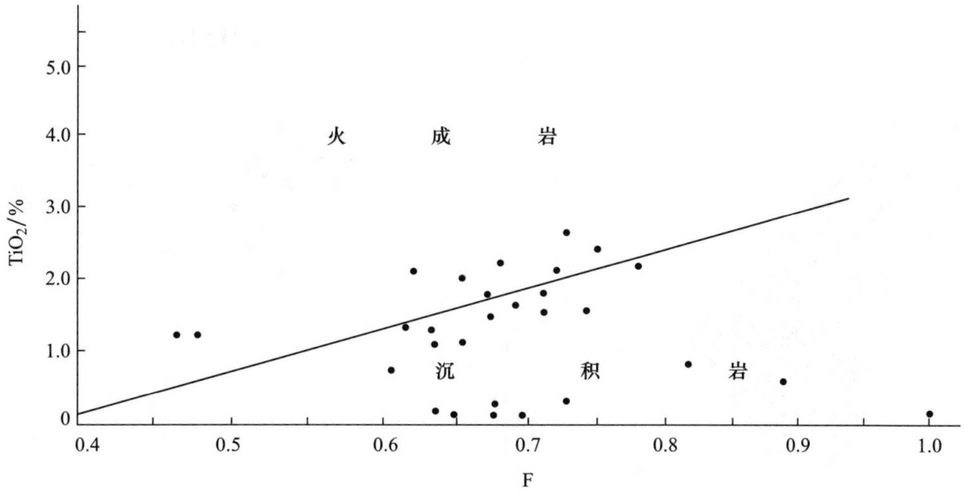

图 13-4　米斯拉 TiO_2-F 图解

Fig. 13-4　Misra's TiO_2-F diagram

4. 周世泰 K-A 图解

20 个角闪质类岩石样品投点有 17 个落在玄武岩区（表 13-6；图 13-5～图 13-7），2 个落在泥质-粉砂质岩石区，1 个落在沉积岩分类图解 3 个圈定的岩石区之外；黑云斜长片麻岩及 1 个浅粒岩样品投点落在流纹岩区；1 个浅粒岩样品投点落在粗面岩区的外侧；绿帘阳起片岩样品投点落在泥质-粉砂质岩石区；白云母石英岩、白云母石英片岩样品投点全部落在砂质岩石区，含石英大理岩样品投点全部落在磷酸盐岩石区。

5. 尼格里（al-alk）、c 关系图解

29 个样品投点（3 个含石英大理岩未在该图上投点）除个别点较分散外，呈较明显的负相关关系，显示出沉积岩的相关特征（图 13-8）。

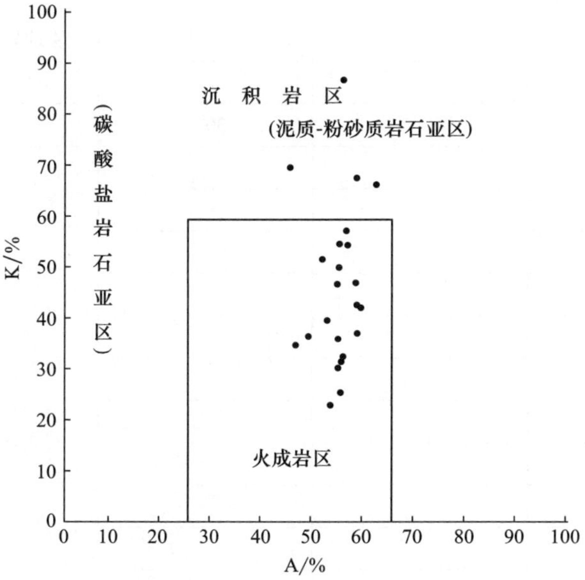

图 13-5　周世泰 K-A 图解

Fig. 13-5　Zhou's K-A diagram

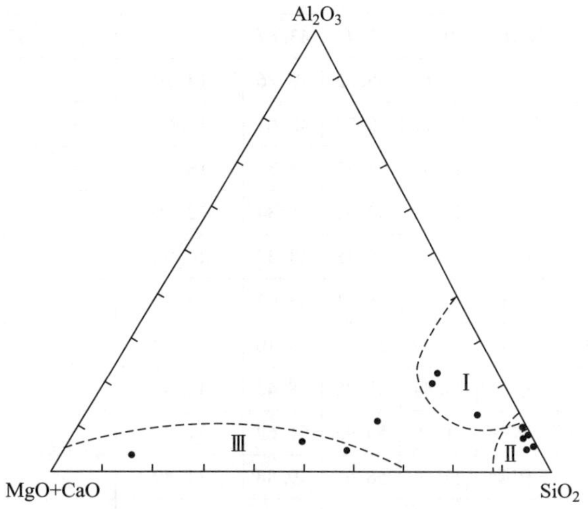

图 13-6　周世泰沉积岩分类图解

Fig. 13-6　Zhou's sedimentary rock classification diagram

Ⅰ—泥质-粉砂质岩石区；Ⅱ—砂质岩石区；Ⅲ—碳酸盐岩岩石区

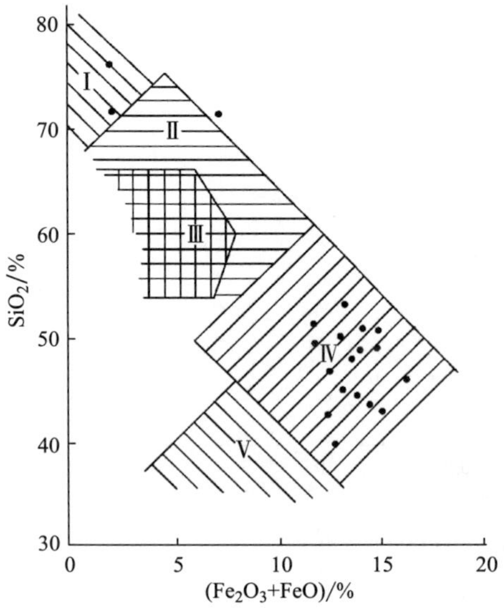

图 13-7　周世泰岩浆岩分类图解

Fig. 13-7　Zhou's magmatic rock classification diagram

Ⅰ—流纹岩；Ⅱ—粗面岩；Ⅲ—安山岩；

Ⅳ—玄武岩；Ⅴ—辉橄-橄榄岩

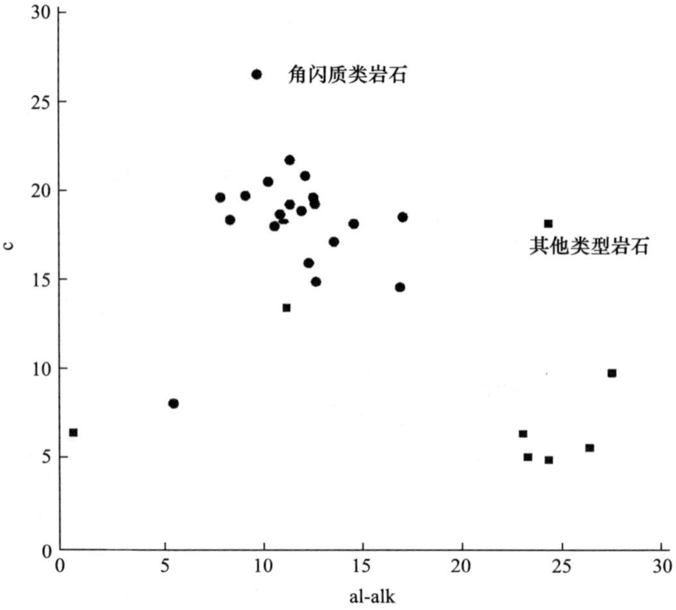

图 13-8　尼格里［al-alk］-c 关系图解

Fig. 13-8　Niggli's［al-alk］-c relation diagram

表 13-6　K-A 图解法有关参数及投影结果

Table 13-6　Parameters and projection results of the K-A diagram method

序号	地点	K/%	A/%	氧化物含量/%		$Al_2O_3/(SiO_2+Al_2O_3+MgO+CaO)$	$SiO_2/(SiO_2+Al_2O_3+MgO+CaO)$	$(MgO+CaO)/(SiO_2+Al_2O_3+MgO+CaO)$	原岩
				SiO_2	Fe_2O_3+FeO				
1	三林塘	65.87	62.82	53.04		20.9	65.1	14.0	泥质-粉砂质岩石
2	三林塘	67.46	58.34	51.57		18.8	64.6	16.6	
3	赵家宅	31.61	56.22	50.38	12.83				玄武岩
4	沈庄	51.85	52.53	46.96	12.38				
5	肖塘	37.29	48.99	42.29	13.01				
6	肖塘	70.23	45.45	43.87		12.0	57.3	30.7	
7	肖塘	31.15	55.32	49.26	14.79				玄武岩
8	肖塘	25.29	55.27	48.67	13.61				
9	肖塘	42.67	58.72	45.32	16.03				
10	康家宅	54.62	55.52	46.34	12.99				
11	康家宅	46.76	55.18	43.32	14.96				
12	康家宅	36.66	58.91	43.62	14.44				
13	康家宅	55.09	55.48	39.10	12.68				
14	康家宅	35.84	55.28	48.42	13.52				
15	三林塘	31.18	55.95	51.62	11.46				
16	三林塘	29.89	56.66	49.88	11.49				
17	康家宅	34.59	46.94	44.89	13.75				
18	康家宅	22.71	53.84	53.62	13.03				
19	天花庵	42.18	60.30	50.96	14.57				
20	天花庵	39.69	53.11	50.85	13.91				

续表

序号	地点	K/%	A/%	氧化物含量/%		$Al_2O_3/(SiO_2+Al_2O_3+MgO+CaO)$	$SiO_2/(SiO_2+Al_2O_3+MgO+CaO)$	$(MgO+CaO)/(SiO_2+Al_2O_3+MgO+CaO)$	原岩
				SiO_2	Fe_2O_3+FeO				
21	沈庄	57.69	57.07	71.04	2.23				流纹岩
22	沈庄	86.45	56.18	64.58		13.1	78.3	8.6	泥质-粉砂质岩石
23	康家宅	50.00	55.30	75.30	1.73				流纹岩
24	赵家宅	46.64	58.92	70.80	7.15				靠近粗面岩
25	赵家宅			88.16		5.7	92.7	1.6	砂质岩石
26	华泾			81.98		8.0	89.9	2.1	
27	华泾			89.21		5.0	93.7	1.3	
28	华经			84.77		7.1	91.0	1.9	
29	华泾			84.16		7.1	90.8	2.1	
30	吕巷			40.46		5.96	55.74	38.30	碳酸盐岩
31	吕巷			30.69		7.51	45.46	47.03	
32	吕巷			7.99		3.33	14.38	82.29	

综合上述研究成果，并合地质产状、岩石自然共生组合等特征，对金山群变质岩的主要原岩类型推断如下。

1. 角闪质类岩石的原岩

角闪质类岩石包括斜长角闪岩、角闪斜长片麻岩、斜长角闪片麻岩、角闪片岩、角闪岩等，兼具玄武质或安山玄武质岩浆岩和沉积岩岩石化学特征，经有关图解投影，有67.5%的点投在火成岩区；Sr-Ba、Cr-Ni成对化学元素的比值与变质岩中辉绿岩脉的Sr/Ba、Cr/Ni值接近；95%的样品，其氧化铝的分子数小于碱加钙的分子数，H. A. 叶利谢夫认为这个系数对火成岩是十分典型的，这一原则有时遭到破坏是由于交代作用造成的。

上述特征说明角闪质类岩石的原岩属岩浆成因。角闪质类岩石的氧化率最低为49，最高97，平均71（表13-7）。Hyndman（1972）认为，由玄武岩变质生成的斜长角闪岩氧化率低，约为30，由玄武质凝灰岩变质生成的斜长角闪岩的氧化率高，约为68。据此推断金山群b岩性段的角闪质类岩石的原岩为玄武质凝灰岩。此外，根据AFM图解投影结果判断，其原岩主要为拉斑玄武岩系列，次为钙碱性玄武岩系列（图13-9）。

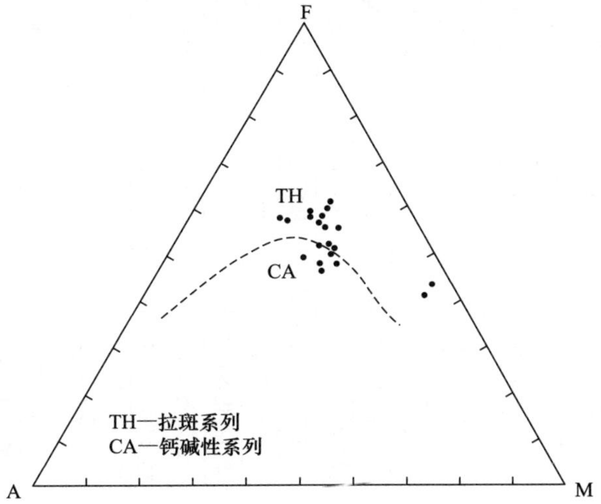

图13-9 角闪质岩石AFM图解

Fig. 13-9 AFM diagram of amphibolic rocks

表 13-7　金山群原岩恢复结果

Table 13-7　Primary rock restoration results in the Jinshan Group

序号	岩性	地点	可能原岩					
			尼氏四面体图解	A. 西蒙南[(al+fm)-(c+alk)]-Si 关系图解	S. N. 米斯拉 TiO_2-F 图解	周世泰 K-A 图解法	氧化率	综合结果
1	黑云母化斜长角闪岩	三林塘	黏土质沉积物	厚层泥岩	沉积岩	泥质-粉砂质岩	77	玄武质凝灰岩
2	黑云母化斜长角闪岩	三林塘	岩浆岩	厚层泥岩	沉积岩	泥质-粉砂质岩	78	玄武质凝灰岩
3	斜长角闪岩	赵家宅	岩浆岩	火山岩	沉积岩	玄武岩	97	玄武质凝灰岩
4	斜长角闪岩	沈庄	岩浆岩	火山岩	沉积岩	玄武岩	72	泥质灰岩和钙质页岩
5	含绿帘斜长角闪岩	肖塘	岩浆岩	厚层泥岩	岩浆岩	玄武岩	55	玄武质凝灰岩
6	蚀变斜长角闪岩	肖塘	岩浆岩	厚层泥岩	岩浆岩	玄武岩	49	玄武质凝灰岩
7	绿帘斜长角闪岩	肖塘	岩浆岩	厚层泥岩	沉积岩	玄武岩	73	玄武质凝灰岩
8	绿帘斜长角闪岩	肖塘	岩浆岩	火山岩	沉积岩	玄武岩	72	玄武质凝灰岩
9	含绿帘斜长角闪岩	肖塘	岩浆岩	厚层泥岩	沉积岩	玄武岩	95	玄武质凝灰岩
10	斜长角闪岩	康家宅	岩浆岩	厚层泥岩	岩浆岩	玄武岩	54	玄武质凝灰岩
11	绿泥石化斜长角闪岩	康家宅	岩浆岩	火山岩	岩浆岩	玄武岩	61	玄武质凝灰岩
12	绿泥石化斜长角闪岩	康家宅	岩浆岩	厚层泥岩	岩浆岩	玄武岩	61	玄武质凝灰岩
13	绿泥石化斜长角闪岩	康家宅	岩浆岩	厚层泥岩	岩浆岩	玄武岩	89	玄武质凝灰岩
14	斜长角闪岩	康家宅	岩浆岩	火山岩	岩浆岩	玄武岩	53	玄武质凝灰岩
15	绿泥石化斜长角闪岩	三林塘	岩浆岩	火山岩	沉积岩	玄武岩	82	玄武质凝灰岩
16	角闪斜长片麻岩	三林塘	岩浆岩	火山岩	沉积岩	玄武岩	89	玄武质凝灰岩
17	绿泥石化斜长角闪岩	康家宅	岩浆岩	火山岩	岩浆岩	玄武岩	54	玄武质凝灰岩
18	黑云母化斜长角闪岩	康家宅	岩浆岩	火山岩	沉积岩	玄武岩	66	玄武质凝灰岩
19	角闪片岩	天花庵	黏土质沉积物	厚层泥岩	岩浆岩	玄武岩	85	玄武质凝灰岩
20	角闪片岩	天花庵	岩浆岩	火山岩	沉积岩	流纹岩	67	玄武质凝灰岩
21	黑云斜长片麻岩	沈庄	岩浆岩	火山岩	沉积岩	泥质-粉砂质岩		中酸性火山岩
22	绿帘阳起片岩	沈庄	岩浆岩	厚层泥岩	沉积岩	流纹岩		泥质-钙质页岩或粉砂质页岩
23	浅粒岩	康家宅	岩浆岩	厚层泥岩	沉积岩	粗面岩?		流纹岩
24	浅粒岩	赵家宅	岩浆岩	厚层泥岩	岩浆岩	砂质岩石		流纹斑岩
25	白云母石英岩	赵家宅	残余沉积	厚层泥岩	岩浆岩	砂质岩石		含泥质石英砂岩
26	白云母石英片岩	华泾	残余沉积	砂岩	沉积岩	砂质岩石		泥质石英砂岩
27	白云母石英片岩	华泾	残余沉积	砂岩	沉积岩	砂质岩石		泥质石英砂岩
28	白云母石英片岩	华泾	残余沉积	砂岩	沉积岩	砂质岩石		泥质石英砂岩
29	白云母石英片岩	华泾	残余沉积	砂岩	沉积岩	砂质岩石		泥质石英砂岩

2. 黑云斜长片麻岩的原岩

根据黑云斜长片麻岩的岩石化学特征及其样品在各种图解上的投影，推测其为中酸性火山岩。

3. 云母片岩类的原岩

白云母片岩、绢云母片岩等云母片岩与大理岩、绿片岩等呈整合互层状产出，层理清晰。推测其原岩系正常沉积成因的黏土质岩石。

4. 白云母石英岩、白云母石英片岩的原岩

该类岩石的化学特征及其样品在有关岩石化学图解上的投影，均反映出原岩具有沉积岩特征。这些岩石的氧化铝分子数均较碱加钙的分子数要大。据此推测它们的原岩分别为含泥质石英砂岩及泥质石英砂岩。

5. 绿片岩类的原岩

a、c 岩性段中发育的斜长绿泥片岩、阳起片岩、钠长阳起片岩、绿帘阳起片岩等绿片岩类岩石，与大理岩、白云母片岩等呈整合互层状产出。除在尼格里四面体图解上样品投点落于火成岩区域外，其余均落在沉积岩区城，推测原岩为泥质-钙质页岩或粉砂质页岩等正常沉积岩。

6. 变粒岩的原岩

所见变粒岩不多，夹于 b 岩性段角闪质类岩石之间，推测系中酸性火山岩变质而成。

7. 浅粒岩及长石石英岩的原岩

1）b 岩性段中与角闪质类岩石共生的浅粒岩与长石石英岩，以含少量绿泥石、黑云母及角闪石为特征，偶含少量石榴子石（铁铝榴石?），推测系酸性火山碎屑岩变质而成。少数浅粒岩在周世泰 K-A 图解上样品投点落在流纹岩区，其氧化铝的分子数小于碱加钙的分子数，推测系流纹岩所变。

2）a、c 岩性段中所见浅粒岩与长石石英岩基本不含暗色片柱状矿物，而普遍含白云母或绢云母，且与白云母石英岩、白云母石英片岩等共生，推测系含泥质长石石英砂岩变质而成。但根据偶尔见于浅粒岩中的个别钾钠长石残斑及长石石英分别聚集成条带相间排列的现象，也不应排除其为流纹斑岩变质而成的可能性。

8. 大理岩类的原岩

大理岩普遍含石英，有时含少量蛇纹石、硅灰石、透闪石及长石、白云母、绿泥石等矿物，并与钠长阳起片岩、斜长阳起片岩、含磁铁阳起片岩、白云母片岩等片岩类岩石呈整合状互层或互为夹层产出，推测其为含硅、铝、镁杂质的石灰岩，即由泥质灰岩或泥质白云质灰岩变质而成。

根据变质岩石自然共生组合及原岩恢复结果，金山群 a 岩性段原岩建造类型为由石英砂岩、长石石英砂岩、页岩、灰岩等岩石组成的砂质、泥质-碳酸盐型；c 岩性段出现较多的泥质-钙质页岩、黏土质岩石等细碎屑岩；b 岩性段的原岩建造以基性火山岩建造为主，夹少量中酸性及酸性火山岩或火山碎屑岩，其中基性火山岩建造以拉斑玄武岩系列的玄武质凝灰岩为主，钙碱性玄武岩系列的玄武质凝灰岩次之。

（二）惠南板岩

由泥质、微细鳞片状绢云母、绿泥石、微粒状方解石及长英质粉砂组成。其中粉砂质点仍保留棱角—次棱角状沉积特征，无明显重结晶现象。绿泥石、绢云母则由泥质成分经变质结晶而成。镜下能见到较清晰的变余泥质结构及变余粉砂状结构。从岩石的矿物组分及残余结构可知，其原岩应为沉积成因的含钙质、粉砂质泥岩类岩石。

四、大地构造环境分析

金山群内部花岗斑岩和朱泾花岗岩代表了中元古代初期（1781±16Ma）、晚期（1082.4±7.9Ma）

不同的构造环境（图 13-10），前者为破坏性活动板块边缘或同碰撞构造环境。区域上，1.77Ga 前大规模的板内玄武岩浆活动很可能与区域伸展构造有关，并引发了广泛的地壳物质重熔（李献华，1993）。后者推测与不成熟的岛弧或洋壳相关，形成幔源型花岗岩。与朱泾花岗岩年龄相当的金山群变质流纹岩、流纹斑岩和中酸性火山岩形成于造山带环境且向碱性、富碱性岩石过渡，个别与现代日本岛弧吻合（图 13-11）。可能分别与哥伦比亚超大陆和罗迪尼亚超大陆的相关活动大陆边缘体系有关，尤其是，在中元古代晚期演化为洋内弧，具有洋壳的弧后盆地伸展环境。与张永康（2008）认为江南地区在中元古代为弧后盆地的认识较一致。赣东北铁砂街组的双峰式火山岩（1172±10Ma～1132±8Ma），具板内特征，可能形成于早期的裂谷环境下（高林志，2013；张恒，2015）。至后期，位于弧前或弧后环境的赣东北和皖南蛇绿岩的形成（王孝磊，2017），都表明区内中元古代末期至晋宁运动Ⅰ幕发生之前，一直处于持续伸展的构造环境。青白口早期，为相对稳定的弧后浅海—次深海环境。

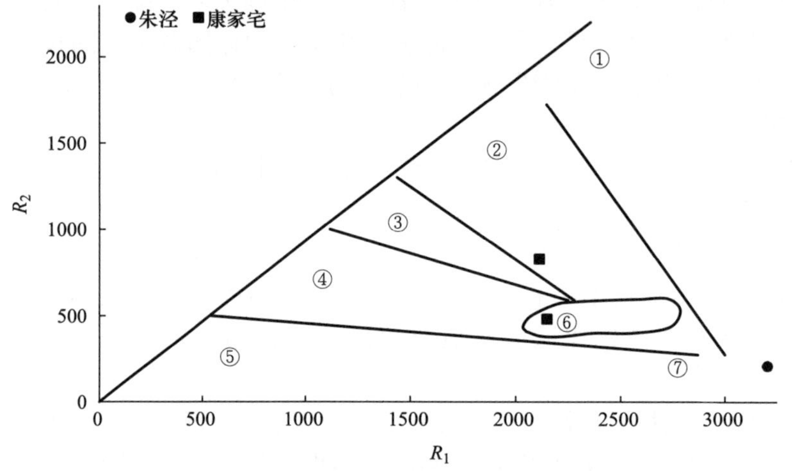

图 13-10　上海不同阶段闪长岩和花岗岩构造环境判别图解

Fig. 13-10　Tectonic environment discrimination diagram of the different periods of diorite and granite in Shanghai

①地幔斜长花岗岩；②破坏性活动板块边缘（板块碰撞前）花岗岩；③板块碰撞后隆起期花岗岩；④晚造山期花岗岩；⑤非造山区 A 型花岗岩；⑥同碰撞（S 型）花岗岩；⑦造山期后 A 型花岗岩

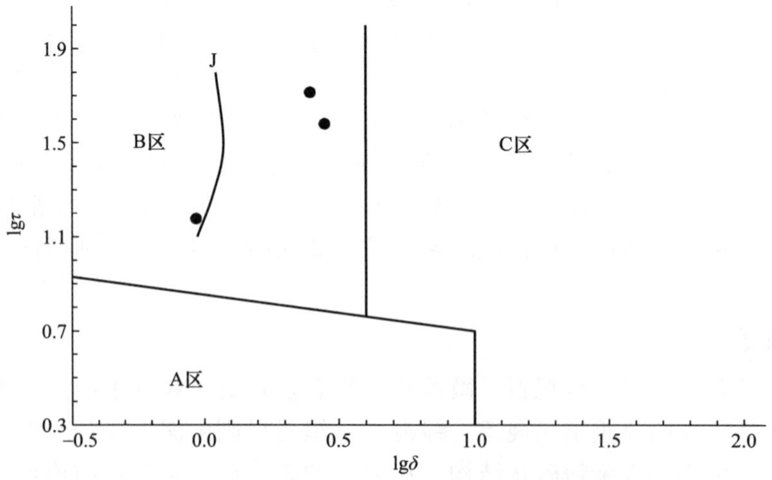

图 13-11　金山群中酸性火山岩构造环境图解

Fig. 13-11　Tectonic environment diagrams of the intermediate-acidic volcanic rocks in the Jinshan Group

A 区—非造山带火山岩；B 区—造山带火山岩；C 区—A 区、B 区派生的碱性、富碱性岩；J—日本火山岩

在上述大地构造环境中，中元古代区内主要形成一套具有类复理石建造特征的浅海相沉积，伴随有间歇性的（海底）火山活动。由早期向晚期，海水似有加深趋势，并且岩浆作用似有加剧特征。青白口早期，沉积了一套细粒碎屑岩。

第三节　变质矿物共生组合和变质相划分

一、金山群

根据矿物共生组合及主要变质矿物特征，金山群可分为低绿片岩相和高绿片岩相。

1. 低绿片岩相

低绿片岩相与a、c岩性段的分布一致，相当于一般的绿片岩相（不包括绿帘角闪岩相），常见的矿物共生组合有：

1）白云母（绢云母）+绿泥石+石英；

2）白云母（绢云母）+斜长石+石英；

3）钠长石+阳起石+绿帘石+绿泥石+石英；

4）斜长石+绿帘石+绿泥石+石英；

5）方解石+斜长石+绿帘石+透闪石+石英。

1)~4）为变质泥质岩的主要矿物共生组合，5）为变质钙质岩所具有的矿物组合。上述组合中以组合3）最为典型。阳起石、钠长石、绿帘石、绿泥石、白云母（绢云母）等为该变质相的特征变质矿物。阳起石的存在标志着金山群的绿片岩相为阳起石绿片岩相。这种阳起石绿片岩相存在于变质的砂质、泥质-碳酸盐型及泥质-碳酸盐型原岩建造中，在泥质-钙质页岩或粉砂质页岩与碳酸盐岩互层并发生变质时，由于富CaO和相对贫Al_2O_3，阳起石的出现是完全可能的。阳起石绿片岩相代表相对低的P_{co_2}。

2. 高绿片岩相

高绿片岩相分布上与b岩性段一致，常见的矿物共生组合有：

1）普通角闪石+斜长石+绿帘石+石英；

2）普通角闪石+斜长石+透辉石+绿帘石+石英；

3）普通角闪石+斜长石+铁铝榴石+黑云母+石英；

4）普通角闪石+斜长石+黑云母+绿帘石+石英；

5）斜长石+钾长石+铁铝榴石+石英；

6）斜长石+钾长石+黑云母+白云母+石英；

7）斜长石+黑云母+绿帘石+石英。

组合6)、7)、8)、9）为变质基性岩矿物共生组合，也是本区的典型矿物共生组合，其中普通角闪石+斜长石+绿帘石+石英为特征矿物组合。按照Ramberg将与绿帘石处于平衡的An=30斜长石成分作为划分绿帘角闪岩相和角闪岩相界线的原则，及普通角闪石N_g的颜色判断，金山群b岩性段应属角闪岩相。在角闪石Al^{IV}-Al^{VI}变异图（图13-12）上，Df_4、Df_5落在绿帘石-角闪岩相，Df_6落在绿帘石-角闪岩相与绿片岩相交界线上，Df_7落在角闪岩相，在角闪石Al′-Si变化关系图（图13-13）上，Df_4与Df_7落在角闪岩相。

根据上述典型矿物共生组合、斜长石An的含量、普通角闪石N_g方向的颜色、在有关变异图及变化关系图上的投影位置以及普通角闪石部分组分含量等特征综合考虑，认为本区金山群b岩性段所处变质相为变质基性岩的高绿片岩相（相当于绿帘石角闪岩相，图13-14）。

二、惠南板岩

岩石变质程度低，矿物成分简单，矿物重结晶不完全，岩石中能见到较多的泥质及少量隐晶状硅质成分。变质矿物绿泥石、绢云母为细小的鳞片状，绢云母带有很淡的绿色。由于惠南板岩中缺少变质基性岩，对其变质相的属性尚很难确定，根据绿泥石+绢云母组合，初步认为惠南板岩所属变质相相当于低绿片岩相的下部。其变质作用类型相当于董申保的区域动力变质作用中的板岩千枚岩变质作用。

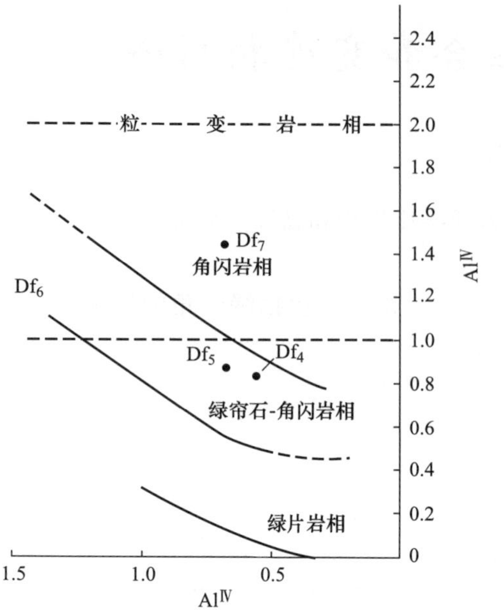

图 13-12　萨克路特金角闪石 $Al^{Ⅳ}-Al^{Ⅵ}$ 变异图

Fig. 13-12　King's hornblende $Al^{Ⅳ}-Al^{Ⅵ}$ variogram

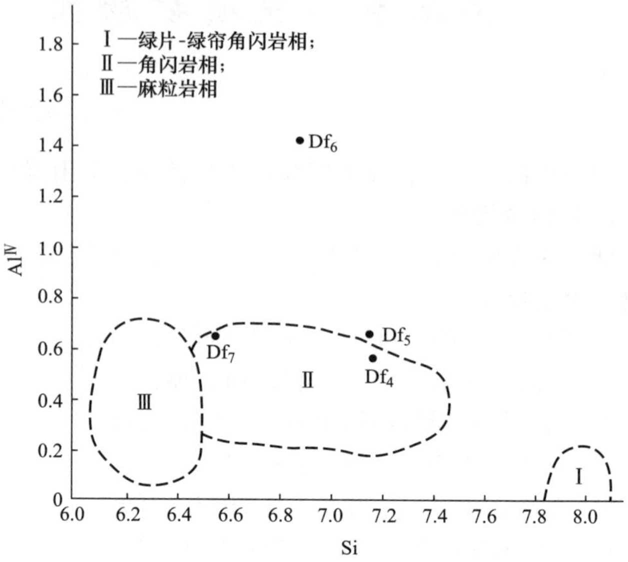

图 13-13　角闪石 $Al^{Ⅳ}$-Si 变异图

（据崔文元，1980）

Fig. 13-13　Hornblende $Al^{Ⅳ}$-Si variations

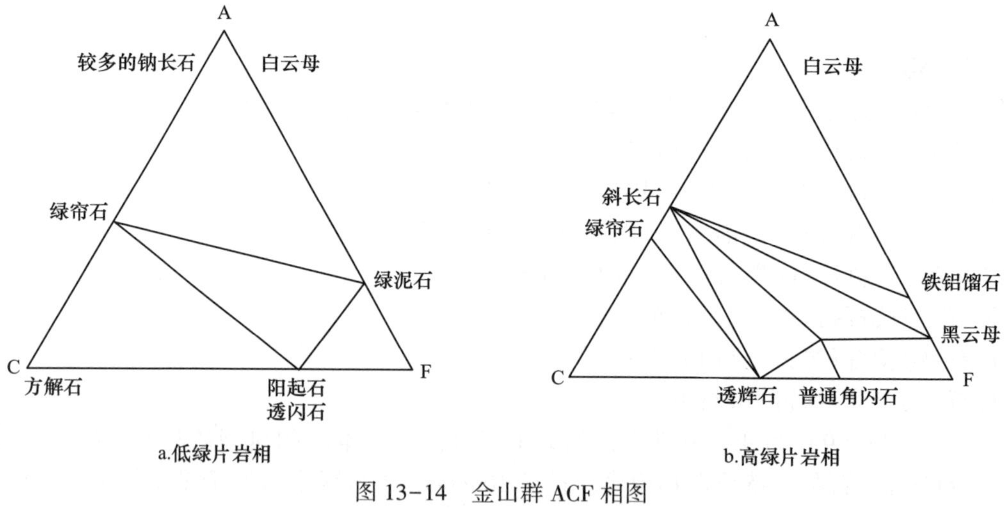

图 13-14　金山群 ACF 相图

Fig. 13-14　ACF phase diagram of the Jinshan Group

第四节　温压条件估算

角闪石 N_g 方向的颜色与 TiO_2 的含量有关，所以随着温度的升高，颜色显示相对有规律的变化：在低温下（通常为绿帘角闪岩相—低角闪岩相）重结晶的变质基性岩中，角闪石 N_g 方向的颜色通常呈蓝绿色（淡绿蓝色或淡蓝绿色），在较高温度下（通常为角闪岩相）为绿色（没有淡蓝的色调），在更高的温度下为棕色。金山群变质基性岩中大量存在的普通角闪石，N_g 方向的颜色为淡绿—深绿色及带有蓝色的绿色，没有淡蓝及棕色色调。

如前所述，b 岩性段变质基性岩的特征矿物组合成分中的斜长石主要为更长石（An=15~28），部分为中长石（An=32~38），这不同于日本三波川带高压变质基性岩和苏格兰高地巴罗式中压变质基性岩中斜长石的成分（三波川带及苏格兰高地变质基性岩中的斜长石 An=5），而与日本领家带和阿武隈高原中部低压变质基性岩中 An=20~30 的斜长石成分较接近。说明金山群具有低压变质特征。惠南板岩变质程度远低于金山群，所以形成的温压条件更低。

根据都城秋穗“变质相的温度和压力”关系图，推测区内变质岩形成的温、压条件为：低绿片岩相形成于 350~450℃，压力为 2~4kbar❶；高绿片岩相（绿帘石角闪岩相）的形成温度为 450~570℃，压力为 3~6kbar。

第五节　变质期次和时期讨论

金山群与惠南板岩在岩石自然共生组合、变质程度、褶皱强度、原岩类型等方面明显不同。金山群岩石类型复杂，惠南板岩岩石类型单一；金山群达高绿片岩相，惠南板岩仅为低绿片岩相；金山群褶皱强烈，挠曲发育，倾角多达 60°~70°，惠南板岩板理面平直，褶皱微弱，板理面倾角大多小于 20°。上述特征表明，两者应该为不同期次的变质作用建造，金山群的变质作用发生时间应早于惠南板岩原岩形成时间。

金山群全岩 Rb-Sr、Sm-Nd 等时线年龄分别为 1123±27Ma、1364±113Ma，前者被认为是变质年龄，后者为原岩年龄。本志在斜长角闪岩、糜棱岩化花岗斑岩中分别获得 1078±8.9Ma、1781±16Ma 的岩浆锆石 U-Pb 年龄，指示其原岩年龄主体应为古—中元古代。最晚的原岩年龄与晋宁运动Ⅰ幕即四堡运动时间等相当，所以金山群的变质作用应该在中、新元古代之交，属于晋宁中期变质岩。前述 Rb-Sr、Sm-Nd 等时线年龄为一种混合年龄，受样品可能不同源、后期蚀变等因素影响，其地质意义并不明确。角闪石 Ar-Ar 法获得的 1259±30Ma、585.3±15Ma、496±13Ma、379.9±10Ma、303.5±8.4Ma、238.2±6.7Ma、223.5±6.3Ma、190.3±9.0Ma、140.1±4.1Ma、124.4±3.6Ma、118.2±3.4Ma、112.2±3.3Ma 等视年龄表明金山群后期受到了多期构造热事件作用的叠加影响。

惠南板岩原岩为青白口纪早期沉积岩，且在上覆河上镇群中获得 774±12Ma、801±13Ma 的锆石 U-Pb 年龄，显示惠南板岩应该属于晋宁晚期变质岩，是晋宁运动Ⅱ幕的产物。

---

❶ 1bar=10^5Pa。

第十四章　接触变质岩及变质作用

接触变质岩主要形成于燕山晚期，分布上明显受燕山期岩体及围岩性质控制。目前已知规模最大的接触变质岩分布在上海西南金山境内，范围为西起界山，东经时家、百家村至戚家浜，向北倾斜。已知上覆地层为劳村组长石石英砂岩、流纹质凝灰岩、流纹斑岩或直接伏于松散层之下，深部与燕山期石英闪长岩接触。主要岩石类型为角岩、大理岩和矽卡岩。

20 世纪 70 年代，《上海市金山县张堰铁铜矿地质特征的初步认识》一文确认界山、张堰变质岩为接触变质岩。《上海市某铁铜矿的地质矿床特征》认为张堰变质岩主要由一套早古生代泥质粉砂质岩经热变质而成，也可能有火山岩成分。《华东地区区域地层表江苏省及上海市分册》（1978）将张堰的一套黑云斜长角岩与矽卡岩组合的变质岩划归"寒武系"。《上海市金山县张堰铜铁矿北接触带阶段性普查评价报告》（1979）认为百家村地区变质岩的原岩可能为一套砂、泥质夹钙镁质成分组成的沉积岩系，经受过区域变质及较广泛的动力变质后，其中部分经后期岩浆活动产生了接触和热液蚀变而形成的一套变质岩系，并将其地层时代划归前泥盆纪。

第一节　主要岩石类型及特征

一、变质岩自然共生组合

区内自上而下有以下 3 种变质岩自然共生组合：

1）上部黑云斜长角岩、黑云堇青角岩、斜长角岩组合。

2）中上部为透辉石榴矽卡岩、石榴子石矽卡岩、绿泥金云石榴矽卡岩、绿帘石榴矽卡岩等各种矽卡岩与大理岩、黑云斜长角岩、透辉斜长角岩形成共生组合；中下部为透辉斜长角岩、黑云斜长角岩夹少量绿泥石化、碳酸盐化石英石榴矽卡岩组合，局部地区靠近下部有角闪斜长角岩。

3）下部为黑云斜长角岩、角闪斜长角岩夹少许透辉斜长角岩与含磁铁矿角闪斜长角岩组合（该组合有时可缺失）。

组合 3）与石英闪长岩体接触；当组合 3）缺失时，组合 2）的下部则直接与深部石英闪长岩体接触。

二、主要变质岩岩石学特征

（一）角岩

1. 黑云斜长角岩

黑云斜长角岩呈灰黑色，具典型的角岩结构、显微等粒状变晶结构、斑状变晶结构，块状构造及不明显的条带状构造，部分呈片状构造，揉皱现象明显，有时见有扁豆状、肠状石英脉体沿条带贯入。主要矿物成分为斜长石、黑云母，含有少量角闪石及微量磁铁矿、磷灰石、绿帘石、锆石及绿泥石。斜长石为显微粒状变晶，近等轴状，大者粒径在 0.7~0.5mm，小者约为 0.01mm，为更长石及中长石；黑云母呈微细鳞片状，长一般为 0.35mm，宽为 0.1mm，星散状或集中呈条带状分布，条带中单个黑云母鳞片的长轴方向与其集合体条带间往往有 10°~20°的夹角，深部近岩体处，岩石条带构造变得模糊不清，矿物颗粒变粗，呈粒状花岗变晶结构。岩石黄铁矿化较为普遍，局部见透辉石化、绿泥石化，致密坚硬，节理裂隙发育。

2. 角闪斜长角岩

角闪斜长角岩呈灰绿—灰黑色，具显微纤状花岗变晶结构，微细条带状构造。主要成分有角闪石、中—更长石，次要成分有黑云母、石英、磁铁矿等，此外尚有极少量的榍石、方解石、黄铁矿等。角闪石、黑云母、磁铁矿等暗色矿物常相对集中呈微细条带分布于长英质浅色矿物中，形成条带状构造。角闪石有两种明显不同的粒级，粗粒者呈粗大的板柱状，晶面弯曲，细粒者呈细小柱状，为两个不同世代。部分角闪石被阳起石交代。岩石致密坚硬，节理裂隙发育，裂隙中为碳酸盐矿物、黄铁矿、绿泥石等充填。角闪斜长角岩多见于本变质岩系下部组合中，与黑云斜长角岩一样，近石英闪长岩体处岩石中的条带构造变得极不明显。

3. 透辉斜长角岩

透辉斜长角岩呈浅绿—灰绿色，具显微花岗变晶结构、显微等粒粒状变晶结构，微细条带状构造并有揉皱现象。矿物成分有中长石、透辉石，少量石英、方解石、绿泥石、榍石等。有不同程度的绿泥石化、碳酸盐化、黄铁矿化。

4. 黑云堇青角岩

黑云堇青角岩呈灰黑—黑褐色，具斑状变晶结构，片状、条带状、块状构造。矿物成分有堇青石、黑云母及少量石英、磁铁矿、绿泥石、金红石等，有时见斜长石及少量矽线石。堇青石一般为粗大变斑晶或变聚斑晶，粒径在0.5~1.0mm，最大达10mm，近等轴卵状、短柱状，常为绢云母鳞片集合体交代，仅保留堇青石假象。黑云堇青角岩多见于该变质岩系的最上部。

5. 红柱石正长角岩

红柱石正长角岩具微粒花岗变晶结构，隐约可见残余的粉砂碎屑结构。主要矿物含量：正长石在40%左右，红柱石在20%~25%，石英在10%左右。次要矿物含量：磁铁矿在10%~15%，呈浸染状，绢云母+斜长石在10%~15%，黑云母小于5%，磷灰石等小于0.1%。红柱石为无色—略带红色，干涉色为一级灰，部分已绢云母化。石英呈粉砂状。

（二）大理岩

大理岩呈灰白—乳白色，细—中粒花岗变晶结构，块状构造，部分具条带状构造。矿物成分以近等轴粒状方解石为主，白云石次之，其他矿物有石榴子石、透辉石、符山石、蛇纹石、硅灰石等，故有石榴透辉大理岩、符山石大理岩、蛇纹石大理岩及硅灰石大理岩之分。常呈透镜体夹于矽卡岩之中。分布范围不大，厚度不等，最厚62.77m，最薄0.30m。

（三）矽卡岩

矽卡岩呈灰白、灰棕、浅绿、黄绿、灰绿等色，不等粒变晶结构，块状构造，部分呈条带状构造。矿物成分有石榴子石、透辉石、硅灰石、绿帘石、石英、绿泥石、方解石、透闪石、金云母、滑石、蛇纹石等。根据所含主要矽卡岩矿物的不同，可分为透辉石榴矽卡岩、石榴子石矽卡岩、石榴透辉矽卡岩、绿泥金云石榴矽卡岩、赤铁矿方解石石英石榴矽卡岩、绿帘石榴矽卡岩、硅灰石榴矽卡岩等多种矽卡岩，以透辉石榴矽卡岩为主。矿物大部分为自形—半自形晶，粗粒矽卡岩矿物间常有交代残留矿物（方解石）存在。岩石具黄铜矿化、磁铁矿化、赤（镜）铁矿化、闪锌矿化，局部具辉钼矿化，为该变质岩系的主要赋矿层位。其中黄铜矿、磁铁矿、闪锌矿已富集成工业矿体。值得指出的是：凡Cu含量达工业品位要求时，Ag含量也都达到工业品位要求，两者呈明显同步增长的趋势。

矽卡岩与大理岩关系极为密切，均分布于岩体附近的接触带上。

三、变质矿物特征

（一）斜长石

斜长石为他形近等轴粒状晶，粒径一般为0.05~0.1mm，大者0.7mm，大部分重结晶现象明显。

似由先后两期生成，早生成者表面布有较多的尘状物而显混浊状，晚生成者表面干净明亮。黑云斜长角岩中斜长石的光轴角 $+2V=80°\sim90°$，斜长石单矿物经 X 光粉晶法测定（表 14-1），其有序度为 0.9~1，牌号 An=22~37。

表 14-1　张堰黑云斜长角岩 X 光粉晶斜长石有序度测定结果

Table 14-1　X-ray powder method for the measurement of order parameter in meplagioclase of the Zhangyan biotite plagioclase hornstone

样品号	测定的 2Q 值			测定结果			
	面网	2Q/(°)	ΔZQ/(°)	名称	有序度	牌号	形成温度/℃
Df_1	131	31.575	1.67	更长石	1.0	An=22	500
	$1\bar{3}1$	29.903					
	$\bar{2}04$	50.700	1.54				
	400	49.164					
Df_2	131	31.453	1.10	更长石	1.0	An=24	500
	$1\bar{3}1$	30.354					
	$\bar{2}04$	51.352	1.20				
	400	50.017					
Df_3	131	31.545	1.74	更长石	0.9	An=37	500~650
	$1\bar{3}1$	29.810					
	$\bar{2}41$	35.853	0.71				
	$24\bar{1}$	35.142					

（二）黑云母

黑云母呈红棕、绿褐色，多色性明显，N_p 为淡黄色，N_g 为略带绿色色调的深褐色，有时被阳起石、绿泥石交代。黑云母为黑云斜长角岩的主要组成矿物之一。

（三）透辉石

透辉石通常为不规则粒状、燕麦粒状和放射柱状体，最长达 8~12mm。有浅灰色、微带淡绿的灰色及褐色等。在（010）切面上测得 $C\wedge N_g=37°\sim40°$，$2V=40°\sim45°$，两组解理交角近 86°。本变质岩系除 4 号样品相当于多勃列佐夫等辉石分类表的 6a 少钙透辉石外（表 14-2），其余 6 个样品均相当于 8a 透辉石。含铁度 f 值（仅 1 个样品为 14%，其余均小于 10%）及阳离子系数值均与接触交代成因镁矽卡岩中辉石的平均 f 值及阳离子系数平均值接近，为接触交代成因类型的透辉石。这和与其密切共生的石榴子石的接触交代成因特征互相呼应。

（四）石榴子石

石榴子石有致密块状和晶粒状两种。常见的晶形为菱形十二面体及菱形十二面体与四角三八面体的聚形，粒径在 0.2~0.5mm，最大可达 1.2mm。具红、棕、黄绿、棕褐等色。有均质和非均质两种，后者具很好的环带构造和扇形双晶，并呈现二轴晶、负光性。石榴子石常被磁铁矿、透辉石和石英交代。19 个石榴子石样品全部为钙铁榴石系列石榴子石（表 14-3），成分以钙铁榴石为主，平均值占石榴子石总成分平均值的 76.59%，钙铝榴石次之，占 19.52%，两者共占 96.12%；镁铝榴石占 2.30%，锰铝榴石占 1.34%，铁铝榴石仅占 0.01%，钛榴石占 0.23%。根据石榴子石的成分及其组成岩石的产状，其成因类型为矽卡岩型。

表 14-2 张堰变质岩中透辉石化学成分、标准晶体化学式及其类型一览表

Table 14-2 Chemical compositions, standard crystal chemical formula and types of the diopside in the Zhangyan metamorphic rocks

序号	氧化物含量/%						标准晶体化学式	以 6000(O) 为基准换算的阳离子数					含铁度 f	颜色	形态	类型
	SiO_2	Fe_2O_3	Al_2O_3	CaO	MgO	FeO		Ca	Fe^{2+}	Fe^{3+}	Mg	Al^{VI}				
1	51.94	1.11	1.49	24.75	16.24	2.22	$Ca_{0.99}(Mg_{0.91}, Fe^{2+}_{0.07}, Fe^{3+}_{0.03}, Al_{0.01})_{1.03}[(Si_{1.94}, Al_{0.06})_{2.00}O_6]$	992	69	31	906	60	10	褐色	块状	透辉石
2	45.28	0.06	0.45	25.40	18.34	1.32	$Ca_{1.11}(Mg_{1.12}, Fe^{2+}_{0.05})_{1.17}[(Si_{1.85}, Al_{0.02})_{1.87}O_6]$	1110	45	2	1115	20	4	灰白色	放射长柱状	透辉石
3	54.37	0.13	0.76	25.17	17.17	2.03	$Ca_{0.98}(Mg_{0.93}, Fe^{2+}_{0.06}, Al_{0.01})_{1.00}[(Si_{1.98}, Al_{0.02})_{2.00}O_6]$	984	62	4	934	20	7	浅绿灰色	放射柱状	透辉石
4	52.90	0.63	0.72	22.09	18.40	2.12	$Ca_{0.88}(Mg_{1.03}, Fe^{2+}_{0.07}, Fe^{3+}_{0.02}, Al_{0.01})_{1.13}[(Si_{1.98}, Al_{0.02})_{2.00}O_8]$	884	66	18	1025	20	8	浅绿灰色		少钙透辉石
5	53.97	0.31	0.76	26.92	14.88	3.88	$Ca_{1.06}(Mg_{0.81}, Fe^{2+}_{0.12}, Fe^{3+}_{0.01},)_{0.94}[(Si_{1.97}, Al_{0.03})_{2.00}O_8]$	1057	119	8	813	30	14	白灰色		透辉石
6	51.52	0.12	0.51	25.09	18.06	1.67	$Ca_{1.01}(Mg_{1.01}, Fe^{2+}_{0.05})_{1.06}[(Si_{1.94}, Al_{0.02})_{1.96}O_8]$	1013	53	4	1014	20	5	浅灰色	放射长柱状	透辉石
7	49.53	0.60	0.81	25.90	21.68	1.47	$Ca_{1.039}(Mg_{1.19}, Fe^{2+}_{0.05}, Fe^{3+}_{0.02})_{1.20}[(Si_{1.83}, Al_{0.04})_{1.87}O_8]$	1025	45	17	1194	40	5	绿灰色	放射长柱状	透辉石

注：含铁度 f=（$Fe^{2+}+Fe^{3+}$）/（$Fe^{2+}+Fe^{3+}+Mg$）。据华东地质科学研究所上海小组，《上海金山张堰铁铜矿床地质特征的初步研究（岩石部分总结）》，1977。

表 14-3　张堰变质岩中石榴子石化学成分、标准晶体化学式及石榴子石成分

Table 14-3　Chemical compositions, standard crystal chemical formula and compositions of the garnet in the Zhangyan metamorphic rocks

序号	氧化物含量/%								阳离子系数								标准晶体化学式	石榴子石成分/%						类型
	SiO_2	Fe_2O_3	Al_2O_3	CaO	MgO	FeO	TiO_2	总和	Si	Ti	Al	Fe^{3+}	Fe^{2+}	Mn	Mg	Ca		Sch	Andr	Pyp	Mgn	Gro	Ald	
1	34. 33	29. 64	1. 88	31. 75	0. 08	0. 27	0. 07	99. 84	2. 911	0. 004	0. 198	1. 891	0. 019	0. 124	0. 010	2. 884	$(Ca_{2.884}, Mg_{0.010}, Mn_{110.124}, Fe^{2+}_{0.019})_{3.037}(Al_{0.109}, Fe^{3+}_{1.891}, Ti_{0.004})_{2.004}(Si_{2.911}, Al_{0.089}O_{12})$	0. 15	97. 16	0. 35	2. 34	0. 00	0. 00	钙铁榴石系列
2	34. 78	29. 55	1. 34	32. 28	0. 29	0. 51	0. 05	100. 0	2. 943	0. 003	0. 134	1. 882	0. 036	0. 086	0. 037	2. 926	$(Ca_{2.926}, Mg_{0.037}, Mn_{0.086}, Fe^{2+}_{0.036})_{3.085}(Al_{0.077}, Fe^{3+}_{1.882}, Ti_{0.003})_{1.962}(Si_{2.943}, Al_{0.057}O_{12})$	0. 11	95. 69	1. 24	2. 92	0. 04	0. 00	
3	37. 20	18. 11	7. 72	33. 06	1. 36	0. 53	0. 46	98. 96	3. 026	0. 028	0. 740	1. 109	0. 036	0. 036	0. 165	2. 881	$(Ca_{2.881}, Mg_{0.165}, Mn_{0.036}, Fe^{2+}_{0.036})_{3.118}(Al_{0.740}, Fe^{3+}_{1.109}, Ti_{0.028})_{1.877}(Si_{3.000}, Al_{0.000}O_{12})$	1. 01	58. 95	5. 95	1. 29	32. 80	0. 00	
4	35. 83	27. 60	1. 29	31. 41	0. 42	1. 29	0. 03	99. 03	3. 039	0. 002	0. 129	1. 762	0. 091	0. 083	0. 053	2. 854	$(Ca_{2.854}, Mg_{0.053}, Mn_{0.083}, Fe^{2+}_{0.091})_{3.081}(Al_{0.129}, Fe^{3+}_{1.762}, Ti_{0.002})_{1.893}(Si_{3.000}, Al_{0.000}O_{12})$	0. 07	93. 11	1. 87	2. 94	2. 01	0. 00	
5	35. 06	28. 68	1. 17	32. 81	0. 36	0. 46	0. 05	98. 86	2. 985	0. 003	0. 117	1. 838	0. 033	0. 019	0. 046	2. 993	$(Ca_{2.993}, Mg_{0.046}, Mn_{0.019}, Fe^{2+}_{0.033})_{3.091}(Al_{0.102}, Fe^{3+}_{1.838}, Ti_{0.003})_{1.943}(Si_{2.985}, Al_{0.015}O_{12})$	0. 11	93. 89	1. 56	0. 66	3. 78	0. 00	
6	34. 84	28. 95	0. 69	32. 84	0. 39	0. 42	0. 04	98. 56	2. 984	0. 003	0. 070	1. 866	0. 030	0. 028	0. 050	3. 014	$(Ca_{3.014}, Mg_{0.050}, Mn_{0.028}, Fe^{2+}_{0.030})_{3.122}(Al_{0.054}, Fe^{3+}_{1.866}, Ti_{0.003})_{1.962}(Si_{2.984}, Al_{0.016}O_{12})$	0. 09	96. 31	1. 72	0. 97	0. 91	0. 00	
7	34. 72	28. 58	1. 02	32. 73	0. 51	0. 30	0. 02	98. 29	2. 977	0. 001	0. 103	1. 844	0. 022	0. 030	0. 065	3. 007	$(Ca_{3.007}, Mg_{0.065}, Mn_{0.030}, Fe^{2+}_{0.022})_{3.124}(Al_{0.080}, Fe^{3+}_{1.844}, Ti_{0.001})_{1.925}(Si_{2.977}, Al_{0.023}O_{12})$	0. 05	94. 66	2. 23	1. 02	2. 04	0. 00	

续表

序号	氧化物含量/%								阳离子系数								标准晶体化学式	石榴子石成分/%						类型
	SiO_2	Fe_2O_3	Al_2O_3	CaO	MgO	FeO	TiO_2	总和	Si	Ti	Al	Fe^{3+}	Fe^{2+}	Mn	Mg	Ca		Sch	Andr	Pyp	Mgn	Gro	Ald	
8	35.48	28.03	0.81	32.45	1.23	0.36	0.03	98.79	3.013	0.002	0.081	1.791	0.026	0.029	0.156	2.952	$(Ca_{2.952}, Mg_{0.156}, Mn_{0.029}, Fe^{2+}_{0.020})_{3.165}(Al_{0.081}, Fe^{3+}_{1.791}, Ti_{0.002})_{1.874}(Si_{3.000}, Al_{0.000}O_{12})$	0.07	95.60	4.33	0.00	0.00	0.00	钙铁榴石系列
9	37.48	28.16	1.39	32.39	0.85	0.46	0.02	98.43	2.970	0.001	0.140	1.810	0.033	0.027	0.108	2.964	$(Ca_{2.964}, Mg_{0.108}, Mn_{0.027}, Fe^{2+}_{0.033})_{3.132}(Al_{0.110}, Fe^{3+}_{1.810}, Ti_{0.001})_{1.921}(Si_{2.970}, Al_{0.030}O_{12})$	0.04	92.78	3.70	0.94	2.54	0.00	
10	34.88	29.47	0.92	32.62	0.19	0.28	0.03	98.73	2.980	0.002	0.093	1.895	0.020	0.02	0.024	2.986	$(Ca_{2.986}, Mg_{0.024}, Mn_{0.025}, Fe^{2+}_{0.020})_{3.005}(Al_{0.073}, Fe^{3+}_{1.895}, Ti_{0.003})_{1.970}(Si_{2.980}, Al_{0.020}O_{12})$	0.07	95.27	0.81	0.83	3.02	0.00	
11	37.56	11.51	13.78	33.96	0.30	0.64	0.11	98.39	2.996	0.007	1.295	0.967	0.043	0.036	0.036	2.902	$(Ca_{2.902}, Mg_{0.036}, Mn_{0.036}, Fe^{2+}_{0.043})_{3.017}(Al_{1.291}, Fe^{3+}_{0.691}, Ti_{0.007})_{1.989}(Si_{2.996}, Al_{0.004}O_{12})$	0.22	34.56	1.20	1.20	62.63	0.19	
12	37.03	11.67	13.25	33.86	0.29	1.10	0.16	97.99	2.981	0.001	1.257	0.707	0.704	0.045	0.031	2.921	$(Ca_{2.921}, Mg_{0.031}, Mn_{0.045}, Fe^{2+}_{0.074})_{3.071}(Al_{1.238}, Fe^{3+}_{0.7072}, Ti_{0.010})_{1.955}(Si_{2.981}, Al_{0.019}O_{12})$	0.03	35.66	1.06	1.53	61.42	0.00	
13	37.48	16.53	9.70	34.27	0.93	0.87	0.00	100.18	3.000	0.000	0.915	0.966	0.058	0.027	0.111	2.939	$(Ca_{2.939}, Mg_{0.111}, Mn_{0.027}, Fe^{2+}_{0.058})_{3.135}(Al_{0.915}, Fe^{3+}_{0.996}, Ti_{0.000})_{1.911}(Si_{3.000}, Al_{0.000}O_{12})$	0.00	52.11	3.87	0.95	43.07	0.00	
14	37.10	11.98	12.82	33.74	0.56	0.72	0.40	97.81	2.988	0.024	1.217	0.726	0.048	0.033	0.067	2.912	$(Ca_{2.912}, Mg_{0.067}, Mn_{0.033}, Fe^{2+}_{0.048})_{3.060}(Al_{1.205}, Fe^{3+}_{0.726}Ti_{0.024})_{1.955}(Si_{2.988}, Al_{0.012}O_{12})$	0.83	36.54	2.31	1.15	59.17	0.00	
均值																		0.23	76.59	2.30	1.34	19.53	0.01	

注：Sch—钛榴石；Andr—钙铁榴石；Pyp—镁铝榴石；Mgn—锰铝榴石；Gro—钙铝榴石；Ald—铁铝榴石。据华东地质科学研究所上海小组，《上海金山张堰铁铜矿床地质特征的初步研究（岩石部分总结）》，1977。

第二节 原岩类型

根据角岩+大理岩+矽卡岩岩石组合及各类岩石的主要矿物共生组合特征，推测有下列原岩类型存在。

一、泥质岩及富钙泥质岩

泥质岩及富钙泥质岩经低级区域变质作用形成千枚岩及千枚状片岩，复经热接触变质作用发生重结晶，形成近等轴的糖粒状斜长石和石英、片柱状习性较差的黑云母、不规则圆形和椭圆形堇青石以及红柱石等变质矿物，并构成黑云斜长角岩、黑云堇青角岩、红柱石正长（斜长）角岩等具有典型角岩结构的热接触变质岩石。堇青石、红柱石及矽线石等均系泥质原岩接触变质的特征矿物。矽线石较少出现，这可能是因为接触变质作用比低压区域变质作用具有更低的岩压。

此外，本变质岩系普遍见有黑云母，而堇青石仅见于个别层位。这是由于其原岩主要为富钙泥质岩，缺钙或少钙泥质岩较少的缘故。温克勒和冯·普拉顿1960年的试验已经证明：原来存在于沉积岩中的方解石数量决定了由变质作用产生的黑云母和堇青石的数量……，原来沉积岩中方解石的数量增加，堇青石的数量就减少，而黑云母的数量却增加。

二、长英质碎屑岩

含有一定数量泥质、钙质成分的长石砂岩、长石石英砂岩等长英质碎屑岩，早期经区域变质作用曾发生轻微变质，燕山期热接触变质作用又使其进一步发生重结晶，最终形成以含斜长石为主，石英次之，并含一定数量黑云母，具等轴粒状变晶结构、块状构造的斜长角岩。

三、含白云质灰岩

大理岩的主要矿物成分为方解石，次为白云石，并含有少量蛇纹石，说明其原岩为以碳酸钙为主、碳酸镁次之的含白云质灰岩。大理岩及矽卡岩中的石榴子石（钙铝榴石系列石榴子石）、透辉石、符山石、硅灰石等均属矽卡岩型接触交代变质矿物，它们形成时所需的 $A1_2O_3$、SiO_2 等均系矽卡岩阶段的外来成分，并非原岩所固有。

四、基性脉岩

部分角闪斜长角岩中可隐约见到变余辉绿结构，可能是早于石英闪长岩岩体的基性脉岩（辉绿岩）经重结晶变质而成。

综上所述，界山—张堰的燕山期接触变质岩，其原岩主要为一套正常沉积的泥质、砂质细碎屑岩与碳酸盐岩，间有少量基性脉岩。上部、下部主要为泥质岩、富钙泥质岩及长英质碎屑岩，中上部为含白云质灰岩、富钙泥质岩与长英质碎屑岩互层，中下部以富钙泥质岩为主，夹少量白云质灰岩。

第三节 变质作用类型

一、接触热变质作用

业已发生低级区域变质作用的前震旦系金山群变质岩，由于燕山期中酸性岩浆岩的侵入，在接触带及其附近发生重结晶变质作用，形成具花岗变晶结构、等轴粒状变晶结构及块状构造等典型角岩结构的各种岩石。

局部地段，劳村组底部的长石石英砂岩、流纹质凝灰岩等，由于岩体所携热流的影响也发生角岩化现象，形成长英质角岩与或角岩化岩石。这种岩石的原岩结构、构造及矿物成分基本上仍清晰可辨。当原岩为泥质成分含量高的凝灰质或粉砂质泥岩及泥质砂岩时，对热接触变质作用最敏感，常形成含堇青石、红柱石的角岩或角岩化砂岩。

上述各种角岩及角岩化岩石的分布明显受燕山期石英闪长岩及脉岩的控制，岩石的重结晶程度与距离岩体远近关系密切，愈近岩体，岩石的重结晶愈明显，矿物颗粒愈粗大，岩石的条带状构造则变得愈加模糊不清，远离岩体者反之。

二、接触交代变质作用

燕山期石英闪长岩体使围岩受热流影响发生热接触变质并析出挥发分，大量聚集的挥发分沿接触带或围岩破碎带及片理面向围岩运移的过程中，对围岩特别是对碳酸盐岩围岩进行选择交代作用，改变了围岩化学成分，带进 SiO_2、Al_2O_3，带出 CaO、MgO 及 CO_2，形成含方解石、石榴子石、透辉石、硅灰石、绿帘石、阳起石、透闪石、金云母等矿物成分的岩石。钙铁榴石系列石榴子石及透辉石的化学成分均可说明它们为接触交代成因的矽卡岩矿物。继矽卡岩形成之后，富铁溶液于热液阶段在有利部位堆积并形成磁铁矿体。接着进入石英-硫化物-碳酸盐阶段，形成了以黄铜矿为主，闪锌矿、辉钼矿为次并富含银的铜多金属硫化物矿物的富集体。石榴子石矽卡岩即为黄铜矿、闪锌矿及银的贼存层位，银含量与铜含量呈明显的正相关关系。

交代变质作用稍晚于接触变质作用发生时间，这一点可以被一些角岩的矽卡岩化现象所证实。

对界山—张堰变质岩来说，接触变质作用为一种叠加变质作用。在此之前，早已发生过较轻微的区域变质作用。依据有：①千枚岩、千枚状片岩残留体的存在；②角岩及矽卡岩所具有的条带构造的倾角在 40°~60°之间变化，以 50°~60°为多，这与区内前震旦系金山群区域变质片岩类、角闪质类岩石的片理和条带的倾角几近一致。所不同的是，角岩和矽卡岩所具条带构造未能给予岩石以裂开成薄板的趋势，这一点正是接触变质角岩所具条带（继承性条带）与区域变质岩所具条带的根本区别。

在界山—张堰变质岩系发生燕山期接触变质作用之后，又经后期的动力（碎裂）变质作用叠加，以致各类岩石发生碎裂甚至糜棱岩化，可见斜长石双晶纹的扭弯曲或错位。张堰矿床中的磁铁矿破碎，说明在成矿阶段之后仍有动力变质作用发生。

第四节　变质矿物共生组合和变质相划分

角岩类岩石的典型矿物共生组合有斜长石-黑云母-角闪石-石英、斜长石-黑云母-堇青石-矽线石（少量）及红柱石-正长石-斜长石-黑云母-石英组合（表 14-4）。

表 14-4　主要变质岩结构构造及矿物成分

Table 14-4　Textures, structures and mineral compositions of the main metamorphic rocks

岩石类型	结构构造	矿物组成
斜长角岩、黑云母斜长角岩、角闪斜长角岩、透辉石斜长角岩、黑云堇青角岩、红柱石正长(斜长)角岩	粒状变晶结构、变斑状结构、鳞片花岗变晶结构、显微纤状花岗变晶结构、显微花岗变晶结构,条带状构造、片状构造	斜长石(中长石、更长石)、角闪石、黑云母、石英、堇青石、红柱石、正长石、黑云母,堇青角岩中偶见矽线石
大理岩	细—中粒变晶结构,块状和部分具条带状构造	方解石、白云石、石榴子石、透辉石,少量符山石、蛇纹石、硅灰石
石榴子石矽卡岩、透辉石榴矽卡岩、石榴透辉矽卡岩、绿泥金云石榴矽卡岩、绿帘石榴矽卡岩、硅灰石榴矽卡岩	不等粒变晶结构,块状、部分具条带状结构	石榴子石、透辉石、绿帘石、硅灰石、石英、方解石、透闪石、金云母、绿泥石、滑石、蛇纹石

大理岩类的矿物共生组合有方解石-白云石-石榴子石-透辉石-硅灰石（少数情况下可见符山石）组合。

矽卡岩类矿物共生组合有石榴子石-透辉石-硅灰石-绿帘石等。

根据上述矿物共生组合及结构、构造特征判断，界山—张堰变质岩所处变质相至少相当于角闪角岩相，其中的红柱石-正长石-斜长石-黑云母-石英组合已达辉石角岩相。

第五节　变质作用时期讨论

区内发生的接触变质作用时间基本上均系燕山期。

界山—张堰变质岩的下部层位原岩推测为元古宇—下古生界，上部层位原岩为劳村组，变质作用发生的上限应为早白垩世早期。主要热源为石英闪长岩侵位，经锆石 U-Pb 法测得的同位素年龄值为 101.6±1.1Ma。穿插在金山群变质岩中的辉绿岩岩脉，经 K-Ar 法测定，其同位素年龄值为 102.0Ma（角闪石）及 106.0Ma（全岩），均属燕山晚期。黑云斜长角岩的 K-Ar 法年龄值（黑云母）为 147.0Ma 及 121.0Ma，与上述岩体、岩脉同位素年龄值接近，说明见于界山—张堰一带大面积产出的角岩、矽卡岩均系燕山晚期岩浆活动产物。

见于其他岩体周围的接触变质作用与相关的侵入岩体侵位时间相当。

第四篇　地质构造

第十五章　构造单元及主要特征

上海位于江山—绍兴深断裂的北西侧，受地质和地球物理资料不足的制约，其大地构造位置和构造单元的划分并不一致；但通常被划为扬子克拉通的一部分。有关地质构造方面的论述，最早见于20世纪60年代初石油及天然气地质普查报告和图件中。随着石油地质、矿产地质、水文工程地质、地震地质工作的深入开展，特别是钻探、地面物探、航空磁测工作的增多和综合应用，积累了愈来愈多的地质构造方面的基础资料，认识也在逐步深化。

1960年，《上海市石油及天然气地质普查工作报告》将上海划为浙西坳陷，南侧在杭州湾北岸与华夏古陆相邻。境内又可细分为上火山岩系分布带、祝桥—新腾断块隆起带、下火山岩系分布带、嘉定—昆山断块隆起等。1962年，《上海地区矿产地质普查工作总结报告》将上海统称为上海中新坳陷带，细分为张堰—奉贤中新坳陷带、吕巷—祝桥褶断隆起带、青浦—浦东中新坳陷带、嘉定褶断隆起带、庙镇中生代断陷。1980年，《中国大地构造及其演化》显示上海位于扬子准地台浙西、皖南台褶带和下扬子台褶带东缘，崇明—景德镇深断裂及球川—萧山深断裂延经区内。

1988年，《上海市区域地质志》首次对上海进行了4级构造单元划分。将其划为扬子准地台一级构造单元，以湖苏断裂和江绍断裂为界的浙西—皖南台褶带之上海台陷和下扬子台褶带之南通—湖州台拱2个三级构造单元，并将前者划分出青浦—宝山坳褶断束和金山—南汇隆褶断束2个四级构造单元。杨志坚（1989）认为，上海为皖南—浙西构造单元的东北自然延伸部分，为一中元古代形成的裂谷系，可细分为太湖复向斜、江南（上海）复背斜、钱塘复向斜等。陈焕疆（1990）在进行东海北部及邻区基底构造单元划分时，将上海划为扬子准地台，东部水域和杭州湾南部划为海陆衔接带。吕炳全（1991）提出了上海地块的概念，认为上海地块与周边相邻的地块有较大差别，应属外来地体，于中生代早期拼贴于中国大陆的东部。1992年，《上海地区地质构造研究》等将上海及邻区划分出了上海地块、苏州—南通地块、余杭—嘉兴地块和浙东地块等4个一级构造单元，又将上海地块细分为上海—长江口断隆区、角直—青浦断隆区、金山—杭州湾断隆区等3个二级构造单元，庙镇—启东隆起带、浏河—新开河坳陷带、上海—佘山岛隆起带、泥城—牛皮礁坳陷带、阳澄湖—昆山隆起、安角凹陷、青浦隆起、松江—北桥凹陷带、金山—钱桥隆起、杭州湾—小戢山坳陷等10个三级构造单元。1997年，《上海市新构造运动与地震活动》在前述方案基础上进行了修改，将上海及邻区划分出了上海台隆、下扬子台陷、浙西台陷、浙东台隆等4个一级构造单元，并将上海台隆划分出了昆山—佘山岛台凸和金山卫—灶前礁台凸2个二级构造单元。原二级构造单元划为三级，并增加了松江—牛皮礁断陷。原三级构造单元划为四级构造单元。郭玉贵（1996）在东海及邻区构造划分中将上海划归华南岭南地块和扬子京畿地块，并将后者细分为勿南沙隆起区。朱光（1999）认为，上海划分为皖南—苏南坳陷、江南隆起带和钱塘坳陷3个二级构造单元，上海中部隆起是江南隆起带的一部分，通过天目山—白际山剪切带与皖南、江西地区相接。陈沪生（1999）在进行下扬子及邻区构造区划时，将上海划为华东大陆边缘板块之下扬子裂陷盆地区，三级构造单元为苏南隆起。2008年，“上海市三维城市地质调查”项目将区内划分为扬子陆块，以湖苏断裂为界划为上海—浙西地块、苏州—南通地块2个二级构造单元，并进行了三、四和五级构造单元划分。其中，三、四级构造单元与《上海地区地质构造研究》的二、三级构造单元相当。尹延鸿（2008）在对中国东部海区进行编图时，将上海划归扬子地块之苏南-勿南沙-光州隆褶带和华南地块之浙闽隆褶带2个四级构造单元。

值得注意的是，白杨明桂等（1997）提出钦杭结合带的认识后，历经20余年大量地质事实和地球物理资料的检验，钦杭结合带这一构造单元已为国内外地质界广泛认同和接受；界线划分上，东段被广泛接受的划分方案是北以景德镇—崇明断裂为界，南以江山—绍兴断裂带为界；但在板块归属、华南洋特征等方面仍存在不同的认识。

第一节　构造单元划分

根据工作指南建议，结合本区的实际地质情况和区域上构造演化阶段特征，对上海进行了5级综合构造单元划分（表15-1；图15-1）。

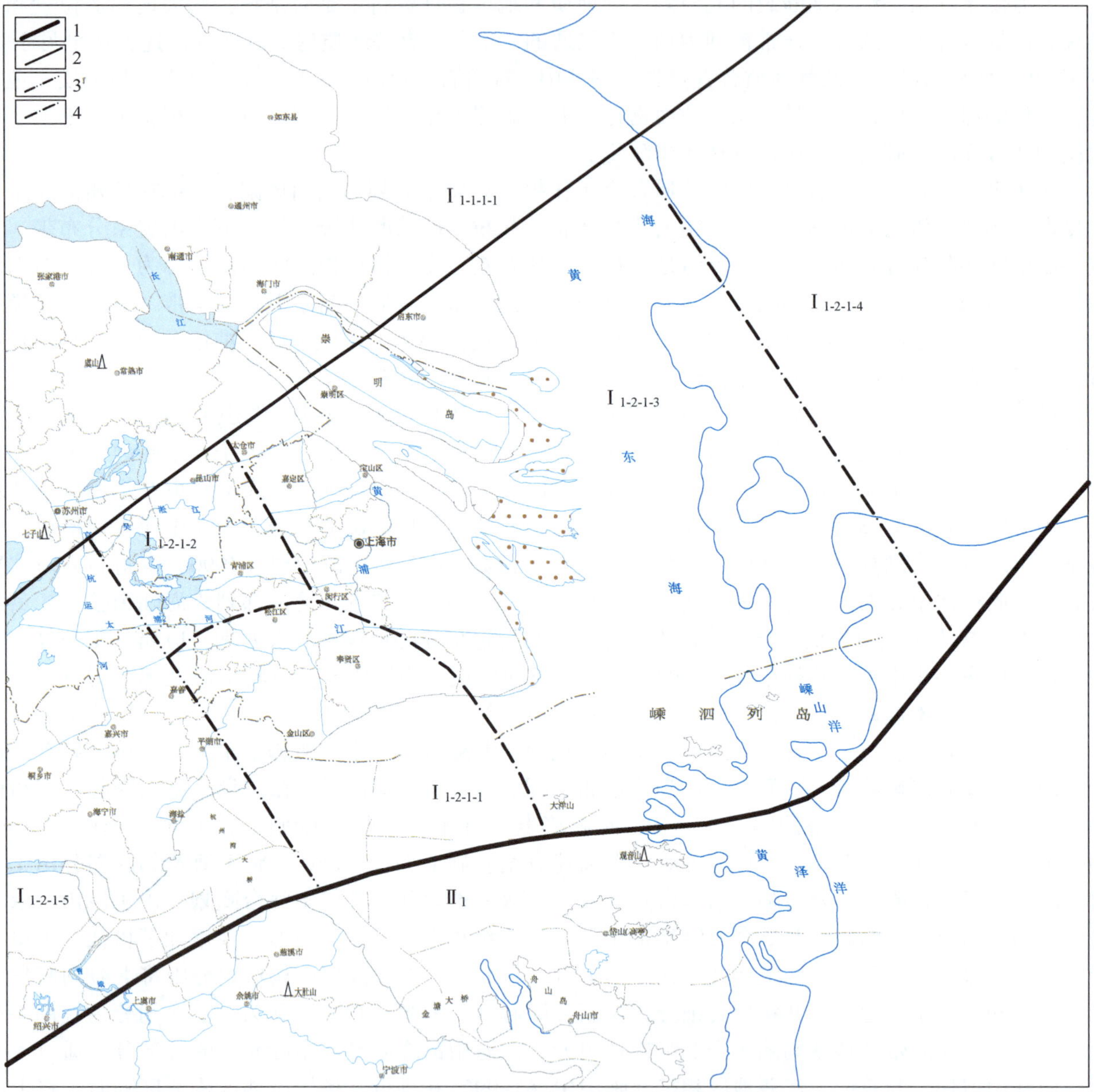

图15-1　综合构造单元区划图

Fig. 15-1　Integrated tectonic unit zoning map

1—Ⅰ级、Ⅱ级分区界线；2—Ⅲ级分区界线；3—Ⅳ级分区界线；4—Ⅴ级分区界线

以江绍断裂带为界，将上海及邻区划分为扬子克拉通和华夏造山系两个Ⅰ级构造单元。下扬子陆块和江绍拼合带两个Ⅱ级构造单元。下扬子陆块是叠置于前南华纪基底之上的一个多旋回沉积复合盆地，经历了多期构造运动的改造，内部构造极为复杂，广泛发育早古生代、晚古生代海相、海陆过渡相沉积。江绍拼合带是一个古华南洋在经历了俯冲消减后，最终发生闭合、两侧陆块（扬子克拉通和武夷地块）发生碰撞拼贴而形成的构造带，以陈蔡俯冲增生杂岩发育为特征。

表 15-1　综合构造单元划分简表

Table 15-1　Integrated tectonic unit division table

<table>
<tr><th>Ⅰ级</th><th>Ⅱ级</th><th>Ⅲ级</th><th>Ⅳ级</th><th>Ⅴ级</th></tr>
<tr><td rowspan="6">扬子克拉通
（Ⅰ）</td><td rowspan="6">下扬子陆块
（I_{1}）</td><td>下扬子克拉通盆地
（I_{1-1}）</td><td>吴江—南通断块区
（I_{1-1-1}）</td><td>南通断隆区（$\mathrm{I}_{1-1-1-1}$）</td></tr>
<tr><td rowspan="5">钦杭结合带
（I_{1-2}）</td><td rowspan="5">I_{1-2-1}
杭嘉沪断块区</td><td>金山—杭州湾断隆区（$\mathrm{I}_{1-2-1-1}$）</td></tr>
<tr><td>佘山—青浦火山隆起区（$\mathrm{I}_{1-2-1-2}$）</td></tr>
<tr><td>上海—长江口断隆区（$\mathrm{I}_{1-2-1-3}$）</td></tr>
<tr><td>长江口东断隆区（$\mathrm{I}_{1-2-1-4}$）</td></tr>
<tr><td>余杭—嘉兴断坳区（$\mathrm{I}_{1-2-1-5}$）</td></tr>
<tr><td>华夏造山系
（Ⅱ）</td><td>江绍拼合带
（II_{1}）</td><td></td><td></td><td></td></tr>
</table>

Ⅲ级构造单元主要依据北东向区划性断裂、前燕山期的建造和变形特征进行划分。具体为以湖苏断裂带、江绍断裂带为界划分为下扬子克拉通盆地和钦杭结合带。钦杭结合带在前南华纪为一系列的岛弧、弧前盆地和弧后盆地，见有蛇绿混杂岩。中—新元古代地层十分发育，青白口系上部以火山碎屑岩-火山岩系建造为主。南华纪以来演化为裂谷盆地、被动陆缘盆地、陆表海盆地。下扬子克拉通盆地基底形成阶段弧盆格局简单，南华纪—早古生代构造格局与江南地区相似，但晚古生代以来陆表海盆地沉积比较发育。

Ⅳ级和Ⅴ级构造单元主要依据北西向断裂构造、滨太平洋活动阶段的火山和断陷盆地建造总体发育特征进行划分。各构造单元在区域重、磁场上具有明显不同的特征。

此外，在Ⅴ级构造单元中，由于燕山期以来断块活动差异和深部构造因素影响，形成一系列断凸和断凹、火山洼地和隆起等。根据其特征可以划分出一系列次级构造单元。不同次级构造单元内，基岩面埋深、基岩地层组合差别较大。

第二节　各构造单元特征

作为扬子克拉通边缘的一小部分，无论是基底与盖层的发育特征，还是陆缘活动阶段的地质构造特征，上海均与下扬子陆块的特征基本一致。同时，作为Ⅰ级、Ⅱ级构造单元的结合部位，区内建造和改造特征具有鲜明的过渡性特征，已发现的下古生界与浙西—皖南发育情况十分相似，苏南地区广布的上古生界在区内暂未发现或缺失。

一、南通断隆区

南通断隆区位于湖苏断裂带以西地区，只包括上海境内崇明岛西北角的极小部分。这是一个晚三叠世以来相对上隆的区域，其中上古生界发育齐全，印支运动使其产生以下古生界为背斜核部，中、下三叠统为向斜核部的北东向褶皱。早白垩世，陆相火山盆地发育；晚白垩世，内陆湖泊相断陷式红盆发育。喜马拉雅期，见内陆湖盆或三角洲相红色碎屑岩沉积，上海境内钻孔揭露资料有限，推测其地质特征与苏锡地区相似。

二、金山—杭州湾断隆区

金山—杭州湾断隆区位于上海西南部，北以卖花桥—鲁汇断裂为界，西抵苏州—嘉善断裂，南至江绍断裂带。发育有大量区域变质岩、接触变质岩和火山岩，局部有断陷盆地发育。重力异常带总体呈向北凸起的弧形，其中部为一片正异常带，南、北两侧均为负异常。钻探揭示正异常带基岩多属前震旦系—奥陶系，部分老地层上覆盖有数百米厚的下白垩统。而负异常带基岩则一般为中、新生代地层构成的断（坳）陷盆地。包括松江—北桥盆地、吕巷断凸、钱桥断凸、杭州湾—小蕺山坳陷、庄行构造火山盆地等次级构造单元。

三、佘山—青浦火山隆起区

佘山—青浦火山隆起区位于上海西部，北为湖苏断裂带，东抵太仓—奉贤断裂，南以练塘—新桥一线为界，西抵苏州—嘉善断裂。为中生代大面积的火山岩分布区，局部有中、新生代断陷盆地沉积，中、新生代地层厚度推测大都超过3000m。在布格重力异常图上，除东北角的昆山—巴城—太仓和东侧重固周围为重力正异常外，其余均为布格重力负异常区。在航磁异常图上主要表现为正磁异常。包括了安角盆地、天马山—青浦火山隆起、重固—凤西断凸等次级构造单元。

四、上海—长江口断隆区

上海—长江口断隆区位于上海东北部，北以湖苏断裂带为界，西至太仓—奉贤断裂，南抵松江—北桥坳陷带北界，东延入海至苏北沿岸断裂。区内前震旦系、下古生界断凸和长江口坳陷带发育，发育大面积的燕山期侵入岩和金桥构造火山盆地，且新生代火山岩发育。在航磁异常图上有大量正磁异常，局部成片分布。在重力异常图上，除长江口北支口外，该区重力场均显示大片重力正异常。包括庙镇—启东隆起、浏河—新开河坳陷、上海—佘山岛隆起和泥城—牛皮礁坳陷、长江北支口坳陷等次及构造单元。

五、长江口东断隆区

长江口东断隆区位于上海东北部海域，北以湖苏断裂带的海域延伸段为界，南以江绍断裂带为界，西至苏北沿岸断裂。在重力异常图上，基本为正异常区。多为重力高，极个别为相对低值，推测发育小型的断凸和断凹。

六、余杭—嘉兴断坳区

余杭—嘉兴断坳区位于上海西南部和邻近的江苏、浙江，北以湖苏断裂带为界，南以江绍断裂带为界，东至苏州—嘉善断裂，上海只有金山枫泾西南部的极少部分地区。在重力异常图上，除嘉兴、海宁发育明显的重力高外，多为明显对重力低，且在长河盆地为明显的负异常区。广泛发育上白垩统，在嘉兴和海宁断凸发育古生界。

第十六章　构造形变

独特的大地构造位置，使得区内构造较为复杂。多旋回、多期次的构造运动在本区地壳岩层中塑造了各种规模级别不等、形态复杂多样、有一定规律可循的构造形迹。但由于区内新生界盖层广厚，构造形迹的识别十分困难。钻孔资料表明，区内褶皱、断裂、节理等韧性和脆性构造都有发育。小褶皱主要在金山群和寒武系软弱层中，呈揉皱变形。节理在从金山群到白龙港玄武岩中都有发育，有的节理面平直光滑，有的弯曲，发育1组、2组甚至3组节理。除明显的共轭节理外，一组通常较陡、甚至直立，一组较缓，表明在地质史上应该存在分别以挤压、剪切和伸展为主的多种构造体制。下古生界构造形态比较简单。寒武系倾角多在40°~60°，少量偏缓或偏陡，10°~20°或70°~90°；奥陶系倾角多在40°~65°，少量偏缓或偏陡，20°~30°或70°~90°。劳村组倾角多在20°~35°，少量偏陡，45°~60°。赤山组倾角多10°左右，局部可达50°，甚至60°~70°，其倾角变化多受盆缘形态控制。

有关基岩褶皱构造的研究较少。通常认为区内基岩断块构造、单斜构造发育，褶皱构造不发育。1988年，《上海市区域地质志》对基底、盖层褶皱进行了简略分析。

断裂构造的认识较为丰富，但多是通过区域重磁解释等地球物理勘查方法获得的，钻探验证不足，不同阶段对同一条断裂认识差别较大。1960年，《上海市石油及天然气地质普查工作报告》在境内划分出苏州—昆山、嘉兴—奉贤两条断裂带。1962年，《上海地区矿产地质普查工作总结报告》将区内断裂分为北北西、北北东和近东西向3组，基本上形成于燕山期和喜马拉雅期。1979年，在编制《上海市基岩地质图》时，罗店—周浦断裂、湖苏断裂、大场—九亭断裂、张堰—南汇断裂、枫泾—川沙断裂、绿华—三星断裂、灯塔—赵家宅断裂、嘉定—昆山断裂等主要断裂的认识已具有雏形。1988年，《上海市区域地质志》基本确定了区内的断裂构造格架，共划分了25条主要断裂。1987年，《上海及邻近地区重磁场特征探究》进一步确认了徐市—钱桥断裂（太仓—奉贤断裂）的存在。1990—1993年，顾彭涛、朱子沾等分别在上述断裂格架的基础上，对区内断裂系统进行了解释，提出了长江北支断裂（新河—牛棚断裂）、常熟—庙镇断裂、天马山—横沙岛（长兴岛）断裂、练塘—芦潮港断裂、奉城—九段沙断裂等构造，并对太仓—二堡镇断裂等个别断裂的展布提出了不同认识。2008年，“上海市三维城市地质调查”项目在《上海市区域地质志》断裂格架的基础上，去除了昆山—嘉定断裂、江口断裂、城桥—新光断裂和天马山断裂，增加了太仓—奉贤断裂。

关于中新生代断陷盆地，在1962年编制的《上海地区矿产地质普查工作总结报告》认为区内发育松江—北桥、角直和新民3个中新生代沉积凹陷。20世纪70年代以来，仅仅保留了松江—北桥、角直两个断陷盆地。20世纪90年代中期，在近岸杭州湾北沿和长江口地区划分出了两个中新生代沉积凹陷，这个认识一直沿用至今。

第一节　褶皱构造

晋宁运动之前，上海是华南洋弧盆体系的一部分。经普宁运动Ⅰ幕，本区基底初次褶皱，金山群形成北东—北东东向的大型隆坳；嗣后，又经过晋宁运动Ⅱ幕，惠南板岩、金山群等基底岩层接受变形改造；最后，在强烈的印支运动影响下，区内褶皱格架基本定型。

一、复式褶皱

（一）朱泾—周浦复背斜

朱泾—周浦复背斜分布于枫泾—川沙断裂和张堰—南汇断裂带之间，核部由前震旦系金山群及惠南板岩组成，推测其轴向为北东60°左右。由于受早白垩世松隐、金桥火山盆地及松江—北桥晚白垩世—古新世断陷红盆的叠加和破坏，已残缺不全，很不连续。复背斜的两翼见有上震旦统灯影组和上寒武统超峰组。地层倾角较陡，一般在65°~70°，甚至接近直立。

（二）昆山—港西复背斜

昆山—港西复背斜位于崇明岛西部，为邻区苏州—吴兴复背斜的北东向延伸部分，受湖苏断裂带控制，形态复杂。根据重力场资料和邻区地层分析，推测在港西地区发育前寒武系，构成复背斜核部，向翼部发育寒武系、奥陶系和志留系。

（三）青浦—宝山复向斜

青浦—宝山复向斜分布于枫泾—川沙断裂以西地区，至今未见基底褶皱岩系。主要由下古生界构成，最老为寒武系超山组，最新为奥陶系长坞组。从其下古生界发育齐全来看，推测当地早古生代为一北东向的坳陷带，至燕山期又被巨厚的火山岩系及红层覆盖。

（四）金山—大团复向斜

金山—大团复向斜位于张堰—南汇断裂带南东侧的杭州湾北岸一带。据航磁资料推测为一坳陷带，已于钱桥及齐贤发现中—上寒武统、奥陶系，在浙西北乍浦一带推测发育有上泥盆统—中三叠统，构成复式背斜核部。

二、其他盖层褶皱

（一）罗店背斜

罗店背斜分布在嘉定—罗店一带，轴向北东60°，向南西倾伏，其东北端被罗店—新场断裂断失。背斜核部为超山组及大陈岭组，翼部由杨柳岗组组成。北西翼被寿昌组超覆不整合覆盖，南东翼为东西向的昆山—嘉定断裂所破坏，现存部分，构成背斜残留的断凸形态。

（二）徐家汇—五角场背斜

根据地层的出露情况，推测原背斜的核部位于徐家汇—五角场一线及其以东。背斜的北西翼由杨柳岗组构成，被断裂构造肢解，显得残缺不全。核部可能由底—下寒武统组成。与此相邻的向斜核部位于酒精一厂—江湾一线，由奥陶系组成。北北东向的凉城断裂，使北新泾杨柳岗组、大场大陈岭组逆覆于奥陶系之上（图16-1），至早白垩世，金桥构造火山盆地形成，静安寺断裂、大场—江湾断裂、淞南断裂发生块断差异升降运动，致使北新泾、大场相对上隆呈断凸形态，并引起北新径—大场重力高。

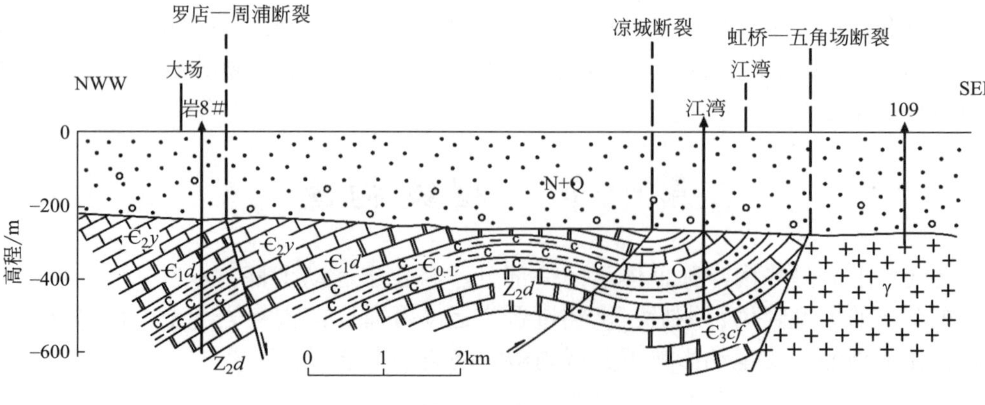

图16-1　大场—江湾地质剖面示意图

Fig. 16-1　Geological section of Dachang-Jiangwan

N+Q—新近系+第四系；O—奥陶系；ϵ_3cf—超峰组；ϵ_2y—杨柳岗组；ϵ_1d—大陈岭组；$\epsilon_{0-1}c$—超山组；Z_2d—灯影组；γ—花岗岩

第二节　断裂体系

自晋宁期以来，本区形成一系列规模不等、性质不同的断裂构造。按其走向主要分为北东、北北东、北西-北北西、近东西向 4 组，以北东向断裂最为发育（图 16-2）。其中，北东、近东西向断裂一般属早期构造，北北东、北西-北北西向断裂为晚期构造，时代相对较新。形成于晋宁期或更早的北东向断裂，持续活动，对区内建造具有明显的控制作用。

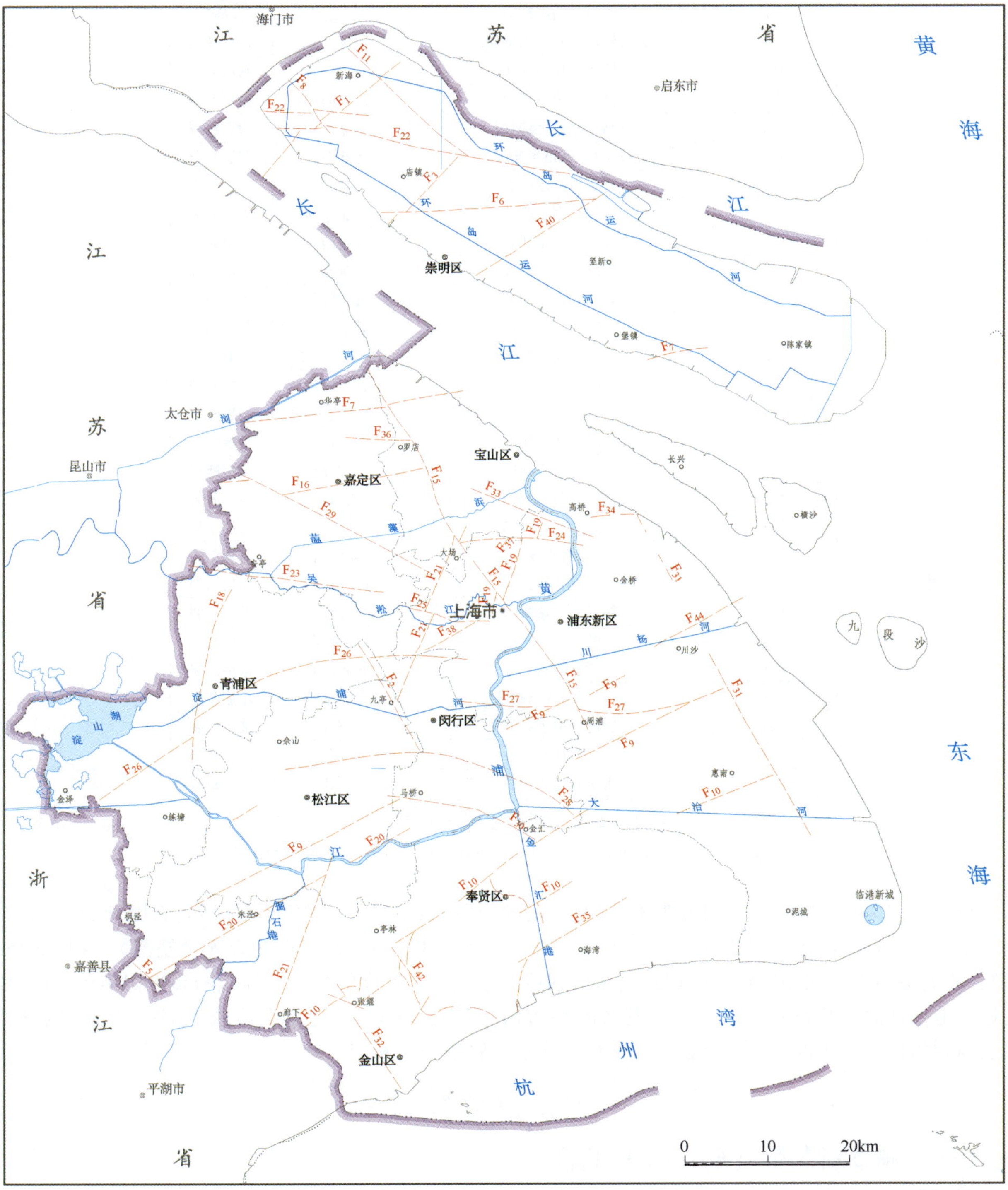

图 16-2　上海断裂构造分布图

Fig. 16-2　Distribution of fault structures in Shanghai

编号	名称	方向	分类
F_1	沙溪—吕四断裂	北东向	基底断裂
F_2	太仓—奉贤断裂	北西向	基底断裂
F_3	浮桥—长征断裂	北东向	深断裂
F_5	苏州—嘉善断裂	北西向	基底断裂
F_6	徐市—大兴断裂	近东西向	盖层断裂
F_7	太仓—二堡镇断裂	近东西向	盖层断裂
F_8	跃进农场—育新断裂	北西向	盖层断裂
F_9	枫泾—川沙断裂	北东向	基底断裂
F_{10}	张堰—南汇断裂	北东向	深断裂
F_{11}	新海农场—长征农场断裂	北西向	盖层断裂
F_{15}	罗店—周浦断裂	北西向	基底断裂
F_{16}	嘉定—罗店断裂	近东西向	盖层断裂
F_{18}	白鹤—姚家港断裂	北北东向	盖层断裂
F_{19}	虹桥—五角场断裂	北北东向	盖层断裂
F_{20}	兴塔—泖港断裂	北东向	盖层断裂
F_{21}	大场—廊下断裂	北北东向	盖层断裂
F_{22}	绿化—三星断裂	近东西向	盖层断裂
F_{23}	千灯—黄渡断裂	近东西向	盖层断裂
F_{24}	大场—江湾断裂	近东西	盖层断裂
F_{25}	丰庄—静安寺断裂	近东西向	盖层断裂
F_{26}	青浦—龙华断裂	北东—近东西向	基底断裂
F_{27}	梅陇—邓镇断裂	近东西向	盖层断裂
F_{28}	卖花桥—鲁汇断裂	北西—近东西向	盖层断裂
F_{29}	葛隆—南翔断裂	北西向	盖层断裂
F_{30}	马桥—金汇断裂	北西向	盖层断裂
F_{31}	灯塔—赵家宅断裂	北西向	盖层断裂
F_{32}	张堰—金山卫断裂	北西向	盖层断裂
F_{33}	淞南断裂	北西向	盖层断裂
F_{34}	高东断裂	近东西向	盖层断裂
F_{35}	钱桥—奉城断裂	北西向	盖层断裂
F_{36}	永胜—施庙断裂	近东西向	盖层断裂
F_{37}	凉城断裂	北北东向	盖层断裂
F_{38}	愚园—动物园断裂	北北东向	盖层断裂
F_{40}	浏河—前进断裂	北东向	基底断裂
F_{42}	朱行断裂	北西向	盖层断裂
F_{44}	唐镇—合庆断裂	北东向	盖层断裂

图 16-2 上海断裂构造分布图（续）

Fig. 16-2 Distribution of fault structures in Shanghai (continued)

根据切割深度，将本区断裂分为深断裂、基底断裂和盖层断裂。深断裂包括岩石圈断裂和地壳断裂，指切穿岩石圈或地壳的断裂。参照《活动断裂调查与区域地壳稳定性调查评价规范》（DD 2015—02），岩石圈断裂指切穿岩石圈或切穿莫霍面到达上地幔一定深度的断裂；地壳断裂指切割至硅镁层内或达上地幔顶面（莫霍面）的断裂；基底断裂指切穿基底顶部，并进入上地壳下部—中地壳的断裂；盖层断裂指切割沉积盖层及火成岩体一定深度或达基底顶面的断裂。

一、深断裂

（一）湖苏断裂带（F_1、F_3 和 F_{40}）

断裂带总体走向北东 50°，由湖州，经苏州、阳澄湖、崇明岛西端一线分布，向东北延入黄海。向西至江西省内，又称崇明-景德镇断裂带。区内崇明岛段仅是其中一小部分，长约 20km。推测由浏河—前进断裂（F_{40}）、沙溪—吕四断裂（F_1）和浮桥—长征断裂（F_3）3条断裂组成，被绿华—三星东西向断裂带切割，向西南延入长江南支水域。其中，推测浮桥—长征断裂为主干断裂，切割较深，为深断裂。沙溪—吕四断裂和浏河—前进断裂为两条重要的分支断裂，为基底断裂。深部倾向南东，浅部倾向北西或南东，断裂深部可能沿中地壳或深变质岩系底部发生拆离滑移。

该断裂带为一条综合物探、地质资料确定的隐伏断裂构造。区内崇明呈现为重力异常梯度带，梯度带呈北东向，多宽缓，在庙镇和港西之间稍密；在航磁异常平面图上处于正、负磁异常交变带附近。断裂带沿线和南东侧见前震旦系、震旦系及下古生界，下白垩统火山岩分布广泛，上白垩统为断陷盆地型堆积；北西侧可见泥盆系—三叠系，下白垩统为局部的盆地型沉积，上白垩统分布广泛。唐市以南和堡镇的花岗岩体受其控制，为燕山晚期活动的产物。CSAMT显示，在阳澄湖和东太湖地区浅部由2~3条断裂组成，在崇明绿华—新海农场以西地区浅部由3~4条断裂组成，构成地堑构造。

浏河—前进断裂在崇明40孔见糜棱岩化花岗岩。江苏吴江吴3孔中，在780~815m井段见断层角砾岩，镜下见胶结物中绢云母鳞片多呈定向排列，泥质物中可能为硅质隐晶物。岩石挤压破碎强烈，且见扭曲、擦痕，裂隙中还见有方解石或方解石与硅质所组成的细脉充填。孔内所见断裂上盘下白垩统火山岩与火山碎屑岩覆盖在下盘下二叠统堰桥组细砂岩夹粉砂岩之上，显示了正断层性质。但由于断裂带中的断层角砾岩和旁侧的堰桥组粉砂岩均见定向排列的挤压片理，从而说明该断裂早期可能为压性，后期表现为张性的特点。

断裂带推测于晋宁期开始活动，主要形成于印支晚期，燕山早期复活，燕山晚期活动强度加剧，喜马拉雅期仍继续活动。断层性质较为复杂，早期为逆断层，晚期为正断层。

（二）张堰—南汇断裂带（F_{10}）

张堰—南汇断裂带大致沿钱圩—张堰—朱行—奉贤—惠南一线展布，走向北东60°。朱行至奉贤之间被庄行构造火山盆地覆盖，形迹不显。奉贤以东分为南、北两支，北支沿齐贤桥—航头—新场—盐仓一线分布，南支沿光明—曹家宅—塘桥—惠南一线分布。区内断续出露长40余千米，向东延入海至九段沙浅滩，向西南延入浙江。主断面倾向南东，倾角为45°~50°，浅部多破碎带，以倾向北西为主，倾角较陡。断裂两侧地层发育有差异，金山群、惠南板岩及灯影组分布于断裂带的西北侧，而黄尖组主要分布于断裂带南东盘或断裂带上。沿断裂带多个钻孔中均见挤压片理、动力变质岩或构造破碎带，热液蚀变现象普遍。如张堰普遍见角岩、矽卡岩及构造破碎现象，朱行发现英安质角砾岩破碎带和多处碎裂安山岩。齐贤上寒武统上部与下部岩层的倾角变化很大。朱行钻孔中曾发现厚为200m的构造破碎带，新场见断层角砾岩，惠南钻孔中见有厚为25m的断层破碎带，三灶一钻孔中3个井段见断层破碎带，错动面倾角为65°~70°，甚至直立，肖塘金山群中亦已发现糜棱岩化现象等。

断裂带有明显的控岩、控矿作用，对构造火山盆地也有明显控制作用，如钱圩—查山火山隆起、松隐火山盆地、庄行构造火山盆地及钱圩和张堰石英闪长岩、荷花池石英闪长玢岩、齐贤辉石闪长岩等均受其控制。张堰矽卡岩型铜铁矿床即产于该断裂带中，此外坦直也见有多种矿化迹象，说明该断裂为区内的一条主要的控矿构造。

沿断裂带出现多个呈串珠状分布的航磁异常，如钱圩、张堰、朱行、庄行、奉贤、齐贤、新场等地。已经钻探验证的大多为浅成中酸性侵入体，异常展布方向与断裂走向一致。断裂北支有明显的重力梯度反映，南支位于正、负航磁异常交变带或彼此的转折带，并沿边部通过。

断裂带形成较早，推测形成于晋宁期，加里东、印支期有活动迹象，燕山期强烈复活，为主要活动期。早期为逆冲断层，晚期为正断层。

（三）江绍断裂带

该断裂带在浙江境内展布清楚，但从绍兴东南入海后的走向存在不同认识。本次结合区域地质资料，根据区域重磁场特征，认为该断裂带进入海域后控制长河盆地东南缘，在庵东东北部进入杭州湾海域，并逐渐向北东东偏转，于衢山岛北侧通过。在杭州湾中部被通过上海中部的北西向断裂切割。

杭州湾及邻区布格重力异常图上，该断裂构造带处于重力梯级带上，以其为界，北西侧重力场显示为高背景场，西部有小块负异常，东北部分布有密集带状分布的正异常圈闭。自杭州湾向东依次分布有长河重力低、杭州湾中部重力低和嵊泗列岛重力低，为中新生代断陷盆地。南东侧则为低背景场区，向台州、宁波、绍兴市东南等地延伸，为大面积负异常区，构成一个完整的布格异常低值区。在浙江余姚牟山、慈溪坎墩和衢山岛附近发育相对重力高，为变质基底出露或抬升区。

杭州湾及邻区航磁异常图上，断裂沿线表现为磁异常分界线。沿断裂带少量弱的串珠状异常，延伸方向为北东和北东东向。南东侧为平静的磁场区，局部正负相间剧烈变化。磁异常值都较低，幅值变化也不大，磁异常大小的变化范围多在-100~100nT。在总体平缓的背景上局部零星分布有几个相对高的磁异常圈闭；西北侧总体为正磁异常背景区，正负异常跳动剧烈，轴向杂乱。磁异常大小的变化范围多在-300~300nT。等值线北东、北东东向延伸，越往东南向延伸，异常值就越低。伴生的负异常范围相对较小，走向取北东向或北西向。航磁异常随着上延高度不断加大，在上延 10km 航磁异常图和 135°水平导数图上可以看出，航磁异常呈北东向线性梯度变化，磁场等值线变化特征明显不同。卫星照片上线性特征明显，基本上与地表断裂排列组合相吻合。

二、基底断裂

（一）太仓—奉贤断裂（F_2）

断裂自江苏福山长江边，经常熟支塘、太仓和上海外岗、诸翟、莘庄、奉贤西南入杭州湾，走向北西 330°，深部倾向北东，倾角上陡下缓，呈犁式，长约 140km，宽数千米。断裂的性质与深度有严格的分段性，其主体部分在支塘至新桥之间，在北桥断陷带和杭州湾被掩盖。切割深度在 10km 以上，支塘以北和新桥以南后期伸展的浅层断裂发育。

断裂明显制约新近纪以来地貌、古地理的演变以及沉积物的发育。新近纪和第四纪初期，断裂以西除安角断陷外，均属于剥蚀低山区；断裂以东地区基岩埋藏骤然加深。基岩起伏走势在断裂带附近发生改变，沿断裂带分布有一些沟谷，在断裂两侧发育的某些高地和洼地延至断裂带突然截断或终止。断裂带附近已断续发现一些串珠状分布的渐—中新世碱性玄武岩体，其中在常熟支塘至昆山城区附近揭露岩体厚度可达 70m，另在北桥也有分布；总体上，基性火山岩多分布于断裂以东地区，且越近滨岸地带分布越多，岩体规模越大，喷溢时代亦越新。断裂以西，很少发现。

在各种比例尺的布格重力等值线图上，均可见断裂沿断续分布的重力异常梯度带或等值线扭曲部位通过。东侧为浮桥-钱桥正异常区，有一系列重力正异常，如狼山、仪桥、浮桥、嘉定北新泾—大场、莘庄—坦直、钱桥等异常；西侧为白茆-松江负异常带，分布有绵丰—乐余、白茹、花桥、松江、胡桥等重力负异常或重力低值区。在 1：20 万、1：50 万重力异常上延至 30km 等值线图上，除陆域部位有重力异常梯度带显示外，至杭州湾水域呈 150°方向继续延伸，直达滩浒山附近，构成坦直—钱桥—三墩重力高区长江口水域部分的西部边界。沿断裂有串珠状斜列的局部航磁异常分布，有支塘、直塘、南翔、七宝、肖塘青树、钱桥等局部异常。区内两条主要的北东向航磁异常带在该断裂附近即被中断，呈不连续或收敛，致使两侧异常带宽度、局部异常轴向发生变化。

断裂主要活动时期为燕山期-喜马拉雅期。早期逆冲挤压，晚期转化为张扭性。HQ-13 剖面反映该断裂深切至上地壳下部，构成深、浅变质岩系之间的滑移界面。

（二）枫泾—川沙断裂（F_9）

断裂自枫泾经车墩、周浦至川沙一线断续分布，由数条北东向断裂组成，北支在龙华、杨思一带。东北段被金桥构造火山盆地覆盖，形迹亦不显。马桥—颛桥之间，行迹亦不显。区内断续出露约 40km，往西南延入浙江，向东北经白龙港附近延入东海海域。走向北东 60°，深部倾向南东，浅部倾向北西，倾角在 75°以上，切割深度至少可达 10km。

断裂对两侧的岩性、岩相有一定控制作用。金山群和惠南板岩仅见于断裂东南侧；震旦系镁质碳酸盐岩建造在南东侧较西北侧分布广泛；下古生界海相类复理石建造常见于断裂西北侧；劳村组广泛分布于东南侧基岩顶部。沿断裂见有中新世玄武岩流局部喷溢。断裂控制了金桥构造火山盆地和天马山构造火山盆地的发育，并直接控制了中性、酸性岩浆的喷溢和侵入，如涂泾花岗岩体、关港花岗岩体、三林塘—赵家宅等地的石英闪长岩体等。沿断裂带，构造破碎、动力变质、热液蚀变现象强烈。松江钻孔中见有构造角砾岩及褪色现象，构造、热液活动迹象明显。沈家浜、车墩、曹行及陈行、题桥的钻孔中见有构造破碎带、构造角砾岩与断层泥，天花庵、三林塘、赵家宅、康家宅一带的钻孔中蚀变、构造破碎及糜棱岩化现象极为普遍。其糜棱岩化作用仅见于金山群，未波及惠南板岩、灯影组和下古生界。

在布格重力异常图上，枫泾-车墩段为重力正异常，断裂北西侧为重力负异常，在宽达 4km 范围内梯度变化可达 $4\times10^{-3}\sim5\times10^{-3}$（$cm\cdot s^{-2}/km$）。关港—周浦段及周浦—川沙段，基本上沿等值线方向延伸，东南侧为正异常，西北侧异常带相对低缓。车墩—颛桥段，因白垩纪红盆的覆盖，重力图上无显示，按总体延伸方向在红盆下应于北桥、新桥两重力低的鞍部通过。在航磁 ΔT 平面图上，断裂的东南侧多为强正异常，北西侧则为低缓正异常或负异常，彼此排列方向与断裂带方向基本一致。频率电磁测深证实，黄浦江以东部分（以西未做工作）断裂带破碎，含水充分，呈低阻异常反应。

推测断裂形成于晋宁期，加里东期和印支期均有活动，燕山中晚期强烈复活，喜马拉雅期仍有活动。早期为压扭性逆冲断裂，早白垩世以后表现为张扭性正断层。

（三）罗店—周浦断裂（F_{15}）

断裂由江苏浮桥经上海罗店、大场至周浦、新场、四团一带分布，是上海横切一系列北东向构造的一条重要的北西向断裂。断裂走向 325°~330°，深部倾向北东，呈陡倾斜。浅部断层组合特征差别较大，多由 3~4 条断裂组成地堑构造。具有张性及右行走滑性质。

断裂对早期断裂、岩体起改造和破坏作用。某些地段构成断凸与断凹的边界，或切过它们。据物探资料推断，罗店背斜在月浦附近被盖于火山岩之下。在沈庄、周浦一带中偏基性辉石闪长岩或中性闪长岩分布广泛，且周浦一带常有矿化显示。在宝山罗店及大场的钻孔中均见有挤压破碎带，在同一钻孔中上、下地层间的倾角变化很大。

断裂位于断续分布的北西向重力梯度密集带上，方向大体与重力等值线方向一致。罗店—周浦段西盘为重力正异常带，由嘉定—罗店重力高、大场—北新泾重力高、六里桥—周浦相对重力高等组成，东盘为盛桥—凌桥相对重力低。周浦—四团段的东盘为坦直—三墩重力高带，西盘为相对重力低带。在六里桥至周浦之间一系列紊乱的航磁负异常背景上，出现局部航磁正异常，呈边幕式排列，方向为 315°左右，与断裂在该段的位置基本一致。断裂线东侧为低缓航磁负异常（0~250nT），西侧为 250~1000nT 正异常。

断裂主要活动于燕山晚期，在新生代仍有活动。沿断裂带近代有地震活动，如 1624 年的南市 5 级地震、近年崇明岛西端长江水域的 1~2 级小震，可能均与该断裂的活动有关。

（四）青浦—龙华断裂（F_{26}）

断裂主要沿青浦莲盛—沈巷—青浦镇—徐泾—龙华一带分布，呈北东—近东西向。向东可能切过金桥火山盆地，延入长江口地区，向西推测与马金—乌镇断裂相连。推测为基底断裂，切割深度推测可达中下地壳，深部断面南倾，倾角在 50°左右。浅部由 2~3 条断层组成地垒构造。

断裂构成了天马山构造火山隆起的北部边界。北侧发育寿昌组火山碎屑岩、沉积岩，南侧为黄尖组中性火山岩系。对第四系厚度有一定的控制作用。断裂附近脉岩多见，并发育有构造破碎带，如在青浦钻孔中岩芯破碎成棱角状，徐汇龙华钻孔中见糜棱岩等。

断裂有较为明显的重力梯度显示，金泽—青浦段为重力高与重力低的转换带，等值线宽缓，但明显向北东方向延伸。青浦—龙华段，北侧为重固—北新泾重力高，南侧为泗泾相对重力低，梯度变化约为 3.5mGal/km。断裂南侧多为航磁正异常，其轴向亦为北东—北东东—东西向，并略显弧形展布。

推测断裂主要活动于早白垩世，延续至更新世。

三、盖层断裂

（一）太仓—二堡镇断裂（F_7）

太仓—二堡镇断裂位于太仓—盛桥—向化—二堡一线，走向近东西，长江南支水域以东转为北东东向，推测向西与江苏境内的望亭—太仓东西向断裂相接，东部可能继续向东延伸穿过崇明岛东南角潜入长江口海域。西段由多条断裂组成，倾向北，其总体落差在120m以上。区内长度大于75km。属张扭性正断层。

断裂控制了白垩纪以来的岩浆作用及唐行断凹的形成，如陆渡桥的玄武岩、崇明燕山晚期花岗岩体、崇明陈家钻孔所见石英斑岩、安山玢岩，北侧的唐行断凹和南侧的嘉定断隆等。东段两侧新生界松散层沉积厚度存在明显差异，北侧厚在420~480m，南侧一般在350m之内。

布格重力等值线西段南侧为嘉定—罗店重力高，北侧为唐行—陆家宅重力低，重力异常呈东西向展布；东段长江南支水域北侧为重力低，在崇明陈家附近，断裂于牛棚重力高与陈家镇重力高的鞍部通过，布格重力异常轴向近东西，梯度变化在4~5mGal/km。西段局部航磁正异常呈东西向串珠状排列，东段断裂北侧为北东东向的正异常，南侧则为负异常区。

推测断裂形成于燕山早期，延续至燕山晚期，喜马拉雅期仍有活动。

（二）昆山—嘉定断裂（F_{16}）

昆山—嘉定断裂沿江苏唯亭、昆山以南、蓬朗及上海外冈、刘行一线分布，蓬朗以东被北西向葛隆—南翔断裂错开，向南平移至外岗一带，东端又为北西向的罗店—新场断裂所截。区内外岗—刘行段是嘉定断凸的南界断裂，长约19km。走向北东东—东西向，倾向为南（图16-3），倾角为65°。属张性—张扭性正断层。

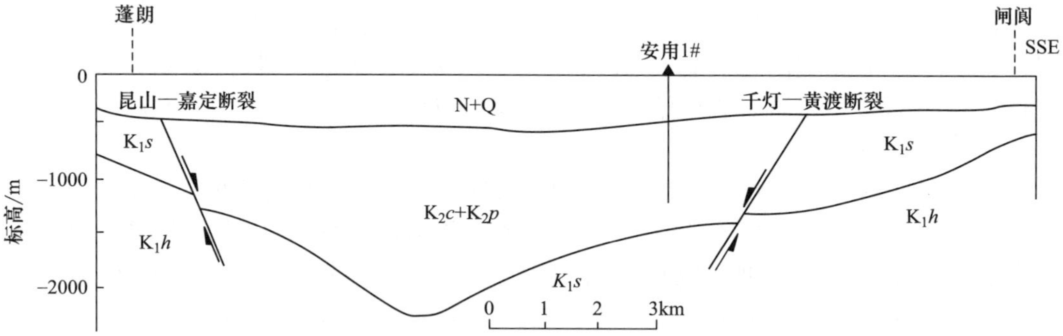

图16-3　安角断陷盆地电测深解释剖面示意图

Fig. 16-3　Electric sounding interpretation section of the Anlu faulted basin

N+Q—新近系+第四系；K_2c—赤山组；K_2p—浦口组；K_1s—寿昌组；K_1h—黄尖组

断裂控制了嘉定断凸和南侧中新生代断凹的发育。北侧主要为下白垩统火山岩系，见有寒武系—奥陶系；南侧西段主要为上白垩统浦口组和赤山组，东段嘉定以南寿昌组分布广泛。安角断陷盆地内，历史上地震频繁。1731年昆山以南的5.0级地震可能与该断裂有关。

该断裂西段，布格重力、人工地震、电测深资料及地质依据都很充分，其北侧为阳澄湖、马鞍山、罗店重力高带，单个重力异常轴向及排列均近东西向。断裂南侧为重力负异常带和广福—顾村相对重力低带。正负异常长轴方向与断裂延伸方向一致，断裂位于重力等值线密集的梯度带上，变化达5mGal/km。断裂两端见有轴向近东西向的航磁异常展布。

推测该断裂形成于早白垩世末期，晚白垩世活动强烈，古新世至现代仍有活动。

（三）白鹤—姚家港断裂（F_{18}）

白鹤—姚家港断裂分布于白鹤经青浦城西至练塘一带，长约25km，走向北东25°，倾向南东，倾角为70°。

断裂处于凹陷地带，分隔东以佘山为中心、西以淀山附近为中心的两个北东向火山隆起区。对第四系厚度有一定的控制作用。在李墟村航磁异常区，钻孔揭露的安山岩、安山质角砾熔岩中见厚10余米的构造破碎带。断裂附近发生过多次小震活动，如1975年青浦西南部1.0级地震、1978年白鹤2.3级地震、

1991 年练塘 3.0 级地震等。据 1978 年 1 月 5 日白鹤 2.3 级地震的震源机制解释，该断裂似为其发震构造。

沿断裂有密集的重力梯度带显示，其等值线近南北向延伸。

推测断裂形成于燕山期，喜马拉雅期持续活动，主要呈张性特征。

（四）虹桥—五角场断裂（F_{19}）

虹桥—五角场断裂自江湾五角场，经北站西向南延至虹桥附近，走向北东 30°，倾向南东，长约 17km。断裂分别为大场—江湾断裂、罗店—周浦断裂所截。

该断裂位于金桥破火山口西缘，早白垩世曾是破火山口环状断裂的一部分，与破火山口的放射状断裂互为交切关系。断裂西侧多为下古生界，东侧一般以花岗岩为主。沿断裂分布的多个钻孔中均发育有构造破碎带，岩石似糜棱岩化，蚀变也较强烈。

在布格重力图上呈较为宽缓的重力梯度带，断裂与重力等值线方向基本一致。

推测断裂形成于印支期，燕山早期被火山构造利用，成为破火山口的环状断裂，活动时间至少延至早白垩世，为正断层。

（五）大场—廊下断裂（F_{21}）

大场—廊下断裂沿廊下、松隐、七宝、北新泾、大场一带呈北东 25°方向断续分布。向西南延入浙江省，越杭州湾，可能与浙东的丽水—上虞断裂相连，大场以北形迹不显，至苏北启东又复出露。区内断续延伸约 67km，断裂带宽为 2~3km。切割一系列北东向断裂构造，并分别为罗店—周浦断裂、丰庄—静安寺断裂所截。

断裂大场—九亭段倾向北西，倾角约 50°，切割重固—北新泾早古生代断凸（图 16-4），北西侧地层埋深相对加深，落差大于 400m。东北段东南侧出现下古生界碳酸盐岩、碎屑岩断凸，北西侧为晚白垩世红盆或火山构造盆地火山碎屑沉积岩系。在北新泾及大场—江湾断凸上，于断裂东侧发育有平行于主干断裂的逆冲断层，致使寒武系逆覆于奥陶系之上。在北新泾钻孔中，有 4 个井段都见到断层角砾岩，大场钻孔中也见到 40m 的构造破碎带。松隐—廊下段断裂倾向南东，倾角约 50°。构造形迹比较清晰，东北为枫泾—川沙断裂所截。东南侧为松隐火山构造盆地和河上镇群，广泛分布劳村组火山碎屑沉积岩系；西北侧发育金山群。区内燕山晚期主要的酸性侵入体，如虹桥、市南、市北、吴淞、崇明花岗岩均沿该断裂带分布。岩体本身也多见构造破碎现象。

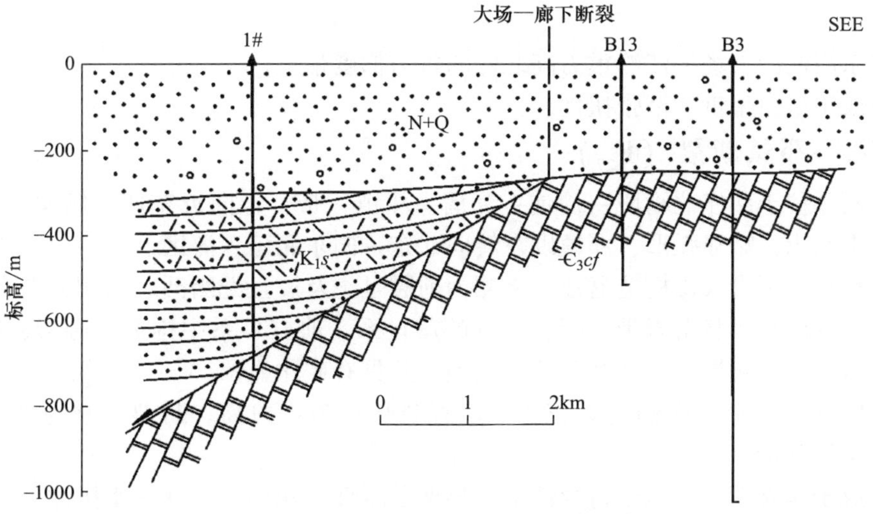

图 16-4 北新泾—化工院地质剖面示意图

Fig. 16-4 Geological section of Beixinjing-Institute of Research Institute of Chemical Industry

N+Q—新近系+第四系；K_1s—寿昌组；ϵ_3cf—超峰组

断裂松隐—廊下段有较明显的重力梯度反应，西侧为吕巷重力高，东侧为松隐—亭林重力低，断裂位于重力梯度带上，梯度变化达 4mGal/km，延伸方向基本与重力等值线相一致。断裂东北段位于断续分布的重力梯度密集带上。

断裂推测形成于印支期，燕山期仍继续活动。印支期为压扭性，燕山期为张扭性。

（六）绿华—三星断裂（F_{22}）

绿华—三星断裂又称三星—新光断裂。断裂位于崇明西北部，沿绿华、三星北、新光一线分布，向东延入长江北支。走向近东西，绿华一带由3条阶梯状断层组成，倾向北。

断裂为海桥断凹与庙镇断凸的分界线，两侧基岩面埋深差别较大。北侧为海桥中、新生代沉积，南侧由下古生界组成断隆。海桥断凹是早白垩世以来长期发育的继承性盆地，其形成和发育与该断裂的活动有一定关系。

断裂南北航磁特征差异极大，北部均为区域性平缓正异常区，南部则为区域性负异常背景上发育有零星局部正异常。断裂所处地域，航磁负异常圈闭的长轴方向均为北东东—东西向。重力场显示，南侧为庙镇重力高带，北盘为相对重力低带，断裂沿重力梯度带延伸。

断裂推测形成于早白垩世，西段曾频繁发生弱小地震。主要为张扭性。

（七）千灯—黄渡断裂（F_{23}）

千灯—黄渡断裂西起江苏吴江，经千灯延至上海的黄渡附近。上海内长约18km，走向北东东—东西向，断面北倾，倾角50°左右。为张性—张扭性断层。

断裂构成了安角断凹的南界，南侧为寿昌组、黄尖组火山岩、火山碎屑岩等分布区，北侧是由浦口组、赤山组及夏驾桥组组成的断陷红盆。安深1井中，于井深3046.5m处见浦口组与寿昌组上段中、酸性火山岩及火山碎屑岩呈超覆不整合接触关系。

在布格重力异常图上，千灯—黄渡一线有明显的重力梯度密集带，电测深图上也显示为一等值线密集带，在人工地震磁带剖面上也有反应。

断裂推测形成于早白垩世末，晚白垩世—古近纪、新近纪仍有活动。呈张性—张扭性。

（八）大场—江湾断裂（F_{24}）

大场—江湾断裂位于宝山大场—虹口江湾一线，长约13km，走向东西，断层面北倾。

该断裂原为金桥破火山口的放射状断裂，与淞南断裂、丰庄—静安寺断裂构成破火山口西缘的放射状断裂组，西端为北西向的罗店—周浦断裂所截。断裂以北为寿昌组、黄尖组，南部为震旦系—下古生界断凸。

断裂位于呈东西向的大场—江湾重力高北侧的重力梯度带上。

断裂推测形成于早白垩世，呈张扭性。

（九）梅陇—邓镇断裂（F_{27}）

梅陇—邓镇断裂位于莘庄北梅陇—周浦—瓦屑—邓镇一带，走向近东西，倾向北，全长约30km。周浦—瓦屑段向南凸出，略显弧形，邓镇以东转向北东。向东北延入海区。

断裂北侧发育早白垩世火山构造盆地，南侧为前震旦纪及震旦纪断凸。两侧分布有中基性侵入体，组成金桥破火山口的南部环状杂岩带。断裂附近的震旦系、前震旦系中构造角砾岩、糜棱岩化岩石十分普遍、中酸性岩脉穿插频繁、热液蚀变也很强烈，并见有矿化显示。

断裂东段北侧为重力低，南侧（天花庵—坦直断凸）为重力高。断裂北侧为紊乱的航磁负异常带，南侧为航磁正异常带。

断裂推测形成于印支期，白垩纪持续活动。早期呈压性，晚期呈张性—张扭性。

（十）卖花桥—鲁汇断裂（F_{28}）

卖花桥—鲁汇断裂沿松江富林—卖花桥—闵行颛桥—鲁汇一线展布，西段走向北东东，东段走向北西西，倾向南，长约35km。断裂总体呈向北凸出的弧形，切割枫泾—川沙断裂带。

断裂构成松江—北桥断陷盆地的北界，卖花桥以东地段，断裂由大地电磁资料确定。曾于卖花桥钻孔中发现该断层，见有厚54m的构造破碎带。西段有多次小地震记载。断裂两侧的地热异常差异较大。两侧所见地层不同，松江断凹北侧为黄尖组火山岩系，北桥断凹北侧为劳村组火山-沉积岩系，松江断凹南侧为上白垩统赤山组，北桥断凹南侧为古新统北桥组。西段与枫泾—川沙断裂带联合控制了松江断陷盆地的发育，东段与马桥—金汇断裂联合控制了北桥地堑式断陷盆地的发育。

断裂位于向北凸出的弧形布格重力等值线密集带上，其南侧航磁负异常呈带状展布。

断裂推测形成于燕山期，主要活动期为燕山晚期至古新世，为正断层。

（十一）葛隆—南翔断裂（F_{29}）

葛隆—南翔断裂自蓬朗东葛隆，经外冈、南翔附近延至桃浦一带，区内长约 32km，走向北西 300°，倾向南西。北西延入江苏省内，经石牌延至横泾附近。

区内断裂构成安角断凹的东部边界，断裂两侧的地层不同，北侧为寿昌组分布区，南侧为浦口组、赤山组分布区。两侧基岩埋深不一，断裂东北侧一般为 300m 左右，而南侧一般为 350m 以上，安亭一带更深达 470m。断裂附近钻孔中，岩芯破碎，有蚀变。赤山组砂泥岩中见擦痕。

断裂沿线呈明显的重力梯度密集带，梯度变化最大可达 6mGal/km。东北侧为嘉定—罗店重力高，西南侧为安角重力低。外冈至马陆之间重力等值线呈北东方向凸出的弧形。

断裂推测形成于早白垩世末，晚白垩世—古近纪、新近纪仍有活动。呈张性—张扭性。

（十二）马桥—金汇断裂（F_{30}）

马桥—金汇断裂沿车墩北—马桥—金汇一线分布，走向由西到东由近东西向转为北西向，呈弧形展布。倾向北东，长约 20km。该断裂大致与卖花桥—鲁汇断裂相平行，切割枫泾—川沙断裂带。

马桥—金汇断裂与卖花桥—鲁汇断裂共同组成北桥地堑式断陷盆地。断裂北侧为北桥组、夏驾桥组等新生代沉积，南侧为劳村组火山角砾岩。

断裂位于北桥断凹南缘的重力梯度带上。

断裂推测形成于燕山早期，早白垩世活动明显，可延至古新世，为正断层。

（十三）灯塔—赵家宅断裂（F_{31}）

灯塔—赵家宅断裂自浦东高东镇东长江口边灯塔，沿曹路、机场、赵家宅、老港一线分布，走向北西 335°，倾向北东，陆上全长大于 40km。属张扭性正断层。

断裂控制了金桥破火山口、金桥构造火山盆地，杨家宅火山杂岩体的发育。据钻探资料，在杨家宅施工的钻孔中均见有构造破碎带和断裂错动的擦痕。

在祝桥—东海段断裂具有较密集的重力梯度显示，断裂于坦直重力高的东缘通过；祝桥以东的航磁异常相对于周浦—瓦屑之间的航磁异常北移；小营房—江镇段断裂为中新世玄武岩覆盖，北段的重力等值线也产生向北西方向的扭曲。

断裂推测形成于燕山早期，喜马拉雅期仍有活动迹象。

（十四）永胜—施庙断裂（F_{36}）

永胜—施庙断裂位于宝山罗店北—施庙—永胜一线，走向北东东 75°，倾向北。推测与太仓—二堡镇断裂构成一套近东西向断裂带，东端为罗店—周浦断裂所截。

断裂控制了罗店下古生界断凸的形成，其南东侧为震旦系—寒武系含泥砂质碳酸盐岩，北侧为寿昌组火山碎屑岩。所处位置为重力等值线密集带。

断裂推测形成于早古生代以后，早白垩世以前，为正断层。

（十五）张堰—金山卫断裂（F_{32}）

张堰—金山卫断裂自张堰西至金山卫，向南推测延入杭州湾。断裂长大于 15km，走向北西 322°，倾向南西，陡倾角。切割张堰—南汇断裂带，为钱圩—查山火山穹窿发育稍后的放射状断裂。

断裂控制了万春桥、钱圩、查山等寄生火山锥的发育，并分割了查山凸起和金山凹陷。断裂东侧第四系等厚线呈北东向分布，厚度在 100~140m；西侧第四系等厚线呈北西向分布，沉积厚度向西迅速变厚，最厚可达240m。

断裂西南侧为相对布格重力低，东北侧为相对重力高，断裂于重力梯度稀疏平缓处通过，与重力等值线延伸方向基本一致。断裂带及其东侧为航磁正异常，航磁从+220nT直升至+1500nT。重力在10km延拓后已不再显示。

断裂推测形成于早白垩世，晚白垩世、古近纪、新近纪持续互动，呈张性。

此外，区内还划分有徐市—大兴断裂（F_6）、新海农场—长征农场断裂（F_{11}）、愚园—动物园断裂（F_{38}）、凉城断裂（F_{37}）、高东断裂（F_{34}）、淞南断裂（F_{33}）、丰庄—静安寺断裂（F_{26}）、天马山断裂（F_{25}）等，延伸较短或者依据不充分。

第三节　断陷盆地

一、松江—北桥断陷盆地

（一）分布范围

松江—北桥断陷盆地位于上海中部的古松—松江—北桥—闸港一带，呈一略向北突出的狭长弧形，由松江断陷盆地和北桥断陷盆地组成（图16-5）。松江断陷盆地总体上沿枫泾—川沙断裂呈北东向延伸，北桥断陷盆地呈北西西向。松江—北桥断陷盆地是晚白垩世发育的一个较大型的断陷红盆，其中北桥断陷盆地形成期可延续到古新世。从区域重磁场特征看，该断陷盆地似可延伸至泥城—牛皮礁地区。

（二）重力场特征

松江—北桥断陷盆地位于呈弧形带状展布的重力低上，由西向东有古松、三家桥、新桥、北桥、闸港等重力低。松江断陷盆地的南东侧、北侧及北桥断陷盆地的南、北两侧都有极为明显的重力梯度反映，最大梯度变化可达5mGal/km。

（三）地质背景和构造控制条件

松江断陷盆地的形成与枫泾—川沙断裂带密切相关，上白垩统厚度在松江城区附近最大，而南西段甚小。西南部及西北部与下白垩统呈超覆不整合关系。北桥断陷盆地是受卖花桥—鲁汇断裂和马桥—金汇断裂共同控制的地堑式盆地，东端为区域性的张堰—南汇断裂带所截切。

（四）物质组成

松江断陷盆地广泛分布上白垩统赤山组，为一套紫红—灰紫色的粉砂质泥岩、泥质粉砂岩夹中细砂岩组成的韵律层，偶见粗砂岩。碎屑形态多为棱角状，磨圆度及分选性都较差。薄层状和脉状石膏多见。厚度大，已见厚度大于771.9m，局部钻孔见其不整合于寿昌组凝灰质角砾岩之上，厚度仅为148m。北桥断陷盆地内北桥组分布广泛，岩性以棕红、灰褐色泥岩、粉砂质泥岩夹粉细砂岩和玄武岩为主。已见厚度为289.4m，南侧厚度锐减或尖灭。揭露的赤山组与下伏劳村组呈超覆不整合接触，与松江盆地相比，碎屑组分较粗，层理不发育，分选性差，且无石膏夹层。重力低与盆地深凹形状较吻合。其中古新统发育，说明新生代以来盆地的沉陷活动仍显著。

二、安角断陷盆地

（一）分布范围

安角断陷盆地位于江苏苏州以东的吴中角直至上海西北部的花桥—安亭一带。南北分别以吴江—黄渡断裂（区内千灯—黄渡断裂是其东段）和昆山—嘉定断裂为界，西界是北东向的湖苏断裂带，东部为北西向的葛隆—南翔断裂所限制。为一周边受断裂构造控制，呈北东东向展布的中、新生代隐伏断陷盆地。盆地大部分在江苏境内，区内仅是其东段的一小部分，呈南东东向展布。

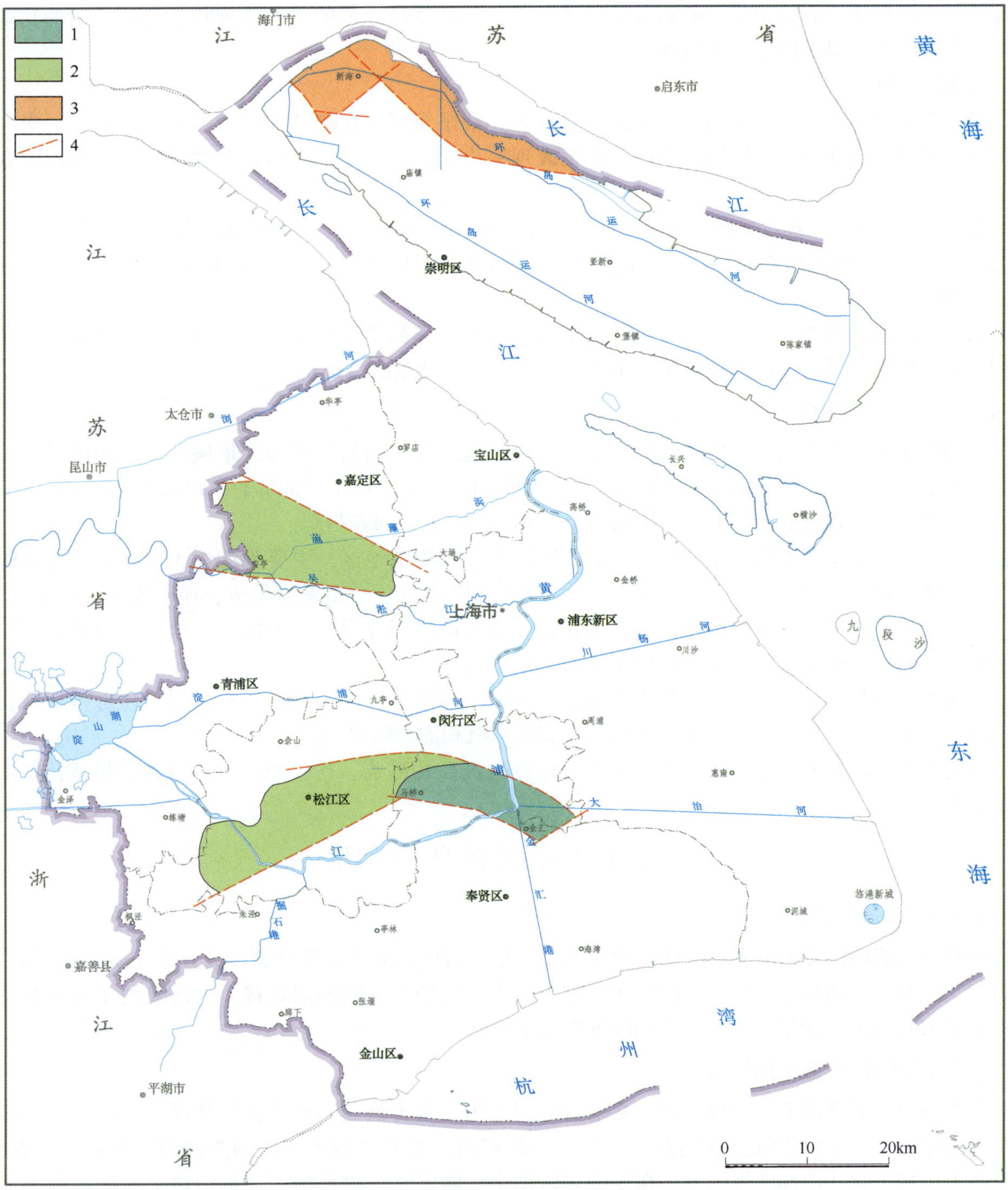

图 16-5　上海中、新生代断陷盆地及地层分布图

Fig. 16-5　Distribution map of the Meso-Cenozoic fault basins and formations in Shanghai

1—古近系；2—下白垩统；3—下白垩统—古近系；4—断裂

（二）地球物理场特征

1. 重力异常

在布格重力异常图上，安角断陷盆地显示为一个北东东向的负异常带，周边均为重力高。北侧有阳澄湖重力高、马鞍山重力高及嘉定—罗店重力高，它们呈近东西向的串珠状分布。东侧的北新泾—大场重力高、西侧的七子山重力高则为北东向，南侧为成茂、千灯相对重力高。

断陷盆地北侧为重力等值线的密集带，梯度变化达5mGal/km，江苏昆山蓬朗附近更大。南侧重力梯度平缓，反映为一斜坡地带。断陷盆地东端外岗、西端横泾一带的重力梯度带，是北西向的葛隆—南翔断裂和北东向湖苏断裂带的反映。

2. 磁力异常

安角断陷盆地内，尤其是角直、花家桥为磁力异常平缓区，推测是红层覆盖较厚的地区。南部斜坡带的二次异常分布零星，特别在千灯、成茂表现突出，是下白垩统火山岩系埋深较浅，火山岩系磁化不均匀所致。

断陷盆地北部的间断性串珠状正磁异常，为侵入及喷出岩的反映。昆山磁力高带及浏河磁力高带内，均见有喜马拉雅期喷发的玄武岩。至于断陷内部安深1井所见的玄武岩，因埋深较大，厚度薄而无反映。

3. 电测深

安角断陷盆地两侧出现等值线密集带，重力、人工地震测量结果相一致。图上也反映断凹有两个沉积中心，但它所反映的断凹的范围比重力、人工地震异常范围略小。

（三）物质组成

安角断陷盆地中有两个东西向沉积中心，且被凸起隔开。凸起东侧为天福庵凹陷，西侧为角直凹陷。安角3孔于430m见古新统夏驾桥组，503m见上白垩统赤山组；安深Ⅰ井井深456m处直接见到上白垩统赤山组，1551m处见上白垩统浦口组，3046.5m处钻遇寿昌组。

寿昌组组成了安角断陷盆地的基底，见于安深1井和周围地区，主要为酸性火山熔岩与火山碎屑岩，由球泡流纹岩、流纹斑岩、英安流纹岩、含角砾熔结凝灰岩、凝灰岩组成，厚度大于326.9m。

浦口组分布于断陷盆地的南部，区内小村、安亭及黄渡都曾发现，为棕红—紫红色杂色砾岩，间夹砂砾岩及粉—细砂岩。安深1井所揭厚度为1405.2m。

赤山组分布最广，安亭、南翔都曾发现，为砖红—棕红色粉砂岩、泥质粉砂岩、含砾粉细砂岩夹棕色泥岩及棕杂色砂砾岩薄层，由泥质与钙质胶结，沉积韵律明显。

夏驾桥组仅见于江苏昆山夏驾桥，主要由棕红色泥质粉砂岩、粉砂质泥岩、粉砂岩组成，厚度大于101.9m。

三、新村—兴隆断陷盆地

（一）分布范围

新村—兴隆断陷盆地分布于崇明岛西段北缘，呈北西向沿北支上段分布，大致位于崇明跃进农场、新村、兴隆和江苏的启东北新、海门临江一带。向东北可能与麒麟凹陷相接。上海境内主要由新海农场—长征农场断裂、绿华—三星断裂和湖苏断裂带控制。

（二）地球物理场特征

在布格重力异常图上，新村—兴隆断陷盆地明显为一重力低，向西延伸至麒麟。该重力低值区有两个低值中心，一个位于红星农场东，一个位于麒麟西，尤其是后者。两侧为新海农场—长征农场、临江—万安两个宽缓的梯度带围限。在航磁异常图上，断陷盆地表现为明显的负磁异常区，在盆地东南缘兴隆沙发育一明显磁力高值。

（三）物质组成

控制钻孔较少，崇明新海LQ24孔于基岩面以下揭露了一层厚达30m的玄武岩。玄武岩下部为一套砂泥岩沉积，461.4m以浅以棕红、灰色、淡紫色为主，下部以紫红色为主。玄武岩K-Ar年龄为31.6±1.31Ma~38.57±2.05Ma，为始新世末期—渐新世初期，相当于三垛组。红星农场P41孔揭露地层主要为赭红色泥钙质砂砾岩，砂砾状结构，胶结差，胶结物占1/4，碎屑占3/4，碎屑成分为各种火成岩，以玄武岩、安山岩、流纹岩为多，并有一些花岗闪长岩、花岗斑岩等，矿物都经受不同程度的次生变化（多为黏土化）和铁染。428~442m井段见明显擦痕。永安HYQD01孔揭露基岩为上白垩统血红色砂质泥岩。

第十七章　构造阶段和变形机制

地壳和岩石圈构造的发展演变表现出相对平静、和缓、漫长的量变期与相对集中、短暂、强烈活动的质变期的交替。每个交替期具有相对稳定和独特的建造和改造特征，即为构造发展的阶段性。构造阶段的划分既要考虑地壳发展的渐进性，又要承认地壳发展过程中的突变性，而上、下地层间的重大角度不整合（构造事件）就是这些突变事件的表现。区内构造阶段的划分是随着对区内地层发育和区域构造演化特征的认识不断完善的。1960年，《上海市石油及天然气地质普查工作报告》将区内构造阶段划分出加里东期、燕山期、古近纪、上新世—第四纪。1962年，《上海地区矿产地质普查工作总结报告》中进一步将区内划分为太古宙—古元古代、新元古代、古生代、中生代、新生代等构造阶段。1988年，《上海市区域地质志》将区内构造阶段划分为前晋宁期、晋宁期、加里东期、燕山期和喜马拉雅期，这一划分方案基本被广泛接受，并沿用至今。1991年，吕炳全将区内划分为前震旦纪、早古生代、新生代3个阶段。

第一节　构造阶段划分和建造特征

根据本次工作指南、上海及邻区相关地质资料，区内的地质构造演化经历了活动—稳定—再活动的多阶段螺旋式发展过程，中元古代以来，总体上可划分为基底形成、盖层形成、滨太平洋陆缘活动等3个大阶段和晋宁期、扬子—加里东期、华力西—印支期、燕山期和喜马拉雅期等5个构造阶段（表17-1）。

一、晋宁期

晋宁期指中元古代—青白口纪这一构造阶段，是区内最早的大规模地壳建造发育和演化阶段。根据区内地质特点，可进一步划分为晋宁早—中期和晋宁晚期。该阶段形成的碳酸盐岩夹黏土质、长英质碎屑岩、泥质碎屑岩、基性火山岩、火山碎屑岩夹少量酸性火山岩建造，经过晋宁运动Ⅰ幕和Ⅱ幕改造发生区域变质作用，形成区内的基底，其中，晋宁运动Ⅱ幕作用使晋宁早—中期构造层发生退变质作用。

晋宁早—中期建造由金山群组成，为晋宁运动Ⅰ幕的产物，见于枫泾—川沙大断裂与张堰—南汇大断裂之间，原岩为火山岩建造、碎屑岩-碳酸盐岩建造，重、磁场呈高密高磁特征。曾经过多期变质作用的叠加，动力变质、热接触交代变质作用普遍可见，是区内变质程度最深，构造变形最强烈的构造层。a岩性段见于周浦赵家宅一带，b岩性段广布于周浦、康家宅、肖塘及朱行一带的钻孔中，片理、片麻理倾角多在60°~70°之间，最缓为40°~45°，最陡者达85°。c岩性段见于沈庄、华泾、厍里等地，岩性与a段相似。

晋宁晚期建造由惠南板岩和河上镇群组成。惠南板岩为晋宁运动Ⅱ幕的产物，分布局限，岩性单一，主要为钙质绢云母板岩与粉砂质绢云母板岩。原岩为粉砂质泥岩及钙质泥岩。变质程度浅，构造变形比金山群简单得多。区域上，河上镇群为在造山晚期至后造山阶段形成的一套滨海和陆相的碎屑岩、火山岩建造。区内目前仅揭露火山岩层位，锆石U-Pb年龄为774±12Ma。

表 17-1 构造阶段划分及建造、应力场特征

Table 17-1 Division of tectonic stages and characteristics of formation and stress field

地质年代				岩石地层	构造阶段		构造运动	岩浆作用	变质作用	沉积作用	构造应力场
新生代	第四纪	Q			滨太平洋陆缘活动阶段	喜马拉雅期			接触变质岩 角闪岩相	以陆相湖泊和河流沉积为主	
	新近纪	N_2	上新世	崇明组N_2c			喜马拉雅运动Ⅱ	基性岩，呈席状			NEE-SWW向挤压
		N_1	中新世	白龙港玄武岩N_1b							近SN向挤压
	古近纪	E_3	渐新世	?			喜马拉雅运动Ⅰ				
		E_2	始新世								NWW-SEE向挤压
		E_1	古新世	北桥组E_1b 夏驾桥组E_1x				基性岩夹层			
中生代	白垩纪	K_2	晚白垩世	赤山组K_2c 浦口组K_2p		燕山期		基性岩夹层			
		K_1	早白垩世	? 寿昌组K_1s 黄尖组K_1h 劳村组K_1l			燕山运动Ⅱ			以陆相火山碎屑沉积为主	NE-SW向挤压 ↑ NWW-SEE向挤压
	侏罗纪	J_3	晚侏罗世					以中酸性岩为主，偶见基性岩		以陆相湖泊和河流沉积为主	
		J_2	中侏罗世	?			燕山运动Ⅰ				
		J_1	早侏罗世								
	三叠纪	T_3	晚三叠世		盖层形成阶段	华力西—印支期	印支运动			陆表海盆地滨浅海沉积(?)	NW-SE向强烈挤压
		T_2	中三叠世								
		T_1	早三叠世								
古生代	二叠纪	P_2	晚二叠世	?							
		P_1	早二叠世								
	石炭纪	C_2	晚石炭世								
		C_1	早石炭世								
	泥盆纪	D_3	晚泥盆世							被动大陆边缘盆地滨浅海沉积	
		D_2	中泥盆世								
		D_1	早泥盆世								
	志留纪	S_4	末志留世			扬子—加里东期	加里东运动				
		S_3	晚志留世								
		S_2	中志留世	?							
		S_1	早志留世	康山组S_1k 河沥溪组S_1h							近SN向挤压
	奥陶纪	O_3	晚奥陶世	长坞组O_1c 黄泥岗组O_3h							
		O_2	中奥陶世	牯牛潭组$O_{1-2}g$							
		O_1	早奥陶世	红花园组O_1h 仑山组O_1l							
	寒武纪	$Є_3$	晚寒武世	超峰组$Є_3cf$							
		$Є_2$	中寒武世	杨柳岗组$Є_2y$							
		$Є_{0-1}$	底—早寒武世	大陈岭组$Є_1d$ 超山组$Є_{0-1}c$							
新元古代	震旦纪	Z_2	晚震旦世	灯影组Z_2d							
		Z_1	早震旦世	?							
	南华纪	Nh_2	晚南华世								
		Nh_1	早南华世								
	青白口纪	Qb_2	晚青白口世	河上镇群Pt_2H			晋宁运动Ⅱ		区域变质岩 低绿片相		NW(W)-SE(E)向挤压
		Qb_1	早青白口世	惠南板岩Pt_3h	基底形成阶段	晋宁期	晋宁运动Ⅰ	酸性侵入岩、中酸性和基性火山岩	区域变质岩 低绿片相—高绿片岩相	弧后盆地浅海和次深海沉积	
古—中元古代				金山群$Pt_{1-2}J$							近SN向挤压

注：构造应力场据涂荫玖(1993)和万天丰(1993)等。

二、扬子—加里东期

该阶段指南华纪—志留纪这一构造阶段。区域上，通常将南华纪到底寒武世划为扬子期。因为区内南华系、下震旦统暂未发现，未进行细分。

该阶段区内上震旦统—志留系发育齐全，表现为浅海—滨海相的连续沉积，组成了区内第一套沉积盖层，地层整体上倾角较小。暂未发现该阶段的岩浆作用和变质作用记录。

上震旦统灯影组主要分布于坦直及罗店、大场，为镁质碳酸盐岩建造。寒武系—志留系在枫泾—川沙大断裂以西地区发育较好。其中超山组为黑色页岩、碳质页岩建造；大陈岭组为含镁碳酸盐岩建造，以具条带状构造及富含黄铁矿为特征。杨柳岗组为碳酸盐岩建造，见缝合线构造。超峰组为镁质碳酸盐岩建造，岩性以白云岩及角砾状白云岩为主。奥陶系见于市区西部地区，沉积连续，黄泥岗组及其以下层位为碳酸盐岩建造，具硅质、泥质团块及豹皮状构造。长坞组为类复理石建造，由灰黑、灰色粉砂质泥岩、泥质粉砂岩及粉砂岩组成韵律层。志留系仅在崇明庙镇一带发现，上部是以紫红色为主的石英砂岩，含泥砾，中下部为灰黑、灰绿色石英砂岩、粉砂质泥岩及泥质砂岩互层。

从沉积岩岩相分析，区内该阶段建造有以下几个特点：

1）为被动大陆边缘盆地沉积，上震旦统—志留系构成一个大的由海进到海退的沉积旋回。沉积环境以海相、浅海相为主，已见累计总厚近2000m。

2）构造层的发育程度、岩相、厚度变化，受晋宁运动所形成的基底隆、拗的控制。枫泾—川沙大断裂以西发育较好，厚度也大，东南部隆起带上分布零星或缺失。

3）奥陶系发育有类复理石建造。

三、华力西—印支期

该阶段指泥盆纪—中三叠世末这一构造阶段，区域上为陆表海沉积阶段。印支运动是我国境内一次划时代的构造运动，经过印支运动，我国东部进入陆相沉积阶段。由于区内迄未发现泥盆系—侏罗系，所以推测区内的主要构造运动幕发生在中三叠世晚期—晚三叠世早期。

四、燕山期

该阶段指晚三叠世—白垩纪末期这一构造阶段。以区域上的燕山运动Ⅰ幕不整合面为界，将该阶段的沉积建造划分为燕山早—中期和燕山晚期。燕山晚期是本区建造记录的壳幔活动最为剧烈的阶段，其间断裂发育，并伴有大规模的陆相岩浆活动，动力变质及热接触交代变质作用发育，同时形成各种火山构造和断陷盆地，造成地层间的假整合和不整合接触，成为区内地质构造史上最具特色的发展阶段。

燕山早—中期仅发现中侏罗世侵入岩建造，中—下侏罗统沉积建造在区内暂未发现，但后者在邻近的浙江、江苏都有分布。如浙西北的王沙溪组和同山群、江苏的象山群，江西的多江组、林山组、漳平组等。主要岩石类型为泥岩、粉砂岩、砂岩、粉砂质泥岩、含砾砂岩、砾岩等，局部夹煤层或炭质条带。

燕山晚期建造在区内分布十分广泛，由下白垩统劳村组、黄尖组、寿昌组火山碎屑-沉积和上白垩统浦口组、赤山组断陷红盆沉积构成。上白垩统和下白垩统之间为明显的角度不整合接触关系。所见构造形变简单，地层倾角一般在25°~30°

劳村组下段为复陆屑、火山碎屑岩及火山岩建造，分布于枫泾—川沙断裂带的东南侧，其中松隐火山盆地及其外围发育最好。劳村组上段为复陆屑、火山碎屑岩建造，主要见于松隐盆地及张堰一带。劳村组上、下段之间呈整合接触关系。劳村期晚期火山活动渐趋平息，火山盆地积水成内陆湖盆。在这一火山运动间隙后，区内最为强烈的黄尖期火山活动又告开始。黄尖组主要为火山岩、火山碎屑岩建造，主要为安山岩及安山质火山碎屑岩、英安岩、流纹英安岩建造，广泛分布于枫泾—川沙断裂带以北、庄行构造火山盆地。寿昌组下段为紫红、青灰、灰黑色粉细砂岩、砂岩、砾岩组成的多个韵律层，局部夹黑色炭质条带和煤线，偶夹凝灰岩，并出现多层泥灰岩，层理发育，为河湖—沼泽相沉积，中段为火山碎屑、复陆屑杂色岩建造，为含火山物质的内陆湖泊相或湖盆陆缘相沉积，上段为火山岩、火山碎屑岩建造，分布局限。寿昌组主要分布于青浦—龙华断裂以北地区，是在黄尖期构造火山盆地背景上发育起来的盆地沉积。

浦口组和赤山组为红盆沉积，夹有玄武岩夹层，分布局限，仅分布在松江—北桥断陷盆地和安角断陷盆地之中。浦口组为山麓相类磨拉石建造，由砾岩间夹砂砾岩和粉细砂岩组成。赤山组由粉砂质泥岩、粉砂岩、泥质粉砂岩、含砾粉细砂岩夹棕色泥岩及杂色砂砾岩组成，与下伏浦口组呈整合过渡关系。

五、喜马拉雅期

该阶段指古近纪以来的这一构造阶段。由于古近纪和新近纪建造差别较大，对应可分为早期和晚期两个阶段。古近纪继承晚白垩世沉积格局，仅分布在局部断陷盆地之中。沉积物以细粒为主，尤其是北桥组。中、上新统为河湖相杂色碎屑岩，间有大范围的玄武岩喷溢。第四系广覆全区，由黄褐、灰、暗绿与杂色黏土、粉砂质黏土、黏土质粉砂、砂、砾石层等组成，沉积环境复杂，多为浅海—河湖相沉积。

第二节　构造体制和变形机制

区内中、新元古代构造格局的形成演化是在哥伦比亚超大陆和罗迪尼亚超大陆聚合、裂解的构造背景中演化的。晋宁早—中期，受近南北向的挤压作用影响，区域上的湖苏断裂带、张堰—南汇断裂带、枫泾—川沙断裂、江绍断裂带、太仓—奉贤断裂等北东向和北西向断裂雏形、近东西向的基底褶皱于此时形成。断裂性质主要呈压扭性。布格重力异常的向上延拓表明，该阶段形成的深部北西向和北东向构造带对后期北东向和北西向断裂构造具有了明显控制作用。晋宁晚期，在雏形断裂基础上，湖苏断裂带等超壳断裂开始明显活动，尤其是陆内裂谷背景下，各断裂持续活动。在北西西-南东东向区域主压应力场的作用下，青白口早期地层发生褶皱作用，形成了北东—北北东向隆、坳相间的基底褶皱构造。

扬子—加里东期，在南北向挤压主应力场作用下，早期断裂持续活动。在形成东西向构造形迹的同时，派生了部分较大规模的北西向断裂构造。湖苏断裂带对区内沉积已具有明显的控制作用，构成台地、台地边缘、斜坡、盆地、浅海、潮坪等沉积相的分界。由于加里东运动主要呈造陆特征，所以本期构造变形强度相对较弱，加里东期建造主要表现为宽缓的褶皱构造。

印支期是区内主要构造形迹形成的重要时期。在北西-南东向构造挤压应力场作用下，印支运动奠定了区内以北东走向的线性构造为主的构造格架。印支运动在浙西被认为是晋宁运动以来最为强烈的一次褶皱运动，使基底及盖层产生剧烈的褶皱和逆掩构造，形成走向北东的线性断褶带。推测区内震旦系、下古生界的紧密线性褶皱，是在加里东宽缓褶皱的基础上经过印支运动而得到进一步加强和定型的。

燕山构造期欧亚板块与太平洋板块发生斜向碰撞，区内进入强烈的陆内构造改造阶段。早期表现为挤压-走滑，晚期转为拉张-碰撞的大地构造背景。燕山早中期，持续的水平挤压，使湖苏断裂带等北东向断裂带转换为大规模的壳内拆离断层，滑移面为上地壳下部或者中地壳上部。至燕山晚期，在北东-南西向主压应力场背景下，我国东部转化为以伸展为主的构造体制。在该体制背景下，区内活动加剧，北东向断裂浅部转化为张性，并派生出次级正断层，局部构成地堑构造，发育了广布的白垩系火山碎屑岩、火山岩类和断裂带控制的断陷红盆沉积。北北东向断裂开始明显活动，主要表现为张扭性。北西向和近东西向断裂主要呈压性、压扭性。

喜马拉雅期构造运动主要表现为早期构造基础上的晚期构造及盆地叠加和活化，尤其是北东向、北西向、东西向断裂的活化。北东向断裂与东西向断裂为主要控盆断裂，北西向断裂起主要以切割、位移及破坏为主的作用，部分断裂则亦具一定控盆作用。

第十八章　新构造运动特征

新构造运动直接影响和决定着现代地貌的发育过程及其形态特征，影响着现代地壳的稳定性。区内新构造运动研究主要集中在地震活动和活动性断裂、区域地壳稳定性评价方面。20 世纪 80—90 年代为区内地震地质研究的活跃阶段，研究内容涉及控震构造、地震地质背景、构造应力场、地震活动特征和活动趋势、基本烈度等。在地震活动规律上，提出了地震分布的丛集性、周期性和迁移性等认识；但通常采用有感及以上规模的地震，且研究范围较大，以地震区、带或长江中下游地区、发震频度较高的南黄海地区为主。主要成果有：

1983 年，华东师范大学开展了上海邻近地区活动性断裂与地震活动的初步研究。郑家欣（1986）论述了苏南（包括上海）新构造运动的基本特征，提出了新构造运动标志，划分了新构造分区。1989 年，上海大学开展了上海地震活动性研究。刘若愚（1991）讨论了石灰华对区内新构造隐伏活动断裂的指示意义，评价了 6 条隐伏断裂的活动性。1990 年，上海市地震局完成了上海地震危险性分析与基本烈度复核研究，在地震地质背景研究基础上，对地震活动性和危险性、地震基本烈度等内容进行了系统研究。同年，上海市环境地质站开展了上海及邻近地区区域地震地质研究，重点分析了区内地震活动规律，进行了潜在震源区划分。1992 年，上海市环境地质站初步开展了新构造运动特征研究，进行了新构造分区，总结了地震活动特征，提出了初步的地壳稳定性分区方案。地质矿产部第一综合物探大队（1992）通过浅层地震勘查，认为枫泾—川沙断裂、大场—廊下断裂、灯塔断裂和赵家宅断裂仅切割基岩面，罗店—周浦断裂上断点埋深 150m 左右。秦大正（1994）对区内枫泾—川沙断裂、罗店—周浦断裂、张堰—南汇断裂、昆山—嘉定断裂的新构造期活动性进行了论述，认为前两者最新活动时间在早更新世末期，张堰—南汇断裂未切割松散层，昆山—嘉定断裂最晚活动时间为新近纪。沈永盛（1984）、顾志文（1997）分别讨论了新生代沉积与活动性断裂的关系。1995 年，上海市岩土地质研究院开展了区内地震地质背景分析，提出了新的构造单元划分方案，总结了新构造运动特征、活动性断裂、盆地构造与地震的关系，进行了稳定性区划等；1997 年，上海市城市地质研究院进一步开展了上海新构造运动与地震活动特征研究。王卫平（1997）对上海及邻区的区域地壳稳定性进行了分区，认为总体上属于基本稳定区，而且浦东稳定性最好，崇明次之，青浦稍差。

2004 年，上海市地震局在上海陆域和海域范围开展了全面的活动性断裂调查，并出版专著《上海市隐伏断裂及其活动性》对断裂活动性进行了详细论述，认为区内断裂最新活动时间差别较大，最新可至晚更新世。2008 年，“上海市三维城市地质调查”成果从更大范围内对上海及邻区的地震活动规律、新构造分区等进行了研究，开展了潜在震源区划分，评价了地震危险性及活动断裂对城市建设的影响。

第一节　新构造运动标志和基本特征

在中国新构造区划上，上海处在华北新构造区的南缘（李祥根，2003），接近与华南新构造区的分界处。两者的分界约在杭州湾。新构造运动相对较活跃，表现形式多样，在空间及运动形式上既有继承性，又有新生性。

根据工作指南，本志中的新构造运动一般指上新世以来的构造运动，活动构造指 15 万年以来活动的构造。

一、运动标志

（一）地震活动

据不完全统计，1500~2017 年间，在 N30°~32°40′、E120°30′~123°30′范围内发生地震超过 1750 次，其中有感地震 79 次、中强震 28 次、强震 4 次。最高为 1984 年南黄海 $M_s=6.0$ 级地震。上海陆域地震 96 次，最高为 $M_s=4.75$ 级（或 $M_s=5.0$ 级）。

（二）断裂活动

浅层人工地震勘查显示，区内新构造断裂较发育。至目前，上海陆域和近岸水域共发现新构造断裂约 37 条，其中，东西向 7 条、北东东向 11 条、北北东向 5 条、北北西向 9 条、北西西向 4 条、南北向 1 条。

（三）差异沉降和夷平面

上新世以来，区内继承了早期差异性沉降特征，伴随的差异运动则形成了由隆起区向坳陷区的阶梯状过渡带。中更新世以来，总体上转入相应的加速沉降时期，显示相对升降的地壳运动趋于衰退状态。全区基岩埋深变化大体出现三级夷平面：较高一级埋深为 180~220m；中间一级埋深为 320~350m；较低一级在东部长江三角洲沉积区河口沙岛分区，埋深大于 400m。三级夷平面间的落差比较均匀，为 100~120m。根据磁性地层推算，塑造夷平面的地质年代较低的一级形成于 3.60Ma B.P. 左右，中间的一级形成于 3.00Ma B.P. 左右，较高的一级形成于 1.50Ma B.P. 左右。据此揭示，区内晚新生代地壳沉降与堆积速率，由新近纪的约 17cm/ka 降至早更新世早、中期的 8cm/ka 左右。

（四）节理和擦痕

在埋深大于 150m 的下更新统—上新统内，可以见到含铁锰质结核和钙质结核、网脉的黏土被节理切割（钙质结核、网脉被切割明显可辨），局部显示微小的错动。顺节理淋滤形成一定宽度的黄绿、灰绿或蓝灰色网纹。

（五）地热活动

浅部第四承压含水层水温和区内地温梯度表现出沿某些断裂呈带状高值区的特征，并且含水层有石灰华分布的基岩断裂带上的地温，较区域平均地温值一般偏高 1~2℃，与我国山区沿新构造活动断裂带多见有零星分布的温泉（如西藏等地）并相伴有石灰华沉淀的特征情况类似（刘若愚，1991）。地热异常的长轴方向大致与北西向、东西向张扭性断裂有关（黄焕忠，1990）。

（六）垂向变形

区域上，北北东向的茅山断裂带，在溧阳附近测得在断裂以西地区向东北方向的位移（马杏桓，1986）。据中国大陆构造环境监测网络，佘山 GNNS 基准站（编号：JZ08）近 20 年的位移时间序列拟合显示，区内垂向的变形速率为-1.38 ± 0.27mm/a。上海佘山台伸缩仪记录的东西向形变速率平均为 1mm/a（宋先月，2003）。

二、基本特征

据上述识别标志，总结区内新构造运动总体上具有以下特征：

1）新构造期主压应力为北东东—近东西向，河谷分布的优势方向显示主压应力场方向为 96°左右。北北东向断裂主要表现为右旋压扭，北东向断裂以张扭居多，东西向主要为张性正断层，北西向断裂则多表现为左旋张扭（朱履熹，1990）。

2）断裂活动有很强空间和时间差异性，各条断裂在不同时期其活动强度是不一样的，甚至同一条断裂的不同区段在同一时期内其活动性也有很大差异。北西—北北西向断裂自西向东活动性减弱或分段性不明显。北北东向断裂则反之。北东向断裂一般西段的活动性要较东段的稍强。断裂活动性相比，由强至弱，依次为北北西—北西向、北北东向、北东向。

3）断裂最新活动时间一般为早—中更新世，北北西向断裂的最新活动时间更新些，一般为中更新世，有的可延续至晚更新世。一般来说，东西向、北东向断裂活动时间略早，北北东向次之，北西西向、北北西向较晚。

4）断裂的最大断距一般为5~8m、几十米，在断点剖面上显示，往上断距变小，有时代越新幅度趋小的特点。第四系底界断距在10~30m；中更新统底界断距在12~13m。陆域断裂平均位移速率以太仓—奉贤断裂为最，在0.01~0.028mm/a之间。

5）区域上，地震在空间上具有很强的丛集性、重复性、分块性和条带性。在时间上表现出相对平静期和显著活动期的交互特征，呈现出一定的周期性。震源深度多在5~20km，存在5~10km、15~20km两个优势层。弱震震级与震源深度无相关性，其他地震震源深度与震级总体呈一定线性关系。自北西向南东方向明显减少，总体呈北强南弱、东强西弱（或海域强、陆地弱）的特征。

6）绝大部分地震受第四纪活动断裂，尤其是晚更新世活动断裂的控制。在多组断裂的交汇处、新生代断陷盆地的边界断裂，尤其是北北东（北东）向和北西向断裂、北东（北北东）向与东西向断裂交叉部位是主要发震部位。中强地震在条带相交区域的频度与强度相对较高。此外，地震与新生代盆地构造关系也很密切。历史上超半数的中强地震都发生在早新生代继承性断陷盆地边缘和内部的次级隆、坳交界处，而在中生代断陷盆地很少发生。大型盆地边缘往往有6~7级地震发生。那些既控制盆地沉积中心，又是盆地边缘的断裂带有较大地震。

7）区域上，沉积厚度的大小普遍受断裂构造的影响，其中尤以北西或北北西向断裂的影响更甚（顾志文，1997）。在北东向和北西向断裂联合控制下，区内形成了北西向分带、北东向分块的基岩面起伏形态。松散层厚度自西南和东南岛屿的浅埋区向崇明岛方向变厚。

8）上新世时，在邻近浙江等地区有火山作用。由于受块断构造活动影响，区内出现剥蚀低山与堆积盆地相间分布的格局，形成山前洪积平原和缀布其间的扇间洼地。第四纪以来，伴随着地壳的持续沉降、气候和海平面波动，区内逐渐全面转化为沉降盆地。

第二节　断裂活动性

新构造期以来断裂构造的活动性，主要是通过浅层人工地震勘查方法获得的。陆域主要断裂都表现出一定的活动性，其中晚更新世断裂2条，中更新世断裂3条。

一、晚更新世断裂

（一）太仓—奉贤断裂

断裂北段（沪宜公路剖面）断面倾向北东，断裂上延突破T_3反射波阻界面，为中更新世断层；断裂南段（申兴路、黎安路等）由两条相距250~500m，倾向向背的正断层构成（图18-1）。断裂突破T_2反射波阻界面一直进入上更新统青灰色细砂层中，上断点埋深约80m，断距10~15m。古地磁结果显示，86.7~88.8m的地层年代为10.8×10^4~11.4×10^4a B.P.（布莱克反极性事件），因而南段最新活动时间约10×10^4 a B.P.。垂直位移平均速率为0.01~0.03mm/a。

断裂明显制约了新近纪以来的地貌、古地理演变，是重要的地貌界线，且和相当于海侵范围边界的“冈身带”基本一致。沿断裂北桥、莘庄等处有地热异常。1990年太仓4.9级地震、1992年奉贤西北部的2.1级地震的等震线长轴及余震分布方向均与断裂走向一致。

（二）张堰—金山卫断裂

该断裂可分成两段，呈雁行排列。断裂北段走向北西330°~335°，倾向南西，倾角为60°，上断点埋深在160m左右，为早更新世断裂。往南至张堰附近，表现为由阶梯状的两条断裂构成，宽为60m，倾向北东，倾角为55°~65°，基岩面落差约为30m，上断点埋深在90m左右，属中更新世断裂。断裂

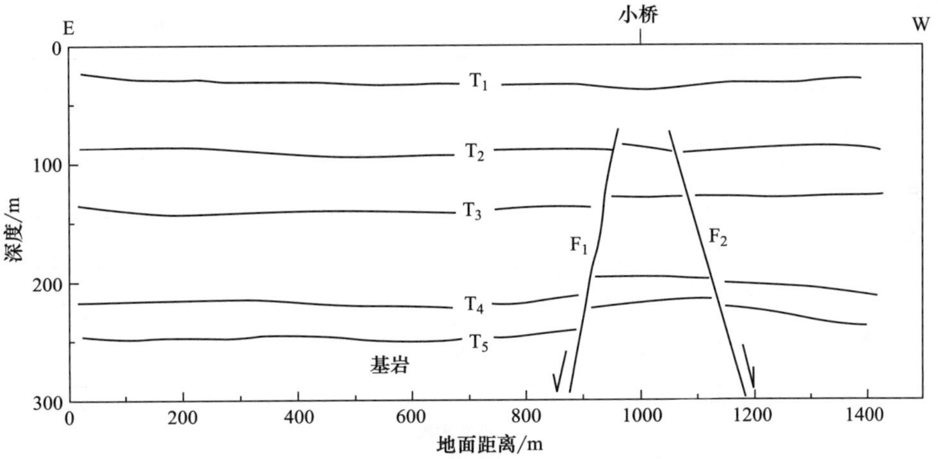

图 18-1　黎安路测线地震深度剖面图

Fig. 18-1　Seismic depth section of the Li'an Road survey line

南段（金山卫城南部）走向为北西 325°，倾向南西，倾角为 70°，基岩面落差达 12.5m，断裂上断点突破 T_2 反射波阻界面，终止于上更新统底部，埋深为 75m，属晚更新世断裂（图 18-2）。

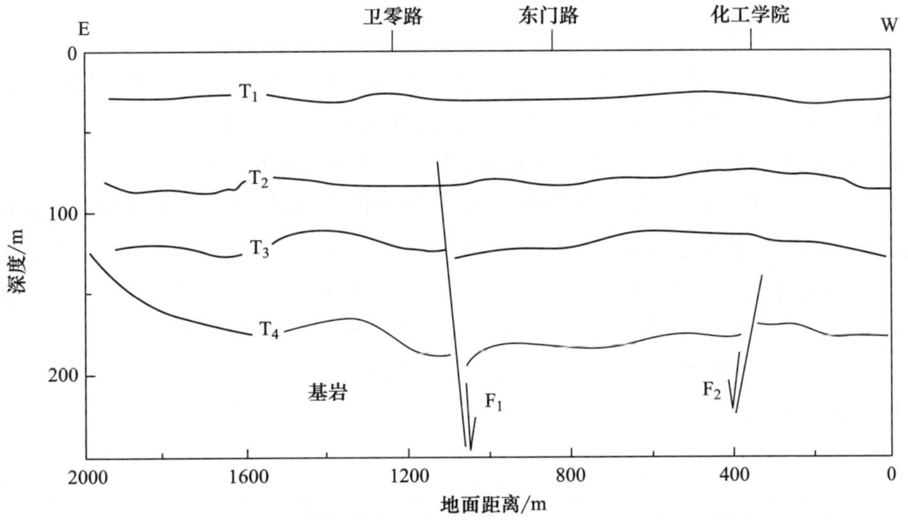

图 18-2　金山卫测线地震深度剖面图

Fig. 18-2　Seismic depth section of the Jinshanwei survey line

南段上更新统顶部的暗绿色黏性土层等厚线横跨断裂，呈近东西向延伸，未出现畸变，说明晚更新世后期以来未受断裂影响。

二、中更新世断裂

（一）枫泾—川沙断裂

断裂为由两条倾向相反的断裂组成的地垒构造，宽为 0.5~2km，有分段性，呈西强东弱的特点。

断裂西南段（枫泾—车墩），在松江城南基岩破碎带宽为 50~60m，基岩面落差为 30m，185~195m 深处落差为 10m，160~165m 深处落差为 5m。断层进入中更新统下部，为中更新世断裂。

断裂中段（莘庄南，经朱行、长桥，至黄浦江边）北支在申兴路、沪闵路等地基岩破碎带宽为 20~60m，基岩面落差为 13~30m，倾向北西；南支基岩破碎带宽为 20~60m，落差为 24~44m，倾向南东（图 18-3）。上断点均突破 T_3 反射波阻界面，埋深为 130~150m，被蓝灰色粉质黏土（$23.09\times10^4\pm1.87\times10^4$a）覆盖，最新活动时间为中更新世。莘庄地温梯度为 4.4℃/100m，属地热异常地段。

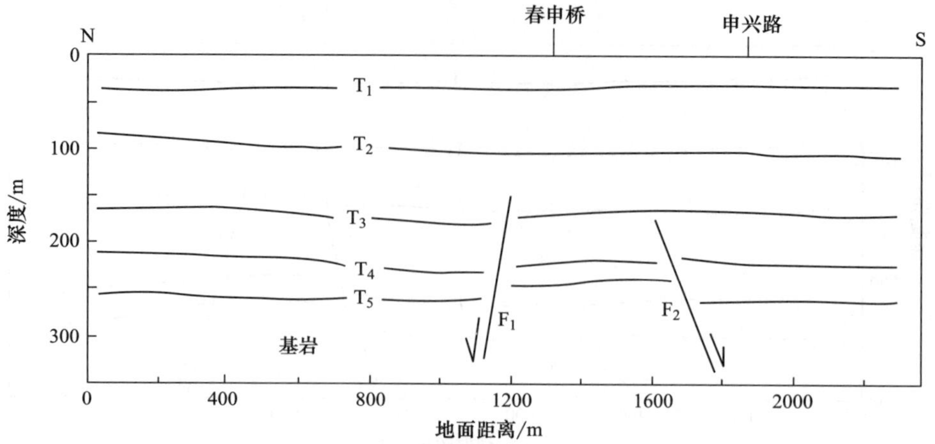

图 18-3　沪闵路测线地震深度剖面图

Fig. 18-3　Seismic depth section of the Humin Road survey line

断裂东段，南支在杨思南断层带宽为 50m，基岩面落差为 22m，仅上延至第四系底部。北支在北蔡等地基岩断层带宽为 20~50m，基岩面落差为 11~13m，穿越 T_4 反射波阻界面，属早更新世断裂。

(二) 罗店—周浦断裂

在上海市内大致可分为 3 段：

断裂北段（江苏浏河—大场）在宝山罗店等地由 3 条断层组成，表现为复式地堑构造，宽 610 余米，北西两支断裂相向倾斜，倾角均在 75°以上。基岩面断距在 12~18m 不等，且均突破 T_3 反射波阻界面，上断点埋深在 145~155m，进入中更新统底部，为中更新世早期断裂。

断裂中段（大场—黄浦江边）由 4 条断层组成，西侧两条倾向北东，东侧两条倾向南西，构成复式地堑构造（图 18-4）。4 条断裂自东向西，F_1 从大场经甘泉新村，过吴淞江在常熟路北京路口被北东向断裂所截，于打浦路复兴路西侧又复显现至老南火车站江边。F_2 自大场北，经侯家宅、海防路东端、江宁路延安路口、斜土路，至黄浦江边。F_3、F_4 自彭浦始经蒙自路东西两侧至南浦大桥。4 条断裂皆为正断层，其中 F_1 仅穿越 T_4 反射波阻界面，而东侧 F_2、F_3、F_4 则突破 T_3 反射波阻界面，断距一般为 7~10m，断层已影响到中更新统顶部。

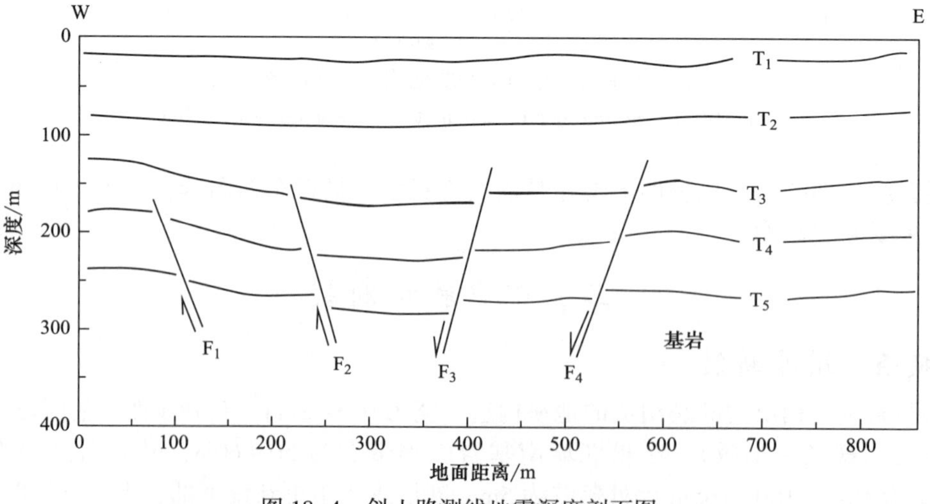

图 18-4　斜土路测线地震深度剖面图

Fig. 18-4　Seismic depth section of the Xietu Road survey line

断裂南段（浦东六里—新场）在东方路等地由 2 条断裂雁列而成，走向北西 330°~340°，倾向南西，倾角为 70°。基岩面断距在 12~30m，自北向南断距逐渐减小。上断点埋深在 160~190m，亦呈向南逐渐加深的趋势。向上错断 T_3 反射波阻界面，为中更新世断裂。

1615 年南通 5 级地震及 1624 年上海 4 $\frac{3}{4}$级地震在该断裂各分段断裂的端部，等震线长轴方向与断裂一致。此外，沿断裂在浮桥、浏河、罗店、北蔡、新场等地还有小地震发生。

（三）朱行断裂

断裂在朱行、护塘等地基岩面断距为 5m 左右，错断 T_4 反射波阻界面，上断点埋深在 115~120m，为中更新世早期断裂。

三、早更新世断裂

（一）张堰—南汇断裂带

浅层人工地震剖面揭示，浅部以倾向为北西者多，陡倾角。断裂在金山张堰松金路（图 18-5）、奉贤南桥北沪杭公路等地都有显示，F_4 上断点切割下更新统，为早更新世断裂。

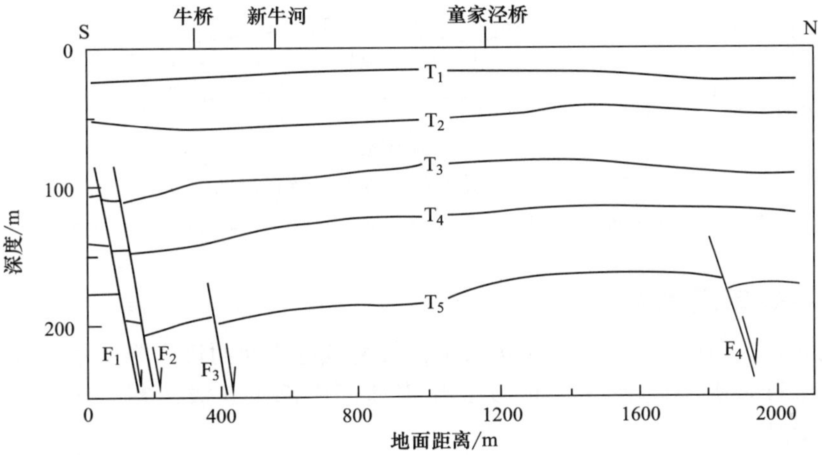

图 18-5　松金路测线地震深度剖面图

Fig. 18-5　Seismic depth section of the Songjin Road survey line

（二）灯塔—赵家宅断裂

断裂在顾高路、灵通路、闻居路等地均见进入上覆第四系；北段显示活动时间为新近纪末，灵通路和闻居路都反映为突破 T_4 反射波阻界面，应为早更新世断裂（图 18-6）。

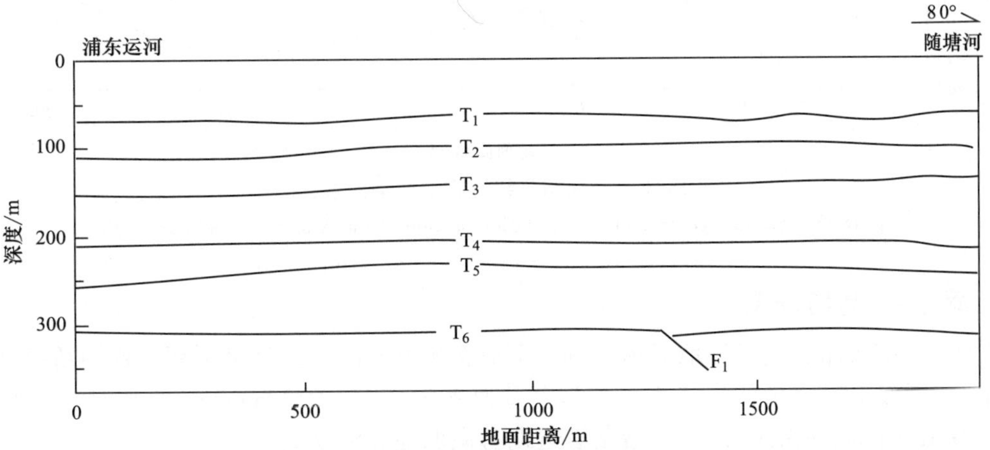

图 18-6　顾高路测线地震地质解释剖面图

Fig. 18-6　Seismic geological interpretation section of the Gugao Road survey line

（三）卖花桥—吴泾断裂和马桥—金汇断裂

卖花桥—吴泾断裂南北两侧基岩反射层特征不同（图 18-7 之 F_1），上盘为古近系，其上第四系厚为 295~328m，下盘为下白垩统，上覆 265~284m 的松散堆积。马桥—金汇断裂两侧的第四系厚度也有所不同，上盘第四系厚为 258~275m，下盘第四系厚为 223~227m。两条断裂都从基岩进入第四系后突破 T_4 反射波阻界面，为早更新世断裂。

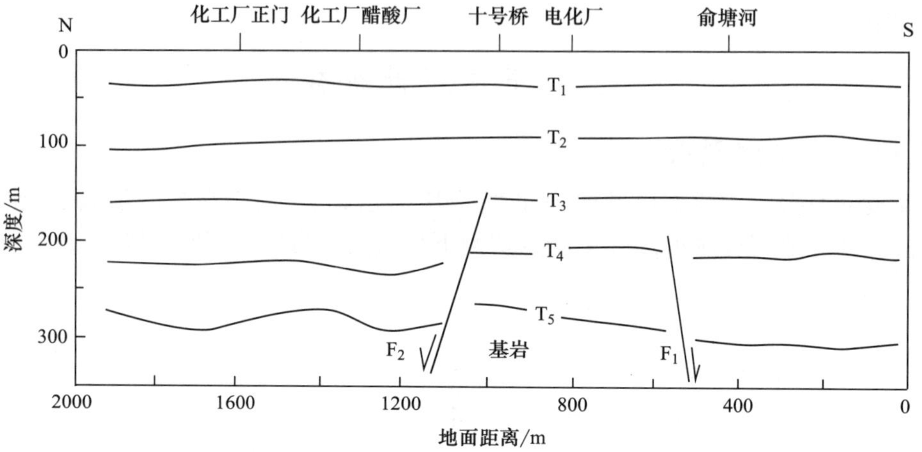

图 18-7　龙吴路测线地震地质解释剖面图

Fig. 18-7　Seismic geological interpretation section of the Longwu Road survey line

（四）白鹤—姚家港断裂

断裂在青浦城北香大路（图 18-8）断面东倾，仅突破基岩面而进入第四系底部，属早更新世早期断层。断裂附近发生过多次小震活动，如 1975 年青浦西南部 1.0 级地震、1978 年白鹤 2.3 级地震、1991 年练塘 3.0 级地震等。

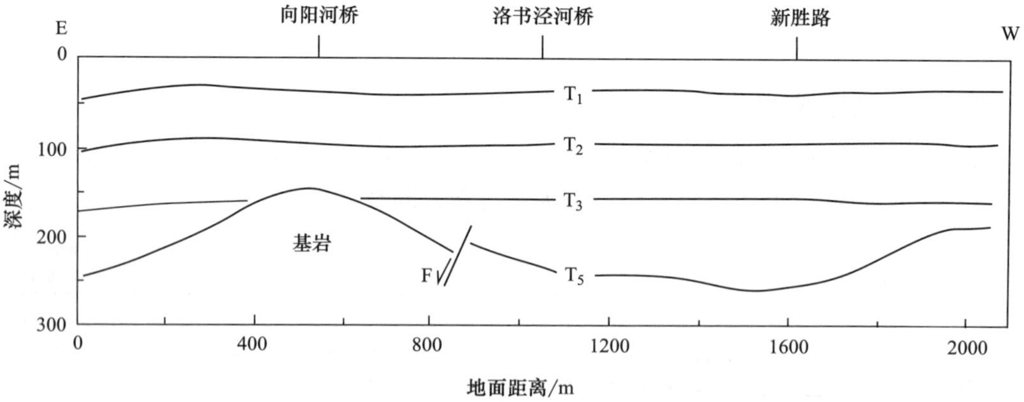

图 18-8　香大路测线地震地质剖面解释图

Fig. 18-8　Seismic geological interpretation section of the Xiangda Road survey line

（五）廊下—大场断裂

断裂在七宝、虹桥路、松隐等地显示切割至下更新统上段底部，为早更新世晚期断裂（图 18-9）。断裂带两侧基岩重复水准测量初步认为：东盘上升速率为 0.35~0.39mm/a，西盘桃浦、华漕则下降，速率分别为-0.13mm/a、-3.56mm/a。沿断裂近期有微弱地震活动。

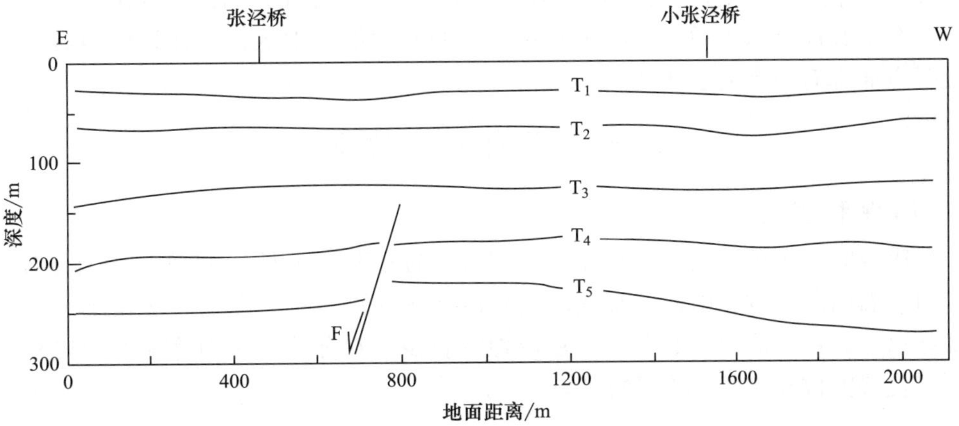

图 18-9 松隐测线地震地质剖面解释图

Fig. 18-9 Seismic geological interpretation section of the Songyin survey line

（六）昆山—嘉定断裂

断裂在沪嘉辅道及胜辛路等地显示由两条断层组成，基岩面落差在15~17m和21~30m，断层视宽度分别为15~50m和20~50m，构成地垒形式。断裂上延入第四系被T_3反射波阻界面（中更新统底界）整合覆盖，为早更新世断层。

（七）青浦—龙华断裂

断裂在青松公路、赵重公路、黄方公路等地显示为由2~3条断层组成断裂带，宽为0.7~1km，呈地垒形式出现，基岩面断距在4~26m不等，由西向东断距增大。西段断层突破T_5反射波阻界面，东段进入第四系后还突破T_4反射波阻界面，判断其西段为早更新世早期断裂、东段为早更新世晚期断裂；对第四系厚度有一定的控制作用（图18-10）。

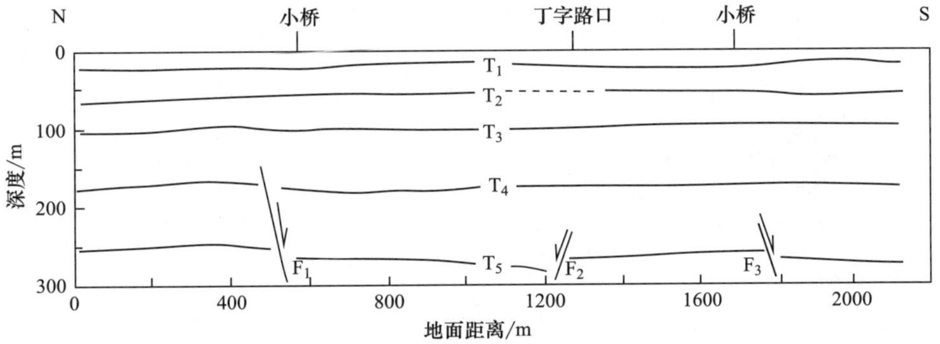

图 18-10 赵重公路测线地震地质剖面解释图

Fig. 18-10 Seismic geological interpretation section of the Zhaozhong Highway survey line

除上述断裂外，葛隆—南翔断裂、大场—江湾断裂、吴江—黄渡断裂、湖苏断裂带（上海段）在第四纪都表现出一定的活动性。绿华—三星断裂、太仓—二堡镇断裂和永胜—施庙断裂在上新世呈明显活动性。

第三节 地震活动性

区内地震活动大致呈现北强南弱、东强西弱（或海域强、陆地弱）的态势。据《中国地震动参数区划图》（GB18306—2015），上海及邻区主要位于华北地震区东南边缘地段，进一步划分为华北地震区的长江下游—南黄海地震带。自有地震记载（公元1475年）以来，据不完全记载，上海曾遭受过

160 余次中强地震袭击，其中发生在今上海行政区及其东侧海域内的地震有 70 余次。主要分布在北至南黄海震源区，东至长江口东震源区，西至常熟，南至舟山群岛地区，上述地区地震活动在空间、时间、震源深度等方面都有一定的规律性。

一、空间分布特征

（一）不均性和重复性

区域上，地震空间分布极不均匀，有几个比较明显的密集区（图 18-11）。分别分布在南黄海、长江口东、长江口、衢山岛东侧、南通、常熟以北福山和张家港等地。在这些区密集内的地震活动次数占研究区的半数以上。杭州湾南部的浙西北陆域地震较少。在苏州—昆山等地的历史和现代地震分布较均匀，无明显优势方位和丛集、重复特征。

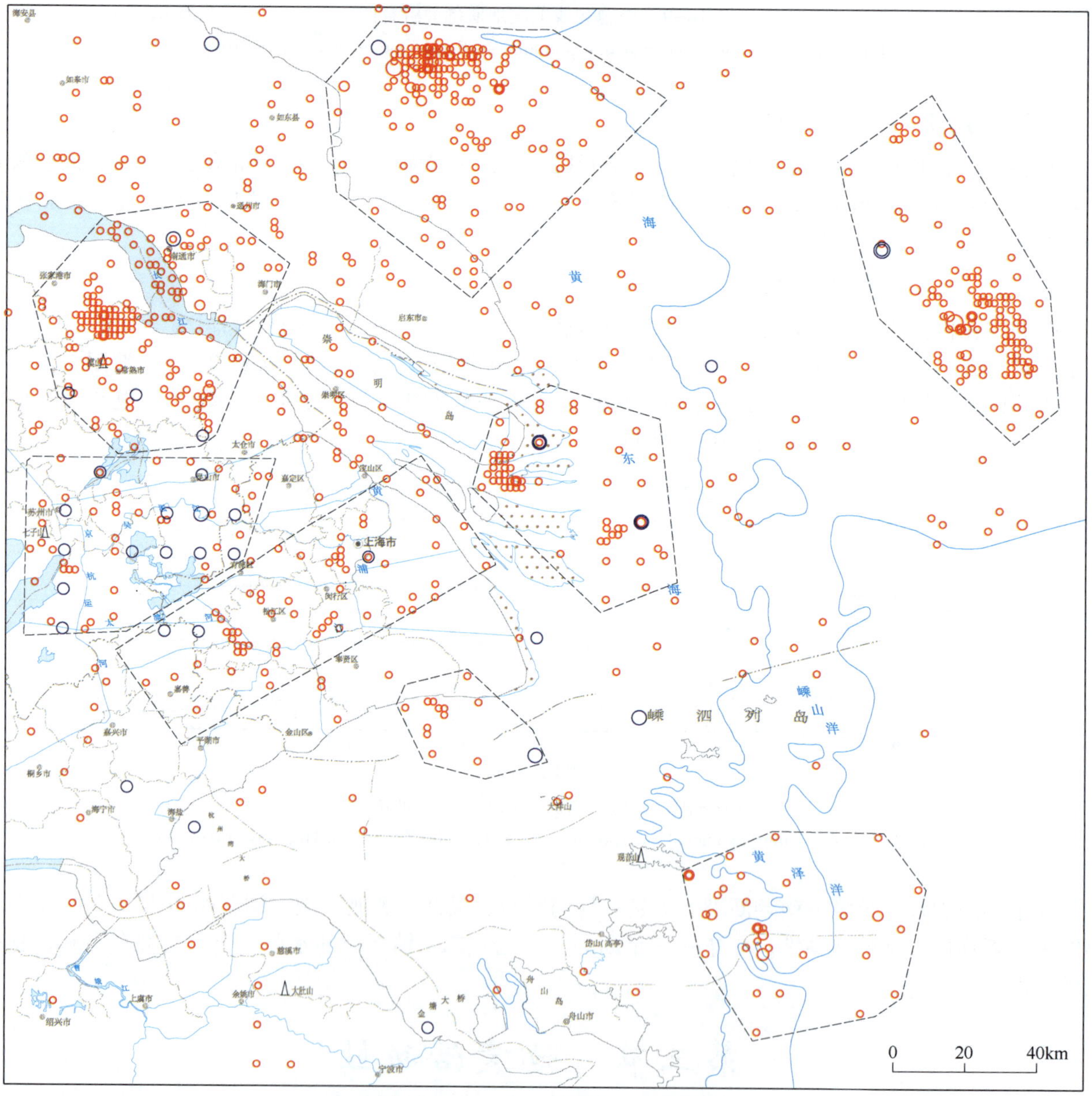

图 18-11　地震活动震中分布特征

Fig. 18-11　Distribution of the epicenters of seismic activity

历史地震：○6.0~6.9 级；○5.0~5.9 级；○4.0~4.9 级。

1970 年以来地震：○6.0~6.9 级；○4.0~4.9 级；○3.0~3.9 级；○小于 3.0 级

不同构造单元中分布特征具有明显差异。在下扬子地块和钦杭结合带内部分布较多。华南造山带分布较少，地震活动水平低、频度小、强度弱，属弱震、少震区。分布的具体构造部位多处于不同次级构造单元的结合带、多组断裂的交汇处。主要有勿南沙隆起带中部、南通断块西南缘、江阴—昆山断块中部、昆沪地块中部、崇明东凹陷带周缘等。

强震分布区呈北西向分布在长江口到江苏盐城附近的南黄海海域。主要强震有 1994 年 6.3 级、1996 年 6.1 级等地震，其中 1996 年 6.1 级地震就位于上海附近，曾造成上海市某些建筑物的损毁，上海市有感强烈。弱震、有感地震与大多数中强震—强震在空间上分布具有很强的一致性。在地震丛集区，中强震、强震发生的前后，弱震—有感地震明显增多。

（二）条带性和分块性

中强震、强震通常伴随有地震条带，异常条带的展布范围一般与大断裂带相一致（刘蒲雄，1989）。区内弱震等虽具有弥散状分布的特征，但以丛集区为中心，仍表现出一定分块性，有长江口东、南黄海、长江口、舟山群岛东、南通—常熟—沙溪。将中强震、强震分布区的几个地震群与零散的地震活动贯穿起来，分布方向同样具有一定的优势性，表现出一定的条带性，主要条带沿北东和北西向分布。总体上显示，地震活动受北西向断裂带和北东向断裂带的联合控制。

北东向主要为上海市中部—长江口—长江口东、张家港—南通—勿南沙、白茆—启东等地；北西向主要为张家港—白茆、勿南沙东部—长江口、长江口东部等地。

二、时间分布特征

历史地震资料显示，区内地震活动在时间上分布也是不均匀的，表现出地震活动的相对平静期和显著的活动期，呈现一定的规律性、重复性（图 18-12，图 18-13）。

1）大于 3 级的地震自 1839 年以来呈现出加强的趋势，尤其是自 1960 年以来。这种趋势也有可能是早期地震记录的缺失造成的。在 1630 年、1850 年等强震的前后比较密集，多呈现出 5～10a 的活动间隔期。

2）1500 年以来该带经历有 3 个地震活跃期（1500—1679 年、1719—1761 年和 1839 年以来），最后一个活跃期地震活动又经历有 3 个活跃幕，每个活跃幕 40 余年。目前，该带正处于第二活跃期的第三个活跃幕（1970 年以来）。估计未来百年内该带地震活动处于活动阶段的后期，但仍存在发生 6～7 级地震的可能。

3）1970 年以来的全部地震活动特征显示，区内弱震频发，呈现出明显的峰、谷特征，呈 1～2a 的周期。在一次中强震、强震发生后，均有活动减弱的特征。总体上，自 20 世纪中叶以来，地震活动频次具有加强的趋势。

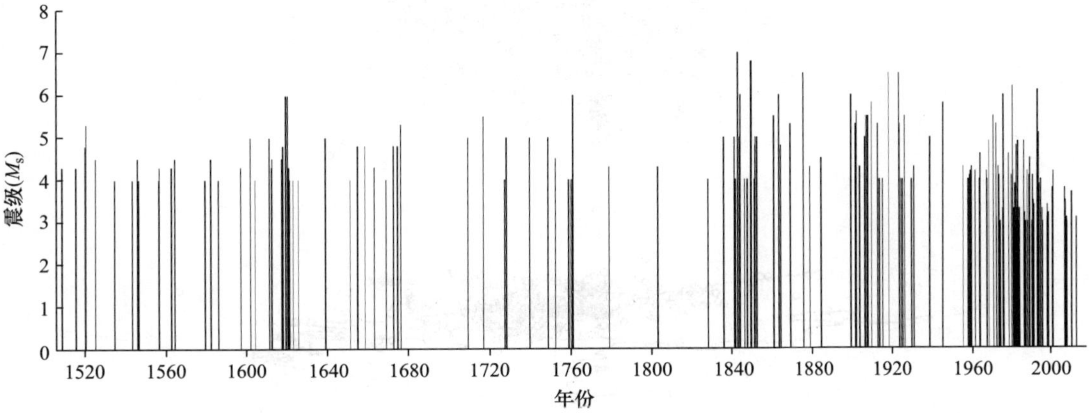

图 18-12　1500 年以来研究区大于 3 级地震 $M-t$ 关系图

Fig. 18-12　$M-t$ relations of earthquakes greater than magnitude 3 in the study area since 1500

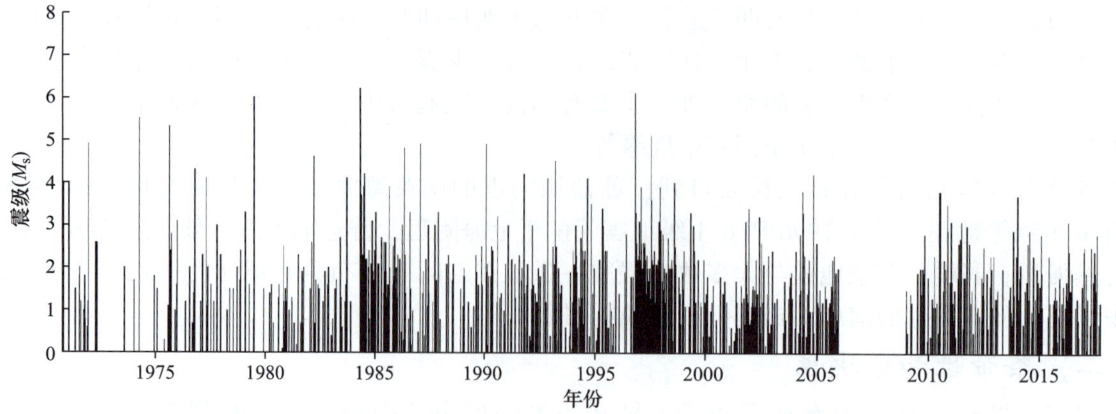

图 18-13　1970 年以来研究区全部地震 $M-t$ 关系图

（2006—2008 年缺数据）

Fig. 18-13　$M-t$ relations of all earthquakes in the study area since 1970

（other than 2006-2008）

三、震源深度分布特征

本次收集的研究区的地震震源深度在 1～37km 之间，$H\leqslant5$km 的地震次数约占 10.7%，$5<H\leqslant10$km 的地震次数约占 37.8%，$10<H\leqslant15$km 的地震次数约占 20.2%，$15<H\leqslant20$km 的地震次数约占 25.0%。大于 20km 的地震次数约占 6.0%。主要的震源深度总体分布在上地壳下部—中地壳范围内，多为 5～20km，存在 4～11km、15～21km 两个优势层（图 18-14）。

沿纬度方向上，下地壳以深的震源深度集中在 31°～33°之间；沿经度方向上，地壳以深的震源深度集中在 120.5°～122°、123°～123.5°之间（图 18-15）。南黄海地区震源深度分布范围较大，平均为 12km，向北部深度加大。南通—常熟地区的震源深度变化较大，在 1～31km 都有分布，集中在 20km 以浅，平均为 11.3km。长江口东地区震源深度分布在 1～28km，平均深度约为 13.4km；上海的震源深度较浅，分布范围为 1～22km，平均深度为 9.5km。杭州至太仓以西深度在 1～20km，平均为 10.4km，长江口海域深度多在 1～18km，平均约为 11.1km，舟山群岛地区震源深度为 8～20km，平均为 11.1km。总体表现为海域平均深度深、陆域浅的特征。

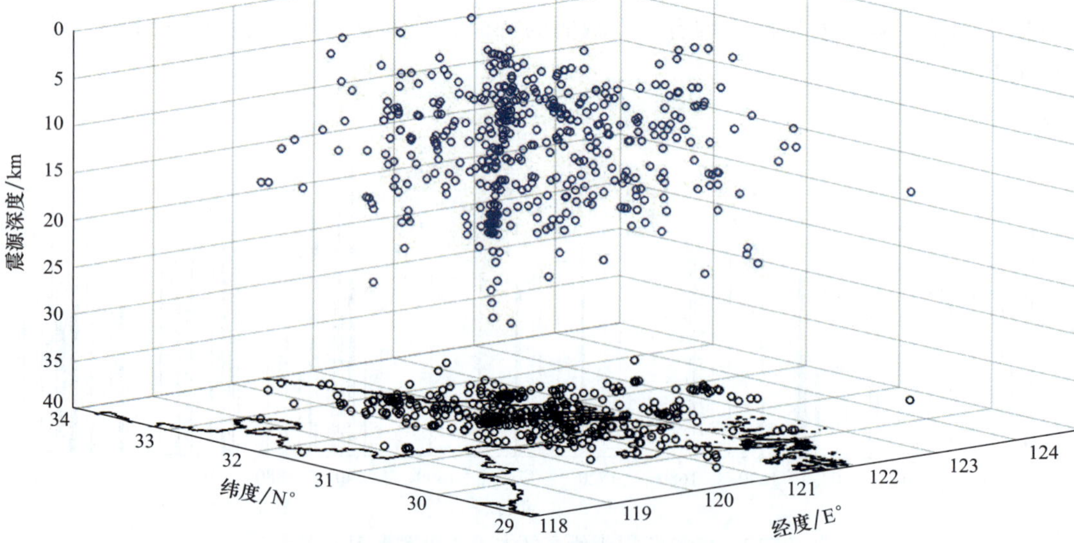

图 18-14　研究区地震震源深度三维空间分布图

Fig. 18-14　3D spatial distribution of the earthquake focal depths in the study area

统计表明，区内地震震级与震源深度关系不大，有感地震、中强震和强震震源深度集中在中地壳，少量在上地壳，总体上表现为一定的线性关系（图18-16）。弱震的震级与震源深度基本没有关系，各个震级的震源深度在整个地壳范围内都有分布。

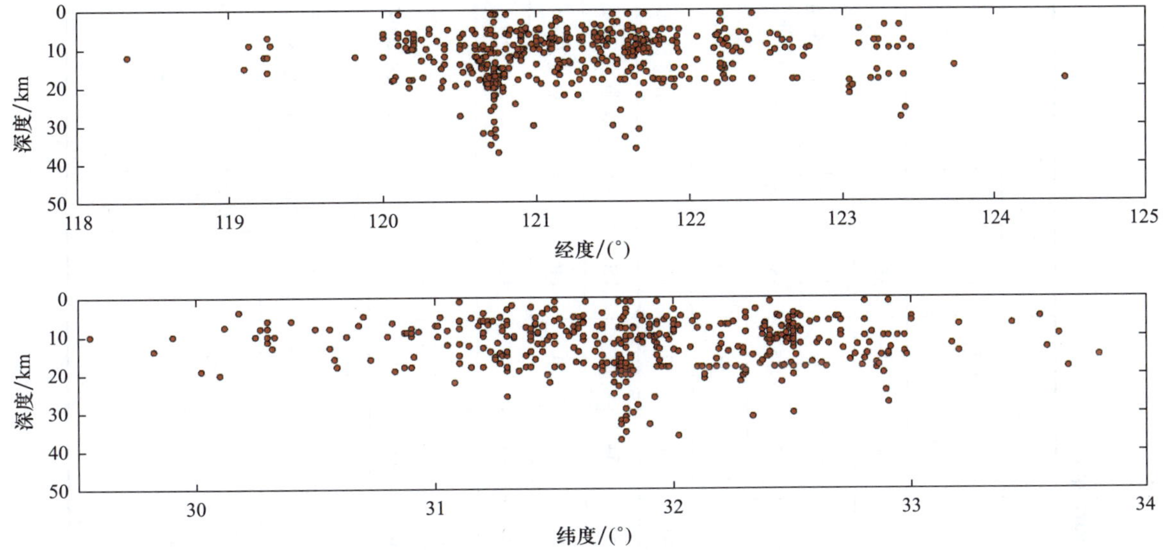

图18-15　地震震源深度剖面分布图

Fig. 18-15　Sections of earthquake focal depths

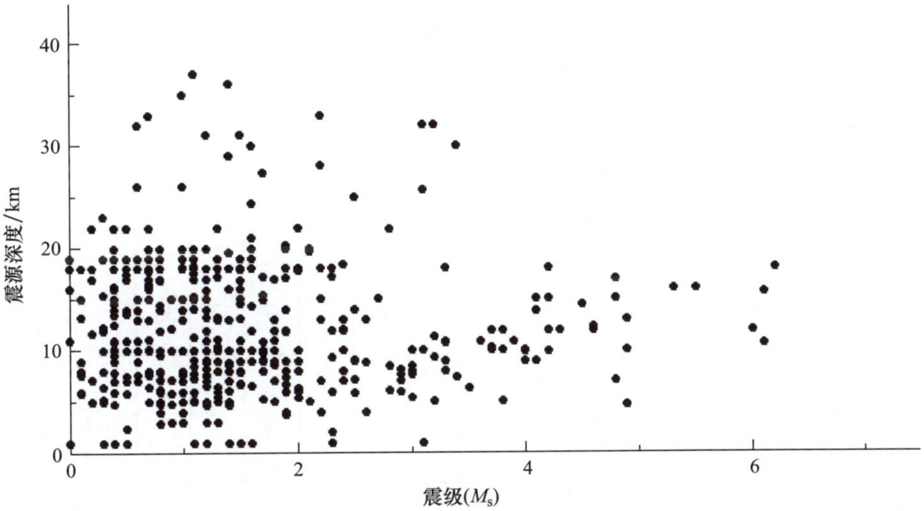

图18-16　地震震源深度与地震震级关系图

Fig. 18-16　Earthquake focal depth versus earthquake magnitude

四、主要地震震源机制解

震源机制解显示（表18-1），区内主压应力轴为北东东—近东西向，优势方位角约为78°，平均仰角为23°，主张应力轴方位角为349°，平均仰角为25°。南黄海地区的主压力方向平均为50.7°，长江口和南通等地中强震资料明显显示轴向南偏。大致以南渡—金坛断裂为界可以分为两个区，中等主应力倾角为60°~65°。太仓—奉贤断裂附近地区的P轴、T轴方位较为凌乱；湖苏断裂带两侧区域P轴方位集中在北东向，部分北西向（孙业君，2015）。

表 18-1　不同地震震源机制解

Table 18-1　Focal mechanism solutions of different earthquakes

序号	发震时间（年．月．日）	地点	震级（M_s）	震源深度/km	节面 I			节面 II			P 轴		T 轴		资料来源
					走向/（°）	倾向	倾角/（°）	走向/（°）	倾向	倾角/（°）	方位/（°）	仰角/（°）	方位/（°）	仰角/（°）	
1	1971. 12. 30	长江口	4. 9	10	30	NW	81	295	NNE	47	247	21	354	36	《上海及邻近地区区域地质地震地质研究报告》，1990
2	1975. 9. 2	南黄海	5. 3	30	93	NNE	76	176	SWW	64	223	7	317	29	
3	1978. 1. 5	白鹤	2. 3	3	28	NW	55	317	NE	64	82	84. 53	356	46. 4	
4	1984. 5. 21	南黄海	5. 7	17	350	NE	85	77	NNW	60	37	25	299	10	
5	1985	舟山岛	3. 2	22	15	SEE	59	102	SSW	79	51	12	156	27	周春喜，1990
6	1990. 2. 10	常熟	5. 1	3. 9	2	SEE	55	111	SSW	65	54	6	150	45	王争争，2001
7	1991. 2. 17	练塘	3. 0	8. 5	22	SE	41	82	NNW	65	38	54	146	13	章振铨，2004
8	1996. 11. 9	长江口东	6. 1	15. 7	219		85	128		81	84	10	353	1	郑月军，2006
9	1997. 8. 3	长江口	3. 6		65		62	99		33	168	15	300	68	章振铨，2004
10	1997. 8. 6	长江口	2. 5		41		3	106		76	355	26	228	51	
11	1997. 9. 19	长江口东	2. 4		41		37	75		57	208	73	334	10	
12	2009. 7. 9	南黄海	3. 7	8	298		87	29		74	252. 3	13. 4	344. 5	9. 1	康清清，2013
13	2001—2011	南通市	综合震源机制	195		81	54		11	97	53	291	36	盛书中，2013	
14	2009	松江	综合震源机制	10		73				67		324		王小平，2010	

两个节面通常呈北东和北西向，少部分呈近东西向，节面产状中等到近垂直，倾角集中在 33°～87°。南黄海地区一组节面走向在 273°～356°，一组在 29°～77°。南通—常熟地区一组节面走向为北西 291°，一组在 2°～54°。长江口地区一组节面在 30°～65°，一组在 279°～295°。长江口东一组在 39°～41°，一组约在 308°。在此应力场背景下，北东向和北北东向断裂呈右行走滑兼张性特征，北东东向和近东西向断裂呈张性特征。北西向和北西西向断裂以压性兼左行走滑特征。根据地震破裂方式，区内以走滑型错动为主，倾滑型错动占比较少，滑动角在-30°～0°（孙业君，2015）。这与在现有应力场下断裂的主要活动性质认识较一致。

第四节 区域应力场特征

上海及邻区的现代构造应力场处于北东东向水平压应力与北北西向水平张应力下，在构造应力的

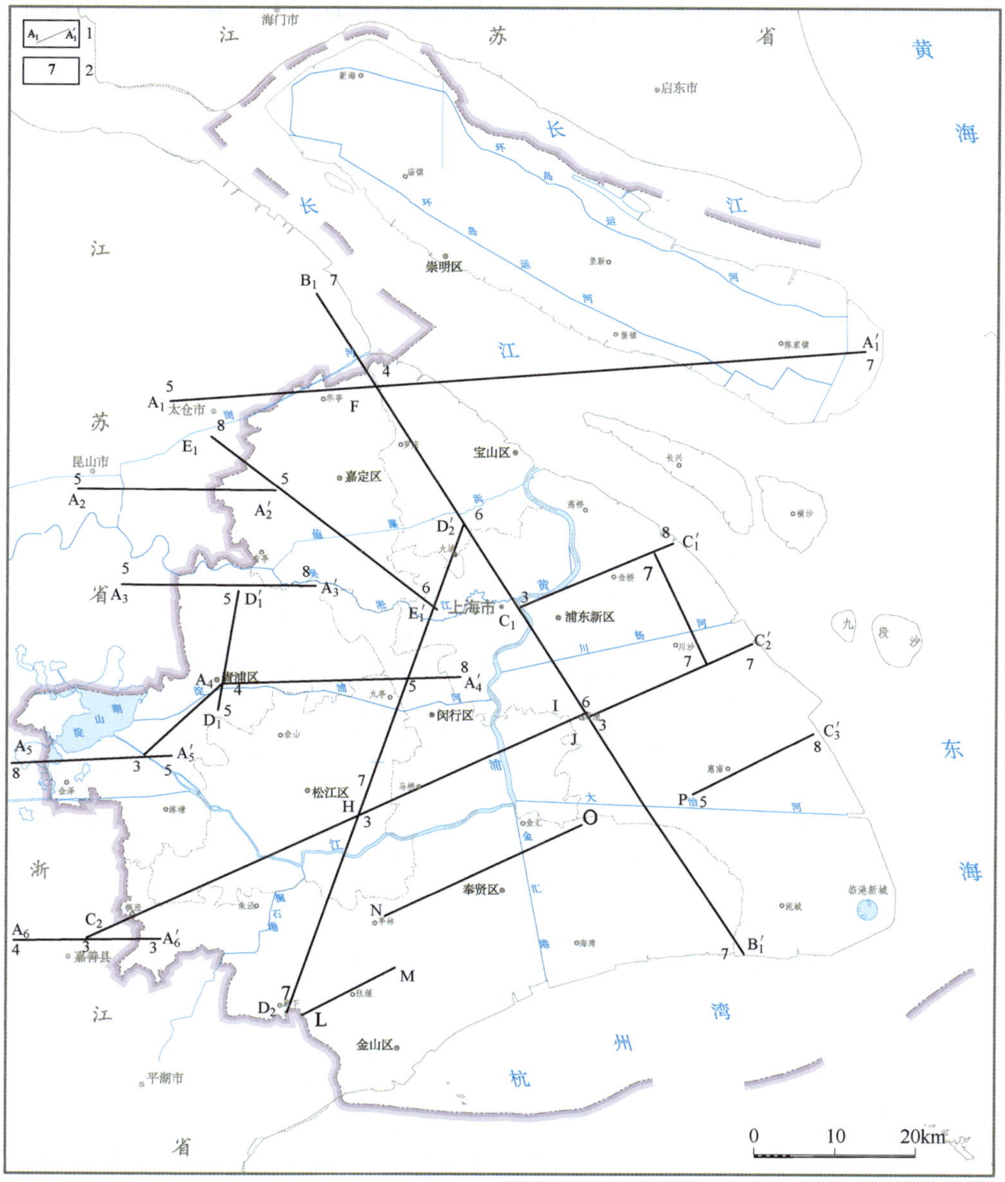

图 18-17 上海光弹性模拟模型及应力集中系数

（图中数字为应力值的 2 倍）

Fig. 18-17 Photoelastic simulation model and stress concentration coefficient in Shanghai

(Numbers in the Figures are twice the value of stress)

1—简化（断裂）构造；2—应力集中系数

作用下，由于地壳复杂的构造格局以及地壳固体介质的不均匀性，区域应力场和应变场的分布呈现不均匀状态。

上海应力场特征的光弹性模拟显示，最大剪应力方向约为北东 55°和北西 35°。围岩内的最大剪应力较低但均匀，一般在 0. 5~1. 0 间。沿断裂两侧，最大剪应力值大都为 1. 5，有些地段如松江—七宝段东侧、松江城东、高桥东和川沙附近应力值较高，为 2. 5~3. 0。高和较高的最大剪应力点全在各断裂点和几处交而不汇的部位，最大剪应力值在 1. 5~4. 0 间。太仓南部、安亭东南部、龙华、芦墟西部、高桥东部、廊下东南部等地可达 4. 0。在周浦西北部、七宝和安亭北部等 3 处断裂交汇区最大剪应力值也较高，在 2. 5~3. 0。HID_2'三角形断块区域和高桥—川沙之间的应力集中系数很高。其他地区、断裂不甚发育的地区，最大剪应力大都在 0. 5~1. 0 之间。最大主应力方向约为北东 80°，应力值分布特征与最大剪应力相似（图 18-17）。

根据应力模拟结果，判断太仓南部、安亭东南部、龙华、芦墟、高桥东部、廊下东南部、惠南等 7 处的应力集中系数最大，是围岩的 8 倍，最易集聚高应变。HID_2'三角形断块区域和高桥—川沙之间为危险区域。

第五篇

岩石圈结构构造

第十九章　区域地球物理场特征

随着板块构造学说的发展和大陆动力学、地幔柱理论、地震成因研究的深入，人们逐渐认识到研究地壳和上地幔横向结构的重要性。但是现在的研究手段和方法有限，主要限于地球物理方法、岩石地球化学探针及大陆深钻等。深部地球物理方法是区内岩石圈结构研究的主要方法，自20世纪50年代以来，主要工作进展和成果有：

为寻找油气的物探普查工作（1958—1962年），开展了电法、1∶5万重力（沿公路自由网布点）及1∶20万重力和航磁测量，完成了油气孔65口、煤田孔1口的电测井。

寻找固体矿产资源的物探普查工作（1962—1982年），开展了1∶2.5万航磁测量，发现142个航磁异常。另外，1963年在全市范围内完成1∶20万的6条电法剖面。1972—1979年，完成金山张堰金属电测井、三分量磁测井，1981年又补做电测井工作，共完成56口井的测井工作。

为配合地震地质研究（1982—1991年），开展了全市1∶5万重力普查、1∶10万~1∶20万航磁测量、金山-常熟100km大地电磁测深。

1991—1995年，地质矿产部第六物探大队在下扬子地区开展了区域性MT深部探测，HQ13、HQ26剖面通过上海西南部地区，并以前者为基础编制了奉贤—灵璧地学断面，详细论述了沿线的地球物理场特征及岩石圈结构。

2004—2005年，上海市地震局沿上海南部（沿浙江湖州—上海奉城一线）开展了以深地震反射方法为主，并结合高分辨折射、地震宽角反射P折射和大地电磁测深的联合深部探测。解释了上海南部的岩石圈结构。

近几十年来，部分学者通过区域重磁反演、远震接收函数、天然地震层析成像等方法对上海及邻区岩石圈结构构造进行了研究（顾斌，1987；黄兆熊，1989；王小平，2006；等），讨论了地壳的速度结构，深部构造与地震的关系等。

第一节　岩石圈速度结构

上海10个地震台站下方的v_p/v_s值在1.65~1.73，地壳各向异性分裂参数大致为67.7°及0.26s，上地幔各向异性分裂参数为66.6°及0.25s（王小平，2006，2007），表明区内岩石圈内部各向异性特征明显。在HQ13地学断面、泥城—湖州深地震反射和折射剖面上，区内岩石圈表现出如下速度特征。

一、HQ13地学断面

该断面东起安徽灵璧，西至上海奉贤，横跨下扬子地区。

实测资料显示，不同时代地层和岩性的P波速度变化较大，第四系为1700~2000m/s，新近系、古近系和上白垩统的陆源碎屑岩分别为2200~2500m/s、2500~3500m/s、3800m/s，下白垩统陆源碎屑岩和火山岩为4000~4250m/s，三叠系陆源碎屑岩和碳酸盐岩分别为3200~5000m/s、5600m/s，二叠系陆源碎屑岩及煤系、碳酸盐岩分别为2500~4100m/s、6200m/s，石炭系碳酸盐岩为5700~6700m/s，泥盆系碎屑岩为3700~5000m/s，志留系它生陆源碎屑岩为4000~4200m/s，奥陶系—震旦系碳酸盐岩为5400~6200m/s，前震旦系浅变质岩、深变质岩系分别为5700~6000m/s、6300m/s，下地壳上部为6000m/s，下地壳下部为6800~7600m/s，岩石圈上地幔为8000~8200m/s。

深地壳地震测深剖面取得了6个地震界面，分别对应于中新生代陆相层、震旦纪—中生代海相层、深—浅变质岩系、塑性层（低速层）和下地壳下部、岩石圈上地幔的顶部。全地壳反射剖面进一步识别出了中生界顶部、浅变质岩系底部和深变质岩系内部3个界面。岩石圈的速度结构从表层的4200m/s，向下总体上呈递增，至下地壳下部达6800m/s，岩石圈上地幔在8000~8200m/s。据此概括出如下速度结构模型（图19-1）：

第一层：纵波速度5000m/s，为扬子地块上地壳上部的盖层；

第二层：纵波速度5800m/s，为扬子地块上地壳下部的浅变质岩系；

第三层：纵波速度6000m/s，为扬子地块中地壳上部的深变质岩系；

第四层：纵波速度6300m/s，为扬子地块中地壳下部的深变质岩系；

第五层：纵波速度6000m/s，为扬子地块下地壳顶部的壳内低速层；

第六层：纵波速度6800m/s，为扬子地块下地壳下部的深变质岩系；

第七层：纵波速度8200m/s，岩石圈上地幔。

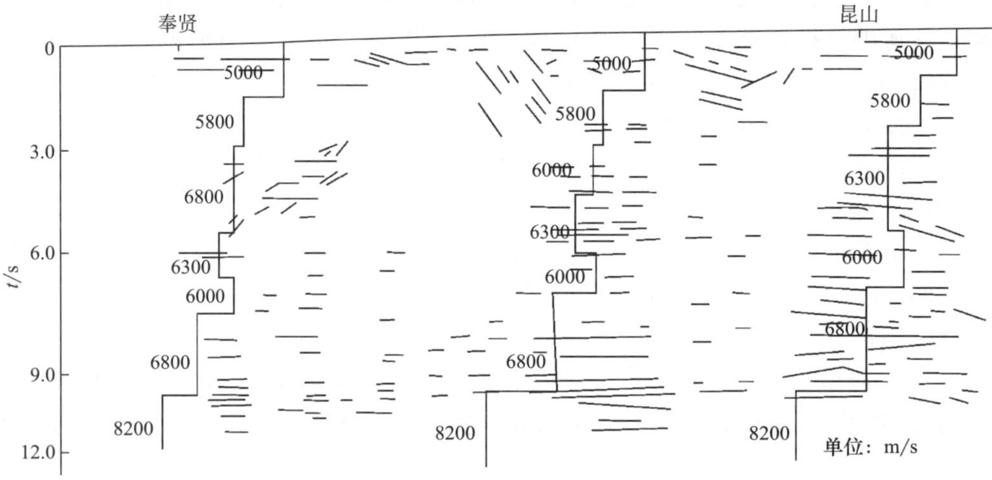

图19-1　地震反射同相轴与速度结构图

（据陈沪生，1993）

Fig. 19-1　Seismic reflection events and velocity structure diagram

二、泥城—湖州深层地震反射/折射剖面

该剖面东起浦东书院东的海滩，西至浙江湖州，综合开展了深地震反射、深地震宽角反射/折射、高分辨率折射等地震勘查（姚保华，2005，2007），在人工地震深部探测剖面上，区内岩石圈表现出如下速度特征（图19-2）：

第一速度层：底界双回程反射时间在200~400ms，150~320m，为松散层。

第二速度层：底界深部一般在200m左右、600~700m，局部区域上要深一些，泥城凹陷最深达1.2km。纵波速度为2.69~4.27km/s，该层中的断裂特征不明显。

第三速度层：底界埋深一般在1.3~1.8km，横向起伏变化相对较大，最浅处埋深约600m。层内纵波速度在4.27~5.12km/s之间，明显高于第一层，各断裂带在该层的底部附近都有显示。在深地震反射剖面上，又有两个内部反射界面，双回程反射时间分别为550~750ms，710~1060m，连续延伸；800~1000ms或750~900ms，1150~1500m或1030~1350m。

第四速度层：底界埋深一般在2.6~4km，局部区域上要深一些，界面自西向东倾伏（见图19-2）。该层内的速度结构具有明显的横向非均匀性，P波速度范围为5.12~5.97km/s。剖面上部分垂向低速带在深度约4.5km处汇合。

第五速度层：至上地壳底部，顶界横向变化十分强烈，在金山断凸下方为一明显的高速隆起区。在宽角地震折射/反射剖面上基本上是一个比较平坦的界面，可以被连续对比追踪。地震波速度为

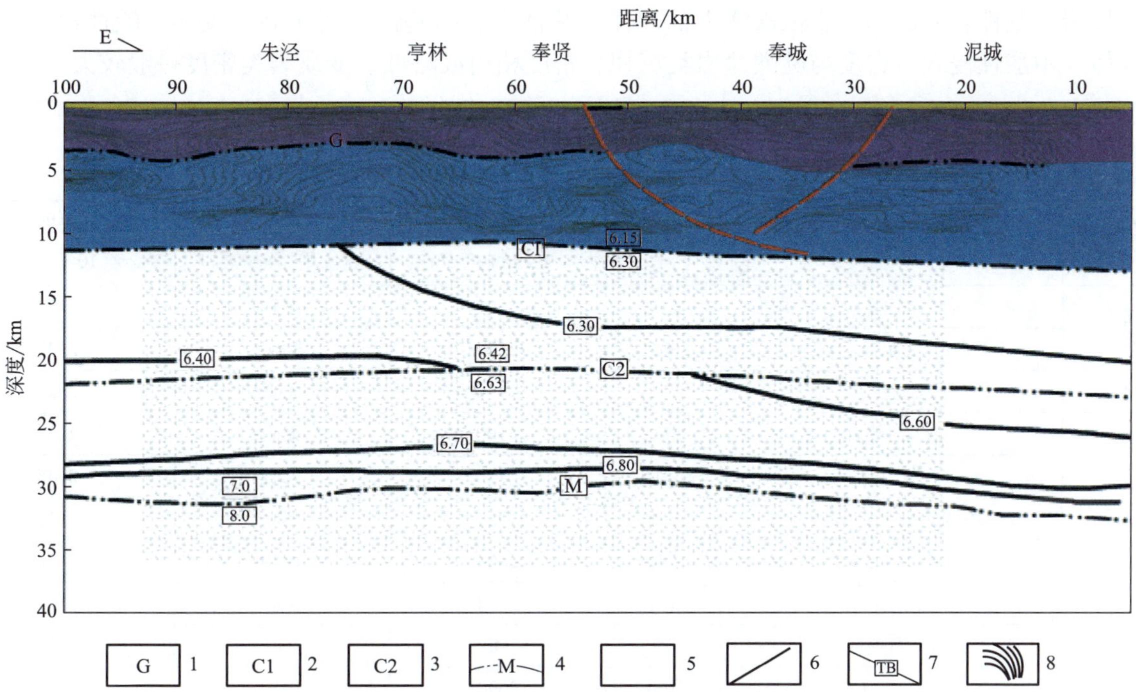

图 19-2 速度结构和深部探测综合解释图

（据姚保华，2007）

Fig. 19-2 Velocity structure and integrated interpretation

1—盖层界面；2—上地壳底面；3—中地壳底面；4—莫霍面；5—反射带；6—断裂；
7—速度等值线（深地震反射，km/s）；8—速度等值线（高分辨地震反射，km/s）

5. 80~6. 20km/s，在靠近 G 界面的顶部附近，其速度为 5. 80~6. 00km/s，其下在 5~7km 的厚度范围内是一个较弱的速度梯度变化层，其速度变化范围为 6. 00~6. 10m/s，C1 界面的速度跳跃差在 0. 1~0. 2km/s 之间。该层平均速度为 5. 68~5. 98km/s。

第六速度层：中地壳，顶界面双回程时间在 4~4. 5s，11. 2~12. 8km；该层西段上部为一弱的速度梯度层，靠近底部界面速度梯度较强，该层东段上部有一厚约 7km 且速度为 6. 30km/s 的匀速度层，向下至底部界面是速度梯度层，界面速度跳跃差为 0. 15~0. 2km/s。该层平均速度为 5. 95~6. 25km/s。

第七速度层：下地壳，顶界面双回程时间在 7~7. 6s，21~22. 6km。上部有厚度约为 6km 的正弱速度梯度层，梯度为 0. 012km/(s · km)，靠近莫霍面存在厚度约为 4km 的正强速度梯度带，梯度为 0. 088km/(s · km)。下部地壳的层速度由顶部的 6. 63km/s 左右增大到底部的 7. 05km/s 左右，平均速度为 6. 18~6. 34km/s。

第八速度层：岩石圈地幔，顶界面双回程时间在 9. 5~10. 5s，29. 5~32. 5km。

第五层之上的速度结构基本上是由一个强速度梯度层所组成，由上部的 2. 50~4. 00km/s 变化到基底之上的 5. 20~5. 60km/s。剖面东端浅部地层较低的速度结构表现较为明显，显然是受沉积盖层的影响所致。

第二节 区域重磁异常

一、岩石密度和磁性特征

区内大部分区域为第四系覆盖，岩石物性测试资料较少。已有资料显示各类岩石密度变化范围较大（表 19-1，表 19-2）。其中沉积岩属于低密度岩类，岩石密度按黏土岩—粗碎屑岩—灰岩依次增大，也随埋深和成岩固结程度增加而增大。侵入岩密度略大于喷出岩。岩浆岩的密度值具有从酸性岩—中性岩—基性岩—超基性岩依次增大的规律。基性岩、超基性岩属于高密度体，酸性侵入岩属于低密度体。中酸性侵入岩密度与碳酸盐岩类沉积岩密度相当或略低。变质岩类密度一般较大。

表 19-1 上海地层岩石物性参数一览表

Table 19-1 Petrophysical parameters of the strata in Shanghai

地质年代	岩石名称	标本数	磁化率 $k/(4\pi\times10^{-6}SI)$	剩磁 $J_r/(10^{-3}A\cdot m^{-1})$	密度 $\sigma/(10^3kg\cdot m^{-3})$
新生代	第四系	73	<100	无磁	2.13
	玄武岩	69	2150	3060	2.47
	红色砂岩	7	<100	无磁	2.29
中生代	泥岩、砂岩	78	<100	无磁	2.32
	砂岩	31	<100	无磁	2.46
	中酸性凝灰岩、熔岩	325	1080	240	2.66
	安山岩、英安岩	780	2860	1040	2.66
	流纹斑岩	171	1470	490	2.62
	长石、石英砂岩	29	740	740	2.67
	辉石安山岩	166	2690	1720	2.75
早古生代	灰岩	17	0	无磁	2.78
	大理岩	200	<100	<100	2.79
	微晶灰岩	150	<100	<100	2.83
	粉砂质泥岩	40	<100	<100	2.83
	砂岩	22	<100	无磁	2.66
中元古代	大理岩	134	0	无磁	2.73
	白云质灰岩	124	0	无磁	2.81
	石英片岩	27	0	无磁	2.63
	绢云母片岩	19	0	无磁	2.75
	黑云母角闪片岩	337	5510	6160	2.89
	变粒岩	26	5120	11260	2.72
	长石角闪片岩	12	15240	17000	2.97
	角岩	188	5790	1350	2.75

表 19-2 上海侵入岩岩石物性参数一览表

Table 19-2 Petrophysical parameters of the intrusive rocks in Shanghai

样品点位	岩石名称	标本数	磁化率 $k/(4\pi\times10^{-6}SI)$	剩磁 $J_r/(10^{-3}A\cdot m^{-1})$	密度 $\sigma/(10^3kg\cdot m^{-3})$
查山朱行	石英斑岩	47	0	无磁	2.50
查山松江	正长斑岩	123	1380	160	2.60
康家宅	花岗斑岩	102	2230	850	2.65
崇明	花岗岩	67	1240	250	2.62
界山肖塘	花岗闪长岩	682	3700	410	2.68
奉贤	闪长岩	350	3700	500	2.69
坦直、川沙	石英闪长岩	176	4320	350	2.73
周浦	辉石闪长岩	497	6480	2740	2.81

续表

样品点位	岩石名称	标本数	磁化率 $k/(4\pi\times10^{-6}SI)$	剩磁 $J_r/(10^{-3}A\cdot m^{-1})$	密度 $\sigma/(10^3kg\cdot m^{-3})$
周浦	辉长岩	164	7610	3740	2.88
张堰	角闪玢岩	31	5860	1570	2.90
界山	辉绿岩	83	4970	850	2.80
界山、张堰	磁铁矿	106	128520	62650	3.60
张堰	含铁矽卡岩	11	44850	33710	3.10

（一）密度特征

根据岩石密度数据和不同时代地层的岩石类型，区内地层主要存在3个密度界面：

1）新生界及上白垩统沉积岩与下白垩统火山岩间密度差为 $0.3\times10^3\sim0.4\times10^3kg/m^3$。

2）新生界及上白垩统沉积岩与三叠系及古生界、变质岩基底之间密度差为 $0.4\times10^3\sim0.5\times10^3kg/m^3$，是本区内最大的密度层界面。

3）下白垩统火山岩与三叠系—古生界、变质岩基底间密度差为 $0.15\times10^3\sim0.2\times10^3kg/m^3$。

另外，古生界与元古宇之间密度差约为 $0.1\times10^3kg/m^3$。

第1）、第2）密度层界面是造成本区重力异常的主要因素。

（二）磁性特征

岩石磁性数据显示，不同岩石类型磁性特征差别较大。

区内各类沉积岩基本上为无磁性岩石。

变质岩类一般为中磁性岩石，但磁性变化大，可从片岩、板岩、石英岩、大理岩的无磁性变至各种斜长片麻岩的强磁性。感磁与剩磁的比值也变化较大。

火山岩普遍具有磁性，而且磁性变化较大，一般有较强剩磁。中基性火山岩要比酸性火山岩有较强的磁性，能引起较强磁异常。中基性火山岩及次火山岩可引起强度大、范围小的尖峰异常。

侵入岩类磁性从基性超基性岩—中性岩—酸性岩由强变弱。中酸性侵入岩有磁性，一般会引起几百纳特（nT）的航磁异常，呈规则形状。

从重磁参数来看，区内浅部地壳层存在4个明显的物性层：前震旦系变质岩属“高密高磁”层；震旦系和下古生界属于“高密低磁”层；下白垩统属于“中密中磁”层；上白垩统和新生界属于“低密低磁”层。分别与晋宁期构造层、加里东期构造层、燕山期构造层第二亚层及燕山期构造层第三亚层、喜马拉雅期构造层对应。新生代基性火山岩及广泛分布的中生代火山岩类、元古宙变质岩和中酸性、中基性侵入岩是引起区内磁异常的主要因素。

二、区域重力场特征

区内重力场强度变化较大，除松江、青浦西部和崇明堡镇一带外，基本上全部为正异常区。总体上，呈现有两条重力高值区、一环状重力高和一低负值区（图19-3）。

阳澄湖—崇明重力高值带异常轴向为北东向和北东东向，岳王、陆渡以东异常为连续且以多峰出现，以西则以断续异常出现。嘉兴—惠南高值带总体走向为北东东，但嘉兴异常轴向近东西，昆山—真如—吕巷高值带以淀山湖为中心，呈3面的环状展布。角直—松江重力低值区又可分为近东西向的3个带，角直—花桥低值带，为区内重力最低区，松江—金汇重力低带和淀山湖—凤溪相对重力低。

局部重力高发育明显，主要有庙镇—港西重力高、吕巷重力高、嘉定—罗店重力高、宝山西重力高、重固重力高、北新径重力高、坦直—祝桥重力高，钻探结果显示大多与前中生代地层的出露有关，个别可能与下部这套地层的上隆有关。局部发育重力低，如庄行重力低、唐行重力低、松江重力低，钻探揭示前两者为一中生代构造火山盆地，后者为断陷盆地。

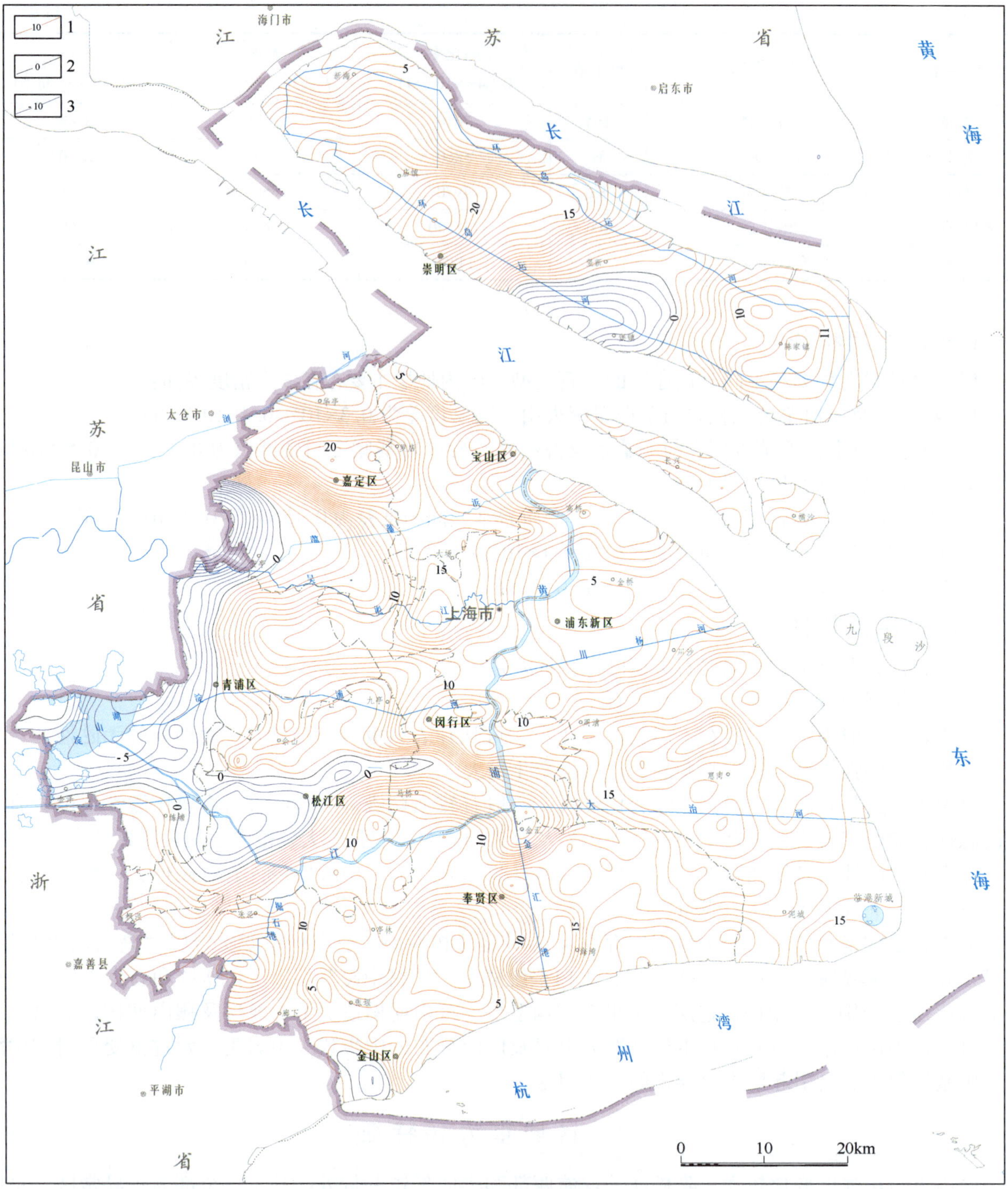

图 19-3　上海布格重力异常特征

Fig. 19-3　Bouguer gravity anomalies in Shanghai

1—正值等值线（$10^{-5}m/s^2$）；2—零值等值线（$10^{-5}m/s^2$）；3—负值等值线（$10^{-5}m/s^2$）

重力高值区和低值区通常对应于隆起区和坳陷区，一般由不同级别的梯度带分割或之间形成线状过渡带。重力高值区的产生原因与前白垩系的上隆有关，重力低值区的产生原因除浅部的密度低的花岗岩体外，大多数与中新生代断陷和火山盆地有关。

三、区域航磁场特征

航磁 ΔT 异常平面等值线图显示航磁异常强度变化在-499~1066nT 之间。区域磁场以北东-南西向占优势，明显呈带状分布，总体构成醒目的东西向异常带（图 19-4）。自北向南，区内有 3 条醒目的北东东向航磁异常带，即宝山—崇明磁异常带、天马山—顾路磁异常带和张堰—惠南磁异常带。其中，中带由中酸性火山岩所引起，北带和南带由中酸性火山岩及侵入岩所引起。规模较大的低（负）异常区有角直、重固—长兴、庙镇和崇明东南部等地，其中角直为一中、新生代断陷盆地，重固—长兴、庙镇与下古生界隆起有关。

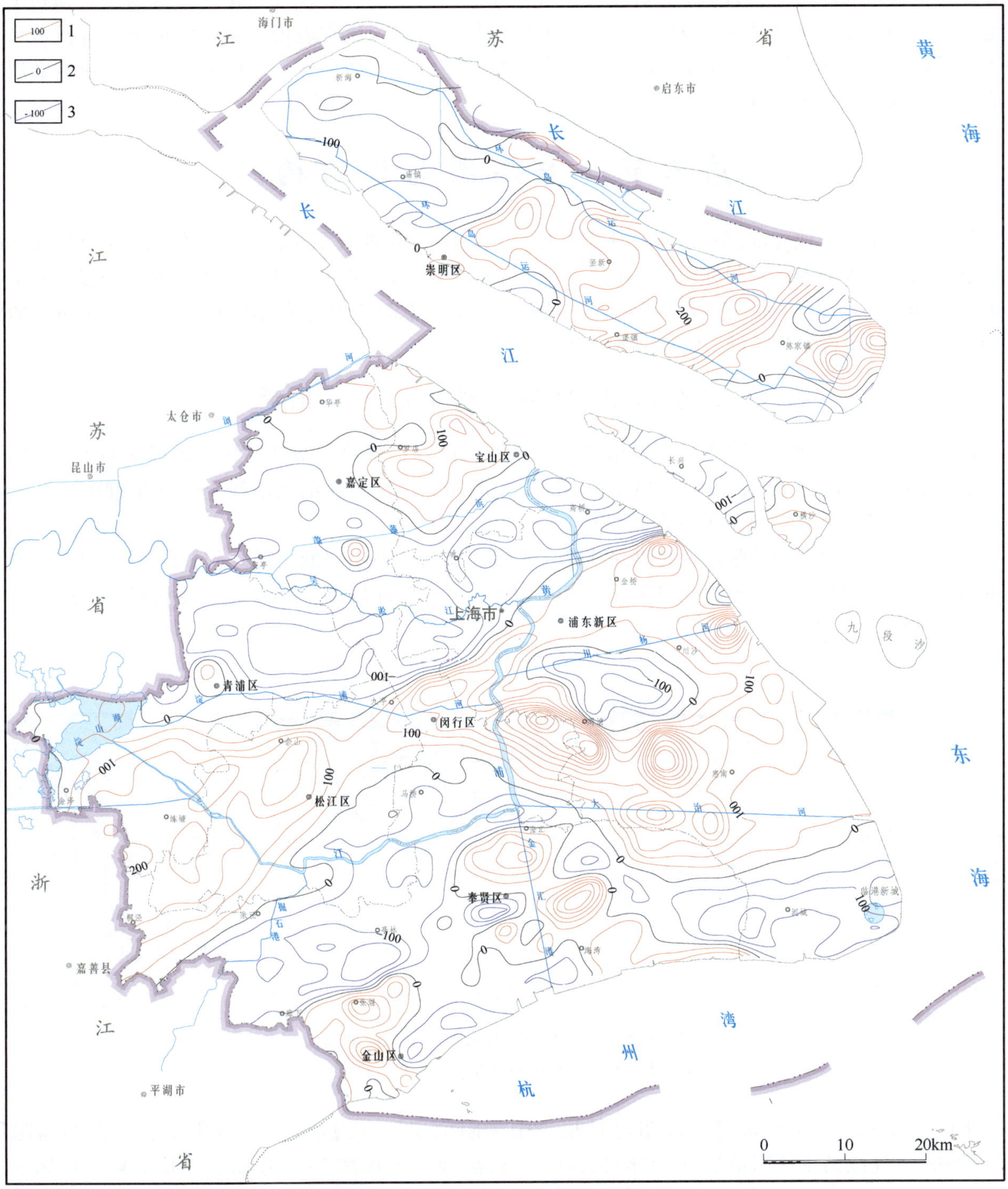

图 19-4　上海航磁（ΔT）异常特征

Fig. 19-4　Airborne magnetic（ΔT）anomalies in Shanghai

1—正值等值线（nT）；2—零值等值线（nT）；3—负值等值线（nT）

在航磁异常图上，局部正磁异常区为基性、中性（或中酸性）火山岩、侵入岩，有时是构成火山锥的浅成—超浅成—喷出相火山杂岩区，或为基底角闪质岩所组成的块断隆起。其中侵入岩、角闪质岩反映的磁异常较规则，火山岩则显得紊乱。负低磁异常区通常对应于陆源碎屑岩类的浅埋区。因受倾斜磁化的影响，磁异常与磁性地质体在空间上有明显位移。

综合重、磁场特征，区内基底浅埋区通常呈现高重力异常、高磁力异常或低磁力异常特征；下古生界浅埋区通常表现为高重力异常和低磁力异常特征；下白垩统通常表现为相对平缓的重力场和磁力场特征，局部呈相对的高异常区或低异常区。上白垩统—古新统通常表现为明显的负重力低异常和负磁力低异常特征。基性侵入岩呈明显的正磁异常或相对高的重力高异常，而中酸性侵入岩通常表现为重力低异常和高磁力异常特征。

第三节　岩石圈电性结构

从电参数来看，地壳浅部具有4个典型的电性层：Q+N松散沉积层，电阻率值为4～$n\times10\Omega\cdot m$；白垩系火山岩、火山碎屑岩和沉积岩，电阻率值为$n\times10^2\Omega\cdot m$；Z-Pz_1以碳酸盐为主的地层，电阻率值可达$n\times10^3$～$n\times10^4\Omega\cdot m$；前震旦系变质岩和侵入岩类，电阻率值为$n\times10^3\Omega\cdot m$。区内4条大地电磁测深剖面上的电性结构不同，特征分别如下：

一、金山卫-常熟大地电磁测深剖面

该剖面自金山卫开始，向西北经过松江、青浦和昆山，至常熟东南部结束。各测深点的ρ_{TE}曲线呈AAK、KHAK、HK、KHK、AK、HAK等不同类型的断面显示，在剖面沿线上壳幔呈现出横向上被低阻带分块，纵深上分层的5~6层层块结构。

横向上被深达岩石圈的张堰以北纵深陡倾低阻带、安角盆地中部低阻带分割，2号测点和10测点视电阻率小于200Ω·m，7号测点电阻率小于400Ω·m。35km以下，断面的三高和三低特征十分清楚（图19-5）。

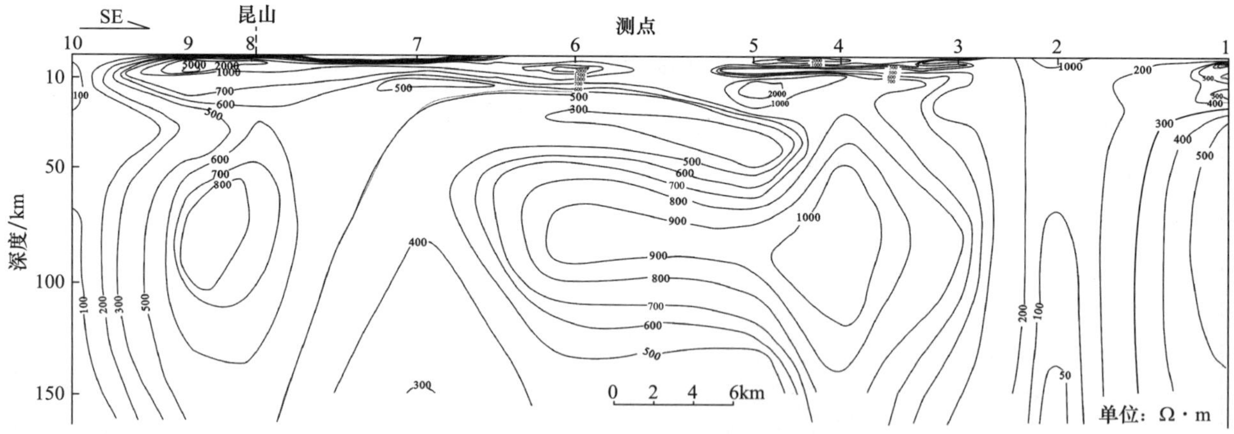

图19-5　金山卫—常熟大地电磁测深剖面

Fig. 19-5　Jinshanwei-Changshu MT Section

垂向上分为地壳（冷块层）高阻电性层、中等阻值层块（过渡层）、低阻层、巨厚的高阻层块（冷块）、连续的中等阻值层（冷热过渡块）、软流圈。尤其是3~7号测点形成高阻—低阻—高阻—低阻—高阻南倾叠瓦式电性结构断面。其中5~6号测点25km左右处有一厚约8km的壳内低阻层，电阻率小于250Ω·m。7~9号测点3~10km处有一近水平展布的高阻电性层，电阻率约500Ω·m，间断于10号测点南侧，趋薄于7号测点的4~5km深处。在4~5号测点之间、9号测点附近的中上地壳明显具有两个低阻层和两个高阻层，与HQ-13大地电磁测深断面反映的中上地壳结构近乎一致。

二、HQ-13 地学断面

该断面大地电磁测深曲线代表性的类型为 $K_1H_1K_2H_2K_3$ 型。深部，青浦一带的下地壳及以下的岩石圈为高阻体，电阻率高达 10000Ω · m，为区域上团块状高阻层（一般为 500~700Ω · m）的一部分，相当于下地壳下部的中性麻粒岩和岩石圈上地幔部分。在中下地壳交界地带，大部分地区发育低阻层，顶界埋深在 18km 左右，底界埋深在 22km 左右，电阻率多在 12~40Ω · m，同时纵波波速自上部的 6.3km/s 下降为 5.9~6.0km/s，为典型的高导低速层。松江—奉贤段为电阻率低值区，一般为 100~200Ω · m。岩石圈下部与软流圈的过渡带（上地幔高导层）电阻率多小于 40Ω · m（图 19-6）。

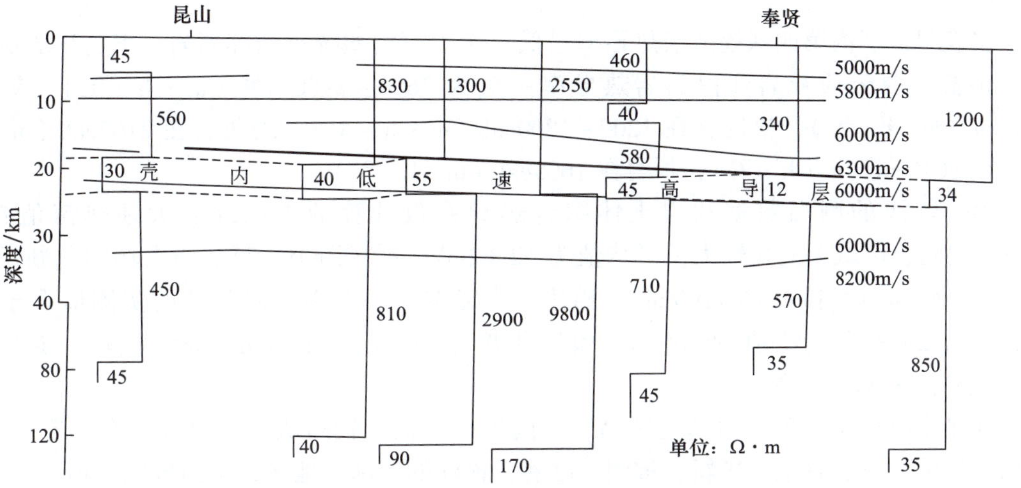

图 19-6　HQ-13 地学断面地震界面图与电性结构图

Fig. 19-6　Seismic interface and electric structure diagram of the HQ-13 geoscience transect

电性界面和密度界面在划分上具有等效性，浅部地壳从下向上各层间的界面多对应于晋宁运动以来多次构造运动形成的构造-岩性界面，不同物性层相互叠置形成一系列不同程度变形的构造层次。

三、泥城—湖州大地电磁测深剖面

该剖面上测点的曲线类型西段、中段和东段不同（于鹏，2008）。西段以 HKHK 型为主，在 1.0Hz 附近出现低阻极值；中段以 HK 型为主，深部低频的高阻极值点对应频率大致在 0.1Hz 之后，幅值高，一般大于几百欧姆，高者过千欧姆，曲线形态连续性较好，说明高阻块体稳定且厚度巨大。东段以 HKH 型为主，深部低频的电阻率幅值高，往往超过 1000Ω · m，说明高阻块体稳定且厚度巨大，中深层构造较为稳定，分层的特点不明显，地下电性结构横向变化小。甚低频出现的特征，说明高阻巨厚沉积之下有一定深度的低阻层分布。

在剖面西侧发育纵深低阻带，两侧电性结构具有一定差异。浅部剖面上，剖面东侧高阻基底以上出现 2 个相对低阻层，东侧基底下部岩石层电阻率值较高且较均匀，没有发现壳内高导层。剖面西侧高阻基底以上出现 3 个相对低阻层。基底下部岩石层电阻率值相对偏低，于 13~15km 埋深处出现西低东高、厚度约为 2km 的壳内高导层，其落差在 215km 左右。深部，在剖面东侧、与软流圈顶界面相差 26km 左右的地方出现厚度约为 4km 的低阻层，由东往西埋深加大，最后消失在软流圈的隆起部位。

四、HQ-26 大地电磁测深剖面

HQ-26 断面的 28~30 测点在上海境内青浦以南呈北东走向。地表为火山岩与中生代陆相碎屑岩所覆盖，ρ_s=50~80Ω · m，总厚度为 500m~1.5km；下部可能为下古生界或隐伏侵入岩，ρ_s=500Ω · m，顶面埋深约 1km。

第四节 区域地温场

区内变温带深度在9.0~17.0m之间变化，平均值为13.3m；恒温带底界在17.0~27.0之间变化，平均值为23.0m，以下为增温带。200m以浅地层增温率在2.31~3.73℃/100m，平均值为3.04℃，总体上呈西南部较高、东北部略低的变化趋势。与邻区相比，明显低于苏北盆地，和苏南地区相当，比浙西北地区要高。

一、岩石热导率和大地热流值

实测数据表明，区内第四系松散层热导率最低，次为下白垩统火山碎屑岩，下古生界碳酸盐岩和断裂破碎带最高。白云质灰岩和白云岩热导率一般为2600~2900cal[1]/(m·h·℃)，灰岩一般为1960~2120cal/(m·h·℃)，松散层在1200~1800cal/(m·h·℃)，凝灰岩在2162cal/(m·h·℃)，断层破碎带在3636cal/(m·h·℃)，板岩在1605cal/(m·h·℃)。

根据150m以浅的测温数据和岩土体综合热导系数计算的本区地表大地热流值在40.67~68.37mW/m^{2}之间，区域上变化较大，平均值为52.62mW/m^2（图19-7）。与大场和北新泾测得的平均热流值（64.26mW/m^{2}和65.64mW/m^2）相比，明显偏小。大场和北新泾的实测值介于苏北白垩纪—古近纪盆地的平均热流值68mW/m^{2}和华北热流值61.5±13.4mW/m^{2}之间。与苏南地区的1.53HFU平均热流值基本一致。

地表大地热流值整体表现出西部地区高，东南地区较低。较高值区主要分布在嘉定外岗、安亭、马陆、南翔、江桥，青浦赵屯、华新、重固、赵巷，松江的大部分地区，金山城区、枫泾、朱泾等地；较低值主要分布在浦东临港新城，闵行三林、浦江，奉贤新城、金汇、奉城，崇明西北部的红星农场、长征农场、庙镇、港西、东风农场、城桥、横沙岛等地。

二、地温梯度特征

地温在垂向上变化特征明显，随着深度的增大，地温总体上处于增大趋势。地温梯度受岩性、构造等因素影响，不同地层变化较大。下古生界平均地温梯度为2.1~3.4℃/100m，变化幅度最大，上寒武统白云岩在陆家楼井中为0，其他井在1℃/100m左右；下白垩统火山碎屑岩为1.8~2.36℃/100m；第四系松散层为3.1~4.6℃/100m。50~100m、100~150m、150~200m间区域地温场的平均垂直梯度值分别为3.14℃/100m、3.06℃/100m、3.10℃/100m。屈家桥B21孔300~500m井段下白垩统火山碎屑岩的平均地温梯度达3℃/100m。

地温梯度在横向上也具有一定的变化。在正向构造单元和基岩浅埋区地温梯度通常较高，莘庄的地温梯度可达4.4℃/100m；负向构造单元和基岩深埋区地温梯度通常较低，上海西部和安角断陷盆地两侧、坦直—周浦明显较高，崇明地温梯度多小于3.0℃/100m（图19-8）。

三、岩石圈温度结构

据上海市地矿工程勘察院（2014）资料，区内1000m深度地温一般在35.6~51.9℃，平均为45.2℃。2000m深度地温一般在45.9~80.3℃，平均为63.3℃。3000m深度地温一般在56.2~108.7℃，平均为78.3℃。4000m深度地温一般在66.6~137.1℃，平均为91.3℃。高值区主要分布在区内西南部和西部。

朱介寿（2005）总结了华南及东海地区的岩石圈热结构，显示区内大地热流值在65~70mW/m^2。70km以上的岩石圈温度递增明显，10km温度为230~240℃，20km温度为425~525℃，30km温度为

---

[1] 1cal=4.18J。

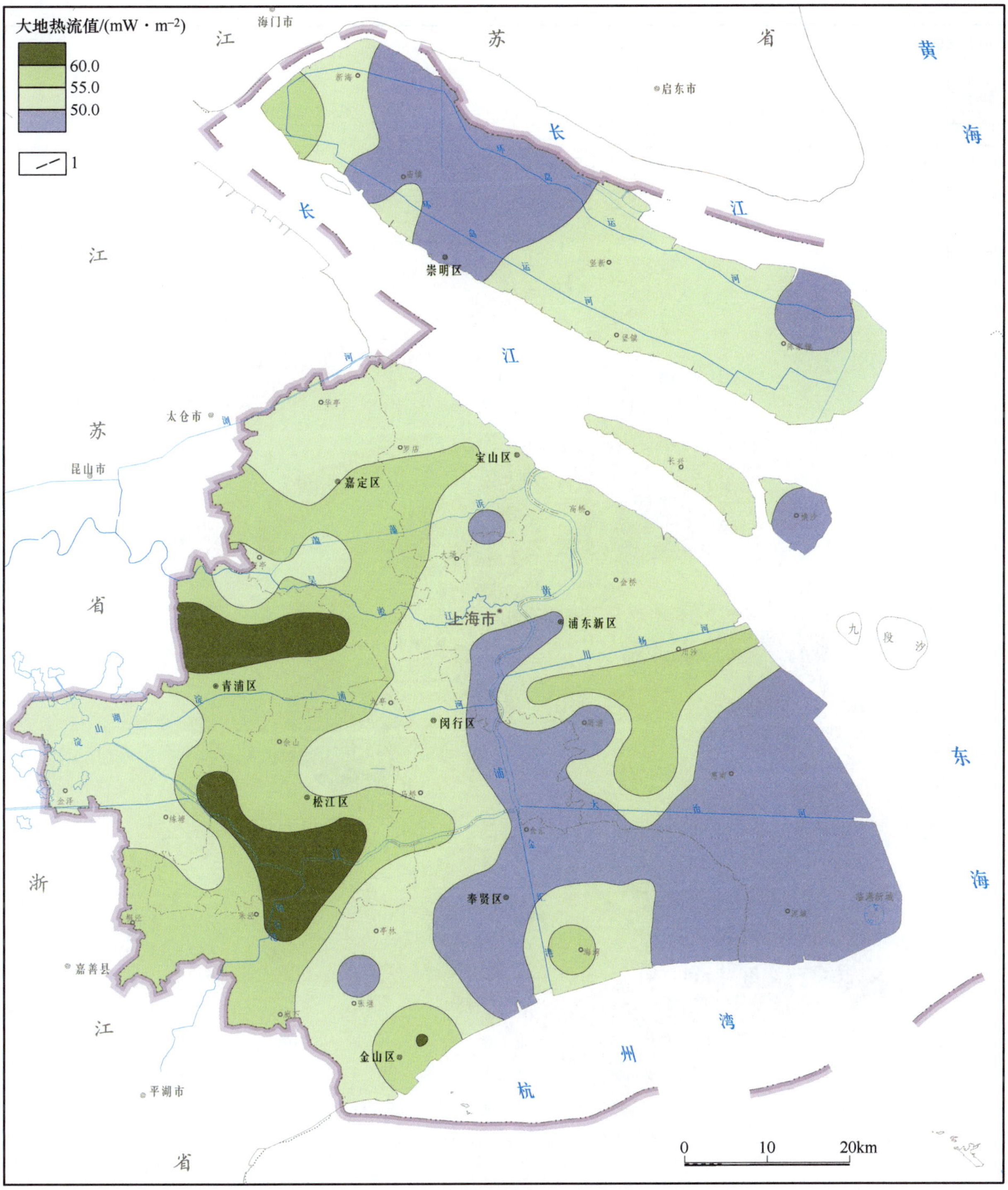

图 19-7 上海地表大地热流值分布图

Fig. 19-7 Distribution of surface terrestrial heat flow values in Shanghai

1—大地热流等值线（mW/m^2）

600~720℃，40km 温度为 750~900℃，50km 温度为 900~1000℃，60km 温度为 1050~1200℃，70km 温度为 1150~1250℃，80~120km 温度基本上都在 1200~1250℃，莫霍面的温度在 680~780℃。100℃等温面深度在 3.8~4.5km，600℃等温面温度在 23~30km。等值线方向多呈北东-南西向。

四、第四承压含水层水温特征

区内第四承压含水层水温平面上变化幅度较小，在平面上呈现出几个明显的北东向条带状高值

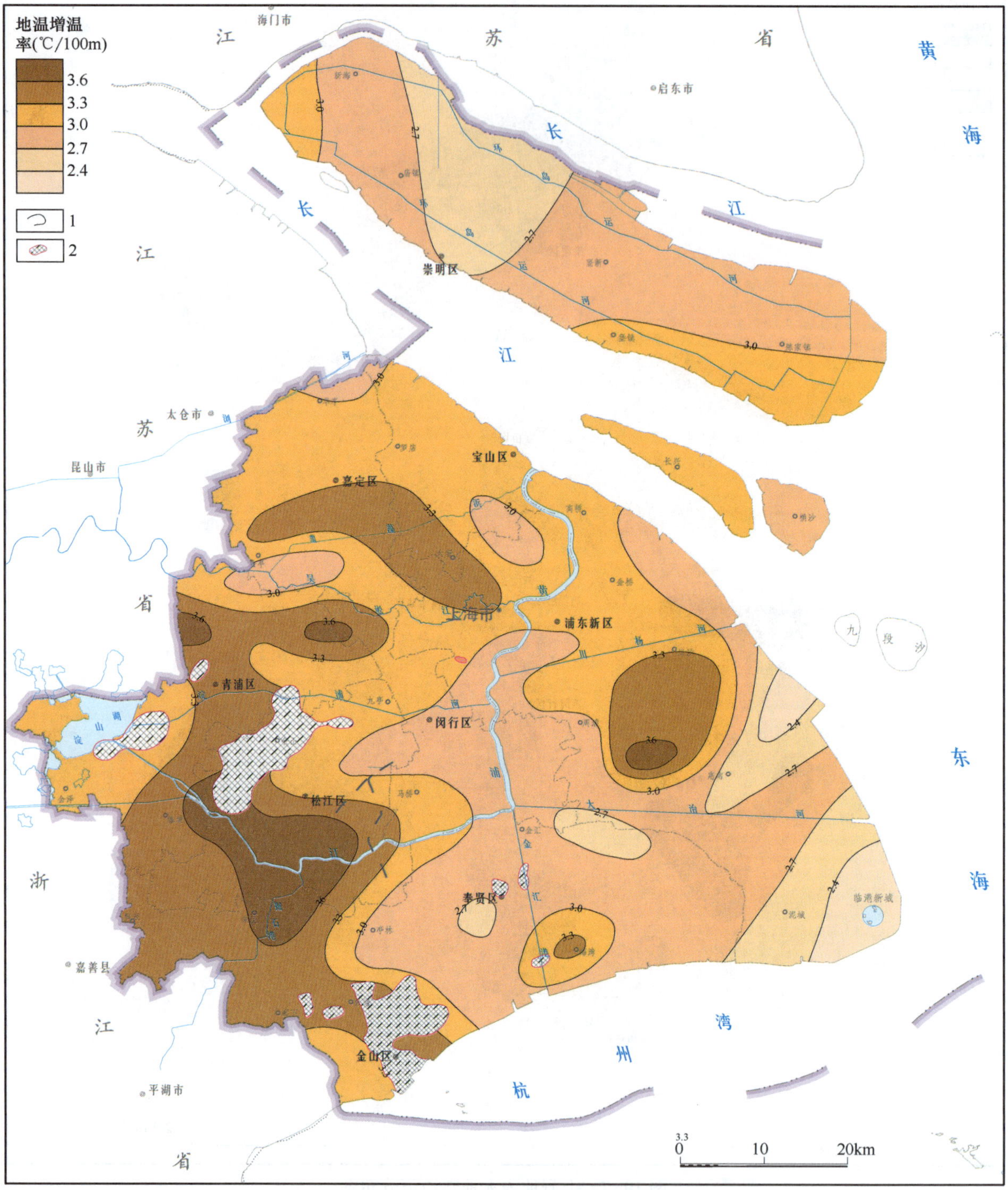

图 19-8　上海 150m 以浅地温梯度图

Fig. 19-8　Geothermal gradient above 150 m in Shanghai

1—地温增温率等值线（℃/100m）；2—第四系厚度小于 150m

区（图 19-9）。以太仓—奉贤断裂和罗店—周浦断裂为界大致划分为 3 个区，太仓—奉贤断裂以西温度比罗店—周浦断裂以东略高，但与两断裂带之间的区域相比略低。3 个分区中都可以划分出具有温度略高的局部异常。这种特征除受本身埋藏深度的影响外，与基岩埋深、岩性、深部构造等因素密切相关。

局部高值区主要呈北东向，个别呈北西—北西西向，主要集中在太仓—奉贤断裂和罗店—周浦断裂两条北西向断裂之间。主要包括安角盆地北缘、青浦—北新泾—静安寺、庙镇—唐行、顾村—高

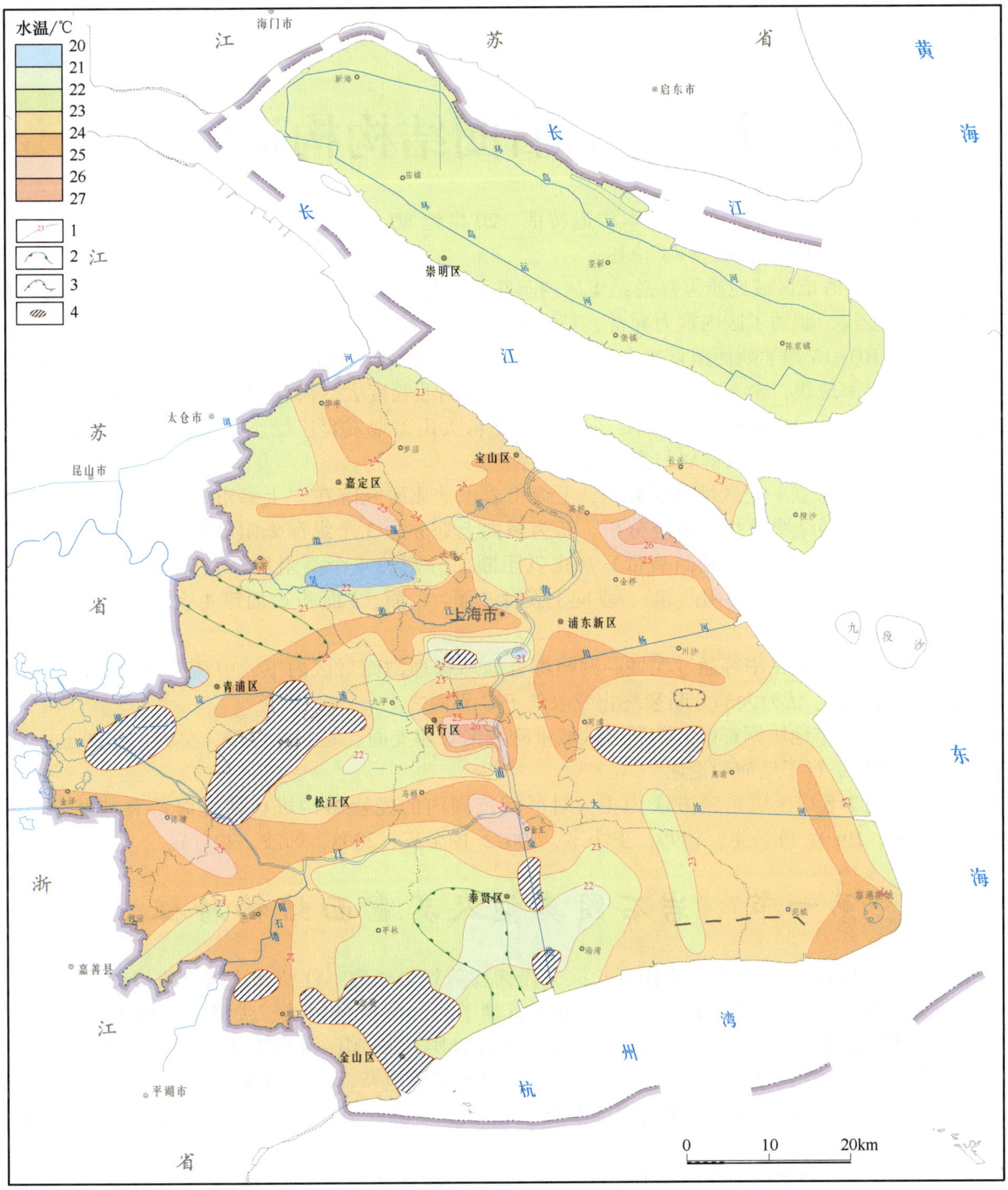

图 19-9　上海第四承压含水层水温等值线图

Fig. 19-9　Water temperature contour IV map of the Quaternary confined aquifers in Shanghai

1—水温等值线（℃）；2—第三、四承压含水层沟通区；3—第四、五承压含水层沟通区；4—基本缺失区

行—长兴、北桥盆地、练塘—石湖荡、枫径—马桥、书院—临港新城、朱泾—廊下、莘庄—周浦和三林—川沙高等。上述高值区在布格重力异常上多表现为沿布格重力异常梯度带和局部重力高异常分布。如吕巷重力高、北新泾重力高、长兴重力高、莘庄重力高、枫泾—北桥重力异常梯度带、葛隆—南翔重力异常梯度带、嘉定—罗店重力异常梯度带等。在航磁异常上多表现为负磁异常带或正、负磁异常的变化带，如周浦负磁异常区、北新泾—青浦负磁异常区、嘉定—南翔的正、负磁场过渡带等。

第二十章　岩石圈结构构造特征

层块结构构造特征是岩石圈的基本构造特征，20 世纪 80 年代以来，部分区内、区域上的地质、地球物理研究成果对相关特征进行了论述。

1988 年，《上海市区域地质志》总结了区内莫霍面的起伏特征，认为区内莫霍面自西、西南向东、东北方向总体变浅，制约了区内重力异常、基岩面的起伏变化。

1993 年，HQ-13 地学断面图首次划分了区内奉贤以西地区的岩石圈结构。以昆山—吴江断裂为界划分为昆山—上海地块，岩石圈相对较厚，燕山期抬升剥蚀幅度大，海相中、古生界复式褶皱构造层保存差，浅变质岩系因推覆而突出增厚，以及早白垩世火山盆地发育广泛。垂向上发育 9 大地质构造界面和 8 大构造层，并将金山群划归为浅变质岩系。

1995 年，基于最新一轮区域重磁数据，上海市岩土地质研究院在“上海市地震地质背景分分析”项目中对区内的壳幔结构进行了系统分析。认为横向上划分为 3 个纵深低阻断面（沙溪—吕四断裂、昆山—嘉定断裂和松江—牛皮礁断裂带）和由低阻断面分割的 4 个高阻块体（无锡—南通凸起、昆山—嘉定凸起、青浦凸起和山阳凸起），垂向上发育 S 面、G 面、J 面和 C 面等 4 个滑移面，金山群为区域上深变质岩系的上部。

2005—2007 年，姚保华等通过泥城—湖州综合地球物理剖面对区内下、中和上地壳进行了划分，并识别出 12 条断裂，认为大部分断裂终止或收敛于 G 界面之上。

2008 年，上海市地质调查研究院通过区域重磁数据对印支面、基底面埋深、莫霍面和居里面等深部构造-岩性界面进行了反演和论述。

此外，不同学者以区域重磁方法为主，对区域范围内的壳幔结构进行研究时都有涉及（顾斌，1987；黄兆熊，1989；陈沪生，1999；王小平，2007；杨金玉，2008；韩波，2011）。

第一节　岩石圈类型及莫霍面结构特征

据朱介寿（2005），区内位于嘉山—长兴壳幔韧性剪切带以西，江绍碰撞缝合带以北，属于下扬子强减薄型碎块状岩石圈。岩石圈自下而上分为石榴子石二辉橄榄岩层、尖晶石石榴子石二辉橄榄岩层、尖晶石二辉橄榄岩层、太古宙陆核（基性麻粒岩层）、火山-沉积变质岩层、台地性碎屑-碳酸盐岩沉积岩层、陆相中酸性火山-侵入岩系、陆相伸展盆地及盆缘玄武岩层。岩石圈厚度变化较大，在青浦一带为 120~150km，在松江—奉贤一带为 50~70km，在南部金山—泥城一带多为 100km 左右（图 20-1）。

不同学者对区内莫霍面和居里面埋深进行了反演。1986 年的《上海市区域地质志》中计算的莫霍面在 32.5km 左右。顾斌（1987）年计算的莫霍面在 29km 左右。黄兆熊（1989）计算的莫霍面在 30km 左右。韩波（2011）计算的莫霍面在 29km 左右。在 HQ-13 地学断面上，莫霍面埋深一般在 30~32km，嘉定一带可达 33.5km。在泥城—湖州地震剖面上，莫霍面埋深为 29.5~32.5km，自东向西呈逐渐抬升的趋势，且表现为一个带，厚度在 3.5~2km，自西向东变薄。尽管具体数值有所差别，但总体相似且变化趋势基本一致，主要特征有：①自西向东、自西南向东北呈现出变浅的趋势。②海域、陆域，北部和南部之间都表现出一定的差异。③莫霍面等值线和极值长轴方向与枫泾—川沙断裂带和张堰—南汇断裂带走向相似。④以上海中部的北西向构造带（太仓—奉贤断裂和罗店—周浦断裂之间）为界，莫霍面埋深特征表现出一定的差别。

黄兆熊（1989）计算的居里面在 17~21km。“上海市地震地质背景分分析”（1995）成果报告中为 25km 上下。2008 年，“上海市三维城市地质调查”成果中计算为 28~32km。韩波（2011）计算的居里面为 16~22km，且浅埋区沿上海—长兴岛分布。尽管居里面埋深的差别较大，但都表现出区内居里面起伏较大，且其总体变化特征受北东向主构造线和上海中部北西向构造带的控制。

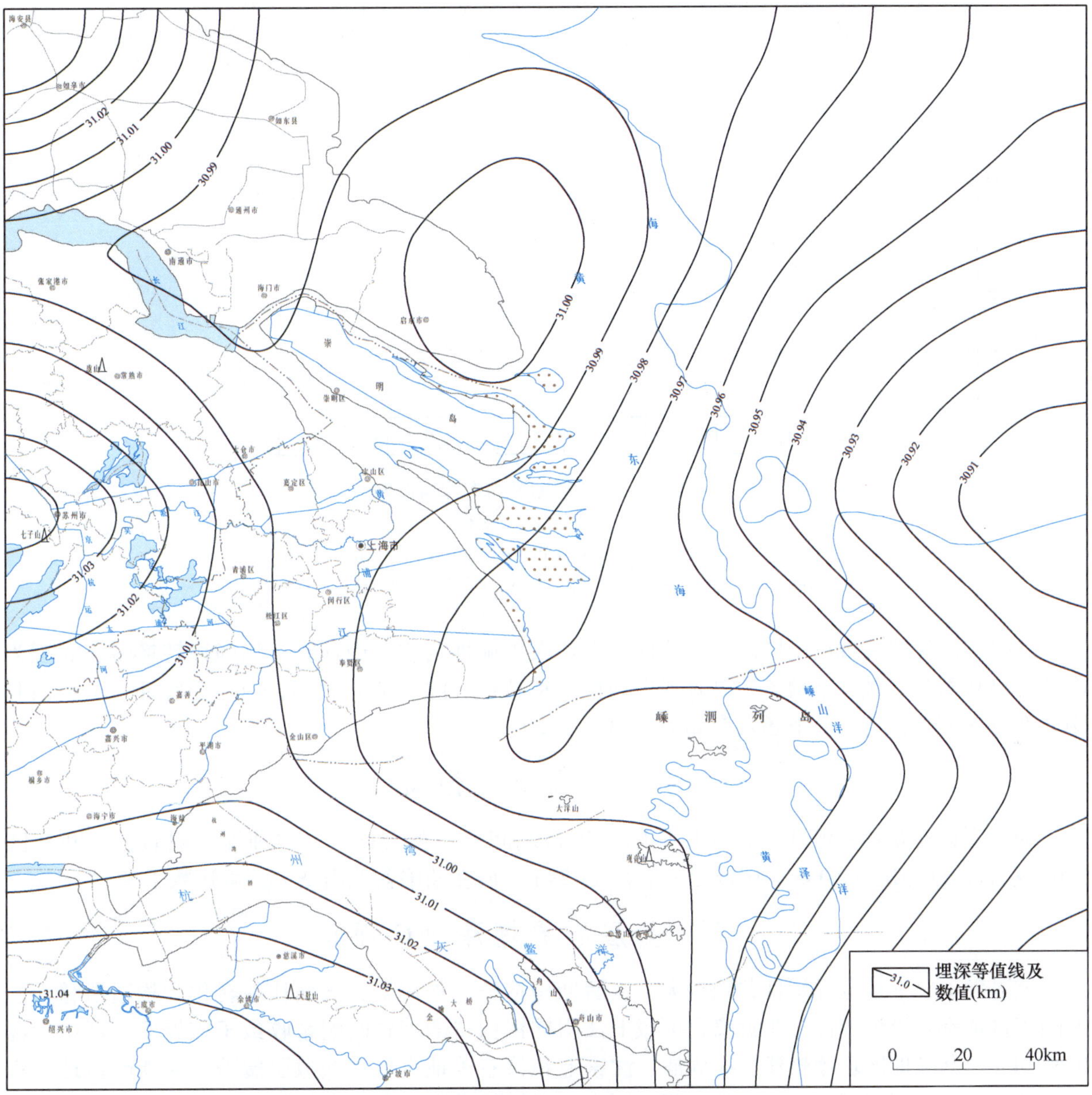

图 20-1　上海及邻区莫霍面埋深等值线图

Fig. 20-1　Moho depth contour map of Shanghai and its neighboring areas

第二节　地壳结构及厚度变化

天然地震层析成像和人工综合地球物理勘查成果都表明，区内地壳属大陆型结构，分上、中、下 3 层（图 20-2）：上地壳厚为 12~14km，底界有自西向东逐渐加深的趋势；中地壳厚度约 10km，埋深在 22~24km，呈现出明显的地层物性或岩石结构的变化；下地壳厚为 10~11km，与岩石圈地幔具有厚 2~3.5km 的壳幔过渡带。在青浦—松江一带，埋深 25km 左右处具有一明显的厚约 8km 的“壳内低阻带”。

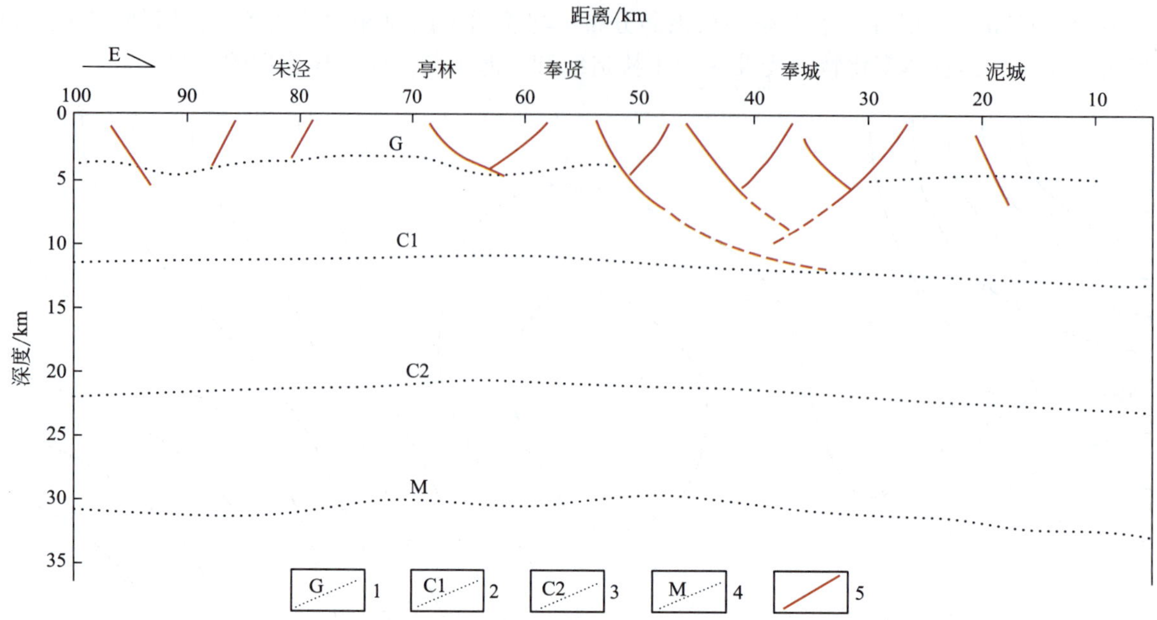

图 20-2　主要断裂与深部构造界面

（据姚保华，2007）

Fig. 20-2　Interfaces of deep structures and main faults

1—盖层界面；2—上地壳底面；3—中地壳底面；4—莫霍面；5—断裂

地壳不同深部表现出不同的变形特征，上地壳以脆性为主，中地壳为韧脆性转换带，表现为多层次的拆离和部分熔融，下地壳主要表现出韧性流动特征。除上地壳底部、中地壳底部、莫霍面和岩石圈底部等界面外，自上而下还发育如下界面（表 20-1）。

一、印支侵蚀面

为印支运动导致的海相沉积与后期陆相沉积之间的不整合面。区内埋深一般较浅，断隆区多在 300m 左右，局部接近 500m。区域上一般在 1~2km，而在断陷盆地内增大，可达数千米。

二、盖层底界埋深（G 面）

该界面上、下地层弹塑性差别较大，在后期构造挤压作用下，该面往往转化为构造滑移面。角直凹陷等盆地都是沿该界面拉张拆离，区域上江阴推覆构造的犁式断裂系收敛于该界面，区内埋深多在 0. 3~5km。断隆地区埋藏较浅，在周浦、沈家宅和吕巷等地直接被松散层覆盖，在断陷盆地和崇明岛西部埋藏较深。据泥城—平湖地震剖面，区内大多数断裂截至或收敛于该界面。

三、浅变质岩系顶界埋深

区内浅变质岩系顶界指区内金山群的顶部，与上覆地层之间岩性和变形程度差别较大，同样是区内部分断裂的滑移面。埋深差别较大，埋深最大区域主要位于长河盆地和安角盆地内。

四、浅变质岩系底界埋深（J 面）

区内浅变质岩系底界为金山群底部，与下部地层岩性差别较大，弹塑性悬殊。太仓—奉贤断裂、罗店—周浦断裂和松江推覆带通常收敛于该面（图 20-3）。埋深一般在 5~10km，区域上，在浙西北地区浅变质岩系较厚，可能至 13~14km。

表 20-1　上海岩石圈垂向结构简表

Table 20-1　Summary of vertical structures of the lithosphere in Shanghai

垂向分层	地质界面	界面深度/km	物性特征	主要岩性岩相	流体特征	变形特征与机制
松散层			地层(中)低密、(中)低磁、(中)低阻、(中)低速;岩浆岩从基性到酸性特征不同	冲洪积、河流和海相	低温富水	断裂作用、碎裂作用和摩擦滑动导致的脆性破裂等线性应变,韧性变形局部可见
	基岩面	0~0.48				
陆相层				陆相碎屑岩		
	印支侵蚀面(S面等)	0.3~3.2				
海相层			高密低磁,中高物性互层	海、海陆过渡相陆源碎屑岩与碳酸盐岩		
	盖层底面(G面)	0.3~5.0				
浅变质基底 惠南板岩			相对低密、低磁、低阻、低速,金山群局部高密高磁	低绿片岩相板岩类		
浅变质基底 金山群				高绿片岩相片岩、片麻岩类		
	浅变质岩系底面(J面)	5~10				
深变质基底			高密、高磁、高阻、高速	角闪岩相、麻粒岩相		
	上地壳底面(C_1)	12~14				
中地壳上部			西部为弱速度梯度,东部匀速度梯度	部分熔融变质岩、岩浆岩	高温含CO_2,富集于活动带	压溶作用、位错和粒间滑动的脆韧性变形
	中地壳上部底面	16~18				
中地壳下部			西部速度梯度增强,东部变成梯度层			
	中地壳底面(C_2)	22~24				
下地壳			高速(6.8km/s)、高阻、高密度,发育低阻层	榴辉岩、麻粒岩	高温、$H_2O+CO_2+CH_4$流体	位错和扩散蠕变导致的固态、超塑性流变等非线性应变
	莫霍面(M)	29.5~32.5				
岩石圈地幔			波速8.2km/s,密度$3.2\times10^3kg/m^3$,电性不均匀	冷热不均匀团块		位错蠕变和水压破裂导致的脆韧性变形
上地幔高导层			电阻率小于40Ω·m			

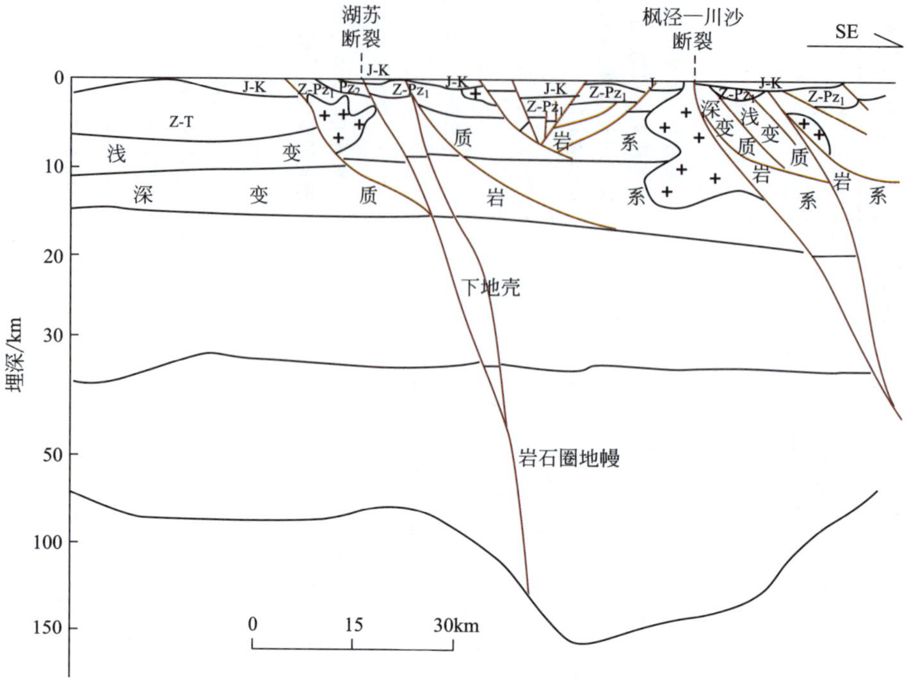

图 20-3　区域构造剖面示意图

（据华东石油局，有修改）

Fig. 20-3　Regional structural section diagram

五、下古生界软弱面

下寒武统底部：区内以寒武系超山组为主，主要为一套黑色碳质（含灰质、黄铁矿、磷质）泥岩、灰黑色碳质页岩组合。邻近地区涉及荷塘组、幕府山组（大陈岭组）下部层位的硅质或碳质泥岩夹层。

志留系：区域上以下志留统高家边组（霞乡组）为主，主要为一套页（泥）岩，局部可能涉及上志留统、上奥陶统的部分层位。

这些界面，区域上埋深区域变化较大，区内埋深多在数百米至4km左右，断陷盆地埋藏较深，推测可至6~7km。

第三节　深部地质作用及其对资源环境的约束

深部地质作用是内动力地质作用主要组成部分。地质历史和现阶段的深部地质过程对地形地貌、资源分布和地质环境条件具有明显的制约作用。地球表层地貌的形成、演化常与断裂构造、山脉隆升和盆地拉张密切相关，并进而受到地壳运动与壳幔相互作用控制。金属矿产多半是来自地幔的金属元素通过岩浆活动在上地壳表层的富集，而地壳减薄是诱发岩浆活动的重要因素。这种影响在区内主要表现在以下几个方面：

一、岩浆作用与成矿、石材资源

中—新元古代、中生代、新生代，区内都有岩浆作用发生。尤其是燕山晚期，在区域岩石圈大规模减薄与软流圈上涌的背景下，壳幔作用十分强烈，形成了大量的侵入岩和火山岩。

在张堰—南汇断裂带侵位的石英闪长岩体北侧形成矽卡岩型张堰铜矿。矿区位于金山张堰附近，东起戚家浜（朱行镇团结村）、西至界山（张堰镇界山村）、北自赵家宅（张堰镇高桥村）、南达兰花墙（张堰镇光明村），其范围占张堰大部分、朱行一部分，东西长约6km，南北宽（平均）约为1km。矿区由西部的界山铜矿、中部张堰铜矿和东北部的戚家浜铜锌矿组成。其中，张堰铜矿分百家村、时家村与东城隍庙3个矿段，东西长约2200m，南北宽约1200m。百家村矿段是唯一开展过详查的地段，控制程度较高，资源量最丰富，品位最高。矿种以铜为主，共（伴）生铁、银、金、锌、镉等。主要矿体的埋深在200~400m之间。矿体按产状与围岩性质分似层状、透镜状与脉状3类，按规模大小分主矿体、次要矿体与零星矿体。共有工业矿体75个，其中主矿体7个，次要矿体21个，零星矿体47个。表外矿体59个，由11个次要矿体和48个零星矿体组成。

佘山、天马山和辰山等西南部的基岩露头主要岩性为安山岩，其次为流纹岩，有少量的正长斑岩、花岗斑岩、闪长斑岩、细晶岩和火山角砾岩等。其强度一般在100MPa以上，可以作为石材使用；但由于节理发育，大型石材适用性差。除此之外，适用于作为混凝土砼料和一般玻璃原料。

此外，最近几年江苏省地质调查研究院在邻近长江北支的启东进行区域地质调查时，在松散层覆盖的火山岩、火山碎屑岩系上部发现了大型膨润土矿。

二、深部地质作用与地热

自晚白垩世开始，地壳的大规模拉张和软流圈的上隆导致岩石圈明显减薄，同时伴随有区域性断裂的构造反转和局部断陷盆地的发育。在中新生代盆地内部，形成盆地传导型地热资源，在深大断裂沿线形成局部热流异常和对流型地热资源。区内张堰—南汇断裂带、湖苏断裂带、枫泾—川沙断裂带和安角、松江—北桥、新村—兴隆等断陷盆地形成和演化过程即与上述深部地质作用过程密切相关。在这些断裂带和断陷盆地周缘通常有地热异常显示，有一定的水热性地热资源潜力。

中生代大规模的岩浆作用，在市区、崇明岛中部、青浦—松江和金山西南部等地形成大量的中酸性岩体，且规模较大。此外，在上海中部和南部地区，结晶基底推测埋深较浅，从岩石圈温度结构看，这些区域在3km以下具有形成干热岩资源的潜力。

三、深部地质作用与活动断裂

浅部构造形态是对壳幔深部构造的一种响应，深部构造对浅部区域构造的生成与展布有明显的制约作用。区内新近纪以来的新构造运动特征表现出了明显的继承性和新生性，特别是在沉积盆地演化上表现得相当明显。中新世玄武岩多沿古近纪断陷盆地边缘或深断裂与北西向断裂的交汇区分布。松散层的沉积厚度继续受北东向老断裂控制，但新生的北西向断裂系统的作用越来越突出，呈现出比较明显的分带和分块特征。基岩断隆区基岩面埋深较浅，断陷盆地区基岩面埋深较大。

第四纪以来的活动断裂，主要表现为老断裂的复活，且活动时期和强度具有明显的分段性。受基岩断裂带控制，走向与基岩断裂走向基本一致，在基岩浅部及上部松散层中多表现为一组倾向一致或相反的断裂。北西向断裂活动较晚，北东向或近东西向断裂活动较早。太仓—奉贤断裂带沿线在全新世中期为上海长期海岸线的分布位置。

四、深部地质作用与地震活动

大陆地震的孕育、发生与地球深部构造具有密切关系，深部构造环境是强震区孕震背景研究的重要内容（王椿镛，2016）。上海及邻近区域内，震中分布的条带性特征和震源深度沿经度、纬度方向的变化相似特征，都表明北东和北西向断裂为区内的主要控震、发震构造。在北东向的江绍断裂带、张堰—南汇断裂带、枫泾—川沙断裂带、湖苏断裂带，北西向的苏北沿岸断裂带、太仓—奉贤断裂带和罗店—周浦断裂带，以及北北东向的镇海—鸡骨礁断裂带等主要断裂交汇处，及其他断裂与这些主要断裂的交汇处是上海及邻区的地震易发部位。

区内主要震中密集区与居里面同样具有密切的关系（黄兆熊，1989），地震一般发生在居里面隆拗的结合过渡带、扭曲带和转折端附近等构造薄弱区域。主要有居里面等深线密集带，尤其是密集带等深线发生扭折的地段；居里面凹凸走向近于90°交汇地段；居里面凹凸交替发育的地段等。

地震的两个优势震源深度（4~11km、15~21km）与区域上的构造-岩性界面、不同规模断裂的拆离、滑脱界面相一致，尤其是寒武系和盖层底部、浅变质岩系底部和深变质岩系底部。浅部高角度张性断层与深部缓倾角拆离滑脱构造呈现了地壳不同构造层次的构造变形特征差异，它们之间具有密切的成因联系，其未连通空间构成了区域上地震的发震区域。

区内大于3级以上的地震主要分布在中地壳的中部，与中地壳的壳内高导层深度相当。壳内高导低速层上方一定深度范围内往往容易诱发地震，是一个构造活跃层。这一高导低速层可能与区内深大断裂在这个深度形成的拆离面有关。

在现今应力场作用下，以北东向和近东西向断裂易于剪切滑动而积累不起较高的地震能量，但由于不同层次滑移面的发育，尤其在苏南地区，所以导致弱震频发，震源深度变化较大。而北北东向和北北西向断裂带以压性活动为主，积累了较多的应变能量。一旦发震，震级可以较高。

第六篇
第四纪地质与地貌

第二十一章　第四纪地层

上海第四纪地层十分发育，除少数基岩残丘以外，均有第四纪沉积物覆盖，覆盖厚度主要在200~280m，局部下伏新近纪松散沉积物。受基岩面起伏控制区内第四系底界起伏由西向东大体呈阶梯状降低的趋势，在松江天马山、佘山、金山卫、北新泾、坦直、罗店、奉贤南桥等构造火山隆起和前古生界断凸地区，埋深较浅，多小于200m，在嘉定、宝山、崇明、浦东沿江以及西南部的金山枫泾等区域第四系厚度埋深较深，为260~280m（图21-1）。

对上海基岩以上疏松沉积层的研究可以上溯到20世纪20年代，大规模的第四纪地质调查研究是在中华人民共和国建立以后，随着水文、工程地质和矿产资源勘查的全面开展而持续进行的。相应的调查研究重点与研究深度，随着时代的进展，大致可以分为3个阶段。

早期，20世纪的五六十年代，可称为奠基阶段。通过上海至长江三角洲地区区域性的广泛实地调查，为阐明区域第四纪地质与地质环境特征奠定了基础。该阶段建立了早期的岩石地层单位和年代地层划分框架。

中期，20世纪的70—90年代中期，可称为夯实阶段。随着区域性水文、工程地质和环境地质普查的蓬勃开展及点上专门研究的启动，第四纪地层与沉积环境研究的基础得到了夯实。如1987年，吴标云和李从先的专著《长江三角洲第四纪地质》出版，书中按多重地层划分方法将第四纪下限划在高斯与松山极性时的分界，确定从早更新世中期以来，包括上海在内的现长江三角洲地区出现5次海侵。1991年，上海市环境地质站应用多时相图像结合土壤、历史等资料的验证，完成了覆盖上海与江、浙邻近地区的“地貌和第四纪地质遥感调查”。1991—1992年，上海市环境地质站与浙江省地质矿产局水文地质队合作完成的浦东，以及上海市环境地质站单独完成的上海城区1∶2万工程地质详查，对埋深80m以上的工程地质层做了详细的划分，根据地层接触关系，指出浦东等地埋深常达40m的工程地质⑤$_4$层硬土，为工程地质⑥层暗绿色硬土被侵蚀后、全新世海侵前形成的“高台深谷”沉积物。1996年，上海海洋地质调查局黄慧珍、唐保根、杨文达等的著作《长江三角洲沉积地质学》出版，作者根据长江水下及滨岸区大量物探、地质和钻探取样等的调查结果，阐述了长江水下三角洲现代沉积特征和沉积环境，应用层序地层和沉积相分析，探讨了晚更新世晚期以来长江水下三角洲的分期。1996年，中国地质科学院林景星的英文专著 *Quaternary Environment in the Eastern China* 出版。作者根据对诸多钻孔松散沉积层中有孔虫化石等的综合研究结果，提出上海及长江三角洲地区在距今约2.1Ma以来的第四纪曾发生8次海侵，分别为早更新世2次、中更新世3次、晚更新世2次和全新世1次。

近期，从20世纪90年代后期以来，引入了在我国黄土高原先行开展的磁化率、磁组构等环境磁学与古土壤母质层形成环境的研究，采用了包括热释光（TL）、光释光（OSL）、电子自旋共振（ESR）等定年新技术。上海第四纪研究逐渐进入深化阶段，开始针对不同时期的沉积古地理演化、古气候演化、第四纪海侵海退、长江贯通入海、河流泥沙源汇过程、全新世相对海平面重建、末次冰期下切河谷演化及长江三角洲形成发育等区域重大地质问题进行了深入探讨，在区域第四纪地质事件和地层界线划分等方面都取得了不同程度的进展。1998年，李从先、汪品先编著的《长江晚第四纪河口地层学研究》出版。作者根据微古化石、磁性地层、同位素测年、浅层地震等多重信息，建立了长江河口地区晚第四纪地层层序框架，分析了距今60000年来长江河口地区晚第四纪沉积环境的演化。2007年，邱金波和李晓编著的《上海市第四纪地层与沉积环境》，总结了近30年的上海第四纪地层调查成果，依据大量的^{14}C同位素、热释光、光释光、电子自旋共振、铀系等测年技术，结合磁性地层学、微体古生物学、元素地球化学、矿物学等多种学科手段，重新厘定了上海第四系重要地层界线和层序特征。

2012年至今，在已有认识和工作基础上，上海市地质调查研究院依托全国“三维地质调查试点”

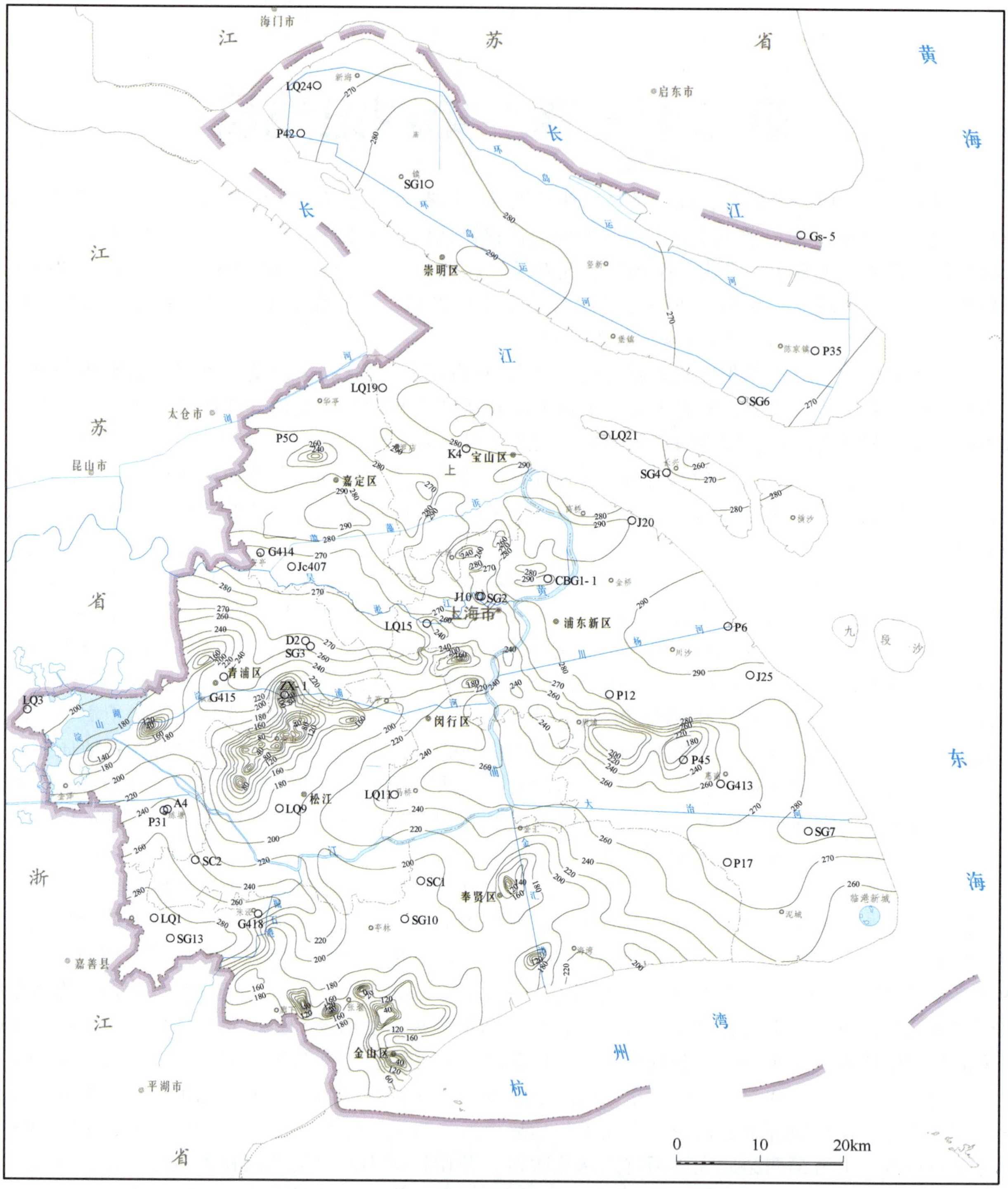

图 21-1　上海第四系厚度等值线和典型钻孔位置图

Fig. 21-1　Quaternary thickness contour map and typical boreholes of Shanghai

子项目“长江三角洲重点地区三维地质调查（上海）”对上海陆域三维地质结构模型进行了补充完善，并联合华东师范大学对第四纪年代地层框架和沉积物来源进行了针对性研究，以专题报告和论文的形式报道了所取得的最新认识。其中，磁性地层采取了比以往更高分辨率的样品采集，并采取热退磁和交变退磁相结合的方法辨识磁性事件，用于检验或试图完善原有古地磁记录确定的第四系重要地层界线（谢建磊等，2017；Liu et al.，2018；Yue et al.，2019）。然而，由于上海第四纪沉积相变频繁，地层很难保持理想的连续性，尤其是砂砾层中很难获取可靠的古地磁记录，一些学者开始尝试新的手段探讨第四纪年代地层划分。例如，Yang et al.（2006）最早通过浦东机场 PD 孔地球化学元素揭示的上新世以来的化学风化变化讨论了第四纪以后气候发生的重大变化，结果显示新近系与第四系的气候地层界线与磁性地层界线基本一致。谢建磊等（2017）尝试在光释光测年和磁性地层基础上，以频率磁化率为信号源、ETP 为靶曲线，通过轨道调谐方法对横沙岛 LZK1 孔开展了天文年代标尺研究，建立了年代序列框架。结果显示，M/B 界线埋深约为 143.0m、Ga/M 界线埋深约为 219.0m、Gi/Ga 界线埋深约为 296.6m。岳伟等尝试利用敏感重矿物指标区分第四系与新近系，研究观点认为进入第四纪，长江三角洲地区气候、物源、地貌同步发生了显著变化，稳定矿物成分可用作区分第四系和新近系的指标（Yue et al.，2019）。上述两个研究成果显示上海局部第四系底界需要上调 10 余米甚至 50 余米，这种结论有待今后更多工作进一步验证。最近 10 余年，随着光释光（OSL）测年精度和方法的提升，上海市调查研究院、华东师范大学调查和研究表明，上海上更新统年代地层划分的传统观点（邱金波和李晓，2007）面临挑战。例如，王张华等（2008）尝试 U 系测年和 OSL 相结合的方法对上海浦东南汇上更新统进行了氧同位素阶段划分，并通过地层对海平面的响应关系进行了佐证，提出传统的上更新统下部砂砾层的年代应归属为中更新世。赵宝成、王张华等相继研究了上海西部湖沼平原上更新统（Zhao et al.，2008；Wang et al.，2013），研究结果均显示上海西部上更新统底界埋深在 60m 左右。2012 年以来，上海市地质调查研究院在开展长江三角洲重点地区三维地质调查（上海）过程中在上更新统布置了大量的光释光测年，进一步揭示西部湖积平原上更新统底部埋藏深度较以往的认识相对较浅。赵宝成（2015）对上海河口沙岛、滨海平原、湖沼平原获取的钻孔及测试数据进行对比，发现上海上更新统在西部湖积平原和东部三角洲平原差异明显，认为 MIS5 阶段以来湖积平原保留了相对完整的暖期高海平面沉积记录，海相和河漫滩（古土壤）交替分布，而三角洲平原则多保留了冷期低海平面沉积记录，不同期次的古河道相较为发育，暖期海相地层多被剥蚀。

第一节　地层分区

在全国第四纪地层区划图中，上海第四纪地层分区总体应属于华北地层大区的华北东部地层区。

考虑基岩面起伏、海平面变化、沉积物来源、现代地貌及沉积作用，根据区内第四纪沉积环境演变及沉积地层结构整体差异，首先将上海第四纪地层划分为长江三角洲沉积区（Ⅰ）和太湖平原沉积区（Ⅱ）两大分区（图 21-2），两者的分界为上海境内湖积平原和滨海平原过渡地带“冈身带”的西界。再依据晚第四纪地质结构差异，长江三角洲沉积区进一步划分为三角洲主体河口沙岛沉积小区（$Ⅰ_1$）和三角洲侧翼滨海平原沉积小区（$Ⅰ_2$）；太湖平原沉积区进一步划分为东部小区（$Ⅱ_1$）和西部小区（$Ⅱ_2$）。

一、长江三角洲沉积分区（Ⅰ）

（一）三角洲主体河口沙岛沉积小区（$Ⅰ_1$）

该分区基岩埋深多大于 320m，第四系厚度在 260~290m。

早更新世地层发育齐全，层底埋深多在 150~290m，层厚在 107~157m，自下而上由以泛滥湖泊—浅湖相杂色黏土（夹粉细砂层）变为以边滩—河床相砂、含砾砂为主。

中更新世地层层底埋深为 132~159m，层厚为 13~45m，主要为中更新世早期地层，而中更新世晚期地层多被晚更新世早期河流下切剥蚀而缺失。崇明岛域内多数地点该时期地层以河床相砂砾、砂层为主，而在崇明西北部、东部以及长兴岛、横沙岛则发育杂色黏土夹砂或以细砂、含砾砂为主的河漫滩、心滩及边滩沉积。

晚更新世早期地层底界深度达到 120~130m，大都由下部的河床相含砾中粗砂，中上部的河流边滩相中细砂和粉细砂夹河床相砾砂组成，局部夹有泛滥平原相青灰色粉土—粉质黏土层，或为河流边滩相中细砂和河床相含砾中粗砂的互层，为摆动不定的曲流河道沉积。

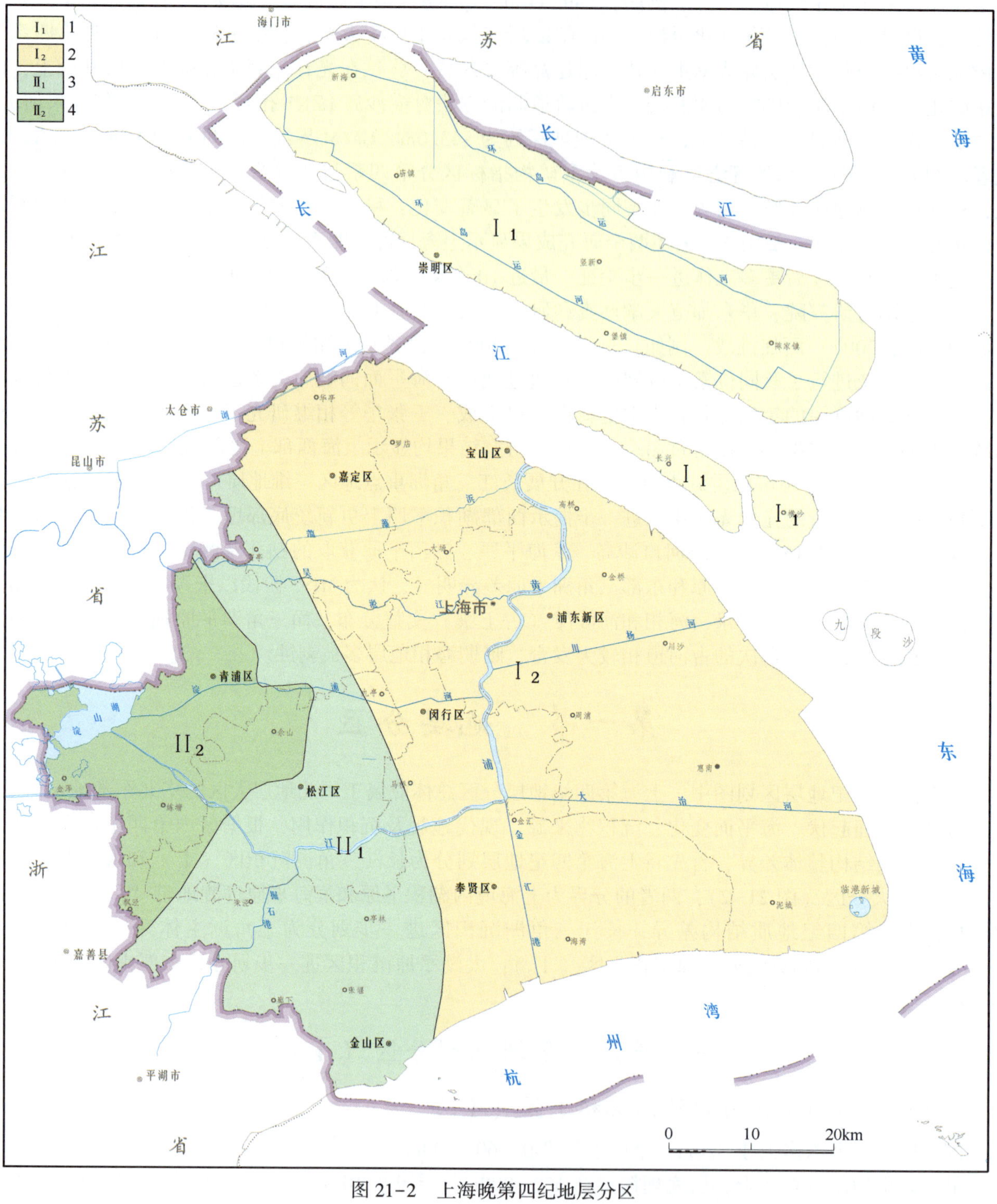

图 21-2 上海晚第四纪地层分区

Fig. 21-2 Stratigraphic division of the Late Quaternary in Shanghai

1—长江三角洲沉积区河口沙岛分区；2—长江三角洲沉积区滨海平原分区；3—太湖沉积区东部分区；4—太湖沉积区西部分区

晚更新世晚期地层，只保存相当于末次冰期间冰阶沉积的河口湾相灰色细砂夹砂质粉土和上覆的滨海相粉细砂和粉土互层。相当于早、晚冰阶的陆相堆积已侵蚀殆尽。

全新世地层在河口分区最为发育，底界深度普遍达到 50～60m，呈由西而东趋深；由下部的溺谷—河口湾相灰色粉细砂和粉土、粉质黏土，中部的滨岸浅海相深灰色淤泥质黏性土与滨海相灰色砂质粉土夹粉质黏土，以及上部的河口相粉土、粉砂和地表的三角洲平原相褐黄、灰黄色粉质黏土组成，三分性明显，构成完整的冰后期海侵旋回。

(二) 三角洲侧翼滨海平原沉积小区（I_2）

该分区基岩埋深多为240~320m，第四系厚度为240~260m。北部嘉定和东南部滨岸一带凹陷区内基岩面和第四系厚度与东部河口沙岛分区相似。

早更新世地层层底埋深为229~299m，厚度为90~160m，总体特征南北差异明显。北部嘉定、宝山及浦东北部地层沉积体系自下而上堆积了一套以杂色黏土（夹砂层）为主的泛滥湖泊（浅湖）相地层和一套以砂砾、砂层为主的河流相地层，而南部闵行—南汇新城一带自下而上堆积了一套以杂色黏土为主，含黏土砂砾、砂为特征的冲洪积扇相地层和一套以砂砾、砂层为主的河流相地层。东部一带早更新世地层中上部发现微弱海侵迹象。

中更新世地层层底埋深为133~159m，厚度为90~160m，主要发育河流相地层，局部地层发育受微弱海侵影响，常见岩性包括含砾砂、砂及杂色黏土（夹砂层），在嘉定—南汇新城和闵行—南汇新城形成两片古河道区，含砾砂、砂较为发育。中更新世晚期地层由于晚更新世早期河流剥蚀作用而局部发育。

晚更新世早期北部的嘉定、宝山经历前期的河道至中、后期以河流边滩为主的曲流河道环境演变，至市区及其东南的浦东、闵行、奉贤进入明显受海水影响的溺谷沉积环境，在浦东的滨岸地带还曾出现河口滨海环境。岩性基本为含砾中粗砂和中细砂、粉细砂组成的沉积序列；在市区—南汇新城的部分地区，中部地层内见有泛滥平原相青灰色粉质黏土层。该分区的上更新统底界起伏较大，大体上由西而东加深，至浦东滨岸地带已下切到中更新统中部。

晚更新世晚期北部为海湾，中、南部为浅海环境，相应堆积了北部的灰色粉质黏土夹粉砂、砂质粉土纹层；中部市区、浦东的粉质黏土与粉细砂薄互层，即俗称的“千层饼”；南部闵行、奉贤、浦东东南部地区的灰、灰黄色粉细砂与砂质粉土互层夹粉质黏土。其底部，断续出现相当于末次冰期早冰阶泛滥平原堆积的灰绿、深灰色黏土或黏质粉土等陆相层；其上部，则在埋深40m左右的滨浅海相“千层饼”层之上，超覆一套基本为河流相的锈黄色粉细砂、黄色砂质粉土，其顶部大都保存末次冰期晚冰阶泛滥湖泊堆积的褐黄—暗绿色硬黏土，成为常见的更新统顶部的标志层。而在下切河谷内，该标志层多被侵蚀而局部分布，其底部多发育泛滥湖泊和废弃河道堆积，完整的剖面由暗绿、灰绿色硬黏土与上覆含多少不一腐殖物的灰、深灰色粉质黏土夹粉砂的序列组成，暗绿色硬黏土以上沉积时间大致与末次冰期末“新仙女木突冷事件”一致。由于区内晚更新世晚期海、陆变迁频繁，地层中形成了中、底部之间及中、上部之间的两三个沉积间断。

全新世地层在滨海平原分区发育一套完整的海侵旋回。由下而上出现河口滨海—滨岸浅海—河口滨海相的变化；相应沉积序列为灰色砂质粉土、灰黑色含腐殖物黏土—深灰色淤泥质黏土、淤泥质粉质黏土和砂质粉土—灰色粉质黏土、砂质粉土互层和褐黄色粉质黏土，显示相对粗—细—粗的岩性变化，其岩相、岩性的三分性明显，这与湖沼平原区仅有海陆过渡，甚或未受海水侵进迥然不同。另外，在中—晚全新世，出现流经嘉定、青浦接壤区，穿越市区后流向宝山、浦东的吴淞江故道切割地带，在埋深约10m以浅堆积了三角洲支流河道—河口相青灰色砂质粉土和粉砂，上覆河口滨岸相褐灰色粉质黏土。

二、太湖沉积分区（Ⅱ）

该分区多位于基岩面凸起区，基岩埋深多小于240m，松散沉积物主要为第四系。局部诸如金山枫泾、青浦淀山湖一带为基岩凹陷区，基岩面埋深为240~320m，第四系厚度约为260m。

早更新世地层层底埋深为218~265m，多在220~240m，多缺失早更新世早期地层，岩相以泛滥平原—湖泊相的杂色黏土（夹砂层）为主。枫泾、淀山湖一带基岩凹陷区早更新世地层发育齐全，下部、中部发育冲洪积扇相含黏土砂砾。

中更新世地层发育齐全，层底埋深为130~150m，厚度为21~64m，以河流相含砾砂、砂、杂色黏土为主，局部地层发育受微弱海侵影响。中更新世末，全为泛滥湖泊环境，普遍沉积了蓝灰色黏土。

（一）西部沉积小区（$Ⅱ_2$）

在西部沉积小区，包括青浦大部及与松江、金山的接壤区：

晚更新世早期，通常形成下部的河床相与上部的沙坝相组成的曲流河道沉积序列，其下部以含砾砂为主夹粉细砂，上部常为粉细砂夹含砾砂和黏性土。

晚更新世晚期，以一套泛滥平原—湖泊相与滨海—潟湖相的交替沉积为特征，这是滆湖组典型发育的地区。在其下部（埋深40~60m）、中部（深20~30m）、上部（深约5m）各出现一层泛滥平原—湖泊沉积的暗绿、灰绿—褐黄色硬黏土，其形成年代由下而上分别为75~55ka、45~35ka和20~13ka。其间多为滨海—潟湖沉积的粉土、粉细砂夹粉质黏土，以及至上部的硬黏土之下则为河流边滩沉积的黄色砂质粉土。表明滆湖海侵曾在55~20ka侵入西部分区，但其表现较弱，并在45~35ka退出，于是在这一分区内形成了常称的上海西部“第二硬土层”。

全新世早—中期，西部小区为未受海侵波及的剥蚀硬土台地；全新世晚期沦为湖沼，堆积了灰黄、灰黑色黏土，局部夹有泥炭。

（二）东部沉积小区（$Ⅱ_1$）

在东部沉积小区，大体包括金山、松江和青浦东北部地区：

晚更新世早期，通常也为曲流河道沉积。其下部以河床相含砾砂为主，间有沙坝相粉细砂、中细砂。上部常以粉细砂、中细砂为主，常夹含砾砂，还有砂质粉砂和粉砂质黏土，反映为沙坝与河床交替，间或为泛滥平原沉积；微体古生物（简称微古）分析显示，在个别钻孔中发现属种、个数均很少的广盐性半咸水有孔虫和半咸水介形虫，伴有偶见的陆相介形虫及淡水腹足类化石，反映为有时受海水淹没的河口湾或感潮河道沉积。

晚更新世晚期，基本也是一套泛滥平原—湖泊相与滨海—潟湖相的交替沉积，为滆湖组比较发育的地区。在埋深15~65m之间出现厚5~10m不等的4层灰绿、灰黄杂蓝灰色硬黏土，通常不含海相微古化石，但可见较多的中生、水生草本植物及环纹藻等孢粉和藻类化石，为泛滥平原—湖泊沉积；其上、中部两层硬黏土可与西部分区相应层位的两层硬土对比，下部两层相当于西部的下层硬土。夹于硬土层间的为潟湖或滨海相的灰、灰黄色（粉质）黏土与粉砂或砂质粉土薄互层，通常含属种、个数均少的半咸水微古化石。表明滆湖海侵在这里比西部分区多出现一次进退。

东部沉积小区（$Ⅱ_1$）的东南侧部分如金山亭林孔揭示，上更新统横向上发生相变，变为中段地层内无陆相硬土层出现为特征的南汇组。上更新统中段以上，包括与末次冰期末的“新仙女木突冷事件”相当地层以至上覆的全新世沉积地层，其岩性、岩相序列，与长江三角洲滨海平原地层剖面可以对比，唯上全新统一般为不含海相微古化石的青浦组湖相沉积。

第二节　年代地层划分

参照全国地层委员会方案，主要依据古地磁、放射性碳法、光释光，并少量热释光、电子自旋共振及铀系方法，将上海更新统划分出下更新统、中更新统和上更新统3个阶段，全新统划分出下全新统、中全新统和上全新统3个阶段。

一、磁性地层及第四纪极性事件

（一）古地磁测量

上海古地磁测量材料均为第四纪钻孔，由于钻取的样品为非定向的，因此只能给出磁倾角及天然磁化率数据。目前区内积累了17个古地磁测试钻孔，其中1980—2008年第四纪钻孔古地磁测量采用交变退磁方法，样品布置间距多在1m以上，2008年以后采用交变退磁和热退磁相结合的方法，样品布置间距多在20~30cm。区内第四纪沉积物古地磁测量难点在于松散砾砂层较厚，测量结果可信度相对较低，因此在极性柱上出现了较多缺样的磁性不清段，在一定程度上影响了第四纪极性柱的完整性。

区内第四纪极性时、亚时的确立参照国际标准极性和国内不同第四纪沉积物揭示和验证的极性事件。美国学者Cox. A（1969）提出了4.5Ma以来的古地磁极性年表即常称的“考克斯年表”，嗣后又经Mackinen（1979）等学者的修正，为全球第四纪磁性地层的划分对比提供了很有价值的科学依据。对古地磁极性柱中一些极性亚时（亚幕）或极性漂移事件的争议较多，通过对海洋、湖泊等沉积物的古地磁研究，目前趋向于被多数学者认识、验证的极性事件，在布容（Brunhes）正极性时（幕）内，有拉尚（Laschamp）负极性亚时（约35ka）、布莱克（Blake）负极性亚时（约115ka）、牙买加（Jamaica）负极性亚时（约195ka）、Big lost Emperor负极性事件（约554ka）、Delta负极性事件（约690ka）等（王永等，2004）；还有历来在长江三角洲钻孔晚更新世末陆相层顶部经常出现的哥德堡（Gothenburg）负极性亚时（13.75~12.35ka）（吴标云和李从先，1987；黄慧珍等，1996）。在松山（Matsuyama）负极性时（幕）内，据阳原—蔚县盆地大道坡和洞沟两个泥河湾组地层剖面的研究（袁宝印等，1996），有贾拉米洛（Jalamillo）正极性亚时（原定年0.97~0.90Ma，后修正为1.07~0.99Ma）、奥尔都威（Olduvai）正极性亚时（1.95~1.77Ma）、留尼昂（Reunion）正极性亚时，由Ⅰ、Ⅱ两个正极性事件组成（2.14~2.01Ma）。在加密采样的洞沟剖面上发现，奥尔都威正极性亚时由两次正极性和一次负极性事件组成，留尼昂正极性亚时由3次正极性和2次负极性事件组成。在洞沟剖面留尼昂正极性亚时以下，接近高斯（Gauss）/松山（Matusyama）（G/M）界限处，记录到一次短暂的正极性事件（X事件，约为2.37Ma）；在该剖面等地的贾拉米洛与奥尔都威正极性亚时之间还记录到一次正极性事件，被称之为前贾拉米洛（Pre-Jalamillo）事件。还有前中一晃等人在陕西洛川黄土剖面的松山（Matusyama）/布容（Brunbes）（M/B）界限与贾拉米洛正极性亚时之间，也记录到一次正极性漂移事件，为0.82~0.80Ma（据朱日祥等，1995），本次以“后贾拉米洛事件（Later-Jalamillo）”称之。在高斯正极性时（幕）内，有凯纳（Kaena）负极性亚时（原定年3.01~2.92Ma）、马莫斯（Mammoth）负极性亚时（原定年3.15~3.05Ma）。

（二）极性时

1. 布容正极性时

布容正极性时分布于区内第四纪地层剖面的上部，古地磁极性柱上总体显示为正向极性（图21-3），间夹若干负向极性事件（表21-1）。布容正极性时包含了长江三角洲地区第四纪全新统、上更新统和中更新统3个年代地层。

布容正极性时与其下的松山负极性时的分界为M/B界限，地质年代为0.78Ma。区内M/B界线埋深为130~150m，大致出现在中更新统底部砂层与下伏下更新统顶部杂色硬土层之间的假整合界面附近。

（1）哥德堡负极性亚时

哥德堡负极性亚时赋存在晚更新世末沉积的陆相标志层暗绿—褐黄色硬土层的顶部，接近与上覆全新统的假整合面附近。在河口沙岛、滨海平原下切河谷内，该事件也可在砂层中识别。其分布深度随该标志层的起伏而升降，已记录到的埋深为10~40m不等，通常为20~30m。

（2）拉尚负极性亚时

拉尚负极性亚时赋存在上更新统上部海相层灰色粉质黏土、粉细砂互层或海陆过渡相灰、灰黄色砂层内。综合上海和毗邻省区诸多钻孔资料分析，其赋存深度通常为40~50m，相当于上海和长江三角洲地区“滆湖海侵”（形成区内第Ⅱ海进层）中期的低海平面层位或上海青浦—太湖平原地区所见第二层陆相暗绿—褐黄色硬土的顶部；但在现河口岛屿上则深达80~90m。记录到这一负极性亚时的地层厚度常为3~4m，已知其年龄约为35ka。

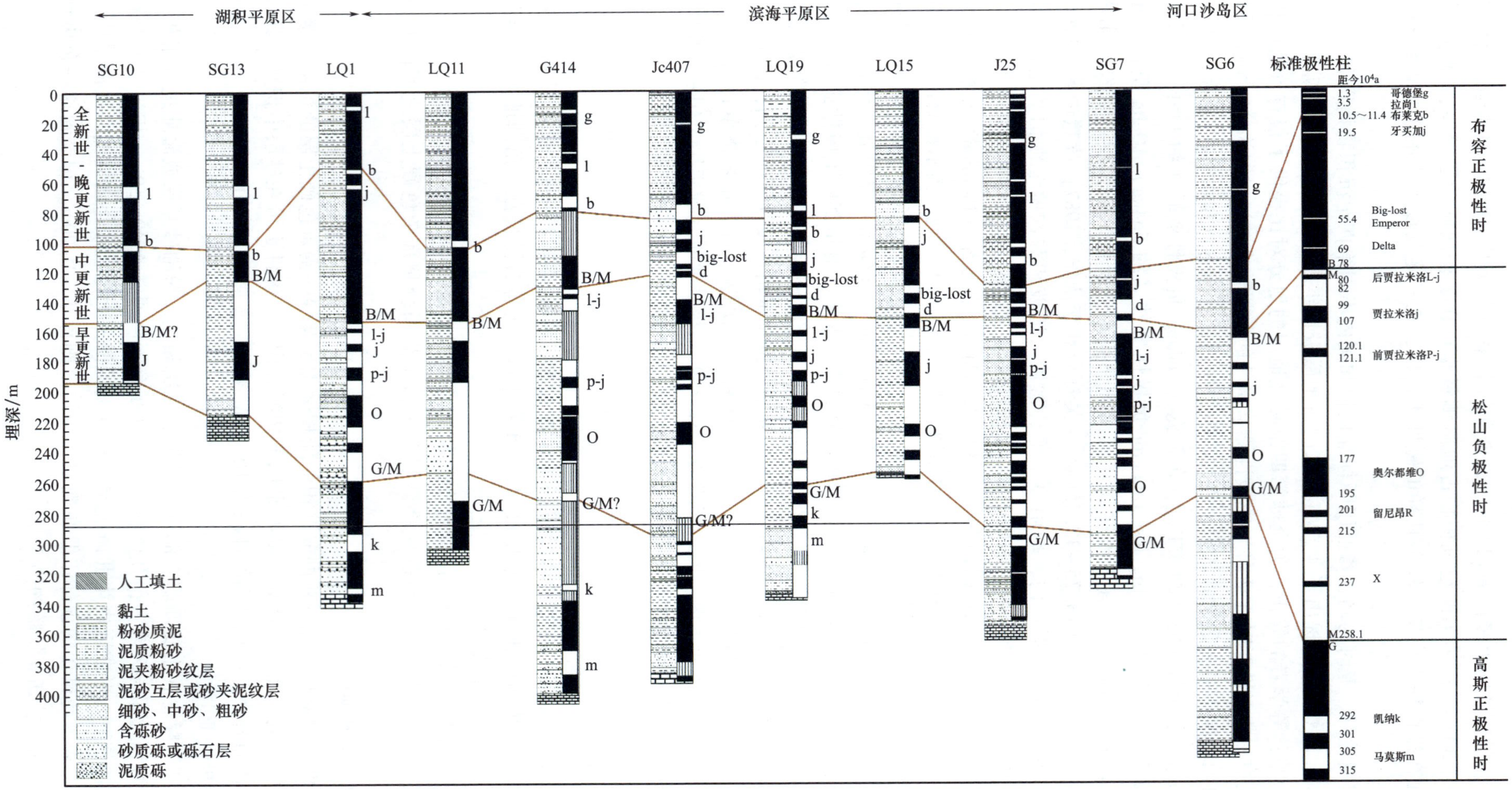

图 21-3 上海第四纪磁性地层划分

Fig. 21-3 Magnetic stratigraphic division of Quaternary in Shanghai

表 21-1　上海第四纪钻孔所见极性事件埋深一览表

Table 21-1　Burial depths of polarity events revealed by boreholes in Shanghai

地貌单元		湖积平原				河口沙岛			滨海平原							
极性事件	距今年龄/ka	金山枫泾 LQ1/m	金山朱泾 G418/m	松江新城 LQ9/m	青浦商塌 LQ3/m	崇明前哨 P35/m	崇明新海 LQ24/m	崇明庙镇 SG1/m	闵行马桥 LQ11/m	城区北新泾 LQ15/m	嘉定黄渡 Jc407	嘉定娄塘 P5/m	嘉定安亭 G414/m	宝山罗泾 LQ19/m	浦东机场 J25/m	浦东临港 SG7/m
Gothenburg	约 13	—	—	—	—	—	59~65	—	—	—	22(?)	41~47	13.5	29~33	28~31	—
拉尚 (Laschamp)	约 35	9~12	—	—	10	—	—	87~90	—	—	—	—	38~42	77~79(?)	56~57	53
布莱克 (Blake)	约 115	52~55	65~71(?)	44~49	42	—	80	125~128	115~119	79~82(?)	75~85	—	72~80(?)	90~93(?)	111~112	99~102
Jamaica	约 195	62~65	—	64~72(?)	61~64	107~112	—	—	—	105~112(?)	95~98	101.5~103	—	108~114	—	120~123
Big lost Emperor	约 554	—	108~115	88~90(?)	100~103	122~126	—	141	—	127~133	109~113	117~125	—	123~128(?)	131~132	—
Delta	约 690	—	—	110~118(?)	119~127	129~132	—	—	—	139~146	118~120	—	132~136	131~137(?)	142	140~150
松山 (Matsuyama)/布容 (Brunhes) (M/B)	780	150	146	129	150	147	143	164	—	155	125	149	138	150	151	155
后贾拉米洛 (Later-Jalamillo)	820~800	154~156	151~154	137~140	163~167	151~153	—	—		—	140~150	—	—	160~164	153~156	—
贾拉米洛 (Jalamillo)	1070~990	164~168	161~164	149~161	170~180	157~160	176~184	—	171~175(?)	168~192	184~193	—	—	174~181	174~180	193~198(?)

续表

地貌单元		湖积平原				河口沙岛			滨海平原							
前贾拉米洛（Pre-Jalamillo）	1211~1201	184~192	—	170~172	—	177~182	191~203	—	—	—	195~200	183~191	192~198	187~199	—	—
奥尔都威（Olduvai）	1950~1770	194~216	—	196~201	203~207（?）	184~229	232~240	170~210	216~219（?）	218~225	220~235	205~241	211~240	203~211	192~235	260~268（?）
留尼昂（Reunion）	2150~2010	—	—	—	210~212（?）	231~248	—	239~267	—	235~240	—	247~257	244~247	—	251~268	—
X 事件	2370	—	—	—	220~224（?）	—	—	270. 5~271	—	—	—	—	—	—	274~276	—
高斯（Gauss）/松山（Matusyama）（G/M）	2581	251	261	—	237	262	275	280	255	—	300?	269	264	261	287	290
凯纳（Kaena）	3010~2920	285~295	—	—	—	—	—	300~303	—	—	—	339~345	328~333	274~283	—	—
马莫斯（Mammoth）	3150~3050	—						324~328						290~305（?）		
高斯（Gauss）/吉尔伯特（Gilbert）G/G	3400	—	—	—	—	—	—	—	—	—	—	—	379	—	—	—

（3）布莱克负极性亚时

布莱克负极性亚时多见于上更新统底部的砂或砂砾层中，其正常的分布深度应小于等于100m，因河口滨海地区上更新统底界深切，赋存这一极性事件的层位也随之下移。记录的厚度常为2~3m，已知其年龄为114~108ka。如上海东北部的浦东J25孔、庙镇SG1孔孔深102~128m所见的三角洲河流—浅滩相灰色含砾砂或灰色粉砂夹粉质黏土层记录到这一极性事件。布莱克负极性亚时在西部枫泾LQ1孔、滨海平原西部北新泾LQ15孔、宝山罗泾LQ19孔以及闵行马桥LQ11孔均有记录，埋深主要在80~100m，但LQ11孔该事件埋深更浅，约为60m。

（4）牙买加负极性亚时

该极性事件已知年龄约为195ka。在娄塘P5孔孔深101.5~103m、前哨P35孔孔深107~112m等处出现，记录的地层厚度为2~5m，赋存于中更新统上部，为灰绿色黏土层或蓝灰色黏土与粉土层过渡处，孢粉和微古分析表明为温凉气候下的陆相堆积。滨海平原西部北新泾LQ15孔、宝山罗泾LQ19孔以及闵行马桥LQ11孔牙买加负极性亚时埋深在100~105m，西部枫泾LQ1孔该事件埋深约80m。

（5）Big lost Emperor负极性事件

该极性事件已知年龄约为554ka。在诸多钻孔地层剖面上出现，埋藏深度大致介于110~130m，记录的地层厚度多为1~4m。赋存于中更新统下部，普遍为蓝灰—褐黄色硬黏土，也为温凉气候下的泛滥平原—湖泊之类陆相沉积。滨海平原西部北新泾LQ15孔、宝山罗泾LQ19孔记录该事件埋深约为120m。

（6）Delta负极性事件

该极性事件已知年龄约为为690ka。在诸多钻孔地层剖面上出现，埋藏深度大致介于130~140m，记录的地层厚度在1~4m不等。赋存于中更新统下部，较典型的如安亭G414孔、浦东J25孔为泛滥平原—湖泊堆积的褐黄杂蓝灰色细网纹状硬黏土；而崇明岛上则为河流浅滩沉积的灰色中细砂夹砂质粉土层。滨海平原西部北新泾LQ15孔、宝山罗泾LQ19孔记录该事件埋深接近140m。

2. 松山负极性时

分布于上海第四纪地层剖面的下部，上海及长江三角洲地区G/M界限的深度为250~285m，松山时磁性地层厚度为100~130m，平均近于120m，仅包含下更新统一个年代地层。目前G/M界限年龄为2.58Ma。上海更新统与上新统分界大都在G/M界限之下2~5m，由此推算区内第四系底界年龄为2.65~2.61Ma，即大致为2.60Ma。

（1）后贾拉米洛正极性事件

在浦东J25等孔下更新统顶部出现，深度为151~156m，记录厚度为2~3m，均赋存于褐黄色硬黏土层中，据朱泾G418孔微古、孢粉分析，该黏土层仅含淡水介形虫，主要出现苔藓和少量草本孢粉，反映为凉干气候下的湖泊堆积。已知其年龄为0.80~0.82Ma。西部枫泾LQ1孔、滨海平原西部北新泾LQ15孔、宝山罗泾LQ19孔以及闵行马桥LQ11孔该事件埋深约160m。

（2）贾拉米洛正极性亚时

在浦东J25等孔下更新统上部出现，深度为174~180m，记录厚度为6m，赋存于河流边滩相灰色砂质粉土中。朱泾G418孔出现深度为161~164m，也属下更新统上部，为河流边滩相灰色粉砂—泥质粉土；据孢粉分析，该层主要含落叶阔叶、部分针叶及中生、水生草本孢粉，微古分析未检出化石，反映为温暖湿润气候下的沉积物。已知其年龄为0.99~1.07Ma。滨海平原西部北新泾LQ15孔、宝山罗泾LQ19孔以及西部枫泾LQ1孔该事件埋深约为170m。

（3）前贾拉米洛正极性亚时

在娄塘P5、安亭G414等孔地层剖面上出现，埋藏深度为180~195m，记录厚度为5~8m。其层位属于下更新统中部上段地层，赋存于富含钙质结核或夹有钙板层的褐黄、褐色硬黏土中。据分析孢粉贫乏，见少量的落叶阔叶和禾本科等孢粉，反映为温凉气候背景下短期温暖波动形成的湖泊沉积。已知其年龄为 1. 211~1. 201Ma。

（4） 奥尔都威正极性亚时

这是松山负极性时内主要的正极性亚时之一，在钻孔地层剖面上普遍出现，埋藏深度介于 180~240m，记录厚度通常为 30~40m。其层位属于下更新统中部下段地层。典型的赋存地层剖面如安亭 G414 孔所见：下部为河床—砂坝相黄灰色砾砂夹含砾细砂，上部为泛滥—湖泊相杂色黏土，显示为具二元结构的曲流河道沉积序列；孢粉分析反映其经历了早期温暖湿润—晚期温凉干燥的气候变迁。已知其年龄为 1. 95~1. 77Ma。据庙镇 SG1 等孔观测知，这一正极性亚时往往由两个正极性事件夹一负极性事件（厚 1m 多）组成，岩性多为灰黄色网纹状黏土、青灰色黏土，普遍含多少不一的钙质结核和网脉。滨海平原西部北新泾 LQ15 孔、宝山罗泾 LQ19 孔以及西部枫泾 LQ1 孔该事件埋深在 220~240m。

（5） 留尼昂正极性亚时

这是松山负极性时内另一个主要的正极性亚时，在钻孔地层剖面上普遍出现，埋藏深度大都介于 240~260m，记录厚度大都为 10 ~ 20m。其层位属于下更新统下部上段地层。已知其年龄为 2. 15~2. 01Ma。

此正极性亚时通常由上、下两个正极性事件夹中间一个负极性事件（厚 1. 5~4m 不等）组成，如浦东 J25、庙镇 SG1、娄塘 P5 等孔均出现。典型极性柱如庙镇 SG1 孔孔深 239~267m 所见地层：中下部为河流边滩相灰黄色粉细砂—砂质粉土夹粉质黏土层，其下伏层为河床相含砾砂层；上部为泛滥湖泊相灰黄色黏土层，富含钙质网脉。

（6） X 正极性事件

这一正极性事件只在近年完成的如浦东 J25、庙镇 SG1 两孔地层剖面上出现，记录深度介于 271~275m，厚度仅为 0. 5~2m。以厚度略大的 J25 孔地层剖面所见，这一极性事件赋存在泛滥平原相褐黄色粉质黏土层内，底部夹一薄钙板层，下伏河床相含砾砂；极性柱显示，在上、下均为负极性的背景上出现一个正极性事件，磁倾角 23°。SG1 孔测得此极性事件所在处磁化率降低，为 $22\times10^{-8}m^3/kg$。已知这一极性事件的年龄为 2. 37Ma。

3. 高斯正极性时

分布于区内松散覆盖层的深部，其埋深在 260~280m 以下，古地磁极性柱显示基本为正向极性，间夹个别负向极性事件。所属地层为第四系下伏的上新世地层。区内钻孔揭露的高斯正极性时磁性地层所占厚度大都小于等于 80m，揭露深度最大的为安亭 G414 孔。该孔高斯（Gauss）正极性时与吉尔伯特（Gilbert）负极性时界限即 G/G 界限的深度为 379m；出现 G/M 界限的深度为 264m。宝山罗泾 LQ19 孔以及闵行马桥 LQ11 孔以及西部枫泾 LQ1 孔 G/M 界线出现在 250~260m。

凯纳负极性亚时和马莫斯负极性亚时：

在娄塘 P5、安亭 G414、庙镇 SG1 等孔地层剖面上出现，埋深大致在 300~330m，记录厚度为 5~9m 不等。在极性柱完整的 SG1 和 P5 孔中均可见上、下两个负极性事件夹一个正极性事件，如庙镇 SG1 孔极性柱揭示，在孔深 324~328m 之间由下而上出现：下部，厚 4m，磁倾角 −35°，磁化率 $7\times10^{-8}m^3/kg$，马莫斯负极性亚时；中间夹一正极性事件，厚 11m，磁倾角 35°~80°，磁化率 10×10^{-8}~$20\times10^{-8}m^3/kg$；上部，厚 3m，磁倾角 −22°，磁化率 $18\times10^{-8}m^3/kg$，凯纳负极性亚时。已知其原年龄：凯纳负极性亚时为 2. 92~3. 01Ma，马莫斯负极性亚时为 3. 05~3. 15Ma。极性事件出现在上新世地层上部，为湖泊堆积的青灰、褐黄色黏土。

4. 吉尔伯特负极性时

在磁性地层剖面研究深度最大的安亭 G414 孔底部揭露。该孔深 379m 处灰褐色泥质砾石层底部记录到吉尔伯特/高斯两极性时的分界，距今年龄约为 3. 40Ma，孔深 387~390m 灰色泥质中粗砂内记录到科奇蒂（Cochiti）正极性亚时，距今年龄为 3. 92~3. 95Ma，其下至孔深 397. 36m 与下伏基岩分界之间的砂砾层因未能采取古地磁样而致极性不清。

二、地质测年及晚第四纪年代地层划分

依据放射性碳法、热释光或光释光等样品的定年结果（表 21-2）确立晚更新世地层地质年龄，确立了 12×10^4a 以来的年代地层框架。

（一）青浦组/如东组

亚大西洋期，河口滨海相粉土、粉砂与泛滥湖泊相褐黄色黏性土，下限年龄为 3~4ka。

表 21-2　上海晚第四纪地质测年与年代地层

Table 21-2　Geological dating and chronological stratigraphy of the Late Quaternary in Shanghai

岩石地层	地层代号	下限年龄 ka B. P.	定年依据	气候期及岩相
青浦组/如东组	Qh_3q/Qh_3r	4~3	青浦 SG13 孔深约 1. 65m 处蓝灰色黏土光释光年龄为 3. 5±0. 2ka；金山亭林 SG10 孔深约 3. 25m 处褐黄色黏土光释光年龄为 3. 3±0. 2ka	亚大西洋期，河口滨海潮滩相粉土、粉砂与泛滥湖泊相褐黄色黏性土
			崇明奚家港 SG6 孔深约 3. 2m 处深灰色黏质粉土 ^{14}C 年龄为 3950±200a B. P.	
上海组	Qh_2s	8	崇明庙镇 SG1 孔深约 25. 77m 处淤泥质黏土 ^{14}C 年龄为 7740±190a B. P. ；面粉厂 SG2 孔深约 16. 75m 处淤泥质黏土 ^{14}C 年龄为 8400±200a B. P. ；金山亭林 SG10 孔深约 15. 85m 处灰色粉质黏土光释光年龄为 6. 6±0. 2ka，与深约 24. 55m 处灰色粉质黏土光释光年龄为 10. 6±1. 0ka（两样内插）；青浦赵巷 ZNC2 孔深 12. 98m 处泥炭 ^{14}C 年龄为 7980±50a B. P. ；城区北新泾 LQ15 孔深 16m 处灰色淤泥质泥中贝壳 ^{14}C 年龄为 6890±125a B. P.	亚北方期—大西洋期，滨浅海相淤泥质黏性土
娄塘组	Qh_1l	10~11	长江北支 GS-5 孔-52. 87m 处砂层中木块 ^{14}C 年龄为 10200±150a B. P. ；青浦凤溪 SG3 孔深约 13. 19m 处灰色黏土光释光年龄为 11. 18±0. 95ka；亭林 SG10 孔深约 24. 55m 处光释光年龄为 10. 6±1. 0ka，与深约 25. 25m 处暗绿色黏土光释光年龄 13. 2±1. 1ka（两样内插）；崇明奚家港 SG6 孔深约 52m 处灰色粉细砂光释光年龄为 11. 8±0. 71ka；城区北新泾 LQ15 孔深 22m 处灰黑色有机质植物碎屑 ^{14}C 年龄为 8502. 5±7. 5a B. P.	北方期—前北方期，河口滨海潮滩相灰色黏土和粉土、粉砂
南汇组/滆湖组上段	Qp_3n^3/Qp_3g^3	20	嘉定娄塘 P5 孔深 44~46. 5m 处泥炭 ^{14}C 年龄为 12155±115a B. P. 为 12565±115a B. P.	“新仙女木事件”11500~12900a，湖沼相泥炭
			青浦凤溪 SG3 孔深约 15. 96m 处暗绿色粉质黏土光释光年龄为 17. 41±0. 88ka，与孔深约 22. 04m 处褐黄色粉砂光释光年龄 22. 72±2. 43ka 两样内插	末次冰期晚冰阶，泛滥湖泊相暗绿—褐黄色黏性土（古土壤）
南汇组/滆湖组中段	Qp_3n^2/Qp_3g^2	23	青浦凤溪 SG3 孔深约 2204m 处褐黄色粉砂光释光年龄为 22. 72±2. 43ka（下推）；南汇临港新城 SG7 孔深约 25. 2m 处灰黄色黏土光释光年龄为 18. 2±1. 6ka，与深约 40. 8m 处灰黄色砂质粉土光释光年龄为 59. 3±3. 5ka（两样内插）；金山枫泾 LQ1 孔 7. 62m 处褐灰色黏土光释光年龄为 24. 9±1. 0ka	末次冰期间冰阶，河流—河口相黄色粉细砂夹粉土
		35	青浦练塘 A4 孔深 19. 2m 处深灰色粉质黏土 ^{14}C 年龄为 28600±1900a B. P. ；面粉厂 SG2 孔深约 38. 95m 处灰黄色粉细砂光释光年龄为 32. 35±2. 91ka；金山枫泾 LQ1 孔深 15. 65m 处黄灰色泥质粉砂光释光年龄为 41. 0±2. 6ka	末次冰期间冰阶，滨浅海相灰色粉细砂及灰色粉质黏土与粉细砂薄互层
		45	青浦练塘 A4 孔深 22. 5m 处灰绿色黏土热释光年龄为 42. 70±2. 80ka	青浦西部“二硬”灰绿色硬土
		70	青浦凤溪 SG3 孔深约 59. 55m 处深灰色黏质粉土光释光年龄为 50. 05±2. 80ka；亭林 SG10 孔深约 70. 5m 处青灰色粉砂光释光年龄为 69. 0±4. 1ka，与 75. 6m 处草黄色中细砂光释光年龄 75. 2±6. 1ka 两样内插 . ；城区北新泾 LQ15 孔 49. 8m 处深灰色黏土夹粉砂纹层光释光年龄为 58. 9±2. 5ka	末次冰期间冰阶，滨浅海相灰色粉细砂、粉质黏土互层和河口滨海—河流相灰色粉细砂—含砾砂

续表

岩石地层	地层代号	下限年龄 ka B. P.	定年依据	气候期及岩相
南汇组/淀湖组下段	Qp_3n^1/Qp_3g^1	75	青浦凤溪 SG3 孔深约 62.05m 处灰色粉细砂光释光年龄为 72.60±7.10ka（上推）；金山亭林 SG10 孔深约 70.5m 处与 75.6m 处年龄为两样内插值	末次冰期早冰阶，泛滥湖泊相深灰、灰绿色粉质黏土、粉土
川沙组	Qp_3c	128	青浦凤溪 SG3 孔深 99.65m 处灰色含砾中细砂光释光年龄为 113.55±11.26ka；下伏孔深约 100.05m 处灰蓝色粉质黏土光释光年龄为 129.95±4.88ka	末次间冰期寒冷阶段，河流相灰—灰黄色细砂、含砾砂
			青浦 LQ3 孔深 51.09m 处灰色粉砂质泥夹细砂光释光年龄为 128.3±5.8ka	末次间冰期温暖阶段，河口滨浅海相泥砂互层

（二）上海组

亚北方期—大西洋期，滨浅海相淤泥质黏性土，下限年龄约为 8ka。

（三）娄塘组

北方期—前北方期河口滨海相灰色黏土和粉土、粉砂，下限年龄为 10～11ka。

（四）南汇组/淀湖组上段

末次冰期晚冰阶，泛滥湖泊相暗绿—褐黄色黏性土，下限年龄约为 20ka B. P.。其中，“新仙女木事件”为 11.5～12.9ka。

（五）南汇组/淀湖组中段

末次冰期间冰阶，河流—河口相黄色粉细砂夹粉土，下限年龄为 23ka。

末次冰期间冰阶，滨浅海相灰色粉细砂及灰色粉质黏土与粉细砂薄互层，下限年龄为 35ka。

青浦西部“二硬”灰绿色硬土，下限年龄为 45ka。

末次冰期间冰阶，滨浅海相灰色粉细砂、粉质黏土互层和河口滨海—河流相灰色粉细砂—含砾砂，下限年龄约为 70ka。

（六）南汇组/淀湖组下段

末次冰期早冰阶，泛滥湖泊相深灰、灰绿色粉质黏土、粉土，下限年龄约为 75ka。

（七）川沙组

末次间冰期寒冷阶段河流—河口相灰色细砂、含砾砂和温暖阶段滨浅海相泥，下限年龄约为 128ka。

第三节　岩石地层划分与区域对比

区内第四系岩石地层共划分为 9 个组。其中，下更新统为安亭组，中更新统为嘉定组，上更新统下部为川沙组，上更新统上部因区域空间差异划分有淀湖组和南汇组，其中前者分布在西部湖积平原沉积区，后者分布在长江三角洲平原沉积区。下全新统为娄塘组，中全新统为上海组，上全新统因区域空间差异而划分出青浦组、如东组，前者分布在西部湖积平原沉积区，后者分布在长江三角洲平原沉积区。

一、安亭组

（一）层型剖面

嘉定安亭 G414 孔剖面，为原建组地层剖面所在。在孔深 153.44～269.00m 间揭露，与上、下地层以假整合接触。3 段地层发育齐全，总厚 115.56m（图 21-4）。

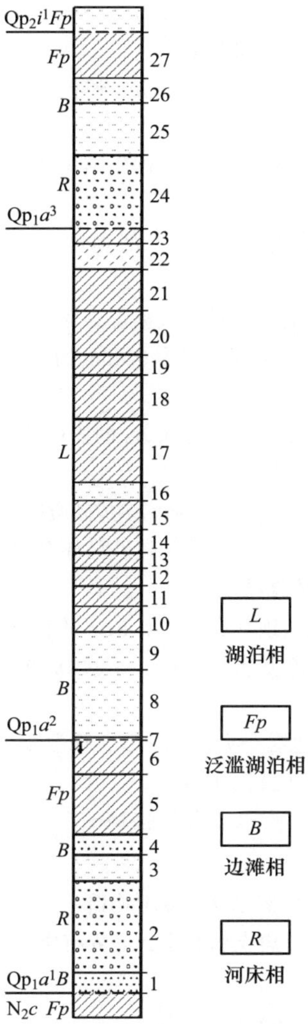

图 21-4　G414 孔安亭组地层剖面

Fig. 21-4 Stratigraphic section of the Anting Formation per borehole G414

上覆地层：中更新统嘉定组灰色细砂、含砾中细砂

－－－－－假　整　合－－－－－

安亭组（Qp_1a）　总厚 115.56m

上段（Qp_1a^3）　厚 23.66m

27. 以褐黄色为主，杂有灰绿、灰白色网纹状黏土，底部有厚 0.35m 的钙板层—钙质粉砂岩。下部个别样品见很少的有孔虫　5.56m

26. 灰色细砂，松散，顶部 0.9m、底部 0.3m 含较多泥质，底部并有深灰色腐殖物碎屑纹层聚集，呈不规则水平层理　3.00m

25. 灰色细砂，松散，顶部厚 0.7m 为砂质粉土夹粉质黏土，具水平层理；　6.24m
中部有厚 0.6m 灰色砂质粉土夹富腐殖物碎屑纹层，具斜波状层理；
下部为中细砂，底部见石英细砾

24. 灰色含砾中粗砂，含砾 10%～15%，次圆—浑圆状，砾径一般为 2～5mm，大者 2～2.5cm，大都为石英砾　8.86m

－－－－－侵蚀接触－－－－－

中段（Qp_1a^2）　厚 61.60m

23. 褐黄杂浅蓝灰、灰白色网纹状、斑杂状粉质黏土，上部夹一厚为 0.6m 的砾石层　1.83m

22. 褐黄杂灰白色砂质粉土　3.07m

21. 咖啡色黏土，含大小、多少不一的钙质结核，一般为 2~3cm，大者 7~9cm，底部 0.4m 为褐色杂灰绿色斑杂状黏土，有钙质物沿裂隙充填 5.00m

20. 锈黄色黏土，穿插大量灰白色斑纹，含大量钙质结核，粒径多在 2~3cm，含少量铁锰质结核，粒径为 2~3mm 5.18m

19. 深褐黄色黏土，间或有蓝灰色杂斑，上部含姜黄色钙质结核和少量铁锰质结核，中部夹厚为 5cm 的钙质砂岩 2.52m

18. 灰白色黏土，有锈黄色斑纹，含少量钙质结核，上部垂直裂隙发育，为灰白色钙质条带充填 5.31m

17. 以褐黄色为主，杂浅蓝灰、灰白色斑杂状、网纹状黏土，含多少不一的钙质结核，粒径在 2~5cm 不等，间或含较多铁锰质结核，中部有厚为 0.45m 的灰白色黏土 7.59m

16. 灰黄色细砂，松散 2.20m

15. 褐黄色黏土，有蓝灰色杂斑，坚硬，底部有厚为 0.9m 的浅褐黄色钙质砂岩 3.44m

14. 以灰绿色为主，杂褐黄色黏土，具硬塑性 2.86m

13. 以深褐黄色为主，杂灰白色网纹状黏土，向下含砂量增多成砂质粉土，中下部含较多钙质结核 1.88m

12. 以灰白色为主，杂褐黄色斑纹状黏土，含有钙质结核，局部夹有褐黄色砂质粉土 2.07m

11. 褐黄色黏土，上部有厚为 0.4m 的钙质结核层 2.44m

10. 以褐黄色为主，杂灰白色网纹状黏土，顶部有厚为 0.51m 的灰黄色砂质粉土 3.11m

9. 灰色细砂，松散、上部 1.9m 为青灰色泥质粉细砂，下接 0.24m 细砂、粉质黏土薄互层，局部见深灰色腐殖物碎屑纹层 4.60m

8. 灰色细砂，松散，顶部厚 0.8m 为灰色黏土夹薄层粉砂，现微波状层理 8.08m

7. 灰色砂质粉土夹粉质黏土，含朽木碎块及钙质结核 0.42m

----- 侵蚀接触 -----

下段（Qp_1a^1） 厚 30.30m

6. 蓝灰色粉质黏土，杂有灰白色黏土，局部见揉皱变形构造 4.00m

5. 以褐黄色为主，杂灰白、蓝灰色网纹、斑杂状黏土，底界有腐木碎块 7.30m

4. 灰色中细砂，松散 2.38m

3. 浅绿灰色泥质细砂，稍密。底部厚 0.37m 为灰色黏土 3.19m

2. 浅灰黄色含砾粗中砂，松散，上部含砾略多，次浑圆状，砾径一般为 3~5mm，大者 1~3cm，下部含砾少见 11.05m

1. 锈黄色中细砂，含泥质，顶部厚 0.2m 为蓝灰色黏土 2.38m

----- 假 整 合 -----

下伏地层：上新统崇明组褐杂绿灰色黏土夹粉砂纹层

上列地层剖面揭示，整个安亭组由含砾砂层与黏土层交替构成，存在 3 个沉积旋回，相应构成 3“段”地层，其间为侵蚀接触。微古分析唯在上段上部黏土层内发现很少的有孔虫，其他层段均未见及。表明每“段”地层为曲流河道、沙坝沉积的含砾砂和中细砂层打底，至泛滥湖泊堆积的黏土层结束的河流二元结构。

据孢粉分析，下段含孢粉较少，主要有 *Cupressaceae*、*Pinus*、*Gramineae* 和苔藓类（*Bryophyta*），反映进入第四纪初期，古气候骤变为寒冷干燥。中段下部（第 7~9 层）含孢粉较丰，其中阔叶植物占 58%~67%，具 *Quercus*（*deciduous*）-*Carpinus*-*Q. glauca*-*Gramineae* 组合，还有些 *Abies*、*Picea*；中部（第 10~18 层）含孢粉较少，组合面貌与顶部相同；上部（第 19~21 层）含孢粉较丰，其中木本植物花粉占 45%，草本植物花粉占 40%，蕨类较少，具 *Quercus*（*deciduous*）-*Ulmus*-*Gramineae* 组合；顶部（第 22~23 层）含孢粉较少，以草本植物 *Gramineae*、*Ranunculaceae* 为主，还有少量的木本植物 *Quercus*（*deciduous*）、*Castanea*、*Ulmus* 及 *Taxodiaceae*、*Cupressaceae* 花粉。反映从中段下部至顶部地层沉积期间，古气候经历了温暖湿润—温凉干燥—温暖稍干—温凉干燥的变化。上段中下部（第 24~26 层）含孢粉较丰，几乎全为木本植物组成，其中阔叶植物花粉即占 51%，具 *Pinus*-*Quercus*（*de-*

ciduous）-*Q. glauca*-豆科（Legumjnosae）组合；上部（第 27 层）孢粉贫乏，主要有 *Cupressaceae*、*Pinus*、*Artemisia*、阴地蕨属（*Botrychium*）。反映上段沉积期间，古气候经历了温暖湿润—寒冷干燥的变化。

（二）对比剖面

1. 金山朱泾 G418 孔剖面

该孔可作为西南部枫泾凹陷内的代表剖面。在孔深 153.3～265.2m 揭露，总厚度为 111.9m，与上、下地层均为假整合接触。3 段地层发育较全（图 21-5）。所见地层如下：

上覆地层：中更新统嘉定组褐黄色砂质粉土

----- 假　整　合 -----

安亭组（Qp_1a）	总厚 111.90m
上段（Qp_1a^3）	厚 34.70m
27. 灰绿色粉质黏土，坚硬。顶部厚 0.2m 呈铁锈色，含有铁质结核和铁锰质胶膜，上部夹有厚 0.7m 的褐黄色砂质粉土，下接厚为 1.9m 的灰褐色粉质黏土，再下为厚 2m 的灰绿色粉质黏土和灰白色粉砂薄互层。本层具水平层理、微波状层理，偶见斜波状层理	5.70m
26. 灰褐色粉质黏土夹粉砂。上部为两者互层，砂中可见炭质碎屑细纹层，下部变为泥质粉土。具不规则水平层理、波状层理	4.90m
25. 灰色细砂，松散	3.10m
24. 灰色中细砂，松散	3.30m
23. 灰色含砾中细砂，松散，含钙质砂岩砾，占 5%～10%，次棱角—次圆状	2.70m
22. 青灰色细砂	0.90m
21. 黄灰色细中砂，含有 2%～3%的石英质细砾（砾径为 2～6mm）	0.70m
20. 灰色细砂，松散	1.20m
19. 灰色含砾粗中砂，松散。砾石占 10%～15%，砾径为 2～6mm，次棱角状，成分主要为石英、燧石	6.00m
18. 灰色中细砂。中部有厚 0.3m 含砾（砾径为 0.5～2cm）15%～20%，并见炭化木碎块	5.80m
17. 粗砂砾石层	0.20m
----- 侵蚀接触 -----	
中段（Qp_1a^2）	厚 58.00m
16. 黄绿灰色中细砂，稍密，夹有灰绿色黏土。具波状层理	4.80m
15. 褐灰色中细砂，松散。上部厚 1.5m 含砾 5%左右，顶部含砾 25%左右，砾径为 0.2～2cm，成分较杂	4.20m
14. 绿灰色粉质黏土，夹有灰白色粉细砂，含泥砾	0.90m
13. 褐灰色砂质粉土，夹有暗绿灰色粉质黏土	3.90m
12. 黄绿色细砂，含泥质，稍密。上部 1.20m 为含细砾中细砂，具不明显板状斜层理	3.20m
11. 褐灰色中细砂，松散。上部厚 1m 含细砾 3%～5%，成分以石英、燧石、凝灰岩较多。下部 2.1m 为含砾中细砂，含砾约 20%，次圆、次棱角状，砾径一般为 0.5～1cm，大者达 3cm	5.70m
10. 灰褐色砾质粗中砂，松散。砾呈次圆、次棱角状，砾径大者达 5cm，成分以石英、燧石居多，含量为 20%～25%	1.50m
9. 黄绿色中细砂，松散。中部有厚 0.7m 含砾中细砂，砾径为 0.2～2cm，含量为 15%～20%；其下接厚 0.3m 的黄绿色黏土夹砂质粉土，具水平层理	5.80m
8. 褐黄色粉质黏土	2.80m
7. 灰白、灰色含砾泥质粉细砂，坚硬，均匀层理。上部夹 1.2m 钙板层，中、下部夹有厚分别为 0.9m 与 0.8m 的泥质砂砾层	20.20m
6. 以褐黄色为主，杂灰白色泥质砂砾层，砾呈次棱角、棱角状，占 50%～60%，砾径一般为 2～3cm，大者 7～8cm，成分有石英、燧石、砂岩等，为泥质胶结	5.00m

----- 侵蚀接触 -----

下段（Qp_1a^1） 厚19.20m

5.	浅黄绿色粉质黏土，杂有少许褐黄色斑纹	1.50m
4.	以褐黄色为主，杂灰黄色斑杂状粉质黏土，含较多黑色铁锰质结核，有灰白色钙质网脉穿插	4.40m
3.	褐黄色粉质黏土，有大量钙质网脉穿插，含多量铁锰质结核，见有铁锰质假化石出现	3.10m
2.	杂色砂质粉土，下部0.6m砂质含量增加	1.20m
1.	砂砾石，松散，砾呈次棱角状，砾径为0.3~10cm，成分有石英、燧石、凝灰岩、泥砂岩等。上部厚1.1m呈锈黄色，其下2.9m为浅黄灰色泥质砂砾石，再下3m为浅黄灰色中细砂砾石，其底含砾可达80%	9.00m

----- 假　整　合 -----

下伏地层：上新统崇明组浅黄灰杂锈黄色斑杂状粉质黏土

上列地层剖面显示，与安亭孔相似，朱泾孔安亭组的3个“段”，也大体存在3个下粗、上细的沉积旋回，中段厚度也可占总厚度的1/2强。但与安亭孔比较：朱泾孔的砂层累厚，尤其是中、上段砂层厚度所占比例较大，黏土层厚度明显较小；二是砂层含砾远比安亭孔为多，砾石成分复杂，磨圆度也差；三是中段由3个小旋回组成，不同于安亭孔中段系单一旋回；四是下段和中段下部砂砾层泥质含量较高，形成河流冲洪积相泥质砂砾石与河间心滩相粉质黏土层交替堆积；中段上部和上段才出现河床相含砾砂与沙坝相中细砂交替，并过渡为边滩—泛滥平原堆积的砂质粉土或粉质黏土夹粉砂—粉质黏土层。总之，朱泾孔安亭组的沉积系长期处于距物源相对较近、摆动多变的河道环境，下段和中段下部仍属辫状河流，其上才演化为曲流河道环境。

据微古分析，唯在上段上部26~27层内含有淡水介形虫土星介（*Ilyocypris*）-玻璃玻璃介（*Condona*（*Condona*））-丽星介（*Cypria*）组合，以下没有见及，说明该孔所见安亭组为陆相沉积。

据孢粉分析，下段（1~5层）含孢粉贫乏，仅见苔藓类（*Bryophyta*）和草本植物*Gramineae*花粉，反映古气候寒冷干燥。中段中下部（6~14层）木本植物和草本植物花粉含量大致相等，以落叶阔叶树种花粉居多，具*Quercus*（*deciduous*）-*Pinus*-*Gramineae*-*Ceratopteris*组合，反映古气候温暖湿润；上部（15~16层）以草本植物花粉占绝对优势，含量达77.8%，具麻栎（*Q. acutissima*）-*Betula*-*Gramineae*-*Artemisia*组合，反映古气候温凉干燥。上段中下部（17~26层）木本花粉占绝对优势，含量达62.9%~74.4%，以落叶阔叶树种花粉居多，也见常绿阔叶成分；草本植物以中生草本*Gramineae*为主，还有一些水生草本植物花粉，具*Q. acutissima*-*Ulmus*-*Pinus*-*Gramineae*组合，反映古气候温暖湿润；上部（27层）孢粉贫乏，主要见大量的苔藓类（*Bryophyta*）孢子，以及少量的*Salix*、*Celtis*、*Tsuga*和*Gramineae*花粉，具*Salix*-*Gramineae*-*Bryophyta*组合，反映古气候温凉干燥。

2. 浦东机场J25孔剖面

揭露的下更新统“安亭组”，可作为南东部三墩凹陷内的代表剖面。该孔在孔深153.70~289.33m段钻遇下更新统“安亭组”，总厚度为135.63m。如同上述两孔一样，所见“安亭组”也出现3个沉积旋回，构成下、中、上3个“段”，相应厚度为49.43m、46.10m、40.10m（见图21-5）。其地层结构与处于同一凹陷带西南部枫泾凹陷内的朱泾孔相似：砂层（包括少部分粉土）累厚所占比例达4/5，黏土层所占厚度明显较少；砂层大都含砾，砾石成分也较复杂；但砾径除下段和中段底部外，总体比朱泾孔减小，磨圆度趋好；砂层一般松散，含泥质也少。这些特征表明，早更新世，机场孔所在地区也长期为古河道所占据，堆积的含砾砂经历了较远距离的搬运，因此其成熟度比朱泾孔提高；再者与朱泾孔“安亭组”有所差异的是，机场孔唯下段为由6个含砾砂与粉质黏土的韵律构成的旋回，砂中时见泥砾，粉质黏土内可见泥裂；中段基本为以含砾中粗砂为主与含砾粉细砂的互层；上段由下而上出现含砾砂—粉细砂—砂质粉土—粉质黏土序列。反映下段仍属河道分、合多变的辫状河流沉积；中段变为河道与砂坝频繁重现的低弯度曲流河道沉积；至上段已演化为较稳定的高弯度曲流河道沉积。

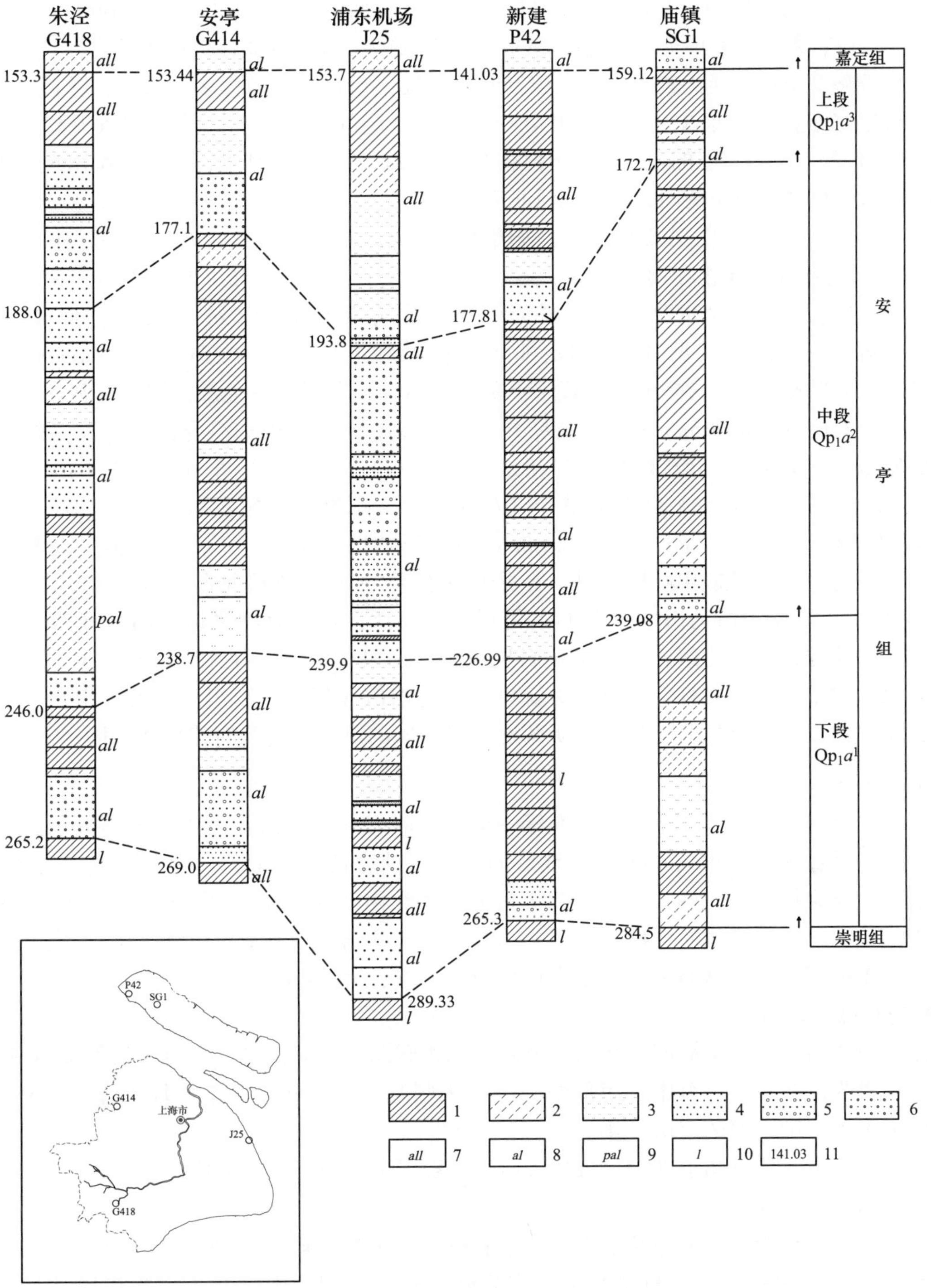

图 21-5　下更新统安亭组地层对比图

Fig. 21-5　Stratigraphic correlation of the Lower Pleistocene Anting Formation

1—黏土、粉质黏土；2—砂质粉土、黏质粉土；3—粉砂、细砂；4—中砂、粗砂；5—含砾砂；6—砂砾石；7—泛滥平原—泛滥湖泊相；8—河流（河床—边滩）相；9—冲洪积相；10—湖泊相；11—深度（m）

机场孔“安亭组”唯在上段上部的粉质黏土层内发现个别的淡水介形虫 *Ilyocypris* sp.、*Condona* sp.，以下岩层内没有微古化石检出，表明属于陆相沉积。

3. 崇明庙镇 SG1 孔剖面

揭露的下更新统“安亭组”，可作为安亭—崇明凹陷东北部的代表剖面。该孔在孔深 159.12～284.50m 段钻遇下更新统“安亭组”，总厚度为 125.38m。由下而上 3 个“段”的厚度分别为 45.42m、

66.38m、13.58m（见图21-5）。其地层结构与处于同一凹陷西北部的安亭G414孔相近：中段厚度同样占全组厚度的一半强；黏土层累厚近于全组厚度的2/3，砂质粉土和砂层累厚仅占1/3强，与安亭孔砂和粉土厚度所占比例略少于黏土层比较，本孔沉积物明显趋细；砾石成分同样多为石英、燧石之类，但砂层含砾大减，砾径趋小，泥质也少，反映本孔处于下游，砂层的分选性和成熟度有所提高。从地层剖面上（见图21-5）看，每段打底的含砾砂以下段较厚，中段减薄，上段则无，往上依次出现细砂—粉土—黏土序列，下段还出现两个小旋回。反映同属曲流河道沉积，晚期河流下切侵蚀作用已显著减弱。

庙镇孔“安亭组”也未发现微古化石，表明属于陆相沉积。但在其西南邻近的新建P42孔“安亭组”上、下两段内发现含陆相介形虫化石：其一在孔深251.94~255.55m段、“安亭组”下段中偏下部的褐色黏土层内发现含少量化石，主要为柯氏土星介相似种（*I.* cf. *cornae*），其次为奇异白花介（*Leucocythere mirabilis*），并见白花介未定种（*L.* sp.）和浪游土星介（*Ilyocypris errabundis*）；其二在孔深141.03~167.05m段、“安亭组”上段中上部的褐黄杂蓝灰或黄绿色黏土层内发现含丰富化石，为苏星小玻璃介（*Candoniellasuzini*）-浪游土星介（*Ilyocypris errabundis*）组合，并有较多的舒滨小玻璃介（*C. schubinae*）、背瘤白花介相似种（*L.* cf. *dorsotuberosa*）、开封土星介（*I. kaifengensis*），还有少量的洁净小玻璃介（*C. albicans*）、柯氏土星介（*I. cornae*）、丽星介未定种（*Cypria* sp.）、近球形金星介（*Cypris subglobosa*）、弯玻璃介（*Candonasinuosa*）、玛纳斯土星介（*I. manasensis*）及轮藻化石。化石组合表明“安亭组”为陆相河湖沉积。所见介形虫化石组合与河北阳原盆地井儿洼村钻孔孔深107.8m以下的下更新统泥河湾组所含介形虫化石可以对比；古地磁测量证明两地均落在松山极性时内，只是阳原井儿洼村泥河湾组含介形虫化石层位跨松山极性时的加拉米洛极性亚时下限（1.07Ma）至留尼昂亚极性时下限（2.15Ma），而崇明新建孔发现相似介形虫化石组合的安亭组上—下段则跨松山极性时顶限至留尼昂亚极性时下限。证明两地下更新统的沉积环境和时代大体相同。

（三）总体特征

下更新统安亭组在全区广泛分布，其埋深一般在150~290m之间，由西南太湖沉积区向东北的河口沙岛区埋深逐渐加大。厚度约占第四纪地层的一半。其岩性为以褐黄、褐色为主，有灰绿、黄绿、蓝灰色网纹、斑纹穿插的杂色网纹状、斑杂状黏土，大致由下而上呈坚硬—硬塑—可塑状变化，以普遍含有钙质结核、网脉和铁锰质结核为特征，与浅灰、灰黄等色含砾砂和砂砾石间互成层。一般存在粗、细相间3个沉积旋回，从而可将之相应划分为下、中、上3“段”。

1. 安亭组下段（Qp_1a^1）

安亭组下段主要分布在上海南、北两个第四纪沉积凹陷区，即安亭—崇明凹陷和金山—三墩凹陷，以前一凹陷内最为发育；在上海中、南部两个凸起区则基本无沉积。其层底埋深介于259~296m，厚度在23~60m不等，随基底的起伏变化。

这套地层大体上形成下粗上细一个沉积旋回，一般可分为上、下两亚段。

（1）下亚段

下亚段主要为冲洪积—河流边滩堆积的灰、灰白、灰黄色含砾砂、砂砾和砾石层，夹有泛滥湖泊堆积的褐黄、灰黄及杂色黏土、粉质黏土，或为砂、黏土互层。中上部夹有多层厚达逾1m的钙板层（或钙质砂岩），下部多砂砾层。下亚段厚度在1~59m不等，大都为10~30m，以在崇明岛和浦东滨岸地带最为发育。在南部的枫泾凹陷区分布的含砾砂层一般分选较差，风化岩屑、长石和泥质含量较高，含砾较多，大小不一，磨圆较差，多次棱角状，成分也较复杂，有石英、燧石、凝灰岩、砂岩、钙积石等。如金山朱泾G418孔所见即是。由于砂砾层的结构和成分成熟度较差，因此这一地区分布的该含水层富水性稍差。而在河口沙岛崇明和浦东滨岸地带分布的下段地层，其堆积厚度较大，含砾砂层结构松散，含泥质很少，砾石大都为次圆状，成分多为石英、燧石之类，说明沉积物经历了较远距离的搬运，砂的分选性和成熟度较高。

（2）上亚段

上亚段以褐黄、灰黄和褐色为基色，杂黄绿、灰绿、蓝灰、紫灰等色的网纹状、斑杂状黏土、粉质黏土、黏质粉土为主，呈坚硬—硬塑状，多钙质网脉和结核，局部形成钙积层，顶部多为黄绿、蓝灰色黏土，夹姜黄、灰色粉细砂、砂质粉土，局部夹有含砾中粗砂、中细砂。黏土层内常见一般具压性擦痕的节理。砂中含岩屑较多，常为泥质弱胶结。上亚段厚度为10~48m。

上亚段的岩性、岩相在区域分布上有些变化。在凹陷边部和凸起部位即上海的大部分地区通常为泛滥湖泊沉积的黏土、粉质黏土夹有一些粉土、粉细砂；在凹陷中心地带多半出现黏土、粉质黏土与砂砾、含砾砂或细砂的互层，砂砾层中时而还见有泥砾，反映为辫状河流沉积；局部出现洪积相的含砾黏土和泥质砂砾石互层，如南部凹陷的浦东机场J25等孔，本亚段厚度多为20~40m。

据古地磁测量，下更新统安亭组中、下段界线略居于松山负极性时奥尔都维正极性亚时下限之下，界线年龄约为2.00Ma。由此确定安亭组下段的年龄为2.60~2.00Ma。

2. 安亭组中段（Qp_1a^2）

安亭组中段通常也出现下粗上细的变化，存在又一个沉积旋回。其埋深一般在190~240m之间，厚度介于26~66m，大都可达45~60m，在下更新统内本段厚度常可占一半，尤以北部凹陷为著。本段也可分为上、下两个亚段。

（1）下亚段

下亚段岩性、岩相变化较大，因所在构造部位及古地貌的差别而异。在嘉定—市区—浦东坦直弧形凸起以东的宝山滨岸带—浦东张桥、川沙、机场—临港滨岸地带，出现早期浏（河）—南（汇）古河道流经场所，为当时区内的主干河流所在，堆积了一套浅灰、灰黄、草黄色砂砾、含砾中粗砂和中细砂，大都结构松散，厚度达到20~50m，如浦东国际机场J25、外高桥J20等孔所见。在北部的安亭—崇明凹陷大部分地区和中、南部凸起区内基本为褐黄、杂色黏土、粉质黏土夹粉细砂的泛滥湖泊沉积，含水砂层不发育。其中部分地带如安亭—宝山城、崇明新河—前哨农场一带以及南部凹陷的闵行北桥—新场—卢潮港一带，为支流河道和边滩沉积的粉细砂夹粉质黏土或砂质粉土；还有在西南部枫泾凹陷内朱泾等地、崇明凹陷西北隅红星农场等地堆积的砾砂、中细砂，夹泥质砾石、泥质含砾粉细砂和粉质黏土，为摆动不定的辫状河流沉积，含水层发育较差。这些含水砂层不发育或发育较差的地区，下亚段厚度略小，一般为10~30m。

（2）上亚段

上亚段岩性、岩相比较单一。基本由泛滥湖泊沉积的黏土、粉质黏土组成，以褐黄、褐色为主，杂有灰绿、蓝灰等色，顶部常为单一的黄绿、灰绿、蓝灰色，中、下部夹杂紫褐、赭色，具网纹、斑杂状构造，普遍含较多的钙质结核和网脉，中下部夹有1~3层钙板层，并含少量的铁锰质结核。局部夹有灰、褐黄色粉细砂、中细砂透镜体。其厚度在5~45m不等，与下段成反消长关系。本段黏土通常具硬塑状态，其中也时而见到节理，大都为具擦痕的压性或压扭性节理。

安亭组中段地层内动物化石贫乏，唯在浦东东侧滨岸地带曾发现少量基本为广盐性的有孔虫及海相腹足类。如浦东海塘P6孔深206~226m段灰色细砂夹含砾细中砂内见及少量有孔虫，有轮虫类的*Ammonia tepida*、*A. koeboeensis*，浮游类的*Globigerina* spp.、*Epistominella naraensis*，平旋壳类的*Nonion* sp.、*Cribrononion* sp.、*Elphidium* sp.和*Florilus* sp.等，并见窄盐性成分*Bolivina* sp.等，伴有轮藻。浦东高桥“长江一号”孔，在大致相当于下亚段、孔深260m处黄灰色中细砂内发现多种海相腹足类，有雕刻友螺（*Nassarius*（*Phromtis*）*cealatula*）、囊螺（*Retusa*）、露齿螺（*Odostomia*）等及陆相腹足类沼螺（*Parafossarulus* sp.）；20世纪60年代初《上海市油气普查报告》记载，曾在川沙邓镇孔深261.3m处见及少量轮虫类有孔虫和海相介形虫。后两个钻孔所见海相层位大致相当于安亭组中段下亚段或略下，属于三角洲河流沉积，唯海塘孔所见可达河口相。

在上亚段黏土内偶见淡水介形虫布氏土星介（*I. bradyi*）和轮藻，及淡水蚌和腹足类介壳碎片，属于陆相河湖堆积。

据古地磁测量，下更新统安亭组中、上段的界线介于松山反极性时贾拉米洛正极性亚时与前贾拉米洛正极性事件之间，界线年龄约为1.20Ma。由此确定安亭组中段的年龄为2.00~1.20Ma。

3. 安亭组上段（Qp_1a^3）

安亭组上段岩性通常也具下粗上细的变化，因此同样可分为上、下两段。其埋深一般在150～190m之间，厚在24～62m不等。

（1）下亚段

下亚段岩性、岩相变化较大。在大部分地区，下部为灰、灰黄、褐黄色含砾中粗砂、砾质中粗砂夹细砂，底部常为砂砾石或砾石层；中上部为灰、褐黄等色细砂夹褐、褐黄色粉质黏土、粉土及杂色网纹状粉质黏土，富含钙质结核，顶部常见厚0.1～1.5m不等的钙板层（钙质砂岩）。该段河流冲积相含砾砂层即第Ⅳ上含水层，在上海陆域分布广泛。主要分布于浏—南古河道流经的嘉定华亭—宝山城—市区—浦东国际机场—东海农场宽达10km的曲形地带内，堆积的含砾砂层较粗，其底界深度及厚度也较大。其次在南部凹陷内的枫泾—朱泾—闵行—鲁汇—奉城—芦潮港一带，惯称为枫（泾）—奉（城）古河道流经地带，分布以褐灰、灰绿色中细砂为主，下部有褐灰、灰黄色含砾中粗砂。在这两条古河道之间的市区与闵行、松江接壤区为河间带滩地堆积的粉细砂夹粉质黏土；在崇明岛、长兴岛等现河口地区则广泛分布边滩—泛滥湖泊相黏性土夹粉细砂或两者的互层，这些地区含水层发育明显较差。还有在中、南部两个凸起地区分布泛滥平原—湖泊沉积的黏性土夹粉细砂层，含水层就不发育了。下段厚度为12～56m不等。

（2）上亚段

上亚段岩性、岩相比较单一，普遍为褐黄、褐色黏土、粉质黏土和褐黄杂蓝灰或黄绿色的杂色黏土、粉质黏土，具网纹状、斑杂状构造。上部常为蓝灰、灰绿色黏土、粉质黏土，中下部夹有姜黄、褐黄色粉细砂和砂质粉土。本段普遍含较多的钙质结核或网脉，并且中上部含少量铁锰质结核，中下部常出现个体较大的淡水螺壳及丽蚌碎片。表明一般为泛滥湖泊相堆积。本段黏土具硬塑—可塑状，其中可见具压性、偶见旋卷擦痕的节理。本段厚度为16～44m，其厚度变化主要与下段成反消长关系，并因上覆地层的起伏而有所增减。

在下更新统上段下亚段内，大部分地区产较丰富的淡水软体动物化石，有纹沼螺（*Parafossarulus striatulus*）、长沼螺（*P. longiconis*）、田螺（*Viviparus* sp.）、河兰蚬（*Corbicula fluminea*）、短沟蜷（*Semisulcospira* sp.）、盘螺（*Valvata* sp.）等，表明大部分地区下亚段为河流冲积的。只是在浦东的滨岸地带有数个钻孔，如横沔P12孔（孔深172～178m灰黑色细砂夹砂质粉土）、浦东海塘P6孔（孔深171～191m灰、灰黄色细砂）发现少量有孔虫和海相介形虫等。有孔虫有具毕克卷转虫（*Ammonia beccarii* var.）-奈良小上口虫（*Epistominella naraensis*）组合；或以*A. tepida*等轮虫类为主，其次为*Elphidium* sp.、*Florilus* sp.、*Globigerina* spp.，并见有深水种太平洋霍氏虫（*Hopikinsina pacifica* Cushman）、瓶虫（*Lagena* sp.）、判草虫（*Brizalina* sp.）等窄盐性成分，伴生半咸水的东台新单角介（*Neomonoceratina dongtaiensis*）、淡水的洁净小玻璃介（*Condoniella albicans*）以及水蕨孢子、轮藻受精卵膜等陆相生物成分。还有在海潮P17孔（孔深175～177m灰色细砂）内分析出少量钙质超微化石大洋桥石（*Gephyrocapsa ocenica*），为上海钙质超微化石出现的最低层位。这些钻孔所见化石表明，当时浦东滨岸地带一度处于三角洲河流趋近河口的环境，由此将之定为区内的第Ⅵ海进层。

在相当于上覆的安亭组上段上亚段地层内，以常见个体大而完整的淡水腹足类化石田螺、角状环棱螺（*Bellamya angularis* Muller）和丽蚌（*Lamprotula* sp.）为特征，并产较丰富的陆相介形虫，主要有浪游土星介、布氏土星介（*I. bradyi*）、洁净小玻璃介（*Candoniella albicans*）、舒滨小玻璃介（*C. schbina*）、苏星小玻璃介（*C. suzini*）、湖花介未定种（*Limnocythere* sp.），其次有玻璃介（*Candona* sp.）、丽星介未定种（*Cypria* sp.）、金星介未定种（*Cypris* sp.）、背瘤白花介（*Leucocythere* cf. *dorsotuberosa*），尚见较少的近球形金星介（*C. subglobosa* Sowerby）、开封土星介（*I. kaifengensis*）、苏北丽星介（*Cypriasubensis*）、斗星介未定种（*Cypridopsis* sp.）、线星介（*Lineocypris* sp.）和锥星介（*Subulacybris* sp.）等，均属陆相属种。再者，20世纪五六十年代之交的上海油气普查期间，曾在浦东唐家宅孔深178.26～183.25m发现海相软体动物化石，在瓦屑孔深160～

163m 分析出海相介形虫。还有在 20 世纪七八十年代于青浦练塘 P31 孔（孔深 137~173.8m 杂色、浅褐色黏土和灰黄色粉细砂内）断续检出个别的有孔虫；在上海面粉厂 J10 孔（孔深 148.10~159.29m 杂色黏土夹黄褐色细砂内）发现少量的有孔虫有螺旋壳类 *Ammonia* sp.、平旋壳类的 *Elphidium* sp. 和 *Florilus decorus* 等广盐性有孔虫；浦东海塘 P6 孔（孔深 146~164m）、横沔 P12 孔（孔深 149~174m），还有安亭 G414 孔（孔深 158.2m 处）的杂色、褐黄色黏土内断续分析出很少的有孔虫，三灶 P45 孔（孔深 151~153m）褐色粉质黏土内偶见钙质超微化石 *Gephyrocapsa oceanica* 等。这表明，上亚段一般属于淡水河湖相，唯主要在南部凹陷，曾偶尔处于略受海水影响的微咸水泻湖环境。

据古地磁测量，中、下更新统的界线恰在布容正极性时与松山反极性时的分界上，说明下更新统顶界的年龄为 0.78Ma。由此确定安亭组上段的年龄为 1.20~0.78Ma。

二、嘉定组

（一）层型剖面

嘉定娄塘 P5 孔剖面，为原建组剖面所在。该孔在孔深 95~149m 钻遇“嘉定组”，上、下两段地层发育齐全，总厚度为 54m。与上、下地层均为假整合接触（图 21-6）。现列述如下：

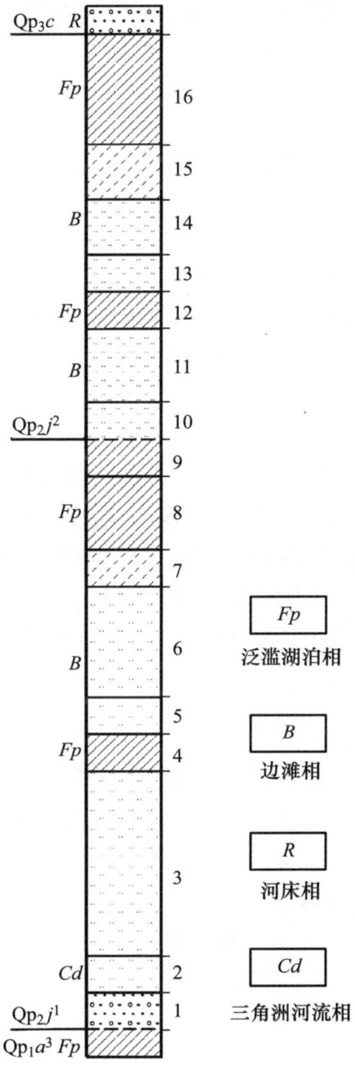

图 21-6　P5 孔嘉定组地层剖面

Fig. 21-6　Stratigraphic section of the Jiading Formation per borehole P5

上覆地层：上更新统川沙组浅灰黄色含砾粗砂

----- 假　整　合 -----

嘉定组（Qp_2j）	总厚 54.0m
上段（Qp_2j^2）	厚 22.0m
16. 蓝灰色黏土，硬塑状。下部厚 2m 为蓝灰杂褐黄色斑纹状黏土，含粒径 1~5mm 的钙结石 5%~10%；中部 2m 约含有粒径为 2~10mm 的石英砾 25%	6.0m
15. 棕灰色砂质粉土，可塑状。向下渐粗。隐约显微状层理	3.0m
14. 灰色粉细砂，松散	3.0m
13. 黄灰色泥质粉砂，稍密	2.0m
12. 棕灰色粉质黏土，夹灰色粉细砂，形成厚 0.2~3mm 的纹层，水平薄互层理	2.0m
11. 灰色细砂，松散，向下渐粗，下部 2m 含石英质细砾 3%~5%，偶见小螺壳及碎屑	4.0m
10. 青灰色含泥质粉砂，稍密	2.0m
----- 侵蚀接触 -----	
下段（Qp_2j^1）	厚 32.0m
9. 淡蓝灰色粉质黏土，硬塑，含粒径 0.2~2cm 的钙结石约 10%	2.0m
8. 黏土，上部灰色，下部褐黄色，硬塑，普遍有顺微裂隙形成的宽 1~3mm 根系状灰色胶泥细脉穿插；下部含粒径 1~3mm 的豆粒状褐铁矿结核 20%~25%	4.0m
7. 暗棕色砂质粉土，可塑，含较多云母细鳞片。具水平—微波状层理	2.0m
6. 黄灰色粉砂，上部松散，下部略含泥质而稍密	6.0m
5. 灰色粉细砂，松散	2.0m
4. 棕灰色粉质黏土，可塑	2.0m
3. 灰色粉细砂，松散。上部 2m 富含腐朽植物细屑富集纹层及云母鳞片；下部 4m 夹单层厚 0.5~2cm 的砂质粉土纹层	10.0m
2. 灰色细砂，松散	2.0m
1. 浅黄灰色含砾细中砂，松散。含砾 5%~8%，次圆、浑圆状，砾径为 2~8mm，成分以脉石英为主，次之花岗岩、凝灰岩等。偶见高约 5mm 的纹沼螺（*Parafosarulus striatulus*）	2.0m

----- 假　整　合 -----

下伏地层：下更新统安亭组上段蓝灰色黏土

上列地层剖面揭示，整个嘉定组地层由下而上出现砂层夹砂质粉土或粉质黏土—砂质粉土—黏土正韵律构成的两个沉积旋回，相应组成上、下两“段”地层，其间为侵蚀接触。

据微古分析，唯在下段 1~3 层内发现数个属种、数量很少的有孔虫，有 *Ammonia beccarii*、*Pseudononionella variabilis*、*Rosalina bradyi*，均为浅水广盐性属种，结合所见岩性序列分析，当属海相性很弱的陆上三角洲河流沉积；其他层段未检出微古化石，应属陆相的河流—泛滥湖泊沉积。

（二）对比剖面

1. 上海面粉厂 J10 孔和 SG2 孔剖面

因系长江三角洲地区最早依据较详细的微古、孢粉分析建立的第四纪地层系统而著名。该孔在孔深 109.5~148.1m 间钻遇中更新世地层，总厚为 38.60m（图 21-7）。现将剖面列述于下：

上覆地层：上更新统川沙组青灰色含砾细砂

----- 假　整　合 -----

嘉定组（Qp_2j）	总厚 38.60m
上段（Qp_2j^2）	厚 8.91m
6. 蓝灰色粉质黏土	2.23m
5. 灰色粉质黏土	2.97m
4. 灰、黄褐色细砂夹粉砂	3.71m

----- 侵蚀接触 -----

下段（Qp_2j^1） 厚29.69m

3. 黄褐色黏土 6.50m
2. 灰黑色砂质粉土 8.29m
1. 青灰色细砂 14.90m

----- 假 整 合 -----

下伏地层：下更新统安亭组上段杂色粉质黏土

上列地层剖面揭示，J10孔为由两个正韵律构成的两个沉积旋回，按岩性、岩相特征及其埋深，基本可与娄塘的嘉定组对比。

据微古分析，下段第1层含少量的有孔虫，有 *Ammonia beccarii* var.、*Elphidium* sp.、*Florilus decorus*，均为浅水广盐性属种；并见介形虫，均为淡水—微咸水属种，有土星介（*Ilyocypris* sp.）、小玻璃介（*Candoniella* sp.）和线星介（*Lineocypris* sp.），反映为海相性很弱的三角洲河流沙坝沉积；第2~3层未检出有孔虫，只是在第2层内见有淡水—微咸水介形虫，为浪游土星介（*Ilyocypris errabundis*）、纯净小玻璃介（*Candoniella albicans*）、舒滨小玻璃介（*C. schubinae*）和玻璃介（*Candona* sp.），反映为基本未受海水影响的河流边滩—泛滥平原沉积。上段第4层唯见少量有孔虫，第5~6层未发现微古化石，反映早期为受海水泛及的河流沙坝，中晚期演变为泛滥平原沉积。

为弥补原来J10孔无定年资料的不足，2003年3月在该孔原址施工了一个系统采取光释光、^{14}C定年兼微古、孢粉分析样品的SG2孔，孔深以达到中更新统顶部标志层蓝灰色粉质黏土为止。结果在孔深112m处钻及此标志层顶界，于孔深112.35m处蓝灰色粉质黏土内获得的光释光定年值为116.40±11.25ka处说明属中更新统上段；同期于青浦凤溪施工的SG3孔在孔深100m处钻及此标志层顶界，于孔深100~100.1m处的蓝灰色粉质黏土内获得的光释光定年值为129.95±4.88ka（见图21-7）。这些是上海首次取得的中、上更新统界线定年结果，为更新统的内分及中更新统顶部标志层的确立提供了可靠的依据。

据J10孔孢粉分析，第1层具麻栎（*Quercus acutissima*）-榆属（*Ulmus*）-禾本科（Glamineae）组合；第2~6层具松属（*Pinus*）-落叶砾（*Quercus deciduous*）-榆属（*Ulmus*）-冷杉属（*Abies*）组合，反映由下而上古气候经历了温暖略干—凉冷湿润的变化。

在崇明、长兴等河口沙岛区和浦东滨岸带分布的中更新统嘉定组因遭受侵蚀而残缺不全，顶部标志层往往缺失，但仍大体保存由两个正韵律组成的上、下两段地层，如崇明庙镇SG1孔和长兴岛SG4孔揭露的嘉定组地层剖面；有的只保留下段地层，如浦东机场J25孔所见（见图21-7）。

2. 长兴岛SG4孔剖面

孔深124.65~144.33m（终孔）揭露嘉定组，总厚度大于19.68m。上覆上更新统川沙组灰色含砾中砂。上、下段分界在141.30m处。所见嘉定组地层由上而下依次为：

嘉定组（Qp_2j）

上段（Qp_2j^2） 厚16.65m

6. 以浅棕色粉质黏土为主，与灰色含砾粗中砂互层，中间夹一段厚为0.15m的蓝灰色粉质黏土夹中细砂团块 1.23m
5. 灰色细砂，松散 2.02m
4. 黄灰色含砾中粗砂，松散，含次圆、次棱角状的石英、燧石、钙结石，砾径0.2~2cm，底部有厚为0.25m的砂砾层 8.80m
3. 上段的灰色细砂，松散，底部有厚为0.25m含次棱角状石英细砾的细砂，底界可见起伏的冲刷面。在孔深141.15~141.25m（距本层底界以上0.05~0.15m处）段样品的光释光定年值为254.81±23.61ka 4.60m

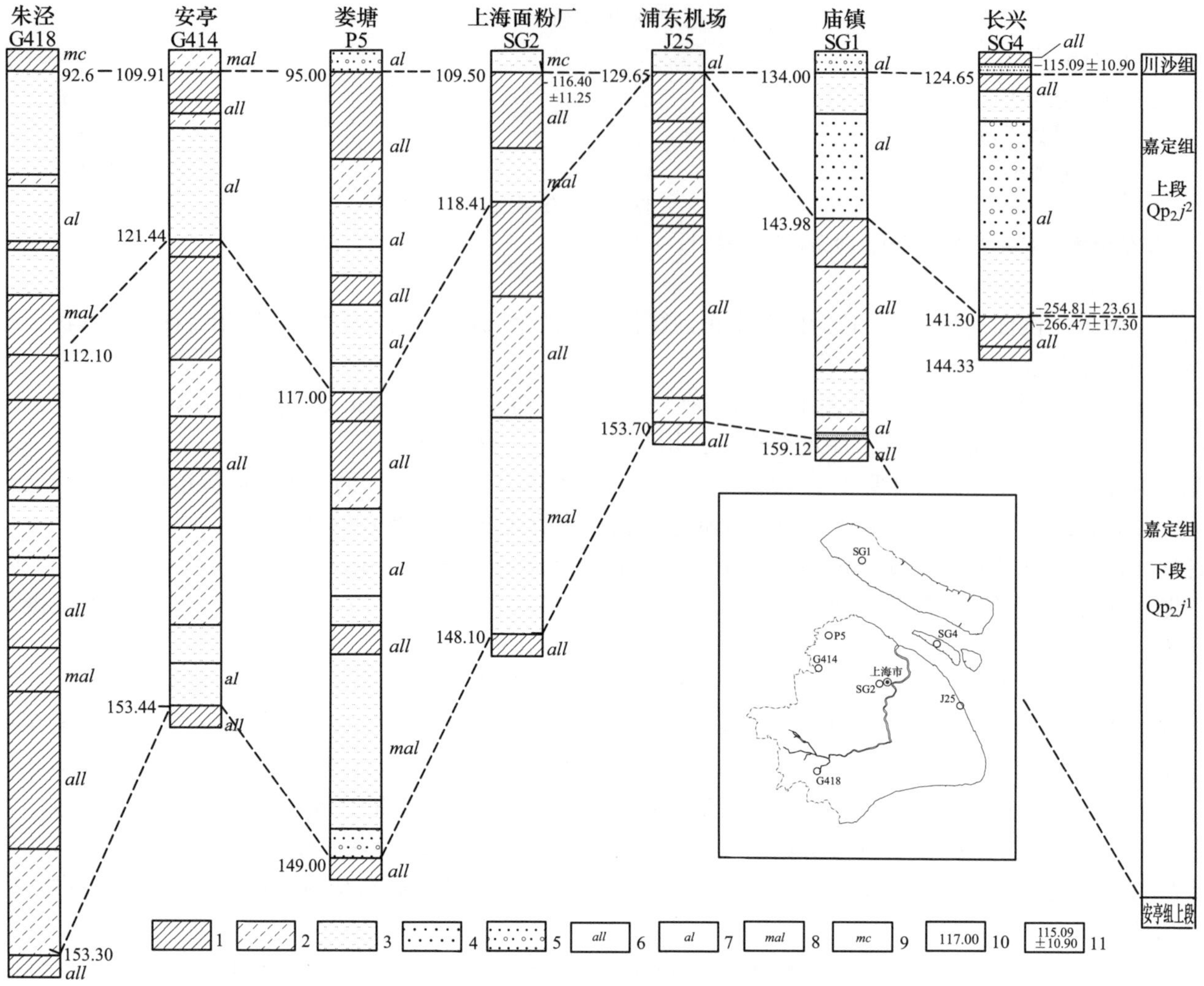

图 21-7　中更新统嘉定组地层对比图

Fig. 21-7　Stratigraphic correlation of the Middle Pleistocene Jiading Formation

1—黏土、粉质黏土；2—砂质粉土、黏质粉土；3—粉砂、细砂；4—中砂、粗砂；5—含砾砂；6—泛滥平原—泛滥湖泊相；7—河流（河床—边滩）相；8—三角洲平原河湖相；9—河口相；10—深度（m）；11—光释光法定年值（ka B. P. ）

－－－－－ 侵蚀接触 －－－－－

下段（Qp_2j^1） 厚度>3. 03m

2. 蓝灰色黏土，可塑—硬塑状，含粒径为 1～6mm 的钙质小结核 5%～10%，在孔深 141. 4～141. 5m（距本层顶界以下 0. 1～0. 2m 处）样品的光释光定年值为 266. 47±17. 30ka 2. 05m

1. 黄褐杂蓝灰色细网纹状黏土 >0. 98m

在孔深 123. 25～123. 35m（本组顶界以上 1. 3～1. 4m 处的浅棕色粉质黏土内）光释光样品的定年值为 115. 09±10. 90ka，证明为晚更新世地层。

据微古分析，本孔嘉定组内未发现化石。上列嘉定组层序反映：下段所见系泛滥湖泊沉积；上段基本为沙坝与河床交替的曲流河道沉积，顶部出现泛滥平原与串沟河流沉积，见有顶部标志层的冲刷残留泥砾堆积。

3. 庙镇 SG1 孔剖面

在孔深 134～159. 12m 钻遇嘉定组，总厚度为 25. 12m。下伏下更新统安亭组上段具硬塑状的青灰、灰黄色细网纹状黏土；上覆上更新统川沙组青灰色含砾中细砂。孔深 143. 98m 处为上、下段分界。所见嘉定组地层由上而下依次为（见图 21-7）：

嘉定组（Qp_2j）

上段（Qp_2j^2）		厚 9.98m
7.	青灰色粉细砂，松散，含腐木和云母碎片	2.78m
6.	灰黄色中粗砂，松散。上部夹有细砂，近底部夹有厚为 0.18m 的腐木碎屑富集和石英细砾层	7.2m
	----- 侵蚀接触 -----	
下段（Qp_2j^1）		厚 15.14m
5.	青灰色黏土，硬塑，具斜波状层理。中部有厚为 0.8m 呈灰黄色，下部含钙质结核，向下增多	3.31m
4.	青灰色砂质粉土。上中部有 6m 夹有粉质黏土，见斜波状层理；下部有 0.8m 呈褐灰色，含腐木细屑，稍有揉皱变形构造	7.13m
3.	灰色细砂，松散	3.03m
2.	褐灰色黏质粉土	1.25m
1.	灰色砾质中粗砂，松散，含砾为 20%~30%，砾径为 0.2~3cm，磨圆差	0.42m

庙镇 SG1 孔所见上、下两段地层为一套河床—沙坝—泛滥平原沉积序列构成的曲流河道沉积，其岩性、岩相及分界深度大体可与长兴岛 SG4 孔对比，只是后者下段地层未钻穿而已。

4. 浦东机场 J25 孔剖面

在孔深 129.65~153.70m 钻遇嘉定组。下伏下更新统安亭组上段褐黄色粉质黏土；上覆上更新统川沙组灰色含砾细砂、灰黄色砾砂。所见嘉定组唯保留下段（见图 21-7）

嘉定组下段（Qp_2j^1）

8.	蓝灰色黏土，下部为有绿色斑纹的灰色黏土，顶部 0.4m 为蓝灰色黏质粉土，具均匀层理	3.35m
7.	黄绿色粉质黏土，含钙质斑点、小结核，顶部见垂直的植物根系	1.40m
6.	灰色黏土	2.40m
5.	灰色黏质粉土。下部变为灰黄色砂质粉土，底部见腐殖物碎屑集成斜波状纹层，中部水平层理，上部具包卷层理	1.65m
4.	绿灰色黏土	1.00m
3.	杂色黏土，有灰绿色网纹，向下渐变为黄绿色黏土，具均匀层理	0.75m
2.	褐黄色黏土，有少量灰绿、蓝灰色斑纹，含钙质、铁锰质结核，粒径在 3mm 左右，向底部渐多，局部含蚌壳碎片，具均匀层理	11.80m
1.	草黄色砂质粉土，有铁锰质斑点，底界见冲刷面	1.70m

据微古分析，唯在 4 层内发现淡水至微咸水介形虫，优势种为 *Ilyocypris bradyi*，次之有双摺土星介（*I. biplicata*）、玛纳斯土星介（*I. manasensis*），还有个别的玻璃介（*Candona* sp.）、小玻璃介（*Candoniella* sp.）、真星介（*Eucypris* sp.）。反映 J25 孔所见嘉定组为陆相的泛滥湖泊沉积。

据古地磁测量，该孔嘉定组底界在布容/松山两极性时界限（年龄 0.78Ma）之下 3m 处，顶界在 Big lost Emperor 负极性事件（年龄 0.554Ma）以上，表明属于嘉定组下段。

（三）总体特征

1. 嘉定组下段（Qp_2j^1）

嘉定组下段在全区普遍分布，其埋深在 120~160m 之间，大体由下粗上细的一套松散岩层组成，据此可相应分为上、下两个亚段。

（1）下亚段

下亚段岩性、岩相变化较大，受中更新世早期继续发育的浏—南、枫—奉两条古河道分布格局的控制。在崇明岛及陆域的宝山—市区东部—浦东—芦潮港这一浏—南古河道发育的宽广地带，还有南部的枫—奉古河道展布地带，堆积的为河床相黄灰或灰色含砾砂和砂砾层，其中枫—奉古河道内沉积物略细；在宝山、浦东的滨岸地带及长兴、横沙岛上则堆积河流心滩相灰色、褐黄杂灰绿、蓝灰色网纹状黏土与粉土、粉砂互层。这两个古河道发育地带（浏—南古河道流域宽达三四十千米，枫—奉古河道流域也宽达一二十千米）即成为区内第Ⅲ下含水层的主要分布地带。其他地区即大体为中、南部两个凸起及其附近，包括青浦、松江、嘉定的大部、市区西部以及金山南部和奉贤东南部均为河流边滩—泛滥湖泊堆积的灰、褐黄和杂色黏土夹粉土和粉砂或互层，在这些地区含水层不发育。本段厚度在 8～36m 不等。在河道带下亚段较厚，其底界也较深；在河间地带则底界略浅，厚度也较小。

（2）上亚段

上亚段以蓝灰色黏性土为特征，并有部分灰、蓝灰与褐黄杂色细网纹、斑杂状黏土、粉质黏土，含少量的钙质、铁锰质结核，并可见多少不一的淡水螺、蚌壳及碎片。在上部常夹有灰、褐黄色细砂和粉土。为一套泛滥平原—湖泊堆积。本段厚度在 8～24m 不等，一般与下伏地层厚度呈反消长关系。在上述凸起区附近上亚段较厚；而在崇明岛和陆域的宝山至芦潮港间下亚段为河床相地层发育的地区，上亚段很薄以至缺失，使上更新统川沙组砂层（即第Ⅱ下含水层）超覆于中更新统嘉定组下段砂层（第Ⅲ下含水层）之上，导致上、下含水层之间沟通。

在长江三角洲区如上海面粉厂、浦东、嘉定娄塘等地诸多钻孔的中更新统嘉定组下段的下亚段砂层内可见及少量有孔虫，有 *Ammonia beccarii*、*Pseudononionella variabilis*、*Nonion* sp.、*Rosalina bradyi* 等广盐种；近南汇嘴出现 *Epistominella naraensis* 和其他浮游有孔虫及箭头虫（*Bolivina* sp.）等窄盐种，间或伴有少量的钙质超微化石大洋桥石（*Gephyrocapsa oceanica*）与陆相介形虫 *Ilyocypris errabundis*、*Candoniella albicans*、*C. schubinae*、*Candona* sp.、*Ilyocypris* sp.、*Candoniella* sp.、*Lineocypris* sp. 共生，还出现淡水软体动物纹沼螺（*Parafosarulus striatulus*）、田螺（*Viviparus* sp.）、丽蚌（*Lamprotula* sp.）等化石，表明属弱海陆过渡的三角洲平原河流沉积。而在西部太湖沉积区，一般见淡水介形虫，只出现个别的有孔虫，唯青浦练塘 P31 孔孔深 133～135m 浅黄灰色砂质粉土夹粉细砂内出现结缘寺卷转虫角状亚种（*Ammonia ketienziensis angulata*）-毕克卷转虫（*Ammonia beccarii*）有孔虫组合，前一属种分布于现南黄海水深 50m 左右处水域，并见具瘤先希望虫（*Protelphidium tuberculatum*）、冷水面颊虫（*Buccelafrigida*）等示冷性有孔虫，复合分异度达 2.92；与之共生半咸水属种斑纹三原介（*Sanyuania psaronion*）和常在具一定盐度水体中生活的陆相介形虫 *Limnocythere* sp.、*Candoniella* sp.；又在金山朱泾 G418 孔孔深 132.2～139.5m 黄褐色黏土内也见及个别的有孔虫。说明在太湖沉积区一般属于略受海水影响的三角洲平原河湖，可能由于海面波动一度沦为开放性海湾环境。其上亚段为一套具特征的蓝灰色和带灰绿色细网纹或皱纹状的杂色黏土，一般无有孔虫，但其中出现的淡水软体动物化石在区内第四纪地层中最为丰富，在该段黏性土及所夹粉细砂薄层内频繁可见。一般为个体较小的腹足类，有角状环棱螺（*Bellamya angularis*）、纹沼螺相似种（*Parafossarulus* cf. *striarulus*）、中华沼螺（*Parafossarulus sinensis*）、短沟蜷未定种（*Semisulcospira* sp.）、小旋螺未定种（*Gyraulus* sp.）、光滑狭口螺（*Stenothgroglabra*），还有丽蚌（*Lamprotula* sp.）和珠蚌（*Unio* sp.）的碎片。相伴出现少量的陆相介形虫 *Ilyocypris errabundis*、*Candoniella albicans*、*C. schubinae* 和 *Candona* sp.，还有轮藻、鱼牙等，表明为河流泛滥平原—洼地湖泊堆积。

据长兴岛 SG4 孔中更新统嘉定组上、下段分界处的光释光定年值 254.81±23.61ka（孔深 141.15～141.25m）和 266.47±17.30ka（孔深 141.40～141.50m），研究区内嘉定组下段的顶界年龄约为 0.26Ma，其底界大都恰好落在极性柱布容正极性时与松山负极性时界限上，年龄为 0.78Ma，据此确定嘉定组下段的年龄介于 0.78～0.26Ma。

2. 嘉定组上段（Qp_2j^2）

嘉定组上段分布范围基本为宝山及南部地区的闵行、奉贤诸区及其以西的陆域中西部地区。大致沿芦潮港农场—航头—市区中山北路—外高桥所围限的陆域东南部地区及崇明、长兴、横沙等河口沙岛区，由于为晚更新世早期河流的侵蚀，上段地层大都残缺不全。

嘉定组上段埋深在95~120m之间，大体也呈下粗上细的岩性变化，据此也可相应分为上、下两个亚段。

（1）下亚段

下亚段岩性、岩相由东向西具有以下规律性变化。

东部，在宝山、嘉定及与市区中西部接壤区，还有浦东外高桥、张江、浦东南部地区与奉贤的东部地区，基本分布河道及河间心滩和边滩相的灰、黄灰色含砾中粗砂、含砾中细砂及部分的粉细砂，有的底部可见砂砾层，为上海第Ⅲ上含水层的主要分布场所。其埋深在100~120m，厚度一般为15~25m。

西南部的天马山凸起附近及濒临杭州湾的秦皇山凸起附近地区，多为泛滥平原—洼地湖泊堆积的灰黄、褐黄、灰绿等色粉质黏土夹粉细砂，厚度一般不超过10m。

介于上述两者之间地区多为河流边滩相的灰、青灰等色粉细砂和粉土、粉质黏土的互层，局部有支流河道堆积的粉细—细中砂夹含砾砂层。砂中常见腐木碎块、淡水螺壳及泥砾。厚度多在5~15m之间。

（2）上亚段

上亚段以蓝灰、灰绿色黏土、粉质黏土为主，部分为灰及蓝灰与褐黄杂色细网纹状、斑杂状黏性土夹浅灰色粉砂，含少许钙质、铁锰质结核，局部产淡水软体动物丽蚌等化石，并富含腐烂植物碎屑和根茎，属于泛滥平原—湖泊堆积。埋深一般介于95~100m，厚度约为5m。

嘉定组上段在滨海平原东部滨岸地带和河口沙岛地区大都因被侵蚀而缺失，其分布基本限于太湖平原分区和滨海平原分区的中西部。在上段的下亚段内，三角洲沉积区个别钻孔（市区J10孔孔深115.19~118.41m灰、黄褐色粉细砂）内检出少量有孔虫；在太湖沉积区个别钻孔（青浦练塘P31孔孔深93~107m灰绿、棕灰色粉质黏土和青灰色细砂）内分析出 *Ammonia beccarii-A. convexidorsa* 有孔虫组合，还见有 *Epistominella naraensis*、*Pseudoeponides andersen*、*Nonion shansiensis*、*Cribrononion subincertum*、*C. vitreum* 等有孔虫，与之共生的为半咸水的中华丽花介（*Sinocytheridea* spp.）组合，还见有 *Leguminocythereis* sp.、*Loxoconcha* sp. 等半咸水属种，并见 *Ilyocypris errabundis*、*Candoniella* sp.、*Candona* sp. 等陆相介形虫；偶尔见及钙质超微化石 *Gephyrocapsa oceanica*。表明当时基本为河流沉积，偶尔受到海水的影响。其上亚段则未发现微体古生物，仅见及陆相软体动物方形环棱螺相似种（*Gyraulus albus*）等，表明为陆相的河泛平原—洼地湖泊堆积。

据孢粉分析，中更新统嘉定组上段下亚段内普遍含孢粉较丰，上亚段内孢粉贫乏。

2003年3月取自青浦凤溪SG3孔样品的光释光定年结果（见图21-7）：孔深99.6~99.7m上更新统川沙组底部灰色含砾中细砂为113.55±11.26ka；孔深100~100.1m中更新统嘉定组顶部灰蓝色粉质黏土为129.95±4.88ka。据此推算，该孔深100m处的中、上更新统分界年龄应为127.9ka，即约为128ka，这与国内所定的中、上更新统界线标准年龄相同。嘉定组上段底界年龄即为上述嘉定组下段顶界年龄，由此可知，区内嘉定组上段的年龄介于0.260~0.128Ma。

再者，中更新统嘉定组的“杂色硬黏土”以具有下述特征，而与下更新统的“杂色硬黏土”有所区别：

1）嘉定组及以上硬黏土内未发现因挤压错动而产生的节理，而在下更新统“杂色硬黏土”内则屡见不鲜。唯一的例外是在外高桥J20孔孔深132.8m处的棕黄色黏土内发现一条∠35°的张性节理，具下滑擦痕，切穿灰绿色细网纹及钙质结核。

2）本组硬黏土内所含钙质结核粒径较小，一般仅为数毫米至2cm，含量也少，无钙质网脉和胶结成岩的钙板层出现；铁锰质结核出现较少，通常其颜色因含铁质略多而偏于褐色，与下更新统硬土内常见含锰较高而偏于黑色者有所差异。

3）本组杂色黏土所具网纹、斑杂状构造较下更新统者细小，并常出现细条纹状构造。

4）本组顶部通常有较纯净、鲜艳的蓝灰色黏土出现，成为中更新统嘉定组顶部的标志层，不同于下更新统安亭组顶部硬黏土以灰绿、黄绿、褐黄色居多，间或有蓝灰色出现。

5）本组内出现淡水腹足类化石最丰，但个体明显较小，并见多呈碎片的淡水蚌，而与安亭组上段黏土内常见较大、完整的淡水腹足类和淡水蚌化石有别。反映前者沉积环境常动荡不定，后者环境则较平静稳定。

三、川沙组

（一）层型剖面

横沔 P12 孔剖面为 1984 年列举的川沙组层型剖面。该孔于孔深 81~121m 段钻遇川沙组，总厚度为 40.0m。所见地层为（图 21-8）：

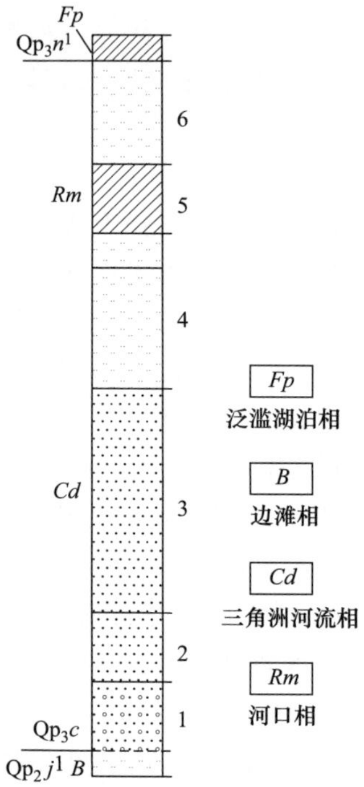

图 21-8　P12 孔川沙组地层剖面

Fig. 21-8　Stratigraphic section of the Chuansha Formation per borehole P12

上覆地层：上更新统南汇组（Qp_3n）灰色粉质黏土

－－－－－ 假　整　合 －－－－－

川沙组（Qp_3c）	厚 40.0m
6.　浅灰色细砂	6.0m
5.　灰色粉质黏土，可塑，含少量泥钙质结核	4.0m
4.　灰色细砂、粉细砂	9.0m
3.　灰色中细砂、浅灰色中粗砂	13.0m
2.　灰色中细砂	4.0m
1.　灰色粗砂砾石层	4.0m

－－－－－ 假　整　合 －－－－－

下伏地层：嘉定组下段（$Qp2j^1$）灰黑色细砂

经有孔虫分析，1~4 层中下部含少量再沉积的有孔虫，4 层上部~6 层出现圆形卷转虫（*Ammonia*

dominicana）-缝裂希望虫（*Elphidium magellanicum*）组合，并有奈良小上口虫（*Epistominella naraensis*）、五叶抱球虫（*Globigerina quinqueloba*）。反映前者为海相性很弱的三角洲河流沉积，后者成为海陆过渡的河口沉积。

（二）对比剖面

1. 浦东机场 J25 孔剖面

该孔于孔深 81.95~129.65m 间钻遇川沙组，总厚度为 47.7m（图 21-9）。

上覆地层：上更新统南汇组（Qp_3n）灰色粉细砂夹粉质黏土

-----假　整　合-----

统川沙组（Qp_3c）	厚 47.7m
11. 灰色粉砂夹砂质粉土，局部沿层面有腐殖物碎屑聚集	3.05m
10. 灰黄色砾砂，砾石多棱角、次棱角状，少有扁圆状，砾径向下渐大可达 2cm，具向上变细的韵律层理	0.40m
9. 灰色粉砂夹粉质黏土，形成纹层厚 2~10mm 的薄互层，具不规则水平层理、斜波状层理	4.85m
8. 灰黄色砾砂，砾石呈次棱角状，大者可在 2cm 左右，由上而下渐多	0.25m
7. 灰色粉砂夹粉质黏土，距底界 1.6m 处见冲刷构造。底部 0.4m 含粒径为 3~5cm 的泥砾，局部含较多贝壳碎屑	9.50m
6. 灰色粉细砂	2.00m
5. 灰色粉砂夹粉质黏土薄层，具水平层理	1.50m
4. 灰色粉细砂夹有不连续的粉质黏土层，下部偶含细砾	14.62m
3. 灰黄色砾砂，砾呈次圆状；底为扁圆状砾石层，最大砾径在 3~4cm	2.48m
2. 灰黄色砾砂，砾呈次棱角状，最大砾径在 0.5~1cm	6.20m
1. 灰色粉砂，含云母较多，显水平层理。底部为含砾细砂，砾呈粒径为 0.5~3cm 的次棱角状	2.85m

-----假　整　合-----

下伏地层：中更新统嘉定组下段（Qp_2j^1）蓝灰色黏土

据微古分析，大都未见化石，唯在 7 层内发现个别的有孔虫，有毕克卷转虫（*Ammonia beccarii*）、压扁卷转虫（*A. compressiuscula*）、亚三刺星轮虫（*Asterorotalia subtrispinosa*）；还有个别介形虫皮桑双角花介（*Kejiella bisanensis*）；5 层内见个别的日本库士曼介（*Cushimanidea japonica*）。所见均为海相介形虫和多盐—真盐潮间带有孔虫，未见陆相介形虫；因此结合岩性特征分析，可以判定中部 5~7 层属于高能砂质潮间带的滨海沉积；下部 1~4 层及上部 8~11 层均为河床相与边滩相交替的曲流河道沉积。

2. 上海面粉厂 SG2 孔剖面

在孔深 73~112m 钻遇川沙组，总厚度为 39m（见图 21-9）。大致可分 3 段：下段（孔深 99.5~112.0m），青灰色粉细砂、含砾细砂，含粒径为 2~5mm 的石英、燧石质细砾，局部夹黏质粉土，厚 12.5m；中段（89.0~99.5m），青灰、褐灰色粉质黏土夹粉细砂，局部为薄互层，时见冲刷构造、泥砾，上部含少量的碎屑螺、贝壳碎屑及腐殖物碎屑，厚为 10.5m；上段（73~89m），青灰色、部分为灰黄色含砾中粗砂、含砾中细砂夹青灰色细砂，砾石多次棱角—次圆状，粒径为 2~7mm 的石英、燧石质细砾，断续见含少量贝壳碎屑及腐殖物碎屑，厚为 16.0m。这一岩性序列与浦东机场 J25 孔所见相似，只是未发现微古化石，很可能为河流沉积；也不能排除“砂质潮间带微古化石通常贫乏、甚至缺失”的可能，因此，也可能如同 J25 孔一样为河流—滨海—河流沉积。

总之，川沙组基本为一套河流—三角洲河流相灰色、部分灰黄色的含砾中粗砂、细砂互层，局部如浦东机场等地的川沙组中上部夹有滨海相的灰色粉细砂夹粉质黏土。川沙组砂层在崇明岛、长兴岛等河口地区十分发育（见图 21-9），如崇明庙镇 SG1 孔在孔深 67.8~134m 钻遇厚达 66.2m 的灰、灰黄色含砾中粗砂、中细砂互层构成的河流沉积；长兴岛 SG4 孔孔深 70.5~124.65m 钻遇厚为 54.15m 的

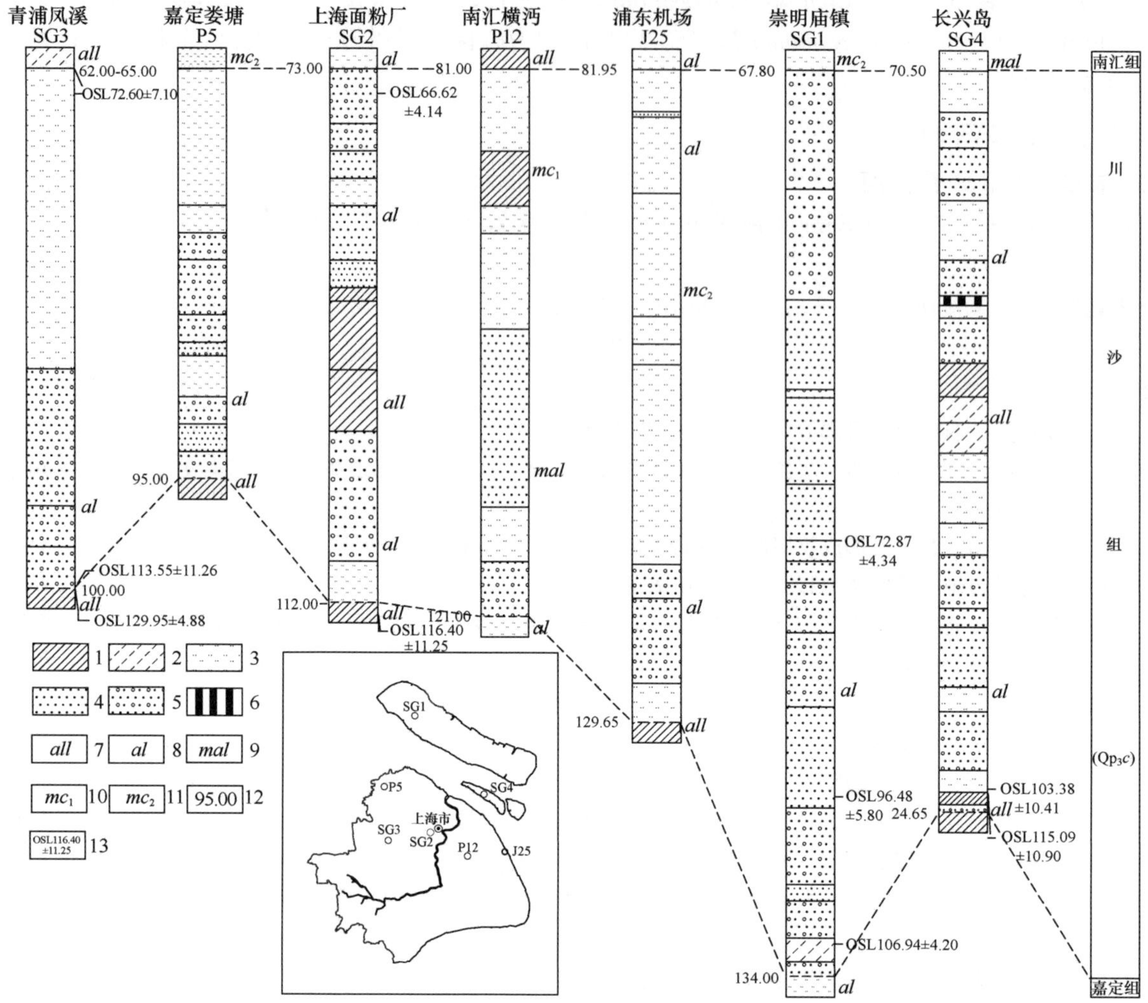

图 21-9　上更新统川沙组地层对比图

Fig. 21-9　Stratigraphic correlation of the Upper Pleistocene Chuansha Formation

1—黏土、粉质黏土；2—砂质粉土、黏质粉土；3—粉砂、细沙；4—细中砂、中粗砂；5—含砾砂；6—腐木层；7—泛滥湖泊相；8—河流相；9—三角周河流相；10—河口相；11—滨海相；12—深度（m）；13—光释光法定年值（ka B. P.）

河流相灰色、黄灰色细砂和含砾中粗砂互层为主，中部夹泛滥平原相青灰色粉土—粉质黏土层。两孔的川沙组内均未发现微古化石。

关于川沙组的年龄，其下限即与前述中更新统嘉定组上限年龄一致，为 128ka；其上限，据青浦凤溪 SG3 孔样品光释光定年结果（见图 21-9）：孔深 59.5～59.6m 段上更新统南汇组底部深灰色黏质粉土夹粉砂的光释光定年值为 50.05±2.80ka；孔深 62.0～62.1m 段上更新统川沙组顶部灰色粉细砂夹薄层粉质黏土的光释光定年值为 72.60±7.10ka。还有，在上海面粉厂 SG2 孔深 74.55m 处、川沙组顶界下 1.55m 处的青灰色中细砂内获得的光释光定年值为 66.62±4.14ka，与凤溪 SG3 孔川沙组顶部定年值相近。

据凤溪 SG3 孔定年结果推算：该孔深 62m 处川沙组与上覆南汇组分界的年龄应是 71.698ka，即约为 72ka。由此可知，区内川沙组的年龄通常介于 128～72ka 之间，其上限与国内通常划定的末次冰期下限一致。

但是需要指出，在崇明岛、长兴岛等河口地区所见的川沙组上部砂层年龄向上穿时现象明显。据庙镇 SG1 孔系列定年结果，以孔深 101.8m 处分界，在川沙组上部（孔深 67.8～101.8m）取得 2 个光释光和 5 个 ESR 定年结果，定年值一般为 33.95±1.55ka B. P. ～35.4ka，相当于末次冰期间冰阶后段；

下部（孔深 101.8～132m）取得 3 个光释光和 8 个 ESR 定年结果，定年值一般为 59.9～106.94±4.20ka，大致相当于通常的川沙组年龄。其上、下的定年值出现跳跃式突变，其间可能存在一个沉积间断，缺失了相当于末次冰期间冰阶前段和早冰阶地层。

（三）总体特征

川沙组在上海全区分布广泛，基本为一套河流—弱海陆过渡的三角洲河流相细砂和含砾砂。其埋深由研究区西南向东北逐渐加大：陆域西部埋深一般介于 65～95m 之间；陆域东部至崇明岛，其顶界埋深在 75～90m，底界埋深在 100～130m。其岩性、岩相由东北至西南具有以下变化：

在东北部的崇明、长兴、横沙等河口沙岛上，沉积物最粗，为河流堆积的灰、灰黄色砾质中粗砂、含砾粗中砂和含砾中细砂，中上部夹有粉细砂和砂质粉土。厚度常达 50～60m。这里是第Ⅱ含水层发育最好的地区。

在陆域东部，大致沿嘉定外冈—北新泾、漕河泾及其以南至燎原农场一线以东地区，基本以河流—三角洲河流堆积的浅灰、黄灰色含砾细中砂—含砾中粗砂为主，夹粉细砂或呈互层，中、上部夹有河口滨海相浅灰色粉细砂、砂质粉土和粉质黏土。厚度一般在 30～40m。这里是第Ⅱ含水层发育良好的地区。

在陆域西部即嘉定外冈—燎原农场一线以西地区，大部分为三角洲河流边滩堆积，一般为灰、棕灰色粉细砂和砂质粉土夹含砾砂；在与浙江毗邻的青浦、金山、松江接壤区，为河口湾堆积的灰、棕灰色粉质黏土与浅灰色粉砂、砂质粉土互层，厚度大都在 30m 左右。这里是第Ⅱ下含水层发育较差的地区。

据微古分析，在长江三角洲区的川沙组内一般无微古化石检出，在浦东滨岸地带—南汇嘴地区的川沙组中上部可见少量的螺旋壳类、平旋壳类有孔虫，如圆形卷转虫（*Ammonia dominicana*）、毕克卷转虫（*A. beccarii*）、缝裂希望虫（*Elphidium magellanicum*）和冷水面颊虫（*Buccella frigida*），有处尚出现奈良小上口虫（*Epistominella nareansis*）、五叶抱球虫（*Globigerina quinqueloba*）等浮游有孔虫；还有在全组砂层内断续见及个别的钙质超微化石大洋桥石（*Gephyrocapsa oceanica*）、赫氏艾氏石（*Emiliania huxleyi*）、瓦氏卷舟颗石（*Helicopontosphaera wallichii*）和非颗石型的胸甲球石（*Thoracosphaera*）。如外高桥 J20 孔就在川沙组内始零星出现 *Gephyrocapsa oceanica*。表明东北部长江三角洲沉积区的川沙组基本为河流沉积，其中上部断续出现受海侵泛及的三角洲河流沉积，浦东—南汇嘴地区的川沙组中部则间或出现河口滨海沉积。

太湖沉积区大部也无微古化石检出，唯在练塘、朱泾地区断续见有孔虫，主要有 *Ammonia beccarii*、*A. convexidorsa*，还见 *Epistominella nareansis*、*Elphidiella kiangsuensis*、缝裂假上穹虫（*Pseudoeponides anderseni*）、山西九字虫（*Nonion shansiensis*）、亚易变筛九字虫（*Cribrononion subincetum*）等，均为广盐性属种。介形虫主要为半咸水的中华丽花介（*Sinocytheridea* sp.），还有霍纪豆艳花介（*Leguminocythere hodgii*）、新单角介（*Neomonoceratina* sp.）、棘艳花介（*Echinocythereis* sp.）和淡水种 *Ilyocypris errabundis*、*Candona* sp.，显示河口湾微古化石组合面貌。

上述化石组合面貌反映，进入晚更新世早期，本区主要仍为陆相河流环境，其中、晚阶段出现微弱—弱海陆过渡的三角洲河流环境，局部进入强海陆过渡的河口滨海环境，形成区内的第Ⅲ海进层。

孢粉分析表明，上更新统川沙组普遍含孢粉较丰。

据光释光定年可知，区内川沙组的年龄通常介于 72～128ka，其上限与国内通常划定的末次冰期下限一致。

四、南汇组/滆湖组

（一）南汇组层型剖面

海潮 P17 孔剖面为 1984 年列举的南汇组层型剖面。该孔于孔深 23～64m 间钻遇南汇组，总厚度为 41m，据岩性、岩相变化可划分为 3 段，相应代表陆相—海相—陆相沉积，构成一个完整的海侵旋回。所见地层为（图 21-10）：

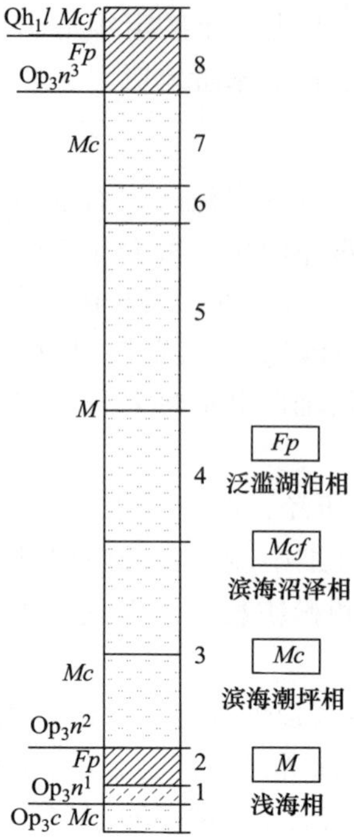

图 21-10　P17 孔南汇组地层剖面

Fig. 21-10　Stratigraphic section of the Nanhui Formation per borehole P17

上覆地层：全新统娄塘组（Qh_1l）灰黑色淤泥质黏土

----- 假　整　合 -----

南汇组（Qp_3n）	总厚 41.0m
上段（Qp_3n^3）	厚 3.0m
8. 暗灰绿粉质黏土、姜黄色黏土，可塑，有铁锰质斑点	3.0m

———— 整　合 ————

中段（Qp_3n^2）	厚 35.0m
7. 浅黄色粉砂夹砂质粉土，含褐黄色粉质黏土碎块	7.0m
6. 黄灰色粉砂，含少量褐黄色粉质黏土碎块	2.0m
5. 浅黄灰色粉细砂，偶见贝壳碎屑	8.0m
4. 青灰色粉细砂，含贝壳碎片及半腐烂植物碎屑	7.0m
3. 浅灰黄、浅灰色细砂，上部夹有半腐烂植物碎屑细纹层	11.0m

----- 假　整　合 -----

下段（Qp_3n^1）	厚 3.0m
2. 灰黑色粉质黏土	2.0m
1. 灰绿色砂质粉土	1.0m

———— 整　合 ————

下伏地层：上更新统川沙组（Qp_3c）浅灰色粉细砂

据微古分析，下段未发现微古化石，属湖泊堆积。中段的 3 层下部未检出微古化石，3 层中上部至 4 层下部含丰富的有孔虫，为 *Elphydium advenum*-*Ammonia beccarii*-*Florulus decorus*-*Globigerina* sp. 组合，并多 *Pseudorotalia*、*Asterorotalia* 等暖水种；4 层上部含有孔虫较丰，主要有 *Ammonia beccarii*、*Elphydium advenum*、冷水面颊虫（*Buccela frigida*），其次为浮游类 *Globigerina* sp.，窄盐性的箭头虫（*Bo-*

livjna sp.），并见少量的钙质超微化石 *Gephyrocapsa oceannica*；3层中上部至4层内并见很少的半咸水介形虫 *Sinocytheridea latiovata*、*Leguminocythereis hodyi*，反映3层下部可能为滨海沉积；3层中上部至4层为滨浅海（河口淡化浅海）沉积。中段中上部的5～6层含有孔虫较丰，主要有 *Ammonia beccarii*、*A. convixidorsa*、*A. ketienziensis*、*Elphydium advenum*，并有较多的暖水种 *Pseudorotalia schroeteriana*，还普遍可见少量的钙质超微化石 *Gephyrocapsa oceannica*，但未检出介形虫，表明属滨岸浅海相；上部的7层含少量的有孔虫 *Ammonia beccarii*、*Elphydium advenum*，表明属滨海相。上段偶见砂杆虫（*Ammobaculites* sp.）、筛九字虫（*Cribrononion* sp.），都属半咸水属种，前者属分布于水深仅1m多的属种，并偶见半咸水介形虫豆艳花介（*Leguminocythereis* sp.），表明为曾受海泛影响的泛滥湖泊沉积。在区内上段暗绿色黏土内检出半咸水微古化石者，这是个别特例之一。

据孢粉分析，下段含孢粉贫乏，只见少量的蒿属（*Artemisia*）及数粒落叶栎（*Quercus deciduous*）、枫香属（*Liquidambar*）花粉，反映古气候可能为温暖偏干；中段含孢粉贫乏，仅5层局部见两粒 *Chenopodiaceae*；上段以 *Quercus diciduous*、*Pinus*、*Polypodiaceae*、*Ceratopteris* 为主，还有少量的 *Cupressaceae*、*Betula*。反映古气候由下段—中段—上段，经历了温暖偏干（?）—海洋温暖湿润（?）—温凉湿润的变化。

（二）南汇组对比剖面

1. 市区杨浦教育局 CBG1-1 孔剖面

该孔于孔深25.2～66.9m间钻遇南汇组，总厚度为41.7m，3段地层发育齐全（图21-11），现列述如下：

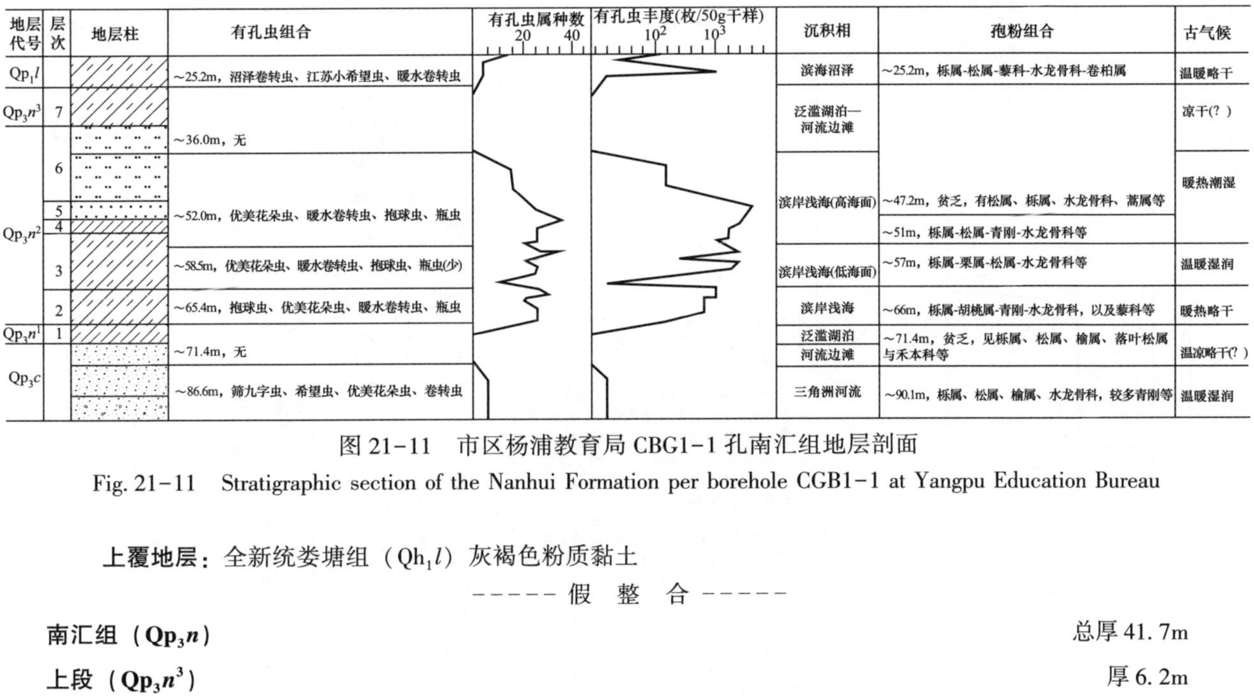

图21-11　市区杨浦教育局 CBG1-1 孔南汇组地层剖面

Fig. 21-11　Stratigraphic section of the Nanhui Formation per borehole CGB1-1 at Yangpu Education Bureau

上覆地层：全新统娄塘组（Qh_1l）灰褐色粉质黏土

-----假　整　合-----

南汇组（Qp_3n）	总厚41.7m
上段（Qp_3n^3）	厚6.2m
7. 墨绿—褐绿—灰黄色粉质黏土，含少量铁锰质结核	6.2m

————整　合————

中段（Qp_3n^2）	厚34.0m
6. 灰黄、锈黄色粉细砂，松散。顶部1m多泥质纹层，下部4m为粉细砂和细砂薄互层	12.6m
5. 青灰色细砂，松散	3.2m
4. 灰褐色黏土夹薄层粉砂，多虫孔构造，并有较多有机物条带	2.0m
3. 深灰色粉质黏土、粉细砂薄互层	9.3m
2. 灰色粉质黏土夹细砂，见脉状、透镜状层理	6.9m

----- 假 整 合 -----

下段（Qp_3n^1） 厚 1.5m

1. 灰褐色黏土，切面致密光滑，质纯 1.5m

———— 整 合 ————

下伏地层： 上更新统川沙组（Qp_3c）青灰、灰色细砂

由上述剖面可知：下段无微古化石检出，为温凉略干气候下泛滥湖泊沉积。中段含有孔虫属种大都有二三十种，丰度大都达每 50 克干样含有孔虫数百至上千枚，唯 3 层下部有孔虫简单分异度（属种数）和丰度均显著下降；所含有孔虫主要为 *Florilus decorus*-*Ammonia tepida*-*Globigerina* sp. 组合，并出现不少深水种 *Lagena* sp.，反映为滨岸浅海相；古气候总体上暖热湿润，只是中间（孔深 51~57m）3 层沉积时气候转为温暖，海平面有所下降，处于低海面时期。中段上部和上段沉积期间古气候变为凉干，沉积物内也无有孔虫检出，反映演化为河流边滩（6 层上部）—泛滥湖泊（7 层）沉积。

2. 上海面粉厂 SG2 孔剖面

在长江三角洲滨海平原沉积区内广泛分布的南汇组大都只有中、上两段，缺失下段，如上海面粉厂、外高桥、娄塘、安亭等孔均是，唯国际机场孔例外。现以上海面粉厂 SG2 孔南汇组剖面为代表，该孔于孔深 26.5~73.0m 间钻遇南汇组，总厚度为 46.5m，包括中、上两段地层（图 21-12），所见地层如下：

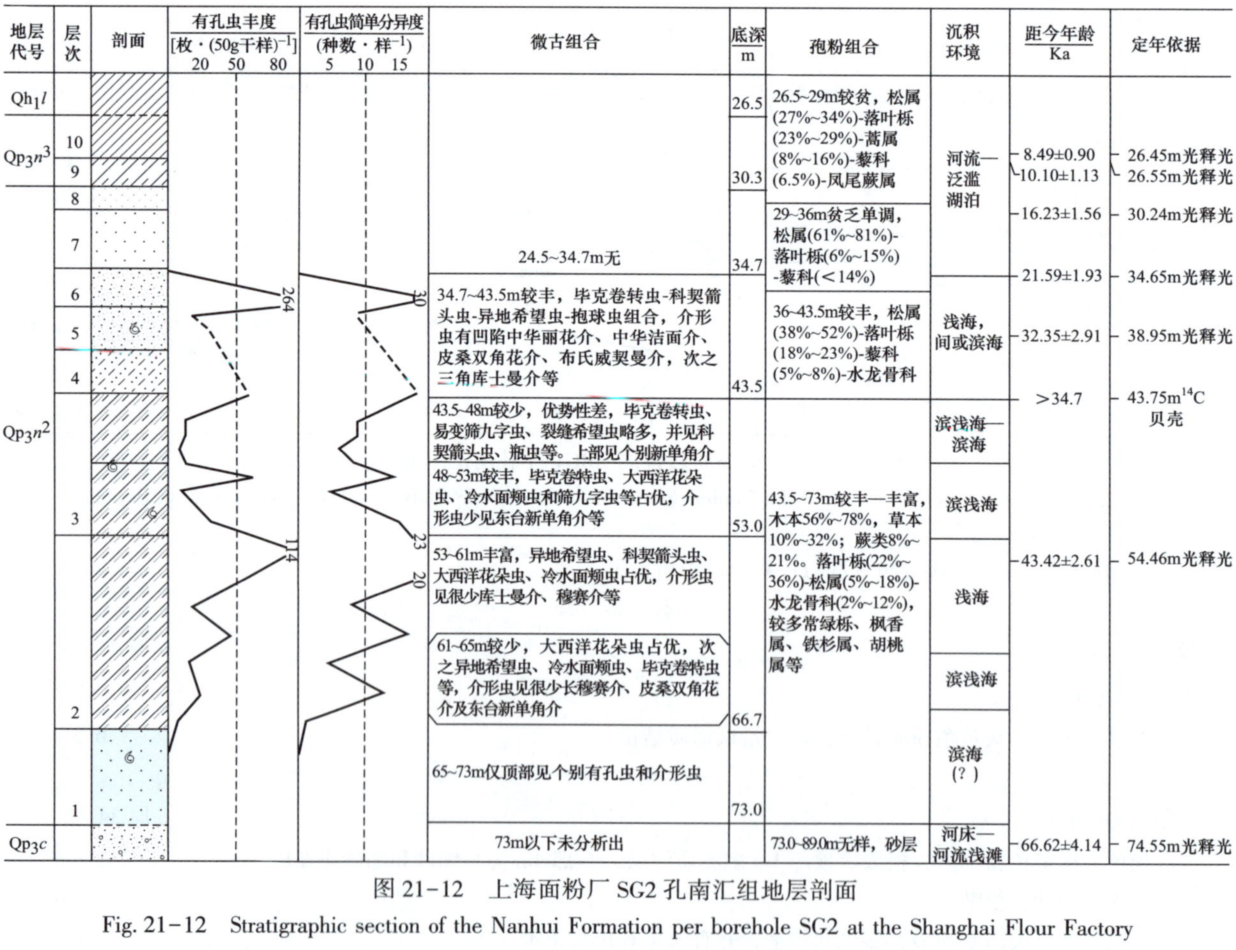

图 21-12 上海面粉厂 SG2 孔南汇组地层剖面

Fig. 21-12 Stratigraphic section of the Nanhui Formation per borehole SG2 at the Shanghai Flour Factory

上覆地层：全新统娄塘组（Qh_1l）褐灰色黏土

----- 假 整 合 -----

南汇组（Qp_3n）	总厚 46.5m
上段（Qp_3n^3）	厚 3.8m
10. 暗绿色黏土，致密，可塑，见有黄褐色锈斑，具细网纹状构造，沿之为灰黑色铁锰质充填，含粒径为 1~3mm 白色的钙质小结核。隐约稍显揉皱变形的斜波状层理	2.3m
9. 褐黄色粉质黏土，致密，可塑，见有宽小于等于 1mm 的灰黑色铁锰质充填的细网纹状构造，含少量粒径为 1~2mm 的铁锰质结核。具揉皱变形的斜波状—微波状层理	1.5m

———— 整 合 ————

中段（Qp_3n^2）	厚 42.7m
8. 褐黄色砂质粉土，上部夹少量粉质黏土	1.2m
7. 褐黄色粉砂，松散—稍密，显斜波状、微波状层理，有较多水平状铁染细纹	3.2m
6. 暗黄灰色细—粉砂，以粉砂为主，松散—稍密，发育斜层理	2.0m
5. 暗灰色粉细砂，局部见贝壳碎片，隐约显斜波状层理。局部夹有砂质粉土和粉质黏土纹层	2.3m
4. 青灰色细砂夹砂质粉土。砂中见冲刷构造，有粒径为 2~10mm 的泥砾。近底部显波状交错层理	4.5m
3. 青灰色粉质黏土夹浅灰色黏质粉土，细层厚为 1~10mm，局部为两者薄互层。少数稍厚的粉土内含细小泥砾。具不规则水平层理、微波状层理；近顶部富集贝壳碎片及腐殖物，还含有粒径为 1~1.5cm 的燧石质砾，显波状、波状交错层理，并见有延深近 1cm 的陡倾斜虫孔构造	9.5m
2. 灰色粉质黏土、粉细砂薄互层，以前者为主，呈 1~5mm 厚的纹层，与后者厚 1~15mm 的纹层构成薄互层；有前者在砂中呈脉状。多具不规则水平—微波状层理，粉细砂具斜波状层理	13.7m
1. 灰色粉细砂夹薄层粉质黏土，上部为两者薄互层。砂单层厚为 1~20mm，粉质黏土厚为 1~4mm，具微波状层理，偶见脉状的粉质黏土	6.3m

----- 假 整 合 -----

下伏地层：上更新统川沙组（Qp_3c）青灰色含砾中细砂

上述剖面揭示，SG2 孔缺失南汇组下段。孔深 34.7~73m 的第 1~6 层组成南汇组中段中下部，超覆于川沙组之上，总体含有孔虫较丰，除大都出现广盐性有孔虫 *Ammonia beccarii*、*Florulus atlanticus* 外，还普遍见正常浅海相的 *Elphydium advenum*、深水种 *Bolivina cochei* 和瓶虫类，与半咸水和正常海相介形虫共生，综合孢粉资料分析，为暖湿气候下的滨岸—浅海或为滨海沉积；据定年资料推算其形成年龄介于 64.83~21.72ka。孔深 30.30~34.7m 的 7~8 层组成中段上部，为向温凉偏干气候过渡环境的河流边滩沉积，形成于 21.72~16.30ka。孔深 26.50~30.30m 的 9~10 层组成南汇组上段，为温凉干燥气候下泛滥湖泊堆积的褐黄—暗绿色硬土，即更新统顶部的标志层，推算形成于 16.30~9.94ka。顶界的年龄值显然偏小，这可能与上段地层沉积后又经历了暴露风化的过程有关，致使其定年结果比地层沉积的年龄减小。

3. 浦东机场 J25 孔剖面

J25 孔是滨海平原东南滨岸地带钻遇完整的南汇组地层剖面钻孔之一。该孔在孔深 29.4~81.95m 间见南汇组，总厚度为 52.55m，3 段地层齐全。所见地层如下：

上覆地层：全新统娄塘组（Qh_1l）灰色黏土夹粉砂

----- 假 整 合 -----

南汇组（Qp_3n）	总厚 52.55m
上段（Qp_3n^3）	厚 3.00m

12.	暗灰绿色黏质粉土，夹有粉砂，具水平层理、斜层理	0.51m
11.	草黄色黏质粉土，局部与粉砂薄互层，并具水平层理	2.49m

———— 整　合 ————

中段（Qp_3n^2） 厚38.31m

10.	灰黄色砂质粉土，具灰绿、锈黄色网纹，含较多顺层锈黄色条纹及云母富集形成的波状条纹，局部夹有粉砂、粉质黏土纹层，含泥砾、淡水螺壳碎片。具不规则水平—波状层理	9.51m
9.	灰黄色粉细砂，中上部断续见有饼状钙结石，具断续微波状层理、水平层理	6.59m
8.	灰黄色粉砂、粉质黏土薄互层，具水平层理、微波状层理	3.41m
7.	草黄色粉砂，具斜波状层理。中偏上部有0.6m砂质粉土、粉细砂薄互层，具水平层理；下接2.1m粉细砂夹粉质黏土，具水平—斜层理	9.29m
6.	灰黄色细砂，底界为冲刷面，与下伏层位突变接触	2.45m
5.	灰黄色砂质粉土，隐约显水平层理	6.45m
4.	灰黄色砂砾，含砾径为0.2~2.5cm的砾石，底部较大。含沼螺及碎片，多沿层面富集	0.61m

-----假　整　合-----

下段（Qp_3n^1） 厚11.24m

3.	灰色黏土夹黏质粉土，局部含钙质结核	2.61m
2.	灰色黏土、粉砂薄互层，显水平层理。底部为0.27m含砾粉细砂，分选差，砾呈扁圆状，径大者有2cm	5.60m
1.	灰色粉细砂夹粉质黏土纹层，上部偶含次圆、次棱角状砾。具不规则水平—波状层理	3.48m

———— 整　合 ————

下伏地层：上更新统川沙组（Qp_3c）灰色粉砂夹砂质粉土

据微古分析，下段无化石检出，为陆相泛滥平原沉积，间或在洪水期有串沟河道沉积的砾砂层。中段4~6层唯在5层中下部见少量螺旋壳、平旋壳类有孔虫，有*Ammonia beccarii*、*A. compressiuscula*、*Elphydiumadvenum*、*Florilus* sp.、*Cribrononion* sp.，还见个别的淡水介形虫*Candoniella albicans*完整个体，为溺谷沉积。7~8层出现分异度（9~17种）、丰度（数十枚/20g样）较低的有孔虫，优势种有*Ammonia beccarii*、*Florilus* sp.、*Globigerina* sp.、*Bolivina cochei*，其次较多*Elphydium advenum*，中下部还常见瓶虫类透明瓶虫（*Lagina perlucida*）、亚线纹瓶虫（*L. substriata*）；列式壳类卡纳利葡萄虫（*Uvigerina canariensis*）、西方三棱虫（*Trifarina occidentalis*）等深水种；介形虫断续出现，属种、个数波动较大，较多出现日本穆赛介（*Munseyella japonica*）、东台中华花介（*Sinocythere dongtaiensis*）、皮桑双角花介（*Keijella bisanensis*），其次为滨海弯贝介（*Loxoconcha binhaiensis*）、中华洁面介（*Albileberis sinensis*），还出现三角库士曼介（*Cushmanidea triangulata*）、三浦翼花介（*Cytheropteron miurense*）等海相种；7层下部兼有淡水种泡丽星介（*Cypria bulloides*）、湖花介（*Limnocythere* sp.）；反映7~8层为滨岸浅海相，7层下部受河口淡化海水影响。9~10层下部有有孔虫断续出现，属种（最多不过8种）、个数（最多20~30枚/20g样）均很少，主要有*Ammonia beccarii*、*Elphydium advenum*、*Florilus* sp.，还见有*Globigerina* sp.、*Bolivinacochei*，*A. compressiuscula*及*Cribrononion* sp.等；以9层上部较多，相伴出现*Keijella bisanensis*、*Cushmanidea triangulata*和小海花介（*Pontocythere parva*）等海相种，未见陆相介形虫，反映为滨海沉积；10层中部（孔深35.8~38.8m）有孔虫分异度（12~13种）、丰度（60~100枚/20g样）均略高，优势种有*Ammonia beccarii*、*Florilus* sp.、*Globigerina* sp.，还有较多的*A. compressiuscula*、*Bolivina cochei*、*Elphydium advenum*和富尔先科虫（*Fursenckoina* sp.）等深水种；相伴断续出现属种、个数均少的半咸水和海相介形虫，有凹陷中华美花介（*Sinocytheridea depssulum*）、布氏威契曼介（*Echinocythereis bradyi*）、中华洁面介（*Albileberis sinensis*），还见有长库士曼介（*Cushmanidea longa*）、三浦翼花介（*Cytheropteron miurense*）、半尾花介（*Hemicytherura* sp.）等海相种，未见陆相介形虫，反映为滨岸浅海沉积；10层上部无微古化石，可能演化为滨海环境。上段均无微古化石，结合岩性特征分析，为泛滥平原沉积。

据孢粉分析，下段具 *Pinus-Quercus deciduous-Betula-Artemisia-Pteris* 组合，以木本植物花粉占绝对优势，还有较多 *Ulmus*，局部出现较多冷杉属（*Abis*），蕨类次之，草本花粉较少，反映古气候温凉湿润。中段一般含孢粉很少，可能因多与滨海、滨浅海砂质沉积物有关；局部孢粉丰富，以木本花粉占绝对优势，落叶阔叶植物为主，针叶植物次之，草本、蕨类花粉很少，具有 *Quercus deciduous-Pinus-Ulmus-Picea*、*Abies-Gramineae-Pteris* 组合，还见有较多 *Liquidambar*，反映古气候温暖湿润。上段孢粉贫乏，见有很少的 *Pinus*、*Quercus deciduous*、*Castanea*，草本的 *Artemisia*、蕨类的卷柏属（*Selaginella*），反映古气候温凉稍干（?）。

4. 庙镇 SG1 孔剖面

在长江三角洲河口沙岛沉积区内分布的南汇组一般只保存中段地层，其上、下段地层均缺失。以庙镇 SG1 孔剖面为例，该孔在孔深 50.25~67.80m 间钻遇南汇组，唯保存中段地层（图 21-13）。所见地层如下：

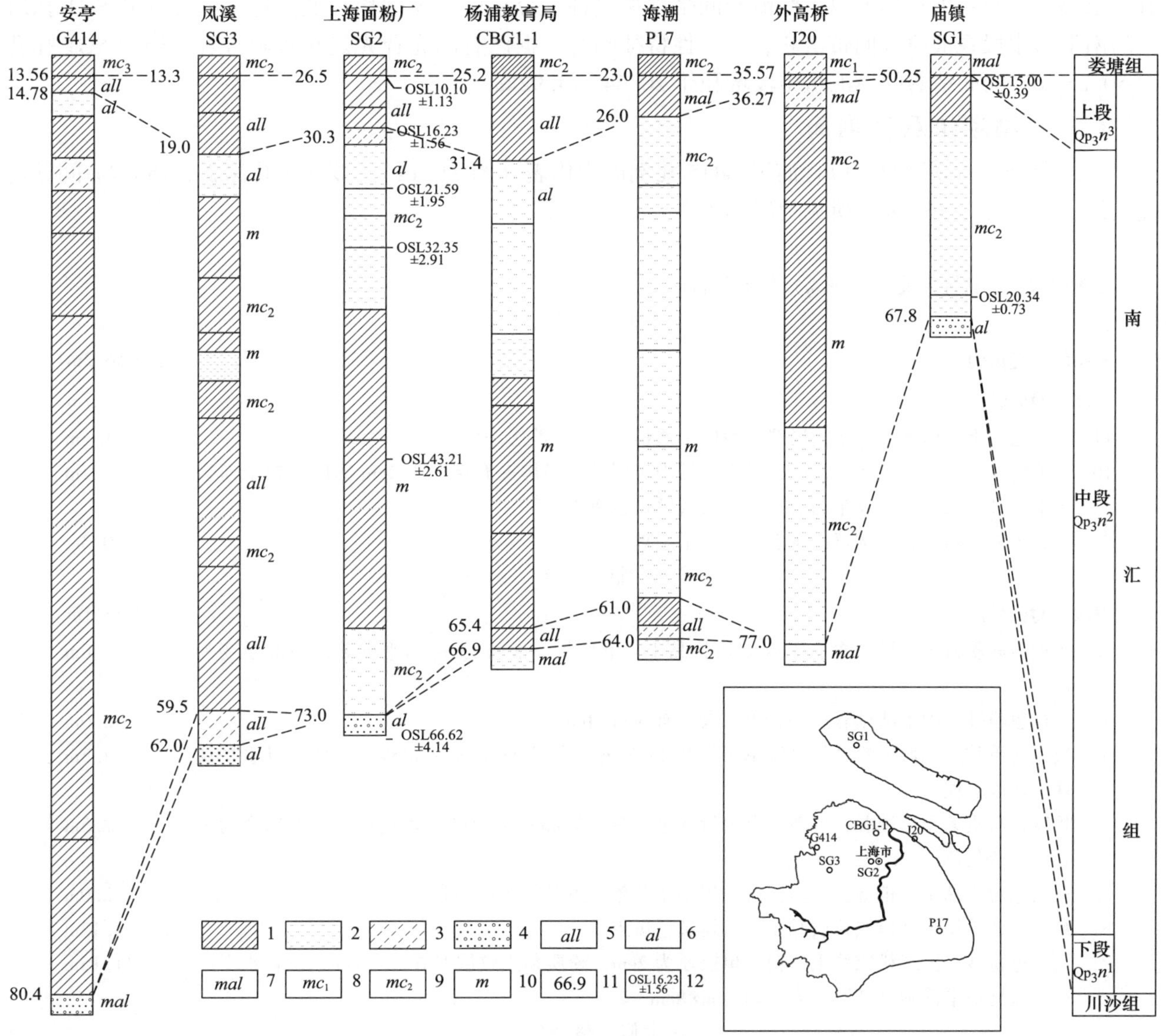

图 21-13 上更新统南汇组地层对比图

Fig. 21-13 Stratigraphic correlation of the Upper Pleistocene Nanhui Formation

1—黏土、粉质黏土；2—粉砂、细砂；3—砂质粉土、黏质粉土；4—含砾砂；5—泛滥平原—泛滥湖泊相；6—河流相；7—三角洲平原河湖相；8—河口相；9—滨海沼泽、潮坪、海湾相；10—滨岸浅海相；11—深度（m）；12—光释光法定年值（ka B. P.）

上覆地层：全新统娄塘组（Qh_1l）灰色黏质粉土与灰白色粉砂互层

----- 假 整 合 -----

南汇组中段（Qp_3n^2）		厚 17.55m
3.	深灰色粉质黏土与浅灰色粉砂互层，以粉质黏土为主，粉砂细层厚度一般小于 3mm。具波状、斜波状、透镜状层理	3.35m
2.	浅灰色粉砂与深灰色黏质粉土。中部细粉砂层厚者有 15~20cm，局部含粉砂云母较多；下部砂层略粗，局部变为粉细砂。具脉状、羽状层理，见冲刷构造	12.64m
1.	浅灰色粉砂与深灰色黏质粉土互层	1.56m

----- 假 整 合 -----

下伏地层：上更新统川沙组（Qp_3c）黄灰色含砾中粗砂

据微古分析，2 层内断续检出少量有孔虫和个别半咸水介形虫，1、3 层内未发现微古化石。所见有孔虫属于少盐—中盐的河口、滨海潮间带常见属种，分异度、丰度均很低。这一特征反映 SG1 孔所见的南汇组中段系滨海潮间带沉积，与岩性特征吻合。据已有的光释光定年资料推算，庙镇 SG1 孔孔深 50.25~67.80m 钻遇的南汇组中段堆积于 20.64~14.92ka。

（三）滆湖组层型剖面

区内所分布的滆湖组可以青浦镇 G415 孔剖面为代表（图 21-14）。该孔于孔深 2.5~58.7m 间钻遇滆湖组，总厚度为 56.2m。所见地层如下：

上覆地层：全新统青浦组（Qh_3q）灰褐色黏土

----- 假 整 合 -----

滆湖组（Qp_3g）		总厚 56.2m
上段（Qp_3g^3）		厚 5.7m
11.	暗绿色黏土，稍湿，可塑，局部有粉砂小透镜体，具均匀层理	1.7m
10.	褐黄色黏土，上部黄绿色，湿，可塑，具均匀层理；局部夹有层厚为 2cm 的砂质粉土而显水平层理。见有垂向分布的蓝灰色条纹及锈黄色斑点	3.7m
9.	褐黄色粉质黏土，有锈黄色斑点，具水平层理	0.3m

———— 整 合 ————

中段（Qp_3g^2）		厚 33.1m
8.	褐黄色砂质粉土，局部含有不规则块状黏土，下部见一 3cm×3cm 扁平砂姜，底部渐变为粉砂	3.8m
7.	褐黄色粉砂，中下部见有扁平砂姜，大者有 3cm×4cm	2.0m
6.	灰色粉质黏土夹薄层粉砂，粉砂单层厚在 1~3mm，偶见 1cm×3cm 粉砂透镜体。具水平层理、透镜层理	7.5m
5.	暗绿色粉质黏土，稍湿，可塑，有少量灰色条带，局部有不规则锈黄色斑点，下部变为黄绿色夹蓝灰色	2.5m
4.	黄灰色粉质黏土，稍湿，可塑，有不规则锈黄色斑点及褐黄色条带	2.0m
3.	灰黄色砂质粉土夹薄层粉质黏土，有锈黄色斑点	4.15m
2.	灰黄色粉砂夹薄层粉质黏土，粉砂单层厚为 2cm，粉质黏土纹层厚在 2~3mm，发育水平层理。偶见扁平浑圆钙结石，大者为 4cm×8cm	11.15m

----- 假 整 合 -----

下段（Qp_3g^1）		厚 17.4m
1.	灰色粉质黏土夹薄层粉砂或砂质粉土，湿，可塑，粉砂也有呈不规则团块状的。上部含有大小不一的扁平钙结石，介于 1cm×2cm~5cm×7cm；局部含黄色泥钙质结核及黑色有机物。粉砂中含少量贝壳碎片	17.4m

———— 整 合 ————

下伏地层：上更新统川沙组（Qp_3c）灰色细砂

据微古分析，1 层未发现化石，为河漫滩沉积。2 层下部见属种、个数均少的有孔虫和介形虫，大都为广盐性半咸水有孔虫，也见个别深水种 *Lagina hispida* 和浮游有孔虫 *Globigirina falconensis*；介形虫仅见个别海相种，未见陆相种，应属滨海沉积。3～5 层未发现化石，为河流边滩—泛滥湖泊沉积。6 层见较少属种、个数的有孔虫和介形虫，大都为广盐性半咸水有孔虫，少数浮游有孔虫 *Globigirina falconensis*、个别深水种 *Bolivina cochei*；介形虫也多为半咸水属种，有很少的海相种 *Cytheropteron kiangsuensis*、*C. miurense*，未见陆相种，属滨海沉积。再在 8 层底部见较少属种、个数的有孔虫和介形虫，大都为广盐性半咸水有孔虫，有少数浮游有孔虫 *Globigirina* sp. 及深水种 *Lagina spicata*、瓷质壳类 *Quinqueloculina sabulosa*；介形虫也为半咸水属种，偶见海相种 *Munseyella* sp.，未见陆相种；其下伏的 7 层及上覆的 9～11 层均无微古化石出现，加之 7、8 层呈渐变，而与 6 层岩性突变等沉积特征分析，7～11 层当大体属于陆相的河流边滩（7～8 层）—泛滥湖泊（9～11 层）沉积，唯 8 层底部沉积时曾一度沦为溺谷。

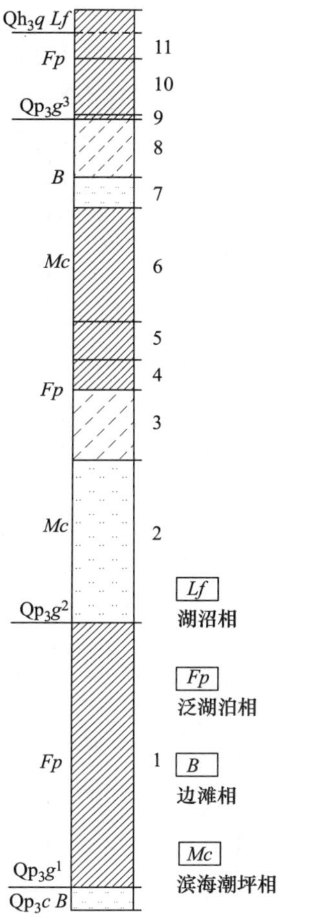

图 21-14　G415 孔滆湖组地层剖面

Fig. 21-14　Stratigraphic section of the Gehu Formation per borehole G415

据孢粉分析，1 层含 *Pinus-Polypodiaceae-Pteris* 组合，木本松属和蕨类孢粉占明显优势，还含其他针叶植物、少量落叶阔叶和个别常绿阔叶成分，草本花粉贫乏，反映气候温凉湿润；2 层含 *Quercus diciduous-Q. glauca-Castanea-Glamineae-Pteris* 组合，阔叶植物花粉占绝对优势，反映气候暖热湿润；3～5 层无孢粉，相应为不含微古化石的陆相层，可能为凉干气候环境；6 层下部含 *Chenopodiaceae-Pinus-Quercus diciduous-Polypodiaceae* 组合，草本藜科占明显优势，反映气候温凉干燥；6 层中上部至 8 层含 *Polypodiaceae-Quercus diciduous-Pinus-Chenopodiaceae* 组合，以阔叶成分及蕨类孢粉略占优势，草本减少，但藜科含量仍较高，反映气候温暖偏干；9～10 层无孢粉，可能为凉干气候环境；11 层含 *Chenopodiaceae-Polypodiaceae-Pinus* 组合，以草本藜科花粉占绝对优势，蕨类及木本花粉较少，反映为凉干气候环境。

根据与青浦镇 G415 孔邻近的青浦镇 A5 浅孔（孔深 24.8m）的 ^{14}C 和热释光（TL）定年资料：

1）孔深 5.5m 处褐黄色黏土，TL 定年值为 14630±1440a，相当于 G415 孔 10 层上部；

2）孔深 9.1m 处褐黄、灰黄色粉砂夹黏土，TL 定年值为 21250±1530a，相当于 G415 孔 8 层上部；

3）孔深 19.3m 处深灰色黏土，^{14}C 定年值为 29600±1800a B. P.，相当于 G415 孔 6 层下部；

4）孔深 22.6m 处灰黄杂灰色粉质黏土，TL 定年值为 49840±3020a，相当于 G415 孔 4 层顶部。

据上述定年资料推算：G415 孔滆湖组上段成岩年龄为 15.37～10.03ka，与上述 SG2 孔的南汇组上段成岩年龄（16.30～9.94ka）基本相同；上海西部的“第二硬土层”，即滆湖组中段出现的暗绿色黏土（上述 G415 孔 5 层）成岩年龄在 44.93～32.67ka。上述 G415 孔微古分析资料揭示，中部陆相层（3～5 层）将滆湖组中段的海相层分隔为上、下两个亚层。现由定年资料推算：上亚层形成年龄介于 33～20ka；下亚层形成年龄约在 45ka 以前。

（四）滆湖组对比剖面

青浦练塘 P31 孔（20 世纪 80 年代初施工的水文地质深孔，浅部 24m 利用 2002 年 4 月原址附近施工的 A4 号浅孔资料），在孔深 3.9～59m 间钻遇滆湖组，总厚度为 55.1m。所见地层剖面（图 21-15）如下：

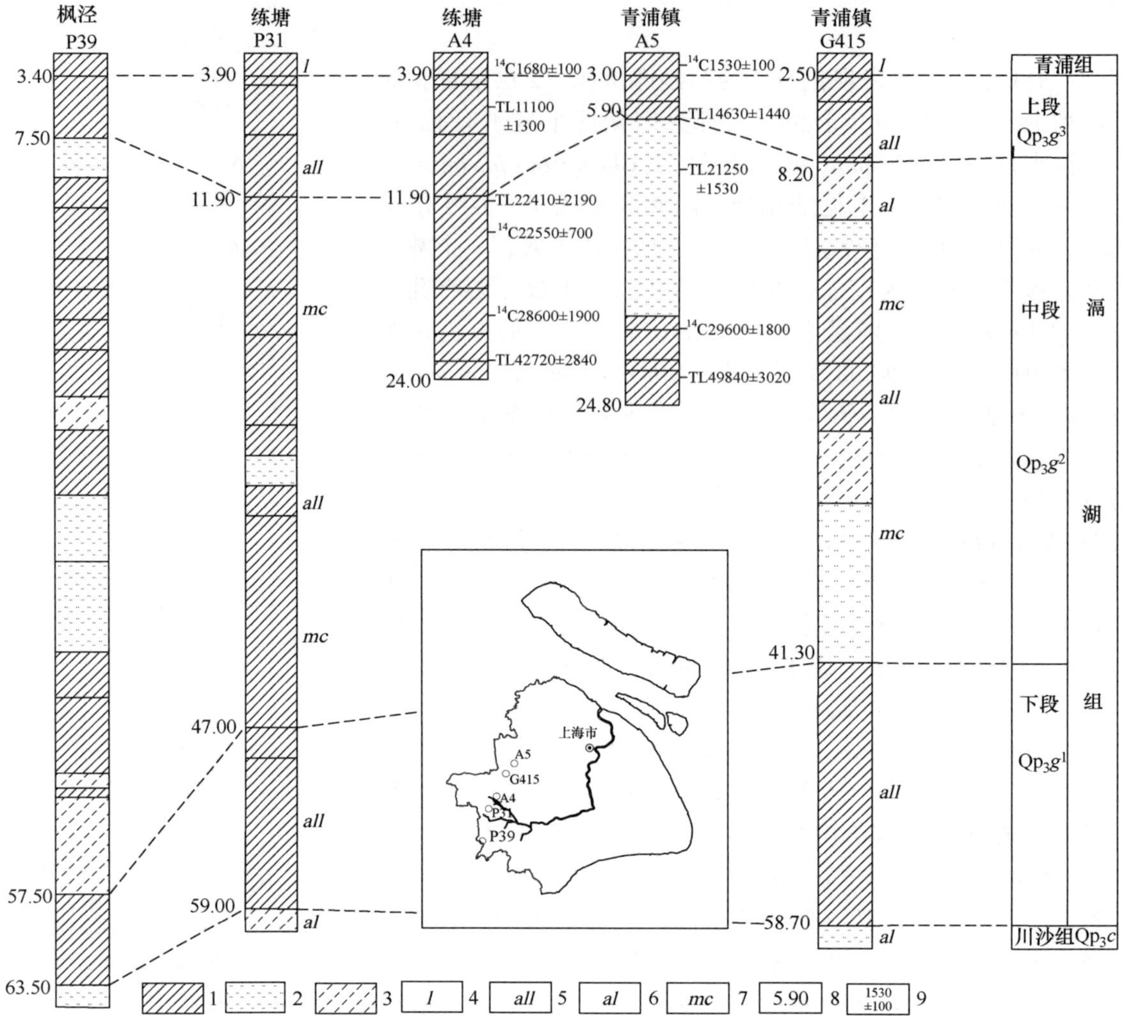

图 21-15　上更新统滆湖组地层对比图

Fig. 21-15　Stratigraphic correlation of the Upper Pleistocene Gehu Formation

1—黏土、粉质黏土；2—粉砂、细砂；3—砂质粉土、黏质粉土；4—湖泊相（据微古分析，下同）；
5—泛滥平原—泛滥湖泊相；6—河流相；7—滨海沼泽、滨海、潟湖相；8—深度（m）；
9—定年值（a B. P.，TL 为热释光法，^{14}C 为放射性碳法）

上覆地层：全新统青浦组（Qh_3q）深灰色淤泥质黏土夹泥炭，距层底 0. 5m 处^{14}C 定年值为 1680±100a B. P.

----- 假　整　合 -----

滆湖组（Qp_3g） 总厚 55. 1m

上段（Qp_3g^3） 厚 8. 0m

13. 暗绿色黏土，可塑。仅见陆相的刺盒虫、圆盘形硅藻，为湖泊相 0. 6m

12. 褐黄杂灰色黏土，有灰色斑纹沿裂隙分布，含少量铁锰质结核和薄膜。见个别广盐性半咸水有孔虫 *Ammonia beccarii*、*Cribononion poeyanum* 及陆相的刺盒虫、圆盘形硅藻，为偶受海水泛及的泛滥湖泊相。距层顶 1. 2m 处热释光定年值为 11. 10±1. 30ka 3. 3m

11. 棕黄、土黄色粉质黏土，可塑，含较多铁锰质薄膜和少量结核，湖泊相。距层底 0. 4m 处的热释光定年值为 22. 41±0. 19ka 4. 1m

----- 假　整　合 -----

中段（Qp_3g^2） 厚 35. 1m

10. 深灰色淤泥质黏土，切面光滑，偶见螺贝壳。见 *Cribononion poeyanum*-*Pseudoeponides compressum* 半咸水有孔虫组合及陆相的水蕨孢子，为潟湖相。距层顶 2. 1m 处的^{14}C 定年值为 22550±700a B. P. 6. 1m

9. 深灰色淤泥质粉质黏土，单层厚在1~3mm，水平层理发育。有优势性极强的*Ammonia beccarii-Sinocytheridea latiovata*组合，还有陆相的水蕨孢子、轮藻、有壳变形虫，为滨海沼泽相。距层顶1.2m处的^{14}C定年值为28600±1900a B.P. 3.0m

8. 深灰、灰绿色黏土，切面光滑，可塑，具均匀层理。无微古化石，为湖泊相。距层底0.3m处的热释光定年值为42.72±2.84ka 1.8m

7. 灰绿夹杂褐黄色粉质黏土，以灰绿色为主，夹有顺层的褐黄、锈黄色纹层，为泛滥湖泊相 4.2m

6. 黄色粉砂。见少量*Ammonia beccarii*、*Cribononion poeyanum*、*Pseudononionella variabilis*等半咸水有孔虫及陆相的水蕨孢子、刺盒虫等，为滨海相 2.0m

5. 黑色黏土，含腐殖物较多。见个别有孔虫及陆相的水蕨孢子、刺盒虫等，为滨海沼泽相 2.0m

4. 以蓝灰色为主，与浅棕、褐黄色间互吴皱纹状黏土，可塑。见个别有孔虫及陆相的水蕨孢子、刺盒虫等，为潟湖相 2.0m

3. 杂色皱纹状黏土，以褐黄色为主，与蓝灰、棕色纹层间互呈皱纹状，可塑。见优势性极强的*Ammonia beccarii*有孔虫及*Leguminocythereis hodyii-Sinocytheridea latiovata*组合，还有陆相介形虫*Candona* sp.及水蕨孢子、小双壳类，为潟湖相 14.0m

----- 假 整 合 -----

下段（$\mathbf{Qp_3g^1}$） 厚12.0m

2. 暗棕灰色粉质黏土，为暗灰、棕色间互呈皱纹状，可塑。见个别有孔虫及水蕨孢子、刺盒虫，为偶受海水泛及的湖泊相 2.0m

1. 浅蓝灰色粉质黏土，可塑。见个别有孔虫，还有陆相的水蕨孢子、刺盒虫，为偶受海水泛及的湖泊相 10.0m

———— 整 合 ————

下伏地层：上更新统川沙组（Qp_3c）灰色砂质粉土

（五）南汇组/漏湖组地层总体特征

如上所述，南汇组/漏湖组为末次冰期期间形成的陆相—海相/海、陆相交替—陆相沉积序列，相应分为下、中、上3段地层。

1. 南汇组下段（$\mathbf{Qp_3n^1}$）/漏湖组下段（$\mathbf{Qp_3g^1}$）

其埋深大体上由西南向东北加大，底界埋深在52~90m，厚度在0~17m不等。岩性、岩相与厚度变化较大。

该段地层在上海西部的太湖湖沼平原区内连续分布；在滨海平原区内，浦东、宝山滨岸地带和嘉定、闵行、市区等地的部分地段缺失，浦东南部、奉贤等地基本连续分布；而在河口沙岛地区因上覆砂层的下超而全部缺失。其典型岩性为泛滥平原—湖泊堆积的灰绿、灰黑、深灰色及局部的杂色黏土、粉质黏土和黏质粉土；局部夹有砂质粉土或粉砂层。含钙质结核、腐木碎块，见有陆相软体动物化石碎片。

分布于西南部青浦与松江、金山接壤区的漏湖组下段（Qp_3g^1）埋深略浅，埋深介于41~59m，厚度为5~17m。如：金山枫泾P39孔孔深44.5~51.5m钻遇该段；松江镇G417孔孔深49.7~53.7m钻遇；青浦练塘P31孔孔深49~59m钻遇；青浦镇G415孔孔深41.3~58.7m钻遇。岩性主要为暗绿、灰绿，部分为深灰、棕灰、蓝灰、灰黄色黏土、粉质黏土，呈可塑性硬土，局部含钙结石和泥灰质结核；夹有砂质粉土或粉砂纹层。所含化石有陆相微体化石水蕨孢子、刺盒虫，偶见个别的有孔虫。表明为偶尔受海水泛及的泛滥平原—湖泊沉积。

除青浦与松江、金山接壤区以外，天马山以东的太湖湖沼平原和滨海平原区分布的为南汇组下段（Qp_3n^1），其与漏湖组下段（Qp_3g^1）的岩性、岩相相同，年代相当，只是地层发育程度及分布连续性较差，地层底界埋深起伏及厚度变化稍大。该段地层埋深在56~90m之间，通常介于56~73m；厚度在1.5~12m不等，大都在2~4m。所见岩性主要为暗绿、灰绿色，部分深灰、褐灰和杂色黏土、粉质黏土，多呈可塑性硬土；夹有灰绿、深灰色砂质粉土或粉砂。通常无微古化石，局部如浦东横沔P12孔偶见个别有孔虫和陆相介形虫*Ilyocypris* sp.；在金山朱泾，本段上部出现半咸水介形虫*Sinocythe-*

ridea longa、*Leguminocythereis hodgii*、*Neomonoceratina dongtaiensis*，也见陆相化石轮藻，还见到淡水软体化石珠蚌（*Unio*）、河兰蚬（*Corbicula*）、田螺（*Viviparus*），表明通常为泛滥平原—湖泊沉积，局部受海水影响。

在河口区的崇明、长兴、横沙诸岛上，南汇组下段均被侵蚀而缺失，只保存有埋深约在75m以上的中段地层。

2. 南汇组中段（Qp_3n^2）/滆湖组中段（Qp_3g^2）

该段地层在全区普遍分布，厚度占上更新统上部地层的大半。其埋深由西南向东北逐渐加大，底界埋深在41~90m，顶界埋深在10~60m。厚度在13~61m不等，主要因上覆全新统底界的起伏而增减。

根据本段地层在全区出现的岩性、岩相变化，可分为以下3片。

在陆域东部的现滨海平原区，大致包括嘉定、宝山、市区、浦东等地，本段的埋深居中，厚度较大。埋深大都介于30~70m，厚度为15~61m不等，大都为30~40m。其北部（大概在浦西—张桥—顾路一线以北）地区基本为浅海—海湾相灰色粉质黏土频夹浅灰色细砂、砂质粉土纹层或呈薄互层，上部常为河口滨海相灰、灰黄色粉细砂夹砂质粉土，如上海面粉厂J10、外高桥造船基地J20及嘉定安亭G414等孔所见；其以南地区为滨海—浅海相灰、灰黄色粉细砂、砂质粉土互层夹粉质黏土，如浦东国际机场J25等孔所见。

在陆域西部的湖沼平原区及中南部的闵行、奉贤等地，本段埋深相对较浅，顶界埋深为7~30m，底界埋深为45~65m，厚度为25~40m。大致在天马山残丘群西北的青浦及与之接壤的松江、金山部分地区分布的为滨海—潟湖相与河流边滩、泛滥平原—湖泊相交替的暗灰、褐黄和杂色黏土、粉质黏土，时夹黄色粉砂、砂质粉土或呈互层，即为“滆湖组”所在。其标志岩性为深20~25m间赋存的陆相暗绿—褐黄色黏土层，历来称为西部“第二硬土层”。以往根据特征岩性等标志，将之与市区相当埋深的硬土层对比；现经热释光定年证明，这是40ka左右一次寒冷气候事件导致西部地区海退而形成的陆相堆积。

在崇明等河口沙岛区本段埋藏较深。顶界埋深一般在45~55m，底界埋深在65~75m，厚度多在20m左右。其中下部通常为灰、黄灰色中细砂、粉细砂夹粉土或呈互层，底部或出现砂砾层；上部多为青灰、灰色粉细砂夹褐灰色粉质黏土、粉土或呈互层。为河流—溺谷和滨海堆积。

3. 南汇组上段（Qp_3n^3）/滆湖组上段（Qp_3g^3）

本段地层在西部湖沼平原区分布普遍；在中东部的滨海平原区内，除主要沿金山廊下—黄浦江西岸—吴淞口—宝山滨岸—上海北侧与江苏接壤区这一逶迤伸展地带，因受古河道的穿插侵蚀而缺失外，大部分地区可见本段地层分布。而在崇明等河口沙岛上，因受晚更新世末—全新世早期古河道的深切，本段地层已荡然无存。

本段地层埋深由西南向东北逐渐加大，顶界埋深在3~42m，底界埋深在6~48m。其埋深空间变化显示由西向东依次下降的3级阶地，下降坡度约为0.6‰（图21-16）。

较高一级，在松江五厍—佘山东麓—嘉定安亭一线以西的太湖平原地区，本段顶界埋深在3~14m，地层也呈连续稳定分布。

中间一级，大致在上一界线与下一界之线间的滨海平原与太湖平原接壤地区，本段地层分布连续稳定，顶界埋深在14~26m。

较低一级，大致在廊下—虹桥—宝山一线以东的滨海平原中、东部地区，本段顶界埋深在26~35m，埋深起伏较大，因上覆层的下切使本段地层断续分布。

在三者分界地带，尤其是中、西部两级阶地之间，出现本段地层埋深降幅及厚度变化比较显著的阶梯。

组成本段的岩层，下部通常为姜黄、褐黄色粉质黏土，局部夹砂质粉土或粉细砂；上部为暗绿、灰绿或黄绿色黏土和粉质黏土，厚度在0~9m不等，以西部地区较厚，中部较薄，东部地区厚薄不一。就岩性观察，下部的姜黄色砂质粉土夹粉砂层具波状交错层理和微波状层理，显示为河流边滩沉积；

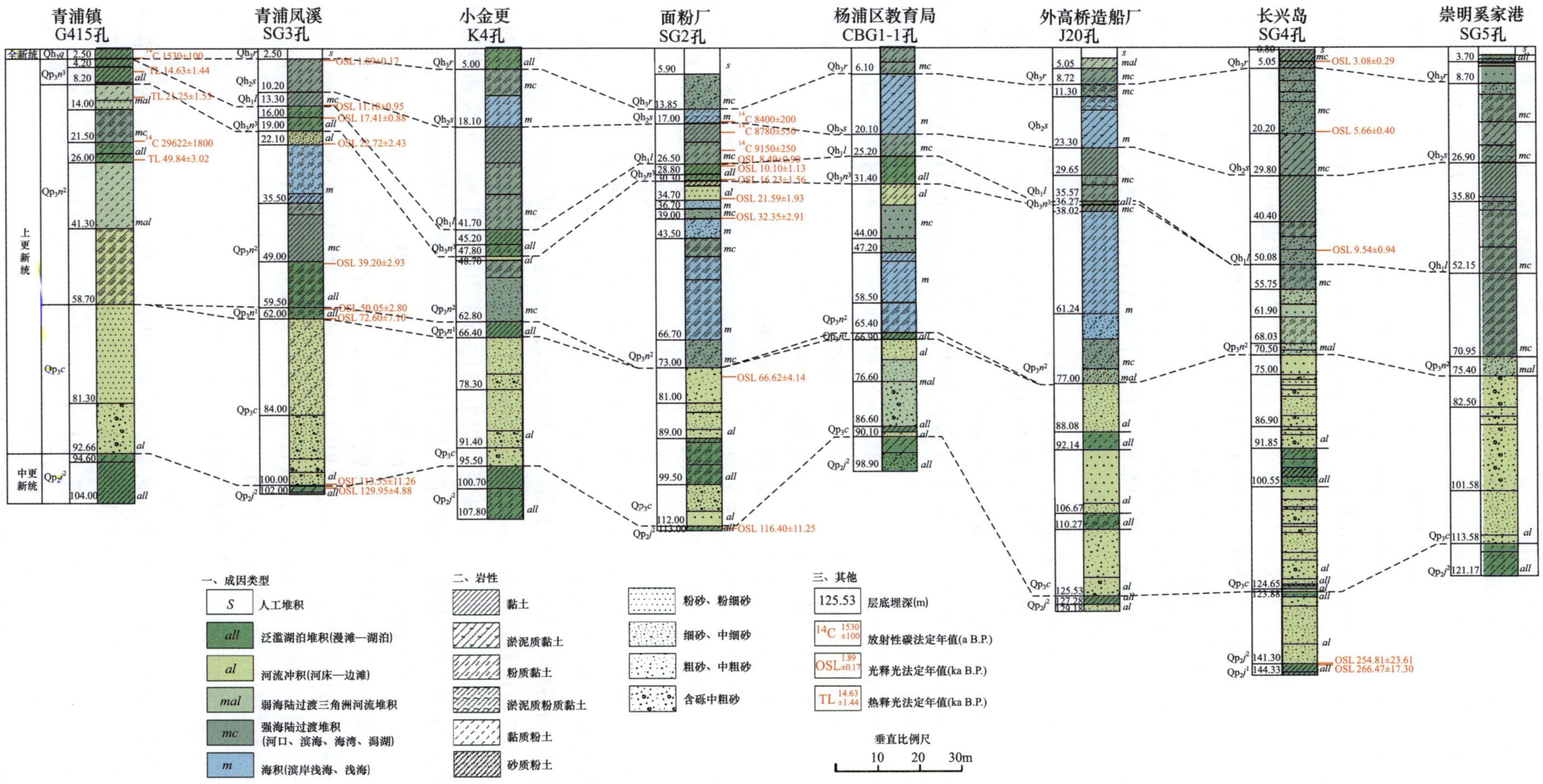

图 21-16　上海北部地区晚第四纪地层对比图

Fig. 21-16　Stratigraphic correlation of the Late Quaternary in northern Shanghai

其上的褐黄色粉质黏土和暗绿色黏土层具揉皱变形的斜波状层理及为铁锰质胶膜充填细微裂隙形成的细网纹构造，表明为泛滥平原—湖泊沉积，之后经历了成陆风化的过程。

据古生物分析，在上海南西部及南汇嘴地区的个别钻孔上段地层内，发现个别的广盐性有孔虫，如青浦练塘 P31 孔见 *Ammonia beccarii*、*Cribononion poeyanum* 和陆相的刺盒虫、圆盘形硅藻；海潮 P17 孔见广盐性有孔虫 *Ammobaculites* sp.、*Cribrononion* sp.、半咸水介形虫 *Leguminocythereis* sp.。在陆域中、北部及现滨海、河口区一般见很少的陆相介形虫，有 *Ilyocypris errabundis*、*Candoniella sulacensis*、*Cypria* sp.、*Candona*sp. 等，局部可见半咸水介形虫 *Alocopocythere transcendens*、*Sinocytheridea longa*、*Cushimanidea japonica* 等，偶见个别有孔虫及大洋桥石藻。反映在总体为泛滥平原—湖泊堆积的背景下，主要在上海南部区域有时还受到海水的影响；这与市区 CY 孔本段地层孢粉研究中发现海水生沟鞭藻，从而得出其在“沉积过程中曾受到海水影响”的结论是一致的。

中段地层以普遍含丰富的海相生物化石为特征，构成广布全区、海相性最强的第Ⅱ海进层（见图21-16）。按其沉积相大体分为两类地区：其一是西部太湖平原及河口沙岛区分布的海相性弱的滨海、潟湖与陆相的河流—泛滥湖泊沉积；其二是东部滨海平原区分布的海相性较强的滨海—浅海沉积。

后者如陆域东部的浦东机场孔，中段地层由下而上出现以下变化：下部（孔深 70.71~61.20m）以含属种、个数均少的广盐性有孔虫为主，并含很少的窄盐种以及个别的淡水介形虫，表明为河口湾内环境；中部（孔深 61.20~48.50m）含微古化石属种、个数有所增多，兼有广盐性、窄盐种和浮游有孔虫，与种数、个数波动颇大的咸水、半咸水介形虫共生，近其底部处出现数种淡水介形虫，表明演化为河口淡化浅海—浅海环境。上部（孔深 48.50~32.4m）含微古化石种数、个数大都减少，并断续出现个别的海相介形虫，在其顶部（深 32.4~35.8m）和近底部（深 44.8~46.8m）处则无化石出现；局部（孔深 35.8~38.8m）兼有略多的广盐性、窄盐种、浮游有孔虫，与断续出现的少量咸水、半咸水介形虫共生，未见淡水介形虫，表明大体为滨海环境，局部为滨岸浅海环境。

在河口沙岛区分布的中段，唯在其上部的灰色粉砂、粉质黏土互层中见少量有孔虫和个别半咸水介形虫，其中、下部砂和含砾砂中无微古化石检出。说明除上部为滨海沉积外，其余为河流沉积。

在西部太湖平原区分布的中段地层，以含种数、个数均少、优势性强的广盐性有孔虫，与很少的窄盐性、浮游有孔虫和半咸水介形虫、陆相微古化石共生的岩层为主，间有无微古化石的夹层。反映区内经历下部的滨海，中部的泛滥湖泊，上部的滨海与泛滥湖泊交替的环境演化。揭示中段地层沉积时，海水在上海西部地区曾有 3 次进退。其中主要的是中部（孔深 20 多米）陆相层即西部“第二硬土层”，可能与市区杨浦教育局 CBG1-1 孔孔深 52~58m 微古分析揭示的低海面层位相当。

上段，即广泛分布于市区等地深约 20m 的暗绿、褐黄色硬土层内，如上所述，可见鱼骨碎片、轮藻受精卵膜等淡水生物化石，通常不含有孔虫等海相生物化石，但在南汇嘴、市区和青浦练塘等上海南部区域的钻孔中陆续发现个别的海相生物化石，揭示海水曾时而泛及当时的泛滥平原—湖泊环境。

孢粉分析表明，南汇组下段地层沉积时植被面貌为针叶、落叶阔叶混交林—草原，古气候温凉偏干。中段地层沉积时，植被面貌先后出现含常绿阔叶落叶阔叶混交林—落叶阔叶、针叶混交林—含常绿阔叶落叶阔叶混交林植被的演化，反映古气候相应经历了暖湿—温湿—暖湿的更替；其上部出现较多指示海侵环境的刺球藻（*Baltisphaeridium*）、极管藻（*Bipolaribucima*）孢子，其中部温和气候的出现与中段低海面层位是相对应的。上段地层沉积时显示以草本植物花粉占优势，并有较多的苔藓类和指示湖泊环境的环纹藻（*Concentrycystis*）等组成的疏林—草原植被，反映古气候凉冷干燥，西部地区湖泊广布。

光释光定年值证明，青浦、练塘等孔深 20m 处所见的陆相层“第二硬土层”，将“滆湖海侵”形成的第Ⅱ海进层分隔为上、下两个亚层：下亚层形成于 45~55ka；上亚层形成于 20~35ka。

据市区和青浦多个光释光定年值表明，南汇组上段上部暗绿色硬土层的形成年龄为 11~17ka，下部褐黄色粉质黏土的形成年龄为 17~20ka；下伏黄色粉细砂层的形成年龄为 20~23ka。

在这一暗绿色硬土层被古河道切割而缺失的地带内，可见一两层埋深在 35~50m 的灰绿色硬土层。

后者在市区、浦东等地多处钻遇，如市区西部小金更 K4 孔孔深 41.7~47.8m 所见即是。上海工程地质界称为“⑤$_4$层”，其土力学指标与暗绿色硬土（称为“⑥层”）相近。“⑤$_4$层”之上时或出现灰黑色粉质黏土夹泥炭层，即“⑤$_3$层”，如嘉定娄塘 P5 孔孔深 41~47m 所见，经^{14}C 定年，表明其形成年龄为 11.0~12.9ka，相当于“新仙女木事件”的同期产物。至于⑤$_4$层的形成年龄，可以推定介于⑤$_3$层与⑥层之间，即在冰后期海侵之前，按目前国内将全新世下限倾向于划在 10~11ka 的方案，⑤$_4$、⑤$_3$层的地层年代无疑应归属晚更新世。

五、娄塘组

（一）层型剖面

1. 嘉定娄塘 P5 孔剖面

嘉定娄塘 P5 孔为命名剖面所在剖面。该孔在孔深 23~41m 钻遇娄塘组，所见地层剖面如下（见图 21-17）：

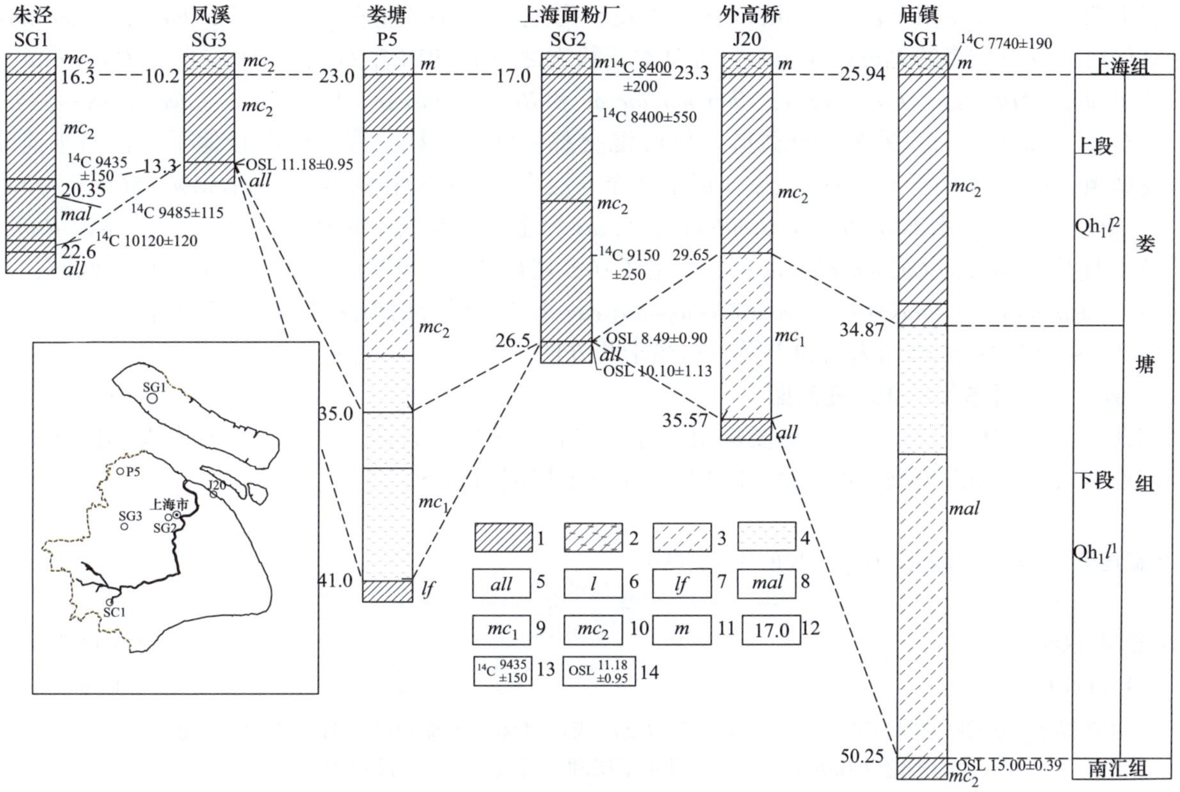

图 21-17　全新统娄塘组地层对比图

Fig. 21-17　Stratigraphic correlation of the Holocene Loutang Formation

1—黏土、粉质黏土；2—淤泥质黏土；3—砂质粉土、黏质粉土；4—粉砂、细砂；5—泛滥平原、泛滥湖泊相；6—湖泊相；7—湖沼相；8—三角洲河湖相；9—河口相；10—滨海潮坪、沼泽、海湾相；11—滨岸浅海相；12—深度（m）；13—放射性碳法定年值（a B. P.）；14—光释光法定年值（ka B. P.）

上覆地层：全新统上海组（Qh_2s）棕灰色砂质粉土

—————— 整　合 ——————

娄塘组（Qh_1l）	总厚 18.0m
上段（Qh_1l^2）	厚 12.0m
5. 棕灰色粉质黏土夹粉砂。具水平薄互层理	2.0m
4. 棕灰色砂质粉土，含云母较多	8.0m
3. 浅灰色粉砂夹砂质粉土，含云母较多，松散。具脉状层理	2.0m

下段（Qh_1l^1） 厚 6.0m

2. 灰色粉细砂夹棕色粉质黏土，含较多云母及腐朽植物细屑集成的细痕。具脉状层理、包卷层理 2.0m

1. 浅灰色粉砂、棕色砂质粉土薄互层，粉砂呈细纹层或透镜体。具水平层理、透镜层理 4.0m

-----假　整　合-----

下伏地层：上更新统南汇组上段（Qp_3n^3）灰黑色粉质黏土

据微古分析，娄塘组下段 1、2 层含有孔虫种数、个数较多，具 *Ammonia dominicana*-*Pseudogyroidina sinensis*-*Nonion usbekistanensis* 组合，并含较多的 *A. beccarii*、*Florilus decorus*、*Turborotalia quinqueloba*、*Cribononion* sp.、*Nonion* sp.、*Globigerina* sp.，顶部出现少量冷水有孔虫，反映属河口相。娄塘组上段 3、4 层含少量有孔虫，有属种 7~16 个，具 *Ammonia beccarii*-*Nonion usbekistanensis* 组合，其次为 *Globigerina* sp.，反映属滨海相。5 层含种数、个数均很少的有孔虫，有 *Ammonia dominicana*、*Nonion* sp. 等广盐种，反映属滨海相。

据孢粉分析，娄塘组下段 1、2 层以含木本花粉为主（占 67%左右），其中以 *Quercus*（*deciduous*）占绝对优势，并有较多的 *Pinus*、*Betula*；含草本花粉次之（占 30%左右），其中多半为 *Gramineae*，另含一些 *Typha*、*Artemisia*、*Compositae*、*Chenopodiaceae* 和 *Ranunculaceae*，具 *Quercus*（*deciduous*）-*Pinus*-*Betula*-*Gramineae* 组合，反映为落叶阔叶、针叶混交林—草原植被，古气候温暖偏干，相当于前北方期；娄塘组上段 3~4 层下部孢粉贫乏；4 层上部至 5 层含木本花粉比前减少（占 50%~56%），草本增多（占 20%~40%）。木本以 *Quercus deciduous*、*Pinus* 为主，并有较多的 *Betula*、*Tilia*；草本以 *Gramineae* 占绝对优势，其次为 *Chenopodiaceae*；蕨类孢子在 5 层内略多，多为 *Polypodiaceae*，往下剧减，具有 *Quercus*（*deciduous*）-*Pinus*-*Gramineae*-*Chenopodiaceae* 组合，反映仍为落叶阔叶、针叶混交林—草原植被，包括 3~5 层的古气候均为温和干燥，相当于北方期。

2. 浦东外高桥造船厂 J20 孔剖面

该孔在孔深 23.30~35.57m 钻遇娄塘组，地层发育齐全，其岩性、岩相在上海中南部区域具有较强代表性，可作为辅助层型剖面（图 21-17）。所见地层剖面如下：

上覆地层：全新统上海组（Qh_2s）灰色淤泥质黏土

———— 整　合 ————

娄塘组（Qh_1l） 总厚 12.27m

上段（Qh_1l^2） 厚 6.35m

3. 灰色黏土，软塑，夹少量厚为 1~3mm 的粉砂纹层或透镜体，上部粉砂中含贝壳碎片，偶见直径约为 1cm 的樱蛤（*Tallina*）壳片。具水平层理、透镜层理、微波层理 6.35m

下段（Qh_1l^1） 厚 5.92m

2. 灰色砂质粉土夹粉质黏土。除顶部 0.3m 两者形成薄互层外，以下夹层较少。具包卷层理、脉状层理 2.16m

1. 浅灰色砂质粉土与灰色粉质黏土互层，前者层厚为 1~5mm，后者为 1~3mm。以水平—微波状层理为主，较少透镜层理 3.76m

-----假　整　合-----

下伏地层：上更新统南汇组上段（Qp_3n^3）黄绿色粉质黏土

据微古分析，1~3 层下部含有孔虫属种、个数多寡不一，优势种为 *Elphidium magellanicum*（占 8.9%~50%，局部无）、*Cribrononion vitreum*（大都 9%~29.6%，部分 2.8%~0），次之抱球虫科（1 层占 30%~40%，2~3 层下部占 0~25%）及其他平旋壳类，并见少量列式壳类、瓷质壳类窄盐性种；介形虫见较多半咸水种 *Sinocytheridea latiovata*，次之为 *Leguminocythereis hodgii*，并见 Cytheropteron sp.、

Munseyella japonica 等海相种，反映为河口沉积。3 层中上部含有孔虫优势种为广盐性属种 *Ammonia beccarii*（占 8%~27.8%）和 *Cribrononion vitreum*（27.1%~38.5%，底部无），介形虫唯见半咸水种 *Sinocytheridea latiovata*，为海湾沉积。总之 J20 孔娄塘组为河口—海湾沉积。

孢粉分析：1、2 层具 *Pinus-Quercus*（*deciduous*）-*Gramineae-Polypodiaceae* 组合，尚见不少 *Picea*、*Abies*、*Quercus*（*evergreen*）、*Larix*；草本花粉的十字花科（Cruciferae）和水生草本 *Cyperaceae*、百合科（Liliaceae）、*Typha*；蕨类有 *Pteridium*、*Pteris*、槐叶苹属（*Salvinia*），反映为针叶、落叶阔叶混交林植被，古气候温凉湿润，气候期大致相当于前北方期。3 层具 *Pinus*-*Quercus*（*deciduous*）-*Quercus*（*evergreen*）-*Polypodiaceae* 组合，还出现不少 *Larix*、*Tsuga*、*Castanea*；草本花粉主要为 *Cruciferae*、*Gramineae*；蕨类尚见较多的 *Pteridium* 及苔藓类（*Bryophyta*）孢子，反映为含常绿阔叶落叶阔叶、针叶混交林植被，古气候温暖湿润，气候期大致相当于北方期。

（二）对比剖面

1. 上海面粉厂 SG2 孔剖面

该孔孔深 17.0~26.5m 揭露娄塘组，有多个定年结果，结合岩性、岩相特征对比，可确切厘定地层时代归属，系相当于外高桥 J20 孔娄塘组上段（见图 21-17）。所见地层剖面如下：

上覆地层：中全新统上海组（Qh_2s）褐灰色淤泥质黏土

—————— 整　合 ——————

娄塘组（Qh_1l）　　厚 9.5m

2. 褐灰色黏土，含灰白色泥钙质结核，往下减少，上部局部富腐殖物根茎并见贝壳碎片　　4.5m

1. 褐灰色黏土，含钙质结核。遍见针孔状气逸构造，具水平层理　　5.0m

----- 假　整　合 -----

下伏地层：上更新统南汇组上段（Qp_3n^3）暗绿色黏土

据微古分析，1、2 层均见广盐性有孔虫，以 *Ammonia beccarii* 为主（其中 2 层内占绝对优势），次之为 *Cribrononion incertum*、*Rosalinabradyi*、*Elphidium* sp. 等；介形虫为半咸水属种，以凹陷中华丽花介（*Sinocytheridea impressa*）为主，次之为中华刺面介（*Spinileberis sinensis*）等。1 层下部 2m 无微古化石。反映基本属滨海沼泽相。

孢粉分析：1、2 层含孢粉丰富，为 *Quercus*（*deciduous*）（占 27%~37%）-*Pinus*（占 13%~26%）-*Castanea*（占 2%~10%）-*Artemisia*（占 4%~9%）-*Pteris*（占 0~5%）组合，还出现不少 *Quercus*（*evergreen*）、*Castanopsis*、*Juglans* 等喜暖热的木本及水生草本花粉，反映为含常绿阔叶落叶阔叶、针叶混交林植被，古气候温暖湿润。

SG2 孔娄塘组内及其上下系列测年结果：*2 层顶界以上 0.25m（孔深 16.75m）处淤泥质黏土，^{14}C 定年值为 8400±200a B. P.；*2 层顶界以下 1.6m（孔深 18.6m）处黏土，^{14}C 定年值为 8780±550a B. P.；*1 层顶界以下 1.95m（孔深 23.45m）处黏土，^{14}C 定年值为 9150±250a B. P.；*1 层底界以上 0.05m（孔深 26.45m）处黏土，光释光定年值为 8.49±0.90ka；*1 层底界以下 0.05m（孔深 26.55m）处暗绿色黏土，光释光定年值为 10.10±1.13ka。据测年结果推算，SG2 孔所见娄塘组顶、底界年龄分别为 8.45ka 与约 9.4ka，其气候期应相当于北方期。

2. 金山朱泾金山气象站 SC1 孔剖面

该孔于孔深 16.3~22.6m 钻遇娄塘组，地层较全，在上海西南部地区较有代表性。加之有多个 ^{14}C 定年样品控制，可据以厘定地层时代的归属（见图 21-17）。所见地层剖面如下：

上覆地层：中全新统上海组（Qh_2s）灰黑色黏土

—————— 整　合 ——————

娄塘组（Qh_1l）　　总厚 6.30m

上段（Qh_1l^2） 厚 4.25m

5. 灰黑色粉质黏土，含大量植物根叶，并含锈黄色泥灰质结核，见有拟蟹守螺（*Cerithidea* sp.）等碎片。一般具均匀层理，偶见不规则水平层理。本层含介形虫个数多，但种数少，优势度高，半咸水种 *Sinocytheridea latiovata* 和 *Spinileberis sinensis* 即占 86%，还有较多的 *Leguminocythereis hodgii* 及很少的 *Loxoconcha binhaiensis*、*Echinocythereis* sp.、*Pontocypris*，反映属滨海沼泽相 3.70m

4. 灰褐色粉质黏土，含少量植物叶片，顺层分布，并含锈黄色泥灰质结核。具水平层理。见很少半咸水介形虫 *Sinocytheridea latiovata* 和 *Spinileberis sinensis*，反映属滨海沼泽相 0.55m

下段（Qh_1l^1） 厚 2.05m

3. 灰黑色粉质黏土，含较多植物叶片，上部尤多，叶片长者近 6cm。偶见半咸水介形虫 *Sinocytheridea latiovata* 和 *Spinileberis sinensis*，属受海水影响的湖沼相 1.10m

2. 灰褐色粉质黏土，含少量植物碎屑及贝壳碎片。具滑塌构造。无微古化石，属湖沼相 0.55m

1. 黑色黏土，含泥灰质结核，见有植物根茎呈缓倾或直立。无微古化石，属湖沼相 0.40m

----- 假　整　合 -----

下伏地层：上更新统南汇组上段（Qp_3n^3）暗灰绿色黏土

SC1 孔娄塘组测年结果：*5 层底界以上 0.8~0.4m（孔深 19.2~19.6m）处灰黑色粉质黏土，^{14}C 定年值为 9435±150a B.P.；*3 层顶界至其下 0.6m（孔深 20.55~21.15m）处灰黑色粉质黏土，^{14}C 定年值为 9485±115a B.P.；*1 层顶至底界（孔深 22.20~22.60m）黑色黏土，^{14}C 定年值为 10120±120a B.P.。

综上表明，SC1 孔所在金山娄塘组的沉积环境经历了两段演化：下段（1~3 层）为湖沼环境，晚期略受海水影响，其气候期大致相当于前北方期，形成年龄为 10100~9500a；上段（4~5 层）遭海水侵进，成为滨海沼泽，其气候期大致相当于北方期，形成年龄小于等于 9500a。两个阶段的晚期均呈现明显的沼泽化，致使植物大量繁盛。

3. 崇明庙镇 SG1 孔剖面

河口沙岛区的全新世地层，包括下全新统娄塘组的厚度最大。庙镇 SG1 孔孔深 25.94~50.25m 钻遇娄塘组，总厚为 24.31m（见图 21-17）。所见地层剖面如下：

上覆地层：全新统上海组（Qh_2s）深灰色淤泥质黏土

———— 整　合 ————

娄塘组（Qh_1l） 总厚 24.31m

上段（Qh_1l^2） 厚 8.93m

3. 深灰色粉质黏土与灰白色粉砂互层，粉砂薄层厚多为 0.5~1cm，见虫孔构造，具波状层理、揉皱变形层理

下段（Qh_1l^1） 厚 15.38m

2. 灰白色粉砂与灰色黏质粉土互层，粉砂薄层厚为 2~3mm，具斜波状、波状层理 4.58m

1. 灰色黏质粉土与灰白色粉砂互层，底部 0.75m 粉砂含量明显增多 10.80m

----- 假　整　合 -----

下伏地层：上更新统南汇组中段（Qp_3n^2）深灰色粉质黏土与粉砂互层

据微古分析，娄塘组下段 1、2 层一般无微古化石检出，唯 1 层近顶部处含有孔虫，基本为广盐性属种，以 *Ammonia beccarii* 占优，其次有 *Cribononion* sp.、*Elphidium nakanokawaense*、*Protelphidium* sp.、*Pseudononionella* sp.、*Rosalina* sp.、*Globigerinoides ruber*；所含介形虫为少量半咸水属种，有中华花介（*Sinocythere* sp.）、中华丽花介（*Sinocytheridea* sp.）、中华洁面介（*Albeleberis sinensis*），还见淡水种玻璃介（*Candona* sp.），反映娄塘组下段 1、2 层可能为三角洲河流相，近顶部处为河口湾相。娄塘组上段 3 层中下部含少量广盐性有孔虫，以 *Ammonia beccarii* 占优，次之为 *Cribononion* sp.、*Florilus* sp.、

Protelphidium sp. 等，并见少量半咸水属种介形虫 *Sinocythere* sp.、*Sinocytheridea* sp. 等；3 层上部 *Ammonia beccarii* 占优，并有较少的平旋壳类和抱球虫，还见有少量的 *Sinocythere* sp.、花花介（*Callistocythereruggieri*）等半咸水介形虫，反映娄塘组上段为滨海沉积。

据孢粉分析，娄塘组下段 1、2 层具 *Quercus*（*deciduous*）-*Pinus*-*Chenopodiaceae*-*Pteris* 组合，木本花粉含有较多的 *Castanea*、*Betula*、*Ulmus*，1 层内尚出现一些 *Quercus*（*evergreen*）、*Juglans* 及 *Picea*、*Abies*；草本花粉含有较多的 *Artemisia* 及一些水生草本 *Typha*、Cyperaceae；蕨类孢子含有一些 *Polypodiaceae*、*Lygodium*，反映为落叶阔叶、针叶混交林—草原植被，古气候温暖湿润，气候期相当于前北方期。娄塘组上段 3 层具 *Pinus*-*Quercus*（*deciduous*）-*Artemisia*-*Pteris* 组合，木本花粉尚见较多的 *Castanea*、*Betula*、*Abies*、*Picea*、*Ulmus*；草本尚有较多的 *Chenopodiaceae*，也见一些水生草本等花粉，反映为针叶、落叶阔叶混交林—草原，古气候温和偏干，气候期相当于北方期。

测年结果：在与下伏地层分界以下 0.21m 处光释光定年值为 15.00±0.39ka；在 3 层顶界以上 0.19m 处的上覆深灰色淤泥质黏土层内，^{14}C 定年值为 7740±190a B.P.。据测年成果推算，SG1 孔娄塘组顶界年龄约为 7.8ka。

（三）总体特征

综合上述地层剖面，可将娄塘组特征归纳如下。

1）地层剖面发育完整者，可分为上、下两段。

下段：长江三角洲沉积区多为河口相，在崇明岛西部为溺谷相。岩性常为灰色砂质粉土、黏质粉土与灰白色粉砂互层，部分如外高桥孔为灰色砂质粉土与粉质黏土互层。普遍含有多少不一的广盐性有孔虫、少量浮游种和窄盐种，还有少量半咸水和海相介形虫，偶见陆相介形虫。太湖沉积区如金山朱泾、青浦赵巷等孔为湖沼堆积的灰黑、灰褐色粉质黏土，偶见半咸水介形虫，常含有较多的植物碎片，局部形成泥炭或碳质黏土层。

上段：普遍为滨海堆积。大都为深灰、褐灰色（淤泥质）黏土、粉质黏土，其特征标志是可见大量的植物茎叶与黄色泥灰质结核，局部形成泥炭或碳质黏土层。广泛分布于金山、奉贤、浦东南部等地和市区、宝山等地的部分地段。其中可检出少量广盐性有孔虫和很少的半咸水介形虫，其优势度高，表明属于滨海沼泽或潟湖沉积。部分为深灰、褐灰色粉质黏土夹灰白色粉砂或成互层，其中含少量广盐性有孔虫，还有很少的浮游种，属种比前者略多，其优势度较低；也含少量的半咸水介形虫，属潮坪或海湾堆积。

2）地层厚度具有西南部薄、东北部厚的变化规律，至现长江河口地区厚度明显加大。太湖沉积区厚度多为 3~6m，长江三角洲沉积区介于 4~32m 不等，在河口沙岛地区普遍厚达 30m。

3）娄塘组与下伏的上更新统南汇组为假整合接触，在上海陆域大都覆盖在南汇组上段标志层暗绿色硬土之上；至河口地区全新统底界深切，使娄塘组超覆于南汇组中段海相层之上。

4）孢粉组合通常与地层分段对应分为上、下两带，相应的植被面貌为由早期的针叶、落叶阔叶混交林演变为晚期的落叶阔叶、针叶混交林—草原，反映古气候由温凉稍湿变为温和偏干，分别相当于 A. Blytt-R. sernander 气候分期的前北方期和北方期。

5）根据钻孔中取得的 20 多个娄塘组地层测年资料统计，娄塘组下段年龄介于 10300~9500a，上段年龄介于 9500~8000a。

六、上海组

（一）层型剖面

1. 上海面粉厂 SG2 孔剖面

1984 年厘定“上海组”时依据上海面粉厂 J10 孔，现用原址新施工的 SG2 孔地层剖面予以说明。该孔在孔深 13.85~17m 钻遇上海组（图 21-18），所见地层剖面如下：

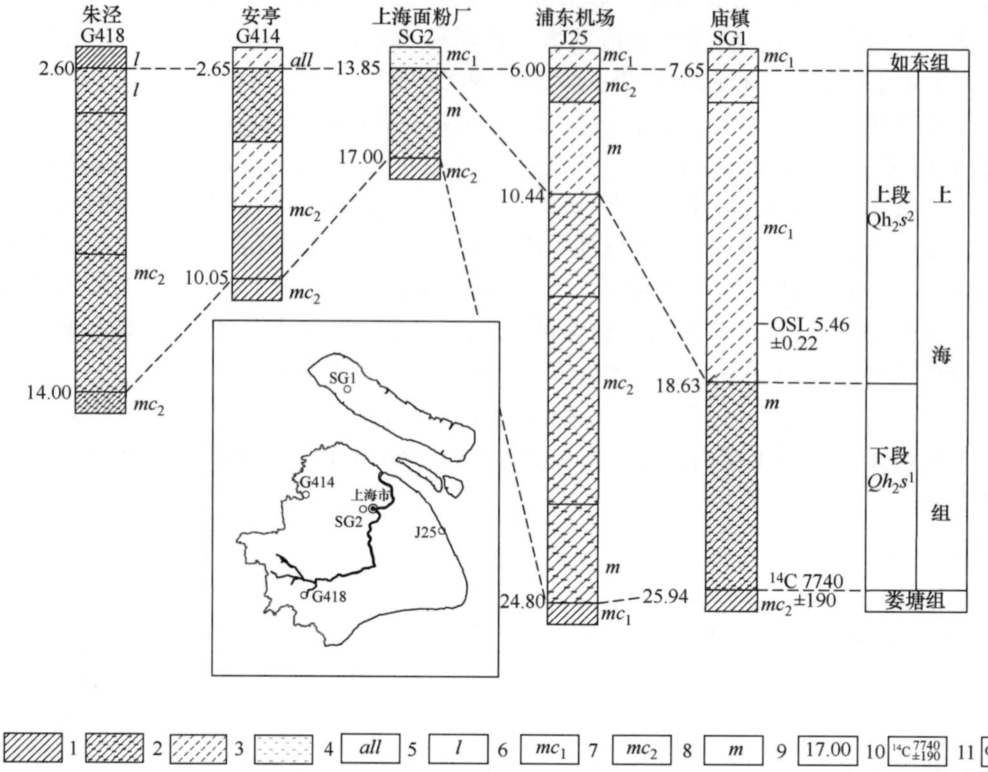

图 21-18　全新统上海组地层对比图

Fig. 21-18　Stratigraphic correlation of the Holocene Shanghai Formation

1—黏土、粉质黏土；2—淤泥质（粉质）黏土；3—砂质粉土、黏质粉土；4—粉砂；5—泛滥平原相；6—湖泊相；7—河口、河口淡化浅海相；8—滨海、沼泽、潟湖相；9—滨岸浅海；10—深度（m）；11—放射性碳法定年值（a B. P.）；12—光释光法定年值（ka B. P.）

上覆地层：上全新统如东组（Qh_3r）青灰色粉砂夹砂质粉土

----- 侵蚀接触 -----

上海组（Qh_2s） 厚 3. 15m

褐灰色淤泥质黏土，局部夹薄层粉砂，近底部含腐殖物根茎 3. 15m

———— 整　合 ————

下伏地层：下全新统娄塘组（Qh_1l）褐灰色黏土

据微古分析，中下部含有孔虫种数、个数较多，以 *Ammonia beccari*、*Elphidium advenum* 为主，其次为 *Florilus atlanticus*、*Globigerina* sp. 等；含少量介形虫，有半咸水的 *Albileberis sinensis*、*Sinocytheridea impressa* 及海相种瞳孔穆赛介（*Munseyella pupilla*）、圆翼花介（*Cytheropteron rotunda*）等，反映为滨岸浅海相。上部未检出微古化石，可能为滨海相。

据分析，含孢粉较丰，具 *Quercus deciduous*（20%～25%）-*Pinus*（16%～27%）-*Castanea*（2%～10%）-*Artimisia*（4%～9%）-*Pteris*（0～5%）组合，反映古气候温暖湿润。

2. 浦东机场 J25 孔剖面

该孔在孔深 6～24. 8m 钻遇上海组，其地层发育比面粉厂孔齐全，岩性、岩相具较强代表性，可作为上海组的辅助层型剖面（见图 21-18）。所见地层剖面如下：

上覆地层：上全新统如东组（Qh_3r）青灰色粉砂夹砂质粉土

----- 侵蚀接触 -----

上海组（Qh_2s） 总厚 18. 80m

上段（Qh_2s^2） 厚 4.44 m

4. 褐灰色黏土夹粉砂、砂质粉土，局部成薄互层。具水平层理、脉状层理 1.20m

3. 灰色黏质粉土，局部夹粉砂或黏土薄层。具水平层理、小型斜层理，底部具冲刷面 3.24m

下段（Qh_2s^1） 厚 14.36m

2. 灰色淤泥质粉质黏土，流塑，层面含粉砂、贝壳碎片。具水平层理 3.60m

1. 灰色淤泥质黏土，偶夹粉砂薄层或团块。含少量贝壳碎片，底部较多，并见雕刻友螺［*Nassarius*（*phrontis*）*Cealatula*］。见虫孔构造，具不规则水平层理 10.76m

———— 整　合 ————

下伏地层：下全新统娄塘组（Qh_1l）灰色黏土

据微古分析，1 层下部含属种、个数均少的有孔虫，优势种为 *Elphidium advenum*，其次有 *Quinqueloculina* sp.、*Brizalina* sp.、*Ammonia annectens* 等窄盐种；下部并见较多 *Ammonia beccarii* 及较少的 *Cribrononion* sp.；含介形虫多寡不一，属种较少，兼有半咸水—海相种，优势种为皮桑双角花介（*Kejiellabisanensis*）、*Sinocytheridea impressa*，其次为 *Albileberis sinensis*，中下部并见三井耳形介（*Aurilamiii*）等，反映为滨岸浅海相。1 层中部微古化石时有时无，属种、个数均少，兼有广盐、窄盐种和浮游有孔虫，有 *Quinqueloculina* sp.、*Nonion* sp.、*Elphidium advenum*、*Rosalina bradyi*、*Spiroloculina* sp. 等；介形虫断续见很少半咸水的 *Leguminocythereis hodgii*、*Kejiella bisanensis*。1 层上部见很少的 *Florilus* sp.、*Quinqueloculina lamarckiana* 等有孔虫和很少的半咸水介形虫 *Neomonoceratina dongtaiensis*、*Kejiella bisanensis*，反映 1 层中上部为滨海相。2 层见很少的有孔虫，主要为广盐种 *Cribrononion* sp.、*Ammonia beccarii*，下部出现窄盐种 *Quinqueloculina* sp.、*Ammonia compressiuscula*；并断续见很少半咸水介形虫 *Neomonoceratina dongtaiensis*、*Kejiella bisanensis*，反映为滨海相。3 层含微古化石较丰，有孔虫优势种有 *Ammonia beccarii*、*A. compressiuscula*、*Bolivina* sp.，并有较多 *Elphidium advenum*、*Globigerina* sp. 等；介形虫优势种为 *Albileberis sinensis*，次之为 *Neomonoceratina dongtaiensis*、*Sinocytheridea impressa*、*Kejiella bisanensis*，尚见 *Munseyella* sp.、*Leguminocythereis hodgii* 等，反映为滨岸浅海相。4 层仅在下部出现较多的 *Albileberis sinensis*，未检出其他微古化石，可能为滨海相。

据孢粉分析，1 层下部具 *Quercus deciduous*（19%～38%）-*Quercus evergreen*（3%～19%）-*Pinus*（3%～22%）-*Castanea*（3%～22%）-*Artimisia*（0.3%～9.9%）-*Pteris*（2.3%～21%）组合；1 层上部具 *Quercus deciduous*（21%～28%）-*Pinus*（5%～20%）-*Castanea*（14%～19%）-*Artimisia*（1.2%～4.5%）-*Pteris*（11%～14%）组合，尚见较多水生草本植物花粉，反映 1 层由下而上为含针叶树、常绿阔叶落叶阔叶混交林—针叶、落叶阔叶混交林植被，古气候由暖热湿润—温暖潮湿，气候期相当于大西洋期；2、3 层同具 *Pinus*（6%～41%）-*Abies*、*Picea*（2%～19%）-*Quercus deciduous*（14%～31%）-*Artimisia*（3%～5%）-*Pteris*（9%～17%）组合，反映为暗针叶、落叶阔叶混交林植被，古气候凉冷湿润。4 层具 *Quercus deciduous*（6%～17%）-*Pinus*（17%～36%）-*Abies*、*Picea*（4%～11%）-*Artimisia*（1%～9%）-*Pteris*（9%～50%）组合，反映为针叶、落叶阔叶混交林植被，古气候温凉潮湿，气候期相当于亚北方期。

（二）对比剖面

1. 崇明庙镇 SG1 孔剖面

该孔在孔深 7.65～25.94m 钻遇上海组，其地层剖面（见图 21-18）在河口沙岛地区具较强代表性，测试资料也较多，可作为上海组地层划分对比的重要依据。所见剖面如下：

上覆地层：上全新统如东组（Qh_3r）深灰色黏质粉土与灰白色粉砂互层

----- 侵蚀接触 -----

上海组（Qh_2s） 总厚 18.29m

上段（Qh_2s^2） 厚 10.98 m

3. 灰色砂质粉土与粉质黏土互层　1.15m

2. 灰色砂质粉土，含腐木碎屑。中下部夹粉砂较多，内可见云母及贝壳碎片，距底1.1m处夹有3cm厚的云母和腐木碎屑富集层。具波状层理，局部见斜层理及针孔状气逸构造　9.83m

下段（Qh_2s^1）　厚7.31m

1. 深灰色淤泥质黏土，局部夹薄层粉砂。具揉皱变形层理及虫孔构造　7.31m

———— 整　合 ————

下伏地层：下全新统娄塘组（Qh_1l）灰色粉质黏土与粉砂互层

据微古分析，1层含 *Ammonia beccarii-Elphidium advenum* 有孔虫组合，并较多其他平旋壳类及 *Bolivinacochei* 等；介形虫见 *Cushimanidae japonica-Neomonoceratina* sp. -*Albileberis sinensis* 组合，反映为滨岸浅海相。2层含 *Ammonia beccarii-Elphidium advenum* 有孔虫组合，还有少量其他平旋壳类、窄盐种和 *Globigerina* sp. 等；介形虫含 *Tanella opima-Neomonoceratina dongtaiensis-Sinocytheridea impressa* 组合；3层有孔虫 *Ammonia beccarii* 占优势，还有少量平旋壳类，介形虫含 *Tanella opima-Neomonoceratina dongtaiensis* 组合，反映均为河口淡化浅海环境。

据孢粉分析，1层具 *Quercus*（*deciduous*）-*Pinus-Artimisia-Pteris* 组合，并见较多阔叶植物 *Castanea*、*Ulmus*、*Juglans*、*Quercus*（*evergreen*）、*Betula*、水生草本植物 *Typha*、*Cyperaceae* 及蕨类 *Polypodiaceae* 等孢粉，局部出现较多 *Picea*、*Abies*，反映古气候暖热潮湿，间或有温凉波动，气候期相当于大西洋期；2、3层具 *Pinus-Quercus*（*deciduous*）-*Artimisia-Pteris* 组合，尚见较多木本植物 *Betula*、*Picea*、*Abies*、*Tsuga*、*Ulmus*、*Castanea*、草本植物 *Chenopodiaceae*、*Gramineae* 花粉及蕨类孢子 *Adiantum*、*Cyathea*、*Polypodiaceae* 等，反映古气候温凉偏干，气候期相当于亚北方期。

庙镇SG1孔测年结果：＊2层底界向上1.95m（孔深16.68m）处粉砂，光释光定年值为5.46±0.22ka；＊1层底界向上0.09~0.25m（孔深25.69~25.85m）处淤泥质黏土，^{14}C 定年值为7740±190a B.P.。按上列测年值推算，SG1孔上海组顶界（孔深7.65m）年龄约为3195a，底界年龄约为7780a。

2. 嘉定安亭G414孔剖面

该孔在孔深2.65~10.05m间钻遇上海组，厚为7.4m（见图21-18）。所见地层剖面如下：

3. 深灰色淤泥质粉质黏土夹粉砂，以不规则水平层理为主，脉状层理次之，见有负荷构造、虫孔构造　厚2.59m

2. 灰色砂质粉土夹粉质黏土纹层，以不规则水平层理为主，局部见包卷层理及垂直虫孔　2.26m

1. 灰色粉质黏土夹粉砂纹层，局部为透镜体，下部1.65m渐变为粉细砂，以不规则水平层理为主　2.55m

据微古分析，*Nonionakitaense* 和 *Ammonia tepida* 两种广盐性有孔虫占绝对优势，并含一些大洋桥石、赫氏艾氏石、瓦氏卷舟颗石等超微化石，反映为潟湖沉积。

据孢粉分析，1、2层主要含 *Quercus*（*evergreen*）、*Quercus*（*deciduous*）、*Castanopsis*、*Pinus* 和水生草本植物花粉，反映为常绿阔叶、落叶阔叶混交林植被，古气候暖热潮湿，气候期相当于大西洋期；3层主要含 *Quercus*（*deciduous*）、*Pinus*、*Quercus*（*evergreen*）和水生草本植物花粉，反映为含常绿阔叶的落叶阔叶、针叶混交林植被，古气候温暖潮湿，气候期趋近于亚北方期。

3. 金山朱泾G418孔剖面

该孔在孔深2.6~14.0m间钻遇上海组，厚为11.4m（见图21-18）。见岩性、岩相单一的弱海相性淤泥质黏土，在上海西南部地区较有代表性。所见地层剖面如下：

3. 灰褐色淤泥质黏土，下部1.25m夹粉砂纹层，并含黑色炭化植物斑点，大都具均匀层理，局部见不规则水平层理　厚6.55m

2. 灰褐色淤泥质粉质黏土，含少量炭化植物碎屑，局部见褐铁矿结核，偶见不规则微波状层理 2.85m

1. 灰黑色淤泥质黏土，普遍含炭化植物叶片、根茎，分布杂乱，并见杏黄色泥灰质结核。含半咸水的中华拟蟹守螺（*Cerithidea sinensis*），具均匀层理 2m

据微古分析，1~3 层中下部含长中华丽花介（*Sinocytheridea longa*）-射阳洁面介（*Albileberissheyangensis*）-近美丽弯贝介（*Loxoconcha subpulchra*）。其中 1、2 层化石含量较少，3 层较多；3 层上部 1.9m 唯见轮藻，反映为滨海沼泽（1、2 层）—潟湖（3 层中下部）—湖沼（3 层上部）沉积。

据孢粉分析，1~3 层下部含孢粉甚丰，具 *Quercus*（*evergreen*）-*Q. acutissima*-*Pinus*-*Chenopodiaceae*-*Polypodiaceae*，反映古气候暖热湿润，气候期相当于大西洋期；3 层中部具 Gramineae-*Typha*-*Pteridium*-*Salix* 组合，反映古气候温和偏干；3 层上部具 *Quercus*（*deciduous*）-*Castanopsis*-*Pinus*-*Castanea* 组合，反映古气候温暖湿润。3 层中上部的气候期相当于亚北方期。

（三）总体特征

综合上述及其他钻孔地层剖面，可将上海组特征归纳如下。

1）按岩性、岩相组合基本可分为两种类型，相应即为两种层型。

一是分布于“冈身带”以东的长江三角洲沉积区的滨岸浅海（下部）—河口滨海堆积（上部）。地层保存完整的剖面可见上、下两段。下段为滨岸浅海堆积的深灰色淤泥质黏土和淤泥质粉质黏土，如面粉厂孔所见和机场孔、庙镇孔剖面的下部地层均是；以含分异度、丰度均较高的有孔虫，与海相、半咸水介形虫共生。上段为滨海堆积的灰色砂质粉土、黏质粉土夹粉砂，其上部常为灰色砂质粉土与粉质黏土互层或为黏土、粉质黏土夹砂质粉土和粉砂，如机场孔、庙镇孔剖面的上部地层即是；大都含属种、个数较少、局部较多的有孔虫，以浅水种为主，也见很少深水种和浮游种，与少量半咸水和很少海相介形虫共生。面粉厂孔显然由于上覆如东组底界的下切而将上海组上段侵蚀。

另一是分布于“冈身带”之西的太湖沉积区的潟湖—滨海沼泽堆积。地层剖面所见通常为较单一的深灰色淤泥质黏土、黏土和淤泥质粉质黏土。以含优势性较强的一两种广盐性有孔虫，与较少的半咸水介形虫共生为特征，顶部演化为唯见轮藻化石的湖沼堆积。其中、下部常含植物茎叶、泥灰质结核及贝壳碎片；中上部频见粉砂、砂质粉土纹层，偶成一两米厚的夹层。此种砂性土夹层的出现，反映时有潮汐水道侵进。

介于上述两者之间、分布于嘉定—七宝—闵行—奉贤南桥一带的“冈身带”，其岩性、岩相特征具东、西两区之间的过渡性质，但大体上与太湖沉积区相近，基本为一套滨海堆积。其中在梅陇以北地带主要为深灰色淤泥质黏土和淤泥质粉质黏土，中上部多夹粉砂纹层，常见贝壳碎片，偶尔形成贝壳砂层；下部常含植物碎片及黄色泥灰质结核，反映早期为滨海沼泽，晚期演化为潮滩、湿地堆积，如这一带的嘉定镇 G412 孔、闵行七宝 G100 孔所见均是。梅陇以南地带通常为灰色粉质黏土夹含贝壳碎片的粉砂层，反映为水动力较强的潮滩堆积。

2）本组的顶、底界深度和厚度，总体上由西南的太湖沉积区向东北的长江三角洲河口区逐渐加大。在西部太湖沉积区，其顶、底界深度通常介于 2.5~15m，厚为 7~12m；东部的长江三角洲沉积区，包括“冈身带”，顶、底界深度通常介于 5~25m，厚度多为 10~18m。

3）孢粉组合大体与地层分段对应分为上、下两带，相应的植被面貌为由早期的常绿阔叶、落叶阔叶混交林演变为晚期的针叶、落叶阔叶混交林，反映古气候由暖热潮湿演变为温凉稍干，分别相当于 A. Blytt-R. sernander 气候分期的大西洋期和亚北方期。

4）根据庙镇 SG1 孔上海组内两个光释光定年值、面粉厂 SG2 孔和长兴岛 SG4 孔的两个上海组底界 ^{14}C 定年值，并综合青浦赵巷等孔的 ^{14}C 定年资料，可以确定上海组形成年龄为 8000~3000a。

5）上海组与下伏娄塘组为整合接触。

七、如东组/青浦组

（一）青浦组层型剖面

命名剖面位于青浦镇青浦气象站SC2孔。孔深3.25m以浅见该组地层，以假整合超覆于上更新统南汇组上段（Qp_3n^3）暗绿色黏土层之上（图21-19）。所见地层剖面如下：

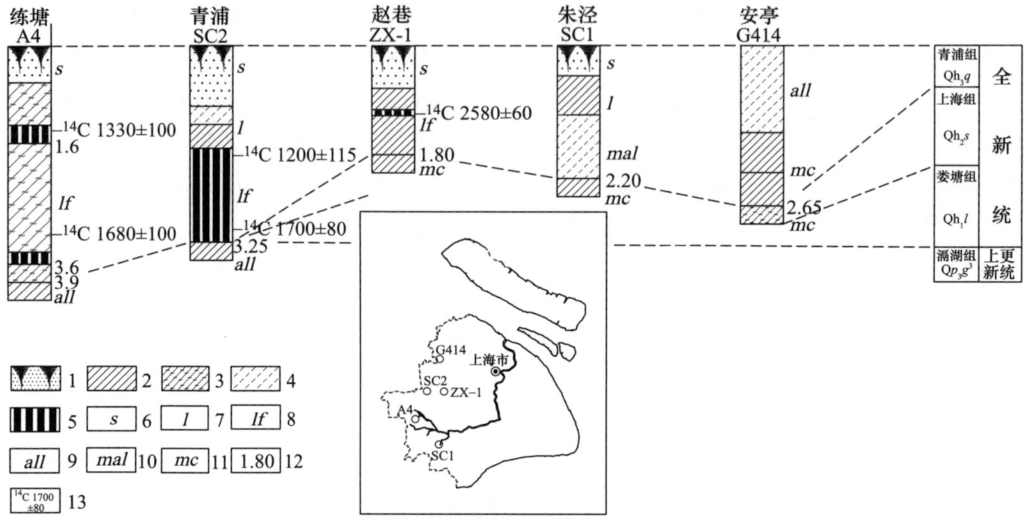

图21-19　全新统青浦组地层对比图

Fig. 21-19　Stratigraphic correlation of the Holocene Qingpu Formation

1—耕植土、填土；2—黏土、粉质黏土；3—淤泥质黏土；4—砂质粉土；5—泥炭；6—人工堆积；7—湖泊相；8—湖沼相；9—泛滥湖泊相；10—陆上三角洲湖泊相；11—潟湖相；12—深度（m）；13—放射性碳法定年值（a B. P.）

青浦组（Qh_3q）		厚3.25m
4.	人工填土	1.00m
3.	灰褐色砂质粉土	0.30m
2.	灰褐色粉质黏土。与上层均具微波状—不规则状水平层理	0.40m
1.	黑色泥炭层，由黏土、粉砂和植物碎屑组成。具均匀层理。顶部有0.06m为褐灰色粉质黏土与泥炭薄互层	1.55m

----- 侵蚀接触 -----

下伏地层：上更新统南汇组上段（Qp_3n^3）暗绿色黏土

据微古和孢粉分析：1层未见微体化石。孢粉以草本植物花粉为主，木本植物花粉次之，形成*Gramineae-Quercus*组合，并含不少水生草本植物、青刚栎、栗属、栲属等花粉，反映处于湖沼环境，附近为含常绿阔叶落叶阔叶林—草原植被，古气候温暖偏干。2、3层检出淡水介形虫，有少量*Ilyocypris* sp.，并见*Candoniella* sp.、*Cypria yangchensis*。在2层底部偶见有孔虫。反映为湖泊—滨湖环境，偶尔受海水影响。孢粉以木本花粉为主，草本次之，具*Quercus-Altingia-Cupressaceae-Gramineae*组合，常绿阔叶成分比前略增，反映古气候温暖湿润。

^{14}C测年：1层顶界向下0~0.4m（孔深1.7~2.1m）泥炭定年值为1200±115a B. P.；1层底界向上0.1~0.45m（孔深2.8~3.15m）泥炭定年值为1700±80a B. P.。

（二）青浦组对比剖面

1. 青浦练塘A4孔剖面

孔深3.9m以浅见青浦组，地层发育良好，并有定年结果，是一处很有价值的地层剖面（见图21-

19）。所见地层剖面为：

2. 耕植土　厚0.60m
1. 深灰色淤泥质黏土，其上、下部各夹一层泥炭，厚为0.3m和0.2m。也以假整合超覆于上更新统南汇组上段（Qp_3n^3）暗绿色黏土之上

^{14}C测年：1层顶界向下0.7m（孔深1.3m）处泥炭为1330±100a B. P.；1层底界向上0.5m（孔深3.4m）处泥炭为1680±100a B. P.。

2. 青浦赵巷ZX-1孔剖面（华东师范大学施工）

该孔在孔深1.8m以浅钻遇青浦组（见图21-19）。所见地层剖面为：

2. 耕植土　厚0.7m
1. 棕灰色粉质黏土，有大片锈斑，均匀层理　1.1m

孔深1.1m处夹一5cm厚泥炭层，^{14}C定年值为2580±60a B. P.。

经分析唯发现大量水蕨（*Parkeriaceae*），表明为湖沼环境。其以整合接触覆于中全新统上海组潟湖沉积青灰色粉质黏土之上，这与上两孔青浦组超覆于上更新统之上显然不同。

根据孔深1.1m处泥炭定年结果推算，青浦赵巷ZX-1孔青浦组的底界年龄为2968a。

3. 金山朱泾金山气象站SC1孔剖面

该孔在孔深2.20m以浅揭露青浦组（见图21-19）。所见地层剖面为：

3. 耕植土　厚0.5m
2. 褐灰色粉质黏土夹粉砂，多锈斑，含少量铁锰结核，下部见角状环棱螺（*Bellamya angularis*），具微波状层理，本层偶见舒滨小玻璃介（*Condoniella schbinae*），表明为湖泊沉积　0.65m
1. 褐黄色砂质粉土，夹有粉砂纹层或小透镜体，顶部0.2m为锈黄色粉质黏土，含较多铁质斑点及泥灰质结核，具水平微层理，本层见少量有孔虫，未检出介形虫，表明为受海水影响的陆上三角洲湖滨滩地沉积　1.05m

同在朱泾的G418孔孔深2.60m以浅见青浦组，有孢粉资料可作为SC1孔的补充：

地层以褐黄色粉质黏土为主，杂有蓝灰色细网纹，下部尤多，上部较多铁质结核；顶部厚0.5m为灰黑色黏土；底部0.1m为灰色砂质粉土，具滑塌构造。微古分析唯见轮藻。反映为湖泊沉积，顶部的灰黑色黏土表明晚期趋向于沼泽化。

据孢粉分析，具*Quercus*（*deciduous*）-*Pinus*-*Quercus*（*evergreen*）-*Ceratopteris*组合，反映为含常绿阔叶落叶阔叶、针叶混交林植被，气候暖热潮湿。

4. 嘉定安亭G414孔剖面

该孔在孔深2.65m以浅揭露青浦组（见图21-19），所见地层剖面为：

3. 灰褐色砂质粉土，含有铁锰结核及植物根茎孔隙　厚1.44m
2. 黄褐色粉质黏土夹粉砂薄层，见有软体动物化石河兰蚬（*Corbicula fluminea*），也含铁锰结核，具不规则水平层理　0.66m
1. 褐灰色粉质黏土与灰白色粉砂互层，具水平层理　0.55m

据微古分析，1~2层中下部唯见少量广盐性有孔虫*Nonion akitaense*、*Ammonia tepita*；2层上部至3层无化石检出，反映前期为潟湖沉积，后期演变为河漫滩沉积。

孢粉分析：为同一的*Quercus acutissima*-*Pinus*-*Betula*-*Gramineae*孢粉组合，反映为落叶阔叶、针叶

混交林植被，古气候温暖湿润，气候期相当于亚大西洋期。

（三）青浦组地层总体特征

综合上列地层剖面，可将青浦组特征归纳如下：

1）岩性、岩相有两种类型：主要类型为夹有泥炭层的深灰、棕灰色淤泥质黏土、粉质黏土的湖沼沉积，这是见于青浦西南部及与松江、金山接壤区，超覆于上更新统滆湖组台地（以存在西部“第二硬土层”为标志）之上的特征岩性；次要类型为分布于上述地区以东至“冈身带”之间湖泊沉积的褐灰、褐黄色粉质黏土夹粉砂，唯局部可见泥炭或碳质黏土夹层。

2）分析表明，泥炭层中草本花粉占明显优势，并有大量水生草本花粉和水蕨孢子，表明为湖沼沉积的草本泥炭；再者，青浦组内通常含有淡水软体动物、介形虫、轮藻等陆相生物化石，其下部可偶尔见及有孔虫，在金山朱泾和嘉定安亭则有少量出现。这表明一般为湖泊环境，早期略受海水影响，局部如金山东南部至安亭近“冈身带”区域曾为潟湖环境。

3）青浦组厚度一般为2～4m，大致有从东向西，即从湖盆边缘向湖盆内部加厚的趋势。

4）青浦组底界年龄约为3ka。所夹泥炭层的形成依次有约2.6ka、约1.7ka和1.2～1.3ka等3个时期，有从湖盆边缘向内部逐渐变新的趋势，揭示泥炭是湖盆趋向萎缩期间形成的。

（四）如东组层型剖面

“如东组”在上海中东部的长江三角洲沉积区内广泛发育，尤以河口地区及江南滨岸地带发育最好。现列举代表剖面予以说明。

1. 崇明庙镇SG1孔剖面

该孔在孔深7.65m以浅揭露如东组（图21-20）。所见地层剖面如下：

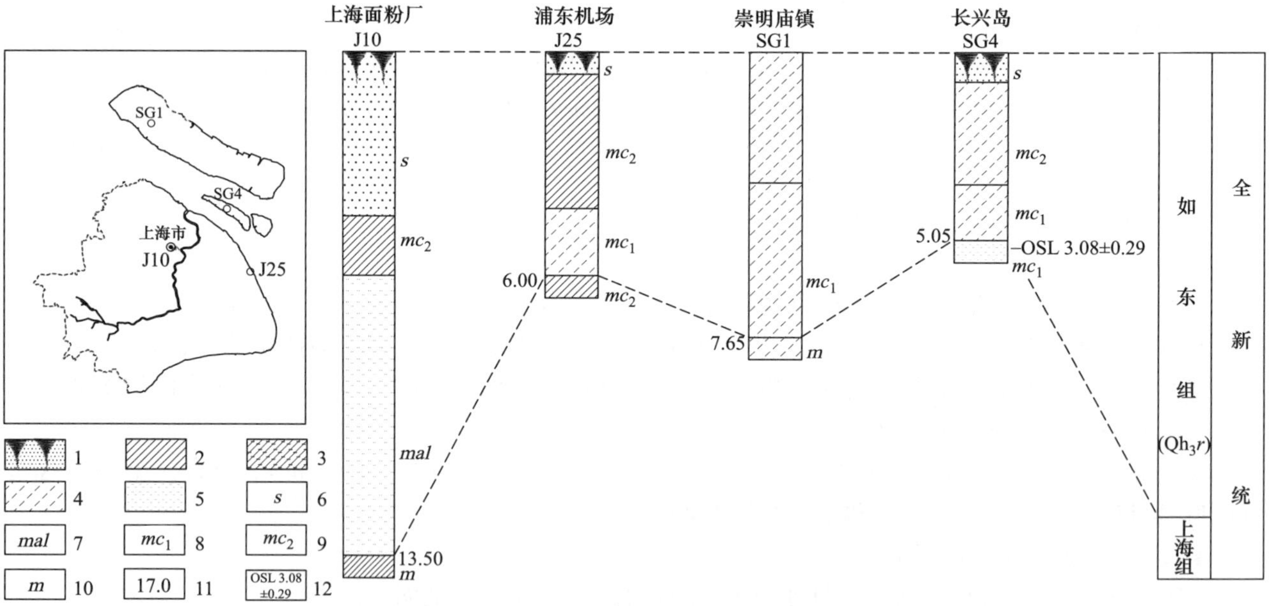

图21-20 全新统如东组地层对比图

Fig. 21-20 Stratigraphic correlation of the Holocene Rudong Formation

1—人工填土；2—黏土、粉质黏土；3—淤泥质黏土；4—砂质粉土、黏质粉土；5—粉砂、细砂；6—人工堆积；7—三角洲支流河道相；8—河口、河口湾相；9—河口滨岸、滨海相；10—滨岸浅海相；11—深度（m）；12—光释光法定年值（ka B.P.）

2. 深灰色砂质粉土　　厚3.50m

1. 以深灰色黏质粉土为主，与灰白色粉砂互层，粉砂多呈厚2～5mm的透镜体，具波状层理、透镜状层理，与下伏上海组（Qh_2s）灰色砂质粉土与粉质黏土互层，为侵蚀接触　　4.15m

据微古分析，1、2 层同具 *Tanellaopima-Albileberis sinensis* 介形虫组合；有孔虫具 *Ammonia beccarii-Cribrononion* sp. -*Florilus* sp. 组合，尚见较多的 *Elphidium magellanicum*、*Lagena* spp.、列式壳类、瓷质壳类及抱球虫类等，反映属河口相。

据孢粉分析，1 层具 *Pinus-Quercus*（*deciduous*）-*Artimisia-Pteris* 组合，还有较多的 *Betula*、*Picea*、*Abies*、*Tsuga*、*Ulmus*、*Castanea* 等木本花粉，*Chenopodiaceae*、*Glamineae* 等草本花粉；2 层具 *Quercus*（*deciduous*）-*Pinus-Artimisia-Adiantum* 组合，还有较多的 *Betula*、*Castanea*、*Juglans*、*Ulmus*、*Tsuga* 等木本及水生草本 Cyperaceae、*Typha* 等花粉，反映气候由早期的温凉偏干转为后期的温暖湿润。

2. 长兴岛 SG4 孔剖面

该孔在孔深 5.05m 以浅揭露如东组（见图 21-20）。所见地层剖面如下：

3.	杂填土	厚 0.8m
2.	灰黄色砂质粉土，含较多云母片及少量腐木碎屑，具微波状—不规则状水平层理	2.75m
1.	浅棕色砂质粉土，含较多云母片，多集成水平细纹层，具不规则水平层理，局部斜波状层理	1.50m

与下伏中全新统上海组浅灰色粉砂夹浅棕色黏质粉土为侵蚀接触。

据微古分析，1 层含有孔虫属种、数量较丰，优势性弱，复合分异度 2.7，具 *Ammonia beccarii-Globigerina bulloides-Florilus atlanticus-Cribrononion incertum* 组合，偶见窄盐种 *Lagena* spp.；介形虫见半咸水属种 *Tanellaopima*、*Sinocytheridea impressa*、*Leptocythere ventriclivosa*，反映为河口沉积；2 层含有孔虫属种、数量均少，优势性弱，复合分异度 1.7，几乎全为广盐性属种，主要见有 *Globorotalia* sp.、*Florilus* sp.、*Nonion* sp.，偶见窄盐种 *Bolivina* sp.；介形虫唯见个别半咸水属种 *Sinocythere dongtaiensis*，反映为河口滨岸沉积。

据孢粉分析，1 层含孢粉较丰，以木本占明显优势，具 *Pinus*（34%～37%）-*Quercus deciduous*（9%～16%）-*Artimisia*（7%～10%）-*Pteris*（6%～9%）组合，尚见较多的木本 *Tsuga*、*Ulmus*、水生草本 *Cyperaceae* 等花粉，还出现 *Quercus*（*evergreen*），反映为针叶、落叶阔叶混交林植被，古气候温和湿润。2 层含孢粉中常，以蕨类占优势，具 *Quercus deciduous*（5%～23%）-*Pinus*（10%～18%）-*Typha*（7%～23%）-*Pteris*（9%-29%）组合，尚见较多蕨类 *Adiantum*、*Selaginella* 及木本 *Ulmus* 等孢粉，反映古气候温暖潮湿。

在 1 层底界以下 0.05～0.15m（孔深 5.1～5.2m）处粉砂的光释光定年值为 3.08±0.29ka，由此推算 1 层底界即该孔如东组底界年龄约为 3ka。

（五）如东组对比剖面

1. 浦东机场 J25 孔剖面

该孔在孔深 6m 以浅揭露如东组（见图 21-20）。所见地层剖面如下：

3.	褐黄色填土	厚 0.60m
2.	褐黄色粉质黏土，见具揉皱的包卷层理	3.60m
1.	灰色黏质粉土，局部有粉砂团块或夹粉质黏土薄层，具水平层理	1.80m

与下伏中全新统上海组（Qh_2s）褐灰色黏土夹粉砂、砂质粉土薄层为侵蚀接触。

据微古分析，1 层含有孔虫属种、个数较丰，优势种为 *Ammonia beccarii*、*A. compressiuscula*、*Bolivina obscura*、*Quinqueloculina* sp.，次之为 *Cribrononion subinsetum*、*Nonion* spp.、*Globigerina bulloides*，还有少量的瓶虫类、螺旋壳类、列式壳类；介形虫属种、个数较少，优势种为 *Albileberis sinensis*，次之为 *Neomonoceratina dongtaiensis*、*Tanella opima*，还见少量的 *Lyguminocythereis hodgii*、*Sinocytheridea impressa* 等半咸水种，出现些淡水种 *Eucypris* sp.，反映为河口沉积。2 层见很少的广盐性有孔虫 *Pseudononionel-*

la variabilis、*Protelphidium tersum*、凹陷圆形虫（*Gyroidina dempressa*）；介形虫仅见个别的 *Sinocytheridea impressa*，反映为滨岸潮滩沉积。

据孢粉分析，2 层上部和 3 层无孢粉资料。1 层和 2 层下部含孢粉丰富，以木本占优势，蕨类次之，草本很少，比下伏层含木本花粉减少，蕨类孢子增加，具 *Pinus*（17%～36%）–*Quercus deciduous*（6%～17%）–*Picea*、*Abies*（4%～11%）–*Artimisia*（1%～9%）–*Pteris*（9%～50%）组合，还有较多的 *Tsuga*、*Ulmus* 木本花粉及蕨类 *Polypodiaceae* 孢子，反映为针叶、落叶阔叶混交林植被，古气候温凉潮湿。

2. 上海面粉厂 J10 孔剖面

这是著名的、上海最早进行系统的微古、孢粉分析的钻孔，在孔深 13.5m 以浅揭露如东组（见图 21–20）。所见地层剖面如下：

3.	人工填土	厚 4.4m
2.	褐灰色粉质黏土	1.6m
1.	青灰色粉砂，距层顶 3～6m（孔深 9～12m）内夹有少量粉质黏土	7.5m

下伏地层为中全新统上海组褐灰色淤泥质粉质黏土。

据微古分析，1 层含有孔虫丰富，大多为细小个体，具 *Epistominella naraensis*–*Ammonia convexidorsa* 组合，与现今长江河口支流河道——吴淞江、黄浦江所见的有孔虫组合成分相同；介形虫较少，主要见半咸水的 *Sinocytheridea latioveta*、*Neomonoceratina dongtaiensis*，陆相介形虫 *Ilyocypris* sp.，还见有海相的海胆刺、海陆过渡的盾形化石、小双壳类、圆形硅藻及陆相的轮藻受精卵膜，与通常河口区所见的组分相似。2 层含有孔虫较少，多为广盐性的 *Ammonia beccarii*，介形虫有现常见于长江河口的 *Neomonoceratina dongtaiensis*，也见陆相介形虫 *Ilyocypris* sp. 以及海相的海胆刺，反映为河口滨岸沉积。

据孢粉分析，1、2 层属同一孢粉带，以木本花粉占优势（55%～71%），草本次之（占 12%～34%），蕨类孢子较少。具 *Pinus*–*Quercus*–*Gramineae* 孢粉组合，以 *Pinus*、*Quercus*（*deciduous*）为主，尚见较多的 *Ulmus*、*Castanea*、*Morus*、*Juglans*、*Cupressaceae* 等，还有 *Quercus*（*evergreen*）以及蕨类的 *Polypodiaceae*，反映为含常绿阔叶、落叶阔叶和针叶混交林植被，古气候温暖湿润，相当于亚大西洋期气候期。

（六）如东组地层总体特征

综合上述，可将如东组地层总体特征归纳如下：

1）岩性、岩相主要类型为河口湾—滨岸潮滩堆积的灰、灰黄色砂质粉土、黏质粉土夹粉砂，其上部普遍为滨岸潮滩堆积的褐黄、褐灰色粉质黏土。次要类型如上海面粉厂孔揭露古吴淞江所在三角洲支流河道堆积的青灰色粉砂和砂质粉土，还有如安亭等孔近地表出现的古吴淞江滨岸堆积的灰褐色砂质粉土。该组主要含有较丰富的广盐性有孔虫、少量浮游有孔虫和半咸水介形虫，也见个别的窄盐海相有孔虫、海相和陆相介形虫。

2）主要类型分布于包括“冈身带”及以东的长江三角洲沉积区，即现上海陆域中、东部的滨海平原和河口沙岛地区，堆积厚度多为 5～8m，局部逾 10m；次要类型为沿吴淞江等三角洲支流河道分布，堆积厚度在 2～10m 不等，具有西薄、东厚的变化规律。

3）所含孢粉组合反映，如东组堆积时古气候总体上温暖湿润，早期略显温凉偏干，其气候期大致相当于亚大西洋期。

4）如东组与下伏中全新统上海组为冲刷接触，其底界起伏较大，致使下伏上海组受到部分侵蚀，如面粉厂孔即是。

5）据长兴岛 SG4 孔所获定年资料确定，如东组底界年龄为 3ka，与青浦组相同，表明两者属同期异相堆积。

全新世地层在上海全区分布广泛，但地层的发育程度及岩性、岩相变化较大。由西向东地层发育程度由差趋好，地层底界埋深及厚度逐渐加大。根据全新世地层发育程度及岩性、岩相的差异，可划为4个全新统分区，现将分区特征列于下表（表21-3）。

表21-3　上海全新统分区特征一览表

Table 21-3　Stratigraphic division of the Holocene in Shanghai

分区	分布范围	沉积区	地层发育	地层结构	厚度/m
陆域西部	青浦西南部与松江、金山接壤区	区内太湖沉积区西部	差	只有 Qh_3q	<4
陆域中部	嘉定、闵行、奉贤及金山、松江东部	区内太湖沉积区东部—“冈身带”	尚好	Qh_3—Qh_1 上部，Qh_2 为滨海相，Qh_1 局部较厚	一般14~≤26，局部40左右
陆域东部	宝山、市区、浦东等地	长江三角洲沉积区滨海分区	较好	Qh_3—Qh_1 上部，Qh_2 为海相，Qh_3、Qh_1 局部较厚	一般26~≤35，局部≤45
河口沙岛	崇明、长兴、横沙三岛	长江三角洲沉积区河口分区	良好	Qh_3—Qh_1 下部，Qh_2 为海相，Qh_3、Qh_1 均厚	>35~60

光释光和^{14}C定年资料表明，区内的全新统底界为一不等时面，大都波动于9~11ka之间，因此可将区内全新统底界年龄厘定为10~11ka，与目前国内外倾向的划分方案一致；中全新统的底界为8ka；上全新统的底界为3ka。

上海面粉厂SG2孔全新统底界与上更新统顶界之间两个紧邻的定年结果表明，在长江三角洲沉积区滨海平原分区内全新统沉积之前，至少经历了1500a以上的沉积间断，而在研究区西部的太湖沉积区内更间断了8000a以上。证明区内上更新统与全新统之间的沉积间断由东向西急剧加大，上更新统顶部的硬黏土层经历了长达数千年之久的暴露风化过程。

第四节　生物地层划分与区域对比

上海第四系生物地层划分主要是依据微体古生物化石开展的，区内已积累了30余个具备全孔分析有孔虫、介形虫、孢粉的第四纪钻孔资料。20世纪70年代，上海面粉厂J10孔为区内最早开展微古分析的钻孔，随后80年代1：20万区域水文地质普查和长江三角洲调查项目、2004—2008年“上海市三维城市地质调查”和2010—2014年“长三角重点地区三维地质调查”项目相继对区内近20个第四纪钻孔进行了有孔虫、介形虫和孢粉、藻类分析，这些钻孔基本控制了不同沉积区，为研究区内第四纪海侵事件提供了扎实的证据，也为恢复区内第四纪植被面貌和建立古气候模式提供了依据。除了有孔虫、介形虫化石，20世纪80年代，上海市地质矿产局曾对嘉定安亭G414、金山朱泾G418、三灶P45孔开展过钙质超微化石和软体动物鉴定，充实了区内海侵认识的证据。

一、第四纪动物群分布特征和区域对比

对区内第四纪钻孔检出的有孔虫、介形虫、软体动物化石按照出现频次特征、生态环境等因素划分为淡水种、半咸水种、广盐滨岸种和盐度要求较高（多盐—真盐）的近岸浅海种和浮游类（汪品先等，1988；李从先等，1998）。

（一）第四纪动物群总体面貌

1. 淡水动物

淡水动物主要指淡水类型的介形虫和软体动物。

介形虫主要种类见 *Candona* sp.（玻璃介）和 *Ilyocypris bradyi*（布氏土星介）。这些属种在我国现代和上新世—第四纪地层中常见于湖泊、河流、池塘和沼泽等淡水环境中。

软体动物化石主要见角状环棱螺（*Bellamya angularis*）、丽蚌（*Lamprotula* sp.）和珠蚌（*Unio* sp.）。

2. 半咸水动物

有孔虫常见 *Elphidiella kiangsuensis*（江苏小希望虫）、*Haynesina germanica*（德国海恩斯虫）、*Helenina anderseni*（缝裂海伦虫）、*Pseudogyroidina sinensis*（中国假圆旋虫）、*Pseudononionella variabilis*（多变假小诺宁虫）、*Rectoelphidiella lepida*（精美直小希望虫）和 *Stomoloculata multangula*（多角口室虫）等。

介形虫常见 *Dolerocypria mukaishimensis*（向岛薄丽星介）、*Spinileberis furuyaensis*（古屋刺面介）和 *Tanella opima*（丰满陈氏介）。

上列有孔虫、介形虫属种在我国东黄海沿岸主要见于盐度低于20‰的潮上带、潮间带和河口等少盐—中盐的半咸水环境。

软体动物常见河兰蚬（*Corbicula fluminea*）、中华沼螺（*Parafosarulus sinensis*）、纹沼螺（*P. striatulus*）、田螺（*Viviparus* sp.）、短沟蜷螺（*Semisulcospira* sp.）。

3. 近岸广盐水动物

有孔虫主要见 *Ammonia confertitesta*（厚壁卷转虫）、*Ammonia tepidai*（暖水卷转虫）、*Cribrononion porisuturalis*（孔缝筛诺宁虫）、*Cribrononion subincertum*（亚易变筛诺宁虫）和 *Elphidium magellanicum*（缝裂希望虫）。

介形虫最常见 *Albileberis sinensis*（中华洁面介）、*Albileberis sheyangensis*（射阳洁面介）、*Bicornucythere bisanensis*（美山双角花介）、*Leptocythere ventriclivosa*（腹结细花介）、*Loxoconcha ocellata*（眼点弯贝介）、*Neomonoceratina chenae*（陈氏新单角介）、*Sinocythere dongtaiensis*（东台中华花介）和 *Sinocytheridea impressa*（典型中华美花介）。

上列有孔虫、介形虫属种对生活的水体盐度有良好的适应性，是我国东海和黄海沿岸广布的广盐种，广见于潮上带、潮间带、潮下带、潟湖和河口的少盐（1‰~5‰）—真盐（30‰~35‰）海水中。广盐滨岸类多是各钻孔数量和种类最多的有孔虫，在多数样品中高于40%以上，构成有孔虫群中的优势成分。

软体动物见中华拟蟹守螺（*Cerithidea sinensis*）、蟹守螺（*Cerithium* sp.）、樱蛤（莫瑞蛤）（*Tellina*（*Moerella*）*jedoensis*）、兰蛤（*Corbula* sp.）、黑龙江兰蛤（*Potamocorbula amurensis*）、雕刻友螺（*Nassarius*（*Phromtis*）*cealatula*）、红带织纹螺（*Nassa succiencta*）和河兰蚬（*Corbicula fluminea*）等。

4. 多盐—真盐近岸浅海动物

多盐—真盐的近岸浅海属种在钻孔中属种数较多，但多数属种的丰度较低。

有孔虫最常见 *Ammonia pauciloculata*（少室卷转虫）、*Elphidium advenum*（异地希望虫）*Florilus atlantica*（大西洋花朵虫）、*Florilus decorus*（优美花朵虫）和 *Protelphidium tuberculatum*（具瘤先希望虫）及瓷质壳类 *Quinqeloculina akneriana*（短五玦虫）、*Quinqueloculina seminula*（半缺五玦虫）和列式壳类 *Bolivina robusta*（强壮箭头虫）。

钻孔中出现的浮游有孔虫的丰度较低，多数在10%以下，浮游有孔虫多数个体细小，可能是由外海经潮流搬运而来的异地埋葬分子。

介形虫最常见 *Cytheropteron* spp.（翼花介诸种）、*Munseyella japonica* + *M. pupilla*（日本穆赛介+瞳孔穆赛介）、*Paijenborchella miurensis*（三浦小佩詹博介）和 *Pistocythereis bradyformis*（布氏形纯艳花介）。

上列有孔虫、介形虫现代主要分布在水深小于50m的近岸陆架和河口地区的前三角洲。

软体动物见有雕刻友螺（*Nassarius*（*Phromtis*）*cealatula*）、樱蛤（莫瑞蛤）（*Tellina*（*Moerella*）*jedoensis*）、求拉立螺（*Gulalius*）、囊螺（*Retusa*）、梯螺（*Scala* sp.）和拟蟹守螺（*Cerithidea* sp.）。相伴产出的有海胆刺、角贝、海相硅藻、蛇尾类和盾形化石等。

在海相动物化石群内还出现较多的钙质超微化石。常见成分有颗石型桥石科的大洋桥石（*Gephyrocapsa oceanica*）和赫氏艾氏石（*Emiliania huxleyi*）。在正常浅海相的上更新统上部海相层内见钙质超微化石最多，除桥石科外，尚见颗石科的瓦氏卷舟颗石（*Helicopontosphaera wallichii*）、透明卷舟颗石（*H. hyalina*）、远洋颗石（*Coccolithus pelagicus*）、弱孔圆颗石（*Cyalococcolithus leptopora*）、锡伯脐球石（*Umbilicosphaera sibogae*）；还见到很少的非颗石型钙质超微化石，有毕氏五边石（*Braarudosphaera bigelowi*）和胸甲球石（*Thoracosphaera* sp.）。

（二）第四纪动物群地层特征

以宝山 LQ19 孔有孔虫为例。

依据有孔虫和介形虫等微体化石组合特征及其所反映的生态环境，在 LQ19 孔地层中划分出 9 个层段，其中在孔深 93m 以浅的上部地层中识别出 4 个海侵层（M1～M4），孔深 93m 之下全为陆相沉积（图 21-21～图 21-23）。

第一层段（孔深 0～0.5m）：无有孔虫和海相介形虫，只见 18 枚陆相介形虫 *Ilyocypris bradyi*（布氏土星介）和 *Candona* sp.（玻璃介未定种）及陆相螺类，推测为浅水的陆相沉积环境。

第二层段（孔深 2.4～39.6m）——第一海相层（M1）：共计分析 37 块样品，连续产有孔虫和海相介形虫。微体化石群中浮游、浅海、广盐和半咸水等不同生态类群落混杂共生，个体偏小，总体上反映为河口沉积特征。主要根据化石群落属种组成上的差异，将该层段进一步划分为 4 小段。

2.1 段（孔深 2.4～3.6m）：该段有孔虫只见底栖的螺旋类和平旋类，不见浮游类、瓷质类、瓶虫类和列式类；海相介形虫只见 2 瓣 *Sinocytheridea impressa*（典型中华美花介）。底栖有孔虫分异度平均为 7.5 种，丰度为 2.0 枚/g。在组成上以广盐类 *Ammonia tepida*（暖水卷转虫）在数量丰度上占优势，平均丰度为 63.6%；第二优势种为 *Cribrononion porisuturalis*（孔缝筛诺宁虫，15.4%）。可称之为 *Ammonia tepida*（暖水卷转虫）-*Cribrononion porisuturalis*（孔缝筛诺宁虫）组合。

海相介形虫群落在属种组成上也是以广盐类占优势，平均丰度为 80.9%，其中以 *Albileberis sinensis*（中国洁面介）为主，平均丰度为 34.0%，其次是 *Neomonoceratina chenae*（陈氏新单角介，20.2%）和 *Sinocytheridea impressa*（典型中华美花介，20.2%）。在下部（孔深 6.2m）样品中见少量的陆相介形虫 *Candona* sp.（玻璃介未定种）。可称之 *Albileberis sinensis*（中国洁面介）-*Neomonoceratina chenae*（陈氏新单角介）-*Sinocytheridea impressa*（典型中华美花介）组合。

该层段有孔虫和介形虫在组合上各个生态类群落属种混生，其中以广盐类为主，并含较多的滨岸浅海类属种和少量的浮游有孔虫，也见少量的半咸水类和个别陆相介形虫，推测为河口边滩环境沉积。

2.2 段（孔深 4.6～16.9m）：该段有孔虫和介形虫个体普遍细小。有孔虫丰度低，平均为 1.7 枚/g，分异度平均为 21 种，优势度 D1 平均为 24.4%。底栖有孔虫以近岸浅海类在数量上占优势，平均丰度为 61.1%；广盐类次之，平均丰度为 36.7%；半咸水类最少，平均丰度为 2.2%。但群落中前 2 位优势种由广盐类 *Ammonia tepida*（暖水卷转虫）和 *Elphidium magllanicum*（缝裂希望虫）构成，平均丰度分别为 14.4%和 14.0%。近岸浅海类最常见的是 *Ammonia pauciloculina*（少室卷转虫，12.7%）、*Bolivina robusta*（强壮箭头虫，10.0%）、*Quinqueloculina akneriana*（短五玦虫，9.7%）和 *Elphidium advenum*（异地希望虫，5.9%），这些种的丰度是各小段中最高的，是该段的主要特征。依据丰度最高的前 2 位优势种，该段命名为 *Ammonia tepida*（暖水卷转虫）-*Elphidium magllanicum*（缝裂希望虫）组合。

海相介形虫丰度和分异度平均分别为 1.1 瓣/g 和 7.9 种，优势度 D1 平均为 43.2%，其中优势度是各小段中最高的。属种组合上与有孔虫不同，海相介形虫以广盐类占绝对优势，平均丰度为 70.4%；近岸浅海类平均丰度为 28.9%，是该层最高值；半咸水类稀少，平均丰度为 0.7%。群落中的数量最多的前 2 位优势种是广盐类 *Albileberis sinensis*（中国洁面介，19.2%）和 *Neomonoceratina chenae*（陈氏新单角介，27.4%），丰度居次的是近岸浅海类 *Pistocythereis bradyformis*（布氏纯艳花介，19.0%）和 *Sinocytheridea impressa*（典型中华美花介，8.0%）。可称之为 *Albileberis sinensis*（中国洁面介）-*Neomonoceratina chenae*（陈氏新单角介）组合。

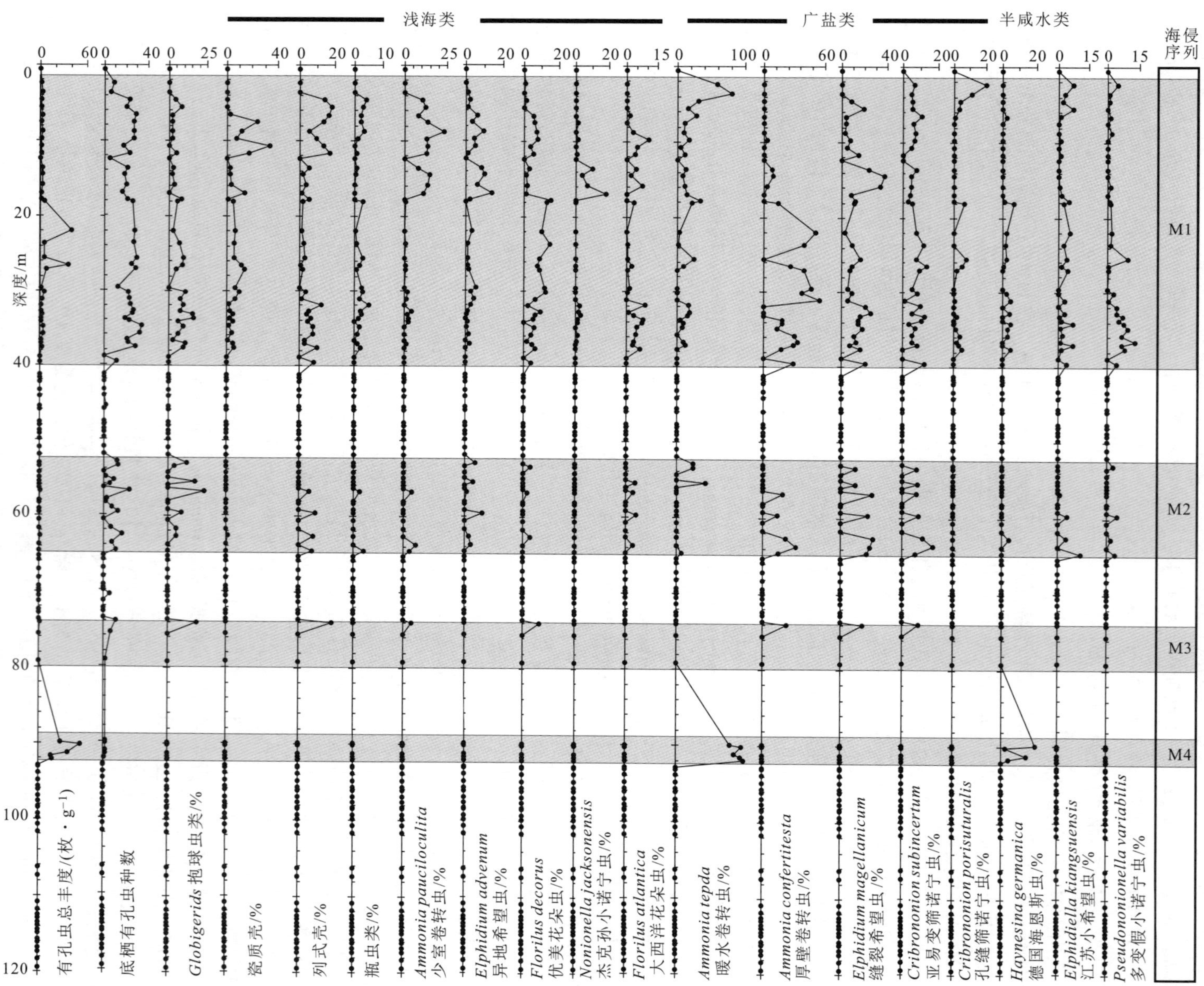

图 21-21　上海罗泾 LQ19 孔有孔虫丰度（枚/g）和含量（%）及海侵地层划分

Fig. 21-21　Foraminifera abundance (pc/g), content(%) and transgressive stratigraphic division per borehole LQ19 in Luojing, Shanghai

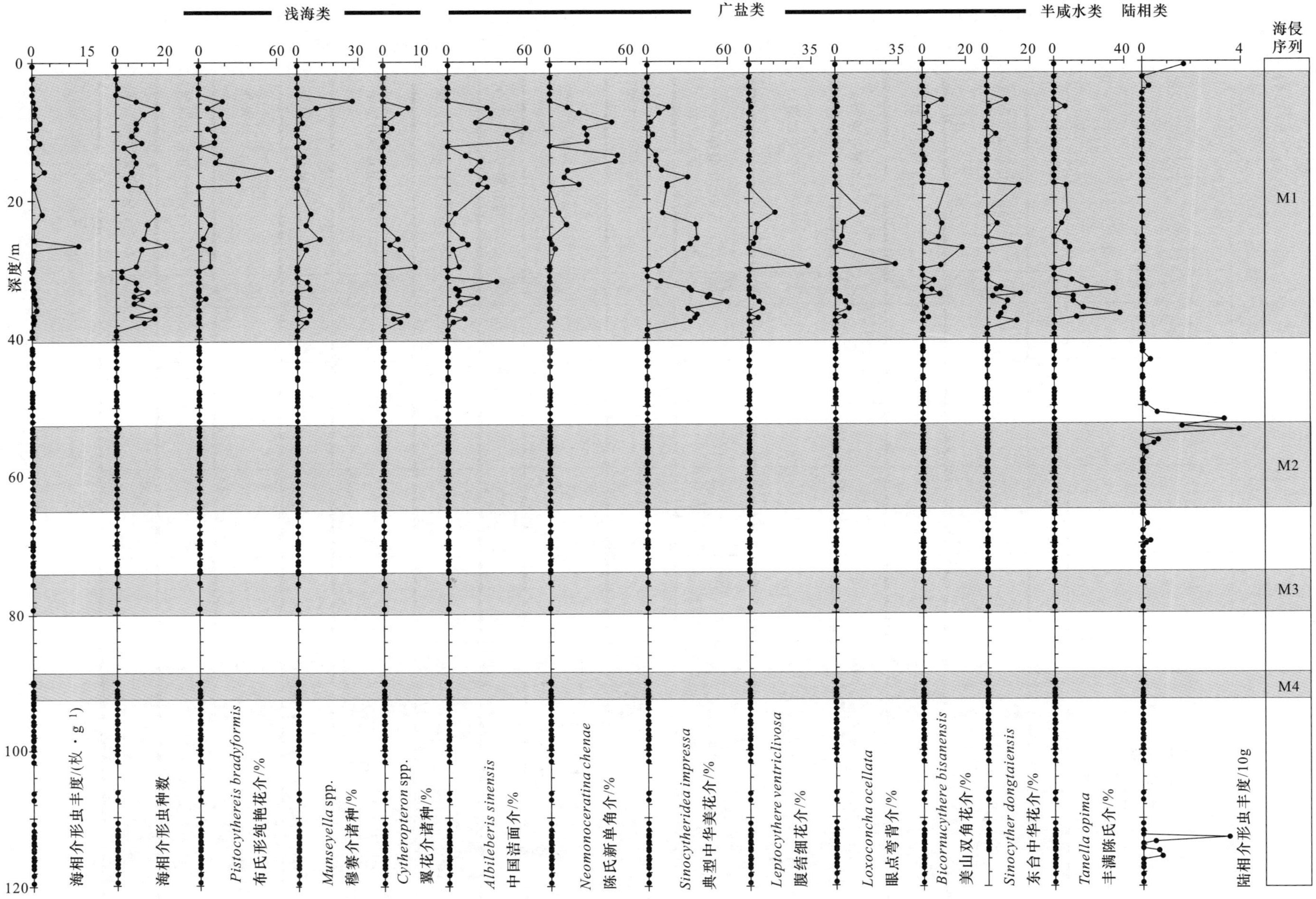

图 21-22 上海罗泾 LQ19 孔介形虫丰度（枚/g）和百分含量（%）及海侵地层划分（一）

Fig. 21-22 Ostracodes abundance (pc/g), content (%) and transgressive stratigraphic division per borehole LQ19 in Luojing, Shanghai (Ⅰ)

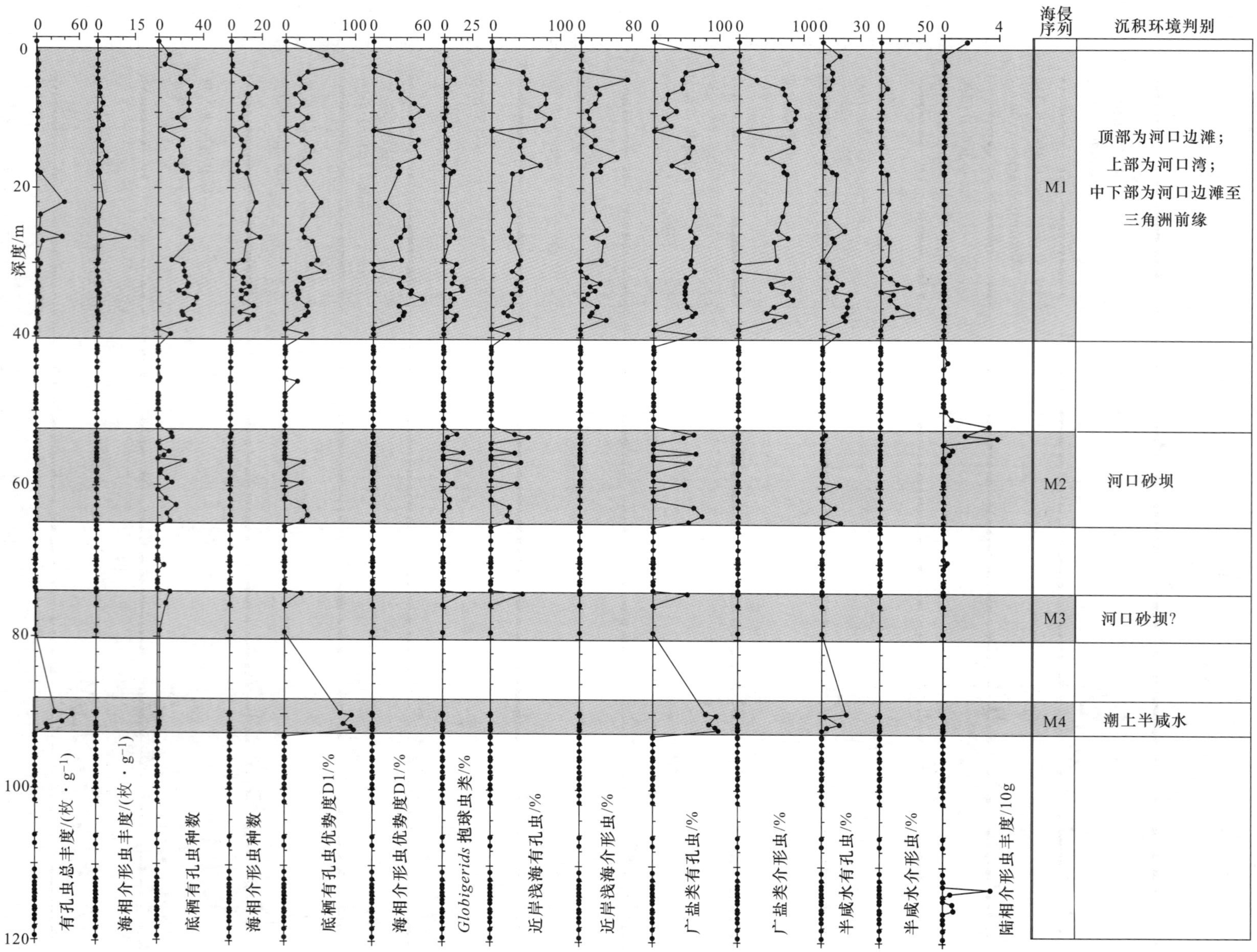

图 21-23　上海罗泾 LQ19 孔介形虫和有孔虫丰度（枚/g）和百分含量（%）及海侵地层划分（二）

Fig. 21-23　Ostracodes and foraminifera abundance (pc/g), content (%) and transgressive stratigraphic division per borehole LQ19 in Luojing, Shanghai (Ⅱ)

该段底栖有孔虫和海相介形虫个体偏小，属种组合上含较多的近岸浅海类，其丰度是各小段中最高的，推测为较开阔的河口湾环境沉积。

2.3 段（孔深 17.8~31.1m）：该段有孔虫和介形虫数量丰度是各小段中最高的，分别平均为 11.3 枚/g 和 1.8 瓣/g；属中组合都以广盐类为主，丰度分别为 58.5%和 66.5%。底栖有孔虫分异度平均为 24.3 种，优势度 D1 平均为 37.6%。浮游有孔虫平均丰度为 6.4%。底栖有孔虫属种组合中以广盐类 *Ammonia confertitesta*（厚壁卷转虫）最丰富，平均丰度为 31.0%。其余常见的依次为浅海类 *Florilus decorus*（优美花朵虫，11.0%）、广盐类 *Ammonia tepida*（暖水卷转虫，9.4%）、*Elphidium magllanicum*（缝裂希望虫，8.6%）和 *Cribrononion subincertum*（亚易变筛诺宁虫，7.3%）。可称之为 *Ammonia confertitesta*（厚壁卷转虫）-*Florilus decorus*（优美花朵虫）组合。

海相介形虫分异度平均为 9.5 种，为各小段最高；优势度 D1 平均为 30.2%，为各小段最低。群落中丰度最高的前 2 位优势种是 *Sinocytheridea impressa*（典型中华美花介，22.9%）和 *Albileberis sinensis*（中国洁面介，12.2%），其余常见种按丰度依次为近岸浅海类 *Pistocythereis bradyformis*（布氏纯艳花介，7.9%）、广盐类 *Bicornucythere bisanensis*（美山双角花介，7.9%）和 *Leptocythere ventriclivosa*（腹结细花介，7.4%）、*Loxoconcha ocellata*（眼点弯贝介，7.4%）。可称之为 *Sinocytheridea impressa*（典型中华美花介）-*Albileberis sinensis*（中国洁面介）组合。

该段底栖有孔虫和海相介形虫分异度和丰度相较高，都以广盐类占优势，同时也含较多的浮游有孔虫（丰度为 6.4%）和半咸水类属种（丰度分别为 8.2%和 5.4%），推测为河口边滩或三角洲前缘环境沉积。

2.4 段（孔深 31.9~39.6m）：该段底栖有孔虫和海相介形虫在数量上仍以广盐类为主，平均丰度分别为 51.0%和 64.2%，但半咸水类丰度显著增加，分别达到 15.7%和 16.1%，为该层的最高值。浮游有孔虫丰度也达最高值，占有孔虫全群的 8.4%。底栖有孔虫分异度平均为 24.4 种，优势度 D1 平均为 23.5%。属种组合中以广盐类 *Ammonia confertitesta*（厚壁卷转虫）和 *Elphidium magllanicum*（缝裂希望虫）丰度最高，平均分别为 17.3%和 16.2%。其余常见种按丰度依次为 *Ammonia tepida*（暖水卷转虫，8.6%）、*Cribrononion subincertum*（亚易变筛诺宁虫，7.3%）和半咸水类 *Pseudononionella variabilis*（多变假小诺宁虫，7.0%）。可称之为 *Ammonia confertitesta*（厚壁卷转虫）-*Elphidium magllanicum*（缝裂希望虫）组合。

海相介形虫分异度平均为 9.9 种，优势度 D1 平均为 38.7%，以广盐类 *Sinocytheridea impressa*（典型中华美花介，36.1%）和半咸水类 *Tanellaopima*（丰满陈氏介，15.2%）为优势种，其余常见种是 *Albileberis sinensis*（中国洁面介，11.3%）和 *Sinocythere dongtaiensis*（东台中华花介，7.1%）。可称之为 *Sinocytheridea impressa*（典型中华美花介）-*Tanella opima*（丰满陈氏介）组合。

该段中有孔虫和介形虫以广盐类居多，含有较多的半咸水类和浮游类，推测为河口边滩环境沉积。

值得注意的是，在该段顶部（17.8m）样品中见 4 枚暖水种 *Pseudorotalia indopacifica*（印度太平洋假轮虫），此种在上海通常见于第二及其之下的海侵层中。其标本保存差，呈黑色或黄褐色，明显经磨蚀，推测为再沉积化石。但需 AMS^{14}C 年龄数据，确定该样品的年代。

第三层段（孔深 41.3~52.0m）：该层段基本不见有孔虫和海相介形虫，仅在孔深 45.6m 处见 2 枚底栖有孔虫，推测可能是在钻进过程中有泥浆带入。在孔深 43.3m 处和 49.8~52.0m 间 3 个样品中见陆相介形虫 *Candona* sp.（玻璃介未定种）和 *Ilyocypris bradyi*（布氏土星介）。在孔深 46.0~49.1m 间的 4 个样品中见陆相螺类。推测为浅水河沼沉积环境。

第四层段（孔深 53.0~64.6m）——第二海相层（M2）：该层段见有孔虫和陆相介形虫，不见海相介形虫。有孔虫丰度低，其中 3 个样品不见有孔虫，6 个样品数量不足 20 枚，最高丰度仅为 1.5 枚/g，多数样品中不足 0.5 枚/g。分异度为 2~24 种，平均为 13.6 种；优势度 D1 平均为 28.8%。底栖有孔虫属种组合以广盐类在数量上占优势，平均丰度为 58.4%；近岸浅海类丰度居次（36.2%），半咸水类最少（5.4%）。群落中丰度最多的 4 个种全是广盐类，它们是 *Elphidium magllanicum*（缝裂希望虫，

19.1%）、*Ammonia confertitesta*（厚壁卷转虫，12.9%）、*Ammonia tepida*（暖水卷转虫，12.6%）和 *Cribrononion subincertum*（亚易变筛诺宁虫，9.0%）。浮游有孔虫丰度较高，平均丰度为9.7%。底栖有孔虫可称之为 *Elphidium magllanicum*（缝裂希望虫）- *Ammonia confertitesta*（厚壁卷转虫）组合。

介形虫只见陆相类 *Candona* sp.（玻璃介未定种）和 *Ilyocypris bradyi*（布氏土星介），产于上部孔深53.0~56.8m间的5个样品中。

该层各个不同生态类群落混杂，个体偏小，数量丰度偏低，数量上以广盐类为主，同时含较多的浮游类和陆相类，推测为河口砂坝或河口内边滩沉积环境。

第五层段（孔深65.3~73.6m）：有孔虫仅在孔深70.5m处见9枚细小个体。不见海相介形虫，在孔深67.1m、69.7m和70.0m等处样品中见个别陆相介形虫 *Candona* sp.（玻璃介未定种）和 *Candoniella albicans*（纯净小玻璃介），推测为河流沉积环境。

第六层段（样品82~84号，孔深74.0~79.1m）——第三海相层（M3）：该层段见少量有孔虫，在每块样品中仅挑得6~26枚壳体，丰度为0.3~0.8枚/g。底栖有孔虫分异度为2~11种，在属种组成上常见广盐类 *Ammonia confertitesta*（厚壁卷转虫）、*Elphidium magllanicum*（缝裂希望虫）、*Cribrononion subincertum*（亚易变筛诺宁虫）等。同时含较多的浮游类，最高丰度达19.2%，推测为河口砂坝或河口内的河道沉积环境。

第七层段（孔深79.1~89.9m）：该层段在上、下样品间隔10.8m内无样品采取，因而很难对该段进行推测。

第八层段（样品85~89号，孔深89.9~92.1m）——第四海相层（M4）：该层段只见底栖有孔虫，丰度较高，平均为30.5枚/g，但属种单调，群落仅由广盐类 *Ammonia tepida*（暖水卷转虫）和半咸水类 *Haynesina germanica*（德国海恩斯虫）组成，以前者居多，平均丰度为91.8%，后者为25.2%。此层海相性程度弱，推测为潮上半咸水沉积环境。

第九层段（样品90~174号，孔深92.9~296.8m）：该层段均不见有孔虫和海相介形虫。在孔深112.9~115.7m间段的4块样品和145.6m、148.6m处样品中见陆相介形虫 *Candona* sp.（玻璃介未定种）、*Candoniella albicans*（纯净小玻璃介）、*Eucypris* sp.（真星介未定种）、*Ilyocypris bradyi*（布氏土星介）和 *Lineocypris* sp.（线星介未定种）等。在孔深112.5m、119.5m、121.1m、122.5m、130.0m、146.2m和157.0m等处样品中见个别陆相螺类或其口盖。推测为河湖相交替沉积环境。

（三）第四纪动物群区域对比

区内第四纪沉积物动物化石分析主要用于揭示海相地层，30余个第四纪钻孔有孔虫、介形虫分布揭示本区第四纪海相动物群主要分布在孔深100m以浅地层中，100~180m间海相化石零星分布。综合大量第四纪钻孔的动物化石分析结果，将区内第四纪动物群综合划分为7个带。

1. 第Ⅰ海相化石带

此化石带位于安亭组中段，分布在孔深200m以下，含有少量基本为广盐性的有孔虫及海相腹足类化石，仅在现今浦东沿江地带局部分布。该化石带主要由20世纪80年代实施的2个钻孔揭示。浦东海塘P6孔，在孔深206~226m灰色细砂夹含砾细中砂内可见轮虫类的 *Ammonia tepida*、*A. koeboeensis*，浮游类的 *Globigerina* spp.、*Epistominella naraensis*，平旋壳类的 *Nonion* sp.、*Cribrononion* sp.、*Elphidium* sp. 和 *Florilus* sp. 等，窄盐性成分 *Bolivina* sp. 等，伴有轮藻。高桥Ch1孔，在大致相当于安亭组中段下亚段、孔深260m处黄灰色中细砂内见海相腹足类雕刻友螺（*Nassarius*（*Phromtis*）*cealatula*）、囊螺（*Retusa*）、露齿螺（*Odostomia*）等及陆相腹足类沼螺（*Parafossarulus* sp.）。2000年以后实施的20余个钻孔均未在200m以下发现海相微体古生物化石。

2. 第Ⅱ海相化石带

此化石带出现在下更新统安亭组上段，为20世纪80年代钻孔揭示，2000年以后未在钻孔该层位中检测出海相有孔虫，多检出陆相介形虫。

安亭组上段下部砂层在浦东的多个钻孔内钻遇，海侵层位属下更新统安亭组上段下部砂层。如在横沔 P12 孔孔深 172~178m 灰黑色细砂夹砂质粉土内见具 *Ammonia beccarii* var. -*Epistominella naraensis* 有孔虫组合，并见有深水种 *Hopikinsina pacifica*、*Lagena* sp.、*Brizalina* sp. 等窄盐性成分；伴生半咸水介形虫 *Neomonoceratina dongtaiensis*、陆相介形虫 *Condoniella albicans* 等以及水蕨孢子、轮藻受精卵膜等陆相生物成分。在浦东海塘 P6 孔孔深 171~191m 灰、灰黄色细砂内可见很少的有孔虫，有 *A. tepida*、*Elphidium magellanicum*、*Epistominella naraensis*、*Globigerinita glutinata*；并见及个别半咸水介形虫 *Cushimanidea japonica*。又如在海潮 P17 孔孔深 175~177m 灰色细砂内分析出少量钙质超微化石 *Gephyrocapsa ocenica*，为上海钙质超微化石出现的最低层位，反映在区内出现的这次海侵只是达到三角洲河流—河口环境。

安亭组上段上部地层内常见淡水软体动物化石，并产较丰富的陆相介形虫，还在少数钻孔中发现海相化石。如在浦东唐家宅孔深 178.26~183.25m 发现海相软体动物化石，在瓦屑孔深 160~163m 间分析出海相介形虫。又如在青浦练塘 P31 孔孔深 137~173.8m 杂、浅褐色黏土和灰黄色粉细砂内断续检出个别有孔虫；在上海面粉厂 J10 孔孔深 148.10~159.29m 杂色黏土夹黄褐色细砂内发现少量的螺旋壳类 *Ammonia* sp.、平旋壳类 *Elphidium* sp. 和 *Florilus decorus* 等广盐性有孔虫；还有在浦东海塘 P6 孔孔深 146~164m、横沔 P12 孔孔深 149~174m、安亭 G414 孔孔深 158.2m 处等的杂、褐黄色黏土内断续分析出很少的有孔虫，三灶 P45 孔孔深 151~153m 褐色粉质黏土内偶见钙质超微化石 *Gephyrocapsa ocenica* 等，表明早更新世末本区转为陆相河湖环境，也还偶尔受到这次海侵后期的微弱影响。

3. 第Ⅲ海相化石带

此化石带局部出现在中更新统嘉定组下段下部灰色细砂层内。如上海面粉厂 J10 孔孔深 133.2~159.5m 青灰色细砂及下伏的黄褐色粉质黏土夹细砂层内见少量有孔虫 *Ammonia* sp.、*Elphidium* sp. 和 *Florirusdecorlus*，与陆相介形虫 *Ilyocypris* sp.、*Candoniella* sp.、*Lineocypris* sp. 共生。嘉定娄塘 P5 孔孔深 141~145m 灰色粉细砂夹砂质粉土中含很少的有孔虫，有 *Ammonia beccarii* var.、*Rosalina bradyi*、*Pseudononionella variabilis* 等。再如在浦东海塘 P6 孔孔深 123~135m 黄灰色中细砂夹灰色粉质黏土中含少量有孔虫，为 *A. tepida*、*Globigerinita* sp. 等。又如在三灶 P45 孔孔深 123.8~145m 灰色细砂、灰黄色含砾砂中含少至很少的有孔虫 *Ammonia beccarii* var.、*Nonion* sp. 和 *Globigerina bulloides*、*Elphidiella kiangsuensis* 等；并偶见钙质超微化石 *Gephyrocapsa oceanica*。另在海潮 P17 孔孔深 108~137m 灰色细砂和含砾中细砂中与横沔 P12 孔孔深 121~149m 灰黑、灰色细中砂中分别发现少量钙质超微化石与再沉积的有孔虫。崇明新海 LQ24 孔孔深 120~140m 中粗砂层中含少量（2~16 枚）浮游和底栖有孔虫，最高丰度为 0.4 枚/g，分异度多数不超过 5 种，不见海相和陆相介形虫。综上所述，长江三角洲沉积区所见常为较粗陆源碎屑沉积物中的少量半咸水有孔虫，或伴有陆相介形虫与钙质超微化石，表明为偶尔受强潮流影响的三角洲平原河流沉积。

在西部太湖沉积区，一般见淡水介形虫，只出现个别的有孔虫，唯青浦练塘 P31 孔孔深 133~135m 浅黄灰色砂质粉土夹粉细砂内出现结缘寺卷转虫角状亚种（*Ammonia ketienziensis angulata*）-毕克卷转虫（*Ammonia beccarii* var.）有孔虫组合，前一属种分布于现南黄海水深 50m 左右处水域，并见 *Protelphidium tuberculatum*、*Buccela frigida* 等示冷性有孔虫，复合分异度达 2.92；与之共生半咸水属种斑纹三原介（*Sanyuaniapsaronion*）和常在具一定盐度水体中生活的陆相介形虫 *Limnocythere* sp.、*Candoniella* sp.；在金山朱泾 G418 孔孔深 132.2~139.5m 黄褐色黏土内也见及个别有孔虫。表明太湖沉积区通常为略受海水影响的三角洲平原河湖，可能一度沦为开放性海湾环境。

4. 第Ⅳ海相化石带

该化石带主要在太湖堆积平原区中更新统嘉定组上段下部出现。如在青浦练塘 P31 孔孔深 93~107m 灰绿、棕灰色粉质黏土和青灰色细砂内分析出 *Ammonia beccarii* var. -*A. convexidorsa* 有孔虫组合，还见有 *Epistominella naraensis*、*Pseudoeponides andersen*、*Nonion shansiensis*、*Cribrononion subincertum*、*C. vitreum* 等有孔虫，伴生半咸水的 *Sinocytheridea* spp. 组合，还见有 *Leguminocythereis* sp.、*Loxoconcha* sp. 等半咸水属种，并见 *Ilyocypris errabundis*、*Candoniella* sp.、*Candona* sp. 等陆相介形虫；偶尔见及钙

质超微化石 *Gephyrocapsa oceanica*。再有在金山朱泾 G418 孔孔深 108~112m 灰色粉质黏土内分析出陆相介形虫 *Ilyocypris* sp.、*Candona*（*Candona*），并见海相介形虫 *Echinocythere isbradei*；还见淡水软体动物化石丽蚌等碎片。这表明当时通常为河流沉积，偶尔受到海水的影响，在太湖沉积区局部出现与现今长江三角洲前缘支流河道相同的有孔虫组合，显示属于三角洲河流堆积。在三角洲沉积区唯市区 J10 孔孔深 115.19~118.41m 灰、黄褐色粉细砂内检出少量有孔虫；其上亚段灰、蓝灰色粉质黏土则未发现微体古生物，仅见及陆相软体动物方形环棱螺相似种（*Gyraulus albus*）等，表明为有时受海侵影响的河流演变为泛滥平原—湖泊洼地环境。

5. 第Ⅴ海相化石带

该化石带分布在上更新统川沙组地层中，除崇明等三岛、宝山和嘉定以外，在上海境内大都有所分布，大体以太湖沉积区偏南部地区的海相性略强。

青浦练塘 P31 孔孔深 73~93m 灰、灰绿色粉质黏土夹灰色粉砂层含 *Ammonia beccarii* var.-*A. convexidorsa* 有孔虫组合，还见 *Epistominella naraensis*、*Pseudoeponides andersen*、*Nonion shansiensis*、*Cribrononion subincertum*、*C. vitreum* 等有孔虫；伴生 *Sinocytheridea* spp.、*Leguminocythereis* sp.、*Loxoconcha* sp. 等半咸水属种和 *Ilyocypris errabundis*、*Candoniella* sp.、*Candona* sp. 等陆相介形虫以及圆盘形硅藻、刺盒虫等陆相化石，反映为陆上三角洲河流环境。金山朱泾 G418 孔此层内断续出现半咸水介形虫和钙质超微化石，其下部（孔深 85.2~92.6m）的灰褐色黏土内主要含 *Sinocytheridea* sp.、*Echinocythereis* sp.、*Neomonoceratina* sp.，并含 *Leguminocythereis hudgii*，上部（孔深 67~71m）的灰色粉质黏土、粉砂互层含少量 *Sinocytheridea* sp.；中部灰色粉质黏土、粉砂互层含少量的钙质超微化石 *Gephyrocapsa oceanica* 及弱孔圆颗石（*Cyalococcolithus leptopora*），表明由下部的河口沉积演变为中上部的滨海沉积。金山枫泾 LQ1 孔孔深 78.40~85.70m 砂层中出现有孔虫和海相介形虫，有孔虫组合为 *Ammonia beccarii-Elphidium magellanicum*，还含较多的 *Elphidium magellanicum*，以及少量的陆架种 *Ammonia compressiuscula*、*Ammonia ketienziensis*、*Florilus decorus* 和 *Protelphidium tuberculatum*。海相介形虫以 *Sinocytheridea impressa* 为主，见少量内陆架种，如 *Neomonoceratina chenae*、*Pontocythere spatiosa* 和 *Pistocythereis bradyformis* 等，反映为弱潮汐的河口前缘沉积，上部海相性减弱，可能为河口内边滩沉积。而在太湖沉积区的青浦北部（如青浦镇 G415 孔）未见有孔虫；唯在嘉定安亭 G414 孔顶、底部的灰色含砾中细砂内检出少量有孔虫，说明两者均为河流沉积，只是后者偶尔受到海水的影响。

在长江三角洲沉积区，此层位大都无或见少量有孔虫，反映通常为河流沉积，间或为受到海水影响的三角洲河流沉积。局部含有孔虫较丰，具河口相组合成分，如市区的面粉厂 J10 孔在此层下部（孔深 92~109.5m）的青灰色含砾细砂夹粉质黏土中含少量有孔虫，具 *Ammonia beccarii* var.-*Epistominella naraensis* 组合，并含 *Florirusdecorlus*、*Protelphidium* sp.，伴生海胆刺、海绵骨针等海相化石，属于河口沉积；其上部（孔深 78~92m）的含砾中细砂中仅含少量陆相介形虫 *Ilyocypris errabundis*、*Candoniella sulacensis*、*C. schubinae*，以及轮藻、鱼牙等陆相生物化石，表明已演变为河流沉积。再如横沔 P12 孔此层中下部（孔深 93~125m）的灰色中细砂、粗砂砾石层仅含少量再沉积有孔虫；上部（孔深 81~93m）含 *Ammonia dominicana-Elphidium magella nicum* 组合，并含 *Epistominella naraensis*、*Globigerina quinqueloba*；表明由中下部的三角洲河流沉积，演变为上部的河口沉积。另如杨浦教育局 CBG1-1 号、浦东外高桥 J20 号、海塘 P6 号、机场 J25 号、三灶 P45 号和海潮 P17 号等孔，均在此层局部见及少量有孔虫或偶见钙质超微化石。如外高桥 J20 孔此层大都为灰色细砂夹粉土和底部的含砾砂，断续见个别的钙质超微化石 *Gephyrocapsa oceanica*，唯顶部（孔深 77~80m）见很少的 *Florirus decorlus*、*Globigerina*，并见 *Epistominella naraensis*、*Ammonia beccarii* var. 及平旋壳类有孔虫，还出现列式壳类深水种，反映通常为间或受海水泛及的河流沉积，晚期演化为三角洲河流沉积（图 21-24）。又如机场 J25 孔中下部（孔深 103.5~129.65m）灰、灰黄色粉细砂、砾砂层无微古化石，其上覆层（孔深 81.95~103.5m）灰色粉细砂夹砂质粉土、粉质黏土的下部（92.6~103.5m）断续见个别的 *Ammonia beccarii* var.、*A. compressiuscula*、*Asterotalia subtrispinosa* 有孔虫及偶见个别的海相介形虫 *Keijellabisanensis*、*Cush-*

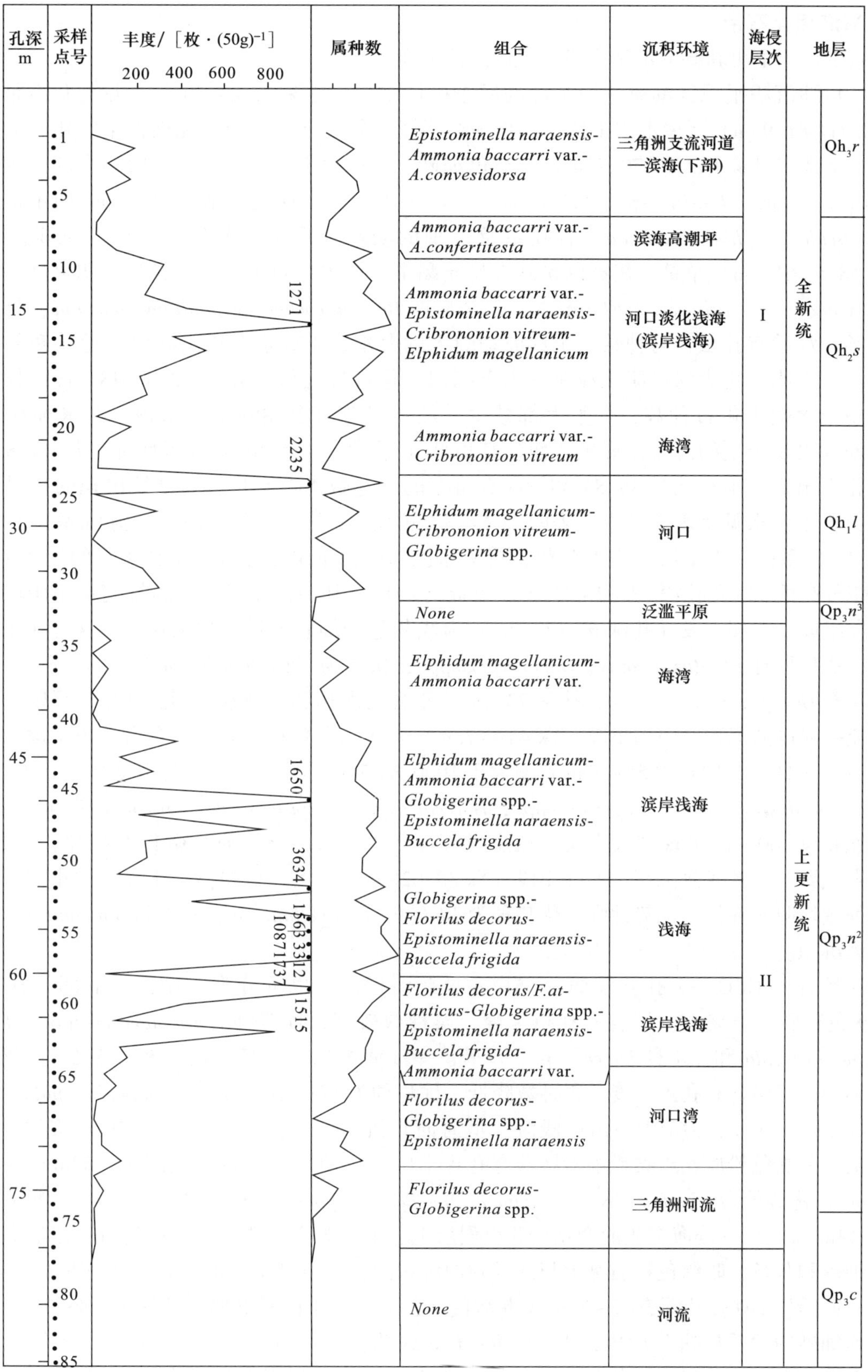

图 21-24　浦东外高桥 J20 孔微古化石组合图

Fig. 21-24　Micro paleofossil assemblages per borehole J20 in Waigaoqiao, Pudong District

imanidea japonica，反映此层通常为河流沉积，只有短期为时而受到海水影响的三角洲河流沉积。

6. 第Ⅵ海相化石带

该化石带位于上更新统南汇组中段，全区广泛分布。

长江三角洲沉积平原区滨海平原东部的浦东机场J25孔孔深32.4~70.71m见此化石带。其下伏层（深70.71~81.95m）系南汇组下段，为由灰色黏土、粉砂薄互层组成无微古生物化石的河漫滩沉积；化石带下部（孔深61.20~70.71m）灰黄色细砂、砂质粉土以含属种、个数均少的半咸水有孔虫*Ammonia beccarii* var.、*Florirus* sp.、*Cribrononion* sp. 为主，还出现代表水深20~50m的正常浅海属种*A. compressiuscula*、*Elphidium advenum*有孔虫及个别的淡水介形虫，表明为河口湾沉积；化石带中部（孔深48.5~61.2m）草黄、灰黄色粉砂夹粉质黏土含微古化石丰度、分异度均明显增高，兼有半咸水属种*Ammonia beccarii* var.、*Florirus* sp.、深水种*Bolivinacochei*、*Elphidium advenum*、瓶虫类、其他列式壳类和浮游有孔虫，与种数、个数波动颇大的海水、半咸水介形虫共生，近其底部出现数种淡水介形虫，表明演化为河口淡化浅海—浅海沉积。化石带上部（孔深32.4~48.5m）灰黄色粉细砂、砂质粉土含微古化石种数、个数大都减少，虽然仍兼有上述的半咸水种、深水种和浮游有孔虫，但只断续出现个别的海相介形虫，而在其顶部（深32.4~35.8m）和近底部（深44.8~46.8m）则无微古化石出现，唯在孔深35.8~38.8m有略多的上述数类有孔虫，与断续出现的少量海水、半咸水介形虫共生，未见淡水介形虫，表明大体演化为滨海沉积，局部为滨岸浅海沉积。上覆层（深29.4~32.4m）系南汇组上段，为由草黄、暗灰绿色黏质粉土组成无微古生物化石的泛滥平原沉积。

在浦东外高桥J20孔孔深38.02~77.00m见此海侵层，其层位属于上更新统南汇组中段。此海侵层由3小段组成：下部小段（孔深67~77.0m）为浅灰色粉细砂夹粉质黏土互层，含有孔虫丰度、属种数均低，优势种有*Florilus decorus*、*Globigerina* spp.，次之为*Elphidium magellanicum*等，并见很少的钙质超微化石*Gephyrocapsa oceanica*，孔深73.5m浅上还有海相和半咸水介形虫共生，反映由下而上为三角洲河流—河口湾沉积。中部小段（深44~67.m）基本为灰色粉质黏土，含有孔虫丰度大都大于等于200枚/50g干样，其中孔深54~60m普遍大于1000枚/50g干样，属种数大都大于等于20个/样，中部优势种有*Florilus decorus*、*Globigerina* spp.，*Buccela frigida*、*Epistominella naraensis*和*Ammonia beccarii* var.，在孔深54~60m间出现列式壳类、瓶虫类等深水种稍多，含钙质超微化石属种、个数也增多，反映为滨岸浅海—正常浅海沉积。上部小段（深38.02~44m）为灰色粉质黏土夹粉砂小团块，有孔虫丰度、属种数均明显降低，优势性强，优势种为*Elphidium magellanicum*，其次为*Ammonia beccarii* var.，反映为海湾沉积。

在杨浦教育局CBG1-1孔孔深36~65.4m见此海侵层，也可分为3小段。下部（孔深58.5~65.4m）灰色粉质黏土夹细砂层含有孔虫丰度、属种数较高，优势种有*Globigerina* spp.、*Florilus decorus*、*Ammonia tepida*和深水种*Lagena* sp.，反映滨岸浅海沉积。中部（孔深52~58.5m）灰黑色粉质黏土、粉细砂薄互层含有孔虫丰度、属种数降低，优势种为*Florilus decorus*、*Ammonia tepida*，并含少量浮游种*Globigerina* spp.，为滨海—滨岸浅海（低海面）沉积。上部（孔深36~52m）灰黄、青灰色粉细砂和灰褐、灰黑色粉质黏土夹粉细砂层含有孔虫丰度、属种数由较高（孔深44~52m）—降低（孔深36~44m），优势种为*Florilus decorus*、*Ammonia tepida*、*Globigerina* spp. 和*Lagena* sp.，反映经历滨岸浅海（深44m以下）—滨海沉积的变化。此海侵层上、下各为不含微体化石的河流—泛滥湖泊沉积的锈黄色粉细砂和灰黄、暗绿色粉质黏土与泛滥湖泊沉积的灰褐色黏土，也构成一个完整的海侵旋回。

在上海面粉厂J10孔孔深30.2~75m见此海侵层，分为两小段。下部（孔深75~43.7m）灰色粉质黏土频夹粉细砂的“千层饼”层内，以*Florilus decorus*为主，并有列式壳、瓷质壳、胶结壳等深水种和示冷性有孔虫*Buccella frigida*出现，复合分异度达2.7~3；与现今见于东海水深大于等于50m的*Munseyella* sp.-*Cythelopteron* sp. 介形虫组合共生，说明属正常浅海沉积。上部（孔深43.7~30.2m）的灰、黄色粉细砂内为*Ammonia beccarii* var. 组合，尚见较少的*Elphidium advenum*、*Florilus decorus*，与*Sinocytheridea latiovata*、*Leguminocythereis hodgii*等半咸水介形虫共生，反映滨岸浅海沉积。

河口沙岛区唯在此海侵层上部的灰色粉砂、粉质黏土互层内见少量有孔虫和个别半咸水介形虫，其中、下部的砂和含砾砂层无微古化石。说明中下部为河流沉积，至上部才演变为滨海沉积。

西部太湖沉积区内此海侵层以含种数、个数均少和优势性强的广盐性有孔虫，与半咸水、陆相介形虫及陆相微古化石共生的层段为主，其与无微古化石的层段交替出现，反映较弱的海陆过渡与陆相交替的沉积演化；揭示这次海侵在上海西部地区曾有两次进退，形成相应的两个亚旋回。以青浦练塘P31孔为例，该孔孔深47~49m为下伏的浅蓝灰色粉质黏土层，出现个别的有孔虫及其他陆相生物成分，说明为间或受海水泛及的湖泊沉积。其上，孔深27~47m为海侵下亚层，由蓝灰、浅棕、褐黄色细纹间互相成皱纹状黏土，上部依次为灰黑色黏土和褐黄色粉砂，其下部出现优势性很强的广盐性有孔虫与陆相、广盐海相介形虫及其他陆相成分共生。中上部出现个数、属种数均少的广盐性有孔虫，仅有其他陆相成分共生，表明基本为潟湖沉积，上部演变为滨海沼泽沉积。再上，孔深23~27m为含褐铁矿结核的灰绿色黏土、浅褐色粉质黏土，即西部的“第二硬土层”，偶见个别有孔虫，表明为偶尔受海水泛及的湖泊沉积。更上，孔深9~23m为海侵上亚层。下部为含植物碎屑的浅棕、灰色（粉质）黏土，含优势性甚强的广盐性有孔虫、个别属种广盐海相介形虫，与其他陆相生物成分共生，反映为滨海沼泽沉积；中上部依次为灰色粉质黏土夹粉砂、褐黄杂有黄绿、蓝灰色斑纹状黏土，含个数、属种数均很少的广盐性有孔虫，与其他个别陆相成分共生，反映为潟湖沉积。上覆层为孔深3.9~9m的褐黄杂黄绿色细网纹状黏土，含有褐铁矿结核，偶见个别的广盐性有孔虫及其他陆相生物成分，表明为偶尔受到海水泛及的湖泊沉积。

7. 第Ⅶ海相化石带

该化石带分布于全新世地层中，全区分布，自下而上化石组合特征揭示了海陆过渡相—海相—海陆过渡相序列，对应娄塘组、上海组和如东组。

在长江三角洲沉积区内，如外高桥J20孔所揭示的（见图21-24），化石带三分性明显，对应于上述3个基本岩石地层单位：

下部（孔深23.30~35.57m）包括孔深29.65~35.57m的灰色砂质粉土夹粉质黏土和孔深23.3~29.65m的灰色黏土两段海陆过渡相地层组成的娄塘组。下段含有孔虫个数和属种数多寡不一，优势种为 *Elphidium magellanicum*、*Cribrononion vitreum*、*Globigerina* spp.，并有其他平旋壳类及少量窄盐种，与较多的广盐海相介形虫 *Sinocytheridealatiovata*、*Leguminocythere ishodgii* 和少量海相介形虫共生，反映为河口沉积；上段有孔虫种数、个数均减少，优势种为 *Ammonia beccarii* var.、*Cribrononion vitreum*，介形虫唯见 *Sinocytheridealatiovata*，反映为海湾沉积。

中部（孔深8.72~23.30m）由下而上依次为灰色淤泥质黏土、灰色砂质粉土夹淤泥质粉质黏土和灰色淤泥质粉质黏土层，这套以淤泥质黏性土为特征的海相地层组成上海组。其中、下两层有孔虫属种、个数均丰，优势种有 *Ammonia beccarii* var.、*Epistominella naraensis*、*Cribrononion vitreum*、*Elphidium magellanicum*，次之为其他螺旋壳类、平旋壳类及抱球虫科，并出现较多的列式壳类、瓷质壳类窄盐性属种，与较多广盐海相介形虫 *Sinocytheridealatiovata*、*Neomonoceratina dongtaiensis*、*Leguminocythere ishodgii* 等共生，还含有较丰富的钙质超微化石，反映为河口淡化浅海（下层）—中潮坪（中层）沉积；上层含有孔虫种数、个数骤减，优势性强，为 *Ammonia beccarii* var. 和 *A. confertitesta*，反映为高潮坪沉积。

上部，孔深8.72m以浅，由下而上为灰色粉质黏土夹砂质粉土及两者的互层，这一海陆过渡相地层组成如东组。其含有孔虫种数、个数较少，优势种为 *Epistominella naraensis*、*Ammonia beccarii* var. 和 *A. convexidorsa*，组合成分与现今长江三角洲前缘支流河道——吴淞江、黄浦江所见有孔虫组分相同。

总之，在长江三角洲沉积区分布的这一海侵层，发育齐全的地层序列通常为：下部包括河口—溺谷堆积的灰色粉土与灰白色粉砂互层和滨海沼泽—海湾堆积的深灰、褐灰色（淤泥质）黏土两段。后者在上海东、南部各区以常见植物茎叶与黄色泥灰质结核，局部形成泥炭或碳质黏土夹层为特征；中部包括滨岸浅海堆积的深灰色淤泥质（粉质）黏土和河口滨海堆积的灰色粉土夹粉砂与灰色粉土、粉质黏土互层；上部为河口堆积的灰、灰黄色粉土、粉砂与滨岸潮滩堆积的褐灰、褐黄色粉质黏土。

在太湖沉积区内如金山和青浦东北部等地所见为陆相—海陆过渡—陆相地层序列，发育齐全的通常具有：下部包括湖沼堆积的灰黑色粉质黏土和滨海沼泽堆积的深灰色（淤泥质）黏土，含较多植物茎叶与黄色泥灰质结核，局部夹泥炭或碳质黏土层；中部为潟湖—滨海沼泽堆积的深灰色（淤泥质粉质）黏土，以含优势性强的一两种广盐性有孔虫与较少的半咸水介形虫共生为特征，其偏下部位常含植物茎叶、泥灰质结核及贝壳碎片，偏上部位时见潮汐水道形成的粉砂、砂质粉土透镜体夹层；上部为湖沼堆积的深灰色黏土夹泥炭与湖泊堆积的褐灰、褐黄色粉质黏土夹粉砂，局部沿古吴淞江分布较厚的灰褐色砂质粉土之类滨岸堆积，常含淡水软体动物、介形虫、轮藻等陆相生物化石，可见少量广盐性有孔虫，显示受到海水的影响。

二、第四纪孢粉化石分布特征和区域对比

（一）孢粉化石主要类型

上海位于北亚热带北缘，受地带性植被影响，孢粉类型既有亚热带分子，也有温带成分，受局地沉积环境影响，常出现滨海及河流湖沼草本花粉、淡水藻类及海相沟鞭藻。第四纪钻孔沉积物中常见孢粉藻类类型如下（王开发等，1998）。

1. 乔木植物花粉

Abies 冷杉、*Cedrus* 雪松、*Keteleeria* 油杉、*Picea* 云杉、*Pinus* 松、*Podocarpus* 罗汉松、Taxodiaceae 杉科、*Tsuga* 铁杉、*Acanthopanax* 五加、*Acer* 槭、*Alnus* 桤木、*Betula* 桦、*Buxus* 黄杨、*Carpinus* 鹅耳枥、*Carya* 山核桃、*Castanea* 栗、*Castanopsis* 栲、*Celtis* 朴、*Corylus* 榛、*Engelhardtia* 黄杞、*Eurya* 柃木、*Fagus* 山毛榉、*Ilex* 冬青、*Juglans* 胡桃、*Liguidambar* 枫香、*Melia* 楝、Menispemaceae 防己科、*Morus* 桑、*Nyssa* 紫树、Oleaceae 木犀科、*Ostrya* 苗榆、*Platycarya* 化香、*Pterocarya* 枫杨、*Quercus* 栎（落叶）、*Quercus* 栎（常绿）、*Rhamnus* 鼠李、*Rhus* 漆树、*Ruta* 芸香、*Salix* 柳、*Tilia* 椴、*Ulmus* 榆、*Cornus* 梾木、*Pistacia* 黄连木。

2. 陆生灌木、草本植物花粉

Ephedra 麻黄、*Elaeagnus* 胡颓子、Ericaceae 杜鹃花科、*Euphorbia* 大戟、Leguminosae 豆科、*Lonicera* 忍冬、Rosaceae 蔷薇科、*Artemisia* 蒿、*Campanula* 风铃草、Chenopodiaceae 藜科、Compositae 菊科、Cruciferae 十字花科、Cyperaceae 莎草科、*Dianthus* 石竹、Gramineae 禾本科、*Humulus* 葎草、Labiatae 唇形科、Liliaceae 百合科、*Malva* 锦葵、*Plantago* 车前、Polygonaceae 蓼科、Ranunculaceae 毛茛科、*Rostellularia* 爵床、Rubiaceae 茜草科、Umbelliferae 伞形科、*Loranthus* 桑寄生。

3. 水生草本植物花粉

Alisma 泽泻、*Myriophyllum* 狐尾藻、*Nymphoides* 荇菜、*Polygonum* 蓼（水生）、Potamogetonaceae 眼子菜科、Sparganiaceae 黑三棱科、*Typha* 香蒲、*Trapa* 菱。

4. 蕨类孢子

Adiantum 铁线蕨、*Ceratopteris* 水蕨、*Cibotium* 金毛狗、Hymenophyllaceae 膜蕨科、*Lycopodium* 石松、Polypodiaceae 水龙骨科、*Osmunda* 紫萁、*Pteridium* 蕨、*Pteris* 凤尾蕨、*Selaginella* 卷柏、*Hicriopteris* 里白。

5. 河湖相藻类

Concentricystes 环纹藻、*Pediastrum* 盘星藻、*Zygnema* 双星藻。

6. 海相沟鞭藻

Lingulodinium machaerophorum、*Operculodinium centrocarpum*、*Selenopemphix quanta*、*S. nephroides*、*Spiniferites hyperacanthus*、*S. ramosus*、*S. bulloideus*、*S. mirabils*、*S.* spp.、*Trinovantedinium applannatum*、*Tuberculodinium vancampoae*。

（二）孢粉化石地层特征

1. 孢粉组合

以滨海平原北新泾 LQ15 孔为例，钻孔所在地点第四系厚度约为 250m。根据地层中孢粉化石浓度、类型等分布特征，自下而上划分为 12 个孢粉组合带。

P1（孔深 227.50~249.50m）：落叶栎-榆-枫香-禾本科组合。该组合中乔木植物花粉占明显的优势，含量为 45.1%~81.3%，其中最为多见的是落叶栎（12.0%~29.4%）、榆（2.4%~20.4%）和枫香（0~26.0%），但后两者的含量变化大。松、常绿栎和朴为常见类型，但含量均不高，很少超过 10%。灌木、草本植物花粉含量为 12.4%~38.7%，其中禾本科含量明显较高，为 4.6%~18.3%。水生草本植物花粉含量为 0.7%~5.9%，香蒲普遍存在。孢子含量为 2.0%~18.6%，大多为水龙骨科。

P2（孔深 130~227.5m）：孢粉贫乏。仅在 177.00m 处见丰富孢粉，组合为落叶栎-禾本科-蒿。组合中乔木植物和灌木、草本植物花粉含量明显高于其他组分，含量均为 43.0%。前者最多为落叶栎（15.0%），后者中含量明显较高者是禾本科和蒿，分别为 15.0% 和 10.0%。组合中水生草本植物花粉和孢子含量低。

P3（孔深 127.00~130.00m）：栎-榆-蒿-莎草科组合。该组合中乔木植物花粉和灌木、草本植物花粉略少，两者含量分别为 31.8%~51.6% 和 29.4%~52.3%。乔木植物花粉中多见落叶栎（6.6%~10.9%）和常绿栎（5.2%~11.1%）。常见类型中榆（2.4%~9.1%）、桦科（3.3%~8.5%）、松（1.3%~8.6%）等含量显著，山毛榉（1.3%~3.9%）、胡桃（0.7%~3.8%）、枫杨（1.3%~2.6%）等含量较低。灌木、草本植物花粉中蒿和莎草科含量明显较高，分别为 5.3%~17.2% 和 6.7%~12.5%，禾本科含量较低，为 5.2%~8.6%。组合中水生草本植物花粉含量低，为 1.2%~3.3%，香蒲较为多见；孢子含量较高，为 13.8%~25.0%，多为水龙骨科。

P4（孔深 111.00~127.00m）：禾本科-莎草科-栎组合。该组合中灌木、草本植物花粉占明显的优势，含量为 12.4%~38.7%，其中禾本科含量明显较高，为 7.1%~39.2%；莎草科和蒿含量较低，分别为 5.3%~30.3% 和 4.4%~20.4%。乔木植物花粉含量为 19.9%~45.2%，其中较为多见的是落叶栎（4.0%~14.8%）和常绿栎（2.6%~16.7%）。组合中水生草本植物花粉含量为 1.9%~9.5%，多为香蒲。孢子含量为 1.9%~10.9%，多为水龙骨科。

P5（孔深 108.00~111.00m）：常绿栎-禾本科-莎草科组合。该组合中乔木植物花粉占优势，含量为 38.0%~64.7%，其中常绿栎含量明显较高，为 10.0%~22.7%。常见类型为落叶栎（6.0%~12.7%）、松（5.9%~13.3%）、榆（5.0%~7.2%）和桦科（1.0%~4.6%）等。灌木、草本植物花粉含量为 24.0%~42.1%，其中禾本科和莎草科含量较高，分别为 6.7%~15.8% 和 6.0%~12.0%。组合中水生草本植物花粉含量低，为 2.0%~3.2%。孢子含量变化较大，为 5.1%~26.0%，大多为水龙骨科。

P6（孔深 108.00）：菊科-莎草科组合。该组合中灌木、草本植物花粉占明显的优势，含量为 65.2%，其中禾本科含量明显较高，为 37.8%；居次的莎草科含量低得多，仅为 10.4%。乔木植物花粉含量为 20.7%，其中较多的是落叶栎和常绿栎，含量均为 5.9%。组合中水生草本植物花粉含量为 3.0%，多为香蒲。孢子含量较高，为 11.1%，绝大部分为水龙骨科。

P7（孔深 66.50~108.00m）：孢粉稀少。

P8（孔深 64.25~66.50m）：莎草科-蒿组合。该组合中灌木、草本植物花粉含量最高，为 69.3%~86.6%，其中以莎草科为主，含量可高达 44.7%~49.0%；而其他常见类型的含量都远不如莎草科，如含量较高的蒿和禾本科分别为 9.6%~21.5% 和 7.9%~11.4%。乔木植物花粉含量为 11.4%~19.3%，其中最多见的是落叶栎，含量为 6.0%~7.4%，其他类型的含量都很低。组合中水生草本植物花粉为零星出现。孢子含量为 1.3%~11.4%，多为水龙骨科。

P9（孔深 36.80~64.25m）：栎-榆-莎草科组合。该组合中乔木植物花粉占优势，含量为 46.7%~

63.4%，其中落叶栎含量明显较高，为10.0%~23.5%；常绿栎和榆含量显著，分别为3.9%~15.1%和0.7%~11.3%。另外的常见类型有松、桦科、枫杨、胡桃等。灌木、草本植物花粉含量为21.7%~40.1%，其中的莎草科含量为4.5%~17.8%，虽较前一组合大幅度下降，但在本组合中仍为含量较高者。组合中水生草本植物花粉含量低或为零星出现，香蒲仍为较多见者。孢子含量为7.1%~23.2%，大多为水龙骨科。

P10（孔深18.00~36.80m）：禾本科-落叶栎组合。该组合中灌木、草本植物花粉含量明显较高，为33.8%~65.6%，其中禾本科含量明显较高，为13.7%~37.9%。常见类型中莎草科和蒿含量较高，但明显不如禾本科。乔木植物花粉含量为15.9%~47.4%，其中较为多见的是落叶栎（6.5%~21.7%）、常绿栎（2.0%~12.1%）和松（0~17.0%），其他类型含量较低，特别是榆，明显少于前述组合。组合中水生草本植物花粉含量为1.9%~17.3%，多为香蒲。孢子含量为1.9%~16.9%，多为水龙骨科。

P11（孔深4.50~18.00m）：栎-松-莎草科组合。该组合中乔木植物花粉占优势，含量为30.7%~61.8%，其中含量明显较高者为落叶栎（7.3%~23.2%）、常绿栎（4.4%~16.1%）和松（2.6%~29.4%），其他类型含量不显。灌木、草本植物花粉含量为23.0%~44.0%，其中莎草科含量较高，为6.6%~19.7%；禾本科和蒿含量较低，分别为1.9%~20.7%和0~9.1%。组合中以香蒲为主的水生草本植物花粉含量为0.6%~7.1%。孢子含量变化较大，为9.4%~30.5%，类型以水龙骨科为主。

P12（孔深4.25m）：以禾本科为主的组合。该组合中以灌木、草本植物花粉为主，含量为77.8%，其中禾本科占绝对优势，含量可达62.1%。其他类型中较多见的是莎草科、蒿、菊科等，但含量不高。乔木植物花粉含量为15.7%，常绿栎（5.2%）、落叶栎（3.9%）和松（3.9%）为含量较显著者。水生草本植物花粉含量低，仅为1.3%。孢子含量为5.2%，所见类型为水龙骨科和蕨。

2. 孢粉记录的古环境信息

以LQ15孔为例，柱状剖面孢粉组合序列总结（表21-4）显示，该钻孔剖面孔深249.50~255.00m、130.00~227.50m和66.50~108.00m间样品内孢粉藻类化石稀少或贫乏，可能由沉积条件不利造成，也可能因气候条件不利所致。

表21-4 LQ15孔柱状剖面孢粉组合序列

Table 21-4 Spora-pollen assemblage of borehole LQ15

<table>
<tr><th colspan="2">样品段/m</th><th>化石分带</th><th>孢粉组合序列</th><th>植被与气候</th></tr>
<tr><td colspan="2">3.21~4.25</td><td colspan="3">孢粉化石稀少、贫乏</td></tr>
<tr><td colspan="2">4.25</td><td>K</td><td>以禾本科为主</td><td>受人类活动影响草甸</td></tr>
<tr><td colspan="2">4.25~18.00</td><td>J</td><td>栎-松-莎草科</td><td>阔叶林-针叶阔叶林、草甸草原，温暖略湿</td></tr>
<tr><td colspan="2">18.00~36.80</td><td>I</td><td>禾本科-落叶栎</td><td>草甸草原、阔叶林，温凉干</td></tr>
<tr><td colspan="2">36.80~64.25</td><td>H</td><td>栎-榆-禾本科-莎草科</td><td>阔叶林、滨岸草甸，温暖略湿</td></tr>
<tr><td colspan="2">64.25~66.50</td><td>G</td><td>莎草科-蒿</td><td>草甸草原，温凉干</td></tr>
<tr><td colspan="2">66.50~108.00</td><td colspan="3">孢粉化石稀少、贫乏</td></tr>
<tr><td colspan="2">108.00</td><td>F</td><td>菊科-莎草科</td><td>草甸草原，温凉干</td></tr>
<tr><td colspan="2">108.00~111.00</td><td>E</td><td>常绿栎-禾本科-莎草科</td><td>阔叶林、草甸草原，温暖略湿</td></tr>
<tr><td colspan="2">111.00~127.00</td><td>D</td><td>禾本科-莎草科-栎</td><td rowspan="2">阔叶林、草甸草原，温暖略干</td></tr>
<tr><td colspan="2">127.00~130.00</td><td>C</td><td>栎-榆-蒿-莎草科</td></tr>
<tr><td rowspan="3">130.00~227.50</td><td></td><td colspan="3">孢粉化石稀少、贫乏</td></tr>
<tr><td>177.00</td><td>B</td><td>落叶栎-禾本科-蒿</td><td>阔叶林、草甸草原，温略干</td></tr>
<tr><td></td><td colspan="3">孢粉化石稀少、贫乏</td></tr>
<tr><td colspan="2">227.50~249.50</td><td>A</td><td>落叶栎-榆-枫香-禾本科</td><td>阔叶林、草甸草原，温略湿</td></tr>
<tr><td colspan="2">249.50~255.00</td><td colspan="3">孢粉化石稀少、贫乏</td></tr>
</table>

P1（227.50～249.50m）的孢粉藻类组合反映，该层段形成时区域性落叶阔叶林的发育和沉积地点附近以禾本科为主草本植物组成的植被存在，指示较为温暖湿润的气候条件。

P2（130.00～227.50m）处于化石稀少或贫乏的样层品段，包含落叶栎-禾本科-蒿组合，表明了该层段形成时发育的区域性植被以落叶阔叶林为多，近沉积地点分布以禾本科、蒿为主的草甸草原，反映的气候条件为温凉略干。

P3（127.00～130.00m）包含栎-榆-蒿-莎草科组合，常绿栎和落叶栎都较多，指示区域性植被为常绿树与落叶树混合的阔叶林，表明气候变得较温暖，但沉积地点附近还发育以蒿、莎草科为主的草本植物。

P4（111.00～127.00m）出现以草本植物为主的组合，表明气候条件趋向于较为干燥。

P5（108.00～111.00m）包含常绿栎-禾本科-莎草科组合，重要特征是常绿栎花粉增多，指示该层段形成时区域性植被的阔叶林扩张，常绿树较为发育，反映气候条件为温暖且较为湿润。

P6（108.00m）获得的菊科-莎草科组合表明无论区域性或是沉积地点附近植被大为改变，以菊科为主的草本植物发育指示气候变干变凉

P7（66.50～108.00m）孢粉匮乏，可能反映了气候冷干。

P8（64.25～66.50m）包含莎草科-蒿组合，是以草本植物为主的组合，表明沉积地点所在地区盛行以莎草科、蒿为主的草甸草原植被，反映温凉略干的气候条件。

P9～P11（4.5～64.25m）孢粉组合经历了栎-榆-莎草科、禾本科-落叶栎、栎-松-莎草科的 3 次变化，反映了气候环境经历了温暖湿润、温和偏干和温暖湿润的变化。

P12（孔深 4.25m）以禾本科为主的组合，反映了沉积地点为受人类活动影响的滨海平原环境。

（三）孢粉地层区域对比

对区内不同地点不同地层单元内的孢粉记录进行了拼接，经对比，上新世以来可以划分出 18 个孢粉组合带。

1 带：落叶栎（*Quercus deciduous*）-常绿栎（*Quercus evergreen*）-莎草科（Cyperaceae）-凤尾蕨属（*Pteris*）孢粉带。此带孢粉大都贫乏，偏上部较丰富。以木本花粉占绝对优势，LQ 系列钻孔（具体钻孔及位置见图 21-1）综合分析表明，上新世孢粉组合以木本为主，并出现喜暖类型。以往调查 J25 孔和 SG1 孔木本花粉含量高达 85%～90%，嘉定安亭 G414 孔（孔深 269～367.9m 见本带）、金山朱泾 G418 孔（孔深 265.2～284.3m 见本带）的木本花粉含量也达 46%～>80%，以麻栎（*Q. acutissima*）、槲栎（*Q. alina*）、柞栎（*Q. dentata*）等落叶栎和青刚栎（*Q. glauca*）、栲属（*Castanopsis*）等常绿阔叶花粉为主，并较多出现一般喜暖热的大戟科（Euphorbiaceae）、胡桃属（*Juglans*）、漆树属（*Rhus*）、枫香属（*Liquidambar*）、木兰科（Magnoliaceae），还有喜温湿的榆属（*Ulmus*）、朴属（*Ciltis*）、桦属（*Betula*）、栗属（*Castanea*）、榉属（*Zelekova*）等。有的层段出现较多喜温凉—寒冷的松属（*Pinus*）、云杉属（*Picea*）、冷杉属（*Abies*）。草本花粉较少，仅占 10%左右，多为喜暖湿的莎草科、茜草科（Rubiaceae），在安亭 G414 孔、朱泾 G418 孔本带内则多禾本科（Gramineae）和一些水生草本香蒲属（*Typha*）。蕨类孢子很少，有凤尾蕨属（*Pteris*）、水蕨属（*Ceratopteris*）等。

2 带：松属-云杉、冷杉属-落叶栎-蒿属（*Artimisia*）孢粉带。该带孢粉大都贫乏或较少，如 LQ1 系列钻孔安亭 G414、朱泾 G418 等孔，通常仅见少量的草本禾本科、苔藓类（*Bryophyta*）和柏科（*Cupressaceae*）、松属花粉，部分钻孔见本带含孢粉较丰，如浦东 J25 孔、庙镇 SG1 孔所见，木本花粉占绝对优势，占 73%～98%，针叶树多于阔叶树花粉，以松属、云杉、冷杉为主，落叶栎次之，还可见较多的榆属、桦属、铁杉属（*Tsuga*）等喜温凉的乔木花粉，也出现常绿栎、栲属等一些喜暖常绿乔木花粉。草本花粉占 0.8%～25.4%不等，常见蒿属、藜科（Chenopodiaceae）、禾本科，较少见及喜暖湿的十字花科（Cruciferae）、莎草科（Cyperaceae），也有喜凉干的麻黄属（*Ephedra*）、菊科（Compositae）等花粉。蕨类孢子仅占 2%左右，有铁线蕨属（*Adiantum*）、凤尾蕨属、水龙骨科（*Polypodiaceae*）等孢子。

3 带：落叶栎-松属-禾本科孢粉带。浦东 J25、娄塘 P5、安亭 G414、朱泾 G418 等孔见本孢粉带。其中上部孢粉贫乏，下部较丰。木本花粉占比大于 50%~95%，其中大都为阔叶树，主要为落叶栎类，局部有较多的常绿栎、栲属、胡桃属、榆属、枫香属、鹅尔枥属（*Carpinus*）等花粉；针叶树主要为松属，还有些冷杉、云杉属花粉。草本占少半至少量，大都为禾本科，部分地段为蒿属、藜科。蕨类孢子很少，见有水蕨属、凤尾蕨属。

4 带：松属-云杉、冷杉属-落叶栎-禾本科-凤尾蕨属孢粉带。浦东 J25、安亭 G414、朱泾 G418、娄塘 P5 等孔见本孢粉带。本带孢粉丰、贫不一。以木本花粉占优势，可占 45%~100%不等，以松属、落叶栎为主，次之为云杉、冷杉属、榆属，局部较多桦属、桤木属（*Alnus*）、铁杉属。草本花粉多寡不一，多处可达 40%~78%，主要为禾本科和蒿属，少见藜科。蕨类孢子很少，以凤尾蕨属为主。

5 带：落叶栎-松属-香蒲属-水龙骨科孢粉带。娄塘 P5、安亭 G414、朱泾 G418 等孔见本孢粉带。以木本花粉占优势，可占 44%~74%，其中阔叶树花粉略多，以落叶栎、松属为主，其次为榆属，并见少量的青刚栎、枫香属、豆科、楮属（*Pasania*）、冬青属（*Ilex*）等喜暖常绿阔叶植物。草本花粉占 14%~30%，以香蒲属、禾本科为主。蕨类占 11%~19%，全为水龙骨科。

6 带：松属-落叶栎-禾本科-水龙骨科孢粉带。娄塘 P5、浦东 J25、安亭 G414、朱泾 G418 等孔见本孢粉带，通常含孢粉较丰富。其中木本花粉占绝对优势，达 67%~96%，大都为松属，其次为落叶栎、云杉、冷杉属，还有桦属、榆属等花粉。草本花粉可占 2%~14%，主要见禾本科，其次为藜科。蕨类占 5%~24%，主要为水龙骨科、凤尾蕨属。

7 带：落叶栎-松属-禾本科-水龙骨科孢粉带。娄塘 P5、浦东 J25、庙镇 SG1、安亭 G414、朱泾 G418 等众多钻孔见本孢粉带，含孢粉丰、贫不一。以木本花粉占优势，通常占 40%~75%，针叶、阔叶树花粉含量相近，以落叶栎、松属为主，其次为桦属、枫香属、榆属，还见有栗属、榛属等，底部出现青刚栎、栲属等常绿阔叶树花粉。草本花粉通常占 23%~36%，以禾本科为主，其次为蒿属、藜科等。蕨类一般占 5%~25%不等，多水龙骨科，还有凤尾蕨属、铁线蕨属等孢子。

8 带：松属-落叶栎-桦属-禾本科-水龙骨科孢粉带。东部滨海平原和河口沙岛区本带地层缺失，在研究区西部有含本带地层保存，如娄塘 P5、安亭 G414、朱泾 G418 等孔见本孢粉带。以木本花粉略多，占 47%~65%，所见主要为松属、落叶栎，其次为桦属、枫香属、榆属，并含少量喜凉冷的铁杉、云杉、冷杉等暗针叶树花粉。含草本多少不一，介于 7%~50%不等，多为禾本科，有的钻孔内尚见较多的蒿属、藜科。含蕨类孢子占 11%~25%，大都为水龙骨科，朱泾 G418 孔则为苔藓所替代。

9 带：落叶栎-松属-枫香属-禾本科-水龙骨科孢粉带。只在研究区西部有含本带地层保存，仅见于娄塘 P5、朱泾 G418 等孔。以木本花粉占优势，可占 60%~81%，其中阔叶树花粉占绝大部分，以落叶栎为主，其次为桦属、枫香属、榆属，还见有栗属、青刚栎、栲属、柃木属（*Eurya*）、楮（*Pasania*）等喜暖热乔木花粉；松属含量则比上带显著减少。含草本花粉 5%~23%，以禾本科为主，藜科次之。含蕨类孢子 13%~19%，几乎全为水龙骨科。

10 带：落叶栎-松属-禾本科-凤尾蕨属孢粉带。广泛见于区内诸多钻孔内。含孢粉丰、贫不一，木本与草本花粉多少不一。通常以木本占优，介于 40%~86%不等，大都占 50%~65%，阔叶树花粉略多于针叶树花粉，以落叶栎、松属为主，其次为枫香属，还有较多的桦属、铁杉属，局部断续出现一些云杉、冷杉、榆属、栗属和青刚栎、栲属。草本花粉大都小于等于 30%，个别钻孔如安亭 G414 孔达 46%~64%，以禾本科为主，其次为蒿属、藜科，主要在本带下部含有多少不一的水生草本眼子菜属、香蒲属、莎草科等花粉。含蕨类孢子占 2.5%~22%不等，以凤尾蕨为主，次之为水龙骨科。

11 带：松属-落叶栎-蒿属-水龙骨科孢粉带。该带广泛见于诸多钻孔内，含孢粉丰富。以木本花粉占绝对优势，通常可占 50%~80%，以松属为主，其次为落叶栎，还常见较多的桦属，局部出现较多的栗属、榆属、铁杉、云杉、冷杉、落叶松属（*Larix*）等花粉。含草本花粉 4%~38%不等，以蒿属为主，藜科次之，禾本科较少。含蕨类孢子 5%~17%，多水龙骨科，次之为铁线蕨、凤尾蕨属。

12 带：广泛见于诸多钻孔内，根据浦东 J25、庙镇 SG1、三灶 P45、安亭 G414 等孔揭示的本带孢粉组分的变化，可划分为 3 个亚带，显示古气候曾经历暖—凉—暖的变迁。

12-1 带：落叶栎-松属-蒿属-铁线蕨属孢粉带。本亚带孢粉丰富。其中以木本花粉占优，含量介于 38%~78%，以落叶栎、松属为主，次之为榆属、桦属、栗属，断续出现一些铁杉、桤木属（*Alnus*）、鹅尔枥属（*Carpinus*）、枫香属、冷杉、云杉属等花粉，在安亭 G414 孔本亚带还含较多的青刚栎。草本花粉介于 6%~54%不等，以蒿属为主，藜科次之，本带上部出现大量香蒲属、莎草科等水生草本花粉。含蕨类孢子占 5%~35%不等，以铁线蕨为主，凤尾蕨属、水龙骨科次之。

12-2 带：松属-落叶栎-香蒲属-凤尾蕨属孢粉带。本亚带孢粉丰、贫不一，以木本花粉占优势，含量介于 43%~87%，大都为 70%~80%，以松属略多，落叶栎居次，还有较多的栗属、榆属、枫香属，三灶 P45 孔本亚带内出现较多的桦属花粉。草本花粉占 10%~20%，多见水生草本香蒲属，也常见盐生草本藜科及蒿属。含蕨类孢子占比大都小于等于 10%，以凤尾蕨属为主，水龙骨科次之。

12-3 带：落叶栎-松属-蒿属-凤尾蕨属孢粉带。本亚带孢粉大体较丰。以木本花粉占绝对优势，含量达 76%~94%，以落叶栎为主，松属次之，还有较多的栗属、桦属、榆属，断续出现云杉、冷杉、常绿栎等花粉。含草本花粉仅占 4%~15%，主要为蒿属、禾本科，其次有水生草本香蒲属、眼子菜属及盐生草本藜科。含蕨类孢子仅占 2%~9%，多为凤尾蕨属。

13 带：蒿属-禾本科-松属-冷杉属-苔藓类（*Bryophyta*）孢粉带。本带地层在河口沙岛区和东部滨海平原部分地区缺失，在研究区西部的娄塘 P5、安亭 G414、朱泾 G418 等孔见本孢粉带。通常以草本花粉最多，占 30%~70%，主要有蒿属、禾本科，也常见旱生草本藜科、麻黄属。其次为苔藓类、藻类及蕨类孢子，含量占 17%~50%，以苔藓类出现最多，蕨类孢子较少，见有水龙骨科，有的地区如朱泾 G418 孔本带内出现大量的环纹藻（*Concentricystis*）。木本植物花粉很少，仅占 10%~20%，常见为针叶树松属、冷杉属；阔叶树花粉较少，见有桦属、落叶栎等。

14 带：松属-落叶栎-禾本科-凤尾蕨属孢粉带。本带在东部滨海平原和河口沙岛区全新世地层发育比较齐，全区域均可见到，含孢粉较丰富。与下伏孢粉带相比，木本含量大增，可占 66%~84%，以松属略多，落叶栎居次，并有较多的栗属、桦属、榆属，断续见喜冷的云杉、冷杉、铁杉，也出现常绿栎、胡桃属、枫香属等喜暖植物花粉。草本花粉占 7%~30%不等，以藜科、禾本科、蒿属为主，并出现较多的水生草本香蒲属、莎草科。蕨类孢子占 8%~17%，以凤尾蕨属为主，水龙骨科次之。

15 带：落叶栎-松属-蒿属-凤尾蕨属孢粉带。本带地层除青浦西北部地区外，在研究区内绝大部分地区广泛分布，含孢粉较丰富。以含木本花粉占优，占 50%~80%不等，其针叶、阔叶树花粉含量相近，以松属略多，落叶栎居次，并有较多的栗属、云杉、冷杉，还常见常绿栎、桦属、铁杉属、榆属等。草本花粉占 10%~40%不等，以蒿属、禾本科为主，藜科次之，有的钻孔如庙镇 SG1 孔本带内见不少水生草本香蒲属、莎草科。含蕨类孢子占 3%~25%不等，通常小于等于 10%，多为凤尾蕨属、水龙骨科。在安亭 G414 孔本带内出现不少刺球藻（*Baltisphaeridium*）。

16 带：落叶栎-常绿栎-松属-禾本科-凤尾蕨属孢粉带。该带在绝大部分地区广泛分布，含孢粉丰富。本带以木本花粉占优，通常占 60%~80%，其中阔叶树花粉占优势，以落叶栎最多，其次为松属、栗属，还常见青刚栎、栲属、桦属、榆属、胡桃属，局部尚多云杉、冷杉、铁杉属。含草本花粉占 10%~25%，见禾本科、蒿属、藜科较多，也出现不少香蒲属、莎草科、眼子菜属等水生草本花粉。含蕨类孢子占 5%~37%不等，以凤尾蕨属为主，水龙骨科次之。

17 带：松属-落叶栎-蒿属-凤尾蕨属孢粉带。该带在绝大部分地区广泛分布，含孢粉较丰富，但其上部有的钻孔含孢粉贫乏。该带以木本花粉占优，通常占 50%~70%，其针叶、阔叶树花粉含量相近，以松属为主，落叶栎次之，并含较多的桦属、云杉、冷杉、铁杉属，还常见栗属、榆属，顶、底部出现较多的常绿栎，或有栲属。含草本花粉占 15%~25%，以蒿属为主，次之为禾本科、藜科。含蕨类孢子通常占 25%~35%，以凤尾蕨属为主，次之为水龙骨科、水蕨属（*Ceratopteris*）。

18 带：落叶栎-松属-禾本科-凤尾蕨属孢粉带。该带在全区均有分布，含孢粉丰富。该带木本花粉占优，通常占 50%~75%，以落叶栎为主，松属次之，并含较多的桦属、胡桃属、栗属、青刚栎、栲属，还常见榆属、云杉、冷杉、铁杉属。含草本花粉占10%~20%，以禾本科、蒿属为主，水生草本香蒲属、莎草科次之，并有少量的盐生草本藜科花粉。含蕨类孢子通常占10%~30%，局部有占孢粉总量1/2左右者，多为凤尾蕨属，次之为水龙骨科、铁线蕨属等。

第五节　化学地层划分与区域对比

上海第四纪沉积物地球化学研究始于20世纪80年代，曾系统采取了包括氧化物分析、微量元素分析和沉积磷酸盐分析样品，主要分布于嘉定安亭G414、金山朱泾G418、崇明新建P42、嘉定娄塘P5等孔和青浦凤溪SG13孔、浦东新港SG7孔、长宁LQ15孔和金山枫泾LQ1孔。其中，较早测试的钻孔样品大都采自黏土层，部分取自粉土和粉细砂，近年来钻孔按照相同间距全孔采样。这些钻孔的元素地球化学分析为探讨区内古气候、沉积物来源等古环境变化提供了一定依据。

一、氧化物特征及古环境意义

（一）氧化物特征

化学分析结果表明，区内松散岩类所含化学成分主要有SiO_2、Al_2O_3、Fe_2O_3、FeO、K_2O、Na_2O、CaO、MgO及少量的P_2O_5、MnO、CO_2等。其中硅、铝氧化物含量占绝大部分，SiO_2占总量的60%~80%，Al_2O_3占10%~20%。其余如Fe_2O_3平均约占3.8%，FeO、K_2O、CaO、MgO均约占2%，Na_2O约占1.5%。

据朱泾G418孔揭示，在纵向地层剖面上，主要的氧化物含量曲线（图21-25）显示下述变化：

1）SiO_2含量波动升高，Al_2O_3含量波动降低，两者含量具反消长关系。

2）Fe_2O_3和FeO含量，尤其是Fe_2O_3与SiO_2含量具明显的反消长关系，而Fe与FeO的含量变化总体上也具反消长关系。

3）CaO、MgO与Fe_2O_3、FeO的含量具同步变化。

4）K_2O、Na_2O与CaO、MgO含量总体上具同步变化。K_2O、Na_2O含量由上而下渐减，两者含量变化大体显示中更新统及以上地层明显具反消长关系。

5）含量较少的P_2O_5、MnO含量曲线起伏，大体上与CaO、MgO及Fe_2O_3含量具相似的同步变化。P_2O_5与MnO含量也大体具正消长关系，与其他氧化物比较，其含量的跳跃变化最大，也显示与上述氧化物相对应的若干阶段性变化。

（二）氧化物含量变化与沉积物的关系

对比分析表明，氧化物含量的变化与沉积物类型及环境变迁具有密切的联系，首先取决于沉积物类型及所含的矿物成分，这与已知的一般规律是相符的。

除了未取样的砂砾层外，SiO_2一般在粉细砂、砂质粉土层中含量较高，在黏土层内则较低，这与石英、长石等富硅矿物为砂性土的主要矿物成分有关；而Al_2O_3在黏土层内含量较高，在粉细砂、砂质粉土层中则含量较低，这与富铝氧化物的伊利石等黏土矿物构成黏土层的主要组分有关；这也揭示硅、铝氧化物含量之所以呈反消长关系的原因。

Fe_2O_3通常在褐黄、赭红等暖色调的硬土层中含量较高（见图21-25），如全新统顶部常称的褐黄色黏土“硬壳层”即是，这与具这些暖色调的沉积物中褐铁矿、假象赤铁矿、水针铁矿等氧化铁、氢氧化铁矿物的相对富集有关，反映其曾经历的陆相沉积-氧化环境；而FeO在灰黑色淤泥质黏土、绿和蓝色黏土、灰绿色粉土和粉细砂中含量略高，这与绿、蓝色黏土、灰绿色砂性土中常含菱铁矿、海相淤泥质黏土中含绿泥石等氧化亚铁矿物有关，反映其所经历的海相或湖相沉积的弱还原环境。

CaO、MgO含量在黏性土中通常较高，而在砂性土中较低。综合朱泾G418孔第四系化学成分变化曲线图和庙镇SG1孔碳酸盐含量曲线图（图21-26），钙、镁氧化物或碳酸盐含量受沉积物类型的制约

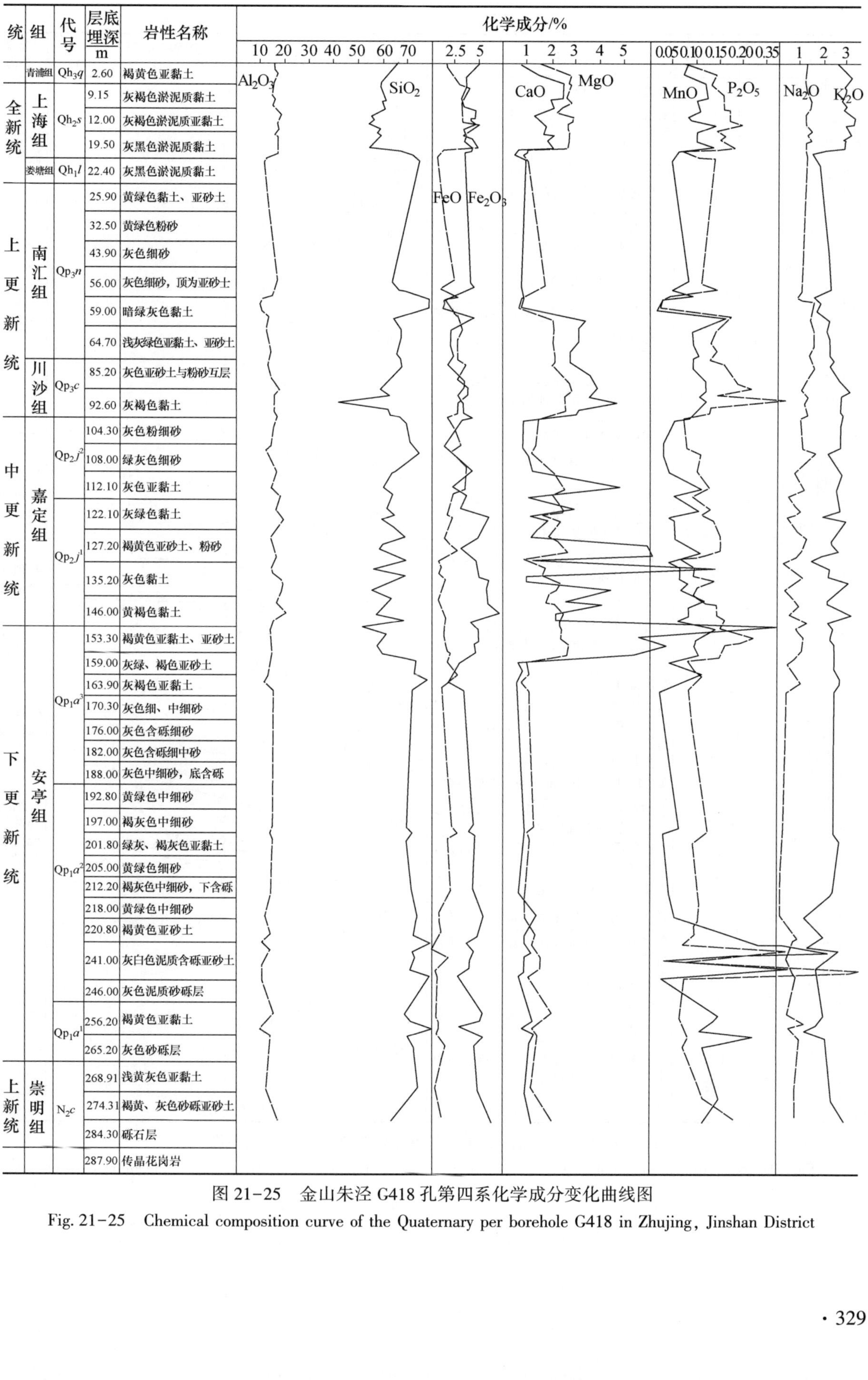

图 21-25 金山朱泾 G418 孔第四系化学成分变化曲线图

Fig. 21-25 Chemical composition curve of the Quaternary per borehole G418 in Zhujing, Jinshan District

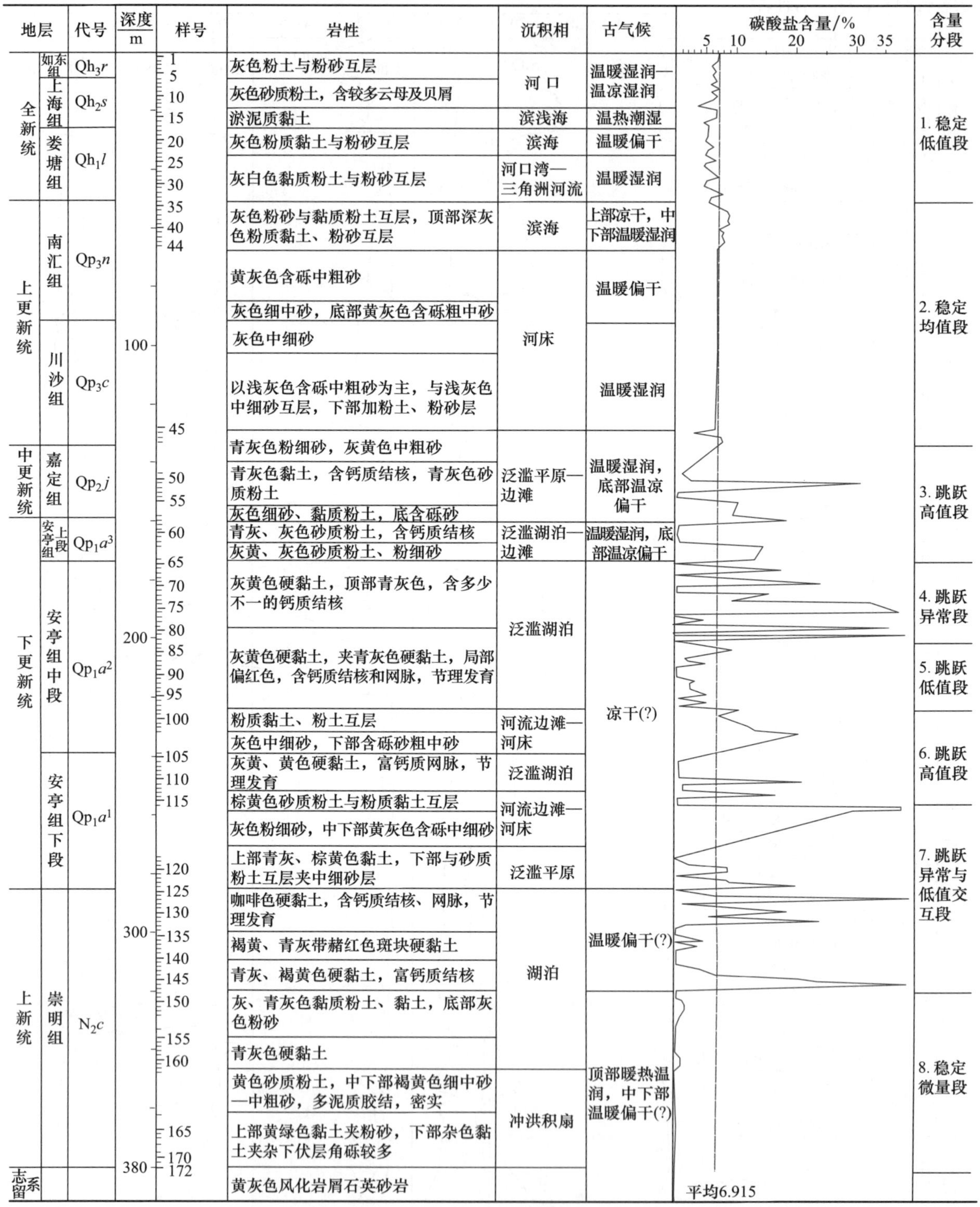

图 21-26 崇明庙镇 SG1 孔第四系碳酸盐含量变化曲线图

Fig. 21-26 Carbonate content curve of the Quaternary per borehole SG1 in Miaozhen, Chongming District

关系相当明显：如朱泾 G418 孔在中、下更新统分界以下，基本由砂性土组成，CaO、MgO 含量剧降；其上则因突变为以黏性土为主，CaO、MgO 含量也骤然升高，并随砂性土层的交替出现而剧烈升降。又如庙镇 SG1 孔在中更新统以上基本由砂性土或粉土组成，碳酸盐含量随之剧降；包括中更新统及以下地层变成基本由黏性土组成，碳酸盐含量因此骤升，而在砂层或粉土夹层出现部位，碳酸盐含量也随之下降。钙、镁氧化物或碳酸盐组分主要与沉积物中所含钙质结核或网脉、方解石，还有菱铁矿、绿泥石及蒙脱石（主要是 Ca-蒙脱石）等矿物有关，这已为重矿物分析和黏土矿物分析所证明。从庙镇 SG1 孔碳酸盐含量曲线图上还揭示，在黏土层内断续出现碳酸盐含量低值段，大都分布在颜色为偏红、赭红，且节理发育的黏土层内或黏土层与上覆砾砂层的接触处，还有上新统崇明组中、下部连续分布碳酸盐含量低值段，这反映了黏土层曾经历湿热气候条件下的化学风化过程，钙、镁氧化物被大量淋失的结果。观察证明，受强烈淋溶的黏土层，往往在与下伏砂层交界处淀积形成钙板层。

碱金属氧化物含量变化与沉积物内含伊利石、黑云母、长石（包括钾长石、斜长石）等含钾、钠硅酸盐矿物的多寡有关。碱金属氧化物总量和 K_2O 含量在中更新统及以上地层内较高，并随黏土、砂层的交替而发生相应的高、低变化；在下更新统以下地层内 K_2O、Na_2O 含量较低，并大致呈同步变化，看来与这部分地层内含长石和蒙脱石（可能部分为 Na-蒙脱石）较多，而伊利石、黑云母等含钾较富的矿物大为减少有关。

化学成分变化曲线图表明，磷、锰氧化物含量比其他氧化物含量低得多。观察证明，在中、下更新统褐黄、杂色硬土层内常含有铁锰质结核；重砂分析表明，主要在上更新统及其以上地层以及下更新统下部沉积物内含有较多的碎屑矿物磷灰石（崇明新建 P42 孔），也发现在中、上更新统与上新统的砂层内有软锰矿的个别高含量样品（崇明庙镇 SG1 孔），显然 P_2O_5、MnO 的含量高峰当与磷、锰矿物的局部富集有关。

综合上述，可据朱泾 G418 孔松散岩层化学成分特征划分出以下 6 个岩石化学带：

1）全新统，低硅高铝，含铁、碱和碱土金属氧化物高值带，以半还原半氧化环境为主，上部为氧化环境，以海陆过渡相为主和部分的陆相沉积，基本由含伊利石、绿泥石等黏土矿物较多的黏性土组成。

2）上更新统上部，硅和铝较高、含铁氧化物较高、碱和碱土金属氧化物较低值带，为以氧化环境为主的陆相—海陆过渡—陆相旋回沉积，大都由粉细砂和部分黏性土组成。

3）上更新统下部，硅和铝双低、含铁氧化物中等、碱和碱土金属氧化物较高值带，基本为半氧化半还原环境下的海陆过渡相沉积，基本由黏性土和粉砂互层组成。

4）中更新统，诸种氧化物含量剧烈跳动带，为以氧化环境为主，上部与半氧化半还原环境更替的陆相沉积，大都由黏性土和部分粉细砂交互组成，时而经受化学风化，致碱和碱土金属氧化物间或被淋失。

5）下更新统中上部，高硅低铝、含铁中等、碱和碱土金属氧化物较低、诸种氧化物含量平稳带，为以氧化条件为主，部分弱还原条件下的河流间夹部分湖泊沉积，大都由砂性土夹部分黏性土组成，反映沉积物中存在大量石英、长石，黏粒相对较少，碱和碱土金属氧化物部分遭受淋失。

6）下更新统下部—上新统，高硅低铝、含铁氧化物较高、锰和磷氧化物奇高、碱和碱土金属氧化物较低、诸种氧化物含量较强波动带，为较强氧化环境下辫状河流间夹心滩—湖泊的陆相沉积，大都由泥质砂砾石夹部分黏性土组成，反映沉积物中存在大量石英、长石和岩屑成分，碱和碱土金属氧化物严重淋失，间或有自生矿物铁锰结核、软锰矿和碎屑矿物磷灰石的局部富集。

（三）氧化物含量变化与风化程度的关系

已知沉积物中某些元素氧化物含量的变化，既主要受沉积物质来源、沉积环境、沉积介质和古气候条件所制约，也受到后来化学风化作用的影响。这里我们以安亭 G414 孔系统取自黏性土和粉细砂层的 130 个样品化学全分析结果为依据，应用岩石风化壳若干风化系数指标的分析方法，探讨沿松散沉积岩层纵向剖面上风化程度的变化及其沉积环境。所计算的风化系数指标包括：反映湿热化程度的硅铝化系数（SiO_2/Al_2O_3）和硅铁化系数（SiO_2/Fe_2O_3）、淋溶系数（$SiO_2/RO+R_2O$）、分解系数（Fe_2O_3/MgO）及风化系数（$R_2O_3/RO+R_2O$）。现将有关风化系数指标及判别结果列于表21-5，并绘成地球化学特征图展示（图21-27）。判别结果与根据孢粉分析及岩性特征所反映的气候条件等基本上是相符的。

全新统的铁氧化物指标显示，除上部的如东组为氧化环境外，中、下部均系还原环境下沉积的。由硅铝化系数和硅铁化系数指标普遍较高表明，其沉积后经受湿热化程度均弱，与依据淋溶、分解和风化程度指标做出的弱—甚弱的判别一致，唯上部地层相对略强些。

上更新统的铁氧化物指标显示，上部的南汇组上段（即暗绿—褐黄色硬土层）为半还原环境下沉积，中段的海相层则为较强还原环境下沉积的。综合各风化系数指标判定，在这部分地层沉积后大体上只经受了中偏弱—弱的化学风化过程。其下至上更新统下部的川沙组属未布样的砾砂层而未予判别。

中更新统的铁氧化物指标显示，嘉定组沉积经历了下段的较弱氧化环境至上段的较强还原环境的变化。综合风化系数指标判定，其沉积后大体上经受了中等—较弱的化学风化过程。

下更新统的铁氧化物指标显示，安亭组是在氧化与还原环境交替、大部分是在较强的氧化环境下沉积的。其沉积后一般明显经历了较强的化学风化过程。

上新统的铁氧化物指标显示，崇明组是在强氧化环境下沉积的。综合各项风化系数指标判定，其沉积后一般经历了强—甚强的化学风化过程。地层剖面上反映，以孔深约340m处为界，存在其下沉积物经受风化作用较强，其上则明显较弱的差异。

二、微量元素特征及古环境意义

（一）微量元素组合及物源指示意义

根据J25孔碎屑物全岩元素地球化学特征并结合聚类分析方法将J25孔元素物源变化划分为Ⅰ～Ⅴ阶段（图21-28，图21-29）。

第Ⅰ阶段（孔深350~287m）为晚上新世—早更新世，稀土元素REE（La、Ce、Nd、Hf、Y、Zr、Nb）、Pb、Th、U、Ba以及Mn含量在整个钻孔里最高，Ti的含量也较高，反映了具有中酸性母岩的特点，研究表明这些类型的岩石在河口地区的基岩中分布非常广泛，而高Ti则说明基岩玄武岩的影响。判别比值方面，f（Cr，Th）和f（K，La）低于0，但是f（La）却是正值。上述标型元素将上新世与第四纪沉积物明显区分出来，体现了河口地区物质来源的改变，推测三角洲在这一时期物源仅仅为局地贡献，区域化的盆地。

第Ⅱ阶段（孔深287~230m）为早更新世早期—中期，Pb、Th、U以及Mn继承了第Ⅰ阶段含量特征，Fe、K、Rb和As的含量上升较快，而高含量的Fe、K、Rb以及As恰是长江中游汉江碎屑物元素地球化学特征。判别比值方面，f（Cr，Th）保持负值，f（La）也降低至0，而f（K，La）却稳定维持在正值。这一时期高含量的Fe、K、As以及Rb指示中游物源进入，推测镇江口被切开，但此时长江上游的物质仍然没有汇入。Mn含量在Ⅰ和Ⅱ阶段明显偏高，指示在中更新世以来该区域没有沉降至低海拔处。前人研究揭示，在第四纪以前朝鲜半岛至福建之间的东海地区广泛存在着浙闽隆升带残留的大量低山。

第Ⅲ阶段（孔深230~135m）为早更新世中期—中更新世，特征元素为Ti、V、Cr、Sr、Sc、Cu、Co、Ni、Mg、Ca、Na以及P含量快速增加，高含量的Ti、V、Cr、Sc是长江上游玄武岩产出碎屑物所具有的标型元素，同时Mg、Ca是上游广泛分布的碳酸盐岩碎屑物的标型元素。判别比值f（Cr，Th）也迅速从负值上升至0，但f（La）和f（K，La）没有明显变化，数值仍然保持在0左右。上述元素及判别比值均显示了长江三角洲物源区的进一步扩展，指示上游峨眉玄武岩以及碳酸盐岩母岩所产出的碎屑物的汇入，推测三峡在这一时期被切穿。

表 21-5　安亭 G414 孔松散岩层风化程度一览表

Table 21-5　Weathering degree of the unconsolidated rock layers per borehole G414 in Anting

地层		硅铝化系数		硅铁化系数		淋溶系数		分解系数		风化系数		还原程度		氧化程度	
		SiO_2/Al_2O_3	判别	SiO_2/Fe_2O_3	判别	$\frac{SiO_2}{(RO+R_2O)}$	判别	Fe_2O_3/MgO	判别	$\frac{R_2O_3}{(RO+R_2O)}$	判别	FeO/%	判别	Fe_2O_3/%	判别
全新统	Qh_3r	5. 30	高	18. 30	较低	8. 23	较弱	1. 72	弱	2. 00	弱	1. 45	弱	3. 69	中常
	Qh_2s	5. 38~5. 68	高	28. 33~42. 24	甚高	6. 92~7. 61	弱	0. 73~1. 00	甚弱	1. 50~1. 65	甚弱	2. 31~3. 75	强	1. 61~2. 33	弱
	Qh_1l	3. 95~4. 91	中常	17. 13~27. 47	较高	6. 90~7. 02	弱	0. 95~1. 30	弱	1. 69~2. 15	弱	2. 52~3. 14	强	2. 33~3. 48	弱
上更新统	Qp_3n^3	3. 64~6. 44	高	13. 73~39. 19	较高	7. 02~12. 26	中常	1. 28~3. 36	较弱	1. 92~3. 89	中常	1. 33~2. 91	较强	1. 90~4. 66	中常
	Qp_3n^2	3. 05~5. 04	较低	11. 76~24. 66	较低	4. 27~10. 19	较弱	0. 25~3. 39	弱	1. 41~1. 81	甚弱	1. 30~5. 68	强	2. 40~4. 34	较弱
	Qp_3n^1	—	—	—	—	—	—	—	—	—	—	—	—	—	—
	Qp_3c	—	—	—	—	—	—	—	—	—	—	—	—	—	—
中更新统	Qp_2j^2	3. 32~3. 89	较低	19. 24~69. 23	高	4. 89~11. 64	较弱	0. 66~1. 06	弱	1. 51~2. 57	弱	3. 67~4. 97	强	0. 98~2. 93	弱
	Qp_2j^1	3. 16~6. 78	中常	11. 03~27. 91	中常	5. 42~15. 88	中常	0. 98~2. 83	弱	1. 41~3. 43	较弱	1. 47~3. 66	较强	1. 31~6. 49	较弱
下更新统	Qp_1a^3	4. 84	中常	15. 56	较低	12. 65	较强	2. 99	较强	3. 43	较强	2. 32	中常	4. 31	较强
	Qp_1a^2	2. 80~6. 12	较低	9. 51~21. 66	较低	6. 50~20. 92	较强	1. 21~4. 61	较强	1. 50~6. 36	强	0. 36~3. 30	较弱	1. 27~6. 43	较强
	Qp_1a^1	4. 25~10. 03	较高	15. 30~43. 88	较高	14. 09~21. 86	强	1. 26~4. 92	较强	2. 34~4. 40	较强	0. 58~2. 38	较弱	0. 59~4. 58	弱
上新统	N_2c	3. 03~6. 44	上部高，中下部低至中常	9. 64~33. 28	上部高，中下部低至中常	10. 04~17. 03	强	2. 70~6. 12	甚强	2. 72~5. 72	甚强	0. 00~2. 22	甚弱	1. 23~7. 16	强
平均值		4. 44		20. 85		10. 23		2. 59		2. 89		1. 81		3. 81	

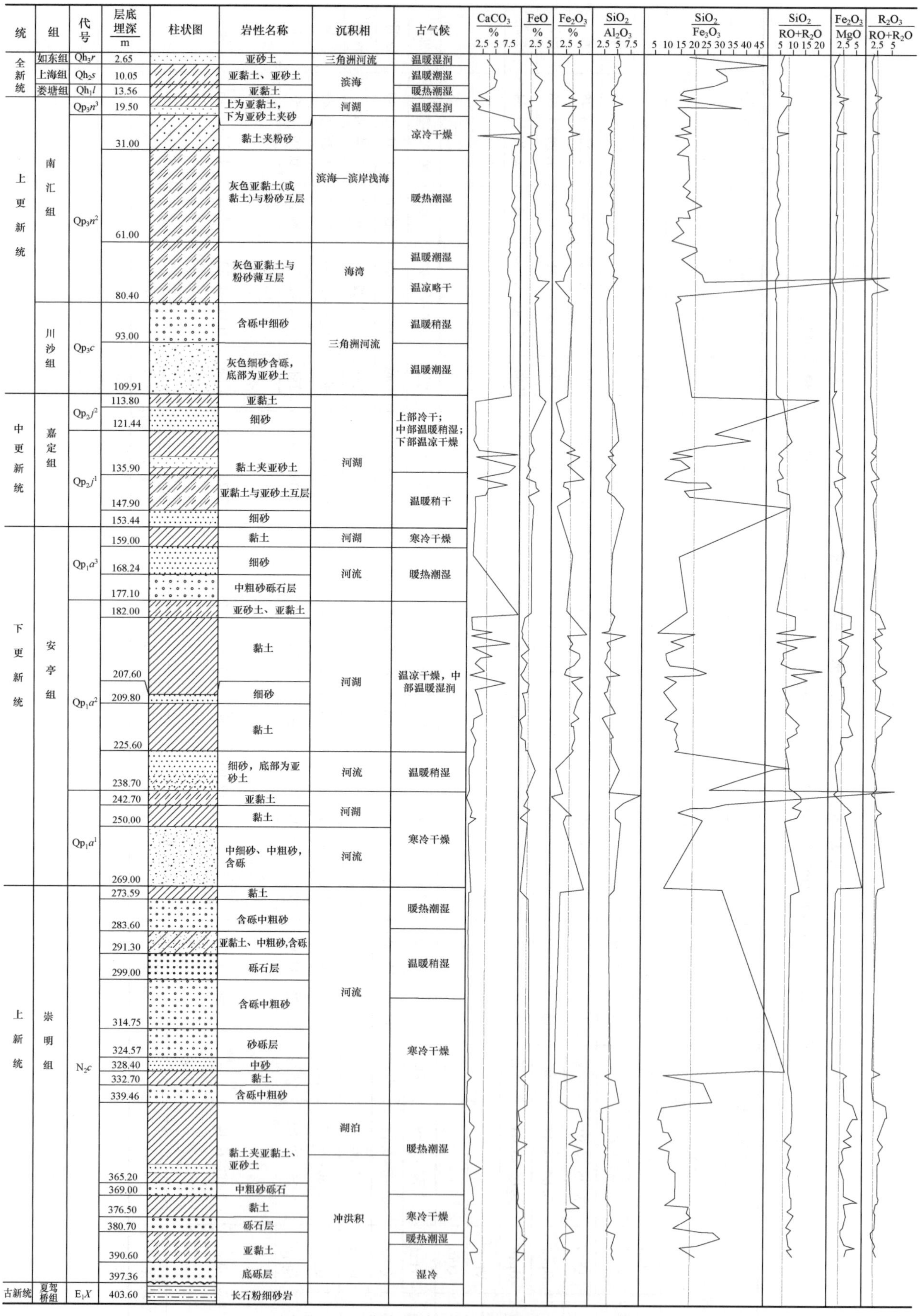

图 21-27　嘉定安亭 G414 孔地球化学特征图

Fig. 21-27　Geochemical characteristic diagram per borehole G414 in Anting, Jiading District

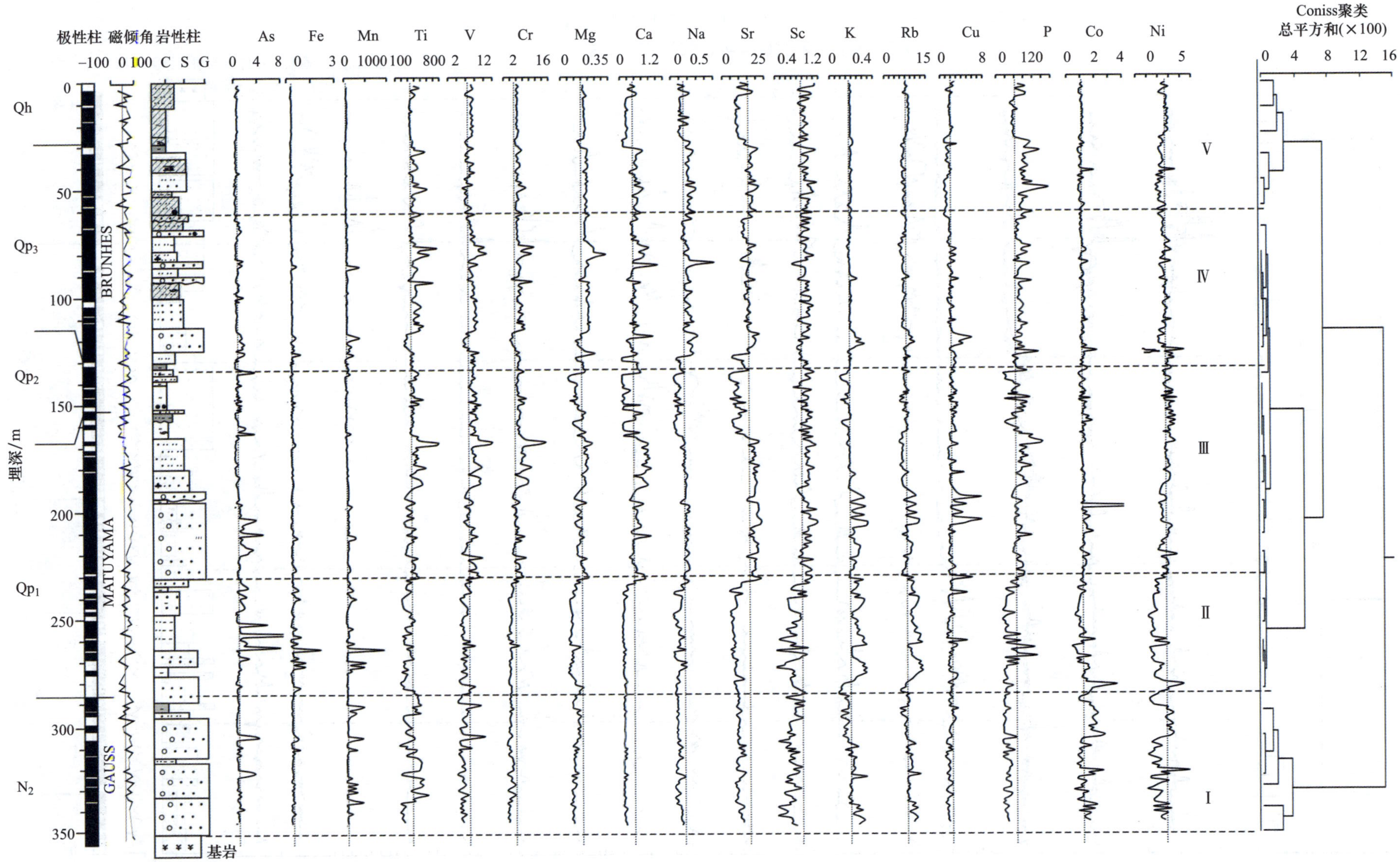

注：所有元素量值经Al归一化处理。

图 21-28 J25 孔元素地球化学垂向分布(一)

Fig. 21-28 Vertical distribution of element geochemistry per borehole J25 (I)

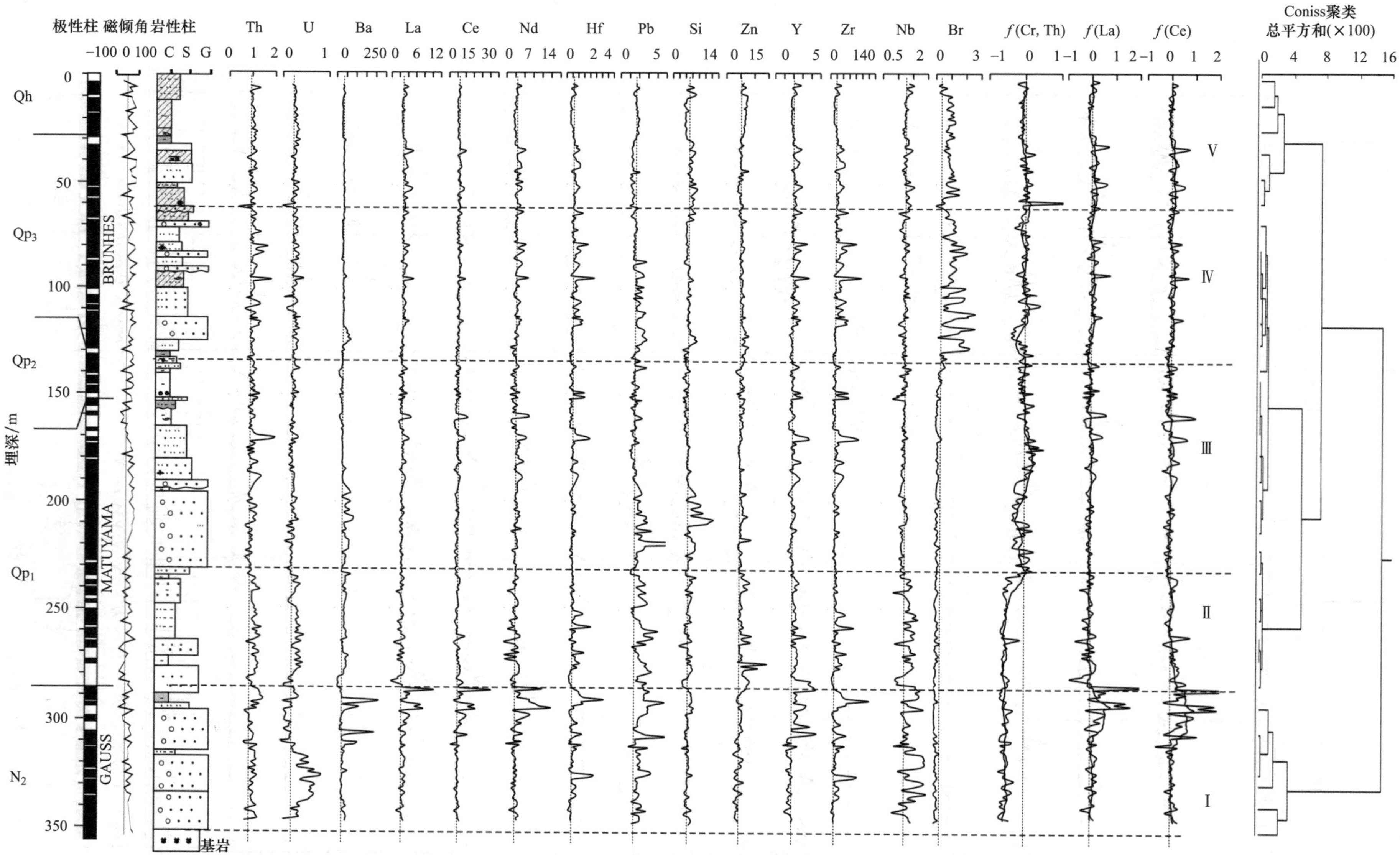

注：所有元素量值经Al归一化处理。

图 21-29　J25 孔元素地球化学垂向分布(二)

Fig. 21-29　Vertical distribution of element geochemistry per borehole J25 (Ⅱ)

第Ⅳ（孔深135~61m）和第Ⅴ阶段（孔深61~0m）为中更新世以来，所有元素地球化学特征同第Ⅲ阶段十分相似，第IV阶段的Mg、Ca和Na进一步增加，而Br含量较前3个阶段出现了突然的大幅度增加。Br是海洋中广泛存在的元素，可以作为指示盐度敏感指标，在河口地区也可以追踪海水的入侵。在第Ⅳ和第Ⅴ阶段Br含量猛增，不仅指示了中更新世以后海水的明显入侵，还间接反映此时期以后地貌类型较先前发生了巨大变化，处在接近海平面的较低位置。第Ⅴ阶段多数元素保持稳定，判别比值f（Cr，Th）、f（La）以及f（K，La）同第Ⅲ阶段相似，这些特征均验证了从第Ⅲ阶段以来（1.0~1.2Ma）长江上游物质的汇入。

（二）微量元素比值及物源指示意义

利用位于枫泾凹陷LQ1孔、三墩凹陷SG7孔、安亭凹陷Jc407孔、天马山凸起SG13孔以及凹陷和凸起过渡区LQ15孔的Nb/Co、Cr/Nb、Th/Cr、Th/Sc、REE物源标志对比了上海不同沉积区的第四纪物源演变。

钻孔对比显示（图21-30），上新世—早更新世中期，4个钻孔元素比值显示来自基性岩的指示元素Nb以及长江口局地物源指示元素Th、U含量较高，反映了区内物源均以局地为主，仅在凹陷区的Jc407孔、SG7孔Cr、Sc含量有微弱增加的迹象，暗示了中期可能有少量长江远源泥沙的输入。早更新世晚期，根据古地磁标定，大约前贾拉米洛事件前后（约1.2Ma B.P.）北部和东部凹陷区的Jc407、SG7孔Cr、Sc含量明显增加，表明了开始稳定接受长江远源泥沙的输入，西部枫泾凹陷LQ1孔元素比值则显示仅在接近贾拉米洛事件前后（约1.0Ma B.P.）枫泾凹陷开始稳定接受长江远源泥沙的输入。然而，位于过渡区的LQ15孔（基岩面埋深约250m）高Nb、Th和REE表明早更新世仍以局地物源为主，仅到中更新世（B/M）以后才稳定接受长江远源泥沙的输入。值得注意的是，位于西部天马上凸起SG13孔（基岩面埋深约220m）早更新世晚期以来物源指标十分稳定，显示了该区域基本不受长江物源的影响。此外，该孔位于全新世最大海侵线以西，再次佐证了其不受长江泥沙影响的可能性。

不同沉积区钻孔第四纪物源演化（图21-31）显示长江泥沙稳定输入上海的时间节点约为1.2Ma B.P.。上海第四纪不同时期物源空间变化概括如下：

约1.2Ma B.P.，长江泥沙影响北部和东部凹陷区；约1.0Ma B.P.，长江物源影响范围可扩大至西部凹陷区；约0.78Ma B.P.以来，长江物源影响范围可抵达凹陷之间的过渡沉积区，这些区域基岩面埋深应不小于220m，对于西部基岩浅埋区（基岩面埋深200m以浅），第四纪物源可能一直以西部天目山近源为主。上海第四纪不同时期物源演变空间差异显示了构造沉降对流域沉积地貌的控制。

三、沉积磷酸盐特征

20世纪80年代对嘉定娄塘P5、崇明新建P42等孔黏土层作过沉积磷酸盐试验，并据之计算了磷酸钙组分和古盐度（表21-6）。结果表明：

1）据尼尔松公式计算的磷酸钙组分和古盐度，虽然其极低值均分布在中、早更新世陆相层中，但是其高、低值在海相、海陆过渡和陆相地层中互有出现，与根据微古分析和沉积特征所确定的沉积相差异较大，其指向价值不如上述的B和B/Ga值。

2）磷酸钙组分高值的断续出现，虽然不能排除采样和化学分析的偶然误差，但主要还是与沉积物中碎屑矿物磷灰石分布的多寡不一有关，这已为重砂分析所揭示，也获得了不少沉积磷酸盐试验实例的证明。

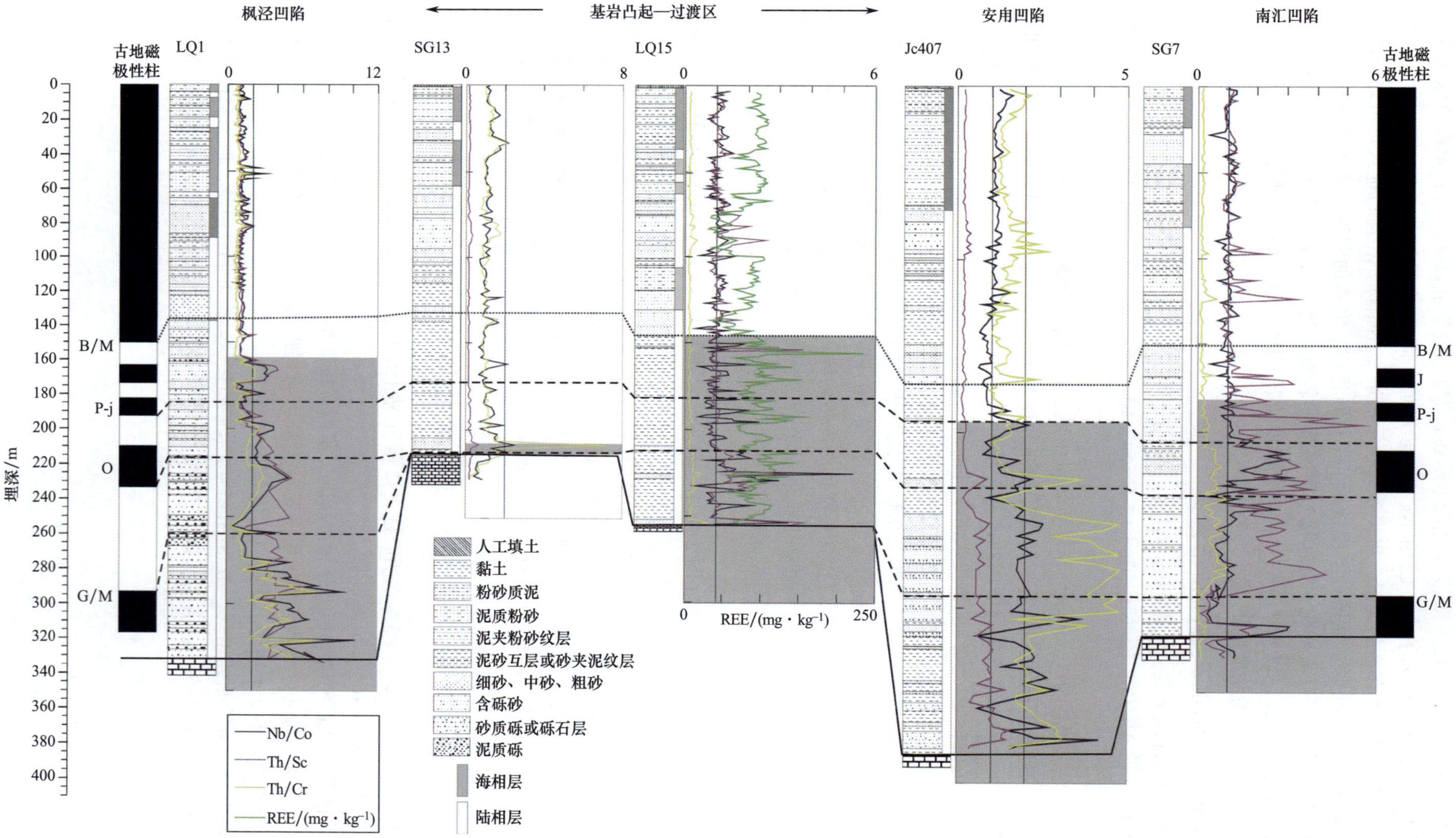

图 21-30　上海第四纪早期不同沉积区物源标型元素指标分布

Fig. 21-30　Distribution of the source standard element indexes among different Early Quaternary sedimentary regions in Shanghai

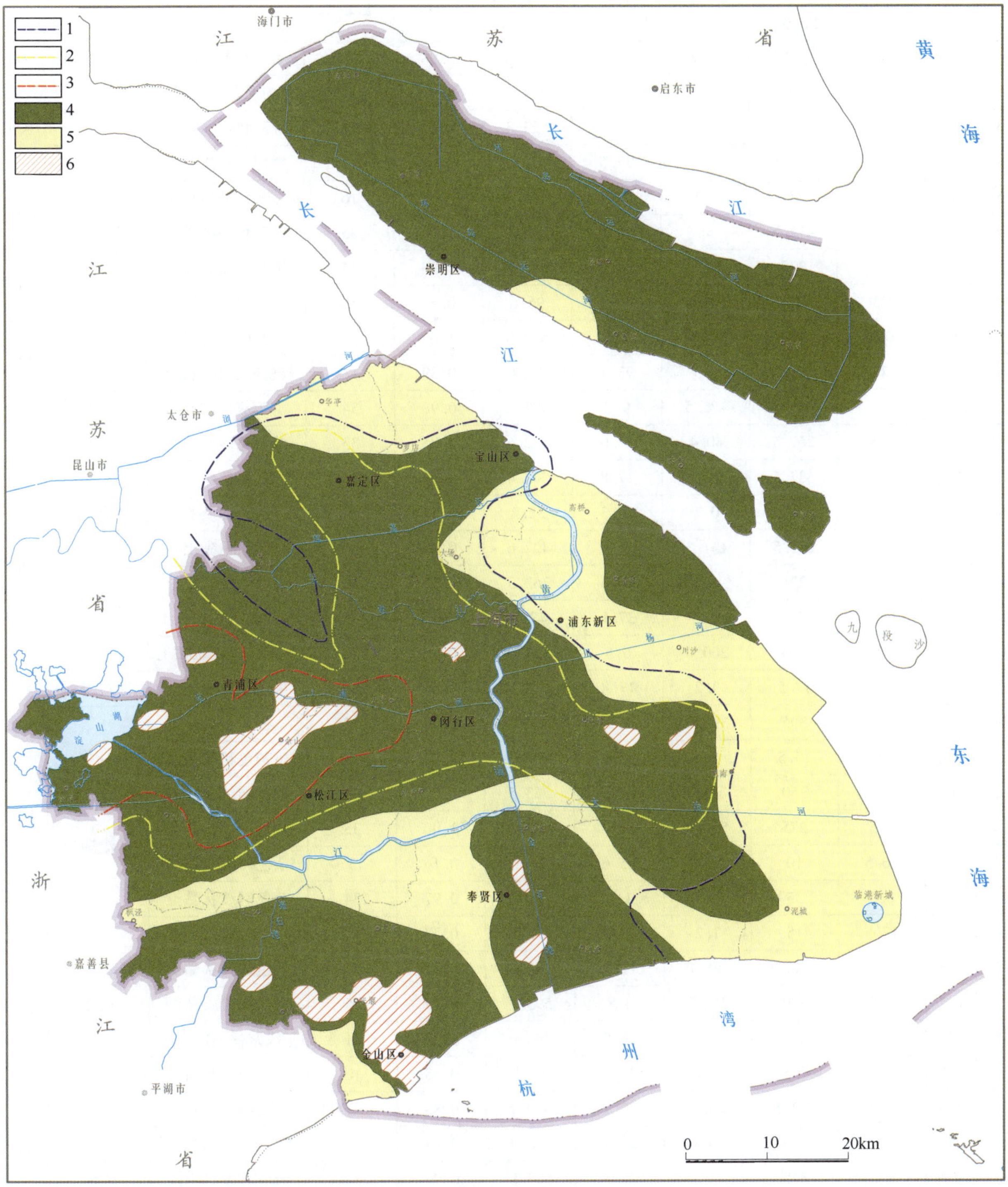

图 21-31　上海第四纪长江物源范围演变

Fig. 21-31　Evolution of the range of the Quaternary Yangtze River provenance in Shanghai

1—1. 2Ma B. P. 长江物源界线；2—1. 0Ma B. P. 长江物源界线；3—0. 78Ma B. P. 长江物源界线；4—第四纪早期河间地；5—第四纪早期古河道；6—基岩浅埋区

表 21-6 嘉定娄塘 P5 孔沉积磷酸盐分析与古盐度计算结果表

Table 21-6 Phosphate analysis and paleosalinity calculations per borehole P5 in Loutang, Jiading District

样号	样深/m	$FePO_4$ 中的 $P/10^{-6}$	$Ca(PO_4)_2$ 中的 $P/10^{-6}$	两者比值	$FePO_4$ 物质的量/10^{-8}(Fe)	$Ca(PO_4)_2$ 物质的量/10^{-8}(Ca)	$\frac{Ca}{Fe+Ca}$	计算盐度/‰	岩性	沉积相	地层
1	2	5.5	450	81.8	17.8	724.5	0.98	34	黄灰色粉质黏土	河口滨岸	Qh_3
2	8	16.5	412.5	25.0	53.3	664.1	0.93	32	灰粉质黏土夹粉砂	河口湾	Qh_2
3	10	88.5	325	3.7	285.9	523.3	0.65	21	棕灰色砂质粉土		
4	20	24	417.5	17.4	77.5	672.2	0.90	31	淤泥质粉质黏土	海湾	
5	28	131.5	275	2.1	424.7	442.8	0.51	16	棕灰色砂质粉土	滨海	Qh_1
6	40	0	412.5	∞	0.0	664.1	1.00	35	灰色粉细砂	河口湾	
7	42	3	375	125.0	9.7	603.8	0.98	34	黑灰色粉质黏土	湖沼	Qp_3^{2-3}
8	44	23	60	2.6	74.3	96.6	0.57	18	灰黑色泥炭		
9	50	0.5	400	800.0	1.6	644.0	1.00	35	灰粉砂、砂质粉土	滨浅海	Qp_3^{2-2}
10	62	29	37.5	1.3	93.7	60.4	0.39	12	灰粉砂、粉质黏土		
11	96	3	367.5	122.5	9.7	591.7	0.98	34	蓝灰色黏土	泛滥湖泊	Qp_2^2
12	102	6.5	262.5	40.4	21.0	422.6	0.95	33	棕灰色砂质粉土	边滩	
13	120	5.5	337.5	61.4	17.8	543.4	0.97	34	褐黄色黏土	泛滥湖泊	Qp_2^1
14	134	29.5	20	0.7	95.3	32.2	0.25	6	棕灰色粉质黏土	泛滥平原	
15	150	2	262.5	131.3	6.5	422.6	0.98	34	蓝灰色黏土	泛滥湖泊	Qp_1^3
16	156	5.5	225	40.9	17.8	362.3	0.95	33	杂色网纹状黏土		
17	178	11.5	7.5	0.7	37.1	12.1	0.25	6	蓝灰色斑纹状黏土		Qp_1^2
18	182	2	400	200.0	6.5	644.0	0.99	35	褐黄色斑纹状黏土		
19	188	5.5	17.5	3.2	17.8	28.2	0.61	20	褐黄色砂质粉土	边滩	
20	192	1	395	395.0	3.2	636.0	0.99	35	褐黄色黏土	泛滥湖泊	
21	196	3	17.5	5.8	9.7	28.2	0.74	25	暗灰色粉质黏土		
22	200	0	350	∞	0.0	563.5	1.00	35	杂色网纹状黏土		
23	206	0.5	275	550.0	1.6	442.8	1.00	35	杂色网纹状黏土		
24	214	26.5	10	0.4	85.6	16.1	0.16	3	杂色斑纹状黏土		
25	222	0.5	35	70.0	1.6	56.4	0.97	34	杂色网纹状黏土		
26	230	12.5	15	1.2	40.4	24.2	0.37	11	褐色黏土		
27	238	4	200	50.0	12.9	322.0	0.96	34	杂色斑纹状黏土		
28	242	1.5	370	246.7	4.8	595.7	0.99	35	斑杂状粉质黏土	边滩	
29	248	30	35	1.2	96.9	56.4	0.37	11	浅紫褐色黏土	泛滥湖泊	Qp_1^1
30	250	7	7.5	1.1	22.6	12.1	0.35	10	杂色斑纹状黏土		
31	256	24.5	372.5	15.2	79.1	599.7	0.88	31	浅紫褐色黏土		
32	264	9	235	26.1	29.1	378.4	0.93	32	浅紫褐色粉质黏土		
33	270	8.5	375	44.1	27.5	603.8	0.96	33	杂色黏土、粉砂	边滩	
34	272	18.5	17.5	0.9	59.8	28.2	0.32	9	斑杂状粉质黏土	泛滥湖泊	
35	276	0	170	∞	0.0	273.7	1.00	35	斑杂状粉质黏土		
36	280	0	55	∞	0.0	88.6	1.00	35	灰绿色黏土	湖泊	N_2
37	288	14.5	15	1.0	46.8	24.2	0.34	10	紫褐色粉质黏土		
38	292	0.5	142.5	285.0	1.6	229.4	0.99	35	褐黄色泥质砾石	洪积扇	

第六节　矿物地层划分与区域对比

一、重矿物特征

（一）重矿物成分

根据多个钻孔的重矿物分析结果，上海第四纪沉积物中共有 30 多种重矿物，其中含量较高的有角闪石、绿帘石、磁铁矿、钛铁矿、褐铁矿、赤铁矿、石榴子石、独居石、锆石等，含量较低的有金红石、白钛石、电气石等，较少见及的有榍石、锐钛矿、黄铁矿、磷灰石、重晶石、菱铁矿、软锰矿等，偶尔出现的有辉石、蛇纹石、透闪石、十字石、硅线石、蓝晶石、刚玉、石棉，还有黄铜矿、方铅矿、闪锌矿、自然铅、自然铜、自然金、白钨矿、辰砂等。在单个重砂样品中出现的重矿物不过一二十种，中更新统以上地层中重矿物种类较少，其下较多。重矿物在重砂样品中所占质量一般约为 1%，在下更新统上部及以上地层中含量较高，其下急剧降低。

在重矿物中，属于稳定和极稳定的矿物有磁铁矿、钛铁矿、褐铁矿、赤铁矿、石榴子石、独居石、锆石、金红石、白钛石、电气石、榍石、锐钛矿、磷灰石、十字石、硅线石、蓝晶石等，非稳定矿物有角闪石、绿帘石、黑云母、绿泥石、透闪石等，自生矿物有黄铁矿、黄铜矿、重晶石、菱铁矿、软锰矿等。在重矿物中，属于非稳定矿物的角闪石、绿帘石和次之的黑云母、绿泥石占绝对优势，平均达 60%左右；其在下更新统上部至全新统中平均可达 80%左右，在下更新统下部及以下砂层中降为 30%~50%。稳定和极稳定矿物与非稳定矿物呈消长关系，由下而上渐趋减少，如在下更新统中稳定和极稳定矿物平均可占 50%~70%，而在中更新统至全新统中稳定矿物和极稳定矿物平均仅占 20%左右。

在区域平面分布上，由西部太湖沉积区至东部长江三角洲沉积区，稳定矿物含量所占比例递增，非稳定矿物则相应减少。而在上海北部与江苏接壤的凹陷中，下更新统中、下部稳定矿物所占比例则远比南部的金山—川沙凹陷为高。

重砂样品中所含的轻矿物主要有石英、钾长石、斜长石、黑云母、白云母、绢云母、绿泥石、方解石、燧石、黏土矿物以及岩屑等。总含量一般可占砂样质量的 99%左右。其中石英、长石、岩屑（包括黏土矿物和其他轻矿物）各占 1/3 左右。

岩屑和黏土矿物在下更新统中部及以下砂层中含量剧增，仅岩屑即可占砂样质量的 30%~50%，在下更新统下部砾砂层内其含量甚至逾半。而在中更新统以上砂层中，岩屑和黏土矿物含量减少，一般仅占重砂样品质量的 5%~20%。

（二）重矿物组合分带

依据崇明庙镇 SG1 孔重砂样品分析结果，由上而下可分为 3 个重矿物组合带（图 21-32）：

（1）角闪石-绿帘石-磁铁矿、钛铁矿-褐铁矿、赤铁矿-独居石带

SG1 孔孔深 84.6~159.1m 的相应层位为上更新统下部—中更新统。孔深 84.6m 以浅的全新统—上更新统上部无分析资料。该带以非稳定重矿物占优势，其中角闪石约占 50%，绿帘石约占 20%；余为稳定重矿物，主要有磁铁矿、钛铁矿和褐铁矿、赤铁矿等。此带的一个重要特点是普遍见有较多的独居石（约≥5%），尤其在孔深 105~121m 间的上更新统下部出现高含量带；其次在该带中部出现石榴子石高含量带，带内还普遍见到少量的锆石。在上部灰色细砂层内可见呈柱粒或六方柱状完好晶形的磷灰石，显示可能为来源较近的碎屑矿物；中部青灰色细砂层内，与所夹腐木碎屑相伴，出现小于 0.1mm 的细小球状、葡萄状、草莓状的黄铁矿集合体，表明为自生矿物；底部灰色细砂的重矿物内见有较多呈暗淡不规则块状的软锰矿，并无其他海相标志矿物或海相生物化石伴生，可能为滨湖沉积的自生矿物。

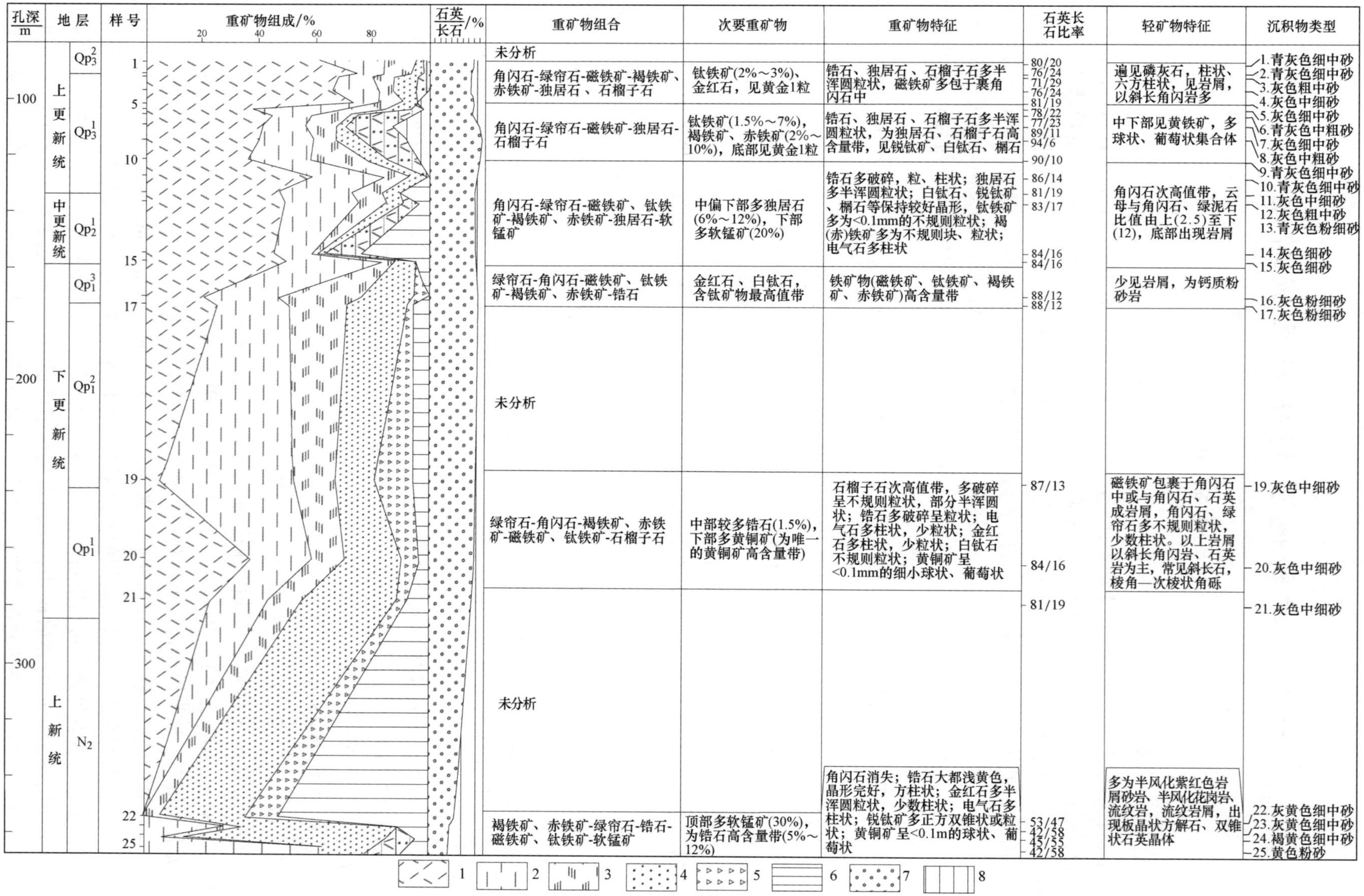

图 21-32　崇明庙镇 SG1 孔重矿物分析图

Fig. 21-32　Heavy mineral composition per borehole SG1 in Miaozhen, Chongming District

1—角闪石；2—绿帘石；3—磁铁矿、钛铁矿；4—褐铁矿、赤铁矿；5—锆石、石榴子石、独居石；6—软锰矿、金属硫化物等；7—石英；8—长石

（2）绿帘石-角闪石-褐铁矿、赤铁矿-磁铁矿、钛铁矿-石榴子石、锆石带

SG1孔孔深159.1~284.5m的相应层位为下更新统。该层位仍以非稳定重矿物占优势，其含量约为55%，但以绿帘石为主，角闪石居次；稳定重矿物含量显著增加，其含量约占40%，有褐铁矿、赤铁矿和磁铁矿、钛铁矿等；还出现5%~15%的锆石、石榴子石等极稳定矿物。在底部的灰色细砂层内出现唯一的黄铜矿高值带，为小于0.1mm的细小粒状和球状、葡萄状黄铜矿集合体构成，表明为自生矿物。

（3）褐铁矿、赤铁矿-绿帘石-锆石-磁铁矿、钛铁矿带

SG1孔孔深284.5m以下的相应层位为上新统。该层位以稳定—极稳定重矿物占明显优势，占50%~90%，其中褐铁矿、赤铁矿占30%~70%，余为锆石和磁铁矿、钛铁矿等，其含量较少。非稳定重矿物绿帘石占比小于10%~60%，而角闪石则已消失。其重矿物组合成分与上述第Ⅳ、Ⅴ重矿物组合带相当。该带顶部灰黄色细中砂的重砂内含软锰矿较多，也无海相生物化石或其他海相标志矿物相伴出现，可能也为滨湖沉积的自生矿物。

（三）重砂矿物特征反映的沉积环境与古气候

根据庙镇SG1孔及其他钻孔的重砂样品分析结果（图21-33），深部（孔深284.5m以下）的上新统内，为全孔最低的重矿物含量带；砂的成分成熟度很差，砂中紫红色岩屑砂岩、石英砂岩岩屑和长石占优势，石英/长石值小于等于1；砾石成分基本为岩屑，以半风化花岗岩最多，还有半风化的紫红色岩屑砂岩、红色钾长石和石英砂岩、石英岩、流纹岩；砂砾的磨圆度差，均呈棱角—次棱角状；重矿物组成中，稳定和极稳定矿物几乎占全部，锆石大都为浅黄色、棱角未磨的正方柱状，少数为长柱状，很少呈磨圆状。这些特征说明，上新世地层为近源堆积，沉积物质主要来源于花岗岩分布区，其次为沉积碎屑岩、变质岩和火山岩分布区，因此砾石成分相对较为简单。在上新统内，非稳定矿物角闪石大都消失，半风化紫红色岩屑大量出现，还有褐铁矿、赤铁矿含量占重矿物组分的明显优势，反映当时系处在湿热的气候条件下，风化分解作用异常强烈，这与孢粉分析揭示的古气候环境可互相印证。

结合区内多个钻孔，在横向分布上，西部地区稳定矿物含量持续占绝对优势，东部滨海平原和河口沙岛区非稳定矿物绿帘石与稳定重矿物组成比率则呈剧烈跳动变化。古地理重建揭示，区内晚新生代沉积盆地形成早期的上新世，西部处在盆地边缘的山前或山间冲洪积扇（裙）地带，由沉积物的沿程分选结果，密度相对较大的稳定重矿物在近源地带先行沉积下来，相对较轻的绿帘石之类非稳定重矿物则被携往距物源相对较远、处于盆地内部的东部地区；加之东部当时处在分合多变、摆动不定的辫状河流环境，在纵向上就相应留下了重矿物组成比率剧烈跳动的记录。

SG1孔孔深284.5~159.12m间为下更新统，见第Ⅱ重矿物带，对应于其他钻孔所见的Ⅱ、Ⅲ重矿物带。由下而上出现：早更新世早、中期以稳定重矿物占优势组分的Ⅲ带，至早更新世晚期剧变为以绿帘石为主的非稳定重矿物具优势组分的Ⅱ带。砂层中常见砂砾形状以次棱角状居多，部分为棱角状；砾石成分以石英岩、钙质胶结的粉砂团块为主，在近第四系底部还出现角闪斜长片麻岩、花岗岩和酸性火山岩等较复杂的成分。砂中石英与长石比率达4.4~7，其成熟度较下伏的上新统砂层明显提高。在下更新统中、下部呈普遍可见的锆石大都呈浅玫瑰色或淡紫红色破碎柱粒状，少数呈半浑圆柱状及个别呈椭圆的状。这反映，在第四纪与新近纪交替之际，构造-剥蚀作用可能较为强烈，沉积物来源因之呈现多样性。早更新世早、中期，上海西南部与西邻的江浙地区仍属连绵山地，沉积物来源仍相对较近，因此密度较大的稳定重矿物依然比较富集。至早更新世晚期，上海西南部山地已演化为分散的低丘，沉积物源远向西邻江、浙山地延伸，此时除浏—南古河道所在的现河口滨岸带有较强水动力携来密度较大的稳定重矿物组分仍占一定比率外，中西部泛滥平原—湖泊广布地区出现的重矿物已几乎全为密度相对较小的非稳定重矿物。

孔深159.12m以浅中更新统以上砂层具有下述几个特点：

1）砾石磨圆度明显趋好，大都呈次圆、浑圆状。

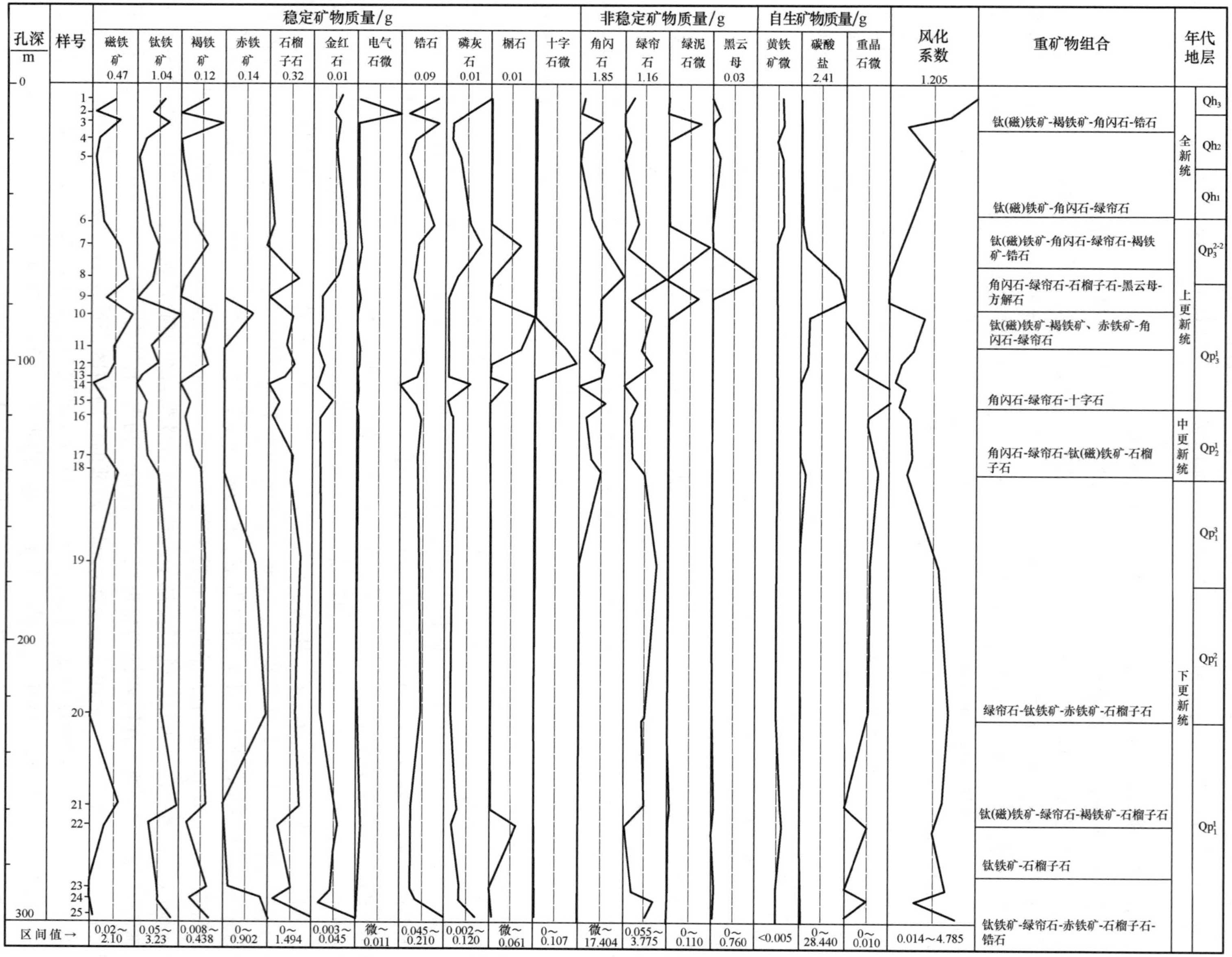

图 21-33　崇明新建 P42 孔重矿物分析图

Fig. 21-33　Heavy mineral composition per borehole P42 in Xinjian, Chongming District

2）砾、砂的成分成熟度比下伏层提高，表现为近源的钙质胶结粉砂团块除该带底部外已消失，花岗岩岩屑比下伏层明显减少，而石英、燧石、石英岩和石英砂岩等石英质砾成为主要成分，砂中石英与长石比率一般达7~9。同时在崇明还普遍出现斜长角闪（片麻）岩、黑云斜长片麻岩、变质石英砂岩等变质岩岩屑。

3）角闪石成为占绝对优势的重矿物组分，与崇明岛普遍出现的斜长角闪（片麻）岩岩屑可互相印证。

4）普遍可见独居石，断续出现独居石高含量带，且大都呈磨圆较好的半浑圆粒状。所含石榴子石大都呈不规则粒状、破碎粒状，少部分呈半浑圆粒状；锆石大都呈浑圆柱状，少数破碎粒状，偶见完好柱状晶形；酸性岩浆岩中常见的副矿物白钛石、锐钛矿、榍石大都呈不规则粒状，也见有较好的晶形，如白钛石的柱状，榍石的板、楔状，唯锐钛矿较多见正方双锥状晶体。

5）稳定和极稳定重矿物占重矿物组成比率剧减，并且在横向上差异较大。如中西部地区平均仅占5%左右，组分仅为单一的磁铁矿、钛铁矿；东部滨岸地带约占20%，组分增加了褐铁矿、赤铁矿；整个陆域稳定重矿物所占比率比较稳定。锆石、石榴子石、独居石等极稳定重矿物仅出现于崇明岛，其和稳定重矿物组成所占比率起伏很大，平均可占35%左右。

综上特征揭示，与早更新世比较，中更新世以后，区内沉积物的来源不仅明显趋于多源化，导致砾砂成分相当复杂；而且源区显然远近不一，与上新世—早更新世沉积物源相对较近的特征迥异。这由上述所含重矿物中较多酸性岩浆岩内常见副矿物及独居石的形态，在岩浆岩和变质岩内都有广泛分布的锆石及常为变质岩特征矿物石榴子石、金红石等重矿物形态特征的显著差异可资证明，反映沉积物来自远近不一的酸性岩浆岩、变质岩和沉积碎屑岩分布区。崇明庙镇孔在晚更新世以后变质岩特征矿物及岩屑的增多，表明相应其间源于变质岩分布区输入的物质呈有增无减的趋势。

中更新世前、后出现独居石及变质岩碎屑有无的显著差异，揭示沉积物来源区域的明显改变。这一改变，与早、中更新世交替期间（0.73Ma）我国曾发生的一次新构造变动事件，从而造成新构造期（0.73Ma）与喜马拉雅期（23.3~0.73Ma）不同的构造地貌特征有关（万天丰，1994）。根据长江三角洲第四纪地质及近年对长江三角洲晚新生代地层独居石年龄谱的研究，揭示长江于早更新世早期贯通，长江主泓在早更新世偏于现长江三角洲北翼的苏北地区流入黄海；至中更新世长江主泓南摆，转为流经现长江河口地区注入东海（吴标云和李从先，1987；范代读等，2004）。有学者（杨子赓，1994）根据南黄海QC2孔第四纪地层重矿物的研究，发现来自长江岩浆岩的榍石、锆石、电气石、石榴子石等重矿物含量明显高于黄河源区的特征，也得出古长江在晚松山极性时曾流向南黄海形成三角洲前缘沉积的认识。崇明庙镇等孔重矿物组成在中/下更新统分界上下发生的急剧变化，揭示了长江主泓可能自中更新世开始以后转而流入东海，导致长江三角洲南翼—河口区域输入物质的多源化。在长江中、下游流经的鄂、赣、皖和苏北等诸省区有较多的太古宇和古元古界角闪岩相—绿片岩相变质杂岩分布（金文山等，1996；沈其韩等，1996），可能由于新构造期强烈的构造-剥蚀作用，改变了长江主泓入海的走向，从而为河口带来了较多变质岩和岩浆岩等多源化组分。由长江水下三角洲沉积物的重矿物组成主要也为角闪石、绿帘石、锆石、石榴子石和钛铁矿（黄慧珍等，1996）的特征表明，新构造期以来形成的构造-地貌格局，为长江三角洲的塑造持续稳定地提供了丰富的陆源碎屑成分。

根据崇明新建P42孔重砂样品分析结果，以稳定重矿物与非稳定重矿物之比计算风化系数，得出了与庙镇孔可相互印证的结果，表明在中、下更新统分界上下，风化系数出现明显的变化：其下风化系数几乎全部处于高值区，反映早更新世研究区处在陆相沉积环境，风化作用比较强烈，非稳定矿物大为减少，很不稳定矿物角闪石消失，只是近第四系底部出现非稳定矿物含量升高的风化系数低值段，与古气候的寒冷变化相对应；分界之上，除上全新统外，几乎全部处于风化系数低值区，表明在中更新世—全新世，研究区通常处在海陆过渡或海相沉积环境下。由于该孔中更新统与上更新统顶部的陆相层均已被侵蚀殆尽，因此总体显示了风化作用较弱，非稳定矿物占明显优势。只是晚全新世以来，崇明演化为陆上三角洲平原，风化作用加强，风化系数才陡升为高值区。

二、黏土矿物特征

（一）黏土矿物种类

主要通过20世纪80年代崇明新建P42等孔的X衍射分析、差热分析和电镜鉴定以及近年浦东国际机场J25孔黏土样品的X衍射分析，查明沉积物中所含的黏土矿物种类主要为伊利石、蒙脱石，其次为高岭石、绿泥石。

1. 伊利石（水云母）

在黏土矿物X衍射图谱上（图21-34，图21-35），未经处理的黏土原样出现9.93~10.04Å、4.96~5Å、3.34Å不同强度的衍射峰；经Mg-甘油饱和处理后，这些衍射峰依然存在，其强度大都与原样相近，原来强度较高的3.34Å峰有所加强。表明在上海第四纪地层中的黏土矿物组成主要为伊利石，尤其在中更新统上部（见图21-34中D2-8号样）占绝对优势。差热分析反映（图21-36），在中温区500℃处普遍显示一中等反应的宽吸热谷，显然为伊利石层间水的脱水反应；另在低温区60~75℃间有一中等—强反应的宽吸热谷，为蒙脱石和伊利石表面附着水的脱水反应。差热分析也反映黏土矿物组成以伊利石为主，印证了X衍射分析结果。电子显微镜鉴定揭示的黏土矿物组成与X衍射分析、差热分析结果一致，即以伊利石为主。扫描电镜下观察，伊利石大致呈边界圆滑的等度片状和细小碎片及其集合体，轮廓清晰，大小小于1~5μm，表明为风化。

碎屑沉积；在透射电镜下，可见伊利石具明显的干涉条纹。

2. 蒙脱石

在X衍射图谱上，未经处理的黏土原样出现弱—中等强度的13.81~14.03Å衍射峰，经Mg-甘油饱和处理后，出现较弱—中等强度的17.33~17.67Å和18.03Å衍射峰；在深部，即下更新统中部以下地层中，经Mg-甘油处理后的黏土样品还出现代表蒙脱石d_{003}特征的5.9~6Å衍射峰和蒙脱石-伊利石混层矿物的8~10Å衍射峰。差热分析反映，在150~160℃之间大都出现一个弱吸热谷，为蒙脱石层间水的脱水引起；另如上述，在60~75℃间有一中—强反应的宽吸热谷，为蒙脱石和伊利石表面附着水的脱水反应。这些分析表明，蒙脱石是本区黏土矿物的较主要组分，并在下更新统中部（见图21-35中D_{2-13}号样）以下地层中渐占优势。电子显微镜鉴定也证实了蒙脱石为区内黏土矿物的主要组分之一。蒙脱石在扫描电镜下呈不规则鳞片状，常凝聚成大小不等的花絮状集合体；在透射电镜下呈边缘半透明的云雾状团块，其轮廓模糊。

3. 高岭石

在X衍射图谱上，未经处理的黏土原样出现7~7.08Å和3.53~3.56Å较弱—中等强度衍射峰；经Mg-甘油饱和处理后，这两个衍射峰依然存在，而且7~7.08Å衍射峰有所加强。差热分析反映，930℃处出现小放热峰，显示高岭石的重结晶；900~1000℃间出现反应微弱的“S”形谷峰，与高岭石再结晶之前的膨胀有关。这些分析显示，黏土矿物中有一定含量的高岭石存在。高岭石在扫描电镜下大都呈边缘受磨损的不规则薄片，部分大致可见假六方片状轮廓，矿物颗粒较蒙脱石、伊利石大，一般小于10μm，大者可大于等于20μm，显示其为碎屑沉积；在下更新统中下部及上新统中，可见到假六方片状高岭石自形晶体，近于等度薄片，但其大小不到0.5μm，看来属自生矿物。

4. 绿泥石

在X衍射图谱上，未经处理的黏土原样，大都可见强度很弱的4.74Å衍射峰，在下更新统上部及以上地层内还普遍出现13.81~14.03Å较弱—中等强度衍射峰；经Mg-甘油饱和处理后的黏土样品，这两个衍射峰仍断续出现，主要在埋深27m以浅（见图21-34中D_{2-6}号样以上）的上更新统上部和全新统这部分海相性较强的地层中较连续反映，但衍射峰强度均弱。差热分析曲线反映不明显，唯在近600℃处出现一微弱的吸热谷。综合这些分析，显示了绿泥石的存在，但其含量较少。

此外，主要在上新统黏土层内，扫描电镜下可见到部分边缘卷曲的埃洛石（也称多水高岭石）；在透射电镜下，明显可见埃洛石特具的管状晶形，也有呈破碎管状晶形者，似为变埃洛石（埃洛石的

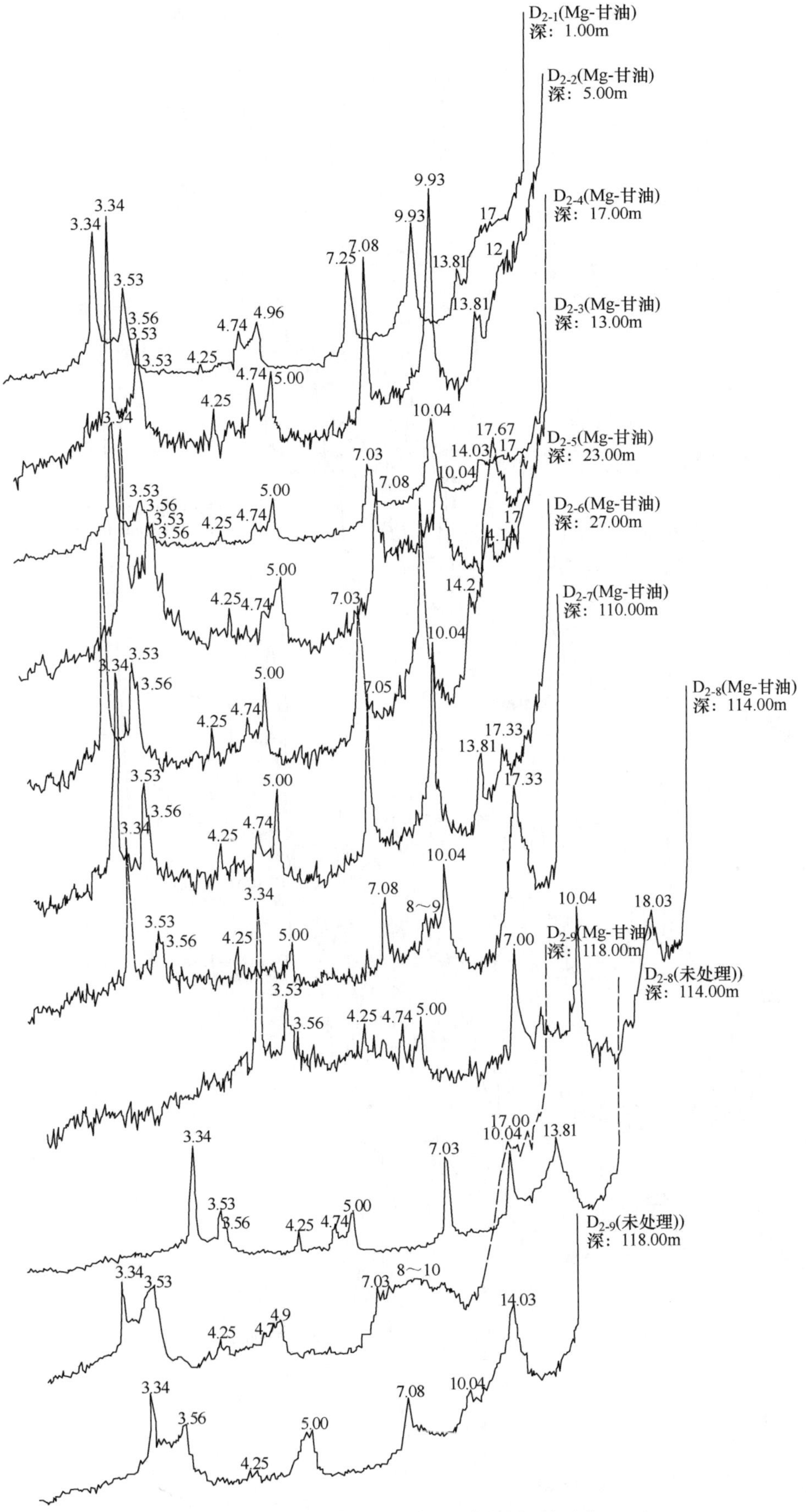

图 21-34　青浦凤溪 D2 孔黏土矿物 X 衍射图谱（一）

Fig. 21-34　X-ray diffraction pattern of clay minerals per borehole D2 in Fengxi, Qingpu District（Ⅰ）

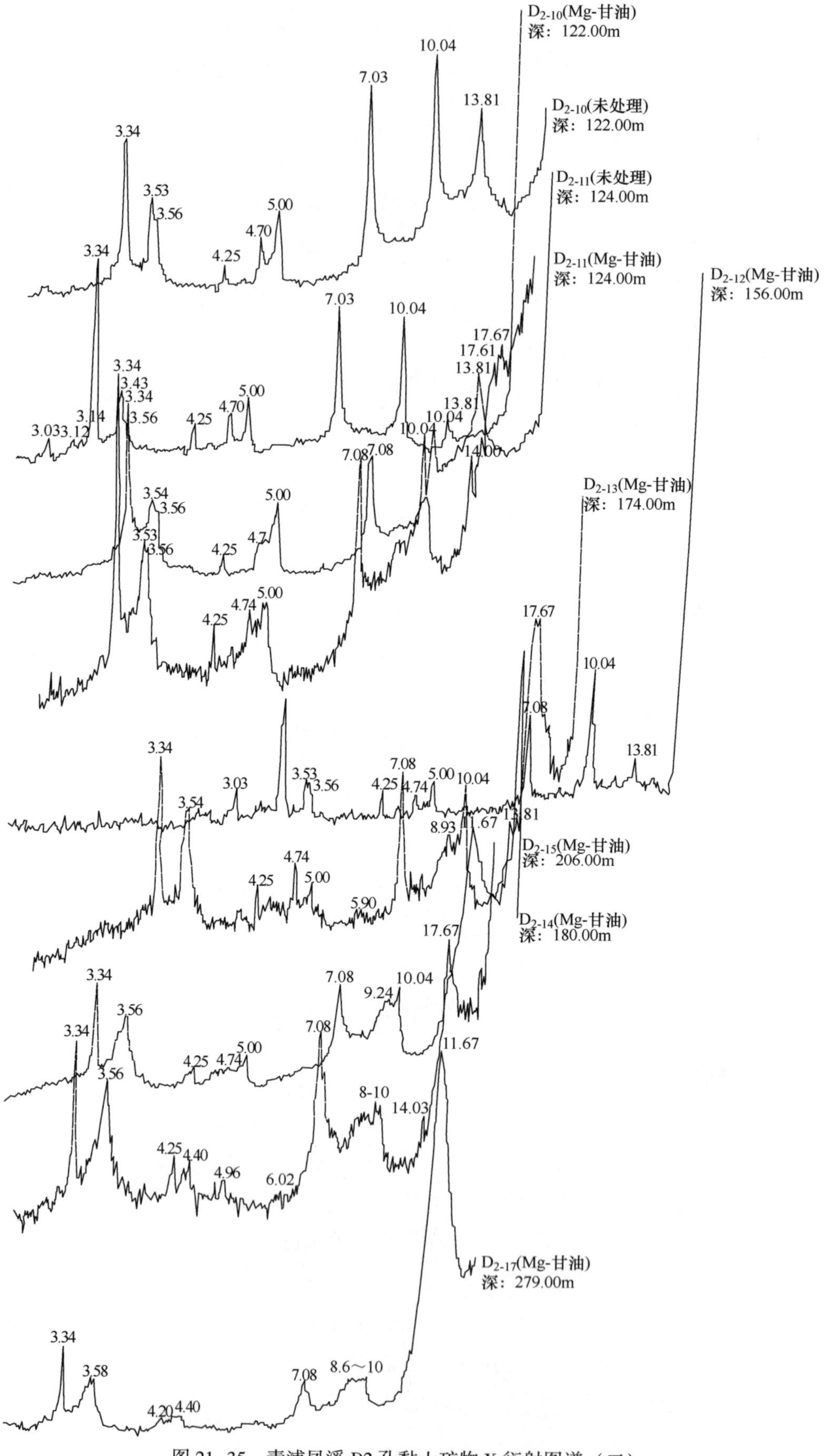

图 21-35　青浦凤溪 D2 孔黏土矿物 X 衍射图谱（二）

Fig. 21-35　X-ray diffraction pattern of clay minerals per borehole D2 in Fengxi, Qingpu District（Ⅱ）

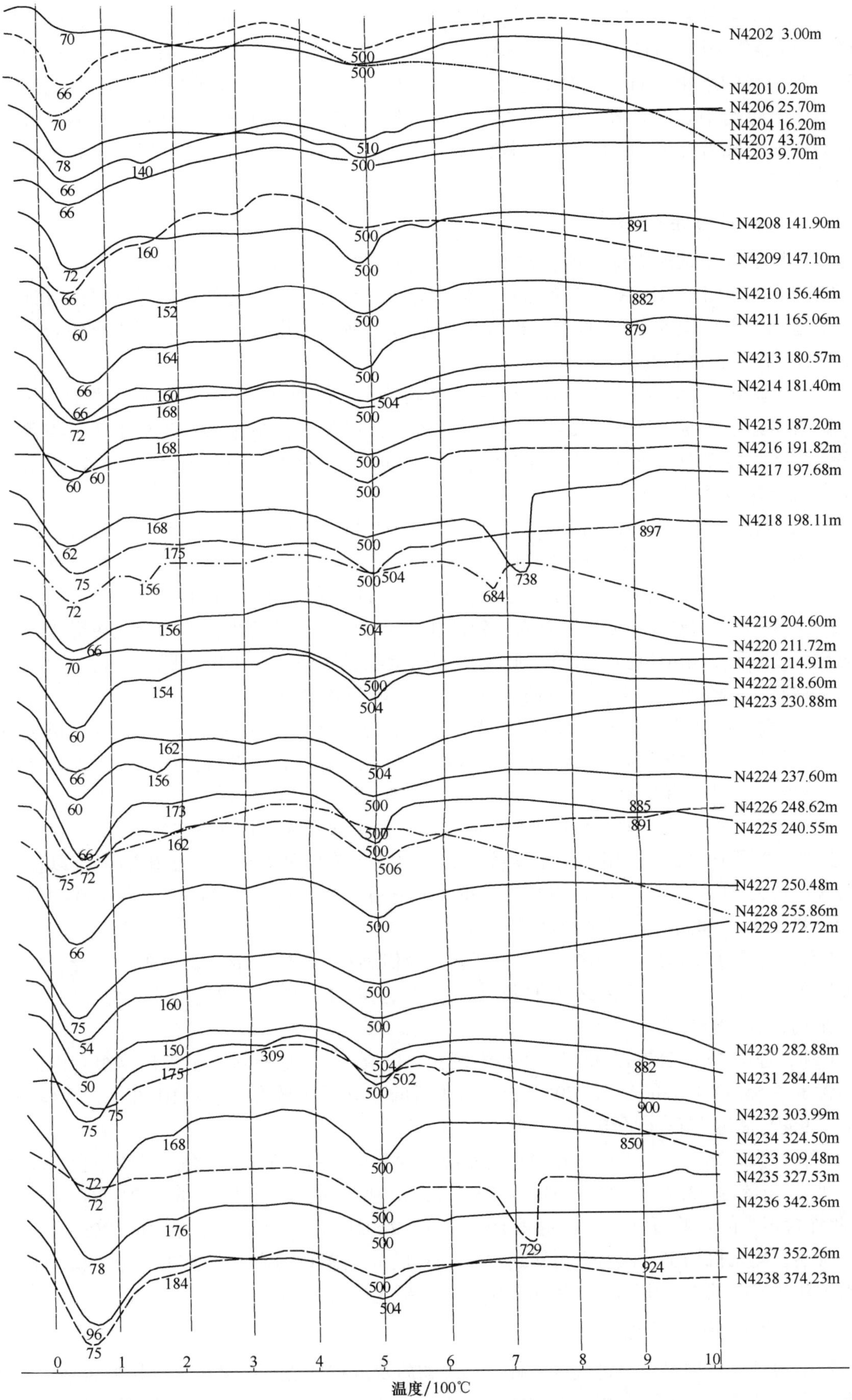

图 21-36 崇明新建 P42 孔差热分析曲线图

Fig. 21-36 DTA curve per borehole P42 in Xinjian, Chongming District

脱水变种)。在深逾 303m 的上新统顶部，差热分析曲线在 309℃ 处出现微吸热谷（见图 21-36 中 N4232 样），可能代表埃洛石层间结构的破坏，也反映有埃洛石的存在。在上新统黏土层上部，在透射电镜下还发现大量的水铝英石，呈对电子束不透明的球粒状集合体，显示为凝胶沉淀，其与埃洛石、变埃洛石共生。这与中酸性火山岩、凝灰岩一般经历的风化过程一样，在淋滤性差的碱性介质环境中易于形成蒙脱石，随着环境 pH 的降低、淋滤作用趋于通畅，经强烈风化，蒙脱石能转变为埃洛石和水铝英石。

（二）黏土矿物组合分带

综合 20 世纪 80 年代所做的崇明新建 P42、惠南 G413、青浦凤溪 D2、嘉定安亭 G414、嘉定娄塘 P5、海潮 P17 和金山朱泾 G418 等 7 个孔的 X 衍射分析结果（前 3 个孔的黏土样品系经甘油水溶液饱和处理，其他则未经该项处理），按本区松散沉积层内的黏土矿物组合可划分为 3 个组合带（图 21-37，图 21-38），由上而下为：

1）上部，埋深 0~145m，相当于全新统—中更新统，为伊利石（水云母）带。本带内伊利石占绝对优势，其含量介于 50%~87%，大体上由下而上增多。其次含蒙脱石 4%~32%不等，与伊利石具反消长关系。再次为绿泥石（介于 1%~24%不等）和高岭石（介于 2%~18%不等），以在全新统内略多，相应蒙脱石含量略减。

2）中部，埋深 145~270m，相当于下更新统，为伊利石-蒙脱石带。两者含量相近，以伊利石略多，一般占 40%~60%，偶尔可达 60%~70%。其次为蒙脱石，大都占 30%~55%，部分为 15%~30%，大体存在由上而下增多的趋势，并含少量的高岭石（通常介于 1%~18%）和绿泥石（介于 2%~11%）。黏土矿物组分的差异与岩性变化有一定的相依关系，如在寒冷期沉积的蓝灰色黏土内伊利石含量占绝对优势，蒙脱石相应减少；而在温暖期形成的紫红、紫褐色黏土中蒙脱石则为主要成分，伊利石减少，高岭石含量略增，尤以白色或褐黄色黏土中高岭石含量较多。

3）下部，埋深约 270m 以下，基本相当于上新统，略跨下更新统底部，为蒙脱石-伊利石带。本带以蒙脱石占优势，其含量通常介于大于等于 50%~70%，愈向下愈占明显优势；伊利石次之，含量介于 20%~40%。此外含有少量的高岭石（1%~5%）和绿泥石（1%~4%）。

浦东机场 J25 孔按黏土矿物组合可划分为 4 个带，由上而下为（图 21-39）：

Ⅰ. 上部，埋深 0~129m，相当于全新统—上更新统，为伊利石带。黏土矿物组合以伊利石占优势，平均含量约为 71%（介于 63%~82%），并含蒙皂石（平均含量仅为 3%，介于 0~10%）、绿泥石（平均含量 16%）和高岭石（平均含量 9%）。

Ⅱ. 中部，埋深 129~154m，相当于中更新统，为伊利石-高岭石带。黏土矿物组合仍以伊利石占优势，平均含量为 63.92%，并含少量的高岭石（平均含量 12.29%）、蒙皂石（平均含量 12.91%）和绿泥石（平均含量 10.88%）。

Ⅲ. 下部，埋深 154~287m，相当于下更新统，为伊利石-蒙皂石带。黏土矿物组合以伊利石占优势，平均含量达 65.11%（介于 44~79%），并含少量的蒙皂石（平均含量为 10.39%）、高岭石（平均含量为 11.39%）和绿泥石（平均含量为 12.63%）。

Ⅳ. 深部，埋深 287~354.50m，相当于上新统，为蒙皂石-伊利石带。黏土矿物组合以蒙皂石占绝对优势，平均含量达 73.7%（介于 51%~100%）；伊利石次之，平均含量为 21.3%（介于 10%~45%）。此外含很少的高岭石（平均含量 4.1%）和微量的绿泥石（平均含量仅为 0.9%）。

综上所述，浦东机场 J25 孔与崇明新建 P42 等孔的黏土矿物组合分带基本对应：机场孔较浅的 3 个带相当于新建等孔的上、中两带；机场孔深部带与新建等孔下部带一致。

（三）黏土矿物古气候意义

沉积物中的黏土矿物组合与古气候条件、母源物质及沉积环境等因素有着密切的关系。20 世纪 80 年代以来，中外学者通过不同沉积盆地和中国黄土中黏土矿物结晶度的研究，发现其与矿物形成时的气候环境条件直接相关，尤其是伊利石和蒙脱石的结晶度变化可用以指示古气候特征。如伊利石在偏

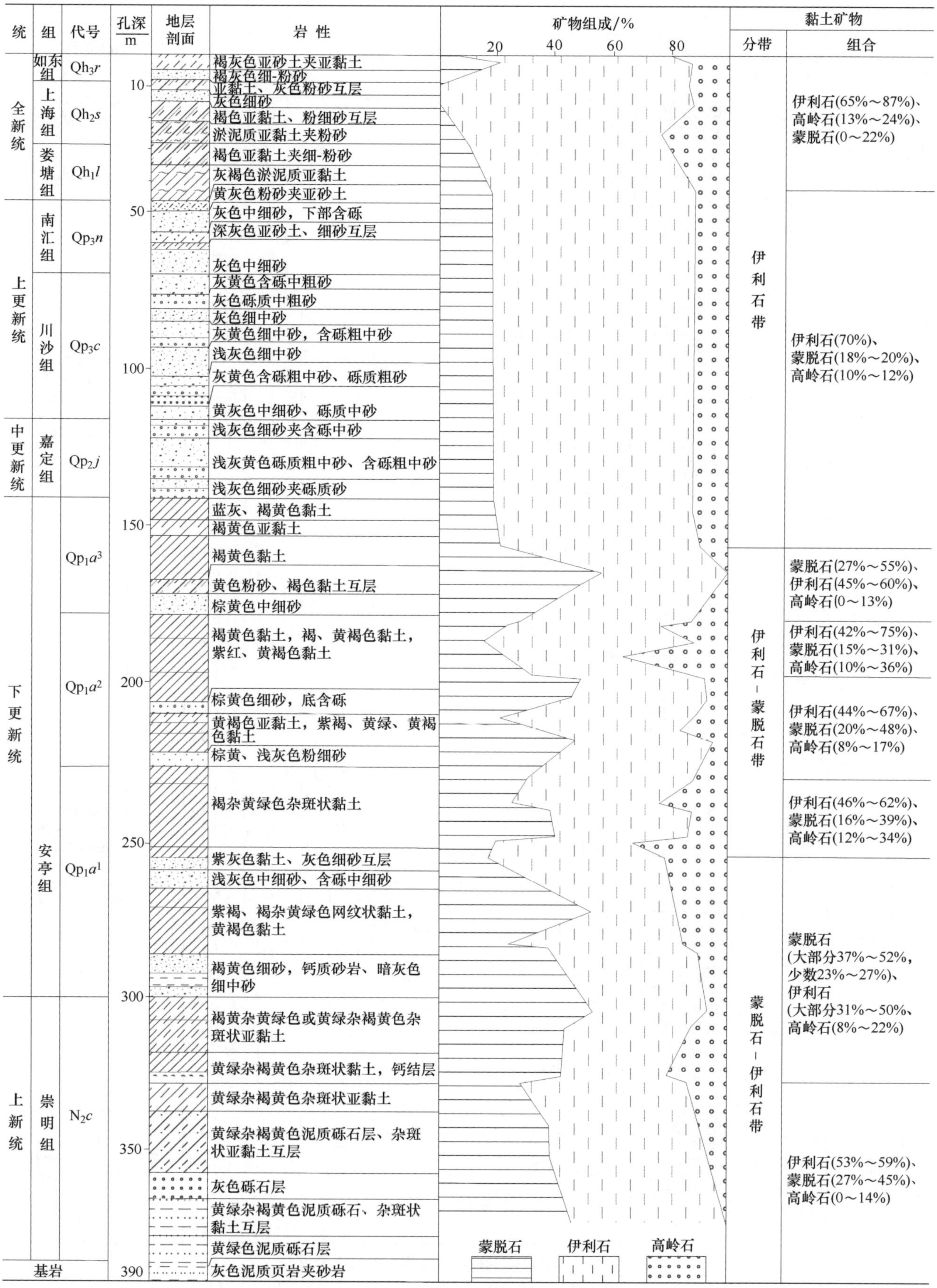

图 21-37　崇明新建 P42 孔第四系黏土矿物组合图谱

Fig. 21-37　Clay mineral assemblage diagram of the Quaternary per borehole P42 in Xinjina, Chongming District

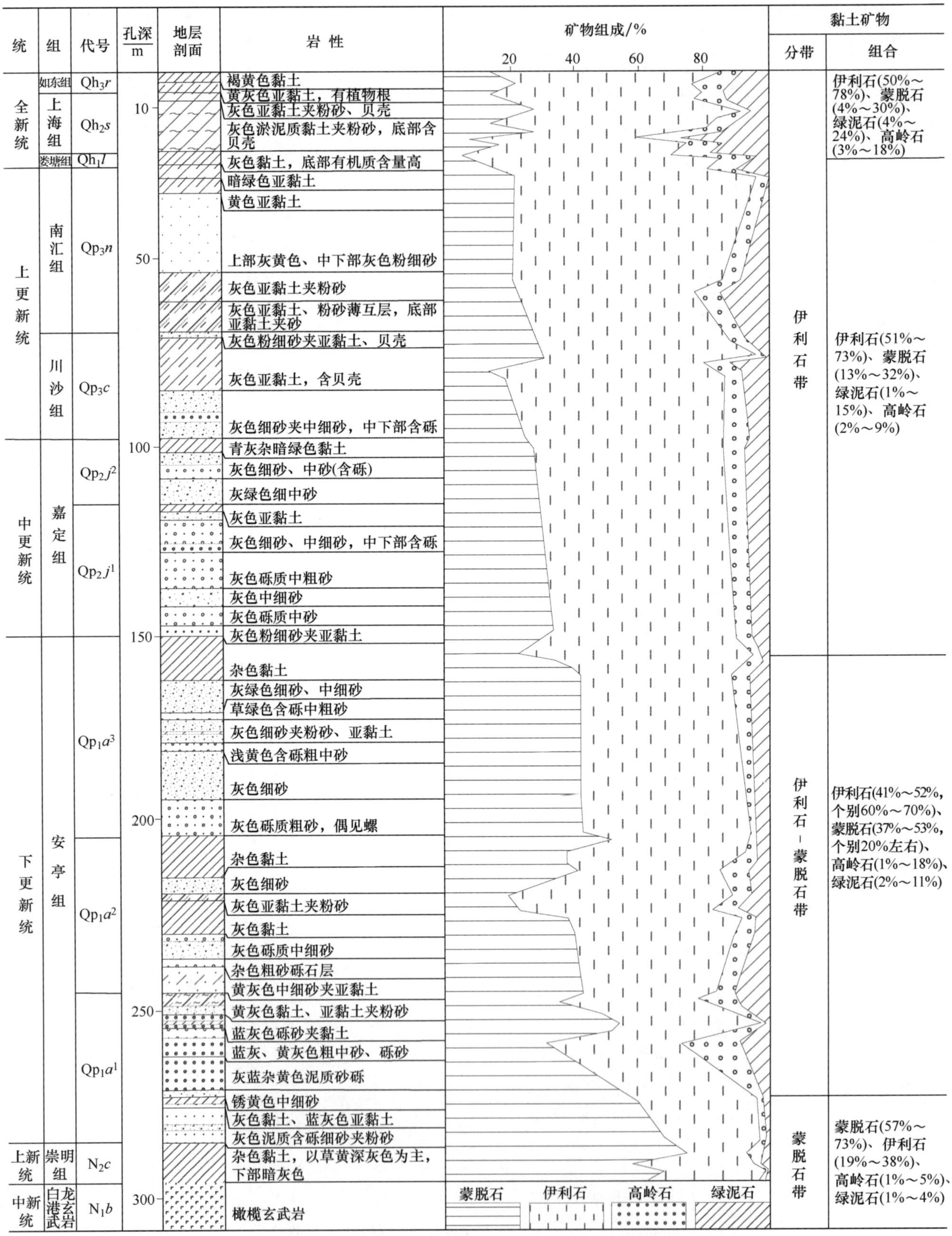

图 21-38 惠南 G413 孔第四系黏土矿物组合图谱

Fig. 21-38 Clay mineral assemblage diagram of the Quaternary per borehole G413 in Huinan

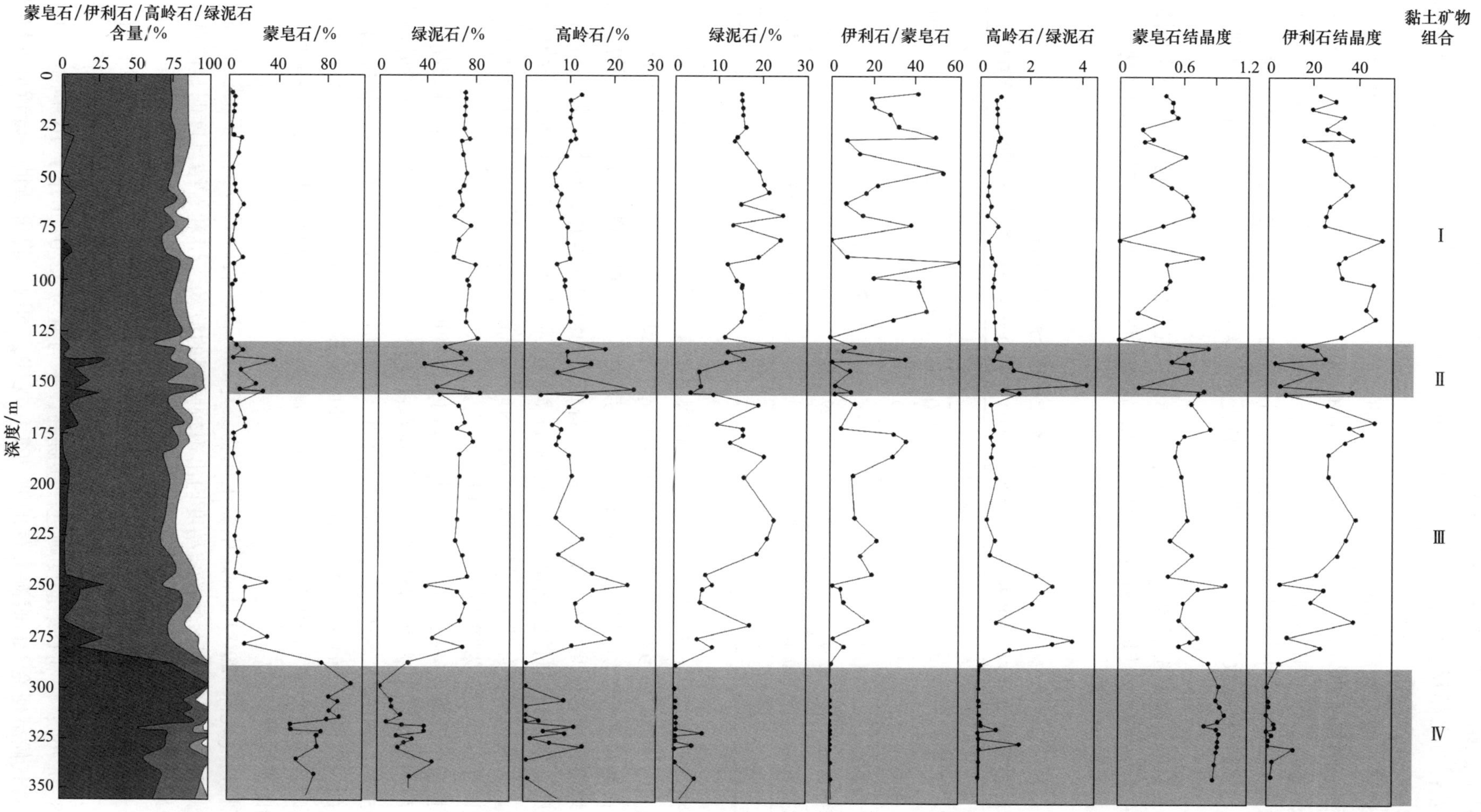

图 21-39 浦东机场 J25 孔第四系黏土矿物组合图谱

Fig. 21-39 Clay mineral assemblage diagram of the Quaternary per borehole J25 at Pudong Airport

湿热的环境下形成开放结构，以“开型指数（Ns）”来表示伊利石（001）衍射峰的开放度，其指数值大，结晶度则差；与之相反，结晶良好的蒙皂石则指示沉积时的暖湿气候环境（Chamley，1975，1983；任磊夫，1984；王诗佾，1987）。

刘东生等（1985）通过对黄土的研究，得出了与 Chamley（1975）一致的结论，提出用 $Hw=H/W$ 反映黄土与古土壤中伊利石的结晶度［式中：H 为伊利石 001（10Å）衍射峰的峰度，mm；W 为 10Å 衍射峰的半高宽，mm］。黄土中伊利石的 Hw 值较大，一般大于 25，即伊利石结晶度高，反映气候干冷；而古土壤和埋藏风化层中伊利石的 Hw 值一般小于 20，即结晶度差，反映气候暖湿。这和 Chamley（1975）的结论是一致的，即伊利石结晶度高，为干燥寒冷气候的反映；结晶度差则反映暖湿的气候。

在浦东机场 J25 孔采取了 63 个黏土样品作 X 衍射分析，现在根据乙二醇饱和片的 X 衍射谱线，采用 Chamley 和刘东生的方法，计算伊利石和蒙皂石的结晶度，以此分析古气候环境的变化。黏土矿物结晶度计算结果列于表 21-7。

表 21-7　浦东机场 J25 孔黏土矿物含量及结晶度

Table 21-7　Clay mineral contents and crystallization levels per borehole J25 at Pudong Airport

黏土矿物组合分带	年代地层	黏土矿物含量/%				矿物结晶度	
		蒙皂石	伊利石	高岭石	绿泥石	蒙皂石	伊利石
Ⅰ	全新统	2.54	71.71	10.79	14.96	0.40	28.06
	上更新统	3.68	71.35	8.24	16.73	0.41	33.57
Ⅱ	中更新统	12.91	63.92	12.29	10.88	0.61	19.15
Ⅲ	下更新统	10.39	65.11	11.39	12.63	0.68	25.25
Ⅳ	上新统	73.70	21.28	4.11	0.90	0.90	2.54

上新世沉积物（孔深 287~354.50m）中黏土矿物含量以蒙皂石占优势。蒙皂石具良好的结晶度，平均值为 0.90（介于 0.80~0.98）；而伊利石结晶度很差，平均为 2.54（介于 0.54~11.66）。高结晶度的蒙皂石含量占绝对优势，伊利石结晶度明显很差，反映了当时较为暖湿的古气候环境，部分伊利石和高岭石产生埃洛石（多水高岭石）化，也说明所曾经受的湿热化作用，这与孢粉分析结果是相符的；同时蒙皂石的富集更与当时沉积物的来源有关。上海及其附近地区基岩广布下白垩统中酸性火山碎屑岩和凝灰岩，其风化产物有利于蒙皂石的生成，因此在晚新生代沉积初期，来自附近山地的冲洪积流就可能形成以富含蒙皂石为特征的快速堆积。

进入早更新世后（孔深 154~287m），由于古气候向干冷的剧变，黏土矿物组合及结晶度也随之产生与上新世截然的差异。黏土矿物组合突变为以伊利石占优势，伊利石的结晶度也陡然升高，平均值为 25.25（介于 5.20~45.74）；蒙皂石结晶度略有降低，平均值为 0.68（介于 0.46~0.85），反映了早更新世比较干冷的气候环境。但在孔深 275m、248m 处的沉积物中黏土矿物含量出现较明显的波动，其中蒙皂石、高岭石含量大增，而伊利石、绿泥石相应减少，反映在早更新世总体干冷的气候条件下，间或出现暂短的暖湿气候波动。

中更新世沉积物（孔深 129~154m）中黏土矿物蒙皂石、高岭石含量较上、下层位略高，而伊利石、绿泥石含量略降。黏土矿物含量及结晶度均出现较强烈的波动，伊利石结晶度明显降低，反映中更新世为温暖湿润的气候环境，古气候曾出现较为强烈的波动。

晚更新世—全新世沉积物（孔深 0~129m）中的黏土矿物含量总体较为稳定，但伊利石、蒙皂石的结晶度及伊利石/蒙皂石比值表现出规律性的波动变化，蒙皂石结晶度进一步降低，平均值为 0.41（介于 0.18~0.76）；伊利石结晶度则进一步提高，平均值为 32.07（介于 16.14~49.38），反映了晚更新世以来古气候呈现冷干与暖湿交替的波动变化规律，同时，晚更新世—全新世沉积物中绿泥石含量比下伏层位明显增多，这与海相沉积中通常绿泥石含量较高的规律相符。

第七节　气候地层划分与区域对比

一、气候地层划分

上海第四纪古气候期的划分主要依据孢粉资料取得，综合孢粉带花粉揭示了第四纪以来古植被演变，进而揭示了气候的冷暖干湿变化。此外，如前所述沉积物中重矿物、黏土矿物和地球化学元素指示的化学风化系数在一定程度上指示了沉积过程中的古气候特点。综合多种指标，上海第四纪古气候冷暖期次变化描述如下。

（一）古植被揭示的古气候波动

孢粉组合成分反映其所代表的植物群面貌及当时的古气候环境。区内18个综合孢粉带相应反映了区内上新世以来所经历的古植被与古气候演变的18个阶段，由下而上依次为：上新世1个，更新世12个，全新世5个。

1. 上新世

第一阶段：此时植被是以麻栎、槲栎、柞栎等落叶栎和青刚栎、栲属等常绿栎为主的常绿阔叶、落叶阔叶混交林，林中杂有喜温湿的胡桃、枫香、榆、桦、栗树，还有喜暖热的大戟科、漆树和茜草科等树种，林下草被为禾本科，其间湿地、湖泊生长着水生草本植物香蒲、莎草和凤尾蕨、水蕨类，附近山地混生有松、云杉、冷杉等暗针叶树。反映气候暖热潮湿，大致相当于现在浙闽地区的中亚热带北部气候，年平均气温比现今上海略高2~3℃。

2. 早更新世

包含2带~6带5个孢粉带，反映相应5个阶段的古植被与古气候演变。

第二阶段：此时植被建群种以松、云杉、冷杉为主，落叶栎居次，构成针叶、落叶阔叶混交林，其间杂生榆、桦、铁杉等。林下中旱生的蒿、藜、禾本科，还有喜凉干的麻黄、菊科等草本植物生长繁茂。当时在东部冲洪积平原上形成针叶、落叶阔叶混交林—草原植被，西部的低山丘陵区则为疏林—草原。气候凉冷干燥，基本相当于现在东北中部的温带气候，年平均温度比现今本区低6~8℃。

第三阶段：以落叶栎类为主，有较多的胡桃、榆、枫香、鹅尔枥等落叶阔叶树（局部有常绿栎、栲等常绿阔叶树间杂），松与一些冷杉、云杉等针叶植物居次，构成含常绿阔叶、落叶阔叶混交林。林下为稀薄的禾本科草地，在西部低山丘陵区，林下则有较多禾本科和一些蒿、藜等草本植物生长，形成含常绿阔叶、落叶阔叶混交林—草原植被。气候温暖稍湿，大致相当于现在淮河流域的北亚热带北缘气候，年平均温度比现今本区低1~2℃。

第四阶段：当时植被建群种以松、落叶栎为主，并有不少云杉、冷杉和榆混杂，其间还杂生桦、桤木、铁杉等喜温凉树种。林下有繁茂的禾本科、蒿等草本植被，构成针叶、落叶阔叶混交林—草原植被，气候温凉干燥，大致相当于现在华北平原—辽宁南部的暖温带气候，年平均温度比现今本区低4~6℃。

第五阶段：当时植被建群种以落叶栎、松为主，间杂有较多的榆树，还有不少青刚栎、枫香、豆科、楮、冬青等常绿阔叶植物，构成常绿阔叶、落叶阔叶混交林。林下普遍生长禾本科植物，湖泊、湿地内出现大量的水生草本香蒲及蕨类。气候温暖湿润，与现今本区所处的北亚热带南缘气候相仿。

第六阶段：当时植被建群种以松为主，其次为云杉、冷杉或柏，间杂有少量的落叶栎、桦、榆、枫香等落叶阔叶植物，构成混有落叶阔叶的针叶林。林下有稀薄的禾本科、藜科等中旱生草本植物生长，林间尚有多种蕨类、苔藓分布。气候凉冷偏干，大致相当于现在东北中部的温带气候，年平均温度比现今本区低6~8℃。

3. 中更新世

包含7带~9带3个孢粉带，反映相应3个阶段的古植被与古气候演变。

第七阶段：当时植被以落叶栎、松为主，间杂较多的桦、枫香、榆、栗、榛等喜温成分，早期还出现不少青刚栎、栲等常绿成分。林下有禾本科、蒿、藜等草本植物生长，林间湖泊、湿地有眼子菜、黑三棱等水生草本植物及多种蕨类密布。形成含常绿阔叶、落叶阔叶混交林，气候温暖湿润，大致相当于现在淮河流域所处的北亚热带北缘气候，年平均温度比现今本区低1~2℃。

第八阶段：当时植被建群种以松、落叶栎为主，间杂较多的桦、枫香、榆，还有一些喜凉冷的云杉、冷杉、铁杉；林下禾本科、菊、蒿等草本植物生长繁茂。在西南部地区建群种多为松、柏；林下多禾本科，并有大量苔藓生长，环境比较阴湿。总体构成针叶、落叶阔叶混交林—草原植被，气候温凉偏干，大致相当于现在华北平原—辽宁南部的暖温带气候，年平均温度比现今本区低4~6℃。

第九阶段：当时植被建群种主要为落叶栎，其间有较多的桦、枫香、榆等落叶阔叶植物，也有一些松、杉等针叶成分，杂有青刚栎、栲、栳木、槠等常绿阔叶成分，形成常绿阔叶、落叶阔叶混交林。林下有较稀薄的禾本科与藜科等草本植被，一些林间湖泊、湿地则有眼子菜、莎草科及蕨类生长。气候暖热湿润，与现今本区所处的北亚热带南缘气候相仿。

另外在第九阶段与第十阶段之间，据西部的娄塘P5号、朱泾G418号等孔中更新统顶部保存较全者揭示，尚有一厚数米、含孢粉贫乏的蓝灰色黏土层，仅发现有个别的禾本科、松属等孢粉颗粒，因此不能单独划为一个孢粉带。这一层位大都因被侵蚀而缺失，它可能代表所曾经历的相当于一期疏林草原甚或荒漠的变迁，相应的古气候似为凉冷干燥。

4. 晚更新世

包含10带~13带4个孢粉带，反映相应4个阶段的古植被与古气候演变。

第十阶段：此时植被建群种以落叶栎、松为主，还有不少枫香，其间杂生较多的桦、铁杉及断续分布的云杉、冷杉、榆等喜温凉成分，也杂有青刚栎、栲等常绿阔叶成分，构成含常绿阔叶、落叶阔叶混交林。林下、河滩、潮滩有禾本科与蒿、藜等草本植被生长，湖泊、湿地内眼子菜、香蒲、莎草等水生草本及蕨类植物生长繁茂，并出现刺球藻、极管藻，反映此时明显有海水侵进。当时气候温暖湿润，大致相当于现在淮河流域所处的北亚热带北缘气候，年平均温度比现今本区低1~2℃。

第十一阶段：此时植被建群种以松、落叶栎为主，间杂不少桦、栗、榆等阔叶成分及暗针叶植物铁杉、云杉、冷杉、落叶松；林下有大量的蒿、藜、禾本科等中旱生草本植物生长，构成针叶、落叶阔叶混交林—草原植被。气候温凉偏干，大致相当于现在华北平原—辽宁南部的暖温带气候，年平均温度比现今本区低4~6℃。

第十二阶段，古植被和古气候曾发生明显变化，可分为早、中、晚3期。

早期：植被建群种以落叶栎、松为主，间杂不少的榆、桦、栗，还杂有一些桤木、鹅尔枥、枫香等落叶阔叶成分和铁杉、云杉、冷杉等暗针叶树，西部地区出现不少的青刚栎之类常绿成分。林下主要生长蒿、藜等草本植物，林间湿地有大量香蒲、莎草等水生草本以及铁线蕨等植物，构成含常绿阔叶的落叶阔叶混交林，气候温暖湿润，大致相当于现在淮河流域的北亚热带北缘气候，年平均温度比现今本区低1~2℃。

中期：植被建群种以松、落叶栎为主，其间杂有较多的栗、榆、枫香，还有桦等落叶阔叶成分，林间屡见湿地内生长的香蒲及凤尾蕨类，也有喜盐草本植物藜、蒿分布，形成落叶阔叶、针叶混交林，气候温和湿润，大致相当于现在华北平原南部—山东的暖温带气候，年平均温度比现今本区低2~4℃。

晚期：植被建群种以落叶栎、松为主，间杂不少的栗、桦、榆，还有常绿栎以及云杉、冷杉之类针叶成分，形成生长茂盛的含常绿阔叶、落叶阔叶混交林。林下生长蒿、禾本科、藜等稀薄草本植被，林间湿地有些香蒲、眼子菜等水生草本植物生长，此时也出现不少的刺球藻、极管藻，表明受到海水侵进的影响。气候温暖湿润，与现今淮河流域的北亚热带北缘气候相仿，年平均温度比现在本区低1~2℃。

第十三阶段：此时植被稀薄，地面植被主要分布蒿、禾本科、藜、麻黄等中、旱生草本植物，其间稀疏生长松、冷杉等针叶树及很少的桦、落叶栎之类落叶阔叶成分，构成疏林—草原植被。当时西部地区出现较多的湖泊、湿地内有大量的环纹藻、苔藓生长。气候凉冷干燥，基本相当于现在东北中部的温带气候，年平均温度比现今本区低6~8℃。

5. 全新世

第十四阶段：此时植被建群种以松、落叶栎为主，杂有较多的栗、桦、榆、枫香，时而出现云杉、冷杉、铁杉等喜冷的暗针叶成分，也有常绿栎、胡桃、枫香等喜暖的阔叶成分。林下有些藜、禾本科、蒿等草本植物分布，较多出现的湿地生长香蒲、莎草等水生草本植物，形成针叶、落叶阔叶混交林，气候温凉稍湿，大致与现今华北平原—辽宁南部的暖温带气候相仿，年平均温度比现今本区低4~6℃。其气候期相当于布列特（A. Blytt）—色尔南德（R. Sernander）气候分期的前北方期。

第十五阶段：此时植被建群种仍以松、落叶栎为主，其间出现不少常绿青刚栎与栗树，也不乏喜凉冷的云杉、冷杉，还有喜温凉的桦、榆、铁杉分布。林下草本植物明显比前一阶段繁茂，多为中、旱生草本蒿、禾本科、藜，构成落叶阔叶、针叶混交林—草原植被。在现崇明、浦东等地有些湖泊、湿地分布，其间香蒲、眼子菜、莎草之类水生草本植物生长茂盛。北部地区出现大量的刺球藻，反映其时明显受到海水的影响。气候温和偏干，大致相当于现在华北平原南部的暖温带气候，年平均温度比现今本区低2~4℃。其气候期与布列特—色尔南德尔气候分期的北方期相当。

第十六阶段：此时木本植物生长茂盛，建群种以落叶栎为主，其次为常绿阔叶乔木青刚栎、栲和栗，杂有不少桦、榆、胡桃等落叶阔叶成分，还有一些柯、槠、冬青、樟、桃金娘、瑞香等常绿阔叶成分，其间杂生较多的松和一些云杉、冷杉和铁杉等针叶成分，构成常绿阔叶、落叶阔叶混交林。草本植被较前减退，林下、海滨有禾本科、蒿、藜等中旱生草本植被分布，林间湿地内有不少的香蒲、莎草、眼子菜等水生植物生长。气候暖热潮湿，大致相当于现在浙闽地区的中亚热带北部气候，年平均气温比现今本区略高2~3℃。

第十七阶段：此时植被建群种以松、落叶栎为主，其间杂生较多的桦、云杉、冷杉、铁杉等喜凉冷乔木，还有不少栗、榆等喜暖落叶阔叶成分；早期有较多青刚栎、栲等常绿成分。林下、海滨均为蒿、藜和禾本科等草本植被，林间阴湿地区密布蕨类，构成针叶、落叶阔叶混交林。气候温凉稍干，大致相当于现在华北平原南部—淮河流域过渡带的暖温带气候，年平均温度比现今本区低2~3℃。其气候期相当于布列特—色尔南德尔气候分期的亚北方期。

第十八阶段：此时植被建群种以落叶栎为主，松居次，其间杂生较多的桦、胡桃、栗、青刚栎、栲、榆等喜温暖的阔叶乔木，还有不少云杉、冷杉、铁杉等喜凉冷的针叶成分。林下河旁、海滨生长禾本科、蒿和少量藜相成的稀薄草本植被，湖泊、湿地有香蒲、莎草、黑三棱、眼子菜等众多水生草本植物和蕨类，形成含常绿阔叶、落叶阔叶混交林。气候温暖湿润，与现今本区气候相仿。其气候期相当于布列特—色尔南德尔气候分期的亚大西洋期。

（二）化学风化指示的古气候情况

如前所述，崇明庙镇SG1孔和崇明新建P42孔稳定重矿物与非稳定重矿物比例指示的风化程度显示第四系相比于下伏的新近系化学风化减弱，古气候相对寒冷。黏土矿物组成成分和结晶度变化显示上新世古气候暖湿，进入早更新世后古气候向干冷的剧变，中更新世气候温暖湿润，古气候波动较为强烈，晚更新世以来古气候呈现冷干与暖湿交替的波动。主量元素氧化物揭示了上新世化学风化十分强烈，而早更新世则相对变弱，第四纪化学风化有逐渐弱化的趋势，可能反映古气候总体有变凉干的趋势。

二、气候地层综合划分及区域对比

上海松散沉积物记录了上新世以来的气候波动，总体而言，上新世以温暖湿润为主，第四纪5次明显的冷暖期交替。

上新世崇明组沉积物记录气候总体暖湿，下部暖热湿润，中间夹一次寒冷干燥期，命名为第一寒冷期，上部暖热湿润—温暖湿润，命名为第一温暖期。上新世气候记录在基岩面凹陷区钻孔上新世沉积物中多有揭示。

进入早更新世，气候开始明显变冷，但有波动，先后经历了3次冷期，中间有两次暖期交替，气候期先后命名为第二寒冷期、第二温暖期、第三寒冷期、第三温暖期、第四寒冷期。

早更新世气候波动划分主要依据孢粉记录获取，区内下更新统孢粉记录差异很大，浦东J25号、娄塘P5号、安亭G414号、朱泾G418号等孔见本孢粉带，但较多钻孔下更新统孢粉稀少，很难区分气候冷暖波动。

中更新世古气候先后经历了一次温暖期和一次寒冷期，分别命名第四温暖期和第五寒冷期，其中第五寒冷期经历了早期温凉偏干、中期暖热湿润、晚期寒冷干燥的变化。

晚更新世128~7ka为一次暖期，命名为第五温暖期，对应MIS5期；70~11ka为寒冷期，命名为第六寒冷期，气候总体温凉偏干，对应MIS3期和MIS2期，在这个阶段大致50~30ka（MIS3期）存在一次间暖期。

全新世为冰后期，气候温暖湿润，气候波动先后经历了温凉稍湿、温和偏干、暖热潮湿、温凉稍干、温暖湿润的变化，对应前北方期、北方期、大西洋期、亚北方期、亚大西洋期。

第八节　多重地层划分与区域对比

依据年代地层、岩石地层、生物地层、化学地层、矿物地层和气候地层综合分析，区内第四系多重地层划分如下（图21-40，图21-41）。

一、安亭组

安亭组分为上、中、下3段，由3个下粗上细沉积旋回组成，每个旋回下部为含砾砂、砂层，上部为杂色黏土，沉积环境为河流—泛滥平原/湖泊交替变化，中下段河流沉积以辫状河沉积为主，上段河流沉积以曲流河沉积为主。

安亭组化学风化强度较下伏上新统变弱，古气候经历了冷—暖—冷—暖—冷波动，安亭组中段和上段局部地点下部砂层见少量海相化石，反映受到微弱海水影响。

安亭组下段、中段沉积物为近源沉积，上段沉积物变为远源沉积，可能为区内最早接受的长江泥沙堆积。

安亭组底部多出现古地磁M/G界限，确定为2.58Ma，安亭组中段下部砂层出现古地磁奥尔都维事件，年代确定为1.95~2.01Ma，安亭组上段下部砂层出现古地磁贾拉米洛事件，年代确定为0.99~1.07Ma。

二、嘉定组

嘉定组分为上、下两段，由2个下粗上细沉积旋回组成，每个旋回下部为含砾砂、砂层，上部为杂色黏土、蓝灰色黏土，沉积环境为河流—泛滥平原/湖泊交替变化，河流沉积以曲流河沉积为主。

嘉定组沉积阶段古气候经历了暖—冷波动，下段局部地点见较丰富的海相化石，上段下部砂层在太湖平原区多出现丰富的海相化石。

嘉定组下段底部多出现古地磁B/M界限，确定为0.78Ma，嘉定组下段顶部OSL年龄约为0.26Ma。嘉定组上段顶部与上更新统接触，年龄为0.26~0.128Ma。

三、川沙组

川沙组是以一套砂、含砾砂为主的沉积层，砂层中偶含或含少量海相化石，体现了受海水影响的河流沉积特征。

川沙组沉积阶段古气候温暖湿润，OSL年龄显示相对于MIS5期，年龄为70~128ka。

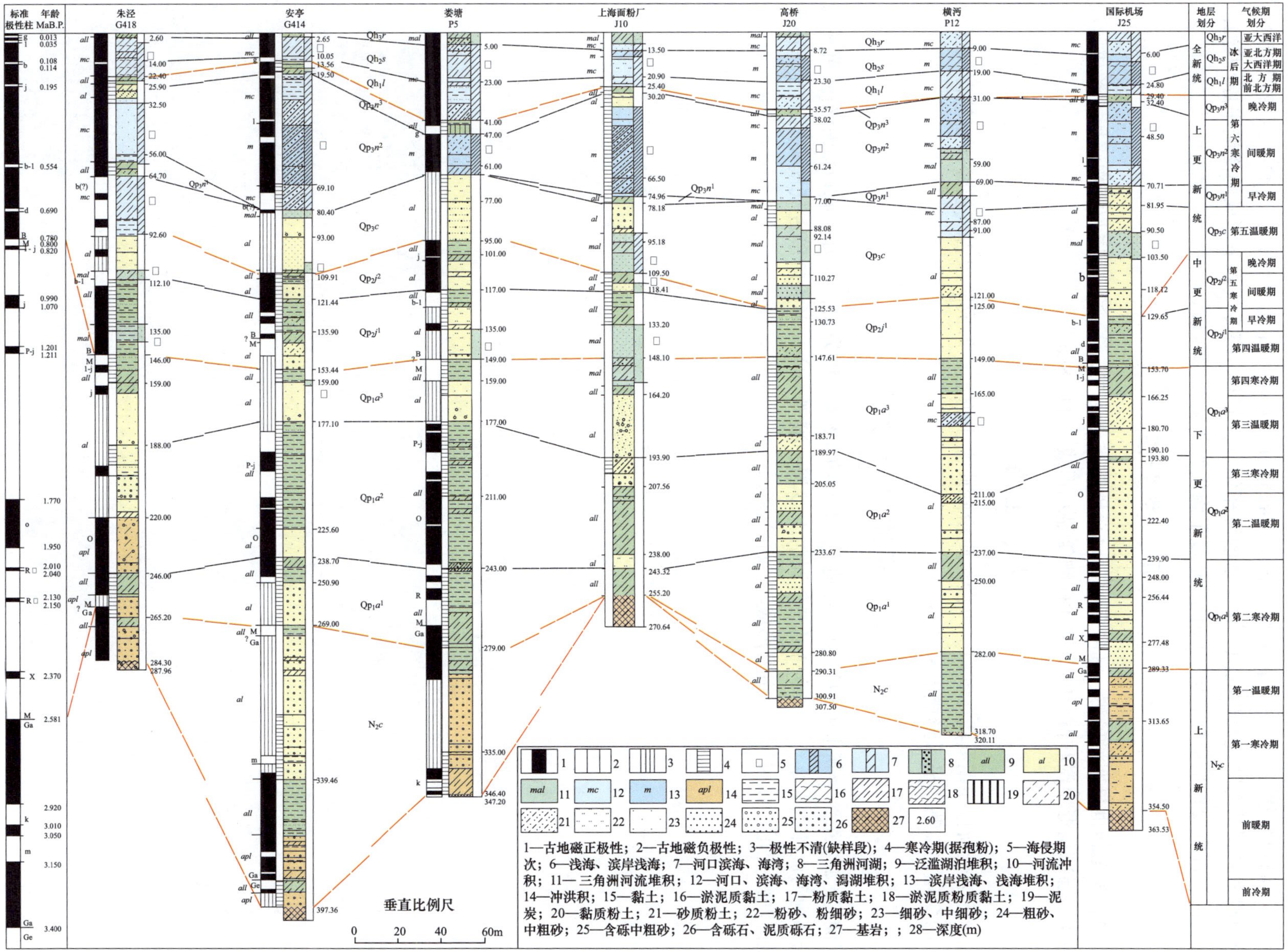

图 21-41 上海第四纪地层综合对比图

Fig. 21-41 Integrated stratigraphic correlation of the Quaternary in Shanghai

四、南汇组/滆湖组

南汇组/滆湖组分为上、中、下3段，下段局部分布，为寒冷气候下形成的泛滥平原/湖泊沉积，中段为间暖期堆积的海陆过渡相沉积，含丰富的海相化石，上段为寒冷气候下形成的泛滥平原/湖泊沉积，为全区浅部重要的地层标志层。OSL测年显示南汇组年龄为70~11ka。

五、娄塘组

娄塘组为全新世早期河口湾、潮滩沉积，以粉砂质黏土为主，多含有机质和植物碎屑，形成阶段古气候经历温凉稍湿至温和偏干的变化，^{14}C测年显示形成于11~8ka。

六、上海组

上海组为全新世中期滨浅海沉积，以淤泥质粉砂质黏土为主，形成阶段古气候经历温暖湿润至温凉稍干的变化，^{14}C测年显示形成于8~3ka。

七、青浦组/如东组

青浦组/如东组为全新世晚期沉积，太湖平原以湖泊沼泽沉积为主，滨海平原以滨岸潮滩沉积为主，局部为河流冲积平原，河口沙岛以三角洲前缘砂坝和潮滩为主，形成阶段古气候经历了温凉稍干至温暖湿润的变化，^{14}C测年显示主要形成于3ka以来。

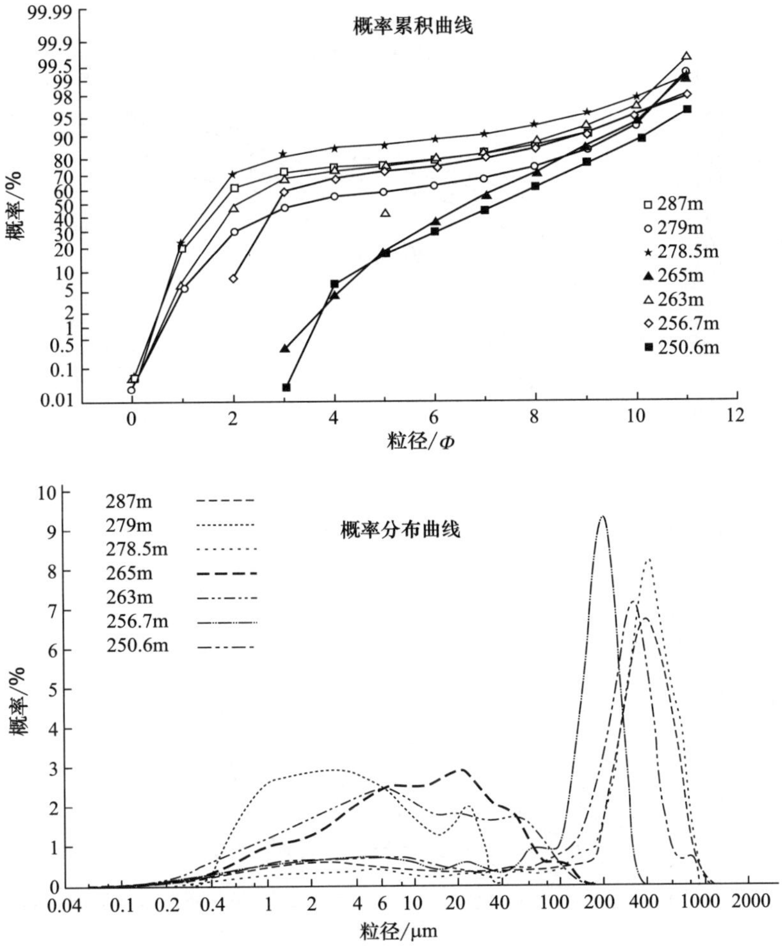

图 23-4 浦东机场 J25 孔第 2 阶段早期沉积物粒度特征

Fig. 23-4 Grain size distribution of the early stage 2 sediments per borehole J25 at Pudong Airport

次总体组成的二段式，以跃移次总体为主，占 30%~55%，斜率大，分选性较好；细截点为 2.8Φ，比早期曲流河沉积细些。由曲流砂坝下部到上部，悬浮总体的含量逐渐增多。

顶部为块状-水平层理的粉砂、粉质黏土，平均粒径为 6.70~7.70Φ（平均 7.22Φ）。标准偏差为 1.77~2.12Φ（平均 1.96Φ），峰度系数为 0.80~1.02（平均 0.86Φ），频率分布曲线为近于正态分布的多峰状（图 23-5）。旋回顶部洪泛平原相细粒越岸沉积物的粒度特征反映出物质来源是多源的，在注入河流混合后经历了较远距离的搬运与改造，沉积物的分选性也较本阶段早期有所提高。

中更新统与下更新统之间可能具有较长时期的沉积间断，浦东机场 J25 孔仅保留了中更新世早期的沉积，岩性为河漫滩—洪泛平原环境形成的具水平层理、波状层理的粉砂、黏土。平均粒径为 4.84~7.67Φ（平均 6.75Φ），标准偏差为 1.71~2.24Φ（平均 1.97Φ），峰度系数为 0.81~1.11（平均 0.95），频率分布曲线为正态分布的单峰或相距较窄的双峰，反映了较好的分选性。从岩石地层、沉积物粒度特征看，中更新世沉积环境仍为曲流河环境，但河床的坡降趋缓，沉积物搬运距离趋远，分选性也趋好。

综上所述，本阶段为曲流河沉积环境，其早期河床的坡降相对较大，为河水深度较浅、水动力强度较大的低弯度曲流河环境；而晚期沉积旋回厚度增大，沉积物分选性变好，表明河流深度加大，曲流河体系发育更加成熟。

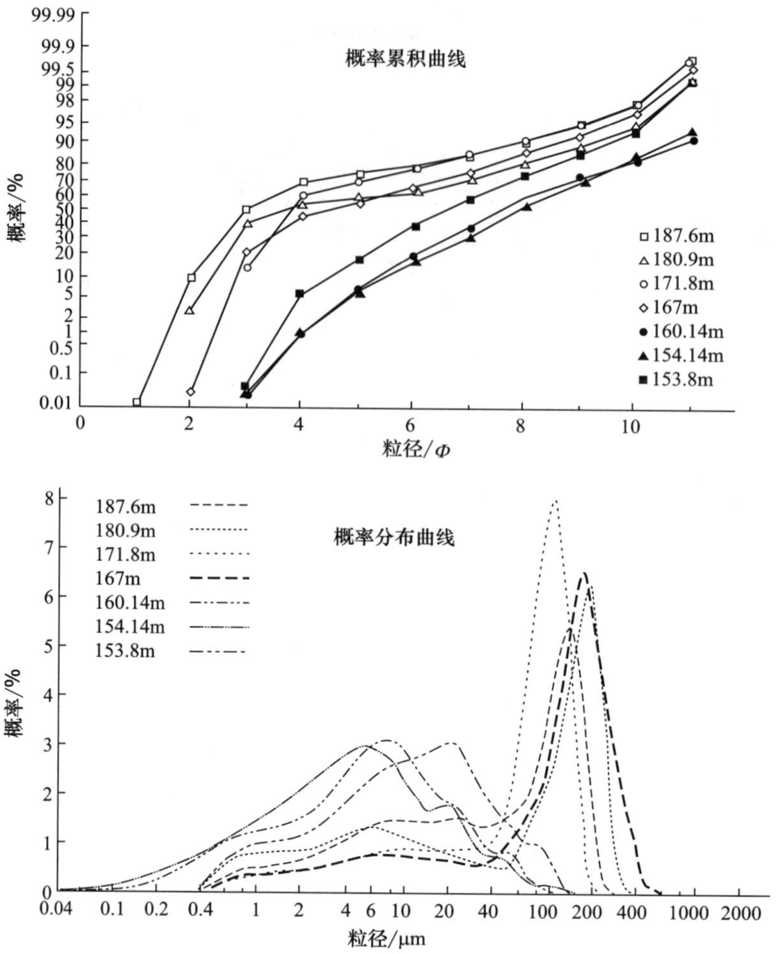

图 23-5　浦东机场 J25 孔第 2 阶段晚期沉积物粒度特征

Fig. 23-5　Grain size distribution of the late stage 2 sediments per borehole J25 at Pudong Airport

三、第 3 阶段：受海水影响的河流环境

晚更新世早中期地层（孔深 129.65~70.71m）由 3 个沉积物粒度向上变细的旋回组成，沉积物粒度表现为向上水动力逐渐变小的河流沉积特征。

下部旋回（孔深 129.65~90.50m）底部为典型的河床相含砾中粗砂，砾呈次圆—扁圆状，较前两阶段砾石的磨圆度提高；向上逐渐变为粉细砂与具水平层理、微斜层理的粉砂。河道沉积物粒度频率分布曲线为带细粒尾的单峰状，平均粒径为 0.67~5.20Φ（平均 4.34Φ），标准偏差为 1.62~2.83Φ（平均 2.48Φ），峰度系数为 0.54~1.60（平均 0.94）。其粒度分布在概率累积曲线图上显示两个次总体，个别为 3 个次总体，以跃移次总体为主（占 75%~90%），中等斜率，分选中等；悬浮次总体占 10%~25%，具低斜率，分选性差。个别出现 3 个次总体者含约 1%的滚动组分。细截点为 1.5~3.0Φ，显示河流水动力条件较强。

中部旋回（孔深 90.50~85.40m）厚度较小，沉积旋回保留不完整。上部旋回（孔深 85.40~70.71m）的底部为灰黄色砾砂，砾石呈棱角—次棱角状，分选差，为典型的河床相沉积。中—上部为河流浅滩沉积的灰色粉砂夹粉质黏土两者的互层，间或夹河床相沉积的灰黄色砾砂，反映当时河流具一定的游荡性。本旋回上部沉积物的平均粒径为 4.39~7.47Φ（平均 6.17Φ），总体呈现向旋回上部粒径逐渐减小的特点；标准偏差为 1.84~2.93Φ（平均 2.31Φ），分选较差—差。频率曲线明显分为两组：中部沉积物为多峰状的正态曲线，沉积物粒度较粗；上部沉积物为带有粗粒长尾的尖峰状曲线。

概率累积曲线表现为两段式向一段式渐变的特点，中部沉积物在概率累积曲线图上显示有两个次总体，悬浮次总体占 60%~80%，跃移次体仅占 20%~40%，细截点为 3.0Φ；上部沉积物概率曲线呈一段式，无跃移次总体（图 23-6）。

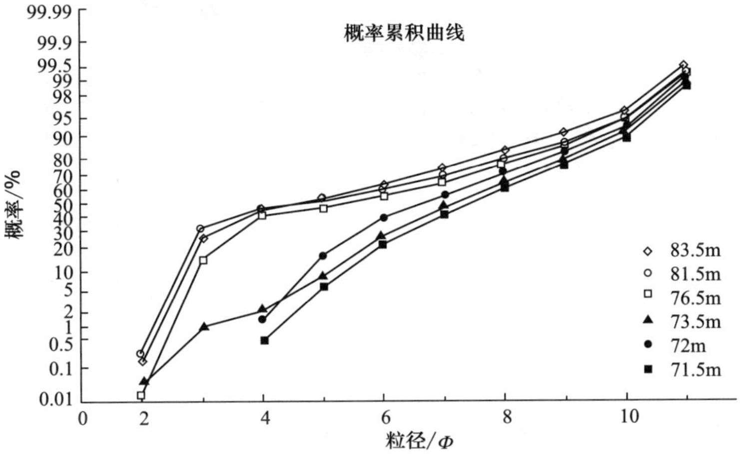

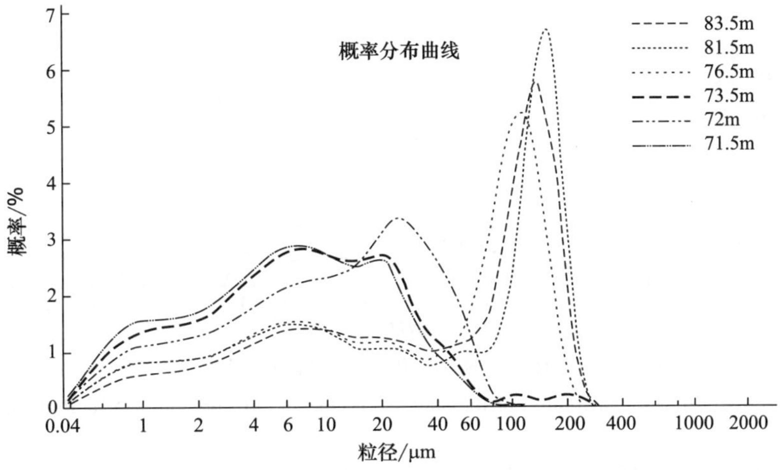

图 23-6　浦东机场 J25 孔第 3 阶段沉积物粒度特征

Fig. 23-6　Grain size distribution of the stage 3 sediments per borehole J25 at Pudong Airport

从上述 3 个沉积旋回沉积物粒度特征分析，河流环境的水动力强度由下而上有逐渐减弱的趋势。在孔深 103.50~92.60m 之间零星分布广盐性有孔虫和半咸水介形虫化石（详见第四、五章），表明晚更新世早期为间或受海水影响的河流环境。

四、第 4 阶段：海岸环境

晚更新世晚期沉积（孔深 70.71~29.4m）表现为一个正旋回特征：底部为灰黄色含砾中细砂，中部以粉砂、砂质粉土为主，至顶部为黏质粉土（图 23-7）。

中下部沉积物平均粒径为 3.72~5.58Φ，中值粒径在 2.26~5.28Φ 之间，标准偏差为 1.80~2.93Φ（平均 2.33Φ），分选程度差。概率累积曲线具有双跃移组分，其跃移组分含量占 90%~95%，悬浮组分占 5%~10%，滚动组分缺失。悬浮与跃移截点在 10.0Φ 左右，双跳跃组分截点在 3~4Φ 之间（见图 23-7）。与典型海滩砂样品对比，其悬浮组分略有增加，反映当时的沉积环境是远离剥蚀区的滨海沙滩，水动力条件较弱。

旋回顶部的沉积物粒度明显减小，平均粒径为 4.69~6.94Φ，中值粒径在 4.11~6.77Φ 之间，标准偏差为 1.85~2.13Φ（平均 1.97Φ），分选程度较差。概率累积曲线为一段式，全为悬浮组分，表现为洪泛平原的沉积特征。

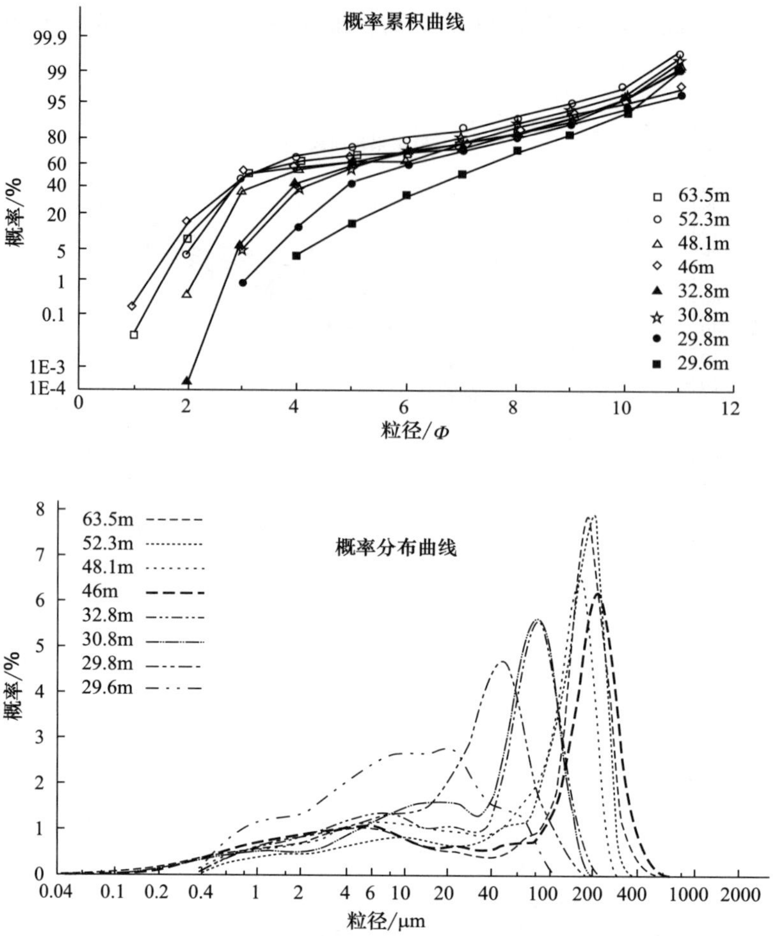

图 23-7 浦东机场 J25 孔第 4 阶段沉积物粒度特征

Fig. 23-7 Grain size distribution of the stage 4 sediments per borehole J25 at Pudong Airport

总之，本阶段早期为河口沉积，之后海平面上升演化为滨浅海，形成海滩砂沉积，晚期随海平面下降而演变为陆相洪泛平原。

五、第 5 阶段：三角洲环境

该阶段为全新世沉积（孔深 29.4~0.6m，0.6m 以上为填土）。按沉积物粒度特征可分为两小段：下段（孔深 29.40~10.44m）平均粒径为 6.11~8.01Φ（平均 7.22Φ），黏粒、粉粒及砂的平均含量分别为 33.11%、63.99%、2.90%，按粒度成分定为黏土，按谢帕德分类为黏土质粉砂，标准偏差为 1.62~2.12Φ（平均 1.83Φ），峰度系数为 0.84~1.10（平均 0.90）；上段（孔深 10.44~0.6m）平均粒径为 6.77~8.45Φ（平均 6.70Φ），黏粒、粉粒及砂的平均含量分别为 19.44%、73.98%、6.58%，按粒度成分定为粉质黏土，按谢帕德分类为粉砂，标准偏差为 1.77~2.12Φ（平均 1.96Φ），峰度系数为 0.80~1.02（平均 0.86）（图 23-8）。

全新统在粒度组成上呈现下细、上粗，分选性略有提高的特点（见图 23-8）。沉积物粒度频率分布曲线特征也具有明显的差异：下段沉积物频率曲线为扁平的多峰状，上段则为在细、粗端各有弱峰出现的正态单峰分布特征，反映全新世早、晚期沉积的搬运营力及物质来源的变化。晚期分选较好的粗颗粒组分的出现，可能反映长江带来物质对本区沉积物的影响，从而塑造了现今的长江三角洲。

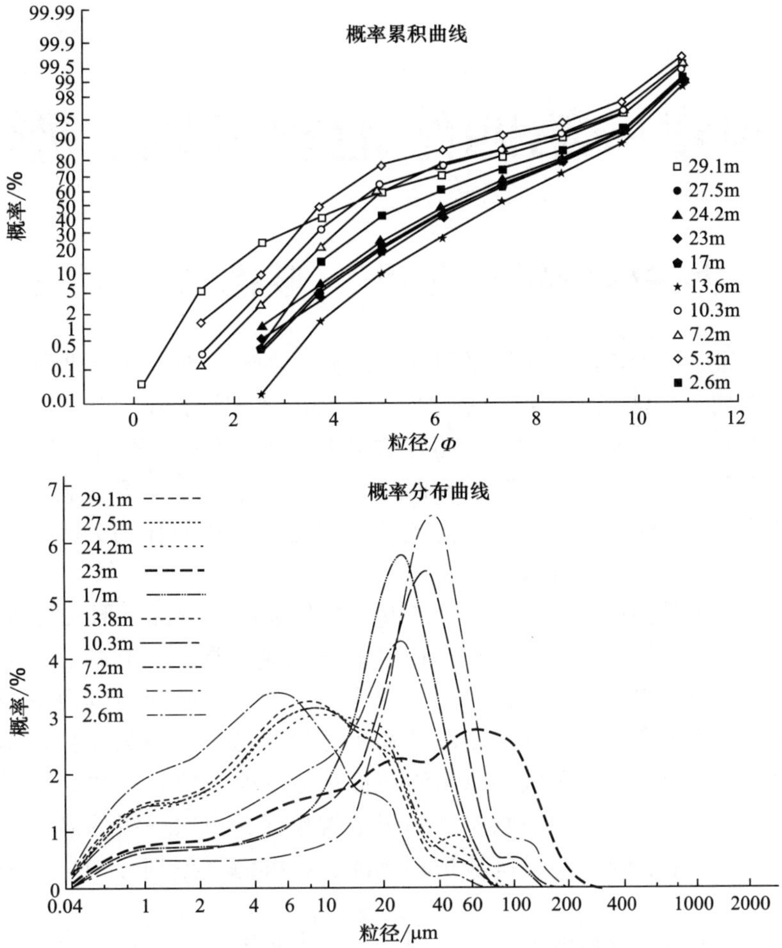

图 23-8　浦东机场 J25 孔第 5 阶段沉积物粒度特征

Fig. 23-8　Grain size distribution of the stage 5 sediments per borehole J25 at Pudong Airport

第二十四章　第四纪地貌成因、形态类型

第一节　第四纪地貌成因、形态类型划分及简述

上海位于长江口，陆域属于长江三角洲平原的一部分，主要位于南翼，地势平坦，海拔一般在2~4m。上海地貌单元主要属于长江三角洲堆积平原区，西部跨太湖堆积平原区，南侧滨杭州湾部分为接钱塘江堆积平原的杭州湾滨海平原区。此外，还有散布于西部的松江、青浦、金山的10多座剥蚀残丘和岛屿。区内的太湖堆积平原区与长江三角洲堆积平原区的分界在江苏太仓—上海外冈—徐泾—邬桥—漕泾一线，即以长江三角洲平原贝壳沙堤构成的“冈身带”西侧作为分界。两者的次一级地貌单元可按地貌的成因与形态差异进一步分级。

据《中国区域地质志工作指南》，本区一、二级地貌分别属于东部平原和中低山、丘陵区（Ⅲ）、长白山山区和山东低山区、丘陵区（$Ⅲ_2$）。根据地貌形态和成因类型，上海陆域进一步可划分为4个三级地貌单元，分别是太湖堆积平原区、长江三角洲堆积平原区、杭州湾滨海平原区、剥蚀残丘和岛屿区（表24-1）。太湖堆积平原区和长江三角洲堆积平原区是区内陆域最主要的地貌类型，其中，太湖堆积平原区分为潟湖积平原、湖沼积低平原和沼积低平原3个四级地貌单元；长江三角洲堆积平原区分为河口沙岛、滨海平原及河流冲积平原3个四级地貌单元，更次一级的微地貌单元可按地貌形成过程将河口沙岛划分为早、中、晚3期河口沙坝和潮滩，将滨海平原划分为早、中、晚3期滨海平原和潮坪。

上海近岸水域可划分为长江三角洲水下三角洲一个三级地貌单元，根据沉积特征，可进一步区分为三角洲前缘区和前三角洲区2个四级地貌单元。

表24-1　上海地貌成因类型划分

Table 24-1　Division of the geomorphologic types in Shanghai

一级	二级	三级	四级	微地貌
东部平原和中低山、丘陵区（Ⅲ）	长白山山区和山东低山区、丘陵区（$Ⅲ_2$）	长江三角洲堆积平原区	河口沙岛	早、中、晚3期河口沙坝和潮滩
			滨海平原	早、中、晚3期滨海平原和潮坪
			河流冲积平原	
		太湖堆积平原区	湖沼积低平原	
			沼积低平原	
			潟湖平原	
		杭州湾滨海平原区		
		剥蚀残丘和岛屿区		
		水下三角洲区	三角洲前缘	分汊河道、拦门沙
			前三角洲	

第二节　第四纪地貌成因、形态类型分布规律及地貌区划

一、陆域地貌单元划分及主要特征

（一）太湖堆积平原区

太湖堆积平原区以江苏太仓—上海外冈—徐泾—邬桥—漕泾一线，即以贝壳沙堤构成的“冈身带”西侧作为分界，西部为湖积平原，属太湖平原东缘。浅表岩性以褐黄色黏土质粉砂为主，局部见灰黑、灰绿色粉砂质黏土夹泥炭，见淡水腹足类壳体及淡水—半咸水介形虫及少量广盐类有孔虫。更次一级的地貌单元可划分为潟湖平原和湖积高地、湖积低地和湖泊洼地（图 24-1；表 24-2）。

（二）长江三角洲堆积平原区

1. 河口沙岛

河口沙岛指三角洲主体部分形成的三角洲平原，3 个河口沙岛位于崇明、横沙、长兴。浅表岩性以潮上带的褐黄色含黏土粗粉砂、黏土质粉砂为主，含淡水腹足类壳体，下伏高潮坪沉积的深灰、褐黄色含黏土粗粉砂、含砂粗粉砂，含半咸水、淡水腹足类幼体壳体及芦苇根茎、钙质结核。按地貌形成过程进一步划分为不同期次的更次一级的地貌单元：将距今约 1400 年以来形成的三角洲河口沙坝、汊道划分为老、中、新 3 期。

2. 滨海平原

滨海平原以江苏太仓—上海外冈—徐泾—邬桥—漕泾一线，即以贝壳沙堤构成的“冈身带”西侧作为分界，东部至河口沙岛之间的范围为滨海平原。浅表岩性以褐黄色含黏土粉砂为主，见淡水腹足类壳体及淡水—半咸水介形虫，嵌布多列贝壳砂带，局部河道发育区粗粉砂含量较高。按地貌形成过程进一步划分为不同期次更次一级的地貌单元：将距今约 6500 年以来形成的三角洲滨海平原划分为古、早、中、晚、新等多期。

3. 河流冲积平原

河流冲积平原指吴淞江、黄浦江两岸，海拔为 1.8～2.9m，沿吴淞江西高东低，市区最低，龙华—得胜港较高，0.5m 以浅为泛滥平原沉积褐黄色含黏土粗粉砂，以下为河床沉积褐黄—褐灰色含砂粗粉砂，吴淞口凸岸为浅滩沉积砂质粗粉砂。

（三）杭州湾滨海平原区

杭州湾滨海平原区指金山、奉贤滨杭州湾岸带，海拔为 2.3～3.1m，向其西北缓倾斜。发育主要为与杭州湾潮流有关的潮上带沉积，由东往西海相性减弱，介于长江三角洲中、晚期滨海平原之间。地表沉积物以黄灰色含黏土粗粉砂为主，含较多广盐性有孔虫，有半咸水、淡水介形虫。

（四）剥蚀残丘和岛屿

剥蚀残丘和岛屿指散布于青浦、松江、金山的佘山、天马山、秦皇山等 17 座火山岩剥蚀残丘以及分布在杭州湾内的大、小金山等，海拔多不足百米，最低仅二三十米。这些残丘岩石覆盖砾质黏土或含砾黏土残积物，陆域内残丘群呈嵌布于平原上的孤丘，海域内残丘可见海蚀崖、初成海蚀柱及砾质滩脊。经过百年采矿，陆域内罗山、甸山、辰山等残丘已遭受破坏而采空夷平或成为矿坑，目前这些矿坑已得到综合治理而成为区内富有特色的植物园、旅游区。

（五）特殊类型

太湖堆积平原与三角洲堆积平原区内的滨海平原交界处以及滨海平原内散布大致 4～5 列古贝壳砂堤，崇明岛东部散布 1～2 列古贝壳砂堤。典型沉积特征为砂中混杂大量完整贝壳和贝壳碎屑，部分地段仅为砂堤，典型地貌特征为砂堤明显高于两侧 1～2m。^{14}C 测年值显示这些砂堤形成于 7000a 以来，其中形成于 7～3.5ka 的 3 列古砂堤集中在湖积平原和滨海平原过渡区域较窄的范围内，称之为“冈身

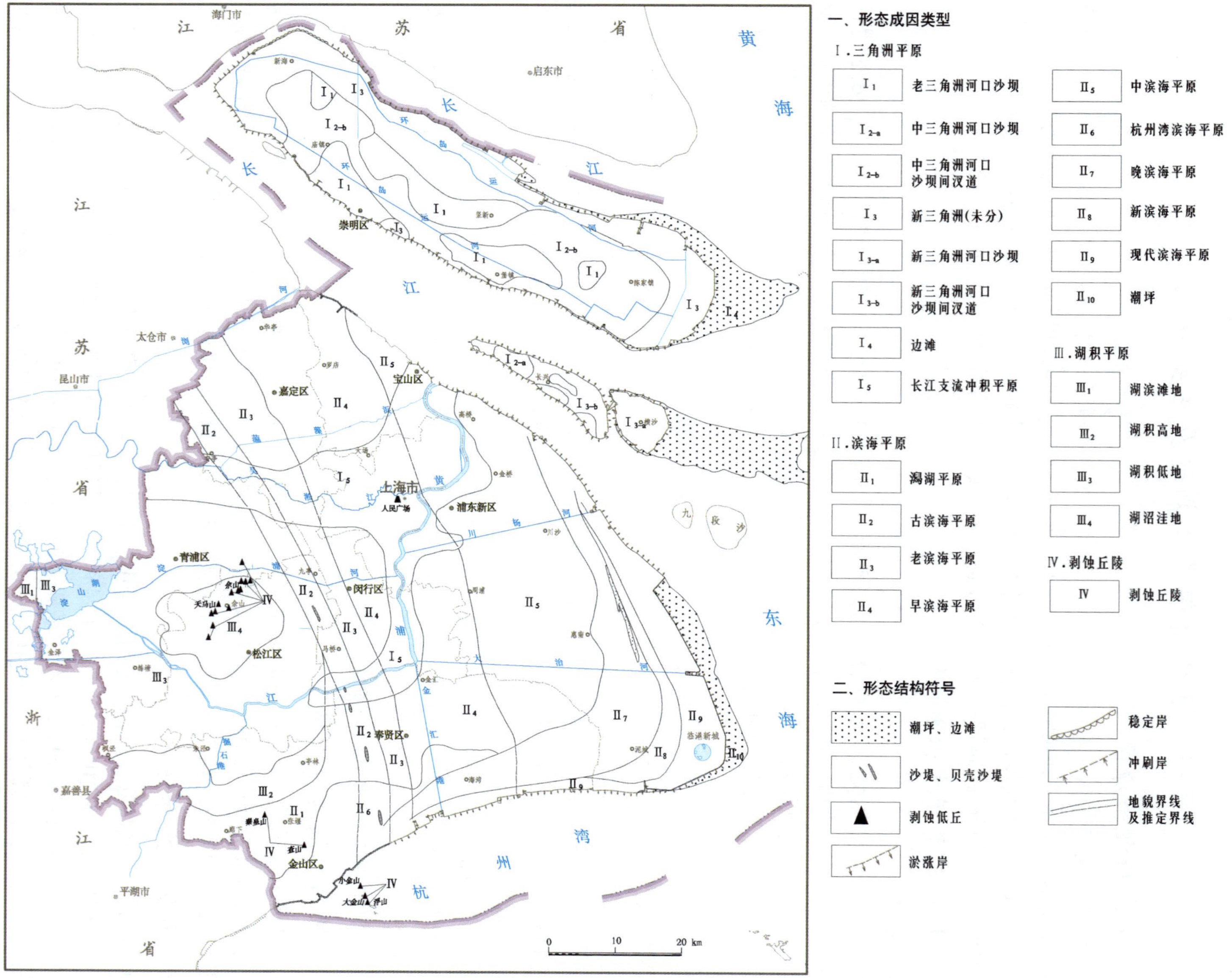

图 24-1 上海陆域地貌类型图

Fig. 24-1 Geomorphologic types in Shanghai

表 24-2　上海陆域地貌类型特征表

Table 24-2　Characteristics of the geomorphologic types in Shanghai

地貌类型		分布地区	地形特征	沉积物特征	其他特征	成陆时间
太湖堆积平原区	潟湖平原	金山张堰、烟墩头—黄浦江附近	地势高亢，海拔（黄海高程，下同）大都为2.5～3m	以褐黄色黏土质粉砂为主，见中华拟蟹守螺、近美丽土星介-布氏土星介，少量毕克卷转虫、平阳卷转虫	张堰以西、干巷、廊下地区地势较低，海拔在2m左右	
太湖堆积平原区	湖沼积低平原	松江、青浦和金山北部	地势低平，河渠、湖荡众多，海拔≤1.5～2.5m，湖滨滩地地势偏高	以褐黄、棕灰色黏土质粉砂、含黏土粗粉砂为主，局部夹碳质黏土、泥炭层或腐烂植物，产少量环棱螺、河兰蚬、珍珠蚌	湖滨滩地沉积砂质粉砂，古松、新五库等地为古泖河低洼带	泥炭，^{14}C定年值为1500～3000a B. P.
太湖堆积平原区	沼积低平原	松江、青浦之间	地势低洼，湖泊、洼淀密布，海拔一般在1.2～1.3m	灰黑、灰绿色粉砂质黏土夹泥炭	此区呈浑圆形洼地，为上海泥炭主要产区	泥炭，^{14}C定年值约为1500a B. P.
长江三角洲堆积平原区 滨海平原	早期滨海平原	嘉定镇—闵行莘庄—奉贤南桥以西的“冈身带”，漕泾贝壳沙堤仍高出附近地表1m	地势高亢，海拔大都为2.5～2.8m，“冈身带”平原梳状水系发育	地表褐黄色含黏土粗粉砂，下为深灰、褐黄色黏土质粉砂，产环棱等淡水腹足类壳体、广盐性和淡水—半咸水介形虫，嵌布多列贝壳砂带	北宽南窄，宽在1～5km。沉积层出现为粗粉砂充填的泥裂、较多芦苇根茎及生物扰动构造，显示为潮上带沉积	西界约6500a B. P.，东界约3000a B. P.
长江三角洲堆积平原区 滨海平原	中期滨海平原	浦东顾路—大团—四团—奉城一线以西地区	地势起伏，海拔一般在2.3～2.8m，嘉定东南部为滨海洼地，海拔多为1.5～2m	以褐黄、灰黄色黏土质粉砂为主，外高桥—川杨河及娄塘—曹行以北地区分别受古吴淞江及古浏河影响，多为含黏土粗粉砂	分布宽度在20～32km。沿盛桥—月浦—北蔡—下沙的沙带为1700a B. P.古海岸线所在	西界约3000a B. P.，东界里护塘约800a B. P.
长江三角洲堆积平原区 滨海平原	晚期滨海平原	1949年以来建的人民塘、八五塘、金汇塘等现代岸堤内	海拔大体在2.2～2.9m，呈向外倾斜，在盐仓—马厂—万祥—奉城一带以西地区为池塘密布的蜂窝状水系滞水区	以灰、浅褐黄色含黏土粗粉砂为主，产半咸水、淡水腹足类，遍有芦苇根茎，显为潮上带沉积。蜂窝状水系带地表下堆积富植物根茎的黑色腐泥，说明为沼泽洼地。沙带为浅灰色细砂、含粉砂细砂，含雕刻友螺、樱蛤等海相软体介壳，为古海岸砂堤遗迹	在南汇嘴突出，最宽处有17km，向北延至外高桥，向西延至漕泾。西侧钦公塘、东部的潘泓有沙堤分布，后者^{14}C定年值约为600a B. P.	约800a B. P.至今
长江三角洲堆积平原区 滨海平原	潮坪	现岸堤以外潮间滩地	高潮滩分布于近海岸带，多被围垦，中潮滩被涨潮淹没		高潮坪多围垦种植芦苇，中潮坪生长藨草和芦苇	

续表

地貌类型			分布地区	地形特征	沉积物特征	其他特征	成陆时间
长江三角洲堆积平原区	河口沙岛	早期河口沙坝	崇明堡镇、新民—港西、城关、海桥、汲浜等地	海拔为2.2~2.9m，地势略高	0.5m以浅为潮上带的褐黄色含黏土粗粉砂、黏土质粉砂，含淡水腹足类壳体，下伏高潮坪沉积的深灰、褐黄色含黏土粗粉砂、含砂粗粉砂，含半咸水、淡水腹足类幼体壳体及芦苇根茎、钙质结核	深0.2~0.5m为褐黄色黏土质粉砂，土质普遍较硬，含瓷器碎片等，以上为含砂量增高的含黏土粗粉砂—耕作层，说明经历成陆—淹没—再成陆的过程	在1400~400a B. P. 形成古沙洲
		中期河口沙坝	崇明岛中南部大部、长兴岛诸古沙洲	与上述相近，在崇明岛沿其分布地带地势偏低	褐黄、棕灰色含黏土粗粉砂，含淡水腹足类幼体介壳较多和有机质较高，为潮上沼泽沉积；崇明东端新桥—牛棚有一宽一二十米的弧形沙带，底部见较多下伏层碎块，为风暴潮在潮上带塑造的滨岸沙堤	长兴古沙洲形成于100~>200a B. P.	约350a B. P. 崇明中南部连片成陆
		晚期河口沙坝	环绕崇明岛西、北、东三侧，长兴岛大部和横沙岛	多辟为农场和水产养殖场，地势偏低	大都为轻盐化含黏土粗粉砂，崇明东北端和横沙外环为盐渍化含砂粗粉砂	横沙出露于1860a A. D.，嗣后近百年向西北迁移达10km	近半个多世纪
		潮滩	以崇明岛东北侧、横沙岛东侧为主	现塘堤外侧，潮滩分带明显，潮沟发育		高潮滩为大片芦苇丛生带，中潮滩多藨草等沼生草本	
	河流冲积平原		吴淞江、黄浦江两岸	海拔为1.8~2.9m，沿吴淞江西高东低，市区最低，龙华—得胜港较高	0.5m以浅为泛滥平原沉积褐黄色含黏土粗粉砂，以下为河床沉积褐黄—褐灰色含砂粗粉砂，吴淞口凸岸为浅滩沉积砂质粗粉砂	吴淞江沉积物较粗，为粗粉砂，黄浦江沉积物略细，为含黏土粗粉砂	
杭州湾滨海平原			金山、奉贤滨杭州湾岸带	海拔为2.3~3.1m，向其西北缓倾斜，地势较高亢	黄灰色含黏土粗粉砂，较多广盐性有孔虫，有半咸水、淡水介形虫，由东往西海相性减弱，为潮流与钱塘江径流交汇堆积，与钱塘江水下沙坝范围一致	发育格状、树枝状水系，主要为与杭州湾潮流有关的潮上带沉积	介于长江三角洲中、晚期滨海平原之间
剥蚀残丘和岛屿			佘山、天马山、秦皇山及大、小金山	剥蚀残丘海拔为23.2~99.8m，大、小金山海拔为32.4~105m，罗山、甸山已采空成为坑、塘	基岩为下白垩统火山碎屑岩，覆盖砾质黏土或含砾黏土残积物	前者呈嵌布于平原上的孤丘，后者东南侧可见海蚀崖、初成海蚀柱及砾质滩脊	

带”，滨海平原东部还散布一列砂堤，^{14}C测年值显示形成年代有900余年，崇明岛东部砂堤贝壳测年同样显示900余年。这些贝壳砂堤是区内全新世古海岸线的标志，见证了全新世最大海侵以来上海的成陆过程。目前，这些砂堤多数遭到破坏，仅在金山曹泾镇沙积村保留一处遗迹。

沿江沿海地带分布数列人工海塘，主要为20世纪60年代以来修建，体现了在人类活动干预下海岸线的变迁过程。建造时间最早可追溯至唐朝，目前仅清雍正时期海塘存有遗址。

二、近岸水域地貌单元划分及主要特征

上海近岸水域地貌类型以长江水下三角洲为主，具体类型划分为三角洲前缘、前三角洲、河口湾前缘、席状砂区、内陆架泥质区（图 24-2）。

（一）三角洲前缘

三角洲前缘分布于近岸 5~10m 以浅的水域范围，可分为分流河道和拦门沙两个地貌单元。

长江口自徐六泾向下，河道分汊，沙洲发育，滩槽相间，南北分支。北支河道沉积物较细，由粉砂和泥组成；南支河道沉积物明显变粗，由砂质粉砂和粉砂质砂组成，其中粉砂质砂主要分布在南支中上段河槽和南港、北港主槽。

拦门沙为长江口 10m 以浅水域堆积的浅滩，包括南汇东滩、九段沙、横沙东滩、崇明东滩、顾园沙。由于水深较浅，波浪和潮汐作用较强，沉积物经长期掀沙簸选，粒径较粗。沉积物主要由砂质粉砂和粉砂质砂组成，其中砂质粉砂主要分布在拦门沙浅滩向陆部分，粉砂质砂主要分布在拦门沙浅滩向海部分。

（二）前三角洲

前三角洲为三角洲前端部分，位于长江口外水深 10~50m 的海域，主要由长江悬浮泥沙堆积而成，表层沉积物以粉砂质黏土和黏土质粉砂为主。前三角洲的发育受科氏力影响而向东南方向延伸。由于夏季冲淡水强盛，水下三角洲泥质沉积可延伸至杭州湾北岸的大、小金山附近，水下三角洲前端可延伸至口外东南 50m 深的水域。前三角洲海域零星分布数个火山岩残丘岛屿。

（三）河口湾前缘

河口湾前缘大致位于大、小洋山至大、小金山一线以西的杭州湾水域，潮流作用强烈，表层沉积物以砂质粉砂、粉砂为主。

（四）残留砂区

位于与前三角洲相邻的扬子浅滩内，表层沉积物以细砂为主，多含贝壳碎片。扬子浅滩主要为形成于晚更新世末—全新世早期海平面快速上升阶段的潮成砂，距今 7000 年以来该区域受旋转潮流改造，形成了准活动性席状潮成砂。

（五）内陆架泥质区

内陆架泥质区位于水下三角洲、河口湾和残留砂以外的内陆架区域，由于受沿岸流携带悬浮泥沙的作用，表层沉积物以粉砂质黏土、黏土质粉砂为主。

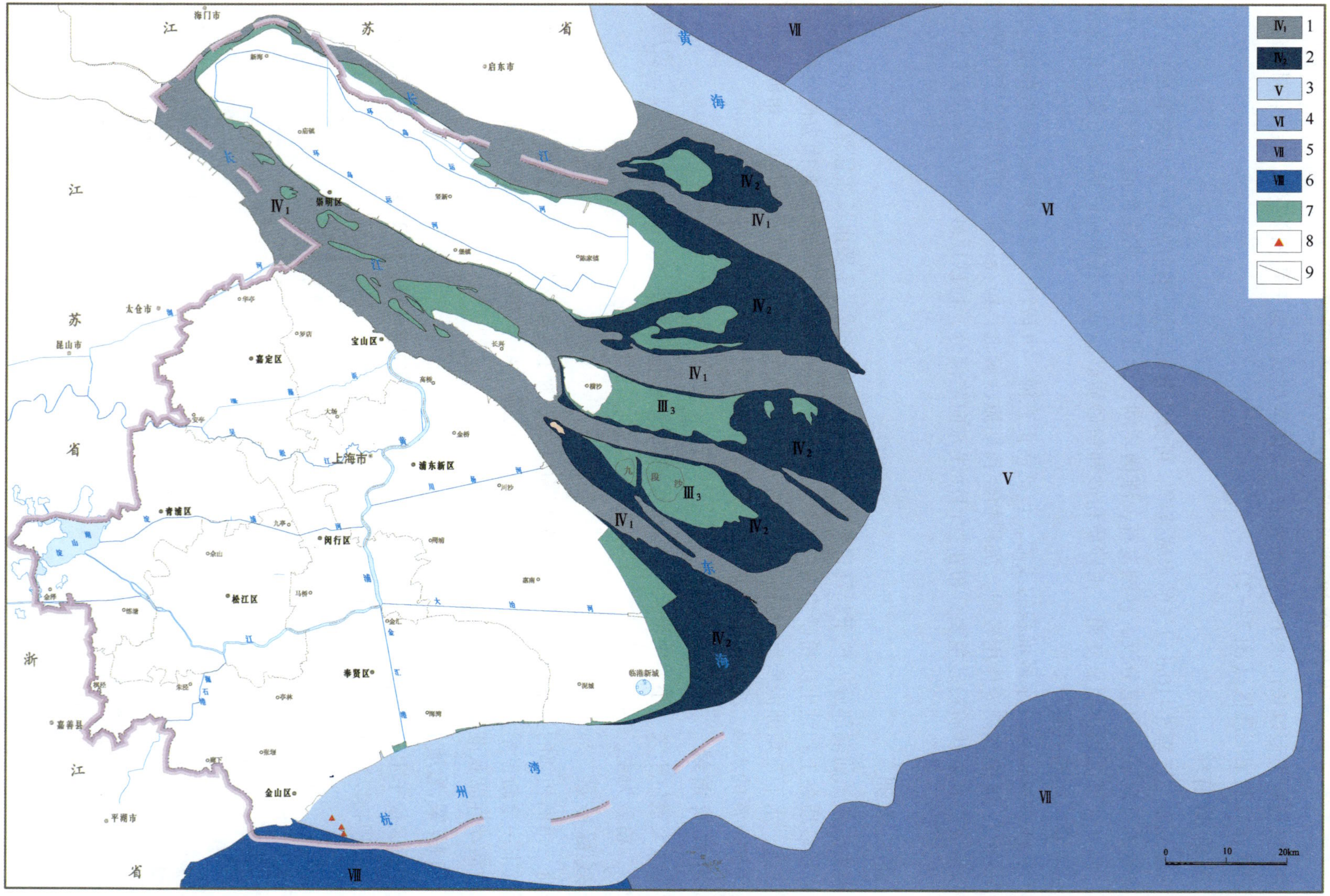

图 24-2 上海地势及近岸水域地貌分区

Fig. 24-2 Terrains and geomorphologic division of nearshore waters in Shanghai

1—三角洲前缘分流河道；2—三角洲前缘拦门沙；3—前三角洲；4—席状沙；5—内陆架；6—河口湾前缘；7—围垦边滩（0m 线以上）；8—剥蚀低丘；9—地貌界线

第二十五章 第四纪海侵海退事件

第一节 海侵期次划分

依据沉积物中微体古生物化石、软体动物化石分布，上海第四纪存在 7 次海侵事件，可与长江三角洲地区第四纪海侵对比（吴标云和李从先，1989；Lin et al.，1989）。

一、如皋海侵——第Ⅶ海侵

如皋海侵为江苏第一水文地质队（1987）根据苏北如皋及南通、启东等地多个钻孔中具河口相有孔虫化石的发现而命名的，海侵层为深逾 220m 的砂层。

上海如皋海侵发现于浦东滨岸地带，海侵层埋深在 200m 以下，位于安亭组中段，含有少量广盐性有孔虫及海相腹足类化石，海侵出现在松山负极性时奥尔都维与留尼昂极性亚时之间，形成年代为 2.00~1.85Ma。这是区内第四纪出现的最早一次海侵踪迹，海侵强度体现为河口—三角洲河流环境（以受海水影响的河流为主），海侵特征可与苏北启东钻孔 220m 以下海侵层位对比。

海侵记录表现：浦东海塘 P6 孔孔深 206~226m 间灰色细砂夹含砾细中砂内所见，有轮虫类的 *Ammonia tepida*、*A. koeboeensis*，浮游类的 *Globigerina* spp.、*Epistominella naraensis*，平旋壳类的 *Nonion* sp.、*Cribrononion* sp.、*Elphidium* sp. 和 *Florilus* sp. 等，窄盐性成分 *Bolivina* sp. 等，伴有轮藻。高桥“长江一号”孔，在大致相当于安亭组中段下亚段、孔深 260m 处黄灰色中细砂内见海相腹足类雕刻友螺（*Nassarius*（*Phromtis*）*cealatula*）、囊螺（*Retusa*）、露齿螺（*Odostomia*）等及陆相腹足类沼螺（*Parafossarulus* sp.）。

二、周浦海侵——第Ⅵ海侵

周浦海侵为上海水文地质队 1984 年在《上海市 1：20 万水文地质普查报告》中提出，1988 年完成的《上海第四纪地层研究报告》中正式命名。这次海侵由南汇嘴入侵，其北端沿浏—南古河道达到浦东海塘—横沔、周浦附近，海侵踪迹在浦东的多个钻孔内发现，海侵层位属下更新统安亭组上段下部砂层，出现于松山负极性时前贾拉米洛极性事件与贾拉米洛极性亚时之间，形成的第Ⅵ海侵层的年代主要为 1.20~1.00Ma，海侵强度体现为河口—三角洲河流环境（以受海水影响的河流为主）。

海侵记录表现：横沔 P12 孔孔深 172~178m 灰黑色细砂夹砂质粉土内见具 *Ammonia beccarii* var. - *Epistominella naraensis* 有孔虫组合，并见有深水种 *Hopikinsina pacifica*、*Lagena* sp.、*Brizalina* sp. 等窄盐性成分，伴生半咸水介形虫 *Neomonoceratina dongtaiensis*、陆相介形虫 *Condoniella albicans* 等，以及水蕨孢子、轮藻受精卵膜等陆相生物成分。浦东海塘 P6 孔孔深 171~191m 灰、灰黄色细砂内见很少的有孔虫，有 *A. tepida*、*Elphidium magellanicum*、*Epistominella naraensis*、*Globigerinita glutinata*，并见及个别半咸水介形虫 *Cushimanidea japonica* 及陆相的轮藻。海潮 P17 孔孔深 175~177m 灰色细砂内分析出少量钙质超微化石 *Gephyrocapsa ocenica*，为上海钙质超微化石出现的最低层位。

此外，在安亭组上段上部以泥层为主的地层内存在零星海相化石。上海面粉厂 J10 孔孔深 148.10~159.29m 杂色黏土夹黄褐色细砂内发现少量有孔虫，有螺旋壳类 *Ammonia* sp.、平旋壳类 *Elphidium* sp. 和 *Florilus decorus* 等广盐性有孔虫；浦东海塘 P6 孔孔深 146~164m、横沔 P12 孔孔深 149~174m、安亭 G414 孔孔深 158.2m 处等的杂、褐黄色黏土内断续分析出很少的有孔虫，三灶 P45 孔孔深 151~153m 褐色粉质黏土内偶见钙质超微化石 *Gephyrocapsa ocenica* 等。这些现象表明早更新世末本区转为陆相河湖环境，也还偶尔受到这次海侵后期的微弱影响。

三、上海海侵——第Ⅴ海侵层

上海海侵为江苏第一水文地质队（1987）命名。这次海侵在上海市区、浦东、嘉定和西南部的青浦、金山等地被广泛发现。

海侵层位属中更新统嘉定组下段下亚段，通常赋存在灰色细砂层内，出现于布容正极性时下界（0.78Ma），终止于Delta（约0.69Ma）与Big lost Emperor（约0.554Ma）两极性飘移事件之间，形成年代为0.78~0.65Ma。海侵程度为偶尔受强潮流影响的三角洲平原河湖、短暂海湾环境。

海侵记录表现：在东部长江三角洲沉积区，上海面粉厂J10孔孔深133.2~159.5m青灰色细砂及下伏的黄褐色粉质黏土夹细砂层内见少量有孔虫，有 *Ammonia* sp.、*Elphidium* sp. 和 *Florirus decorlus*，与陆相介形虫 *Ilyocypris* sp.、*Candoniella* sp.、*Lineocypris* sp. 共生。嘉定娄塘P5孔孔深141~145m灰色粉细砂夹砂质粉土中含很少的有孔虫，有 *Ammonia beccarii* var.、*Rosalina bradyi*、*Pseudononionella variabilis* 等。浦东海塘P6孔孔深123~135m黄灰色中细砂夹灰色粉质黏土中含少量有孔虫 *A. tepida*、*Globigerinita* sp. 等。又如三灶P45孔孔深123.8~145m灰色细砂、灰黄色含砾砂中含少至很少的有孔虫，有 *Ammonia beccarii* var.、*Nonion* sp.，还有 *Globigerina bulloides*、*Elphidiella kiangsuensis* 等，并偶见钙质超微化石 *Gephyrocapsa oceanica*。海潮P17孔孔深108~137m灰色细砂和含砾中细砂中与横沔P12孔孔深121~149m灰黑、灰色细中砂中分别发现少量钙质超微化石与再沉积的有孔虫。上述记录表明长江三角洲沉积区较粗陆源碎屑沉积物中常见少量半咸水有孔虫，或伴有陆相介形虫与钙质超微化石，表明为偶尔受强潮流影响的三角洲平原河流沉积。

在西部太湖沉积区，一般见淡水介形虫，只出现个别的有孔虫，唯青浦练塘P31孔孔深133~135m浅黄灰色砂质粉土夹粉细砂内出现结缘寺卷转虫角状亚种（*Ammonia ketienziensis angulata*）-毕克卷转虫（*Ammonia beccarii* var.）有孔虫组合，前一属种分布于现南黄海水深50m左右处水域，并见 *Protelphidium tuberculatum*.、*Buccela frigida* 等示冷性有孔虫，复合分异度H（s）达2.92，与之共生半咸水属种斑纹三原介（*Sanyuania psaronion*）和常在具一定盐度水体中生活的陆相介形虫 *Limnocythere* sp.、*Candoniella* sp.。在金山朱泾G418孔孔深132.2~139.5m黄褐色黏土内也见及个别有孔虫。表明太湖沉积区通常为略受海水影响的三角洲平原河湖，可能一度沦为开放性海湾环境。

四、嘉定海侵——第Ⅳ海侵层

嘉定海侵为上海水文地质队1984年提出，1988年完成的《上海第四纪地层研究报告》中命名。这次海侵踪迹主要保存在太湖沉积区，其海侵层位属于中更新统嘉定组上段下部，而滨海平原东部滨岸地带和河口沙岛区则因多被侵蚀缺失而不明。依据光释光定年值，海侵层形成年代为0.26~0.20Ma。海侵程度为三角洲平原河流环境。

海侵记录表现：青浦练塘P31孔孔深93~107m灰绿、棕灰色粉质黏土和青灰色细砂内分析出 *Ammonia beccarii* var. -*A. convexidorsa* 有孔虫组合，还见有 *Epistominella naraensis*、*Pseudoeponides andersen*、*Nonion shansiensis*、*Cribrononion subincertum*、*C. vitreum* 等有孔虫，伴生半咸水的 *Sinocytheridea* spp. 组合，还见有 *Leguminocythereis* sp.、*Loxoconcha* sp. 等半咸水属种，并见 *Ilyocypris errabundis*、*Candoniella* sp.、*Candona* sp. 等陆相介形虫，偶尔见及钙质超微化石 *Gephyrocapsa oceanica*。金山朱泾G418孔孔深108~112m灰色粉质黏土内分析出陆相介形虫 *Ilyocypris* sp.、*Candona*（*Candona*），见海相介形虫 *Echinocythereisbradei*，还见淡水软体动物化石丽蚌等碎片。这表明当时通常为河流沉积，偶尔受到海水的影响，在太湖沉积区局部出现与现今长江三角洲前缘支流河道相同的有孔虫组合，显示属于三角洲河流堆积。在三角洲沉积区唯市区J10孔孔深115.19~118.41m灰、黄褐色粉细砂内检出少量有孔虫，其上亚段灰、蓝灰色粉质黏土则未发现微体古生物，仅见及陆相软体动物方形环棱螺相似种（*Gyraulus albus*）等，表明为有时受海侵影响的河流演变为泛滥平原—湖泊洼地环境。

五、太湖海侵——第Ⅲ海侵层

太湖海侵为江苏第一水文地质队（1987）命名，认为这次海侵达到了太湖的西界。这次海侵除崇明等三岛、宝山和嘉定以外，在上海境内大都有所分布。其海侵层位属于上更新统川沙组，光释光定年值显示形成年代为0.07~0.128Ma，海侵程度为间或受到海水影响的三角洲河流，偶尔在局部可达三角洲前缘的河口环境。第Ⅲ海侵层虽然在上海境内分布较广，但是仍同其前数个海侵层一样，属于海相性较弱的海侵层。

海侵记录表现：太湖沉积区偏南部地区的海相性略强。如青浦练塘P31孔孔深73~93m灰、灰绿色粉质黏土夹灰色粉砂层含*Ammonia beccarii* var. -*A. convexidorsa*有孔虫组合，还见*Epistominella naraensis*、*Pseudoeponides andersen*、*Nonion shansiensis*、*Cribrononion subincertum*、*C. vitreum*等有孔虫，伴生*Sinocytheridea* spp.、*Leguminocythereis* sp.、*Loxoconcha* sp.等半咸水属种和*Ilyocypris errabundis*、*Candoniella* sp.、*Candona* sp.等陆相介形虫以及圆盘形硅藻、刺盒虫等陆相化石，反映为陆上三角洲河流环境。金山朱泾G418孔此层内断续出现半咸水介形虫和钙质超微化石，其下部（孔深85.2~92.6m）的灰褐色黏土内主要含*Sinocytheridea* sp.、*Echinocythereis* sp.、*Neomonoceratina* sp.，并含*Leguminocythereis hudgii*；上部（孔深67~71m）的灰色粉质黏土、粉砂互层含少量*Sinocytheridea* sp.；中部灰色粉质黏土、粉砂互层含少量的钙质超微化石*Gephyrocapsa oceanica*及弱孔圆颗石（*Cyalococcolithus leptopora*），表明由下部的河口演变为中上部的滨海环境。而在太湖沉积区的青浦北部如青浦镇G415孔未见有孔虫，唯在嘉定安亭G414孔顶、底部的灰色含砾中细砂内检出少量有孔虫，说明两者均为河流沉积，只是后者偶尔受到海水的影响。

长江三角洲沉积区，此层位大都无或见少量有孔虫，反映通常为河流沉积环境，间或为受到海水影响的三角洲河流。局部含有孔虫较丰，具河口相组合成分，如市区的面粉厂J10孔在此层下部（孔深92~109.5m）的青灰色含砾细砂夹粉质黏土内含少量有孔虫，具*Ammonia beccarii* var. -*Epistominellanaraensis*组合，并含*Florirus decorlus*、*Protelphidium* sp.，伴生海胆刺、海绵骨针等海相化石，属于河口沉积；其上部（孔深78~92m）的含砾中细砂仅含少量陆相介形虫*Ilyocypris errabundis*、*Candoniellasulacensis*、*C. schubinae*，还含轮藻、鱼牙等陆相生物化石，表明已演变为河流沉积。再如横沔P12孔此层中下部（孔深93~125m）的灰色中细砂、粗砂砾石层仅含少量再沉积有孔虫；上部（孔深81~93m）含*Ammonia dominicana*-*Elphidium magellanicum*组合，并含*Epistominella naraensis*、*Globigerina quinqueloba*，表明由中下部的三角洲河流，演变为上部的河口沉积环境。如杨浦教育局CBG1-1孔、浦东外高桥J20孔、海塘P6孔、机场J25孔、三灶P45孔和海潮P17孔等孔，均在此层局部见及少量有孔虫或偶见钙质超微化石。如外高桥J20孔，此层大都为灰色细砂夹粉土和底部的含砾砂，断续见个别的钙质超微化石*Gephyrocapsa oceanica*；唯顶部（孔深77~80m）见很少的*Florirus decorlus*、*Globigerina*，并见*Epistominella naraensis*、*Ammonia beccarii* var.及平旋壳类有孔虫，还出现列式壳类深水种，反映通常处于间或受海水泛及的河流环境，晚期成为三角洲河流。又如机场J25孔中下部（孔深103.5~129.65m）的灰、灰黄色粉细砂、砾砂层无微古化石；其上覆层（孔深81.95~103.5m）的灰色粉细砂夹砂质粉土、粉质黏土的下部（92.6~103.5m）断续见个别的*Ammonia beccarii* var.、*A. compressiuscula*、*Asterotalia subtrispinosa*有孔虫及偶见个别的海相介形虫*Keijella bisanensis*、*Cushimanidea japonica*，反映此层沉积时通常处于河流环境，只是短期为时而受到海水影响的三角洲河流环境。

六、滆湖海侵——第Ⅱ海侵层

滆湖海侵为江苏第一水文地质队（1987）命名，指出其分布范围达到宁镇山脉和茅山山麓。这次海侵的强度和淹没范围为长江三角洲区域第四纪历次海侵之最。滆湖海侵遍及上海全区，其海侵层位属上更新统南汇组中段。按其海侵强度大体可分为两类地区：一是西部太湖沉积区及东北部长江三角洲河口沙岛沉积分区分布的海相性弱的滨海、潟湖，间有陆相河流沉积；二是东部长江三角洲滨海平原沉积区分布的海相性较强的浅海—海湾沉积。依据光释光定年值，滆湖海侵——第Ⅱ海侵层的形成年代主要为20~55ka。

海侵记录表现：长江三角洲沉积区滨海平原，以浦东外高桥J20号为例，孔深38.02~77.00m见此海侵层，其层位属于上更新统南汇组中段。此海侵层由3小段组成：下部小段（孔深67~77.0m）为浅灰色粉细砂夹粉质黏土互层，含有孔虫丰度、属种数均低，优势种有 *Florilus decorus*、*Globigerina* spp.，次之 *Elphidium magellanicum* 等，并见很少的钙质超微化石 *Gephyrocapsa oceanica*，孔深73.5m以浅还有海相和半咸水介形虫共生，反映由下而上为三角洲河流—河口湾环境；中部小段（深44~67m）基本为灰色粉质黏土，含有孔虫丰度大都大于等于200枚/50g干样，其中孔深54~60m间普遍为大于1000枚/50g干样，属种数大都大于等于20个/每样，中部优势种有 *Florilus decorus*、*Globigerina* spp.，*Buccela frigida*、*Epistominella naraensis* 和 *Ammonia beccarii* var.，在孔深54~60m间出现列式壳类、瓶虫类等深水种稍多，含钙质超微化石属种、个数也增多，反映为滨岸浅海—正常浅海环境；上部小段（孔深38.02~44m）为灰色粉质黏土夹粉砂小团块，有孔虫丰度、属种数均明显降低，优势性强，优势种为 *Elphidium magellanicum*，其次为 *Ammonia beccarii* var.，反映为海湾沉积环境。此海侵层下伏河流沉积的浅灰色细砂夹含砾砂，上覆泛滥平原沉积的锈黄色砂质粉土和黄绿色粉质黏土构成完整的海侵旋回。上海面粉厂J10孔孔深30.2~75m见此海侵层，分为两小段：下部（孔深75~43.7m）灰色粉质黏土频夹粉细砂的“千层饼”层内，以 *Florilus decorus* 为主，并有列式壳、瓷质壳、胶结壳等深水种和示冷性有孔虫 *Buccella frigida* 出现，复合分异度达2.7~3；与现今见于东海水深大于等于50m的 *Munseyella* sp.-*Cythelopteron* sp. 介形虫组合共生，说明属正常浅海相；上部（孔深43.7~30.2m）的灰、黄色粉细砂内为 *Ammonia beccarii* var. 组合，尚见较少的 *Elphidium advenum*、*Florilus decorus*，与 *Sinocytheridea latiovata*、*Leguminocythereis hodgii* 等半咸水介形虫共生，反映为滨岸浅海沉积。

长江三角洲沉积区河口沙岛区唯在此海侵层上部的灰色粉砂、粉质黏土互层内见少量有孔虫和个别半咸水介形虫，其中、下部的砂和含砾砂层无微古化石，说明中下部为河流沉积，至上部才演变为滨海环境。

西部的太湖沉积区内此海侵层以含种数、个数均少、优势性强的广盐性有孔虫，与半咸水、陆相介形虫及陆相微古化石共生的层段为主，其与无微古化石的层段交替出现，反映较弱的海陆过渡与陆相交替的环境演化，揭示这次海侵在上海西部地区曾有两次进退，形成相应的两个亚旋回。以青浦练塘P31孔为例，该孔孔深47~49m为下伏的浅蓝灰色粉质黏土层，出现个别的有孔虫及其他陆相生物成分，反映了湖泊沉积，间或受海水泛及。其上，孔深27~47m为海侵下亚层，由蓝灰、浅棕、褐黄色细纹间互而成的皱纹状黏土，上部依次为灰黑色黏土和黄色粉砂；下部出现优势性很强的广盐性有孔虫与陆相、广盐海相介形虫及其他陆相成分共生；中上部出现个数、属种数均少的广盐性有孔虫，仅有其他陆相成分共生，表明基本为潟湖沉积，上部演变为滨海沼泽环境。再上，孔深23~27m间为含褐铁矿结核的灰绿色黏土、浅褐色粉质黏土，即西部的“第二硬土层”，偶见个别有孔虫，表明为湖泊沉积，偶尔受海水泛及。更上，孔深9~23m为海侵上亚层，下部为含植物碎屑的浅棕、灰色（粉质）黏土，含优势性甚强的广盐性有孔虫、个别属种广盐海相介形虫，与其他陆相生物成分共生，反映为滨海沼泽沉积；中上部依次为灰色粉质黏土夹粉砂、褐黄色杂有黄绿、蓝灰色斑纹状黏土，含个数、属种数均很少的广盐性有孔虫，与其他个别陆相成分共生，反映为潟湖沉积。上覆层为孔深3.9~9m间的褐黄杂黄绿色细网纹状黏土，含有褐铁矿结核，偶见个别的广盐性有孔虫及其他陆相生物成分，表明为湖泊沉积，偶尔受到海水泛及。

七、镇江海侵——第Ⅰ海侵层

镇江海侵为江苏第一水文地质队（1987）命名，认为这次海侵范围沿长江两岸分布，顶端达到镇江。这次冰后期海侵在上海境内分布广泛，唯佘山、天马山附近的青浦西南部及与松江、金山接壤区于全新世早、中期至晚更新世硬土台地所占据，未受到这次海侵的侵进。

这次海侵在上海境内形成海陆过渡相—海相—海陆过渡相地层序列，构成一个完整的海侵旋回，相应由下而上组成娄塘组、上海组和如东组。海侵层底界埋深介于60~13m不等，以河口沙岛沉积分区最为发育，向西经滨海平原沉积分区至太湖沉积区渐薄以至尖灭。

在长江三角洲沉积区内，如外高桥J20孔所揭示的，海侵层发育完整，三分性明显，对应于上述3个基本岩石地层单位：

下部，孔深23.30~35.57m，包括孔深29.65~35.57m间的灰色砂质粉土夹粉质黏土和孔深23.3~29.65m间的灰色黏土两段海陆过渡相地层，组成了娄塘组。下段含有孔虫个数和属种数多寡不一，优势种为*Elphidium magellanicum*、*Cribrononion vitreum*、*Globigerina* spp.，并有其他平旋壳类及少量窄盐种，与较多的广盐海相介形虫*Sinocytheridea latiovata*、*Leguminocythereis hodgii*和少量海相介形共生，反映为河口相；上段含有孔虫种、数均减少，优势种为*Ammonia beccarii* var.、*Cribrononion vitreum*，介形虫唯见*Sinocytheridea latiovata*，反映为海湾相。

中部，孔深8.72~23.30m，由下而上依次为灰色淤泥质黏土、灰色砂质粉土夹淤泥质粉质黏土和灰色淤泥质粉质黏土层组成，这套以淤泥质黏性土为特征的海相地层组成上海组。其中、下两层含有孔虫属种、个数均丰，优势种有*Ammonia beccarii* var.、*Epistominella naraensis*、*Cribrononion vitreum*、*Elphidium magellanicum*，次之为其他螺旋壳类、平旋壳类及抱球虫科，并出现较多的列式壳类、瓷质壳类窄盐性属种，与较多广盐海相介形虫*Sinocytheridea latiovata*、*Neomonoceratina dongtaiensis*、*Leguminocythere ishodgii*等共生，还含有较丰富的钙质超微化石，反映为河口淡化浅海（下层）—中潮坪（中层）；上层含有孔虫种、数骤减，优势性强，为*Ammonia beccarii* var.和*A. confertitesta*，反映为高潮坪环境。

上部，孔深8.72m以浅，由下而上为灰色粉质黏土夹砂质粉土及两者的互层，这一海陆过渡相地层组成如东组，其含有孔虫种、数较少，优势种为*Epistominella naraensis*、*Ammonia beccarii* var.和*A. convexidorsa*，组合成分与现今长江三角洲前缘支流河道——吴淞江、黄浦江所见有孔虫组分相同。

总之，在长江三角洲沉积区分布的这一海侵层，发育齐全的地层序列通常为：下部包括河口—溺谷堆积的灰色粉土与灰白色粉砂互层和滨海沼泽—海湾堆积的深灰、褐灰色（淤泥质）黏土两段。后者在上海东、南部各区以常见植物茎叶与黄色泥灰质结核，局部形成泥炭或碳质黏土夹层为特征；中部包括滨岸浅海堆积的深灰色淤泥质（粉质）黏土和河口滨海堆积的灰色粉土夹粉砂与灰色粉土、粉质黏土互层；上部为河口堆积的灰、灰黄色粉土、粉砂与滨岸潮滩堆积的褐灰、褐黄色粉质黏土。

在太湖沉积区内如金山和青浦东北部等地区所见为陆相—海陆过渡—陆相地层序列，发育齐全的通常具有：下部包括湖沼堆积的灰黑色粉质黏土和滨海沼泽堆积的深灰色（淤泥质）黏土，含较多植物茎叶与黄色泥灰质结核，局部夹泥炭或碳质黏土层；中部为潟湖—滨海沼泽堆积的深灰色（淤泥质粉质）黏土，以含优势性强的一两种广盐性有孔虫与较少的半咸水介形虫共生为特征，其偏下部位常含植物茎叶、泥灰质结核及贝壳碎片，偏上部位时见潮汐水道形成的粉砂、砂质粉土透镜体夹层；上部为湖沼堆积的深灰色黏土夹泥炭与湖泊堆积的褐灰、褐黄色粉质黏土夹粉砂，局部沿古吴淞江分布较厚的灰褐色砂质粉土之类滨岸堆积，常含淡水软体动物、介形虫、轮藻等陆相生物化石，可见少量广盐性有孔虫，显示受到海水的影响。

依据光释光和放射性^{14}C定年值，上海镇江海侵——第Ⅰ海侵层的形成年代始于11ka。在8~5ka进入海侵最盛期，形成上述上海组下段的淤泥质黏土。区域性最大海侵时期形成最大海泛面在6.5ka左右，为上海所见分划长江三角洲堆积平原与太湖堆积平原的首条贝壳沙堤——“沙冈”的构筑时期。之后进入高水位体系域，陆续形成上海组上段—如东组的多粉土和粉砂、相对较粗的陆源碎屑堆积。

第二节　海侵分布特征

一、海侵分布范围

上海第四纪不同阶段海侵范围有所差异，并不是每次海侵都淹没全区。这种现象与海平面波动和岩相古地理背景密切相关。总体而言，早更新世和中更新世4次海侵踪迹在现今的滨海平原、湖沼平原局部地点发现，河口沙岛并未发现。

早更新世两期海侵微弱，如皋海侵踪迹仅在现今的浦东滨海局部地点发现。早更新世晚期的周浦海侵略有变强，海侵踪迹不仅在浦东个别地点发现（如P6孔、P12孔），还可在青浦和金山（如P31孔、LQ1孔）有发现。

中更新世早期周浦海侵范围明显较早更新世两次海侵范围明显变大，但仍局部分布，除了在浦东以及金山枫泾一带出现海侵踪迹，还在北部的嘉定出现（例如P5孔）。中更新世晚期嘉定海侵的海相性程度又明显变高，海相地层发育也较厚，但仅在现今湖沼平原发现，青浦LQ3孔、松江LQ9孔以及枫泾LQ1孔均发现。

总体而言，区内晚更新世以来普遍发生海侵（图25-1，晚更新世两期海侵在全区分布）。

晚更新世早期的太湖海侵虽遍及全区，但海侵地层仅在西部湖沼平原发育，在滨海平原和河口沙岛却明显变差。

晚更新世中晚期海侵遍及全区，在滨海平原最为发育，其次为湖沼平原，但在河口沙岛多受到剥蚀。

全新世海侵范围最大，至湖沼平原和滨海平原的交界处，发育了古贝壳砂堤，在河口沙岛最强，其次在滨海平原，湖沼平原受到了海侵的微弱影响。

二、成因分析

上海早—中更新世4次海侵的分布明显与构造沉降控制下的地貌格局有关，总结区内具备微古分析的钻孔，发现海侵地点均发生在地势低洼的古河道内（见图25-1），反映了第四纪早期区内地势相对较高，地形较为崎岖。随着不断的构造沉降，区内地势开始变缓，以至于从早更新世至中更新世海侵逐渐变强且范围逐渐变大。结合磁性地层等地层记录，至中更新世晚期，全区地势已变得平缓，可能海侵能够广泛影响全区。然而，中更新世晚期海侵仅保存在西部，其现象可能与晚更新世早期海侵类似。结合区域地质资料，滨海平原可能在中更新世以来多发育河带，海侵地层可能在海平面波动的不断作用下，受到了后期海退阶段下切河道的剥蚀而不复存在。

区内晚更新世以来的3次海侵分布特点可总结为3次海侵均为大规模海侵，海侵地层的发育与海平面波动控制下的地貌演变密切相关。根据区内晚更新世地质年龄，湖沼平原保留了较好的海侵地层形成于晚更新世早期暖期高海面阶段，而晚更新世早期滨海平原、河口沙岛为河流相沉积，可能为晚更新世早期冷期低海面阶段形成，强烈的河流下切导致海侵地层保留不完整，这反映了晚更新世早期区内地势起伏仍相对较高，河带范围较大。

晚更新世中晚期海侵地层能够发育较好，与地势更加平坦有关，河带范围明显变小。晚更新世中晚期海侵地层在河口沙岛缺失是晚更新世末期长江下切河谷剥蚀的结果。

全新世海侵范围与晚更新世末期长江下切河谷深切产生的河谷地貌限制有关，海侵沿下切河谷逐渐扩大，在下切河谷内发育了诸如崇明岛、长兴岛、横沙岛这种三角洲河口砂坝，在河谷阶地发育了滨海平原。

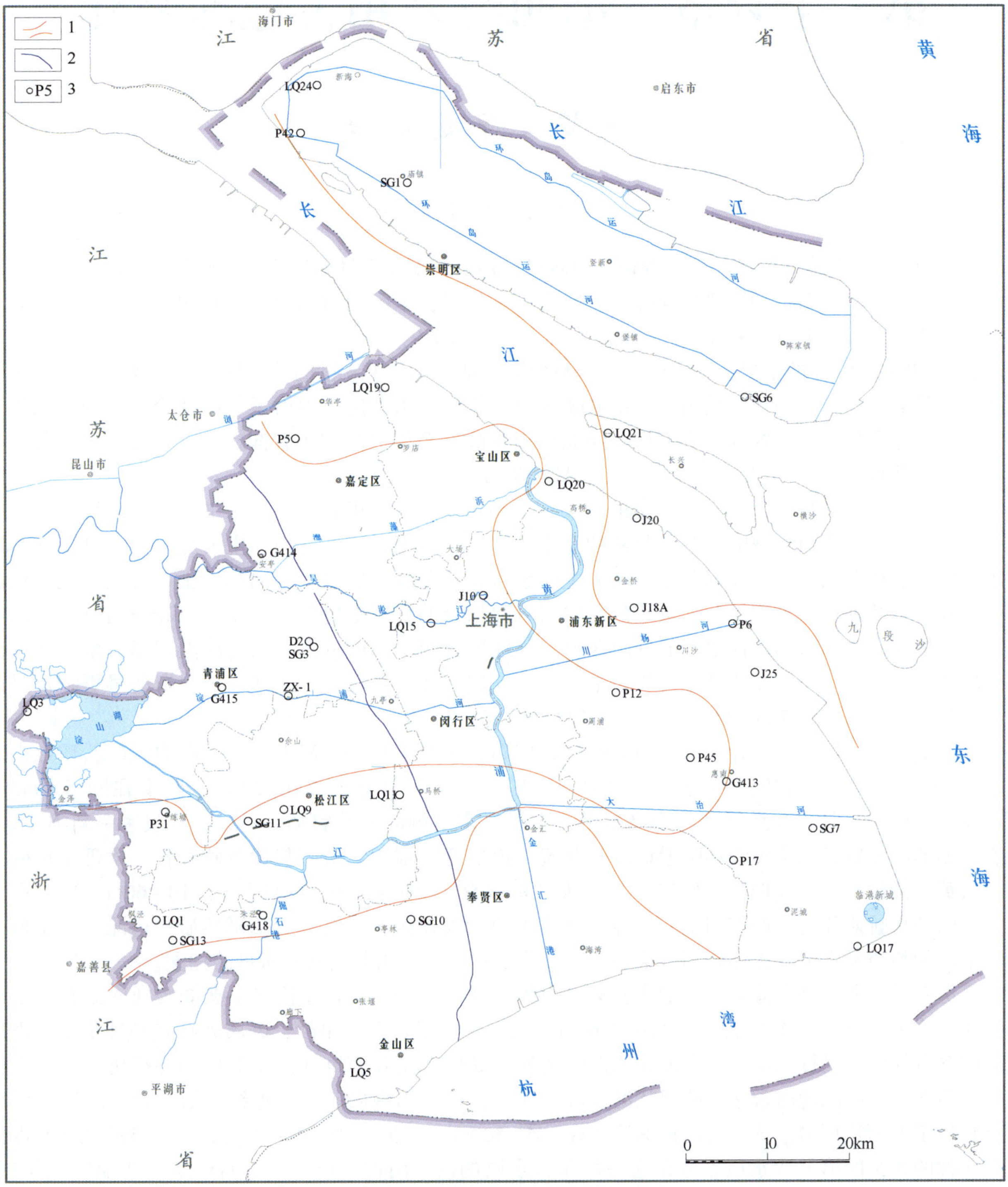

图 25-1 上海第四纪海侵分布范围

Fig. 25-1 Distribution of the Quaternary transgressions in Shanghai

1—早更新世—中更新世海侵范围（古河道）；2—全新世海侵最大范围界线（西界）；3—钻孔位置及编号

第二十六章　第四纪岩相古地理演化

第一节　早更新世

一、早更新世早期

在2.60Ma左右地磁场极性发生倒转，地磁极性期由高斯正极性时转变为松山负极性时，古气候发生显著的转换，由上新世的湿热气候骤变为凉冷干燥气候，本区周边山地针叶林明显扩张，长江流域化学风化强度明显减弱（Yang et al.，2006；Yue et al.，2019）。

在2.60~2.00Ma的早更新世早期，区内新构造运动继承了上新世的差异性升降运动特征，原来上新世上海陆域广泛分布的连绵低山仍占据着陆域西南部的广大地区，但由于气候变冷，侵蚀基准面下降，本区普遍发生溯源侵蚀和河流下切，因此低山丘陵的范围已收缩到青浦、松江、金山、闵行、奉贤境内，山势也已降低（Chen et al.，1995；邱金波和李晓，2007）。

基岩面埋深分布显示，现市区西南的大场—小闸镇—三林—坦直、三灶一带，在湖滨地带内分布着一串剥蚀低丘。当时在现陆域东、北部的嘉定、宝山、市区和浦东境内，主要为一片湖盆，其中堆积了具网纹、斑杂状构造，以褐黄、褐色为主间杂灰绿、黄绿、紫灰等色的黏性土与姜黄、灰等色粉细砂的互层；滨湖地带堆积棕黄、灰等色的中细砂和部分的含砾中粗砂夹砂质粉土和粉质黏土。现在陆域浦东的东南部与金山枫泾附近的低山区内，分别形成两个规模较大的山间冲洪积扇与扇上湖泊洼地，堆积了一套冲洪积相泥质砂砾石与黏土、粉土互层。在现陆域东、北部地带分布的早更新世早期沉积物的埋深通常介于230~290m之间，其厚度大都为30~50m。

在早更新世早期，当时崇明基本为河间滩地广布，主要堆积一套漫滩—泛滥湖泊相的粉质黏土和黏土夹粉土，间或夹有边滩相的粉细砂和细中砂，埋深一般介于220~280m，厚度在40~60m不等（图26-1）。只是在现崇明岛的周缘地带，主要如东南侧的前哨农场—奚家港和侯家—新河附近，还有北侧的红星农场和富民农场附近地段有曲流河道发育，堆积了黄、灰黄色砂砾石、含砾中粗砂与灰、灰黄色细砂互层。其埋深、厚度差异较大，埋深介于240~300m，厚度介于10~60m。当时在长兴、横沔两岛普遍经历了曲流河道—边滩—泛滥平原的演化过程，有河床相砾质砂—边滩相粉细砂—泛滥平原相粉质黏土夹粉土和粉细砂堆积，其埋深、累厚与崇明岛上通常所见的相同。

早更新世早期，现在浦东滨岸的龚路东—海塘—浦东机场—祝桥东渔场—老港及迤南地带，出现一条浦（东）—南（汇）古河道。其中在祝桥以北，堆积了含砾中粗砂、砾质砂频夹黏性土层，由含砾砂内含有泥砾、黏性土层内见有泥裂构造以及砂与黏性土呈频繁重现韵律等可知，祝桥以北属于河道分合多变、河间滩地时常暴露的辫状河流沉积环境；以南出现砂砾石的连续厚层堆积，表明已演化为比较稳定的河道环境。这一古河道的埋深在245~300m，厚度在30~60m不等，所蕴含的地下水即为历来所称的狭义的第Ⅴ含水层。分布这一承压含水层的浦—南古河道带、崇明环岛一些地段和长兴、横沔岛成为探采这一层地下水的有利地区。

由于河流溯源侵蚀和下切作用，早更新世早期出现了新的沉积物源，黏土矿物中蒙皂石含量有所下降，而扇中或河床相沉积物中Fe、Ti、Mg等元素含量显著下降，反映原先出露的白龙港玄武岩此时已基本剥蚀殆尽，中酸性火成岩物源加入本区。

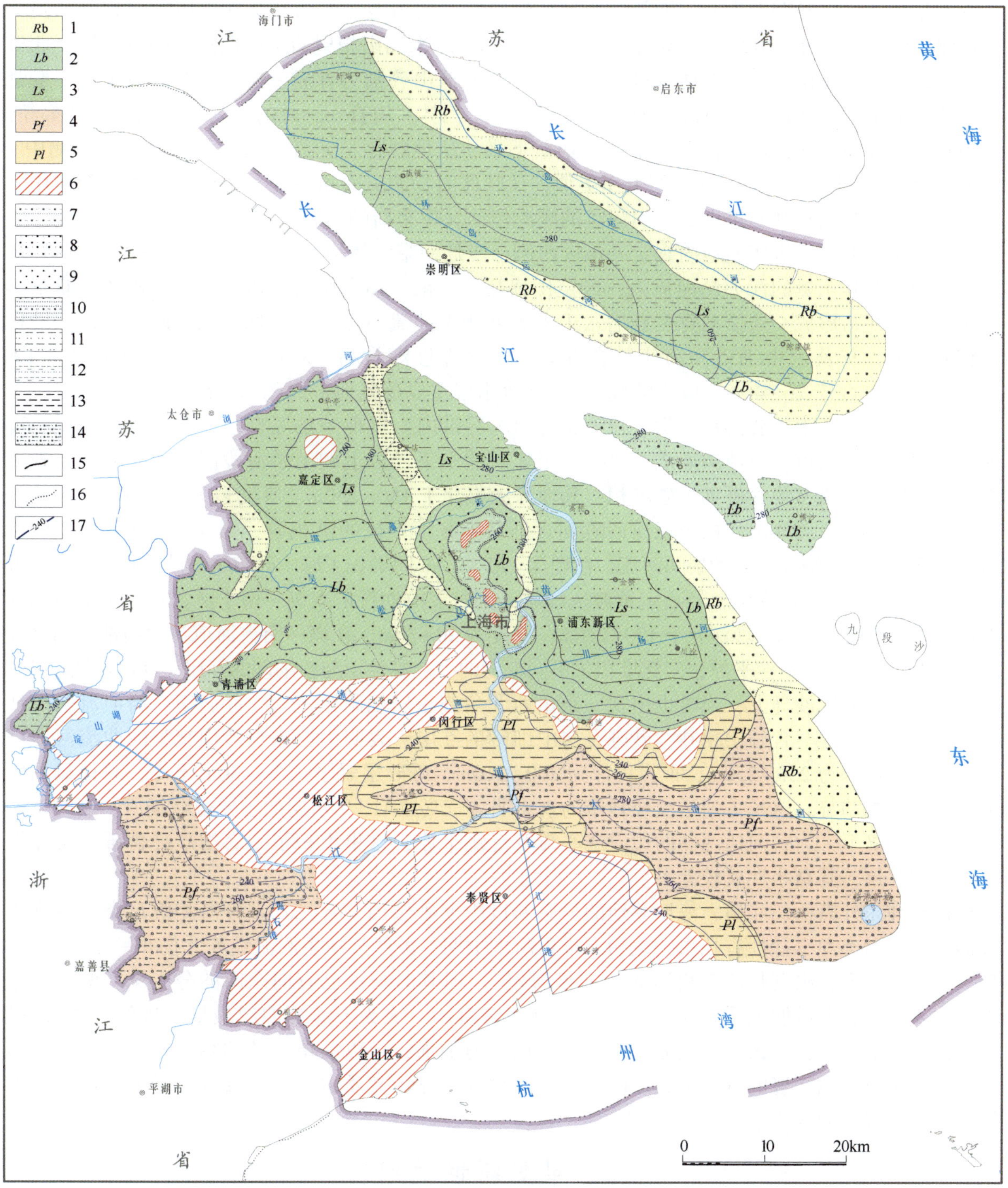

图 26-1 上海早更新世早期岩相古地理图

Fig. 26-1 Lithofacies paleogeographic map of the early Early Pleistocene in Shanghai

1—河流相；2—河流边滩—滨湖相；3—泛滥湖泊—浅湖相；4—洪积扇（裙）相；5—扇上、扇间湖泊相；6—基岩剥蚀区；7—中细砂与含砾中粗砂互层；8—砂砾石或砾质砂；9—细—中砂；10—细砂夹含砾砂；11—黏性土与中细砂互层；12—黏性土夹粉细砂；13—黏性土，间或含砾；14—泥质（砂）砾石；15—岩相分区界线；16—岩性分区界线；17—层底等高线（黄海高程，m）

二、早更新世中期

受到早更新世中晚期发生的一期新构造运动的影响，区内地壳振荡性沉降作用较前有所加强，进入了缓慢沉降阶段。在2.00~1.20Ma的早更新世中期，原来隔山相望的金山枫泾与闵行马桥、北桥迤东的两条山谷经溯源侵蚀近于沟通，将原来连绵分布的低山分割为青浦—松江接壤区与金山—奉贤接壤区的两片低山，山势进一步下降。原来分布于大场—三林—坦直、三灶的一列弧形低丘已大都湮没，只有坦直、三灶与市区西侧虹桥等数个低丘残留下来。从地球化学元素分布来看，此时沉积物中Fe、Ti、Mg等元素含量有所上升，与基岩分布类似，反映此时物源又有所变化，可能是由于沉降作用，本地的中酸性火成岩已经基本被埋藏，而河流的溯源侵蚀又带来新的物源。

早更新世中期，受沉积盆地内嘉定—市区西侧—三灶一带大体呈北西-南东向延展的反“S”形凸起带的控制，在其东侧的江苏浏河流入，经宝山盛桥、吴淞—浦东张桥、川沙—浦东机场—浦东东南部滨岸地带，开始形成一条摆动宽度有七八千米的低弯度曲流河道，即常称的浏—南古河道，堆积了沙坝相的浅灰、灰黄色中细砂和河床相含砾砂，其底界埋深常为230~240m，顶界埋深常为190~200m，厚度大都介于20~40m，成为陆域第$Ⅳ_3$下含水层主要蕴藏地带（图26-2）。

其西南方向，在南、北两片低山、丘陵夹持的沟谷带内，出现一条源自松江和闵行、北桥，流经下沙、大团的支流河道，至芦潮港汇入浏—南干流。浦东下沙以上的上游河段沉积物略粗，为砾质砂和细中砂；以下河段多为含砾细中砂堆积，其埋深大都介于180~230m，厚度多为30~40m。在两片低山之间的垭口及东西两侧山谷内为泥质砂砾和含砾黏土充填形成的洪积扇、裙和扇上湖泊，西侧山谷内出现一条经枫泾向浙江流去的扇间河流，在埋深170~210m间堆积了砾质砂与中细砂的互层。在北部低山区北侧也可能出现一股支流，经嘉定黄渡、马陆，流向宝山刘行汇入浏—南干流河道，为细砂和含砾砂夹粉土、粉质黏土堆积，埋深介于190~220m，厚度为20~30m。在浏—南古河道与西南部低山之间的广阔地带为河流边滩和泛滥平原，广泛堆积了不同粒度的砂及黏性土与粉细砂的互层，其富水性较差。

在现崇明等河口沙岛上，当时也出现一条由西北角红星农场附近流入，经庙镇、新河流向横沙岛的古河道，堆积了一套含砾中粗砂、砂砾石与中细砂的互层，表明为河床与沙坝交替的曲流河道沉积环境，其底界埋深为230~250m，顶界埋深为200~210m，砾砂层累厚在20~50m不等，总体显示向下游加深、加厚的趋势。这一古河道流经地带为现崇明等三岛第$Ⅵ_3$含水层的主要蕴藏场所；当时崇明岛的大部和长兴岛上，为古河道两侧的边滩或心滩及泛滥平原，这里堆积了一套细砂夹含砾砂及黏性土与粉细砂的互层，其富水性就较差了。

根据在浦东外高桥、邓镇、海塘等地的早更新世中期砂层内，曾见及多种海相腹足类及少许广盐性底栖类、浮游类和窄盐性有孔虫，表明在早更新世中期，虽然长江主泓流经苏北扬州—泰州—海安、如皋一带注入黄海，出现长江三角洲的“如皋海侵”，但其时长江已有向东南的分支浏—南古河道流经本区注入东海，形成明显受海水影响的三角洲河流沉积，成为上海第四纪最早出现的第Ⅶ海进层。这一海侵踪迹的出现，是区内地壳振荡性沉降作用较前加强，由以前单一的陆相沉积环境演变为陆相间或有海陆过渡沉积环境的重要标志。同时，重矿物分析显示变质岩物源的加入，反映长江的泥沙有可能在此时已经影响到本区。

三、早更新世晚期

进入1.20~0.78Ma的早更新世晚期，陆域西南部的两片低山经长期剥蚀、收缩，已成为散落于冲积平原上的丘陵。原来在现陆域滨岸地带分布的浏—南古河道此时向西摆动，河道摆动宽度较早更新世中期拓宽，弯度也有所加大，其西界到达嘉定—宝山月浦—市区真如—周浦—芦潮港一线，河流摆动宽度达到10~15km。堆积了一套砾质砂、含砾中粗砂与中细砂的互层，反映为河床与沙坝交替的曲流河道沉积，其埋深通常在165~205m之间，厚度大都为30m左右（图26-3）。在浦东受到当时区内较早一次海侵的影响，形成了间或出现弱海陆过渡的溺谷河流沉积，成为上海的第Ⅵ海进层。

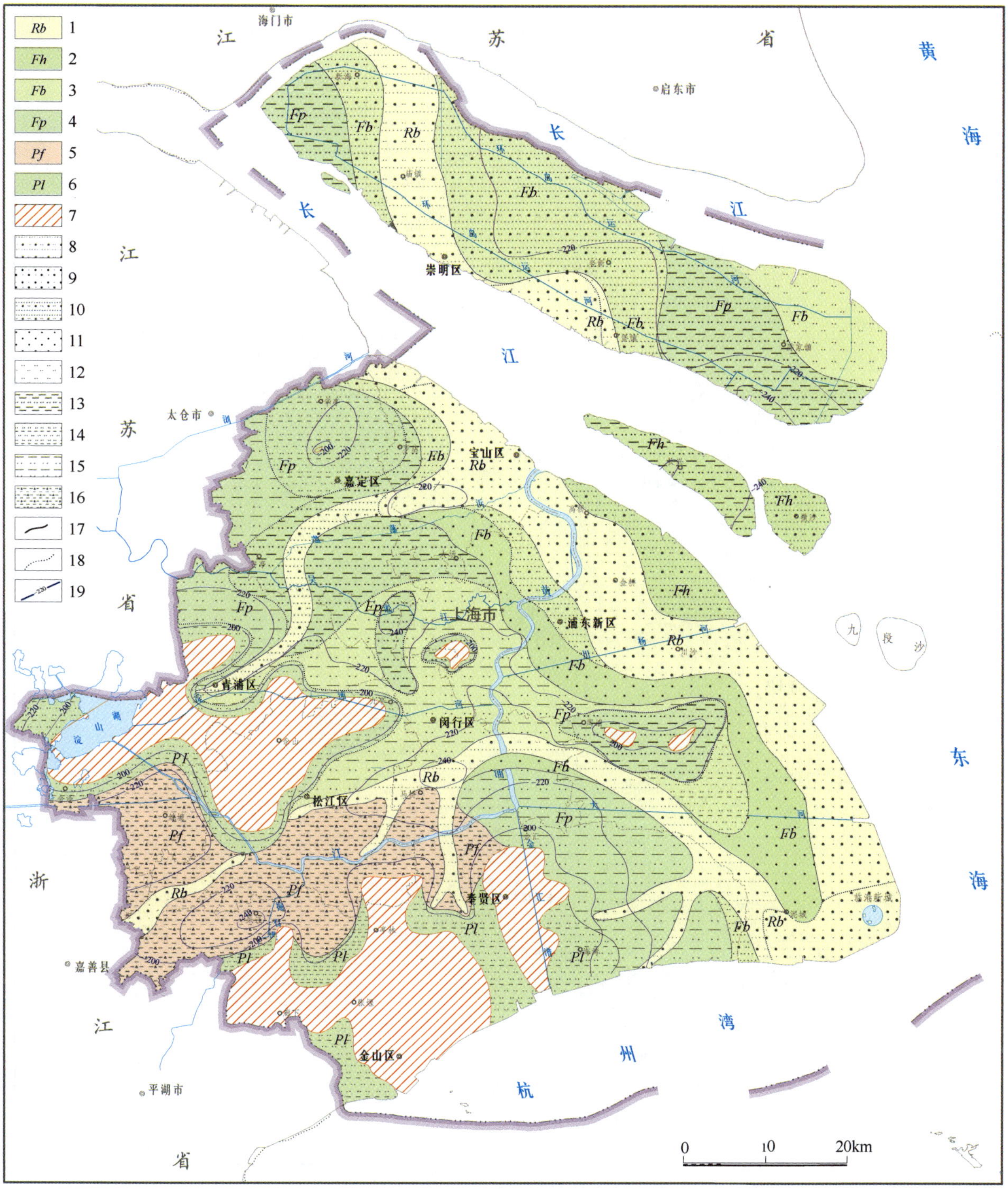

图 26-2　上海早更新世中期岩相古地理图

Fig. 26-2　Lithofacies paleogeographic map of the middle Early Pleistocene in Shanghai

1—河流相；2—河流心滩相；3—河流边滩、天然堤相；4—河漫滩—湖泊相；5—洪积扇（裙）相；6—扇上、扇间湖泊相；7—基岩剥蚀区；8—中细砂与含砾中粗砂互层；9—砂砾石或砾质砂；10—细砂夹含砾砂；11—细—中—粗砂；12—粉细砂和砂质粉土；13—粉细砂夹黏性土；14—黏性土，或夹粉细砂；15—黏性土与粉细砂互层；16—泥质砂砾和含砾黏土；17—岩相分区界线；18—岩性分区界线；19—层底等高线（黄海高程，m）

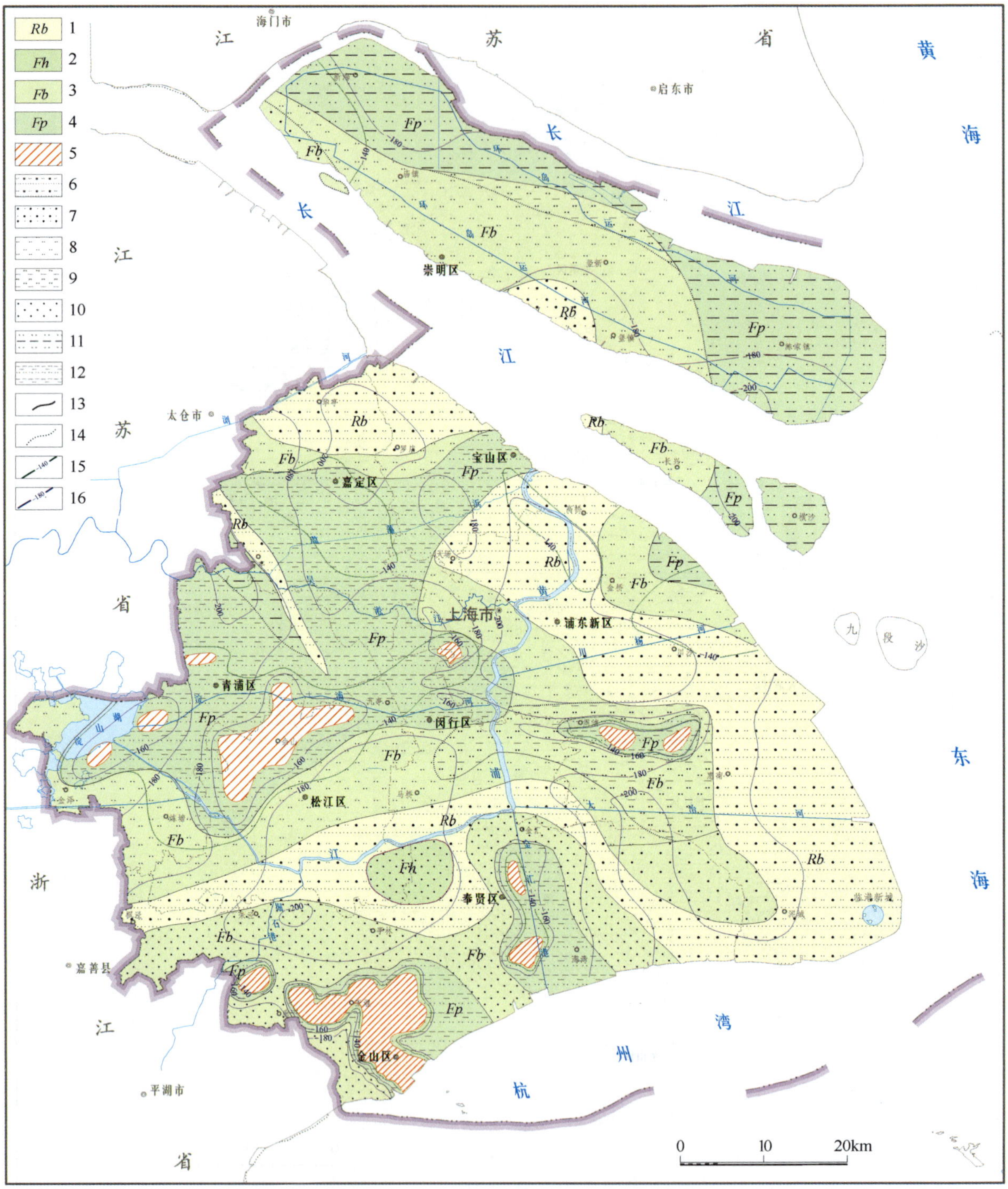

图 26-3　上海早更新世晚期岩相古地理图

Fig. 26-3　Lithofacies paleogeographic map of the late Early Pleistocene in Shanghai

1—河床相；2—心滩相；3—边滩、天然堤相；4—泛滥平原—湖泊相；5—基岩剥蚀区；6—含砾砂与中细砂互层；7—砂砾石；8—粉细砂夹粉土或两者互层；9—粉细砂夹粉质黏土；10—细砂、中细砂，或夹粉土；11—黏性土与粉土、粉细砂互层；12—黏性土夹粉细砂；13—岩相分区界线；14—岩性分区界线；15—层顶等高线（黄海高程，m）；16—层底等高线（黄海高程，m）

原来早更新世中期在南部低山丘陵区流出的一条比较顺直、狭窄的支流河道，此时明显向西南方向摆动、拓宽，在向西溯源侵蚀作用下，使原为松江—叶榭垭口阻隔的东、西沟谷贯通，将原经枫泾西流的河流袭夺，形成陆域南部的枫—奉古河道，至芦潮港附近与浏—南古河道汇合。当时的枫—奉古河道为一时有心滩缀布、摆动宽度达到5~10km的曲流河道，通常具河流沉积的二元结构，其下部堆积河床—沙坝相的灰、灰黄色含砾中粗砂、含砾细中砂与细砂互层，埋深在180~200m，厚度约为20m；上部堆积河漫滩相黏性土夹粉细砂层，埋深为150~180m，厚度介于20~30m。这两条古河道分布地带成为上海第$Ⅳ_1$含水层的主要蕴藏场所，但枫—奉古河道带沉积物的粒度及厚度均不如浏—南古河道。

在两条古河道的两侧分布中细砂、粉细砂夹粉土和粉质黏土的边滩沉积；在环绕低山丘陵的附近地区则为黏性土夹粉细砂组成的大片泛滥平原—湖泊洼地；在现崇明等三岛上也为河流边滩和泛滥平原—湖泊洼地广布，大多堆积粉细砂夹粉土及黏性土与粉土、粉细砂的互层。在这些河间地带，尤其是泛滥平原—湖泊洼地上分布的地层，其富水性就大为减弱了。

粒度分析也显示，此时沙体的分选明显变好，而泥质沉积物中往往夹分选较差的粉、细砂层，反映泛滥平原上的决口扇沉积。可见，此时在构造沉降和风化剥蚀共同作用下，本区的地势已经明显变缓，河流的摆荡性也因此加大。

另外，由于气候的频繁变化，早更新世本区发生了多次河流下切，在气候寒冷期发育河道滞留相的粗颗粒沉积，气候温暖期则发育泛滥平原相细颗粒沉积，且由于河流的下切作用，暖期沉积往往被侵蚀，从而留下厚层砂、薄层泥记录。

第二节 中更新世

一、中更新世早期

在0.78Ma时，地磁场极性发生倒转，地磁极性期由松山负极性时转变为布容正极性时。在0.78~0.26Ma的中更新世早期，区内古气候进入了新的温暖期。据研究，可能由于从中更新世开始的新构造期构造活动，使长江主泓转向南摆，取道扬州—南通—启东一带注入东海，在这一带形成中更新世底界下切较深、厚度较大的三角洲河流沙质沉积。当时上海境内经早更新世末期的堆积及成陆剥蚀后，陆地表面已趋于准平原化。中更新世，区内地壳以平均7cm/ka的速率缓慢沉降，与早更新世平均6.7cm/ka的沉降速率相近，表明此时仍处于缓慢沉降阶段，总体的古地理格局及其演化仍保持与早更新世晚期相似的特征。此时原来在西南部地区分布的秦皇山与天马山两片丘陵，其山势与范围都进一步缩小，成为散布于泛滥平原—湖泊洼地上的剥蚀残丘群；原在坦直、三灶附近残存的低丘则已消失。同时，分布于陆域东部的浏—南古河道向东拓展到崇明岛的中西部地区，成为心滩与边滩均比较发育的曲流河道，主要堆积了一套含砾中粗砂与中细砂的互层，反映了河床与沙坝交替的沉积环境，其埋深大都为120~145m，砂层累厚为20~30m（图26-4）。

南部的枫—奉古河道仍大体承袭早更新世晚期的分布格局，只是河间心滩明显向西伸展，其与浏—南古河道的汇合口则已向北摆动至浦东新场附近。其时枫—奉古河道沉积物仍具二元结构：下部以中细砂为主夹含砾中粗砂，埋深大都在120~155m，厚度约在30m，沉积物粒度较浏—南古河道细些；上部过渡为杂色黏性土夹粉细砂层。这两条古河道分布地带，构成了上海境内层位稳定、分布较广、厚度也较大的第$Ⅳ_3$含水层，并以含水砂层分布范围及厚度、粒度均较大的浏—南古河道的富水性较好。在古河道的两侧及河间带普遍有边滩或心滩分布，堆积了粉细砂与粉土互层、粉细砂和粉土夹黏性土或黏性土与粉细砂互层；在枫—奉古河道南、北两片丘陵的周围地区仍为泛滥平原—湖泊洼地环绕，堆积了黏性土与粉细砂的互层，其富水性较差。

在中更新世早期，主要于浏—南古河道流经的嘉定以南地带发现有孔虫化石，如嘉定娄塘、上海

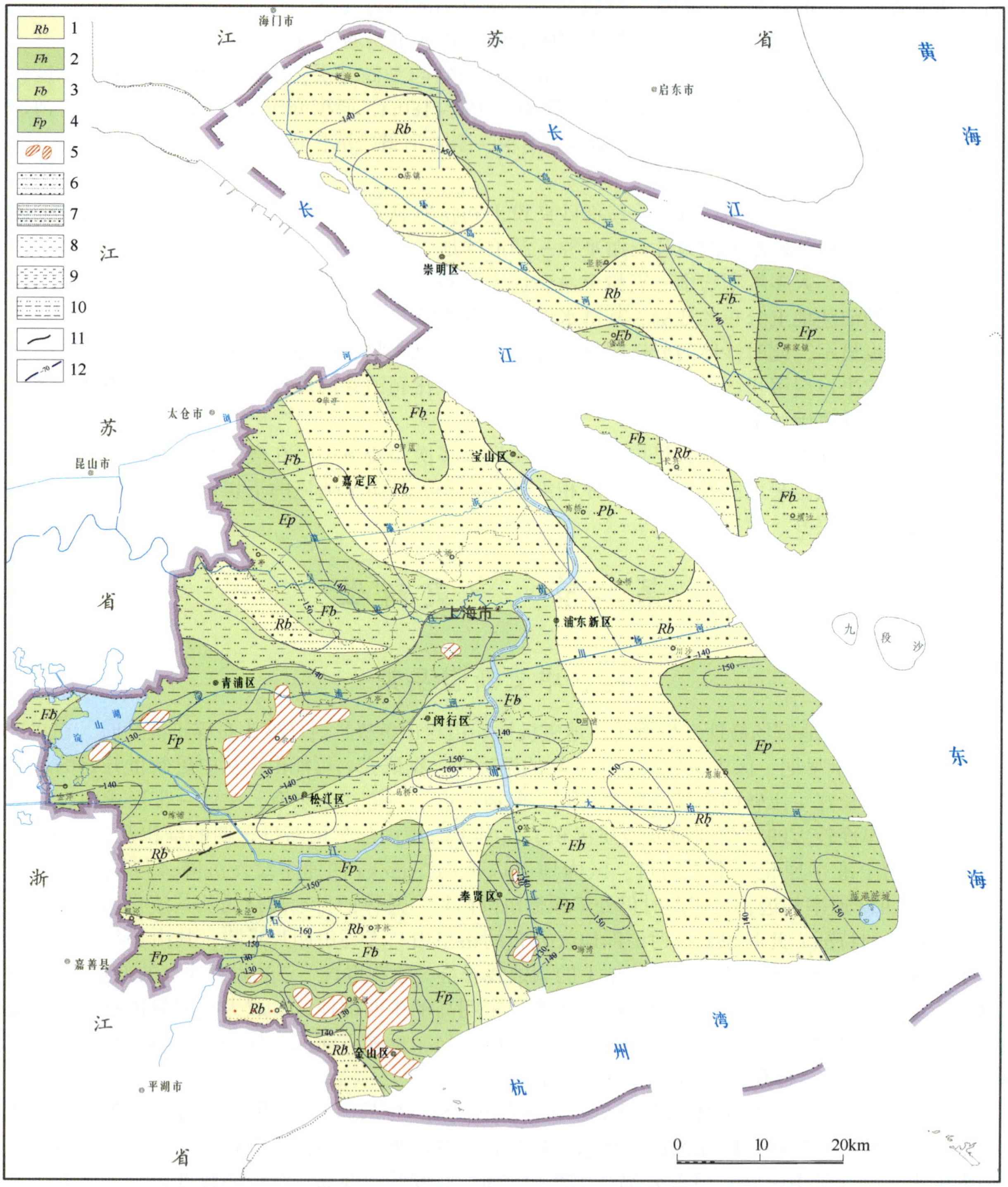

图 26-4 上海中更新世早期岩相古地理图

Fig. 26-4 Lithofacies paleogeographic map of the early Middle Pleistocene in Shanghai

1—滨海相；2—心滩相；3—边滩、天然堤相；4—泛滥平原—湖泊相；5—基岩剥蚀残丘；6—含砾中粗砂与中细砂互层；7—含砾中细砂夹含砾中粗砂；8—粉砂、粉土夹粉质黏土；9—粉细砂、粉土夹黏性土；10—黏性土夹粉细砂或两者互层；11—岩相分区界线；12—层底等高线（黄海高程，m）

面粉厂、横沔、三灶、海潮和浦东海塘等处所见。含化石层位通常出现在该期地层下部的河流相黄灰、灰色含砾中细—粉细砂内，一般为少量的半咸水底栖有孔虫，多为浅水广盐性属种；在南汇嘴地区还出现浮游有孔虫及窄盐种，间或伴有少量的钙质超微化石。与之共生的有陆相介形虫，还有数量丰富、个体较小的淡水软体动物化石，表明属于弱海陆过渡的陆上三角洲平原河流环境，至南汇嘴地区趋近于河口，形成了区内的第Ⅳ海进层。在西部太湖沉积区，一般仅见淡水介形虫，偶尔出现有孔虫，唯在青浦练塘曾发现现分布于南黄海较深水域的底栖有孔虫，与浅水、示冷性属种有孔虫以及半咸水、陆相介形虫共生，表明其时太湖沉积区一般属于陆上三角洲平原河湖环境，可能由于海面波动一度沦为开放性的海湾。

二、中更新世晚期

根据对晚更新世早期河流强烈侵蚀而残留的地层分析，在0.26~0.13Ma的中更新世晚期，古地理面貌仍大体承袭前期的基本格局。在现陆域东部的嘉定镇—上海市区—南汇嘴一带仍为浏—南曲流河道广布，为区内的主干河流所在。在河流西北侧边缘地带广泛镶布着北宽、南窄的边滩—泛滥平原。在陆域西南部，由浙江境内流来，经枫泾、朱泾、松江、亭林等地的两股支流河道构成的枫—奉古河道，过奉贤后，仍在南汇嘴汇入浏—南古河道。当时，天马山与秦皇山两片丘陵的分布范围与山势继续收缩，缀布于周围的泛滥平原—湖泊洼地上。

在中更新世晚期的较早阶段，海水曾泛及区内的河流，因之在西南部的青浦练塘等地嘉定组上段下部的青灰、灰绿色细砂、黏土层内出现与现今长江三角洲前缘支流河道相同的半咸水底栖有孔虫组合，相伴出现少量的陆相介形虫及轮藻，表明为三角洲河流或偶受海水微弱影响的河流环境，成为区内的第Ⅳ海进层位。浦东机场第四纪钻孔沉积物该阶段出现指示火山岩的蒙皂石高峰，与此时西太平洋火山频繁喷发吻合，推测是海侵作用将洋流携带来的火山喷发物质带至本区。临港新城和青浦凤溪第四纪钻孔还显示，自此时沉积物中卤族元素的含量明显上升，可能为海水的入侵作用。较晚阶段堆积的地层仅见陆相腹足类动物化石，表明已演化为陆相的河泛平原—湖泊洼地环境。

由于后来晚更新世早期河流的深切，使中更新世晚期地层在市区东部、浦东及河口三岛上分布残缺不全，保存比较完整的有宝山、嘉定、市区中西部、闵行、奉贤及以西的滨海平原沉积区和太湖沉积区。所见地层为河道和边滩堆积的灰、黄灰色砂、含砾砂夹粉土和黏性土，埋深在100~110m，厚度约为10m，为上海第$Ⅲ_1$含水层的主要分布地区。在当时上海西南部的丘陵区周围仍为一片泛滥平原—湖泊洼地环绕，沉积物为较弱水动力条件下堆积的黏性土与粉细砂、粉土的互层，其富水性较差。

元素分析显示，自早更新世中期以来，本区物源已经不再是前期的单一物源，显示多种母岩的混合，但在气候寒冷期，往往显示中酸性火成岩的加强，可能反映由于气候寒冷期河流下切和溯源侵蚀，带来较多西部茅山、天目山的物质。

第三节　晚更新世

一、晚更新世早期

进入晚更新世以后，中国大陆的差异性升降运动愈趋强烈，我国西部的山地和高原进一步隆升，东部滨海平原转入了强烈沉降阶段。区内陆地遭受一次深切侵蚀，由西南而东北落差逾50m。盆地的沉积速率急剧增大，据磁性地层测算，晚更新世上海的平均沉积速率由其前的7cm/ka增至64cm/ka。这种变化在地层层序上体现为晚更新世早期出现了河道滞留相和溺谷相的厚层含砾砂（图26-5）。

在128~75ka的晚更新世早期，古气候显著回暖。其间，上海所在的沉积盆地进入强烈沉降阶段，古地理面貌随之发生显著改观，这由沉积物孢粉中开始出现刺球藻、极管藻之类海侵指相标志生物化石，海陆过渡沉积环境的广泛、持续出现等可资证明。这次海侵范围达到太湖地区和现长江三角洲的

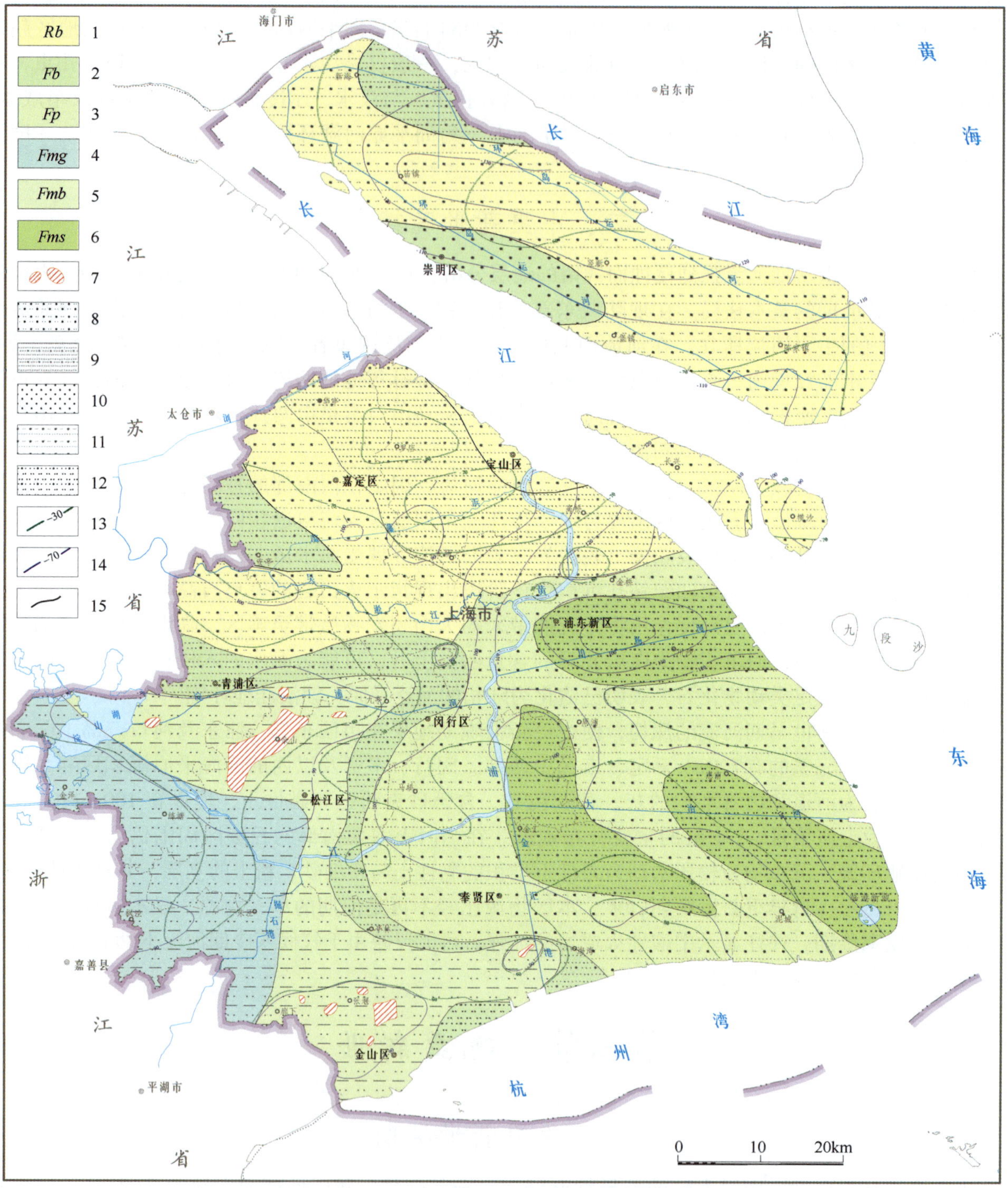

图 26-5　上海晚更新世早期岩相古地理图

Fig. 26-5　Lithofacies paleogeographic map of the early Late Pleistocene in Shanghai

1—河流相；2—边滩、心滩相；3—泛滥平原相；4—河口湾相；5—三角洲分流河道；6—三角洲河流心滩相；7—基岩剥蚀残丘；8—砂质粉土粉细砂与粉质黏土互层；9—细砂夹含砾砂；10—细砂；11—含砾中粗砂、砂砾夹中细砂或成互层；12—粉细砂夹含砾砂、砂砾；13—层顶等高线（黄海高程，m）；14—层底等高线（黄海高程，m）；15—岩相分区界线

南北两翼，即常称的“太湖海侵”。当时，原来分布于天马山—虹桥与秦皇山附近的两片丘陵经长期剥蚀，使虹桥、青浦西岑及奉贤附近分布的孤丘已告消失，成为基本局限于松江北部和金山南部散布的两片基岩残丘。

在大致沿嘉定南界—虹桥—闵行马桥—奉贤钱桥一线以东的陆域东部和崇明等三岛上，当时为河流广布，其中在市区以南的陆域东部地区为由分流河道与朵叶间心滩相间组成的三角洲平原沉积环境。在河道带内堆积了浅灰、黄灰色含砾砂、砂砾石与中细砂互层，河间心滩上堆积细砂、粉细砂夹含砾砂及黏性土。其埋深底界大致起伏于90~130m之间，顶界埋深大致起伏于60~80m之间，厚度一般为20~30m。据有孔虫分析揭示，当时嘉定安亭已偶尔受到海水的泛及；在市区、浦东的砂层内断续发现少量的半咸水和浮游有孔虫，偶尔伴有钙质超微化石，显示为三角洲河流环境；在太湖沉积区的青浦练塘、金山朱泾等地的粉质黏土与粉砂互层内也断续出现稍多的半咸水浅水种有孔虫，伴生半咸水介形虫和陆相介形虫，反映当时介于南、北两片丘陵之间的地区处于河口湾沉积环境。这一其间所形成的弱海陆过渡沉积即为区内的第Ⅲ海进层。而在这一河口湾与东北部的三角洲河流之间的丘陵附近地区则为河流边滩和泛滥平原广布的陆相沉积环境。

二、晚更新世中晚期

在75~11ka，本区进入了相当于末次冰期的中晚期。开始在75~70ka的早冰阶期间，上海处在类似现今暖温带的温凉气候下，演化为泛滥平原—淡水的环纹藻繁殖的泛滥湖泊环境，堆积了暗绿、灰绿和部分深灰、浅蓝灰、灰黄色的（粉质）硬黏土，夹砂质粉土或粉细砂纹层，大都厚为2~4m，组成区内的南汇组下段或滆湖组下段。其中含陆相的介形虫、轮藻等微古化石，还有淡水软体动物，偶尔见及有孔虫和半咸水介形虫化石。反映南部的南汇嘴、金山距海域较近，偶尔回升的海水曾泛及那里的泛滥平原—湖泊。

在70~20ka，本区进入末次冰期间冰阶期间，上海复转入相当于北亚热带的温暖湿润气候环境。当时，散落于松江北部和金山南部的两片基岩残丘已削弱到与现今相似的孤丘形态。其间，随着沉积盆地的强烈下沉，导致区内出现第四纪以来规模和强度最大的一次海侵，海侵范围广泛波及长江三角洲和太湖地区，其前锋抵达江苏滆湖附近，由此被称为“滆湖海侵”。这次海侵淹没了上海全境，堆积了海相或海陆交替相地层南汇组中段或滆湖组中段，形成本区海相性最强的第Ⅱ海进层。这次海进使市区中西部—浦东大部及奉贤等东南部区域沦为滨岸浅海，堆积了黄、灰色粉细砂夹粉质黏土或两者的互层。沉积物中含有属种、数量丰富的微古化石，有孔虫主要为底栖类，并有较多浮游种及多种窄盐种；介形虫主要为内陆架浅海属种，还有外陆架深水属种，反映相当于南黄海水深20~50m的陆架浅海环境。其埋深介于30~80m之间，厚度一般为40~50m。其北部的嘉定、宝山为开放性海湾—滨岸浅海环境，堆积了以灰色粉质黏土为主与粉砂或砂质粉土的互层，即常称的“千层饼”层（图26-6）。一般含微古化石较少，有孔虫主要为半咸水浅水种，伴有半咸水和淡水介形虫。其埋深通常介于20~75m，厚度大都为40~60m。

当时的崇明等三岛处于滨海环境，大都堆积以灰色粉细砂和粉土为主夹粉质黏土，断续见有很少的有孔虫，为半咸水浅水种，偶见窄盐种和浮游种，伴有个别的半咸水介形虫，反映为滨海沉积环境。其埋深顶界因末次盛冰期河道的侵蚀而大为降低，顶、底界埋深通常介于45~65m，厚度仅为15~20m。

在松江、金山与闵行、奉贤的接壤区以及青浦的东北部，当时处于河口湾相环境，主要堆积了灰、灰黄色粉细砂夹粉质黏土；在佘山、天马山与秦皇山这两片残丘附近地区堆积的为粉细砂与粉质黏土互层。主要在下部含有很少的半咸水浅水有孔虫和半咸水介形虫，伴有陆相的轮藻及淡水软体动物化石。其埋深介于20~65m，厚度大都为30~40m。

在佘山、天马山残丘群附近的青浦西南部及与金山、松江接壤区，当时处在潮上带潟湖与淡水湖泊—泛滥平原频繁更替的海陆交替环境，堆积了一套颜色纷杂、灰色与灰绿、灰黑、褐黄、蓝灰等色间互的海相软土、粉砂、粉土与陆相硬土的互层，构成了夹有“西部第二硬土层”等的滆湖组中段。

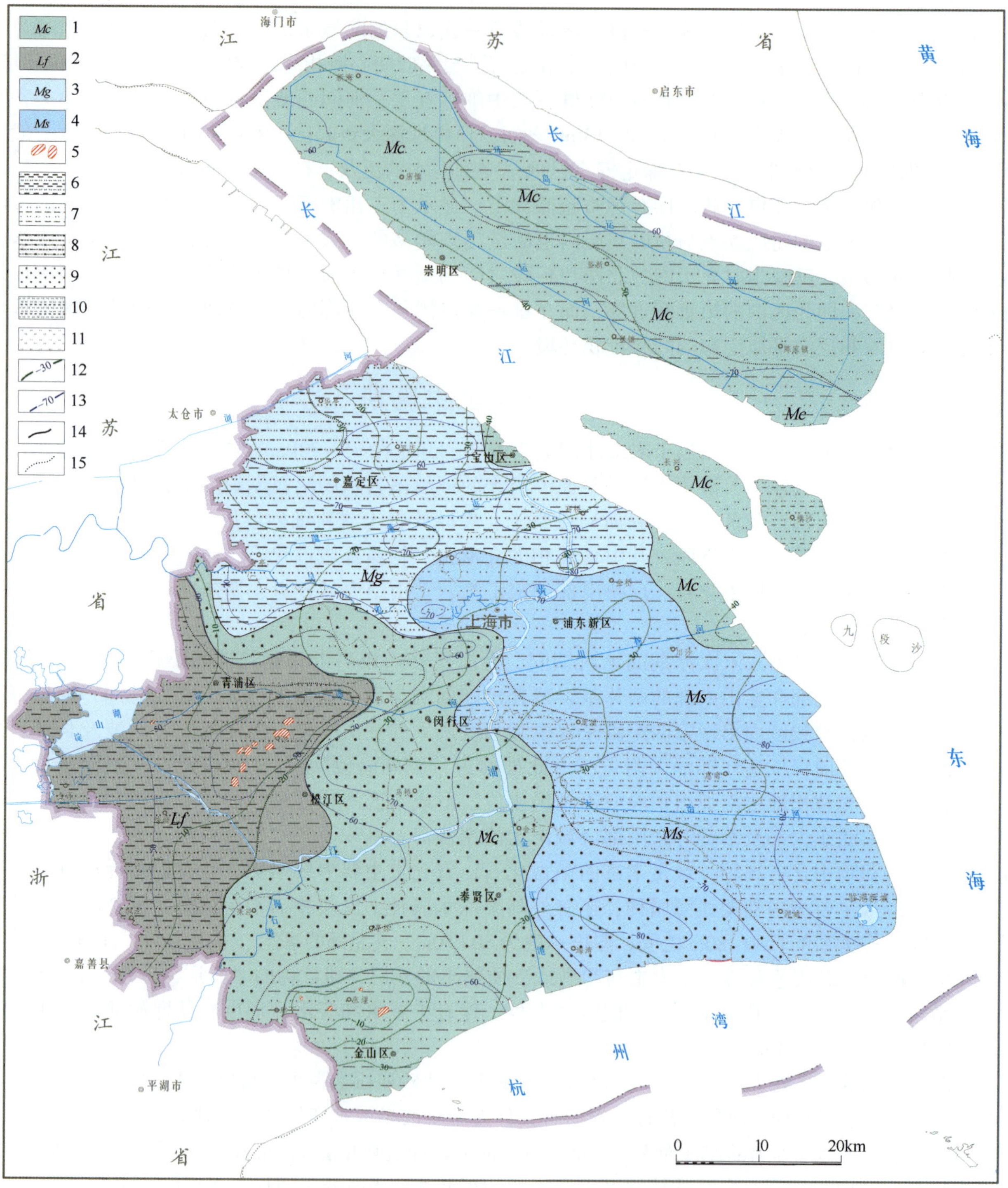

图 26-6　上海晚更新世晚期岩相古地理图

Fig. 26-6　Lithofacies paleogeographic map of the late Late Pleistocene in Shanghai

1—滨海相；2—潟湖相；3—海湾—浅海相；4—浅海相；5—基岩剥蚀残丘；6—粉质黏土夹粉砂、粉土；7—粉细砂与粉质黏土互层；8—粉砂、粉土夹粉质黏土；9—粉细砂、细砂，偶夹粉质黏土；10—粉细砂夹粉质黏土；11—粉细砂夹粉土；12—层顶等高线（黄海高程，m）；13—层底等高线（黄海高程，m）；14—岩相分区界线；15—岩性分区界线

其中断续见有少量半咸水浅水有孔虫，伴生半咸水和陆相介形虫化石。其埋深介于7~65m，厚度为40~50m。

在末次冰期间冰阶期间，随着古气候出现温暖湿润—温凉略干的周期性交替变迁，引起海水相应的升降、进退变化。根据岩石地层和微古化石综合分析，在70~20ka的间冰阶期间发生的“滆湖海侵”，在太湖沉积区东部分区偏西的松江新城附近曾出现3次海水进退，海进期分别为63~60ka、55~46ka和24~20ka；而在太湖沉积区西部分区的青浦等地则显示两次海水进退，海进期分别为50~45ka和35~23ka，其间的海退发生于45~35ka，形成青浦西部等地埋深20余米的灰绿—灰黄色硬土陆相层，即常称的“西部第二硬土层”。

三、晚更新世末期

在23~20ka，本区进入了末次冰期间冰阶后期，由于海面开始下降导致海水逐渐向东海陆架退去，原来上海境内的浅海和海湾演化为滨海—三角洲平原河流环境，堆积了南汇组中段上部的褐黄、锈黄色的砂质粉土和粉砂、粉细砂；海水处于进退不定、经常暴露氧化的浅水环境，因此沉积物中常可见锈黄色铁质浸染斑纹及偶尔见及的泥裂构造。其埋深和厚度明显由西向东倾斜、加大。

当时在青浦及与松江、金山的接壤区及嘉定、宝山部分的上海西北部地带，原来的滨海、潟湖与陆相交替环境，演化为泛滥平原—湖泊的陆相环境，堆积了褐黄、锈黄色粉砂夹黏土或单纯的土黄色黏土，其埋深大致介于5~20m，厚度在5~10余米不等。在其以东的嘉定东部和宝山、市区、浦东、闵行、松江、金山及奉贤东部等地的上海中部地带，此时演化为呈网状穿插的陆上三角洲平原分流河道和分流间滩地或边滩环境，沉积了褐黄色细砂、粉细砂和粉砂、砂质粉土夹黏性土，其中偶见有孔虫，反映偶尔受到海水上侵的影响。其埋深、厚度均比其西边的泛滥平原区增大，通常顶界埋深介于15~25m，底界埋深介于25~45m，厚度通常为10~20m。杭州湾北岸的奉贤及其以东地区处在滨海环境，堆积了浅灰黄、黄灰色粉细砂夹砂质粉土，埋深通常介于25~45m，厚度通常为10~20m，其中断续出现少量的广盐性和窄盐性有孔虫，局部伴有很少的半咸水、海相介形虫。

在崇明等三岛及浦东、宝山局部地区，此期地层已被侵蚀殆尽而未能恢复其时的沉积环境。

在20~11ka全球进入末次冰期晚冰阶期间，本区古气候变为总体上凉冷干燥，海面曾急剧下降至最低位置。据研究，在20~15ka的末次冰期最盛期时，东海海面比现今海面降低130~150m，海岸带远退至现今海下的陆坡带，比现代海岸线后退逾500km。当时，原来局限于上海西北部发育的泛滥平原—湖泊已扩展到除现河口区外的上海整个陆域，广泛沉积了上更新统顶部的褐黄色粉质黏土和具特征的暗绿色黏土层（图26-7），组成南汇组上段或滆湖组上段，其中常见陆相介形虫、轮藻受精卵膜、鱼骨碎片等淡水生物化石，还出现大量的环纹藻以及同生菱铁矿结核之类的湖泊指相成分，表明了当时所处的陆相沉积环境；同时在上海南西部及南汇嘴地区曾陆续检出含半咸水浅水有孔虫和半咸水介形虫和海水沟鞭藻，这反映在海面总体下降的背景下，当时东南部地区还受到时而回溯的海水影响。

光释光定年值表明，南汇组上段/滆湖组上段下部的褐黄色粉质黏土形成于17~20ka，上部的暗绿色黏土层形成于11~17ka。由于较长时间的低海面，褐黄—暗绿色黏性土逐渐脱水成陆，经历了风化成壤的过程。

在13~11ka的晚冰阶末期，由于海平面尚未回升至本区，现上海陆域中东部地带分布的褐黄—暗绿色硬土台地受到频繁变迁、分合河网的侵蚀切割，在相当于新仙女木突冷事件侵袭本区时，在河道切割形成的内叠阶地或废弃河道内，堆积了如市区、浦东等地深40余米的暗灰绿—褐灰色黏性土及嘉定娄塘、上海外滩等地所见的泥炭层。沉积物孢粉中普遍可见环纹藻、原球藻，还有较多的旱生草本植物花粉，反映为温凉偏干气候下的淡水泛滥湖泊沉积。据市区外滩和嘉定娄塘深约45m泥炭层的^{14}C定年值，其为形成于12.6~11.5ka的沉积物，表明与20世纪80年代在欧洲苏格兰和挪威西海岸等地发现的“新仙女木突冷事件”为同期的产物。

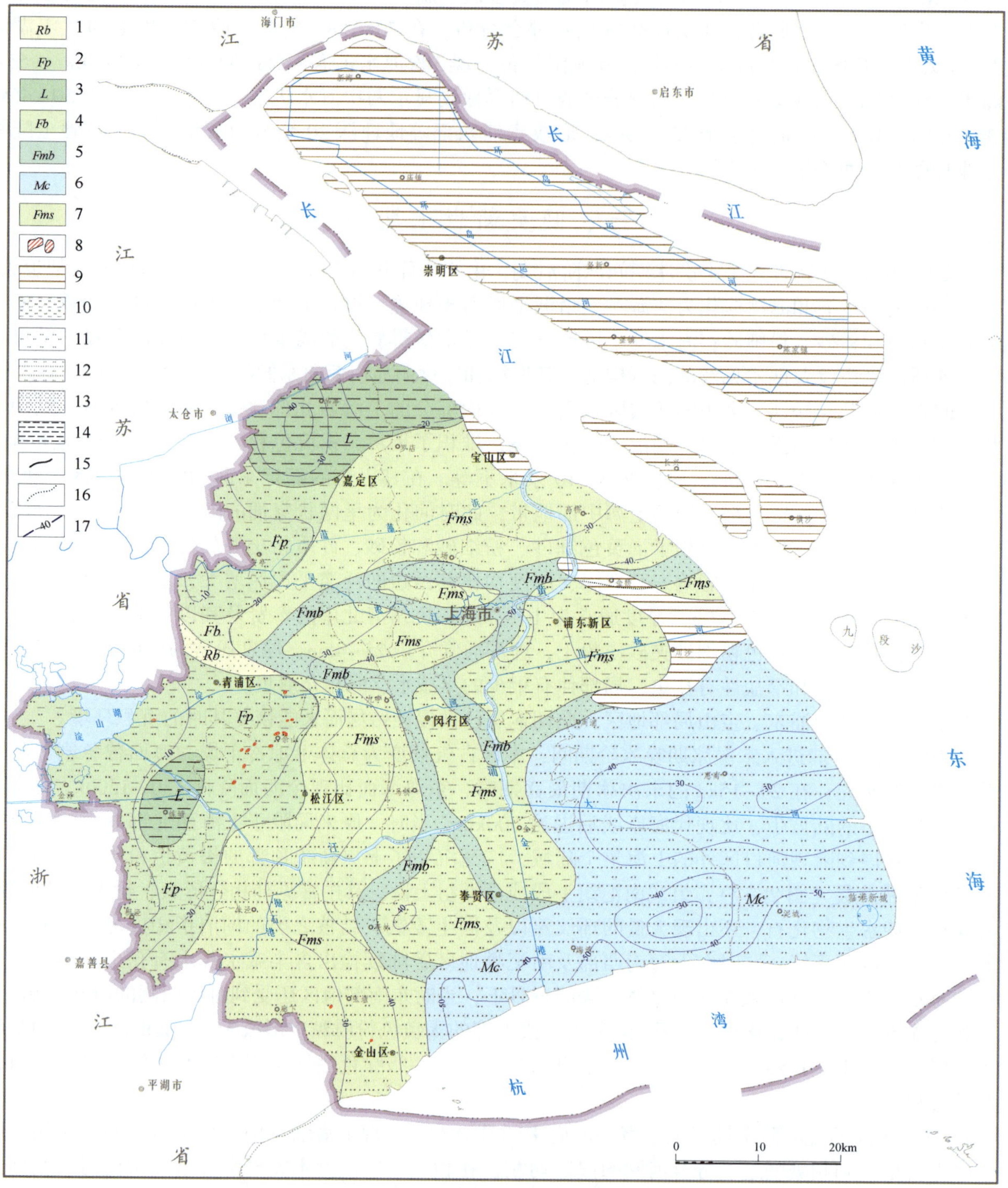

图 26-7　上海晚更新世末期岩相古地理图

Fig. 26-7　Lithofacies paleogeographic map of the end Late Pleistocene in Shanghai

1—河流相；2—泛滥平原相；3—湖泊相；4—边滩相；5—三角洲平原分支河道相；6—沙质潮坪相；7—分流间滩地或边滩相；8—基岩残丘；9—侵蚀缺失区；10—粉砂、粉土夹黏性土；11—粉土为主和粉砂；12—粉细砂；13—细砂和粉细砂；14—黏性土，偶夹粉砂；15—岩相分区界线；16—岩性分区界线；17—层底等高线（黄海高程，m）

第四节　全新世

一、全新世早期

约 15ka，随着全球气候的回暖，大陆冰盖逐渐融化，进入了常称的“冰消期”，导致包括中国东部陆架海在内的全球海平面的回升。依据本区取得的光释光和^{14}C定年值，冰后期海侵开始波及上海在12~11ka，其前锋首先抵达现崇明岛东端河口区，其时海面达到的高度约为现黄海高程-50m 处。

在 10.3~9.5ka，气候期进入相当于 A. Blytt—R. Sernander 气候分期的北方期，当时上海处于大致为现淮河流域的北亚热带北缘气候条件下，古气候趋于温暖稍湿。其时东部陆架海面大幅上升，初期海水沿现河口区—浦东滨岸地带及原陆域内切割较深的支流河谷侵进，当时海平面达到的高度约为现黄海高程-35m 处，形成河口湾及溺谷堆积，主要为灰色砂质粉土、黏质粉土与灰白色粉细砂互层（图 26-8）；在浦东滨岸地带多为灰色砂质粉土与粉质黏土互层。其中普遍出现多少不一的广盐性有孔虫、少量浮游种和窄盐性有孔虫，还有少量半咸水和海相介形虫及个别的陆相介形虫。这部分地层构成了娄塘组下段，厚度可达 10~20m。

在 9.5~8.0ka，相当于 A. Blytt—R. Sernander 气候分期的北方期，古气候趋向温和偏干，转为相似于现华北平原南部的暖温带气候，海平面上升趋于缓慢以至相对停顿，其时海面达到的高度为现黄海高程-15~-20m 处，侵进的海水淹没了包括金山、松江、闵行、宝山及其以东地区，使现今上海的广大陆域沦为滨海沼泽和潮滩，堆积了富含芦苇根茎和黄色泥灰质结核的灰黑色黏土和泥炭夹层，其中可检出少量但优势度高的广盐性有孔虫和很少的半咸水介形虫；在现崇明等三岛和浦东则为河口湾和潮滩环境，堆积了深灰色粉质黏土夹灰白色粉砂或成互层，含有少量的广盐性有孔虫和半咸水介形虫及很少的浮游有孔虫，含属种略多而优势度较低。这个时期的河口湾堆积物构成了娄塘组上段，厚度多为 5~10m。

佘山、天马山残丘群以西包括青浦、嘉定及与之接壤的金山、松江、宝山的部分地区，则基本为连片裸露的晚更新世暗绿色硬土台地，凸出于滨海沼泽之上，未受冰后期海侵的侵进。

二、全新世中期

在 8.0~5.0ka，相当于 A. Blytt—R. Sernander 气候分期的大西洋期，进入冰后期气候的最适宜时期。上海处在相当于现浙闽地区的中亚热带北部气候条件下，古气候趋向暖热潮湿，其间海平面再度显著上升，成为全新世海侵的最盛期，海侵范围直抵江苏镇江，由此被名为“镇江海侵”。随着海侵向西推进，原来全新世早期广布于现上海西北部的晚更新世暗绿色硬土台地，此时已退缩至佘山、天马山附近的青浦及与松江、金山的接壤区。包括宝山、市区、浦东和闵行东部的现陆域东部地带和崇明等三岛所在的河口地区沦为滨岸浅海环境，堆积了深灰色淤泥质黏土。在海侵最盛的中、后期，即在 6.5~5.0ka，海岸带长期稳定滞留于嘉定中部—闵行、奉贤两区的西部地带，先后构筑了紧邻的两道断续相连的贝壳砂堤“沙冈”（北部称“吴塘冈”）和“竹冈”（北部称“外冈”），与相邻的滨海滩涂组成“冈身带”。其时，与“冈身带”毗邻的西部金山、松江及青浦、嘉定的部分地区演化为潟湖、沼泽，堆积了普遍含植物碎片和褐铁矿结核的灰色粉质黏土（图 26-9）。

在 5.0~3.0ka，相当于 A. Blytt—R. Sernander 气候分期的亚北方期，上海转入相当于华北平原南部与淮河流域过渡的暖温带气候条件下，古气候趋向温凉稍干，致使海面略有下降，海岸线随之略为向东后撤，又陆续在嘉定城至颛桥一带构筑了两道贝壳沙堤。当时在“冈身带”以东的长江三角洲沉积区演化为河口滨海环境，主要堆积了灰色粉土与粉质黏土互层和部分的淤泥质粉质黏土；在“冈身带”之西的太湖沉积区则由前期的潟湖演化为后期的湖沼，堆积了深灰色（淤泥质）粉质黏土夹粉砂或粉土。这一时期与上述海侵最盛期形成的以淤泥质黏性土为特征的海相地层组成了上海组，其埋深

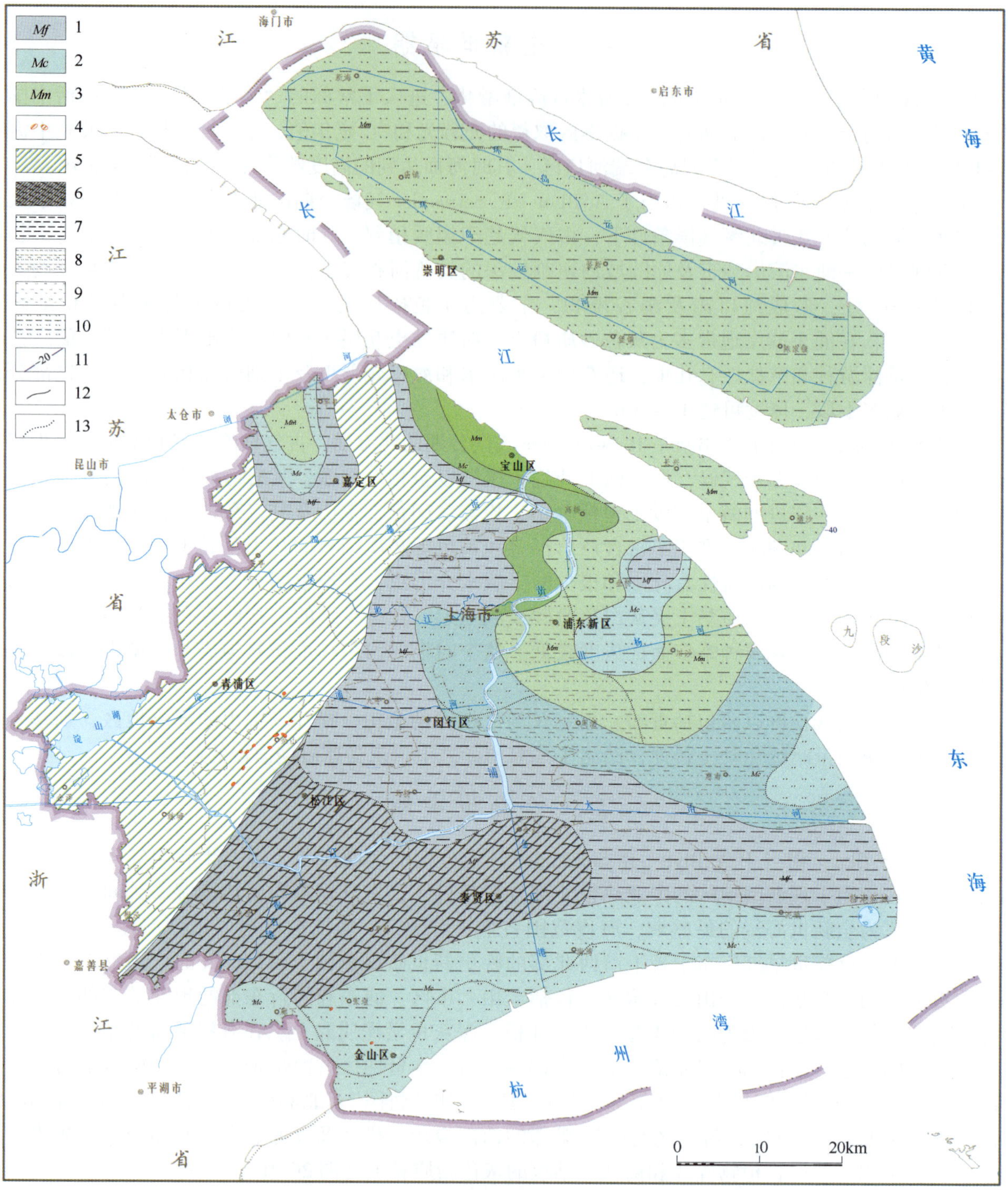

图 26-8　上海全新世早期岩相古地理图

Fig. 26-8　Lithofacies paleogeographic map of the early Holocene in Shanghai

1—滨海沼泽；2—潮滩；3—河口湾；4—基岩残丘；5—硬土台地；6—淤泥质黏土；7—黏性土；8—黏性土夹粉细砂；9—粉细砂、粉土，或夹有黏土；10—黏性土与粉细砂互层；11—层底等高线（黄海高程，m）；12—岩相分区界线；13—岩性分区界线

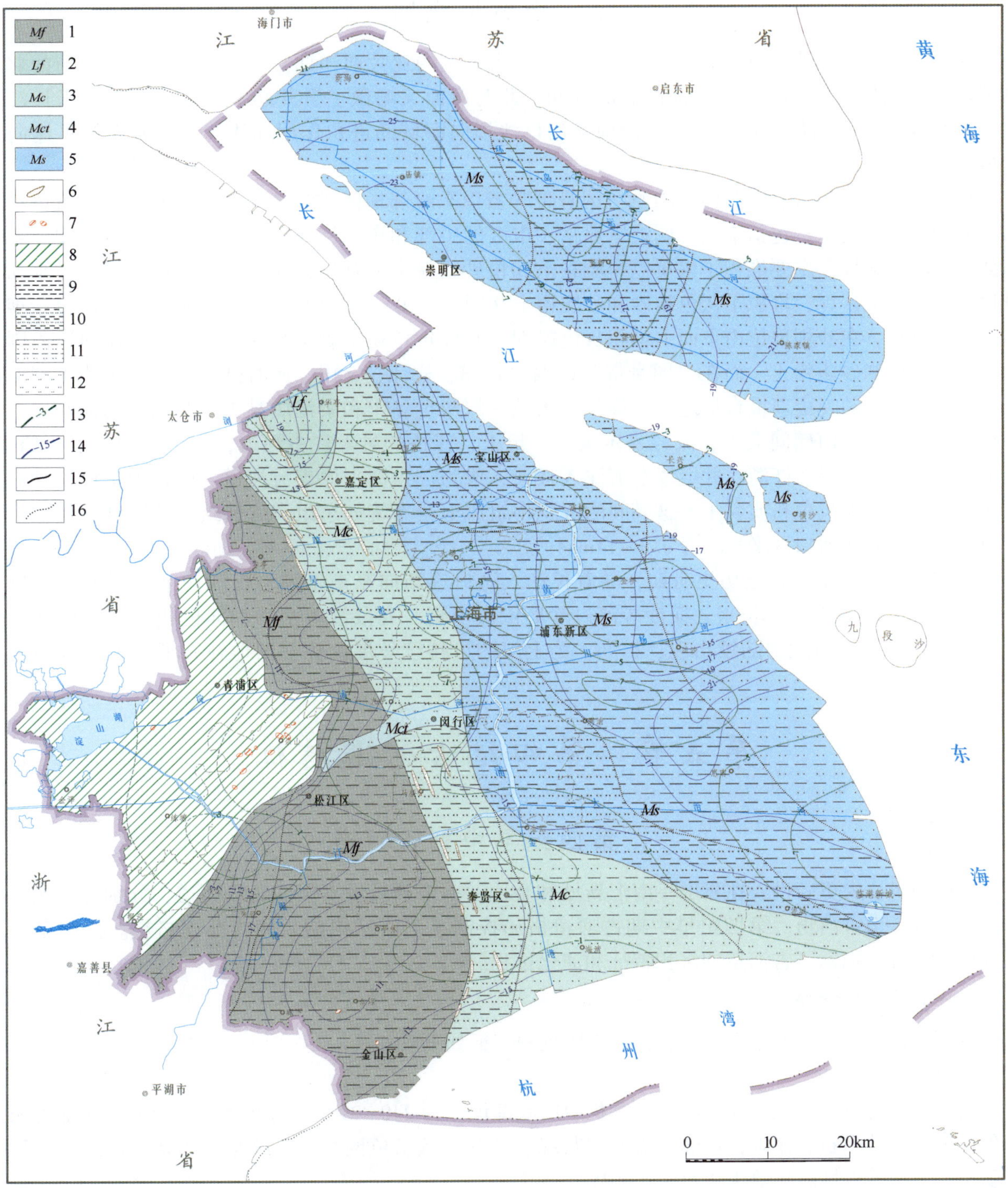

图 26-9 上海全新世中期岩相古地理图

Fig. 26-9 Lithofacies paleogeographic map of the middle Holocene in Shanghai

1—滨海沼泽相；2—潟湖相；3—潮坪相；4—潮汐水道相；5—滨岸浅海—河口湾相；6—贝壳沙堤；7—基岩剥蚀残丘；8—硬土台地；9—淤泥质黏性土；10—（淤泥质）黏性土夹粉砂、粉土；11—（淤泥质）黏性土与粉砂、粉土互层；12—粉砂；13—层顶等高线（黄海高程，m）；14—层底等高线（黄海高程，m）；15—岩相分区界线；16—岩性分区界线和厚度均由西南太湖区向东北的长江三角洲沉积区增大，在太湖沉积区（包括“冈身带”）埋深介于2.5~15m，厚度通常介于7~12m；至长江三角洲沉积区埋深介于5~25m，厚度多为10~18m。

三、全新世晚期

约3ka以来，相当于A. Blytt—R. Sernander气候分期的亚大西洋期，上海转入与现今相仿的北亚热带南缘的气候条件下，古气候回复温暖湿润。此时海面又略微上升并趋于稳定，但是由于长江中上游携来泥沙显著增加，使长江三角洲前缘河口沙坝向海域的进积作用加快，导致海岸线迅速向东后撤。

嘉定、宝山、闵行、奉贤诸区的原“冈身带”及以东毗邻地区成为潮上泥坪，堆积了褐黄色粉质黏土夹粉砂层；其东，包括宝山东侧滨岸地带、市区的中部和东部、浦东北部则为潮间带沙泥混合坪和部分的潮上泥坪，主要堆积了褐黄色粉质黏土和粉砂或粉土的互层；在南汇嘴附近地区则在水动力较强条件下形成沙坪，堆积了灰色粉砂和砂质粉土（图26-10）。在崇明等河口三岛上，为河口沙坝与汊流河道分布，在沙坝部位由下而上堆积灰色粉细砂和棕黄色粉质黏土与粉砂互层，汊流河道内则堆积灰色粉细砂和砂质粉土。而在“冈身带”以西的青浦、松江、金山，此时已基本演化为淡水湖泊沼泽，超覆于原来的晚更新世台地之上，堆积出褐黄—深灰色黏土和淤泥质黏土以及浅部的泥炭层。另外从西侧的安亭、白鹤间流入，流经嘉定黄渡、纪王、江桥的吴淞江古河道，至市区后扩展为扇形，南抵虹桥、北至大场、江湾，并继续东流注入长江，成为三角洲平原支流河道，堆积了灰色粉砂夹砂质粉土和表部的褐黄色粉质黏土。这其间的沉积物组成上全新统如东组和青浦组，其厚度总体上也由西南太湖区向东北河口区增大，介于1~14m不等。

约在1.2ka以来的挽近历史时期，随着长江入海口逐渐向东南推移，长江输来的泥沙在河口不断进积，使上海的崇明、长兴、横沙等河口沙岛依次出现，九段沙等水下沙洲也相继出水成陆。随着河口沙坝和侧翼滩地的陆续出现、扩展，为上海的可持续发展源源不断地补充了弥足珍贵的土地资源。

第五节　第四纪古河道特征

一、第四纪早中期古河道

早更新世、中更新世，在嘉定—南汇新城和金山—南汇新城发育了两条古河道，分别称为浏—南古河道和枫—奉古河道。第四纪早中期古河道格局控制了当时区内的岩相古地理格局。典型钻孔对比（图26-11）揭示了古河道的沉积特征和演化过程。

（一）沉积特征

古河道区下更新统安亭组下、中、上3段沉积特征如下：

1. 下段

浏—南古河道区和枫—奉古河道区沉积地层底部埋深为240~290m，厚度为30~50m，底部为灰黄等色含砾砂和砂砾石、泥砾，顶部覆盖薄层灰绿、黄绿、蓝灰色网纹、斑纹穿插的杂色网纹状、斑杂状黏土。古地磁极性时年龄为2.60~2.00Ma。

2. 中段

枫—奉古河道区该段沉积地层底部埋深为180~220m，沉积特征与下段相似，仍为灰黄等色含泥砾砂和砂砾石，以泥砾为主，顶部覆盖薄层斑杂状黏土，但沉积物粒径总体较下段变细。浏—南古河道区该段沉积地层底部埋深为190~240m，沉积特征为一套下粗上细的旋回，底部不再以泥砾为主，而以含砾砂或砂砾为主，上部变为灰绿、黄绿、蓝灰色网纹、斑纹穿插的杂色网纹状、斑杂状黏土，局部夹砂层。古地磁极性时年龄为2.00~1.07Ma。

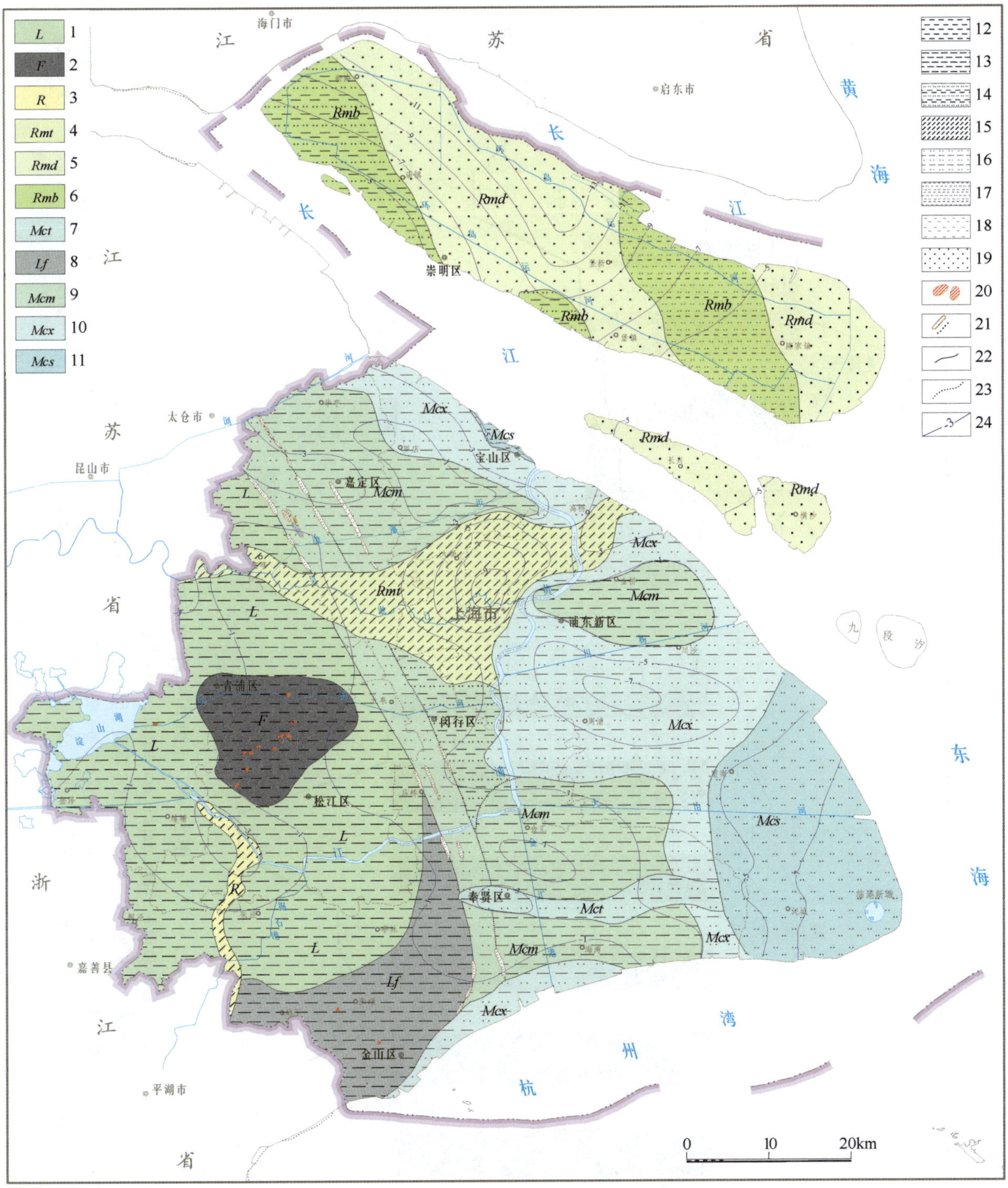

图 26-10　上海全新世晚期岩相古地理图

Fig. 26-10　Lithofacies paleogeographic map of the late Holocene in Shanghai

1—湖泊相；2—沼泽相；3—河流相；4—支流河道—河口相；5—河口汊流河道相；6—河口沙坝相；7—潮道相；8—潟湖相；9—潮上泥坪相；10—潮间混合坪相；11—潮下沙坪相；12—黏土夹泥炭；13—淤泥质土、黏土，偶夹粉砂；14—粉质黏土夹粉土、粉砂；15—粉土为主和粉砂或夹粉质黏土；16—粉质黏土与粉土、粉砂互层；17—粉细砂、粉土夹粉质黏土；18—粉砂为主和粉土；19—粉细砂和粉土；20—基岩残丘；21—中全新世贝壳沙堤；22—岩相分区界线；23—岩性分区界线；24—层底等高线（黄海高程，m）

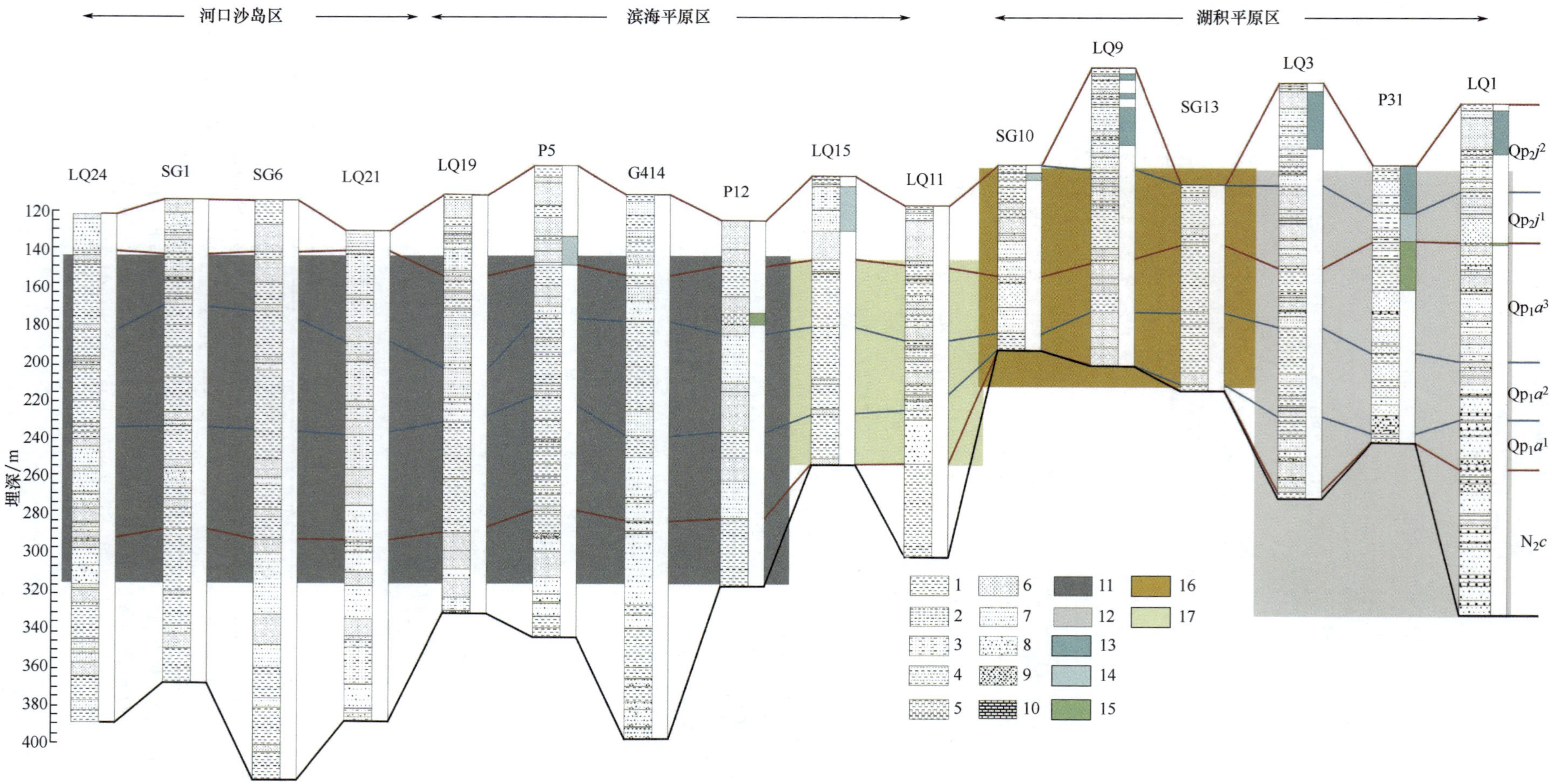

图 26-11　上海早更新世地层对比

Fig. 26-11　Stratigraphic correlation of the Early Pleistocene in Shanghai

Qp_2j^2—中更新统嘉定组上段；Qp_2j^1—中更新统嘉定组下段；Qp_1a^3—下更新统安亭组上段；Qp_1a^2—下更新统安亭组中段；Qp_1a^1—下更新统安亭组下段；N_2c—上新统崇明组。
1—黏土；2—粉砂质泥；3—泥质粉砂；4—泥夹粉砂纹层；5—泥砂互层或砂夹泥纹层；6—细砂、中砂、粗砂；7—含砾砂；8—砂质砾或砾石层；9—泥质砾；10—基岩；11—浏南河带；12—枫泾河带；13—第四海相层（M4）；14—第五海相层（M5）；15—第六海相层（M6）；16—第四纪早—中期台地/阶地（第一级）；17—第四纪早—中期台地/阶地（第二级）

3. 上段

浏—南古河道区和枫—奉古河道区沉积地层底部埋深为 140~190m，地层沉积特征均为一套下粗上细的旋回变化，底部以含砾砂或砂砾为主，上部变为灰绿、黄绿、蓝灰色网纹、斑纹穿插的杂色网纹状、斑杂状黏土，局部夹砂层。古地磁极性时年龄为 1.07~0.78Ma。

（二）演化过程

浏—南古河道和枫—奉古河道下更新统 3 个阶段均分别在嘉定—南汇新城和金山—南汇新城内的凹陷内发育，总体范围没有大的变动，但河道的类型却发生了演化。

早更新世早期，两条古河道均没有完全形成，均为继承上新世的洪积扇沉积体系，分别分布在枫泾凹陷和三墩凹陷（图 26-12），两个沉积扇体中间为松江和金山之间的基岩剥蚀区所隔。枫泾 LQ1 孔和浦东机场 J25 孔沉积物组成及其中的地球化学元素指示该期间两地物源为局地物源。早更新世早期古河道仅在崇明及浦东沿江局部分布。

早更新世中期，嘉定—南汇新城明显形成了由北而南的浏—南古河道，在河道两侧形成了河流边滩带，河道的走向依然受基岩隆起的限制，主要在浦东沿江一带，上海市区则为泛滥平原—湖泊。相比之下，枫—奉古河道还没完全形成，仅在金山—南汇新城一带出现了狭窄的古河道，形成了河道的雏形（图 26-13）。

该时期，金山、松江一带仍为洪积扇体系，并且扇体的范围较早期更大，松江和金山之间的基岩剥蚀区已沦为沉积扇发育场所。钻孔物源分析显示浏—南古河道已经可以接受少量的可能来自长江中游的远源泥沙的堆积，微古化石显示浏—南古河道可能遭受十分微弱的海侵影响。相比之下，枫—奉古河道未发现任何海侵踪迹，同时，枫泾地区沉积物源仍为局地近源沉积。

早更新世晚期，由北而南的浏—南古河道范围进一步拓宽，表现在河流边滩进一步向上海市区延展，而此时的枫—奉古河道也完全形成，金山—南汇新城之间的基岩凸起完全被河流溯源侵蚀切开，东西贯通（图 26-14）。金山、临港及浦东机场处钻孔物源分析表明，沉积物来源发生了重大的调整，长江中上游的远源泥沙能够较多的输入浏—南古河道，并可能代入枫—奉古河道。

中更新世早期，古河道发育继承了早更新世晚期的古河道，但范围更大，发育更具规模，浏—南古河道边滩拓展至崇明等三岛大部分区域，枫—奉古河道进一步展宽，在金山、奉贤、松江交界地带产生心滩。中更新世古河道格局体现了区内地貌准平原化特征。

二、第四纪晚期古河道

晚更新世晚期，由于海平面大幅度下降，河流强烈下切侵蚀形成了长江三角洲地区的古河谷和古河间地（阶地）（图 26-15）。长江主河谷宽度可达 70km，下切深达 80m。长江河谷两翼古河间地广泛分布暗绿色硬黏土，古地势表现为三级阶地特征，埋深分别小于 5m、5~15m 和 20~30m。古河间地分布数条支流古河道（Li et al.，2002；Wang et al.，2018）。河口沙岛地区属长江主下切河谷；滨海平原分布支流河谷，这些区域第一暗绿色硬黏土层被剥蚀殆尽，可能由于一直以侵蚀作用为主，并未明显堆积河流滞留堆积物。末次冰期结束后，随着海平面快速回升，河口迅速后撤，下切古河道遭受海侵从河流环境演化为河口湾，河口湾堆积物充填了下切河谷。

上海古河道和古河间地典型钻孔对比揭示了区内晚更新世末期下切河谷的演化过程。晚更新世末下切河谷区沉积层序依次为晚更新世末的河道滞留沉积、末次冰消期潮汐河道沉积、全新世河口湾相海侵充填沉积、三角洲相海退沉积（图 26-16），体现了末次冰消期海平面上升控制下沉积环境的演化特征。

三、晚更新世末期的河道滞留沉积

该沉积层仅在河口沙岛可以识别，由一套厚度可达 30m 的含砾砂层组成，底部埋深最大可达 80~90m，地质年龄为 30~20ka B.P.，底部见侵蚀面。滨海平原分布下切河谷内一般不见河道滞留沉积，底界埋深相对较浅，一般为 30~40m。

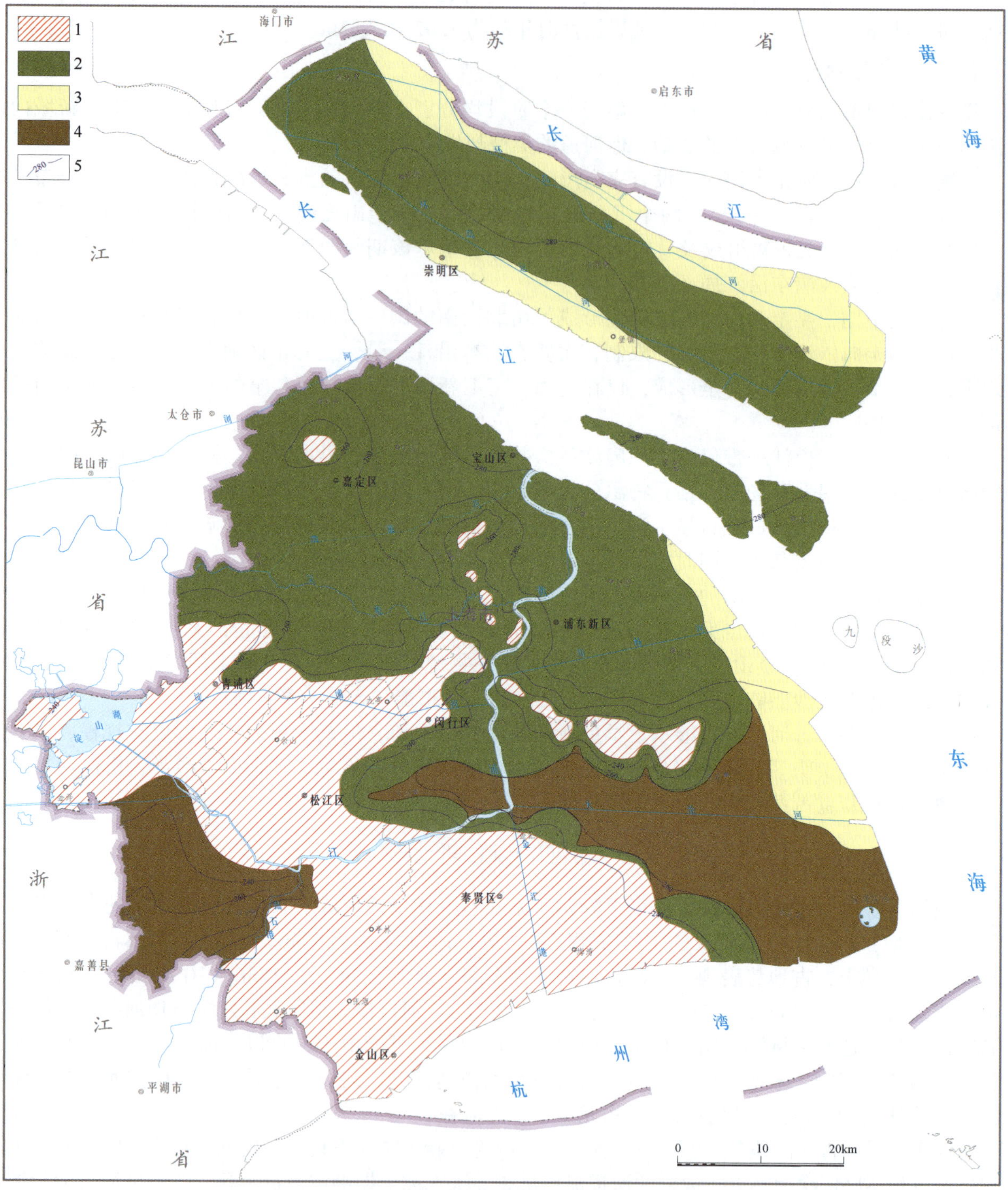

图 26-12　上海早更新世早期古河道分布

Fig. 26-12　Distribution of the early Early Pleistocene paleochannels in Shanghai

1—基岩剥蚀区；2—河间地（泛滥湖泊、河漫滩、边滩）；3—河床；4—冲积扇；5—下更新统下段层底埋深等值线（m）

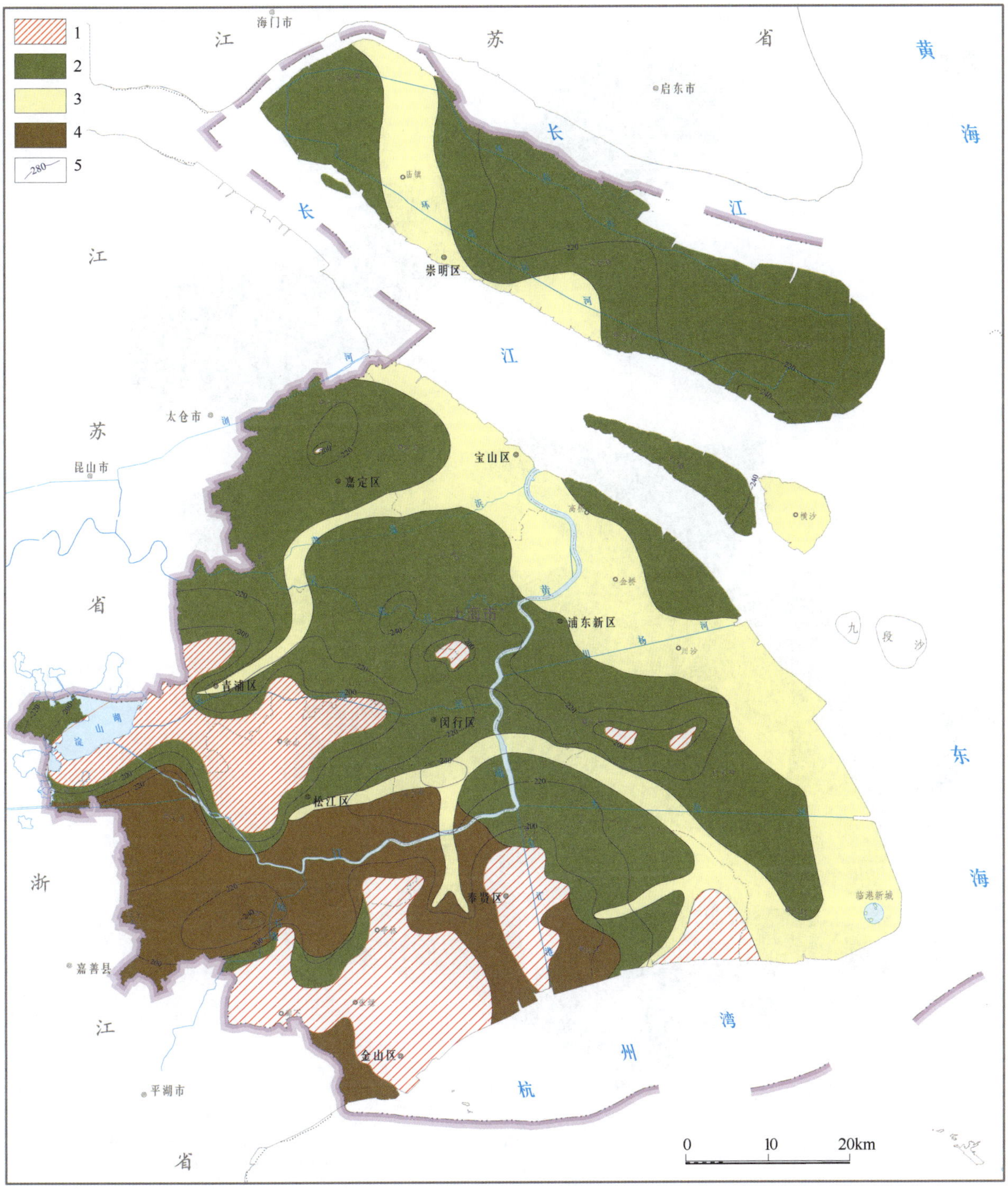

图 26-13　上海早更新世中期古河道分布

Fig. 26-13　Distribution of the middle Early Pleistocene paleochannels in Shanghai

1—基岩剥蚀区；2—河间地（泛滥湖泊、河漫滩、边滩）；3—河床；4—冲积扇；5—下更新统中段层底埋深等值线（m）

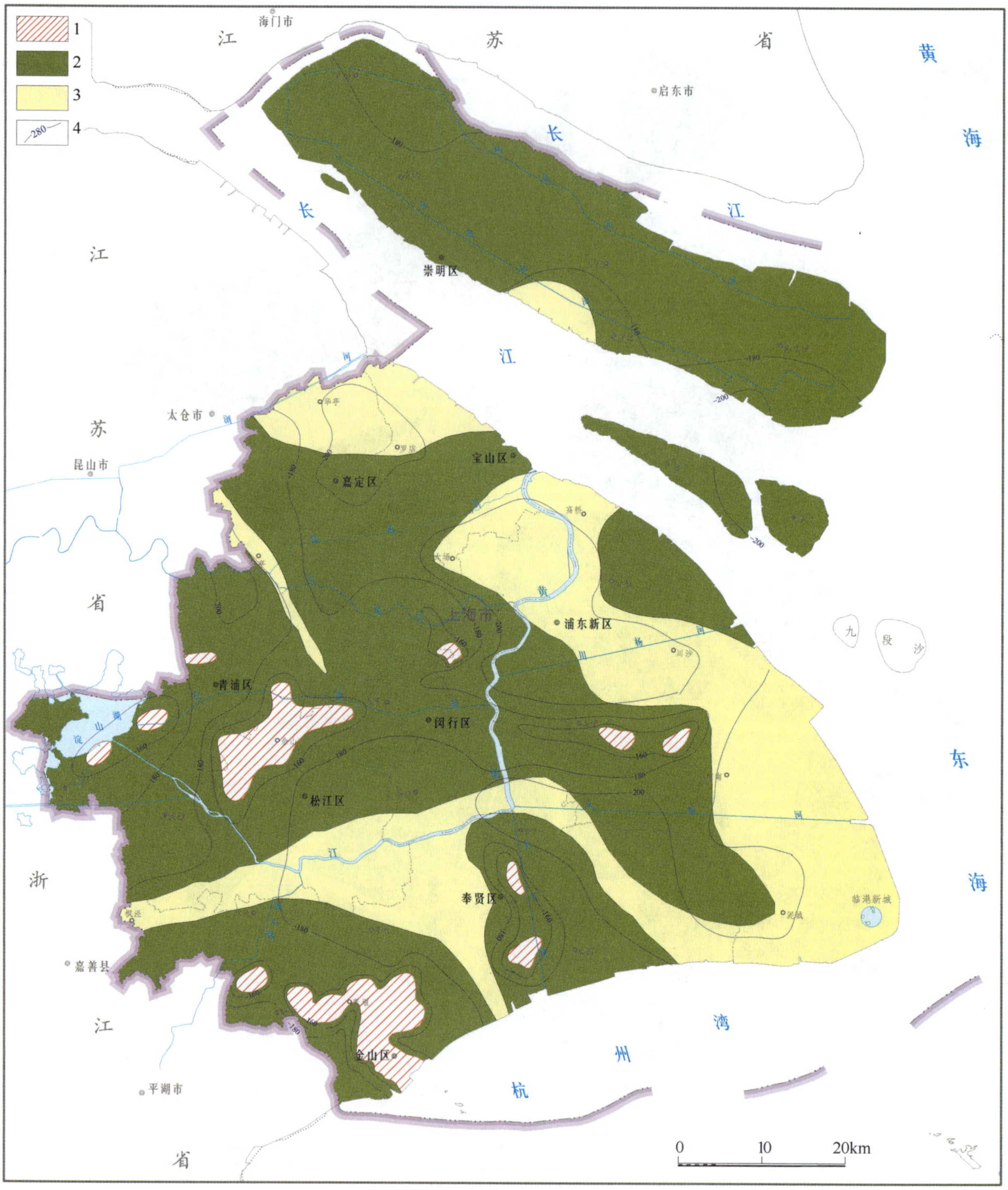

图 26-14　上海早更新世晚期古河道分布

Fig. 26-14　Distribution of the late Early Pleistocene paleochannels in Shanghai

1—基岩剥蚀区；2—河间地（泛滥湖泊、河漫滩、边滩）；3—河床；4—下更新统上段层底埋深等值线（m）

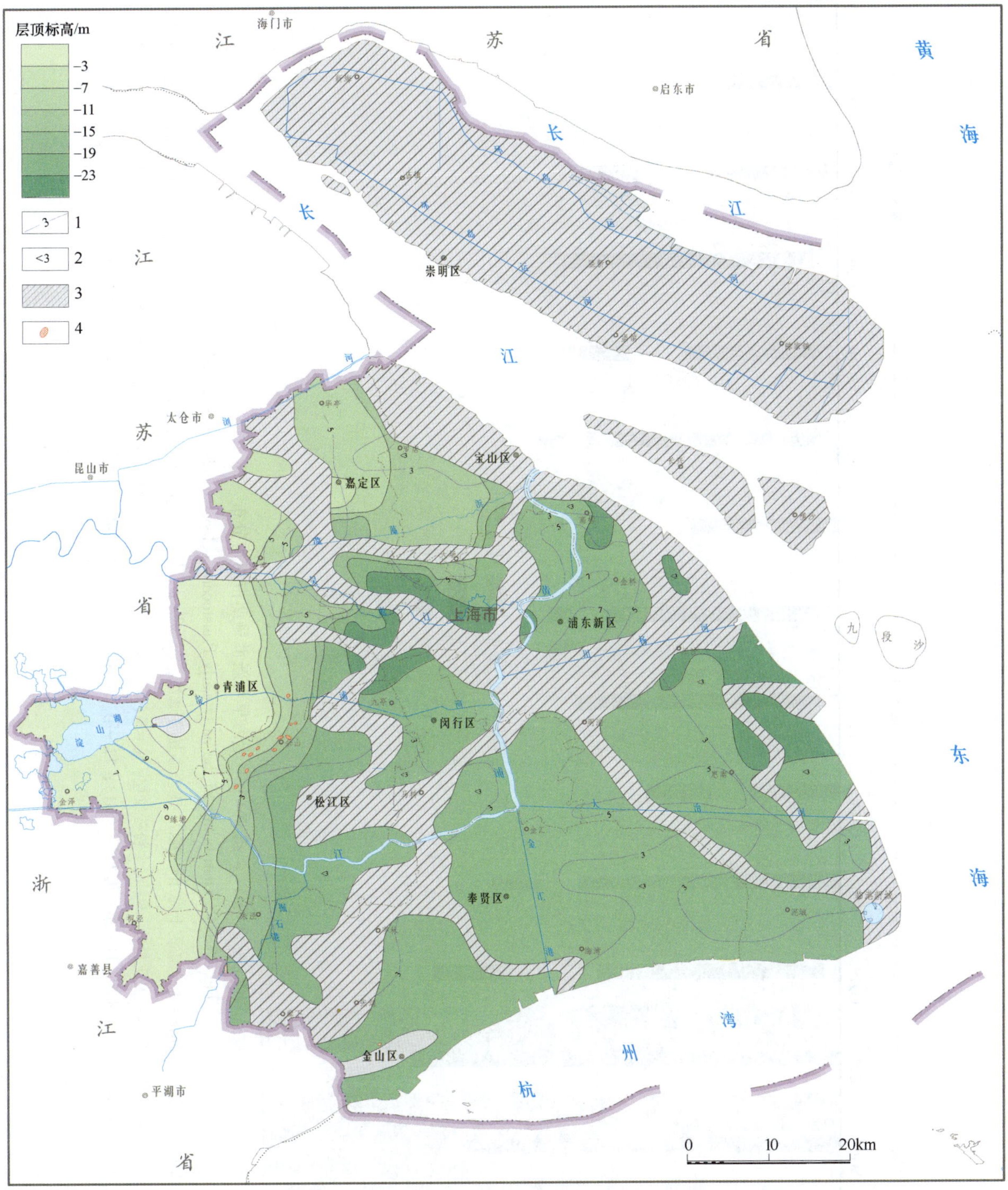

图 26-15　上海晚更新世末期暗绿色硬黏土分布图

Fig. 26-15　Distribution of dark green hard clay formed in the end-stage Late Pleistocene in Shanghai

1—层厚等值线（m）；2—层顶标高范围（m）；3—缺失区；4—基岩区

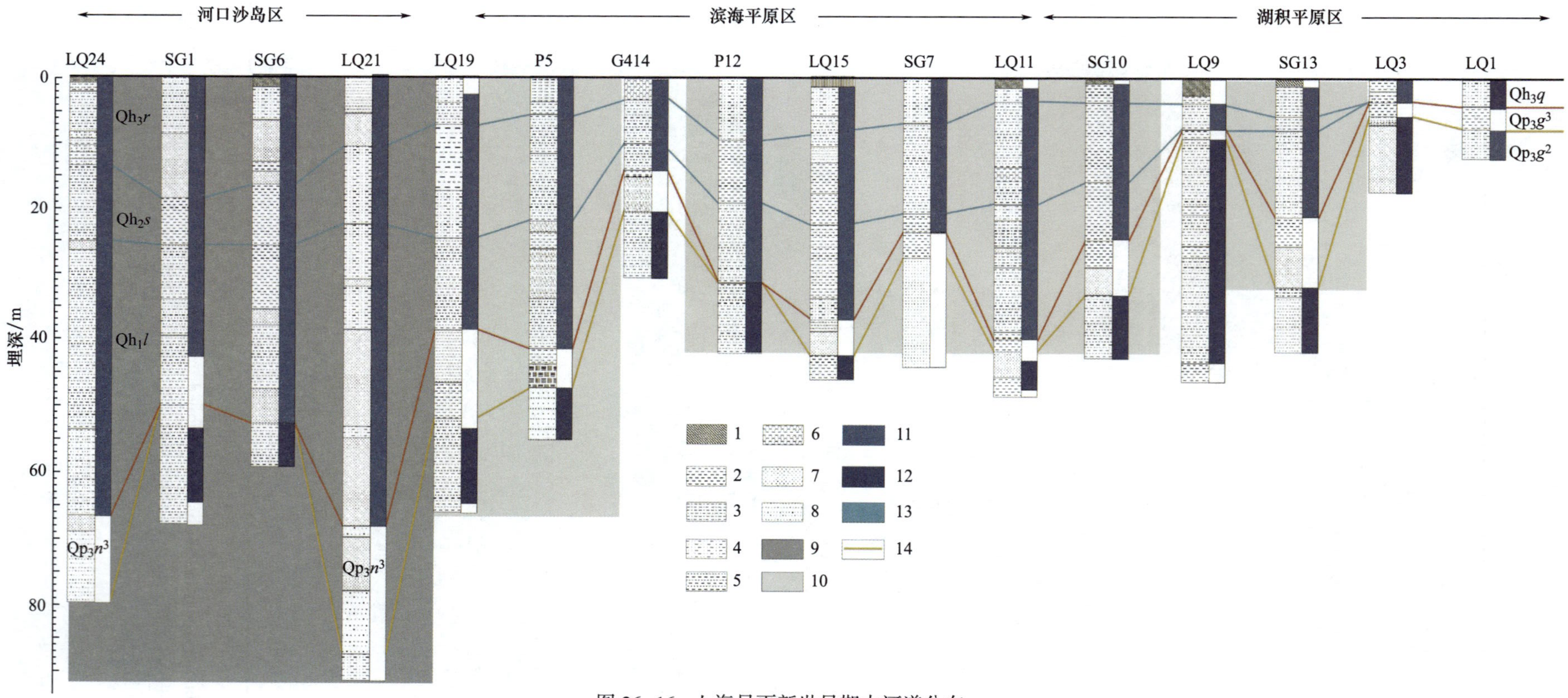

图 26-16 上海早更新世早期古河道分布

Fig. 26-16 Distribution of the early Early Pleistocene paleochannels in Shanghai

Qh_3r/Qh_3q—上全新统如东组/青浦组；Qh_2s—中全新统上海组；Qh_1l—下全新统娄塘组；Qh_3n^3/Qh_3g^3—上更新统南汇/漏湖组上段；Qh_3n^2/Qp_3g^2—上更新统南汇/漏湖组中段

1—人工填土；2—黏土；3—粉砂质泥；4—泥质粉砂；5—泥夹粉砂纹层；6—泥砂互层或砂夹泥纹层；7—细砂、中砂、粗砂；8—含砾砂；9—河口沙岛长江下切主河谷；10—三角洲平原支流河谷；11—第一海相层（M1）；12—第二海相层（M2）；13—第三海相层（M3）；14—侵蚀面或间断面界线

四、末次冰消期潮汐河道沉积

该沉积层仅分布在河口沙岛主河谷区，埋深为40~60m，由砂和砂泥互层沉积为主，微古化石显示遭受海侵的影响，地质年龄为18~11ka B.P.。滨海平原支流河谷一般不保留末次冰消期早期沉积，但在娄塘P5孔埋深43m处见新仙女木事件泥炭沉积。

五、全新世早期海侵沉积

该沉积地层埋深在25~40m，地质年龄为10~9ka B.P.，底部为泥砂互层潮滩沉积，上部为灰色泥夹粉砂纹层河口湾沉积。

六、全新世中晚期最大海泛——海退沉积

该沉积层埋深在10~25m，地质年龄为8~3ka B.P.，在河口沙岛主河谷区为以深灰色均匀泥为主的浅海—前三角洲沉积，在滨海平原支流河谷则为以深灰色均匀泥为主的浅海—潮滩沉积。

七、全新世晚期海退沉积

该沉积层在河口沙岛主河谷区由三角洲前缘相砂组成，埋深在10m左右，在滨海平原则由灰黄色潮下带砂和褐黄色潮上带泥组成，埋深多在5m以浅。

第六节　长江演化沉积记录

长期以来，关于长江演化问题开展了大量研究，并取得了丰硕的成果，但对于“长江东流水系形成于何时”这一关键科学问题却一直存在重大争议，有前新生代、第三纪早期（古近纪）、第三纪晚期（新近纪）、更新世早期和更新世晚期等多种观点（张信宝等，2018）。过去几十年，区域调查和科研项目也在区内开展了长江演化的研究。陆志坚（1989）最早基于上海大量水文地质钻孔岩性资料分析指出，区内松散层深部埋藏两条古河道，分别沿嘉定—南汇新城南北向分布和枫泾—南汇新城东西向分布，分别命名为浏—南古河道和枫—奉古河道。区内磁性地层学研究也表明这两条古河道在早更新世和中更新世一直发育（邱金波和李晓，2007）。依据“杭州幅1∶250000区域地质调查”“上海市幅1∶250000区域地质调查”“长江三角洲重点地区三维调查（江苏）”和“长江三角洲重点地区三维调查（上海）”对调查区第四纪岩相古地理的恢复，浏—南古河道属于长江古河道的分支，而枫—奉古河道是钱塘江古河道的变迁范围。最近10年，一些学者相继从物源角度在上海开展了长江演化和长江贯通入海的沉积记录研究。范代读等（2004）利用浦东机场PD-99号孔独居石年龄谱，认为第四纪是青藏高原隆升对东中国海沉积产生直接影响的开端，暗示着长江贯通入海发生自第四纪；Chen et al.（2009）基于浦东机场PD孔重矿物组合数据指出晚更新世长江贯通入海；Gu et al.（2014）依据浦东机场PD孔微量元素组合，认为长江贯通入海的时间节点是1.0~1.2Ma；2015年，“长江三角洲重点地区三维调查（上海）”对上海湖积平原SG13孔、LQ1孔和滨海平原SG7孔、PD孔和LQ15孔的指示酸性岩和基性岩矿物比例的微量元素含量比值进行了对比，发现基性岩明显的影响开始于1.0~1.2Ma，暗示着长江泥沙明显输入上海的时间发生自1.0~1.2Ma，结合岩相古地理分析，进一步发现构造沉降控制了长江物源的影响范围时空演化规律：1.0~1.2Ma以后长江泥沙开始扩散至古河道区域，0.78Ma以后长江泥沙扩散扩大至古河间地；Yue et al.（2016）对浦东临港新城SG7孔碎屑磁铁矿地球化学元素进行了分析，认为异常高含量Cr_2O_3的磁铁矿可作为示踪长江沉积物入海的指标，也认为长江贯通入海发生在1.0~1.2Ma；Yue et al.（2018）进一步对长江流域干流和支流表层沉积物和崇明岛LQ24孔沉积物重矿物和碎屑磁铁矿地球化学进行综合分析比对，发现1.2Ma以后出现高含量的辉石和富含钛、铁、钒、铬的磁铁矿，表明沉积物物源由近距离转移到远距离，暗示长江贯通入海。

Liu et al.（2018）对宝山 LQ19 孔沉积物中磁性参数进行了系统的分析，发现 145m 以上沉积物磁性矿物含量是下伏沉积物的 5~10 倍，并且与长江上游受峨眉玄武岩物源影响的表层沉积物中磁性矿物含量相似，据此认为长江贯通入海时间发生在 1.0~1.2Ma。最近，Yue et al.（2018）对江苏区内 SK10 孔、上海区内 LQ11 孔、PD 孔和 LQ24 孔重矿物以及长江流域表层沉积物中重矿物进行分析和对比，发现第四纪开始稳定矿物组分发生了显著变化，认为长江贯通入海时间节点发生在 2.58Ma。

总结上述研究成果，类似现代的长江水系及长江贯通入海的时间依然因指标体系不同而存在争议，但总体上偏向于早更新世。值得注意的是，目前区内关于长江入海的认识多是基于磁性地层年代标尺，但如二十一章第二节所述，本区磁性地层框架存在问题，因此，今后年代地层的准确厘定是精确揭示长江泥沙扩散至本区的关键。

第七篇　城市地质

概　　述

城市地质工作，是在城市及其周围地区或潜在城市化地区的特定空间范围内，综合考虑各种地质要素，研究其对城市发展所提供的资源、所施加的约束条件以及城市发展对其产生的影响，为城市规划、建设和管理服务的地质工作。调查和研究内容具有高度的综合性和渗透性，涉及众多的学科领域，有基础地质、水文地质、工程地质、环境地质、矿产地质、地球化学、遥感地质等许多地质学科。新中国成立以来，区内地质工作先后开展了金属、非金属和油气资源，以城市规划和供水为主的水文地质、工程地质，以地面沉降防治为目标的地质环境，综合性的城市地质等地质调查和研究工作。尤其是，20 世纪 90 年代以来，开始开展中大比例尺的水文地质、工程地质、环境地质调查及城市地质综合调查和研究，并专门成立了城市地质研究机构。代表性项目有：1991—1993 年，上海市环境地质站开展了“黄浦江以西、规划外环线以内地区的上海市城区工程地质详查（1：20000）”；1996—1998 年，开展了“各区县 1：2 万~1：10 万水文地质调查”；1998—1999 年，在上海市 20 个县(区)水文地质调查和地下水资源开发区域规划工作的基础上，上海市城市地质研究所开展了“上海市区域水文地质调查”；1996—1999 年，上海市城市地质研究所开展了“上海市环境地质调查（1：20 万）”；1999—2002 年，上海市地质调查研究院开展了“长江三角洲地区上海市地下水资源与地质灾害调查评价”；与此同时，为参与全国水资源评价工作，上海市环境地质站开展了“上海市地下水资源评价工作”；2002 年，上海市地质矿产局组织编制出版了《上海市地质环境图集》；2004—2005 年，上海市地质调查研究院开展了“上海市浅层地下水有机污染调查评价”。2006—2007 年，上海市地质调查研究院开展了“上海市地下水资源可持续利用及应急水源地建设可行性调查研究”。2007—2009 年，上海市地质调查研究院开展了“上海市基本农田环境质量动态监测与预报方法研究”。随着工作的不断推进，工作区也逐渐从陆域逐渐扩展到了近岸海域，相继开展了海岸带地区的多目标地球化学调查、滩涂资源演化规律研究、综合地质调查等工作。

2004 年开始，作为全国首个城市地质工作的试点城市，原国土资源部与上海市政府合作开展了“上海城市地质调查”，这标志着上海城市地质工作迈入 1.0 时代。该项目对区内历史资料进行了系统收集和整理，共设置了 4 个一级课题、两个示范调查区调查和 11 个二级子课题，一级子课题和示范区调查分别为“上海市三维地质结构调查与地下空间开发适应性评价”“上海市区域地球化学调查”“上海市地质环境及其对城市安全影响调查与对策研究”“上海市环境地球化学调查与评价”“上海市城市地质基础信息管理与服务系统”“中国 2010 年上海世博会会址区地质环境调查与安全性评价”“临港新城规划区地质结构调查及对工程建设影响评价”。

项目构建了三维地质结构模型，实现了控制深度达百米的全市域、不同地质结构的“透明化”显示（图Ⅶ-1，图Ⅶ-2）。掌握了土壤和浅层地下水地球化学环境背景条件及近 30 年土壤重金属元素含量变化趋势，基本建成了覆盖上海的全地类土地环境质量动态监测网络，研发了三维可视化的上海城市地质信息管理和服务系统。开展了上海城市地质环境容量评价研究，重点进行了地下水环境容量、地球化学环境容量和城市建筑容量评价尝试。通过“上海城市地质调查”项目，建立了自 1917 年以来的海底地形数据库，对滩涂资源变化规律和后备土地资源潜力进行了综合研究，结合同期实施的“上海市近岸海域多目标区域地球化学调查”等项目，初步实现了区内地质工作由陆向海的跨越。

2009 年，原国土资源部和上海市人民政府签订了《关于共建保障和促进上海科学发展国土资源管理新机制的合作备忘录》，开启了上海城市地质工作 2.0 时代。通过实施“上海地质资料信息集群化产业化”等项目，建立了完善的地质资料汇交、管理和共享机制，出台了市政府规章《上海市地质资料管理办法》，建立了商业性地质资料的汇交管理机制。此外，建成了上海市地质资料数据中心，形成了

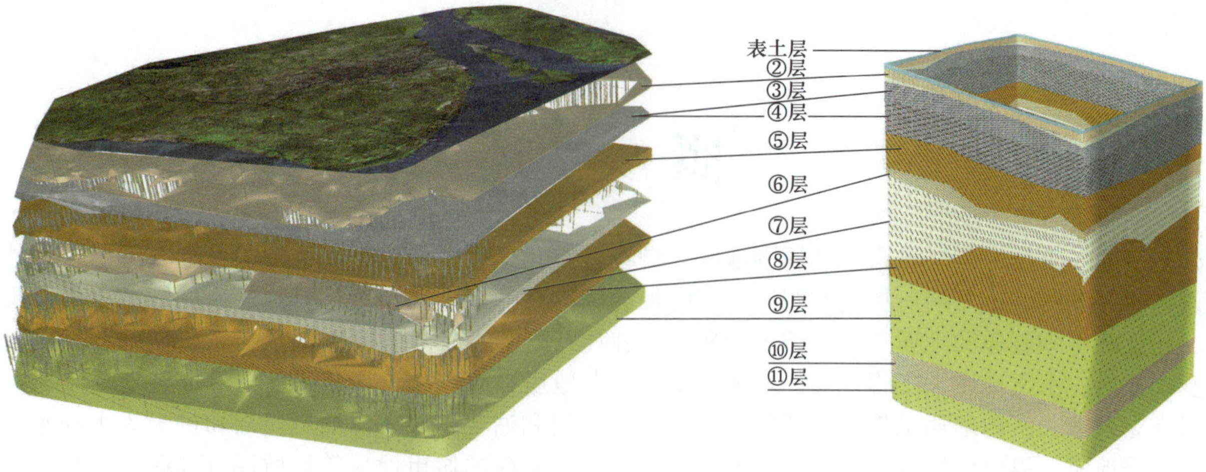

图Ⅶ-1 上海工程地质结构模型图

Fig. Ⅶ-1 Engineering geological model diagram of Shanghai

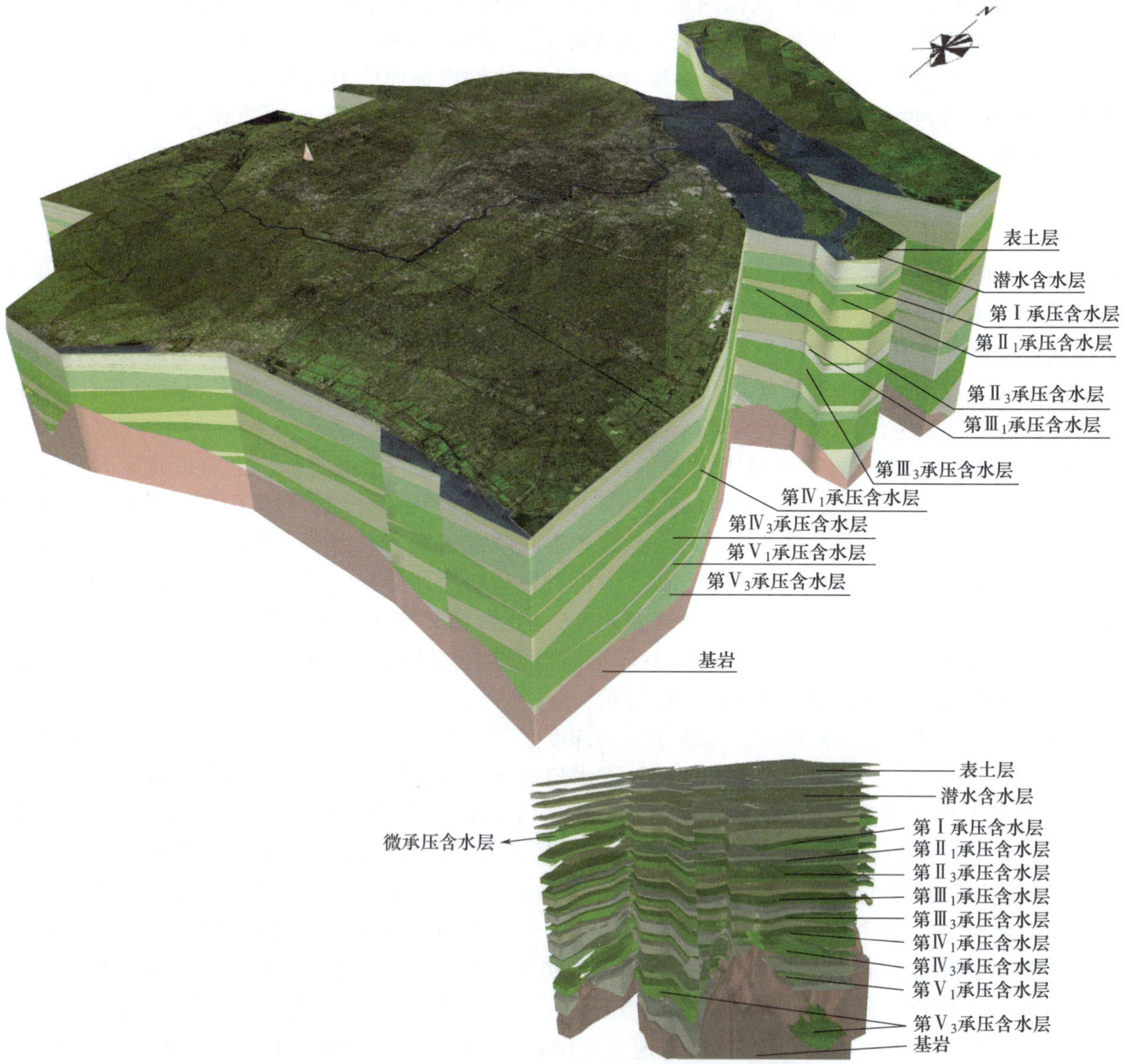

图Ⅶ-2 上海三维水文地质结构图

Fig. Ⅶ-2 3D Hydrogeological structural diagram of Shanghai

覆盖上海市和各地质专业领域、跨越多比例尺和多时间尺度的城市地质数据库体系。同时，构建了面向政府、企事业和公众的“1+X”应用服务体系。2011 年，中国地质调查局与上海市规划和国土资源管理局合作开展的“上海市海岸带地质调查与监测预警示范”项目正式获得立项。该项目拟开展上海近岸海域 1：25 万及重点地区 1：5 万地质调查、海岸带地质环境监测、专题研究等；但受经费限制主要开展了 1：25 万地质调查和近岸海岸带监测工作。通过该项目，初步掌握了海陆一体的地质结构和表层沉积物环境质量状况，构建了海岸带地质环境监测体系，开展了上海后备土地资源潜力和重大海岸带工程安全评价，全面开启了全市海陆一体化的地质调查、地质环境监测、地质信息集成和地质成果应用等工作。

2014 年，为更好地支撑服务上海城市规划、建设、运营和管理，围绕后工业化时期城市转型发展的需要，上海将城市地质工作全面纳入政府管理主流程，标志着上海城市地质工作进入以地质信息多元化服务为特色的 3.0 时代。先后出台了《上海市地面沉降防治管理条例》《地质工作进一步服务规划国土资源管理实施方案》《重大市政工程地面沉降预警机制》。此外，形成了系统的地质信息多元服务技术标准和规范体系，为地质科技成果转化和共享提供支撑。结合后工业化时期上海城市发展特点，相继开展了地下空间、区域地壳稳定性和潜在震源区地震地质条件等调查究工作，评价了地下空间资源量和开发利用适宜性（图Ⅶ-3）。

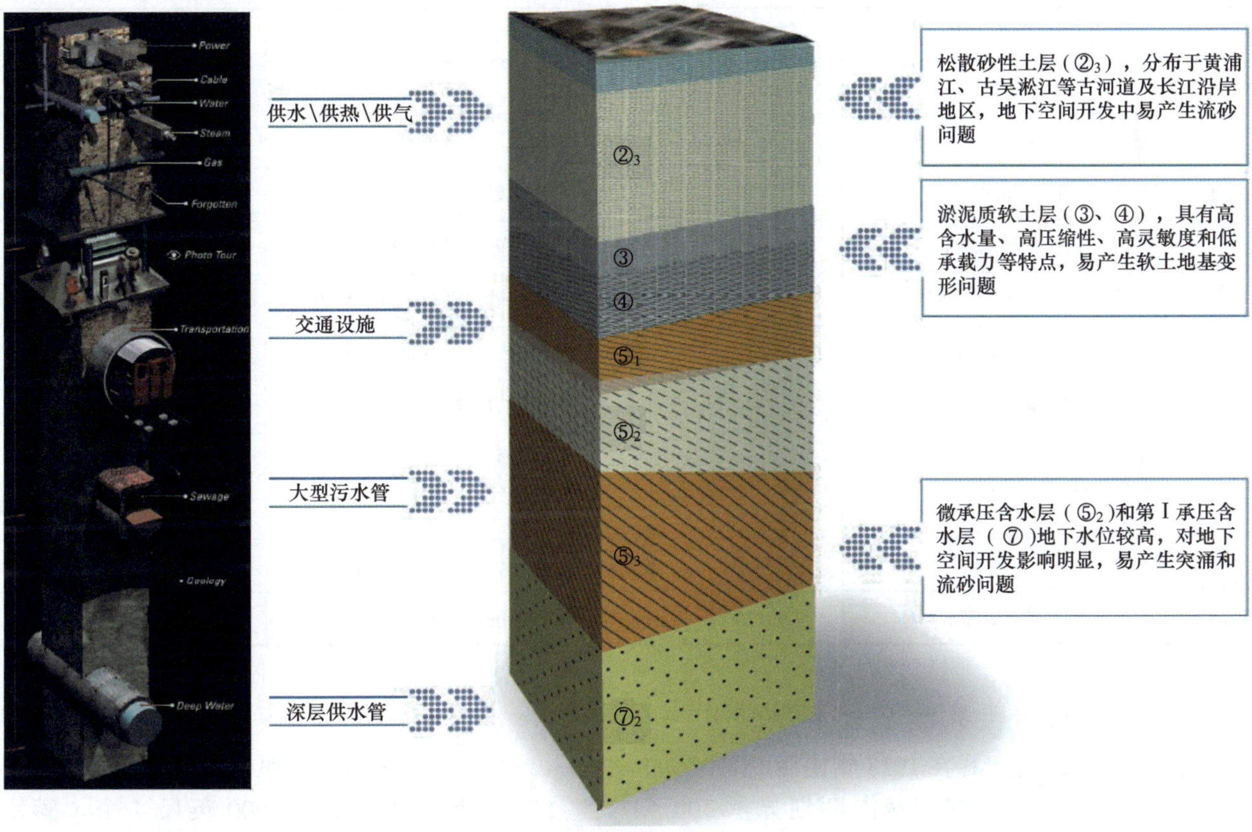

图Ⅶ-3　上海地下空间利用示意图

Fig. Ⅶ-3　Underground space utilization diagram of Shanghai

第二十七章　水文地质条件

一、以往含水层划分结果

（一）含水层划分总体思路

上海系统性水文地质普查工作始于 20 世纪 50 年代，对松散岩类、基岩水文地质条件开展了不同精度区域性调查和专门研究。依据地下水赋存介质岩性和空间类型，将地下水划分了 4 种类型：松散岩类孔隙水（第四系）、碎屑岩类孔隙水（新近系、古近系和白垩系）、碳酸盐岩类裂隙溶洞水（奥陶系、寒武系、震旦系）和基岩裂隙水（前上新统）（表 27-1）。

表 27-1　上海地下水类型及含水层划分一览表

Table 27-1　Groundwater types and water-bearing formations in Shanghai

地下水类型		含水层		分布
大类	亚类	地层时代代号	主要含水岩性	
松散岩类孔隙水	潜水	Qh、Qp_3	粉细砂、粉土、粉质黏土	除基岩裸露区外均有
	上更新统承压水（Ⅰ组）	Qp_3	中粗砂、细中砂、砾石粗砂、粉细砂	平原区
	中更新统承压水（Ⅱ组）	Qp_2	中粗砂、细中砂、砾石粗砂、粉细砂	平原区
	下更新统承压水（Ⅲ组）	Qp_1	中粗砂、细中砂、砾石粗砂、粉细砂	平原区
	上新统承压水（Ⅳ组）	N_2	细砂、中粗砂、含砾及半胶结砂	平原区
碎屑岩类孔隙水		N_1、E_1、K_2	砾岩、含砾砂岩、砂岩	平原区
碳酸盐岩类裂隙溶洞水		O、ϵ、Z	灰岩、白云质灰岩、白云岩	丘岗碳酸盐岩断凸及其附近隐伏区
基岩裂隙水	碎屑岩、火山岩、变质岩类裂隙水	前 N_2	砂岩、石英岩、砂页岩、火山岩系	丘岗与平原
	侵入岩类裂隙水	γ、λo 等	以燕山期为主的花岗岩、二长岩等	零星分布

（二）松散岩类孔隙水含水层划分

美国人 G. H. 克里赛在 1928 年出版的 The *China Journal of science and Arts* 文献中，将上海的含水层描述成“蓄水池”，根据埋藏深度划分了 300 英尺❶、500 英尺、700 英尺、900 英尺 4 个“蓄水池”（相当于第Ⅱ、第Ⅲ、第Ⅳ、第Ⅴ承压含水层）。

1957 年，上海市政工程设计院钟淳昌、沈家锡，把上海的地下含水层划分为 1 个无压潜水层和 4 个承压含水层。

1958 年，苏联水文地质专家阿拉诺维奇在《供水水文地质勘测》一书中对含水层的划分与钟淳昌、沈家锡的划分一致（表 27-2）。

❶ 1 英尺 = 30.48cm。

表 27-2　阿拉诺维奇的含水层划分方案（1958 年）

Table 27-2　Aranovic's aquifer division (1958)

含水层编号	埋藏深度/m	厚度/m	含水层的成分
第一层	70~95	10~15	粗砂和砾石
第二层	128~155	10~20	中砂和细砂
第三层	190~220	10~20	中砂和细砂
第四层	245~280	12~20	细砂

1961 年，何塘在《上海地区地质工作总结报告》中认为，长江三角洲的疏松层呈透镜体，其厚度各地不一，含水层互相交叠，其规律是由西而东，层次是逐渐增加的。为此，将含水层划分为潜水和 3 个承压含水岩组（表 27-3）。

表 27-3　何塘的含水层划分方案（1961 年）

Table 27-3　Hetang's aquifer division (1961)

含水岩组	埋深/m	岩性	含水层厚度/m	静水位/m	涌水量 $[L\cdot(s\cdot m)^{-1}]$	渗透系数 $[m\cdot(24h)^{-1}]$	德国度
潜水	0~3						28.55
第Ⅰ承压含水组	-40~-80	粉砂、细砂、中砂	70~90	-8~-10	3~4	7~23	8.25
第Ⅱ承压含水组	-180~-260	细砂、中砂、粗砂	40~60	-10~-12	12~20	5~15	
第Ⅲ承压含水组	-220~-250	中砂、粗砂、砾石	20±	-10	12	7.7~7.8	

1965 年，包曼芳等把上海含水层进一步划分成 1 个潜水层、5 个承压含水层（表 27-4）。此后，含水层的划分基本统一。

这一划分原则一直沿用至今。

表 27-4　上海松散岩类孔隙水含水层划分表

Table 27-4　Division of unconsolidated rock pore aquifers in Shanghai

含水层次	埋藏深度/m	厚度/m	岩性	分布
1	35~50	5~15	以粉砂为主	杨浦、普陀
2	70~100	25~35	以粗砂砾石为主	全市
3	110~150	20~30	以细砂为主	全市
4	180~240	40~60	以粗砂为主	全市
5	260~280	10~20	以细中砂、砾石为主	市区中东部

注：1965 年，包曼芳等的划分表。

二、近期含水层划分结果

依据上海松散岩类孔隙水含水层划分原则，结合近几十年来水文地质调查和研究成果，进一步细化了含水层的划分。

潜水含水层：为冰后期堆积地层，以河口—滨海相为主，伴有河流、湖泊相堆积。含水砂层主要沿分支古河道、当代河口和滨海岸分布。由于在亚北方期大规模海进，区内堆积了较厚的淤泥质黏土和粉质黏土，使其分为上部潜水含水层和下部微承压含水层。

中、上更新统承压含水层：为河口—滨海—浅海相堆积，可分别划分为一个含水层和两个含水层，即第Ⅰ承压含水层和第Ⅱ、第Ⅲ承压含水层。对于含水层，进一步根据岩性、彼此之间水力联系程度的不同，将含水层划分为上部含水层、下部含水层及含水层内弱透水层，如第Ⅱ承压含水层进一步划分为Ⅱ$_1$、Ⅱ$_2$、Ⅱ$_3$，其中Ⅱ$_1$、Ⅱ$_3$分别代表上、下部承压含水层，Ⅱ$_2$为第Ⅱ承压含水层内弱透水层。

下更新统承压含水层：为河口—滨湖相（晚期为河口—滨海相）堆积，且以湖泊相为主，可划分为两个含水层，即第Ⅳ、第Ⅴ承压含水层。对于含水层，进一步根据岩性、彼此之间水力联系程度的不同，将含水层划分为上部含水层、下部含水层及含水层内弱透水层，如第Ⅳ承压含水层进一步划分为Ⅳ$_1$、Ⅳ$_2$、Ⅳ$_3$，其中Ⅳ$_1$、Ⅳ$_3$分别代表上、下部承压含水层，Ⅳ$_2$为第Ⅳ承压含水层内弱透水层。

上海松散岩类孔隙水含水岩组划分结果详见表27-5。

表27-5 上海松散岩类孔隙水含水层划分结果一览表

Table 27-5 Division of unconsolidated pore water-bearing formations in Shanghai

<table>
<tr><th colspan="2">地质年代</th><th colspan="3">水文地质层填图单位</th><th colspan="2">岩性特征</th><th colspan="2">主要分布范围</th><th colspan="2">成因类型</th></tr>
<tr><td rowspan="4">全新世</td><td rowspan="2">晚期 Qh_3</td><td rowspan="2">潜水含水层</td><td colspan="2">表土层 A_1</td><td rowspan="2">黄褐色粉质黏土、褐灰色淤泥质粉质黏土夹泥炭</td><td>褐黄色黏性土、灰黄色黏性土、淤泥质黏性土</td><td colspan="2">全区分布</td><td rowspan="2">湖沼</td><td>河口滨海</td></tr>
<tr><td colspan="2">A_2</td><td>灰色粉土、粉砂</td><td colspan="2">河口砂岛、东部海岸带、冲积平原</td><td>河口—溺谷</td></tr>
<tr><td>中期 Qh_2</td><td>第一隔水层</td><td></td><td>N_1</td><td colspan="2">灰色淤泥质粉质黏土、黏土</td><td colspan="2">东部滨海平原普遍分布</td><td colspan="2">河口滨海</td></tr>
<tr><td>早期 Qh_1</td><td>微承压含水层</td><td></td><td>B</td><td colspan="2">灰色砂质粉土、粉砂</td><td colspan="2">东部滨海平原局部分布</td><td colspan="2">溺谷相</td></tr>
<tr><td rowspan="10">更新世</td><td rowspan="6">晚期 Qp_3</td><td>第二隔水层</td><td colspan="2">N_2</td><td colspan="2">灰、暗绿—褐黄色黏性土、草黄色粉性土</td><td colspan="2">⑤$_3$、⑤$_4$分布于古河道区，⑥层广泛分布，唯古河道缺失</td><td colspan="2">洪泛平原</td></tr>
<tr><td>第Ⅰ承压含水层</td><td colspan="2">Ⅰ</td><td colspan="2">黄灰、灰色砂质粉土、粉砂</td><td colspan="2">广泛分布，但在河口砂岛及古河道深切区缺失</td><td colspan="2">河流—河口、滨海—潟湖</td></tr>
<tr><td>第三隔水层</td><td>西部 N_3</td><td>东部 N_3</td><td>与灰色粉质黏土夹粉砂互层，夹暗绿、灰黄色硬塑状的粉质黏土</td><td>灰色粉质黏土夹粉细砂或为薄互层</td><td>西部湖沼平原</td><td>东部滨海平原</td><td>滨海—洪泛平原</td><td>滨浅海</td></tr>
<tr><td rowspan="3">第Ⅱ承压含水层</td><td colspan="2">上部Ⅱ$_1$</td><td colspan="2">灰和灰绿色细砂、含砾中粗砂、含砾中细砂</td><td colspan="2">普遍分布</td><td colspan="2">滨岸浅海、浅海，局部地区中下部为河湖</td></tr>
<tr><td colspan="2">第Ⅱ承压含水层内隔水层Ⅱ$_2$</td><td colspan="2">蓝灰、褐黄、杂色粉质黏土夹细砂</td><td colspan="2">分布较普遍</td><td colspan="2">河湖</td></tr>
<tr><td colspan="2">下部Ⅱ$_3$</td><td colspan="2">黄灰色粉细砂、中细砂、砂砾石</td><td colspan="2">较普遍分布</td><td colspan="2">河口滨海—河流</td></tr>
<tr><td rowspan="4">中期 Qp_2</td><td>第四隔水层</td><td colspan="2">N_4</td><td colspan="2">灰绿、蓝灰色黏土</td><td colspan="2">与上覆含水层沟通区缺失</td><td colspan="2">湖泊</td></tr>
<tr><td rowspan="3">第Ⅲ承压含水层</td><td colspan="2">上部Ⅲ$_1$</td><td colspan="2">灰、灰黄色中细砂，含砾砂、砂砾石</td><td colspan="2">普遍分布</td><td colspan="2">河口湾—河流</td></tr>
<tr><td colspan="2">第Ⅲ承压含水层内隔水层Ⅲ$_2$</td><td colspan="2">灰、蓝灰杂色黏土、粉质黏土</td><td colspan="2">分布较普遍</td><td colspan="2">河湖</td></tr>
<tr><td colspan="2">下部Ⅲ$_3$</td><td colspan="2">灰、灰黄色粉细砂，底部砾质砂、砂砾石</td><td colspan="2">分布较普遍</td><td colspan="2">河流—河口滨海</td></tr>
</table>

续表

地质年代		水文地质层填图单位		岩性特征	主要分布范围	成因类型
更新世	早期 QP_1	第五隔水层	N_5	灰绿、褐黄、杂色网纹黏性土夹粉细砂	普遍分布	湖河、洪泛平原
		第Ⅳ承压含水层	上部$Ⅳ_1$	黄、淡绿色细砂、中粗砂、砂砾石	普遍分布，东部滨海地区与下层沟通	河流
			第Ⅳ承压含水层内隔水层$Ⅳ_2$	杂色网纹、斑质状黏土、粉质黏土	分布较普遍	湖河、洪泛平原
			下部$Ⅳ_3$	浅灰、灰黄色中粗砂、含砾砂，夹黏性土	杭州湾北岸及西部基岩隆起区缺失	河流
		第六隔水层	N_6	黄绿、褐黄、紫红杂色黏性土，含钙质结核	分布于陆域北部至河口三岛屿	湖河、洪泛平原
		第Ⅴ承压含水层	上部$Ⅴ_1$	灰白、灰黄色含砾中细砂和中粗砂	分布于陆域北部至河口三岛屿	河流
上新世	N_2		第Ⅴ承压含水层内隔水层$Ⅴ_2$	灰绿、黄色黏土、粉质黏土，夹砾石	分布于陆域西北部及河口三岛屿	洪泛平原—湖泊
			下部$Ⅴ_3$	含泥质中粗砂及砂砾层	河口三岛屿发育	冲洪积
		第七隔水层	N_7	灰绿、黄色黏土、粉质黏土，夹砾石	河口三岛屿发育	残坡积、湖积、冲洪积
	基岩		J			

三、本志水文地质参数应用有关说明

含水层渗透性：以渗透系数评价结果表征；

含水层富水性：以单井涌水量评价结果表征；

地下水矿化度：以矿化度评价结果表征；

地下水质量：以单因子或多因子综合评价结果表征；

地下水污染：以典型有机、无机组分评价结果表征；

地下水水化学类型：以舒卡列夫分类及定名结果表征；

地下水位埋深：指设施测口至地下水位埋深面之间高度；

地下水位标高：指将地下水位埋深值以吴淞高程基准换算后的地下水位；

浅部承压水：指第Ⅰ承压含水层地下水；

深部承压水：指第Ⅱ、第Ⅲ、第Ⅳ、第Ⅴ承压含水层地下水。

第一节　地下水赋存条件与分布规律

上海地下水主要赋存于松散岩类孔隙中，尤其是历次暖期堆积之粗碎屑颗粒间蕴藏有丰富的地下水资源，其次赋存于碳酸盐岩类岩溶-裂隙和基岩裂隙之中。地下水的赋存条件与分布规律，一般受控于区域地貌、地层岩性及厚度，以及地质构造等因素。而松散岩类孔隙承压水亚类则主要受疏松地层内粗碎屑颗粒大小及其发育厚度所制约。

按第四纪地质分区，地下水赋存条件与分布规律表现为：

一、西部湖沼平原区

天马山、秦皇山凸起地段：该地段自第四纪以来长期处在陆地或高漫滩的沉积环境，所以松散地层较薄，沉积物厚度介于0～100m，以黏性土为主，偶夹砂质粉土、粉砂或细砂。含水层一般缺失，或呈薄层状产出，地下水主要赋存在黏性土或泥质砂性土孔隙和基岩裂隙之中，渗透性和导水性差，为上海贫水地段。

安定、金山凹陷地段：该地段自第四纪以来受分支河道堆积控制，松散沉积物厚度一般介于150～250m，中心部位可达300～400m，砂层频繁交互，含水层发育。金山凹陷砂层累计厚度为60～150m，安亭凹陷累计厚度为80～160m，分别约占松散地层的50%，并由凹陷中心至两侧凸起带厚度递减，含水层岩性由含砾中粗砂渐变为粉细砂。因此，地下水较集中赋存于凹陷的中心地带，然而不同时期亦随分支河道南北迁移而摆动。

在湖沼平原区，除赋存松散岩类孔隙水外，且在青浦重固—凤溪、金山松隐—闵行等隐伏古生界凸起处，基底分布的碳酸盐类岩层中，推测可能赋存有裂隙溶洞水。

二、东部滨海平原区

除中心城小闸镇和浦东新区坦直地区基底凸起松散地层略薄外，通常第四系发育，厚度一般为280～300m，这是南部分支河流入湖或入海的主要堆积场所。砂层迭次发育，累计厚度通常为150～200m，最厚处达240m左右，且揭示含水层齐全。因此，东部滨海平原区是区内地下水最富集的区段。

三、河口砂岛区

河口砂岛区主要指长江河口的崇明三岛（崇明岛、长兴岛、横沙岛），也是上海疏松地层堆积最厚的地区。除第四系外，新近系也发育。疏松地层厚度一般为320～400m，凹陷中心部位可达500m。砂层累计厚度一般为150～200m，但在向化、崇明城桥等凹陷区可达250m，含水层发育齐全，也是地下水富集地区之一。

河口砂岛区由于受后期长江流水冲刷，所以第Ⅰ承压含水层普遍缺失，但新近系碎屑岩类孔隙含水层较发育。含水层厚度及岩性在东西方向有明显的差异，西部以含砾中粗砂夹中细砂为主，东部则为中细砂夹细砂多见，偶夹中粗砂。西部第四系砂层累计厚度比东部薄，但东部新近系砂层累计厚度大，因此，存在东部砂层累计厚度虽比西部大，但总的富水性较西部弱的分布特征。

综上所述，上海地下水赋存条件与分布规律表现为：地下水主要赋存在含砾中粗砂、中细砂或砂砾石层孔隙间，其次赋存于灰岩裂隙、溶洞空间及其构造带裂隙之中。地下水资源主要赋存在东部滨海平原区、河口砂岛地区及西部湖沼平原区之断块凹陷地带，东部滨海平原区地下水资源最为丰富。

第二节　水文地质特征

20世纪中叶以来长期的水文地质勘查工作基本查明了潜水含水层和承压含水层埋藏、富水性、渗透性、地下水质量、水化学类型、地下水位（详见本章第四节）等分布现状。同时通过建立地下水动态监测网，客观地反映了地下水质量（矿化度、质量、水化学类型）、地下水位（详见本章第四节）等动态变化特征。

一、潜水含水层

（一）含水层埋藏特征

潜水含水层为全新世（Qh）河口—滨海相沉积，分布广泛。受地形地貌、成因不同影响，含水层具黏性土与粉性土构成的二元结构特征，和由相对单一的黏性土构成的一元结构特征。

陆域滨海平原的东部、南部沿江海地带主要为二元结构分布区，岩性上部为砂质粉土、粉质黏土，下部为粉细砂夹薄层砂质粉土或粉质黏土；河口砂岛区含水层结构也为二元结构区，中下部岩性为砂质粉土、粉质黏土与粉细砂互层，上部为砂质粉土、粉质黏土夹粉砂；滨海平原北部、西部以及西部湖沼平原地区，呈现一元结构特征，岩性以淤泥质黏性土夹粉砂为主。潜水含水层底面埋深有较大差异，层底埋深在 3~25m 不等。粉性土厚度差异较大，二元结构区粉性土厚度普遍大于 3m，陆域东南部沿江海地带及河口砂岛区粉性土厚度普遍大于 10m（表 27-6；图 27-1）。

表 27-6　潜水含水层埋藏特征对比表

Table 27-6　Burial conditions among different phreatic aquifers

地区	吴淞江、黄浦江沿岸	“冈身带”沿线	河口砂岛及陆域东部沿岸	南汇嘴地区
所属第四纪层（组）	如东组	上海组、如东组	如东组	如东组
沉积环境	三角洲河流	滨海	河口	滨海
空间变化特征	上部均有 A_1 层分布，埋深一般为 2~4m			
	砂层厚度变化较大，平均厚度约为 6m	砂层厚度有一定变化，平均厚度约为 3m	砂层厚度大，但变化不大，平均厚度约为 11m	砂层厚度较大，由西向东厚度加大，平均厚度约为 9m
岩性变化特征	以砂质粉土为主，但黏粒含量较多，且土质不均	大部分地区以砂质粉土为主，部分地区以黏质粉土为主	垂向岩性有一定变化，一般顶部为砂质粉土（②$_{3-1}$），中部为黏质粉土（②$_{3-2}$），下部为砂质粉土或粉砂（②$_{3-3}$）	以砂质粉土为主，颗粒较均匀，具水平层理，土质较均匀

（二）含水层渗透性与富水性

潜水含水层渗透性与富水性的强弱，受含水层岩性、结构等因素影响明显。

调查结果显示，潜水含水层岩性主要为黏性土、粉性土，区域上含水层渗透系数、单井涌水量等相关水文地质参数均表现出较小的特征。由于潜水含水层一元结构、二元结构分区较为明显，空间上含水层富水性、渗透性总体表现为二元结构分布区较一元结构分布区相对好的特征。河口砂岛、陆域东南部紧邻东海和杭州湾北岸的浦东—奉贤粉性土发育的二元结构区，渗透性能相对好，渗透系数总体大于 1m/d，富水性也相对好，单井涌水量略大于 $10m^3/d$（口径 500mm，降深 2m 时）；广大的一元结构分布地区，渗透性相对较差，渗透系数普遍小于 0.5m/d，富水性也相对较差，单井涌水量普遍小于 $1m^3/d$（图 27-2）。

（三）地下水矿化度

1. 陆域地区

普遍分布矿化度小于 1g/L 的淡水，约占全市总面积的 55%；但宝山西南部、松江大部、闵行南部、金山西南大部、浦东东南部—奉贤沿江海地带分布微咸水，矿化度为 1~3g/L，累计分布面积约占全市总面积的 20%；东南沿江地带局部分布咸水，矿化度为 3~10g/L，分布面积约占全市总面积的 4%。

2. 河口砂岛区

崇明庙镇—建设—前进农场一线以北地区、港沿—中兴一线以东地区、城桥—新河沿江地带，长

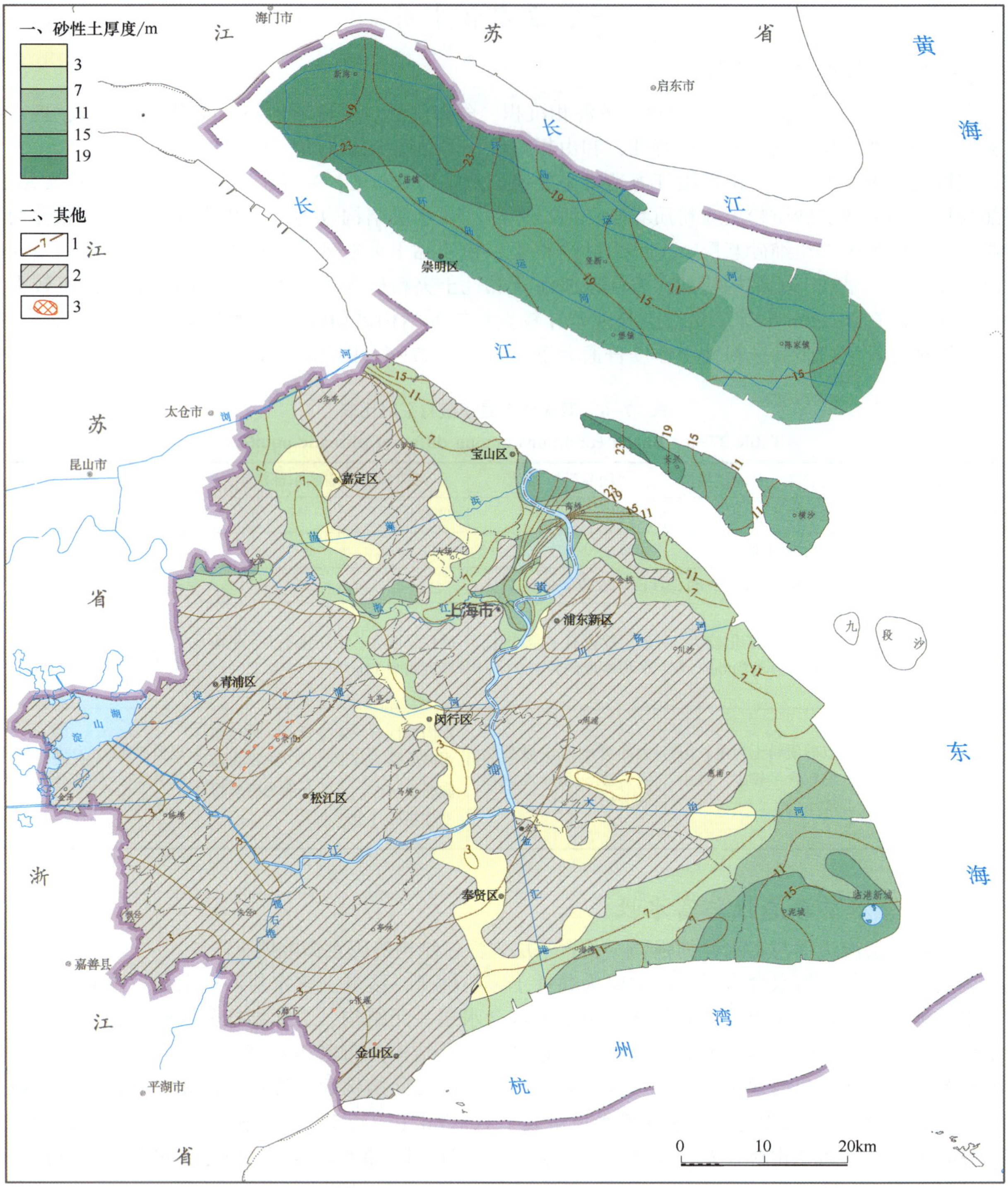

图 27-1　上海潜水含水层埋藏分布图

Fig. 27-1　Buried depth distribution of phreatic aquifers in Shanghai

1—层底界面等值线（m）；2—黏性土分布区；3—基岩分布区

兴岛中西部，普遍分布矿化度为 1～3g/L 的微咸水，约占全市总面积的 14%；城桥局部分布咸水，分布面积约占全市总面积的 1%；除上述以外地区分布淡水，分布面积约占全市总面积的 6%（图 27-2）。

3. 水质稳定性

潜水与大气降水、地表水体等水交替作用较为强烈，地下水中常量组分、微量元素、重金属、污染组分等受之影响明显，最为明显的以矿化度大小表征的咸淡水，其含量在不同阶段有所变化，总体

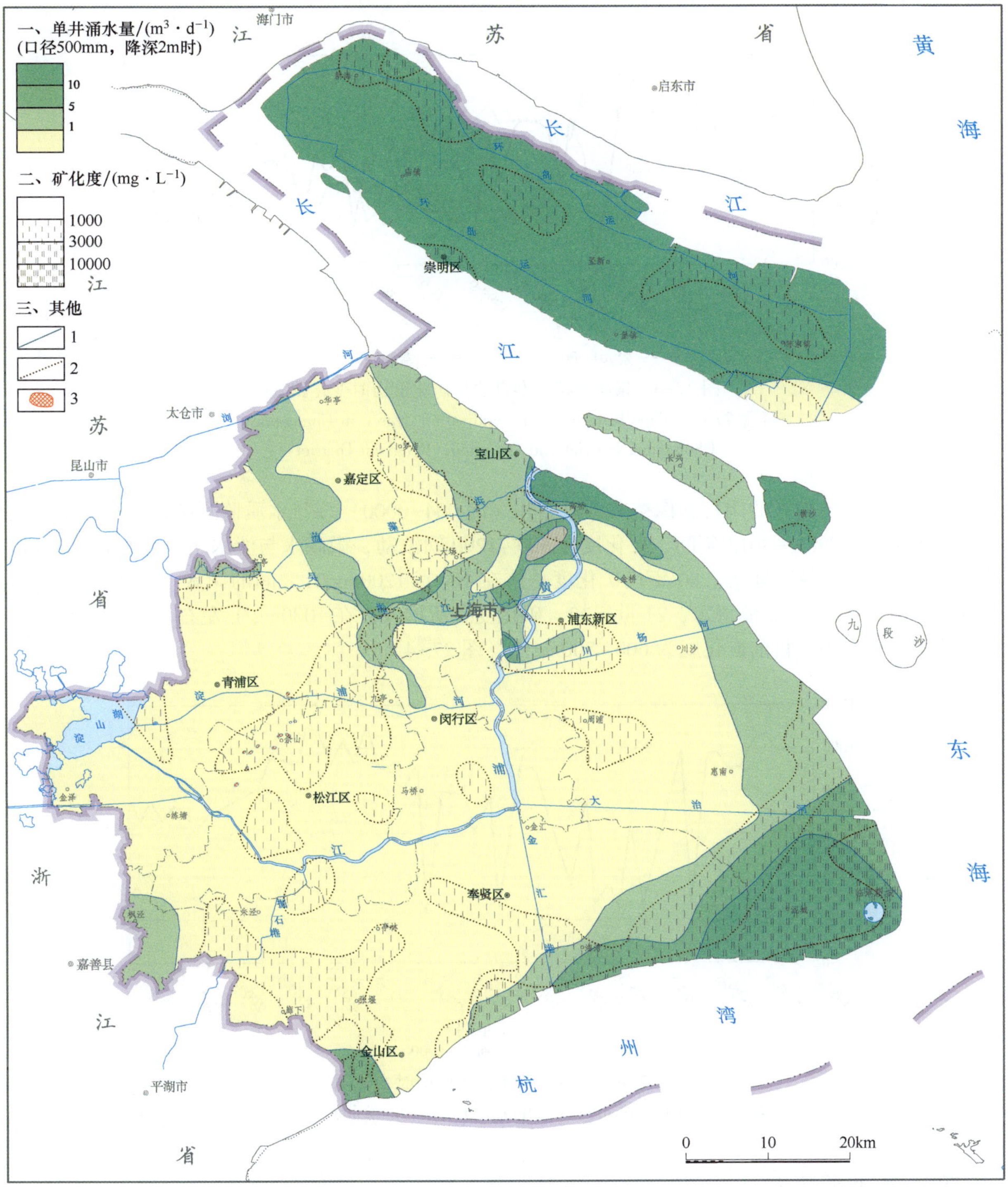

图 27-2　上海潜水含水层富水性、矿化度分布图

Fig. 27-2　Water abundance and mineralization degree map of phreatic aquifers in Shanghai

1—富水性分界线；2—矿化度分界线；3—基岩分布区

表现出沿江沿海地区原生咸水水质淡化趋势明显，而内陆原生淡水水质总体在小范围内波动的变化特征。

位于河口砂岛二元结构区的崇明岛西北部三星典型监测井（崇 001-0000 井），以及位于南部沿海一元结构区的奉贤钱桥典型监测井（奉 005-0000 井）多年水质监测结果显示，矿化度呈持续减小特征。20 世纪 80—90 年代，潜水矿化度在 1000~1300mg/L 之间波动，随后矿化度呈持续减小趋势，由微咸水逐渐转变为淡水，目前矿化度在 600~800mg/L 间波动（图 27-3）。

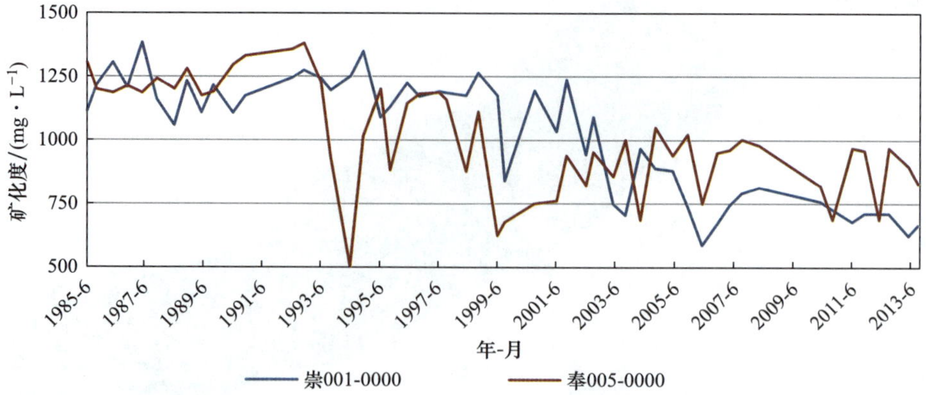

图 27-3　崇明城桥、奉贤钱桥矿化度历时变化图

Fig. 27-3　Time histories of mineralization degree in Chengqiao, Chongming District and Qianqiao, Fengxian District

位于内陆一元结构区的闵行召稼楼典型监测井（海 004-0000）多年水质监测结果显示，受大气降水、地表水影响，1985—2013 年间，矿化度在 300~1200 间波动，微咸水与淡水之间存在转化的过程，趋势性上升或下降持续时间短。80 年代矿化度总体在 1000~1200mg/L 之间的微咸水，90 年代显示为矿化度小于 300~800mg/L 的淡水，21 世纪初，矿化度有所增大，在 1000mg/L 左右，2008 年至今，再次呈现阶段淡化态势，目前矿化度为 600~800mg/L（图 27-4）。

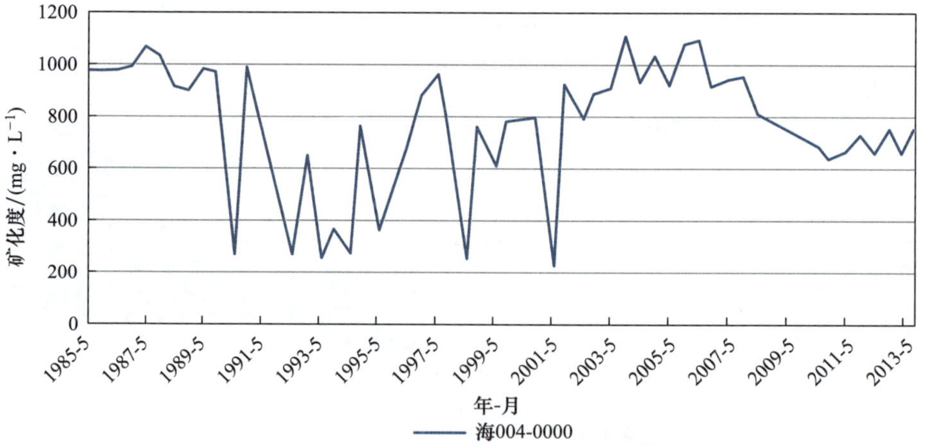

图 27-4　闵行召稼楼矿化度历时变化图

Fig. 27-4　Time history of mineralization degree in Zhaojialou, Minhang District

（四）地下水质量

20 世纪 80 年代初潜水质量评价结果显示，潜水因受外界综合因素影响小，地下水质量基本反映了原生地下水质量状况。陆域地区潜水质量总体较好，约占全市面积 70%的地区分布Ⅱ类、Ⅲ类水，仅浦东东南部—奉贤南部沿江地带，地下水质量相对差，为Ⅳ类、Ⅴ类水。河口砂岛主要分布Ⅲ类水，约占全市总面积的 18%，仅崇明岛、长兴岛、横沙岛的北部沿江地带质量稍差，为Ⅳ类、Ⅴ类水。

20 世纪 80 年代—21 世纪初，在城市加速发展的过程中，区域上地表水质量有逐渐下降的迹象，潜水与地表水在长期交替作用下，受地表水质量下降影响，潜水质量也呈下降特征。潜水质量调查结果显示：全市天然河道Ⅱ类、Ⅲ类水基本不存在（仅浦东北部、金山南部局部地区分布Ⅲ类水，分布面积不足全市面积的 1%），而主要为Ⅳ类、Ⅴ类水，劣Ⅴ类水也占相当权重。

水质稳定性：潜水与大气降水、江河海水补排关系密切，为此大气降水、地表水（江河海水）质量将直接影响着潜水质量。长期监测结果显示，局部原生咸水分布地区，地下水质量呈现转优的迹象，而遭受自然和人为因素（污水排放、农药等）影响明显的局部淡水分布区，潜水质量呈现出下降现象。

影响原生地下水质量的主要组分主要为矿化度、锰、亚硝酸盐、铵、高锰酸盐指数、总铁、总硬度和硝酸盐等。

（五）地下水水化学类型

区域上，潜水水化学类型较为复杂。

1. 陆域地区

浦东—奉贤—金山沿江海咸水和微咸水分布地带，主要分布氯-钠型水和氯·重碳酸-钠型水；中心城、闵行、青浦北部、松江、金山东部、奉贤北部局部、浦东西部局部，主要分布重碳酸·氯-钠·钙（钙·镁·钠、钙·钠）型水，其间零星分布重碳酸·硫酸型水；其他地区主要为重碳酸-钙·钠（钙·镁·钠、钠·钙·镁）型水（图 27-5）。

2. 河口砂岛区

河口砂岛区主要分布重碳酸-钠（钙·镁）和重碳酸·氯-钠（钙·镁）型水，仅北部前卫生态村地区、崇明东滩湿地、长兴岛东部、横沙岛东部，分布氯·重碳酸-钠型水和氯-钠（钠·钙、钠·镁）型水。

二、微承压含水层

（一）含水层埋藏特征

1. 区域含水层埋藏特征

上海将全新统、具有一定承压性、呈透镜体状分布的饱水含水层，统称为微承压含水层（$⑤_2$层）。以$⑤_2$层表征的微承压含水层，因含水层分布局限，供水价值低，多年来水文地质条件调查程度较低。

调查成果显示，该含水层（$⑤_2$层）为全新世早期（Qh_1）溺谷相沉积物，主要分布于全新世早期古分支河道地带。陆域地区，在嘉定安亭—市区北新泾—龙华、淀山湖—松江泗泾—龙华—浦东蔡路、闵行—奉贤庄行—金山朱泾、奉贤庄行—南桥—柘林、浦东三林—新场—临港新城等多条全新世早期古分支河道地带呈零星透镜体状分布。

河口砂岛区$⑤_2$层相对发育，且分布相对连续。含水层岩性以砂质粉土、粉细砂为主，其顶面埋深为 15～22m，厚度在 2～15m 不等（图 27-6）。

2. 中心城含水层埋藏特征

针对浅部地下水对工程设计、施工影响明显及由此造成的灾害频发现象，21 世纪以来以工程建设活动相对强烈的中心城为重点调查地区，加大了微承压含水层水文地质条件调查力度。

中心城除对$⑤_2$层进行调查外，对具承压性的$③_a$层、$④_a$层也进行了全面调查。调查反映，$③_a$层、$④_a$层和$⑤_2$层较为发育，空间分布也较为复杂。$③_a$层空间分布比较零散，与当时沉积环境无明显的相关性，但与当时海侵时携带物质中砂土的多少有关，因此，在空间分布上具有随机性和不确定性，表现在每处分布地区，厚度、面积较小，相对独立，且中心城均有零星分布；$④_a$层空间分布与当时沉积时的地理环境有一定关系，一般地势低洼地区砂层容易沉积，表现在中心城西南部局部地势相对低洼处容易沉积呈层砂层，分布面积小、零散；$⑤_2$层一般分布于古河道切割区或附近地区，空间分布往往不连续，但西南部局部地区呈连续分布，且分布面积相对较大（图 27-7）。

一般$③_a$层层顶埋深为 3～12m，厚度大多在 1～3m，局部在 5m 左右；$④_a$层层顶埋深为 6～20m，厚度大多在 3m 左右，莘庄局部在 11m 左右；$⑤_2$层层顶埋深为 16～30m，厚度在 5～20m 不等。$③_a$、$④_a$层岩性以灰色砂质粉土或夹薄层黏性土为主，$⑤_2$层岩性以灰色砂质粉土和粉砂为主。

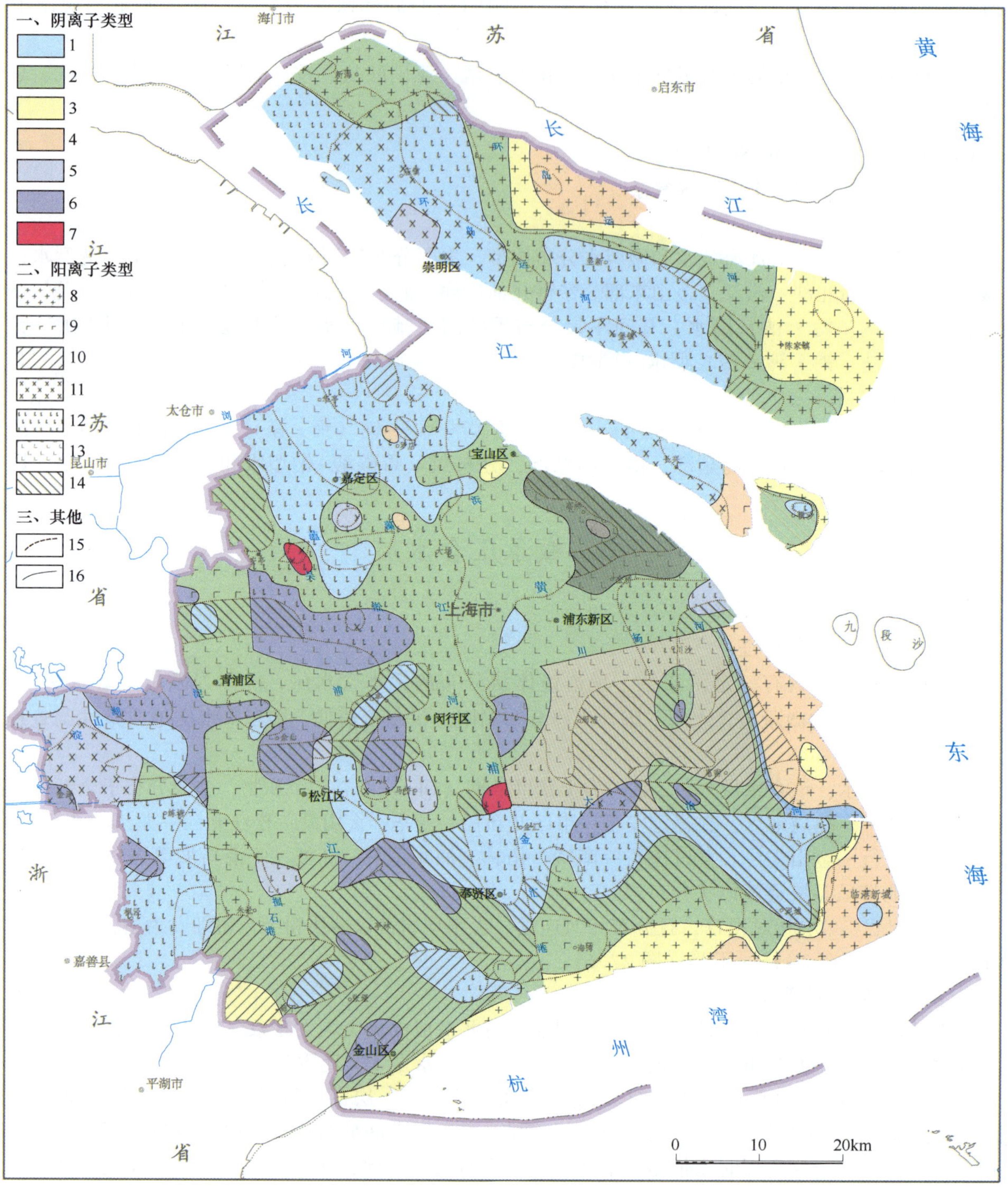

图 27-5　上海潜水水化学类型分区图

Fig. 27-5　Hydrochemical type zoning map of phreatic water in Shanghai

1—重碳酸根；2—重碳酸根・氯；3—氯・重碳酸根；4—氯；5—硫酸根、硫酸根・氯、硫酸根・重碳酸根；6—重碳酸根・硫酸根、重碳酸根・氯・硫酸根；7—硝酸根・重碳酸根、重碳酸根・硝酸根；8—钠；9—钠・钙、钠・镁；10—钠・钙・镁、钠・镁・钙；11—钙；12—钙・镁；13—钙・钠；14—钙・镁・钠、钙・钠・镁；15—阳离子分界线；16—阴离子分界线

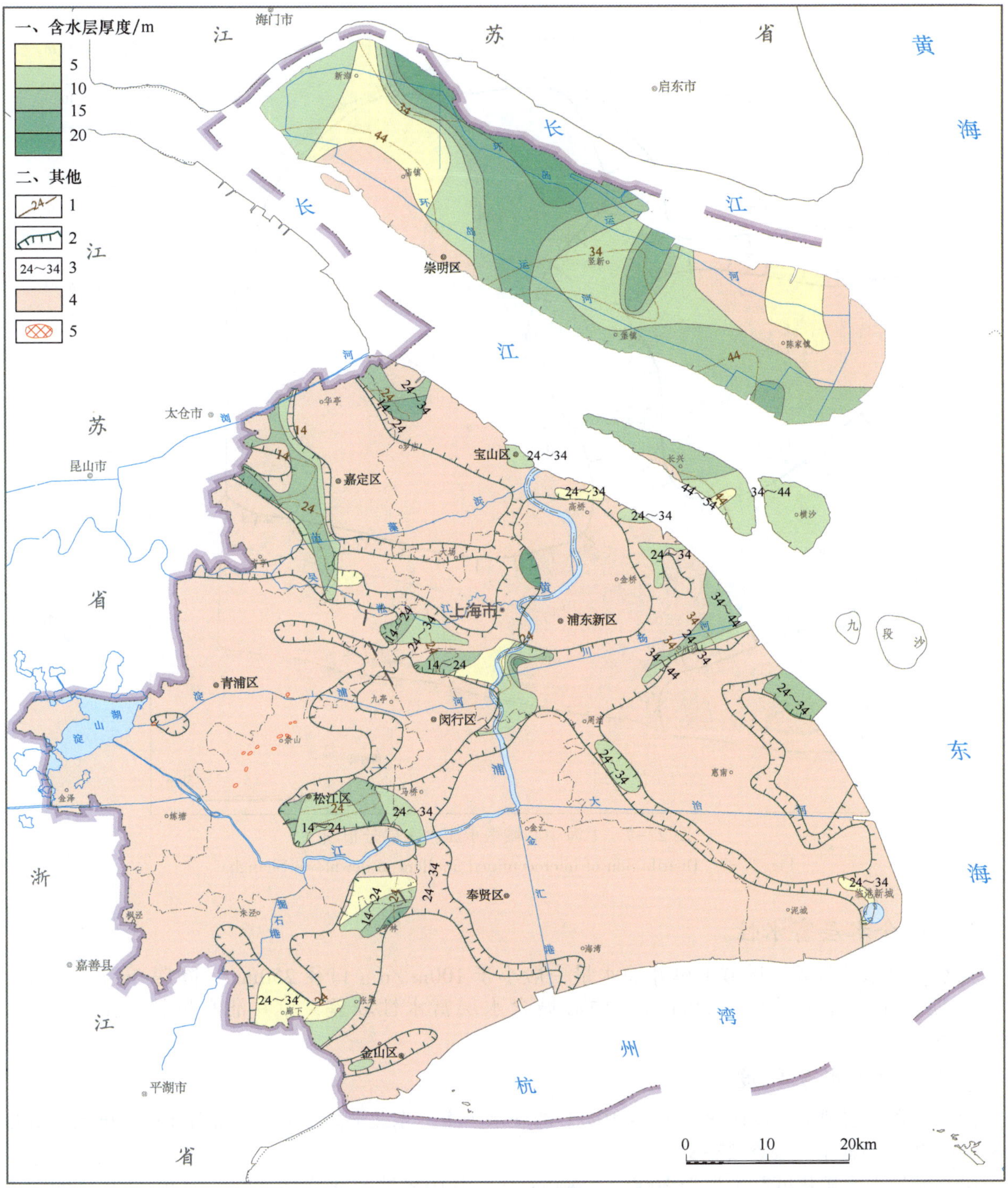

图 27-6　上海微承压含水层（⑤$_2$）埋藏分布图

Fig. 27-6　Buried depth distribution of microconfined aquifers（⑤$_2$）in Shanghai

1—层顶埋深等值线（m）；2—古河道界线；3—层顶埋深等范围（m）；4—含水层基本缺失区；5—基岩区

（二）含水层渗透性

含水层渗透性较差，区域上渗透系数一般小于 3m/d。中心城渗透系数基本在 0.11～2.85m/d，渗透系数与古河道主流向有一定关系，粉细砂分布地区相对大，砂质粉土分布地区相对小。

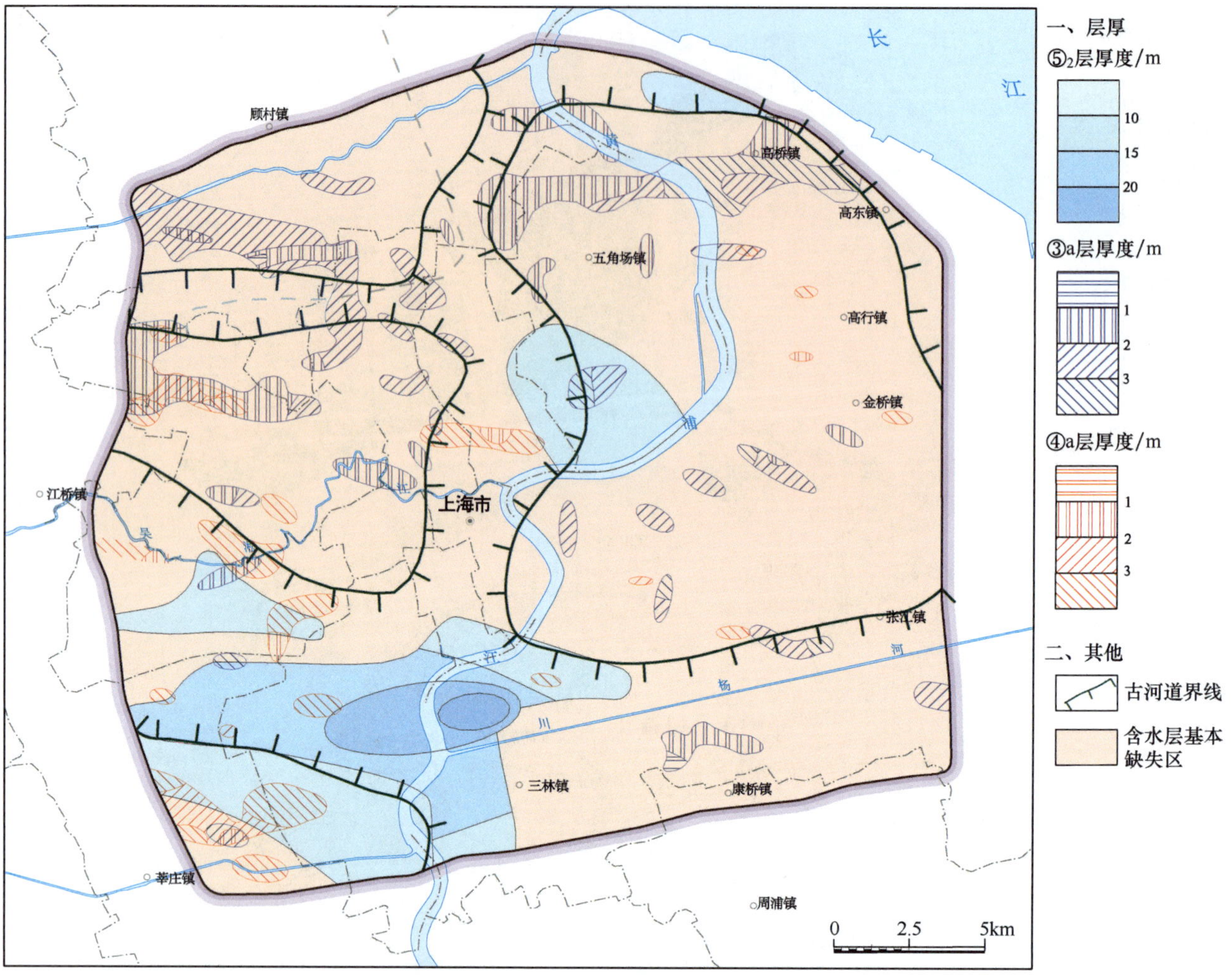

图 27-7　上海中心城微承压含水层分布图

Fig. 27-7　Distribution of microconfined aquifers in downtown Shanghai

（三）含水层富水性

含水层富水性较差，区域上单井涌水量一般小于 100m^3/d（口径 250mm，降深 5m 时，下同），但大部地区单井涌水量小于 50m^3/d。中心城含水层富水性较差，单井涌水量基本在 1～20m^3/d（图 27-8）。

（四）地下水矿化度

微承压含水层地下水以矿化度为 3～10g/L 的咸水为主，东北部高桥地区局部分布矿化度大于 10g/L（18.650g/L）的盐水（见图 27-8）。

中心城地下水矿化度基本界于 3000～10000mg/L。其中杨浦矿化度在 3800～7140mg/L，世博园区矿化度在 7343～9220mg/L，徐汇—闵行莘庄一带矿化度在 4696～4703mg/L 之间。地下水中氯离子含量界于 1000～12000mg/L。

（五）地下水质量

依据地下水常量组分、微量组分等测试结果，微承压含水层地下水质量级别总体为Ⅴ类，属于“极差”水。导致原生地下水质量差的因子主要为溶解性总固体、氯化物、铁（Fe）、锰（Mn）、亚硝酸盐、氨氮等。

（六）地下水水化学类型

据地下水常量组分测试结果，阴离子以氯离子为主，阳离子以钠离子为主，地下水水化学类型以

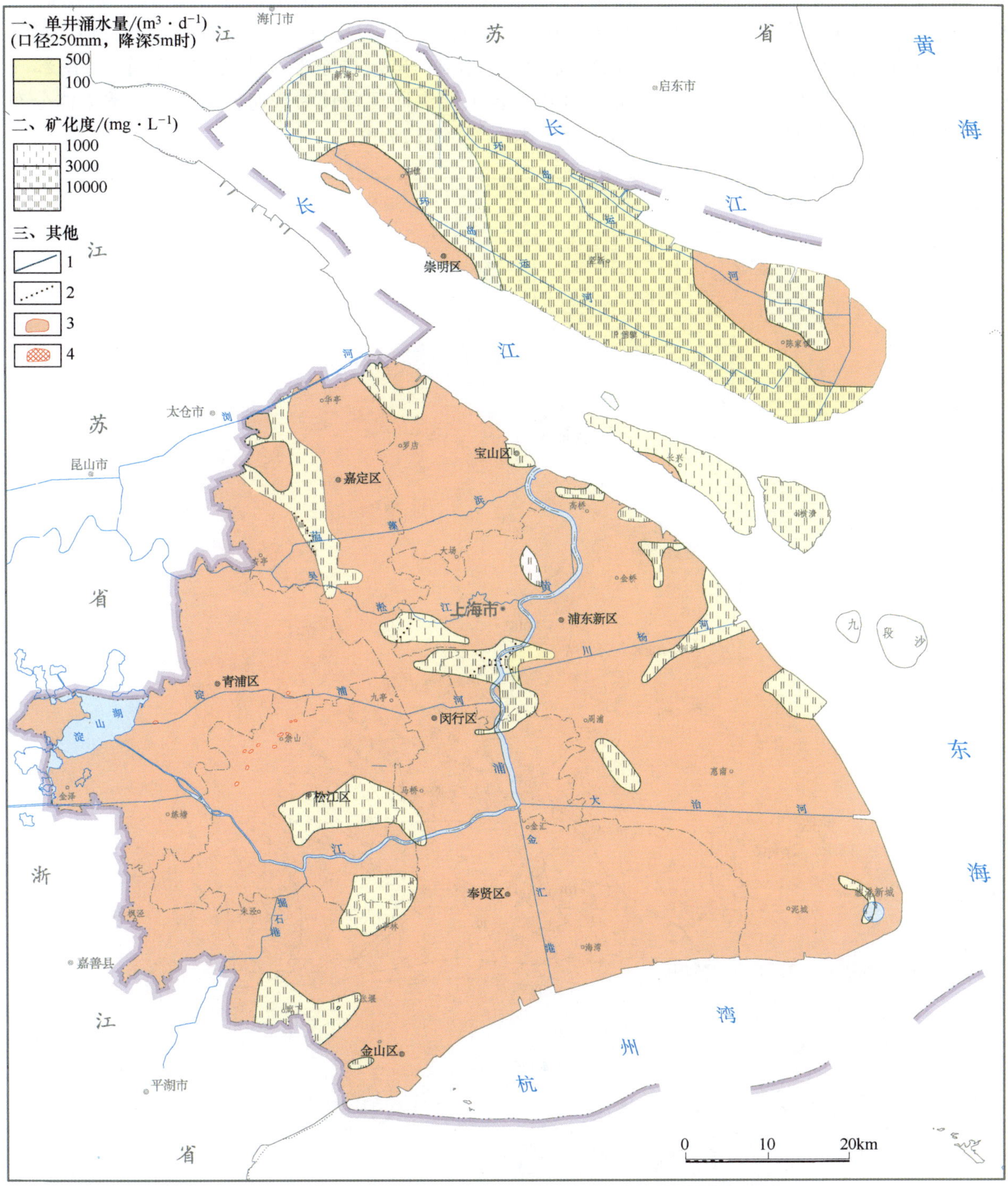

图 27-8　上海微承压含水层富水性、矿化度分布图

Fig. 27-8　Water abundance and mineralization degree map of microconfined aquifers in Shanghai

1—富水性界线；2—矿化度分界线；3—含水层基本缺失区；4—基岩分布区

氯-钠型为主。

三、第Ⅰ承压含水层

（一）含水层埋藏特征

含水层为晚更新世晚期滨海—潟湖相沉积物。西部佘山—天马山、西南部秦皇山—查山等剥蚀残

丘-基岩浅埋区，以及宝山—嘉定—青浦大部和河口砂岛等地区含水层基本缺失，累积面积约2007km²，占全市总面积的29%。含水层分布地区其空间分布较为复杂，一般呈单层分布，但西部湖沼局部地区发育上、下两层。受古地理沉积环境、海进影响，陆域中东部（中心城、浦东）、南部（金山、奉贤）地区，第Ⅰ承压含水层与下伏第Ⅱ承压含水层沟通现象较为普遍，沟通区呈透镜体分布特征，单处沟通区面积由数平方千米至数百平方千米不等（图27-9）。

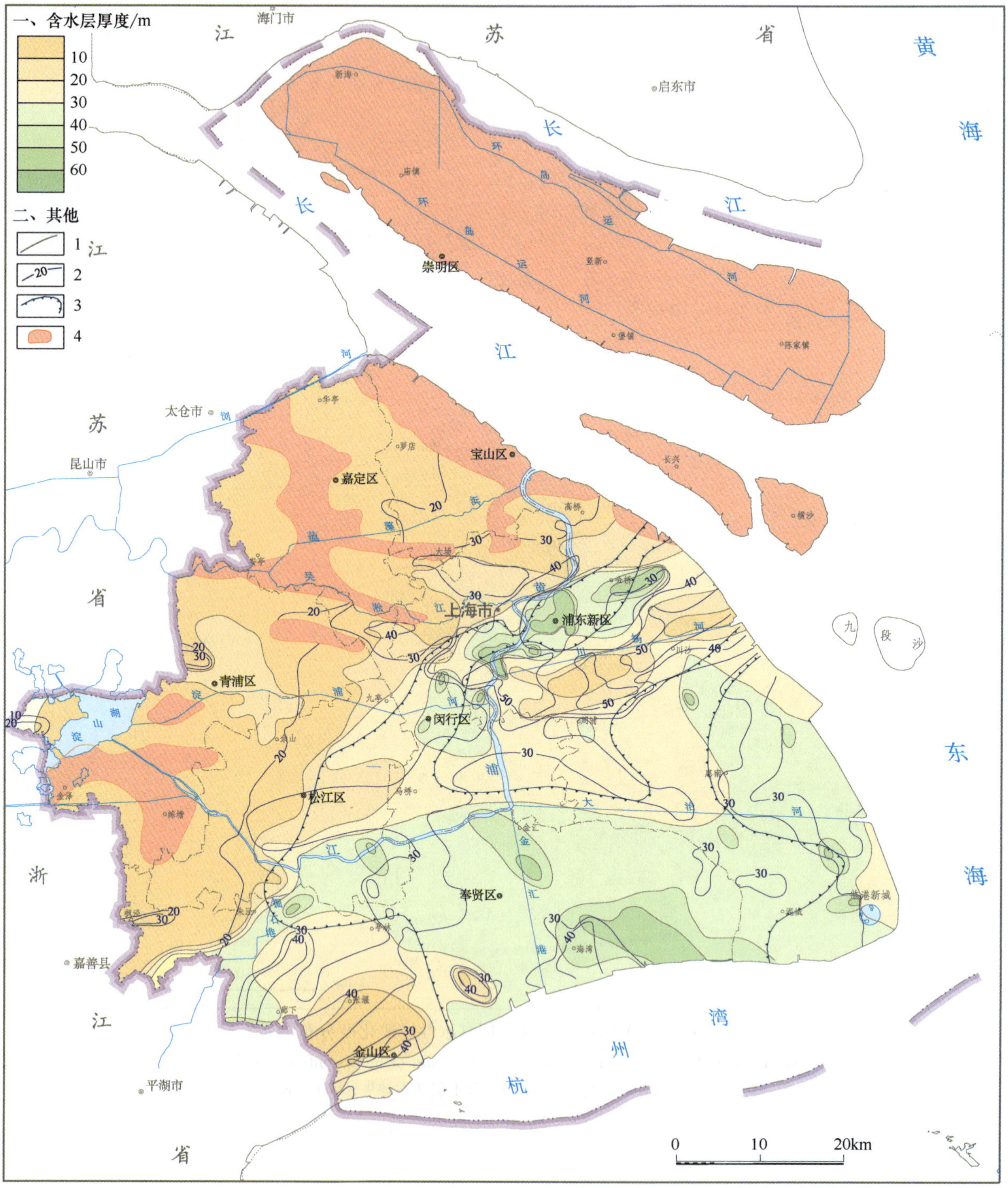

图27-9 上海第Ⅰ承压含水层埋藏分布图

Fig. 27-9 Buried depth distribution of confined aquifer Ⅰ in Shanghai

1—厚度等值线（m）；2—顶面埋深等值线（m）；3—第Ⅰ、Ⅱ承压含水层沟通区；4—含水层基本缺失区

含水层砂颗粒总体较细，分选性尚好。一般自上而下由草黄、灰黄色粉细砂，向下过渡至灰色粉细砂、细砂，偶夹砂质粉土、粉质黏土薄层。含水层顶面埋深一般介于20~40m，西部湖沼地区顶面埋深在10m左右。含水层底面埋深一般介于40~60m，沟通区局部底面可达80m左右。含水层厚度在数米至逾20m不等，沟通区及邻近地区含水层厚度在30m左右，而上述提及的含水层基本缺失区附近地区，厚度大多小于10m。

（二）含水层渗透性

含水层分布地区以渗透系数大小表征含水层的渗透性差异，滨海平原区较湖沼平原区相对好。

浦东浦江—周浦—新川沙、奉贤柘林—南桥新城—青村—海湾含水层渗透性能相对好，渗透系数大于20m/d，除此以外的大部分地区，渗透性能相对差，渗透系数大多小于10m/d，尤其是松江、金山基岩浅埋地区以及北部青浦、嘉定、宝山含水层基本缺失区附近地区，渗透性更差，渗透系数小于1m/d。

（三）含水层富水性

含水层分布地区以单井涌水量表征含水层的富水性，滨海平原区较湖沼平原区相对好。奉贤柘林—南桥新城—海湾，浦东浦江局部、六灶局部及世博园地区富水性较好，单井涌水量大于1000m^3/d，其邻近地区为500~1000m^3/d；除此以外的大部分地区富水性相对差，单井涌水量小于500m^3/d；尤其是松江、金山基岩浅埋地区以及北部青浦、嘉定、宝山含水层基本缺失区附近地区，富水性更差，单井涌水量小于100m^3/d，局部小于10m^3/d（见图27-7）。

（四）地下水矿化度

含水层矿化度在空间上分布较为简单，以咸水、微咸水、盐水为主，局部为淡水。

宝山、嘉定、青浦、松江、金山、奉贤及浦东国际机场—老港、大团—临港新城，普遍分布矿化度为3~10g/L的咸水，面积约占全市总面积的52%（图27-10）。

中东部中心城、浦东矿化度分布复杂，可能遭受后期河流冲刷、补给影响，除分布有咸水，尚分布有淡水、微咸水和盐水。其中，矿化度小于1g/L的淡水零星分布于花木、金桥、曹路局部地区，约占全市总面积的0.1%；矿化度为10~50g/L的盐水主要分布在国际机场—朝阳农场—老港地区，约占全市总面积的2%；其余地区广泛分布矿化度为1~3g/L的微咸水，约占全市总面积的17%。

矿化度稳定性。历史上，该含水层因水质差，地下水位波动对地质环境影响大，开发利用价值极低，为此，相对的水质长期监测覆盖面小，且监测程度也非常低，故区域上地下水质连续性较好的长期监测数据较为缺乏，郊区局部地区处于监控盲区。

据位于中心城国棉二厂地下水开采（回灌）井（普026-06）20世纪60—70年代水质监测结果，第Ⅰ承压含水层矿化度比较稳定。60年代初期，地下水为矿化度在5700g/mL左右的咸水，近8年的持续监测，矿化度基本保持在5700~6200mg/L间波动，相对稳定，地下水质基本保持了原生水质分布状态（图27-11）。

（五）地下水质量

区域上分布有Ⅱ类、Ⅲ类、Ⅳ类、Ⅴ类地下水，但以分布Ⅴ类地下水为主。

不同质量的地下水空间分布相对简单，区域上大部分地区分布Ⅴ类、Ⅳ类水，仅浦东局部地区分布有Ⅲ类和Ⅱ类水。

全市占含水层分布面积的90%的咸水、微咸水分布地区主要分布Ⅴ类水。在中心城的闸北、黄浦、徐汇，浦东康桥—周浦、金桥、临港新城，闵行梅龙，奉贤金汇局部微咸水分布地区主要为Ⅳ类水，占含水层分布面积的8%左右。其中，在中心城的闸北、闵行梅龙Ⅳ类水分布范围内小面积分布有Ⅲ类和Ⅱ类水。

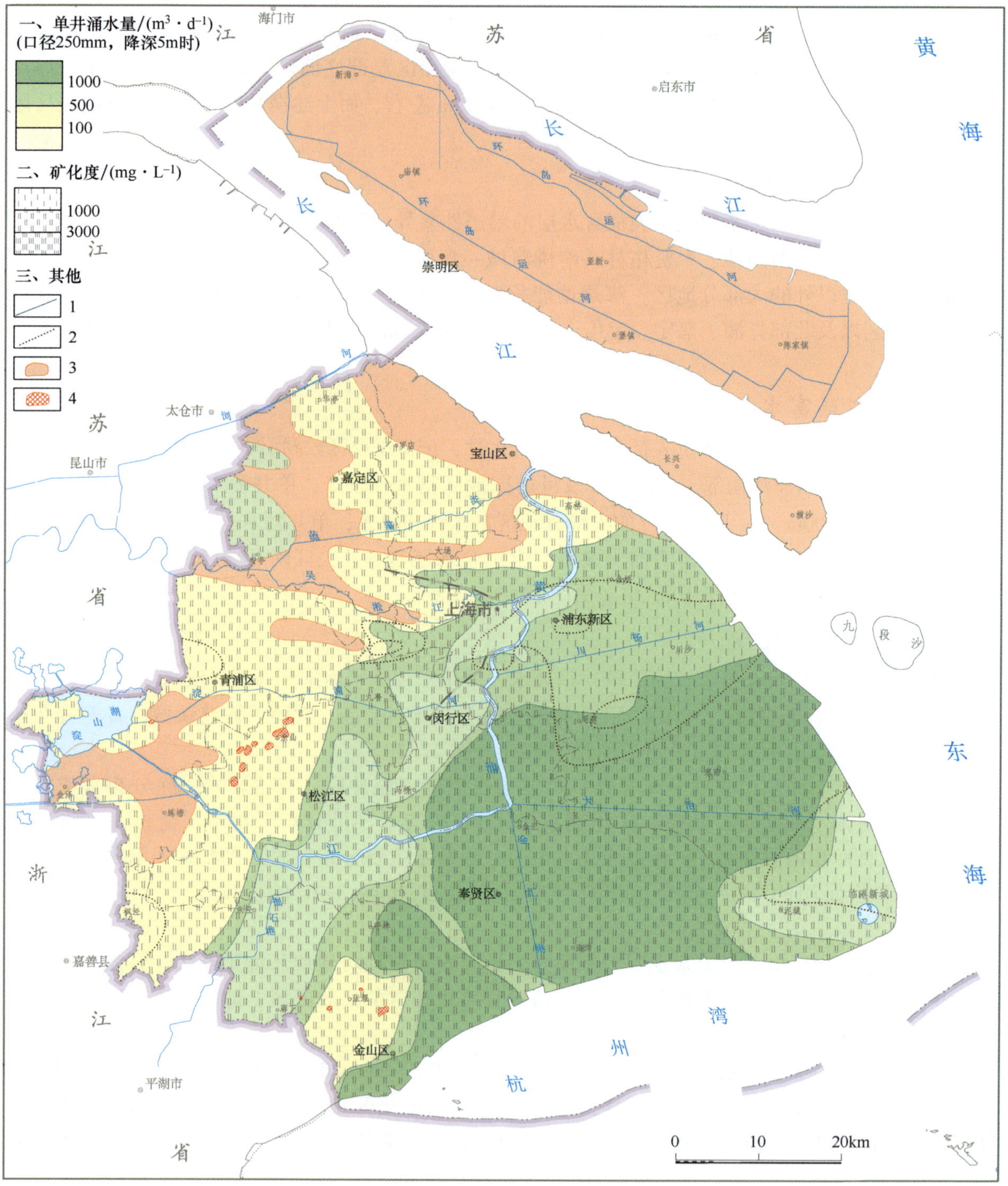

图 27-10　上海第Ⅰ承压含水层富水性、矿化度分布图

Fig. 27-10　Water abundance and mineralization degree map of confined aquifer Ⅰ in Shanghai

1—富水性界线；2—矿化度分界线；3—含水层基本缺失区；4—基岩分布区

（六）地下水水化学类型

地下水水化学类型相对简单。区域上，浦东周浦—川沙以及嘉定东部局部地区分布氯·重碳酸-钙·钠及氯·重碳酸-钠·钙（钙）型水，其他地区广泛分布氯型水，其中闵行—松江—奉贤—金山大部，以及惠南、芦潮港地区分布氯-钠型水，其余大部地区主要分布氯-钠·钙型水（图 27-12）。

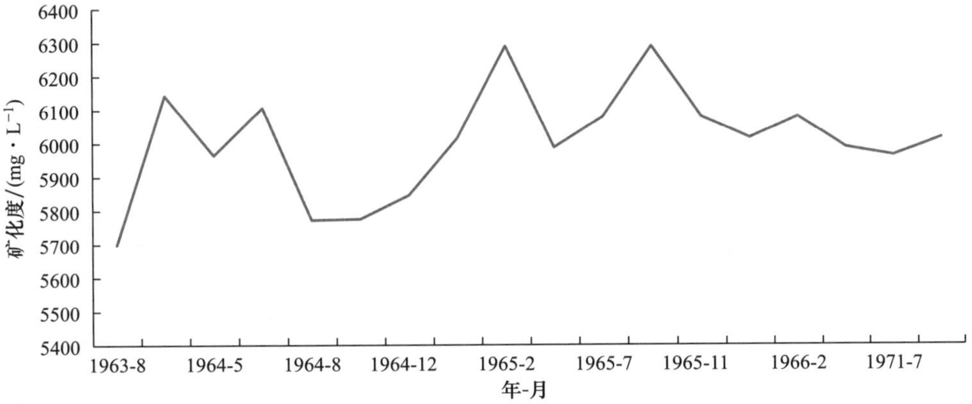

图 27-11　上海中心城第Ⅰ承压含水层典型地下水监测井（普 026-06）矿化度历时变化图

Fig. 27-11　Time history of mineralization degree of confined aquifer Ⅰ in downtown Shanghai per typical groundwater monitoring well（Pu026-06）

四、第Ⅱ承压含水层

（一）含水层埋藏特征

含水层为晚更新世早期（Qp_3^{1-2}）河口滨海—河流相沉积物。除西部佘山—天马山、西南部秦皇山—查山等剥蚀残丘地区含水层基本缺失外，广泛分布。由于大面积遭受海进，广泛接受碎屑物沉积，所以含水层分布最为广泛。伴随气候波动出现两个沉积旋回，形成上、下两个含水层次，在东部滨海平原区，两层汇聚成一层。上层基本为滨岸浅海、浅海相沉积，局部地区中下部为河湖相沉积，气候温暖潮湿；下层为河口滨海—河流相沉积物，气候温凉潮湿。受前、后期古地理沉积环境、海进海退、剥蚀等影响，含水层空间分布较为复杂，陆域中东部（中心城、浦东）、南部（金山、奉贤）地区，即陆家嘴—世纪公园—金桥、迪斯尼乐园—祝桥、吴泾—航头、金汇—青村、宣桥—大团—惠南、庄行—漕泾—柘林、奉城—海湾国家森林公园、临港新城，第Ⅱ承压含水层与上覆第Ⅰ承压含水层沟通现象较为普遍，同时，中心城大虹桥地区和杨浦局部，浦东三林—康桥、惠南、朝阳农场、临港新城局部地区，第Ⅱ承压含水层与下伏第Ⅲ承压含水层呈沟通格局。沟通区呈现透镜体分布特征，单处沟通区面积由数平方千米至数百平千米不等。第Ⅱ承压含水层与第Ⅰ承压含水层沟通区累积面积约占全市面积的14%，约占大浦东地区（中心城、浦东、金山、奉贤）面积的37%，第Ⅱ承压含水层与第Ⅲ承压含水层沟通区累积面积约占全市面积的5%，约占大浦东地区面积的11%。

含水层岩性自上而下呈现由细变粗的沉积韵律。上层岩性以灰色中细砂、细砂为主，频夹砂质粉土和粉质黏土薄层；下层岩性，古河道地带以含砾中粗砂为主，夹中细砂，其他地区以中细砂为主，夹少量中粗砂或砂质粉土，含贝壳碎屑。层间弱透水层岩性主要为灰、灰绿、杂色粉质黏土。

含水层埋藏深度在区域上有一定差异。陆域地区，上述提及的第Ⅱ承压含水层与第Ⅰ承压含水层沟通地区，含水层顶面埋深在30m左右，其附近小范围含水层顶面埋深较浅，在40~50m；除此以外的地区，含水层埋藏深度普遍在60~70m，但在西部朱家角—赵巷—佘山—练塘—金泽、南部金山的廊下—张堰—石化基岩出露或浅埋区、浦东大部、嘉定镇—中心城局部地区，含水层顶面埋深大多大于70m。河口砂岛区，崇明岛北部绿华—三星—庙镇—港西—前进农场一线以北地区，含水层顶面埋深在40~50m，该线以南、东地区，含水层埋藏深度普遍大于60m，仅新河镇局部地区在40~50m，陈家镇南部沿江局部地区，含水层顶面埋深大于70m。长兴岛中部凤凰镇地区含水层顶面埋深小于60m，向东、西两侧逐渐加深至70m左右。横沙岛含水层顶面埋深在60~70m。

含水层厚度在区域上变化也较大，总体呈现东部滨海平原区和河口砂岛区较西部湖沼平原区大。陆域地区，在上述提及的西部松江、金山含水层基本缺失区邻近的石湖荡—佘山—小昆山、吕巷—廊

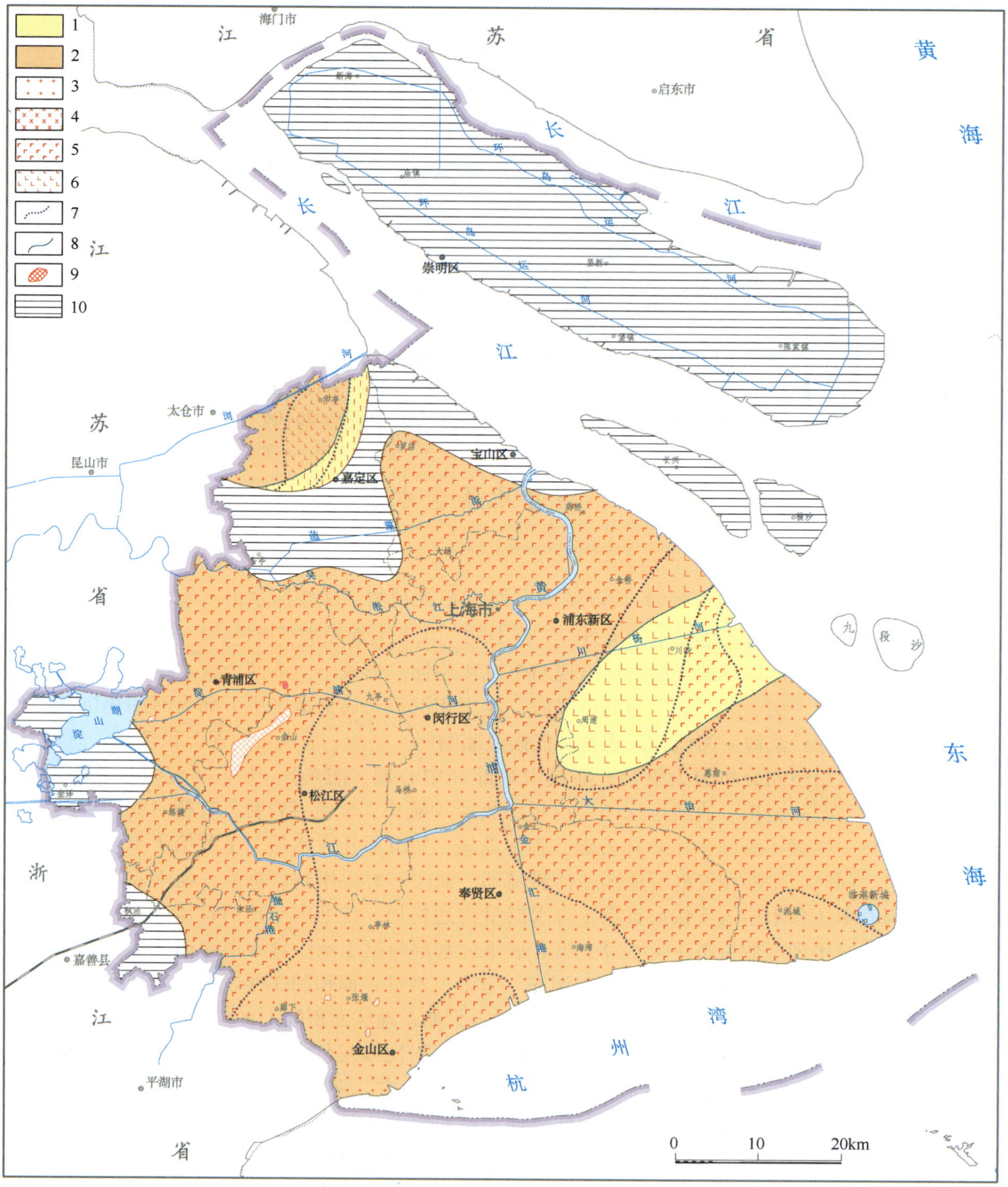

图 27-12　上海第Ⅰ承压含水层水化学类型分区图

Fig. 27-12　Hydrochemical type zoning map of confined aquifer Ⅰ in Shanghai

1—氯・重碳酸根；2—氯；3—钠；4—钙；5—钠・钙；6—钙・钠；7—阳；8—阴离子分界线；9—基岩区；10—含水层基本缺失区

下—金山卫镇—张堰—山阳局部地区，厚度总体小于 10m。第Ⅱ承压含水层与第Ⅰ承压含水层沟通区，及第Ⅱ承压含水层与第Ⅲ承压含水层沟通区，含水层厚度普遍大于 50m，局部在 80m 左右，沟通区邻近地区含水层厚度一般大于 30m，除上述以外的地区含水层厚度普遍在 10~30m。河口砂岛区，崇明岛该含水层厚度总体较大，除西北部新村乡—红星农场局部地区含水层厚度小于 30m 外，其余地区普遍在 30~50m，其中绿华局部、长江农场—前进农场局部地区，含水层厚度大于 50m。长兴岛凤凰镇以西地区，含水层厚度总体大于 30m，长兴岛凤凰镇以东地区及横沙岛，含水层厚度总体小于 30m（图 27-13）。

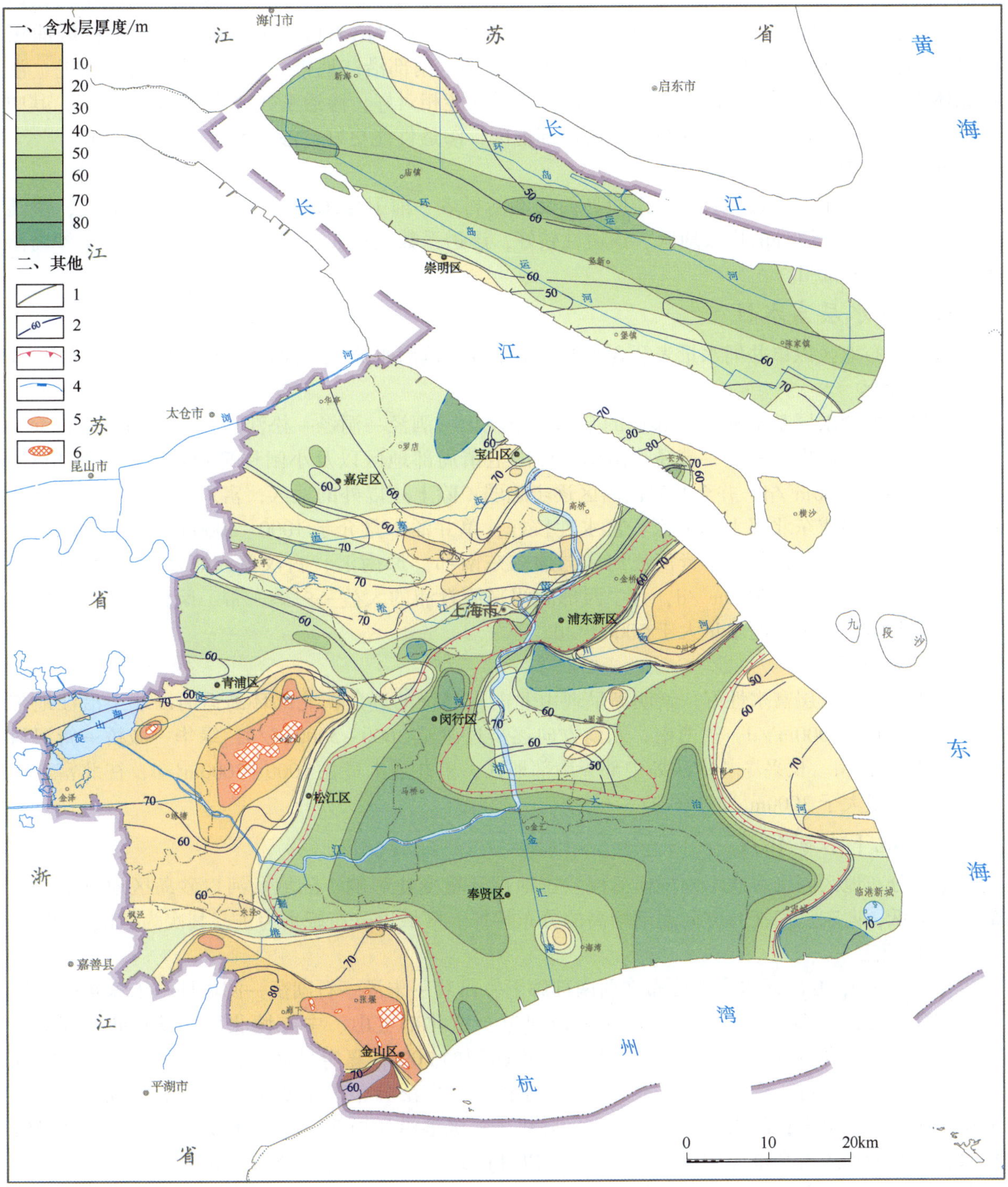

图 27-13　上海第Ⅱ承压含水层埋藏分布图

Fig. 27-13　Burial depth distribution of confined aquifer Ⅱ in Shanghai

1—厚度等值线；2—顶面埋深等值线（m）；3—第Ⅰ、Ⅱ承压含水层沟通区；4—第Ⅱ、Ⅲ承压含水层沟通区；5—含水层基本缺失区；6—基岩分布区

（二）含水层渗透性

含水层分布地区以渗透系数表征含水层的渗透性，滨海平原区、河口砂岛较湖沼平原区相对好。

1. 陆域地区

除基岩出露区邻近地区含水层渗透性较差外，大部分地区含水层渗透性总体较好。佘山基岩出露区邻近的佘山—小昆山—泗泾—洞泾—松江城区—石湖荡—新浜—枫泾—练塘—金泽—商榻镇，张堰基岩出露区邻近的张堰—漕泾—山阳—金山卫镇，市区小闸基岩出露区邻近地区，滨海平原区浦东北部（原川沙县）以及朝阳农场—老港—书院—东海农场东南部沿江地带，含水层渗透性相对差，渗透系数总体小于20m/d；除上述以外的广大地区，渗透性能相对好，渗透系数普遍在20~50m/d，其中惠南、新场、闵行—南桥、白鹤、徐行—罗店局部地区，渗透性能更好，渗透系数大于50m/d。

2. 河口砂岛区

河口砂岛区含水层渗透性较好。崇明岛西北部新村乡—前卫生态村地区，渗透性相对差，渗透系数小于20m/d，除此以外的广大地区，渗透性较好，渗透系数大多在30~60m/d；长兴岛、横沙岛渗透系数大多在30~40m/d。

（三）含水层富水性

含水层分布地区以单井涌水量表征含水层的富水性，滨海平原区、河口砂岛较湖沼平原区相对好。

1. 陆域地区

上述提及的含水层基本缺失区邻近的佘山—小昆山—泗泾—洞泾—松江城区—石湖荡—新浜—枫泾—练塘—金泽—商榻，张堰—漕泾—山阳—金山卫镇局部地区以及小闸局部地区，富水性相对差，单井涌水量小于1000m^3/d，并至基岩出露区逐步趋弱；除上述以外的地区，富水性较好，西南部金山汊流河道及东北部浦东北部大部地区，富水性较好，单井涌水量介于1000~3000m^3/d，北西-南东向展布的自嘉定、青浦向闵行—奉贤汊流河床与浦东东南部（原南汇）古汊流入海地段，富水性更好，单井涌水量大多在3000~5000m^3/d，其中西部湖沼平原青浦的大盈—白鹤一带、闵行城区—南桥、青村—海湾局部地区，单井涌水量大于5000m^3/d。

2. 河口砂岛区

河口砂岛区含水层富水性好。崇明岛呈现由北向南趋好的迹象，北部环岛运河以北至沿江地带单井涌水量在1000~3000m^3/d，往西南、东南方向逐步过渡至3000~5000m^3/d，绿华—城桥—新河沿江地带大于5000m^3/d。长兴岛、横沙岛中部—南部地区，单井涌水量在1000~3000m^3/d，往北富水性趋好，单井涌水量大于3000m^3/d（图27-14）。

（四）地下水矿化度

含水层分布地区因受多次不同规模海侵、古地理环境变化影响，陆域及河口砂岛区地下水均呈现西部矿化度低，向东、南江海方向逐渐增高的分布特征。区域以分布咸水为主，其次为微咸水和淡水。

1. 陆域地区

矿化度小于1g/L的淡水主要分布在西南部廊下—吕巷—朱泾—石湖荡—泖塔圩—朱家角一线以西与浙江交界地带，北部青浦白鹤自西向东叉流河道至东北部宝山、中心城北部，以及浦东周浦—北蔡—张江—金桥—高行—高桥，分布面积约占全市面积的20%；嘉定城区—竹桥—葛隆，以及松江东部、中心城南部、闵行、金山东部、奉贤、浦东东南部，广泛分布矿化度为3~10g/L的咸水，累计咸水分布面积约占全市面积的32%，其中浦东国际机场—老港沿江局部地区，分布有矿化度为10~50g/L的盐水，分布面积约占全市面积的2%；除上述以外的地区较为广泛地分布矿化度为1~3g/L的微咸水，分布面积约占全市面积的22%。

2. 河口砂岛区

崇明岛西南部绿化—城桥—新河沿江地带、长兴岛潘石以西地区分布矿化度小于1g/L淡水，分布面积约占全市面积的2%；崇明岛北部环岛运河至沿江地带分布矿化度为10~50g/L的盐水，分布面积约占全市面积的5%；崇明岛邻近淡水区的绿华—三星—庙镇—港西—建设—新河—堡镇一带、横沙岛，分布矿化度为1~3g/L的咸水，分布面积约占全市面积的6%；除上述以外的地区广泛分布矿化度为3~10g/L的咸水，分布面积约占全市面积的8%（见图27-14）。

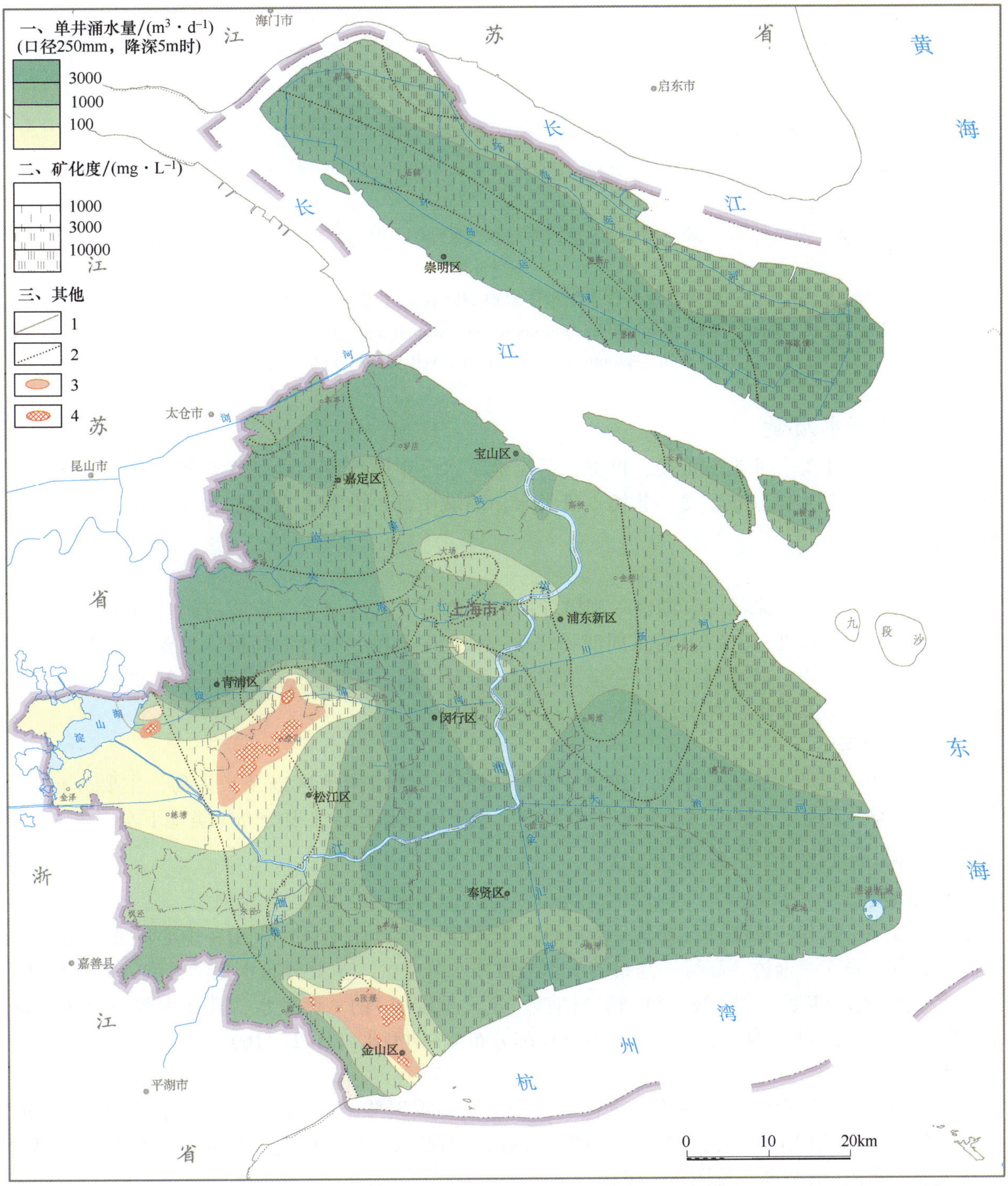

图 27-14　上海第Ⅱ承压含水层富水性、矿化度分布图

Fig. 27-14　Water abundance and mineralization degree map of confined aquifer Ⅱ in Shanghai

1—富水性分界线；2—矿化度分界线；3—含水层基本缺失区；4—基岩分布区

3. 矿化度稳定性

历史上，该含水层为上海地下水主要开发利用层次，水质长期监测程度相对较高，区域上地下水矿化度基本维持原生状态。据位于中心城上海硫酸厂地下水开采井（宝 021-02）20 世纪 60 年代—21 世纪初水质监测结果，20 世纪 60 年代末期，地下水矿化度在 300mg/L 左右，为淡水，近 38 年的持续监测，矿化度基本保持在 300mg/L 上下波动，水质较为稳定，保持了原生地下水质量状

态（图 27-15）。

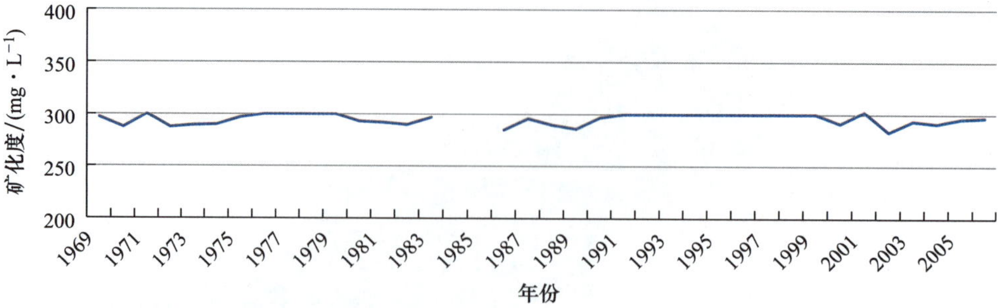

图 27-15　上海中心城第Ⅱ承压含水层地下水典型监测井（宝 021-02）矿化度历时变化图

Fig. 27-15　Time history of mineralization degree of confined aquifer Ⅱ in downtown Shanghai per typical groundwater monitoring well (Bao021-02)

（五）地下水质量

区域上分布有Ⅰ类、Ⅱ类、Ⅲ类、Ⅳ类、Ⅴ类地下水。不同级别的地下水空间分布相对复杂。总体上陆域西部和北部主要分布Ⅰ类、Ⅱ类、Ⅲ类水，陆域中部—南部、东南部和河口砂岛主要分布Ⅳ类和Ⅴ类水。

1. 陆域地区

西南部廊下—吕巷—朱泾—石湖荡—泖塔圩—朱家角一线以西地区，青浦白鹤自西向东叉流河道至东北部宝山、中心城北部，以及浦东周浦—北蔡—张江—金桥—高行—高桥，主要分布Ⅰ类、Ⅱ类地下水；松江东部—中心城南部—闵行—金山东部—奉贤—浦东东南部，广泛分布Ⅴ类地下水；除上述以外的地区主要分布Ⅲ类、Ⅳ类地下水。

2. 河口砂岛区

崇明岛绿华—三星—庙镇—港西—建设—新河—堡镇一线以南地区以及长兴岛，主要分布Ⅱ类、Ⅲ类地下水；该线以北、东的广大地区以及横沙岛，地下水质量差，主要分布Ⅴ类地下水。

（六）地下水水化学类型

1. 陆域地区

西南部廊下—吕巷—朱泾—石湖荡—泖塔圩—朱家角一线以西地区、宝山—中心城北部地区、浦东周浦—北蔡—张江—金桥—高行—高桥的淡水与微咸水分布区，主要分布水化学类型为重碳酸-钙·钠型、重碳酸-钙·镁型、重碳酸·氯-钙·钠型、氯·重碳酸-钠·钙型水，其他大部分地区主要分布氯-钠·钙型和氯-钙·钠型水，闵行、奉贤局部分布氯-钠型水（图 27-16）。

2. 河口砂岛区

崇明岛城桥—新河沿江以及绿华，水化学类型主要为重碳酸·氯-钙·钠型、氯·重碳酸-钙·钠（钠·钙）型，其他地区广泛分布氯-钠型和氯-钠·钙型水。长兴岛主要分布氯·重碳酸-钠型水。横沙岛主要分布氯-钠型水（见图 27-16）。

五、第Ⅲ承压含水层

（一）含水层埋藏特征

含水层为中更新世早中期河流—河口滨海相沉积物。具有与第Ⅱ承压含水组相似的古地理沉积环境和分布规律，但海进程度明显减弱，海水后撤，而河口段显著东迁。推测大致沿嘉定—南翔—莘庄—齐贤一线以东地区，应为广袤的口外沉积环境，接受碎屑物质堆积，西部基底凸起处脱水成陆，凹陷内为古分支河流。为此，在松江、青浦、金山、奉贤、中心城局部基岩出露区或浅埋区，含水层基本缺失，其余地区普遍分布。根据成因、古气候条件不同，含水组发育有上、下两层，但大部分地

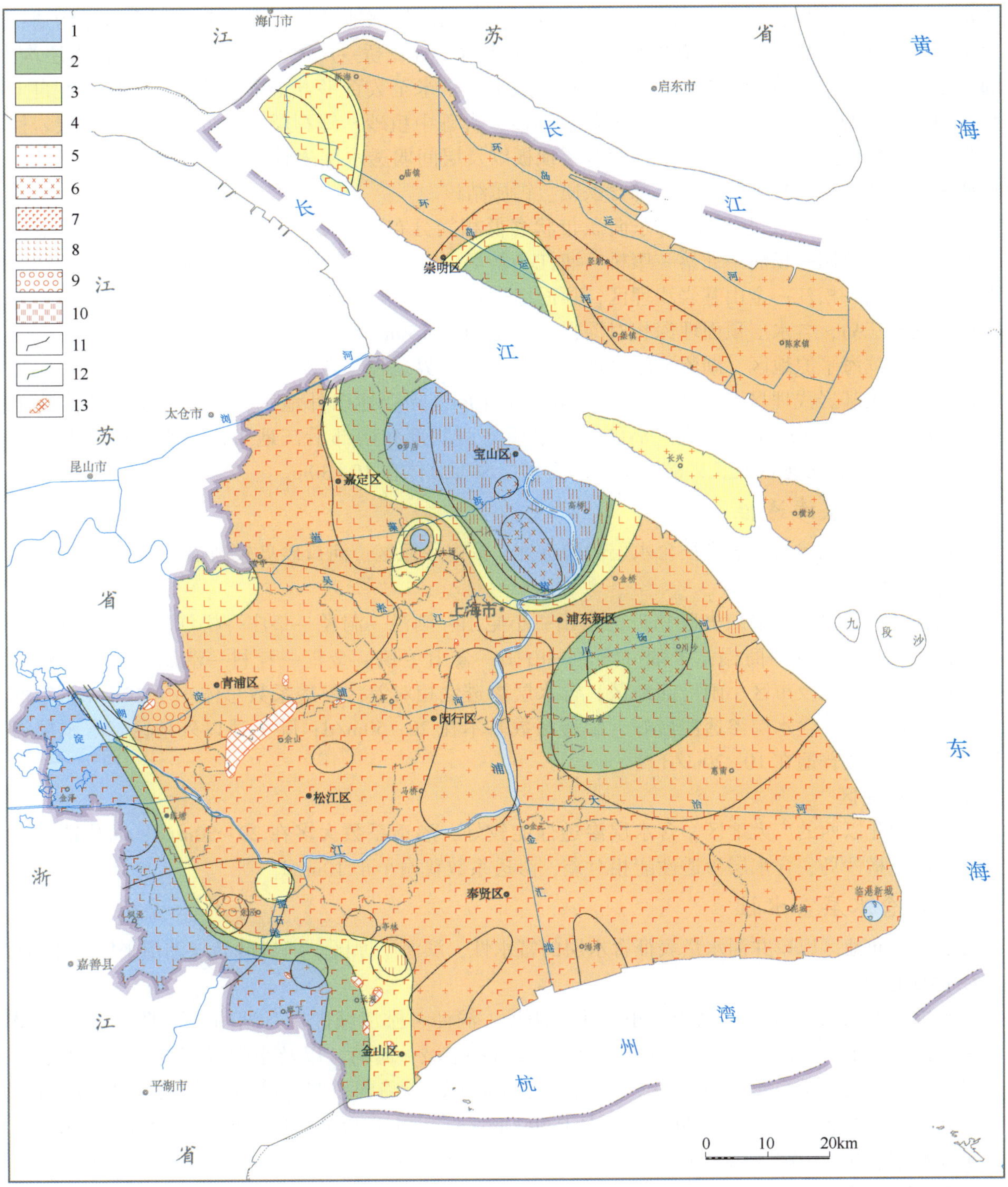

图 27-16　上海第Ⅱ承压含水层地下水水化学类型分区图

Fig. 27-16　Hydrochemical type zoning map of confined aquifer Ⅱ in Shanghai

1—重碳酸；2—重碳酸根·氯；3—氯·重碳酸根；4—氯；5—钠；6—钙；7—钠·钙；8—钙·钠；9—钙·镁·钠；10—钙·镁；11—阳离子分界线；12—阴离子分界线；13—基岩及含水层基本缺失区

区含水砂层呈单层分布，上层基本为河口湾—河流相沉积，气候暖热潮湿，下层为河流—河口滨海相沉积，气候暖热潮湿。

陆域地区，含水层埋藏总体具有湖沼平原区浅—东部滨海平原区最深—河口岛屿地区浅的分布规律，毗邻基岩浅埋区和三角洲前缘含水砂层逐渐尖灭。含水层空间分布较为复杂，局部地区与上、下含水层有沟通现象，宝山月浦—盛桥，中心城五角场—杨浦公园，浦东杨思—三林—御桥—孙桥—黄楼局部地区、朝阳农场、三灶、黄路、临港新城，奉贤四团局部地区与上覆第Ⅱ承压含水层沟通；在奉贤庄行—新寺—胡桥—漴缺—柘林及青浦赵屯—大盈—香花桥—凤溪与下伏第Ⅳ承压含水层沟通。沟通区呈现透镜体分布特征，单处沟通区面积由数平方千米至数百平方千米不等。含水层岩性自上而下呈现由细变粗的沉积韵律。上层岩性为灰色细砂、含砾中粗砂、砾石层、含贝壳碎屑，下层岩性为灰色粉细砂、粉砂、中细砂、砂砾石。含水层顶板以及层间弱透水层岩性主要为蓝灰、灰、褐黄、杂色粉质黏土夹灰、灰褐色细砂，西南部含铁质、钙质结核。

含水层埋藏深度大致呈北西-南东向条带状高低相间分布（图 27-17）。陆域东部大致以宝山罗泾—罗店—顾村公园—杨浦—北蔡—康桥—浦江—新场—青村—海湾国家森林公园一线以东地区，以及西部比邻青浦—松江—金山基岩浅埋区，顶面埋深较浅，一般在 90~110m；中部嘉定—闵行—奉贤一带以及河口三岛，含水层顶面埋深相对深，一般在 110~130m。

陆域东部滨海平原地区、湖沼平原区西南部与浙江交界地带，含水层厚度普遍大于 30m，比邻青浦—松江—金山基岩浅埋区，含水层厚度一般小于 10m，除上述以外地区，含水层厚度总体在 10~30m，基岩出露区含水层基本缺失。岛屿区，崇明岛西北部和南部沿江地带，含水层厚度普遍大于 30m，除此之外大部分地区及长兴岛、横沙岛含水层厚度总体在 10~30m。

（二）含水层渗透性

含水层分布区以渗透系数表征含水层的渗透性，滨海平原区、河口砂岛较湖沼平原区相对好。

1. 陆域地区

上述提及的含水层基本缺失区邻近的佘山—小昆山—泗泾—洞泾—莘庄—松江城区—石湖荡—新浜—枫泾—练塘—金泽—商榻，张堰—漕泾—山阳—金山卫镇，金汇—青村—奉城—海湾，嘉定城区—南翔—安亭，以及市区小闸局部地区，含水层渗透性相对差，渗透系数总体小于 10m/d；东部宝山、浦东沿江地带，含水层渗透性最好，渗透系数基本在 20~50m/d；除上述以外地区，含水层渗透性较好，渗透系数普遍在 10~20m/d。

2. 河口砂岛区

崇明岛向化以西的广大地区，含水层渗透性能较好，渗透系数大多在 20~50m/d；向化镇以东地区及长兴岛、横沙岛，含水层渗透性能相对差，渗透系数普遍小于 10m/d。

（三）含水层富水性

含水层分布地区以单井涌水量表征含水层的富水性，滨海平原区、河口砂岛较湖沼平原区相对好。

1. 陆域地区

含水层基本缺失区邻近的佘山—小昆山—泗泾—洞泾—松江城区—石湖荡—练塘—金泽—商榻，张堰—漕泾—山阳—金山卫镇局部地区，金汇—青村—奉城—海湾，以及小闸局部地区，含水层富水性相对差，单井涌水量小于 $1000m^3/d$，并至基岩出露区逐步趋弱；东部滨海平原地区含水层富水性好，单井涌水量总体大于 $3000m^3/d$，局部大于 $5000m^3/d$。除上述以外的广大地区，含水层富水性也较好，单井涌水量介于 $1000\sim3000m^3/d$。

2. 河口砂岛区

崇明岛向化以西的广大地区，含水层富水性好，单井涌水量大于 $1000m^3/d$，西北部绿化含水层的单井涌水量大于 $5000m^3/d$；向化镇以东地区及长兴岛、横沙岛，含水层富水性相对差，单井涌水量小于 1000m/d（图 27-18）。

（四）地下水矿化度

因受多次不同规模海侵、古地理环境变化影响，地下水总体呈现西部、北部矿化度低，向东、南江海方向逐渐增高，河口砂岛最高的分布特征。区域上以分布淡水、盐水、咸水为主，其次为微咸水。

1. 陆域地区

矿化度小于 1g/L 的淡水主要分布在宝山、嘉定、青浦、松江、金山等西部、北部地区，分布面积

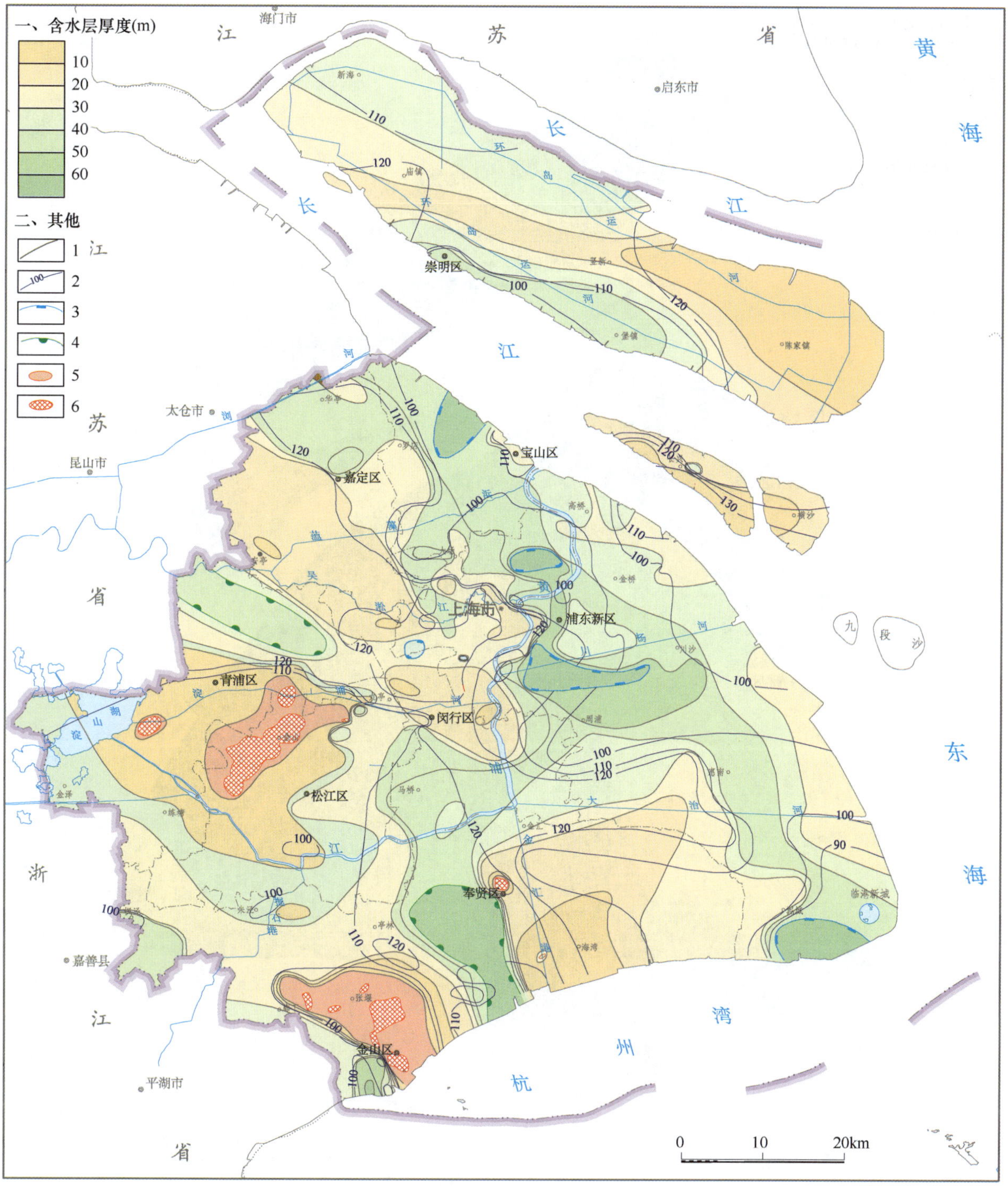

图 27-17　上海第Ⅲ承压含水层埋藏分布图

Fig. 27-17　Burial depth distribution of confined aquifer Ⅲ in Shanghai

1—厚度等值线；2—顶面埋深等值线（m）；3—第Ⅱ、Ⅲ含水层沟通区；4—第Ⅲ、Ⅳ含水层沟通区；5—含水层基本缺失区；6—基岩分布区

约占全市面积的 37%；嘉定城区—竹桥—葛隆，以及松江东部、中心城南部、闵行、金山东部、奉贤、浦东东南部，广泛分布咸水，矿化度为 3~10g/L，分布面积约占全市面积的 19%，其中浦东国际机场—老港镇沿江局部地区，分布有矿化度为 10~50g/L 的盐水，分布面积约占全市面积的 10%；除上述以外的地区广泛分布微咸水，矿化度为 1~3g/L，分布面积约占全市面积的 11%。

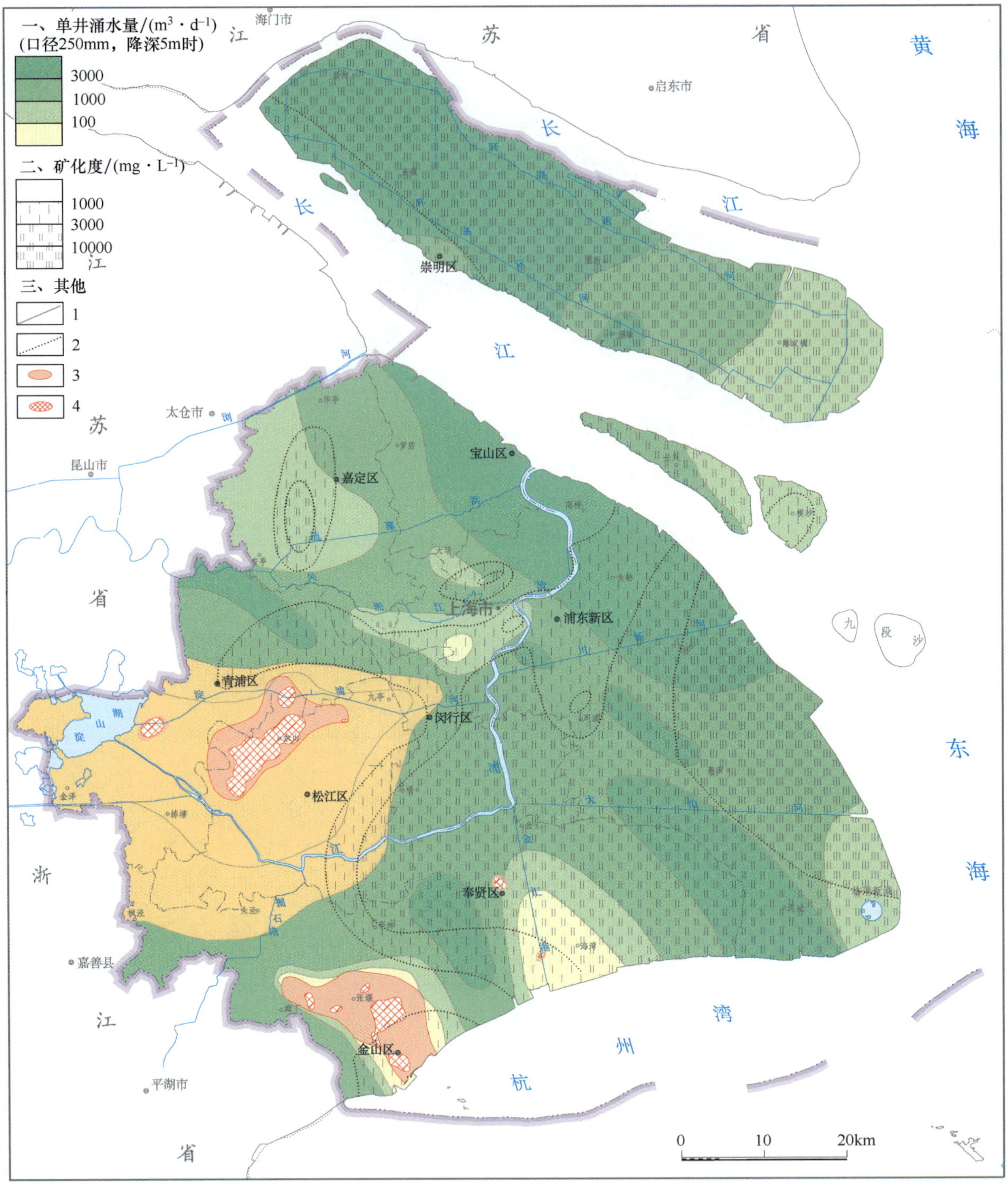

图 27-18　上海第Ⅲ承压含水层富水性、矿化度分布图

Fig. 27-18　Water abundance and mineralization degree map of confined aquifer Ⅲ in Shanghai

1—富水性分界线；2—矿化度分界线；3—含水层基本缺失区；4—基岩分布区

2. 河口砂岛区

河口砂岛区主要分布盐水，矿化度为10~50g/L，分布面积约占全市面积的17%；仅崇明岛西部绿华、横沙岛中部地区分布咸水，矿化度为3~10g/L，分布面积约占全市面积的4%（见图27-18）。

3. 矿化度稳定性

历史上，该含水层为上海地下水主要开发利用层次，水质长期监测程度较高，区域上地下水矿化度基本维持原生状态。位于中心城东部浦东高桥水厂地下水开采井（浦 009-02）20 世纪 70 年代—21 世纪初水质监测结果显示：20 世纪 70 年代初期，地下水矿化度在 400mg/L 左右，为淡水，经过近 38 年的持续监测，矿化度基本在 350～450mg/L 间波动，水质相对稳定，保持了原生地下水质量状态（图 27-19）。

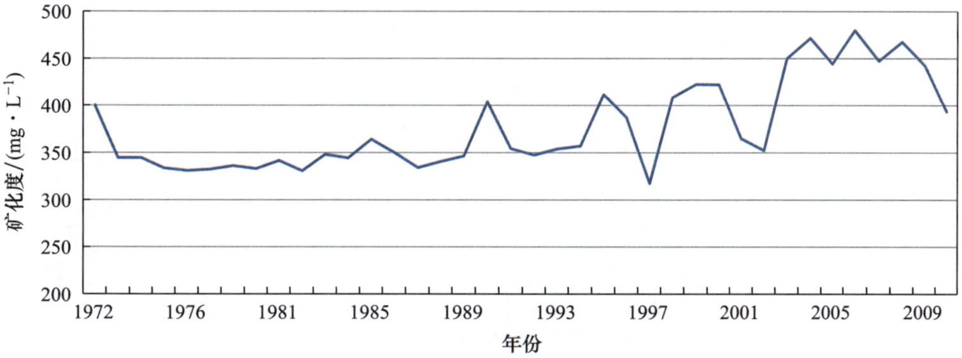

图 27-19　上海中心城第Ⅲ承压含水层典型监测井（浦 009-02）矿化度历时变化图

Fig. 27-19　Time history of mineralization degree of confined aquifer Ⅲ in downtown Shanghai per typical monitoring well (Pu009-02)

（五）地下水质量

区域上分布有Ⅰ类、Ⅱ类、Ⅲ类、Ⅳ类、Ⅴ类地下水。不同级别的地下水空间分布相对复杂。总体陆域西部和北部主要分布Ⅰ类、Ⅱ类、Ⅲ类地下水，陆域中部—南部、东南部及河口砂岛区，主要分布Ⅳ类和Ⅴ类地下水。

1. 陆域地区

奉贤、闵行莘庄以南地区、浦东川沙新城以南地区，广泛分布Ⅳ类及Ⅴ类地下水，嘉定城区局部分布Ⅲ类、Ⅳ类地下水；除上述以外的地区，广泛分布Ⅰ类、Ⅱ类地下水，并以Ⅰ类地下水居多。

2. 河口砂岛区

崇明岛、长兴岛及横沙岛东部沿江地带，普遍分布Ⅴ类地下水仅横沙岛西南部地区分布Ⅱ类、Ⅲ类地下水。

（六）地下水水化学类型

地下水水化学类型分布比较复杂，以重碳酸型水和氯型水为主。

1. 陆域地区

宝山、嘉定、青浦、松江、金山等西、北部淡水、微咸水分布区，水化学类型主要为重碳酸-钙型水、重碳酸-钙·钠（钠·钙）型水，重碳酸·氯-钙·钠（钠·钙）型水；嘉定城区—竹桥—葛隆，以及松江东部、中心城南部、闵行、金山东部、奉贤、浦东东南部咸水分布区，水化学类型主要为氯-钠·钙型水，氯-钠型水，以及氯·重碳酸-钠型水、氯·重碳酸-钠·钙型水（图 27-20）。

2. 河口砂岛区

河口砂岛区主要分布盐水，水化学类型主要为氯-钠型水，仅横沙岛局部分布氯·重碳酸-钠型水。

六、第Ⅳ承压含水层

（一）含水层埋藏特征

除青浦莲盛—泗联、金山吕巷—金山嘴、奉贤齐贤—星火农场和北新泾—小闸等 5 块基底凸起区外，全区第Ⅳ承压含水层分布较为广泛。形成于早更新世晚期—中晚期，可分为上、下两个承压含水层。上、下两个承压含水层间唯江苏浏河—上海浦东东海农场以及闵行—东海农场宽 8～4km 的范围内

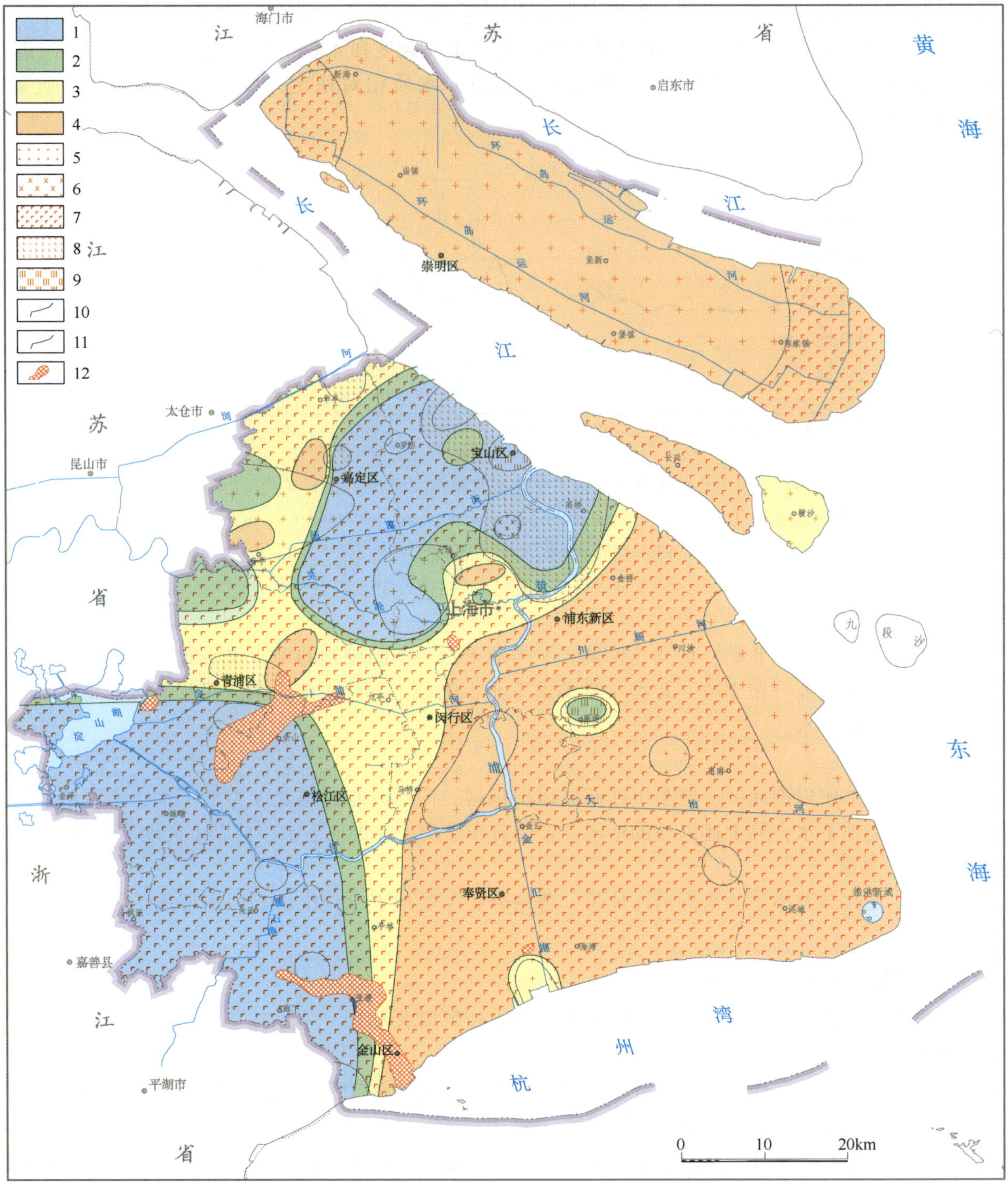

图 27-20 上海第Ⅲ承压含水层地下水水化学类型分区图

Fig. 27-20 Groundwater hydrochemical type zoning map of confined aquifer Ⅲ in Shanghai

1—重碳酸根；2—重碳酸根·氯；3—氯·重碳酸根；4—氯；5—钠；6—钙；7—钠·钙；8—钙·钠；9—钙·镁；10—阳离子分界线；11—阴离子分界线；12—基岩及含水层基本缺失区

彼此沟通外，其他地区一般均为杂色硬黏土分隔，而且上层在陆域地区较下层发育，在河口砂岛则相反。

1. 第Ⅳ承压含水层上部（$Ⅳ_1$）

第$Ⅳ_1$承压含水层为早更新世晚期河流、河口相沉积物。陆域西南部的两片低山（青浦—松江、金山）剥蚀残丘区含水层基本缺失，横沙岛含水层基本缺失。陆域滨岸地带分布的浏—南古河道向西摆动，河流宽度达到 10~15km，河流南段的原南汇区境内当时已受到海水的波及。早更新世中期在南部低山丘陵区流出的一条比较顺直、狭窄的支流河道，为枫—奉古河道，至芦潮港汇入浏—南主干河道，宽度达到 5~10km，成为上海第$Ⅳ_1$承压含水层的主要分布地带。在两条古河道之间以及崇明等现在的岛屿上，则为大片的河流边滩和泛滥平原—湖泊洼地，大多堆积了砂、粉土夹黏性土或两者的互层。

毗邻含水层基本缺失区，含水层埋深一般大于 170m。浏—南古河道分布区，含水层顶面埋深一般为 160~180m，在枫—奉古河道分布区，含水层顶面埋深一般为 130~160m。毗邻含水层基本缺失区，含水层厚度小于 10m。浏—南古河道、枫—奉古河道分布区，含水层厚度总体大于 30m，局部大于 50m；除此之外厚度大多在 10~30m。

崇明岛第$Ⅳ_1$承压含水层顶面埋深为 150~190m。东部堡镇—富民农场一线以东，含水顶面埋深在 164~196m，中部含水层顶面埋深在 154~195m，西部在 160~188m。第$Ⅳ_1$承压含水层厚度在 4~27m 不等，较厚地段分布于中部的南门县城—堡镇城区一带。长兴岛含水层顶面埋深在 150m 左右，厚度在 4~16m 不等（图 27-21）。

2. 第Ⅳ承压含水层下部（$Ⅳ_3$）

第$Ⅳ_3$承压含水层为早更新世中期湖河相沉积物。青浦、松江接壤区，金山、奉贤境内的两片低山区，以及市区虹桥—小闸、南汇坦直、三灶等地散布若干天马山余脉的剥蚀残丘区，含水层基本缺失，嘉定西北部地区大部含水层基本缺失。在宝山盛桥—市区东部—浦东张桥、川沙—机场—南汇东侧滨岸地带，开始出现一条宽达七八千米的曲流河道，即常称的浏—南古河道，成为陆域第$Ⅳ_3$承压含水层的主要分布地带。其西侧则分布较多分支，为源自西部低山丘陵区的顺直河流，其中主要的在南部有从闵行、北桥流出，流向南汇新场—大团—芦潮港的支流，在汇入浏—南主河道的现南汇嘴地区形成一泥质砂砾石与含砾砂夹黏土的冲积扇堆积；在北部，有从江苏花桥流来，流经安亭—刘行后汇入浏—南主河道的一股支流。在现崇明、长兴等沙岛上当时也大部为冲积平原，只是在崇明岛东南部的侯家—新河—向化—前哨一带及西北角的红星农场附近地区各有一条近于东西流向的河流，堆积了含砾砂层，成为崇明岛上第$Ⅳ_3$承压含水层的主要分布地带。

陆域毗邻含水层基本缺失区，含水层顶面埋深在 200m 左右。沿浏—南古河道发育方向，含水层层面大体上顺古河道方向倾斜，埋深在 200~210m。东南部古河道交汇地区，含水层埋藏在 210~230m。毗邻含水层基本缺失区，含水层厚度小于 10m，浏—南古河道分布地带，厚度普遍大于 30m，其他地区大多在 20m 左右（图 27-22）。

崇明岛第$Ⅳ_3$承压含水层顶面埋深 200~220m 不等。东西向展布的中部地带，厚度在 10m 左右，该带两侧厚度有所增大，厚度逾 20m。长兴岛含水层顶面埋深 220~230m，厚度一般为 7~35m 不等。横沙岛含水层顶面埋深 200~210m，厚度界于 5~42m。

（二）含水层渗透性

含水层分布地区，以渗透系数表征的含水层渗透性，滨海平原区较湖沼平原区、河口砂岛相对好。

1. 陆域地区

毗邻含水层基本缺失区，含水层渗透性相对差，渗透系数总体小于 10m/d；东部宝山、浦东浏—南古河道地带，渗透性最好，渗透系数基本在 30~50m/d，南部枫—奉古河道地带，渗透性也较好，渗透系数普遍在 20~30m/d。除此以外的地区，含水层渗透性一般，渗透系数在 10~20m/d。

2. 河口砂岛区

含水层渗透性总体较差，渗透系数大多在 10m/d 左右。

（三）含水层富水性

含水层分布地区，以单井涌水量表征的含水层富水性，滨海平原区较湖沼平原区、河口砂岛相对好。

1. 陆域地区

毗邻含水层基本缺失区，含水层富水性相对差，单井涌水量总体小于 $100m^3/d$；宝山、浦东浏—

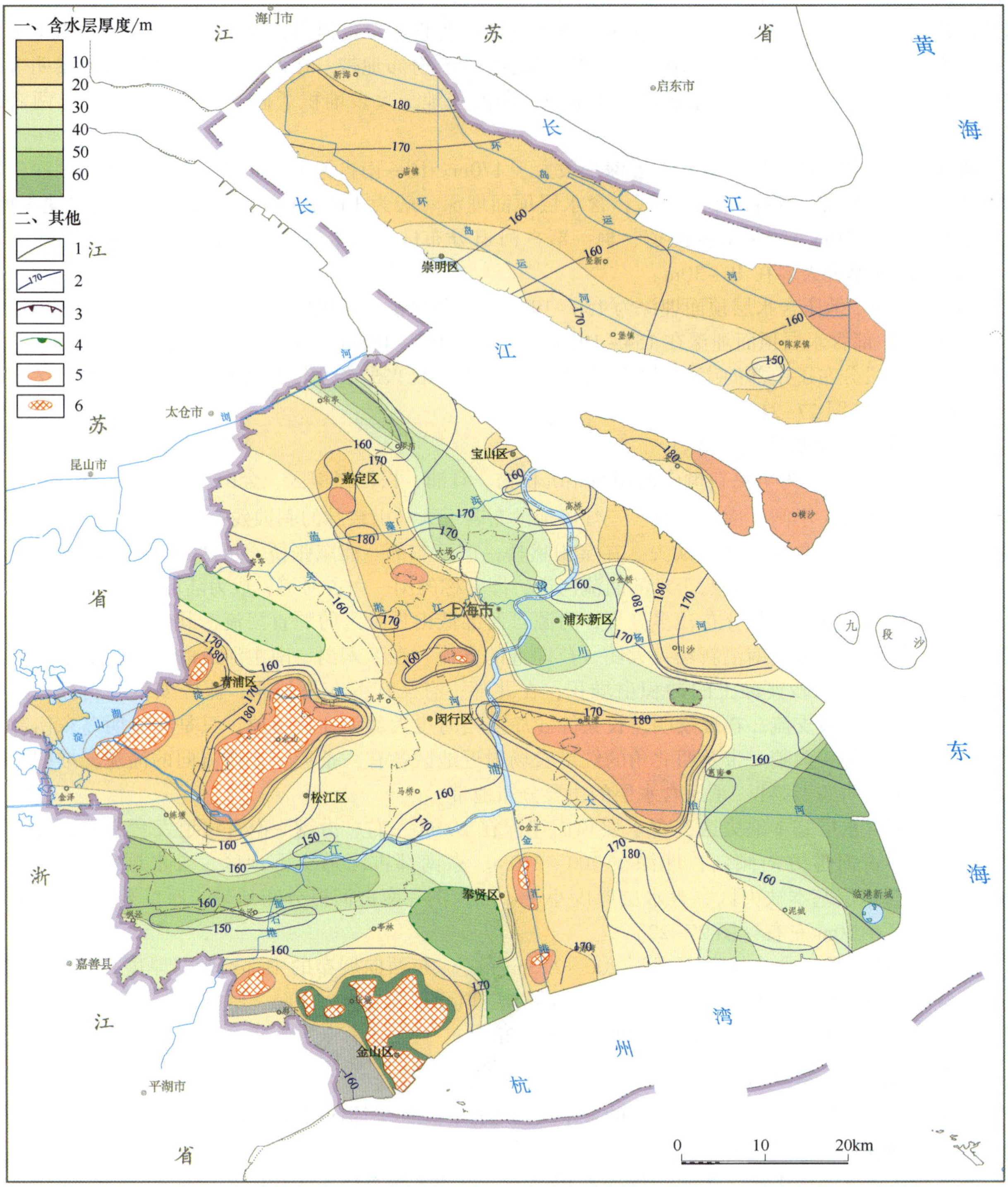

图 27-21　上海第Ⅳ$_1$承压含水层埋藏分布图

Fig. 27-21　Buried depth distribution of confined aquifer Ⅳ$_1$ in Shanghai

1—厚度等值线；2—顶面埋深等值线（m）；3—第Ⅳ、Ⅴ含水层沟通区；

4—第Ⅲ、Ⅳ含水层沟通区；5—含水层基本缺失区；6—基岩分布区

南古河道地带，富水性最好，单井涌水量总体大于 3000m^3/d；南部庄行—柘林—漕泾、青浦白鹤—重固，富水性好，单井涌水量总体大于 3000m^3/d；其他地区，含水层富水性也较好，单井涌水量基本在 1000~3000m^3/d（图 27-23）。

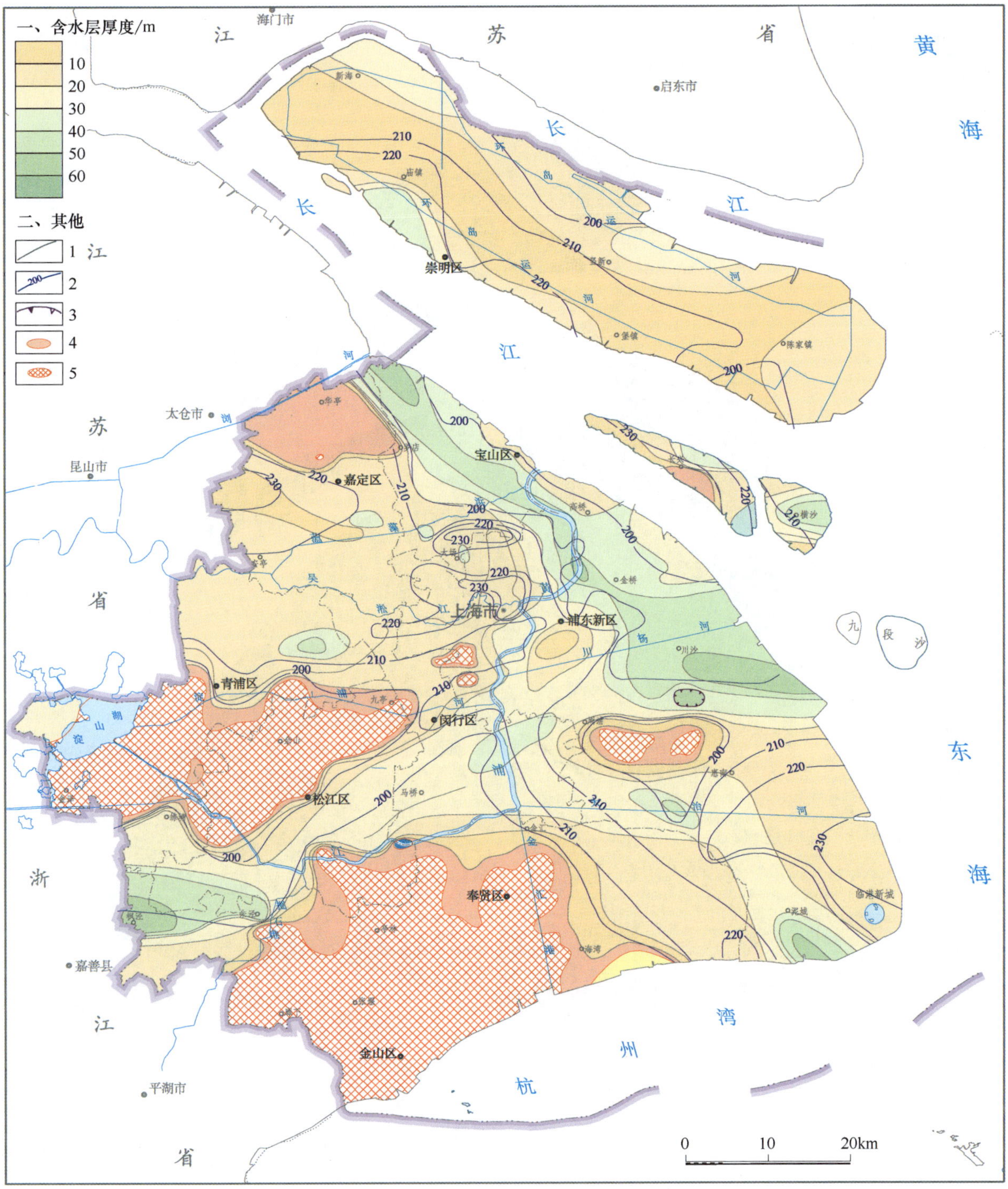

图 27-22　上海第IV_3承压含水层埋藏分布图

Fig. 27-22　Buried depth distribution of confined aquifer IV_3 in Shanghai

1—厚度等值线；2—顶面埋深等值线（m）；3—第Ⅳ、Ⅴ含水层沟通区；4—含水层基本缺失区；5—基岩分布区

2. 河口砂岛区

含水层富水性总体较差，单井涌水量总体小于 1000m^3/d。

（四）地下水矿化度

含水层受沉积环境变化影响，区域上地下水主要分布淡水，其次为微咸水、咸水。

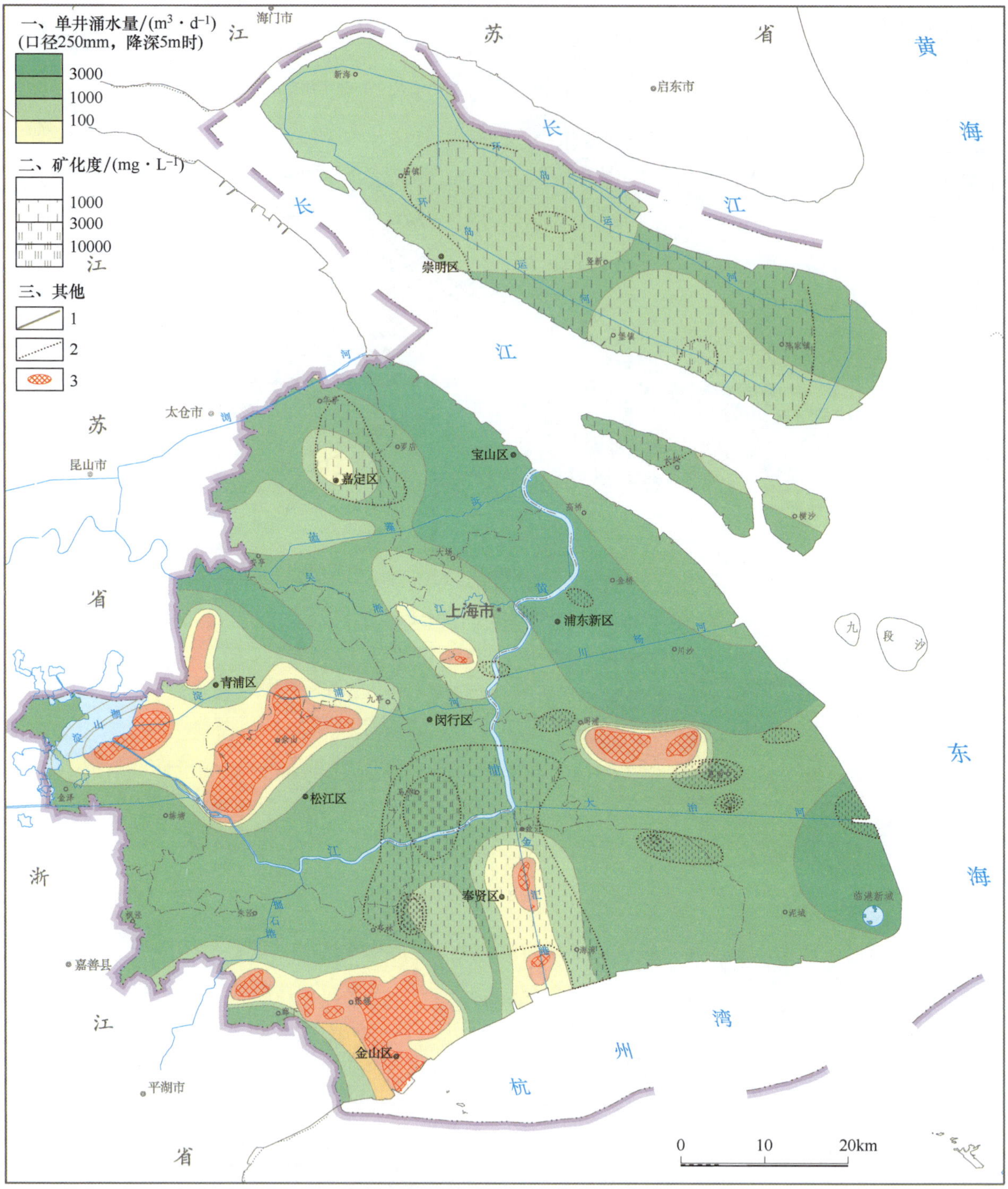

图 27-23　上海第Ⅳ承压含水层富水性、矿化度分布图

Fig. 27-23　Water abundance and mineralization degree map of confined aquifer Ⅳ in Shanghai

1—富水性分界线；2—矿化度分界线；3—基岩分布区

1. 陆域地区

陆域地区普遍分布矿化度小于 1g/L 的淡水，约占全市面积的 61%；微咸水、咸水零星分布于闵行老城区—奉贤南桥，奉贤头桥、嘉定城区、浦东惠南局部地区，分别占全市面积的 10%和 2%。

2. 河口砂岛区

河口砂岛区主要分布淡水和微咸水。崇明岛城桥—东风农场一线以西地区和东部前哨农场局部地区，分布矿化度小于 1g/L 的淡水，分布面积约占全市面积的 10%；崇明岛长江农场和向化局部、长兴岛东部和横沙岛主要分布咸水，分布面积约占全市面积的 2%；其余地区分布矿化度为 1~3g/L 的微咸水，分布面积约占全市面积的 9%（见图 27-23）。

3. 矿化度稳定性

历史上，该含水层为上海地下水主要开发利用层次，水质长期监测程度较高，区域上地下水矿化度基本维持原生状态。位于浦东东南部（原南汇）下沙水厂地下水开采井（汇 021-01）20 世纪 70 年代—21 世纪初水质监测结果显示：70 年代初期，地下水矿化度在 620mg/L 左右，为淡水，经过近 26 年的持续监测，矿化度基本在 610~630mg/L 间波动，相对稳定，保持了原生地下水质量状态（图 27-24）。

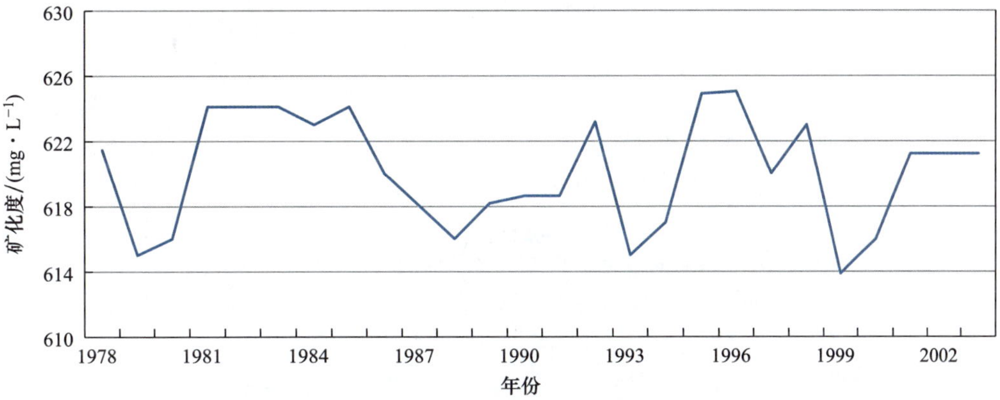

图 27-24　上海第Ⅳ承压含水层典型监测井（汇 021-01）矿化度历时变化图

Fig. 27-24　Time history of mineralization degree of confined aquifer Ⅳ in Shanghai per typical groundwater monitoring well (Hui021-01)

（五）地下水质量

区域上分布有Ⅰ类、Ⅱ类、Ⅲ类、Ⅳ类、Ⅴ类地下水。不同级别的地下水空间分布相对简单。总体上，陆域大部分布Ⅰ类、Ⅱ类、Ⅲ类地下水，仅陆域南部局部和崇明岛，主要分布Ⅲ类和Ⅳ类地下水。

1. 陆域地区

闵行城区—奉贤南桥新城、庄行咸水区，分布有Ⅳ类及Ⅴ类地下水，其邻近地区分布Ⅲ类地下水，同时嘉定城区局部分布Ⅲ类地下水；除上述以外的淡水分布区，广泛分布Ⅰ类地下水。

2. 河口砂岛区

崇明岛港西—建设—新河—竖新—港沿—堡镇—向化微咸水分布地带，普遍分布Ⅳ类地下水，东滩湿地局部地区分布Ⅰ类地下水，西滩湿地局部地区分布Ⅱ类地下水；除上述以外的地区，主要分布Ⅲ类地下水。长兴岛及横沙岛，主要分布Ⅰ类和Ⅲ类地下水。

（六）地下水水化学类型

（1）第$Ⅳ_1$承压含水层地下水水化学类型

陆域地区以重碳酸型水、重碳酸·氯型水为主，河口砂岛以氯·重碳酸型、氯型水为主。

A. 陆域地区

北部嘉定城区—宝山，主要分布重碳酸·氯-钠型水、氯·重碳酸-钠型水；闵行南部—奉贤—浦东东南部（原南汇），主要分布重碳酸·氯-钠型水、氯·重碳酸-钠型水，局部分布氯-钠型水。

除上述以外的地区，广泛分布重碳酸-钠型水、重碳酸-钠·钙型水和重碳酸-钙·钠型水。

B. 河口砂岛区

崇明岛城桥—长江农场一线以东地区主要分布氯-钠型水，该线以西地区主要分布氯·重碳酸-钠型水，西北部绿化主要分布重碳酸·氯-钠型水。

长兴岛、横沙岛主要分布氯-钠型水（图 27-25）。

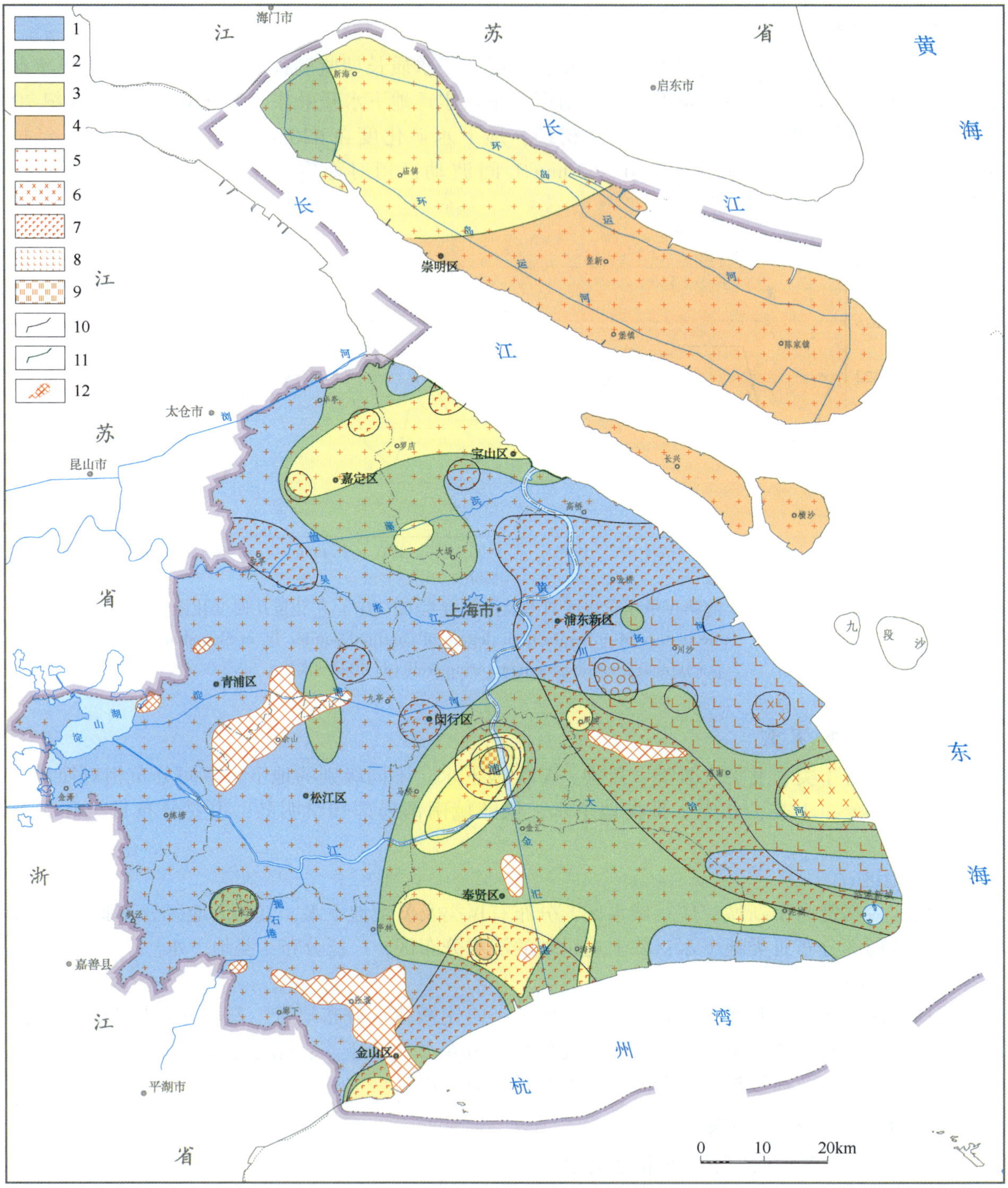

图 27-25 上海第Ⅳ$_1$ 承压含水层水化学类型分区图

Fig. 27-25 Hydrochemical type zoning map of confined aquifer Ⅳ$_1$ in Shanghai

1—重碳酸根；2—重碳酸根·氯；3—氯·重碳酸根；4—氯；5—钠；6—钙；7—钠·钙；8—钙·钠；9—钙·镁；10—阳离子分界线；11—阴离子分界线；12—基岩及含水层基本缺失区

（2）第Ⅳ$_3$ 承压含水层地下水水化学类型

陆域地区以重碳酸·氯型水、重碳酸型水为主，河口砂岛以重碳酸·氯型水为主。

A. 陆域地区

青浦南部—金山西部地区，以及浦东川沙新城以北地区，主要分布重碳酸-钠型水、重碳酸-钙·钠（钠·钙）型水。嘉定、中心城南部局部、浦东惠南—大团及临港新城局部地区，主要分布氯·重碳酸-钠型水、氯·重碳酸-钙·镁型水。

除上述以外的地区，广泛分布重碳酸·氯-钠型水、重碳酸·氯-钠·钙型水和重碳酸·氯-钙·钠型水。

B. 河口砂岛区

崇明岛长江农场主要分布氯·重碳酸-钠型水、氯-钠型水；西北部新海局部地区分布重碳酸-钠型水；其余广大地区主要分布重碳酸·氯-钠型水。

长兴岛东部主要分布重碳酸·氯-钠型水，西部主要分布氯·重碳酸-钠型水、氯-钠型水。横沙岛主要分布重碳酸-钠型水（图27-26）。

七、第Ⅴ承压含水层

（一）含水层埋藏特征

含水层为早更新世早期河流、河口—滨湖相沉积物。含水层在青浦大盈—香花桥—沈泾—徐泾—虹桥—长宁夷定路—人民广场—西藏路—浦东龙阳路立交桥—六里—惠南—大团一线以南地区，嘉定昌桥—永新，宝山场中路以北—江杨南路以西一带基本缺失外，北部、东部的沟谷凹陷区分布较为广泛。岩性在古河道和河口地段以砾质中粗砂或粗砂卵石层为主，夹中细砂，毗邻基岩边滩则为细砂或粉细砂。含水组通常被泥质、钙质弱胶结，且以砂砾石分选性和磨圆度较差、成分复杂为特点。河口砂岛由2~3个含水层组成，往深部，富水性减弱，局部呈半成岩状态。

毗邻含水层基本缺失区，含水层顶面埋深在230m左右；嘉定陆渡—曹王—宝山盛桥一带、高镜—闸殷路一带，顶面埋深在270~290m之间，由此向南过渡至250~260m，近含水层缺失区，顶面埋深在240~250m；中心城以东地区，顶面埋深基本在250~270m。崇明岛中部近东西向展布地带，含水层顶面埋深在250m左右，由此向两侧逐渐加深至大于270m。长兴岛、横沙岛含水层顶面埋深一般在260~280m。

陆域毗邻含水层缺失区，含水层厚度大多小于10m，与江苏交界地带及浦东机场—东海农场一带，含水层厚度大于30m。崇明岛含水层厚度普遍大于50m，长兴岛、横沙岛含水层厚度总体在10~30m（图27-27）。

（二）含水层渗透性

含水层分布地区，以渗透系数表征的含水层渗透性，河口砂岛较滨海平原区相对好。

1. 陆域地区

毗邻含水层基本缺失区，含水层渗透性总体较差，渗透系数大多在5~10m/d，仅东部浦东高桥、川沙新—国际机场，西部白鹤—安亭—黄渡，渗透性相对好，渗透系数在10~20m/d。

2. 河口砂岛区

含水层渗透性总体较好，渗透系数大多在10~20m/d，仅绿化—新河沿江地带相对差，渗透系数在5m/d左右。

（三）含水层富水性

含水层分布地区，以单井涌水量表征的含水层富水性，河口砂岛较滨海平原区相对好。

1. 陆域地区

毗邻含水层基本缺失区，含水层富水性总体较差，单井涌水量总体小于100m^3/d，东部浦东高桥、唐镇—川沙新城—国际机场，西部白鹤—安亭—黄渡，富水性相对好，单井涌水量在1000~3000m^3/d。其他地区单井涌水量在100~1000m^3/d。

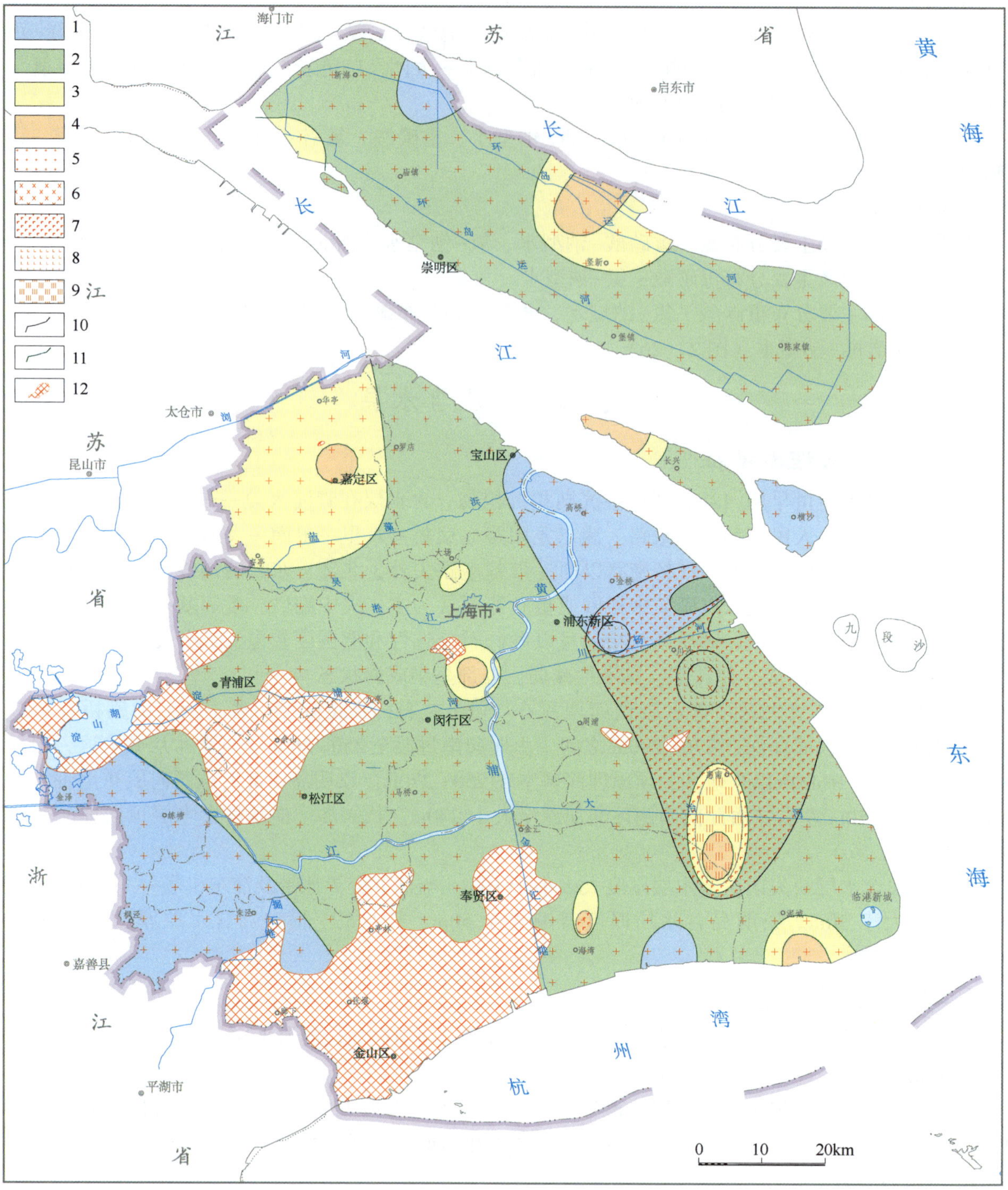

图 27-26　上海第Ⅳ$_3$承压含水层水化学类型分区图

Fig. 27-26　Hydrochemical type zoning map of confined aquifer Ⅳ$_3$ in Shanghai

1—重碳酸根；2—重碳酸·氯；3—氯·重碳酸根；4—氯；5—钠；6—钙；7—钠·钙；8—钙·钠；9—钙·镁；10—阳离子分界线；11—阴离子分界线；12—基岩及含水层基本缺失区

2. 河口砂岛区

含水层富水性总体较好，单井涌水量大多在 1000~3000m^3/d。仅绿化—新河沿江地带相对差，单井涌水量在 100~1000m^3/d（图 27-28）。

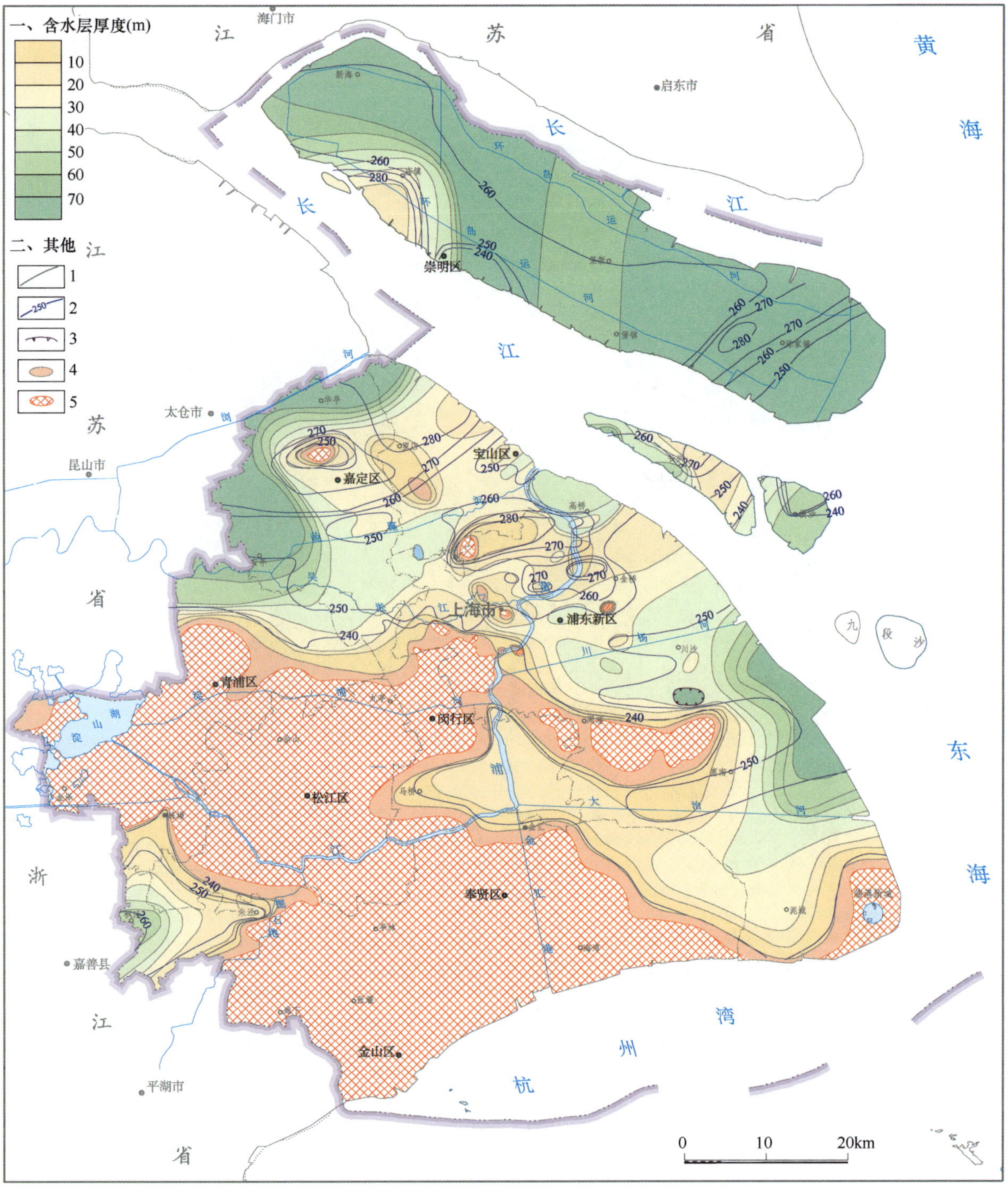

图 27-27　上海第Ⅴ承压含水层埋藏分布图

Fig. 27-27　Buried depth distribution of confined aquifer Ⅴ in Shanghai

1—厚度等值线；2—顶面埋深等值线（m）；3—第Ⅳ、第Ⅴ承压含水层沟通区；4—含水层基本缺失区；5—基岩分布区

（四）地下水矿化度

受沉积环境影响，以分布淡水、微咸水为主。

1. 陆域地区

南部吴泾—航头—新场—奉城—泥城—书院—东海农场地带，主要分布咸水，占全市面积的12%；惠南、嘉定城区—宝山城区一线以北地区，主要分布微咸水，占全市面积的9%；除上述以外的广大地

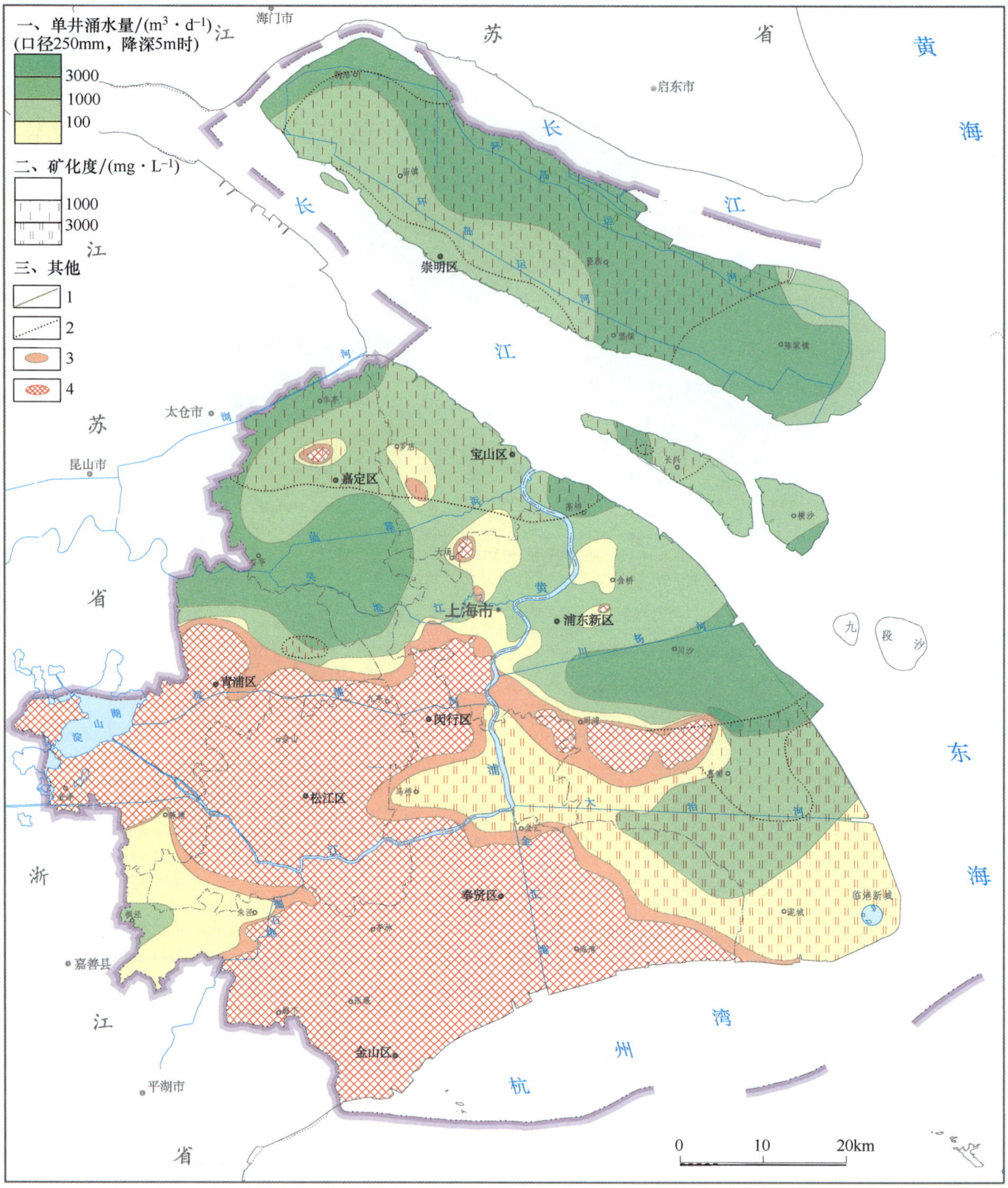

图 27-28 上海第Ⅴ承压含水层富水性、矿化度分布图

Fig. 27-28 Water abundance and mineralization degree map of confined aquifer Ⅴ in Shanghai

1—富水性分界线；2—矿化度分界线；3—含水层基本缺失区；4—基岩分布区

区分布矿化度小于 1g/L 的淡水，约占全市面积的 22%。

2. 河口砂岛区

主要分布微咸水和淡水。崇明岛周围沿江地带、长兴岛东部和横沙岛，主要分布矿化度小于 1g/L 的淡水，分布面积约占全市面积的 6%；除上述以外的地区分布矿化度为 1~3g/L 的微咸水，分布面积约占全市面积的 15%。

3. 矿化度稳定性

历史上，该含水层为上海地下水主要开发利用层次，水质长期监测程度较高，区域上地下水矿化度基本维持原生状态。位于郊区青浦白鹤地下水开采井（青 005-01）20 世纪 70 年代—21 世纪初水质监测结果显示：20 世纪 70 年代初期，地下水矿化度在 540mg/L 左右，为淡水，经过近 40 年的持续监测，矿化度基本在 530~560mg/L 间波动，相对稳定，保持了原生地下水质量状态（图 27-29）。

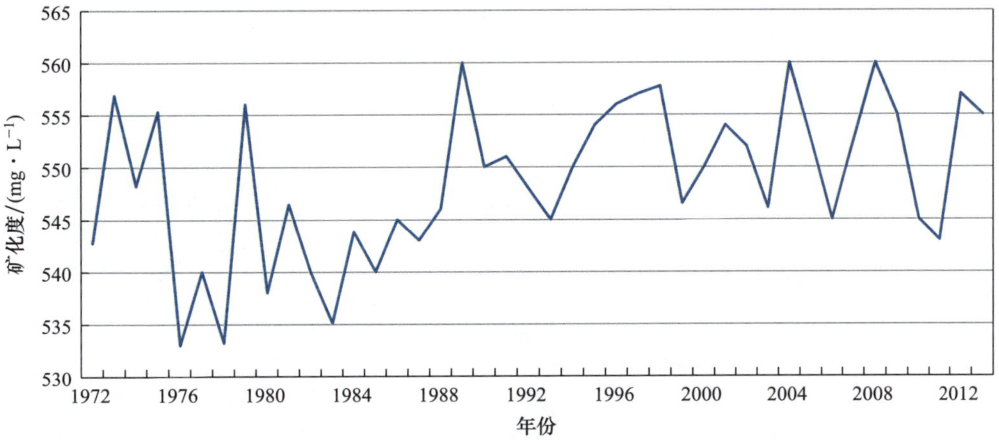

图 27-29　上海第Ⅴ承压含水层地下水典型监测井（青 005-01）矿化度历时变化图

Fig. 27-29　Time history of mineralization degree of confined aquifer Ⅴ in Shanghai per typical groundwater monitoring well (Qing005-001)

（五）地下水质量

区域上分布有Ⅰ类、Ⅱ类、Ⅲ类、Ⅳ类、Ⅴ类地下水。总体上，陆域北部、河口砂岛分布Ⅰ类、Ⅱ类地下水，陆域南部局部地区，分布Ⅲ类、Ⅳ类和Ⅴ类地下水。

1. 陆域地区

青浦—嘉定—宝山—中心城—浦东北部（原川沙县）以及金山枫泾淡水分布区，地下水普遍为Ⅰ类、Ⅱ类；南部闵行吴泾—浦东东南部（原南汇区）为微咸水和咸水分布区，主要分布Ⅲ类、Ⅳ类及Ⅴ类的地下水。

2. 河口砂岛区

地下水以淡水和微咸水为主，故地下水质量较好，普遍分布Ⅱ类地下水，其中城桥沿江地带、陈家—东滩湿地、长兴岛东部和横沙岛，分布Ⅰ类地下水。

（六）地下水水化学类型

地下水水化学类型较简单，以重碳酸·氯型水及氯·重碳酸型水为主。

1. 陆域地区

嘉定北部—宝山—中心城北部，主要分布氯·重碳酸-钠型水；浦东东南部（原南汇），主要分布氯-钠型水；金山枫泾主要分布重碳酸-钠型水。除上述以外地区（除缺失区），主要分布重碳酸·氯-钠型水（图 27-30）。

2. 河口砂岛区

崇明岛新河—前进农场一线以东地区，主要分布重碳酸·氯-钠型水，该线以西地区主要分布氯·重碳酸-钠型水。

长兴岛主要分布重碳酸·氯-钠型水，东北部局部分布氯·重碳酸-钠型水。

横沙岛主要分布重碳酸·氯-钙·钠型水（见图 27-30）。

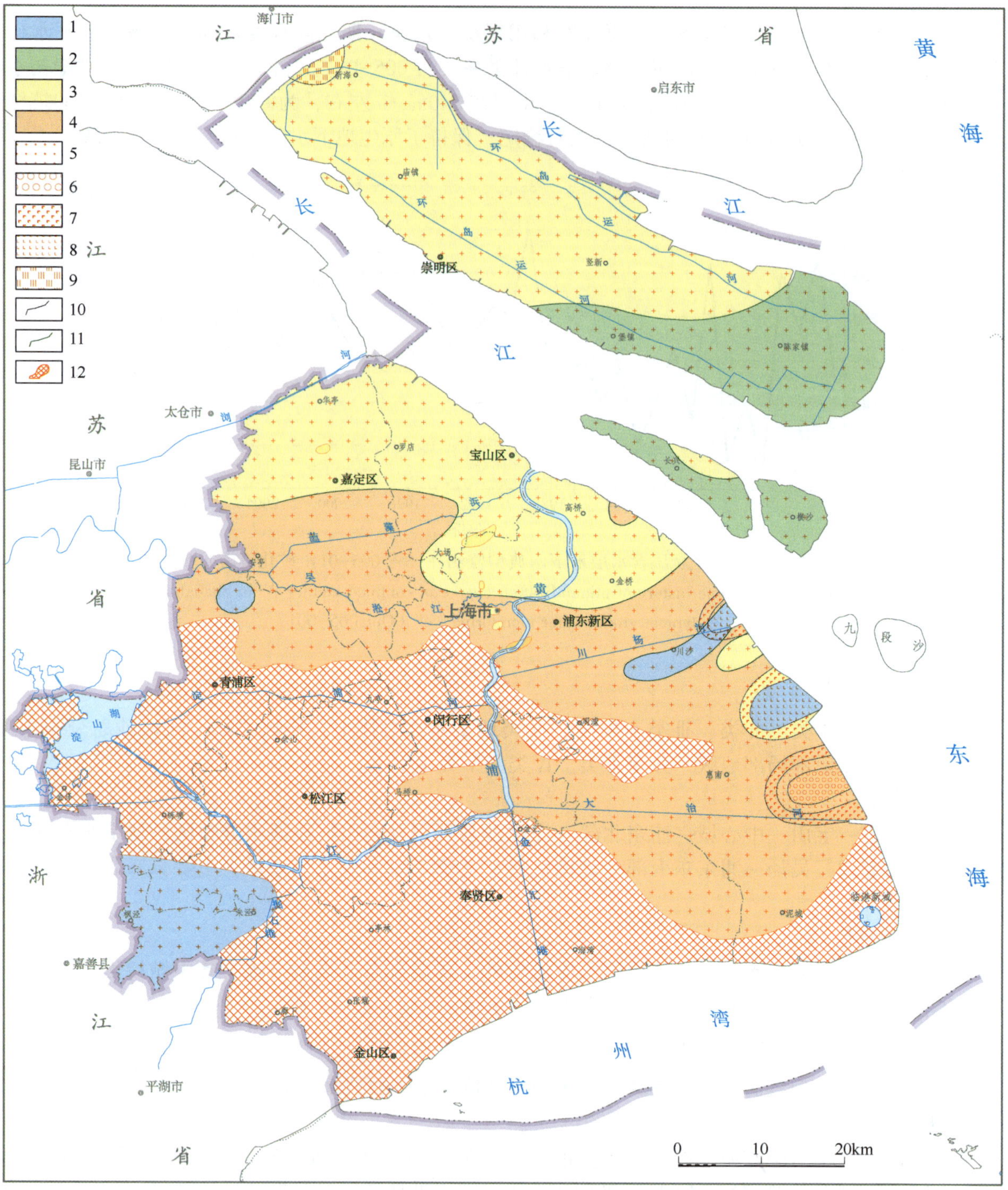

图 27-30　上海第Ⅴ承压含水层水化学类型分区图

Fig. 27-30　Hydrochemical type zoning map of confined aquifer Ⅴ in Shanghai

1—重碳酸根；2—重碳酸根·氯；3—氯·重碳酸根；4—氯；5—钠；6—钙·镁·钠；7—钠·钙；8—钙·钠；9—钙·镁；10—阳离子分界线；11—阴离子分界线；12—基岩及含水层基本缺失区

第三节　水文地质条件分区

上海隶属于长江三角洲水文地质单元。受古地理环境约束，不同含水层的水文地质条件均有其自身分布特点。为此，结合上海地貌单元，以及含水层埋藏、富水性、水质、水位等分布总体特征，进行水文地质亚区分区，反映不同块段水文地质特征（表27-7）。

表27-7　上海水文地质分区表

Table 27-7　Hydrological geological zoning table of Shanghai

区	亚区		段		水文地质条件
	名称	代号	名称	代号	
长江三角洲平原水文地质区	河口—滨海平原水文地质亚区	Ⅰ	淡水水量极丰富地段	I_1	含水层齐全，砂层极其发育，厚度大，岩性以中粗砂含砾为主；地下水极丰富，各层累计单井涌水量为7000~10000m^3/d，地下水质量除第Ⅱ承压含水层在嘉定浏河局部块段为微咸水外，其余层和地段均为淡水
			上咸下淡水量极丰富地段	I_2	水文地质条件与I_1相似，唯第Ⅱ、第Ⅲ承压含水层为微咸—咸水，各层累计单井涌水量为7000~8000m^3/d
	剥蚀残丘-叉流河道水文地质亚区	Ⅱ	剥蚀残丘水量贫乏地区	II_1	基岩残丘零星分布，含水层不发育或基本缺失
			西淡东咸水量中等—较丰富的漫滩地段	$\mathrm{II}_{2\text{-}a}$	含水层不齐全，缺失第Ⅴ承压含水组，砂层不发育，厚度薄，岩性以粉细砂和中细砂为主；各层累计单井涌水量为100~3000m^3/d，地下水质量在亭林镇以西各层主要为淡水，其东侧为微咸水—半咸水取代
			西淡东上咸下淡水量中等至较丰富的漫滩地段	$\mathrm{II}_{2\text{-}b}$	含水层不齐全，缺失第Ⅴ承压含水层，砂层厚度较薄，岩性以中细砂为主；各层累计单井涌水量为1000~3000m^3/d，地下水质量在青浦城区—练塘一线以西为淡水，其东第Ⅱ、第Ⅲ承压含水层为微咸水—半咸水，第Ⅳ承压含水层为淡水
			西淡东咸水量丰富—极丰富的河床地段	$\mathrm{II}_{3\text{-}a}$	含水层较齐全，砂层发育，厚度较大，以中细砂中粗砂含砾为主；各层累计单井涌水量为3000~8000m^3/d，由西往东递增。地下水质量在朱泾以西各层全为淡水，其东主要为微咸水—半咸水
			淡水水量极丰富的河床地段	$\mathrm{II}_{3\text{-}b}$	含水层较齐全，砂层发育，厚度较大，以中粗砂中细砂含砾为主；各层累计单井涌水量为6000~7000m^3/d。地下水质量除东部第Ⅱ承压含水层为微咸水外，其余皆为淡水。

（据《上海地质矿产志》，1999）

第四节　区域地下水位动态

上海地下水（潜水和承压水）因遭受不同程度的各种自然因素和人为作用影响，天然地下水流场也发生了不同程度变化。长期来，通过建设不同含水层地下水监测设施以及长期监测，基本掌握了各含水层地下水位动态变化规律与区域分布特征。

一、潜水含水层

潜水水位埋藏较浅，一般在地表以下0.5~2.0m。潜水因与大气降水、地表水等水交替强烈，多年来水位呈现动态变化的特征，区域上地下水位标高总体在2.5~4.0m（吴淞高程，下同）间波动，在日常间断性降水补给的情况下，水位趋势性上升或下降现象不明显。极端高、低水位分别为4.6m（2001年5月）和2.0m（1998年12月）（图27-31）。

受地势、包气带岩性、含水层厚度不一等综合因素影响，不同地区潜水水位波动特征略显差异。湖沼平原地区，潜水水位相对低，一般在2.5~3.0m；滨海平原地区，潜水水位相对高，一般在3.0~3.7m；河口砂岛区潜水水位波动幅度较大，一般在2.5~3.6m（见图27-31）。

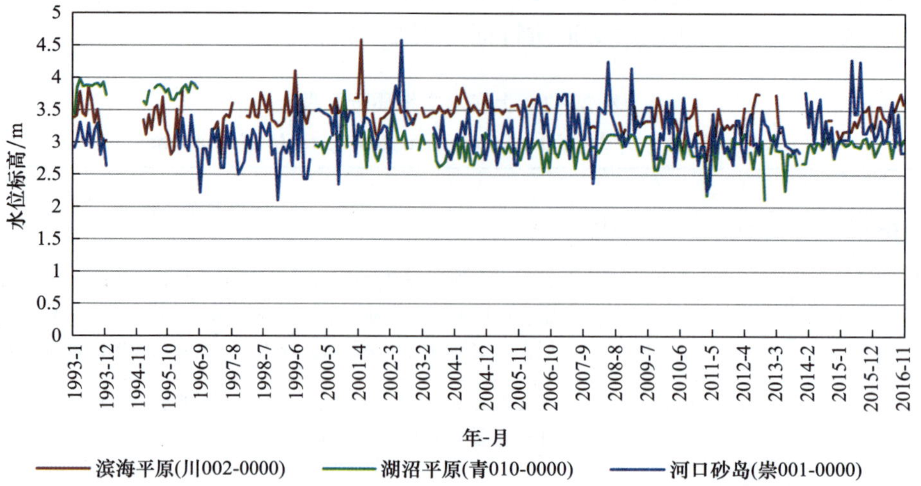

图 27-31　上海不同地貌地区典型潜水含水层地下水位动态变化图

Fig. 27-31　Groundwater table dynamics of typical phreatic aquifers in Shanghai by geomorphologic region

二、微承压含水层

区域上，微承压含水层地下水位系统的监测工作始于 21 世纪初期，以往基本处于监控盲区。自然状态下的微承压含水层地下水位基本处于正标高状态。

中心城典型监测井地下水位监测数据反映：2010 年前后，为城市建设活动强烈时期，由于受工程建设过程中减压降水等因素综合影响，对局部地下水位影响较为明显，地下水位处于相对低的分布状态，水位标高总体在-1～-3m；2012 年以后，随着工程减压降水活动的减弱，地下水位逐渐回升，至 2016 年间，地下水位基本在 0～1m 间波动，趋于相对稳定（图 27-32）。

郊区（青浦、松江、金山等）典型监测井地下水位监测结果显示：不同地区地下水位略显差异，但总体埋藏较浅，地下水位基本处于天然状态，水位标高总体在 0～2m，且多年来相对稳定（见图 27-32）。

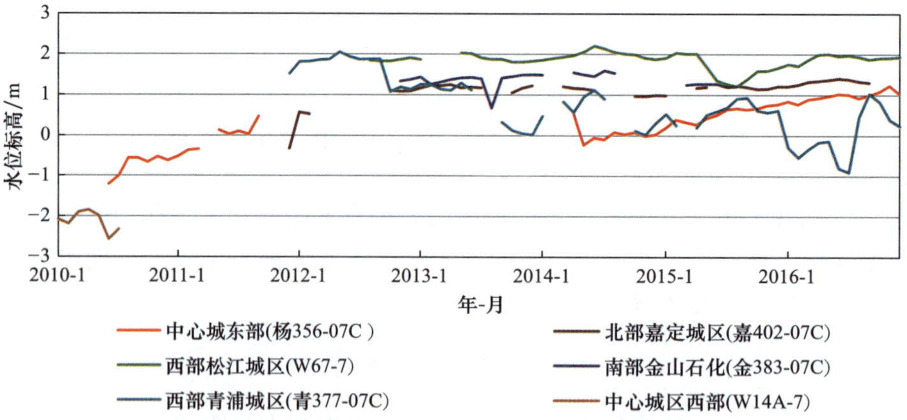

图 27-32　上海不同地区典型微承压含水层地下水位动态变化图

Fig. 27-32　Groundwater table dynamics of typical microconfined aquifers in Shanghai by district

三、第 I 承压含水层

自然状态下的第 I 承压含水层地下水位基本处于正标高状态。但后期由于受短期开采、或沟通区受下伏含水层地下水位波动、或工程建设过程中减压降水等因素综合影响，地下水位呈现出特定的动态变化特征。

（一）区域地下水位动态

该含水层地下水位系统的监测工作，中心城始于20世纪60年代中期，郊区大部分地区始于80年代初期。据中心城和郊区典型监测井监测数据，1965—2015年区域地下水位总体呈现埋藏较浅（非沟通区）的特征。地下水位标高总体在2~-2m间波动，呈现相对稳定—缓慢下降—缓慢上升变化特征（表27-8；图27-33）。

表27-8　上海中心城、郊区第Ⅰ承压含水层地下水位变化特征

Table 27-8　Groundwater table variations of confined aquifer Ⅰ in downtown and suburban Shanghai

地区 \ 时间		1965年	1966—1969年	1970—1981年	1982—1997年	1998—2015年
中心城（杨013-01SG）		-3~-4m，低水位时期	-1~2m，迅速回升	0~2m，高水位相对稳定	0~-2m，缓慢下降	-1~2m，相对稳定至缓慢上升
郊区	北部地区（宝02A-04W）					-1.4~1.3m，缓慢上升
	南部地区（金021-01SG）				0.3~-0.8m，相对稳定，1992年之后缓慢下降	-1.1~-0.1m，相对稳定，缓慢上升
	西部地区（青377-01C）					-0.7~0.1m，相对稳定

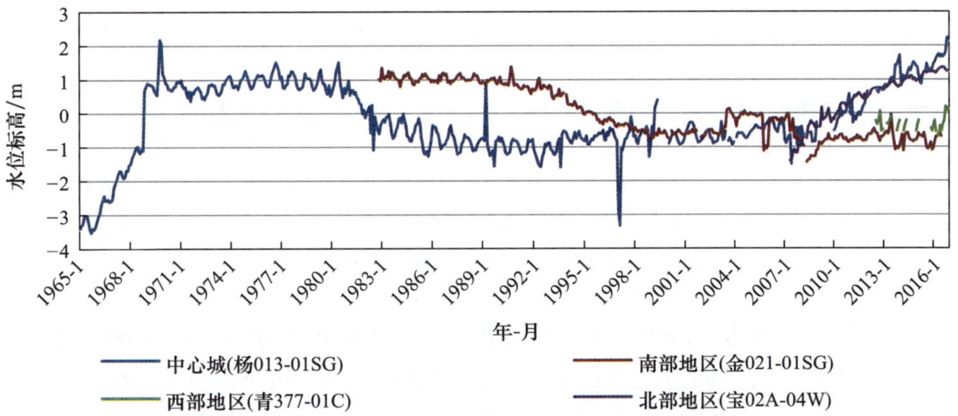

图27-33　上海中心城、郊区第Ⅰ承压含水层区域地下水位历时变化曲线图

Fig. 27-33　Groundwater table time histories of confined aquifer Ⅰ in downtown and suburban Shanghai

1. 中心城

地下水位总体呈现低—高—低—高的分布特征。因市区于1961—1966年间进行过短期的、少量的开采，地下水位下降至-3~-4m；1966年停采后，地下水位迅速回升，至1969年水位最高上升至2.0左右；1970—1981年地下水位基本在0~2m之间波动；1982—1997年地下水位再次处于缓慢下降状态，总体在0~-2m间波动；1998—2015年地下水位相对稳定，并逐步缓慢上升，至2016年地下水位在2m左右。

2. 郊区（含水层非沟通区）

地下水位总体呈现高—低—高的分布特征。地下水位变化特征与中心城在时段上有一定差异，主要是20世纪80年代—21世纪初期，郊区地下水位由相对稳定转而小幅持续下降，而中心城处于相对低水位分布状态。2005 年之后变化趋势基本与中心城一致。

南部金山典型监测井监测结果显示，1982—1992 年，郊区地下水位基本处于天然状态（市区处于下降状态），在 1m 上下波动；1993—1997 年，受区域水位下降影响，地下水位缓慢下降至-0.5m 左右（市区处于相对稳定状态）；1997 年后至今，南部、西部地下水位均处于相对稳定状态，地下水位在 0~-1m 间波动，北部地区 2010 年以来逐步缓慢上升，2016 年上升至 1m 左右，与天然状态水位分布基本一致（表 27-8；见图 27-33）。

（二）沟通区地下水位动态

中心城、浦东、奉贤第Ⅰ承压含水层与下伏第Ⅱ承压含水层沟通现象较为普遍，这些特定地质环境区，地下水位动态（变化趋势、幅度等）受下伏地下水采灌活动较为强烈的第Ⅱ承压含水层地下水位波动影响非常明显。

位于中心城含水层沟通区（浦 015-03F）及邻近地区（杨 005-01SG）典型监测井地下水位长期监测数据显示：第Ⅰ承压含水层沟通区及邻近地区水位与第Ⅱ承压含水层地下水位动态变化规律具有较好的一致性（图 27-34）。

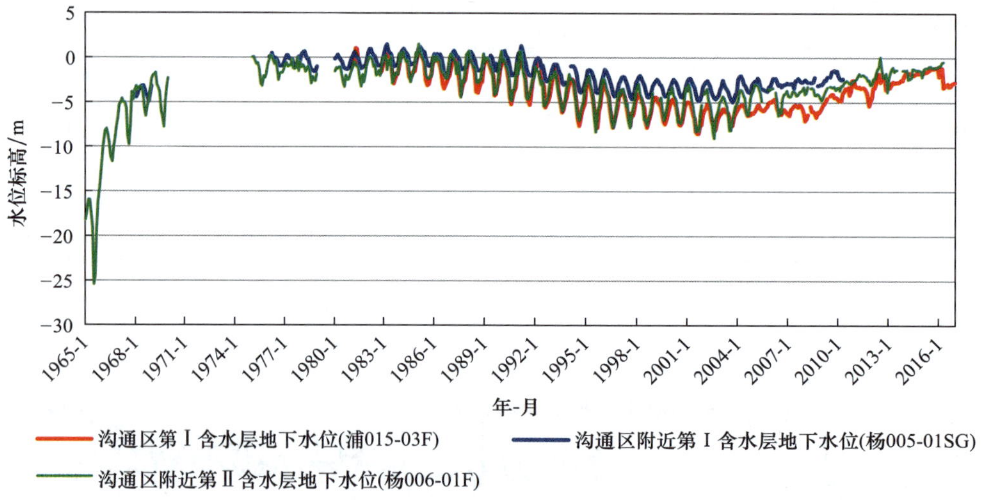

图 27-34　上海中心城第Ⅰ承压含水层沟通区及邻近地区地下水位动态变化曲线图

Fig. 27-34　Groundwater table dynamics of confined aquifer Ⅰ in and around the aquifer-connected regions in downtown Shanghai

20 世纪 60 年代末期（1968—1969 年），受第Ⅱ承压含水层地下水位下降影响，第Ⅰ承压含水层地下水位也处于历史上较低状态，沟通区邻近地区的水位也在-2~-6m 间波动；1970—1985 年地下水位基本稳定在 0~-2m；1986—2001 年地下水位处于持续小幅下降态势，沟通区及邻近地区再次呈现低水位分布状态，最低地下水位分别下降至-8m 左右及-5m 左右；2001 年至今呈现缓慢上升态势，至 2015 年地下水位上升至-2m 左右，与区域水位基本一致，沟通区低水位现象逐步消失（与第Ⅱ承压含水层水位分布基本一致）。

地下水位年内有一定的变化幅度，其变化幅度大小也与下伏第Ⅱ承压含水层地下水位变化幅度相对大小有密切关系。在第Ⅱ承压含水层地下水“三集中”采灌特征相对明显阶段（1980—2002 年），地下水位年度波动幅度可达 2~5m，“十一五”以来，当地下水“三集中”采灌格局逐渐转变为日常采灌格局的情况下，年内地下水位波动幅度明显减小，上下波动幅度与当时回灌强度相对大小有关，大多呈现 0.5m 左右的不规则波动特征。

局部地下水位受工程降水影响也较明显。如 2010 年以来，在区域地下水位总体呈现上升态势的过程中，中心城局部地区地下空间开发施工降水后地下水位短期剧降现象较为明显，如短期极端降至-5.3m（2011 年 11 月）和-3.3m（2016 年 3 月）（见图 27-34）。

四、第Ⅱ、第Ⅲ承压含水层

自然状态下的第Ⅱ、第Ⅲ承压含水层地下水位处于正标高状态。20 世纪初期以来，尤其是新中国成立后，受持续的地下水开采（回灌）等人为活动影响，区域上不同地区的地下水位发生了动态变化，并随着地下水开采（回灌）格局的变化而呈现出特定的变化特征。

（一）中心城地下水位动态

资料显示，自 1918 年开始，第Ⅱ、第Ⅲ承压含水层地下水率先在中心城被少量、间断性地开发利用，新中国成立后，尤其是 20 世纪 60 年代初期至今，表现为大量、持续性开采（回灌）格局。

几十年来，第Ⅱ、第Ⅲ承压含水层地下水位与地下水采灌格局调整等方面较为相似的背景下，地下水位呈现急剧下降至历史最低水位—快速上升—相对高水位期—相对低水位期—缓慢持续上升这一动态变化特征，但第Ⅲ承压含水层最低水位略低于第Ⅱ承压含水层（表 27-9；图 27-35，图 27-36）。

表 27-9 上海中心城第Ⅱ、第Ⅲ承压含水层地下水位变化特征

Table 27-9 Groundwater table variations of confined aquifers Ⅱ and Ⅲ in downtown Shanghai

含水层 \ 标高/m \ 阶段	迅速下降（1962—1965 年）	快速上升（1966—1970 年）	高水位状态（1971—1986 年）	小幅下降状态（1987—1996 年）	缓慢上升（1997 年—至今）
第Ⅱ承压含水层	-25.0~-34.0	-12.0~1.5	-6.5~3.0	-9.6~2.7	-8.5~-1.7
第Ⅲ承压含水层	-30.0~-37.0	-12.0~-0.6	-10.5~1.3	-12.0~1.7	-9.5~-1.7

注：-25.00~-34.00m，为 1962—1965 年水位标高波动区间。

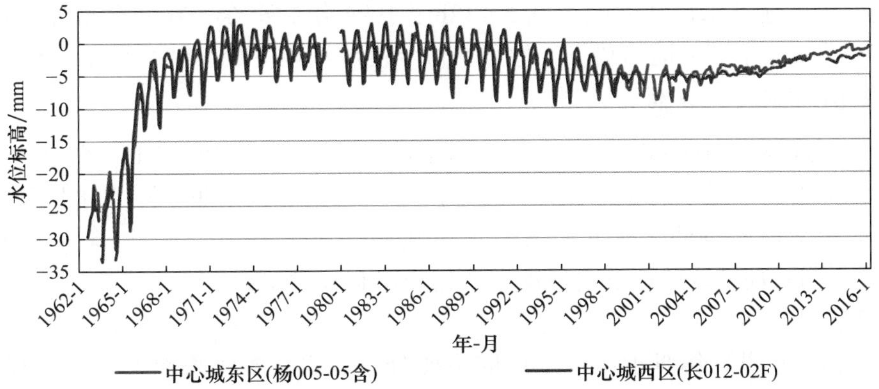

图 27-35 上海中心城第Ⅱ承压含水层地下水位动态变化图

Fig. 27-35 Groundwater table dynamics of confined aquifer Ⅱ in downtown Shanghai

历史最低期：20 世纪 60 年代中期之前，在地下水持续大量开采的情况下，至 1965 年地下水位急剧下降至历史最低水平，第Ⅱ、第Ⅲ承压含水层地下水位分别在-34m、-37m 左右。地下水位的急剧下降，导致市区地面沉降灾害明显，并伴随较多次生灾害。

快速上升期：1966—1970 年，为缓解地面沉降，采取了地下水大幅调减措施，同时起步实施地下水人工回灌措施，第Ⅱ、第Ⅲ承压含水层地下水位分别迅速回升至 1.5m、0m 左右。

相对高水位期：1971—1986 年，在地下水持续维持较低强度开采且采灌基本平衡的格局下，第Ⅱ、第Ⅲ承压含水层地下水位总体在-6.50~3.00m、-10.50~1.30m 间波动，地下水位处于相对高的稳定状态；

相对低水位期：1987—1996 年，在地下水开采量微量增大的过程中，地下水位再次出现持续小幅下降态势，第Ⅱ、第Ⅲ承压含水层地下水位进入相对低水位状态，最低水位分别下降至-9.60m、

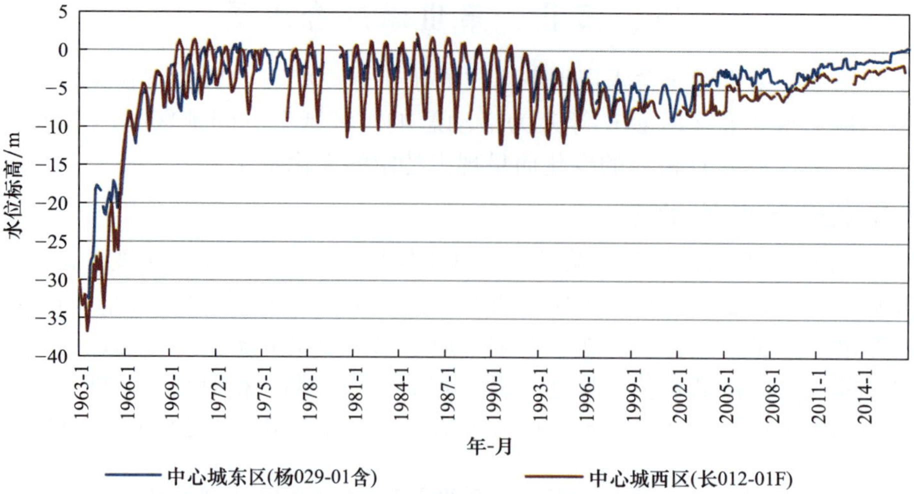

图 27-36　上海中心城第Ⅲ承压含水层地下水位动态变化图

Fig. 27-36　Groundwater table dynamics of confined aquifer Ⅲ in downtown Shanghai

-12.00m 左右。

缓慢持续上升期：1997—2016 年，第Ⅱ、第Ⅲ承压含水层开采量持续大幅调减，同时，第Ⅱ承压含水层地下水采灌比由 1∶1 逐渐上升至 1∶29，第Ⅲ承压含水层地下水采灌比由 1∶0.15 逐渐上升至 1∶18，这一采灌格局下，地下水位呈现缓慢、持续上升态势，目前水位均已上升至-1.70m 左右。这一阶段中心城地面沉降基本控制在阶段目标内。

年度变化幅度：1966—1996 年间地下水开采量大于回灌量阶段，受地下水人工回灌（夏灌冬用、冬灌夏用）影响，年度地下水位波动幅度达 5～12m；1996 年后至今，在回灌量逐渐增大至大于开采量，以及季节性回灌现象不明显的格局下，地下水位年度波动幅度较小，一般在 0.5m 左右。

（二）郊区地下水位动态

郊区大致划分为近郊区和远郊区。以中心城为中心，邻近中心城的宝山、嘉定、闵行、浦东为近郊区，其余崇明、金山、奉贤、松江、青浦为远郊区。郊区第Ⅱ、第Ⅲ承压含水层地下水位的动态变化，也主要受当地水文地质条件、地下水采灌及区域水位变化等因素影响。监测资料显示，远郊崇明（岛屿区）、金山与其他郊区的地下水位动态变化规律有较大差异。

1. 远郊岛屿区

远郊崇明岛屿区，第Ⅱ、第Ⅲ承压含水层因主要分布咸水，多年来地下水开发利用（人工回灌）程度低，其中第Ⅱ承压含水层 1963 年开始实施地下水开发利用至今，年开采量一般小于 $150\times10^4m^3/a$，1983 年开始实施地下水人工回灌，回灌量在 10×10^4～$280\times10^4m^3/a$，且 20 世纪 80 年代以来一直处于回灌量大于开采量格局，第Ⅲ承压含水层基本处于停采状态，为此，第Ⅱ、第Ⅲ承压含水层地下水位总体处于相对稳定的正标高状态，分别在 0.50～3.30m、0.50～2.85m 间波动，且年内波动幅度一般小于 0.50m（图 27-37）。

2. 远郊金山

优质地下水资源较为丰富，20 世纪 70 年代开始实施地下水开发利用，但开发利用程度较低，一般年开采量小于 $100\times10^4m^3/a$，极端最大年开采量在 $450\times10^4m^3/a$（第Ⅲ承压含水层），第Ⅱ、第Ⅲ承压含水层分别于 2010 年、2015 年先后停采。区域地下水位受区内外开采综合影响。监测资料显示，区域地下水位总体处于相对高的状态，总体在 1～-3m。

西南部枫泾，地下水位长期处于低水位分布状态，调查认为主要受区内外地下水开发利用综合影响，且邻省浙江地区地下水位下降（上升）影响更为明显。监测资料显示，20 世纪 70 年代末时，第Ⅱ、第Ⅲ承压含水层地下水位分别在-2.0m、-7.0m 左右，随后呈现持续下降态势，至 2010 年已下降

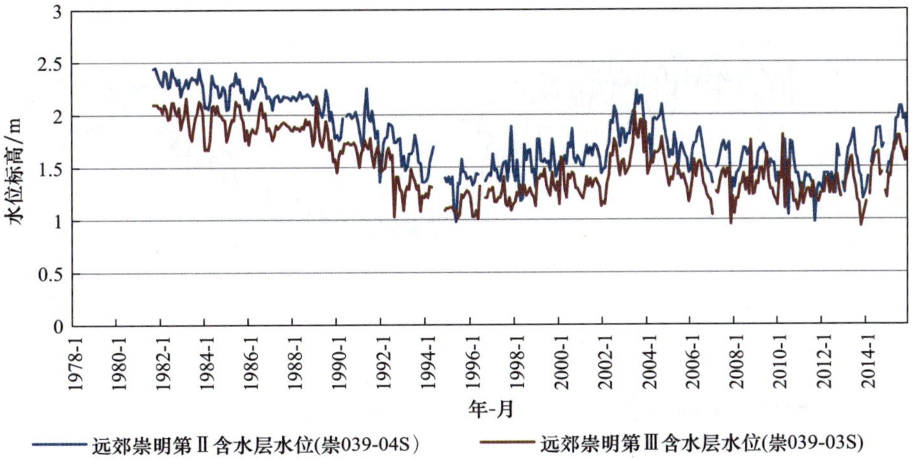

图 27-37 上海河口砂岛第Ⅱ、第Ⅲ承压含水层地下水位动态变化图

Fig. 27-37 Groundwater table dynamics of confined aquifers Ⅱ and Ⅲ in the estuarine sand islands of Shanghai

至-20m、-39m 左右；2011 年以来，在邻区浙江地下水大量调减、地下水位不断上升过程中，枫泾地下水位也呈现持续上升态势，2016 年均已上升至-11m 左右，并保持着进一步上升的态势（图 27-38）。目前枫泾第Ⅱ、第Ⅲ承压含水层地下水位仍是金山乃至全市的最低分布地区。

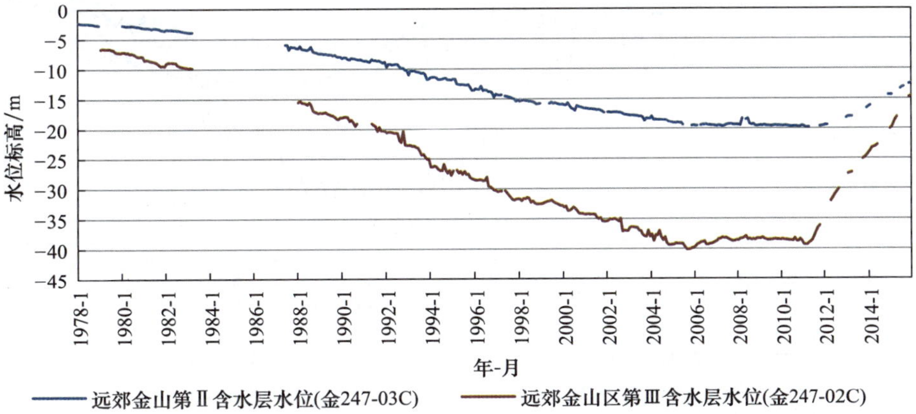

图 27-38 上海远郊金山第Ⅱ、第Ⅲ承压含水层地下水位动态变化图

Fig. 27-38 Groundwater table dynamics of confined aquifers Ⅱ and Ⅲ in the outer suburbs' Jinshan District of Shanghai

3. 其他郊区

地下水位同样受当地地下水开采（回灌）、区域同层次地下水位变化等综合影响。第Ⅱ、第Ⅲ承压含水层区域地下水位动态变化规律较相似，总体呈现相对稳定—持续下降—缓慢上升的动态变化特征。

第Ⅱ承压含水层不同地区同期最大下降幅度略有差异。20 世纪 70 年代末—80 年代中期，区域地下水位基本在 1～-3m，宝山较其他地区略低；1987—1997 年，地下水位缓慢下降至-5～-8m，宝山、嘉定相对低；1998 年以来，地下水位呈现相对稳定—缓慢持续上升态势，2015 年已上升至-2～0m，嘉定、浦东相对低（图 27-39）。

第Ⅲ承压含水层不同地区同期最大下降幅度也有差异。多年来，因嘉定地下水开发利用程度相对高，地下水位总体呈现相对低的分布格局。20 世纪 70 年代末—80 年代中期，区域地下水位基本在 1～-6m，嘉定较其他区低；1987—1997 年，地下水位缓慢下降至-4～-14m，嘉定相对低；1998 年以来，地下水位呈现相对稳定—缓慢持续上升态势，2015 年区域水位已上升至-4～0m（图 27-40）。

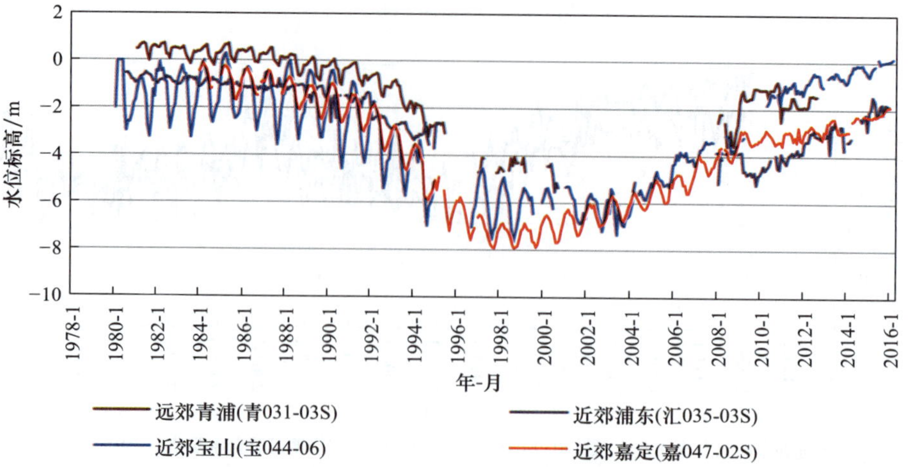

图 27-39　上海远郊青浦和近郊浦东、宝山、嘉定第Ⅱ承压含水层地下水位动态变化图

Fig. 27-39　Groundwater table dynamics of confined aquifer Ⅱ in the outer suburbs′ Qingpu District and the near suburbs′ Pudong, Baoshan and Jiading Districts of Shanghai

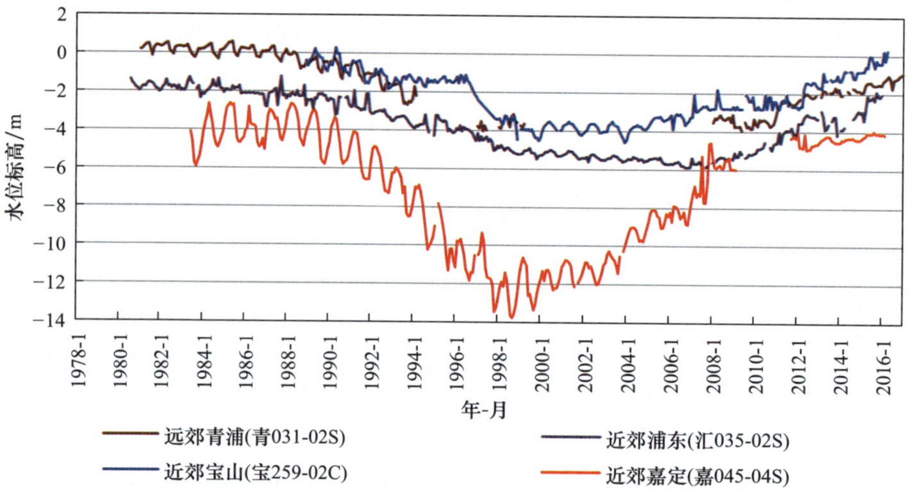

图 27-40　上海远郊青浦和近郊浦东、宝山、嘉定第Ⅲ承压含水层地下水位动态变化图

Fig. 27-40　Groundwater table dynamics of confined aquifer Ⅲ in the outer suburbs′ Qingpu District and the near suburbs′ Pudong, Baoshan and Jiading Districts of Shanghai

五、第Ⅳ、第Ⅴ承压含水层

自然状态下的第Ⅳ、第Ⅴ承压含水层地下水位处于正标高状态。尤其受 20 世纪 60 年代以来持续、大量的地下水开采（回灌）等人为活动影响，区域上不同地区的地下水位发生了变化，并随着地下水开采（回灌）格局的变化而呈现出特定的动态变化特征。

（一）中心城地下水位动态

资料显示，中心城 1918 年、1949 年先后开始对第Ⅳ、第Ⅴ承压含水层地下水进行间断、少量的开发利用，但 20 世纪 60 年代至今，表现为持续、大量的开采（回灌）状态。几十年来，第Ⅳ、第Ⅴ承压含水层地下水位在地下水采灌格局调整等方面较为相似的背景下，地下水位动态变化规律亦较为相似，均呈现急剧下降—快速上升—持续下降至历史最低水位—缓慢持续上升的动态变化特征（表 27-10；图 27-41，图 27-42）。

表 27-10 上海中心城第Ⅳ、第Ⅴ承压含水层地下水位变化特征

Table 27-10 Groundwater table variations of confined aquifers Ⅳ and Ⅴ in downtown Shanghai

阶段 / 标高/m / 含水层	迅速下降（1962—1965 年）	快速上升（1966—1970 年）	持续下降（1971—1998 年）	缓慢上升（1999 年—今）
第Ⅳ承压含水层	-32.0~-34.0	-15.0~-4.0	-4.0~-38.0	-34.0~-18.0
第Ⅴ承压含水层	-18.0~-20.0	-23.0~-4.0	-10.0~-57.0	-53.0~-13.0

注：-32.00~-34.00m，为 1962—1965 年水位标高波动区间。

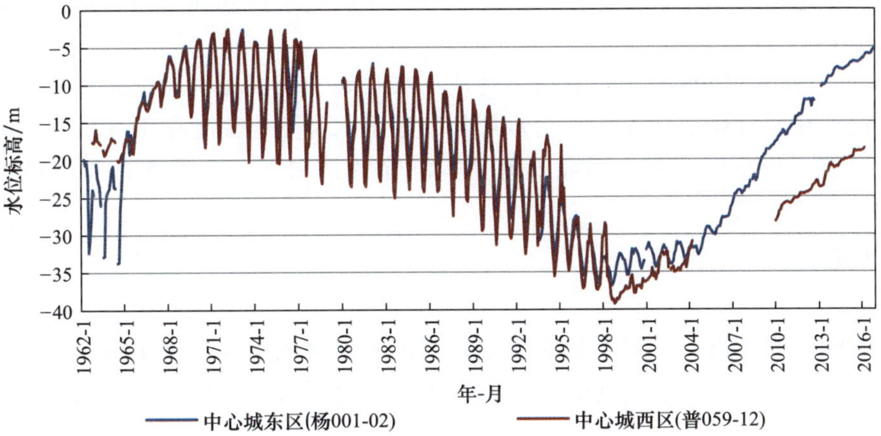

图 27-41 上海中心城第Ⅳ承压含水层地下水位动态变化图

Fig. 27-41 Groundwater table dynamics of confined aquifer Ⅳ in downtown Shanghai

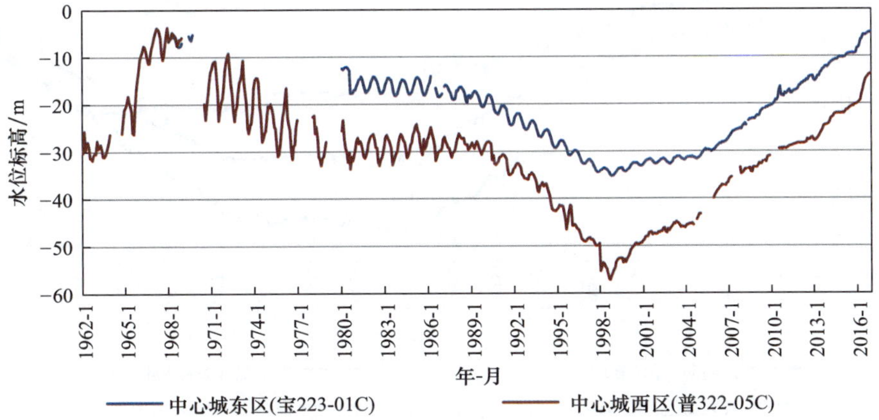

图 27-42 上海中心城第Ⅴ承压含水层地下水位动态变化图

Fig. 27-42 Groundwater table dynamics of confined aquifer Ⅴ in downtown Shanghai

急剧下降：新中国成立前夕—1965 年，在地下水持续大量开采的情况下，至 1965 年地下水位快速下降至历史较低水平，第Ⅳ、第Ⅴ承压含水层地下水位分别在-34m、-31m 左右。地下水位的迅速下降，导致市区地面沉降灾害明显，并伴随许多次生灾害。

快速上升：1966—1970 年，为缓解地面沉降，采取了地下水开采大幅调减以及人工回灌措施，第Ⅳ、第Ⅴ承压含水层地下水位迅速上升，最高水位均上升至-4m、-5m 左右。

历史最低水位：1971—1998 年，由于城市需水量的不断增长，在确保第Ⅱ、第Ⅲ承压含水层适量开采、回灌的同时，第Ⅳ、第Ⅴ承压含水层成为需水增量的主要层次，为此，自 20 世纪 70 年代开始至 90 年代中期，相对漫长的一个时期，地下水开采量基本逐年少量增加，为此，地下水位也缓慢持续下降，至 1998 年，地下水位呈现历史最低，第Ⅳ、第Ⅴ承压含水层地下水位分别降至在-37.00~-38.00m、-35.00~-57.00m，并在-28.00~-37.00m，-30.00~-49.00m 间波动。这一阶段区域地面沉降再次出现加剧现象。

缓慢持续上升：1999—2016 年，为控制微量增大的地面沉降速率，加大了地下水开采调减和地下水人工回灌力度，实施第Ⅳ、第Ⅴ承压含水层年度开采量持续大幅调减的同时，逐渐提高回灌强度，使得第Ⅳ承压含水层地下水采灌比由 1：0.03 逐渐上升至 1：8.50，第Ⅴ承压含水层地下水采灌比由 1：0.003 逐渐上升至 1：3.55，在这一采灌格局下，地下水位呈现缓慢、持续上升态势，2016 年第Ⅳ、第Ⅴ承压含水层地下水位已分别上升至-6.00~-18.00m、-5.00~-13.00m，并保持了进一步上升态势。这一阶段中心城地面沉降速率逐年有所降低，2005 年后基本控制在阶段目标（全市年度平均沉降速率小于 7mm/a）内。

年度变化幅度：1966—1998 年间，为季节性地下水开采、回灌相对明显阶段，第Ⅳ承压含水层年度地下水位变化幅度在 8~18m，第Ⅴ承压含水层年度地下水位变化幅度在 5~10m；1999 年后至今，在回灌量逐渐少量增加，季节性开采、回灌现象基本消除的格局下，地下水位年度波动幅度一般小于 1m（见图 27-41，图 27-42）。

（二）郊区地下水位动态

1. 岛屿区

历史上，岛屿区（崇明岛、长兴岛和横沙岛）第Ⅳ承压含水层地下水未曾开发利用，地下水位主要受区域地下水位波动影响。多年监测结果显示，岛屿区地下水位变化规律基本一致，仅水位标高差异较大（图 27-43）。

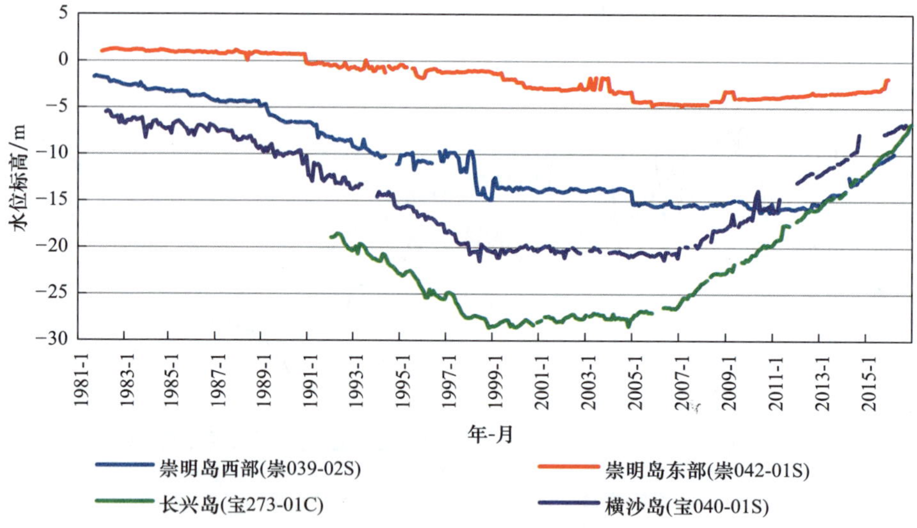

图 27-43　上海岛屿区第Ⅳ承压含水层地下水位动态变化图

Fig. 27-43　Groundwater table dynamics of confined aquifer Ⅳ in the island areas of Shanghai

崇明岛东西部地区地下水位差异较大，西部地区从 1981 年的-2m 左右，逐渐下降至 1999 年的-15m 左右，至 2012 年基本稳定在-14~-16m，2013 年后呈现上升态势，2016 年在-10m 左右；崇明岛东部地区地下水位比较高，变化比较平缓，1981 年在 1m 左右，至 2016 年在-3m 左右。

长兴岛地下水位从 1992 年的-19m 左右，逐渐下降至 1999 年的-28m 左右，至 2007 年基本稳定在-26~-28m，2008 年后呈现上升态势，2016 年上升至-7m 左右。

横沙岛地下水位从 1982 年的-5m 左右，逐渐下降至 1999 年的-21m 左右，至 2007 年基本稳定在-20~-21m，2008 年后呈现上升态势，2016 年上升至-7m 左右。

历史上，岛屿区第Ⅴ承压含水层是地下水主采层，地下水资源长期作为当地生活用水补充水源，地下水位动态变化主要受开采（回灌）影响。

岛屿区第Ⅴ承压含水层地下水位系统性监测时间起步较早，先后从20世纪70年代末和80年代初开始保持至今。多年监测结果显示，岛屿区地下水位变化规律基本一致，仅局部地区水位标高有一定差异（图27-44）。

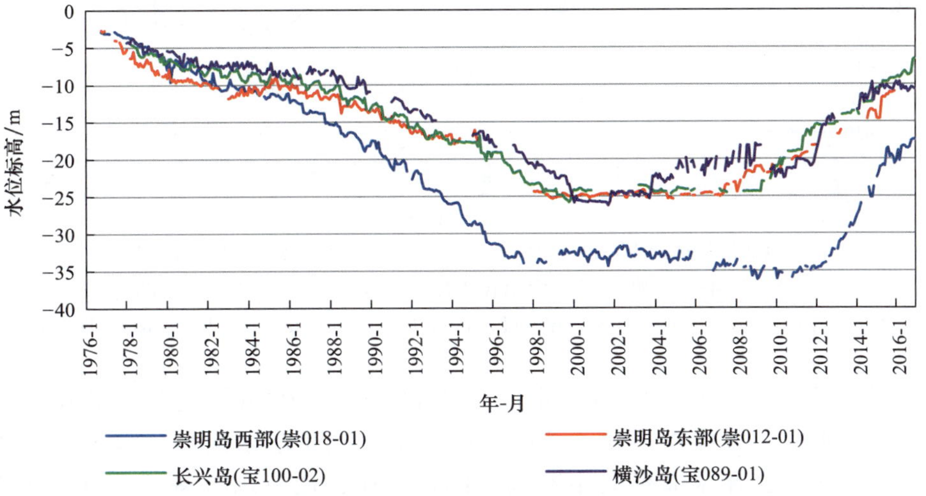

图27-44　上海岛屿区第Ⅴ承压含水层地下水位动态变化

Fig. 27-44　Groundwater table dynamics of confined aquifer Ⅴ in the island areas of Shanghai

崇明岛东、西部地区水位差异较大，西部地区从1976年的-3m左右，逐渐下降至1997年的-34m左右，至2012年基本稳定在-33～-36m，2013年后呈现上升态势，2016年在-17m左右。

崇明岛东部地区、长兴岛、横沙岛，地下水位变化特征比较一致，从1976年的-3m左右，逐渐下降至1999年的-25m左右，至2010年基本稳定在-20～-25m，2011年后呈现上升态势，2016年上升至-7～-10m。

2. 远郊金山

远郊金山第Ⅴ承压含水层基本缺失。

第Ⅳ承压含水层区域地下水位动态变化较为复杂，东部地区和西部地区地下水位标高差异大。

第Ⅳ承压含水层地下水开采井主要集中在金山的中部、东部地区。自1974年开始至今有少量的开发利用，开发利用程度相对较低，除2002～2006年年开采量在150×10^4～$400\times10^4m^3$外，其他年开采量基本小于$50\times10^4m^3$；同时阶段性实施了地下水人工回灌措施，1983～1990年年回灌量小于$10\times10^4m^3$，2007年以来，年回灌量在100×10^4～$150\times10^4m^3$。东部地区监测资料反映，20世纪80年代地下水位在-4.50～-6.00m，90年代至今总体在-5.00～-6.50m间波动。西部地区地下水位埋藏较深，据调查，除受区内开采影响外，主要与邻近的浙江大量开发利用、地下水位波动影响密切相关。金山枫泾、浙江嘉善监测资料显示，区域动态变化规律基本一致，嘉善地下水位长期处于相对低的状态，影响范围波及相邻的金山枫泾。1980年，金山枫泾、浙江嘉善地下水位分别在-7m、-15m左右，2006年、2003年分别下降至-37m、-52m最低分布状态；之后随着区内外地下水开采格局的不断调整，地下水位持续上升，2015年两处地下水位基本在-15m左右波动（图27-45）。

3. 其他地区

宝山、嘉定、闵行、青浦、奉贤、浦东是上海第Ⅳ、第Ⅴ承压含水层地下水开发利用程度较高的地区。这些地区第Ⅳ承压含水层地下水位变化规律大致一致，均呈现相对稳定—持续下降—持续上升的变化规律。但受各地地下水开采强度不同影响，水位下降幅度有较大差异。20世纪60年代中期，地下水位普遍在-2～-15m，近郊的宝山略低；至90年代末，区域地下水位持续下降，近郊的宝山、

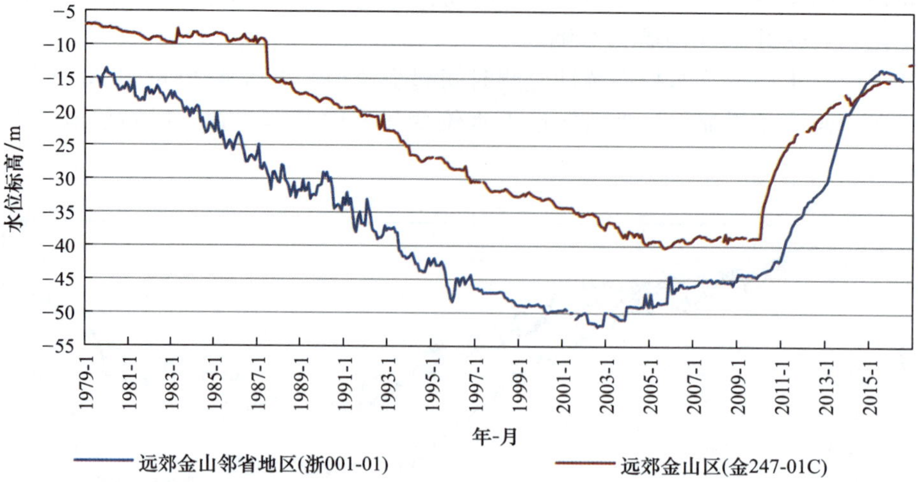

图 27-45　上海远郊金山第Ⅳ承压含水层地下水位动态变化图

Fig. 27-45　Groundwater table dynamics of confined aquifer Ⅳ in the outer suburbs' Jinshan District of Shanghai

嘉定、闵行，地下水开采强度相对大，地下水位下降幅度大，1998 年闵行华漕最低水位降至-68m 左右，宝山、嘉定、青浦最低水位在-40～-50m；远郊奉贤、浦东东南部（原南汇）最低水位降至-22～-12m。1999 年至今，在采灌格局不断调整的过程中，地下水位持续上升，2015 年已上升至-20～-4m，西部青浦相对低，东部浦东东南部（原南汇）、南部奉贤相对高（图 27-46）。

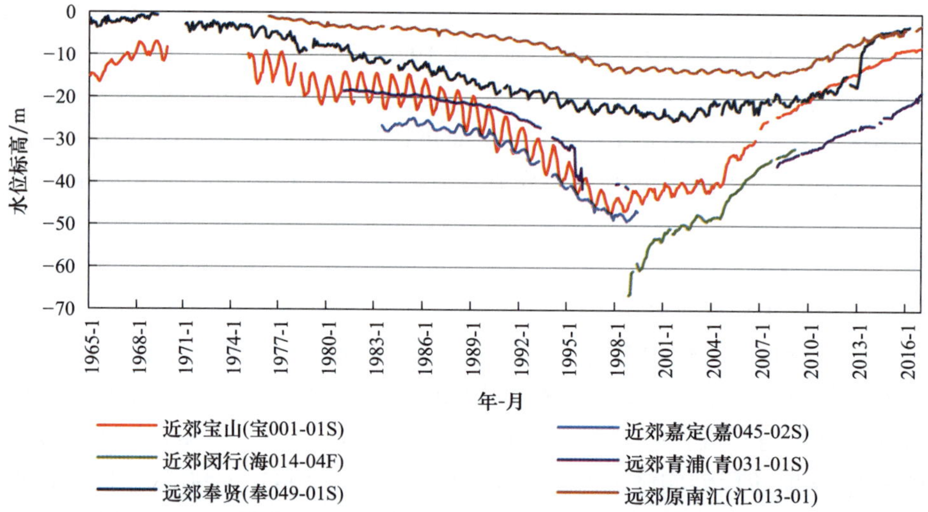

图 27-46　上海宝山、嘉定、闵行、青浦、奉贤、浦东（原南汇）第Ⅳ承压含水层地下水位动态变化图

Fig. 27-46　Groundwater table dynamics of confined aquifer Ⅳ in Baoshan, Jiading, Minhang, Qingpu, Fengxian and Pudong (formerly Nanhui) Districts of Shanghai

调查显示，这些地区第Ⅴ承压含水层地下水位变化规律大致一致，均呈现相对稳定—持续下降—持续上升特征。但受各地地下水开采强度不同影响，水位下降幅度有较大差异。地下水开发利用强度低的浦东东南部（原南汇），地下水位分布比较高，几十年来基本在 0～-12m 间波动，20 世纪 90 年代末为历史最低期。宝山、嘉定、青浦、闵行地下水开发利用强度相对大，70 年代中期，地下水位普遍在-25～-40m，闵行略低；至 90 年代末，区域地下水位持续下降，1998 年最低水位普遍降至-58～-62m。1999 年至今，在采灌格局不断调整的过程中，地下水位持续上升，2015 年已上升至-20m 左右，西部青浦略低（图 27-47）。

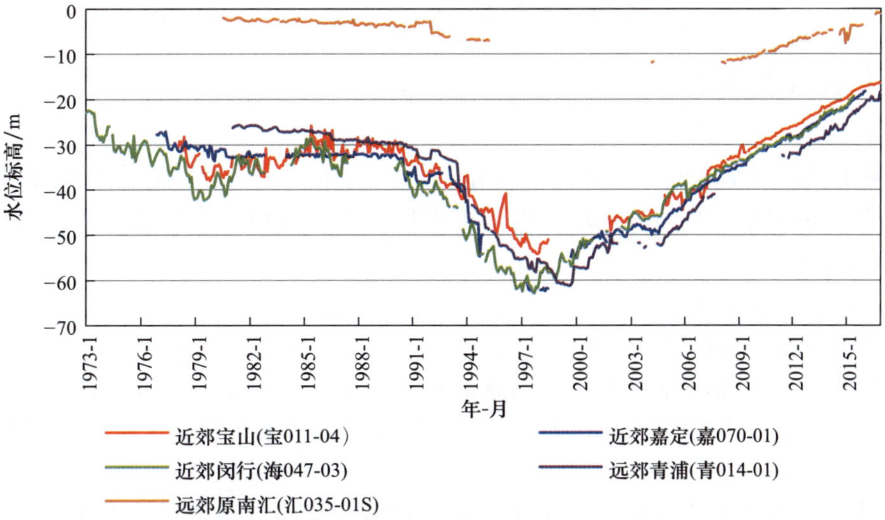

图 27-47　上海宝山、嘉定、闵行、青浦、浦东（原南汇）第Ⅴ承压含水层地下水位动态变化图

Fig. 27-47　Groundwater table dynamics of confined aquifer Ⅴ in Baoshan, Jiading, Minhang, Qingpu and Pudong (formerly Nanhui) Districts of Shanghai

第五节　地下水补径排条件

一、潜水含水层地下水补径排条件

（一）补给条件

潜水含水层岩性主要由黏性土和砂性土构成，土层具有一定的渗透能力，具备了良好的储水空间。调查显示，大气降水、地表水、农田灌溉水等是潜水含水层的主要补给水源。

上海气候湿润，雨量丰富，多年年降水量常年保持在1100～1200mm/a。充沛的降水量形成了丰富的地表径流，从而为潜水含水层提供了大量的入渗水源。

大气降水为潜水的主要补给水源，体现在潜水中氢同位素、水位动态变化趋势与大气降水变化规律具较好的相似性。潜水中同位素^{3}H含量与大气降水^{3}H含量接近。一般枯水期水位埋深相对深，丰水期水位埋深相对浅，水位变化幅度大小与降水强度、持续时间长短基本呈正相关关系。

潜水含水层与地表水体直接接触地带，当地表水体水位高于邻近潜水水位时，含水层将接受地表水源的补给。

农田灌溉区，大量的灌溉水透过上覆黏性土层、粉性土层下渗至潜水含水层中，成为潜水的补给水源之一。

（二）径流条件

上海除局部地区分布基岩残丘外，地势较为平坦，东部滨海平原、河口砂岛区地势略高于西部湖沼平原区。全市河网纵横交叉，分布普遍。含水层存在一元结构区和二元结构区并存的特征。潜水含水层受其岩性影响，渗透能力总体差。受地势、纵横密布的地表水体切割影响，潜水含水层形成局部径流系统，径流途径较短及水力坡度小。径流方向受所在地表水体控制，当潜水水位高于地表水体时，潜水向地表水体方向径流。因潜水以蒸发、蒸腾作用为主要排泄方式，故潜水以垂向径流为主。

（三）排泄条件

潜水排泄方式多样。以天然条件下的蒸发、蒸腾、同地表水体侧向排泄为主要排泄形式。农村地区的家庭洗涤用水、基坑施工过程中的潜水局部疏干，也是潜水的人为排泄形式。

二、承压含水层地下水补径排条件

（一）第Ⅰ承压含水层

1. 补给条件

1）与现代降水及地表水之间不存在补给关系。第Ⅰ承压含水层总体埋藏在地表30m以下。地下水年龄大于1万年，说明承压水为封存水，尚未得到现代降水或地表水等的补给，与现代降水及地表水之间不存在补给关系。

2）无地下水人工补给来源。

3）工程降水引发相邻含水层地下水的越流补给。深大基坑实施第Ⅰ承压含水层地下水降水过程中，有产生相邻黏性土层中地下水极为缓慢的越流补给现象；在第Ⅰ、第Ⅱ承压含水层沟通地区，当基坑工程实施第Ⅰ承压含水层地下水降水时，将直接接受与之接触的第Ⅱ承压含水层地下水的补给。

2. 径流条件

第Ⅰ承压含水层岩性主要为黄色、灰黄色、灰色粉砂、细砂。渗透系数大多小于10m/d，渗透性能总体较差，地下水径流条件总体较差。

地下水径流方向受区域水位分布控制。依据2015年度水位分布现状，总体北部高南部低，由北向南径流；西南部金山枫泾、东部周浦—惠南为地下水位低分布区，其附近表现为向低水位区径流的特征。

3. 排泄条件

工程降水为第Ⅰ承压含水层地下水的主要排泄方式。

（二）第Ⅱ承压含水层

1. 补给条件

1）与现代降水及地表水之间不存在补给关系。第Ⅱ承压含水层总体埋藏在地表60m以下。地下水年龄大于2万年，说明承压水为封存水，未得到现代降水或地表水等的补给，与现代降水之间及地表水之间不存在补给关系。

2）地下水开采引发相邻含水层地下水的越流补给。第Ⅱ承压含水层在地下水开采过程中，有产生相邻黏性土层中地下水极为缓慢的越流补给现象；在第Ⅱ、第Ⅲ承压含水层沟通地区，当实施第Ⅱ承压含水层地下水开采时，将直接接受第Ⅲ承压含水层地下水的补给。

3）地下水人工补给。20世纪60年代初至今，上海持续实施地下水回灌。为此，地下水人工回灌为含水层人工补给来源。

2. 径流条件

第Ⅱ承压含水层岩性主要为灰色细砂、中砂、粗砂，局部含细砾石，含水层总体渗透性能较好，径流条件较好。

地下水径流方向受区内外地下水开采格局影响。2015年度地下水位显示，河口砂岛区地下水总体由西北向东南方向径流。陆域地区，总体表现为中心城以北地区、东南部临港新城相对高，地下水总体由北向南径流。但新场—惠南、枫泾地区局部为低水位分布区，地下水由其附近分别向低水位区径流。

3. 排泄条件

抽汲地下水为第Ⅱ承压含水层地下水的主要排泄方式。

（三）第Ⅲ承压含水层

1. 补给条件

1）与现代降水及地表水之间不存在补给关系。第Ⅲ承压含水层埋藏在地表100m以下。第Ⅲ承压含水层地下水年龄大于2万年，说明承压水为封存水，未得到现代降水或地表水源等的补给，与现代降水之间及地表水之间不存在补给关系。

2）地下水人工补给。20 世纪 60 年代初至今，上海持续实施地下水回灌。为此，地下水人工回灌为含水层人工补给来源。

2. 径流条件

第Ⅲ承压含水层岩性主要为粉砂、细砂、中砂，局部含细砾石，渗透性能较好，径流条件较好。

地下水径流方向受区内外地下水开采格局影响。2015 年度地下水位显示，岛屿区地下水总体由西北向东南方向径流。陆域地区总体表现为东部较西部高，地下水总体由东向西径流。但白鹤、枫泾地区局部为低水位分布区，地下水由其附近分别向低水位区径流。

3. 排泄条件

抽汲地下水为第Ⅲ承压含水层地下水的主要排泄方式。

（四）第Ⅳ承压含水层

1. 补给条件

1）与现代降水及地表水之间不存在补给关系。第Ⅳ承压含水层埋藏在地表 150m 以下。第Ⅳ承压含水层地下水年龄大于 3 万年，说明承压水为封存水，未得到现代降水或地表水源等的补给，与现代降水之间及地表水之间不存在补给关系。

2）地下水开采引发相邻含水层地下水的越流补给。上、下相邻承压含水层在地下水开采过程中，当产生一定的水头差的情况下，下部含水层地下水通过弱透水层而产生越流补给现象；在第Ⅳ、第Ⅴ承压含水层沟通地区，当开采第Ⅳ承压含水层地下水时，将直接接受第Ⅴ承压含水层地下水的越流补给。承压含水层间越流补给是承压水开采期间获得补给的方式之一。

3）地下水人工补给。20 世纪 60 年代初至今，上海持续实施地下水回灌。为此，地下水人工回灌为含水层人工补给来源。

2. 径流条件

第Ⅳ承压含水层岩性主要为细砂、中砂、粗砂，局部含细砾石，渗透性能总体较好，径流条件较好。

地下水径流方向受区内外地下水开采格局影响。2015 年度地下水位显示，岛屿区地下水总体由东南向西北方向径流。陆域地区总体表现为东部较西部高，地下水总体由东向西径流。但白鹤、枫泾、华亭地区局部为低水位分布区，地下水由其附近分别向低水位区径流。

3. 排泄条件

抽汲地下水为第Ⅳ承压含水层地下水的主要排泄方式。

（五）第Ⅴ承压含水层

1. 补给条件

1）与现代降水之间及地表水之间不存在补给关系。第Ⅴ承压含水层埋藏在地表 250m 以下。第Ⅴ承压含水层地下水年龄大于 4 万年，说明承压水为封存水，尚未得到现代降水或地表水源等的补给，与现代降水之间及地表水之间不存在补给关系。

2）地下水人工补给。20 世纪 70 年代初至今，上海持续实施地下水回灌。为此，地下水人工回灌为含水层人工补给来源。

2. 径流条件

第Ⅴ承压含水层岩性主要为粉砂、细砂，局部含细砾石、胶结物。渗透性能相对差，径流条件较差。

地下水径流方向受区内外地下水开采格局影响。2015 年度地下水位显示，岛屿区地下水总体由东南向西北方向径流。陆域地区总体表现为东部较西部高，地下水总体由东向西径流。

3. 排泄条件

抽汲地下水为第Ⅴ承压含水层地下水主要排泄方式。

第二十八章　工程地质条件

上海工程地质条件总体变化不大，工程建设影响范围内（100m以浅）主要为晚更新世以来海陆交替相、以海相为主的松散堆积物（严学新，2009），岩性主要以黏性土和砂、粉性土为主，且软黏性土分布较为广泛。

目前，根据沉积时代和土层物理力学指标特征共划分出11个工程地质层，即第①层填土、第$②_1$层褐黄—灰黄色粉质黏土，第$②_3$层粉性土层（浅部砂层），第③、④层灰色淤泥质黏性土层（软土层），第⑤层灰色黏性土层（局部夹有第$⑤_2$层砂、粉性土层），第⑥层暗绿—草黄色黏性土层（第一硬土层），第⑦层草黄—灰色砂、粉性土层，第⑧层黏性土夹砂层（局部夹第$⑧_{2-1}$层暗绿—草黄色黏性土硬土层），第⑨层灰色砂性土、第⑩层褐灰—蓝灰色黏性土层和第⑪层灰色粉细砂层。

不同工程地质层具有不同的工程地质特性，易产生不同的工程地质问题。其中，第$②_1$层为上海俗称的"硬壳层"，可作为轻型建（构）筑物的天然地基持力层；第$②_3$层为浅部砂层，埋藏浅，易产生震动液化和渗流液化现象；第③、④层为典型的软土层，是天然地基变形主要层次，也为上海的第一压缩层；第⑤层为一般黏性土层，亦为第二压缩层，其中第$⑤_2$层为微承压含水层，基坑开挖时应注意由其引发的流砂、管涌等问题；第⑥层为第一硬土层，埋深从西部向东部逐渐加深，该层为更新统和全新统分界的标志层，在上海广受关注，其土性较好，对控制地基变形较为有利，可作为一般建（构）筑物的桩基持力层；第⑦层为下部砂、粉性土层，土性好，为中高层建筑和市政桥梁工程的良好桩基持力层，该层亦为第Ⅰ承压含水层，地下工程施工时易产生流砂和基坑突涌问题。

第一节　松散层工程地质层划分与依据

在20世纪80年代以前，上海主要依据市区地面沉降研究需要划分工程地质层，主要按照地层的成因类型、地质年代、埋藏深度、土的物理学性质、土层变形和强度特征等因素综合考虑。之后，又利用气候地层学和生物地层学方法对上述土层进行验证，实际上把相同地层年代、成因类型、岩性岩相、地质结构和相近物理力学性质的土层划为同一工程地质层；同一工程地质层又按岩性、结构、构造、矿物成分、粒度成分、埋藏条件及天然状态等划分若干亚层。根据上述划分原则，将上海市区基岩面以上土层划分为表土层、第一压缩层、第二压缩层、暗绿色黏土层、第Ⅰ承压含水层、第三压缩层、第Ⅱ承压含水层、第Ⅲ承压含水层、杂色黏土、第Ⅳ承压含水层、第Ⅴ承压含水层共11个大层。20世纪80年代后特别是90年代中后期，上海工程地质层划分主要依照岩土工程勘察规范，并将上海划分为4个地貌类型区，即湖沼平原区、滨海平原区、河口砂岛区和潮坪区。每个地貌类型区其地层层序表不一致，未进行工程地质层统一，无法进行工程地质层对比。

2008年结束的三维地质调查项目，对部分土层的形成时代及成因环境得出了新的成果，并对上海工程地质层的划分有了更进一步的认识。依据第四纪沉积时代及成因、岩性，结合上海的特点及工程地质研究需要，提出了整个上海统一的工程地质层划分方案（表28-1），即100m以浅共划分出11个工程地质层和40个亚层及次亚层。同时，总结了不同工程地质层物理力学指标特征（表28-2）。

表 28-1　上海工程地质层层序及特征

Table 28-1　Stratigraphic sequence and characteristics of engineering geological strata in Shanghai

地质时代		土层层号	土层名称	层厚/m	层面标高/m	颜色	湿度	状态	密实度	压缩性	土层描述
		①$_1$	填土	$\frac{0.1\sim3.2}{1.1}$							人工获得产物，农田地区为素填土，城市化地区为杂填土，以黏性土为主，含植物根茎及碎石砖块等
		①$_2$	浜填土								主要为暗浜土
		①$_3$	冲填土	$\frac{0.4\sim6.0}{3.4}$	$\frac{5.46\sim0.81}{3.67}$	灰	饱和		松散—稍密	中偏高	以粉性土为主，含云母及贝壳碎片，部分地区以淤泥质黏性土为主，含有机质染斑
全新统	晚期 Qh_3	②$_1$	黏土、粉质黏土	$\frac{0.4\sim4.7}{2.0}$	$\frac{5.56\sim0.27}{2.92}$	褐黄—灰黄	湿	可塑—软塑		中等	见有植物根茎及铁锰质、浸染斑点或小结核
		②$_2$	泥炭质土	$\frac{0.2\sim0.9}{0.4}$	$\frac{4.04\sim1.06}{2.40}$	灰黑	饱和	流塑		高	含黑色有机质，夹棕红色半腐殖物根茎及碎屑，有臭味
		②$_3$	粉土、粉砂	$\frac{0.6\sim22.6}{6.8}$	$\frac{4.30\sim-1.70}{0.87}$	灰	饱和		松散—稍密	中等	含云母及染斑，夹薄层粉质黏土，偶见贝壳碎片。河口三岛区垂向岩性有所差异，上部一般为砂质粉土，中部为黏质粉土，下部为粉砂
	中期 Qh_2	③	淤泥质粉质黏土	$\frac{0.5\sim10.9}{4.6}$	$\frac{3.38\sim-7.31}{-0.29}$	灰	饱和	流塑		高	分布于滨海平原区，层内夹薄层粉砂，粉砂夹层厚为 0.5～1.0mm，具有波状水平层理，局部见有交错层理，层内气孔发育，多呈圆形，孔径 0.5～1.0mm，孔内无充填物，含贝壳碎屑
		③	灰色黏土、粉质黏土	$\frac{0.5\sim12.7}{5.4}$	$\frac{3.54\sim1.83}{0.79}$	灰	很湿	软塑		高	分布于湖沼平原区，含灰白色团块及有机质染斑，夹蓝灰色条纹，偶夹 0.3cm 左右的腐殖物根茎，土质较均匀。部分地区颜色呈蓝灰色，含灰白色泥质条纹及少量腐殖物
		③$_a$	砂质粉土	$\frac{0.5\sim3.7}{2.5}$	$\frac{1.56\sim-7.77}{-1.31}$	灰	饱和		稍密	中等	含云母，夹黏性土薄层，土质不均
		④	灰色淤泥质黏土	$\frac{0.9\sim17.7}{7.2}$	$\frac{0.40\sim-22.42}{-6.87}$	灰	饱和	流塑		高	含云母、有机质，横断面见有鱼鳞状构造，鳞片排列紧密，直径大小以 1～2mm 多见，局布夹贝壳碎片
		④$_a$	砂质粉土	$\frac{1.5\sim14.0}{5.8}$	$\frac{-2.30\sim-18.38}{-8.31}$	灰	饱和		稍密	中等	含云母，夹黏性土薄层

续表

地质时代		土层层号	土层名称	层厚/m	层面标高/m	颜色	湿度	状态	密实度	压缩性	土层描述
全新统	早期 Qh_1	⑤$_{1-1}$	黏土	0.70~19.9 6.4	-0.30~-27.00 -14.02	灰	很湿	软塑—流塑		高	含云母、有机质，见灰白色泥钙质结核及半腐芦苇根茎，夹薄层黏土，土质较均匀
		⑤$_{1-2}$	灰色粉质黏土	0.5~28.1 7.6	-2.20~-37.39 -17.97	灰	很湿	可塑—软塑		中偏高	夹较多砂质粉土或黏质粉土，在金山局部地区岩性为黏质粉土夹粉砂，土质不均
		⑤$_2$	砂质粉土、粉砂	1.8~24.3 9.0	-4.00~-48.47 -26.07	灰	饱和		中密	中等	岩性以灰色砂质粉土夹粉砂为主，局部地区为粉砂，含有机质、钙质结核、腐殖物等，具交错层理，层内气孔发育
		⑤$_3$	粉质黏土	1.0~33.5 12.4	-7.99~-51.17 -28.50	灰	湿	可塑—软塑		中偏高	层内富含有机质，局部地段夹粉砂层，甚至为互层，夹层处见有斜层理或波状水平层理，该层由于在封闭环境中沉积，故一般含天然气
上更新统	上段 Qp_3^2	⑤$_4$	粉质黏土	0.8~9.6 3.4	-20.20~-55.46 -34.42	灰绿	稍湿	可塑		中等	部分地区为黏土，颜色有暗绿、灰绿、草黄色等，含氧化铁斑点、有机质，见铁锰质结核，偶夹粉砂薄层
		⑥	黏土、粉质黏土	0.8~12.5 4.1	1.76~-28.70 -16.49	暗绿—褐黄	稍湿	硬塑—可塑		中等	含铁锰质侵染斑点、结核及少量植物残体，颜色由上至下由暗绿色逐渐过渡为褐黄色、草黄色
		⑦$_1$	黏质粉土、砂质粉土	1.0~17.0 5.0	-3.21~-31.13 -18.57	草黄、灰	饱和		中密	中等	西部湖沼平原区颜色多为灰色，而“冈身带”及滨海平原区颜色多为草黄色，略具水平层理，夹较多黏性土薄层，含少量贝壳碎片
		⑦$_2$	粉砂	1.3~38.0 19.2	-13.1~-56.42 -34.07	灰黄—灰	饱和		密实	中偏低	具水平层理，矿物成分以石英为主，暗色矿物少量，局部云母片富集，土质较均匀
		⑧$_{1-1}$	黏土	1.4~23.8 9.6	-3.93~-53.76 -25.71	灰	很湿	软塑		高	含云母、腐殖物，土质较均匀
		⑧$_{1-2}$	粉质黏土	1.0~35.5 9.8	-4.80~-54.74 -25.23	灰	很湿	可塑—软塑		中偏高	含云母，夹粉砂薄层，夹层单层厚一般为0.1~0.5cm，粒度成分以粉砂粒为主，层内富含动物化石
		⑧$_{2-1}$	粉质黏土	0.9~21.5 6.9	-12.53~-61.50 -25.37	暗绿—褐黄	稍湿	硬塑—可塑		中等	含铁锰氧化物条纹、腐殖物，夹薄层粉土、粉砂，偶见鱼骨碎片

地质时代		土层层号	土层名称	层厚/m	层面标高/m	颜色	湿度	状态	密实度	压缩性	土层描述
上更新统	上段 Qp_3^2	⑧$_{2-2}$	砂质粉土	$\frac{4.0\sim25.0}{13.2}$	$\frac{-17.82\sim-61.19}{-37.72}$	草黄—灰	饱和		中密—密实	中等	含云母、腐殖物，夹薄层黏性土。在河口砂岛区，该层基本为粉砂夹砂质粉土层和粉砂夹粉质黏土层
		⑧$_{2-3}$	粉质黏土夹砂	$\frac{2.0\sim40.0}{13.6}$	$\frac{-30.63\sim-69.22}{-48.24}$	灰	湿	可塑		中等	含云母、腐殖物，层内见有螺贝壳，在滨海平原区该层主要为粉质黏土夹砂和粉质黏土与粉砂互层，层间有间隙，层面间有鱼鳞状构造，具明显的水平或单斜层理，呈“千层饼”状，局部夹少量钙质结核及半腐烂植物，土质不均，湖沼平原区基本为粉质黏土夹粉砂
		⑧$_{3-1}$	粉质黏土	$\frac{2.0\sim10.0}{5.1}$	$\frac{-33.46\sim-66.60}{-46.54}$	暗绿—蓝灰	稍湿	硬塑—可塑		中等	含氧化物条纹，夹薄层粉砂
		⑧$_{3-2}$	砂质粉土	$\frac{1.9\sim12.5}{5.4}$	$\frac{-35.00\sim-79.78}{-53.09}$	褐黄	饱和		中密	中等	含云母、腐殖物，夹薄层黏性土
		⑧$_{3-3}$	粉质黏土	$\frac{1.1\sim13.1}{8.7}$	$\frac{-48.76\sim-72.50}{-57.08}$	蓝灰—灰	稍湿	可塑		中等	含云母、腐殖物，夹薄层粉砂
		⑧$_4$	粉质黏土	$\frac{8.6\sim13.2}{11.6}$	$\frac{-60.00\sim-74.68}{-65.04}$	灰	稍湿	可塑		中等	含云母，夹薄层粉砂，土质不均
	Qp_3^1	⑨$_1$	粉砂	$\frac{1.5\sim22.8}{9.0}$	$\frac{-48.63\sim-71.24}{-63.60}$	青灰—灰	饱和		密实	中偏低	矿物成分以长石、石英、云母为主，次为暗色矿物，分选性较好，部分地区顶部夹黏性土薄层，见有水平层理，夹少量贝壳碎片
		⑨$_2$	含砾中砂	$\frac{11.3\sim26.0}{12.9}$	$\frac{-63.76\sim-78.64}{-73.63}$	青灰	饱和		密实	低	上部一般为细砂，至下逐渐过渡为中粗砂、砾石层，矿物成分以长石、石英、云母为主，次为暗色矿物。局部夹有薄层黏性土透镜体
中更新统	Qp_2	⑩	粉质黏土	$\frac{0.6\sim7.2}{4.0}$	$\frac{-75.89\sim-94.94}{-87.36}$	蓝灰	稍湿	硬塑—可塑		中偏低	含铁、锰质结核及黑色有机质
		⑪	粉细砂	未钻穿	$\frac{-90.39\sim-96.30}{-94.32}$	灰黄—灰	饱和		密实	低	矿物成分以长石、石英、云母为主

注：层厚数值为$\frac{最小值\sim最大值}{平均值}$；层面标高数值为$\frac{最大值\sim最小值}{平均值}$。

表 28-2　上海各工程地质层物理力学指标

Table 28-2　Physical-mechanical parameters of the engineering geological strata in Shanghai

土层层号	土层名称	层厚 m	层面标高 m	含水量 W %	重度 γo KN/cm^3	孔隙比 e	塑性指数 I_P	液性指数 I_L	直剪固快（峰值） 黏聚力 c KPa	直剪固快（峰值） 内摩擦角 φ (°)	压缩系数 $\alpha_{0.1-0.2}$ MPa^{-1}	压缩模量 $Es_{0.1-0.2}$ MPa	标准贯入 $N_{63.5}$ 击	比贯入阻力 ps MPa
②$_1$	褐黄—灰黄色黏土、粉质黏土	0.4~4.7 1.7	5.62~0.27 3.00	25.0~44.5 32.0	17.8~19.9 18.5	0.67~1.06 0.91	11.2~16.95 14.2	0.28~1.25 0.81	8.0~29.0 15	12.2~24.6 23.0	0.24~0.63 0.38	3.17~5.32 5.18		0.70~1.45 0.85
②$_3$	灰色砂质粉土	0.7~22.6 6.4	4.12~-1.70 1.30	26.5~38.9 30.8	18.0~19.4 18.9	0.78~1.07 0.84			3.0~15 4	12.1~34.0 29.0	0.20~0.47 0.25	4.40~8.95 6.5	4.6~15.9 7.3	1.96~3.47 3.41
③	灰色淤泥质粉质黏土	0.5~10.9 4.6	3.38~-7.31 -0.29	35.8~47.2 42.8	16.9~18.5 17.8	1.01~1.33 1.21	10.2~16.9 15.1	1.01~1.45 1.36	7.0~13.9 12	12.8~24.9 16.5	0.27~1.08 0.78	2.29~4.67 3.12		0.60~1.28 0.68
③$_a$	灰色砂质粉土	0.5~3.7 2.5	1.56~-7.77 -1.31	27.8~34.8 31.9	17.6~18.9 18.5	0.79~1.03 0.92			4~7 5	26.0~35.5 29.5	0.13~0.34 0.25	5.96~13.51 8.43	4.5~10.0 7.5	0.98~2.69 1.88
④	灰色淤泥质黏土	0.6~17.7 8.1	0.40~-14.80 -5.81	38.7~58.9 49.6	16.3~18.1 16.8	1.05~1.49 1.42	17.1~25.8 21.3	1.09~1.48 1.25	10.1~14.9 11	7.2~17.5 11.4	0.42~1.65 1.09	1.32~3.68 2.23		0.41~0.74 0.63
⑤$_{1-1}$	灰色黏土	1.1~17.9 7.0	-5.85~-18.72 -13.51	39.1~53.4 40.5	16.3~17.9 17.6	1.01~1.25 1.10	17.7~24.2 18.2	0.72~1.34 1.02	12~17 13	10.0~18.0 15.5	0.58~1.16 0.72	2.12~3.90 3.21		0.58~1.04 0.76
⑤$_{1-2}$	灰色粉质黏土	1.0~17.3 6.9	-5.94~-22.6 -16.48	31.7~39.1 33.9	17.2~18.5 18.0	0.90~1.19 0.99	12.0~21.7 14.7	0.67~0.99 0.96	8~13 11	20.0~27.0 23.5	0.28~0.74 0.42	2.88~6.87 4.74		0.85~1.32 1.05
⑤$_2$	灰色砂质粉土	1.8~22.5 8.9	-7.93~-44.75 -21.15	26.7~36.9 32.1	17.7~19.6 18.2	0.72~1.05 0.91			0~13.6 6	21.4~33.0 30.0	0.12~0.54 0.31	4.15~10.24 6.90	10.3~17.8 16.0	2.58~7.96 3.68
⑤$_3$	灰色粉质黏土	2.0~28.0 13.2	-16.91~-51.17 -26.31	26.9~39.8 34.3	17.5~19.6 18.2	0.74~1.13 0.95	10.2~16.8 14.3	0.74~1.12 0.92	9.1~25.0 17.1	13.0~22.6 18.2	0.21~0.56 0.39	5.01~6.35 5.14		1.15~2.78 1.41
⑤$_4$	灰绿色粉质黏土	0.5~8.7 3.4	-19.62~-60.52 -35.63	18.2~26.9 24.3	18.5~21.0 19.8	0.51~0.89 0.68	9.8~16.8 14.6	0.21~0.72 0.49	27.1~35.0 33.0	14.1~27.0 20.0	0.14~0.36 0.26	5.18~7.96 7.01		1.54~2.81 2.34

土层层号	土层名称	层厚 m	层面标高 m	含水量 W %	重度 γo KN/cm^3	孔隙比 e	塑性指数 I_P	液性指数 I_L	直剪固快（峰值）		压缩系数 $\alpha_{0.1-0.2}$ MPa^{-1}	压缩模量 $Es_{0.1-0.2}$ MPa	标准贯入 $N_{63.5}$ 击	比贯入阻力 ps MPa
									黏聚力 c KPa	内摩擦角 φ （°）				
⑥	暗绿—褐黄色粉质黏土	0.5~12.6 3.8	−3.66~−29.80 −19.62	20.1~26.9 25.1	19.0~20.9 19.6	0.60~0.80 0.75	10.8~17.9 16.5	0.18~1.04 0.39	38.6~53.0 45.0	14.2~19.6 15.0	0.14~0.35 0.23	5.48~7.89 7.68		1.75~3.24 2.25
⑦$_1$	草黄色砂质粉土	1.2~12.5 6.0	−9.06~−31.13 −23.33	20.1~34.8 27.9	18.1~20.5 19.1	0.51~0.99 0.86			0~10 4	25.0~38.0 30.0	0.08~0.36 0.17	6.28~18.84 11.74	15.8~38.7 25.9	3.58~13.46 8.58
⑦$_2$	灰黄—灰色粉砂	1.3~38.0 16.8	−17.20~−56.42 −37.29	17.8~34.2 25.6	18.2~20.8 19.5	0.54~0.92 0.76			0~6 3	24.8~39.0 32.5	0.08~0.27 0.15	7.86~22.58 13.06	35~>50 46.0	9.80~26.98 16.42
⑧$_{1-1}$	灰色黏土	2.0~23.8 12.4	−15.30~−52.36 −36.21	31.0~43.4 37.4	17.2~18.9 18.2	0.96~1.18 0.95	15.5~23.0 17.9	0.53~0.97 0.92	10~20 16	13.0~20.0 16.3	0.43~0.76 0.47	2.76~4.93 3.56		1.08~2.12 1.58
⑧$_{1-2}$	灰色粉质黏土	4.6~35.5 11.6	−13.76~−54.74 −40.68	29.7~41.4 35.0	17.2~18.9 18.6	0.78~1.02 0.88	12.7~19.1 15.5	0.69~0.99 0.92	7.0~20.0 16.5	18.0~27.0 23.5	0.27~0.61 0.38	3.51~7.09 5.30		1.70~2.40 1.89
⑧$_{2-3}$	灰色粉质黏土夹粉砂	4.0~40.0 13.0	−35.25~−69.22 −51.20	21.8~38.2 33.0	18.1~20.3 18.9	0.58~1.08 0.88	7.8~20.7 15.8	0.29~0.74 0.59	7.8~18.9 17.0	16.7~25.8 22.5	0.14~0.48 0.32	4.28~8.56 6.84		
⑨$_1$	青灰色粉细砂	4.9~22.8 8.0	−49.08~−71.24 −63.40	19.4~33.0 25.4	18.2~21.5 19.7	0.41~0.86 0.65			0~8 3	34.0~38.0 33.0	0.07~0.32 0.13	9.28~23.65 16.26		
⑨$_2$	青灰色含砾中粗砂	15.2~26.0 10.0	−63.76~−78.64 −70.45	13.5~29.3 21.2	18.5~22.0 20.0	0.39~0.79 0.64			0~2 0	27.5~44.0 36.0	0.06~0.24 0.12	10.25~26.84 12.88		
⑩	蓝灰色粉质黏土	0.6~7.2 4.0	−75.89~−94.94 −87.63	25.0	19.5	0.72	14.6	0.31	28.0	26.0	0.25	7.03		

注：表中数值为$\frac{最小值\sim最大值}{平均值}$。

第二节　工程地质结构分区

一、工程地质结构分区依据与原则

区域工程地质结构分区主要依据是全新世晚、中期及现代沉积环境对工程地质层结构、特征进行组合分区。由于地貌类型能较好地体现区内地基土层的变化特征，因此，首先根据上海地貌类型进行土体结构大区划分，其次，根据古地貌，亦即是否受有古河道以及近期河道切割等进行工程地质结构亚区划分：湖沼平原区以暗绿—褐黄色硬土层作为分区依据；滨海平原区以浅部砂层（②$_3$）及暗绿色硬土层（⑥）分布状况为依据；河口砂岛区则由于受长江古河道切割，因此无工程地质结构亚区。

二、工程地质结构分区及特征

根据上述分区原则，上海陆域共划分出 3 个工程地质结构区：湖沼平原区（第Ⅰ工程地质结构区）、滨海平原区（第Ⅱ工程地质结构区）和河口砂岛区（第Ⅲ工程地质结构区）。根据亚区划分原则，湖沼平原区划分出 3 个工程地质结构亚区（I_1、I_2、I_3），滨海平原区划分出 4 个工程地质结构亚区（II_1、II_2、II_3、II_4），河口砂岛区未进行工程地质结构亚区划分。全区共划分出 8 个工程地质结构亚区，即 I_1、I_2、I_3、II_1、II_2、II_3、II_4、Ⅲ。

各工程地质结构区具有不同的工程地质层组合特征、天然地基和桩基条件，易引发的工程地质问题也不相同。主要工程地质层分布情况见表 28-3、图 28-1 和表 28-4。

表 28-3　上海各工程地质结构分区的地层组合（硬土层及砂、粉性土层）

Table 28-3　Stratigraphic combinations (hard soil and sandy/silty soil layers) in Shanghai by engineering geological structural region

工程地质结构分区			40m 以浅硬土及砂、粉性土分布		备注
区	亚区	段	有分布	缺失	
湖沼平原区（Ⅰ）	I_1		⑥、⑧$_{2-1}$		正常沉积区，②$_3$ 层零星分布
	I_2	I_{2-1}	⑥	⑧$_{2-1}$	第二硬土层缺失，②$_3$ 层“冈身带”分布较普遍
		I_{2-2}	⑧$_{2-1}$	⑥	第一硬土层缺失，②$_3$ 层“冈身带”分布较普遍
	I_3			⑥、⑧$_{2-1}$	第一、第二硬土层均缺失，古河道切割深度大
滨海平原区（Ⅱ）	II_1		②$_3$、⑥		正常沉积区，仅局部地区零星分布有第二硬土层，⑦层均有分布
	II_2		⑥	②$_3$	
	II_3		②$_3$	⑥	古河道切割区，仅局部地区零星分布有第二硬土层，⑦层均有分布
	II_4			②$_3$、⑥	
河口砂岛区（Ⅲ）			②$_3$	⑥、⑦	长江古河道切割区

表 28-4　上海各工程地质结构分区及评价

Table 28-4　Zoning and assessment of engineering geological structure in Shanghai

工程地质分区			工程地质特征及问题				
区名	亚区名	块	工程地质层组成	天然地基条件	桩基条件	浅部砂土液化	其他地质问题
湖沼平原区（Ⅰ）	Ⅰ$_1$：有⑥和⑧$_{2-1}$分布		基岩露头分布，岩性以黏性土为主，赋存2个硬土层，软土层不发育，无浅部砂层	表土层以黏土为主，下卧层有泥炭层分布，天然地基条件较差	第二硬土层（⑧$_{2-1}$）和第二（⑧$_{2-2}$）、第三（⑨）砂层等桩基持力层有分布，桩基条件较好	基本无浅部砂层分布	局部有工程地质条件不良的泥炭层分布
	Ⅰ$_2$：有⑥和⑧$_{2-1}$分布	Ⅰ$_{2-1}$：有⑥无⑧$_{2-1}$	岩性以黏性土为主，第一硬土层有分布，第二硬土层缺失，软土层有发育	表土层以黏土为主，局部地区有软土层下卧层，天然地基条件较差	第二砂层（⑧$_{2-2}$）分布不稳定，第三砂层（⑨）埋藏较深，桩基条件较差	无浅部砂层分布	局部有泥炭层和软土层分布，有轻度地面沉降
		Ⅰ$_{2-2}$：无⑥有⑧$_{2-1}$	岩性以黏性土为主，第一、第二硬土层缺失，软土层有发育	表土层以褐黄色黏性土为主，一般无泥炭层分布，下部有软土层分布，天然地基条件一般	第二砂层（⑧$_{2-2}$）和第三砂层（⑨）均有分布，桩基条件一般	无浅部砂层分布	存在软土地基变形及轻度地面沉降问题
	Ⅰ$_3$：无⑥和⑧$_{2-1}$分布		岩性以黏性土为主，第一及第二硬土层均缺失，软土层发育	表土层以黏土为主，下部软土层发育，天然地基条件较差	第二砂层（⑧$_{2-2}$）大部缺失且不稳定，第三砂层（⑨）埋藏较深，桩基条件差	无浅部砂层分布	存在软土地基变形及轻度地面沉降
滨海平原区（Ⅱ）	Ⅱ$_1$：有⑥和②$_3$分布		浅部砂层（②$_3$）、第一硬土层有分布，第二硬土层缺失，软土层（③、④）均有分布，第二砂层均有分布	表土层以黏性土为主，下卧层为砂，天然地基条件较好	⑥层埋深浅，⑦层厚度小于7m地区桩基条件差，其余地区桩基条件好	浅部砂层发育，有震动液化及渗流液化问题	存在软土地基变形及地面沉降
	Ⅱ$_2$：有⑥无②$_3$分布		浅部砂层（②$_3$）缺失，第一硬土层有分布，第二硬土层缺失，软土层发育且厚度较大	表土层以黏性土为主，软弱下卧层发育且厚度较大，天然地基条件一般	⑥层埋深浅，⑦层厚度小于7m地区桩基条件差，其余地区桩基条件好	一般无浅部砂层分布	存在软土地基变形及地面沉降
	Ⅱ$_3$：无⑥有②$_3$分布		浅部砂层（②$_3$）均有分布，第一、第二硬土层缺失，软土层发育，局部第二砂层缺失	表土层以黏性土为主，下卧层为砂性土层，天然地基条件较好	受古河道切割，桩基条件变化较大，桩基条件差，⑤$_2$层分布且厚度较大地区条件一般	浅部砂层发育，有震动液化及渗流液化	存在软土地基变形问题，地面沉降发育，极易产生不均匀沉降
	Ⅱ$_4$：无⑥无②$_3$分布		浅部砂层（②$_3$）缺失，第一、第二硬土层缺失，软土层发育且厚度大，局部第二砂层缺失	表土层以黏性土为主，软土层发育，天然地基条件一般	受古河道切割，桩基条件变化较大，桩基条件差，局部⑤$_2$层分布且厚度较大地区条件一般	无浅部砂层分布	存在软土地基变形问题，地面沉降发育，极易产生不均匀沉降
河口砂岛区（Ⅲ）			岩性以砂性土夹黏性土为主，浅部砂层发育，软土层有分布，第二砂层缺失	表土层下有巨厚砂层分布，天然地基条件好	无⑦层分布，但⑤$_2$层分布较为稳定，因此有⑤$_2$层分布区桩基条件一般，无该层分布区条件差	浅部砂层广泛分布，易产生震动液化及渗流液化	轻度地面沉降，易产生不均匀沉降和海潮水侵袭

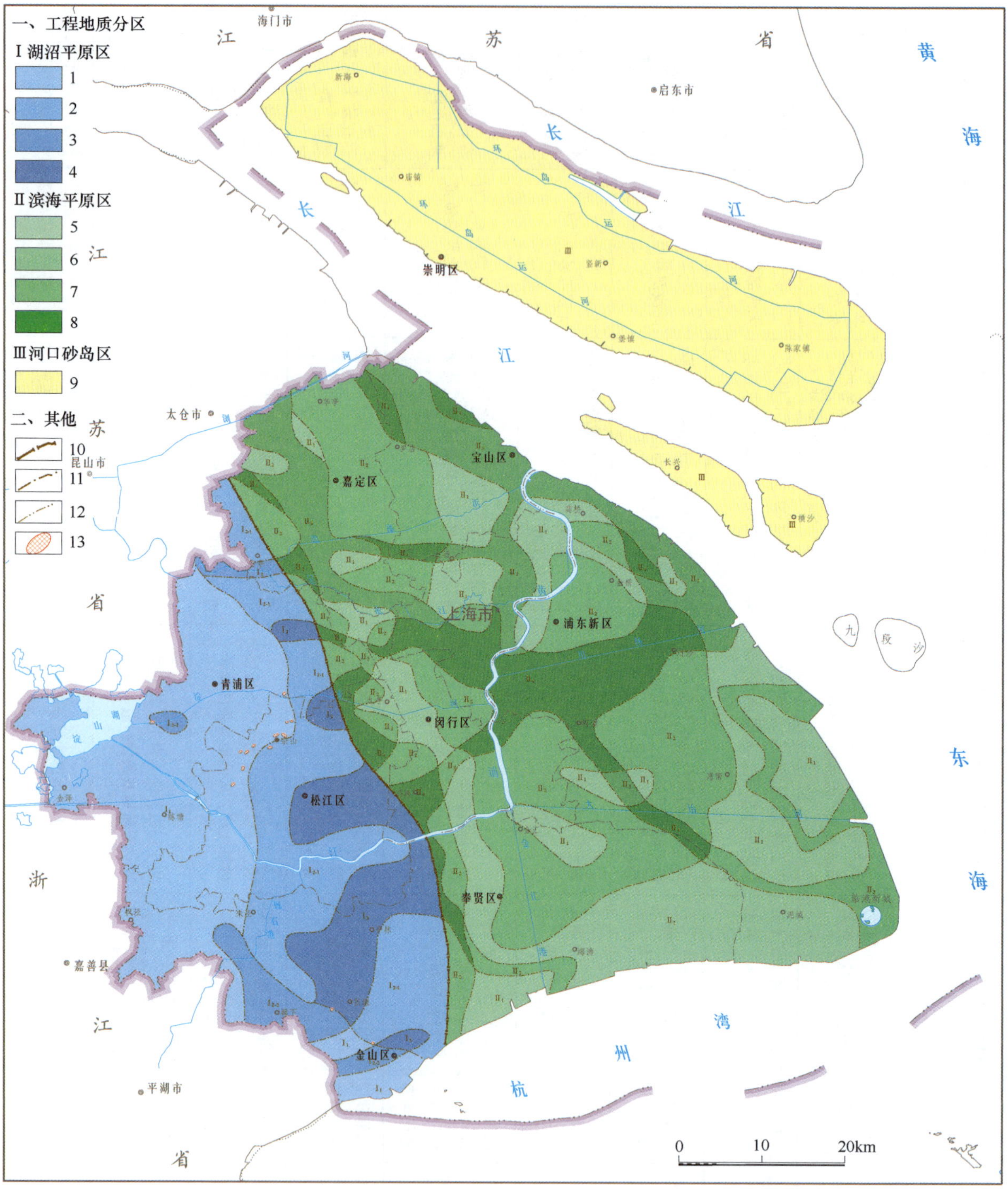

图 28-1 上海工程地质结构分区示意图

Fig. 28-1 Zoning map of engineering geological structure in Shanghai

1—Ⅰ$_{1}$ 有⑥和⑧$_{2-1}$ 分布区、Ⅰ$_{2}$ 有⑥和⑧$_{2-1}$ 分布区；2—Ⅰ$_{2-1}$ 有⑥和⑧$_{2-1}$ 分布区；3—Ⅰ$_{2-2}$ 有⑥和⑧$_{2-1}$ 分布区；4—Ⅰ$_{3}$ 有⑥和⑧$_{2-1}$ 分布区；5—Ⅱ$_{1}$ 有⑥和②$_{3}$ 分布的第Ⅱ$_{1}$ 亚区；6—Ⅱ$_{2}$ 有⑥无②$_{3}$ 分布的第Ⅱ$_{2}$ 亚区；7—Ⅱ$_{3}$ 有⑥无②$_{3}$ 分布的第Ⅱ$_{3}$ 亚区；8—Ⅱ$_{4}$ 有⑥无②$_{3}$ 分布的第Ⅱ$_{4}$ 亚区；9—河口砂岛区；10—区界线；11—亚区界线；12—地块界线；13—基岩区

第三节　工程地质结构特征

一、第②$_1$层——表土层

该层是上海的“硬壳层”，可作为天然地基的持力层，因此广受关注。

（一）分布特征

其分布较为普遍，仅在黄浦江、吴淞江两岸及沿江、沿海局部地区缺失，缺失区呈条带状，其中临港新城东部及崇明东滩由于为新成陆区，均缺失该层。

（二）埋深、厚度变化特征

该层分布于填土层之下，埋深、厚度变化均不大，埋深为0.5~2m，厚度则在0.5~4m之间，市区及其西部大于沿江、沿海地区。

（三）岩性变化特征

该层的岩性在平面展布上存在着差异，河口砂岛区为粉质黏土夹砂质粉土，具水平层理，滨海平原区东部海岸地带地区则以粉质黏土为主，中心城以黏土为主，湖沼平原区以粉质黏土为主。此外，大部分地区该层上部土性较好，以可塑的粉质黏土为主，而下部土层以淤泥质粉质黏土和黏土为主，土性较差，因此具有上硬下软特性。

（四）工程地质层评价

该层是区内表土层，土性较好，为上海俗称的“硬壳层”，总体上3个工程地质结构区内其物理力学指标特征变化不大，其含水量为32.0%（平均值，下同），孔隙比为0.91，液性指数为0.65，压缩系数为0.42MPa^{-1}，压缩模量为5.16MPa，内聚力为27kPa，内摩擦角为25°，从指标亦可看出该层含水量、压缩性较低，强度相对较高，静探比贯入阻力为0.70~1.45MPa（平均值为0.85MPa），为浅基础的良好持力层，但由于该层具有上硬下软特性，因此基础宜浅埋，同时还须注意该层厚度的变化，作为天然地基持力层时，应进行强度验算。

二、第②$_3$层——浅部砂层

（一）分布特征

该层在第Ⅲ工程地质结构区，亦即崇明、横沙、长兴三岛区均有分布（图28-2），长江沿岸地区呈条带状分布，其中在南汇嘴地区分布较为广泛，主要是由于该地区为新近成陆区。黄浦江两岸在上海市区北部均有分布，以南缺失，吴淞江两岸也均有分布，但局部地区不连续。此外，“冈身带”亦有分布，呈条带状，为古海岸线。湖沼平原区该层不发育。该层在不同地区特性具有一定差异（表28-5）。

（二）埋深、厚度变化特征

该层埋深总体变化不大，三岛区埋藏较浅，层顶标高一般均大于2m，仅中部南门港地区层顶标高在0~2m之间。长江沿岸局部地区层顶标高大于2m，其余大部分地区均在0~2m之间。

该层厚度变化较大，在1~27m之间。河口砂岛区厚度最大，一般均大于11m，其中在崇明岛西北端厚度为最大。第Ⅱ工程地质结构区南汇嘴地区厚度一般在7~15m之间，主要集中在临港新城规划区内，吴淞江、黄浦江两岸厚度亦较大，一般为6m左右，其余地区厚度一般小于5m。

（三）岩性变化特征

平面上该层岩性变化不大，以砂质粉土为主，仅黄浦江两岸及三岛局部地区为黏质粉土。垂向上该层在滨海平原区基本以砂质粉土为主，变化不大，但在三岛区具有明显变化。从上至下一般有3种

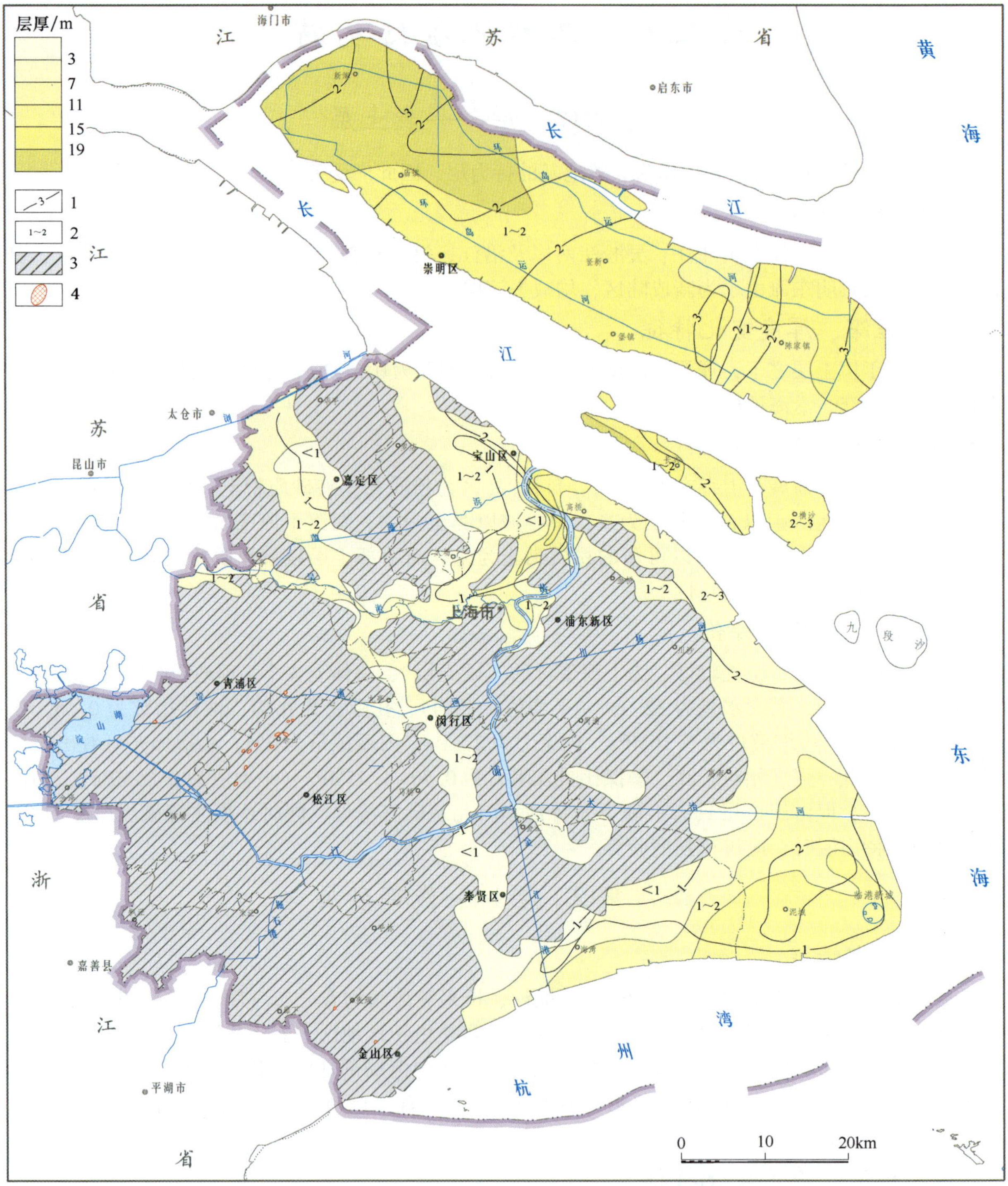

图 28-2　上海第②$_3$ 工程地质层埋藏分布示意图

Fig. 28-2　Burial depth distribution of engineering geological stratum②$_3$ in Shanghai

1—层顶标高等值线（m）；2—层顶标高范围（m）；3—缺失区；4—基岩区

岩性，上部一般为砂质粉土，中部为黏质粉土，局部为淤泥质粉质黏土，下部为粉砂夹砂质粉土。

（四）工程地质层评价

该层为上海的浅部砂层，一般为可液化土层，应注意由其引起的砂土震动液化及渗流液化问题。

表 28-5 上海第②$_3$层工程地质特性对比

Table 28-5 Engineering geological properties of stratum②$_3$ in Shanghai

项目				吴淞江、黄浦江沿岸	“冈身带”沿线	崇明三岛区及宝山、浦东沿岸			南汇嘴地区
地层编号				②$_3$	②$_3$	②$_{3-1}$、②$_{3-2}$、②$_{3-3}$			②$_3$
测年成果及所属层组				如东组	6.5ka B.P.（嘉定“冈身带”测年），上海组	3.2ka B.P.（7.65m），5.5ka B.P.（16.68m）（SG1 孔），如东组（②$_{3-1}$，②$_{3-2}$），上海组（②$_{3-3}$）			1172 年以来沉积（据里护塘海堤），如东组
沉积环境				三角洲河流	滨海	河口			滨海
空间变化特征				该层在上海埋深变化不大，位于表土层（②$_1$）之下，埋深一般为 2～4m					
				厚度变化较大，平均厚度约为 6m	厚度有一定变化，平均厚度约为 3m	厚度大，但变化不大，平均厚度约为 11m			厚度较大，由西向东厚度加大，平均厚度约为 9m
岩性变化特征				岩性以砂质粉土为主，但黏粒含量较多，且土质不均	岩性大部分地区以砂质粉土为主，部分地区以黏质粉土为主	垂向岩性有一定变化，一般顶部为砂质粉土（②$_{3-1}$），中部为黏质粉土（②$_{3-2}$），下部为砂质粉土或粉砂（②$_{3-3}$）			岩性以砂质粉土为主，颗粒较均匀，具水平层理，土质较均匀
物理力学指标变化特征	含水量	w	%	34.4	30.3	②$_{3-1}$	②$_{3-2}$	②$_{3-3}$	29.6
						30.1	33.1	31.2	
	重度	γo	kN/m^3	18.0	18.6	19.1	18.8	19.0	18.6
	孔隙比	e_0		0.98	0.79	0.87	1.01	0.89	0.84
	黏粒含量	ρ_c	%	9.4	6.1	6.5	12.1	7.3	5.3
	压缩系数	$\alpha_{0.1-0.2}$	MPa^{-1}	0.30	0.17	0.22	0.40	0.28	0.17
	压缩模量	$Es_{0.1-0.2}$	MPa	7.13	10.93	8.69	6.30	7.70	11.79
	内聚力	c	kPa	7	5	3	8	5	4
	内摩擦角	ϕ	(°)	29.5	32.0	31.2	22.0	29.5	34.0
	比贯入阻力	ps	MPa	1.48	2.13	2.65	0.98	3.65	3.41
	标准贯入	击	次	6.2	9.9	8.3	5.6	12.5	11.6

三、第③+④层

（一）分布特征

该两层在上海分布较为广泛（图 28-3），湖沼平原区的青浦和金山西部缺失，河口砂岛区的崇明岛西北端由于受到后期古河道的切割亦缺失。其中第③层在第Ⅲ工程地质结构区和沿江、沿海地区由于受到后期古河道切割而缺失，而在湖沼平原区普遍分布，第④层则在西部湖沼平原区基本缺失。

（二）埋深、厚度变化特征

软土层埋深总体由湖沼平原区向滨海平原区、河口砂岛区逐渐加深，湖沼平原区和滨海平原区（除黄浦江和吴淞江沿岸局部地区）总体变化不大，层顶标高一般在 3.5～-1m 之间，仅处于南汇嘴地区的临港新城在-1～-9m 之间。三岛区埋藏较深，在-9～-16m 之间，崇明岛西北端埋深最大。

厚度由西到东逐渐加大，西部地区包括嘉定西部、青浦、金山西部、闵行西部，厚度为 6～10m。紧邻东部的上海市区、原南汇西部、奉贤、金山东部和宝山厚度过渡为 10～14m，浦东部分地区厚度

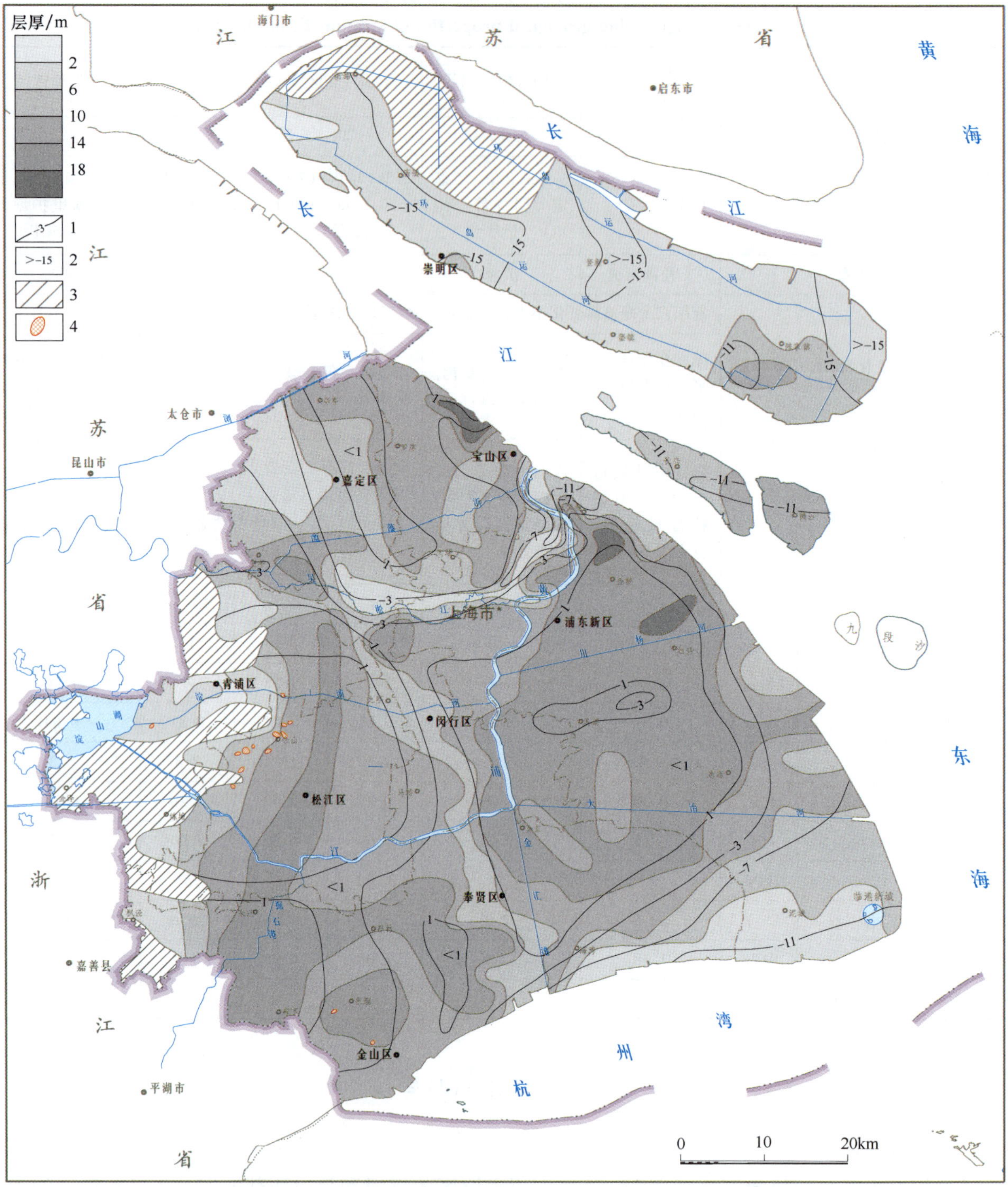

图 28-3 上海第③+④工程地质层（软土层）埋藏分布示意图

Fig. 28-3 Burial depth distribution of engineering geological strata③+④ (soft soil layers) in Shanghai

1—层顶标高等值线（m）；2—层顶标高范围（m）；3—缺失区；4—基岩区

最大，在 14~20m 之间。而南汇嘴地区由于浅部砂层厚度较大，使得软土层厚度变薄，一般为 2~6m。第Ⅲ工程地质结构区厚度最小，一般为 1~6m。

（三）岩性变化特征

软土层以淤泥质黏性土为主，其中第③层以淤泥质粉质黏土为主，第④层基本为淤泥质黏土。③层土由东向西由滨海沼泽相向湖沼相沉积过渡，因此，③层土在岩性上有所变化，第Ⅰ工程地质结构区一般黏性土层和淤泥质土层均有分布，且一般黏性土层占大部分，含水量较小，压缩性较低（表28-6，表28-7），夹粉土薄层少，工程性质较好，而滨海平原则属水动力条件更弱的滨海—浅海相沉积，③层普遍为淤泥质粉质黏土，且普遍夹有粉土、粉砂薄层，工程特性差。④层土由于均为浅海相沉积，湖沼平原区不发育，主要分布在滨海平原区，以淤泥质黏土为主。

表 28-6　上海第③层工程地质特性对比

Table 28-6　Comparison of engineering geological properties of stratum③ in Shanghai

项目				湖沼平原区（Ⅰ）		滨海平原区（Ⅱ）		河口砂岛区（Ⅲ）
				$Ⅰ_1$	$Ⅰ_2$	$Ⅱ_1$	$Ⅱ_2$	
工程地质层编号				③	③	③	③	缺失
测年成果及所属层组				距今约 2000 年，上海组	距今约 2000 年，上海组	距今约 2000 年，上海组	距今约 2000 年，上海组	
沉积环境				滨海	滨海	滨海—浅海	滨海—浅海	
空间分布特征				平均埋深约为 2.5m	平均埋深约为 3m	黄浦江、吴淞江两岸、长江沿岸及南汇嘴地区大部分缺失，埋深较大，厚度较小		
						平均埋深约为 3.5m	平均埋深约为 4.5m	
				平均厚度约为 4.9m	平均厚度约为 6.0m	平均厚度约为 5.8m	平均厚度约为 4.4m	
岩性变化特征				以黏土为主	以黏土为主，东部以淤泥质粉质黏土为主	以淤泥质粉质黏土为主	以淤泥质粉质黏土为主	
物理力学指标变化特征	含水量	w	%	34.6	37.2	43.3	41.9	
	重度	γo	kN/m^3	18.1	17.8	17.6	17.9	
	孔隙比	e_0		1.01	1.07	1.25	1.16	
	塑性指数	I_p		15.0	15.2	15.6	14.6	
	液性指数	I_l		0.96	1.13	1.38	1.22	
	压缩系数	$\alpha_{0.1-0.2}$	MPa^{-1}	0.66	0.67	0.82	0.73	
	压缩模量	$Es_{0.1-0.2}$	MPa	3.64	3.33	3.27	5.32	
	内聚力	c	kPa	12	13	13	7	
	内摩擦角	ϕ	（°）	16.5	17	16.2	14	
	比贯入阻力	ps	MPa	0.54	0.50	0.48	0.54	

（四）工程地质层评价

③层在第Ⅰ工程地质结构区以黏土、粉质黏土为主，在第Ⅱ工程地质结构区为淤泥质粉质黏土，其物理力学指标在不同工程地质结构区亦有相应差异，但基本为软土。

④层为灰色淤泥质黏土，土质更软。

从指标可以看出，该两层软土层具有含水量高、孔隙比大、压缩性高、强度低等不良工程地质特性，为地基沉降的主要层位，因此在高层建筑和路基工程施工过程中极易产生变形。此外，该两层还具有流变和触变特性，在基坑开挖和隧道盾构施工过程中，易引起边坡失稳或地面沉降。

表 28-7　上海第④层工程地质特性对比

Table 28-7　Comparison of engineering geological properties of stratum ④ in Shanghai

项目				湖沼平原区（Ⅰ）		滨海平原区（Ⅱ）		河口砂岛区（Ⅲ）
				I_1	I_2	II_1	II_2	
工程地质层编号				缺失	④	④	④	④
测年成果及所属层组					上海组	上海组	上海组	上海组
沉积环境					浅海	浅海	浅海	浅海
空间分布特征					局部分布，平均埋深约为 9m	广泛分布，平均埋深约为 8.5m	广泛分布，平均埋深约为 10m	广泛分布，平均埋深约为 17m
					平均厚度约为 5.9m	平均厚度约为 7.9m	平均厚度约为 9.0m	平均厚度约为 5.9m
岩性变化特征					该层为浅海相沉积，岩性变化不大，基本以淤泥质黏土为主，河口砂岛区则夹有较多粉性土			
物理力学指标变化特征	含水量	w	%		47.5	49.8	49.5	46.3
	重度	γo	kN/m^3		16.7	16.7	16.8	17.5
	孔隙比	e_0			1.38	1.42	1.40	1.30
	塑性指数	I_p			20.10	22.10	21.10	19.6
	液性指数	I_l			1.21	1.21	1.16	1.10
	压缩系数	$\alpha_{0.1-0.2}$	MPa^{-1}		1.12	1.20	1.01	0.84
	压缩模量	$Es_{0.1-0.2}$	MPa		2.20	2.18	2.46	2.75
	内聚力	c	kPa		12	13	11	10
	内摩擦角	ϕ	(°)		12.5	11.5	11.0	8
	比贯入阻力	ps	MPa		0.59	0.50	0.65	0.72

四、第⑤层

（一）分布特征

该层为滨海、沼泽相沉积地层，主要分布在滨海平原区（第Ⅱ工程地质结构区）及河口砂岛区（第Ⅲ工程地质结构区），湖沼平原区（第Ⅰ工程地质结构区）由于沉积间断而局部分布，此外，滨海平原区嘉定中部、宝山南部部分地区亦有缺失。

（二）埋深、厚度变化特征

该层位于软土层之下，埋深变化不大，总体由第Ⅰ工程地质结构区向第Ⅲ工程地质结构区逐渐加深，嘉定西部和宝山局部地区层顶标高为-6～-9m，市区及奉贤、金山层顶标高变化在-9～-15m，长江沿岸地区在-15～-18m，在三岛区埋深达最大，层顶标高在-18～-24m之间。

厚度上，该层由于古河道的切割使得厚度变化最大，在1～43m不等。正常沉积区厚度一般为3～8m，而在古河道切割区厚度则骤然增大，滨海平原区古河道内该层厚度一般为12～28m，河口砂岛区则最大，一般均大于28m，最厚可达35m。

（三）岩性变化特征

由于受古河道切割，该层岩性变化较大，尤其垂向上变化更大，岩性从上至下依次为灰色黏土层（⑤$_{1-1}$）、灰色粉质黏土层（⑤$_{1-2}$）、灰色砂质粉土、粉砂层（⑤$_2$）、灰色粉质黏土夹粉土层（⑤$_3$）和灰绿色粉质黏土层（⑤$_4$）。

其中⑤$_1$ 层黏性土为滨海相沉积，埋深、厚度变化不大，厚度为5~8m，上海市区及奉贤、松江等地岩性以黏土为主，三岛区及宝山、浦东沿海地区岩性以粉质黏土夹粉土为主，至金山沿海地区以砂质粉土夹粉质黏土为主。

⑤$_2$ 层砂质粉土、粉砂为溺谷相沉积，分布于古河道切割区，第Ⅲ工程地质结构区普遍分布，而滨海平原区则局部分布，且不连续。该层岩性以砂质粉土为主，部分地区下部为粉砂，埋深、厚度变化大，滨海平原区层顶标高一般为-14~-28m，厚度为2~8m，三岛区埋深逐渐加深、加厚，标高一般为-29~-42m，厚度为3~18m。

⑤$_3$ 为溺谷相沉积地层，分布于古河道底部，岩性以粉质黏土夹砂质粉土为主，部分地区夹粉砂，呈“千层饼”状，颜色以灰色、褐灰色为主。

（四）工程地质层评价

总体上该层岩性变化较大，以黏性土为主，夹有砂质粉土、粉砂透镜体，因此土层总体表现为不均匀特性，且厚度变化大，为荷载较大建筑物的地基压缩层，易产生地基不均匀沉降。其中⑤$_{1-1}$、⑤$_{1-2}$ 层很湿—饱和、软塑—流塑、压缩性较高、强度低，为荷载较大建筑的压缩层，此外，该两层由于埋藏适中，可作为沉降控制复合桩的桩基持力层。⑤$_2$ 层为上海的微承压含水层，在大的基坑开挖工程和隧道工程中有可能揭露该层，应注意该层在地下水压力作用下可能产生的流砂现象。⑤$_3$、⑤$_4$ 层为溺谷相地层，分布在古河道切割区，厚度、埋深变化较大，且土质不均，易引起荷载较大建筑的不均匀沉降。

五、第⑥层

该层为全新统和更新统分界的标志层，因此一直为研究的重点层次。由于受到古河道的切割，该层呈块状分布（图28-4）。

（一）分布特征

西部湖沼平原区（第Ⅰ工程地质结构区）由于成陆时代较早，分布较为普遍。最大的一条古河道为长江古河道，使得河口砂岛区（第Ⅲ工程地质结构区）和宝山、浦东沿江沿海地区均被切割。滨海平原区（第Ⅱ工程地质结构区）亦分布有几条大的古河道，南北一条贯穿上海，经宝山吴淞口、五角场至上海市区，往南分为两支，一支经闵行西部至松江再到金山，另一支经徐汇至奉贤西部再到金山；东西一条由嘉定娄唐经嘉定城区至市区与南北向相交后向东经浦东至长江；另外，在浦东东南部有两条较小的古河道。古河道的切割深度在崇明岛最大，一般在50~60m，在长兴岛、横沙岛其切割深度稍浅，一般在45m左右，至滨海平原区其深度又有所变浅，为35~45m。

（二）埋深、厚度变化特征

前面已提及，根据本次的研究成果，第Ⅰ工程地质结构区第一硬土层与第Ⅱ工程地质结构区浅部硬土层为同一层次，由此该层埋深从西部向东部逐渐加深，西部湖沼平原区和嘉定西部埋深在3~5m，“冈身带”和宝山南部埋深12~16m，至滨海平原区埋深则为22~28m。

厚度上，该层变化不大，从西向东厚度逐渐减小，西部湖沼平原区厚度一般为5~7m，而滨海平原区则一般为3~5m。

（三）岩性变化特征

该层为湖沼相沉积地层，厚度较小，岩性变化不大，大部分地区以粉质黏土为主，部分地区以黏土为主，一般从上至下黏粒含量减少，粉粒含量增加，颜色则从暗绿色逐渐过渡为褐黄色、草黄色。

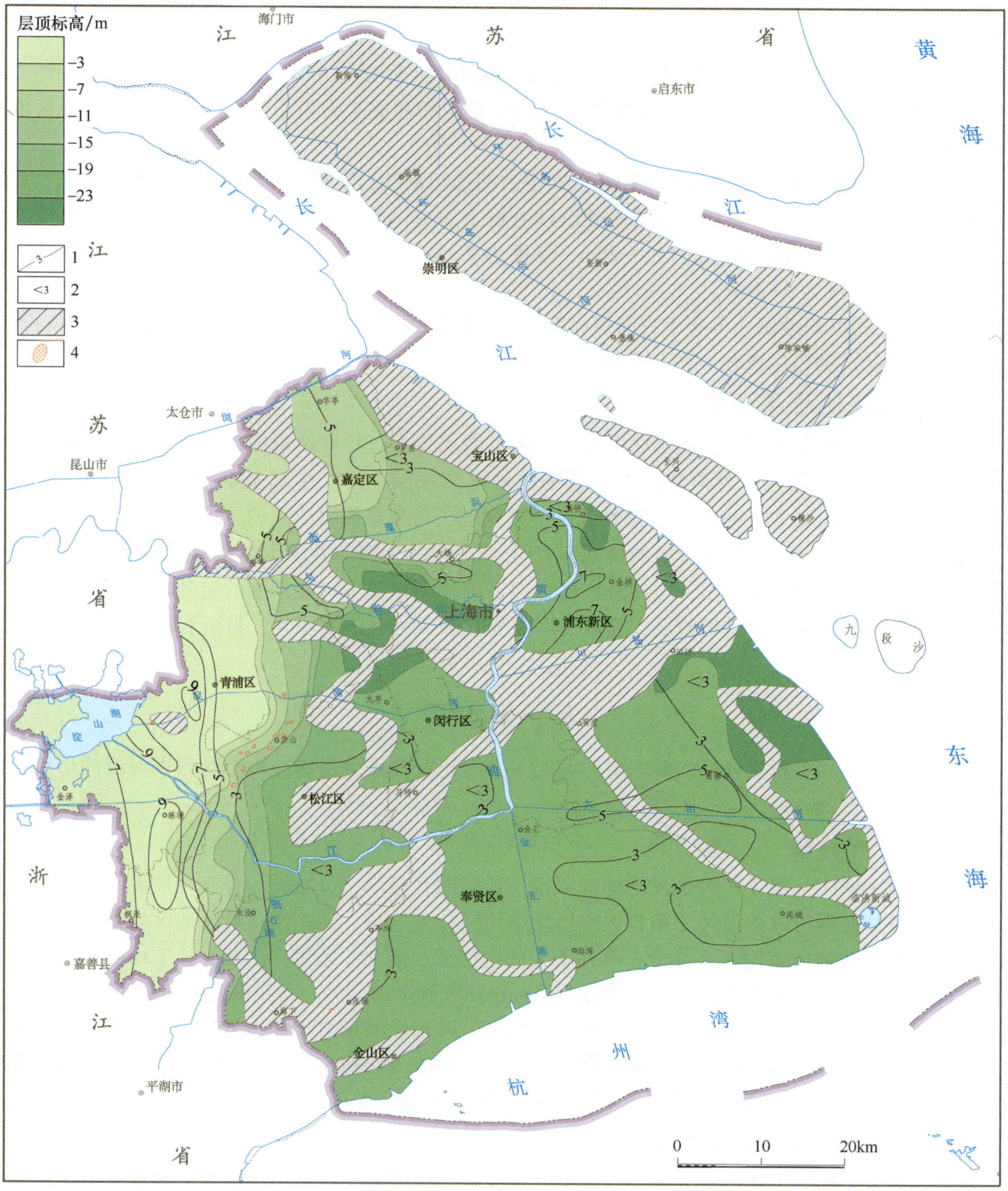

图 28-4　上海第⑥工程地质层（第一硬土层）埋藏分布示意图

Fig. 28-4　Buried depth distribution of engineering geological stratum⑥ (first hard soil layer) in Shanghai

1—层厚等值线（m）；2—层顶标高范围（m）；3—缺失区；4—基岩区

（四）工程地质层评价

该层为湖沼相沉积地层，后期经脱水固结，结构较密实，该层分布较为稳定。不同工程地质结构区其物理力学指标见表28-8。从表28-8可看出，该层具有含水量低、压缩性低、强度高等特点，土性好，该层与下部⑦层联合可作中型建筑物的桩基持力层。此外，该层在控制地面沉降方面，除本身不易压缩外，在一定程度上还能消散或滞后下部应力对上部土层的影响。

表 28-8 上海第一硬土层（⑥）工程地质特性对比

Table 28-8 Comparison of engineering geological properties of the first hard soil layer（⑥）in Shanghai

项目				湖沼平原区（Ⅰ）		滨海平原区（Ⅱ）		河口砂岛区（Ⅲ）
				$Ⅰ_1$	$Ⅰ_2$	$Ⅱ_1$	$Ⅱ_2$	
原工程地质层编号				④	④	⑥	缺失	缺失
新工程地质层编号				⑥	⑥	⑥		
测年成果及所属层组				14.6ka B.P.（青浦A5孔），滆湖组	17.0ka B.P.（青浦SG3孔），滆湖组	10.1~16.2ka B.P.（面粉厂SG2孔），南汇组		
沉积环境				泛滥湖泊	泛滥湖泊	泛滥湖泊		
空间分布特征				埋深 3~5m	埋深 12~16m	埋深 22~28m		
				厚度一般为 5~7m	厚度一般为 5~7m	厚度 3~5m		
岩性变化特征				岩性变化不大，以粉质黏土为主，部分地区以黏土为主，一般从上至下黏粒含量减少，粉粒含量增加，颜色则从暗绿色逐渐过渡为褐黄色、草黄色				
物理力学指标变化特征	含水量	w	%	25.1	26.5	27.5		
	重度	γo	kN/m^3	20.0	19.7	19.4		
	孔隙比	e_0		0.66	0.75	0.79		
	塑性指数	I_p		16.3	16.5	16.6		
	液性指数	I_l		0.24	0.39	0.51		
	压缩系数	$\alpha_{0.1-0.2}$	MPa^{-1}	0.21	0.23	0.28		
	压缩模量	$Es_{0.1-0.2}$	MPa	8.74	7.68	6.82		
	内聚力	c	kPa	52	48	44		
	内摩擦角	ϕ	（°）	14	16	19		
	比贯入阻力	ps	MPa	2.75	2.52	2.25		

六、第⑦层

（一）分布特征

该层为河流相沉积地层，第Ⅲ工程地质结构区和宝山沿岸地区由于受到长江古河道切割，均缺失（图 28-5），滨海平原区（第Ⅱ工程地质结构区）由于切割深度不大而普遍分布，此外，西部湖沼平原区（第Ⅰ工程地质结构区）部分地区和嘉定西部也有缺失。

（二）岩性变化特征

该层岩性平面上变化不大，垂向上有一定差异，从上至下砂粒含量逐渐增加，粉粒含量逐渐减少，局部地区底部夹有砾石，颜色亦由草黄色变化为灰黄色，底部一般为灰色，因此从上至下由粉土（$⑦_1$）过渡为粉砂、细砂（$⑦_2$）。$⑦_1$ 层岩性以砂质粉土为主，西部湖沼平原区（第Ⅰ工程地质结构区）和嘉定、宝山局部地区则以黏质粉土为主，$⑦_2$ 层以粉砂为主，部分地区为细砂，底部夹有砾石，西部地区一般无分布。

（三）工程地质层评价

该层广泛分布，为上海的下部砂层，从上至下该层由砂质粉土逐渐过渡为粉砂、细砂，其中$⑦_1$ 层

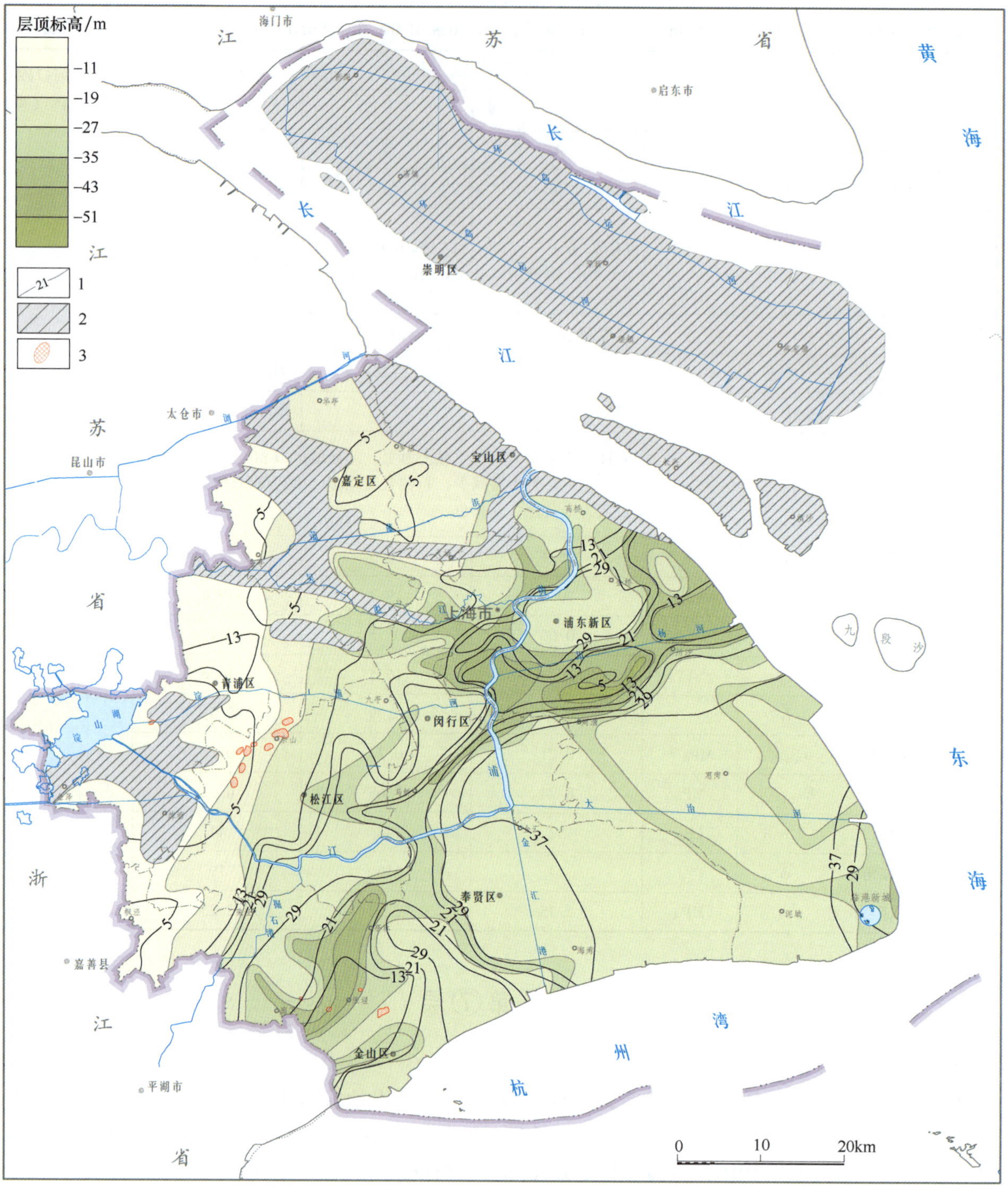

图 28-5　上海第⑦工程地质层层顶埋深示意图

Fig. 28-5　Burial depth of the top surface of engineering geological stratum⑦ in Shanghai

1—层厚等值线（m）；2—缺失区；3—基岩区

压缩性低、强度高（表 28-9），在上海中部地区埋藏适中，可作为大型建筑物的桩基持力层。$⑦_2$ 层可作为大型及重型建筑物的桩基持力层。该层亦为上海的第 I 承压含水层，水头较高，地下空间开发过程中应注意由该层引起的基坑突涌和流砂问题。

表 28-9 上海第⑦层特性对比

Table 28-9 Comparison of the properties of stratum⑦ in Shanghai

项目				湖沼平原区（Ⅰ）			滨海平原区（Ⅱ）			河口砂岛区（Ⅲ）
				$Ⅰ_1$	$Ⅰ_2$		$Ⅱ_1$		$Ⅱ_2$	
工程地质层编号				$⑦_1$	$⑦_1$、$⑦_2$		$⑦_1$、$⑦_2$		$⑦_2$	缺失
测年成果及所属层组				21.2ka（青浦 G415 孔），滆湖组	22.7ka（青浦 SG3 孔），滆湖组		21.6ka（面粉厂 SG2 孔），南汇组		南汇组	
沉积环境				河口—滨海	河口—滨海		河口		河口	
空间分布特征				仅局部分布	$⑦_1$ 分布较广，$⑦_2$ 局部分布		均有分布		$⑦_1$ 基本缺失，$⑦_2$ 广泛分布	
				平均埋深约为 9m	平均埋深约为 14.5m		平均埋深约为 29m		平均埋深约为 48m	
				平均厚度约为 3.5m	平均厚度$⑦_1$约为 3.5m，$⑦_2$约为 10.3m		平均厚度$⑦_1$约为 6m，$⑦_2$约为 19.6m		平均厚度约为 9.1m	
岩性变化特征				以黏质粉土为主，颜色为草黄色，部分为灰色	$⑦_1$ 同$Ⅰ_1$区，$⑦_2$以粉砂为主，颜色为灰黄—灰色		$⑦_1$ 以砂质粉土为主，颜色为草黄色，$⑦_2$ 同$Ⅱ_2$区		$⑦_2$ 以粉砂为主，颜色为灰黄—灰色	
物理力学指标变化特征				$⑦_1$	$⑦_1$	$⑦_2$	$⑦_1$	$⑦_2$	$⑦_2$	
	含水量	w	%	27.5	29.9	24.9	27.9	26.7	25.5	
	重度	γo	KN/m^3	18.5	18.3	18.7	19.1	19.4	19.6	
	孔隙比	e_0		0.80	0.85	0.74	0.86	0.78	0.74	
	黏粒含量	ρ_c	%	11.0	10.5	7.0	5.1	4.1	3.5	
	压缩系数	$\alpha_{0.1-0.2}$	MPa^{-1}	0.29	0.22	0.14	0.17	0.16	0.14	
	压缩模量	$Es_{0.1-0.2}$	MPa	5.43	8.11	13.71	11.74	12.59	13.24	
	内聚力	c	kPa	5.0	3.0	3.0	3.5	3.0	4.0	
	内摩擦角	ϕ	(°)	20.0	31.0	34.2	25.0	27.0	26.0	
	标准贯入	击	次	22.0	26.0	35.7	25.9	45.0	48.0	
	比贯入阻力	ps	MPa	4.85	5.02	5.88	5.68	9.67	9.95	

七、第⑧层

（一）分布特征

该层为滨海相沉积和泛滥平原沉积，西部湖沼平原区（第Ⅰ工程地质结构区）和河口砂岛区（第Ⅲ工程地质结构区）分布较为广泛，滨海平原区（第Ⅱ工程地质结构区）西部和原南汇东部均有分布，浦东西部、中心城南部、奉贤东部和原南汇西部缺失。

该层中有上海的第二硬土层（$⑧_{2-1}$），仅在青浦西部和金山西部有分布，东部地区仅零星分布（图 28-6）。

（二）埋深、厚度变化特征

该层埋深总体由西部向东部逐渐变深，西部湖沼平原区和嘉定、宝山、普陀和浦东北部部分地区层顶标高一般为-8～-24m，“冈身带”一般为-24～-40m，中心城、浦东、金山和原南汇部分地区埋深最大，达-40～-64m，三岛区埋深变化不大，一般为-48～-56m。厚度上，西部湖沼平原区最厚，一

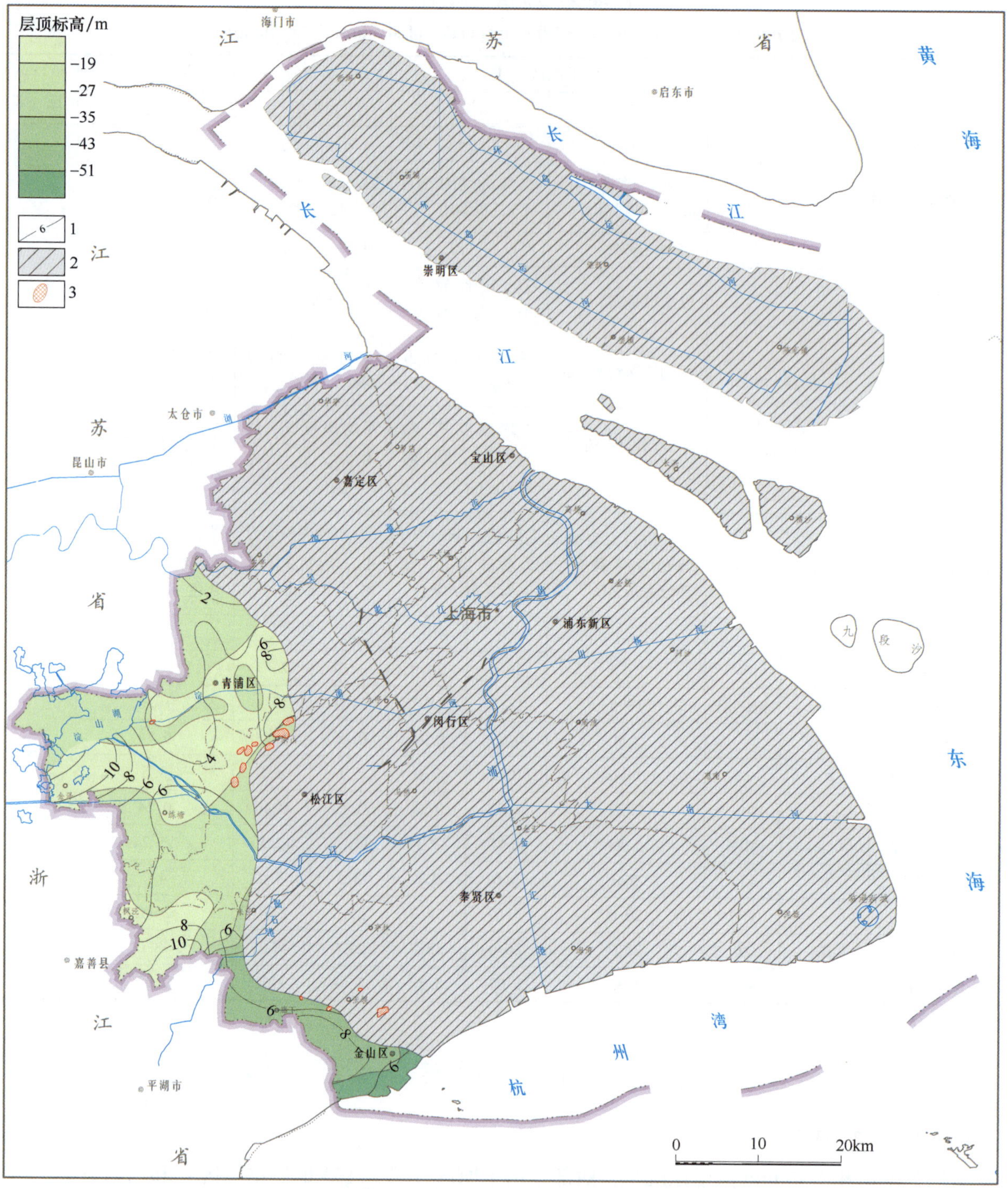

图 28-6　上海第⑧$_{2-1}$ 工程地质层（第二硬土层）埋藏分布示意图

Fig. 28-6　Buried depth distribution of engineering geological stratum⑧$_{2-1}$ (second hard soil layer) in Shanghai

1—层厚等值线（m）；2—缺失区；3—基岩区

般均大于 36m，最厚达 68m，"冈身带"和宝山西部地区为 20~36m，至滨海平原区东部和河口砂岛区逐渐变薄，一般为 4~20m。

该层中的第二硬土层（⑧$_{2-1}$）在上海西部埋深一般均小于 28m，而东部零星分布地区则变化最大，最深达 60m，厚度则变化不大，一般为 4~10m。

（三）岩性变化特征

该层由于沉积成因较为复杂，岩性变化最大，尤其在垂向上，岩性以黏性土为主，夹有较多粉土层和砂层，局部地区呈互层状，从上至下岩性依次为：灰色粉质黏土夹薄层粉砂层（$⑧_1$），第二硬土层$⑧_{2-1}$暗绿—褐黄色硬土，褐黄—灰色砂质粉土、粉砂（$⑧_{2-2}$），灰色粉质黏土夹粉砂层（$⑧_{2-3}$），第三硬土层$⑧_{3-1}$暗绿—褐黄色硬土，褐黄色砂质粉土、粉砂层（$⑧_{3-2}$），蓝灰色粉质黏土夹粉砂层（$⑧_{3-3}$），灰—灰黄色粉质黏土夹粉砂层（$⑧_4$）。其中$⑧_1$层为滨海相沉积地层，岩性上部一般为黏土，下部为粉质黏土，三岛区和长江沿岸局部地区由于长江古河道切割一般均缺失；$⑧_{2-1}$层为上海第二硬土层，为湖沼相沉积，岩性多为粉质黏土，且主要分布在西部，其下部$⑧_{2-2}$层为河流相沉积，湖沼平原区岩性以砂质粉土为主，夹有粉质黏土薄层，局部为粉砂，而三岛区则以粉砂为主，局部为砂质粉土，颜色变化为灰色，滨海平原区该层一般均缺失，下部$⑧_{2-3}$层在湖沼平原区以粉质黏土夹粉砂为主，而在滨海平原区则以粉质黏土和粉砂互层为主，三岛区一般缺失；$⑧_3$层岩性变化特征与$⑧_2$层基本一致，只是该层仅分布在湖沼平原区，滨海平原区和三岛区均缺失，且颜色以灰黄色、蓝灰色为主，土性较$⑧_2$层好。

（四）工程地质层评价

$⑧_1$层以黏性土为主，其含水量为31.5%，孔隙比为0.91，液性指数为0.80，压缩系数为0.41MPa^{-1}，压缩模量为5.47MPa，内聚力为18kPa，土性一般，且土质不均。由于上海桩基持力层一般选择为⑦层，因此$⑧_1$层为桩基的主要压缩层，应注意由该层引起的桩基沉降和差异沉降问题。

$⑧_{2-1}$层为第二硬土层，其性质同第一硬土层，其含水量为29.8%，孔隙比为0.86，液性指数为0.36，压缩系数为0.25MPa^{-1}，压缩模量为7.36MPa，内聚力为51kPa，与第一硬土层基本相近，在西部湖沼平原区该层可与下部$⑧_{2-2}$层作为大型建筑物的桩基持力层。$⑧_{2-3}$层土性较好，中等压缩性，滨海平原区该层分布相对较为稳定，在第⑦层缺失时，可选择该层作为大型和超大型建筑物的桩基持力层。

八、第⑨层

该层为上海的深部砂层，普遍分布，埋深亦较为稳定，一般为65~70m，仅局部地区埋深超过80m。由于埋藏较深，本次所用钻孔大部分未揭穿该层，根据揭穿钻孔揭示，该层厚度一般为15~20m。

该层岩性以砂土为主，从上至下颗粒逐渐变粗，上部一般为粉砂层，中部一般为细砂层（$⑨_1$），往下逐渐过渡为中粗砂（$⑨_2$），底部含有砾石。该层土含水量低，压缩性低，强度高，土性好，为超大型建筑物理想的桩基持力层。

第四节　主要工程地质问题和防治措施

一、地基变形（软土、泥炭土）

上海滨海平原区软土层（第③层淤泥质粉质黏土、④层淤泥质黏土）发育，层顶标高一般在3.5~-1m之间，临港新城埋藏较深，在-1~-9m之间，此外在三岛区埋藏最深，在-9~-16m之间。厚度变化不较大，三岛区一般为1~6m，其余大部分地区为6~14m，浦东部分地区厚度最大，在14~20m之间。在湖沼平原区，特别是I_1亚区，浅部普遍分布有泥炭土，层顶标高一般在1.0~0.5m，层厚一般为0.3~0.6m。根据指标可看出，软土和泥炭土均具有高含水量、大孔隙比、低强度、高压缩性、低渗透性等特点，此外，软土还具有触变性和流变性等不良工程地质特征（史玉金，2009），对工程建设产生不同程度的不良影响。主要表现在：

（一）对地面建（构）筑物的影响

对于采用天然地基的建（构）筑物，在外荷作用下地基压缩变形量大，易产生较大地基沉降和不均匀沉降。而且由于软土为弱透水性土层，排水固结缓慢，沉降会延续相当长的时间。

（二）对地下空间开发的影响

由于软土层抗剪强度低，具有明显的流变特征，基坑开挖时破坏了土体的应力平衡，软土在水、土压力及坑边堆载等外荷作用下易产生侧向变形和剪切破坏，导致支护结构变形或边坡失稳。因此，当基坑开挖施工时，应采取措施，防止坑底土体扰动。

对于天然地基工程应选择合适的基础埋深或采取适当的地基处理措施以减轻地基变形的不利影响；对于基坑工程，应选择合适的基坑围护和基坑开挖施工方案，确保围护墙接缝防渗效果，防止漏水及水土流失。减小地面超载，避免基坑周边振动荷载。

二、砂土液化

（一）砂土震动液化

上海埋深20m以浅的砂土、粉土主要有②$_3$、⑤$_2$、⑦层，其中20m以浅的第⑦层主要分布在西部湖沼平原区，为更新世地层，根据现有规范判定为不液化土层。此外，③层中的透镜体③$_a$层和④层中透镜体④$_a$层也会产生震动液化问题。第②$_3$层岩性主要为砂质粉土，部分地区为粉砂，埋深总体变化不大，层顶标高一般在2m左右，但厚度变化较大，在1~27m之间。三岛区厚度最大，一般均大于11m；滨海平原区南汇嘴地区厚度一般在7~15m之间，主要集中在临港新城规划区内；吴淞江、黄浦江两岸厚度亦较大，一般为6m左右；其余地区厚度一般小于5m。该层在上海一般为可液化土层，液化程度轻微—中等，因此工程建设中应对该层的震动液化问题予以重视。③$_a$层、④$_a$层为透镜体，分布不连续，③$_a$层顶埋深一般在5m左右，④$_a$层顶埋深一般在10m左右，该两层大部分地区岩性以砂质粉土为主，为可液化土层，工程勘察时应对其液化程度进行详细判别。⑤$_2$层分布不连续，为全新世地层，埋深、厚度变化大，滨海平原区层顶标高一般为-14~-28m，厚度为2~8m，根据其指标特征，该层在部分地区亦为可液化土层，在该区进行工程施工时应对其液化程度进行判别，并采取相应的防治措施。三岛区埋深一般均大于20m，因此一般无震动液化问题。

（二）砂土渗流液化

可能产生渗流液化（俗称流砂）的土层主要为砂性土和粉性土，上海工程建设影响范围内可能产生流砂的层次为②$_3$、⑤$_2$、⑦层。其中⑤$_2$层和⑦层分别是上海的微承压含水层和第Ⅰ承压含水层，水头较高，基坑开挖面或盾构施工时易形成较大的水头差，更易诱发流砂灾害。另外，对③$_a$层和④$_a$层的渗流液化问题亦应引起重视。因此，在地下空间开发时应特别重视上述土层。

对于砂土的震动液化，可在查明浅部砂、粉性土的分布范围与深度，并详细判别其液化可能性及场地液化等级的情况下，采取适当的抗液化措施进行防治；对于地下空间开发过程中的渗流液化，可根据情况采取相应的止水及降、排水等措施，确保基坑隔水体系的施工质量，提高围护结构接缝部分的止水效果，避免流砂现象的产生，发现渗水应及时堵漏。

三、水土突涌

上海地下水位高，当基坑下有承压水存在时，开挖基坑减小了含水层上覆不透水层的厚度，当它减小到一定程度与承压水的水头压力不能平衡时，基坑底板被顶裂或冲毁，出现突涌的现象。上海可能产生水土突涌的土层主要为砂性土和粉性土，在浅层和中层地下空间开发过程中可能产生突涌的层次主要为⑤$_2$、⑦层。⑤$_2$层和⑦层分别是上海的微承压含水层和第Ⅰ承压含水层，水头较高，基坑开挖面或盾构施工时易形成较大的水头差，易诱发水土突涌问题。因此，在地下空间开发时应特别重视上述土层。

为防止水土突涌事故发生，应加强勘察工作，查明承压含水层的埋藏分布特征及水文地质条件；设计、施工时应酌情选择降水方案，以满足抗突涌要求，并加强基坑施工过程中承压水水位监测工作；

做好勘探孔、降水孔的封孔工作，确保围护结构施工质量，发挥其对承压水的阻水功能。

四、边坡失稳

上海浅部砂层及软土层均比较发育，在基坑开挖过程中易产生流砂及软土剪切引起的边坡失稳问题。对于底部位于软土层的基坑，应注意软土对基坑边坡稳定性的影响，同时软土抗剪强度低、具有较明显的触变性和流变性，在基坑工程施工过程中，在震动和坑边堆土等上部荷载作用下易产生侧向变形、基坑周围地面沉降及坑底隆起等现象，导致基坑和支护结构变形，严重时会因软土剪切破坏而导致边坡失稳，应采取适当的基底加固措施。对于揭露砂层或粉土层的基坑，应注意由于流砂所导致的边坡失稳，做好加固措施。

五、潜水腐蚀性

潜水含水层为全新世（Qh）河口—滨海相沉积，潜水含水层底面埋深为3~25m，厚度介于2.5~24m之间，长江河口砂岛区厚度较大。根据含水层岩性结构特征可分为两类，一类以单一黏性土（A_1）为介质的孔隙水，全市普遍分布，岩性以填土、褐黄色粉质黏土为主；另一类上部为黏性土（A_1），下部以砂性土（A_2）为介质的孔隙水，A_2层分布范围有限，与工程地质②$_3$层相当，主要分布在滨海平原区以及长江河口沙岛区。潜水含水层富水性差，单井涌水量为1~20m^3/d（口径500mm，降深2m时）。潜水地下水系统天然防护条件较差，直接与大气圈、水圈相接触，易受外界影响。

潜水水位埋深一般离地表面0.3~1.5m，受降水、潮汛、地表水的影响有所变化，年平均水位埋深一般为0.5~0.7m。根据上海经验，当不存在环境水污染源时，潜水一般对混凝土有微腐蚀性；当长期浸水时，潜水对混凝土中的钢筋有微腐蚀性；当干湿交替时，对混凝土中的钢筋有微或弱腐蚀性；潜水对钢结构有弱腐蚀性。

第五节　地面工程建设适宜性

一、天然地基

上海表层一般均分布有②$_1$褐黄色黏性土层，俗称“硬壳层”，一般可作为天然地基持力层，但由于其下部土层具有一定变化，因此天然地基条件存在有一定差异。本次天然地基工程建设适宜性分区主要依照上海表土层（②$_1$）下卧层的不同岩性进行分区（图28-7）。

（一）天然地基条件较好区（A）

“冈身带”、滨海平原区沿江、沿海和河口砂岛区表土层下部为浅部砂层（②$_3$），天然地基条件较好，划分为A区。A区主要分布在河口砂岛区和滨海平原区“冈身带”、临港新城西部和黄浦江、吴淞江两岸，表土层均有分布，其下为②$_3$层灰色砂、粉性土，层顶标高一般为0~2m，厚度变化较大，可与表土层联合作为天然地基持力层，因此天然地基条件较好。但应注意部分地区②$_3$层液化问题，如判别为液化土层，则不宜直接作为天然地基持力层。

（二）天然地基条件一般区（B）

滨海平原区和湖沼平原区东部地区表土层下部为软土层（③或④层），天然地基条件一般，划分为B区。B区广泛分布于滨海平原区，表土层下部分布有厚度较大的软土层（③层淤泥质粉质黏土、④层淤泥质黏土），其厚度在“冈身带”以西一般小于8m，而滨海平原区厚度一般大于8m，由于软土层含水量高、强度低、压缩性高，在附加荷载作用下易产生变形，因此该区为天然地基条件一般区。

（三）天然地基条件较差区（C）

湖沼平原区西部表土层下有泥炭层分布，天然地基条件较差，划分为C区。C区分布于青浦、金

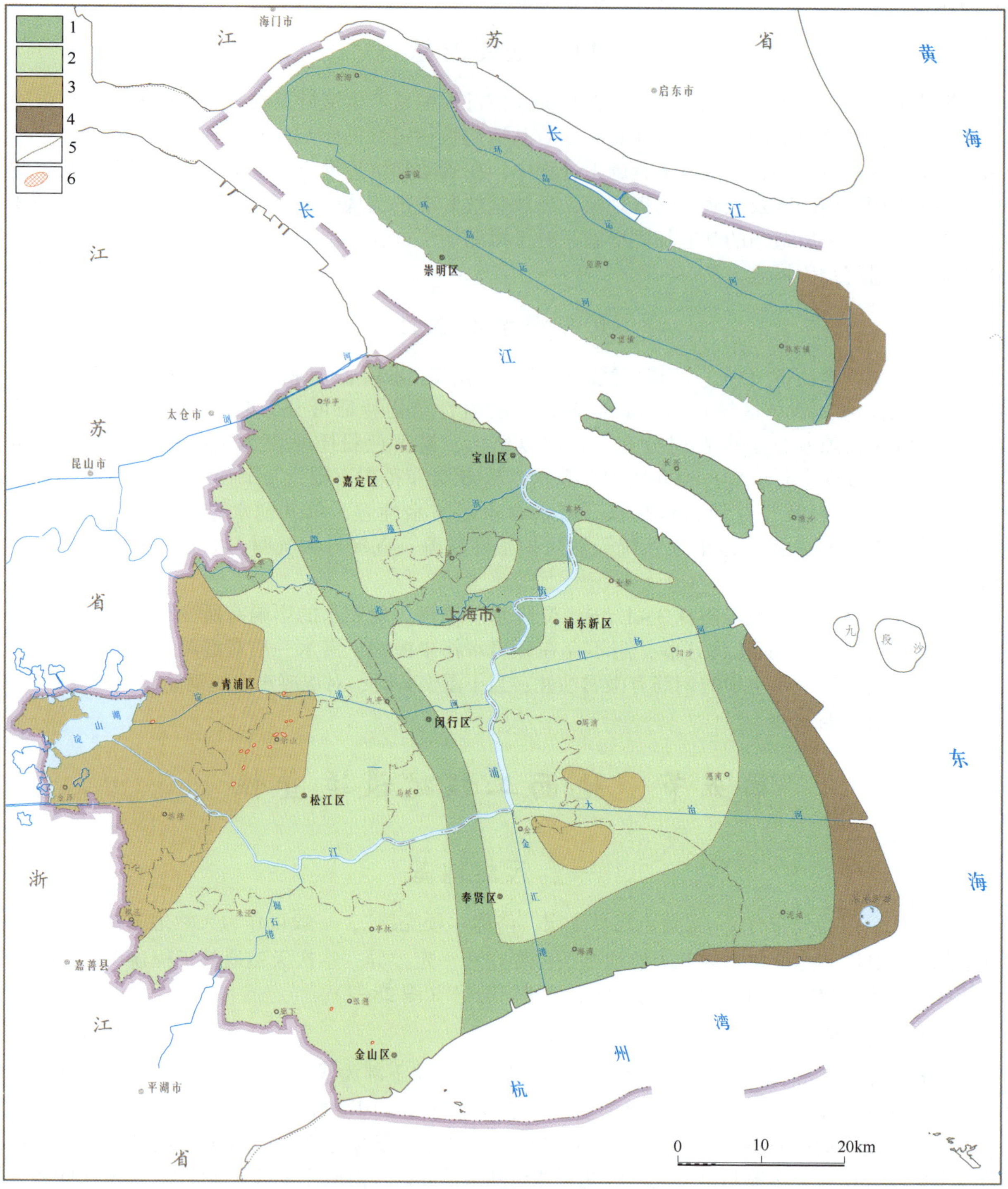

图 28-7　上海天然地基条件分区图

Fig. 28-7　Zoning map of natural foundation conditions in Shanghai

1—天然地基条件好区；2—天然地基条件一般区；3—天然地基条件较差区；4—天然地基条件差区；5—分区界线；6—基岩区

山和松江西部地区，表土层下部分布有泥炭层，厚度一般在 0.3~0.5m 之间，土性极差，易引起地基变形，工程建设时必须进行处理，因此天然地基条件较差。

（四）天然地基条件最差区（D）

沿海地区如临港新城东部、崇明东滩地区无表土层分布，大量分布有欠固结的冲填土层，不能作为天然地基持力层，天然地基条件最差，划分为 D 区。D 区主要分布在新近成陆区，主要有临港新城东部和崇明东滩地区。缺失表土层，广泛分布有冲填土层，岩性一般为砂、粉性土或淤泥质粉质黏土，其形成时代一般小于10年，处于欠固结状态，如不进行处理，不宜作为天然地基持力层，因此该区为天然地基条件最差区。

二、桩基

（一）桩基持力层分布及其稳定性

1. 中部砂层⑤$_2$层

该层灰色砂质粉土、粉砂分布不稳定，因此一般不推荐作为桩基持力层。但在河口砂岛区，由于受到古河道切割第⑥、⑦层均缺失，下部⑧$_{2-2}$层埋藏深，因此在该区内的部分地区如果⑤$_2$层分布稳定，且厚度适中，可作为中型或大型建筑物的桩基持力层。

2. 第⑥层暗绿—褐黄色硬黏性土层

该层土性较好，岩性以黏性土为主，但其埋深变化较大。在西部湖沼平原区和嘉定西部埋深一般为3~5m，埋藏太浅，不能作为桩基持力层，“冈身带”和宝山南部埋深在12~16m之间，其下卧层⑦层粉土、砂土厚度不大，可联合作为一般建筑物的桩基持力层，但不能做中型或大型建筑物桩基持力层。东部滨海平原区埋深则为22~28m，埋藏适中，但厚度较小，一般为3~5m，一般也不作为桩基持力层。

3. 第⑦砂层

该层岩性以砂质粉土、粉砂为主，属低压缩性土层，土质好，其分布与上覆⑥层土密切相关，因此其埋深、厚度亦有一定变化。

湖沼平原区该层厚度较小，且西部缺失，因此一般不宜作为桩基持力层。“冈身带”和宝山南部埋深为15~23m，厚度一般为2~10m，其下部分布有较厚的⑧$_1$层高压缩性土层，易引起桩基沉降，因此该地区该层只能作为一般建筑物桩基持力层。

东部滨海平原区该层由于受到古河道的切割，条件也有所不同，正常沉积区埋深一般为28~31m，厚度一般大于30m，尤其在浦东西部、中心城南部、奉贤东部和原南汇西部其下不压缩层缺失，厚度更大，该层上部为砂质粉土（⑦$_1$层），下部为粉砂（⑦$_2$层），土层中密—密实，平均标准贯入击数大于50击，为良好的桩基持力层，可作为高层建筑、大型和特大型桥梁以及其他重型、超重型建（构）筑物的良好桩基持力层。但在滨海平原区的古河道切割区，第⑦层埋深在35~55m不等，厚度也有较大差异，因此桩基条件差异较大，且基础投入较多。

4. 第⑧$_2$层粉质黏土夹粉土、粉砂层

由于沉积成因较为复杂，该层不同地区岩性差别较大，其桩基条件也有一定差异。

湖沼平原区和嘉定西部该层顶部为第二硬土层⑧$_{2-1}$暗绿—褐黄色粉质黏土，埋深为15~28m，埋藏适中，其下部即为⑧$_{2-2}$灰黄—灰色的砂质粉土、粉砂层和⑧$_{2-3}$灰色粉质黏土夹粉砂层，土质均较好，因此⑧$_{2-1}$层或⑧$_{2-2}$层可以作为桩基持力层。

滨海平原区该层顶部⑧$_{2-1}$和⑧$_{2-2}$一般均缺失，分布有⑧$_{2-3}$层灰色粉质黏土夹粉砂层，部分地区为粉质黏土与粉砂互层，其工程地质特性较好，中等压缩性，亦可作为大型或特大型建筑物桩基持力层。但由于滨海平原区该层埋藏较深，一般不作为桩基持力层。

河口砂岛区⑧$_2$层主要以⑧$_{2-2}$层灰色粉砂为主，埋深一般为50~60m，无软弱下卧层，可作为大型或特大型建筑物桩基持力层，但由于埋藏较深，基础投入大。

5. 第⑨层深部砂层

该层分布稳定，厚度大，属低压缩性土，可作为特大型建筑物桩基持力层。但由于其埋深过大，一般不推荐作为桩基持力层。

综上所述，上海可选择作为桩基持力层土层主要有⑥、⑦、⑧$_{2-1}$和⑧$_{2-2}$层，其分布特征与硬土层的缺失情况及埋深变化特征有密切关系。

（二）桩基工程适宜性分区

1. 大区划分

按照有两层硬土层（湖沼平原区西部）、⑥层（湖沼平原区东部和滨海平原区的正常沉积区）和无硬土层（古河道切割区）进行大区划分，有⑥、$⑧_{2-1}$ 同时存在的为 A 区，可选择的桩基持力层较多，且分布稳定，有⑥无$⑧_{2-1}$ 分布的为 B 区，桩基条件有一定变化，无硬土层分布的为 C 区（图 28-8；表 28-10）。

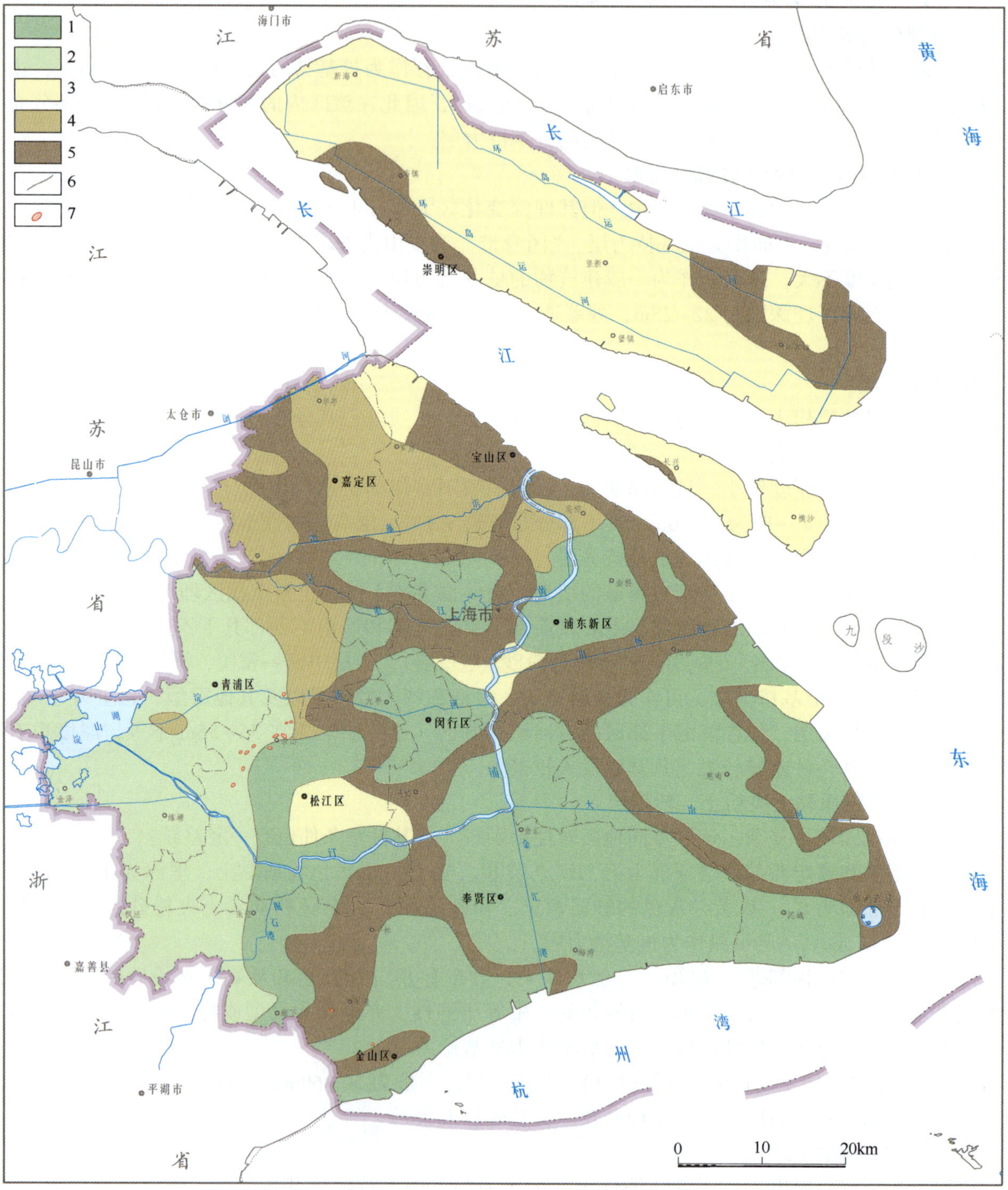

图 28-8　上海桩基工程条件分区图

Fig. 28-8　Zoning map of pile foundation engineering conditions in Shanghai

1—桩基条件好区；2—桩基条件较好区；3—桩基条件一般区；4—桩基条件较差区；5—桩基条件差区；6—分区界线；7—基岩区

表 28-10　上海桩基工程建设适宜性分区及评价

Table 28-10　Suitability zoning and assessment of pile foundation projects in Shanghai

区	适宜性较好区（A）	B		C	
亚区		适宜性好区（B_1）	适宜性较差区（B_2）	适宜性一般区（C_1）	适宜性差区（C_2）
分布地区	青浦、金山及松江西部	上海市区、奉贤、金山、原南汇正常沉积区	宝山、嘉定局部地区	崇明、长兴、横沙三岛区	滨海平原大部分古河道切割区及河口砂岛部分地区
桩基持力层分布特征	有第一硬土层（⑥）、第二硬土层（⑧$_{2-1}$）层分布，下部⑧$_{2-2}$层分布稳定	有⑥层硬土层和⑦层分布，⑦层厚度一般均大于 7m，⑥层层顶标高<-17m	有⑥层硬土层和⑦层分布，⑦层厚度一般均小于 7m，⑥层层顶标高>-17m	无硬土层分布，⑤$_2$层大部分地区分布较为稳定，厚度适中	无硬土层分布，⑤$_2$层分布无规律，大部分地区厚度较小
建筑适宜性评价	地层分布稳定，可选择⑧$_{2-2}$层作为大型或特大型建筑物桩基持力层，且下部土层⑧$_{2-3}$层或⑧$_3$层土性好，产生桩基沉降及不均匀沉降的可能性较小	地层分布稳定，⑦层埋深适中，且厚度大，可作为大型或特大型建筑物桩基持力层，产生桩基沉降的可能性较小	地层分布稳定，⑥层埋藏较浅，下部⑦层相应也较浅，且厚度较小，易产生桩基沉降	⑤$_2$层可作为桩基持力层，但应注意其埋深、厚度变化情况	⑥层、⑦层均缺失，较难选择桩基持力层，必须注意桩基沉降及不均匀沉降问题
地质灾害问题	适当注意区域地面沉降对桩基工程的影响	应特别注意区域地面沉降对桩基工程的影响			应注意区域地面沉降对桩基工程的影响

2. 亚区划分

亚区划分根据桩基持力层埋深、厚度变化情况进行划分。

A 区桩基条件差别不大，因此不进行亚区划分。

B 区则根据⑦层的厚度进行亚区划分，选择⑦层厚 7m 为界，大于 7m 的一般分布在湖沼平原区东部、滨海平原区东部、南部，⑥层顶板标高一般小于-17m，下部⑧$_1$层桩基压缩层埋深大，甚至缺失，桩基条件较好；而⑦层厚小于 7m 的一般分布在“冈身带”、嘉定西部和宝山西部，⑥层顶板标高大于-17m，桩基压缩层埋深浅，厚度大，桩基条件较差，如选择⑦层作为中长桩桩基持力层，则有可能产生较大的桩基变形。由此，在 B 区，⑦层厚度大于 7m 的为 B_1 区，桩基条件好，小于 7m 的为 B_2 区，桩基条件较差。

C 区由于均为古河道切割区，主要分布在河口砂岛区和滨海平原区的古河道切割区，缺失两层硬土层，亚区划分根据⑤$_2$层进行，河口砂岛区有该层分布区一般埋深、厚度较为稳定，划分为 C_1 区，桩基条件一般，滨海平原区的古河道切割区内仅局部地区⑤$_2$层分布稳定，已划分为 C_1 区，其余地区划分为 C_2 区，桩基条件差。

第二十九章 地面沉降

地面沉降是上海主要环境地质问题之一（魏子新，2010），已成为影响城市安全、制约经济发展的主要地质灾害之一。自1921年发现地面沉降以来，已有近百年的历史。从上海总体沉降格局来看，地面沉降漏斗区主要集中在老市区及浦东黄浦江沿线，最大累计沉降量达到3m左右，位于黄浦区西藏北路北京西路附近；另外西部青浦—虹桥也表现为较明显的地面沉降漏斗区。郊区由于监测较晚，累计沉降量一般小于200mm。

上海地面沉降特征与趋势是随着地下水开采与人工回灌格局的调整、城市发展进程而不断发展的，总体上可分为两个时期（图29-1；表29-1），即地面沉降严重发展时期（1921—1965年）、地面沉降

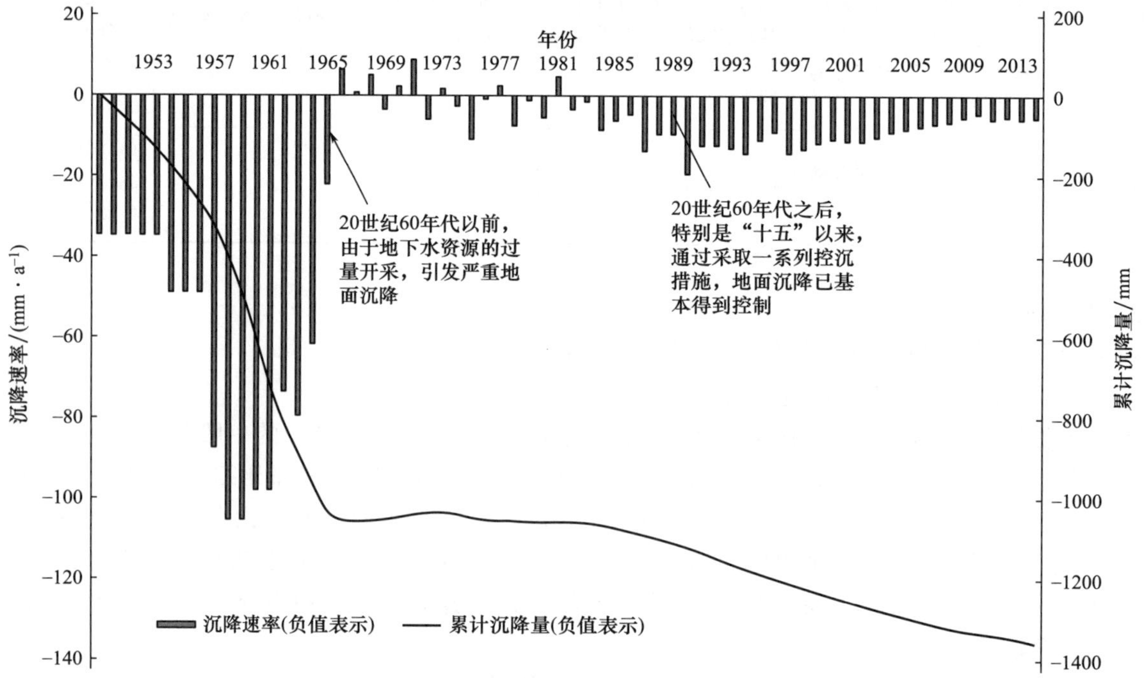

图29-1 地面沉降量历时特征曲线图

Fig. 29-1 Time history Characteristics of Land subsidence

表29-1 典型沉降阶段特征

Table 29-1 Characteristics of typical subsidence periods

地面沉降阶段		沉降特征	年平均沉降量/mm
地面沉降严重时期（1921—1965年）	沉降明显（1921—1948年）	在沪西的静安和黄浦出现两个沉降漏斗中心	22.8
	沉降加快（1949—1956年）	沉降范围继续扩大，在普陀、杨浦工业区出现了两个沉降漏斗中心；在徐家汇和南市城厢出现了新的沉降漏斗中心	40.3
	沉降剧烈（1957—1961年）	沪东、沪西两个沉降漏斗范围迅速向郊区扩展	98.6
	沉降缓和（1962—1965年）	经采取压缩用水措施，地面沉降减缓。西部沉降漏斗基本消失；东部沉降漏斗向东沟一带扩展	59.3

续表

地面沉降阶段		沉降特征	年平均沉降量/mm
地面沉降控制时期（1966—）	微量回弹（1966—1971年）	进一步采取地下水计划用水、人工回灌及调整开采层次等措施，在主要集中灌采区内地面出现微量回弹	-3.0
	微量沉降（1972—1989年）	地下水位呈现周期性变化，土层压缩量周期性地略大于回弹量。地面则呈微量下沉趋势，分布不均	3.5
	沉降加速（1990—2000年）	地下水开采量的急剧增加以及大规模集中工程建设，导致地面沉降速率加大	14.3
	沉降平稳（2001—2005年）	进一步采取人工回灌及调整开采层次等措施，前期年均地面沉降量稍有反复，后期保持持续下降，总体控制态势良好	10.6
	不均匀沉降凸显（2006至今）	区域平均总体沉降逐年明显下降，但不均匀沉降现象凸显	小于7

控制时期（1966年至今）。20世纪60年代前，由于工农业生产对地下水需求量较大，导致了严重的地面沉降，最大年沉降速率达到105mm/a。1966年以来上海中心城地面沉降监测资料显示，随着地下水开采与人工回灌格局变化和城市建设的发展，上海地面沉降在时间上、空间上呈现不断发展的态势。尤其是2006年以来，区域平均总体沉降逐年明显下降，年均沉降量已不足7mm，但不均匀沉降现象凸显。

地面沉降不仅降低了市区防洪能力，增大了防汛墙的压力，还加剧了市区潮灾、涝灾，破坏市政基础设施，危害城市的正常运行。

第一节　地面沉降监测网络

从1961年至今50余年的时间里，结合城市的发展，上海地面沉降监测范围从市区扩大到了全市，逐步形成了由精密水准监测网、地面沉降监测站、地下水动态监测网、重大基础设施地面沉降骨干监测网、GPS地面沉降监测网和InSAR技术地面沉降监测组成的地面沉降监测网络。

一、水准监测网

中心城是上海最早开展地面沉降监测的区域，由于地面沉降发育最为严重，历来是地面沉降重点监测防治区。到1961年，在当时经济能力许可条件下，建立了覆盖市区杨浦、黄浦、普陀等沉降漏斗区的地面沉降水准测线，同时在沉降漏斗分布区建设适量的基岩标、分层标，构成市区地面沉降水准监测网的雏形。20世纪80年代后，随着近郊工业用水量的大幅增加，地面沉降范围开始扩大，水准监测范围扩大至近郊的工业基地。2005年以来，为满足城市建设、规划对地质环境需求，水准监测网已全部覆盖了整个外环线以内的中心城（图29-2），监测网络由2006年以前的一个起算基准自由网，过渡至2008年后由多个基岩标组成的复合网，监测精度得到进一步的提高。

中心城地面沉降水准监测，监测频率为每年1次。监测方法以人工监测为主，兼顾自动化监测手段。

地面沉降对城市生命线工程（地铁、隧道、高架道路等）安全运行已构成隐患，为掌握轨道交通、高架道路等工程沿线地面沉降，及时为工程性地面沉降防治提供依据，以现有地面沉降监测网为基础，在工程沿线加密部署了水准点、分层标、基岩标等设施，形成既相对独立、又与中心城水准网

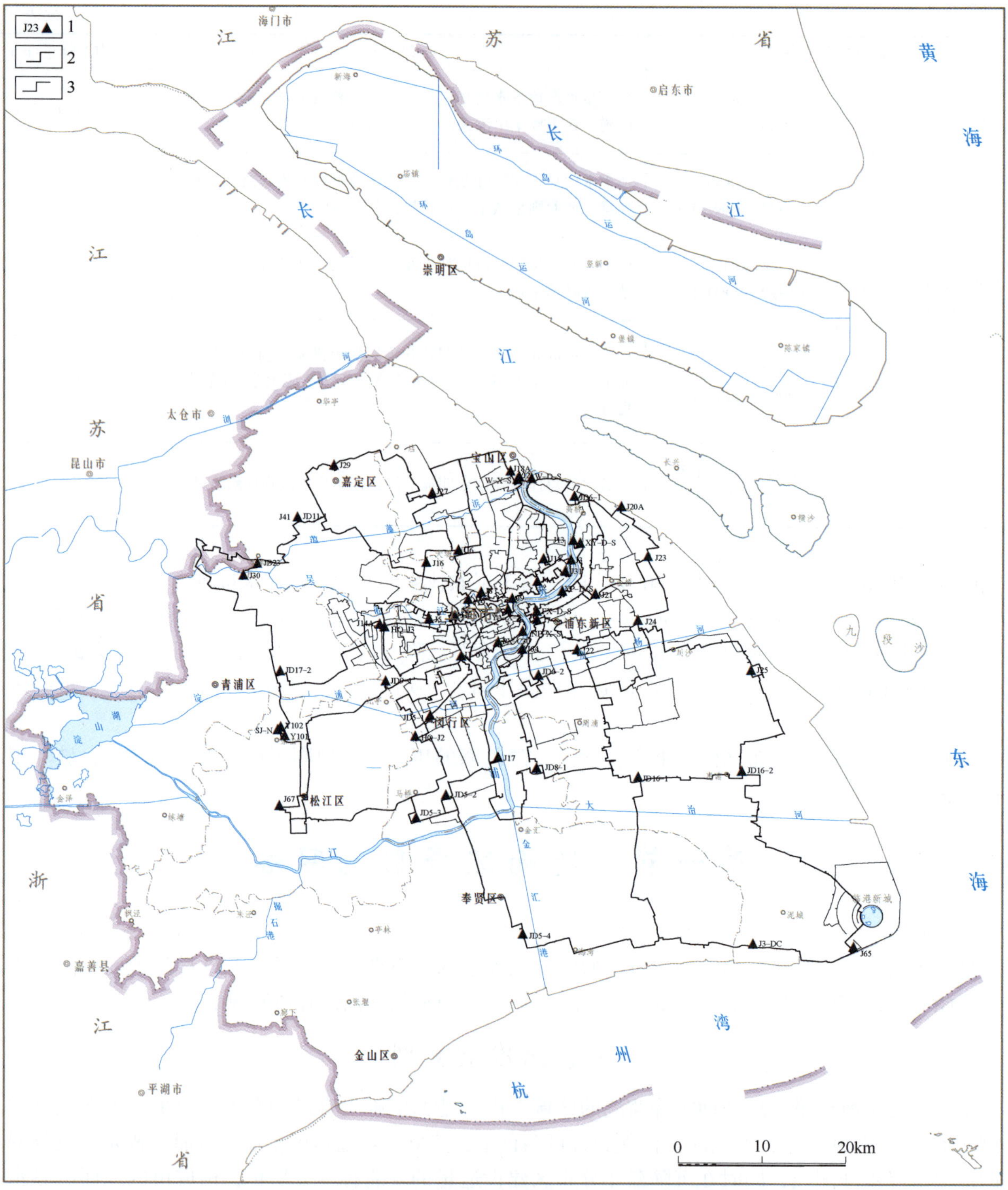

图 29-2　上海水准监测网现状分布图

Fig. 29-2　Distribution of the current level monitoring network in Shanghai

1—基岩标；2——等水准测量线路；3—二等水准测量线路

紧密联系的地面沉降生命线工程骨干网（图 29-3）。同样以人工监测和自动化监测相结合的方式进行。

全市地面沉降监测网由城市水准点、分层标、基岩标等监测设施构成。地面沉降监测站基本覆盖全市不同地质结构区（图 29-4），大大增强了全市及沿江沿海地面沉降实时监控力度。监测网络以佘山新基点为高程起算点，由小闸、北新泾、吴淞和塘桥 4 座基岩标为水准测量高程控制点，并组成水准测量附合网。

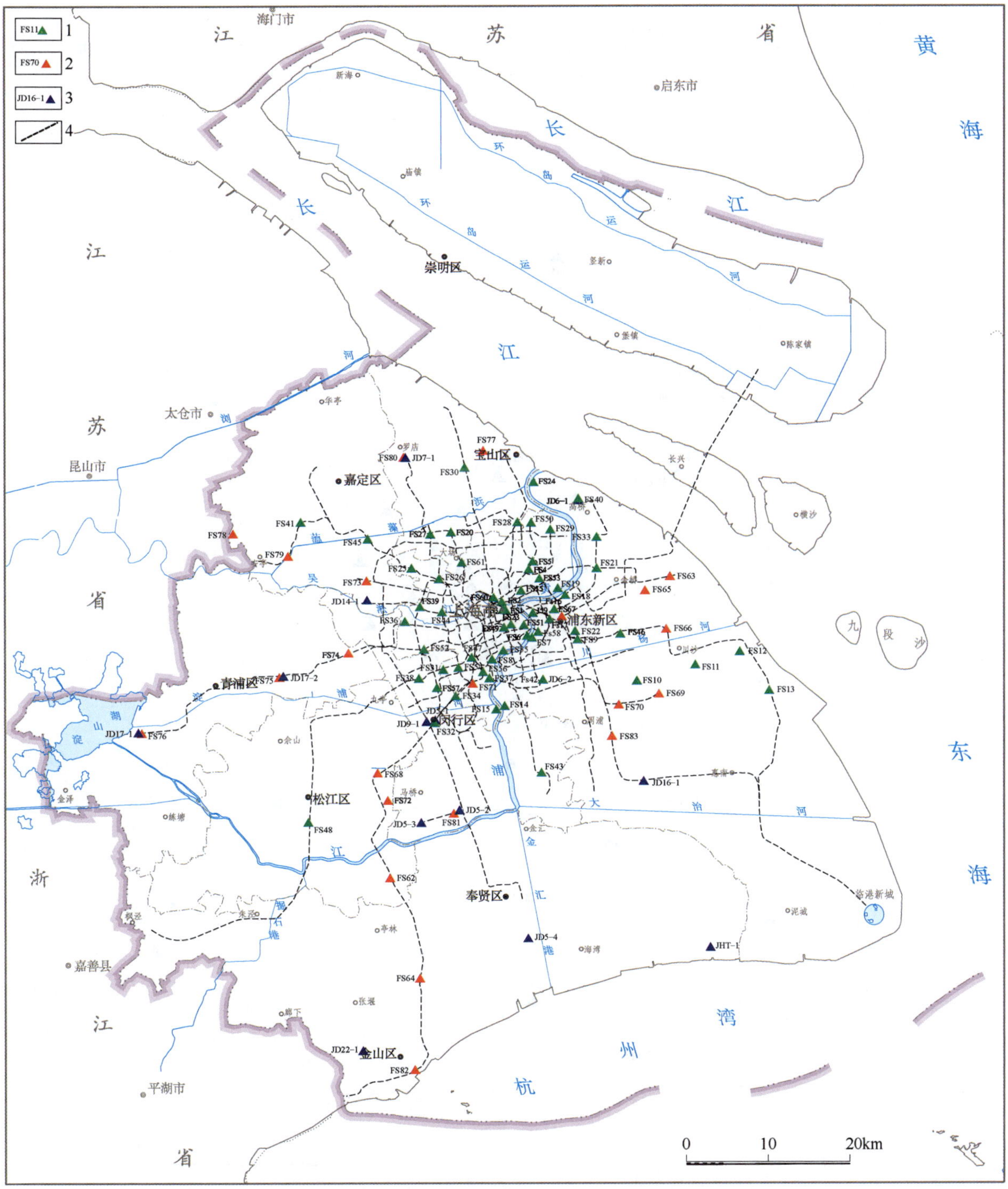

图 29-3　上海生命线工程地面沉降骨干监测网分布图

Fig. 29-3　Distribution of the backbone land subsidence monitoring network for lifeline projects in Shanghai

1—“十一五”及之前建设分层标组及编号；2—“十二五”新建分层标组及编号；

3—“十二五”新建基岩标组及编号；4—已有或规划地铁线

全市实施一、二等精密水准监测，监测频率为每 5 年 1 次。地面沉降监测站基本采用自动化监测方法，进行数据的实时采集和传输，大大提高了数据采集的实效性和快速反应能力。

二、GPS 监测网

2000 年时，上海以中心城杨浦为试验区，开展 GPS 监测技术应用试验研究，至 2005 年，全市基

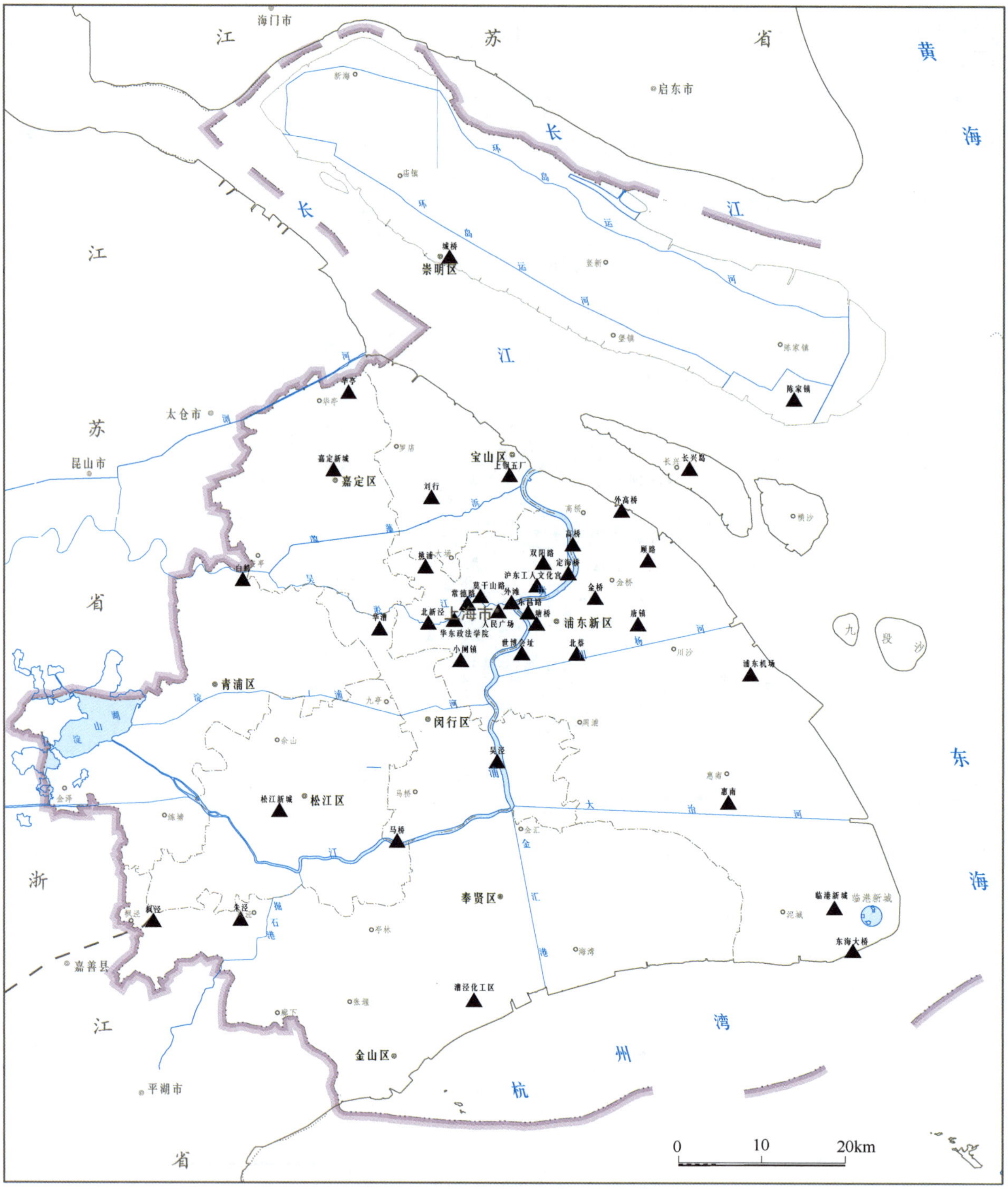

图 29-4　上海地面沉降监测站现状分布图

Fig. 29-4　Distribution of the current land subsidence monitoring stations in Shanghai

本形成由 4 座 GPS 基准站、33 座一级网点、110 座二级网点组成的 GPS 监测网。目前，形成由 4 座 GPS 监测站、55 座一级网点、165 座二级网点构成的 GPS 监测网络（图 29-5）。

三、InSAR 监测

上海市地质调查研究院于 2003 年与武汉大学合作，开展了 PS-InSAR 技术的研究试验。在多年研究试验的基础上，于 2013 年开始按年度进行 InSAR 监测，且年度监测时间和面积水准监测时间同步。

前期的研究主要利用中等分辨率 SAR 影像（如欧洲空间局卫星 ERS-1/2 和 ENVISAT 的 C 波段

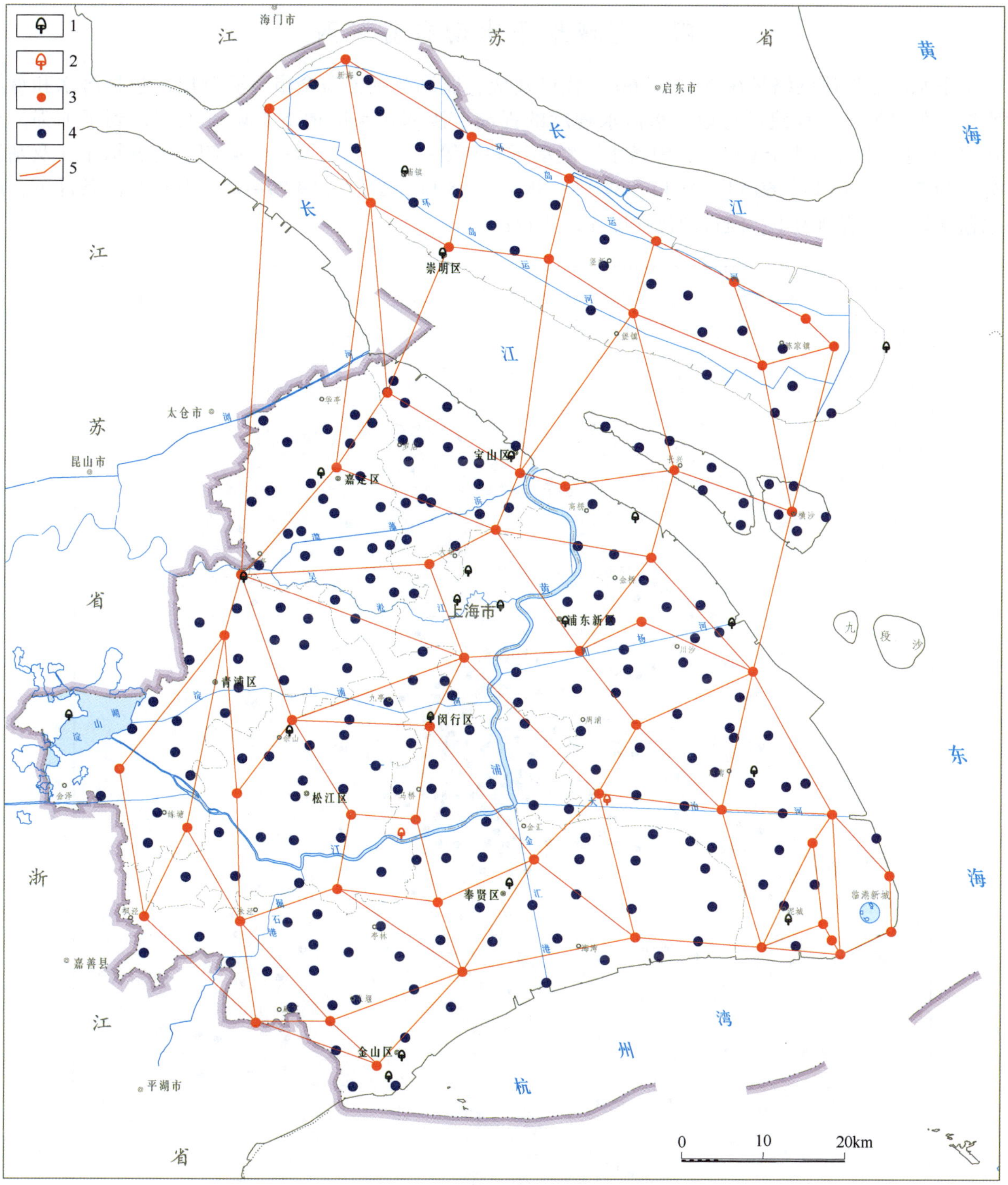

图 29-5 上海 GPS 监测网分布图

Fig. 29-5 Distribution of the GPS monitoring network in Shanghai

1—“十一五”及之前建设 GPS 固定站；2—“十二五”新建 GPS 固定站；3—“十一五”及之前建设 GPS 一级网点；4—“十一五”及之前建设 GPS 二级网点；5—GPS 一级网

SAR 影像)，后期采用了高分辨率 SAR 影像数据和 L 波段的长波长数据进行了大量的对比分析研究。2011 年开始使用具有更强识别能力，且对短周期微小形变更为敏感的德国卫星 TerraSAR-X（X 波段波长为 3.1cm）高分辨率 SAR 影像数据为数据源。

为了提高 InSAR 技术测量精度，2008 年布设了 8 个卫星同步监测水准点，监测频率与卫星过境时间相同。另外，在浦东机场地面沉降监测站内布设了 2 个卫星角反射器。

四、区域地下水动态监测网

地下水动态监测网络早在20世纪60年代已建成，监测范围覆盖全市，监测层次包括潜水和承压含水层；随着城市工程建设对地下水需求程度的提高，浅部含水层地下水调查工作得到了加强。至2016年，地下水监测井451口，其中潜水、微承压井和第一、第二、第三、第四、第五承压井及基岩井分别为50口、16口、60口、93口、87口、88口、55口和2口（图29-6）。现状经济条件下，地下水监测以人工监测为主，但自动化程度在逐步提高。

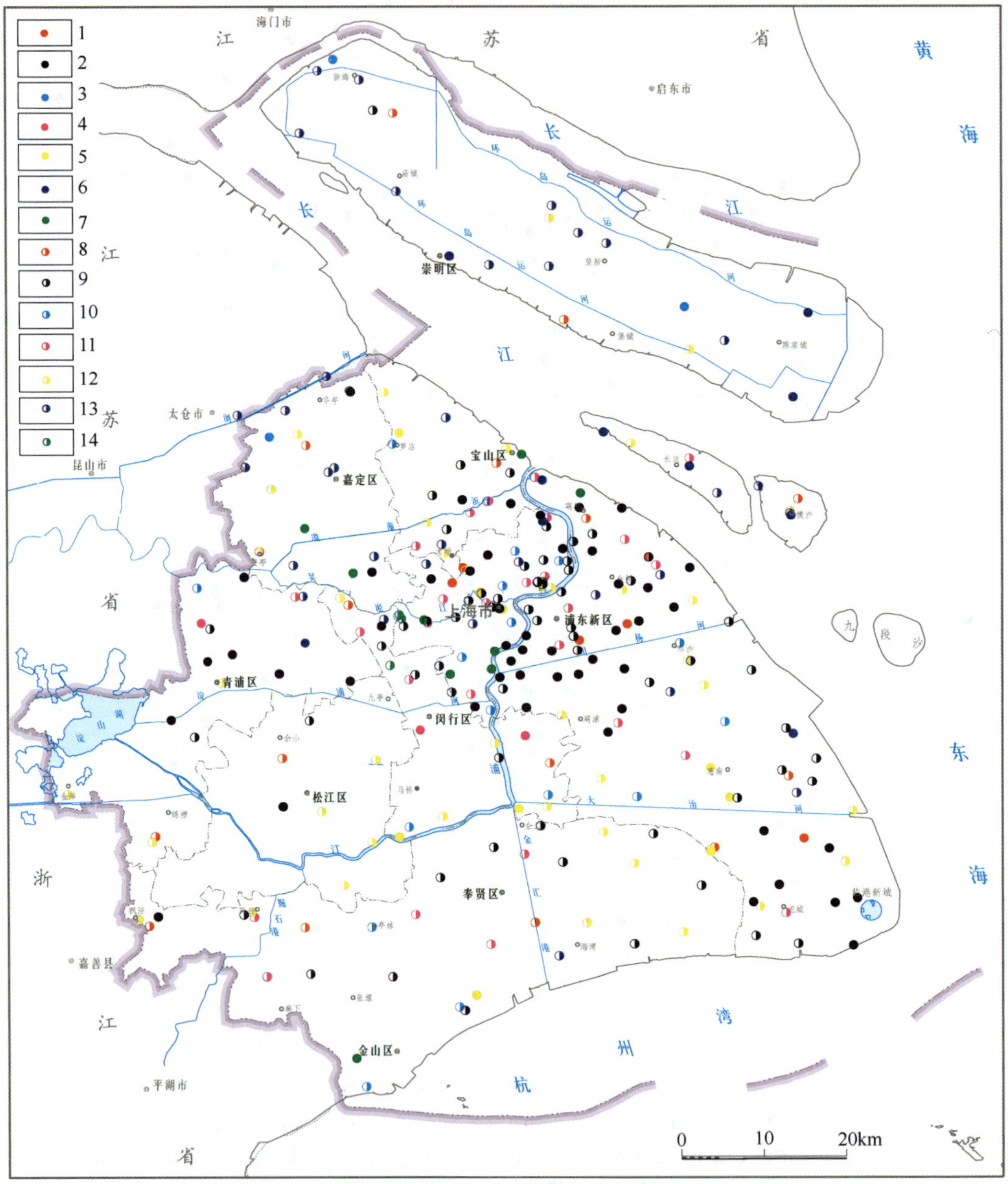

图29-6 上海地下水位监测井现状分布图

Fig. 29-6 Distribution of the current groundwater table monitoring wells in Shanghai

1—潜水含水层地下水监测井（2001年以前）；2—第Ⅰ承压含水层地下水监测井（2001年以前）；3—第Ⅱ承压含水层地下水监测井（2001年以前）；4—第Ⅲ承压含水层地下水监测井（2001年以前）；5—第Ⅳ承压含水层地下水监测井（2001年以前）；6—第Ⅴ承压含水层地下水监测井（2001年以前）；7—微承压地下水监测井（2001年以前）；8—潜水含水层地下水监测井（2001年以后）；9—第Ⅰ承压含水层地下水监测井（2001年以后）；10—第Ⅱ承压含水层地下水监测井（2001年以后）；11—第Ⅲ承压含水层地下水监测井（2001年以后）；12—第Ⅳ承压含水层地下水监测井（2001年以后）；13—第Ⅴ承压含水层地下水监测井（2001年以后）；14—微承压地下水监测井（2001年以后）

上述各种监测方法具有各自优缺点，其适用条件不一致，但具有互补性，通过统一参考基准、数据采集上的一致性与互补性、综合多种数据成图等措施实现了优势互补，提高了地面沉降监测的整体效能。

在地面沉降监测网络建设的同时，制订了加强地下水资源和工程活动管理等防治措施，发布了《上海市地面沉降防治管理办法》《上海市地面沉降防治管理条例》等法规，建立了上海市“两委两局”部门联动、重大市政工程设施沿线地面沉降监测与安全预警、长江三角洲地面沉降区域联防联控机制。

通过综合防治措施的实施，从近年来地面沉降发展趋势来看，地面沉降防治效果明显，地面沉降量及影响范围逐渐减小，但由于地下水开采与人工回灌格局发生了变化，尤其是叠加了深基坑降排水影响，使地面沉降格局更加复杂，中心城及近郊区形成了若干个地面沉降次中心，使地面沉降在空间上的差异性更加显著。

第二节　地面沉降特征

长期的地面沉降水准测量、GPS 测量资料显示，区内地面沉降发育特征如下。

一、全市地面沉降特征

（一）1980—1995 年

全市的地面沉降比较明显，沉降量超过 100mm 的地区有 484.7km^2，超过 200mm 的地区有 2.2km^2。市区是主要沉降漏斗区，市区地面沉降主要分布在黄浦、浦东陆家嘴等区域，向北延伸至虹口—杨浦，平均地面沉降速率为 5.6mm/a，最大沉降量为 200mm。市区西部闵行虹桥—华漕是另一个主要的沉降漏斗区，最大沉降量为 150mm。邻近江苏的嘉定娄塘，邻近浙江的金山枫泾地区，由于邻省地下水开采影响，地面形成了局部的地面沉降中心，嘉定娄塘最大累计沉降量达 200mm，金山枫泾地区最大累计沉降量为 175mm。崇明的地面沉降发展缓慢，平均地面沉降速率只有 3.7mm/a，主要分布在前哨农场和新海农场、红星农场，最大沉降量为 75mm（图 29-7）。

（二）1996—2001 年

全市地面沉降呈加速发展趋势，沉降速率超过 20mm/a 的地区有 1340.0km^2，其中超过 40mm/a 的地区约 127km^2。地面沉降中心影响范围明显扩大，沉降速率明显增加。市区的平均地面沉降速率达到 24.5mm/a，地面沉降影响面积扩大，在南浦大桥和高东镇出现了新的沉降漏斗，最大沉降量为 250mm。市区西部虹桥—华漕地面沉降中心范围向西部、北部扩展，与市区沉降漏斗形成“哑铃”形沉降中心，最大沉降量达 300mm，沉降速率达 60mm/a。邻近浙江省的枫泾地面沉降中心向西部、南部扩展，影响金山西北部地区，最大沉降量达到 200mm。另外，在靠近江苏的娄塘地面沉降中心向南扩展到华亭和罗泾，最大沉降量达到 250mm。全市其他地区沉降速率均有明显增加，如崇明的前哨农场加速沉降，平均地面沉降速率达到 10.1mm/a，最大沉降量达到 300mm（图 29-8）。

（三）2002—2006 年

与 1996—2001 年间相比，2002—2006 年全市地面沉降情况有进一步缓解，地面沉降速率超过 20mm/a 的地区减少至 185.0km^2，超过 40mm/a 的地区仅 13.7km^2。进入 20 世纪以来，由于杨浦、普陀桃浦、宝山大场等工业区的关停并转，地下水开采量随之锐减。

由于中心城地下水开采量的大幅压缩，以及地下水人工回灌主要集中在中心城，这一阶段，中心城的地面沉降基本得以控制，中心城的平均地面沉降速率为 12.7mm/a，中心城大范围地面沉降中心“解体”，分解为孤立分布的沉降点，如杨浦引翔港附近地面沉降影响范围不大，但是最大沉降量达 225mm；虹口凉城公园的沉降漏斗最大沉降量为 150mm。中心城西部虹桥—华漕地面沉降中心已基本

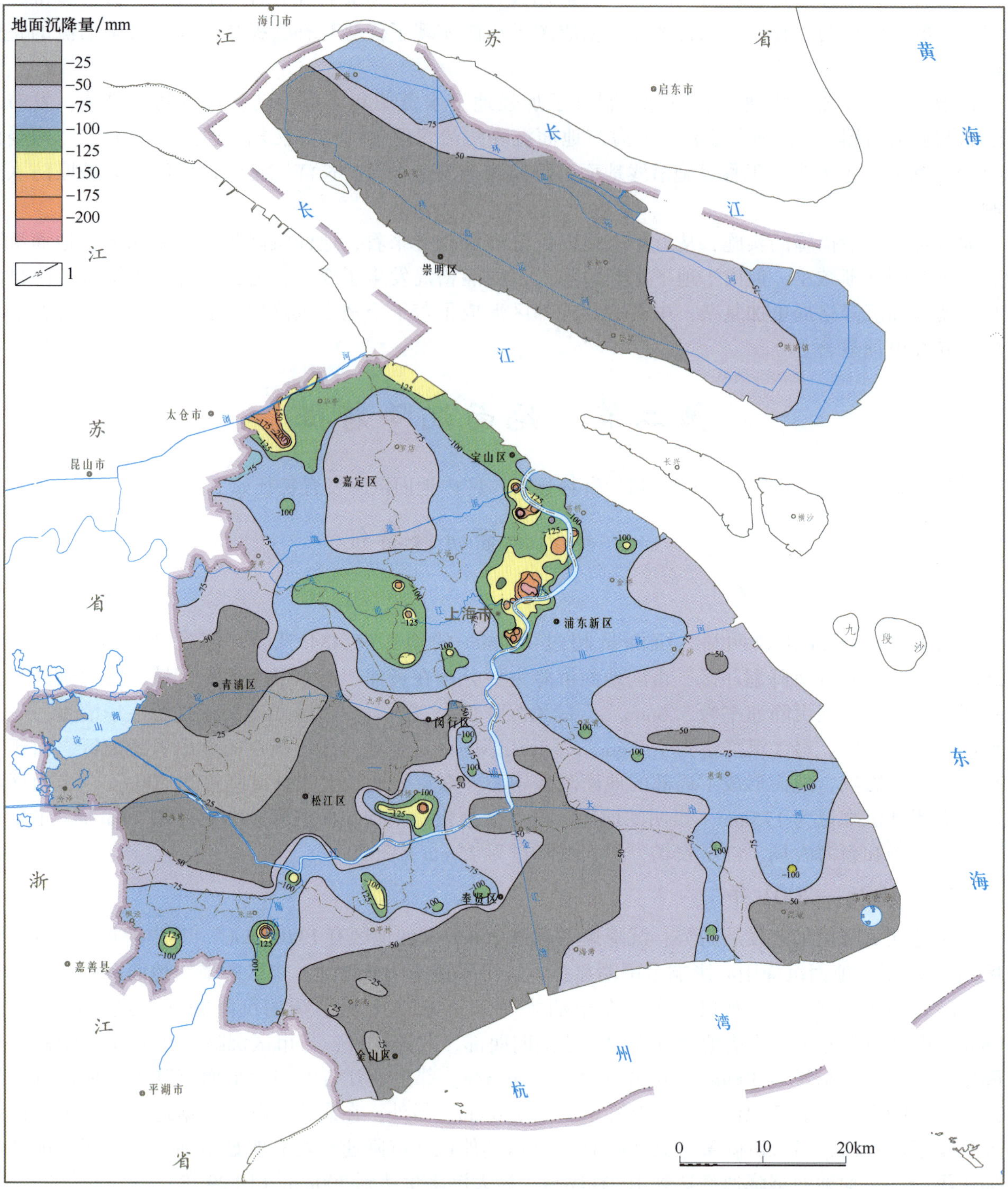

图 29-7　上海地面沉降等值线图（1980—1995 年）（水准复测）

Fig. 29-7　Land subsidence contour map of Shanghai (1980-1995) (leveling resurvey)

1—地面沉降等值线（mm）

消失，地面沉降速率大幅降低为 16.2mm/a。

受苏锡常地区实施地下水禁采措施的影响，与江苏邻近的娄塘一带沉降漏斗规模明显减小，而这一阶段，杭嘉湖地区地面沉降扩展范围和速度增加较快，至 2005 年已波及平原大部分地区，受其影响，邻近浙江省的枫泾镇地面沉降中心仍保持较快的沉降速率（30mm/a）。

崇明地面沉降的发展趋势基本保持不变，平均地面沉降速率为 5.9mm/a，长江农场和绿华镇一带

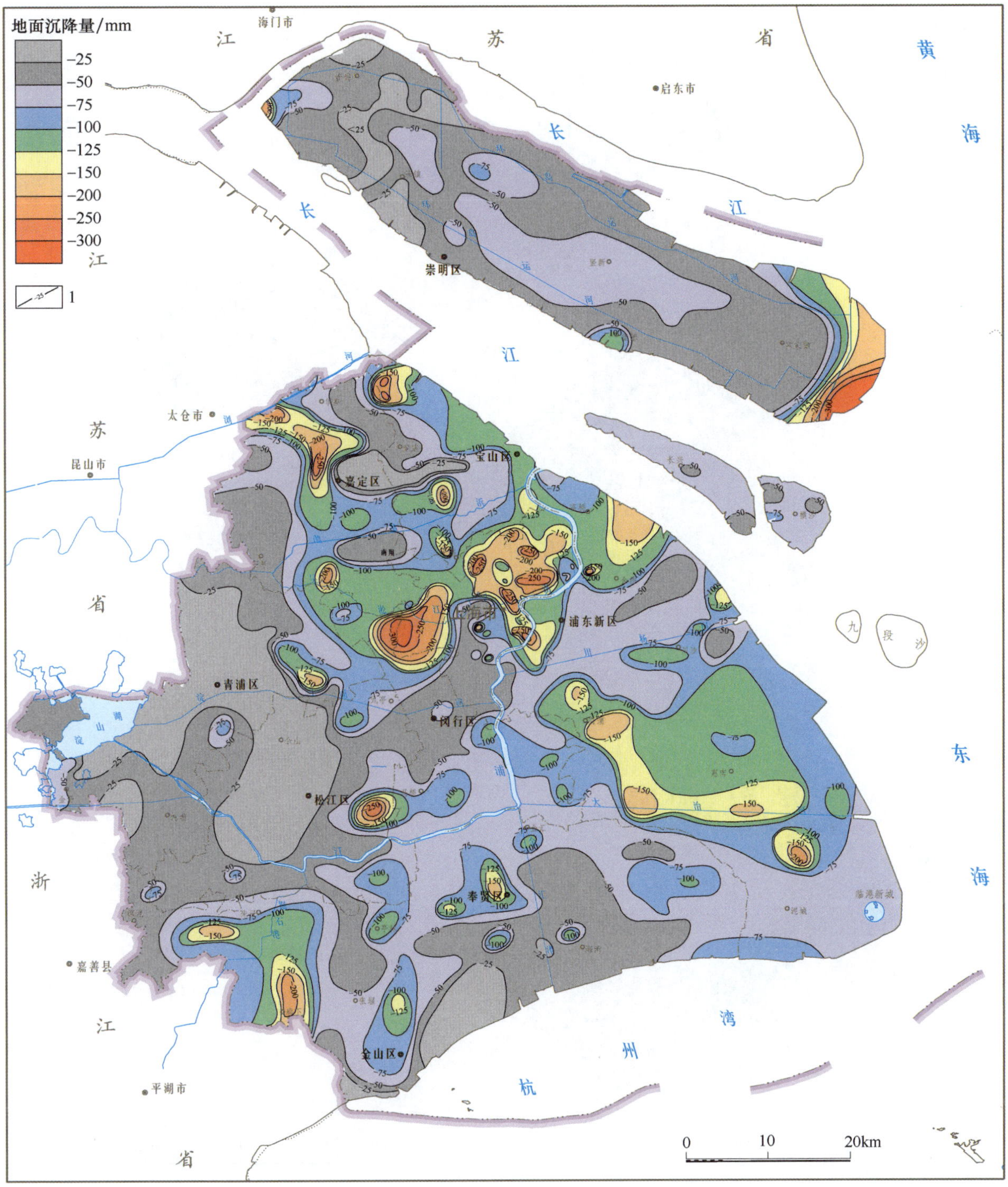

图 29-8　上海地面沉降等值线图（1996—2001 年）（水准复测）

Fig. 29-8　Land subsidence contour map of Shanghai（1996-2001）（leveling resurvey）

1—地面沉降等值线（mm）

出现新的沉降漏斗，最大沉降量为 175mm，其他地区地面沉降的发展趋势基本保持稳定（图 29-9）。

（四）2007—2011 年

除局部地区出现沉降漏斗外，该阶段绝大部分地区累计沉降量较小，在 0~25mm 之间；沉降漏斗主要分布在浦东、闵行、金山、崇明等局部地区，西部、北部局部地区甚至出现地面微量回弹趋势（图 29-10）。

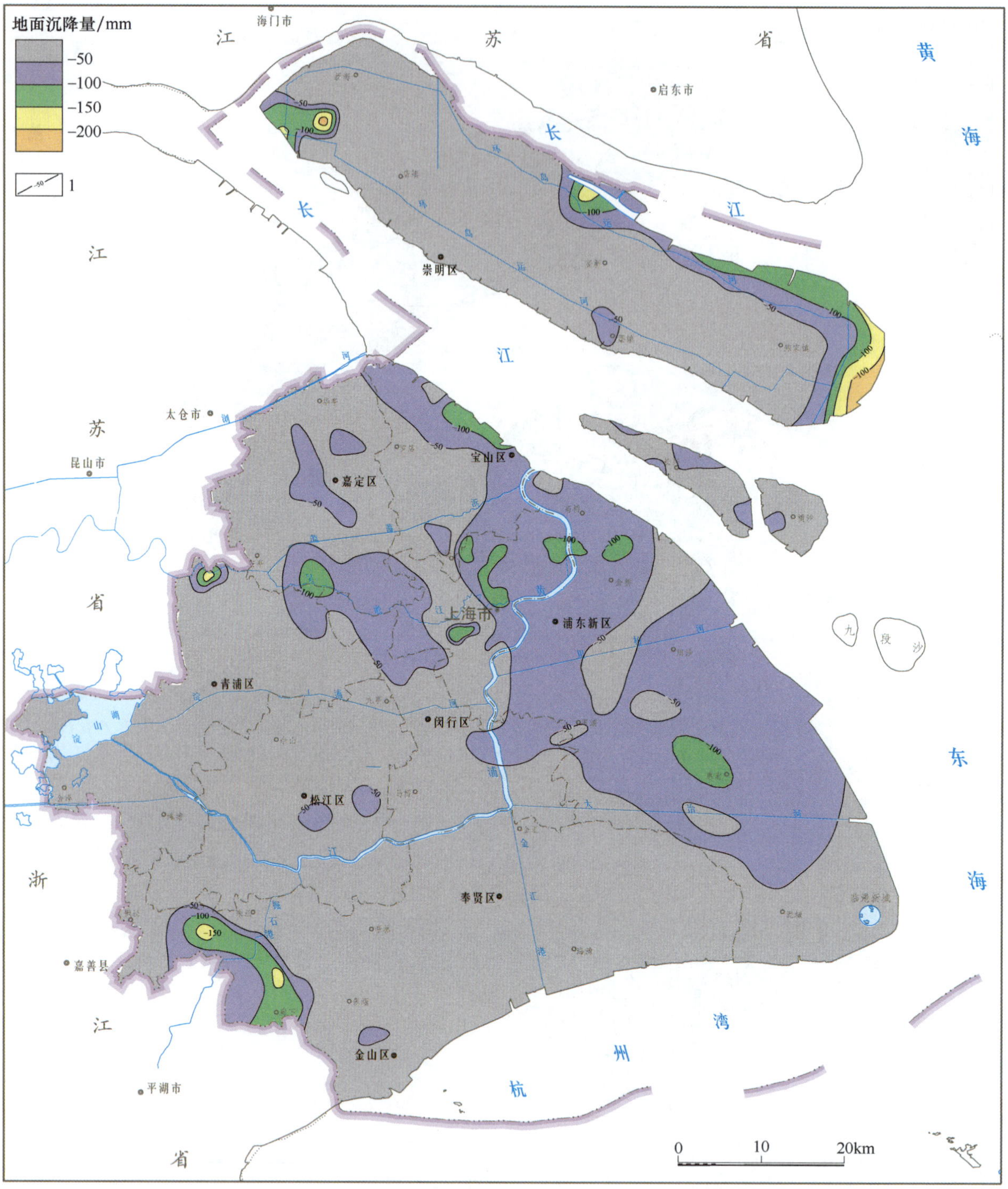

图 29-9 上海地面沉降等值线图（2002—2006 年）（水准复测）

Fig. 29-9 Land subsidence contour map of Shanghai (2002-2006) (leveling resurvey)

1—地面沉降等值线（mm）

从整体看来，崇明东滩和向化沉降最严重，漏斗中心累计沉降量超过 200mm。这一阶段，浦东尚未实现集约化供水，地下水开采仍然是本地区主要的供水来源，受地下水开采影响，该阶段浦东累计沉降量依然较大，分布的区域也较大，全区基本上累计沉降量在 25～50mm 之间。受大规模城市建设活动的影响，其中金桥、张江、航头等地出现沉降漏斗，漏斗中心累计沉降量大于 100mm。

闵行马桥、华漕，嘉定南翔，金山廊下、兴塔等地出现小的沉降漏斗，漏斗中心累计沉降量在

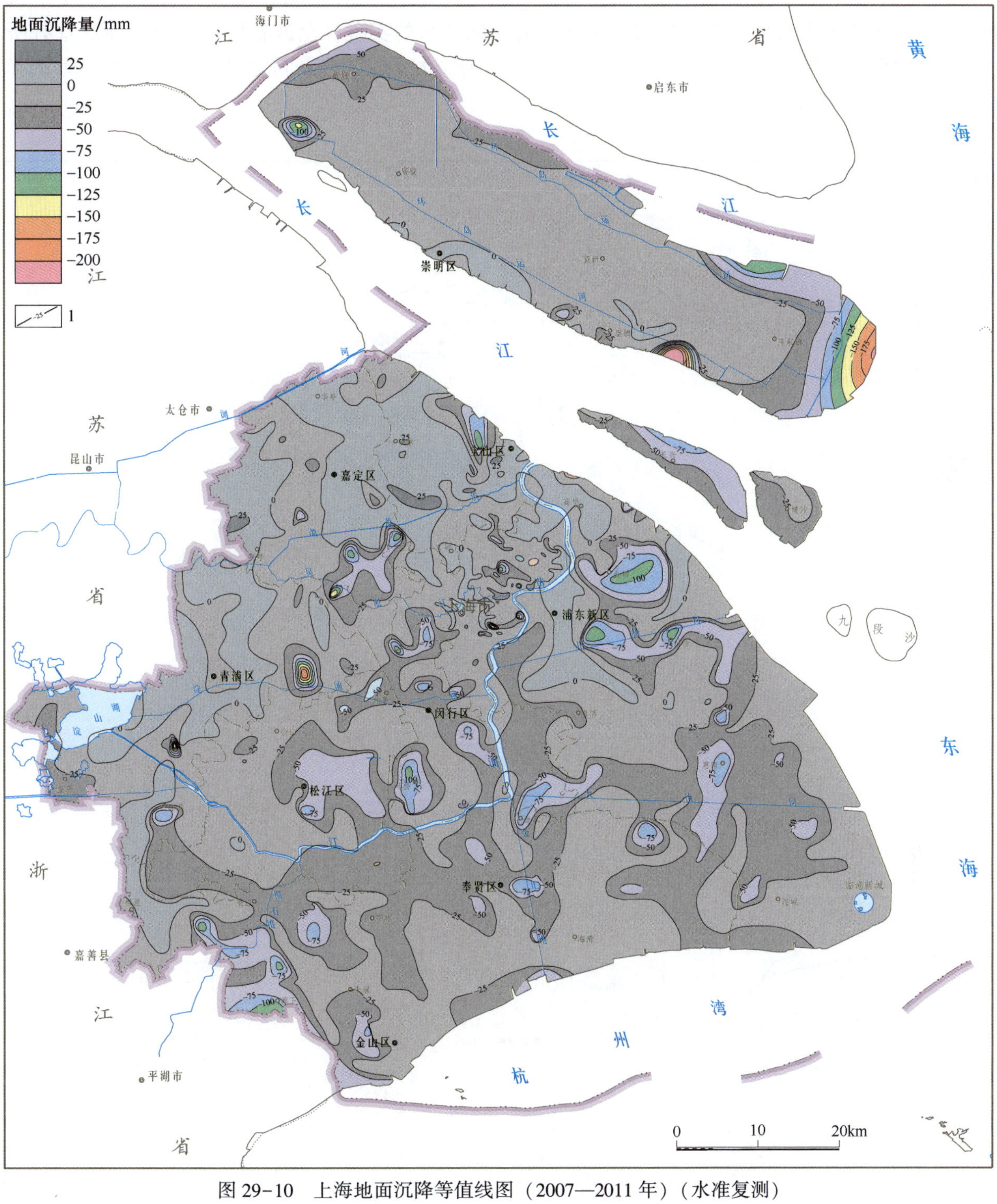

图 29-10　上海地面沉降等值线图（2007—2011 年）（水准复测）

Fig. 29-10　Land subsidence contour map of Shanghai (2007-2011) (leveling resurvey)

1—地面沉降等值线（mm）

75～100mm 之间，青浦赵巷出现沉降量大于 150mm 的沉降漏斗。西部青浦白鹤、嘉定华亭、宝山杨行等局部区域出现小范围地面微量回弹趋势。

（五）2012—2015 年

除局部地区出现沉降漏斗外，绝大部分地区此阶段累计沉降量在 0～30mm。沉降漏斗主要分布在浦东、闵行、金山、崇明等局部地区，西部、北部局部地区甚至出现地面微量回弹趋势。

从整体看来，崇明东滩和南门、新海等地沉降速率较上一阶段明显减小，累计沉降量在30~40mm；这一阶段由于浦东自来水管网的通达，集约化供水的实现，全区整体上地面沉降发育有所缓解，受工程建设活动影响，高桥、周浦—杜行、大治河滨海等地出现沉降漏斗，漏斗中心累计沉降量大于100mm（图29-11）。

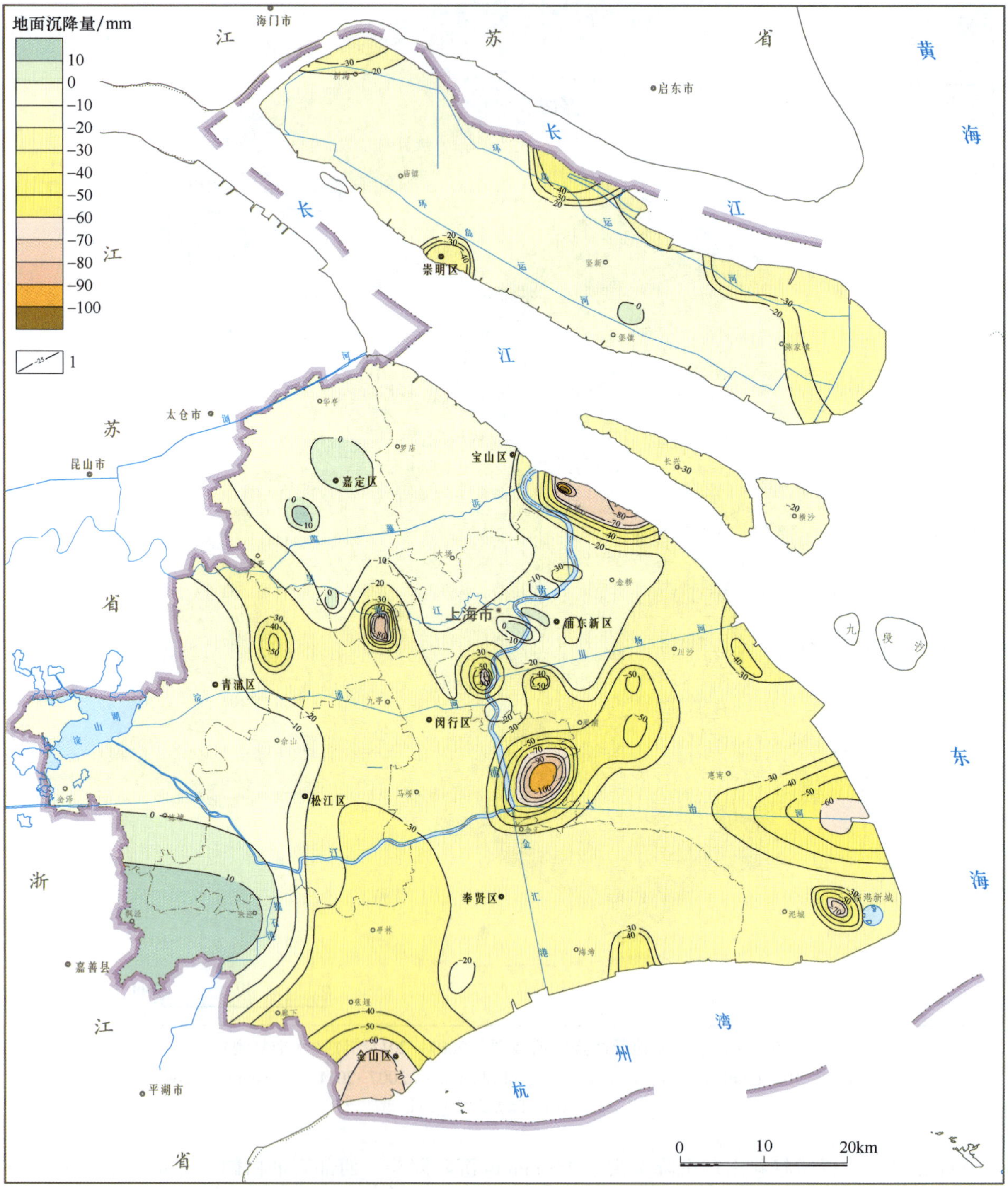

图29-11　上海地面沉降等值线图（2012—2015年）（GPS复测）

Fig. 29-11　Land subsidence contour map of Shanghai（2012-2015）（GPS resurvey）

1—地面沉降等值线（mm）

闵行的大虹桥地区、漕河泾，金山卫等地呈现小的沉降漏斗，漏斗中心累计沉降量在60～80mm之间。

西部青浦蒸淀—松江新浜一带、嘉定城区局部区域，仍出现小范围地面微量回弹迹象，但回弹格局有所变化，范围有所减小。

（六）2002—2016年

从整体上看，该阶段的地面沉降空间发育格局呈现出一些新的变化特征（图29-12）。

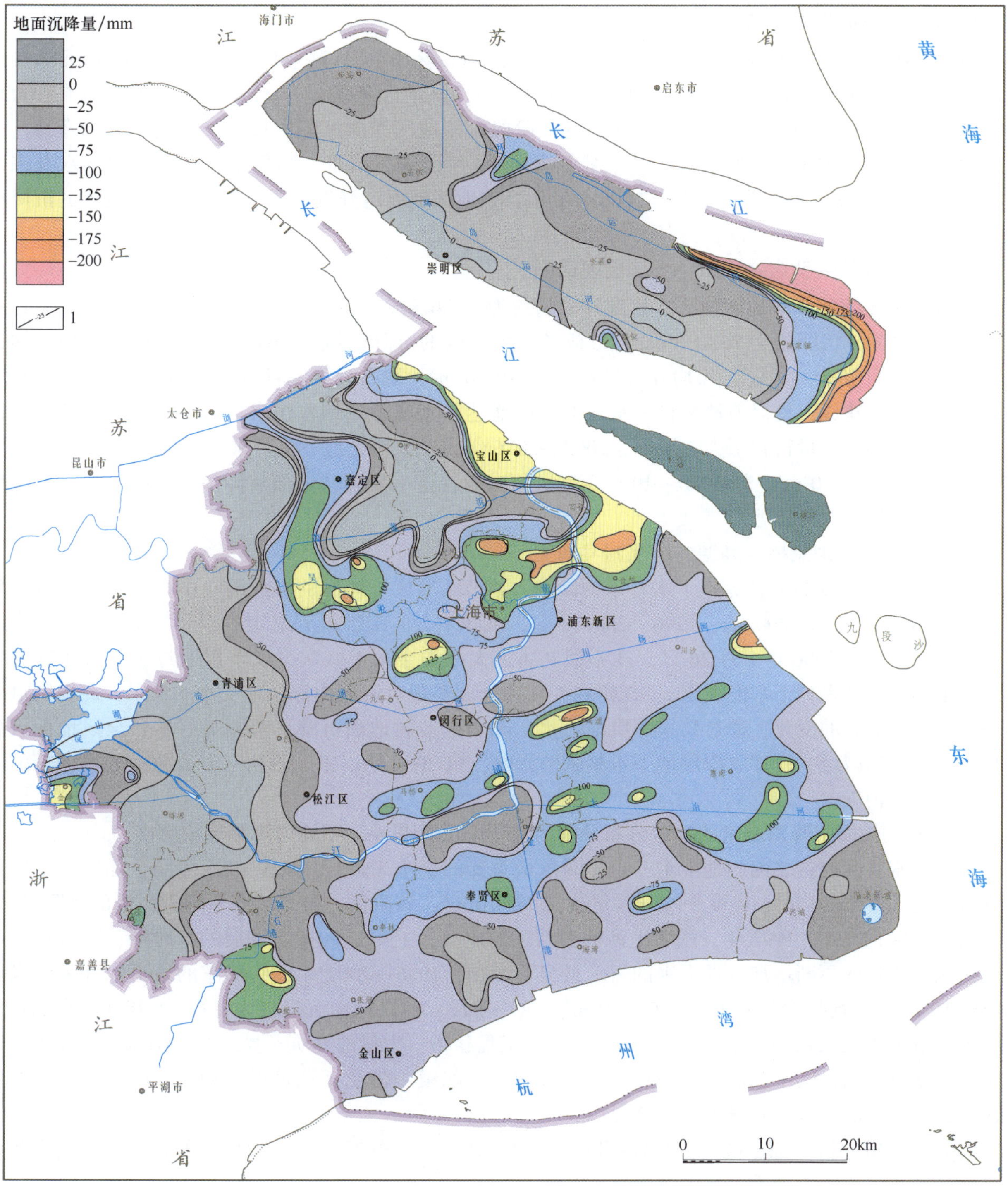

图29-12 上海地面沉降等值线图（2002—2016年）（水准复测）

Fig. 29-12 Land subsidence contour map of Shanghai (2002-2016) (leveling resurvey)

1—地面沉降等值线（mm）

岛域地区基本保持原有的沉降格局，崇明东滩由于广泛分布欠固结的新近纪沉积土和厚层吹填土，自然固结过程尚未完成，同时，在大面积厚层吹填土的附加荷载作用下，下卧土层在相当长的时间内将产生持续的压缩变形，该地区沉降较为突出。

陆域地区，全市仍发育中心城及宝山北部一带和嘉定黄渡—南翔一带两个较大范围的沉降漏斗，西南部金山吕巷—廊下一带发育新的沉降漏斗，省界竹桥—朱家桥—娄塘的沉降漏斗消失，原来闵行马桥的沉降漏斗基本消失，嘉定华亭—徐行—罗店—马陆—宝山刘行—顾村一片出现地面回弹。浦东的沉降范围由原来的片状分布分解为若干个沉降漏斗。

1. 闵行马桥沉降漏斗基本消失

1980—1995 年间该地区地下水累计开采量约为 $18000\times10^4m^3$，这一阶段，该地区最大沉降量达 175~200mm，形成显著的沉降漏斗。1996~2001 年间，该地区年均地下水开采量基本保持在 $1000\times10^4m^3$ 以上，累计开采量约为 $5500\times10^4m^3$，这一阶段，该地区地面沉降发展呈加剧趋势，沉降漏斗中心最大沉降量大于 250mm。21 世纪以来，特别是 2003 年以来，第Ⅲ、第Ⅳ承压含水层地下水开采量大幅减少，15 年间地下水开采量小于 $2000\times10^4m^3$，该地区地面沉降发育明显减缓，原来发育的沉降漏斗已经逐渐消失。

2. 华亭—徐行—罗店—顾村—大场出现回弹

1980—1995 年间，该地区地下水累计开采量约为 $18840\times10^4m^3$，年均开采量达 $1256\times10^4m^3$，这一阶段，该地区累计沉降量为 75~100mm，形成了较大范围的沉降区域。1996—2001 年间，该地区累计地下水开采量为 $7080\times10^4m^3$，年均开采量超过 $1400\times10^4m^3$，地面沉降发育呈加剧趋势。21 世纪以来，随着陈行水库的建成，使得该地区自来水管网得以通达，实现了集约化供水，地下水开采量特别是第Ⅳ承压含水层的开采量得到了大幅压缩，2002—2016 年间，该地区地下水开采量约为 $5100\times10^4m^3$，年均开采量约为 $340\times10^4m^3$，较 1980—1995 年间大幅减少，这一阶段，地面沉降呈现回弹态势，回弹量为 0~25mm，控沉效果显著。

3. 省界竹桥—朱家桥—娄塘的沉降漏斗消失

2000 年以前，受省界江苏地下水开采的影响，该地区形成了局部的沉降漏斗中心，1980—1995 年间，最大累计沉降量达 200mm，1996—2001 年间，沉降中心向南扩展到嘉定华亭和宝山罗泾，最大沉降量达到 250mm。2000 年 8 月 26 日，江苏省九届人大常委会第十八次会议审议通过了《关于在苏锡常地区限期禁止开采地下水的决定》，规定在 5 年内全面实现对区域内深层地下水的禁采，此后区域地质环境发生了显著的变化，通过禁采深层地下水，苏锡常地区地面沉降形势明显好转，全区地面沉降出现不同程度的减缓。受邻省深层地下水禁采的影响，自 2001 年以来，竹桥—朱家桥—娄塘一带的沉降漏斗已逐步消失。

4. 浦东的地面沉降由原来的片状分布分解为若干个沉降漏斗，不均匀沉降现象突出

在自来水管网通达之前，地下水一直是浦东最主要的供水来源，1980—1995 年间，浦东地下水累计开采量达 $56676\times10^4m^3$，年度平均开采量达 $3778\times10^4m^3$，该阶段，整个浦东发育大范围的沉降区域，沉降量在 50~100mm。1992 年，随着上海浦东新区的成立，上海进入振兴发展的新时期，随着城区的不断扩张，人口的不断增加，地下水的需求量显著增加，1996—2001 年间，浦东的地下水开采量高达 $23815\times10^4m^3$，平均年度开采量达 $4763\times10^4m^3$，这一阶段，浦东地面沉降较为发育，5 年间浦东的地面沉降量普遍大于 75mm，建筑工程特别是高层建筑工程猛增，大量的基坑开挖、井点降水，致使不均匀沉降现象渐渐突出，其中周浦镇、新场镇、万祥镇—书院镇一带受工程建设活动影响，发育若干个沉降漏斗，漏斗中心最大沉降量超过 150mm。

2001—2016 年间，浦东整体上地面沉降量在 50~100mm，相较 1996—2001 年间，大范围的沉降漏斗区已经消失，分解为若干个小的沉降漏斗。

5. 大虹桥地区、浦东张江—唐镇—三林—康桥等地局部仍发育沉降漏斗

2010 年，上海市规划委员会召开 2010 年度第一次全体会议，审议并原则同意《关于虹桥商务区规划工作的汇报》和《关于加快推进上海新城规划建设工作情况的回报》，其后，大虹桥商务区开展了大规模的新城建设。大虹桥地区是历史上西部沉降漏斗所在地，据分层标监测数据，自 2011 年以来，大虹桥地区浅部土层呈大幅压缩趋势，可能与该地区近些年大规模的工程建设活动有关。近几年来，受附近工程建设活动的影响，不均匀地面沉降现象较为严重。

在浦东，受地下水开采及工程建设双重影响，三林—康桥、张江—唐镇发育局部范围的沉降漏斗。

二、中心城地面沉降特征

（一）1991—1995 年

1991—1995 年间，城市对地下水的需求量在大幅度的增加，各含水层水位下降幅度明显加大，其中第Ⅳ、第Ⅴ承压含水层水位下降幅度较大。中心城地面沉降的速率相应明显增大，年均沉降速率达 11.9mm。沉降漏斗中心在以往基础上进一步发展，形成了黄浦—闸北不夜城地区—杨浦控江路—五角场面积达 88.0km^2 的地面沉降中心漏斗区，5 年累计沉降量大于 50mm；其中又出现了黄浦（179mm）、闸北不夜城地区（126mm）等数个新的沉降中心（图 29-13）。

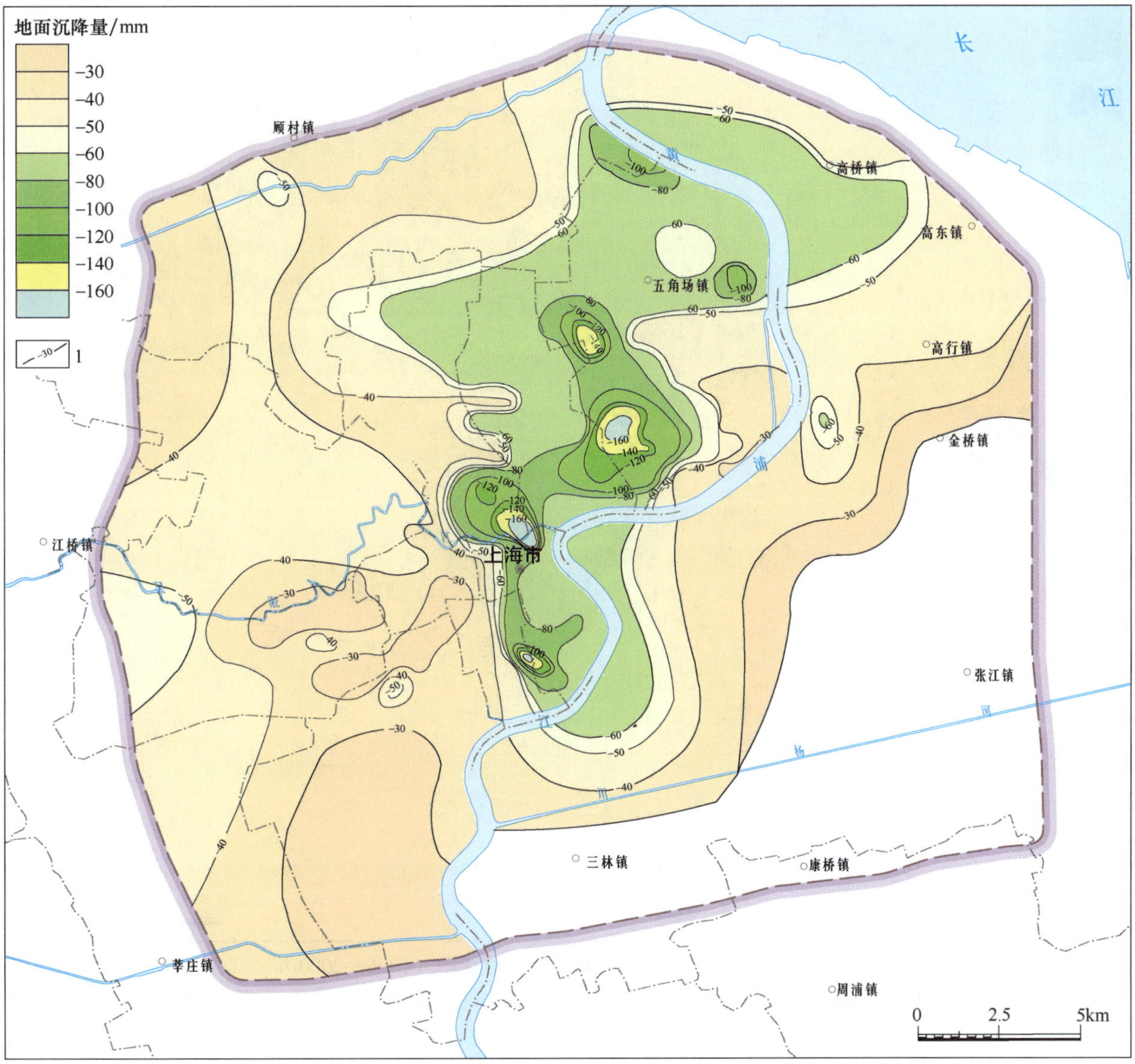

图 29-13　上海中心城地面沉降累积等值线图（1991—1995 年）

Fig. 29-13　Contour map of cumulative land subsidence in downtown Shanghai (1991-1995)

1—地面沉降等值线（mm）

本阶段内，市区地面沉降速率的空间差异显著增大，差异沉降较显著的地区主要分布在杨浦（28~160mm）、黄浦（43~179mm）、虹口（39~141mm）等地。

（二）1996—2000 年

1996—2000 年间，中心城及外围近郊区各含水层地下水位下降幅度虽趋减缓，但城市改造建设强度显著增加，中心城年平均地面沉降速率显著增大，达 20.3mm/a。市区东部地区的地面沉降，在前阶段形成的沉降漏斗区进一步发展，沉降速率加快，影响范围进一步扩大，闸北不夜城地区的最大累计沉降量达到了 280mm；外环线以内的浦东大部分地区累计沉降量由前一阶段的 30~50mm，增加至大于 100mm。市区西部长宁北新泾形成明显的大范围沉降漏斗区，漏斗最大累计沉降量达 200mm，影响范围较上阶段明显扩大。随着本阶段地面沉降速率增大，市区沉降漏斗和西部沉降漏斗的形成，使中心城地面沉降不均匀特点更加突出（图 29-14）。

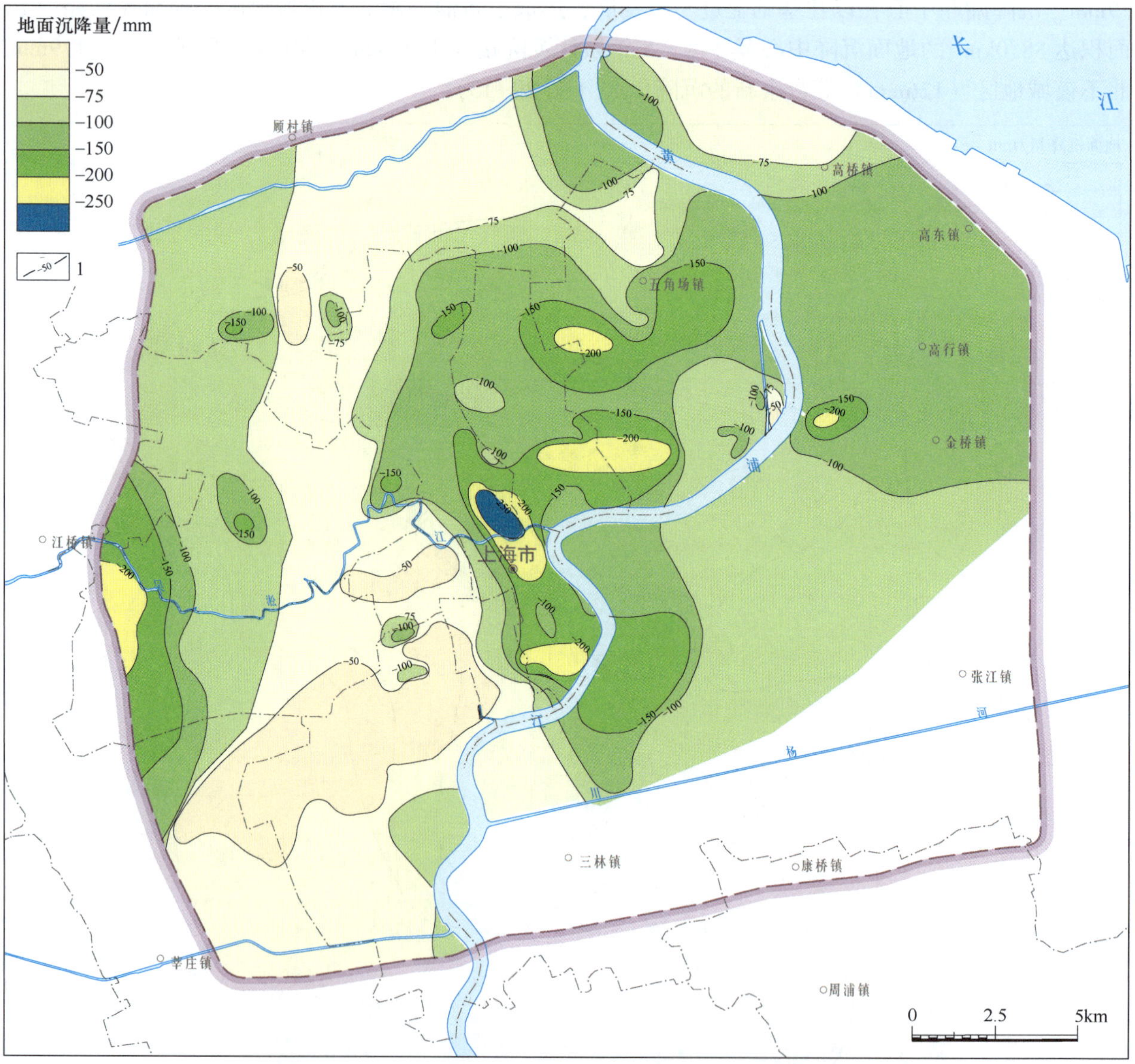

图 29-14　上海中心城地面沉降累积等值线图（1996—2000 年）

Fig. 29-14　Contour map of cumulative land subsidence in downtown Shanghai (1996-2000)

1—地面沉降等值线（mm）

（三）2001—2005 年

2001—2005 年，进一步实施开采量调减方案，各含水层地下水位继续处于上升态势，第Ⅳ、第Ⅴ承压含水层水位上升速率明显加快，使得地面沉降下降速率明显减缓，中心城平均地面沉降速率有所下降，为 15.3mm/a。随着地面沉降速率的减缓，原市区沉降漏斗解体为多个次级沉降中心，但地面沉降的影响范围进一步扩大。北部吴淞地面沉降速率较上 5 年有较明显的下降，最大年均沉降约为 30mm；浦东形成了“六里—陆家嘴”和“东沟—高行—黄浦江”两个地面沉降中心，年沉降速率均超过 20mm。市区西部闵行的虹桥—华漕沉降漏斗基本消失，长宁区北新泾年平均沉降速率由 1996—2000 年的 40mm 减至 22mm，沉降范围明显缩小。中心城地面沉降不均匀的特点仍较为突出（图 29-15）。

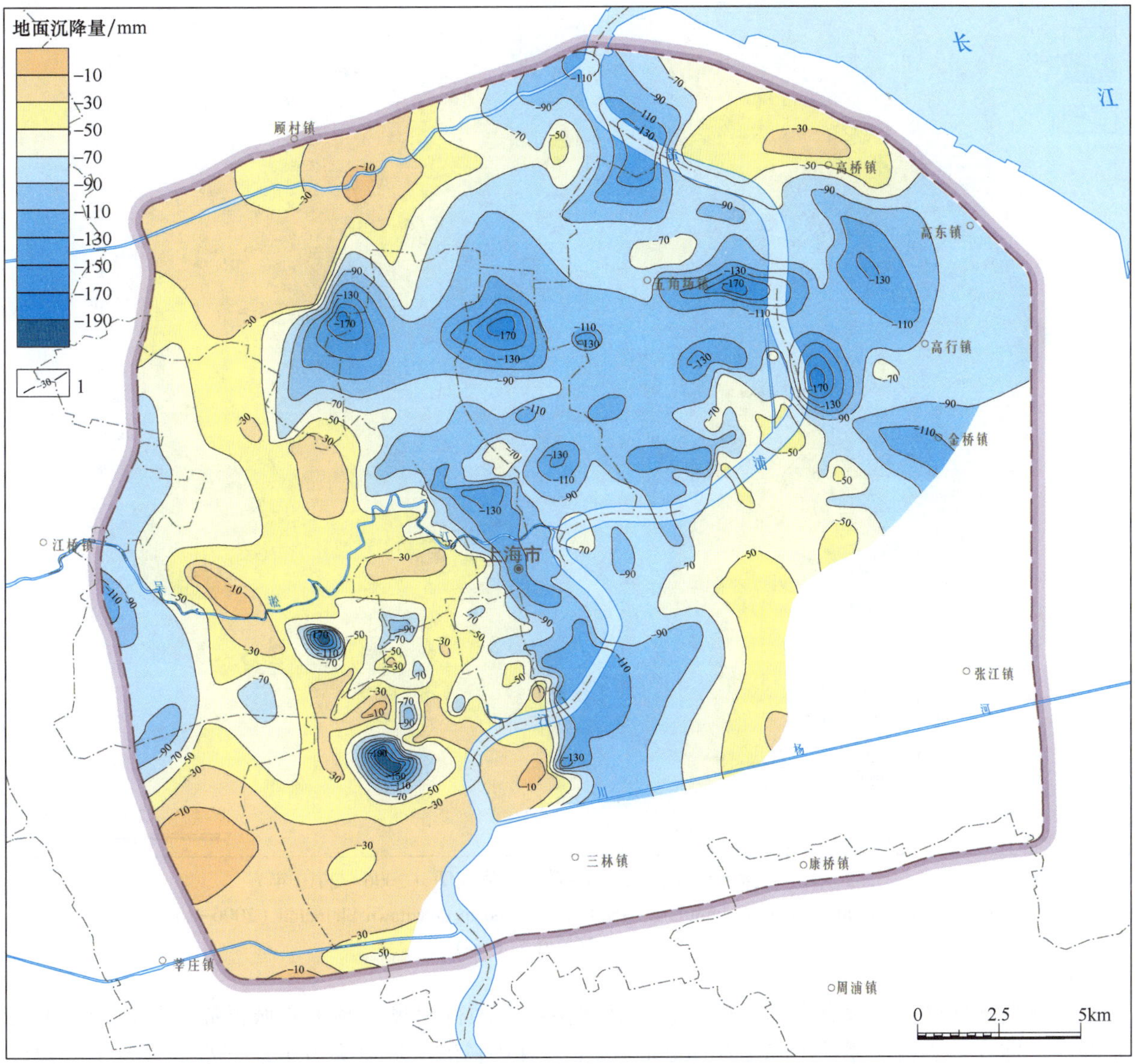

图 29-15　上海中心城地面沉降累积等值线图（2001—2005 年）

Fig. 29-15　Contour map of cumulative land subsidence in downtown Shanghai (2001-2005)

1—地面沉降等值线（mm）

（四）2006—2010 年

从发展趋势来看，地面沉降量及影响范围逐渐减小，但沉降漏斗中心量逐渐增多，沉降漏斗区同等距离上的变化更加剧烈，地面沉降的不均匀性逐年加剧。

从2006—2010年累计地面沉降量看（图29-16），中心城地面沉降总体上仍较发育，累计地面沉降量大于50mm的范围达162km²，地面沉降严重区与地下水位降落漏斗范围相吻合，尤其是叠加了深基坑降排水等因素的影响，地面沉降格局更加复杂，共分解为6个地面沉降影响区，使不均匀地面沉降影响更趋严重。

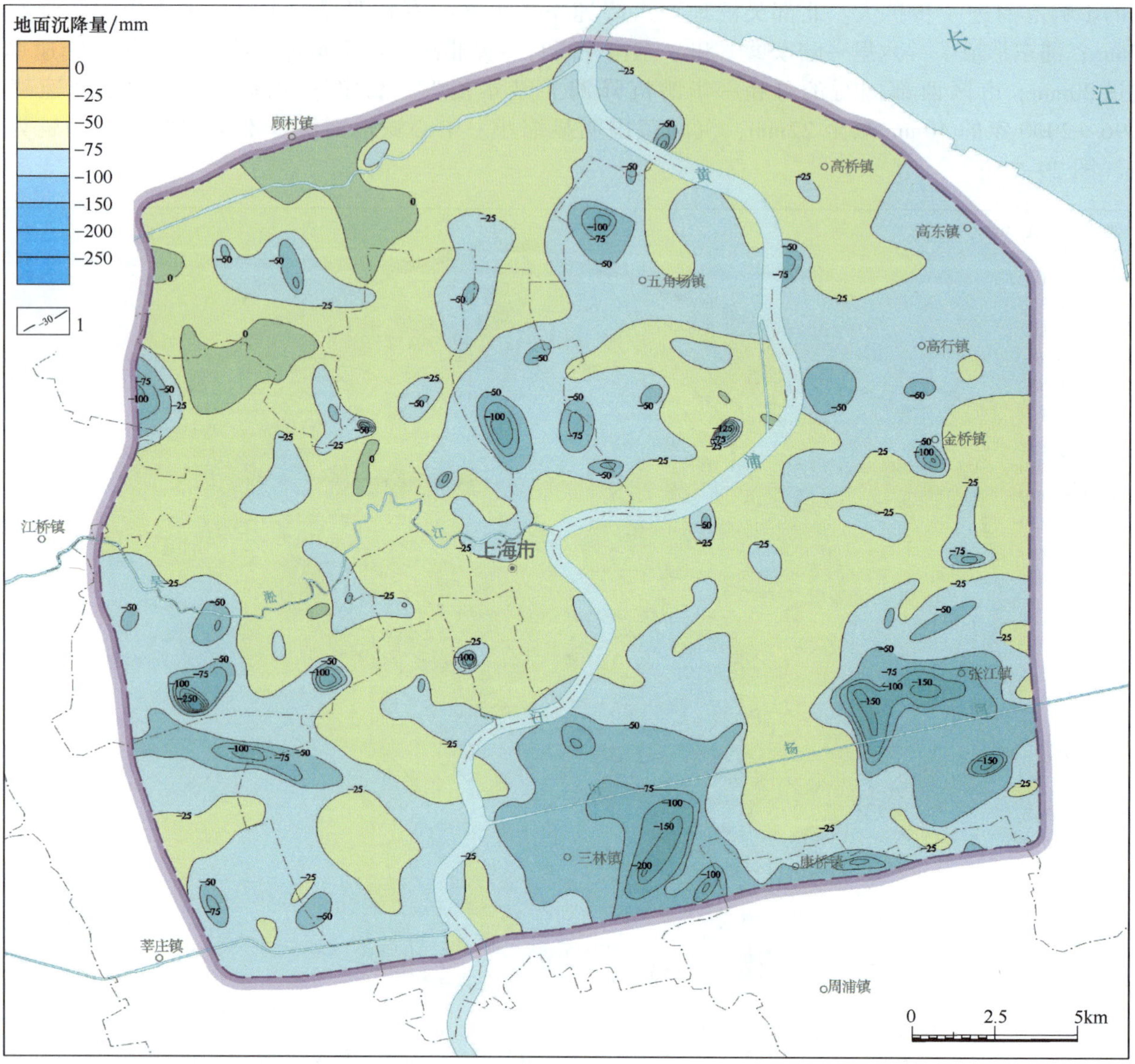

图29-16　上海中心城地面沉降累积等值线图（2006—2010年）

Fig. 29-16　Contour map of cumulative land subsidence in downtown Shanghai（2006-2010）

1—地面沉降等值线（mm）

中心城西部地区：发育有大虹桥地区域、嘉定南翔、闵行梅陇以及中心城杨浦与虹口局部地区等沉降漏斗，漏斗中心极端累计沉降量在300mm以上。其中杨浦与虹口受历史地下水开采及工程建设双重影响，局部地区不均匀沉降现象严重，对区内轨道交通8号线和10号线安全运营影响较大；嘉定南翔为11号线经过地区，以往曾为地下水集中开采区，现为地面沉降中心，现阶段地下水开采量有所压减，地面沉降趋势有一定的缓解；大虹桥地区现阶段为大规模工程建设阶段，受工程建设影响，不均匀地面沉降现象较严重，对包括2号线、9号线、10号线多条轨道交通及高铁等安全运营均有较大影响；在闵行梅陇对5号线影响较大。

中心城东部地区：浦东受地下水开采和工程建设双重影响，成为中心城地面沉降主要发育地区，在张江、三林等发育较大范围的沉降漏斗，漏斗中心最大沉降量在200mm以上，影响范围（累计地面沉降量大于50mm）达100km^2，对轨道交通2号线、6号线、11号线、磁悬浮列车的安全运营及规划在建13号线均会产生较大影响。

第三节　地面沉降规律与成因分析

一、地面沉降规律分析

据分层标组的统计分析结果（图29-17），2001—2015年间，全市浅部土层累计沉降量约为64mm，深部土层累计沉降量约为25mm，浅部土层累计变形量对地面沉降量的贡献率约为72%。与以往各阶段相比（1980—1990年间，浅部土层和深部土层沉降量占地面沉降总量的比例约为68%与32%；1990—2000年间，浅部土层和深部土层沉降量占地面沉降总量的比例约为40%与60%），近15年间，浅部土层的压缩变形已是影响地面沉降的主要因素，深部土层已出现微量回弹趋势。由此可见，控制浅部土层的沉降对控制地面沉降至关重要，尤其是在地下水开采量较小时，深部土层沉降已得到显著缓解（龚士良，2008；刘飞，2002；王寒梅，2015），近几年来，地面沉降几乎全部是由浅部土层沉降所致，深部土层则呈微量回弹的趋势。

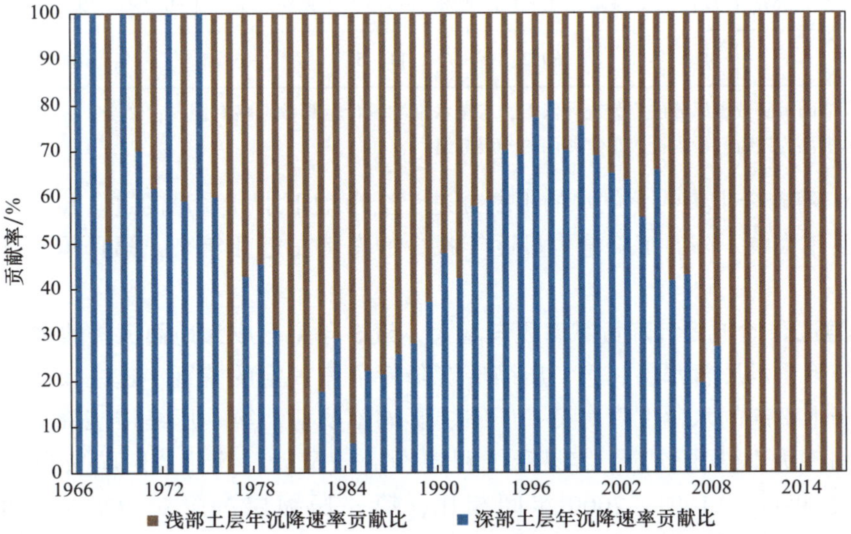

图29-17　深部土层及浅部土层累计沉降量贡献比历时图

Fig. 29-17　Time histories of cumulative subsidence contributions of deep and shallow soil strata

注：“深部土层”指第Ⅰ承压含水层以深土层，“浅部土层”指第Ⅰ承压含水层及以浅土层

从分层标测量结果来看（表29-2），在20世纪60—80年代，由于地下水开采层主要集中在中心城第Ⅱ、第Ⅲ承压含水层，第Ⅱ、第Ⅲ承压含水层地下水位的下降引发了第Ⅰ承压含水层及上部土层沉降；20世纪90年代—21世纪初，随着深部第Ⅳ、第Ⅴ承压含水层成为主采层，地面沉降主要以深部土层沉降为主。

21世纪以来，地下水开采量的大量压缩再加上工程降排水活动的影响，浅部土层成为地面沉降的主要发育层次，且在不同阶段，不同土层表现出不同的变形特征。

1. 第Ⅳ承压含水层沉降逐渐减缓，近5年呈现微量回弹态势

从表29-2可以看出，自1991—2003年间，深部土层的沉降对地面沉降贡献大于浅部土层，这一时期，随着上海浦东新区的成立，城区的不断扩张，人口的不断增加，地下水的需求量逐步增加，加之这一阶段第Ⅳ承压含水层是地下水的主要开采层次，第Ⅳ承压含水层成为最主要的沉降发育层次，对地面沉降的贡献率约在50%。

表 29-2　各土层不同阶段变形量特征

Table 29-2　Deformation rates of the soil strata during different periods of time

土层	1966—1979 年		1980—1990 年		1991—1995 年		1996—2000 年		2001—2005 年		2006—2010		2011—2015	
	累计沉降/mm	贡献率/%	累计沉降/mm	贡献率/%	累计沉降/mm	贡献率/%	累计沉降/mm	贡献率/%	累计沉降/mm	贡献率/%	累计沉降/mm	贡献率/%	累计沉降/mm	贡献率/%
第一压缩层	-23.8	59.65%	-31.9	64.06%	-5.5	9.14%	-18	15.86%	-20.5	31.78%	-7.2	54.55%	-7.39	123.99%
第二压缩层	-9.8	24.56%	-7.7	15.46%	-3.5	5.81%	-6.7	5.90%	-4.5	6.98%	-4.1	31.06%	-3.47	58.22%
第Ⅰ承压含水层	-4.2	10.53%	5.5	-11.04%	-3.8	6.31%	-3.4	3.00%	-3	4.65%	-0.9	6.82%	-1.03	17.28%
第三压缩层	-5.6	14.04%	-6.6	13.25%	-6.5	10.80%	-6.6	5.81%	-5	7.75%	-1.6	12.12%	-0.52	8.72%
第Ⅱ承压含水层	5.6	-14.04%	-2.2	4.42%	-1.5	2.49%	-5.4	4.76%	-0.5	0.78%	0	0.00%	0.62	-10.40%
第Ⅲ承压含水层	2.8	-7.02%	0.8	-1.61%	0.4	-0.66%	-14.1	12.42%	-0.5	0.78%	1.6	-12.12%	1.31	-21.98%
第Ⅳ承压含水层	-4.2	10.53%	-8.8	17.67%	-38	63.12%	-56	49.34%	-31	48.06%	-4.3	32.58%	1.93	-32.38%
第Ⅴ承压含水层	-0.7	1.75%	1.1	-2.21%	-1.8	2.99%	-3.3	2.91%	0.5	-0.78%	3.3	-25.00%	2.59	-43.46%
小计	-39.9	-	-49.8	-	-60.2	-	-113.5	-	-64.5	-	-13.2	-	-5.96	-

注：①本表各阶段地面沉降数据以分层标为统计依据，与按水准测量计量算数据略有差异；②全市分层标组是逐渐建设完善，因此各阶段参与统计的标组均为该时间段内投入正常使用的分层标组。

2003 年以来，随着全市集约化供水的不断推进，全市地下水开采量的进一步减少，第Ⅳ承压含水层地下水开采量逐年大幅下降，沉降量也呈逐年减少的态势，近几年来，第Ⅳ承压含水层已呈现微量回弹的趋势。

2. 第一压缩层沉降对地面沉降贡献率逐渐增大，近 5 年来已成为地面沉降的主要发育层次

第一压缩层与第Ⅰ承压含水层和第二压缩层沉降特征有一定的差别，这主要是由于第二压缩层的阻隔，使得基坑降水引起的水位下降对第一压缩层影响不明显。第一压缩层沉降量较大主要是由于其埋深较浅，土层物理力学性质较差，容易受工程建设活动及动、静荷载影响所致。据分层标组监测结果，2001—2015 年间与 1991—2000 年间相比，第一压缩层沉降对地面沉降的贡献率逐渐增大（图 29-18），已成为地面沉降的主要发育层次，特别是近 10 年来，第一压缩层沉降对地面沉降的贡献已经大于 50%。

3. 第Ⅰ承压含水层及其上覆相邻第二压缩层沉降趋势较为一致，受深基坑降水影响较为明显

第Ⅰ承压含水层与第二压缩层沉降趋势较为接近（图 29-19），表明两者均受基坑降水影响，在第Ⅰ承压含水层水位持续下降的 20 世纪 90 年代第Ⅰ承压含水层和第二压缩层也呈现持续下沉，在第Ⅰ承压含水层水位开始恢复后，第Ⅰ承压含水层和第二压缩层沉降速率有一定的减缓，但不同地区表现出的沉降特征略有差别，这主要是由于地区间发展上的时空差异导致深基坑工程分布的非均匀性，在基坑工程相对密集的城区或者大型开发区第Ⅰ承压含水层和第二压缩层仍持续下沉，甚至出现加速下沉现象。

因此，在浅部土层中以第一压缩层沉降最为显著，而且其影响因素较为复杂；第Ⅰ承压含水层与其上覆相邻第二压缩层沉降趋势较为一致，受深基坑降水影响明显，在空间上表现为受深基坑工程分布影响，在基坑工程密集区域沉降明显，此外，部分地区第Ⅰ承压含水层水位仍低于历史高水位也是其持续沉降的原因之一，因此，控制第Ⅰ承压含水层和第二压缩层沉降需对基坑降水引起的水位下降进行控制。

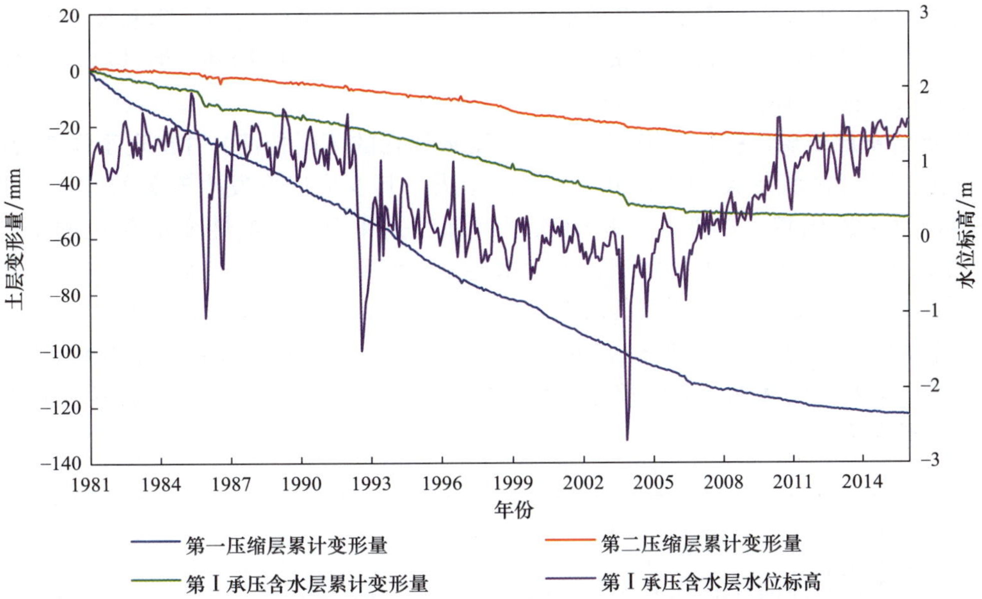

图 29-18　第Ⅰ承压含水层及其以上土层累计沉降量与第Ⅰ承压含水层水位历时曲线

（数据来源：浦东高化分层标 F13、第Ⅰ承压含水层水位监测井浦 218-02C）

Fig. 29-18　Time histories of cumulative soil strata subsidence in and above confined aquifer Ⅰ and water table of confined aquifer Ⅰ (data based on layermark F13 and confined aquifer Ⅰ water table monitoring well Pu 218-02C in Gaohua, Pudong District)

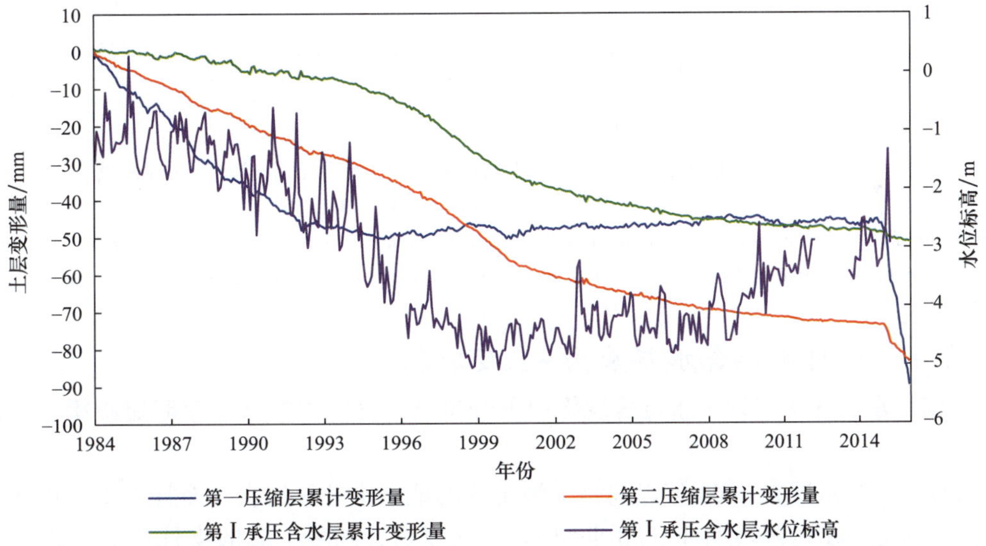

图 29-19　第Ⅰ承压含水层及其以上土层累计沉降量与第Ⅰ承压含水层水位历时曲线

（数据来源：吴泾电化厂分层标 F17、第Ⅰ承压含水层水位监测井海 017-04F）

Fig. 29-19　Time histories of cumulative soil strata subsidence in and above confined aquifer Ⅰ and water table of confined aquifer Ⅰ (data based on layermark F17 and confined aquifer Ⅰ water table monitoring well Hai 017-04F at Wujing Electrochemical Factory)

二、地面沉降与地下水动态相关性

（一）第Ⅰ承压含水层

1. 第Ⅰ承压含水层水位整体抬升，但第Ⅰ承压含水层及其上覆软土层持续压缩变形

第Ⅰ承压含水层地下水由于水质较差，较少被开采利用，地下水排泄以工程降水为主。20 世纪 90

年代开始，由于大规模工程降水活动，出现了第Ⅰ承压含水层水位的大幅下降和该含水层以及上覆第一、第二压缩层的压缩变形。21世纪以来，第Ⅰ承压含水层水位整体抬升，但其以及上覆软土层持续压缩变形，现阶段是主要沉降层（见图29-18，图29-19）。

2. 第Ⅰ承压含水层表现出弹性—弹塑性—塑性变形演化的不同特征，并出现了蠕变现象

随地下水位下降，含水层承受的附加应力逐渐增大，其变形速率发生显著的非线性变化的同时，含水层颗粒结构也会发生较大的变化，从而表现出弹性—弹塑性—塑性变形演化的不同变形特征。2000年前后的平均地下水位基本保持稳定，但含水层累计变形量仍持续增加，甚至有加速的趋势（图29-20），表明此时含水层已出现了明显的蠕变现象；2005年后地下水位迅速回升，而含水层的累计变形量处于微量增加状态，应力-应变形迹表现为在地下水位波动过程中持续微量压缩的波浪式应变路径，说明第Ⅰ承压含水层的附加压力已超过了其弹性极限，第Ⅰ承压含水层随地下水位波动而产生的变形将以塑性变形为主。

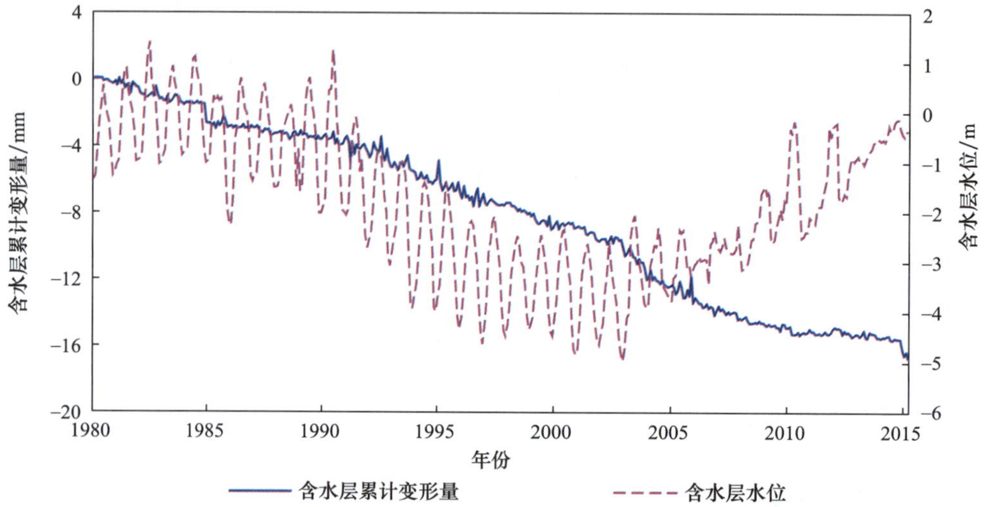

图29-20　第Ⅰ承压含水层土层累计沉降量与第Ⅰ承压含水层水位历时曲线

（数据来源：双阳中学分层标F4）

Fig. 29-20　Time histories of cumulative soil strata subsidence in and above confined aquifer Ⅰ and water table of confined aquifer Ⅰ（data based on layermark F4 at Shuangyang Middle School）

（二）第Ⅱ、第Ⅲ承压含水层应力-应变特征

1. 第Ⅱ、第Ⅲ承压含水层地下水位呈现缓慢上升态势，含水层沉降速率明显减缓，并出现微量回弹态势

第Ⅱ、第Ⅲ承压含水层在1963年以前是上海主要的地下水开采层次，而且集中于市区开采，形成了以市区为中心的地下水位降落漏斗。20世纪90年代以后，随着上海的支柱产业纺织行业大部分企业的关停并转，地下水人工回灌量逐渐减少。这一阶段，第Ⅱ、第Ⅲ承压含水层地下水位显著下降，含水层呈现持续压缩的态势。由于这些含水层的水位较20世纪60年代以前总体抬升了，有效应力一般不超过其前期最大固结压力，土层压缩性小，变形主要以弹性变形为主，土层随水位的升降而膨胀、压缩，残余的塑性变形很小，变形和水位变化是基本同步的。

2001年以来，地下水开采量逐步减少，第Ⅱ、第Ⅲ承压含水层地下水位出现缓慢上升态势，含水砂层沉降速率明显减缓。

2. 第Ⅱ、第Ⅲ承压含水层表现为压缩—回弹的弹性变形特征

第Ⅱ、第Ⅲ承压含水层应力-应变曲线表现为一系统的平行紧闭的应变椭圆（图29-21～图29-24），在地下水位周期性波动的加荷、卸荷作用下，含水层同样表现为压缩、回弹的弹性变形状态，土层加荷段（低水位期）压缩量与卸荷段（高水位期）回弹量基本相等，残余变形微量，显示出较好的弹性变形特征。杨浦劳动公园第Ⅱ承压含水层 35 年的多年残余变形量仅为 4mm，年均残余变形量为 0.11mm，第Ⅲ承压含水层 30 多年来残余变形量几乎为 0。

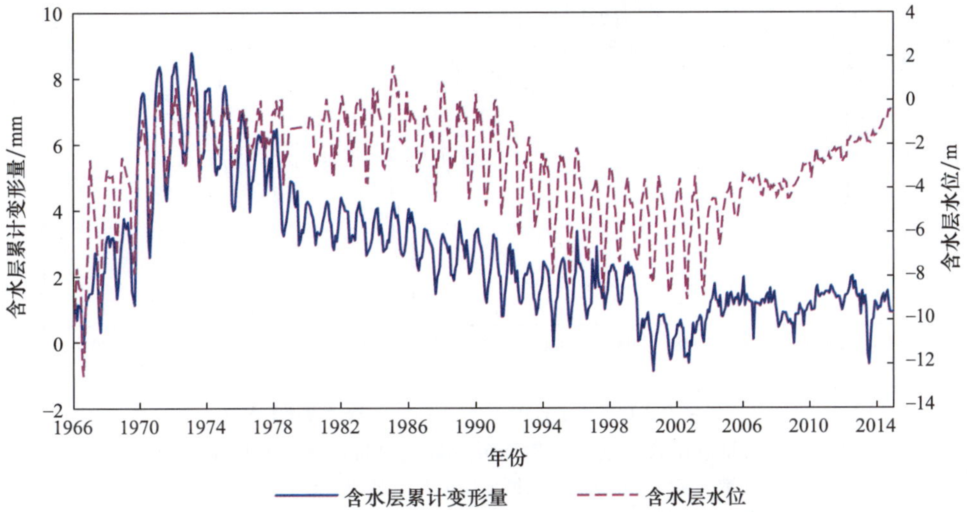

图 29-21　第Ⅱ承压含水层土层累计沉降量与第Ⅱ承压含水层水位历时曲线
（数据来源：杨浦劳动公园分层标 F3，监测井杨 005-05）

Fig. 29-21　Time histories of cumulative soil strata subsidence in and above confined aquifer Ⅱ and water table of confined aquifer Ⅱ (data based on layermark F3 and monitoring well Yang 005-05 at Laodong Park, Yangpu District)

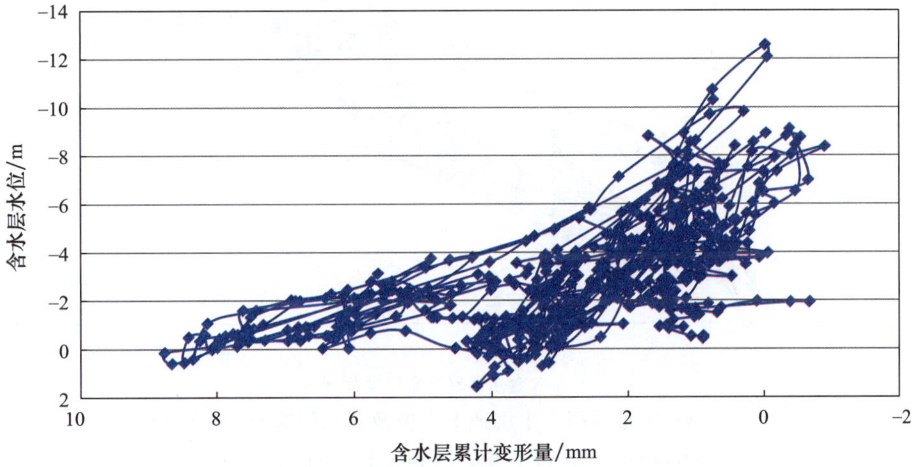

图 29-22　第Ⅱ承压含水层地下水位及含水层应力-应变特征

Fig. 29-22　Groundwater table and stress-strain distribution of confined aquifer Ⅱ

（三）第Ⅳ承压含水层应力-应变特征

1. 第Ⅳ承压含水层地下水位大幅回升，含水层变形速率显著减缓，部分地区第Ⅳ承压含水层出现回弹，扭转了快速沉降的趋势

20 世纪 60 年代后，随着地下水开采的全面调整，第Ⅳ承压含水层成为地下水的主要开采层次，由于第Ⅳ承压含水层在多年来持续、大量开采后，地下水位持续降低，含水砂层表现为压缩态势，且含水层累计变形量曲线与含水层地下水位曲线表现为非线性关系。

自 1998 年加强了对超采区地下水开采量的压缩，使第Ⅳ承压含水层地下水位略有回升，土层变形速率也有所减小，2003 年以后，随着地下水开采量的不断压缩、回灌量的增加，第Ⅳ承压含水层地下水位抬升幅度增大，土层沉降速率明显减缓，部分地区第Ⅳ承压含水层出现回弹，扭转了以往不控沉

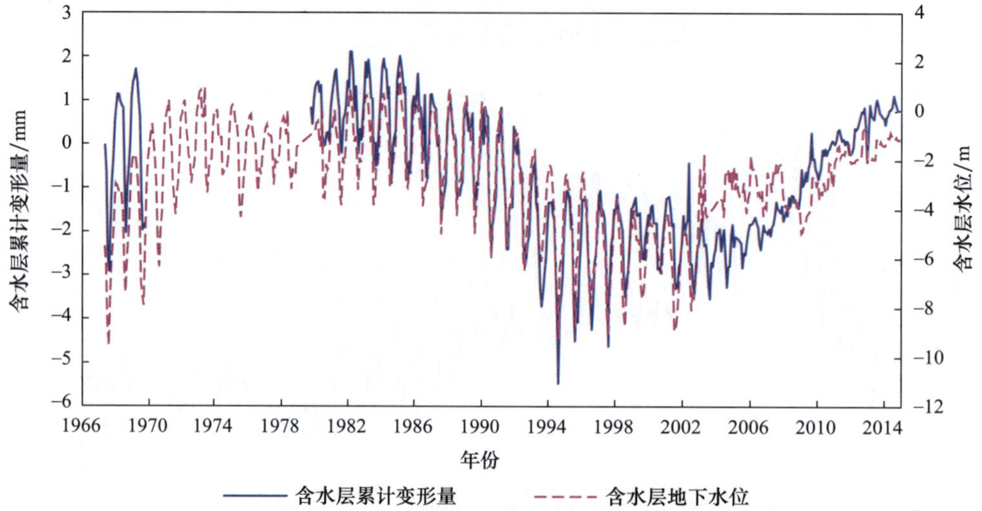

图 29-23　第Ⅲ承压含水层土层累计沉降量与第Ⅲ承压含水层水位历时曲线

（数据来源：杨浦劳动公园分层标 F3，监测井杨 029-01）

Fig. 29-23　Time histories of cumulative soil strata subsidence of soil strata in confined aquifer Ⅲ and water table of confined aquifer Ⅲ (data based on layermar F3 and monitoring well Yang 029-01 at Laodong Park, Yangpu District)

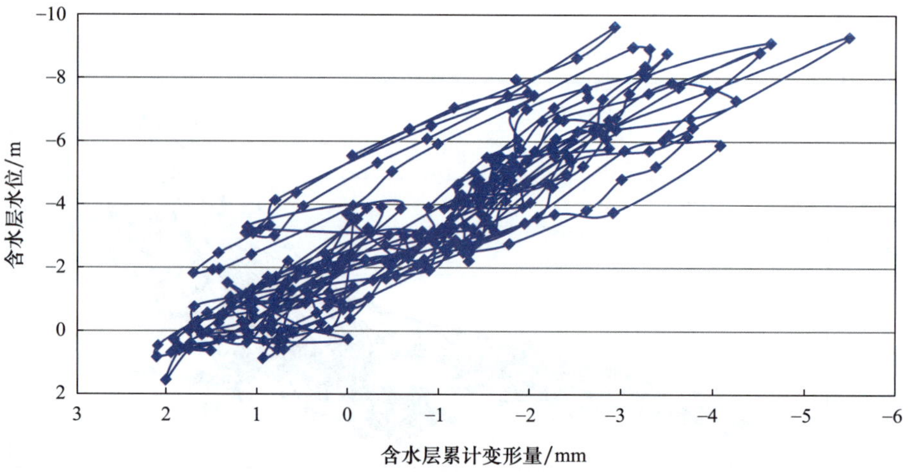

图 29-24　第Ⅲ承压含水层地下水位及含水层应力-应变特征

Fig. 29-24　Groundwater table and stress-strain distribution of confined aquifer Ⅲ

阶段快速沉降的趋势。

2. 弹性变形—弹塑性—塑性变形的变化特征

分层标杨浦劳动公园第Ⅳ承压含水层标组（F3）（图 29-25，图 29-26）及国棉十七厂第Ⅳ承压含水层标组（F6）在地下水变化的不同状态下，第Ⅳ承压含水层变形特征（如应力-应变曲线）明显表现出由弹性变形→弹塑性→塑性变形的变化规律。

20 世纪 90 年代以前，地下水位基本保持稳定，含水砂层表现为与地下水位波动基本同步的周期性的夏沉冬升，每年仅有很小的残余变形，为典型的弹性变形状态。含水砂层的弹性变形主要由季节性低水位期固体颗粒骨架间空隙减小、高水位期孔隙水压力增大使颗粒骨架间隙扩大而表现出来，应力-应变形迹为一系列闭合的应变椭圆。

20 世纪 90 年代初，随地下水位进一步下降，当地下水位降低至-28～-32m 时，含水层变形特征逐渐发生了变化，季节性低水位期的沉降量逐年增大，冬季水位回升期的回弹量逐年减小，表现为弹

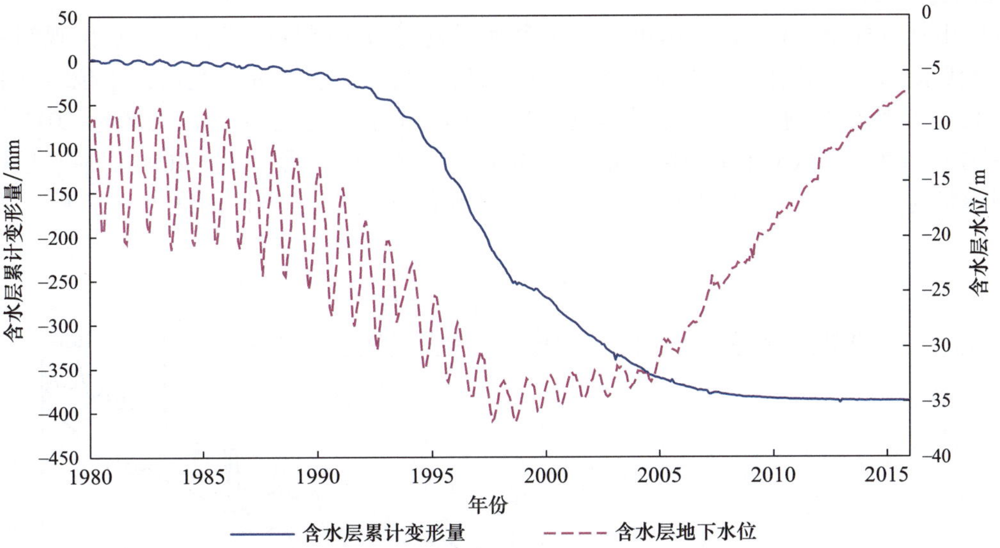

图 29-25　第Ⅳ承压含水层土层累计沉降量与第Ⅳ承压含水层水位历时曲线
（数据来源：劳动公园分层标 F3 标组）
Fig. 29-25　Time histories of cumulative soil strata subsidence in confined aquifer Ⅳ and water table of confined aquifer Ⅰ（data based on layermark F3 at Laodong Park）

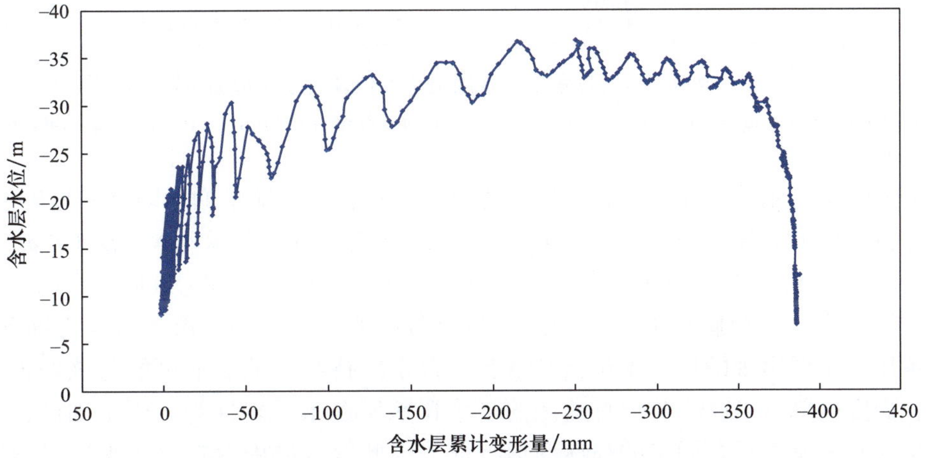

图 29-26　第Ⅳ承压含水层地下水位及含水层应力-应变特征
（数据来源：劳动公园分层标 F3 标组、第Ⅳ承压含水层监测井杨 034-01G）
Fig. 29-26　Groundwater table and stress-strain distribution of confined aquifer Ⅳ（data based on layermark F3 at Laodong Park，confined aquifer Ⅳ monitoring well Yang 034-01G）

性变形向塑性变形过渡。此阶段含水层变形已不仅仅表现为颗粒骨架弹性的压缩与回弹，颗粒排列方式已开始出现重新调整的趋势，应力-应变形迹由闭合应变椭圆向尖峰谷状应变轨迹转变。当地下水位低于此水位以下时，含水层在季节性高、低水位期均表现为持续性的固结压缩，表明含水层此时承受的附加应力已超过了其前期固结压力，呈现颗粒重新排列的破坏型变形，应力-应变形迹表现为在地下水位波动过程中持续压缩的波浪式应变路径。

自 1998 年后加强了对超采区地下水开采量的压缩，并逐步增大回灌量，第Ⅳ承压含水层地下水位开始回升，但土层没有随着水位抬升而回弹，而是存在明显变形滞后和较大残余变形，35 年的多年残余变形量达到 380mm。

20 世纪 90 年代以前，地下水位基本保持稳定，含水砂层形变与地下水位波动基本同步。

3. 线弹性变形特征

分层标宝山刘行标组（F27）显示第Ⅳ承压含水层表现为线弹性变形方式。水位观测井显示此处20世纪90年代水位降深较大，水位在1998年达到最低值，约为-47m，此后，水位缓慢抬升，2004年水位为-41m，年均抬升约为1m，之后迅速抬升，2015年水位达到约-8m，2004—2015年年均抬升3m（图29-27）。1998—2004年土层沉降量约为3mm，有轻微的变形滞后。2004以后土层持续稳定回弹，年均回弹量约为1.25mm。在水位迅速抬升阶段水位与变形有明显的线性相关性，为弹性变形特征。

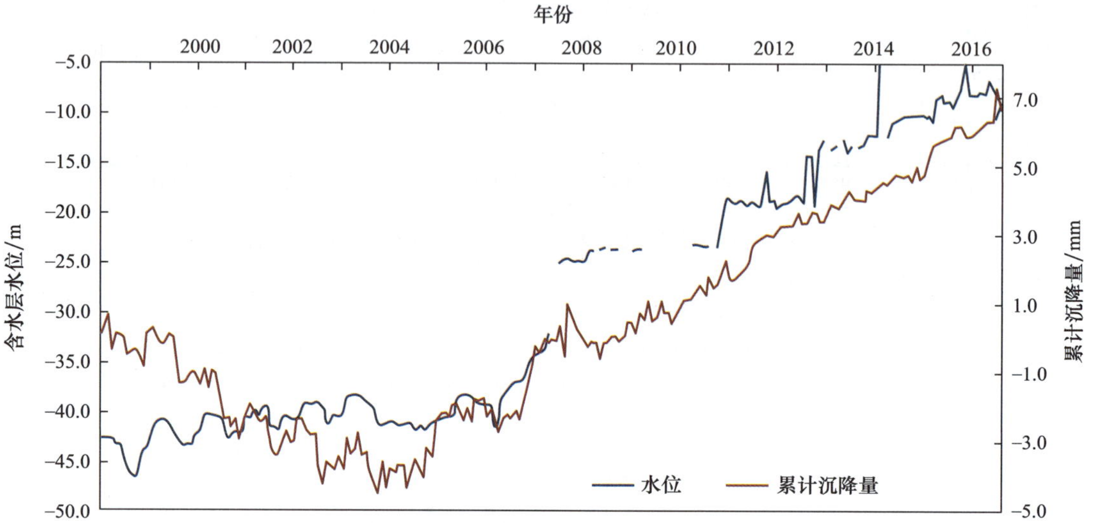

图29-27 宝山刘行第Ⅳ承压含水层地下水位及含水层应力-应变特征

Fig. 29-27 Groundwater table and stress-strain distribution of confined aquifer Ⅳ in Liuhang, Baoshan District

有关试验研究结果揭示，第Ⅳ承压含水层在历史上存在较大压缩量的原因之一是砂土也存在蠕变。此外，对比杨浦劳动公园第Ⅳ承压含水层F3#1分层标与宝山刘行第Ⅳ承压含水层F27#2分层标监测土层的岩性：劳动公园F3#1分层标处的上部为第Ⅳ承压含水层上层，岩性为细砂，中部为第Ⅳ承压含水层中间层为黏土层，岩性为粉质黏土夹细砂，下部为第Ⅳ承压含水层下层，岩性为细砂夹粉质黏土，劳动公园第Ⅳ承压含水层出现的地下水位抬升情况下含水层持续压缩变形可能与该处含水层含有大量粉质黏土有关；而宝山刘行F27#2分层标监测的是第Ⅳ承压含水层岩性为含砾中细砂，含水层颗粒较粗，所以反映出随水位基本保持同步的弹性变形特征。因此含水层中存在黏土夹层对压缩量和变形特征也会有较大影响。另一方面，该地区位于古河道，受前期固结压力较大，临界荷载大，含水层水位下降产生的附加应力没有超过临界荷载，因而随着水位抬升，含水层变形显示出明显的线性相关性，呈现出弹性变形特征。

（四）第Ⅴ承压含水层应力-应变特征

第Ⅴ承压含水层地下水位大幅回升，含水层变形速率显著减缓，部分地区第Ⅴ承压含水层出现回弹，扭转了快速沉降的趋势。

第Ⅴ承压含水层在20世纪60—70年代因开采强度较小，地下水位总体下降幅度较小，地下水位基本在-10~-15m。20世纪80—90年代，第Ⅴ承压含水层地下水开采强度不断加大，沉降漏斗中心地区因开采强度的加大地下水位持续下降。第Ⅴ承压含水层主要分布在上海市区北部和崇明三岛区，1999年前基本没有开展系统的第Ⅴ承压含水层分层沉降监测工作，只能对其后的数据进行分析。

20世纪60—70年代，第Ⅴ承压含水层开采强度较小，开采范围以中心城为主；之后开采范围迅速扩展至郊区青浦、崇明等区（县），90年代开采强度达到最大，闵行—青浦地下水位沉降漏斗中心开采强度大于$60\times10^4m^3/km^2$。根据嘉定新城地下水位、含水层变形监测结果（图29-28），随着第Ⅴ

承压含水层地下水开采强度增大，该含水层地下水位逐年大幅下降，至 1998 年地下水位降至-55m，但由于第Ⅴ承压含水层地下水位、变形的系统监测数据较少，地下水位下降过程中的沉降特点没有监测数据，90 年代末期逐步完善了第Ⅴ承压含水层沉降监测设施。从地下水位、含水层变形曲线可以看出，1998 年后第Ⅴ承压含水层水位大幅回升，至 2015 年嘉定新城地下水位达到-20m 左右，含水砂层表现为微量回弹特征。

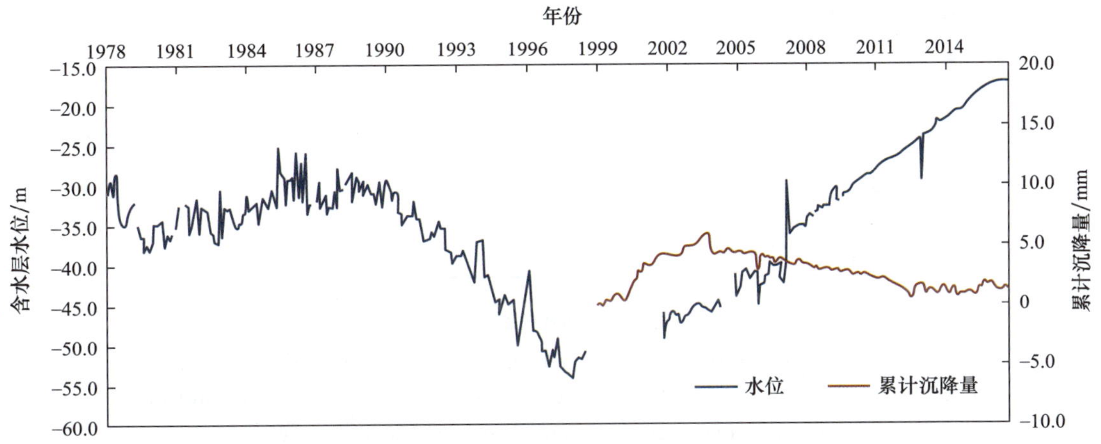

图 29-28　嘉定新城第Ⅴ承压含水层地下水位及含水层应力-应变特征

Fig. 29-28　Groundwater table and stress-strain distribution of confined aquifer Ⅴ in Xincheng, Jiading District

三、工程建设引发地面沉降机理

（一）深基坑减压降水引发地面沉降

目前主要的沉降层以浅部土层为主，其中，以地下空间开发为代表的深基坑工程是引发浅部土层沉降的一个重要因素（骆祖江，2007；杨天亮，2009）。

1. 影响范围与降水方式密切相关，且基坑 3*H* 以外的地面沉降主要受减压降水影响

止水帷幕完全阻断降水目的层的基坑工程，因基坑内外基本没有水力联系，坑内降水对坑外地下水位无影响，对坑外地面沉降影响不大，其影响范围一般在 3*H* 以内（图 29-29，*H* 为 34m，实测范围为 10*H*）（杨天亮，2012），主要地面沉降影响范围在 100m 以内。

对于止水帷幕非完全阻断降水目的层的基坑工程，当采用坑内降水方式时，其影响范围一般在 6*H* 范围内（图 29-30，*H* 为 18m，实测范围为 10*H*，主要影响范围在 100m 范围内）；当采用坑外降水方式时，其影响范围一般可达到 10*H*（叶为民，2009），甚至超过 10*H*（图 29-31，*H* 为 18m，实测范围为 10*H*，主要影响范围达 160m 以外）。

此外，大定海泵站基坑工程现场试验结果（龚士良，2008；王建秀，2009）表明，距基坑 3*H*（*H* 为 19.4m）范围以外区域地面沉降主要受基坑减压降水的影响。减压降水前，距基坑 3*H* 范围以外区域地面沉降不明显（黄鑫磊，2015）；减压降水开始后，坑内水位大幅下降导致坑外距基坑 3*H* 处水位降深达 6.3m，此时 3*H* 范围以外区域地面沉降明显，距基坑 3*H* 处最大沉降量达到 7.9mm 左右，距基坑 10*H* 处最大沉降量达到 5.4mm 左右，可见减压降水对坑外地面沉降影响显著（杨天亮，2018）。减压降水结束后，在地下水位恢复过程中，距基坑 3*H* 范围以外区域地面回弹较为明显（吴建中，2009；杨天亮，2010），回弹量在 1.7mm 左右（图 29-32）。

2. 不同地质结构区典型工程实例试验结果证实，基坑工程减压降水以降水目的层上部软土层沉降变形为主

正常沉积区基坑工程第Ⅰ承压含水层降水运行过程中，以降水目的层和上部软土层沉降变形为主，且在降水运行过程中，上部软土层变形量持续增大，最终沉降量甚至大于降水目的层，降水结束后降

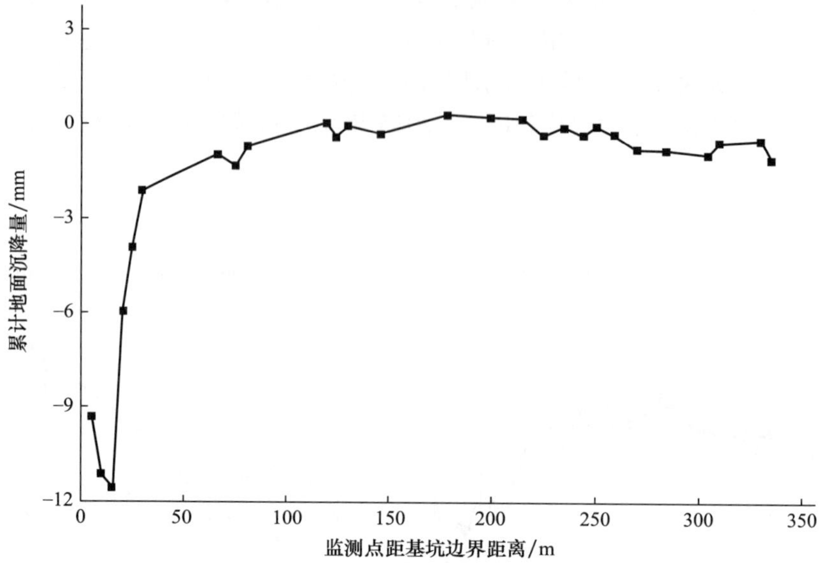

图 29-29　世博变电站基坑累计地面沉降曲线图（止水帷幕完全阻断降水目的层）

Fig. 29-29　Cumulative land subsidence curve of the foundation pit at Expo Substation (with waterproof curtain completely cutting off the dewatering target layer)

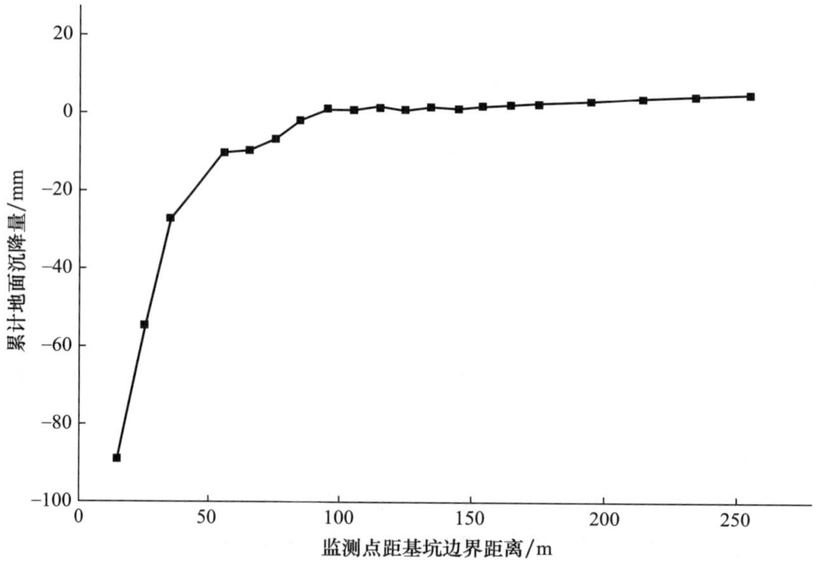

图 29-30　同济大学站基坑累计地面沉降曲线图（止水帷幕非完全阻断降水目的层）

Fig. 29-30　Cumulative land subsidence curve of the foundation pit at Tongji University Station (with waterproof curtain not completely cutting off the dewatering target layer)

水目的层沉降变形有所恢复，上覆软土层沉降变形趋向于收敛。

古河道沉积区基坑工程只降微承压含水层承压水，降水运行过程中地面沉降变形影响范围有限，但微承压含水层减压降水对其上覆软土层沉降层影响显著，其他土层变形量均较小。

古河道沉积区基坑工程第Ⅰ承压含水层降水引发地面沉降的影响较大，降水目的层上部软土层为主要沉降变形土层。

（二）典型建筑密集区地面沉降效应

目前，国内外对于地面沉降的研究主要集中于开采地下水诱发的地面沉降，对于密集建筑群引起的地面沉降研究甚少。

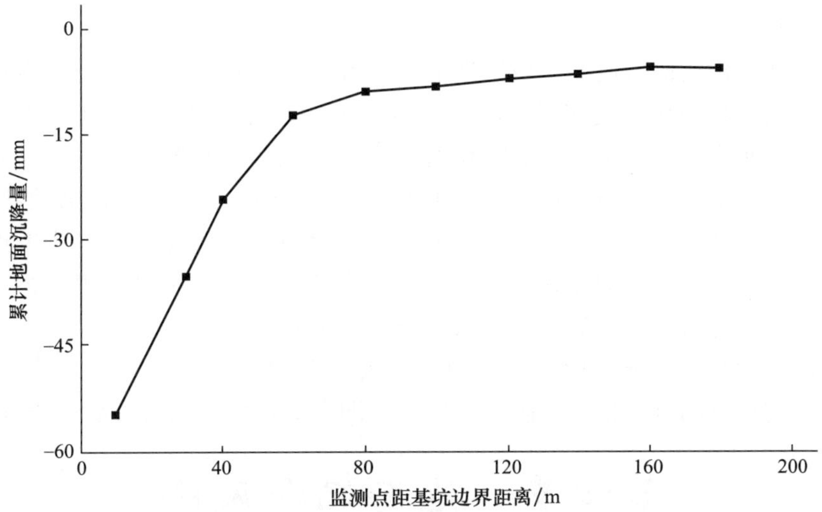

图 29-31　国权路站基坑累计地面沉降曲线图（坑外降水）

Fig. 29-31　Cumulative land subsidence curve of the foundation pit at the Guoquan Road Station (external dewatering)

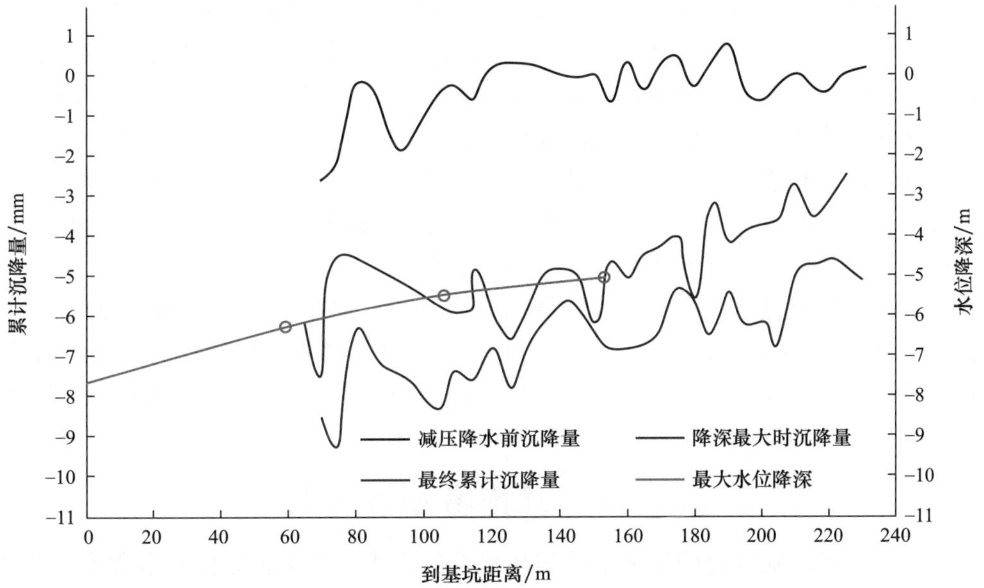

图 29-32　大定海泵站基坑地面沉降与地下水位时空特征曲线图

Fig. 29-32　Characteristics time-space curves of land subsidence and groundwater table of the foundation pit at the Dadinghai Pumping Station

陆家嘴地区是上海典型的高层建筑密集区，研究成果表明：高层建筑群诱发的地面沉降取决于建筑容积率和地质条件；浅部的淤泥质黏土层沉降最大；高层建筑群造成的区域性地面沉降呈不规则的凹凸起伏状；高层建筑群对地面沉降的影响与建筑物施工的先后顺序有关；密集高层建筑群之间存在明显的应力叠加效应。

由监测分析可见，高层建筑群诱发的地面沉降取决于建筑容积率和地质条件，由于建筑基础的约束作用，建筑本身的沉降相对较小，而在离开建筑物一定距离的地面沉降则逐渐增加，最后又逐渐减小，呈现凸凹起伏状。其中第④层淤泥质黏土层和主要持力层（⑥、⑦、⑨）附近沉降较大，说明在特定地质条件下建筑物的荷载叠加效应是造成此现象的原因，可以将其归结为“密集高层建筑群”在特定地质条件下的“工程环境效应”。

模型试验表明，在第⑧层缺失的情况下，第④层淤泥质黏土层虽然层厚不大，但却是引起地面沉降的关键因素，占50%以上。

模型试验表明，高层建筑群造成的区域性地面沉降并不是简单的凹状，而是在总体下沉的趋势下呈不规则的凹凸起伏状。越是靠近建筑群中心（在附加应力作用范围内），地面沉降越大。对于按照实际布置的高层建筑群而言，建筑本身的沉降相对较小，而在离开建筑物一定距离的地面沉降则逐渐增加，最后又逐渐减小，地面沉降最大点出现在单个建筑群的中心，呈现凸凹起伏状。

高层建筑群对地面沉降的影响与建筑物施工的先后顺序有关，离先建建筑物近点的沉降大于离后建建筑物近的点；对荷重相差不大的相邻建筑物而言，先建建筑物的沉降大于后建建筑物。这主要是由于先建建筑物受到后建建筑物的影响，在原来沉降的基础上产生了叠加。

密集高层建筑群之间存在明显的应力叠加效应，表现在超孔隙水压力与土压力均有阶梯形增长，对应于土层的变形与建筑物的沉降也呈现阶梯形增长，即在荷重的施加阶段变形会有明显的突变。

第四节　地面沉降风险

一、地面沉降危险性分区

依据上海地面沉降发育机理、孕育条件、监测成果等，选取地面沉降易发程度、历时灾害强度、预测沉降速率、地势高程作为地面沉降危险性评价的指标，先对各个单项指标进行评价。其中，地面沉降易发区指标评价中将全区划分为地面沉降高易发区、较高易发区、中易发区、较低易发区、低易发区5个级别；地面沉降历史灾害强度指标评价中划分为地面沉降历史灾害大强度区、较大强度区、中强度区、较小强度区、小强度区5个区；地面沉降预测沉降速率评价中将预测沉降速率划分为速率大区、速率较大区、中速率区和速率较小区、速率小区5个等级。地势高程指标评价中划分为低地势区、较低地势区、中地势区、较高地势区、高地势区5个区。

依据单项指标评价结果进行地面沉降危险性的综合评价，结果显示，上海主要处于地面沉降危险性中等区（图29-33），危险性高区主要分布于中心城及与江浙两省交界的嘉定、金山的局部地区，而地面沉降危险性低区主要分布在地质环境发育较好、地面沉降预测速率较低地区，以奉贤及青浦的局部地区为代表。此外，三岛区均属于地面沉降危险性中等区。

二、地面沉降易损性分区

地面沉降易损性是指承灾区遭受地面沉降破坏机会的多少与发生损毁的难易程度。基于行政单元，采用社会经济统计指标代替承灾体具体数据获得了地面沉降的易损性分区。其中，上海地面沉降易损性增加的因素归纳为城市人口密度和单位面积的GDP以及建设用地所占比重，另外，还包括市政设施：防汛墙长度、供水管密度、高架道路密度和轨道交通密度；影响易损性减小的因素为减灾防护的投入，包括单位面积减灾防护投入以及单位面积回灌投入。

其中，将人口因素、单位面积GDP、单位面积建设用地比重3个指标都划分为很大、大、中等、较小、小5个等级。单位面积减灾防护投入、供水管网密度、轨道交通和高架道路等指标划分为1~5共5个等级。

根据上海的各个区的实际指标进行分类，经过模糊综合评价，得到因子权重，最后计算出上海地面沉降易损性评价结果（图29-34）。

上海地面沉降易损性体现了区域经济发展水平和地面沉降发生程度的综合特征。以中心城为代表的经济最发达区同时也是地面沉降程度最为严重区，因此，易损性最为严重；浦东新区北部也是地面沉降比较严重的地区，形成了高桥、张江、塘桥等区域性沉降漏斗，同时单位面积GDP、建设用地比重以及市政设施密度都比较大，受到地面沉降危害程度较高，归为较高易损性区；宝山和闵行工业比

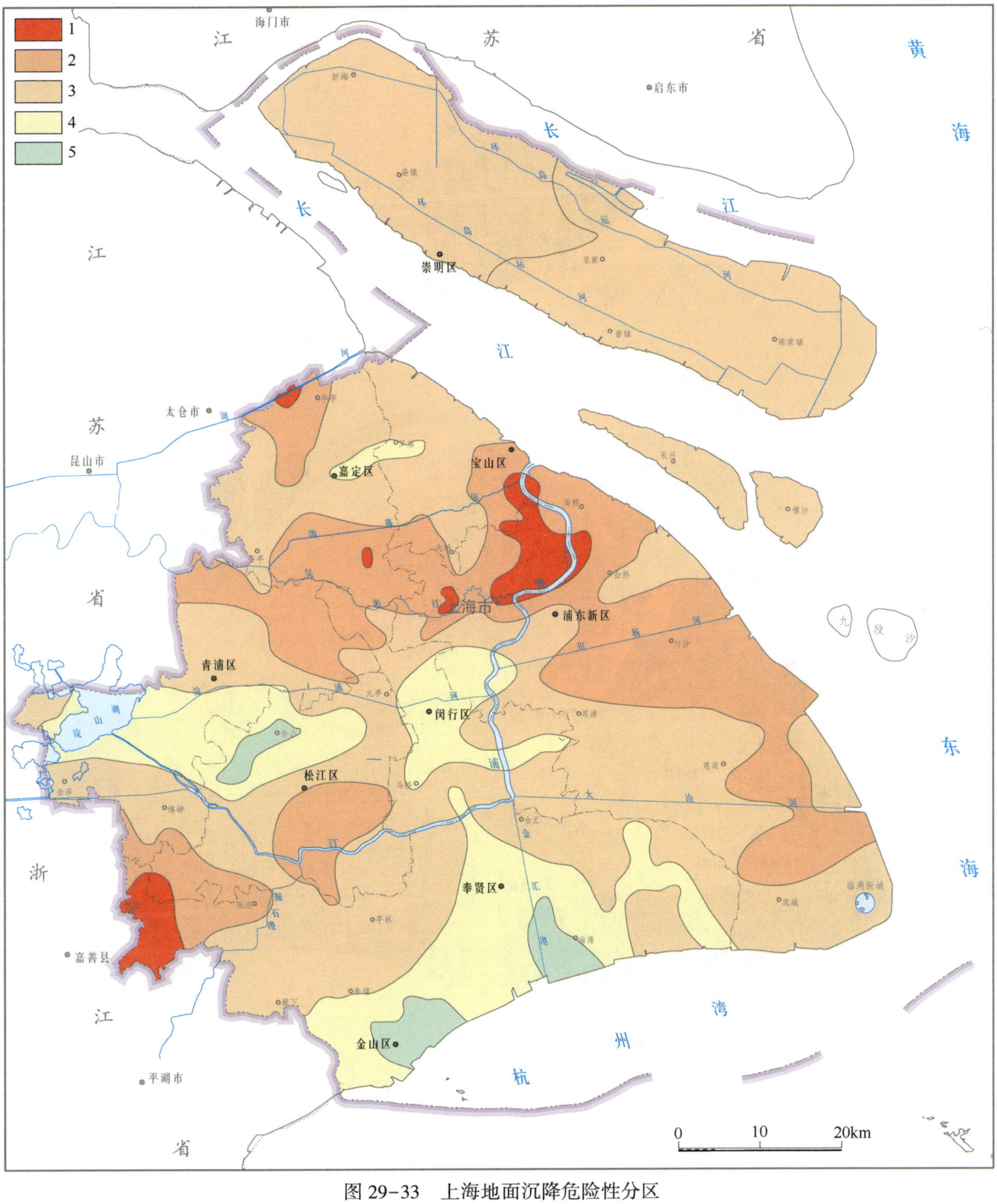

图 29-33　上海地面沉降危险性分区

Fig. 29-33　Land subsidence hazard zoning map of Shanghai

1—危险性高区；2—危险性较高区；3—危险性中等区；4—危险性较低区；5—危险性低区

较发达，建设用地比重比较大，单位面积 GDP 和人口密度、市政设施密度一般，受到地面沉降危害程度中等，归为中等易损性区；松江、浦东新区南部、嘉定建设用地占比中等，人口密度、单位面积 GDP 以及市政设施密度较小，受到地面沉降危害程度较小，归为较低易损性区；奉贤、崇明、金山、青浦人口密度、单位面积 GDP 以及市政设施密度都小，因此，受到地面沉降危害程度小，归为低易损性区；

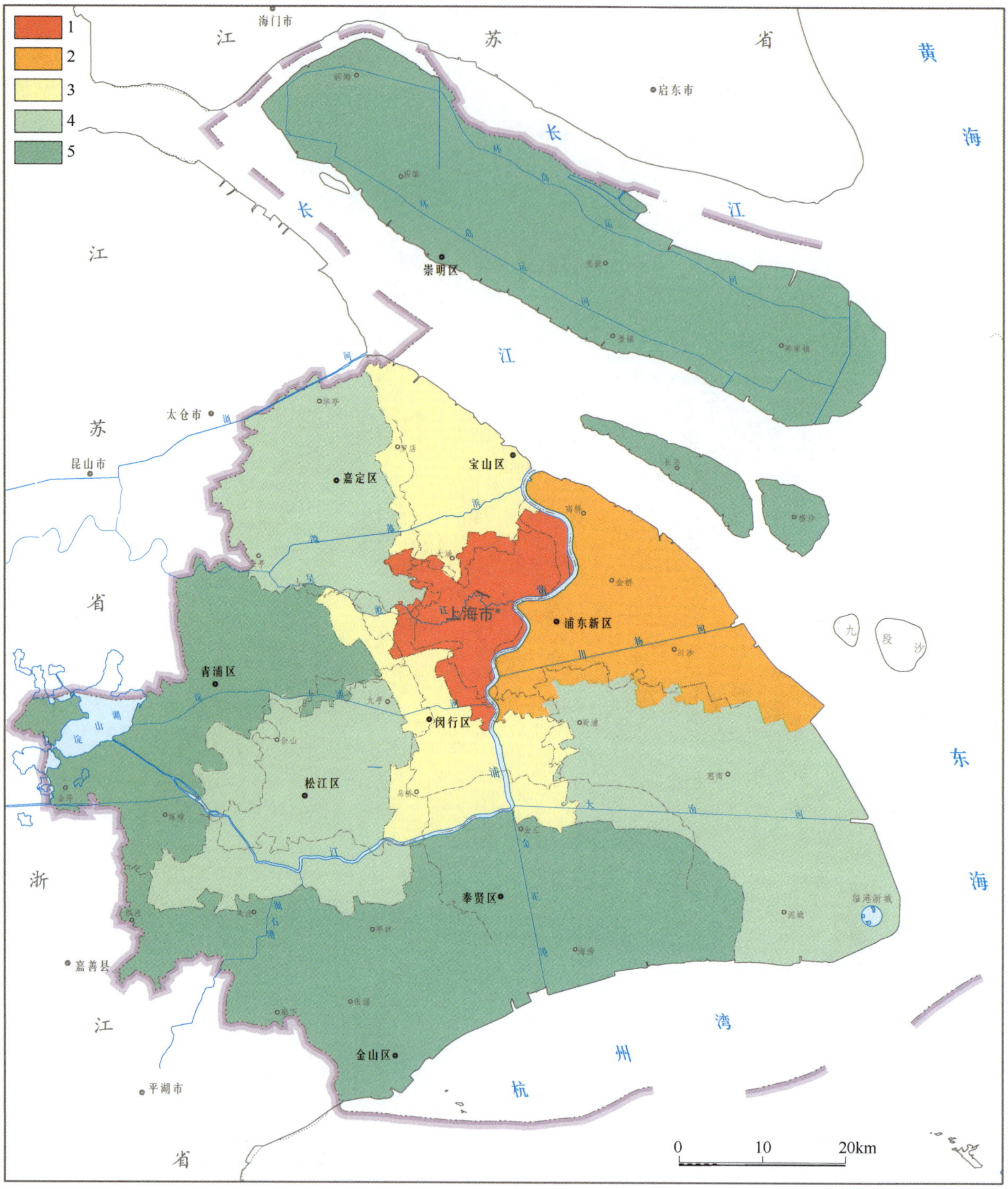

图 29-34　上海地面沉降易损性评价图

Fig. 29-34　Land subsidence vulnerability assessment map of Shanghai

1—易损性高区；2—易损性较高区；3—易损性中等区；4—易损性较低区；5—易损性低区

在上述地面沉降危险性及易损性评价基础上进行地面沉降风险评价（图 29-35），上海地面沉降风险程度在空间上的分布特征为：以中心城为中心形成地面沉降的三级体系，即中心城基本处于地面沉降风险高区，近郊区处于地面沉降较高至中等区，郊区处于地面沉降风险较低区（王寒梅，2013；杨天亮，2014）。

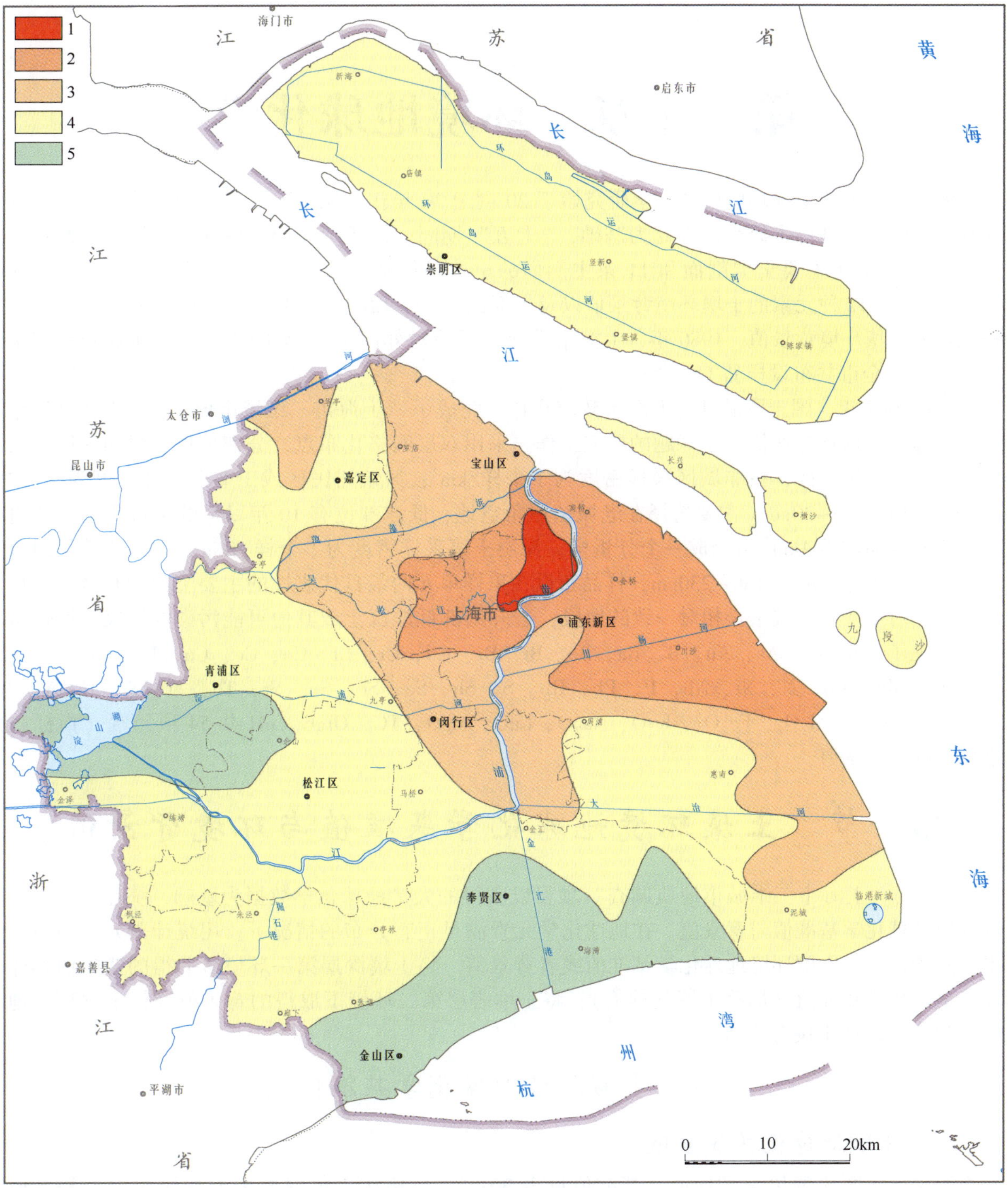

图 29-35　上海地面沉降风险综合评价图

Fig. 29-35　Integrated Land subsidence risk assessment map of Shanghai

1—风险高区；2—风险较高区；3—风险中等区；4—风险较低区；5—风险低区

第三十章 环境地球化学

上海较为系统的土壤环境地球化学研究始于20世纪70年代中期，对主要农业土壤和粮食作物中的9种元素的含量进行了研究。以此为基础，“七五”期间，依托“全国土壤环境背景值调查研究”，上海又开展了20条主要土类剖面和11条主剖面的土壤环境背景值调查研究，提出了主要土类Cu、Pb、Zn、Cd等12种元素的土壤环境背景值以及主剖面的Li、Rb、Cs、Be、Sr、Ba、Sc、Y、La等10余种元素的土壤环境背景值。1980年，上海市农业科学院对郊区10个县的土壤有机氯农药残留情况进行了调查（全市共布设样品112个）。

2004—2008年，国土资源部与上海市政府合作，完成了“上海市三维城市地质调查”“上海市土壤环境地球化学调查与评价”子课题的研究工作。采用双层网格化布点，分别采集表层和深层两层土壤样品。表层土壤采样，在非城区采样密度为1个样/km^2；城镇居民区或工业区采样密度为2个样/km^2，采样深度为0~20cm，主要选择在老城镇屋角空地、低洼部位有10年以上的未被扰动土壤进行采样；所有样品均是4km^2组合成一个分析样。深层土壤采样密度为1个样/4km^2，16km^2等量组合成一个分析样，采样深度为180~230cm，样品布设在采样单元内最具代表性的主要土类分布区，且位于地形起伏相对平缓、植被生长相对一致的地段，避开局部低洼沃土或其他可能污染的地段。通过该项工作，建立了包括Ag、As、Au、B、Ba、Be、Bi、Br、Cd、Cr、Cl、Co、Ce、Cu、F、Ga、Ge、Hg、I、La、Li、Mn、Mo、N、Ni、Nb、P、Pb、Rb、S、Sb、Sc、Se、Sn、Sr、Th、Ti、Tl、U、V、W、Y、Zn、Zr、SiO_2、Al_2O_3、Fe_2O_3、K_2O、Na_2O、CaO、MgO、TC、OrgC、pH共54项元素（指标）的基准值和环境背景值。

第一节 土壤环境地球化学基准值与环境背景值

2008年，“上海市三维城市地质调查”项目以获得的大量地球化学数据为基础，统计计算了上海土壤环境地球化学基准值与背景值。在地球化学元素满足正态分布的情况下，用统计单元的地球化学元素含量均值表示土壤环境地球化学基准值或者背景值。将土壤深层第一环境所取得的背景值称作土壤环境地球化学基准值（简称土壤基准值）；将土壤表层第二环境下取得的背景值称作土壤环境地球化学背景值（简称土壤背景值）。

一、土壤环境地球化学基准值

（一）土壤地球化学基准值

上海深层土壤地球化学平均值、基准值见表30-1，与长江三角洲地区土壤地球化学基准值相比（汪庆华，2007；陈国光，2008），反映出某些组分含量有所差异，但也具有较多的相似性。其中的相似性反映出长江冲积带的共源特点，而差异性则说明了流域内物质组分的迁移性差异和补偿性差异。

（二）不同沉积地质背景的土壤基准值

1. 河口相和河流相沉积母质土壤基准值

早期河口相沉积主要分布在崇明岛近陆的一侧，即崇明—宝山一带；晚期河口相沉积主要分布在崇明岛外侧及长兴岛和横沙岛；河道边滩沉积主要在黄浦江、吴淞江沿岸分布。岩性为砂质粉土、粉质黏土、粉细砂等。

表 30-1　上海深层土壤地球化学平均值、基准值

Table 30-1　Geochemical mean and reference values of deep soil in Shanghai

元素（指标）	平均值（N=427）	标准离差	变异系数	中位数	众数	最大值	最小值	基准值	基准值标准离差	基准值数据个数
Ag	0. 086	0. 027	31. 52	0. 080	0. 100	0. 360	0. 057	0. 080	0. 011	387
Al_2O_3	13. 51	1. 24	9. 18	13. 73	14. 27	21. 35	10. 79	14. 00	0. 72	316
As	7. 7	1. 5	19. 55	7. 5	7. 4	15. 6	4. 7	7. 4	1. 1	377
Au	1. 4	0. 7	45. 69	1. 3	0. 9	8. 4	0. 5	1. 2	0. 2	318
B	64. 1	8. 0	12. 46	64. 0	68. 0	96. 0	42. 0	63. 6	5. 4	366
Ba	464. 3	32. 3	6. 95	462. 0	448. 0	805. 0	394. 0	461. 9	14. 3	335
Be	2. 31	0. 41	17. 76	2. 34	2. 49	6. 60	1. 54	2. 33	0. 26	392
Bi	0. 34	0. 09	26. 81	0. 36	0. 37	0. 85	0. 12	0. 36	0. 06	352
Br	3. 06	0. 87	28. 41	2. 91	2. 70	7. 08	1. 07	2. 80	0. 41	313
C	1. 00	0. 26	26. 11	1. 01	1. 24	2. 11	0. 26	1. 00	0. 21	395
CaO	2. 75	1. 04	37. 87	2. 75	4. 23	4. 69	0. 81	2. 75	1. 04	427
Cd	0. 121	0. 051	41. 99	0. 110	0. 100	0. 800	0. 059	0. 106	0. 018	344
Ce	77. 9	7. 4	9. 56	78. 5	80. 6	113. 0	56. 2	77. 6	6. 3	401
Cl	133. 3	111. 5	83. 61	101. 0	109. 0	970. 0	51. 6	94. 3	14. 1	307
Co	15. 3	2. 1	13. 88	15. 5	15. 9	27. 5	9. 6	15. 3	1. 6	382
Cr	84. 9	9. 3	10. 99	86. 6	88. 5	158. 0	61. 8	85. 3	7. 7	400
Cu	26. 2	5. 2	19. 91	25. 7	24. 7	68. 5	15. 5	25. 8	3. 5	388
F	621. 5	77. 0	12. 39	624. 0	629. 0	839. 0	444. 0	618. 5	66. 2	393
Fe_2O_3	5. 50	0. 74	13. 47	5. 59	6. 09	7. 25	3. 92	5. 49	0. 71	416
Ga	17. 4	2. 2	12. 54	17. 7	16. 4	28. 6	12. 2	17. 5	1. 9	402
Ge	1. 5	0. 3	19. 62	1. 5	1. 6	5. 7	0. 9	1. 5	0. 1	400
Hg	0. 063	0. 038	59. 87	0. 054	0. 047	0. 420	0. 028	0. 051	0. 007	316
I	3. 89	2. 56	65. 72	3. 15	1. 57	14. 60	0. 53	2. 74	1. 04	324
K_2O	2. 51	0. 26	10. 21	2. 56	2. 72	3. 15	1. 89	2. 52	0. 25	421
La	41. 23	3. 81	9. 23	41. 30	41. 30	64. 10	31. 10	41. 31	3. 10	394
Li	48. 10	10. 67	22. 19	49. 90	54. 10	118. 00	26. 10	47. 73	9. 27	409
MgO	2. 08	0. 21	10. 05	2. 11	2. 11	2. 68	1. 10	2. 11	0. 10	330
Mn	774. 2	154. 9	20. 0	751. 0	591. 0	1346. 0	436. 0	756. 3	130. 5	398
Mo	0. 47	0. 17	36. 99	0. 45	0. 42	2. 64	0. 20	0. 45	0. 04	301
N	577	168	29. 04	554	531	1382	268	532	86	334
Na_2O	1. 31	0. 26	19. 97	1. 28	1. 04	1. 83	0. 91	1. 31	0. 26	427
Nb	17. 6	1. 3	7. 48	17. 8	17. 8	26. 6	14. 0	17. 9	0. 6	288
Ni	38. 1	6. 7	17. 47	38. 9	41. 7	64. 2	25. 3	37. 9	6. 4	422
P	615	76	12. 39	615	634	1064	264	614	29	313
Pb	25. 0	4. 3	17. 02	25. 2	25. 9	51. 8	14. 5	25. 3	2. 5	348
Rb	116. 6	16. 9	14. 53	120. 0	126. 0	155. 0	81. 5	116. 7	16. 5	420
S	339	252	74. 32	263	258	2139	60	255	139	356
Sb	0. 58	0. 28	49. 37	0. 54	0. 48	4. 73	0. 31	0. 53	0. 07	351
Sc	13. 11	1. 91	14. 55	13. 40	14. 30	18. 40	8. 86	13. 29	1. 70	401

续表

元素（指标）	平均值（$N=427$）	标准离差	变异系数	中位数	众数	最大值	最小值	基准值	基准值标准离差	基准值数据个数
Se	0.14	0.13	91.44	0.13	0.11	2.10	0.06	0.12	0.03	374
SiO_2	64.17	1.28	2.00	64.12	64.78	68.89	60.05	64.13	0.83	347
Sn	4.1	2.0	49.35	3.8	3.0	32.0	1.9	3.6	0.6	347
Sr	140	32	22.87	137	114	588	103	138	16	419
Th	13.3	1.8	13.59	13.6	14.4	23.3	7.4	13.8	1.0	329
Ti	4828	310	6.43	4919	5009	5789	3911	4997	127	286
Tl	0.6	0.1	16.53	0.6	0.7	1.7	0.4	0.6	0.1	384
U	2.3	0.2	10.59	2.3	2.4	4.8	1.5	2.3	0.2	374
V	99.0	14.8	14.91	101.0	108.0	133.0	59.0	100.5	12.3	394
W	1.9	0.3	15.69	1.9	2.1	3.6	1.1	2.0	0.1	271
Y	28.5	2.6	9.07	29.0	29.0	43.3	20.3	29.7	1.1	285
Zn	93.7	17.1	18.28	91.6	101.0	177.0	57.4	89.7	9.8	342
Zr	243.9	25.4	10.40	240.0	246.0	367.0	183.0	237.5	16.4	360
OrgC	0.53	0.21	39.09	0.49	0.55	1.57	0.24	0.46	0.10	353
pH	8.16	0.28	3.38	8.19	8.22	8.88	6.55	8.21	0.11	325

注：pH 量纲为 1；氧化物和 C、OrgC 的含量单位为%，Au 为 ng/g，其他为 mg/kg。

1）相比较于上海土壤基准值，分布于崇明、长兴、横沙三岛的河口早、晚期沉积相，除极个别元素（Cl）外，基准值无多大变化（表 30-2）。在地球化学组成上，绝大多数元素的基准值偏低于上海基准值，其中 Bi、I、Li、Ni、Sc 相对更低，而 CaO、Cd、C、S、Na_2O、Sr 高于上海基准值。从地球化学分布模式上看，长江河口一带无论早期还是晚期，沉积环境、物质来源，基本变化不大。

表 30-2　上海河流相沉积母质土壤地球化学基准值

Table 30-2　Matrix soil geochemical reference values of fluvial sediments in Shanghai

元素（指标）	晚期河口相（$N=52$）		早期河口相（$N=47$）		河道边滩相（$N=46$）		河流相基准值	上海基准值	中国背景值
	基准值（C_0）	Cv/%	基准值（C_0）	Cv/%	基准值（C_0）	Cv/%			
Ag	0.08	13.9	0.08	11.1	0.09	15.5	0.08	0.08	
Al_2O_3	11.82	3.0	11.85	4.5	13.29	4.2	12.29	12.84	
As	6.54	10.9	6.53	14.7	7.19	13.2	6.74	7.44	11.5
Au	0.97	16.5	0.95	21.9	1.34	18.3	1.08	1.21	
B	58.1	8.8	58.4	10.5	65.7	8.8	60.6	63.6	
Ba	454	1.7	453	1.8	484	3.7	463	462	
Be	1.89	4.0	1.87	8.3	2.29	7.3	2.01	2.33	
Bi	0.22	15.8	0.22	16.5	0.36	12.7	0.26	0.36	
Br	2.43	12.3	2.48	17.8	2.92	12.5	2.60	2.80	
C	1.26	2.5	1.21	5.5	0.94	13.8	1.14	1.00	
CaO	4.18	6.9	3.72	9.9	2.81	14.0	3.60	2.75	
Cd	0.15	12.0	0.14	11.8	0.11	18.3	0.13	0.11	0.084
Ce	67.8	2.4	68.8	3.6	76.9	5.3	71.0	75.9	
Cl	108	20.9	89.0	14.3	92.3	17.8	96.7	94.3	

续表

元素（指标）	晚期河口相（$N=52$）		早期河口相（$N=47$）		河道边滩相（$N=46$）		河流相基准值	上海基准值	中国背景值
	基准值（C_0）	Cv/%	基准值（C_0）	Cv/%	基准值（C_0）	Cv/%			
Co	13.5	5.0	12.9	8.6	15.1	9.0	13.8	14.4	13.4
Cr	73.2	3.4	72.5	5.0	86.8	6.3	77.3	85.3	60.8
Cu	22.5	8.3	21.2	11.6	26.2	11.9	23.2	25.8	23.1
F	526	3.4	553	7.8	644	5.7	572	619	507
Fe_2O_3	4.57	4.8	4.55	5.5	5.44	7.4	4.84	5.49	
Ga	14.7	7.7	15.2	7.2	16.9	6.3	15.5	16.6	
Ge	1.34	3.5	1.38	3.8	1.50	5.4	1.40	1.51	
Hg	0.05	7.8	0.05	13.0	0.06	18.7	0.05	0.05	0.044
I	1.59	20.7	1.79	21.8	2.55	24.5	1.96	2.74	
K_2O	2.18	3.5	2.15	4.2	2.45	5.7	2.25	2.40	
La	36.7	3.8	36.5	3.5	39.6	3.3	37.5	39.8	
Li	34.1	8.3	35.2	7.8	50.5	12.4	39.6	47.7	
MgO	2.09	0.9	2.06	5.1	2.17	5.7	2.10	2.11	
Mn	623	5.4	622	5.7	793	15.1	677	756	597
Mo	0.45	12.4	0.43	12.6	0.46	11.6	0.45	0.45	
N	439	12.2	472	20.8	563	15.6	489	532	
Na_2O	1.69	3.6	1.68	3.8	1.50	7.8	1.63	1.46	
Nb	16.1	5.4	16.2	5.6	18.2	4.8	16.8	17.9	
Ni	28.9	5.9	29.2	6.8	35.5	8.7	31.1	36.2	28.6
P	636	1.6	628	3.5	628	5.1	631	614	
Pb	20.7	8.3	21.0	12.6	26.7	8.2	22.7	25.3	24.7
Rb	94.1	4.8	94.7	6.0	116	8.4	101	110	
S	450	30.6	518	19.8	291	60.1	422	255	
Sb	0.49	7.6	0.46	14.7	0.52	18.1	0.49	0.53	
Sc	10.6	5.1	10.5	8.7	12.9	8.4	11.3	13.3	
Se	0.10	14.7	0.10	14.5	0.14	25.8	0.12	0.12	0.246
SiO_2	63.96	0.8	64.37	0.9	64.89	1.0	64.39	64.13	
Sn	2.91	7.5	2.73	7.1	4.07	16.8	3.22	3.58	
Sr	162	2.6	157	3.1	141	6.2	154	138	
Th	11.0	6.3	11.5	10.2	13.2	10.7	11.9	12.8	
Ti	4378	1.5	4301	2.4	4769	3.1	4477	4850	
Tl	0.52	8.1	0.55	9.8	0.61	11.2	0.56	0.63	
U	2.14	3.9	2.13	7.4	2.14	8.6	2.14	2.19	
V	84.4	6.2	79.3	13.9	95.9	8.1	86.4	97.5	84.3
W	1.59	10.6	1.70	11.6	2.01	5.8	1.76	1.96	
Y	24.6	2.7	24.9	3.1	27.8	3.4	25.7	28.2	
Zn	80.8	8.7	82.1	7.1	105	8.4	88.9	98.8	71.1
Zr	252	6.1	247	7.4	234	5.0	245	237	

注：中国背景值据中国环境监测总站（1990）。氧化物和 C 的含量单位为%，Au 为 ng/g，其他为 mg/kg。

2）与河口相沉积对比，发育于黄浦江、苏州河河道的边滩沉积，绝大多数元素的基准值（Al、As、Au、Bi、Cr、Cu、F、Fe、Li、Mn、N、Sn 等）高于河口冲积相元素的基准值，而河口相中几个含量相对较高的组分（如 Ca、S、Cd、C），在边滩沉积中其基准值则偏低。

将该沉积物 54 种元素（指标）的基准值和上海基准值进行对比，河口相沉积的早期、晚期比值的变化趋势和分布形式几乎一致，反映出共同的沉积环境和相似的沉积地球化学组合特征。河道边滩冲积相图谱曲线在形态上和河口相图谱完全不重叠，反映出该沉积区深部是一个完全不同于河口冲积相的另一个沉积体系。

2. 滨海相沉积母质土壤基准值

滨海相沉积按时间划分为晚期、中期和早期 3 种沉积类型。将不同时期沉积相母质土壤基准值进行统计，结果见表 30-3。

表 30-3　上海滨海相沉积母质土壤地球化学基准值

Table 30-3　Matrix soil geochemical reference values of littoral sediments in Shanghai

元素	晚期（N=32）		中期（N=100）		早期（N=30）		基准值均值	区域土壤基准值	
	基准值（C_0）	Cv/%	基准值（C_0）	Cv/%	基准值（C_0）	Cv/%		上海	中国
Ag	0.08	13.5	0.07	11.2	0.08	11.6	0.08	0.08	
Al_2O_3	11.98	3.6	12.94	1.9	13.42	3.4	12.84	12.84	
As	8.17	13.7	7.40	12.0	7.18	10.8	7.51	7.44	11.5
Au	1.26	20.0	1.22	16.2	1.23	24.7	1.23	1.21	
B	60.6	8.0	65.0	7.9	64.6	3.6	64.1	63.6	
Ba	427	3.9	460	3.6	465	3.0	454	462	
Be	2.13	4.2	2.44	3.3	2.42	5.2	2.38	2.33	
Bi	0.30	12.9	0.38	7.4	0.37	16.0	0.36	0.36	
Br	4.51	13.7	3.08	11.3	2.77	10.5	3.30	2.80	
C	1.16	8.9	0.93	13.9	0.82	19.2	0.95	1.00	
CaO	4.06	5.5	2.74	8.4	2.33	15.9	2.92	2.75	
Cd	0.13	16.0	0.10	8.0	0.10	13.3	0.10	0.11	0.084
Ce	73.1	3.3	79.7	3.3	80.9	5.1	78.6	75.9	
Cl	170	18.5	90.3	10.9	91.8	17.0	106	94.3	
Co	12.9	6.9	15.1	7.5	14.9	5.2	14.6	14.4	13.4
Cr	79.9	3.7	89.7	3.8	88.7	4.9	87.6	85.3	60.8
Cu	23.9	12.7	26.0	7.7	25.3	11.1	25.4	25.8	23.1
F	608	4.3	667	8.0	631	3.6	649	619	507
Fe_2O_3	5.07	7.9	5.85	3.5	5.76	6.8	5.68	5.49	
Ga	14.7	5.9	17.3	6.4	17.8	5.2	16.9	16.6	
Ge	1.41	3.4	1.56	3.5	1.57	4.9	1.53	1.51	
Hg	0.05	15.3	0.05	14.2	0.05	9.8	0.05	0.05	0.044
I	3.57	26.9	6.50	37.6	4.41	38.0	5.54	2.74	
K_2O	2.27	4.2	2.51	2.8	2.55	3.5	2.47	2.40	
La	38.9	5.9	41.4	3.9	42.8	5.7	41.2	39.8	
Li	42.2	6.5	52.8	9.8	51.9	9.7	50.6	47.7	
MgO	2.09	3.8	2.21	3.0	2.17	4.5	2.18	2.11	

续表

元素	晚期（$N=32$）		中期（$N=100$）		早期（$N=30$）		基准值均值	区域土壤基准值	
	基准值（C_0）	Cv/%	基准值（C_0）	Cv/%	基准值（C_0）	Cv/%		上海	中国
Mn	698	9.7	835	12.4	891	7.0	819	756	597
Mo	0.52	12.5	0.43	7.9	0.41	14.6	0.45	0.45	
N	467	13.5	564	15.9	525	14.3	538	532	
Na_2O	1.48	2.5	1.36	3.3	1.39	6.4	1.39	1.46	
Nb	16.7	3.1	18.0	2.5	18.3	4.7	17.8	17.9	
Ni	33.6	6.8	39.5	4.2	39.2	8.2	38.3	36.2	28.6
P	640	3.9	595	4.5	609	2.6	607	614	
Pb	22.4	10.0	25.2	7.4	26.0	7.9	24.8	25.3	24.7
Rb	97.2	7.4	116	4.4	120	5.8	113	110	
S	198	24.1	95.8	17.7	147	44.3	126	255	
Sb	0.62	12.4	0.52	12.3	0.48	12.1	0.54	0.53	
Sc	12.3	5.0	14.2	2.3	13.9	7.1	13.8	13.3	
Se	0.11	16.2	0.12	26.0	0.12	22.9	0.12	0.12	0.246
SiO_2	64.20	1.9	63.63	1.4	64.16	2.0	63.84	64.13	
Sn	3.17	12.5	3.81	13.7	3.86	13.5	3.69	3.58	
Sr	152	2.1	135	3.4	131	3.6	137	138	
Th	12.1	6.5	13.2	2.8	13.3	6.0	13.0	12.8	
Ti	4614	1.6	4866	1.4	4912	2.6	4825	4850	
Tl	0.56	5.6	0.65	4.9	0.65	6.2	0.63	0.63	
U	2.19	2.9	2.21	4.7	2.23	4.3	2.21	2.19	
V	88.5	8.0	103	4.5	103	6.6	99.9	97.5	84.3
W	1.79	4.7	1.92	2.5	2.00	7.7	1.91	1.96	
Y	27.0	3.3	28.2	2.5	28.8	4.6	28.1	28.2	
Zn	88.6	6.9	106	5.1	105	6.2	102	98.8	71.1
Zr	247	11.1	243	8.2	256	11.1	246	237	

注：氧化物和 C 的含量单位为%，Au 为 ng/g，其他为 mg/kg。

1）滨海潮滩沉积相分布特征与河口沉积相俨然不同，除多数元素含量分布存在差异外，集中表现为 I、S 两个完全相反的分布量。滨海沉积相为 I 的高量和 S 的低量分布特征。

2）相比于全国土壤基准值，As、Se 基准值明显偏低，而 Cd、Cr、F、Mn、Ni、Zn 等重金属元素的基准值则明显偏高。

3）对比早、中、晚 3 个不同沉积时期，早、中期沉积相在分布模式上基本相似，而晚期则有明显差异，其中 As、Br、CaO、Cl 基准值明显增高，其他多数元素较早、中期偏低。出现这种不同，初步推断可能是晚期海相沉积母质保留有滨海盐土特征，而早、中期海相沉积母质在退盐化过程中，逐渐降低了上述成分。在退盐过程中，存在 C、Cu、Sr、Sb、As 等元素基准值由晚期、中期至早期呈递降的趋势。

3. 湖沼相沉积母质土壤基准值

湖沼相沉积主要分布于上海西部，包括潟湖相、滨湖相以及湖沼相 3 种沉积类型，统计不同沉积类型母质土壤基准值（表 30-4）。湖沼相沉积母质主要有以下几个特点：

表 30-4 上海湖沼相沉积母质土壤地球化学基准值

Table 30-4 Matrix siol geochemical reference values of limnetic sediments in Shanghai

元素（指标）	泻湖相（N=23）		滨湖相（N=85）		湖沼相（N=12）		基准值均值	C层土壤基准值均值	
	平均值（X）	Cv/%	平均值（X）	Cv/%	平均值（X）	Cv/%		上海	中国
Ag	0.09	14.1	0.08	13.0	0.08	6.7	0.08	0.08	
Al_2O_3	13.41	4.0	13.55	3.0	13.36	3.9	13.50	12.84	
As	8.77	9.0	8.76	14.0	7.69	12.5	8.65	7.44	11.5
Au	1.63	21.8	1.42	18.8	1.37	14.9	1.46	1.21	
B	64.6	7.4	65.9	7.7	63.4	6.2	65.4	63.6	
Ba	465	2.6	472	1.6	472	1.7	471	462	
Be	2.54	4.9	2.50	6.0	2.59	5.2	2.52	2.33	
Bi	0.39	8.6	0.39	10.5	0.41	8.9	0.39	0.36	
Br	3.00	17.4	2.55	14.1	2.55	22.4	2.63	2.80	
C	0.74	13.7	0.79	24.8	0.95	21.8	0.80	1.00	
CaO	1.74	23.2	1.29	22.1	1.50	18.7	1.40	2.75	
Cd	0.10	7.2	0.10	11.6	0.10	8.7	0.10	0.11	0.084
Ce	82.7	3.4	80.0	2.3	76.9	7.4	80.2	75.9	
Cl	94.6	8.0	96.8	14.4	112	5.7	97.9	94.3	
Co	15.7	9.2	15.1	5.9	14.9	11.2	15.2	14.4	13.4
Cr	91.9	2.4	90.3	3.3	91.2	5.1	90.7	85.3	60.8
Cu	28.3	8.1	29.2	8.3	30.1	8.2	29.1	25.8	23.1
F	672	8.5	657	8.1	649	9.3	659	619	507
Fe_2O_3	6.16	6.9	6.15	5.8	6.25	7.1	6.16	5.49	
Ga	17.4	6.8	17.3	7.9	17.0	4.3	17.3	16.6	
Ge	1.58	5.1	1.57	4.4	1.65	3.1	1.58	1.51	
Hg	0.05	19.5	0.05	17.1	0.05	9.4	0.05	0.05	0.044
I	3.88	19.5	2.47	28.2	2.01	22.8	2.70	2.74	
K_2O	2.59	3.7	2.53	4.6	2.45	3.7	2.53	2.40	
La	42.3	3.3	40.3	4.4	40.0	6.6	40.6	39.8	
Li	55.2	6.4	54.8	7.8	55.7	7.4	55.0	47.7	
MgO	2.05	2.5	1.90	8.0	1.94	7.6	1.94	2.11	
Mn	875	11.6	802	14.9	769	19.2	812	756	597
Mo	0.49	18.2	0.45	9.3	0.42	7.5	0.46	0.45	
N	585	13.4	628	14.7	687	18.6	626	532	
Na_2O	1.35	5.2	1.33	3.9	1.42	3.9	1.34	1.46	
Nb	18.1	2.0	18.1	2.3	18.6	2.7	18.1	17.9	
Ni	41.9	8.4	40.5	6.9	39.3	7.9	40.7	36.2	28.6
P	590	5.7	562	9.6	520	10.5	563	614	
Pb	26.2	5.7	26.8	3.5	27.3	4.8	26.7	25.3	24.7
Rb	121	7.0	118	9.8	117	5.8	118	110	
S	160	18.2	206	40.2	257	17.2	202	255	

续表

元素（指标）	潟湖相（$N=23$）		滨湖相（$N=85$）		湖沼相（$N=12$）		基准值均值	C 层土壤基准值均值	
	平均值（X）	Cv/%	平均值（X）	Cv/%	平均值（X）	Cv/%		上海	中国
Sb	0.63	14.4	0.58	6.0	0.56	8.3	0.59	0.53	
Sc	14.7	5.9	14.8	5.5	15.0	6.4	14.8	13.3	
Se	0.14	5.2	0.16	16.9	0.19	17.2	0.16	0.12	0.246
SiO_2	63.90	1.4	64.59	1.5	63.39	0.6	64.34	64.13	
Sn	3.82	11.2	3.89	7.7	3.97	15.2	3.88	3.58	
Sr	122	5.1	117	1.8	118	3.4	118	138	
Th	13.5	5.3	13.3	4.4	12.6	5.7	13.3	12.8	
Ti	4863	1.6	4879	1.6	4858	2.9	4874	4850	
Tl	0.67	7.2	0.67	5.5	0.64	6.8	0.67	0.63	
U	2.22	5.5	2.25	4.3	2.14	5.2	2.23	2.19	
V	105	9.0	105	5.2	103	7.0	105	97.5	84.3
W	2.08	6.8	1.98	4.3	2.06	3.1	2.01	1.96	
Y	28.3	2.8	28.6	2.1	28.8	1.3	28.6	28.2	
Zn	103	6.9	101	4.5	95.4	4.9	101	98.8	71.1
Zr	234	7.9	228	4.1	221	3.4	228	237	

注：氧化物和 C 的含量单位为%，Au 为 ng/g，其他为 mg/kg。

1）Cd、Co、Cr、F、Mn、Ni、V、Zn 等元素的基准值低于中国背景值；

2）潟湖相中的 As、Ca、I、Mn 等元素的基准值高于其他两种类型沉积物的基准值，而 N、S 等元素则低；在湖沼相的 3 种沉积类型中，As、Au、Mn、P 等元素的基准值表现出由湖沼相→滨湖相→潟湖相逐渐递增的趋势，而 C、I、N 等元素则截然相反。

3）3 种沉积相区基本展示出相类似的分布图谱，曲线走势大体一致，和上海基准值对比，典型的是 CaO、MgO、Na_2O、P、Sr 的基准值偏低，而 As、Fe、Li、N、Ni、Cr 等金属元素、非金属元素则偏高。

4）3 种沉积相区之间相互对比，潟湖相又和湖滨相、湖沼相在 I、S 分布上差异较大，而与滨海相 I、S 分布近似。由此认为，潟湖相主体上包含了湖积相地球化学分布特点，但由于其和滨海潮滩相沉积的某种“亲缘”关系，又兼容了滨海潮滩相沉积的主要分布形式。

二、表层土壤地球化学背景值

（一）表层土壤区域地球化学背景值

1. 表层土壤元素区域地球化学变异特性

土壤表层元素分布，总体受区域地质母质环境、土壤表层发生作用、自然生态环境、土壤耕作活动、环境污染以及自身化学性质等复杂因素制约，表现出区域上显著的变异特征。上海受自身特殊的环境要素制约，显示出符合自身特点的地球化学变异特征。对其按照区域变异特性分解，大体划分出以下 5 种类型：

1）均匀分布类型：变异系数（Cv）≤25%，包括硼、铍、铈、钴、氟、镓、锗、镧、锂、锰、铌、镍、铷、钪、锶、钍、钛、铊、铀、钒、铱、锆、硅、铁、铝、钾、钠、镁、pH 等 29 种元素（指标），区域内基本表现出均匀分布，含量变差不大。据含量直方图统计，这些元素中多数元素含量基本服从正态分布，含量区间窄小，变差不大。

在区域分布上，除镍、锂、铷、钾、硼、铝等与某些专属土壤区有简单的对应性分区外，其他元素这种指示作用则并不明显，某些元素只是显示北部河口沉积层区与西部湖积层区及中部滨海潮滩沉积区含量上有所差别。该元素组，基本可以表达了上海平原区原始沉积比较相近的地球化学组成，同时也代表了土壤中化学活动性方面的惰性组分，即便在化学活动性强、甚至强烈人工扰动的土壤表层，亦相应保持着稳定的含量特征。

2）相对分异型：$25\%<Cv\leqslant 50\%$，包括砷、铬、碘、氮、磷、钙、碳、有机碳。这些组分，表现出在全区土壤表层较弱的分异特征，含量有不大的变化范围。含量统计直方图显示，本组元素多数呈对数正态分布，其中基本呈标准对数正态分布的元素有碘、溴、氮、磷、钙诸元素。考察它们的分布，这些元素基本表现为在北部河口沉积区持以低分布量，在西部湖积区持以分布高量，广泛发育的滨海潮滩冲积层持以中间分布量；或者相反，在各土壤单元区均持以有变异性的分布量，仅在局部区域形成高分布量。本组元素基本表现为在土体中具一定的化学活动性，尤其在表生环境条件变化较大的情况下，可产生一定的变异。

3）分异型：$50\%<Cv\leqslant 75\%$，仅包括铋、溴、硫、硒、钨等5种元素。统计显示，铋虽属该类，但分异程度并非很高。在全区含量直方图上，铋、硒表现为二重母体分布的形式。表现为在原始分布上有着铋量两种不同的分布区，即原生背景场和城市人类活动环境污染叠加场。

4）强分异型：$75\%<Cv\leqslant 100\%$，仅包括钼。变异系数为88%，显示出钼相对于铋、溴、硫、硒、钨稍强的区域变异，含量具有更大的变差。在全区含量直方图上，钼基本显示为对数正态分布，带有较强的左偏性质。总体分析，钼、铋、溴、硫、硒、钨等6种元素在上海土壤表层环境较强污染条件下所产生的地球化学变异。

5）极强分异型：$Cv>100\%$，包括银、金、镉、氯、铜、汞、铅、锑、锡、锌等10种元素，变异系数达到200%以上的就有金、镉、氯、铜、锑、锡，其中金达656%，镉达787%，铜达615%。在含量直方图上，10种元素基本表现为二重母体分布，其一母体反映为土体内上述元素的基本分布特征，另一母体为异常母体，反映为区内环境污染作用的叠加特征。在地球化学图上，这种分布特性十分明显，形成的强高量异常散布于中心城上方。

2. 表层土壤区域地球化学背景值

将全区背景值与长江三角洲地区土壤地球化学基准值以及上海的基准值进行对比，可以看出：

1）与长江三角洲平原土壤基准值比较，十分显见两者的分布有着极大的相似性和差异性（表30-5）。

表30-5　上海表层土壤地球化学平均值、背景值

Table 30-5　Geochemical mean and background values of deep soil in Shanghai

元素（指标）	平均值（N=1801）	标准离差	变异系数	中位数	众数	最大值	最小值	背景值	苏锡常背景值	上海基准值	富集系数
Ag	0.137	145.0	0.198	0.100	0.110	4.600	0.043	0.095	0.11	0.080	1.18
Al_2O_3	13.21	6.7	0.89	13.30	13.42	17.06	10.49	13.33	12.22	14.00	0.95
As	8.2	27.8	2.3	7.9	7.9	69.8	3.9	7.8	8.0	7.4	1.05
Au	3.8	656.3	24.8	2.1	1.5	1026.0	0.5	1.9	2.1	1.2	1.54
B	63.7	11.7	7.5	64.0	60.0	113.0	39.0	63.9	63	63.6	1.01
Ba	479.0	9.6	45.8	473.0	472.0	1031.0	370.0	473.0	460	461.9	1.02
Be	2.25	9.7	0.22	2.25	2.24	4.95	1.56	2.26	2.08	2.33	0.97
Bi	0.45	61.4	0.28	0.41	0.40	4.63	0.13	0.40	0.32	0.36	1.12
Br	6.22	56.1	3.48	5.82	5.85	118.00	1.22	5.56	4.61	2.80	1.99
C	1.67	31.4	0.53	1.59	1.32	7.33	0.29	1.55	1.5	1.00	1.55
CaO	2.47	46.7	1.16	2.22	1.20	7.30	0.76	2.42	2.67	2.75	0.88

续表

元素（指标）	平均值（$N=1801$）	标准离差	变异系数	中位数	众数	最大值	最小值	背景值	苏锡常背景值	上海基准值	富集系数
Cd	0.279	786.8	2.192	0.180	0.160	90.900	0.053	0.170	0.183	0.106	1.61
Ce	77.3	17.3	13.3	77.3	79.6	593.0	58.2	77.0	75	77.6	0.99
Cl	230.4	259.0	596.7	115.0	102.0	12146.0	48.9	89.1	108	94.3	0.94
Co	14.7	10.7	1.6	14.7	15.3	29.7	10.4	14.7	12.9	15.3	0.96
Cr	90.0	39.7	35.8	86.7	101.0	1341.0	59.4	86.2	76	85.3	1.01
Cu	41.3	615.0	254.0	31.5	31.0	10779.0	12.4	30.7	27	25.8	1.19
F	656.8	19.2	126.1	643.0	600.0	2656.0	320.0	639.5	567	618.5	1.03
Fe_2O_3	5.28	9.8	0.52	5.27	5.18	9.45	3.53	5.25	4.64	5.49	0.96
Ga	17.2	8.9	1.5	17.3	17.7	31.5	12.2	17.3	15.3	17.5	0.99
Ge	1.5	8.6	0.1	1.5	1.5	4.2	1.1	1.5	1.36	1.5	1.00
Hg	0.171	102.5	0.175	0.140	0.140	3.350	0.021	0.116	0.115	0.051	2.27
I	3.38	48.6	1.64	3.01	2.22	12.50	0.78	2.70	1.93	2.74	0.99
K_2O	2.42	7.7	0.19	2.42	2.40	2.98	1.55	2.42	2.23	2.52	0.96
La	41.49	14.3	5.93	41.40	41.00	269.00	29.30	41.50	40	41.31	1.00
Li	45.57	14.9	6.80	46.10	47.00	86.70	24.70	45.91	36	47.73	0.96
MgO	1.94	13.3	0.26	1.98	2.07	3.37	1.06	1.96	1.97	2.11	0.93
Mn	717.6	18.5	133.0	706.0	683.0	2142.0	437.0	703.4	601	756.3	0.93
Mo	0.68	88.3	0.60	0.57	0.52	13.40	0.24	0.55	0.49	0.45	1.22
N	1351	36.3	491	1304	963	3623	209	1283	1329	532	2.41
Na_2O	1.35	16.5	0.22	1.33	1.12	3.59	0.88	1.33	1.42	1.31	1.01
Nb	17.6	5.7	1.0	17.7	17.5	22.3	12.6	17.8	16.8	17.9	0.99
Ni	37.0	13.6	5.0	36.8	35.4	76.2	23.5	36.8	31	37.9	0.97
P	964	28.4	274	911	890	2718	361	864	964	614	1.41
Pb	38.5	197.7	76.1	30.0	31.2	2521.0	14.7	29.1	25.5	25.3	1.15
Rb	112.1	11.2	12.6	113.0	116.0	144.0	65.5	112.3	98	116.7	0.96
S	419	60.0	251	353	253	2869	77	297	295	255	1.16
Sb	1.42	267.8	3.79	0.82	0.76	55.70	0.30	0.79	0.82	0.53	1.48
Sc	12.60	10.8	1.35	12.70	13.70	18.00	8.34	12.67	11.3	13.29	0.95
Se	0.26	61.1	0.16	0.23	0.21	2.27	0.05	0.23	0.21	0.12	1.86
SiO_2	63.86	3.1	1.99	64.04	63.98	70.04	53.43	64.09	65.81	64.13	1.00
Sn	12.2	264.6	32.3	7.8	11.0	726.0	2.0	6.8	6.86	3.6	1.90
Sr	136	18.6	25	132	121	918	102	131	142	138	0.95
Th	13.0	10.7	1.4	13.2	13.3	22.4	4.9	13.3	12.1	13.8	0.96
Ti	4795	5.3	253	4839	4871	6095	3754	4872	4839	4997	0.98
Tl	0.6	11.0	0.1	0.6	0.7	1.4	0.4	0.6	0.57	0.6	0.99
U	2.3	9.1	0.2	2.3	2.4	3.4	1.7	2.3	2.15	2.3	1.01
V	97.9	12.6	12.3	97.3	101.0	197.0	61.5	97.1	84	100.5	0.97

续表

元素（指标）	平均值（N=1801）	标准离差	变异系数	中位数	众数	最大值	最小值	背景值	苏锡常背景值	上海基准值	富集系数
W	2.5	68.4	1.7	2.2	2.1	38.2	1.2	2.1	1.8	2.0	1.07
Y	28.2	7.1	2.0	28.6	29.5	39.7	19.7	28.5	27	29.7	0.96
Zn	122.7	133.2	163.5	103.0	102.0	6241.0	53.6	99.3	82	89.7	1.11
Zr	241.2	9.2	22.2	238.0	236.0	363.0	170.0	235.9	260	237.5	0.99
OrgC	1.24	42.4	0.53	1.18	1.43	5.78	0.19	1.13	1.07	0.46	2.45
pH	7.61	8.9	0.67	7.84	8.04	8.96	4.56	8.03	7.88	8.21	0.98

注：pH 量纲为 1；氧化物和 C、OrgC 的含量单位为%，Au 为 ng/g，其他为 mg/kg。

2）与上海基准值对比，可以发现表层土壤背景值与基准值之间某些元素含量悬殊差异很大，说明这些元素受后期人为扰动影响比较大。

3）上海土壤表生环境下表层土壤相当多的元素背景值接近于上海的基准值，且其中有铍、铈、氯、钴、镓、锗、碘等26种元素含量偏低于基准值，造成这种情况的原因十分复杂，除基本说明它们在第二环境污染不显著外，最直接的原因可能是土体中存在与环境条件相关的稳定发生层，而这些元素经淋滤易在下部淀积层中富集；而与此相反，对环境敏感、化学活动性大的元素普遍富集，包括银、铋、铜、钼、铅、硫、磷、锑、金、溴、镉、汞、氮、硒、锡、碳、有机碳等，其富集系数均在 1.1 以上（表 30-6），其中金、镉、碳表层富集均为深层的 1.5 倍以上，汞、溴、氮、硒、锡、有机质富集达 1.8~2.45 倍。这种富集集中表现了上述元素在表土环境下较强的活动性和易累积性，另外也显示出农业耕作和经济活动人为污染作用。上海的客观实际也表明，上海城市土壤污染实际上主要为上述元素组合。

表 30-6 上海表层土壤地球化学背景值与基准值比值分组统计

Table 30-6 Soil surface background value to soil reference value ratios in Shanghai by group

R 值	元素及指标	富集情况
0.9~1	铍、铈、氯、钴、镓、锗、碘、锂、锰、铌、镍、铷、钪、锶、钍、钛、铊、钒、铱、锆、硅、铁、铝、钾、钙、镁	淋溶输出
1~1.1	砷、硼、钡、铬、氟、镧、铀、钨、钠	外源成分输入
1.1~1.3	银、铋、铜、钼、铅、硫	
1.3~1.5	磷、锑	
>1.5	金、溴、镉、汞、氮、硒、锡、碳、有机碳	

（二）各沉积单元表层土壤地球化学背景值

按对上海第四系地质单元的划分，以河口—河流相、滨海相和湖沼相沉积物为母质形成的各地质单元表层土壤背景值进行统计，结果列于表 30-7。

1. 晚全新世河口沉积区

1）整个晚全新世河口早、晚两期沉积区表层土壤多元素含量以偏低于区域背景值分布为特征，且大量元素比值在 0.9~1.0 之间。

2）两期沉积最为显著低量的为汞、金、铋、氮、硫、锡、硒、有机碳等，其中汞在晚、早期沉积区背景值为 0.06μg/g 和 0.07μg/g，和上海表层土壤背景值比值仅为 0.5 和 0.6，是所有沉积区最低的，和相应上海的基准值对比也仅高出 20% 和 40%。汞、锡、硫、铋、铅、氮等在表层土壤属于最易污染的元素。由此可见，河口沉积区污染度最低，目前仍基本保持较为纯净的自然生态环境。

表 30-7 上海不同地质单元土壤环境地球化学背景值

Table 30-7 Soil environmental geochemical background values in Shanghai by geological unit

元素（指标）	河流相			滨海相			湖沼相		
	晚期（N=262）	早期（N=198）	河道边滩相（N=202）	晚期（N=147）	中期（N=373）	早期（N=136）	潟湖相（N=87）	滨湖相（N=341）	湖沼相（N=55）
Ag	0.09	0.10	0.14	0.08	0.10	0.11	0.09	0.10	0.12
Al_2O_3	12.7	12.7	12.5	12.5	13.4	13.0	13.4	13.3	13.5
As	8.38	7.94	7.39	7.71	7.72	7.24	7.82	7.89	7.89
Au	1.43	1.52	2.84	1.36	2.02	2.41	1.96	2.32	2.88
B	58.5	63.0	60.9	59.8	66.4	65.1	64.5	65.3	63.0
Ba	416	420	441	387	426	431	424	432	426
Be	2.09	2.06	2.19	2.12	2.38	2.30	2.42	2.35	2.32
Bi	0.32	0.31	0.45	0.35	0.43	0.41	0.42	0.41	0.44
Br	6.53	6.20	4.77	7.02	5.01	4.87	5.79	6.23	7.09
C	1.52	1.46	1.56	1.65	1.50	1.44	1.51	1.73	2.22
CaO	3.98	3.37	2.54	3.52	1.82	1.63	1.25	1.19	1.20
Cd	0.18	0.18	0.22	0.17	0.17	0.18	0.16	0.16	0.18
Ce	71.9	71.9	74.2	73.9	77.6	78.1	78.7	76.3	74.7
Cl	86.5	88.1	101	86.0	86.7	100	128	131	169
Co	14.5	13.7	13.6	13.8	15.2	14.6	15.4	15.0	14.7
Cr	80.3	77.8	87.0	81.7	90.1	88.0	87.1	86.7	89.7
Cu	29.8	28.4	32.5	27.4	31.4	32.1	31.3	31.3	33.3
F	614	614	661	628	675	639	643	617	656
Fe_2O_3	5.14	4.91	5.10	5.07	5.54	5.23	5.55	5.30	5.22
Ga	16.6	16.5	16.6	16.2	17.9	17.5	18.1	17.8	17.7
Ge	1.43	1.44	1.50	1.46	1.55	1.59	1.58	1.55	1.53
Hg	0.06	0.07	0.22	0.08	0.14	0.17	0.15	0.17	0.26
I	2.42	2.79	2.86	4.14	4.40	3.01	3.10	2.20	2.19
K_2O	2.31	2.29	2.25	2.27	2.45	2.33	2.43	2.26	2.23
La	40.8	40.0	41.3	40.5	42.1	42.0	42.5	41.7	40.8
Li	40.0	39.4	44.5	41.7	49.6	46.5	50.6	47.7	48.1
MgO	2.20	2.11	1.82	2.09	2.02	1.86	1.79	1.63	1.66
Mn	722	700	718	686	760	731	688	634	556
Mo	0.57	0.52	0.54	0.53	0.52	0.55	0.53	0.57	0.61
N	885	993	1312	1226	1337	1329	1607	1812	2201
Na_2O	1.20	1.22	1.19	1.53	1.32	1.43	1.44	1.47	1.52
Nb	17.1	17.1	17.5	17.1	18.0	17.9	18.0	18.0	17.9
Ni	34.3	32.8	34.6	32.1	36.8	34.4	36.7	34.6	33.5
P	715	888	1042	994	915	905	854	890	868
Pb	23.9	24.3	31.8	25.0	31.3	32.1	30.5	30.6	32.8
Rb	101	101	103	97.1	114	108	113	109	109

续表

元素（指标）	河流相			滨海相			湖沼相		
	晚期（N=262）	早期（N=198）	河道边滩相（N=202）	晚期（N=147）	中期（N=373）	早期（N=136）	潟湖相（N=87）	滨湖相（N=341）	湖沼相（N=55）
S	200	218	473	279	323	316	330	409	679
Sb	0.77	0.73	0.84	0.74	0.80	0.79	0.80	0.81	0.85
Sc	11.9	11.5	11.5	12.1	13.5	12.9	13.8	13.3	13.1
Se	0.18	0.19	0.28	0.19	0.23	0.27	0.24	0.27	0.31
SiO_2	62.5	63.9	64.1	63.6	64.1	64.8	64.9	65.1	62.9
Sn	3.29	4.88	8.89	4.06	9.07	9.34	8.89	8.22	10.5
Sr	150	142	142	145	127	129	122	119	119
Th	11.6	12.0	12.0	10.8	11.8	11.8	11.70	11.5	11.3
Ti	4764	4562	4592	4781	4922	4873	4978	4975	4841
Tl	0.59	0.61	0.60	0.56	0.64	0.62	0.63	0.62	0.63
U	2.21	2.18	2.28	2.35	2.33	2.34	2.39	2.41	2.40
V	95.1	91.4	88.7	94.6	103	95.1	102	99.4	95.5
W	1.93	1.90	2.23	1.96	2.20	2.23	2.09	2.13	2.27
Y	26.2	26.5	27.5	27.9	29.0	29.6	30.1	29.7	29.1
Zn	86.9	86.8	121	91.5	112	112	105	102	102
Zr	234	235	238	246	238	248	232	236	228

注：氧化物和C含量单位为%，Au为ng/g，其他为mg/kg。

3）该沉积区之内，能够显示为高背景分布的元素主要有溴、钙、镁、锶，其次为砷和锰，从元素组合上反映长江现代冲积带碳酸盐矿物组分的富集，其中钙和区域背景值和基准值的比值分别为1.648、1.395，是诸沉积单元中最高的，表现出这类最新冲积区土壤脱钙程度极低的基本特征。

2. 支流河道边滩沉积区

本区与早、晚期河道相沉积区绝大多数元素含量分布上差异很大，支流河道沉积区表层土壤地球化学背景值分布，既反映了原来自然沉积体系的地球化学特征，又反映了在自然沉积体系基础上出现的强烈人为扰动的结果。具体的特点为：

1）砷、硼、钡、铍、铈、钴、铬、氟、镓、锗、碘、镧、锂、锰、铌、镍、铷、钪、锶、钍、钛、铊、铀、钒、钇、锆、硅、铁、铝、钙、钾、氮等多种元素呈背景性分布。这些元素多属于化学活动性不够活泼的组分。

2）钠、镁、溴呈偏低量分布。

3）氯、磷、铅、钨、碳、银、金、铋、镉、铜、汞、钼、硫、锑、硒、锡、锌、有机碳等呈偏高量和高量分布，其中汞富集度最高，与区域背景值比值在1.8以上，背景值为0.22μg/g，仅次于湖沼相，次为银、金、硫，比值在1.4以上，而镉、磷、硒、锡、锌比值也都大于1.2。

3. 晚全新世滨海潮滩沉积区

晚期滨海潮滩沉积地球化学分布形式与早、中两期基本不同，除去碘在三相分布态势相似外，钙、汞、锡、铅、锌、铜、溴、碳、银、金等元素则呈现出与早、中期完全相反的分布形式，总体具有高钙，碳、溴、碘偏高量，钠、磷和大多数金属元素偏低量，尤以汞、锡、金、银、铅低量典型分布为特征。

该种分布形式和晚期河口相十分一致，除碘外，显示出三岛、江海岸带现代沉积物母质具有基本相似的地球化学组成，表现出晚期潮滩相沉积更多地服从于长江入海口地区现代江海沉积体系，基本分布具有极大的相似性。

4. 湖沼洼地沉积区

湖沼相、滨湖相、潟湖相中土壤元素背景值有所差异，但又具有相似的沉积地球化学体系，其特征有如下几点：

1）低量的钙、镁、锰、碘、钡分布。其中最为显著的是钙低量，三相与表层土壤区域背景值比值为0.497~0.518，仅相当于区域背景值的一半；而碘在潟湖相显示为富集，与表层土壤区域背景值比值为1.146，这是三相分布型式上唯一差异之处，也说明了三相在沉积体系上有所差异。

2）较高量的硫、氯、溴、碳、有机碳、氮、硒分布。含量呈潟湖相—滨湖相—湖沼相递增升高的趋势，以湖沼相表层土壤分布量最高，其中硫背景值达679μg/g，碳2.22%，有机碳2.05%，氮2201μg/g，氯169μg/g，硒0.31μg/g，属于所有沉积相中含量最高的。上述元素是湖积环境下的典型元素组合，表明三相在沉积体系具有一致性，但又受特殊沉积环境的限制。

3）只有少数重金属元素主要在湖沼相中背景值偏高，如汞、锡、银、金、钼、铅等，典型为汞、锡的富集。它们自潟湖相—滨湖相—湖沼相有含量规律的升高，在湖沼相中达到最大，其汞背景值达0.26μg/g，锡10.5μg/g。上述元素的高量分布，认为它们是人为扰动的结果，说明湖泊区已存在重金属污染。其他元素接近于区域背景分布，而一些亲石元素和亲硫元素多呈背景偏高量分布。某些亲铁元素呈背景偏低量分布，这种特征说明了湖泊区富铝高黏颗粒物的专属性吸附作用和湖泊还原环境沉积本质。

（三）不同土壤类型表层土壤地球化学背景值

上海的土壤类型主要为水稻土，潮土和盐土次之，潮土和盐土主要分布于崇明、长兴和横沙三岛区和沿江临海地带。水稻土又分出潜育型水稻土、脱育型水稻土、潴育型水稻土和渗育型水稻土4个亚类。上海按照土壤亚类进行统计，但对潴育水稻土按照成图地质环境进一步分解到土属组合，共划分出10个土壤类别，其统计的表层土壤背景值列于表30-8和表30-9。

表30-8 上海不同土壤类型表层土壤地球化学背景值

Table 30-8 Geochemical background values of surface soil in Shanghai by soil type

元素（指标）	脱潜水稻土	潴育水稻土				渗育水稻土	灰潮土	菜-果园灰潮土（N=56）	盐（化）土（N=280）	城区潮田泥（N=69）
	青紫土（N=228）	青黄泥（N=197）	黄潮泥（N=189）	黄泥（N=229）	潮砂泥（N=108）	黄夹砂（N=324）	旱作黄泥（N=119）			
Ag	0.120	0.130	0.140	0.110	0.130	0.120	0.110	0.120	0.100	0.150
Al_2O_3	13.40	13.18	13.24	13.51	12.78	12.68	12.75	12.36	12.75	12.08
As	8.10	7.40	7.54	7.88	7.17	7.74	8.34	8.64	8.30	9.63
Au	2.42	2.29	2.40	2.08	2.07	1.64	1.38	3.20	1.42	6.83
B	64.93	65.26	65.87	66.40	66.58	63.14	60.94	59.56	59.35	55.05
Ba	432	427	437	421	433	429	411	454	408	489
Be	2.36	2.30	2.35	2.40	2.23	2.09	2.07	2.19	2.10	2.17
Bi	0.42	0.41	0.44	0.44	0.41	0.35	0.34	0.54	0.33	0.68
Br	6.23	6.55	4.99	5.06	4.63	5.51	6.29	4.65	6.94	4.55
C	1.82	1.59	1.47	1.52	1.44	1.56	1.51	1.84	1.52	2.58
CaO	1.17	1.23	1.65	1.71	1.81	3.44	3.45	3.08	3.98	3.32
Cd	0.167	0.167	0.201	0.166	0.180	0.178	0.175	0.316	0.170	0.382
Ce	75.89	76.68	77.39	77.72	76.86	73.91	72.05	73.52	71.95	71.66
Cl	153.25	109.63	94.87	93.25	89.53	85.64	88.38	97.66	87.53	137.28
Co	14.99	14.68	14.93	15.29	14.41	13.99	14.06	13.94	14.25	13.74
Cr	86.51	86.05	88.30	89.95	86.58	82.35	79.43	92.39	80.18	93.52

续表

元素（指标）	脱潜水稻土	潴育水稻土				渗育水稻土	灰潮土	菜-果园灰潮土（N=56）	盐（化）土（N=280）	城区潮田泥（N=69）
	青紫土（N=228）	青黄泥（N=197）	黄潮泥（N=189）	黄泥（N=229）	潮砂泥（N=108）	黄夹砂（N=324）	旱作黄泥（N=119）			
Cu	32.27	30.99	33.14	31.47	30.90	28.40	28.43	37.97	29.54	55.88
F	629	637	658	673	647	626	615	661	618	696
Fe_2O_3	5.35	5.20	5.41	5.62	5.11	5.08	5.04	5.17	5.14	5.12
Ga	17.85	17.46	17.82	18.22	17.11	16.64	16.63	16.46	16.52	16.40
Ge	1.53	1.54	1.57	1.56	1.53	1.48	1.46	1.51	1.44	1.59
Hg	0.175	0.183	0.162	0.141	0.149	0.081	0.067	0.187	0.059	0.335
I	2.26	2.26	3.16	4.57	2.84	2.78	3.08	3.17	2.74	3.16
K_2O	2.27	2.30	2.40	2.46	2.31	2.27	2.31	2.29	2.31	2.22
La	41.56	41.26	41.82	42.14	41.95	41.60	39.95	41.34	40.55	41.50
Li	48.88	47.26	48.17	50.80	44.52	41.18	41.37	44.62	41.49	43.76
MgO	1.62	1.70	1.92	2.02	1.95	2.10	2.12	1.97	2.18	1.81
Mn	641	625	775	722	712	705	697	764	716	746
Mo	0.59	0.53	0.55	0.52	0.52	0.54	0.53	0.70	0.55	1.28
N	1903	1651	1387	1416	1300	1081	1030	1138	934	1382
Na_2O	1.46	1.43	1.35	1.36	1.39	1.23	1.23	1.15	1.21	1.16
Nb	17.94	17.92	18.08	17.94	17.89	17.45	17.13	17.57	17.11	17.10
Ni	34.96	34.28	35.83	37.06	34.21	33.11	33.26	35.48	33.98	34.59
P	889	860	927	902	937	866	918	1054	752	1156
Pb	31.43	30.57	31.76	31.54	31.09	25.82	25.40	43.78	24.35	85.21
Rb	109.78	108.92	111.36	115.08	105.65	101.62	101.57	103.92	100.21	99.41
S	454.65	353.99	326.77	326.61	308.53	270.19	245.89	471.36	208.94	750.74
Sb	0.82	0.80	0.84	0.81	0.79	0.75	0.73	1.28	0.77	2.22
Sc	13.44	12.98	13.20	13.67	12.40	11.80	11.84	11.23	11.98	10.78
Se	0.28	0.26	0.27	0.23	0.26	0.19	0.19	0.31	0.17	0.63
SiO_2	64.77	65.58	64.58	63.72	65.10	64.10	63.81	63.17	62.71	62.24
Sn	8.52	9.24	9.09	9.33	8.54	4.40	4.09	10.53	3.34	26.35
Sr	117.83	121.49	127.13	125.54	130.06	144.10	141.01	148.70	149.32	162.64
Th	11.46	11.48	12.09	11.86	11.47	11.83	11.79	12.38	11.34	11.68
Ti	4965	4892	4875	4938	4822	4751	4626	4524	4755	4358
Tl	0.62	0.62	0.63	0.64	0.62	0.60	0.61	0.60	0.58	0.62
U	2.43	2.35	2.30	2.37	2.30	2.27	2.24	2.20	2.23	2.45
V	99.75	96.57	99.25	105.46	93.65	94.70	92.79	90.49	94.27	82.50
W	2.13	2.09	2.21	2.19	2.22	2.00	1.90	2.42	1.93	3.27
Y	29.69	29.49	29.27	29.19	28.59	27.83	26.53	26.85	26.49	26.61
Zn	103.53	101.96	113.96	112.34	105.41	91.00	90.06	146.03	87.58	247.01
Zr	232.82	237.10	241.49	236.47	243.30	238.85	233.99	232.70	235.24	227.26
OrgC	1.59	1.41	1.24	1.22	1.18	0.99	0.89	1.22	0.79	1.83
pH	6.93	7.31	7.77	7.80	7.80	8.05	8.08	7.99	8.12	7.92

注：氧化物和C、OrgC的含量单位为%，Au为ng/g，其他为mg/kg。

表 30-9 上海不同土壤类型表层土壤元素富集系数

Table 30-9 Element concentration coefficients of surface soil in Shanghai by soil type

元素（指标）	脱潜水稻土	潴育水稻土（N=723）				渗育水稻土	灰潮土	菜-果园灰潮土（N=56）	盐（化）土（N=280）	城区潮田泥（N=69）
	青紫土（N=228）	青黄泥（N=197）	黄潮泥（N=189）	黄泥（N=229）	潮砂泥（N=108）	黄夹砂（N=324）	旱作黄泥（N=119）			
Ag	1.50	1.63	1.75	1.38	1.63	1.50	1.38	1.50	1.25	1.88
Al_2O_3	1.04	1.03	1.03	1.05	1.00	0.99	0.99	0.96	0.99	0.94
As	1.09	1.00	1.01	1.06	0.96	1.04	1.12	1.16	1.12	1.29
Au	2.00	1.90	1.98	1.72	1.71	1.35	1.14	2.65	1.17	5.65
B	1.02	1.03	1.04	1.04	1.05	0.99	0.96	0.94	0.93	0.87
Ba	0.94	0.92	0.95	0.91	0.94	0.93	0.89	0.98	0.88	1.06
Be	1.01	0.99	1.01	1.03	0.96	0.90	0.89	0.94	0.90	0.93
Bi	1.17	1.13	1.22	1.22	1.14	0.96	0.94	1.51	0.90	1.89
Br	2.22	2.34	1.78	1.81	1.65	1.97	2.25	1.66	2.48	1.62
C	1.82	1.59	1.47	1.52	1.44	1.56	1.51	1.84	1.52	2.58
CaO	0.42	0.45	0.60	0.62	0.66	1.25	1.25	1.12	1.45	1.21
Cd	1.51	1.52	1.83	1.51	1.63	1.62	1.59	2.88	1.54	3.47
Ce	1.00	1.01	1.02	1.02	1.01	0.97	0.95	0.97	0.95	0.94
Cl	1.63	1.16	1.01	0.99	0.95	0.91	0.94	1.04	0.93	1.46
Co	1.04	1.02	1.04	1.06	1.00	0.97	0.98	0.97	0.99	0.95
Cr	1.01	1.01	1.04	1.05	1.01	0.97	0.93	1.08	0.94	1.10
Cu	1.25	1.20	1.28	1.22	1.20	1.10	1.10	1.47	1.14	2.17
F	1.02	1.03	1.06	1.09	1.05	1.01	0.99	1.07	1.00	1.13
Fe_2O_3	0.97	0.95	0.99	1.02	0.93	0.93	0.92	0.94	0.94	0.93
Ga	1.08	1.05	1.07	1.10	1.03	1.00	1.00	0.99	1.00	0.99
Ge	1.02	1.02	1.04	1.04	1.01	0.98	0.97	1.00	0.95	1.05
Hg	3.51	3.67	3.24	2.83	2.98	1.61	1.35	3.74	1.18	6.69
I	0.82	0.82	1.15	1.67	1.04	1.01	1.12	1.16	1.00	1.15
K_2O	0.95	0.96	1.00	1.03	0.96	0.94	0.96	0.95	0.96	0.92
La	1.04	1.04	1.05	1.06	1.05	1.05	1.00	1.04	1.02	1.04
Li	1.02	0.99	1.01	1.07	0.93	0.86	0.87	0.94	0.87	0.92
MgO	0.77	0.80	0.91	0.96	0.92	0.99	1.01	0.93	1.03	0.86
Mn	0.85	0.83	1.03	0.96	0.94	0.93	0.92	1.01	0.95	0.99
Mo	1.31	1.17	1.22	1.16	1.16	1.19	1.18	1.56	1.22	2.85
N	3.58	3.10	2.61	2.66	2.44	2.03	1.94	2.14	1.76	2.60
Na_2O	1.00	0.98	0.93	0.93	0.95	0.84	0.84	0.79	0.83	0.79
Nb	1.00	1.00	1.01	1.00	1.00	0.97	0.96	0.98	0.96	0.96
Ni	0.97	0.95	0.99	1.02	0.95	0.91	0.92	0.98	0.94	0.96
P	1.45	1.40	1.51	1.47	1.53	1.41	1.49	1.72	1.22	1.88
Pb	1.24	1.21	1.26	1.25	1.23	1.02	1.00	1.73	0.96	3.37
Rb	1.00	0.99	1.01	1.05	0.96	0.92	0.92	0.94	0.91	0.90
S	1.78	1.39	1.28	1.28	1.21	1.06	0.96	1.85	0.82	2.94
Sb	1.54	1.50	1.59	1.53	1.50	1.42	1.38	2.41	1.45	4.20
Sc	1.01	0.98	0.99	1.03	0.93	0.89	0.89	0.84	0.90	0.81
Se	2.30	2.15	2.25	1.91	2.13	1.61	1.56	2.59	1.44	5.26

续表

元素（指标）	脱潜水稻土	潴育水稻土（N=723）				渗育水稻土	灰潮土	菜-果园灰潮土（N=56）	盐（化）土（N=280）	城区潮田泥（N=69）
	青紫土（N=228）	青黄泥（N=197）	黄潮泥（N=189）	黄泥（N=229）	潮砂泥（N=108）	黄夹砂（N=324）	旱作黄泥（N=119）			
SiO_2	1.01	1.02	1.01	0.99	1.02	1.00	1.00	0.99	0.98	0.97
Sn	2.38	2.58	2.54	2.61	2.39	1.23	1.14	2.94	0.93	7.36
Sr	0.85	0.88	0.92	0.91	0.94	1.04	1.02	1.08	1.08	1.18
Th	0.90	0.90	0.94	0.93	0.90	0.92	0.92	0.97	0.89	0.91
Ti	1.02	1.01	1.01	1.02	0.99	0.98	0.95	0.93	0.98	0.90
Tl	0.99	0.98	1.00	1.01	0.98	0.95	0.97	0.95	0.93	0.99
U	1.11	1.07	1.05	1.08	1.05	1.04	1.02	1.00	1.02	1.12
V	1.02	0.99	1.02	1.08	0.96	0.97	0.95	0.93	0.97	0.85
W	1.09	1.07	1.13	1.12	1.13	1.02	0.97	1.23	0.99	1.67
Y	1.05	1.05	1.04	1.03	1.01	0.99	0.94	0.95	0.94	0.94
Zn	1.05	1.03	1.15	1.14	1.07	0.92	0.91	1.48	0.89	2.50
Zr	0.98	1.00	1.02	1.00	1.03	1.01	0.99	0.98	0.99	0.96
OrgC	1.41	1.25	1.09	1.08	1.05	0.88	0.79	1.08	0.70	1.62

1. 脱潜水稻土——青紫土

脱潜水稻土是西部湖积区形成的一种土壤，属青紫土属，中性土，pH 平均值为6.93。与全区表层土壤背景值比较，表现为一种低钙、镁，高硫、氮、碳，富有机质的土壤。常量组分中硅、铁、铝、钠、钾含量基本和全区背景值持平，全钾量2.27%。唯有钙、镁显示低量，两者背景值分别为1.17%和1.62%，其中钙含量低于全区背景值的一半以下。

该土属中的硫、碳、有机碳是自然土壤中含量最高的，这种分布特点应和湖区沉积的母质环境相关。

微量元素方面，显示有汞、锡、硒、砷、氯的偏高量分布，以及溴、磷、金、铋、锂、锌、铷、铍、铅、硼、钪、钍、钼、钇、钴呈背景偏高量分布，而有益元素碘、锰，有害元素砷、镉、锑呈背景偏低量分布，其余铜、氟、铬等微量元素与表层土壤背景值持平。从总的分布型式上看，反映了原始湖积相母质和表层人为污染复合叠加的土壤地球化学背景。

2. 潴育水稻土——青黄泥+青黄土

青黄泥为一种发育于湖沼平原高地的潴育水稻土，属青黄泥土属。青黄土土属则分布于碟缘地段及江河两侧，多数处于西部湖沼平原区内。中偏碱性，表层pH 平均值为7.31。以地球化学组分分布型式总体界定，本土属为低钙、镁、锰，高碳、有机质、氮、硫、氯的土壤种属。和全区表层土壤背景值比较，本土属常量组分中硅、铁、铝、钾含量略偏低于全区背景值，钠含量略偏高于全区背景值，全钾量为2.31%。钙、镁低量，两者背景值分别为1.23%和1.70%，和全区背景值比值分别为0.567和0.867，即钙含量低于全区背景值的一半以下。亦即，青黄泥和青紫土常量组成方面完全一致。碳、有机质、氮、硫、硒及碘分布亦极为相似，但含量有所差异。碳、有机质、氮、硫、硒含量一般在青黄泥中略低，青紫土中略高，碘则反之，反映出两者基本相同的母质沉积环境。

微量组分方面，存在磷、硒、溴、锌、硼、钴和重金属组分汞、铅、锑、锡、银、金、钇、镧、铷、钨、钪的偏高量分布。其中汞含量达到0.215mg/kg，锡9.24mg/kg，显示本土属区保持着比较丰富的土壤农业有益组分。

3. 潴育水稻土——黄潮泥+沟干泥+沟干潮泥+黄泥头

该类别属老江海潮滩沉积或上覆古长江冲积物母质，包括黄潮泥、沟干泥、沟干潮泥、黄泥头土属，由于分布位置不同致使成土行为有所差别。该类别属弱碱性土，pH 平均值为7.77。在土壤基本组成方面，其地球化学图谱和早期滨海潮滩沉积一致，呈低钙、镁、溴、钡、钍，偏高的氮、硫、碘分布形式。和全区背景值对比，钠、硅、铝、钾背景值略高，铁（Fe_2O_3）的背景均值为5.41%，属偏高量分布；钙、镁背景值偏低，尤其钙背景值和区域背景值比值为0.683，显示该类别土壤全钙量偏低。土壤中钾、硅、铁等含量偏高有利于农作物生长。

微量组分方面，该类别土壤除溴、钼、钡等少数元素属背景偏低量分布外，大多数组分呈背景偏高量分布，其中包括农业上的有益组分钴、硼、钼、锰、磷、碘、铜、硫、氮、锌、硒、有机质等，但幅度都不大；也包括有害元素汞、锡、铅、镉、铊、锑等。其中汞背景均值达0.162mg/kg，镉背景均值0.201mg/kg。

4. 潴育水稻土——黄泥

该类别土属分布在浦东大部分地区，一般处在地势较高部位，成土母质为中期滨海潮滩沉积。土壤呈弱碱性，pH为7.80。常量组分仍以钾、铁、铝比较丰沛，偏低量的钙为特征，其中全钾量为2.469%，铁5.46%，两者均为全区所有土属中最高值；钙1.71%，为区域背景值的70%。硅、镁、钠含量与全区背景值持平或者略低。

本土属土壤表层，作为农业重要养分的氮、磷、硫、有机碳含量略高于区域背景值；而其中以碘的偏高量分布最为典型。在本土属区，碘不仅表层含量丰沛，背景值为4.57mg/kg，为全区土类最高含量，且深层含量也较高。相比之下，区内多数金属元素含量接近区域背景值。呈偏高量分布的重金属元素包括锡、金、汞、银、锌等。

5. 潴育水稻土——潮砂泥

该土属广泛分布于河流两侧，受河流冲积和潮汐影响堆积而成，成土母质为河流冲积物。土壤呈弱碱性，pH为7.80。本土属常量组分与前述的潴育水稻土基本分布型式相同，略有不同的主要是钙含量有所升高，背景值达到1.81%；钾、铁、镁、铝含量和前述黄潮泥等相比之下有所降低；硅含量有所升高，为65.1%。

据统计，本土属的氮、磷、硫、碘等农业有益元素含量分布接近于区域背景值，无明显分布优势。重金属元素含量多数接近于背景值，但仍存在汞、锡、金等的偏高量分布。

6. 渗育水稻土——黄夹砂+砂夹黄

该土类分布于沿海地区，成土母质为晚期滨海潮滩相沉积以及河口江海沉积。土壤呈碱性，pH为8.03。统计显示，本土类区呈现高量的钙、偏高量的镁、镉、锶，大多数元素偏低量的分布特征。在基本常量组分方面，与全区表层土壤背景值对比，钙含量为3.512%，为全区背景值的1.424；镁含量为2.1%，其他元素的含量总体接近或低于全区背景值。钾、钠、硅、铁、铝、镁含量和全区背景值比值在0.92~0.97之间。

微量组分方面，统计显示，除碳、锆、锶、镧、锰、碘、镉含量基本和区域背景值持平外，其他绝大多数微量组分含量低于区域背景值，其比值一般在0.9~0.98间，最低为汞，比值仅为0.694。本土属区汞、锡、镉、铅等有毒组分含量较低，反映出土壤可能属最新形成，尚保持自然状态，受人为扰动较少，但是氮、有机碳的含量偏低，不利于农业生产。

7. 灰潮土——旱作黄泥+旱黄夹砂+旱砂夹黄

本土类为旱作黄泥、旱黄夹砂、旱砂夹黄等土属，分布于崇明岛、长兴岛、横沙岛等地，母质为河口泥沙沉积，是一种旱作熟化土壤，与渗育型水稻土完全一致。土壤呈碱性，pH为8.08。据统计，本类别土壤和黄夹砂在地球化学分布型式上十分相似，所存在的差别仅是基本组分钙含量高，背景值为3.45%；其他钾、钠、镁、铁、铝、硅含量偏低于全区背景值，总的反映出本土类脱钙程度低。该土属磷含量偏高，氮和有机质含量偏低，钾持平，土壤肥力有一定的持续性。

绝大多数微量元素含量偏低于全区背景值。偏高量分布的主要有溴、碘两个元素，但幅度都不是太大，最高的碘含量和区域背景值比值也仅1.138，溴1.13。溴含量较高的特征，也是和黄夹砂土属的不同之处。低量分布的元素主要有汞、锡、金、硫、铋、硒、铅等，其中汞背景值为0.067mg/kg，为所有分类区中的最低值，锡、镉、铅呈偏低量分布。说明本土类区受污染程度最低，基本保持原生自然状态。

8. 灰潮土——菜-果园灰潮土

该土属包括菜园灰潮土和果园灰潮土，以菜园灰潮土为主。成土母质为支流河道沉积，原土可能是灰潮土，也可能是水稻土，但经高度耕作改造，致使独具特点。该土属 pH 为 7.99，呈弱碱性。菜-果园灰潮土区别于其他任何土属，除常量组分、部分惰性组分尚保持潮土原有特征外，那些活动性强、对环境条件敏感的组分在本土属中具不同于本土属的偏高钙含量，和接近于区域背景值的钾、镁、硅，偏高含量的有机质、磷、氮农业养分成分分布，表现出土壤具有较高的肥力，但本土属多种有害元素出现高背景值，极有可能出现生态方面的复合污染效应，对农业生产造成负面影响。

9. 滨海盐化土

滨海盐土主要分布在崇明、长兴、横沙三岛边缘和浦东海沿边滩。盐土形成一是海水直接淹浸，使土壤盐渍化，二是咸潮溯河倒灌，引起河口内侧盐渍化。盐土母质为江海交互沉积而成。表层盐土 pH 平均值为 8.12，最高值为 8.96，为所有土类中最高值。盐土区元素分布的突出特点是，全区唯钙、溴呈偏高背景值分布，其中钙背景均值为 3.98%，溴为 6.94mg/kg，两者都是所有土类中最高的。推断钙的富集与盐土完全未脱钙有关；溴相比氯而言，区内表层土壤氯背景值与全区背景值相当，然而深层土壤中氯含量大大增高，深层与表层土壤氯含量比值为 1∶1.5。这说明，氯、溴可能由于化学性质的差异，氯主要积聚在深层土壤，而溴主要积聚在表层土壤。

该类区元素分布的第二个特点是，绝大多数元素背景值偏低于区域背景值，可能作为最新沉积成土的原因，汞、锡、硫、金、硒、铅、铋、锌易受污染的元素和磷、氮、有机碳农业养分均呈低量分布状态。汞背景值为 0.059mg/kg，为所有土壤区最低值，反映出盐土区原始的自然环境状态。

10. 城区潮田泥

该土属范围仅限于中心城，隶属于灰潮土亚类。土壤表层发育特征为：①土壤基本以原支流边滩冲积层为基础，但又有大量的降尘、城镇工业垃圾和生活垃圾等外来物质混加，表层物质成分受到较大改变；②在本已丰沛的地下水条件下，加之含有大量有机及无机物质的工业废水和生活用水的混染，通过土壤渗滤作用进一步改变了土壤物理化学环境及化学组成；③以污染为诱因，本类区本不完整的土壤剖面趋于更加不完善，或显示出具污染性质的特殊的土壤剖面化学变化，土壤化学成分发生着巨大的改变。据本次测定，本土类区表层土壤 pH 最低值为 7.17，最高值为 8.38，平均值为 7.92，为中性—碱性土。本土类分布型式比菜园土具有更为强烈的震荡。按比值统计，钙、碘、氟、铀、铬、钡、锶、砷、磷、碳、钨、铋、铜、镉、硫、钼、银、锌、汞、硒、铅、锑、金、锡、有机质均属于偏高量和高量分布状态。典型重金属元素背景值：汞达 0.335mg/kg，镉 0.382mg/kg，锡 26.35mg/kg，铅 85.21mg/kg。上述分布型式，部分为上海城区潮土区本身固有的（如钙）特点，而大部分是属于表层富集或城区污染叠加所致。

总体而言，上海中心城土地遭受着大量污染物质的肆虐，人为因素已经是城区地球化学元素富集的主体因素。

第二节　土壤元素空间分布特征

一、造岩元素分布特征

造岩元素包括硅（Si）、铝（Al）、镁（Mg）、钙（Ca）、钠（Na）和钾（K）等，是组成地壳的常量元素，也是土壤组成当中的常量元素。上海表层土壤当中造岩元素总的平均含量为 85.13%（最小值为 74.56%，最大值为 88.88%），深层土壤当中造岩元素总的平均含量为 86.35%（最小值为 82.50%，最大值为 90.93%）。表层土壤与深层土壤中造岩元素平均含量基本一致，这在一定程度上表明，表层土壤与深层土壤成因类型大致相同。

（一）硅（Si）

上海深层土壤当中 SiO_2 的含量范围为 60.05%～68.89%，平均含量为 64.17%（图 30-1）。低含量地区主要为“冈身带”以东地区，高含量地区主要为“冈身带”以西地区，尤其是靠近西部淀山湖附近区域。

表层土壤当中 SiO_2 的低含量地区及高含量地区大致相同，含量范围为 53.43%～70.04%，平均含量为 63.86%。值得提出的是，在黄浦江古河道沉积区及浦东南汇边滩区含量普遍较低。在中心城附近表层土壤当中硅含量普遍较低，这与人类活动使土壤淋浴作用加强以及外来污染物的输入导致硅含量相对降低有关。

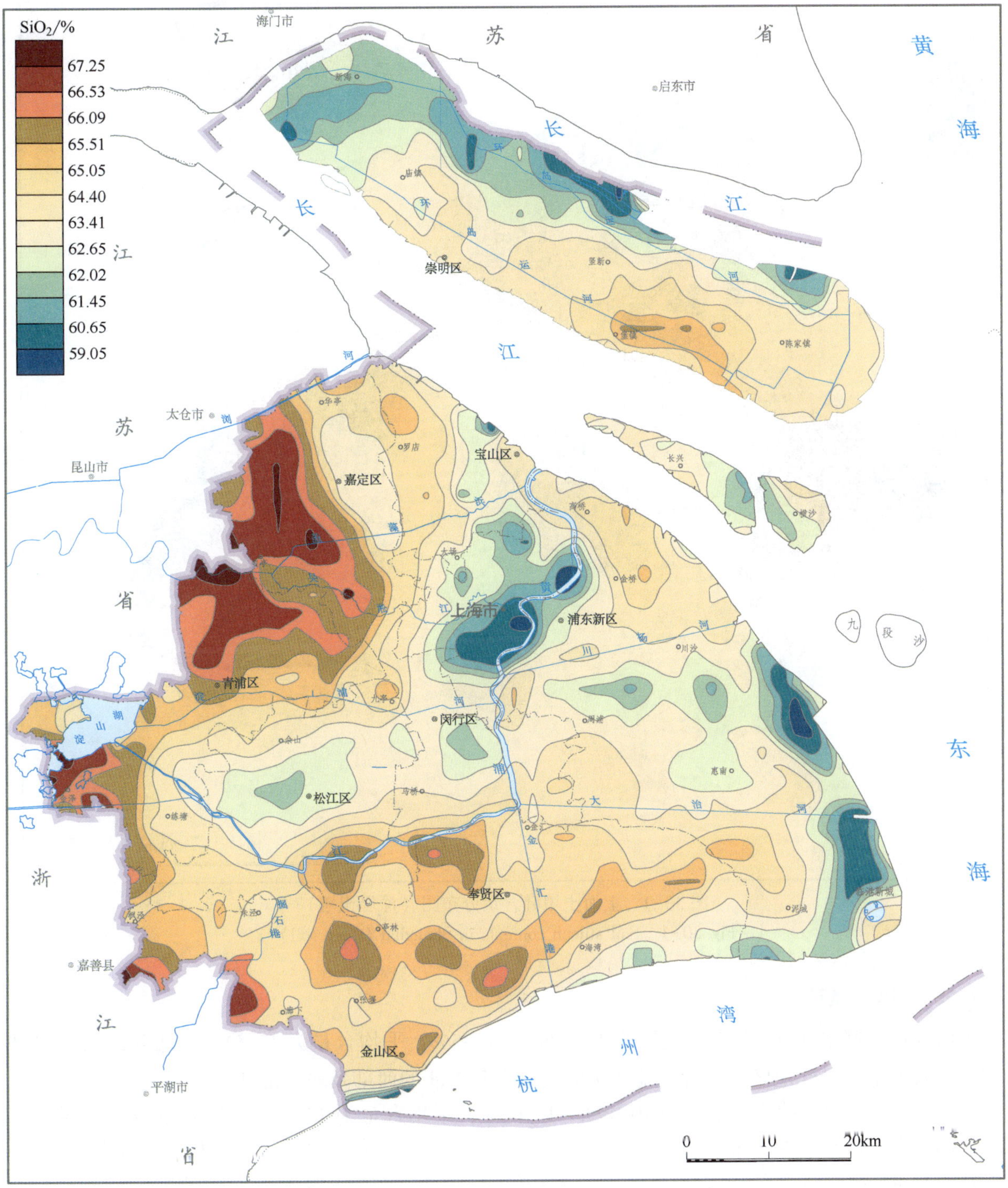

图 30-1　上海土壤硅含量分布图（表层土壤）

Fig. 30-1　Soil silicon contents in Shanghai（surface soil）

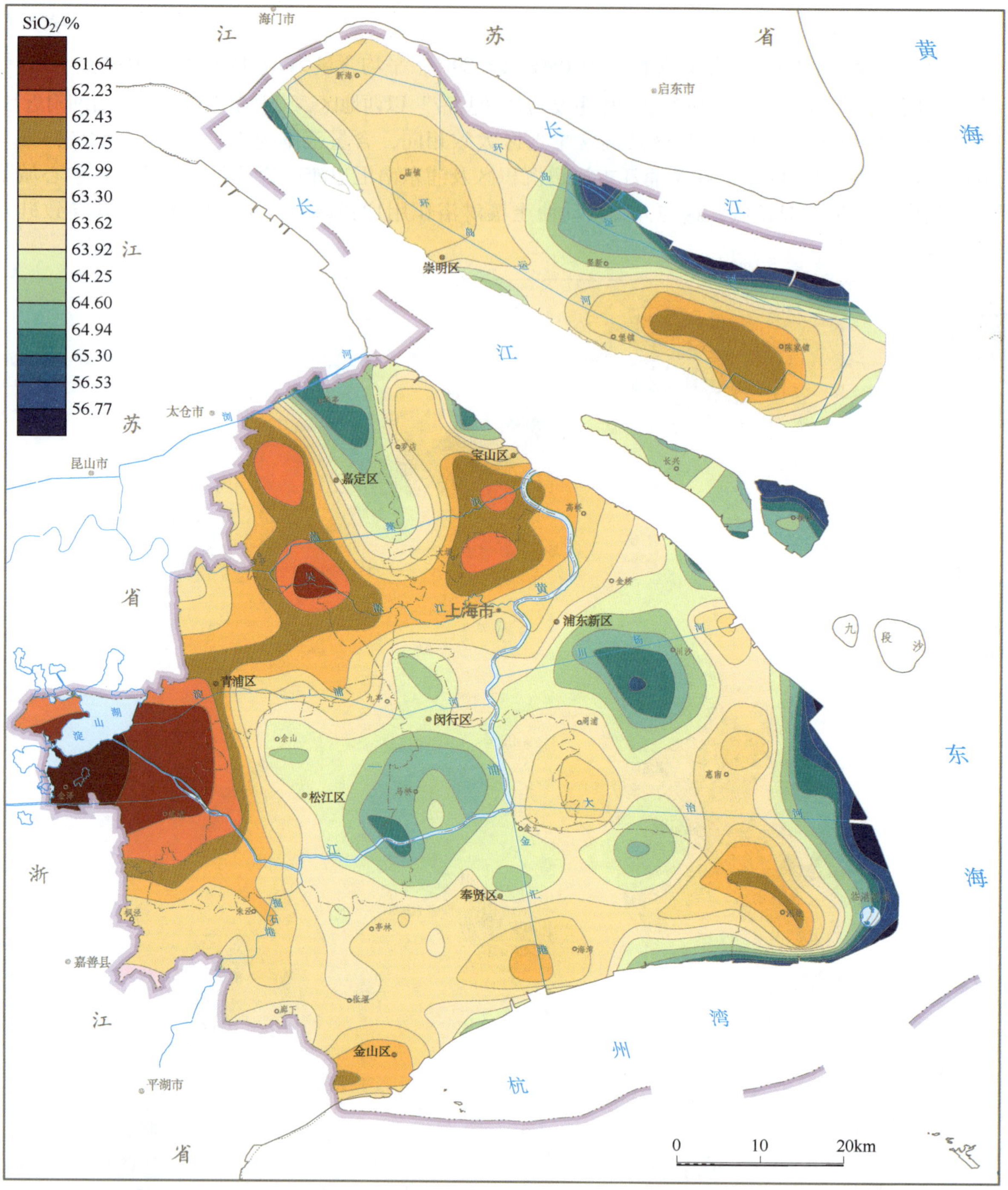

图 30-1　上海土壤硅含量分布图（深层土壤）

Fig. 30-1　Soil silicon contents in Shanghai (deep soil)

（二）铝（Al）

上海深层土壤中 Al_2O_3 的含量范围为 10.79%~21.35%，平均含量为 13.51%。低含量地区主要为崇明三岛及原南汇沿江地区，高含量地区主要为嘉定和青浦的西部及西北部地区。

表层土壤中 Al_2O_3 的含量范围为 10.49%~17.06%，平均含量为 13.21%。低含量地区主要为崇明东滩、原南汇东南部、奉贤南部及中心城等，高含量地区主要为嘉定中部、青浦西部及浦东南部等。

（三）钙（Ca）

上海表层土壤与深层土壤中钙的分布特征与第四纪沉积环境划分极为相似，这主要是由于第四纪沉积物在由海变陆过程中土壤的熟化程度不同造成的。西部地区土壤成陆较早，土壤熟化程度高，土壤脱钙相对较完全。

上海深层土壤中 CaO 的含量范围为 0.81%~4.69%，平均含量为 2.75%，低含量地区主要为西部的青浦和金山，高含量地区主要为崇明和原南汇的沿江地区（图 30-2）。表层土壤当中 CaO 的含量范围为 0.76%~7.30%，平均含量为 2.47%。表层土壤中钙的分布特征与其在深层土壤中分布特征较为相似，值的提出是，中心城也为钙的高含量分布区。

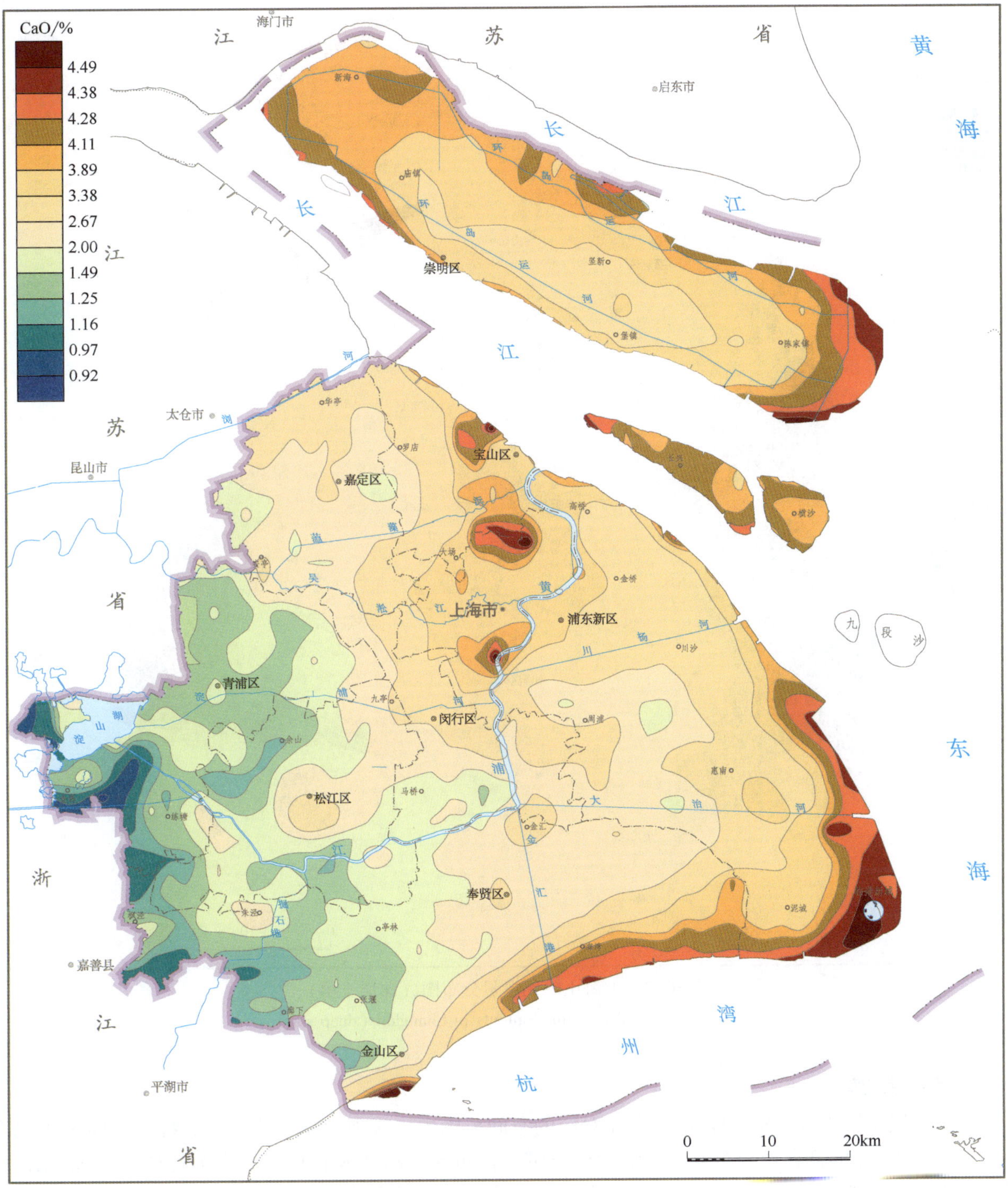

图 30-2　上海土壤钙含量分布图（表层土壤）

Fig. 30-2　Soil calcium contents in Shanghai (surface soil)

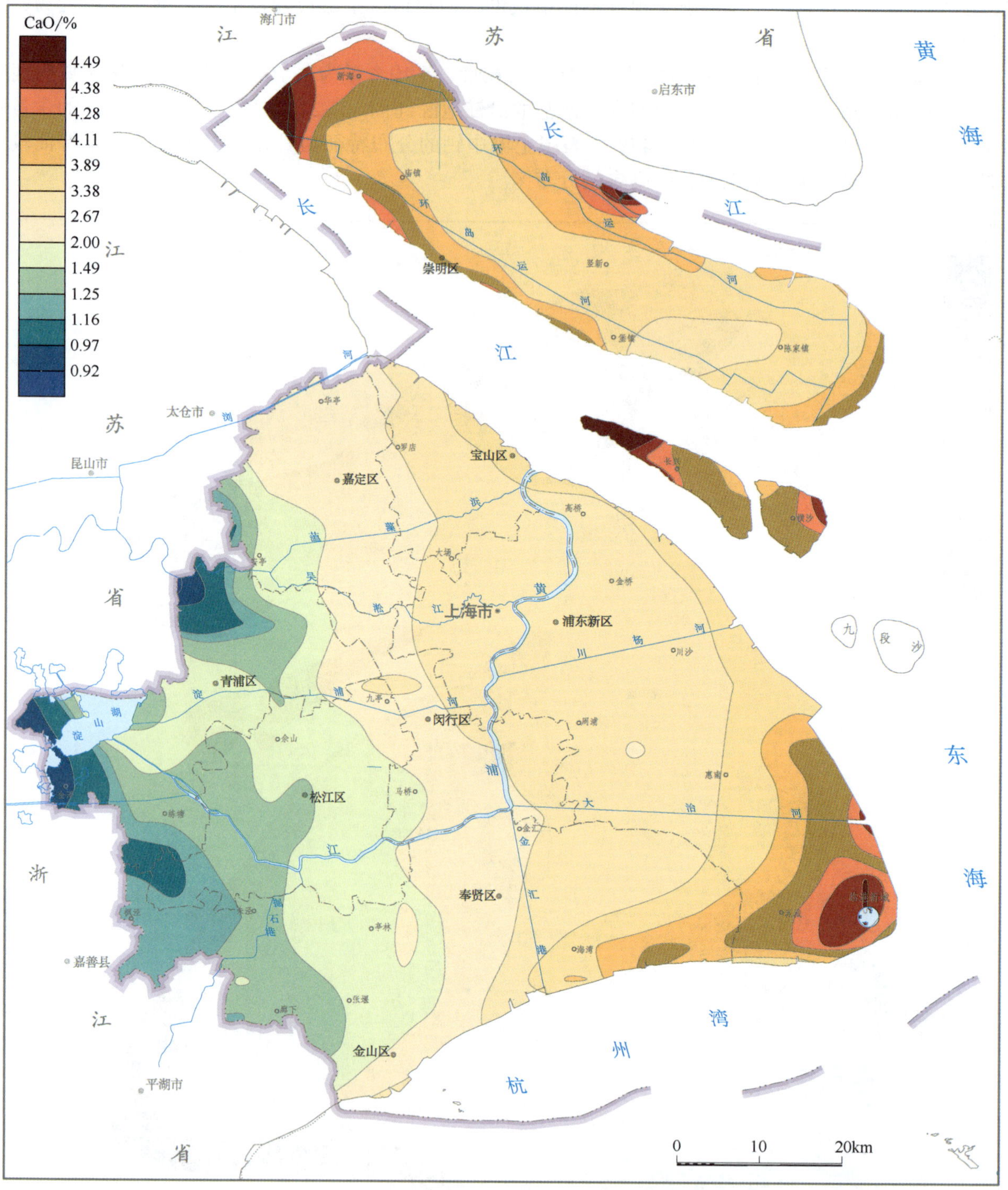

图 30-2　上海土壤钙含量分布图（深层土壤）

Fig. 30-2　Soil calcium contents in Shanghai (deep soil)

（四）其他

上海深层土壤中 K、Na 的含量分布在对应的空间范围内，表现为高、低互应的特点，即 K 的高含量分布区即为 Na 的低含量分布区。如沿“冈身带”附近及三岛区 K 以中低含量为特征，而 Na 则以中高含量为特征。然而，上述特征在以湖积为主的淀山湖地区则不明显。Mg 含量的总体特征为西低东高。由陆向海的过渡区，土壤中 Ca、Mg 含量逐渐增高，其中 Ca 含量特征更为明显。Mg 的高含量地区为嘉定大部分。

上海地层普遍富钾，表层土壤中 K_2O 背景值为 1.55%~2.98%，平均为 2.42%，分布悬殊不是很大。表层土壤 Na_2O 的背景值为 0.83%~3.59%，平均 1.22%。K、Na 的含量分布在对应的空间范围内也表现为高、低互应的特点，即钾的高含量分布区即为钠的低含量分布区。如沿杭州湾—长江口的冲海积沉积物分布带（金山—浦东）K 以中低含量为特征，而 Na 则以中高含量为特征；在嘉定和宝山结合部的早、中期滨海沉积区，K 以中高含量为特征，而 Na 则以中低含量为特征；然而，上述特征在以冲湖积为主的松江则不明显。

二、酸碱度分布特征

上海表层土壤酸碱度范围为 4.56~8.96，平均 7.61。深层土壤酸碱度范围为 6.55~8.88，平均 8.16。表层土壤以碱性和中性为主，在西部的个别地方存在弱酸性土壤。深层土壤则以碱性为主。

表层土壤的酸碱度主要受第四纪沉积环境影响，具体表现为西部地区呈弱酸性，局部地方出现酸性。其中淀山湖周围、泖港靠近黄浦江地段等地方土壤酸碱度总体较低。在上海的中部及东部地区以碱性和弱碱性为主。西部的酸性地区既与该地区地势低洼、排水不畅有关外，还与该地区酸雨频发有一定关系。

上海深层土壤中酸碱度低的地方主要集中分布在西部及中部地区，而在东部及东南部地区则逐渐升高（图 30-3）。在崇明东滩和西北沿江、中央沙、南汇边滩部分地区及上海市化工区外围地方存在高值区，这些地方也是上海土壤盐渍化相对较严重的地区。

三、有机碳分布特征

上海表层土壤有机碳含量范围为 0.19%~5.78%，平均 1.24%。深层土壤有机碳含量范围为 0.24%~1.57%，平均 0.53%。表层土壤中有机碳分布的最大特征是有两个高含量区域，一为中心城，另外一个为西部湖沼相沉积地区。最低值主要集中在崇明东滩、南汇边滩及奉贤、金山杭州湾北岸。西部地区含量高主要与该地区地势低洼和生物残体更易腐烂成有机质有关系。另外也与该地区农业开发历史悠久有一定联系。中心城高含量成因主要与生活垃圾及工业垃圾的长期堆放有关。

上海深层土壤中有机碳含量较高的地方主要集中在西部湖沼地区，而在中部及东部地区明显降低。西部地区有机碳含量高主要有 3 个方面的原因：第一为湖沼地层中生物残体较发育；第二为西部地区深层土壤以黏性土为主，有利于吸附有机质；第三为西部地区成陆较早，土壤熟化程度高，农业开发历史悠久开发强度大。这些条件都有利于有机碳的形成和保存。另外，在宝钢地区也存在一定的高值区，这可能与该地区工业活动有关。在“冈身带”土壤中有机碳含量明显偏低，这与该地区土壤中粉砂含量高有关。砂型土层不利于有机质的保存。

四、主要重金属元素分布特征

主要重金属元素包括 Cd、Hg、Pb、Zn 4 种，被认为是一组地球化学活动性较强的元素，对环境的敏感性较强，因此土壤极易受其污染。上海 4 种元素的土壤基准值分别为 Cd 0.11mg/kg、Hg 0.55mg/kg、Pb 25.3mg/kg、Zn 98.8mg/kg，而表层土壤背景值则分别为 Cd 0.18mg/kg、Hg 0.116mg/kg、Pb 29.1mg/kg、Zn 99.3mg/kg，显示出表层土壤后期污染叠加的幅度。从其分布特征图（图 30-4）看：

1）在上海市区，Cd、Hg、Pb、Zn 分布特征基本相同，即均表现出高量的异常-高背景分布态势，其中以 Hg 最为突出。在城区高量分布，最高含量为 3.35mg/kg，且以高量区为中心，呈现的高背景分布扩散至上海陆域大部地区。在这些高背景区内，又以青浦、松江等城镇为中心，形成局部高量分布异常区。Cd、Pb、Zn 大体相当，在城区形成十分明显的异常区，范围和 Hg 基本一致，最高含量为 Cd 20.0mg/kg、Pb 2521mg/kg、Zn 1094mg/kg；高背景区范围不及 Hg，但均以上海城区为中心，形成似环状的异常-高背景分布区。和 Hg 一样，高背景分布区内亦出现宝山、嘉定、安亭、松江、闵行吴泾等城镇中心上的异常区。由此说明，4 种元素在上海陆域出现的高背景及异常，是城市生态地球化学的基本特点。它们多系人为污染因素所致。城区范围愈大，人口聚集愈多，城市历史愈长，则愈形成这些元素高量的地球化学分布特征。

2）崇明、长兴、横沙三岛及杭州湾滨海地带与上海城区基本可形成相反的对比，在这些地区，4 种元素全部处于低量的分布态势。在地球化学图上，4 种元素以三岛、东部滨岸构成一明显的背景-低量环形带包裹着上海外缘。4 种元素之间也存在差异。其中 Cd，其应是长江镉富集带组成部分，背景

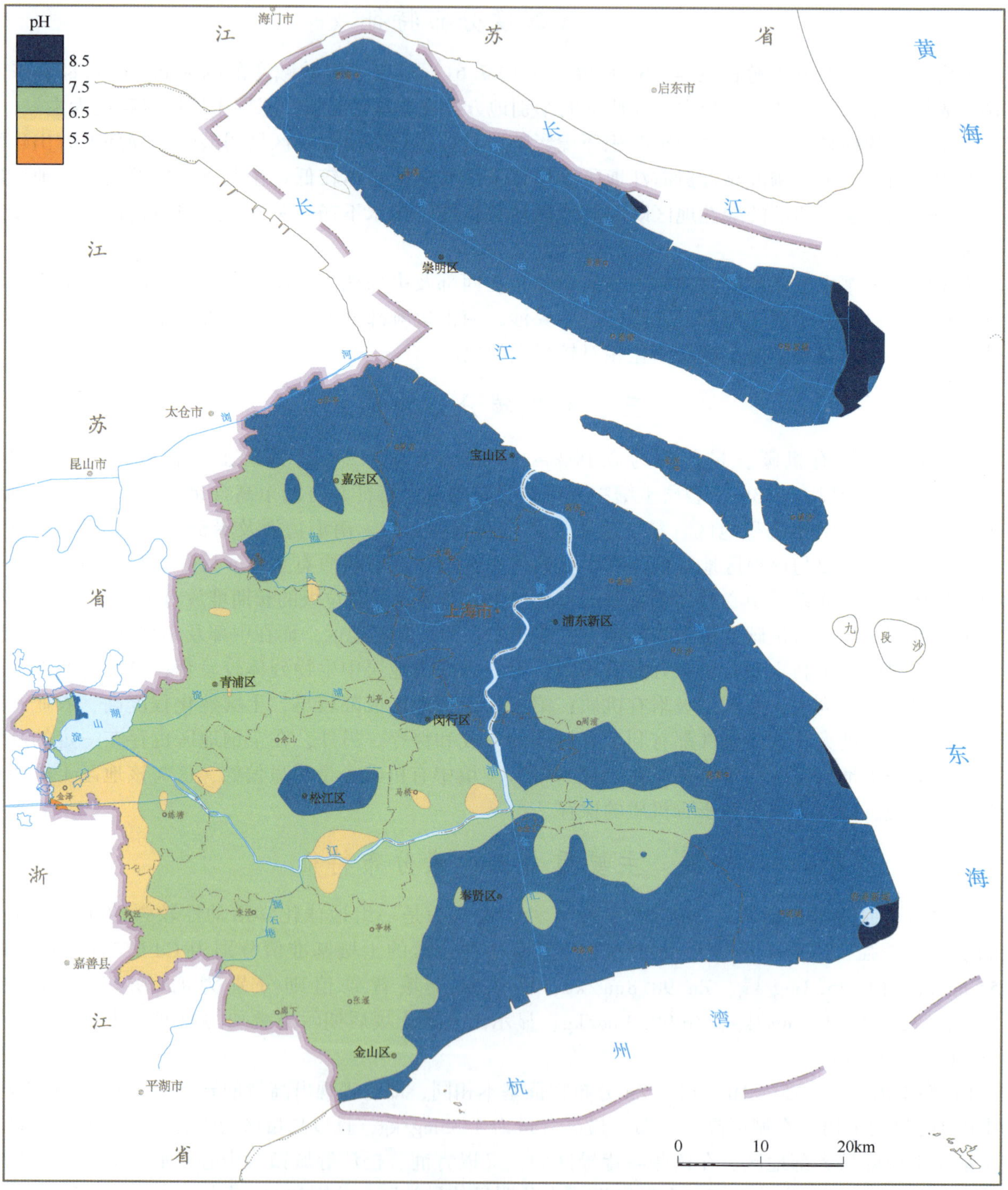

图 30-3　上海土壤酸碱度分布图（表层土壤）

Fig. 30-3　Soil pH distribution in Shanghai（surface soil）

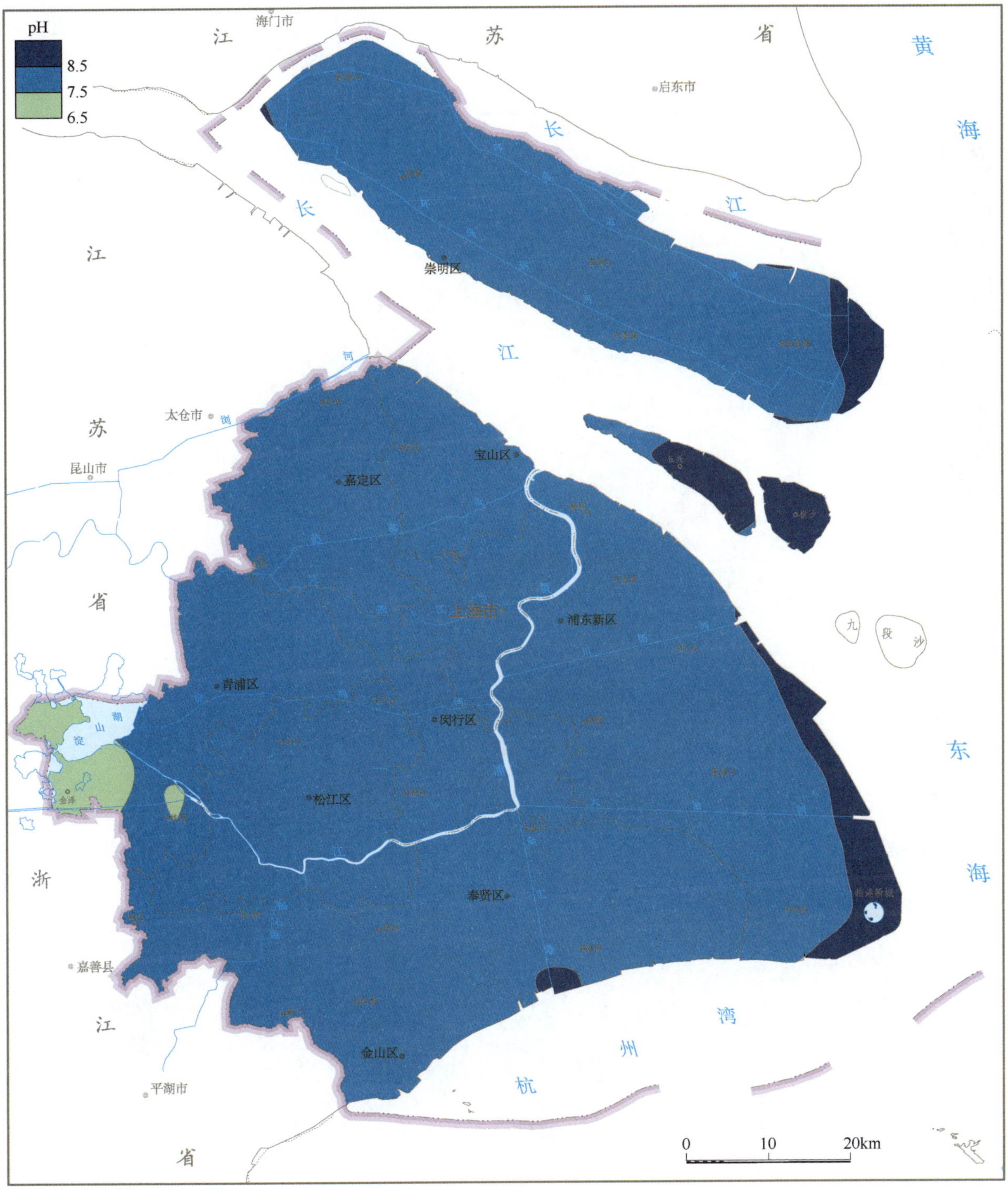

图 30-3 上海土壤酸碱度分布图（深层土壤）

Fig. 30-3 Soil pH distribution in Shanghai (deep soil)

值为 0.18mg/kg，但在城市环境因素的映衬下，这些地区已降为背景-低背景的分布。Hg 环状带不够清晰，即在低背景-低量区内存在零星的高背景及异常点，如崇明城桥、临港新城。这些分布特征表明，这些地区作为最新冲积区尚未受到强烈的人为污染，而之中的 Cd、Hg 具有更为活泼的化学活动性，一旦出现人为因素影响，就十分容易地表现出来。Pb、Zn 低量环形带十分清晰，且三岛的低量分布更为明显。这种分布型式，可以从 Pb、Zn 化学活动性不及 Hg、Cd，从而保持更好的自然分布状态得到解释。

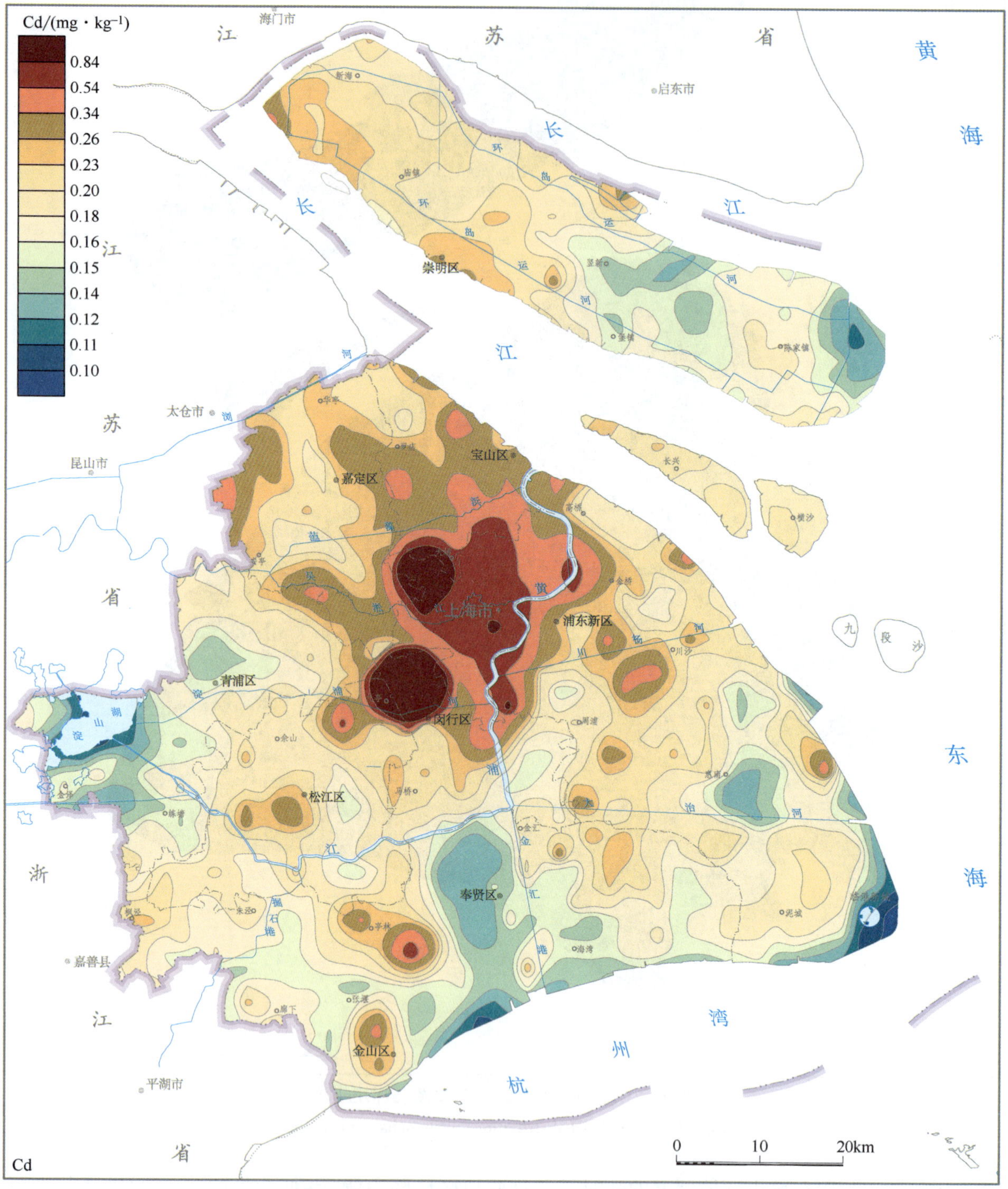

图 30-4 上海表层土壤镉、汞、铅、锌分布图

Fig. 30-4 Surface soil Cd, Hg, Pb and Zn distributions in Shanghai

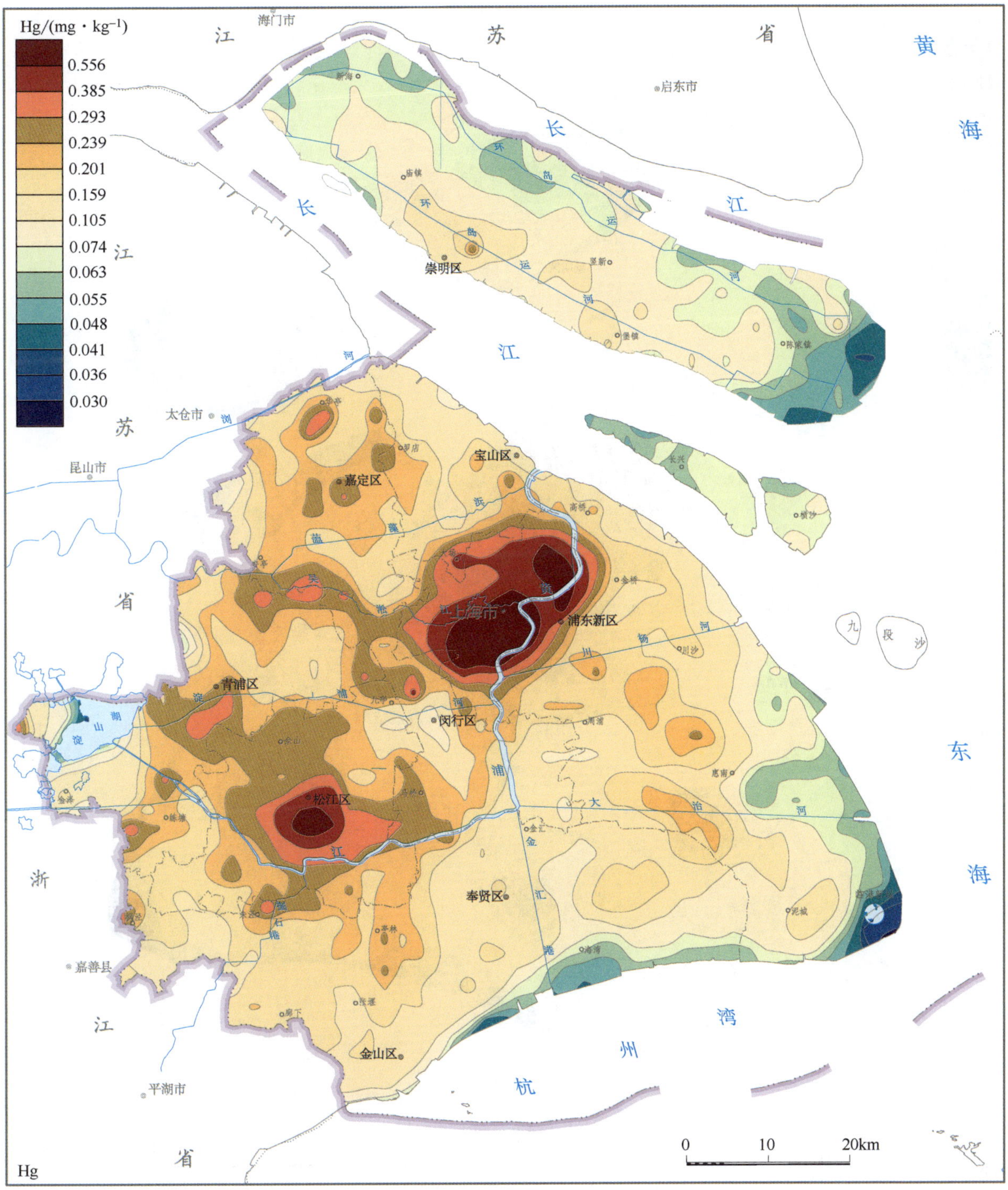

图 30-4 上海表层土壤镉、汞、铅、锌分布图（续）

Fig. 30-4 Surface soil Cd, Hg, Pb and Zn distributions in Shanghai (countine)

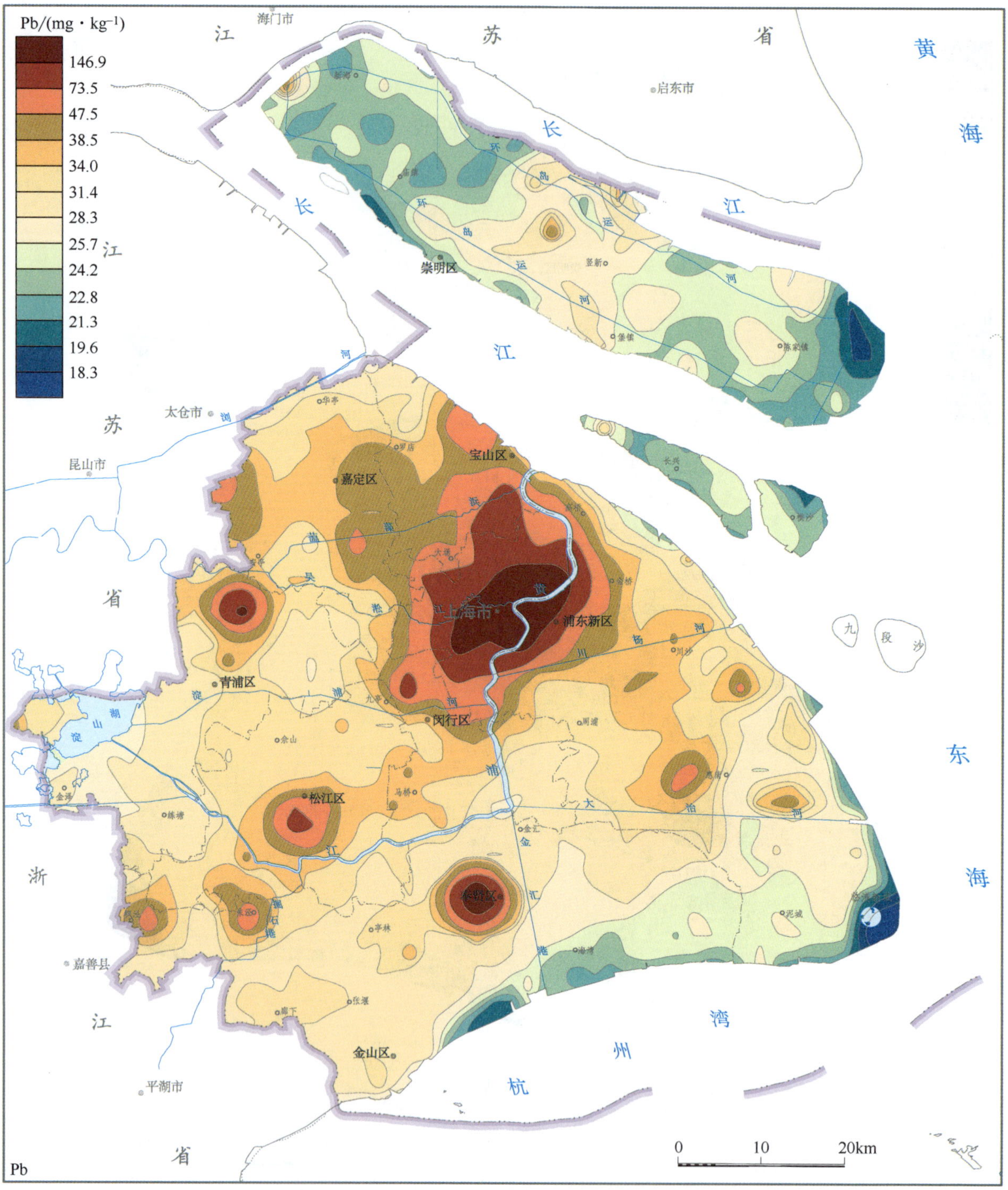

图 30-4 上海表层土壤镉、汞、铅、锌分布图（续）

Fig. 30-4 Surface soil Cd, Hg, Pb and Zn distributions in Shanghai (countine)

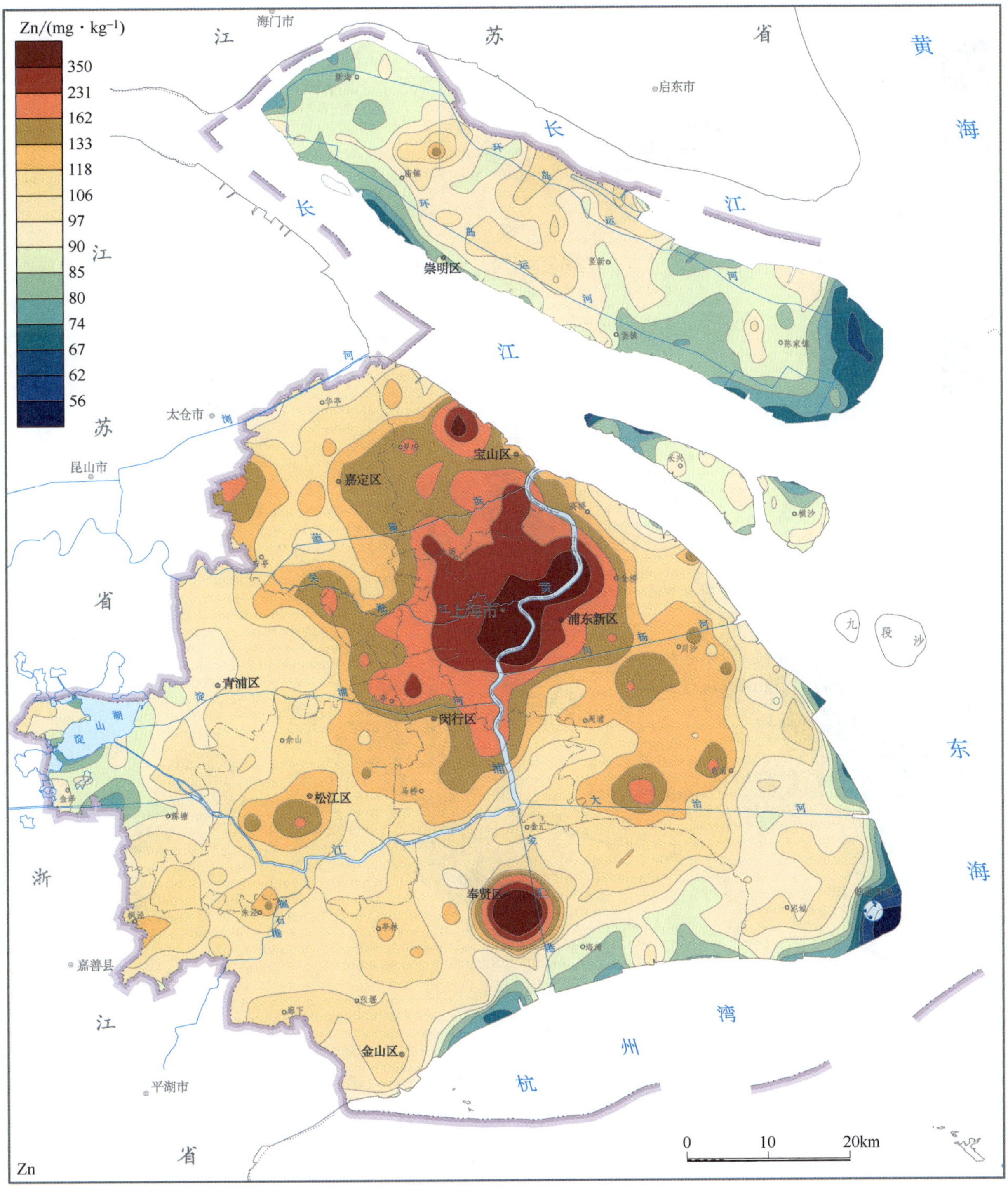

图 30-4 上海表层土壤镉、汞、铅、锌分布图（续）

Fig. 30-4 Surface soil Cd，Hg，Pb and Zn distributions in Shanghai（countine）

3）Cd、Pb、Zn 在上海陆域城区、交通要道之外的地区，如西南部淀山湖、南部局部地区的分布，仍保持着偏低背景分布的态势。

上海深层土壤中镉、汞、铅、锌的分布具有以下特点（图 30-5）。

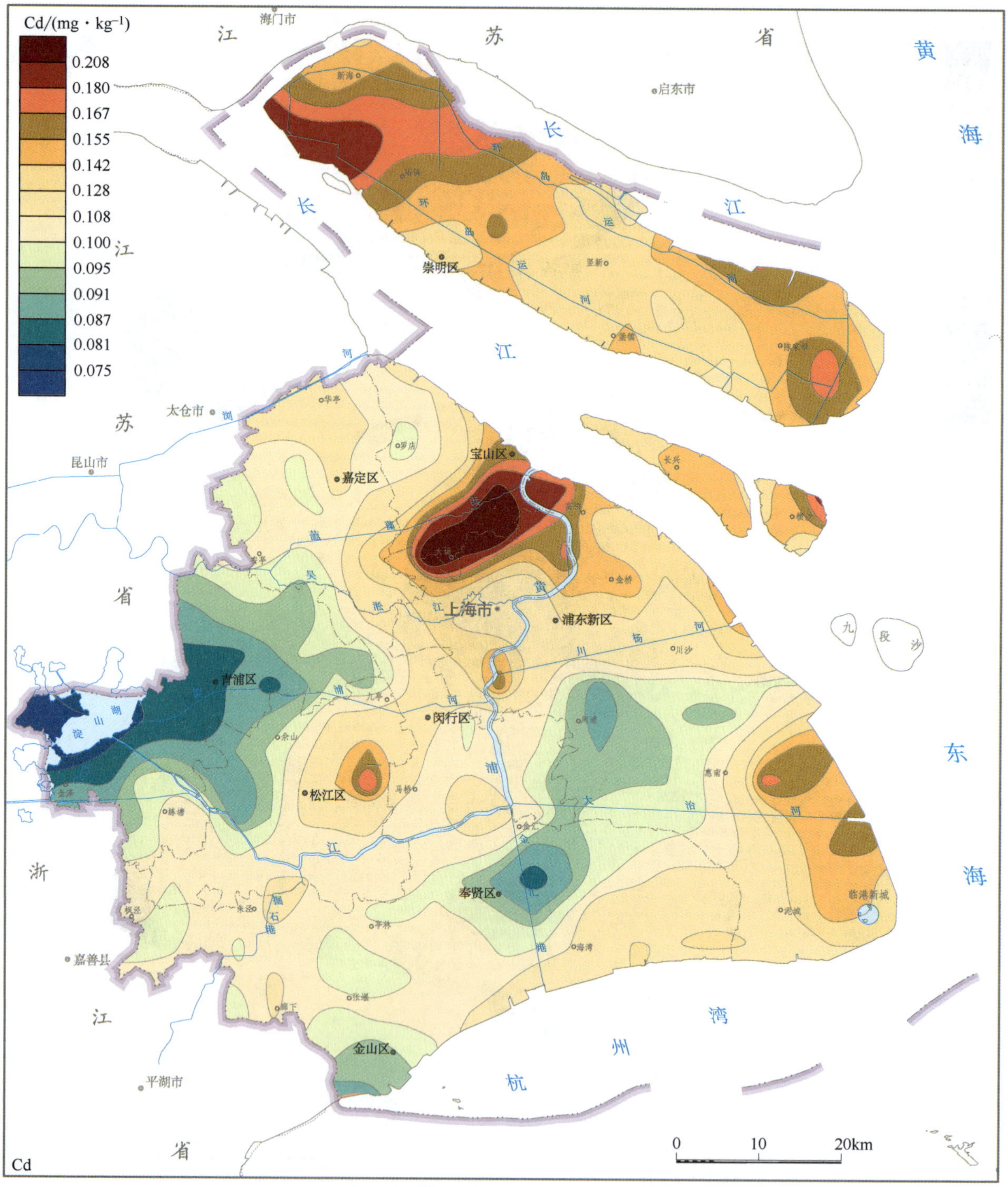

图 30-5　上海深层土壤镉、汞、铅、锌分布图

Fig. 30-5　Deep soil Cd, Hg, Pb and Zn distributions in Shanghai

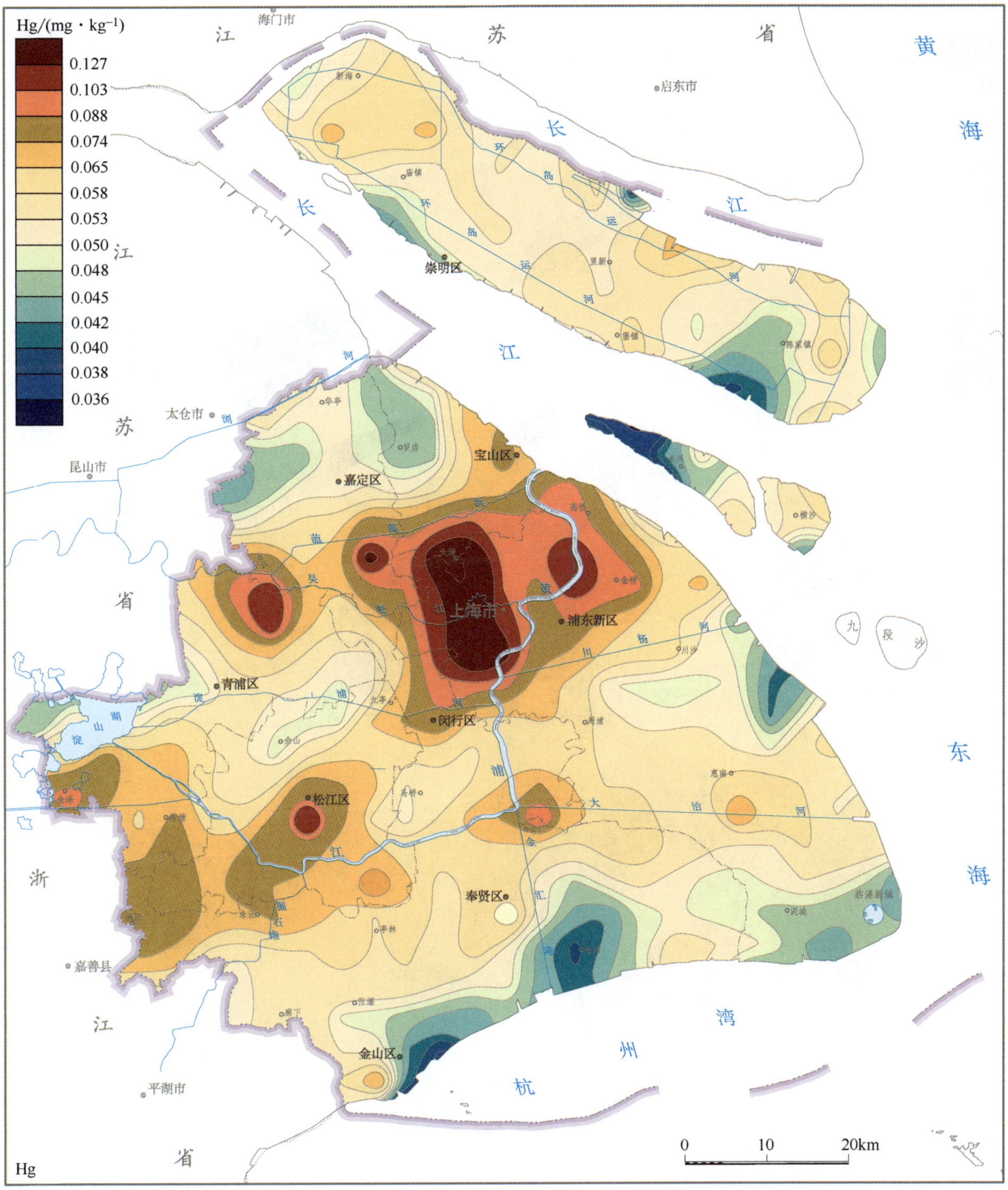

图 30-5　上海深层土壤镉、汞、铅、锌分布图（续）

Fig. 30-5　Deep soil Cd，Hg，Pb and Zn distributions in Shanghai（countine）

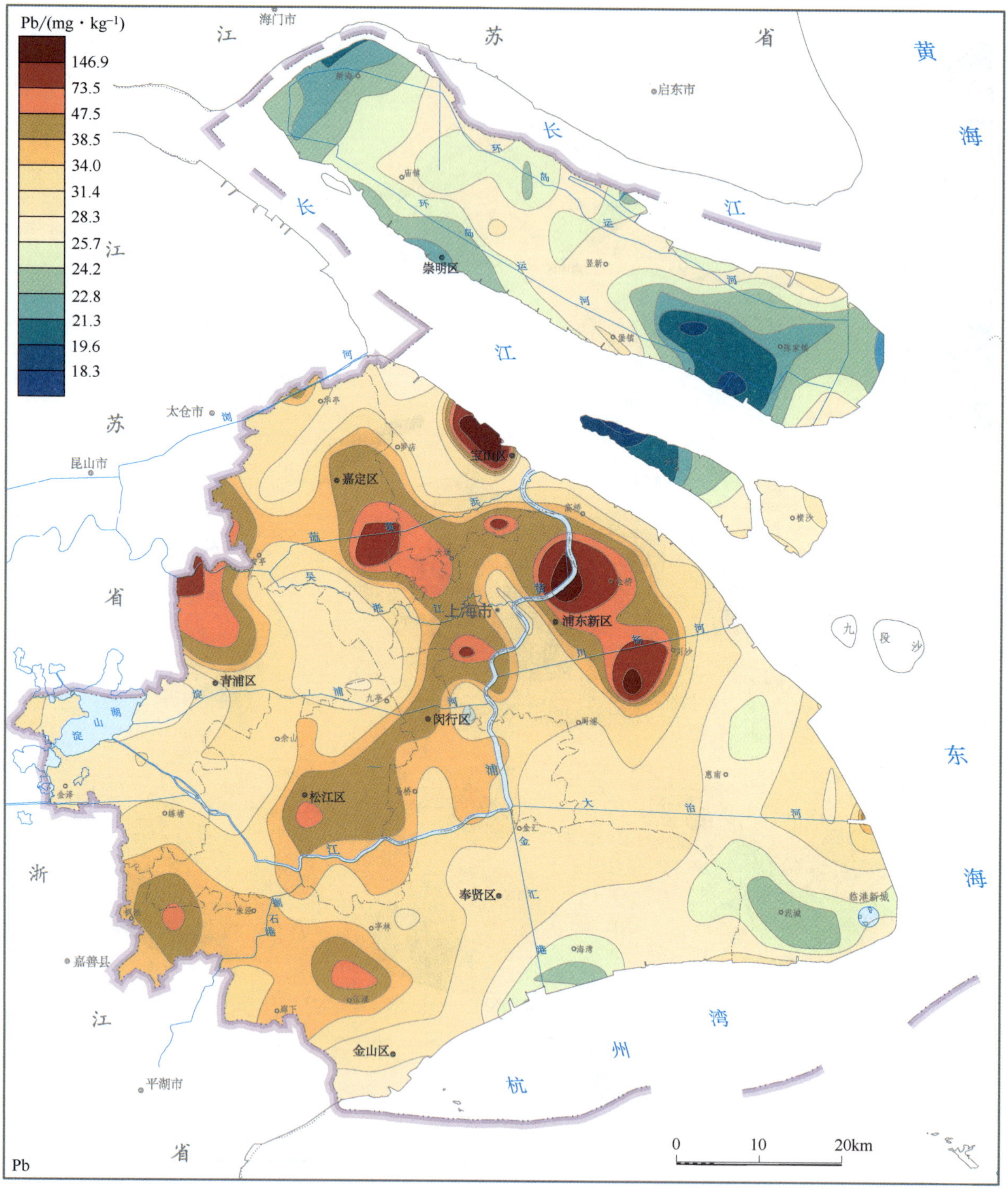

图 30-5 上海深层土壤镉、汞、铅、锌分布图（续）

Fig. 30-5 Deep soil Cd, Hg, Pb and Zn distributions in Shanghai (countine)

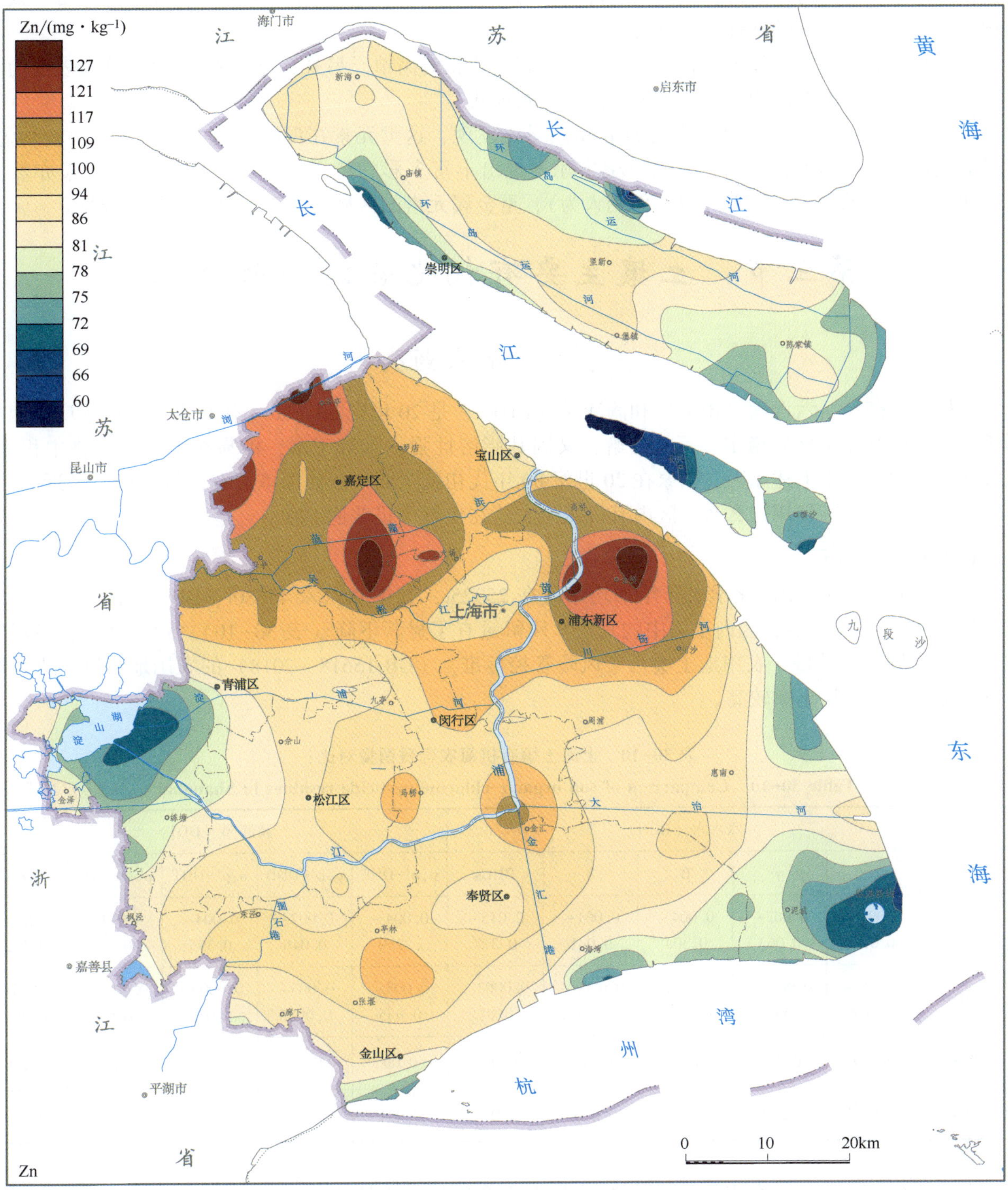

图 30-5　上海深层土壤镉、汞、铅、锌分布图（续）

Fig. 30-5　Deep soil Cd, Hg, Pb and Zn distributions in Shanghai（countine）

1）Cd 在上海东北部地区，即上海市区、宝山吴淞及崇明岛北西段，表现为中高背景分布。以上海市区为中心所呈现的环状地球化学异常是 Cd 在上海分布的一个突出特点。

2）Hg 在上海，除崇明岛、长兴岛、横沙岛及杭州湾沿江地带外，基本上为一个中高背景分布区。在这一背景中，除上海市区外，上海西北部的区域（嘉定、宝山、青浦、松江、金山等）分布着众多的规模不定、形态各异的地球化学异常。

3）Pb、Zn 的分布特征大致相同，但异常分布模式相对简单。另外，Pb、Zn 的含量分布表现为以市区为中心，其浓度向周边地区呈梯度扩散的分布模式。

4）与 Cd、Hg、Zn、Pb 的表层地球化学图对比发现，这类元素在深层土壤中的分布相对较均匀。背景的起伏变化并不大，只是在表层土壤中才表现出丰富多彩的特点来（Cr、Ni、Cu 等元素亦然）。这说明，在不同的作用条件下（自然作用人为），重金属元素的区域分布表现出强烈的差异性。

第三节　土壤主要有机化合物分布特征

一、有机氯农药类

有机氯农药包括六六六（BHC）和滴滴涕（DDT），是 20 世纪 60—80 年代在农业生产中广泛使用的杀虫剂。因其化学性质稳定、不易分解，又因其脂溶性强，难溶于水，极易在土壤及生物中积累，造成环境问题。基于上述原因，国家在 20 世纪 80 年代初就全面禁止使用该类有机氯农药，最近 10 多年的调查表明，尽管停用这么久，区内有机氯类农药在土壤中仍普遍存在。

（一）六六六

1961—981 年间，上海化学农药年平均用量在 25325t，其中六六六 45680t、DDT 18160t。至 2008 年，经近 30 年的降解后，区内土壤中的六六六残留量有了显著下降（表 30-10）。六六六的含量尽管都处在《土壤环境质量　农用地土壤污染风险管控标准》（GB 15618—2018）的农用地土壤污染风险筛选值范围内，但检出率较高。

表 30-10　上海土壤有机氯农药残留量对比

Table 30-10　Comparison of soil organo-chlorine pesticide residues in Shanghai　$w(B)/10^{-6}$

参数	项目	六六六（BCH）					滴滴涕（DDT）				
		α	γ	β	δ	BHCs	p,p′-DDE	p,p′-DDD	o,p′-DDT	p,p′-DDT	DDTs
范围	1978 年	0.008～0.066	0.002～0.063	0.004～0.304	0.001～0.041	0.015～0.377	0.004～0.259	0.002～0.046	0.004～0.395	0.011～0.30	0.009～0.562
范围	2004 年	0.0002～0.0016	0.0002～0.00093	0.0002～0.0063	0.0003～0.00085	0.0002～0.01	0.008～0.055	0.003～0.0085	0.0009～0.013	0.0015～0.07	0.012～0.069
平均值	1978 年	0.029	0.012	0.05	0.008	0.097	0.09	0.01	0.08	0.057	0.248
平均值	2004 年	0.00082	0.0006	0.001	0.0005	0.002	0.021	0.004	0.0028	0.015	0.039

在地理分布上，浦东东南部（原南汇）和嘉定个别地区土壤中六六六残留量相对较高，分别为 0.0046mg/kg 和 0.0037mg/kg；除城区外，崇明土壤中的六六六残留量最低，约为 0.00098mg/kg。

目前，上海表层土壤中 BHC 的含量以农地分布为主要特点，即农用地地区平均含量普遍高于建设用地地区。农用地地区含量相对较高的地方主要集中在嘉定和浦东东南部（原南汇）局部的地区，而在崇明及奉贤等地含量相对较低（表 30-11；图 30-6～图 30-8）。

表 30-11　上海农用地地区土壤六六六（BHC）平均含量

Table 30-11　Soil average BHC contents in agricultural land areas of Shanghai

$w(\mathrm{B})/(\mu g \cdot kg^{-1})$

地区	总六六六		α-六六六	γ-六六六	β-六六六	δ-六六六
	含量范围	均值				
崇明（22）	0.2~2.551	0.793	0.335	0.272	0.403	0.213
宝山（5）	0.2~1.926	0.887	0.2	0.450	0.312	0.325
嘉定（6）	0.707~8.564	3.696	0.566	0.723	2.207	0.334
浦东（20）	0.2~10.282	2.923	0.751	0.611	1.287	0.494
闵行（4）	0.594~3.746	2.364	0.463	0.937	0.907	0.307
奉贤（8）	0.2~4.760	1.341	0.531	0.461	0.426	0.299
松江（7）	0.2~3.485	1.578	0.433	0.470	0.623	0.365
金山（11）	0.43~3.597	1.545	0.384	0.441	0.629	0.291
青浦（10）	0.2~5.270	1.882	0.355	0.394	1.067	0.325
农用地总（93）	0.2~10.282	1.823	0.470	0.476	0.843	0.331
城区对比（11）	<0.20~1.040	0.8450	1.5936	<0.20	<0.20	0.8450

BHC 的 4 种异构体中平均含量最高的为 β-BHC。与其他异构体相比，β-BHC 的所有氯原子都位于 BHC 分子的赤道短轴上，其稳定性高，水溶性和挥发性都非常低，又不易被生物降解，因此土壤及沉积物中 β-BHC 残留量比其他 BHC 异构体都高。

（二）有机氯农药类

土壤中 DDTs 残留量最低小于检出限，最高 736.7μg/kg（图 30-9），与 20 世纪 80 年代初相比，残留量有了显著降低。20 年前平均为 248μg/kg，现在仅为 39μg/kg。按照《土壤环境质量　农用地土壤污染风险管控标准（试行）》（GB 15618—2018），农用地土壤滴滴涕的污染风险筛选值为 100μg/kg，共有 6 件样品超标，分别位于青浦、嘉定、宝山和浦东，其中浦东有 3 个样点超标。这些残留量较高的样点和地区需要被重点关注。除奉贤和浦东区外，不同地区土壤中 DDT 单体组成百分比变化不大（图 30-10）。

二、多环芳烃类

上海土壤多环芳烃（PAHs）含量最低 85μg/kg，最高 24226μg/kg，平均 1037μg/kg。上海多环芳烃含量分布差异明显，崇明、奉贤相对较低，其他区均比较高。按照《土壤环境质量　农用地土壤污染风险管控标准（试行）》（GB 15618—2018），苯并［a］芘的风险筛选值为 0.55mg/kg，在城区的 3 个样点含量均超过了筛选值，值得引起关注。

从多环芳烃各单体的组成百分比来看（图 30-11），浦东东南部（原南汇）和青浦土壤中多环芳烃以 2 环单体为主，2 环含量均超过 55%。多环平均含量最高的是宝山和闵行，这说明上海土壤中多环芳烃的来源具有非常明显的大气沉降的特点，即多环单体近源沉降，2 环及 3 环单体在运送较远距离后才发生沉降。

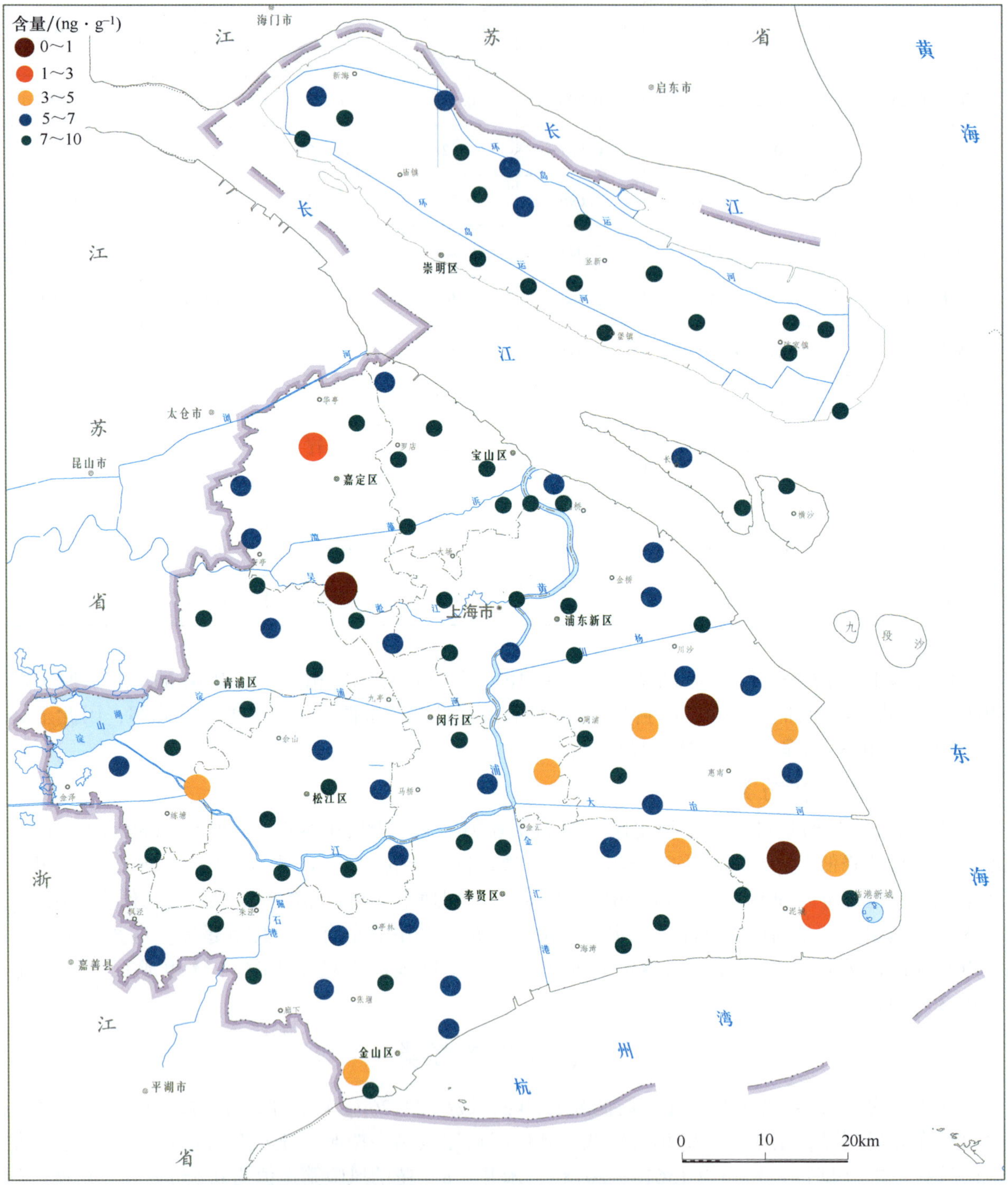

图 30-6　上海表层土壤六六六含量分布示意图

Fig. 30-6　Distribution of surface soil BHC contents in Shanghai

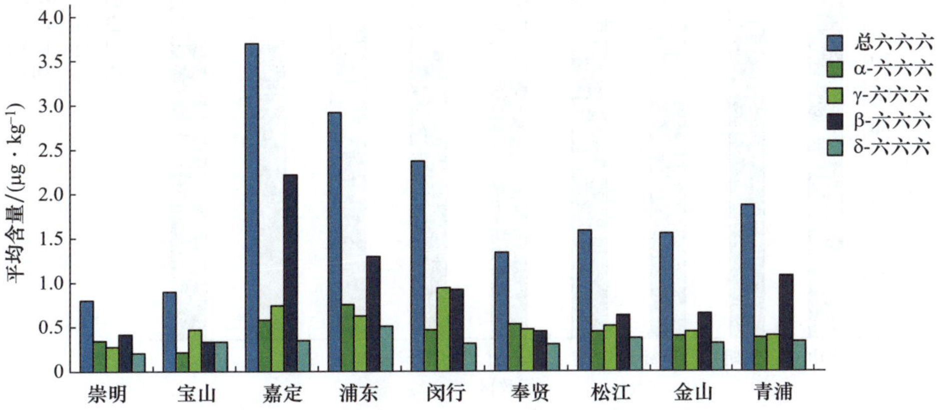

图 30-7　上海各区农用地地区六六六含量对比

Fig. 30-7　BHC distributions in agricultural land areas across different districts of Shanghai

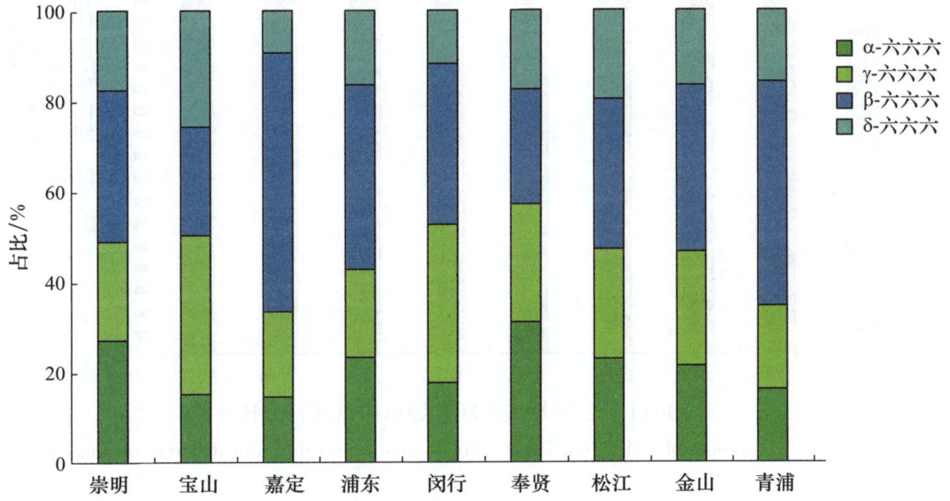

图 30-8　上海农用地地区六六六各单体百分比

Fig. 30-8　% contents among individual BHC components in agricultural land areas of Shanghai

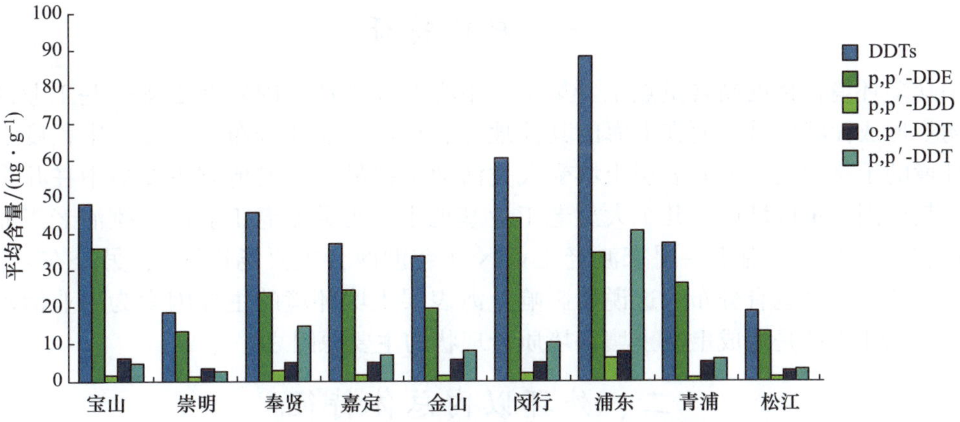

图 30-9　上海各区农用地地区 DDT 含量对比

Fig. 30-9　DDT contents in agricultural land areas across different districts of Shanghai

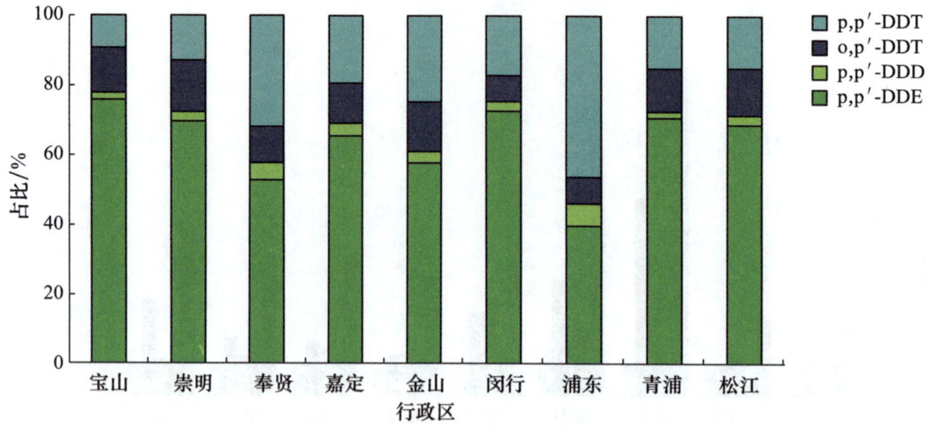

图 30-10 上海农用地地区 DDT 各单体百分比

Fig. 30-10 % contents among individual DDT components in agricultural land areas of Shanghai

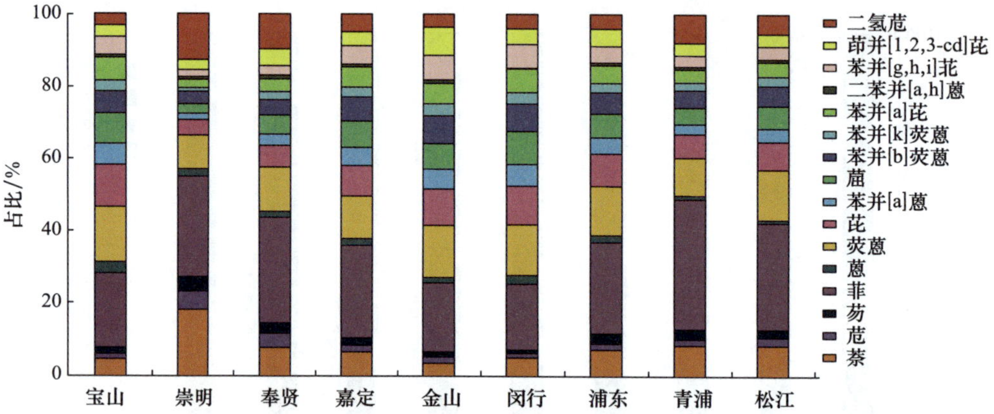

图 30-11 上海各区多环芳烃单体组成百分比

Fig. 30-11 % contents of individual PAH components across different districts of Shanghai

第四节 土壤地球化学异常

一、总体特征

土壤当中众多环境指标的高含量点主要集中在中心城（外环以内）及近郊一些工业区。在外环以内及松江，初步形成面状区域，而在上海的其他地方有局部的点状分布。这在一定程度上反映了人类活动对土壤环境的干扰程度。上海表层土壤受人类活动干扰最严重的地方主要集中在几个大型的老工业基地地区及其周围一定区域内。几个大型老工业基地主要为吴淞老工业区、桃浦老工业区、宝钢老工业区、杨树浦老工业区、南市—周家渡老工业区（今世博会主要场馆区），另外在外高桥工业区、吴泾工业区、辛庄工业区也有分布。这说明影响上海表层土壤环境的主导因素为工业活动，尤其是历史时期的大规模工业生产是造成上海土壤环境质量现状的主要原因之一。

二、外环以内总体特征

上海外环以内以及在外环线附近的宝山吴淞，嘉定江桥，闵行辛庄、七宝及吴泾等地也是城市化较高的近郊地区，表层土壤中平均含量较高的元素分别为 As、Pb、Hg、Cu、Mo、S、Sb、Sn、Se、Zn、OrgC 等。上海中环及内环间表层土壤中平均含量最高的元素为 Ni 等，中环与外环间表层土壤中平均含量较高的元素为 Cd、Cr、F 等（图 30-12，图 30-13）。在 20 世纪的早期及中期内环线以内曾

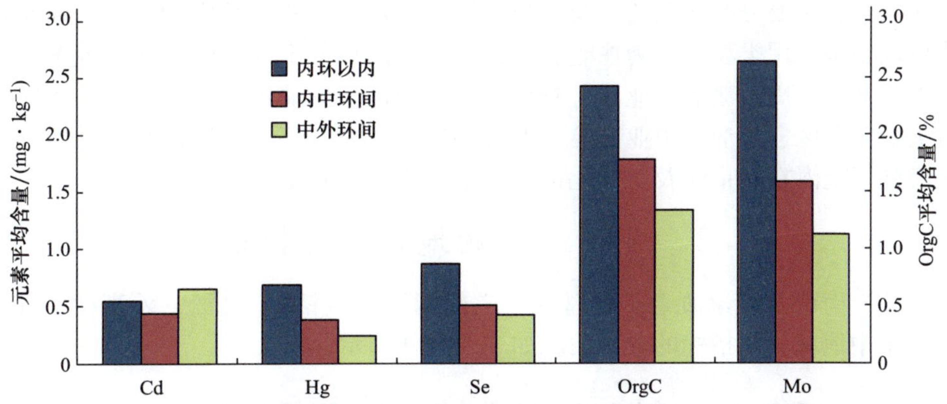

图 30-12 上海外环以内各环线区表层土壤元素平均含量对比图

Fig. 30-12 Average surface soil element contents across different ring line areas within the outer ring of Shanghai

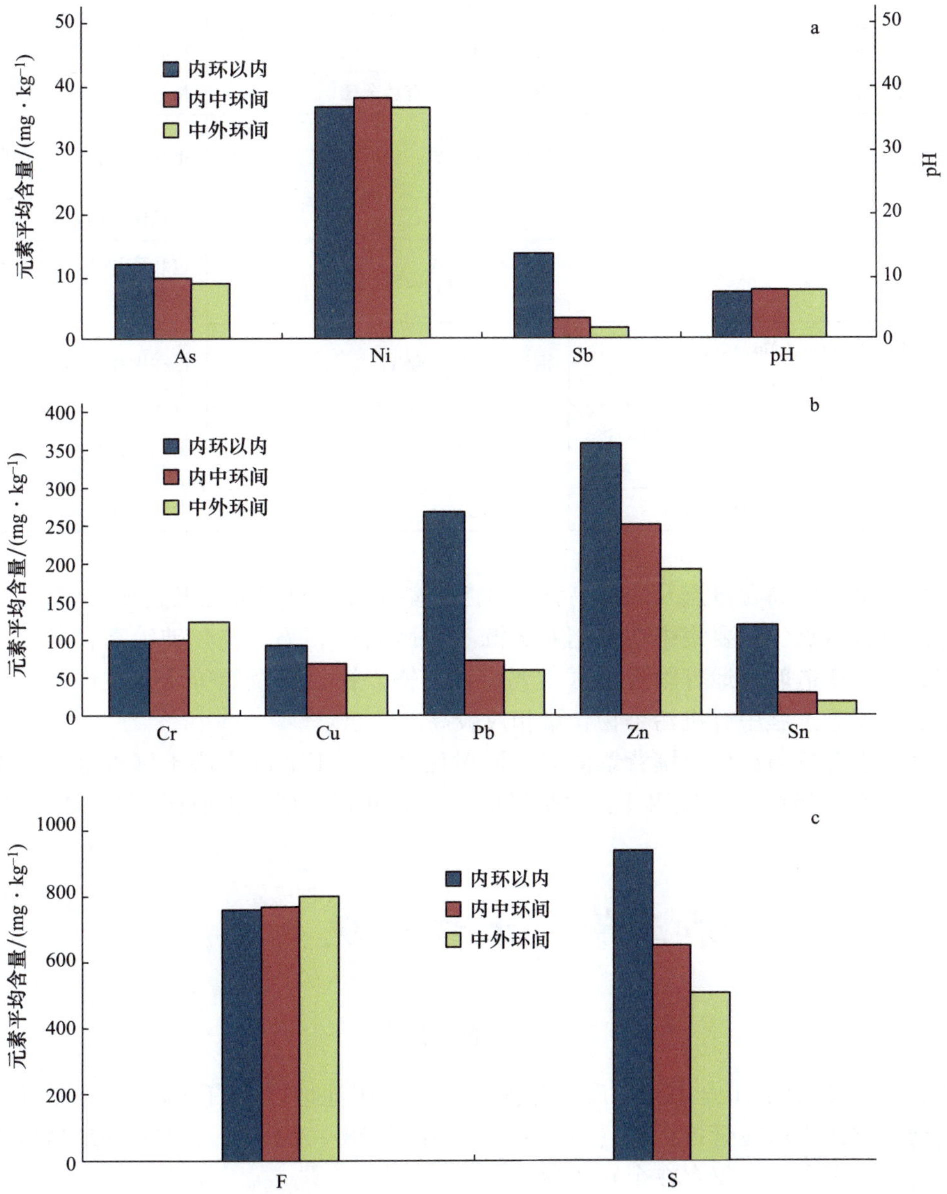

图 30-13 上海外环以内各环线区表层土壤元素平均含量对比

Fig. 30-13 Average surface soil element contents across different ring line areas within the outer ring of Shanghai

有许多大型工业企业，主要分布在黄浦江及苏州河沿线。目前这些区域内的大型工业企业多已拆迁或搬迁，内环线以内土地利用类型主要为住宅和商服用地。内环与中环之间的老工业区主要有北新泾老工业区、南市老工业区、周家渡老工业区、杨树浦老工业区、彭浦市北工业区等。中环与外环间主要的老工业区有吴淞工业区、桃浦老工业区等。上海外环线以内各环线土壤中众多元素含量变化的规律在一定程度上反映了上海在城市化及工业化进程中土壤环境受干扰的程度。

三、不同工业区地球化学特征

目前，杨树浦、南市及周家渡老工业基地，宝山钢铁（宝钢）、吴淞、外高桥、桃浦、吴泾和金山工业基地，土壤中地球化学指标的分布情况如表 30-12 所示。

表 30-12　上海外环线内各环线表层土壤主要地球化学指标对比

Table 30-12　Comparison of main surface soil geochemical parameters across different ring line areas within the outer ring of Shanghai

主要老工业区	平均含量较高指标	历史工业特征企业	现状
宝钢	B、Ge、Mn、Sr、Tl、Fe、Mg	钢铁冶炼、火电、电镀等	精品钢材产业基地、物流中心
吴淞	Cr、W	冶金、化工、有色金属	物流中转基地
杨树浦	Ag、As、Ba、Bi、Cu、Pb、Sb、Sn、C、OrgC	棉纺、柴油机、机床、煤气、电力	正在进行拆迁，规划为滨江旅游
外高桥	pH	化工、农药、石油	大厂区仍旧在生产
桃浦	Cd、Ce、La	化工、制药、染料、化纤	包装印刷中心、物流中心、新居住聚集地
南市、周家渡	Au、F、Hg、Mo、N、Ni、P、S、Se、U、Zn、Ca	钢铁、火电、化工	正在大规模拆迁
吴泾	Be、Co、Ga、I、Li、Nb、Rb、Sc、Th、Ti、V、Y	火电、氯碱、化工	正在进行环境综合整治
金山	Br、Cl、Zr、Si、Al、K、Na	石油、化工	大型石化基地

土壤中众多污染物的高含量点及局部高含量带主要集中在大型工业区及老城区附近。其中汞、铬、铅、钴、钼等的高含量点位主要集中在宝钢及吴淞工业区附近。镉、镍、砷等高含量点位主要集中在桃浦工业区附近。江南造船基地等附近地区主要有镉、铬、铜、锌及铅等含量较高。金山石化、吴泾及外高桥等工业区附近土壤中有机污染物含量相对较高。

从表层土壤元素含量与深层土壤背景值的比值对比来看，工业区表层土壤重金属元素富集程度相对最高，大多数元素（指标）表现明显，并呈局部片状的分布特征。工业区与老城区表层土壤的污染基本连成大面积区域。

第五节　土壤环境质量

一、农用地

上海从 2009 年以来开展了全市土地质量监测工作，农用地地区共有 4462 个监测点。依据《土壤环境质量农用地土壤污染风险管控标准（试行）》（GB 15618—2018），全市农用地地区土壤质量整体良好。大多数可以安全利用，仅存在个别风险点。

（一）As

土壤中 As 总体含量不高，质量情况较好。所有监测点的 As 含量均小于土壤风险筛选值，农用地土壤污染风险低，可以忽略。

（二）Cd

农用地地区土壤 Cd 含量有个别点位（15 个点位）超过了土壤风险筛选值，但是未超过土壤污染风险管控值。这些点位分别分布在宝山、嘉定、青浦、松江、浦东和奉贤，呈孤点状分布，推测人为带入的可能性较大。这些点位对应的地块可能存在农用地土壤污染风险，以及食用农产品不符合质量安全标准等风险，应加强土壤环境监测和农产品质量协同监测，同时应当采取农艺调控、替代种植等安全利用措施。

（三）Cr

农用地地区土壤 Cd 含量有个别点位（8 个点位）超过了土壤风险筛选值，但是未超过土壤污染风险管控值。这些点位分别位于宝山、嘉定、浦东和青浦。其中，宝山的 1 个超标点位位于宝钢附近，且周围的建设用地有多个超标点位，因此推测该点的高含量系周围工业生产污染所致的可能性更大一些；嘉定的超标点位周围全是农用地，2 个超标点位呈孤点状分布，初步推断主要是人为带入的污染可能性更大一些。青浦 1 个超标点位就位于另一个建设用地超标点位的附近，受此影响的可能性比较大。浦东的 4 个超标点位也都呈孤点状分布，农业生产或者其他人类活动的带入可能性比较大。对于超过了土壤污染风险筛选值，但是又低于风险控制值的点位，可能存在农用地土壤污染风险，以及食用农产品不符合质量安全标准等风险，应加强土壤环境监测和农产品质量协同监测，同时应当采取农艺调控、替代种植等安全利用措施。

（四）Cu

Cu 含量较低的时候，属于有益的微量元素，含量过高则属于重金属元素富集，造成土壤污染风险。上海农用地土壤中 Cu 含量超过风险筛选值的监测点有 30 个，除了崇明岛之外，其他郊区均有点位超过了风险筛选值。宝山的几个点位靠近建设用地，其他地区的超标点位均呈分散的孤点状分布，推测人为带入及工业生产导致其含量增高的可能性更大些。

（五）Hg

农用地土壤中的 Hg 含量有 23 个点位超过了土壤污染风险筛选值，除了崇明岛之外，其他几个郊区均有不同程度的 Hg 含量超标情况。宝山的超标点位于宝钢地区，临近有多个建设用地的监测点超标，分析可能是工业生产将其带入了土壤中。松江区的超标点位除了有位于农田区的孤点超标之外，还有临近的两个建设用地的超标点位，工业生产和人类活动是引起超标的主要原因。其他超标点位均呈孤点状分布。

（六）Ni

农用地土壤中的 Ni 含量均在土壤污染风险筛选值范围内，农用地土壤污染风险低，可以忽略，Ni 对农业生产的安全影响较小，比较安全。

（七）Pb

农用地土壤中的 Pb 含量有 2 个调查点超过了风险管控值，分别位于奉贤南桥和四团。这两个点位以及周边一定区域内农用地存在比较高的土壤污染风险，食用农产品不符合质量安全标准，且难以通过安全利用措施改善食用农产品质量的原则上应当采取禁止种植食用农产品、退耕还林等严格管控措施。

（八）Zn

农用地土壤中 Zn 含量有很多超过了风险筛选值，尤以青浦、松江、金山、宝山为甚，表明这些区域的土壤 Zn 可能存在农用地土壤污染风险，应加强土壤环境监测和农产品质量协同监测，同时限制利用方式。

（九）综合质量

在单指标评价的基础上，以监测点对应风险级别较大的污染物确定该点的风险等级，结果表明，全市农用地土壤质量整体较好，除了有2个调查点元素含量超过了土壤污染风险管控值之外，还有部分点位元素（以Cu、Ni、Zn为主）含量超过了土壤风险筛选值，安全利用受到了影响，需要加强土壤环境监测和农产品协同监测，限制其利用方式。

二、建设用地

全市建设用地共有1192个监测点。依据《土壤环境质量建设用地土壤污染风险管控标准（试行）》（GB 36600—2018），按照第一类用地标准，As含量有99.55%的调查点处于风险筛选值范围内，可以安全利用，0.45%的调查点处于风险管制值范围内；Cd含量100%的调查点处于风险筛选值范围内；Cu含量有99.96%的调查点处于风险筛选值范围内，可以安全利用，0.04%的调查点大于风险管制值；Hg、Ni含量分别有99.96%的调查点处于风险筛选值范围内，可以安全利用，0.04%的调查点处于风险管制值内；Pb含量有99.10%的调查点处于风险筛选值范围内，可以安全利用，0.19%的调查点处于风险管制值范围内，0.71%的调查点大于风险管制值。结果表明，全市土地作为一类用地，除个别地块，大多数可以安全利用；作为二类用地，As、Cd、Hg含量均处于筛选值范围内；Cu、Ni、Pb含量基本均处于筛选值范围内，有个别点位大于管制值。

第六节　土壤养分质量丰缺状况

一、土壤养分单指标丰缺分布

（一）有机质

有机质主体成分为有机碳，土壤中的有机碳平均含量为1.24%。在不同类型土壤中，脱育型水稻土中的含量最高，为1.59%；渗育性水稻土中的含量最低，为0.99%；潴育性水稻土有机碳的平均含量为1.26%。灰潮土中的含量在0.89%~1.22%；盐土中的含量在各类土壤中最低，为0.79%。按1.72的系数转换后，便可获得有机质的数据，即水稻土为2.13%、灰潮土为1.81%、盐土为1.36%。土壤的有机质仅处于中等含量水平。其中西部地区、中心城的有机质含量较低，其他地区属于中等含量水平。农田有机质含量减少是一个带有普遍性的问题。

（二）大量养分元素

氮（N）、磷（P）、钾（K）是构成土壤肥力的基础，N、P、K的丰缺直接影响作物的正常生长发育。在《全国第二次土壤普查暂行技术规程》中，依据土壤养分丰缺标准，将N、P、K含量划分出6级（表30-13）。

表30-13　土壤氮、磷、钾（有效态）丰缺分级标准

Table 30-13　Classification standard for soil N, P, K (effective state)

项目	Ⅰ级	Ⅱ级	Ⅲ级	Ⅳ级	Ⅴ级	Ⅵ级
	严重缺乏	缺乏	相对缺乏	适中	丰富	很丰富
氮（N）	<30	30~70	70~100	100~130	130~160	>160
磷（P）	<3	3~6	6~12	12~20	20~40	>40
钾（K）	<30	30~50	50~100	100~150	150~200	>200

1. 氮（N）

区内表层土壤全氮量最低值为209mg/kg，最高值为3623mg/kg，平均背景值1350mg/kg。深层土壤氮含量范围为268~1382mg/kg，平均含量为577mg/kg（图30-14）。

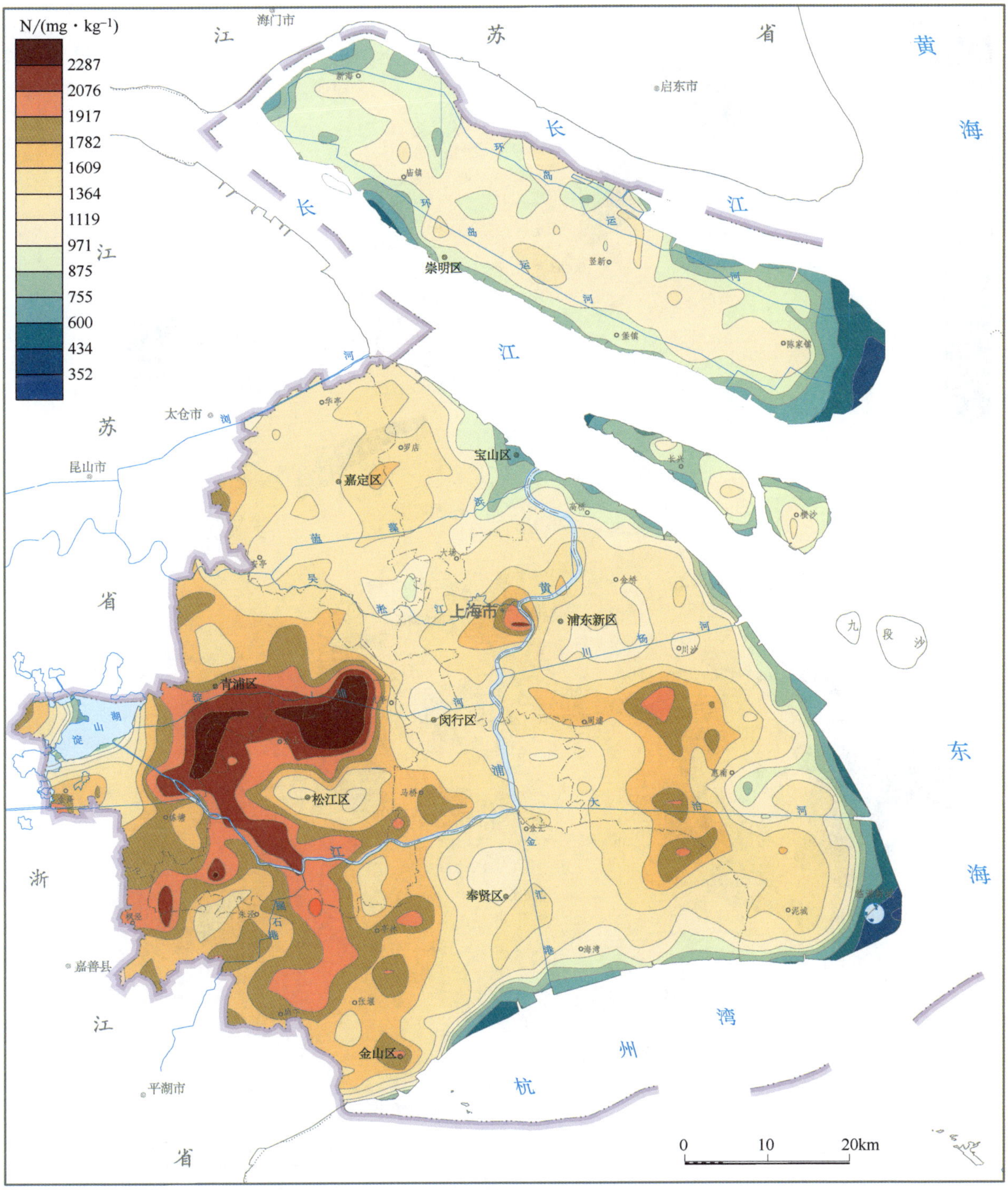

图 30-14　上海土壤中氮含量分布图（表层土壤）

Fig. 30-14　Soil N content distributions in Shanghai (surface soil)

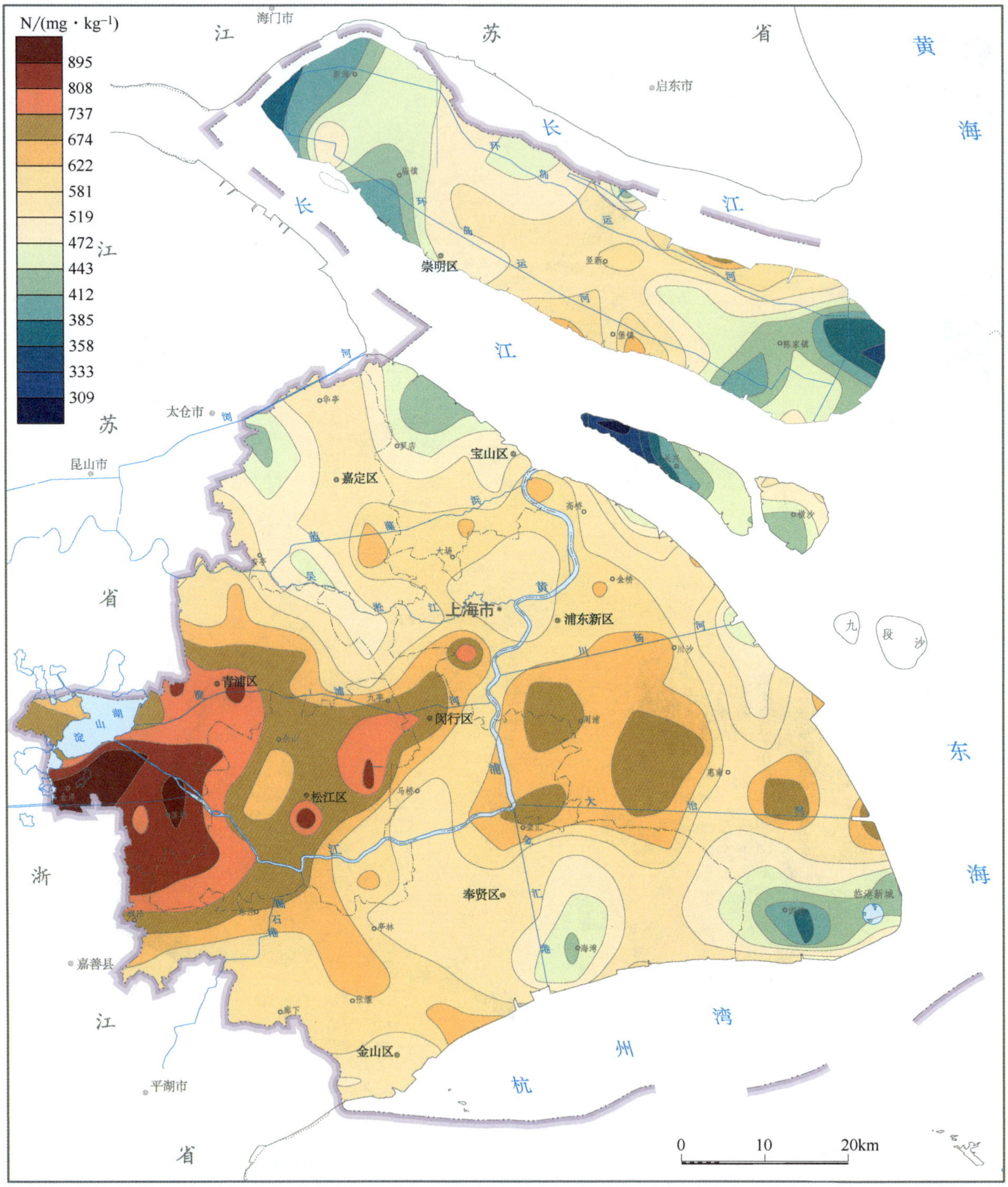

图 30-14　上海土壤中氮含量分布图（深层土壤）

Fig. 30-14　Soil N content distributions in Shanghai (deep soil)

各类土壤中，平均全氮量为934～1903mg/kg，依黄夹砂—旱作黄泥—黄泥—菜园灰潮土—淹育型水稻土—潮砂泥—潮田泥—黄潮泥—青黄泥—青紫土递增，含量差异比较显著。从区域分析，全氮量分布决定于不同的立地条件，决定于区域内有机质的积累和分解作用的相对强度。湖积区青紫土地带，全氮量偏高明显与土壤富有机质有关。滩涂区最新冲淤层区呈低氮量分布，出现沿岛缘边滩明显的低量带。一般存在着壤性化程度低的地区，有机质积累量小，全氮量低；反之，土壤熟化度高，耕作频繁的土壤区，全氮量偏高的规律。

水溶性氮是可被植物利用的有效态氮，是铵态氮、硝态氮、氨基酸、酰胺及易水解的蛋白质的总和。上海北部地区各土类水溶性氮量最低69.68mg/kg，最高162.2mg/kg，平均117.64mg/kg，和全量氮的总平均比率为8.94%。土壤类别中，水溶性氮平均值量出现盐化土<旱作黄泥<黄泥<黄夹砂<潮田泥<黄潮泥<青黄泥<青紫土<菜园灰潮土的分布态势，这种态势与土壤的立地性及环境条件有关。按照土壤酸碱性解析，偏于中性的土壤中含有较多的水溶态氮，偏于碱性土壤则含量较低。

上海土壤中的氮高含量地区主要集中在西部湖沼相沉积区，另外也与该地区的农业开发历史较长有一定关系。在三岛等沿江沿海地区为氮相对低含量地区，其中，崇明岛氮的缺乏是制约高效生态农业发展的不利因素之一。依照氮的丰缺分级指标，土壤存在着氮的缺乏、相对缺乏、适度、富足4个分级区。其中，崇明、长兴、横沙三岛、上海本土沿江、沿海地带隶属于氮的缺乏—相对缺乏区，而以崇明岛的滨海盐土氮素供应最差；而在本土，从城区及周缘向西部湖区，氮素供应由适度—富足过渡。从整个分布趋势上看，有效氮的分布与全氮基本相似。

2. 磷（P）

磷易氧化，在风化作用及水动力条件强的地方，容易从矿物中析出。表层土壤全磷量最低值为361mg/kg，最高值为2718mg/kg，一般在700～1400mg/kg之间，平均964mg/kg。深层土壤磷含量范围为264～1064mg/kg，平均含量为615mg/kg（图30-15）。

区域分布上，表层土壤主要在上海城区及其周缘、西部湖积区显示为偏高含量，基本在1000mg/kg左右；偏低含量分布区基本位于崇明岛、上海本土东部边缘滩涂地带，含量在650～720mg/kg之间。就地域而言，崇明、长兴、横沙等岛屿区总体全磷量偏低，本土区总体偏高，城区最高。城区偏高，是因为有环境污染因素叠加影响所致。从上海土壤总体含磷水平和国内其他地区不同类型土壤比较，属于含量偏高土壤类型。深层土壤中磷的高含量区主要分布在吴淞及宝山工业区。在西部湖沼相沉积区含量普遍较低，其他地区含量中等。

不同土壤类型有效态磷分布量差异很大，有两个含量级次。低含量级为浦东地区的黄泥土属区、崇明岛的黄夹砂土属区、灰潮土旱作黄泥土属区和盐（化）土土属区，分别为22.07mg/kg、14.4mg/kg、14.38mg/kg、12.97mg/kg；偏高量区为黄潮泥、青黄泥、潮田泥、潮砂泥、青紫土、菜园灰潮土等土属区，含量在25.45～93.44mg/kg，以菜园灰潮土为最高。

上海为全磷高、有效态磷也高的分布模式。根据已有成果初步认定，浦东地区的川沙北部和长兴、横沙、崇明为磷的相对缺乏区，西部湖区、城区周缘为磷的富足区。对于三岛区和川沙等缺磷区，农业上以适当补施磷肥为宜。

磷与氮在陆域土壤中的分布特点大致相似，但磷的适量及丰富区范围要明显大于氮，如崇明岛土壤中严重缺氮，而大部分地区土壤中的磷并不缺乏。

（三）钾（K）

土壤中全钾量以 K_2O 计，最低值为1.55%，最高值为2.9%，各类土壤中背景量在2.209%～2.469%，含量差别不大。这一分布特征和上海总的所处的地质环境有关，即基本处在三角洲平原区，普遍接受长江洪冲积为主的沉积，不同之处为少部分地区有湖积或者海积物混合。从分布区域上，嘉定、浦东周浦地区滨海潮滩沉积母质层区钾呈略高量分布，崇明约略偏低，但含量都在2%以上。总体上看，上海全钾量除了南部沿海地区相对缺乏之外，其他地区均较为丰沛。

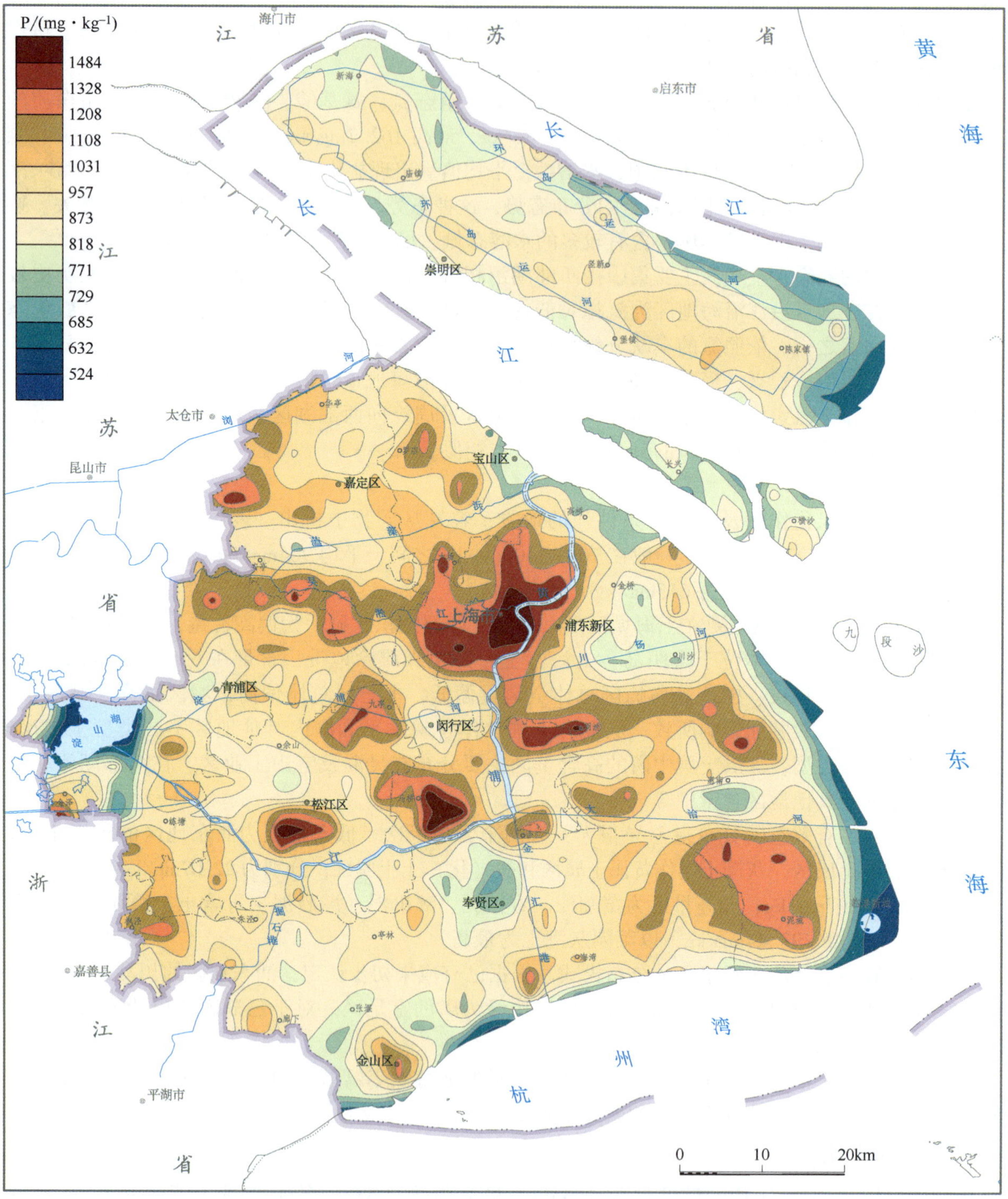

图 30-15 上海土壤中磷含量分布图（表层土壤）

Fig. 30-15 Soil P content distributions in Shanghai (surface soil)

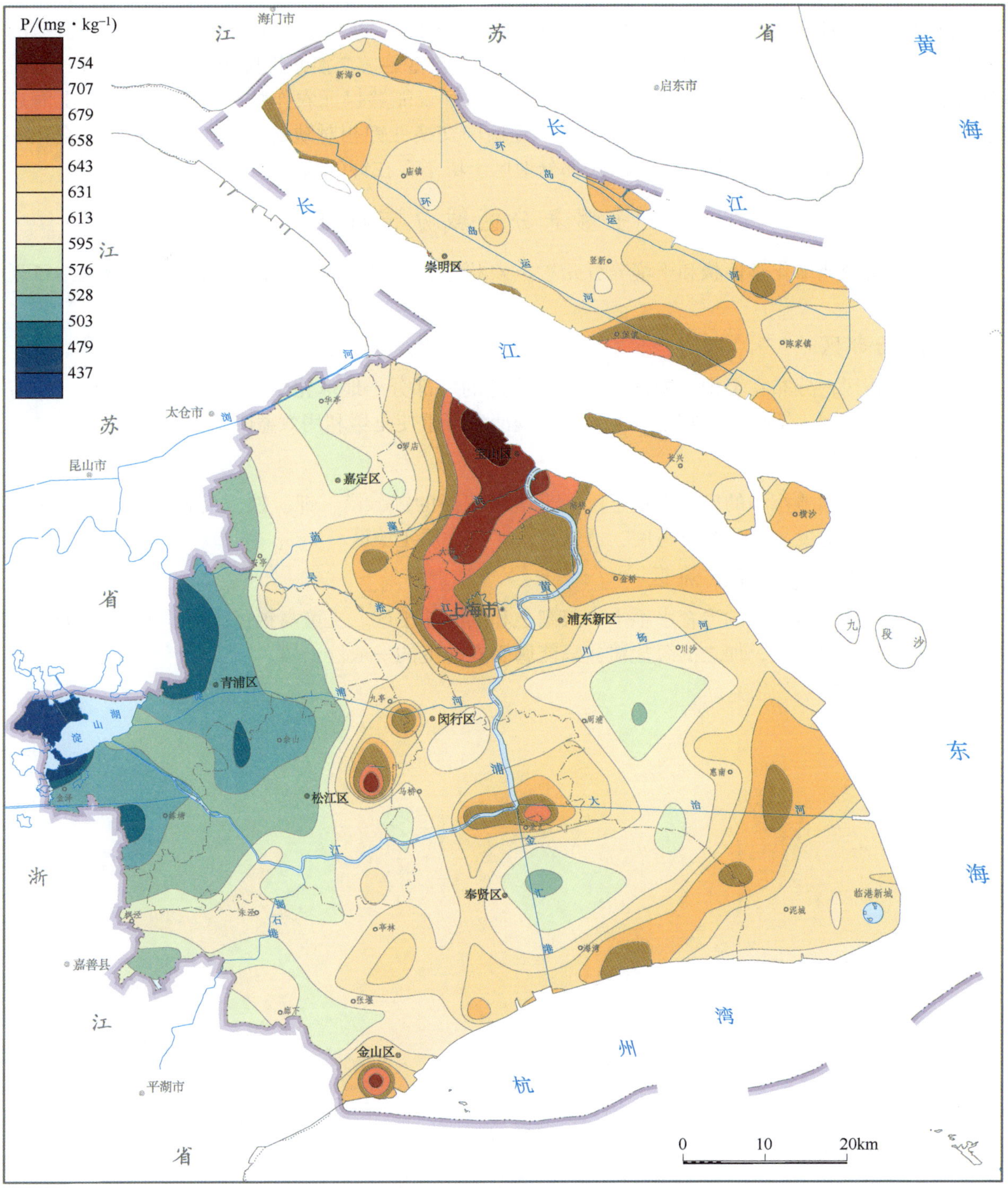

图 30-15　上海土壤中磷含量分布图（深层土壤）

Fig. 30-15　Soil P content distributions in Shanghai (deep soil)

各类土壤中交换性钾最低量平均为117.76mg/kg，最高量平均为329.8mg/kg，总平均量为168.14mg/kg，全钾总比率为0.73%。据统计，交换性钾和全量钾两者之间的相关系数为0.1792，无相关关系。可能反映在全钾基本全为富集区的情况下，交换性钾的含量决定于土壤的其他方面环境因素。但就土壤类型来看，交换性钾偏高量出现在菜园灰潮土、城区潮田泥和脱潜水稻土亚类青紫土土属区，这可能和3种土属有机质含量均偏高，腐殖酸含量较高，有利于钾向溶解性钾转化有关。

按照丰缺标准，表层土壤中全钾平均量在2%以上，较富足。交换性钾按照丰缺分级标准，上海北部及西南部地区土壤基本属于钾素较富足土壤，土壤中钾素能够保证农作物生长需求。而在上海东南部地区（原南汇大部、奉贤和金山沿海地区）的土壤中交换性钾相对缺乏。上海三岛及除沿金山—奉贤—原南汇一线以东的滨海地区外，上海大部分地区土壤不存在钾的缺乏。

二、土壤养分区域分布特征

依据上海土壤地球化学性状和土壤养分元素丰缺状态，上海以农业利用为目标的土壤基本可划分为3类区：

（一）西部区

即西部湖积区，分布在金山、松江、青浦和嘉定的一部分。地球化学上低钙、镁，高硅，富有机质。土壤养分多表现为适度或者富足。但多以土壤渍水严重使土壤利用效益降低。

（二）中部区

即母质为滨海潮滩沉积的广大水稻土区。土壤以偏高的钙、镁、钾、铁、铝和微量碘为特征。因该区土壤养分多为适度，土质黏重适度，透气爽水，为优质土壤区。

（三）东部三岛区及滨海区

属母质为近代河口沉积的渗育水稻土、灰潮土和滨海盐土区。土壤高钙，土壤养分显示为磷、氮、锌、铜、钼、硼相对缺乏，加之滨海盐渍，只能属于农业上的一般性土壤。

第三十一章　岸带冲淤

第一节　重点岸带地区的冲淤演变

一、崇明东滩

本志崇明东滩的范围指崇明东滩围垦区外缘区域，包括北港北沙区域和北港北汊道，北侧为长江口北支，近年来不断束窄淤浅，南侧为北港下段，对岸为横沙东滩大型围垦区。西侧崇明岛，崇明岛在新中国成立后进行了大量围垦工程。2005 年东滩作为国家自然保护区进行保护，外缘为中华鲟自然保护区，东侧为东海海域。

自公元 11 世纪左右开始，崇明岛开始形成。特别是 18 世纪中叶长江主泓走南支以来，崇明岛南岸遭受侵蚀（公元 1894 年，加固堤防才制止坍势），而北岸迅速淤涨，东滩也有所发展，崇明岛面积不断扩大，直到今天还继承着这样的发展趋势。1842 年，崇明东滩呈向东延伸的“鸭舌状”边滩；1907 年，在北港拦门沙段北侧出现阴沙；1960 年，阴沙淤积出露成为心滩明沙，称为团结沙，之后，团结沙“北淤南蚀”，向西侧滩涂并靠发展；1979 年，人工筑坝堵汊，团结沙与东滩连在一起。由于团结沙筑坝堵汊工程的实施，改变了其周围水流特性，原团结沙外侧受涨潮流的冲刷形成第二个离滩沙体，经过多年的发育成为北港北沙，与崇明东滩之间的汊道称为北港北汊道。1842 年至今，崇明东滩不断淤涨，但是近年来淤积速率逐渐趋缓（茅志昌，2014）。

2011—2013 年，崇明东滩出现小幅度冲刷，尤其是在崇明东滩北侧区域，冲刷较为明显，中间北港北汊道局部淤积，而南侧北港北沙区域变化较小，局部冲淤交替，无明显趋势；2013—2015 年，崇明东滩和北港北沙淤积明显，崇明东滩一侧淤积更为明显，北港北汊道略为冲刷；2015—2017 年，崇明东滩继续淤积，南侧淤积大于北侧（图 31-1）。

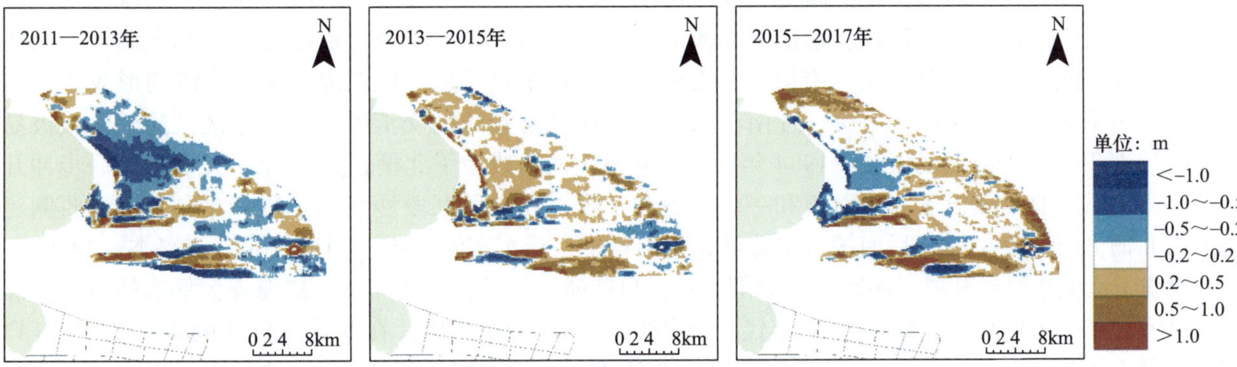

图 31-1　崇明东滩地形冲淤变化

Fig. 31-1　Terrain erosion and deposition in the Chongming East Shoal

二、横沙浅滩

横沙浅滩本志指横沙东滩围垦区外沿 -5m 等深线包络区域，两侧分别为长江口北槽和北港，东侧面临东海，其中北槽建设有深水航道，是长江口主要的水运航道，其北导堤阻挡了调查区与北槽的水沙交换，西侧为横沙东滩围垦区，围垦堤坝建设改变了涨落潮的动力，进而影响局部地形演化，东侧为东海，海洋潮流有风浪大、流速快的特征，北侧受北港涨落潮流影响。

早在1842年，长兴岛、横沙岛、横沙东滩、九段沙和江亚南沙等尚未成陆，零星有滩涂分布，统称为铜沙浅滩；1869年，北港被冲开后，横沙岛开始出露成陆，后长兴岛开始成陆，并跟圆圆沙合并成陆；1954年，长江大洪水将铜沙浅滩冲开，形成北槽，九段沙独立成陆，铜沙浅滩只为横沙岛东侧部分浅滩；20世纪80年代后，改为横沙浅滩，并在滩中间形成一条连通北槽和北港的串沟（王维佳，2014）；1998年后，北槽深水航道北大堤建设，封堵串沟，并在后续时间里将横沙东滩逐渐围垦成陆（陈吉余，2013）。

2011—2013年，横沙浅滩整体为冲刷态势，冲刷尺度在0.2～0.5m之间，局部区域在0m等深线包络区有小范围淤积（图31-2a）；2013—2015年，两侧出现冲刷，中间靠近东侧发生较为明显的淤积，整体淤积面积大于冲刷，北侧变化幅度较大（图31-2b）；2015—2017年，淤积面积继续增加，只在西北侧发生冲刷，中间区域淤积依然明显，淤积厚度甚至超过0.5m（图31-2c）。

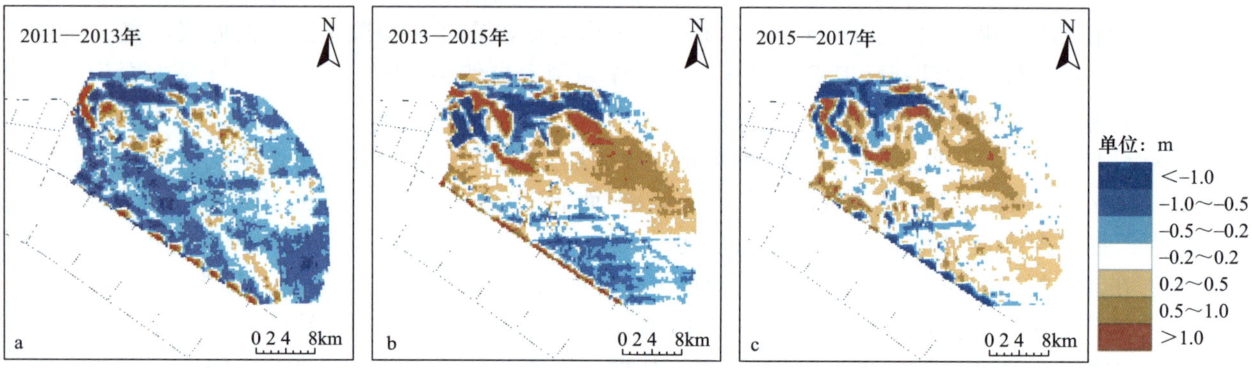

图31-2 横沙浅滩地形冲淤变化

Fig. 31-2 Terrain erosion and deposition in the Hengsha Shoal

三、南汇边滩

南汇边滩在平面形态上为一向海突出的梨头状浅滩地形，是长江入海径流扩散和长江口与杭州湾北岸两股水体涨潮分流、落潮合流共同作用下形成的堆积地貌。通常以石皮勒为界，以北称南汇东滩，以南谓南汇南滩。长期以来南汇边滩为长江河口淤涨速率最快的岸滩之一，1954年，长江流域特大洪水将铜沙浅滩中间冲开一个河槽，形成了北槽的雏形，两侧沙体分别发育成横沙浅滩和九段沙，北槽的形成增加了此河段落潮分流比，南槽逐渐淤积变浅，在南侧形成南汇东滩，南汇东滩的形成主要与科氏力作用下涨落潮流路不同有关，在南汇东滩处形成涨落潮流的交汇处，为缓流区，这个区域极易发生泥沙堆积。由于南槽的淤浅，1984年后，主航道由南槽改为了北槽，1998年，大洪水进一步冲开北槽河道，后在北槽建设了深水航道和一些导堤和丁坝，固定了九段沙北侧沙体，并在九段沙西侧建设了南北槽分流鱼嘴工程。后南槽分水分沙比逐渐增加，南汇东滩也逐渐围垦成陆（郭兴杰，2016）。

历史上南汇边滩虽开展了多次促淤圈围工程，目前滩涂资源仍极为丰富。以海塘为标志的海岸外涨速度显示，自公元8世纪—20世纪60年代，海岸线年平均外涨22.3m。在前后1247年的时间内，长江口南岸海岸线外涨速率不均，713—1172年，年均外延34.85m，1172—1584年间，外涨速度仅为3.4m/a，1584—1884年间，外涨速度增加至24m/a。原因是在此时期，长江主泓由北支入海，导致南侧河道的供沙不足。18世纪后，长江主流经南支入海，南岸边滩供沙丰富，1884—1960年，涨速扩大至41.3m/a。

1997—2002年，南汇边滩大部分区域冲淤幅度在50cm范围内，冲淤基本平衡。仅大治河外侧-2～-5m等深线区域存在一个淤积块状区域，平均淤积厚度在0.5～1m范围，最大淤积厚度为2m左右。

2002—2009年，南汇边滩大部分区域冲淤幅度在50cm内，冲淤基本平衡。在沿岸和-5m等深线区域形成两个淤积条带，特别是大治河以北沿岸区域是淤积集中区域，平均淤积厚度达2m左右。在南汇边滩北部和南部，有小范围的冲刷。

1997—2009 年，除区域北部和南部小范围为冲刷外，大部分区域为冲淤平衡或淤积状态。主要淤积呈带状分布，由大治河以北沿岸斜向东南方向，淤积带宽由沿岸约 3km 逐步扩大为 6km 左右。特别是大治河以北沿岸和大治河以东-5m 等深线左右形成了两条高淤积条带，淤积厚度都在 2m 以上。

四、九段沙

九段沙是长江口第三代新生沙洲，由上沙、中沙和下沙组成。上沙与江南亚沙毗邻，中、下沙南侧为南槽，九段沙北侧为北槽。2000 年 3 月，上海市人民政府批转建立九段沙湿地自然保护区，同年 8 月，九段沙湿地自然保护区管理署成立。2005 年，九段沙湿地晋升为国家级自然保护区，保护区范围包括九段沙、江亚南沙及附近的浅水水域。

由于九段沙与江亚南沙在形成机理和形成时间上的差异，根据 1842—1958 年长江口不同年代的历史海图分析，九段沙从形成到成为独立的沙体，经历了 100 余年的漫长过程。在 1942 年的海图上，铜沙浅滩为一庞大的河口心滩，在其南北两侧为南港和北港。北港 5m 槽向上游封闭，说明以涨潮流作用为主，南港 10m 槽向下游封闭，以落潮流为主，南港为径流入海主汊，北港为支汊。

1860 年，长江流域经历大洪水（宜昌站最大流量为 92500m^3/s，接近 1954 年的洪峰流量），大洪水对河势产生明显作用，南北港河势发生重大变化。北港原 5m 深涨潮槽冲深演变为 10m 深落潮槽，10m 深槽下端延伸至 122°07′E，最大水深达 14.6m；南港由于分流量减少，河槽淤积，10m 槽端点上移了 8000m，北港替代南港成为长江径流入海主汊。1864—1869 年，横沙岛已出露水面，长兴岛沙洲群形成，铜沙浅滩南部冲刷出一条涨潮槽，同时在南港高桥至五号沟河段江中形成木房心滩。

1879—1907 年，木房滩不断下移，抵达 122°06′E，木房滩纵长约为 46km，宽为 2~4km。木房滩上段在北移过程中成为组成长兴岛 6 个沙洲中的圆圆沙。中下段下移北靠铜沙浅滩，故九段沙雏形由木房滩和铜沙浅滩一部分组成。

1931 年、1935 年，长江全流域发生洪水，大通站洪峰量在 7×10^4m^3/s 以上，南港发育成为南支入海主汊，逐渐形成单一河道。

1945 年，九段沙北侧落潮下延 6km，同时在九段沙下段北侧形成北西-南东向的 5m 涨潮槽，水深在 5.8~9.1m 之间。

1949 年、1954 年，长江发大洪水，特别是 1954 年的特大洪水（最大洪峰流量为 92600m^3/s），促使九段沙北侧的落潮槽与涨潮槽贯通，最大水深达 9.4m，北槽形成。至此九段沙成为介于南港南槽和南港北槽之间的独立沙洲。

自 1958 年，九段沙成为独立沙体以来，滩地面积一直处于增长状态。特别是 2000 年以来，受深水航道工程和江亚南沙九段沙促淤工程影响，滩地面积更是加速增长。1997—2002 年，江亚南沙显著淤积，平均淤积 1~2m，部分区域淤积甚至在 2m 以上，而九段沙下沙出现轻微冲刷。2002—2005 年，江亚南沙区继续显著淤积。九段沙中、下沙大范围淤积，淤积幅度呈现阶梯状，南导堤 S4~S5 下平均淤高 1~2m，S5~S6 下平均淤高 0.5~1m，S6~S7 下平均淤高 0~0.5m。2005—2008 年，江亚南沙淤积，九段沙下沙继续大范围淤积，但淤积幅度不大，一般在 0~0.5m 之间。

五、青草沙水库附近水域

青草沙附近海域是上海海岸带地质环境变化最急剧的区域之一。近 10 年来，新浏河沙、中央沙沙头和新桥通道一直处于侵蚀后退状态。南北港分流口控制工程和中央沙、青草沙围垦工程建设后，青草沙附近河势调整明显。

2006—2008 年，新浏河沙、中央沙和青草沙沙头水下斜坡发生侵蚀并向围堤靠近，在新浏河沙护滩潜堤和中央沙青草沙围堤北沿形成带状的侵蚀条带，平均侵蚀深度约为 5m。

2008—2010 年，中央沙、青草沙堤前侵蚀条带继续向东部发展。受青草沙水库围堤合拢影响，北港上段河槽被束变窄，使北港四滧港至奚家港段产生块状侵蚀，平均冲深约为 1.5m。

2010—2011 年，北港地形总体变化趋缓。北港进口段主流南偏，护滩潜堤附近冲刷。北港四滧港至奚家港段由侵蚀转为平衡。长江隧桥断面北段继续侵蚀，河槽加宽变深，主泓偏北。

2011—2012 年，北港地形总体变化幅度较小。北港进口段主流继续南偏，限流潜堤附近发生侵蚀，平均侵蚀深度约为 3.2m。青草沙围堤北沿、长江隧桥断面北段由侵蚀转变为淤积。北港四滧港至奚家港段继续维持较稳定状态。

2012—2013 年，北港冲淤变化格局与 2011—2012 年相比有较大变化，潜堤附近侵蚀基本消失，而限流潜堤附近则发生明显淤积，北港深弘有向南岸偏移。

2013—2014 年，与上年度明显不同的是，北港以显著的淤积为总体特征，根据 2 个年份水深数据统计结果，一年来该区净淤积泥沙 $1.37\times10^8m^3$，其中淤积量 $2.50\times10^8m^3$，侵蚀量 $1.13\times10^8m^3$。横向上，淤积量大的地区主要分布于北港中上段的深泓线附近区域，该区-15m 等深线向深泓线发生显著偏移，偏移幅度达 300～500m。此外，淤积区面积也较侵蚀区偏大，占该区总面积的 57%，侵蚀区占 43%。垂向上，强淤积区和强侵蚀区多分布于北港中段和上段地区，最大侵蚀淤积程度多介于 2～5m 之间。

2014—2015 年，北港较上年度发生较小变化，以弱淤积为主要特征。其侵蚀区面积和淤积区面积相当，分别为 $200km^2$ 和 $201km^2$，淤积量较侵蚀量偏大，分别为 $1.70\times10^4m^3$ 和 $1.49\times10^4m^3$，说明本年度北港延续了上年度以淤积为主的特征，但淤积速率明显减缓。空间上，淤积主要发生在北港上段和北港南部两岸近岸-10m 等深线以浅地区，淤积厚度多在 0.5m 以上，而深泓线附近仍以侵蚀为主，侵蚀深度多大于 1m。青草沙水库北岸岸堤附近呈现弱侵蚀，年度侵蚀深度在 0.5m 左右。横向上，淤积程度的减弱和深水区的侵蚀导致航道向南北两岸总体变宽，其中-10m 等深线在北港中下段向两侧各拓宽 100m 以上。

2015—2016 年，和上年度明显不同，北港以明显的侵蚀为特征，垂向上侵蚀深度多大于 0.5m，侵蚀区面积达 70%～80%。横向上，水道向两侧变化不明显，而沿水道变化程度很大，如-15m 等深线向下段发生明显拓展，南部沿北港水道向岸拓展了约 5km。说明过去一年来北港的水下地形变化以下蚀作用占优势，侧蚀作用相对不明显，导致了区内大面积出现垂向侵蚀，但等水深线的横向变化不显著。这种河势变化以及南北港分岔口控制工程前沿的淤积特征对区内两大工程青草沙水库和上海长江大桥均没有明显不利影响。

2015—2017 年，对青草沙水库提取-2m 和-5m 等深线，结果显示崇明岛北岸-2m 等深线略向南偏移，呈微淤积态势；青草沙水库一侧-2m 等深线也略向南偏移，呈冲刷态势。2015—2017 年间，崇明岛南岸一侧-5m 等深线变化不明显，没有明显的趋势，而青草沙水库北岸一侧-5m 等深线向南偏移，呈冲刷态势。青草沙水库附近海域位于北港上段，2015—2017 年间水库北岸东侧为冲刷态势，冲刷约 2m，而西侧为淤积态势，淤积厚度在 1～2m 之间，青草沙水库的建设束窄了北港 1/3 的河槽断面，减少了北港的分流比，水库一侧来自潮流和径流的冲刷明显，岸线附近冲淤对堤坝的稳定性有较大的影响。

2011—2017 年冲淤图（图 31-3）显示，在北港和新桥通道衔接处青草沙水库北侧发生冲刷，冲刷深度大于 2m，所以对青草沙水库堤坝安全有一定的隐患，被冲刷的泥沙向下游推移，在北港中段堆积形成阴沙。

六、金山深槽

金山深槽为海底侵蚀地貌，始于大、小金山，向西延伸，全长 11300m。其宽度按-15m 等深线测量，在大、小金山之间为 1500m，在戚家墩前为 1700m，在上海石油化工总厂纬三路前为 2200m，过纬三路即收夹为 600m，至纬七路前深槽基本消失。金山嘴附近又有大、小金山及戚家墩凸岸的作用，使深水区靠近岸边，-8m 等深线离岸仅 800m。深槽的底部由一系列深潭串联而成，一般深度为 30～

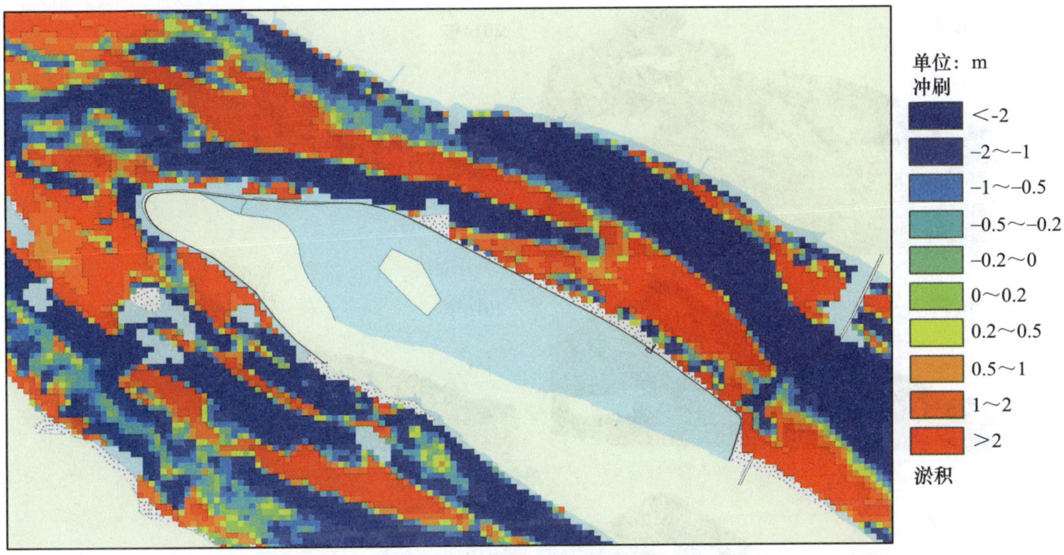

图 31-3　青草沙水库附近 2011—2017 年冲淤图

Fig. 31-3　Erosion-deposition map of the Qingcaosha reservoir nearby area during 2011-2017

40m，以大、小金山附近为最深。

在深槽东部的北侧、靠近小金山西麓有两道东西走向的水下沙堤属堆积地貌；深槽的西端有两个不同方向的深潭，较小的一个指向西北，较大的一个指向正西，表明金山深槽具有活动性。

金山深槽是海潮流侵蚀的结果。12 世纪 80—90 年代（南宋淳熙年间），金山沦入海中以后，最初在大、小金山之间尚保留着一小块陆地。在海潮的不断冲刷下逐步瓦解，到宋末元初，已完全消失，这样在大小金山之间便形成一道海峡，称为“金山门”。

金山门海峡形成以后，原来必须绕过大、小金山的潮流，就可以由金山门向西直冲，汹涌的涨潮流在通过金山门时，由于大、小金山的约束，潮位徒然增高，水流集能，产生强烈的向下淘蚀作用，自金山门以西就逐渐冲刷出一道深水槽。

明嘉靖以前，潮流顶冲点在南汇、奉贤境内，深槽西端不超过今金山嘴镇正南，由淘蚀作用及深槽横向环流所卷起的泥沙就堆积在深槽北侧，形成偏向北边的第一道水下沙堤。

明嘉靖以后，随着南汇嘴延伸，顶冲点逐步西移。金山门以东潮流不断加强，深槽急需向西伸展，至清雍正末年，乾隆初年，即 18 世纪 30 年代后期，顶冲点移至金山嘴一带，引起深槽向南运动，被卷起的泥沙在原有第一道水下沙堤之南堆积，形成偏南的第二道水下沙堤。并在向南运动的同时，迅速向西扩展。

乾隆初年（1736 年）金山深槽已伸展至金山卫南滩地前缘。乾隆《金山县志》卷首海塘图注：“沙滩外有沙堤（指沙滩前缘的贝壳沙堤），堤下深不可测，盖潮激金山，直冲其地故也。与他处以渐而深者不同。”所指“深不可测”之处即金山深槽。1931 年-20m 等深线区已由金山门伸展至戚家墩前，1958 年已越过金山滩嘴（今纬三路），1972 年又向西延伸 1km。

金山深槽存在 3 个比较深的冲刷深坑，其中一个位于大金山和小金山之间，另一个位于大金山和乌龟山之间的深坑，第三个位于大金山西侧，在陆域金山石化公司南侧 1km 处。3 个深坑最大深度都接近 56~48m，前两个深坑与后续一个深坑之间存在一个次级深度的潮流槽。金山深槽存在一些微地貌，如光滑海底、冲沟、凹坑、洼地和沙波等，各类微地貌又存在多种形态，金山深槽微地貌形态复杂，而且逐年也有一些细微的变化（图 31-4）。

2009—2010 年，金山深槽基本保持稳定。

2010 年 5 月—2011 年 6 月，金山深槽南部有一定程度淤积，平均淤积厚度在 0.8m 左右，致使深槽-10m 等深线宽度有所缩小。与 2009 年和 2010 年相比，金山深槽北部近岸一侧基本保持稳定，深槽

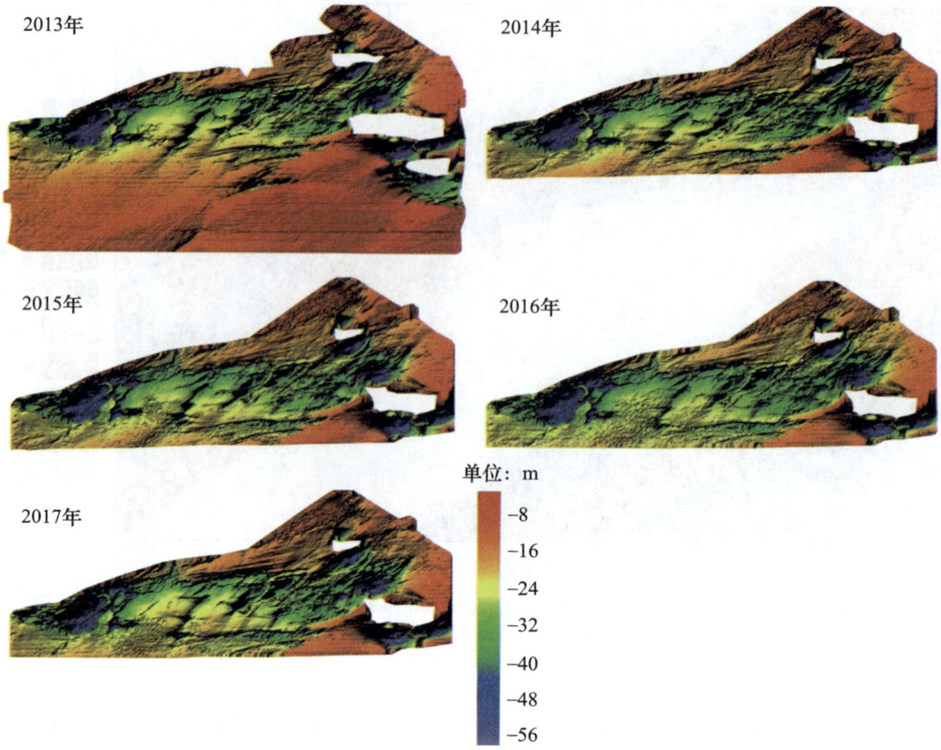

图 31-4　金山深槽地貌图

Fig. 31-4　Geomorphological map of the Jinshan Deep Trough

无明显的向岸摆动趋势。

2011 年 6 月—2012 年 5 月，金山深槽总体保持稳定。北部近岸一侧保持稳定，无明显向岸摆动，与 2011 年相比，-10m 等深线向东有所推展，恢复到与 2010 年相当的位置。

2012 年 5 月—2013 年 5 月，-10m 等深线与-20m 等深线基本保持固定，但其深槽陡坎有继续向陆移动的趋势。

2013—2014 年，金山深槽及其附近地区以侵蚀作用占主导。根据地形叠加对比计算结果，侵蚀量和淤积量分别为 $987\times10^4m^3$ 和 $255\times10^4m^3$，占比分别为 79%左右和 21%左右。面积上，侵蚀区和淤积区面积分别为 $21.5km^2$ 和 $5.2km^2$，占比分别为 81%和 19%。淤积主要发生在深槽中西部深潭附近-25～-30m 等深线之间，淤积幅度多介于 1～5m 之间。深槽近岸和远岸-20m 等深线以浅岸坡侵蚀区面积大，但侵蚀幅度较小，多介于 0.2～0.5m 之间。-20m 等深线以深侵蚀幅度多大于 0.5m，-30m 等深线以深侵蚀幅度多介于 2～5m 之间。横向上，等深线总体向深槽外侧迁移，迁移幅度多介于 20～50m 之间，深槽南部-20m 等深线向南迁移最大达 150m 左右。距离金山石化岸线约 1km 处的深槽岸坡发生显著向岸迁移，-25～-45m 等深线均发生 50～100m 的向岸迁移，-35m 等深线最大迁移幅度近 130m（图 31-5），指示该区进一步发生的海岸侵蚀作用，值得密切关注。

2014—2015 年，金山深槽地区延续上一年度趋势，以侵蚀为主要特征，多波束测量区范围约为 $25km^2$，其中侵蚀区和淤积区面积分别为 $16.0km^2$ 和 $9.46km^2$，占比分别为 63%和 37%，侵蚀量和淤积量分别为 $0.22\times10^8m^3$ 和 $0.16\times10^8m^3$，占比分别为 58%和 42%。淤积区主要分布于调查区东部大金山岛北部地区及深槽南岸地区，垂向淤积厚度多大于 1m。侵蚀区分布广泛，但垂向侵蚀程度较低，多在 0.5～1.0m 之间。金山石化近岸局部地区垂向侵蚀程度大于 2m。相比而言，深槽近岸较远岸岸坡地形变化较大。横向上，远岸向西南方向拓展，最大达约 200m，说明深槽范围趋于增大。金山石化近岸岸坡-30m 等深线向岸蚀退 20～40m（见图 31-5），对沿岸岸堤和工程安全带来不利影响。

2015—2016 年，金山深槽总体仍以侵蚀为主，延续往年变化特征，不同的是，近岸水下岸坡垂向

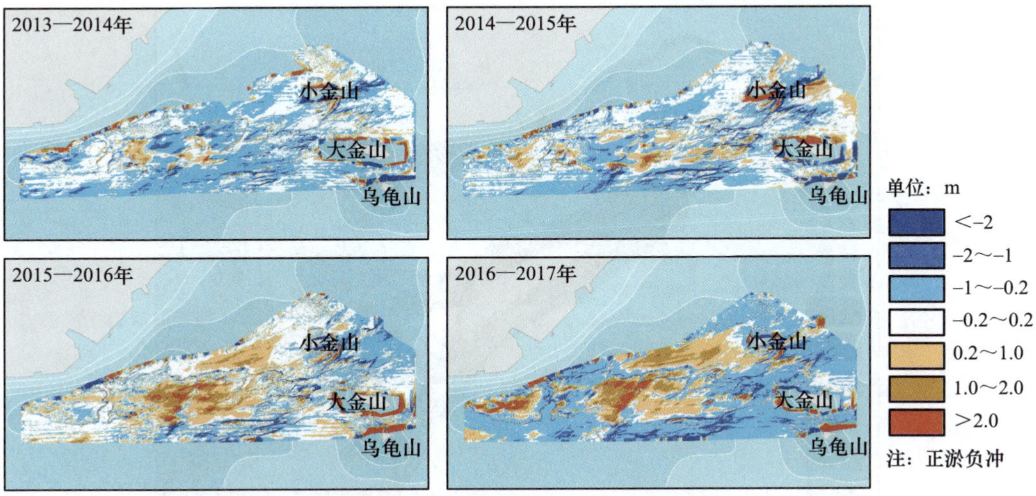

图 31-5　金山深槽 2013—2017 冲淤变化

图 31-5　Erosion-deposition variations of Jinshan Deep Trough during 2013-2017

侵蚀程度较低，远岸侵蚀程度相对较大，多大于 0.5m。但近岸-15m 等深线以浅发生较明显侵蚀，局部侵蚀深度大于 1m。横向上，近岸等深线向岸侵蚀后退不明显，远岸相对较大，最大发生在远岸-15m 等深线，其向远岸侵蚀后退距离多在 100m 以上。此外，南部深槽附近发生大面积淤积，局部淤积厚度在 2m 以上（见图 31-5），推测与水下滑坡有关。

2016—2017 年，金山深槽冲淤态势不明显，基本变化在-5～5m 范围之间，局部区域略有冲刷，在一年间深槽演化较为稳定。等深线微微变化，但是没形成趋势性的演化规律，整体冲刷还是大于淤积（见图 31-5），延续以往变化趋势。根据断面图可知，金山深槽西北侧变化较小，东南侧冲刷明显。

2013—2017 年间，金山深槽之间基本以冲刷为主（见图 31-5），在 2013—2015 年之前基本全区以冲刷为主，冲刷深度在 0.2～1m 之间，局部小的区域有一些淤积，在 2015—2017 年之间也基本以冲刷为主，但是在中间区域有一些淤积趋势，且淤积区在两年间延续。

第二节　岸带冲淤规律和主要影响因素

一、岸带冲淤总体规律

由于近年来上海建造了大量的堤坝工程和航道工程，约束了长江口河槽，且近年来来水通量基本稳定，在来沙量减少的情况下，整体河势比较稳定，局部河势有侵蚀风险。

（一）2011—2017 年区域冲淤和口外-10m 等深线变化特征

从第二级分汊口南北港分流口为界，南港、北港、南槽和北槽以及北港、北槽和南槽口外都以冲刷为主，而南支和北支中上段尽管冲淤变化较大，但是冲淤趋势并不明显；而北支下段、崇明东滩北港口门和横沙东滩基本以淤积为主，局部地区淤积厚度较大。杭州湾北岸水域在南汇南滩外侧出现较大范围的淤积（图 31-6），这可能与南汇东滩大范围围垦有关。

受上游来沙减少，长江口、杭州湾北岸基本还以冲刷为主，一些滩槽如北支下段、崇明东滩和横沙东滩以及南汇南滩外侧基本以淤积为主，这些地区淤积可能与一些工程相关，由于北槽深水航道和横沙东滩围垦工程的建设，隔断了长江口外南北向的海流，导致此区域淤积，南汇南滩口外可能与南汇东滩工程的围垦相关，在南汇南滩外侧形成一个水流滞留点，形成较大范围的淤积。

北支和北港口外-10m 等深线变化很小，基本趋势不变，以局部小幅迁移为主，区域水下地形横向变化总体不显著。而南槽和南汇南滩口外-10m 等深线变化较为明显，南槽口外-10m 等深线向西偏

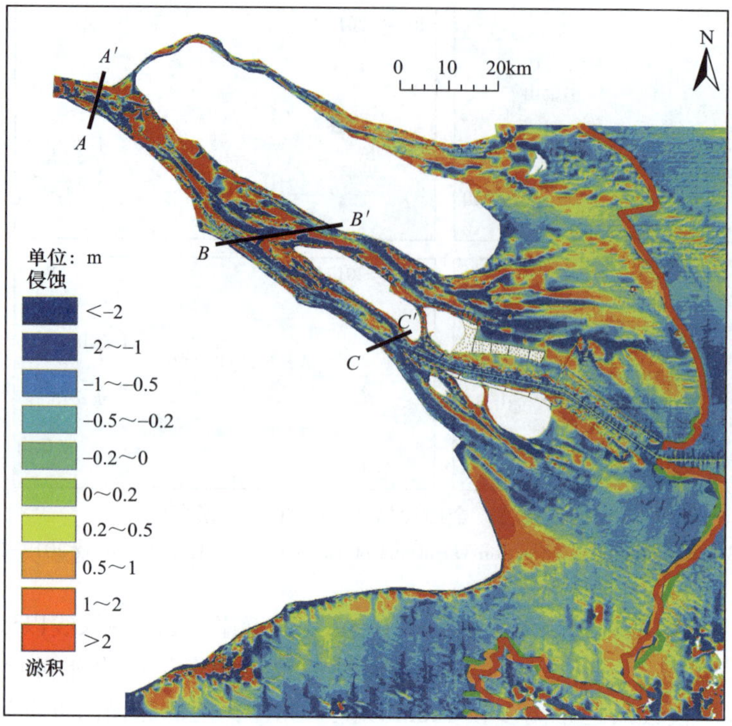

图 31-6 上海海岸带 2011—2017 年冲淤及口外-10m 等深线变化图

Fig. 31-6 Off-the-estuary -10m isobaths and Erosion-deposition variation in the coastal zones of Shanghai during 2011-2017

移，基本以冲刷为主，而南汇南滩外等深线向北偏移，以淤积为主。2011—2017 年间的变化区域主要在南汇边滩和南槽外缘（见图 31-6）。

（二）分阶段区域冲淤变化特征

2011—2013 年，海岸带整体呈“南淤北冲”“冲大于淤”基本特征，南支中上段、北支、崇明东滩、南汇南滩和口外区域整体呈现淤积现象，尤其是北支下段，部分淤积厚度约为 1m，北支河槽继续束窄，口外基本以淤积为主，且淤积程度北侧大于南侧。口内南支下段、南港、北港、南槽和北槽基本都呈冲刷态势，尤其以南支和北港冲刷明显。南港和北槽次之，北槽由于航道疏浚的作用，在河槽中间航道处处于冲刷状态，其余两侧为淤积状态。杭州湾区域除了南汇南滩外侧地区基本呈冲刷状态，由于南汇东滩的围垦作用，南槽冲刷的泥沙随着杭州湾张潮流的作用，在南汇南滩附近出现大范围的淤积，由于东海大桥的作用，大桥所在海域冲刷明显（图 31-7）。口内北支延续淤积态势发展，南支、南港、北港、北槽和南槽冲刷，基本为“北淤南冲”的变化特征。这种一致性的变化态势可能是由于科氏力的作用及北侧区域大量河口工程的建设，这种变化对上海城市安全有较大的不利影响。

2013—2015 年，海岸带整体呈“外淤内冲”基本特征，北支口外、崇明东滩、北港口外、横沙东滩和杭州湾北岸地区以淤积为主，南支、北支、南槽、北槽、南港和北港以及北槽和南槽口外地区以冲刷为主，局部地区有小范围淤积。南支中上段和北支、崇明东滩、南汇南滩与口外区域整体呈现淤积现象，北槽由于航道疏浚的作用，在河槽中间航道处处于冲刷状态，其余两侧为淤积状态。杭州湾区域除了东海大桥水域地区基本呈冲刷状态，其余地区以淤积为主。由于东海大桥的作用，大桥所在海域冲刷明显（图 31-8）。

2015—2017 年，海岸带整体呈“南淤北冲”现象，北支和口外区域整体呈现淤积现象，尤其是北支下段，部分淤积厚度约为 1m，北支河槽继续束窄，口外基本以淤积为主，且淤积程度北侧大于南侧。南支、南港、北港中上段、南槽和北槽基本都呈冲刷态势，尤其以南支和北港冲刷明显。南港和北槽次之，北槽冲淤最不明显。杭州湾区域呈“东冲西淤”态势，由于东海大桥的作用，大桥所在海

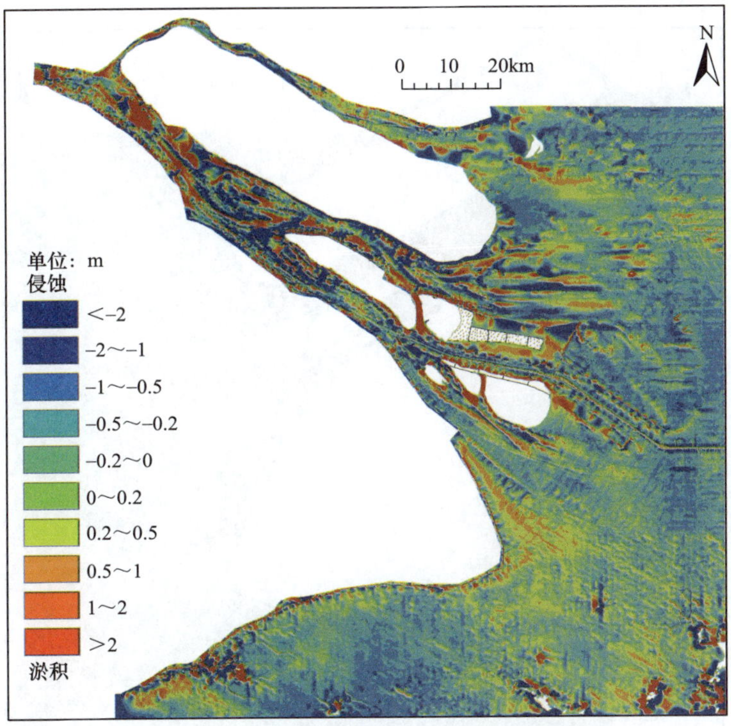

图 31-7　上海海岸带 2011—2013 年冲淤图

Fig. 31-7　Erosion-deposition variations in the coastal zones of Shanghai during 2011-2013

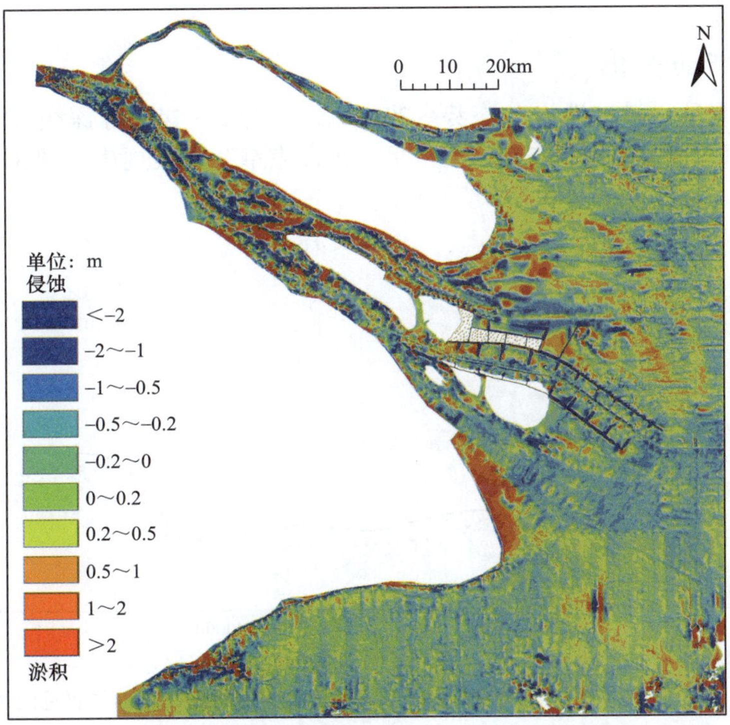

图 31-8　上海海岸带 2013—2015 年冲淤图

Fig. 31-8　Erosion-deposition variations in the coastal zones of Shanghai during 2013-2015

域冲刷明显。2015—2017 年间河槽冲淤变化跟往年监测口外区域略有不同，呈淤积态势，口内北支延续淤积态势发展，南支、南港、北港、北槽和南槽冲刷，基本为“北淤南冲”的变化特征（图 31-9）。这种一致性的变化态势可能是由于科氏力作用及北侧区域大量河口工程的建设，这种变化对上海城市

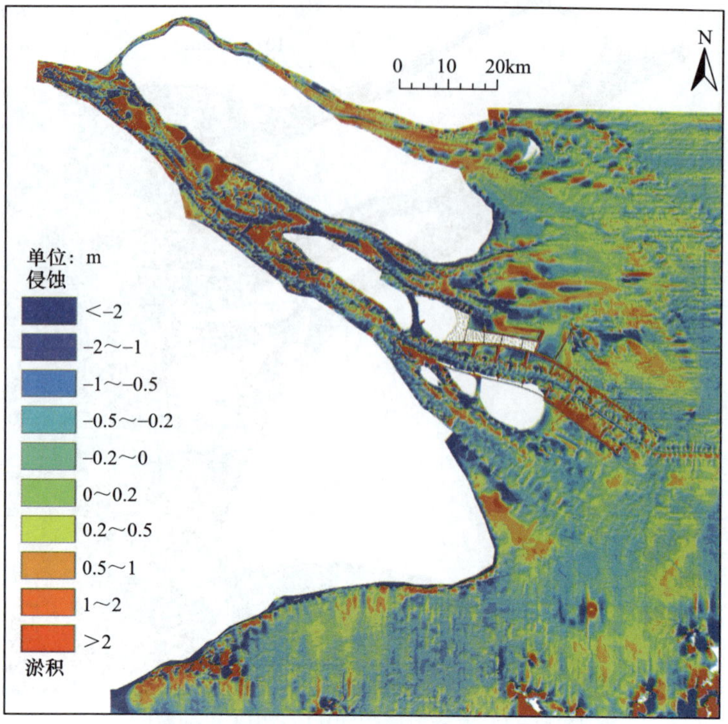

图 31-9　上海海岸带 2015—2017 年冲淤图

Fig. 31-9　Erosion-deposition variations in the coastal zones of Shanghai during 2015-2017

安全有较大的不利影响。

（三）分流口断面变化

南北支分流口河槽呈“槽冲滩淤”态势。2011—2017 年，-14m 等深线以下的河槽冲刷 1～2m，但是在两侧-10m 等深线以上呈淤积态势，河槽呈“加深束窄”发展规律（图 31-10）。

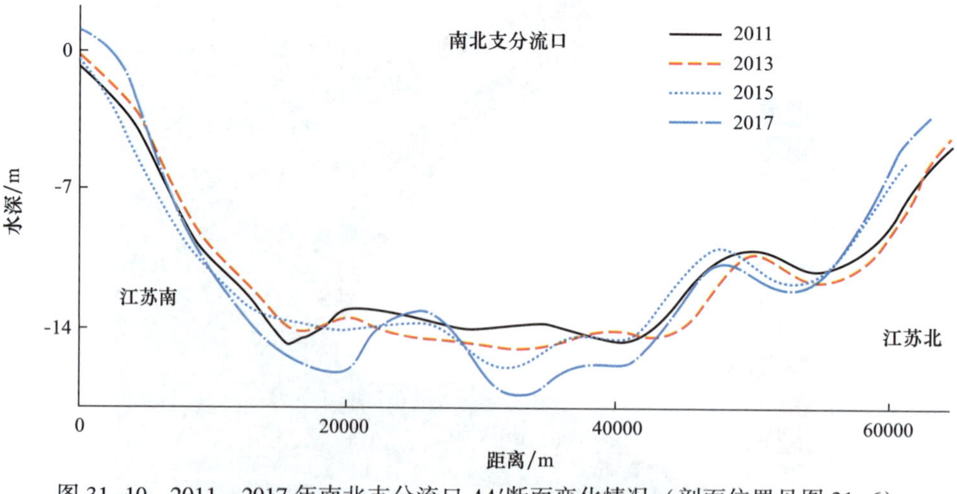

图 31-10　2011—2017 年南北支分流口 *AA'* 断面变化情况（剖面位置见图 31-6）

Fig. 31-10　*AA'* section variations at the South/North Branch bifurcation during 2011-2017 (location shown in fig. 31-6)

南北港分流口由于多个沙体和河槽交错，地形比较复杂。2011—2017 年，南港上段变化较小，只在河槽-16m 等深线以下发生冲刷，冲刷约 3m；中间中央沙区域发生淤积，淤积 3～4m；北港上段崇明岛一侧发生较为明显的冲刷，青草沙水域冲淤现象不明显，河槽底部发生冲刷（图 31-11）。

2011—2017 年，南北槽分流口处南槽一侧变化较小，无明显趋势，河槽-10m 等深线以下冲刷，冲刷约 2m；中间分流口区域也发生明显的冲刷，冲刷约 3m；北槽一侧以-10m 等深线为界，滩处发生明显淤积，淤积厚度约为 3m。但在 2015—2017 年间发生冲刷（图 31-12）。

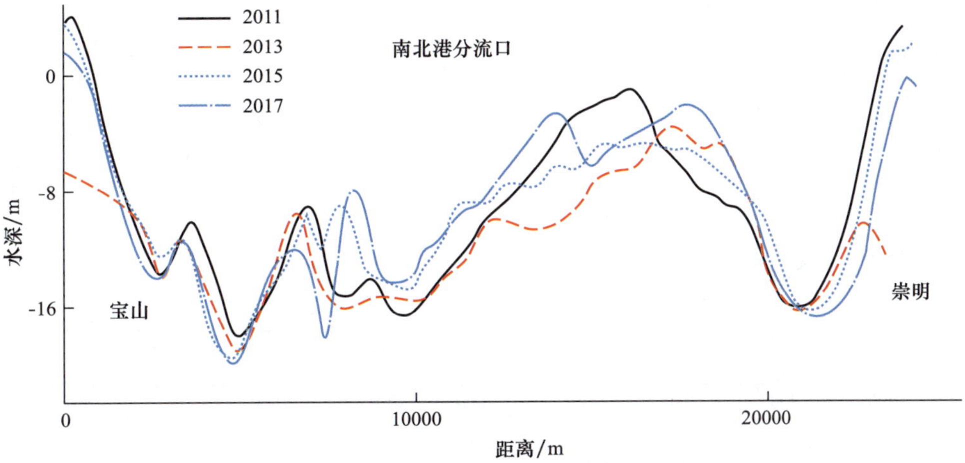

图 31-11　2011—2017 年南北港分流口 *BB′*断面变化情况（剖面位置见图 31-6）

Fig. 31-11　*BB′* section variations of the South/North Port bifurcation during 2011-2017 (location shown in fig. 31-6)

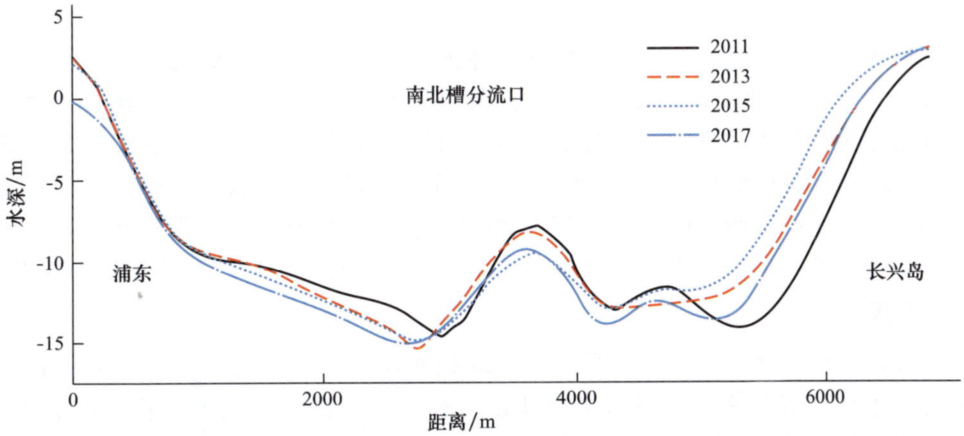

图 31-12　2011—2017 年南北槽分流口 *CC′*断面变化情况（剖面位置见图 31-6）

Fig. 31-12　*CC′* section variations of the South/North Passage bifurcation during 2011-2017 (location shown in fig. 31-6)

二、岸带冲淤主要影响因素

（一）自然因素

长江口形态为“三级分汊四口入海”格局，地形的发育和演变既受到长江径流量、来沙量的影响，也受到东海波浪、潮汐、潮流和输沙的制约，且特殊的地理环境条件如多汊道分水分沙、地转科氏力、悬沙输运、底沙推移、异重流以及盐淡水交汇形成的盐水楔和水动力减弱形成的最大浑浊带和拦门沙，这些因素的变化都对地形的发育和演化有重要影响。

1. 水沙特征

长江口来水来沙可由大通站控制，长江口近年来径流输沙量有所减弱，1960—1986 年之间输沙量与径流量基本保持正相关关系，在 1986 年后大通站径流量变化不大，但是其输沙量开始减少，尤其是 1998 年之后，输沙量减幅增大，2006 年之后输沙量在逐年波动，但是基本趋于稳定，基本维持在 $1.0\times10^8\sim1.2\times10^8$t（图 31-13）。

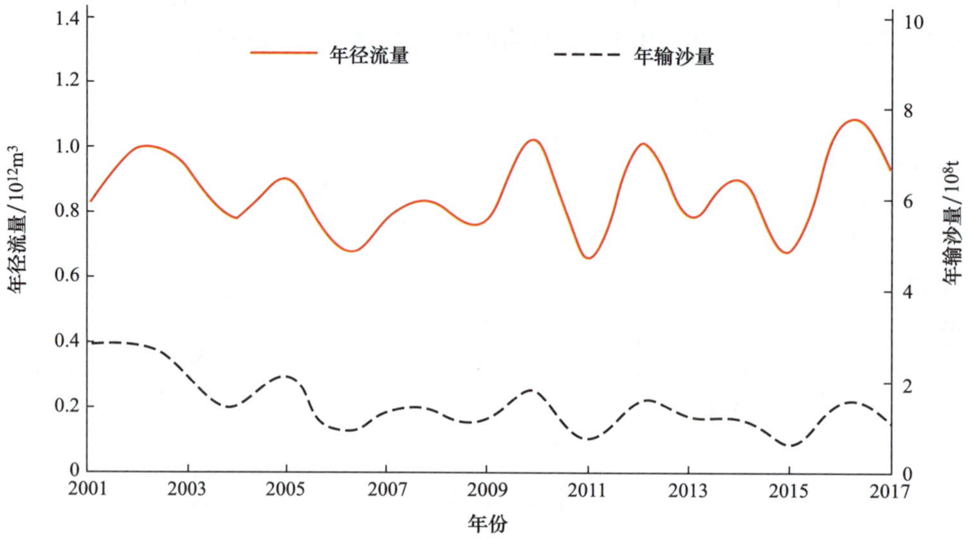

图 31-13　2001—2017 年大通站水沙量统计（根据历年长江泥沙公报统计）

Fig. 31-13　Water and sediment distributions at Datong Station during 2001-2017 (per the "Changjiang Sediment Bulletin" of these years)

2. 风暴潮

风暴潮是沿海主要的灾害之一，可引起局部水流流速和潮位等强烈变化，进而导致局部地形强烈冲淤。台风是导致近海风暴潮的主要原因之一，尤其是风暴潮位与天文潮位相叠加时段，地形的剧烈变化对周围工程设施等有较大的影响。上海市水务局官网防汛系统显示，近年来对上海有影响的风暴潮有 25 个左右，横沙东滩外缘基本接近东海海域，受东海潮动力影响较大，风暴潮对其地形变化较为敏感，本志收集了近年来影响较大的台风，其中 2012 年台风“海葵”对上海影响明显，市区普遍出现 6~8 级、阵风 9 级的东北大风，全市普降暴雨到大暴雨，可能对长江口地形演化影响较为严重。

（二）人为因素

长江口还是长江三角洲经济中心，也是人类活动频繁地带，大量河口和近岸工程的建设同时改变了地形发育的环境条件。近年来，长江口主要建设有长江口北槽深水航道工程、青草沙水源地工程、上海长江大桥和洋山深水港以及配套的东海大桥工程，以及大量的滩涂促淤圈围工程。这些工程的建设改变了局部水流变化特征，进而改变了地形特征。

第三十二章　地下水资源

为控制地面沉降，上海潜水资源开发利用程度极低，系统性评价工作相对薄弱，20 世纪 80 年代以及 21 世纪初进行了相对系统的计算和评价工作。潜水资源计算方法主要依据规范。潜水资源计算内容依据上海特点，主要包括天然资源的大气降水入渗量、地表水体对潜水垂向渗漏量、地表水体对潜水侧向补给量、灌溉水入渗量等。

上海对于承压水可开采资源的计算方法有多种，根据不同阶段水文地质调查精度、地质环境监测资料积累程度，地质环境保护指标要求，以及计算机辅助数据处理能力不断提高等实际情况，针对性地选择传统的相关法、降落漏斗法，以及建立准三维数学模型、真三维数学模型等新方法进行计算。

一、20 世纪 80 年代初期，地质环境普查阶段，主要采用传统计算方法进行概算

20 世纪 60—80 年代，水文地质调查处于 1∶20 万普查阶段，基本查明了各含水层埋藏分布、地下水位、水质等水文地质条件。同时这一阶段市区地下水开发利用程度高，郊区地下水位开发利用程度较低。

1987 年，从供水目的出发，充分总结各含水层水文地质特征、地下水补径排条件、地下水动态变化，应用回归方程及降落漏斗法，以总体水位平均下降 5m 为控制条件，进行了全市地下水开采资源的概算。

地下水资源量计算中，市区地下水可开采资源计算精度相对较高；全市地下水开采资源量，虽然考虑了缓和地面沉降等因素，但采用比拟法计算时，对漏斗区外围开采或扩采所存在的水量、水位和地面沉降的相互关系未予以考虑，未能对水位、地面沉降预报与开采时间的保证率做出适当评价，也无法对上海地下水资源开采补给保证程度做出区域性论证。

二、20 世纪 90 年代以来，当地质环境调查、监测数据和辅助计算功能具备一定条件时，逐渐采用了新方法进行计算

随着水文地质条件调查程度的不断提高，地质环境（水位、水量、水质、地面沉降等）监测数据的大量积累，计算机对于海量监测数据处理能力的不断提升，充分利用计算机运算功能，从 1990 年开始，通过建立不同的数学模型进行地下水可开采资源计算。

1990 年，结合“上海市地下水资源合理开发应用研究”课题的开展，对第Ⅱ、第Ⅲ承压含水层，分别采用水位、水量相关法和降落漏斗法，计算了漏斗区的开采资源，并在缓和地面沉降条件下，计算漏斗区可开采资源，再用比拟法推算全市可供饮用的地下水可开采资源。第Ⅳ、第Ⅴ承压含水层可饮用开采资源，采用平面二维流有限元数值法进行计算。

1995 年，结合“上海市县（区）区域水文地质调查地下水资源计算”课题的开展，根据地下水动力学原理及太沙基一维变形理论，结合地下水流场及各土层变形特征，建立了准三维地下水渗流一维沉降数学模型。对建立的水文地质概念模型采取一类边界拟合、二类边界检验，水文地质参数进行三角剖分，剖分密度体现水文地质调查程度高的地区相对大的原则。依据当时地质环境保护要求，设定了各承压含水层最低水位、总沉降量控制指标，运行检验过的数学模型，计算各承压含水层地下水可开采量。同时对上海地下水资源开采补给保证程度做了区域性论证。

2000 年，结合“上海市地下水资源评价”等项目的开展，以以往建立的准三维地下水渗流一维

沉降数学模型为基础，采用准三维流有限元数值解法进行地下水可开采资源计算。为提高精度，对以往模型进行了改进，剖分、三角形单元数量较前期明显增加，重新进行数值模拟、参数拟合及模型检验。依据地质环境保护要求，重新设定了各承压含水层最低水位、总沉降量控制指标，运行检验过的数学模型，计算各承压含水层地下水可开采量。同时对上海地下水资源开采补给保证程度做了区域性论证。

2005 年、2010 年，结合“上海市城市地质调查”等项目的开展，充分利用水文地质、地质环境监测成果，建立了三维非稳定流地下水运动——维地面沉降非线性耦合数学模型。相对于以往建立的二维流或者准三维流而言，基本实现了真正意义上的三维流。不仅考虑了各个含水层之间的水力联系，也考虑了各含水层间弱透水层中的水位变化；而且对于同一含水层，也考虑到因垂向上岩性的差异和埋深的不同，水位也有相应差异的情况。即将上海整个地下含水层作为一个各向参数均在发生变化的地下含水系统来综合考虑，实现了真正意义上的三维流特征。在此基础上，结合近期城市地质环境保护要求，设定地下水位和地面沉降控制条件，进行地下水可开采资源计算。模型中已对地下水资源开采补给保证程度做了相应界定。

第一节　地下水资源量概况

一、潜水资源

潜水资源主要计算天然资源、调节储存量、补给模数。天然资源由各项补给资源构成。开采资源以调节储存量表征。1987 年、2000 年、2002 年潜水资源计算结果如表 32-1 所示。

计算结果显示：天然资源总量、调节储存量、补给模数，21 世纪初计算结果较 20 世纪 80 年代计算结果偏小。可能与城市化地区有效补给区面积略有减少、各类入渗参数略有调整、评价年降水量差异等因素有关（表 32-1）。

表 32-1　上海潜水资源计算一览表

Table 32-1　Calculations of phreatic water resources in Shanghai

年份	计算方法	天然资源总量 ($10^8 m^3 \cdot a^{-1}$)	调节储存量 ($10^8 m^3 \cdot a^{-1}$)	补给模数 [$10^4 m^3 \cdot (a \cdot km^2)^{-1}$]	评价年
1987	水均衡法	28.266	8.690	31.400	1981
2000	水均衡法	12.938	2.03	24.598	1951—1999
2002	水均衡法	12.060	2.030	23.890	1980—2000

二、承压水可采资源

结合上海不同阶段城市对地下水资源的需求以及地质环境保护目标，分别进行了多轮承压含水层地下水主采层（第Ⅱ、第Ⅲ、第Ⅳ、第Ⅴ承压含水层）可采资源（可开采量）计算。

（一）地下水可开采资源量计算结果

表 32-2 为全市地下水可开采资源量不同阶段的计算结果。随着地质环境保护指标的提高，以及全市不断提高的自来水供给能力的背景下，成功实现了优质地下水资源功能由作为城市工业、生活补充供水水源，逐步向优水优用方向的转变，为此，地下水可开采量从 20 世纪 80 年代的 $4.39\times10^8 m^3/a$，逐渐控制在目前的 $0.250\times10^8 m^3/a$，较好地体现了优质地下水资源在城市发展过程中的重要作用，也实现了几十年来地质环境保护与地下水优水优用相协调的宗旨。

表 32-2 上海承压水可开采资源量计算一览表

Table 32-2 Calculations of exploitable confined water resources in Shanghai

年份	计算方法	地下水位、地面沉降控制条件	含水层地下水可开采量/（$10^8m^3 \cdot a^{-1}$）				全市可开采总量/（$10^8m^3 \cdot a^{-1}$）
			第Ⅱ承压含水层	第Ⅲ承压含水层	第Ⅳ承压含水层	第Ⅴ承压含水层	
1987	比拟法、（相关法、降落漏斗）	地下水位平均下降 5m	1.84	1.13	1.07	0.35	4.39
1990	相关法、二维有限元法	累计沉降量小于 300mm	0.15	0.17	1.50	0.41	2.23
1995	准三维有限元法	地下水位：第Ⅱ、第Ⅲ含水层为-18m、-20m，第Ⅳ、第Ⅴ含水层为-60m；总沉降量小于 250mm	0.21	0.27	0.71	0.26	1.45
2000	准三维有限元法	地下水位：第Ⅱ、第Ⅲ含水层为-18m、-20m，第Ⅳ、第Ⅴ含水层为-60m；总沉降量小于 250mm	0.34	0.28	0.55	0.07	1.24
2005	真三维有限元法	地下水位：第Ⅱ、第Ⅲ含水层为-3m、-5m，第Ⅳ、第Ⅴ含水层为-20m、-25m；区域年均沉降不大于 7mm	0.04	0.05	0.10	0.06	0.25

（二）现状各行政区地下水可开采资源量

以平均年度地面沉降量不大于 7mm，第Ⅱ、第Ⅲ、第Ⅳ、第Ⅴ承压含水层最大地下水位分别不低于-3m、-5m、-20m、-25m，利用最新建立的三维非稳定流地下水运动一维固结耦合数学模型进行地下水可开采资源计算（表 32-3）。浦东地下水可开采量相对最大，中心城、宝山相对最小。

表 32-3 上海各区地下水可开采量计算一览表

Table 32-3 Calculations of exploitable groundwater resources in Shanghai by administrative district

行政区	地下水可开采量/10^4m^3				可开采资源总量/（$10^8m^3 \cdot a^{-1}$）
	第Ⅱ承压含水层	第Ⅲ承压含水层	第Ⅳ承压含水层	第Ⅴ承压含水层	
中心城	0	0	100	0	100
宝山	40	0	50	0	90
浦东	120	100	300	50	570
闵行	0	50	50	0	100
嘉定	40	50	50	50	190
青浦	50	100	50	100	300
金山	50	100	150	0	300
崇明	50	0	0	400	450
奉贤	30	70	150	0	250
松江	50	50	50	0	150
合计	430	520	950	600	2500

第二节 地下水资源开发利用

上海蕴藏着较为丰富的地下水资源，1860 年开始实施了开发利用，尤其自 20 世纪 60 年代初以来，在城市发展的不同阶段，以城市用水需求和地质环境保护为前提，地下水作为工业、生活补充供水水源而被长期大量开发利用。同时通过水环境的长期监测，基本掌握了地下水的开采对区域水环境的影响特征。

一、潜水开发利用

潜水作为资源，其开发利用程度与城市不同阶段自来水供水能力强弱、地表水污染程度的大小、生态环境阶段保护目标等有密切关系，利用程度受到诸多因素的约束。

（一）潜水曾是广大农村地区日常生活用水的重要水源

20 世纪 80 年代之前，生态环境相对较好，当时作为自来水原水的黄浦江、长江、淀山湖等地表水质量较好，以Ⅱ类、Ⅲ类居多，潜水质量也以Ⅱ类、Ⅲ类居多，潜水质量基本符合生活饮用水标准。

受当时全市自来水供水能力限制，自来水受水地区主要为中心城，而在郊区的广大农村地区，当地居民通过开凿大口径潜水井，直接就地利用潜水作为生活、洗涤等主要用水。这一阶段潜水开发利用程度相对高，年度开发利用量较大。据资料记载，当时农村普遍开凿潜水井，累计约有 64 万眼潜水井，年度开采量在 $1.8\times10^8m^3$ 左右。

（二）20 世纪 80 年代之后，生态环境发生了变化，限制潜水的集中开发利用，维持潜水的自然状态是近阶段地质环境保护的权宜之计

20 世纪 80 年代之后，以往一直作为自来水原水的黄浦江、长江、淀山湖等地表水，在城市高速发展的过程中，因受各种客观或人为因素综合影响，大部分地表水受污情况不容乐观，如黄浦江、长江等主要通航河道，作为自来水原水的地表水水质逐渐下降，由以往的Ⅱ类、Ⅲ类逐渐下降至Ⅲ～Ⅳ类，局部甚至达Ⅴ类，其他河道水质也明显下降，导致短期内全市大部分河道水质以Ⅳ类、Ⅴ类居多，部分河道水质为劣Ⅴ类。

“十二五”期间，随着城市生态环境保护力度的不断提高，尤其加强自来水原水的保护措施，部分地区河道水质有所改善。目前长江口、黄浦江上游原水取水口水质有较大好转，总体达到Ⅲ类水；但区域上大部分河道水质仍不容乐观，以Ⅳ类、Ⅴ类居多，部分河段仍为劣Ⅴ类。

全市地表水质量的逐步下降，一定程度上影响到了潜水质量，导致区域上潜水质量也普遍下降，潜水质量也以Ⅳ类、Ⅴ类居多，潜水中对人体有毒有害的许多组分含量大大超标。

另外，农村地区农田灌溉使用污水或受污染的地表水，工业和生活垃圾的不当堆放，长期使用农药、化肥等人为活动，导致局部地区潜水遭受明显污染。据宝山吴淞、浦东张江调查显示，农田灌溉利用城市污水和受污染的地表水后，导致灌溉区潜水遭受明显污染，如潜水中 NO_3^- 含量明显增高，酚、铬、砷、汞等含量远超过饮用水卫生标准。含有高有机质，有毒有害物质的生活垃圾和工业垃圾的随意堆放也是导致潜水中某些物质的含量增高，如江镇垃圾场堆放的生活垃圾对堆场所在地的潜水质量产生了严重的污染，潜水中 COD、NO_3^-、NO_2^-、NH_4^+、TCr、酚等含量远较背景值高，庙行垃圾场堆放的工业有毒废弃物（铬渣），在大气降水的淋溶下，Cr 含量远超饮用水卫生标准，堆放场附近形成铬含量高值区。农药和化肥使用对潜水质量也造成一定影响，潜水中往往 NO_3^-、有机氯含量相对高。

受自来水处理能力限制，根据生活饮用水标准，直接利用潜水或处理不当的潜水作为生活供水水源，居民的身体健康将不能得以保障。为此，从保障人民生活质量出发，原则上潜水不再作为广大农村地区生活用水水源，而仅作为洗涤等一般水源利用。为此，近年来年度利用量非常小或基本未予利用。

“十三五”初始阶段，市政府高度重视生态城市建设，市水务局牵头开展上海水环境治理工作，为还上海一个美丽的生态环境，通过建立市—区—街镇三级河长体系，按照属地管理原则，分类设置河长，实现全市河道河长制全覆盖。同时即将启动苏州河环境综合整治 4 期工程，整治目标是至 2021 年苏州河干支流消除劣类水，干流的堤防工程全面达标，航运功能得到优化，生态景观廊道基本建成，

形成大都市的滨水空间示范区，最终实现安全之河、生态之河、景观之河、人文之河。全市河道环境的好转，为潜水质量的改善创造了良好的条件，但由于潜水水交替能力弱，水环境的改善将是一个漫长的过程，减少或不予对潜水的人为集中开采活动，仍将是近阶段地质环境保护的优选良策。

多年研究表明，在上海各含水层地下水位确保一定埋藏深度的背景下，可较好地保护地质环境，现状潜水未予集中开发利用格局是上海地质环境保护的综合体现。为此，上海根据潜水质量现状及变化趋势，综合考虑保护地质环境，近年来，潜水作为广大农村地区供水水源而开发利用的格局受到控制，遵循潜水自然状态原则。

二、承压水开发利用

上海虽然早在1860年开始利用承压水，但大量的开发利用还是始于新中国成立以后，尤其是20世纪60年代以后，对深部第Ⅱ、第Ⅲ、第Ⅳ、第Ⅴ承压含水层地下水资源实施了大量开发利用，作为国民经济不同发展阶段工业、生活补充水源，而浅部第Ⅰ承压含水层地下水基本维持长期的停采与工程降水并存格局。

（一）浅部承压水长期处于停采状态

1. 地下水质量差，不宜直接作为生活、生产供水水源，长期来处于停采状态

浅部承压水主要是指第Ⅰ承压含水层地下水。调查显示，区域上第Ⅰ承压含水层主要分布于陆域地区，富水性相对弱，地下水矿化度大多大于3g/L，地下水质量差，主要为Ⅴ类水，局部分布Ⅲ～Ⅳ类水。

因地下水质量普遍不能满足生活饮用水标准，难以作为生活供水水源；同时因地下水矿化度、氯离子等含量高，对混凝土有微腐蚀性，对钢筋混凝土结构中钢筋有微腐蚀性，对钢结构有中腐蚀性，也难以直接作为一般的工业生产用水，为此历史上地下水开发利用程度极低，仅19世纪末和20世纪初（1918年之前）曾实施过少量开采，1961—1966年间也进行过少量开采，年开采量在5×10^4～$16\times10^4m^3$，累计开采量仅$70.89\times10^4m^3$。

地下水短期的、少量的开采后，地下水位出现急剧下降现象。受地下水位相对持续的、大幅度下降影响，浅部土层变形量明显增大。

针对该含水层地下水少量开采后地下水位下降明显，及由此引发土层变形显著这一问题，同时考虑到地下水质量差，富水性也较差，不能直接作为生活、工业等用水水源，故从长期、有效保护地质环境角度考虑，1967年后全市对第Ⅰ承压含水层地下水实施停采措施。这一停采格局维持至今。

2. 地下水位埋藏浅，对工程安全施工影响明显，减压降水现象普遍

上海许多工程案例显示：由于第Ⅰ承压含水层地下水位埋藏较浅，工程施工过程中受地下水影响明显，频发基坑突涌、管涌、边坡失稳等地质灾害或诱发的次生灾害，造成工程事故或人员伤亡，为此工程施工、设计单位高度重视，对处于该含水层发育地区施工的深大基坑（或地下空间开发），加强场区和区域地下水调查，优化降水设计方案，确保工程施工期间安全、顺利施工。20世纪90年代—21世纪初，城市建设进入飞速发展阶段，中心城大量的工程降水，导致第Ⅰ承压含水层地下水位明显下降，区域上呈现相对的低水位分布区，中心城与郊区的最大水位差达5～8m。

限于以往工程施工、管理部门与地下水管理部门之间各自为政，工程降水与地下水管理部门之间尚无明确的管理制度和共同执行的法律法规，地下水管理部门对工程施工过程中减压降水水量等计量资料严重缺乏，对该含水层地下水科学管理受到一定影响，同时对地面沉降机理研究等也造成了很大困难。据调查研究，工程降水后一定范围内的土层（主要为目的层以上土层）变形明显，尤其对10倍基坑深度范围内的地面沉降影响明显，3倍基坑深度范围内土层变形更为显著。

近年来，工程设计部门根据工程性质与场区地质、水文地质条件，制定优化降水方案，控制地下水位，以减少工程施工对周边环境的影响。上海将打造最美卓越城市，城市建设将稳步推进，加强对地下空间资源的开发利用已成为趋势。城市地下空间开发过程中采取降水措施将不可避免，科学、合理的降水方案等措施，可较好地实现保护地质环境的目的。

目前浅部第Ⅰ承压含水层地下水的开发利用处于停采与间断少量的工程降水并存格局，且这一格局将在今后一段时期内持续维持。

多年来第Ⅰ承压含水层地下水一般不予开采这一措施的实施，很好地贯彻了上海对于特定的地质环境开展的针对性、有效性地质环境保护宗旨。

（二）深部承压水资源开发利用

1. 地下水开采特征

深部承压水主要指第Ⅱ、第Ⅲ、第Ⅳ、第Ⅴ承压含水层地下水，是上海长期来主要开发利用的地下水资源。1949 年前为初始开发利用时期，1949—1964 年为充分利用时期，1965 年后至今为计划开采时期。其间经历了开采层次、区域、强度等方面的重大调整（图 32-1）。据统计，1960—2016 年全市累计地下水开采量为 50.28$\times10^8$$m^3$。

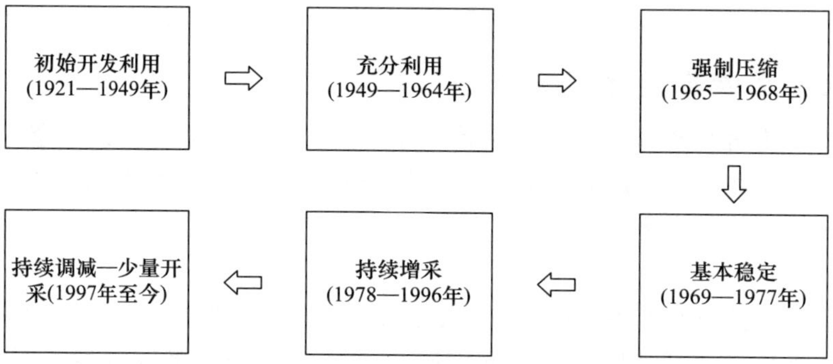

图 32-1　上海地下水开发利用情况沿革

Fig. 32-1　Groundwater exploitation history of Shanghai

（1）初始开发利用时期（1949 年以前）

上海地下水的开发利用始于清咸丰十年（1860 年）。1918 年（民国 7 年），开始实施第Ⅱ、第Ⅲ、第Ⅳ承压含水层地下水资源开采，开采区域主要为市区。

至民国 10 年（1921 年），地下水开采量已从 1$\times10^4$$m^3/a$ 递增至 30$\times10^4$$m^3/a$。

1922—1936 年，深井数量和开采量均呈增长趋势，但开采量增幅不大，平均年开采量在 1500$\times10^4$~2000$\times10^4$$m^3$。

1937—1949 年，上海处于战乱时期，工业发展处于迟缓和停顿状态，开采量较上一阶段减少约 8%。

（2）充分开发利用时期（1949—1964 年）

新中国成立后，上海工业生产迅速发展，地下水开采量急剧增加，第Ⅱ、第Ⅲ承压含水层为主要开采层次，其次为第Ⅳ承压含水层，第Ⅴ承压含水层为少量开采层次。开采区域主要为市区及近郊工业区。

1949 年，地下水开采量有所增大，已达 8000$\times10^4$$m^3$。

至 1964 年，地下水日平均开采量由 24.0$\times10^4$$m^3$ 上升至 55.7$\times10^4$$m^3$，最大年开采量在 2.0$\times10^8$$m^3$ 以上（1960 年和 1963 年）。市区开采量约占全市总开采量的 56.7%，近郊工业区开采量约占郊县开采量的 85.3%~90%。累计开采量大于 6.01$\times10^8$$m^3$。

与此同时，市区出现水位降落漏斗，并在开采区出现与水位漏斗形态基本一致的大面积地面沉降，且 1957—1964 年是市区地面沉降最为严重时期，年沉降量达 60~98mm。

（3）强制大幅压缩时期（1965—1968 年）

这一阶段以城市工业、生活必要用水为前提，在继续强制压缩第Ⅱ、第Ⅲ承压含水层地下水开采量的同时，调整了第Ⅳ、第Ⅴ承压含水层地下水的增采幅度，开采区域也从市区逐步扩展至远郊区。

为缓和、控制地面沉降，从 1965 年开始，对全市地下水采取压缩开采措施，尤其市区压缩力度大。全市开采量由 $13360\times10^4m^3/a$ 减少到 1968 年时的 $5855\times10^4m^3/a$，压缩了近 71%。其中市区开采量由 $4759\times10^4m^3/a$ 减少到 1968 年时的 $831\times10^4m^3/a$，压缩了近 80%（图 32-2）。累计开采量为 $3.40\times10^8m^3$。

（4）基本稳定时期（1969—1977 年）

1969—1977 年，地下水保持相对小的开采强度，年开采总量基本在 $7000\times10^4\sim8000\times10^4m^3$（图 32-2），市区地面沉降得到了有效控制。累计开采量为 $6.51\times10^8m^3$。

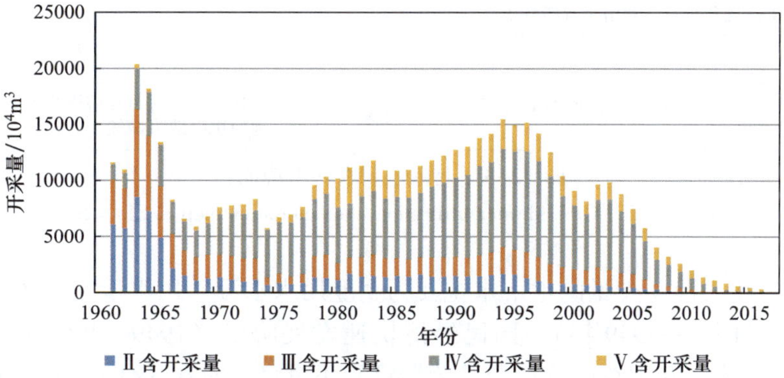

图 32-2　1960—2016 年上海承压含水层地下水年度开采总量历时变化图

Fig. 32-2　Time history of total annual groundwater exploitation of confined aquifers in Shanghai during 1960-2016

（5）持续增采时期（1978—1996 年）

全市第Ⅱ、第Ⅲ、第Ⅳ、第Ⅴ承压含水层仍为地下水主采层，其中第Ⅱ、第Ⅲ承压含水层地下水维持低强度开采；第Ⅳ、第Ⅴ承压含水层地下水开采强度有所增大，地下水增量主要产生在郊区的第Ⅳ、第Ⅴ承压含水层。

这一阶段，城市发展加快，生活、工业等需水量急剧增加，同时自来水供水能力尚处于较低阶段，为此地下水开采量持续呈现微量增长态势，其中，1978—1988 年，开采量增长相对缓慢，基本稳定在 $9000\times10^4\sim11000\times10^4m^3/a$，1989—1996 年增量幅度有所加大，年开采总量从 $12181\times10^4m^3$ 逐步增至 $15098\times10^4m^3$（见图 32-2）。累计开采量为 $23.09\times10^8m^3$。

（6）持续调减—少量开采时期（1997 年至今）

全市第Ⅱ、第Ⅲ、第Ⅳ、第Ⅴ承压含水层仍为地下水主采层，但均处于调减开采状态，减采力度为历史之最。2010 年之后，第Ⅱ、第Ⅲ承压含水层地下水基本处于极低强度开采状态，第Ⅳ、第Ⅴ承压含水层地下水维持少量开采，地下水进入优水优用开采阶段。年度开采总量从 1997 年的 $14895\times10^4m^3$ 逐步减少至 2016 年的 $300\times10^4m^3$ 左右。2016 年各含水层开采量仅为 1997 年时相应层次开采量的 3%、0%、1%、8%（见图 32-2）。累计开采量为 $11.17\times10^8m^3$。

2. 地下水利用特征

上海以公共供水（自来水厂）形式集中抽汲地下水，主要作为城镇居民生活、工业和其他综合性利用水源。对于郊区目前自来水管网尚未到达地区及地表水源水质较差地区，以公共供水形式提供地下水水源，作为居民生活用水、部分企事业单位的日常生活和生产用水水源；部分特殊行业的生产单位自备深井，抽汲地下水作为应急利用、节能冷却、食品生产等必要水源。

1860 年开凿第一口深井至今地下水总体利用特征（图 32-3）显示，地下水利用量大小与城市公共供水能力、城市经济发展速度快慢等紧密相连，城市公共供水能力低、城市经济发展速度相对快的时期，地下水利用量相对大，反之相对小。

（1）1965 年之前，为地下水初始利用—大量开发利用时期

上海自 1843 年 11 月 17 日开埠后进入工业迅速发展时期。1860 年开凿第一口深井，尝试对地下水

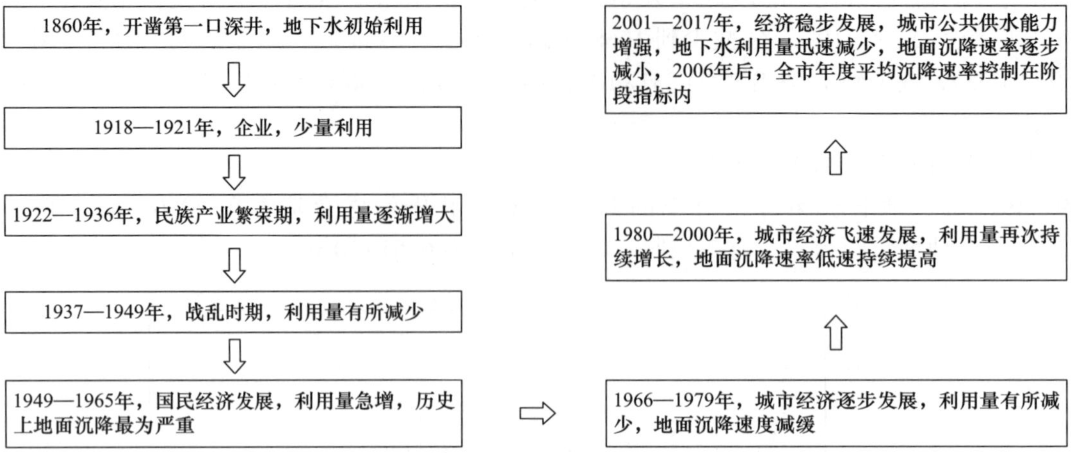

图 32-3　上海地下水利用特征图

Fig. 32-3　Groundwater utilization characteristics map of Shanghai

的开发利用，至 1921 年之前，少量的地下水主要为企业用水水源。用水量在民族产业繁荣期（1922—1936 年）、战乱时期（1937—1949 年）、国民经济快速发展阶段（1949—1965 年）利用量有所增大，但经历了相对大小的变化，其中在民族产业繁荣期日利用量最大时在 $4\times10^4\sim5.5\times10^4m^3$

受当时自来水供水能力影响，大量利用地下水作为轻纺工业的夏季空调降温和冷却用水，部分优质地下水作为生活补充供水水源，其中历史上地下水开采强度最大的年份发生在这一阶段后期（1960 年和 1963 年）。开采区域主要集中在工业相对发达的地区，即中心城和桃浦、宝山、吴泾等 8 大工业区。这一阶段地下水以工业用水（纺织、石化、一般工业、特种行业等）为主，主要为节能冷却和食品加工等利用，少部分作为生活（公共供水、食品饮料等）补充水源。据系统数据记载，利用的地下水总量大于 $74328\times10^4m^3$，地下水用于工业和生活的比例约为 80%和 20%（表 32-4）。

表 32-4　上海地下水利用预估一览表

Table 32-4　Estimations of groundwater utilization in Shanghai

年份	地下水开采总量/10^4m^3	工业用水比例/%	生活用水比例/%	其他用水比例/%
1965 之前	>74328	80	20	
1966—1979	105630	50	50	
1980—2000	257129	25~35	60~75	
2001—2005	43756	22~12	68~80	
2006—2016	21954	10~20	70~85	5~10

（2）1966—1979 年，地下水仍是重要工业、生活利用主要补充水源

受上一阶段大量开采地下水影响，第Ⅱ、第Ⅲ、第Ⅳ、第Ⅴ承压含水层地下水位急剧下降，地面沉降及由此引发的相关次生灾害明显，为减缓市区地面沉降灾害，又力争满足城市工业发展、生活必要的用水需求，市供水部门一方面提高自来水供水能力，另一方面主要通过调整开采层次，最大限度获取地下水资源。

这一阶段，吴泾、桃浦、宝钢等工业区对于地下水需求量增长迅速，纺织、冶金等行业利用含水层具有储冷、储热的功能进行冬灌夏用和夏灌冬用，采集的地下水部分用于调节车间温度，或直接作为生产用水，部分作为居民生活水源利用，为地下水需求量总体较大时期。据统计，这一阶段地下水利用格局有所调整，作为工业用水的比例较上一阶段有所减少，约占总开采量的 50%（表 32-4）。

（3）1980—2000 年，地下水进入新一轮需水高峰

全市仍呈现地表水和地下水同时利用的格局。地表水为工业、生活用水（自来水）主要水源，也

是农业用水主要水源。开采的地下水作为城市生活、工业的补充水源。

据1980—1990年间全市用水量统计结果，地表水作为生活用水水源在$5.1\times10^8\sim8.5\times10^8m^3/a$，地下水利用总量在$1.07\times10^4\sim1.27\times10^4m^3/a$。其间，地下水利用量在增大，年度利用量总体在$1\times10^8m^3$左右，地下水在工业、生活方面的利用比例在不断调整，地下水用于工业和生活的比例预估在25%～50%和60%～75%（表32-4，表32-5）。

表32-5 上海1980—1990年用水量对照表

Table 32-5 Water consumption in Shanghai during 1980-1990

年份	总用水量/10^8m^3	分项用水量/10^8m^3			不同类别用水量/10^8m^3	
		工业	生活	农业	地表水	地下水
1980	80.50	46.70	5.10	28.70	79.43	1.07
1985	89.20	56.00	6.10	27.10	88.12	1.08
1987	95.90	60.10	6.60	29.20	94.77	1.13
1988	122.60	63.20	7.30	52.10	121.40	1.20
1989	97.70	61.50	8.00	28.20	96.48	1.22
1990	107.20	62.90	8.50	35.80	105.93	1.27

（据魏子新，2010）

1990—2000年，受全市地表水质量持续下降影响，地下水需水量进一步有所增加，年度开采量在$1.0\times10^8\sim1.5\times10^8m^3$。据1998年统计资料，地下水用于工业、生活比例分别约为29%和71%。

（4）2001—2016年，供水格局进一步优化，基本实现地下水优水优用目标

市水务局根据供水专业规划，逐步落实，优化调整地下水优水优用格局，利用量逐步向远郊自来水管网未到达区域及部分郊区地表水源水质较差地区的居民用水、国家重大工程用水等方向倾斜。同时加大了自来水集约化供水措施，基本实现了全市相对集中的片区供水目标，为调减地下水开采强度奠定了良好基础。

据2001—2005年统计资料，年度生活用水比例最大，其次为工业和其他方面的利用，从各行业年度趋势性变化来看，生活用水以每年约5%的比例在提高，从2001年的68.17%提高至2005年的80.20%；工业用水比例在逐年下降，由2001年的22.1%减少至2005年的12.30%；其他用途利用量也呈下降趋势，由2001年的9.73%减少至2005年的7.50%，2002—2005年基本稳定在年度开采量的7%～8%（表32-6）。

2006—2016年，地下水开采力度持续、大幅降低，少量开采的地下水主要满足自来水管网难以达到地区的居民生活用水，及特殊工业、食品、矿泉水等必要水源，基本实现了真正意义上的优水优用目标及地下水资源保护。地下水利用于工业和生活的比例为10%～20%、70%～85%（表32-4）。

表32-6 上海2001—2005年地下水利用一览表

Table 32-6 Groundwater consumption in Shanghai during 2001-2005

年份	生活用水		工业用水		其他用水	
	水量/10^4m^3	比例/%	水量/10^4m^3	比例/%	水量/10^4m^3	比例/%
2001	5904	68.17	1912	22.10	845	9.73
2002	7112	73.80	1795	18.60	728	7.60
2003	7763	79.30	1362	13.90	668	6.80
2004	7034	80.40	1083	12.40	635	7.20
2005	5976	80.20	916	12.30	559	7.50

注：生活用水=公共供水+食品饮料，工业用水=纺织+石化+一般工业，其他=非工业+特种行业。资料来源：《上海城市地质》（魏子新，2010）。

3. 地下水保护措施

为确保长期最大限度保持地下水资源合理利用和地质环境保护相协调，地质、水务等相关部门，加强地下水和地面沉降监测，关注地下水采灌格局和地面沉降变化特征，并依据变化特征和规律，及时开展下年度、阶段地下水开发利用方案调整等措施研究。

（1）年度地下水采灌方案调整

20 世纪 60 年代至今，在地下水资源长期开发利用的过程中，因开采强度存在相对大小的周期性动态变化，在此影响下，地质环境也遭受不同程度影响，为此，为实现地下水开发利用和地质环境保护相协调，年度地下水开发利用方案的研制工作已成常态化。

一般基于全市实施上年度地下水开发利用方案实际执行情况及地面沉降变化特征，作为下年度地下水开采（回灌）空间格局调整、方案优化的重要指导依据。如 2015 年地下水采灌方案，要求年度全市地下水开采总量控制在 $800\times10^4m^3$ 以内，地下水人工回灌量达到 $2300\times10^4m^3$，努力形成 $2500\times10^4m^3/a$ 的回灌能力。

（2）阶段地下水开采程度调整

阶段地下水开采程度调整，主要在上阶段地下水超采程度与地面沉降相关性评价结果是否达到相对协调的基础上进行。

21 世纪以来开展的数次地下水超采区评价结果（表 32-7）表明，开采强度的大小基于相应阶段地下水可开采量和实际开采量。2000 年地下水超采区评价结果显示，地下水可开采量为 $1.42\times10^8m^3$，实际开采量为 $9017\times10^4m^3$，年度总开采强度为 64%，第Ⅱ、第Ⅲ、第Ⅳ、第Ⅴ承压含水层开采强度分别为 37%、54%、77%、53%，其中第Ⅱ承压含水层开采强度相对小。但在这一开采格局下，虽然开采强度相对小，但绝对开采量相对较大，导致全市地面沉降控制效果仍不容乐观，尤其中心城，年度地面沉降速率普遍较控制指标偏大，且区域差异沉降明显，局部地区沉降速率几倍于控制指标，开采格局有优化调整的需求。2000 年以后，针对新形势下地质环境保护要求，2005 年再次开展了地下水资源评价工作，基本确定年度地下水可开采量为 $0.25\times10^8m^3$，较上阶段确定结果明显减少。通过 2005 年、2010 年、2015 年地下水超采强度评价，地下水开采强度呈现快速下降态势，开采强度由 2005 年的 298%快速下降至 2015 年的 17%，实际开采量也大幅度减少。反映进入 21 世纪以来，在城市安全对地质环境保护要求越来越高的背景下，对引发地面沉降的主要因子——开发利用地下水资源实施了严格控制，提出了城市水资源利用主要通过自来水集约化措施的加强加以解决，通过阶段地下水开采强度的不断调减，较好地实现了地质环境阶段保护目标。

表 32-7　上海地下水开采强度调整一览表

Table 32-7　Adjustments of groundwater exploitation degree in Shanghai

年份＼含水层	开采程度/%					地下水可开采量（$10^8m^3\cdot a^{-1}$）
	第Ⅱ承压含水层	第Ⅲ承压含水层	第Ⅳ承压含水层	第Ⅴ承压含水层	全市	
2000	37	54	77	53	64	1.42
2005	96	231	475	221	298	0.25
2010	13	44	104	117	79	
2015	4	5	8	51	17	

三、地下水开采引起的环境地质问题

基于上海地下水赋存特点，浅部潜水资源、第Ⅰ承压含水层地下水几十年来基本处于停采状态，而深部第Ⅱ、第Ⅲ、第Ⅳ、第Ⅴ承压含水层地下水资源已有近 160 年的开发利用历史，作为上海工业、生活等不可或缺的补充、优质供水水源。监测显示，在这长期的开发利用过程中，因各种因素直接或间接地引发了诸多环境地质问题，最为显著的是原生水环境发生了变化，表现为区域地下水位下降、局部地区地下水遭受污染及之后的咸水界面迁移等环境地质问题。同时地下水位的变化是引发或加剧区域地面沉降环境地质问题的最为重要因素，且局部地下水污染后，对地下水可持续利用造成一定影响。

（一）引发区域地下水位下降

据上述地下水开发利用与区域地下水位动态变化分析，深部第Ⅱ、第Ⅲ、第Ⅳ、第Ⅴ承压含水层地下水的开采活动直接影响着区域地下水位的变化，天然状态下的地下水位普遍呈现下降特征。

地下水位下降幅度与不同地区开采强度有关。多年资料显示，各含水层均显示出开采量大小与地下水位升降呈正相关性特征，即地下水开采强度相对大的时段，地下水位相对低，地下水开采强度相对小的时段，地下水位相对高。

与地下水开采同期进行的地下水人工回灌活动，对确保地下水位的稳定或减缓地下水位下降速率起到了一定作用。当地下水开采量保持一定强度时，回灌量相对大的时段，地下水位处于相对高的状态，或呈缓慢下降态势（图 32-4～图 32-7）。

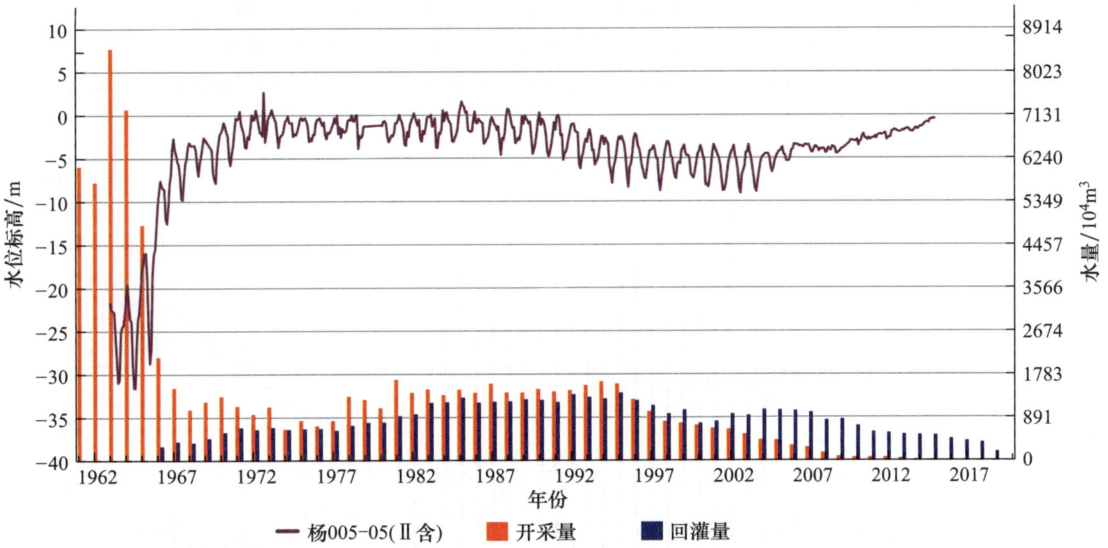

图 32-4 上海第Ⅱ承压含水层地下水开采量与地下水位变化图

Fig. 32-4 Groundwater exploitation of confined aquifer Ⅱ versus groundwater table variation in Shanghai

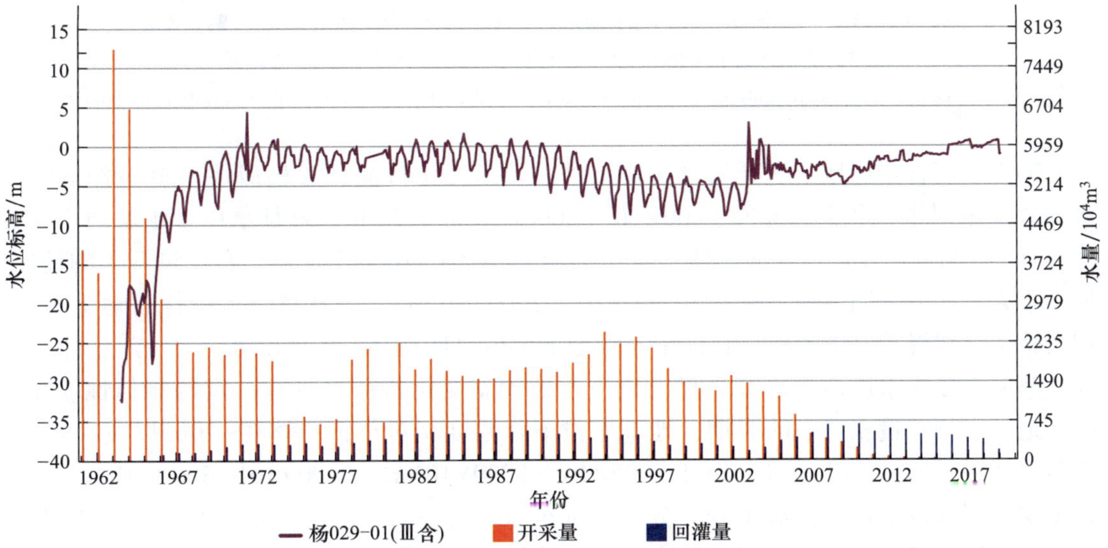

图 32-5 上海第Ⅲ承压含水层地下水开采量与地下水位变化图

Fig. 32-5 Groundwater exploitation of confined aquifer Ⅲ versus groundwater table variation in Shanghai

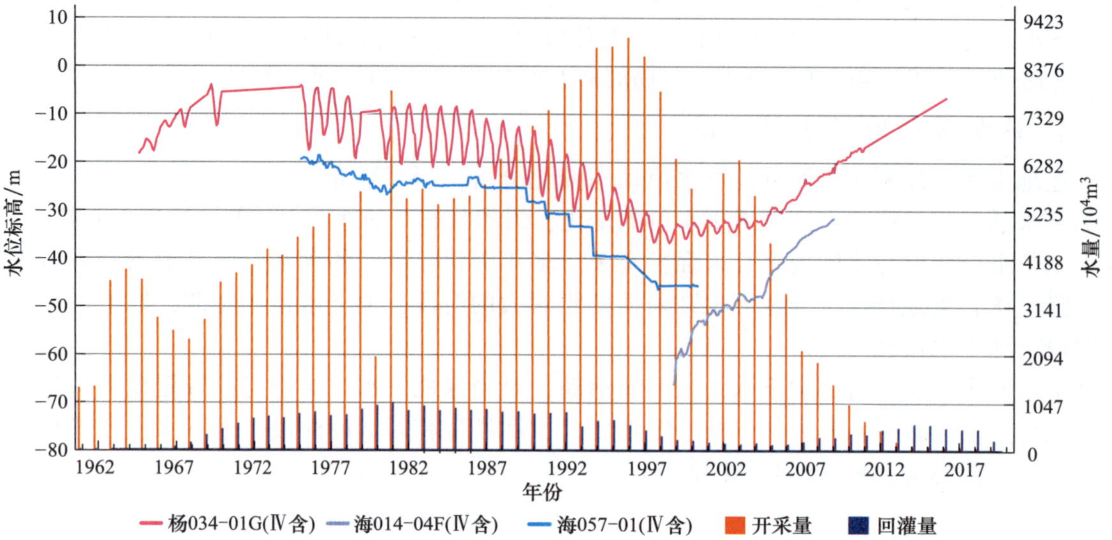

图 32-6 上海第Ⅳ承压含水层地下水开采量与地下水位变化图

Fig. 32-6 Groundwater exploitation of confined aquifer Ⅳ versus groundwater table variation in Shanghai

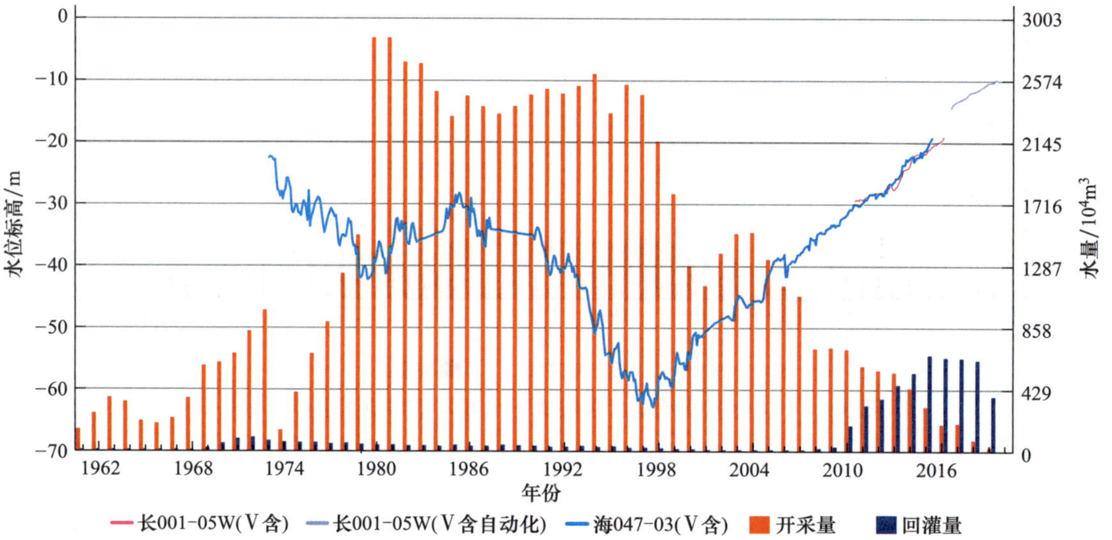

图 32-7 上海第Ⅴ承压含水层地下水开采量与地下水位变化图

Fig. 32-7 Groundwater exploitation of confined aquifer Ⅴ versus groundwater table variation in Shanghai

第Ⅱ、第Ⅲ承压含水层地下水开采量与地下水位动态变化特征具有相似性。20 世纪 60 年代初期为地下水最强开采时期，年开采量在 $5000\times10^4\sim8000\times10^4m^3$，第Ⅱ、第Ⅲ承压含水层地下水位迅速、持续下降，分别下降至-31m、-33m，地下水位为历史最低；80 年代以来，在少量开采的情况下，及持续的地下水人工回灌，回灌强度保持在 $500\times10^4\sim1000\times10^4m^3/a$ 的背景下，确保了地下水位处于相对高、相对稳定、缓慢抬升的态势。目前中心城地下水位总体在 0m 上下波动。

第Ⅳ、第Ⅴ承压含水层地下水开采量与地下水位动态变化特征也具有一定的相似性。20 世纪 80 年代之前，地下水开采量相对小的时期，地下水位下降幅度相对小，总体在-20m 以浅；80—90 年代为地下水最强开采时期，年开采量分别在 $5000\times10^4\sim8000\times10^4m^3$、$2000\times10^4\sim3000\times10^4m^3$，第Ⅳ、第Ⅴ承压含水层地下水位迅速、持续下降，分别下降至-65m、-62m 左右，地下水位为历史最低；2010 年代以来，在保持少量开采的情况下，及持续的地下水人工回灌，回灌强度在 $500\times10^4\sim700\times10^4m^3/a$ 的背景下，确保了地下水位处于持续上升态势。目前中心城第Ⅳ、第Ⅴ承压含水层地下水位总体在

-5m、-12m 左右。

可见，地下水在长期开采、人工回灌等因素综合影响下，地下水环境不可避免地发生了变化，地下水位下降是最为突出的环境地质问题，且地下水开采强度的大小将直接左右着区域地下水位的下降程度。随着区域地下水位下降现象的产生，土体变形也随之发生，地下水位变化已成为引发或加剧地面沉降环境地质问题主要影响因素。

（二）导致局部地区地下水污染

以往地下水调查结果显示，全市深部第Ⅱ、第Ⅲ、第Ⅳ、第Ⅴ承压含水层地下水主采层在多年来的开发利用过程中，总体未受工业“三废”和生活污水等人为活动产生的污染源的直接污染。但因成井质量欠佳及后期维护不当等人为因素，导致局部地区水环境质量发生了一些变化，地下水遭受了一定程度污染，优质地下水资源利用受到一定影响。

1. 凿井止水不当对地下水的污染

20 世纪 60 年代以来，地下水在开发利用过程中，多处第Ⅳ承压含水层开采井因凿井过程中施工质量问题，导致优质地下水遭受上部咸水下渗污染影响。

青浦第十牧场：1964 年该场开凿的第Ⅳ承压含水层深井（青 002-1）因止水不好，导致上部第Ⅱ、第Ⅲ承压含水层咸水向下渗漏，致使该井使用 10 年后，原生地下水中氯化物含量由 120mg/L 上升至 1585mg/L，矿化度由 515mg/L 上升至 2180mg/L，原生地下淡水已受咸化污染，因水质愈趋恶化，1973 年停采。1974 年，距离该污染井 3km 外的国光口琴厂同层次深井（青 009-1）开始进行地下水开发利用，受区域水动力影响，污染井附近形成的咸水界面不断向四周低水位方向迁移，1974—1989 年间，国光口琴厂所在地区地下水质量逐渐受到影响，表现为氯化物含量由 216mg/L 上升至 534mg/L，矿化度由 689mg/L 上升至 1175mg/L，因水质逐渐恶化不能满足生产、生活所需，1989 年亦被迫停采。1989 年预估污染范围已达到 2~5km^2（图 32-8）。目前已经遭受污染地区，处于尚未修复状态。

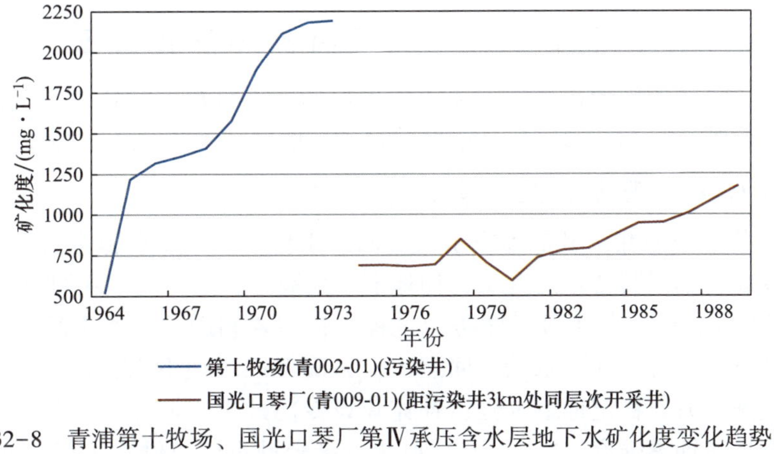

图 32-8　青浦第十牧场、国光口琴厂第Ⅳ承压含水层地下水矿化度变化趋势图

Fig. 32-8　Groundwater mineralization degree variation of confined aquifer Ⅳ at the Tenth Ranch and Guoguang Harmonica Factory，Qingpu District

浦东惠南、奉贤头桥：两处第Ⅳ承压含水层地下水也不同程度地遭受到上部咸水污染，这一现象在近期（2015 年）区域水文地质勘查（上海大浦东地区地质环境调查）过程中被发现，初步认为是以往凿井止水不当造成的。两处地下水矿化度由小于 1g/L，分别上升至 4.58g/L（咸水）、11.15g/L（盐水），地下水污染较为严重，但其动态变化资料同时邻近地区同层次开采井水质监测资料掌握有限。

目前受各种调查条件限制，上述发现的多处地下水污染区，未能掌握具体起始受污时间、污染范围、变化特征等详细信息。建议今后具备条件时开展专项调查，为地下水治理措施的制定提供科学依据。

2. 井管破裂对地下水的污染

据调查，嘉定二中 2 号井（嘉 031-02），该井因钢管接头焊接质量差，在井深 36~60m 处发生裂缝，第Ⅱ承压含水层咸水下渗污染了下伏第Ⅲ承压含水层淡水，导致第Ⅲ承压含水层水质及水化学类型均发生了改变，Cl^-含量由 148.3mg/L 急剧上升到 604.2mg/L，水化学类型也由 HCO_3 · Cl—Ca · Na 型水剧变为 Cl—Na · Ca 型水。20 世纪 90 年代后期预估其污染范围已影响到嘉定整个城中区，影响面积在 $20km^2$ 以上。

3. 维护保养不当造成地下水污染

据调查，1977 年 11 月，高桥海滨油库 200 多吨柴油漏溢，油库旁浦 75-1 号井水水质受到影响。由于未引起足够重视和及时处理，时隔一年，该井附近的两口同层次开采井（浦 75-2、浦 73-2 井）井水水质均受到漏溢柴油的污染，导致短期内两个单位和职工家属区的生产、生活用水遭到严重影响。

另外，许多采灌井井管周围受各类细菌影响，维护保养不当时，对地下水质量也会造成一定影响。硫酸盐还原菌、铁细菌等一类工业微生物生长，尤其在有机污染、人工回灌管路密封不严及开采井（回灌井）疏通、回扬不畅等特定的氧化-还原条件下，采灌井地下水易产生水质变色发臭，导致水质恶化。这些现象在杨浦、普陀多口开采井中均有所反映。

上述提及地区一定范围内地下水质量下降后，受当时各种调查条件限制，污染程度尚未得以进一步掌握，也尚未及时采取修复措施。

第三节　地下水人工回灌

研究表明，地下水的大量开发利用，已成为引发及加剧地面沉降的主要因素，为此，在城市发展过程中对优质地下水开发利用不可避免的背景下，为控制区域地面沉降，从 1966 年开始，在地下水开采的同时，同步实施了地下水人工回灌措施。

一、地下水人工回灌特点和变化特征

几十年来的地下水人工回灌工作贯穿于地质环境保护的全过程，不管是总体回灌格局，还是年度回灌方案、阶段回灌规划等，均体现出与相应阶段地面沉降防治工作相协调的特点。

（一）地下水人工回灌方式呈现多样化特征

上海通过建设管井（回灌井）开展地下水人工回灌工作。就回灌设施本身含义，也是随着不同阶段应用功能的发展变化，辅以其特定的含义，如地下水采灌井、专门地下水回灌井、应急回灌井。

20 世纪 60 年代至今，全市在地面沉降防治过程中，主要利用地下水采灌井，采取地下水人工回灌和开采交替进行的方式，开展地下水人工回灌工作，确保区域地下水位的有效控制；“十一五”以来，为实现短期内快速抬升、稳定地下水位，补充建设专门地下水回灌井，日常发挥地下水人工回灌功能，以进一步加强区域回灌强度；“十二五”以来，结合城市应急供水工程，补充建设应急回灌井，日常主要发挥其回灌功能，而当应急供水情况发生时发挥其应急供水功能。

为此，目前上海地下水人工回灌格局具有由日常采灌结合为主，逐渐转变成日常（应急）采灌、专门回灌等多种形式并存的特点。

（二）年度回灌特征

1. 回灌层次及回灌量

第Ⅱ、第Ⅲ、第Ⅳ、第Ⅴ承压含水层为地下水人工回灌主要层次（图 32-9）。

全市地下水人工回灌总体呈现初期少量回灌—微量增加—相对稳定—历史最强回灌—快速减少—逐年少量增加的动态变化特征。1980—1996 年间，受大量开采地下水影响，区域地面沉降速率出现持续微量加速现象，为有效控制地面沉降，全市加大地下水人工回灌强度，减缓地下水位下降幅度，这一阶段已成为历史上年度回灌量最大时期。1998—2005 年间，受城市建设过程中回灌井遭受破坏严重

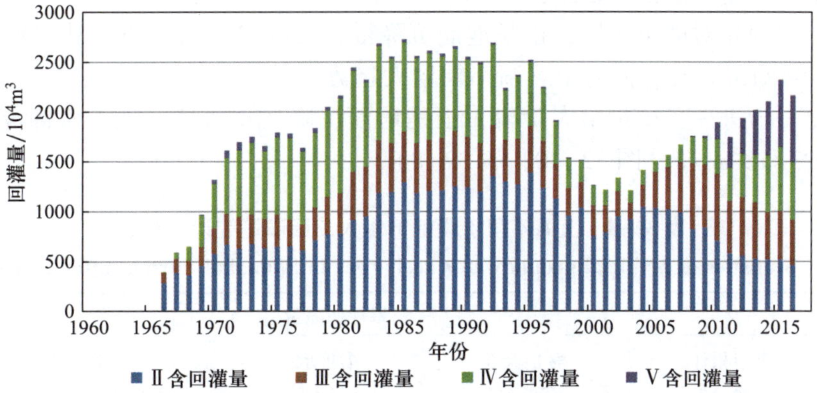

图 32-9　1960—2015 年上海第Ⅱ、第Ⅲ、第Ⅳ、第Ⅴ承压含水层地下水年度回灌总量历时变化图

Fig. 32-9　Time histories of total annual groundwater recharge of confined aquifers Ⅱ, Ⅲ, Ⅳ and Ⅴ in Shanghai during 1960-2015

影响，年度回灌量明显减少，成为历史上年度回灌量较小时期。

第Ⅱ、第Ⅳ承压含水层多年保持较大的回灌权重，其次为第Ⅲ承压含水层。第Ⅳ承压含水层在1999—2009年间回灌强度有明显下降现象，调查发现主要与这一阶段城市建设过程中该含水层采灌井被大量破坏有关；第Ⅴ承压含水层多年维持少量的回灌，2001—2007年呈停灌状态，但2008年以来回灌量增幅较大。2010—2015年，第Ⅱ、第Ⅲ、第Ⅳ、第Ⅴ承压含水层回灌权重相对均衡。

数据统计显示：第Ⅱ、第Ⅲ、第Ⅳ、第Ⅴ承压含水层年回灌量分别在$200\times10^4\sim1400\times10^4m^3$、$100\times10^4\sim600\times10^4m^3$、$100\times10^4\sim1000\times10^4m^3$、$10\times10^4\sim700\times10^4m^3$。1966—2015年，全市累计地下水人工回灌量为$93858\times10^4m^3$。

2. 回灌地区

针对区域地面沉降不同阶段控制目标，结合城市功能定位以及全市地下水开采地区的不断调整，地下水人工回灌地区也随之进行了合理、优化调整。

1966年回灌初始，地下水人工回灌以市区（原市区）为主。随后回灌地区由市区逐步扩展至郊区。这一回灌格局基本延续至今。

21世纪以来，地下水人工回灌工作不断适应新的地面沉降防治规划、城市（应急）供水规划等需求，在确保（电力等企业）地下水采灌结合、（低地下水位区）专门回灌等回灌形式的同时，结合城市地下水应急水源地建设、或地下水应急利用等需求，在满足水文地质条件、人口相对集中（学校、公园、大型社区）地区建设应急回灌井，发挥日常人工回灌及应急利用地下水功能。

（三）阶段回灌特征

纵观阶段回灌强度，是随阶段地面沉降防治目标的不同而进行调整。阶段特征表现为：

1966—1969年，人工回灌初始阶段。因前期地下水的大量开采，地面沉降较为严重，为此启动保护措施，开始实施地下水人工回灌。人工回灌初始阶段回灌量较小，由$387\times10^4m^3/a$增至$968\times10^4m^3/a$。

1970—1979年，为逐渐增大—相对稳定阶段，也是地面沉降得到较好控制的阶段，地下水回灌量基本在$1300\times10^4\sim2000\times10^4m^3/a$间小幅波动。

1980—1989年，为进一步增大—相对稳定阶段，也是地面沉降微量加速阶段。这一阶段地下水回灌量较上一阶段有明显的增加，地下水回灌量基本保持在$2200\times10^4\sim2700\times10^4m^3/a$，是历史上地下水人工回灌强度最大时期。

1990—1999年，为保持相对大的稳定阶段，也是地面沉降微量加速阶段。这一阶段地下水回灌量继续保持上一阶段回灌强度，但后期因设施破坏严重而略有减少。其间地下水回灌量基本保持在

$1500\times10^4\sim2500\times10^4m^3/a$，也是历史上地下水人工回灌强度相对大的时期。

2000—2009 年，为相对减少阶段，也是地面沉降得到较好控制的阶段。其间地下水回灌量基本保持在 $1200\times10^4\sim1700\times10^4m^3/a$，同时呈逐渐小幅增大趋势。

2010—2015 年，为进一步少量增大阶段，也是地面沉降较小阶段。地下水回灌量由 $1892\times10^4m^3/a$，逐步增加至 $2323\times10^4m^3/a$（图 32-10；表 32-8）。

表 32-8　上海不同阶段地下水人工回灌量统计一览表

Table 32-8　Statistics of artificial groundwater recharges in Shanghai during different periods of time

年份	累计回灌量/10^4m^3				
	全市累计回灌总量	第Ⅱ承压含水层	第Ⅲ承压含水层	第Ⅳ承压含水层	第Ⅴ承压含水层
1966—1969	2594	1466	602	512	13
1970—1979	17142	6571	3025	6980	543
1980—1989	25352	11150	5012	8842	289
1990—1999	22069	12092	4204	5589	184
2000—2009	14684	9176	3793	1679	36
2010—2015	12018	3404	3331	2780	2500

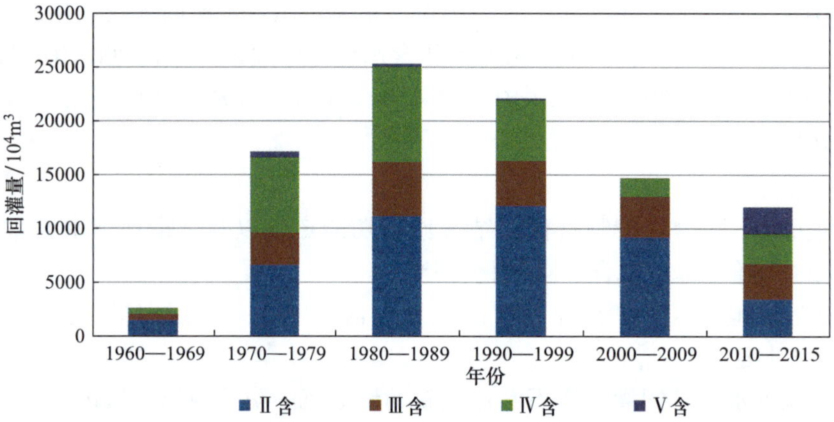

图 32-10　第Ⅱ、第Ⅲ、第Ⅳ、第Ⅴ承压含水层阶段地下水人工回灌量变化特征图（1960—2015 年）

Fig. 32-10　Artificial groundwater recharge variations of confined aquifers Ⅱ, Ⅲ, Ⅳ and Ⅴ during different periods of time (1960-2016)

二、回灌井地下水质量稳定性

地下水人工回灌对水环境影响具有双重效应，在有效抬升区域地下水位或减缓地下水位下降速率的同时，局部原生地下水质量的稳定性也受到了一定程度影响。

（一）回灌水源质量

上海地下水人工回灌水源，一般就近直接利用质量符合饮用水卫生标准的自来水。

上海地表水资源丰富，大小河道纵横交错，黄浦江、长江及其支流纵贯上海东西部和南北部，其水源为上海生产自来水的最为重要的水源。但是地表水质量在不同阶段存在着相当大的变化，往往质量优劣与社会经济发展、生态环境保护力度大小有一定相关性。

20 世纪 80 年代之前，社会经济发展速度相对缓慢，各种污染源的产生程度相对轻微，对地表水质量影响相对小，为此，地表水质量总体较好，以Ⅱ类、Ⅲ类水为主。但 80 年代以来至 2015 年，在城市社会经济迅猛发展的过程中，环境污染也随之有所加剧，地表水质量不断下降，由以往的以Ⅱ类、

Ⅲ类水为主，逐渐转变成以Ⅳ类、Ⅴ类（或劣Ⅴ类）水为主，水体中许多无机组分、有机组分等对人体有害的污染组分含量时有检出或超标。低质量的地表水，在自来水生产过程中，由于受各项技术处理能力的限制，在基本满足饮用水质量的情况下，部分污染组分经常存在检出或超标现象。

区域上，自来水水质测试资料以无机组分为主，有机组分资料基本缺乏。

对反映不同地区（阶段）自来水质量的动态数据掌握程度有限。但据位于中心城上棉八厂（1977—2010年）自来水间断性水质测试结果（表32-9），初步反映了多年来市区作为回灌水源质量的动态变化特征。

表32-9　上海地下水人工回灌水源部分组分统计

Table 32-9　Statistics of some components in the water source used for artificial groundwater recharge in Shanghai

主要组分	饮用水卫生标准限值/（mg·L^{-1}）	自来水中含量/（mg·L^{-1}）
矿化度	1000	220~550（最高1400）
氯离子	250	40~200（最高700）
硫酸根	250	20~100（最高150）
铁	0.3	0.03~1.00（最高10）
锰	0.1	0.005~0.3（最高0.7）
硝酸根	20	0~15（最高27）
氨氮	0.5	0.2~5.0（最高12）
酚	0.002	0.001~0.01（最高0.09）
氰化物	0.05	0.001~0.01（最高0.23）
化学耗氧量	5	2~7（最高12）

中心城上棉八厂部分典型组分资料显示：自来水基本为淡水，矿化度、氯离子含量总体较低，小于饮用水卫生标准限值；硫酸根含量相对较高，但小于饮用水卫生标准限值；铁、锰、氨氮、酚、氰、化学耗氧量等含量，大多符合饮用水卫生标准限值，但时有超标迹象，多年最大值往往是标准限值的数倍（图32-11~图32-13）。

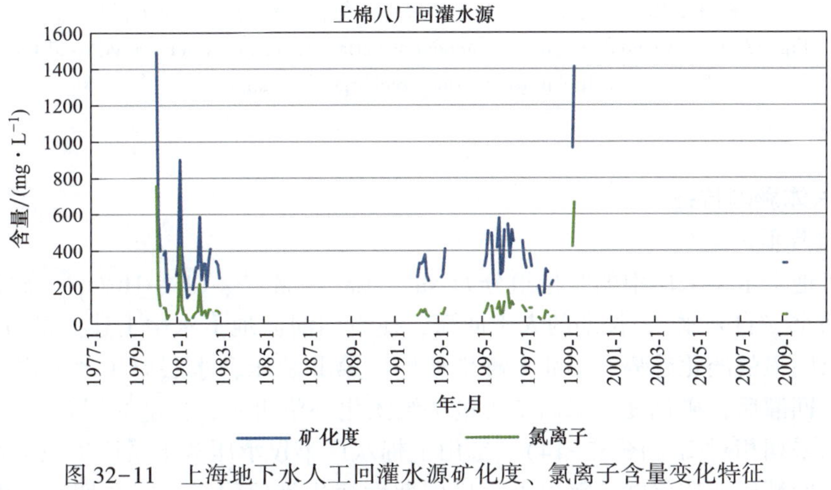

图32-11　上海地下水人工回灌水源矿化度、氯离子含量变化特征

Fig. 32-11　Mineralization degree and chlorine ion variations of the water source used for artificial groundwater recharge in Shanghai

（二）人工回灌对地下水水质总体影响

地下水在几十年来的开采、回灌并举的过程中，区域地下水质量总体稳定，基本维持原生状态。但因受回灌水源质量、回灌方式不同等综合因素影响，回灌井周围一定范围内原生地下水质量发生了

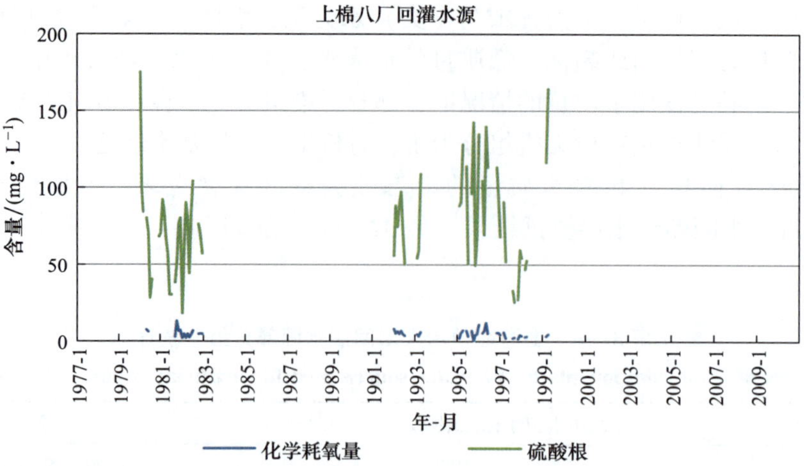

图 32-12　上海地下水人工回灌水源化学耗氧量、硫酸根含量变化特征

Fig. 32-12　Chemical oxygen consumption and Sulfate ion variations in the water source used for artificial groundwater recharge in Shanghai

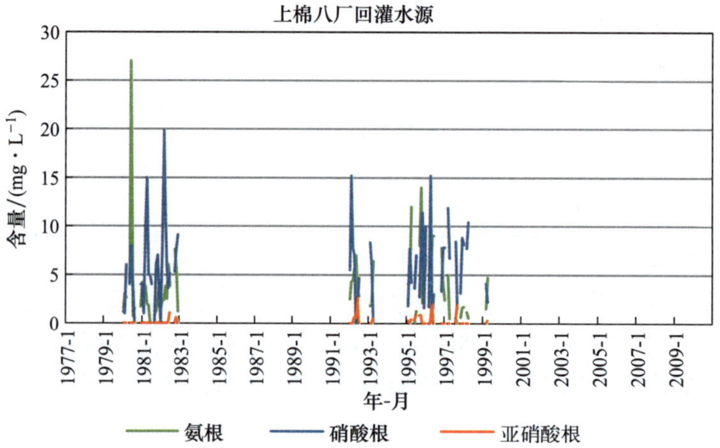

图 32-13　上海地下水人工回灌水源氨氮类组分含量变化特征

Fig. 32-13　Ammonia-nitrogen component variations in the water source used for artificial groundwater recharge in Shanghai

一定变化。

1. 无机组分总体影响特征

（1）指标含量降低的组分

据调查，上海地下水主采层中的常量组分如 Na^+、Ca^{2+}、Mg^{2+}、Cl^-、HCO_3^-、矿化度和硬度指标等含量普遍高于回灌水（自来水），长期的人工回灌，使得回灌后地下水中上述大部分指标浓度被稀释，含量下降，呈现出明显的淡化趋势。如浦东耀华玻璃厂第Ⅱ承压含水层原生地下水为微咸水，日常开展冬灌夏用，人工回灌后，矿化度、氯离子含量持续淡化迹象明显，含量的相对大小也具周期性特点，回灌期相对低，回采期相对高（图 32-14）。宝山上棉八厂第Ⅳ承压含水层原生地下水为淡水，日常开展冬灌夏用，人工回灌后，矿化度、氯离子含量周期性变化迹象明显，回灌期矿化度、氯离子含量进一步降低现象明显（图 32-15）。

（2）指标含量增高的组分

A. 硫酸根

原生地下水中硫酸根含量一般很低，或无，但受长期的人工回灌影响，硫酸根含量上升现象明显，并呈现回灌期高、回采期低的周期性变化特征（图 32-16）。

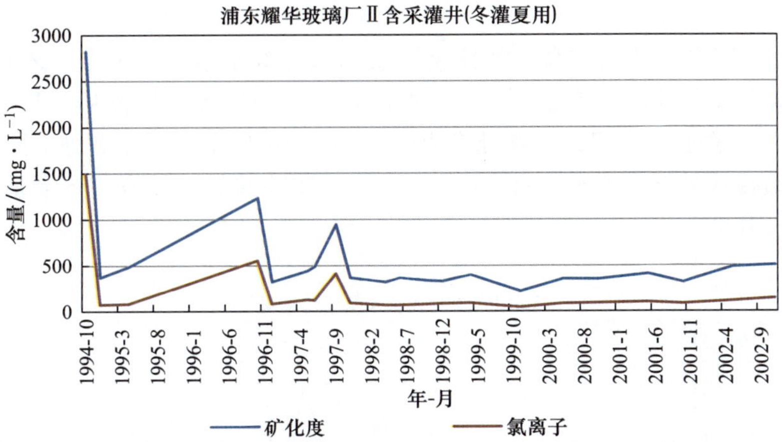

图 32-14　上海第Ⅱ承压含水层典型采灌井（pd023-11）矿化度、氯离子含量变化图

Fig. 32-14　Mineral degree and chlorine ion variations of confined aquifer Ⅱ in Shanghai per typical mining and recharge well (pd023-11)

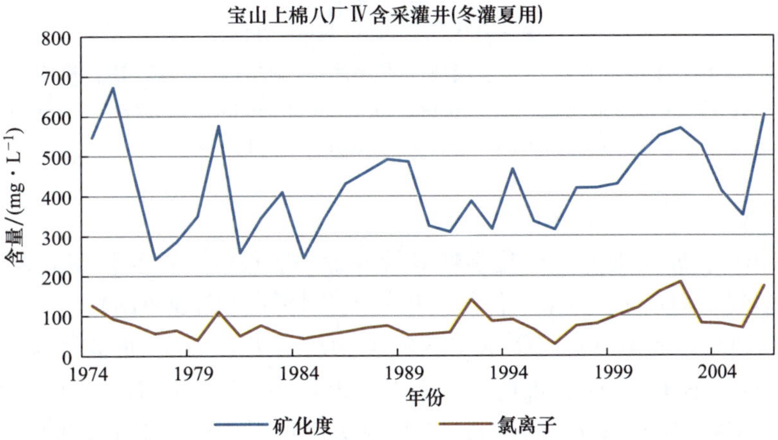

图 32-15　上海第Ⅳ承压含水层典型采灌井（bs024-17）矿化度、氯离子含量变化图

Fig. 32-15　Mineralization degree and chlorine ion variations of confined aquifer Ⅳ in Shanghai per typical mining and recharge well (bs024-17)

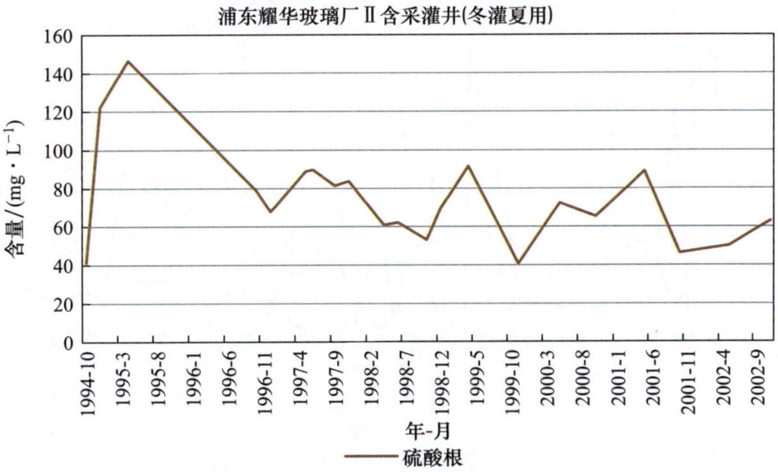

图 32-16　上海第Ⅱ承压含水层典型采灌井（pd023-11）硫酸根含量变化图

Fig. 32-16　Sulfate ion content variation of confined aquifer Ⅱ in Shanghai per typical mining and recharge well (pd023-11)

B. 铁、锰

原生地下水中铁、锰含量普遍较高，往往超出饮用水标准限值。区域上，不同水质分布区，其含量差异较大，一般咸水区含量相对较高，淡水区相对较低。长期人工回灌后，地下水中铁、锰含量较原生水普遍上升，铁、锰处于长期、不断累积状态，如宝山上棉八厂第Ⅳ承压含水层采灌井，地下水（淡水）中铁、锰含量的增加明显（图 32-17）。从回灌井的类型而言，冬灌夏用井地下水中的铁、锰含量普遍较夏灌冬用井高。

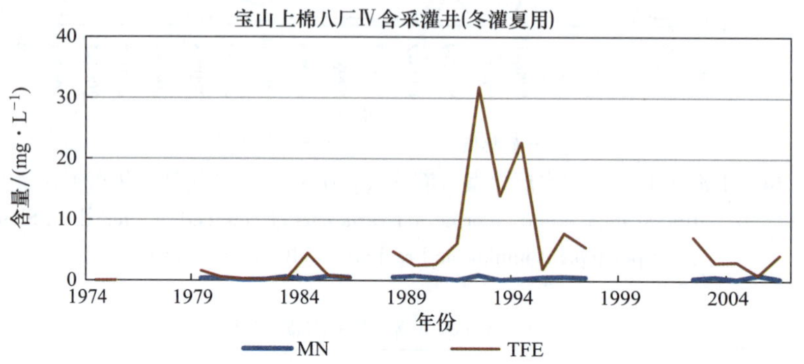

图 32-17　上海第Ⅳ承压含水层典型采灌井（bs024-17）铁、锰含量变化图

Fig. 32-17　Iron and manganese content variation of confined aquifer Ⅳ in Shanghai per typical mining and recharge well（bs024-17）

C. 化学耗氧量、氨氮

化学耗氧量（COD）：原生地下水中化学耗氧量含量一般很低，但受长期的人工回灌影响，其含量上升现象明显，并呈现回灌期高、回采期低的周期性变化特征（图 39-18）。

三氮类（NH_4^+、NO_2^-、NO_3^-）：地下水中三氮类物质主要以 NH_4^+形式存在，NO_2^-、NO_3^-含量均较低，尤其是在淡水区域 NO_2^-、NO_3^-含量多处于未检出状态。人工回灌后，由于改变了地下水中的还原状态，进入了氧化环境，致使地下水中 NO_2^-含量显著增加，NO_3^-含量在不同深度的含水层及不同区域的咸淡水中上升速率不一。NH_4^+的变化则与原生地下水中含量高低有关，原生背景含量高时多呈下降趋势，低时有增高现象（图 32-18）。

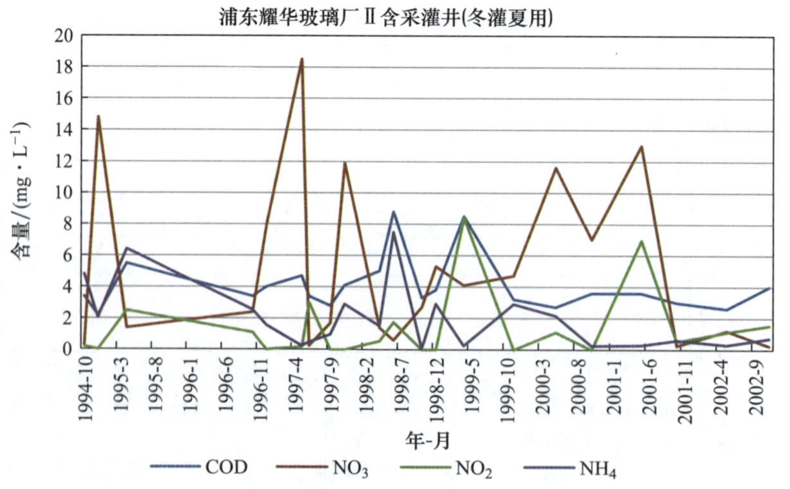

图 32-18　上海第Ⅱ承压含水层典型采灌井（pd023-11）化学耗氧量、氨氮含量变化图

Fig. 32-18　Chemical oxygen consumption and ammonia-nitrogen content variations of confined aquifer Ⅱ in Shanghai per typical mining and recharge well（pd023-11）

D. 微量元素

原生地下水中铜、铅、锌、砷、镉、铬、汞、硒、镍、钴、钼、银等微量元素含量一般均较低，大多符合饮用水标准。但因自来水中上述部分微量元素时有检出或超标，在回灌和开采的交替过程中，回灌井地下水中的微量元素含量表现出一定的规律性变化特征，在一个采灌周期内往往呈现出高—低—高的变化规律，回灌期相对高，回采期相对低。

E. 酚、氰化物

原生地下水中基本不含酚、氰化物。但由于回灌水源中含有酚、氰化物成分，使得回灌井地下水中时常检出有一定量的酚、氰化物。

一般情况下，在回灌期间酚、氰化物时有检出或超标，但通过回扬后，酚、氰化物大部分被抽取和自净，一般不会形成残留性影响，但当回灌水中该类指标含量较高且回灌水大量残留于含水层中时，地下水中有检出（或超标）的可能。

2. 有机组分总体影响特征

原生地下水中基本无卤代烃、苯系物、多环芳烃和有机氯农药类等有机组分，一般处于未予检出状态。自来水中，一般无有机磷农药、多氯联苯、有机氯，但卤代烃、多环芳烃、单环芳烃、有机氯农药指标时有检出（或超标）。由于自来水有时受物理、化学等处理技术限制，部分有机组分仍残留在水中，导致长期周期性回灌、回采后的地下水中卤代烃、多环芳烃、单环芳烃等指标时有检出（或超标）。

3. 回灌井地下水质量变化趋势

地下水采灌方式的不同，影响着地下水质量总体变化趋势

回灌井地下水质量变化趋势与回灌水质、采灌水量大小和采灌时间长短等有着密切的关系。地下水在长期交替采灌过程中，采灌量的相对大小，导致地下水质量变化规律存在明显差异：

当长期呈现回灌量远小于开采量的采灌格局时，回灌井周围一定范围内的地下水质量基本以原生水为主，混合后的地下水中常量组分等含量表现出与原生地下水组分更为接近的特点。

当长期呈现回灌量大于开采量的采灌格局时，回灌井周围一定范围内地下水逐渐被回灌水源水所取代，混合后的地下水中常量组分等含量表现出与回灌水源组分更为接近的特点。

（三）人工回灌影响范围

为切实保障地下水资源安全，先后通过相关试验研究，基本掌握了地下水人工回灌影响范围。试验结果已成为上海制定地面沉降防治措施和地下水资源保护规划的重要依据。

1. 典型地下水人工回灌试验场特点

20 世纪 70 年代以来，地质矿产部所属上海市地质矿产局（现上海市规划和自然资源局），组织技术力量，以满足地质、综合试验研究条件为原则，在地面沉降相对发育的中心城（原市区），以及水文地质条件较好的滨海平原深部古河道分布地区，通过建设（或利用已有处于正常运行的）不同层次人工回灌试验场，先后开展了地下水采灌、专灌条件下对原生地下水质量影响试验研究。

20 世纪 70 年代末—80 年代初，结合上海展览馆地下水采灌井，开展了第Ⅱ承压含水层单个采灌型水质影响试验研究（表 32-10）；结合深部地下水采灌井相对密集、采灌强度大和时间长的杨浦，开展了第Ⅳ承压含水层群井采灌型水质影响试验研究；“十一五”以来，针对深部含水层地下水人工回灌强度增大趋势，开展了深部第Ⅳ承压含水层专门回灌型水质影响试验研究。

试验目的：掌握地下水采灌（单井、群井）、专门回灌（单井）等不同条件下人工回灌对地下水的影响范围。

试验配置：配置主井（采灌井、专灌井）和监测井；回灌水源为直接利用符合生活饮用水卫生标准的自来水；监测原生水、自来水、混合地下水质量；Cl^-（或 SO_4^{2-}、氚）为试验示踪剂。

2. 水质影响范围

上述不同回灌类型试验场试验结果显示，地下水人工回灌后对原生地下水的影响特征以及影响范围有一定差异。主要体现在：

表 32-10　上海典型地下水人工回灌试验场一览表

Table 32-10　List of typical artificial groundwater recharge test grounds in Shanghai

试验场类型	试验目的层	试验地区	试验场设施配置	主要监测内容	回灌水源	原生地下水质量	示踪剂	试验时间/年	监测内容				总体影响特征	试验期间影响范围/m	备注
									现场检测	全分析	有机组分	同位素			
采灌结合型	第Ⅱ承压含水层	上海展览馆	1个主井（采灌井）、5个监测井	原生地下水、自来水、回灌后的混合水水质	自来水	微咸水	Cl^-和SO_4^{2-}	18（1967—1985年）	无	有	无	无	回灌量远大于开采量，主井矿化度、氯离子含量逐渐降低；SO_4^{2-}含量从无到有并有所增多	大于68	①现场检测内容：电导率、溶解氧、氧化还原电位、酸碱度；②有机组分主要指：卤代烃、多环芳烃、有机氯农药；③同位素主要指：氘、氚、^{18}O、^{13}C、^{14}C；④回灌水源为符合饮用水卫生标准（当时的）自来水
采灌结合型	第Ⅳ承压含水层	覆盖杨浦与邻近黄浦江的浦东高桥地区	多个主井（采灌井），2个监测井（开采井）	原生地下水、自来水、回灌后的混合水水质	自来水	淡水	Cl^-与氚	15（1967—1982年）	无	有	有	有	回灌量略大于开采量，主井氯离子含量有所上升，并向外逐渐降低；引起水质明显变化的范围小于1000m	小于1000	
专门回灌型	第Ⅳ承压含水层	宝山顾村	1个主井（专门回灌井）、10个监测井	原生地下水、自来水、回灌后的混合水水质	自来水	淡水	Cl^-	5（2012—2017年）	有	有	有	有	专门回灌，由主井向外围电导率、溶解氧、氧化还原电位、酸碱度由高变低；氯离子、硫酸根、硝酸根含量先升高再降低；铅、汞有毒重金属元素含量未有明显改变；有机氯农药、多环芳烃未检出，卤代烃有检出但未超标；^{13}C浓度降低，^{14}C、^{18}O、^{2}H、^{3}H浓度升高	大于180	

（1）地下水组分总体影响特征

人工回灌后对原生地下水质量有一定影响，且不同背景（淡水、咸水）的原生地下水影响特征或变化趋势有一定差异。

试验结果显示：试验主井（采灌井或专门回灌井）在长期的开采与回灌交替过程中，或在长期的专门回灌条件下，主井地下水中部分无机组分含量发生了变化。如矿化度、氯离子，当原生地下水为咸水时，人工回灌后矿化度总体呈现减小的趋势，当原生地下水为淡水时，人工回灌后矿化度变化趋势不明显；部分常量组分（硫酸根等）含量有明显的增高；三氮组分含量的增高；微量元素含量呈周期变化；部分重金属元素含量的增高；酚、氰化物有检出或含量或增高超标等。部分有机组分卤代烃（二氯甲烷、三氯甲烷、溴二氯甲烷、二溴一氯甲烷）等在地下水中也时有检出。

（2）影响范围

A. 单井采灌条件下影响范围

1967—1985 年，开展了上海展览馆单井采灌条件下对地下水影响范围试验研究。1979 年 12 月后，各监测井水质均发生了明显的变化，距主井最远的监测井（H5 孔）Cl^-含量已出现明显波动，SO_4^{2-}则处于时有时无断续检出状态。至 1984 年 5 月，示踪离子间歇性检出现象完全消失，即使在回采以后也无法恢复到原地下水状态。说明此时回灌水体的最小影响半径已大于 68.1m。试验表明，单井采灌对原生地下水影响范围相对较小。

B. 群井采灌条件下影响范围

1967—1982 年，在黄浦江两岸的杨浦—浦东开展了群井采灌条件下对地下水影响范围试验研究。

1979—1982 年，示踪剂氚同位素监测结果显示，距上棉十七厂约 500m 的华光啤酒厂开采井（杨 19-3）地下水中已有氚同位素的断续检出，检出含量在 6.5～17.0TU 之间，平均含量为 14.16TU，但低于回灌井地下水；而位于黄浦江南岸的 916 仓库开采井（浦 39-1），自 1980 年 10 月以后氚同位素开始有少量零星检出，检出含量在 2.9～8.2TU 之间，平均含量为 2.94TU，反映的是早期灌入含水层中残留的自来水，由此推断试验期间回灌水的最大影响半径在 1000m 左右（图 32-19）。

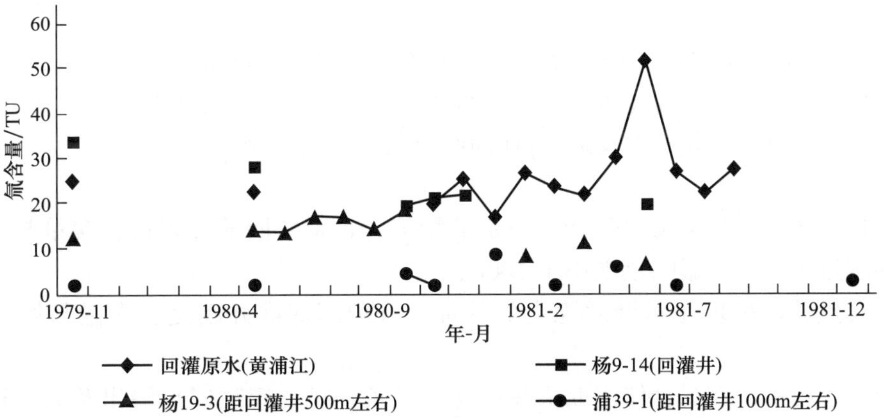

图 32-19　群井采灌条件下距开采井不同距离处氚含量变化

Fig. 32-19　Tritium content variation at different distances from the mining well under group-well mining and recharge

1979—1982 年，示踪剂 Cl^-含量监测结果显示，上棉十七厂采灌井（杨 9-14）地下水中 Cl^-含量变化趋势与回灌水（自来水）Cl^-含量变化趋势基本一致；华光啤酒厂开采井（杨 19-3）地下水中 Cl^-含量在夏用期可高达 60mg/L，仅 Cl^-的峰值略滞后，说明该井地下水已明显受到了回灌水的影响（图 32-20）；而黄浦江南岸的 916 仓库开采井（浦 39-1），虽然氚同位素已检出，但该井地下水中 Cl^-含量未发生明显变化，仍保持原生地下水背景值。

由此推断，以上棉十七厂为中心的集中采灌区，自 1967 年以来大量的开采与回灌交替过程中，对地下水的最大影响边界约为 1000m，但引起水质明显变化的范围小于 1000m。如果保持当时采灌格局，

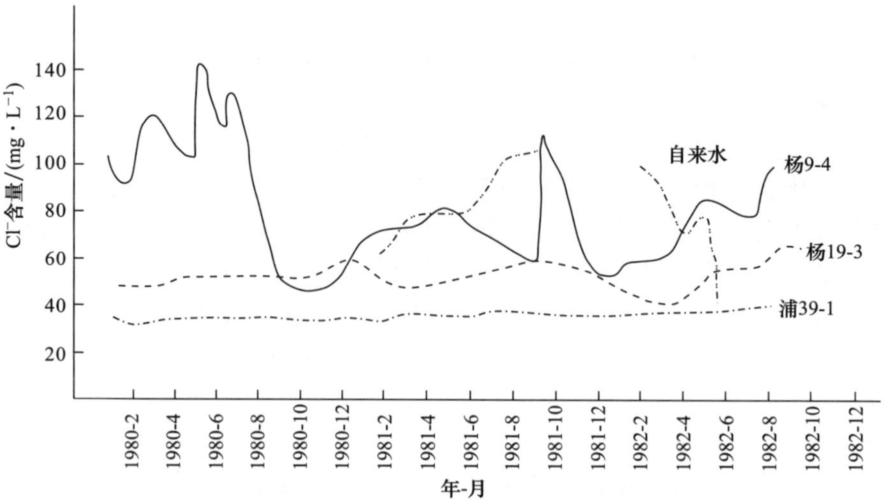

图 32-20 群井采灌条件下距监测井（开采井）不同距离处氯离子含量变化特征

Fig. 32-20 Chlorine ion variation at different distances from the monitoring well (mining well) under group-well mining and recharge

至 2015 年预估影响范围将大于 1.5km，优质地下水资源的可持续利用会存在一定隐患。

试验表明，群井采灌对原生地下水影响范围相对较大。

C. 专门回灌条件下影响范围

2012 年开始，在宝山顾村试验场开展了单井专门回灌条件下对地下水影响范围试验研究。2012—2013 年，现场测试、常量组分、重金属元素、有机组分、同位素测试结果等，初步反映了专门回灌条件下对原生地下水质量影响范围。

1）现场测试。

现场测试包括电导率（Ec）、溶解氧（DO）、氧化还原电位（Eh）、酸碱度（pH），在试验期间，上述因子影响特征为：

电导率（Ec）：距回灌井越远，电导率下降幅度越小，稳定值越高。

氧化还原电位（Eh）、溶解氧（DO）：距离回灌井越远，氧化性逐渐减弱。

酸碱度（pH）：距离回灌井由近及远，pH 曲线降低幅度越来越大。

2）常量组分。

氯离子：含量总体呈现先增高后降低、稳定的特征，且距离回灌井越远，含量与原生地下水含量越接近。连续回灌近 1 年时，影响范围小于 100m。

重碳酸根：含量总体呈现先降低后稳定的特征，且距离回灌井越远，含量与原生地下水含量越接近。

硫酸根：回灌过程中，在混合作用或某些矿物（如黄铁矿、硫锰矿）溶出及氧化还原反应综合影响下，地下水中硫酸根含量呈升高迹象。同时在回灌试验过程中地下水溶解氧含量不断增加时，会将地下水中还原硫氧化为硫酸根，出现混合水中硫酸根含量有高于回灌水源硫酸根含量的现象。

硝酸根：含量表现出先增后急剧下降的特征，且距离回灌井越远含量越低，待氧化还原反应达到平衡时最终稳定在一定水平。

钠离子、钾离子：综合作用后地下水中钠离子含量呈现大幅度降低特征，且距离回灌井越远降低幅度越小。

钙离子：地下水中钙离子含量先降至回灌水源钙离含量以下，而后钙离子含量又逐渐上升并高于水源含量，最终稳定在一定含量，且距离回灌井越远含量越高。

镁离子：回灌初期地下水中镁离子含量急剧降低，此后缓慢降低并最终稳定，且距离回灌井越远镁离子最终稳定含量越高。

总铁：距离回灌主井越近，地下水中总铁含量降低趋势越明显。

总锰：回灌初期地下水中总锰含量降低，连续回灌 30 天，31m 以外区域含量变化不明显。

3）铅、汞有毒重金属。

地下水人工回灌过程中，地下水中铅、汞含量未发生明显改变，趋势性变化不明显，仅呈现窄幅波动态势。

4）有机组分。

有机氯农药（OCPs）均未检出。

多环芳烃（PAHs）均未检出。

卤代烃（挥发性有机化合物，VOC），主要指二氯甲烷、三氯甲烷、溴二氯甲烷、二溴一氯甲烷，试验期间在 60m 范围内和回灌水源中均有检出，但未超标。

D. 同位素

回灌试验前后，对试验场原生地下水、自来水、回灌后混合水同位素指标［氘（^{2}H）、氚（^{3}H）、碳-13（^{13}C）、碳-14（^{14}C）、氧-18（^{18}O）］进行测试。

原生地下水中除碳-13 含量高于自来水（回灌水源）外，其他氘、氚、氧-18 含量均低于自来水。回灌后的混合水中，碳-13 含量降低，氘、氚、氧-18 含量升高，并逐步与自来水相关元素含量一致。且距离回灌井越近，其含量变化幅度越大，距离回灌井越远，其含量变化越小。

试验表明，单井专门回灌对原生地下水有明显影响，且随着时间的延长，其影响范围数倍于单井采灌影响范围。

第四节　地下水开发利用与保护规划

一、地下水开发利用与保护规划

2010 年，进一步围绕地质环境保护条件下的地下水资源开发利用、可持续利用问题，以及满足地面沉降分区精细化管理需要，系统地开展了地下水资源保护规划研究工作。

在充分分析、利用海量地下水位、地面沉降等监测资料的基础上，利用近期建立的真三维有限元法数学模型，模型中辅以全市地面沉降年度平均地面沉降量（不大于 7mm/a）、第Ⅱ、第Ⅲ、第Ⅳ、第Ⅴ承压含水层地下水位最低控制标高为约束条件，运行模型，计算得出全市地下水可开采量为 $2500\times10^{4}m^{3}$。以确定的地下水可开采量指标为依据，综合考虑全市各行政区对地下水资源需求预估，以及地面沉降不同管理分区控制指标等因素，提出了 2010 年上海地下水可持续利用与保护规划方案，圈定了 4 个规划分区，包括地下水资源严格限制开采区、限制开采区、控制开采区和一般监测区（表 32-11）。规划方案奠定了上海此后一定阶段地下水开发利用与保护工作的基础。

地下水资源开发利用规划方案如下：

严格限制开采区（Ⅰ）。中心城主要为外环线以内地区，是历史上地下水开发利用强度最大的地区，处于严重超采状态，也是地面沉降最为严重的地区和地质环境重点防护地区，为最大限度减少地下水开采对地质环境安全影响程度，规划除保留少量特殊用途（矿泉水、特种锅炉用水等）地下水用水外，力争向禁止开采目标努力，以实现长期、有效控制地面沉降的目的。2010 年，规划中心城除保留 $100\times10^{4}m^{3}$ 特殊用途水量外，严禁其他用途的地下水开采活动。

限制开采区（Ⅱ）。近郊的宝山、浦东、闵行、嘉定及远郊的青浦、金山等地地下水开采强度较大，处于严重超采状态，地面沉降较为严重，需要进一步压缩地下水开采规模。2010 年，随着城市供水规划全面推进自来水集约化工程，年总开采量可压缩至 $1550\times10^{4}m^{3}$。

控制开采区（Ⅲ）。远郊的崇明、奉贤、松江等地，地下水开采强度相对较小，处于一般超采状态，地面沉降也明显，但地质环境条件较脆弱。但由于远郊区城市供水能力较低，地下水资源可以适量开发利用。在全市地下水开采总量控制下，对地下水开采量在层次及地区上进行适当调整。至2010年，崇明作为最后一个地区实现自来水集约化工程，届时这些区将严格执行开采量指标，规划开采总量约为$850 \times 10^4 m^3$。

一般监测区（Ⅳ）。主要指某区域松散层中所有含水层原生地下水均为咸水以及地下水资源较为贫乏的地区，这些地区的地下水不能直接作为资源利用，或地下水开发利用价值极低，为此日常基本未予开采，但地下水位受同层次区域地下水位波动影响，地面沉降现象也较为明显，为此对这些地区的地下水，规划日常地下水环境监测为工作重点。

表 32-11　上海地下水资源开发利用保护方案

Fig. 32-11　Groundwater resource exploitation and protection plan of Shanghai 单位：$10^4 m^3/a$

<table>
<tr><th colspan="2" rowspan="2">含水层 / 地区</th><th colspan="4">承压含水层规划开采量</th><th colspan="2" rowspan="2">规划开采总量</th></tr>
<tr><th>第Ⅱ承压含水层</th><th>第Ⅲ承压含水层</th><th>第Ⅳ承压含水层</th><th>第Ⅴ承压含水层</th></tr>
<tr><td>严格限制开采区（Ⅰ）</td><td>中心城</td><td>0</td><td>0</td><td>100</td><td>0</td><td colspan="2">100</td></tr>
<tr><td rowspan="6">限制开采区（Ⅱ）</td><td>宝山</td><td>40</td><td>0</td><td>50</td><td>0</td><td>90</td><td rowspan="6">1550</td></tr>
<tr><td>浦东</td><td>120</td><td>100</td><td>300</td><td>50</td><td>570</td></tr>
<tr><td>闵行</td><td>0</td><td>50</td><td>50</td><td>0</td><td>100</td></tr>
<tr><td>嘉定</td><td>40</td><td>50</td><td>50</td><td>50</td><td>190</td></tr>
<tr><td>青浦</td><td>50</td><td>100</td><td>50</td><td>100</td><td>300</td></tr>
<tr><td>金山</td><td>50</td><td>100</td><td>150</td><td>0</td><td>300</td></tr>
<tr><td rowspan="3">控制开采区（Ⅲ）</td><td>崇明</td><td>50</td><td>0</td><td>0</td><td>400</td><td>450</td><td rowspan="3">850</td></tr>
<tr><td>奉贤</td><td>30</td><td>70</td><td>150</td><td>0</td><td>250</td></tr>
<tr><td>松江</td><td>50</td><td>50</td><td>50</td><td>0</td><td>150</td></tr>
<tr><td colspan="2">总计</td><td>430</td><td>520</td><td>950</td><td>600</td><td colspan="2">2500</td></tr>
</table>

注：地下水规划开采量依据2010年评价结果。

二、地下水应急水源地建设

（一）地下水应急水源地规划

1. 建设地区

“十一五”期间为推进、落实国土资源部关于加强应急水源地勘察工作精神，地质部门（原上海市房屋土地资源管理局）牵头开展过地下水应急水源地建设可行性研究工作，对建设地下水应急水源地的可行性进行了综合研究。初步确定以下3大类地区为地下水应急水源地建设重点部署地区。

（1）第一类是城市中心地区

规划重点是确定大（中）型应急地下水水源地，以应对各种可能发生的供水中断或不足的危机。重点解决紧急状态下的城市居民的生活用水和特殊工业用水。这类水源地一般具供水量大、集中，供水时间相对较短的特点。

（2）第二类是岛域地区

规划重点是应对各种原因引发的，如以地表水源受污染、咸潮时间过长、发生严重干旱等为主的中小型应急地下水供水水源地。重点解决分散岛域居民的生活用水和部分特殊工业用水。这类水源地一般具供水时间相对较长的特点。

（3）第三类是城市规划新城区

规划重点是应对各种原因引发的应急供水事件，建立中小型应急地下水供水水源地。重点是解决中小城镇居民生活用水。这类水源地一般具供水时间相对较长的特点。

2. 建设数量

以含水层满足取水条件和现有开采设施满足需求为基础条件，基于2020年城市总体规划，根据城市规划所确定的城市重点规划建设地区以及人口、产业导向，总体部署中心城、宝山新城、嘉定安亭新城、浦东国际机场、惠南新城、临港新城、奉贤海滨农场、金山新城、松江新城、青浦新城、城桥新城、陈家镇新城、长兴岛、横沙岛等14个地下水应急水源地（图32-21）。

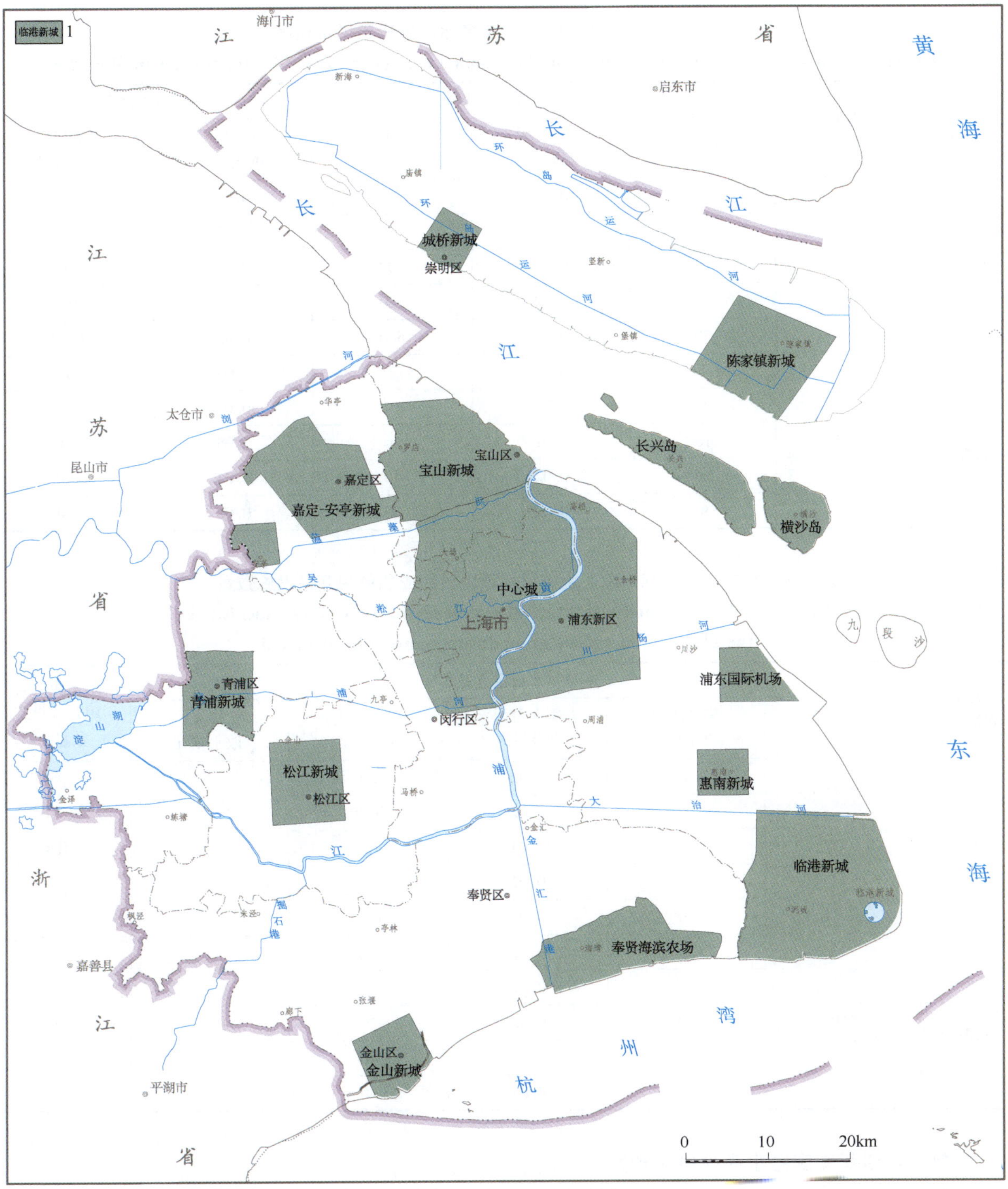

图32-21 上海地下水应急水源地规划部署图

Fig. 32-21 Disposition of the planned emergency groundwater sources in Shanghai

1—应急水源地及名称

（二）地下水应急水源地供水预案

1. 中心城取水预案

预估应急供水人口 1000 万人，当最低人均一天需水量 20L 时，一天总取水量为 $20\times10^4m^3$；当人均一天需水量 50L 时，一天总取水量为 $50\times10^4m^3$。应急供水时间分别按 1d、5d、10d、20d、30d 计。根据上述最低供水条件下确定应急备用开采井数量，若按单井每小时出水量 80t 计，则一天（24h 计）取水量为 1920t（表 32-12），中心城若满足一天应急供水量，则需 104 口开采井满负荷运作即可实现（表 32-13）。

表 32-12 上海规划地下水应急水源地应急供水预案

Table 32-12 Emergency water supply plan for the planned emergency groundwater sources in Shanghai

应急水源地	应急供水人口/万人	应急使用时间/d	应急供水预案 1		应急供水预案 2	
			每人每天供水量/t	应急供水总量/10^4t	每人每天供水量/t	应急供水总量/10^4t
中心城	1000	1	0.02	20	0.05	50
		5	0.02	100	0.05	250
		10	0.02	200	0.05	500
		20	0.02	400	0.05	1000
		30	0.02	600	0.05	1500
郊区规划新城	100	1	0.02	2	0.05	5
		5	0.02	10	0.05	25
		10	0.02	20	0.05	50
		20	0.02	40	0.05	100
		30	0.02	60	0.05	150

表 32-13 上海规划地下水应急水源地预估应急备用开采设施数量

Table 32-13 Estimated quantities of emergency backup mining facilities at the planned emergency groundwater sources in Shanghai

应急水源地	应急供水设定方案					预估应急备用开采设施数量/口
	应急供水人口/万人	应急使用时间/d	每人每天供水量/t	应急供水总量/10^4t	单井每天取水量/t	
中心城	1000	1	0.02	20	1920	104
		5	0.02	100		
		10	0.02	200		
		20	0.02	400		
		30	0.02	600		
郊区规划新城	100	1	0.02	2	1920	10
		5	0.02	10		
		10	0.02	20		
		20	0.02	40		
		30	0.02	60		

2. 郊区规划新城应急水源地供水预案

预估单个应急水源地应急供水人口 100 万人，当最低人均一天供水量 20L 时，一天总供水量为 $2\times10^4m^3$；当人均一天供水量 50L 时，一天总供水量为 $5\times10^4m^3$（见表 32-12）。应急供水时间分别按 1d、5d、10d、20d、30d 计。根据上述最低供水条件下确定应急备用开采井数量，郊区各新城若满足一天应急供水量，则需 10 口开采井满负荷运作即可实现（见表 32-13）。

（三）应急开采对地质环境影响

假设全市同步启动确定的 14 处应急水源地，以中心城水源地应急开采量每天 $20\times10^4m^3$ 为假设前提，其他各水源地应急开采量每天 $10\times10^4m^3$，假设持续开采时间为 3 个月，开展了地面沉降预测（图 32-22）。

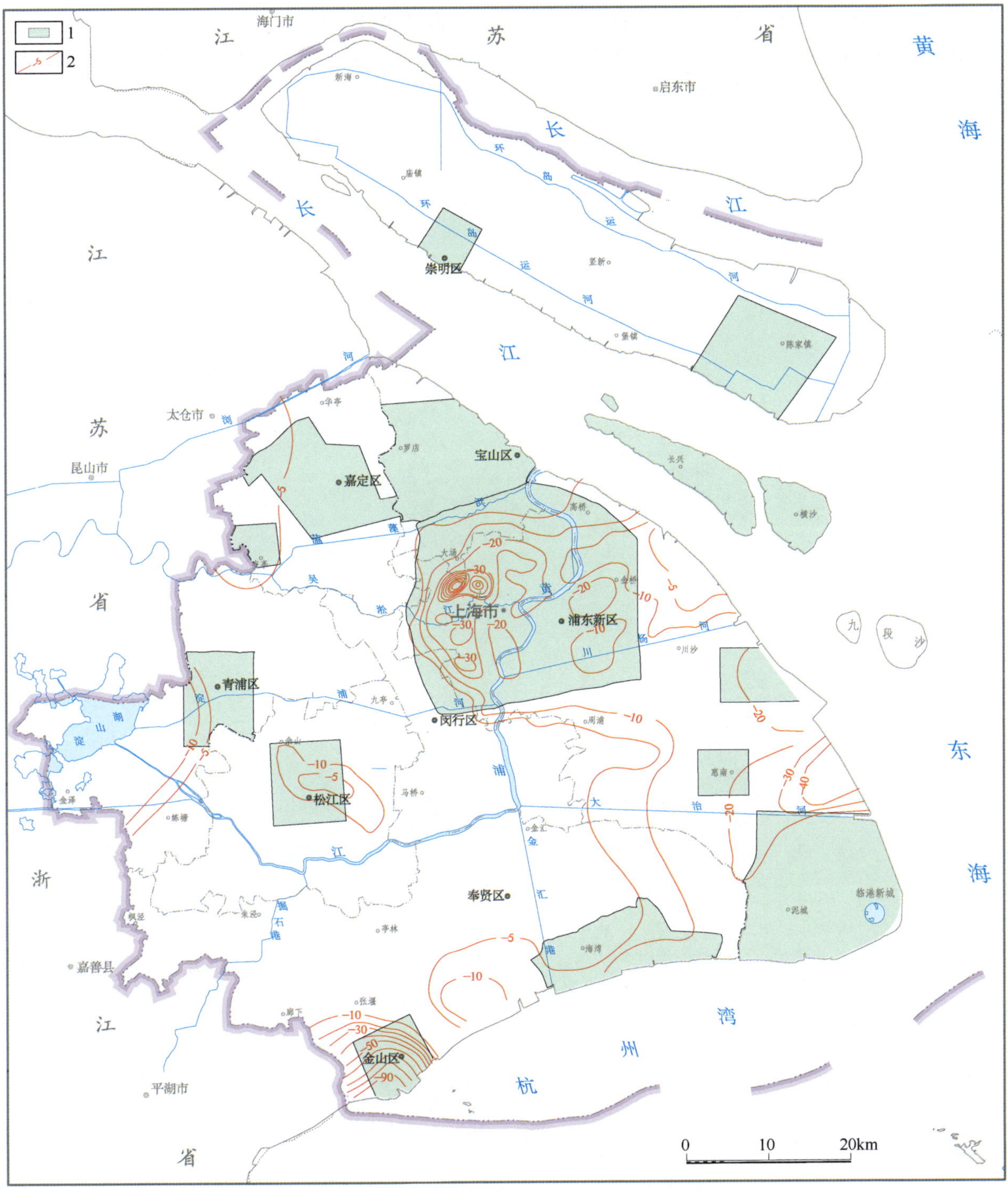

图 32-22　上海地下水应急水源地启用 3 个月后累计地面沉降量预测图

Fig. 32-22　Predicted cumulative land subsidence 3 months after the launch of the emergency groundwater sources in Shanghai

1—应急水源地；2—累计沉降量等值线（mm）

中心城应急水源地因应急开采量为其他水源地的 10 倍，产生的地面沉降现象比较明显，累计沉降量总体在 10~30mm，并且随着时间的推移，呈微量下降态势。

金山新城应急水源地，因含水层水文地质条件相对较差，且其东部基岩隆起含水层侧向补给条件明显削弱，实施开采后，在周边补给条件比较差的情况下，较易产生地下水位的下降，及由此引发的地面沉降现象也较明显，累计沉降量总体大于 30mm/3 个月。

惠南新城、临港新城、松江新城、青浦新城、浦东国际机场、奉贤海滨农场区，有微量的地面沉降发生，一般小于 10mm/3 个月，并随着时间的推移，下降趋势不甚明显。

崇明城桥新城、陈家镇新城区、长兴岛、横沙岛水源地，宝山新城、嘉定安亭新城水源地，地面沉降现象总体不甚明显，基本小于 10mm/3 个月。

第三十三章　地下空间资源

地下空间是城市发展的一个重要部分，与全国其他城市相比，上海的城市地下空间开发利用起步较早。上海城市地下空间开发建设与利用始于1862年的市政管线工程，先后经历了如下3个开发利用阶段。

一、1949年以前——以箱形基础为特征的地下空间开发

20世纪20—30年代，随着多层和高层建筑的出现，作为基础形式的箱形地下室，或打桩再加半层或一层地下室相继在上海出现，如上海第一百货商店、国际饭店、上海大厦、和平饭店、中国银行等。上海是中国近代城市建设中开发利用软土地下空间最有代表性的城市。

二、1949—1990年——平站结合的人防地下空间开发

新中国成立以来，随着上海城市建设的不断发展，地下空间资源的开发利用也得到了快速发展。这一阶段主要发展建设的领域是人防地下工程。1949—1990年间，上海城市地下空间开发利用经历了两次发展高潮。第一次是20世纪60年代初的“深挖洞、广积粮、不称霸”全国全民大搞“防空洞”运动；第二次是80年代中期的全国人防“厦门会议”，以突出“人民防空工程要与城市建设相结合”的平战结合运动。这两次运动，不仅全面推进了上海城市地下空间开发，使得平战结合的人防地下空间超过$100\times10^4m^2$，而且解决了城市旅馆、商业、仓储、停车等设施严重不足的问题，并为社会经济效益和战备做出了积极贡献，从而转变了人们对地下空间资源价值的认识，为后续大规模开发建设打下了良好的基础。

三、1990年以后——以地下轨道交通设施为典型代表的地下空间开发

进入20世纪90年代，浦东的改革开放，给上海带来了千载难逢的发展机遇。城市的发展，也带动了地下轨道交通建设的发展。1993年，上海市中心城地铁1号线的建成，标志着上海已经进入了大规模开发利用地下空间的时代。地下轨道交通设施的建设，不仅其本身的主要设施构筑在地下，而且由于给沿线创造了安全、舒适、便捷的交通条件，使得沿线房地产业得到了飞速发展。尤其是对以轨道交通场站为核心的地下综合体的开发建设，大力推进了上海空间利用立体化的进程。

第一节　地下空间资源利用概况

一、开发利用现状

根据上海市地下空间管理联席会议办公室统计资料，截至2016年，全市已建成地下工程36809个，总建筑面积为$8186\times10^4m^2$，比2011年前增加建筑面积$2487\times10^4m^2$，地下空间开发量以每年约$1000\times10^4m^2$的速度增长。其中，地下生产、生活服务设施占地下工程总量的95%；地下公共基础设施占3%；轨道交通及附属设施占2%；上海地下空间开发规模已位居国际大城市前列。

21世纪初以来，上海地下空间开发数量和规模逐年增加并呈现爆炸性的增长态势，地铁、越江隧道、地下道路、城市地下综合体、大型深埋地下市政设施等地下工程不断增多。

城市轨道交通。截至“十二五”末，上海城市轨道交通网络运行线路总数达14条，总里程逾600km，地铁车站约370座，换乘站51座，城市轨道交通基本网络已全面建成。

地下综合体。以轨道交通站点为重要载体，以提升城市公共活动中心空间品质为目的，形成了以人民广场、静安寺、世博园、徐家汇、花木和五角场副中心等为代表的一批多功能、大规模地下综合体。

地下综合交通枢纽和道路。虹桥枢纽、浦东国际机场、上海火车站、上海南站、十六铺公共交通枢纽等已基本建成。外滩通道、北横通道等地下道路和上中路、西藏南路等越江隧道的建成，为优化城市交通提供了新的渠道。结合绿地、广场、公园建成的地下车库缓解了中心城停车资源紧张的问题。

地下市政设施。市政管线是目前占用城市道路浅层地下空间的主要设施，已形成相当规模。专业管沟和综合管廊建设也正处于探索和推进阶段，已经建成了浦东张杨路综合管廊、松江新城综合管廊、嘉定安亭新镇综合管廊和世博重点配套项目西藏路电力电缆隧道；其他如泵站、变电站等设施也已向地下发展。

民防工程。建成了一批指挥工程、医疗救护工程、防空专业队工程和大型人员掩蔽部等骨干民防工程，进一步完善了民防工程布局和功能配套。

其他特殊工程。据了解，近期上海深层地下空间典型工程主要有苏州河段深层排水调蓄管道系统工程及硬 X 射线自由电子激光装置等工程。

二、开发利用特征

根据上海市地质调查研究院收集和调查的不完全统计资料，全市地下空间利用总体上具有以下明显特征：

从区域分布上看，地下空间开发与土地价值呈现较强的相关性。环线内外地下空间开发规模和功能差异较大，中心城的静安和黄浦地下空间开发强度最大，而郊区的嘉定、青浦、松江、奉贤、崇明开发强度较小。但随着新城的逐步发展，中心城以外的地下空间也将成为今后的开发重点。

在开发深度上，地下空间开发呈现较明显的分层利用特点。目前开发深度主要集中在中浅层（50m 以浅），其中，基坑工程埋深一般小于 30m，隧道工程埋深多在 20~40m。此外，建筑桩基础端部大多在地下 30~70m 之间，在一定程度上占用了地下空间资源。

从开发利用趋势分析，上海地下空间开发已从浅层、中层向深层发展。上海深层地下空间开发深度已超过 50m，如苏州河段深层排水调蓄管道系统（试验段）工程盾构工作井基坑开挖深度超过 60m，围护结构地下连续墙插入深度已达 98~105m，隧道埋深约为 60m；硬 X 射线自由电子激光装置工程盾构工作井基坑开挖深度约为 50m，围护结构地下连续墙插入深度约为 100m；轨交四号线修复工程基坑开挖深度达 41.5m，地下连续墙深度已达 65.5m；上海长江隧桥工程隧道底部埋深最大达江底（55m），正在建设的北横通道工程深度已达 48m，深度大于 40m 的路段长 2.6km。可见，上海深层地下空间（50m 以下）的利用实际上已经进入了起步阶段。

第二节　地下空间资源开发利用影响因素

一、地下水

上海地下水位埋深浅，各种地下工程不但在施工过程中受到地下水的影响，而且建成后还将长期位于地下水位以下（严学新，2004），因此，必须重视地下水对地下空间开发的影响问题。在上海发生的地下工程事故中，如隧道掘进和基坑开挖时的流砂、坍塌、滑坡、坑底涌土等地质灾害的发生，除与土性、施工方法、设计缺陷等有关外，地下水的作用不容忽视。因此，应根据不同地下工程的特点，分析评价地下水的不良影响，并采取必要的防治措施。

（一）地下水水头的不利影响

通过前述潜水含水层、微承压含水层、第Ⅰ承压含水层水文地质条件分析，3 个含水层分别埋藏于地表下不同的深度，从上到下依次介于地下空间分层布局所规划的相应深度范围内，即潜水含水层埋藏深度基本介于0~15m，微承压含水层埋藏深度基本介于15~40m，第Ⅰ承压含水层埋藏深度基本介于36m以下。不同深度的开发空间均处于饱水的含水层中。

上海以往地下空间开发经验表明，在地下水位较高的地区开挖基坑、隧道时，含水层被切割，若设计不当或采取措施不当，地下水进入基坑，容易造成流砂、边坡失稳、地基承载力下降等不良现象，易产生安全事故；基坑开挖至坡底后，由于上部土体的卸载，坑底被承压水顶破而发生涌砂、隆起的危险。有的内部基坑土体发生较大的位移，影响邻近建筑物的安全等异常情况。

地下空间开发利用过程中，地下水为导致基坑、隧道不稳定的最主要的影响因素之一，而粉性土及砂土层是主要含水层，水量丰富，黏性土层为相对隔水层，水量小，地下空间开发过程主要控制粉性土、砂土层中的地下水。

由于各含水层埋藏于地表下不同深度，而不同深度含水层地下水有其自身的水头压力。当不同深度的基坑及隧道开挖过程中，坑底将受到不同类型地下水水头压力的影响。以往经验表明，一般基坑深度小于10m时，主要受潜水地下水影响，当基坑深度大于10m时，除潜水外，还受承压含水层地下水影响，如果坑底所受承压水浮托力大于上覆隔水层的土压力时，说明基坑不稳定，有产生突涌的可能，应降低承压水水头，降低高度应满足设计、施工要求。

中部粉、砂土（第$⑤_2$层）及下部粉、砂土层（⑦层）均为承压含水层，当基坑开挖超过一定深度，坑底下地基土不足以抵抗承压水水头压力时，则有基坑突涌的可能，应降低承压水水头，降低高度应满足设计、施工要求。另外地铁、隧道施工时，当涉及承压含水层时，由于水头具有承压水，如盾构机土舱压力小于承压水水头压力时，则会发生喷水、冒砂事件，将造成很大危害，因此应充分重视。

（二）地下水质量的不利影响

上海中心城潜水含水层地下水中氯离子含量小于5000mg/L，硫酸根含量小于500mg/L，矿化度小于2×10^4mg/L，表明该含水层地下水对混凝土无腐蚀性。第Ⅰ承压含水层地下水对混凝土无腐蚀性，仅在个别点（杨浦明园新村）对钢筋混凝土结构中钢筋为弱腐蚀性外，其余皆为无腐蚀性，对钢结构皆具有中腐蚀性。第Ⅱ承压含水层地下水对混凝土无腐蚀性，对钢筋混凝土结构中钢筋无腐蚀性，中心城的东部、东北部、北部和西北部地区对钢结构具有弱腐蚀性（除个别地区如宝山大场与汶水路），其他地区则为中腐蚀性。

（三）地下水对地下空间开发利用影响分析

对上海地下空间影响最大的是潜水、微承压水及第Ⅰ承压含水层中的承压水。地下水的存在与运移，直接影响地基土物理力学性质的变化，从而会引发一系列的工程问题。地下水对地下空间开发影响主要体现在以下几个方面：

1. 地下水对地下工程的浮力作用

地下水以下的土体和结构物要受到垂直于表面的静水压力所产生的浮力作用。上海地下水位高，一般潜水的高水位埋深平均为地表下0.5m，当地下结构体的底板所受的地下水浮力大于结构荷载时，地下室将会浮起。上海曾发生多起地下水池在平常使用时情况良好，当把水池中水抽干维修时，由于降低了结构荷载，出现水池浮力增大的情况。水底地下工程（如过江隧道）不仅要考虑水土压力大小以便设计衬砌强度，而且还需设计合理的土下埋置深度，以利于用上部土体重力平衡上浮力，防止整个结构物上浮。

2. 施工中地下水的渗透破坏

地下水渗流沿渗透方向产生作用于岩土体骨架上的渗透力。土体在渗透力作用下其颗粒发生移动或颗粒成分及骨架结构发生改变的现象称为渗透破坏，渗透破坏有管涌和流土（砂）两种形式。渗透破坏的发生与渗透力大小及土的颗粒级配、密实度、渗透性等条件有关。

3. 地下水对软弱结构面的影响

地下水运动将软弱结构面中的物质软化或泥化，使结构面的抗剪强度降低。动静水压力抵消了部分法向应力，导致抗滑摩阻力减少。在自重作用下，促使有滑塌趋势的土体沿软弱结构面滑塌。

4. 施工中的涌水和突水

在地下隧道的施工过程中，如防水不当或施工不当，很容易产生涌水或突水事故，对地下工程施工造成影响（表 33-1）。

5. 地下水化学性质对地下空间建筑物的影响

当地下水中含有腐蚀性物质时，将影响衬砌、隔水材料的长期安全性。软水及 pH 为 6.5~7.0 的水对混凝土也是有害的，因为这些水中所溶解的石灰质少，当其流经混凝土裂缝时，将溶解混凝土中的石灰质，使混凝土失去骨料，加大原有裂缝。黏性土中的硫化铁，如果暴露于空气中，经地下水作用后，将分解为硫酸，腐蚀破坏建材及设备，所以土层中含有硫化铁时，开挖后应尽快衬砌，以免暴露在空气中大量分解。若地下水中溶解有有毒气体，对施工中的人员安全会有很大威胁。

6. 地下水埋藏条件对地下工程的影响

上层滞水接近地表且一般水量不大，地下工程施工中应优先考虑采用排水方法处理。

潜水对结构物的稳定和工程施工都有影响，工程选址应选潜水水位深的地带，尽量避免水下施工，当无法避免水下施工时，宜用排水、降低水位、隔离等措施减少潜水对施工的影响。

承压水由于它的压力影响，若地基内存在承压水，开挖基坑时可能会使坑底土层发生隆起甚至破坏。承压水水头高、水量大，施工中若揭穿了承压含水层后再实施排水将十分困难。

表 33-1 涌水和突水的危害性

Table 33-1 Hazards of water inflow and water inrush

项别	产生条件和原因
工作面土体崩溃，埋没隧道	接近涌水带土体自稳定性差，一旦涌水容易坍塌
施工现场积水，淹没设备	开挖引起含水层或含水带被揭穿，发生较大的集中涌水，水量多，流速大
产生地面沉降、地裂缝	大量地表水、地下水向开挖采空地段排泄，水位迅速下降，自重应力、真空吸蚀、冲蚀作用造成土层损失

二、软土

（一）软土层对地下空间开发的不利影响

软土层③、④层以及$⑤_1$层具有高含水量、大孔隙比、低强度、高压缩性等性质，而且还具有低渗透性、触变性和流变性等不良工程特点，在地下空间开发中，软土层对地下空间开发不利影响如下：

1）在外荷作用下压缩变形量大，易产生较大的沉降和不均匀沉降。在上海中心城进行地下空间开发，大部分区域涉及软土层，软土层厚度越大，压缩性越高，沉降量越大。

中心城$Ⅱ_2$、$Ⅱ_4$工程地质结构区软土层埋藏浅、厚度大，工程建设引起的地基变形量要大于其他工程地质结构区。

2）抗剪强度低，基坑开挖时坑边难自立，从而导致边坡失稳。软土层抗剪强度低，基坑开挖坑壁难以直立，因此基坑超过 3m 一般需采取围护措施，受开挖深度及周围环境要求，需设置一道或多道支撑。

3）具有明显的流变特征。对于基坑工程，开挖时易产生侧向变形和剪切破坏，导致支护结构变形或边坡失稳；对于隧道工程，软土流变及次固结变形会导致隧道纵向和横向的长期缓慢变形，且变形收敛时间长。

4）具有明显的触变特性，若土体受到扰动或震动，影响土体结构，会使强度骤然降低，导致土体沉降或滑动，使地下空间施工的安全度降低。

（二）地下空间开发对软土层的不利影响

基坑工程。为确保地下工程顺利进行，需进行打桩、围护以及开挖等工序。如采用预制桩，则将使周围土体产生较大的孔隙水压力，孔隙水压力积聚以及消散的过程，将引起软土强度降低，继而逐渐恢复的过程。基坑开挖，打破原有土体的应力平衡，土体向挖土的一侧位移，造成软土应力重分布，强度降低，待重新固结后强度再逐渐回升。

区间隧道掘进。目前大多采用盾构掘进，盾构机掘进过程中对周围土体将产生一定的扰动，另外隧道管片外还留有一定的孔隙。隧道掘进过程中，将造成土体强度降低以及重新固结。

总之，地下空间开发不可避免地使软土层应力重新分布，强度下降，但随着时间的推移，软土层将逐步固结沉降，强度逐渐提高。

三、砂、粉性土

（一）粉性土、砂土对地下空间开发的不利影响

从上海工程实践看，在粉性土、砂性土分布地区，地下空间开发过程中遇到的工程事故较多，主要为流砂管涌、地基液化以及承压水突涌 3 大类事故。

（1）流砂、管涌问题

上海地下空间开发，经常需要通过降水方式使地下水位达到设计安全施工标高以下，这样就形成了水头差，使地下水的渗流速度加快，这种渗流又往往是竖向的。竖向渗流的作用使土的有效重度发生了变化，向下渗流使有效应力增加，导致土体的渗透压密，增加土的抗剪强度；向上渗流则使有效应力减小，导致土的密实度和强度降低。如果水力梯度增加到使有效应力减少为零的某一数值，此时土中有效应力消失，土粒悬浮，这种渗透液化现象称为流砂。

上海地下水位高，绝大部分基坑开挖均需降水，使坑外地下水位高于坑内，围护墙后侧地下水向下渗流，绕过墙趾后地下水将向上渗流，在地基土中产生自下而上的渗透力。当水头差增大而使水力梯度达到临危梯度时，就会出现流砂现象。当土中渗流的水力梯度小于临界水力梯度时，虽不致诱发流砂现象，但土中细小颗粒仍有可能穿过粗颗粒之间的孔隙被渗流挟带而去，时间长了，在土层中将形成管状空洞，使土体强度降低，压缩性增大，这种现象称为“机械潜蚀”或“管涌”。流砂、管涌形成初期往往不易被发现，随着砂土中细颗粒土不断被带走，在地下形成越来越大的空洞，一般直到地表出现塌方或在坑底出现大量涌砂、涌水才能被施工人员发现，此时已造成大量土体流失，如施工人员不当机立断采取堵漏措施，流砂、管涌形成的空洞将迅速扩大，并相互连通，造成不可估量的损失。

因此，在基坑工程、地铁车站、隧道工程、越江工程等地下空间开挖施工工程中，应特别重视砂土、粉性土层的流砂管涌问题。流砂发生时能造成大量的土体流动，引发滑坡、塌方及塌陷等地质灾害，使周围环境受到严重破坏，特别是当地下工程从江、河等地表水体以下穿越时，流砂造成的管涌可能使大量泥沙和地表水体涌入地下结构物，造成重大财产损失或人员伤亡，地下空间开发时应予以特别重视。

（2）砂土震动液化问题

饱和砂土或砂质粉土在地震力作用下，土中孔隙水压力逐渐上升，部分或完全抵消土骨架承担的有效压力，从而使土体承载力降低甚至完全丧失，发生地基液化。

目前，上海地下空间开发建设中主要包括地下车库、地下商场、地铁、隧道等，地下车库、商场、车站，其埋深 5~20m 不等，隧道最深处达 30m。在这深度范围内将可能涉及吴淞江故道饱和砂质粉土或粉砂和黄浦江两岸新近沉积的砂质粉土（俗称江滩土），上述两种土层在地震力作用下均可能发生液化，给地下空间开发带来不利影响。

根据国家及上海抗震设计规程，上海市区地震基本烈度为 7 度，需对地表下 20m 深度范围内（深

基础）饱和砂质粉土和粉砂进行液化判别。虽然城市地下建（构）筑物深埋地下，四周均有侧限，相对地上建（构）筑物而言，其抗震是有利的；但当建（构）筑物位于液化土层中，将涉及地震发生时地下构筑物地基承载力或侧向力丧失问题、地下构筑物底板地基不均匀沉降问题，上述问题如不采取有效措施，一旦发生较为严重的液化，将引发地下建（构）筑物开裂、渗水，影响安全及使用功能。

（二）地下空间开发对粉性土、砂土的不利影响

基坑工程。为确保地下工程顺利进行，需进行打桩、围护以及土方开挖等工序。如采用预制桩，则将使周围土体产生较大的孔隙水压力，孔隙水压力积聚以及消散的过程，将引起粉性土、砂土松砂挤密，重新固结后，砂土的强度有所提高，并能降低浅部液化土层的液化势。但基坑开挖打破原有土体的应力平衡，土体向挖土的一侧位移，造成粉性土及砂土应力重分布，土体变松，强度降低，待重新固结后强度再逐渐回升；另外地下空间开发需要降低地下水位，引起土体的固结沉降，但地下水抽取过程中，如围护及止水等工程措施不到位，会引发流砂、管涌，使砂土流失，形成空洞，继而土体坍塌，对原状土体造成极大的破坏，继而对周围环境造成极大的影响，并有可能引发次生灾害。

区间隧道掘进。目前大多采用盾构掘进，盾构机掘进过程中对周围土体将产生一定的扰动，另外隧道管片外还留有一定的孔隙。隧道掘进过程中，将造成土体的强度降低以及重新固结。当正常施工条件下，在粉性土及砂土中采取保护泥浆、二次注浆等措施，在粉土及砂土层中掘进引起的地面沉降较软土地基小。但如掘进过程中发生机械故障，则可能引发流砂或管涌，大量砂土及地下水会灌入隧道内，对粉性土及砂土造成极大的破坏。另一方面，如隧道衬砌管片渗水，会造成砂土缓慢流失，日积月累也可能造成粉土及砂土层的空洞或坍塌。

总之，地下空间开发在正常情况将使粉土层密实度有所改变，强度发生变化，但随着时间的推移，粉土及砂土层将逐步固结沉降，强度逐渐恢复。但如在地下空间开发过程中或使用阶段发生流砂、管涌现象，将对粉土及砂土造成极大的破坏性影响。

四、浅层气

（一）第四系浅层天然气地质特征

上海为新生代的沉降平原，以海相为主的第四纪地层由下而上可以划分为：下部滨海河口相沉积层，中部陆相沉积层，上部海相沉积层和近代沉积层 4 个部分；以及本市东部（钦公塘以东地区）和西部（淀山湖区、东与中部接界）3 个沉积古地理环境不同的地区。

上海第四系普遍发育着浅层天然气（吕少伟和唐益群，2000），整个第四系可分为：上部海相层、受中部陆相层控制的沉积层，以及下部滨海河口相沉积层，3 个含气层系。目前资料较多，了解得比较清楚的是上部海相层含气层系（图 33-1）。

（二）浅层天然气的分布特征

浅层天然气按埋深自上而下可划分为 3 个含气层，其分布特征简述如下：

1. 第一含气层

第一含气层是上海最主要的含气层，它与古地理沉积条件有着密切的关系。埋深在 8~30m，通常在 12~25m，分布最广泛，是对地下空间开发影响最大的含气层。主要有 2 个含气层位：20m 以浅的含气层一般呈交互状的扁豆体出现，以贝壳砂、贝壳砂层为主含气层，是本市埋藏最浅的含气层；25m 左右的含气层为砂透镜体。

中心城处于本市最大的浅层天然气分布区——东部滨海平原含气区。含气层埋藏最浅的小于 9m，最深的可达 30m 左右。含气层上部以扁豆体贝壳砂层为主，下部以砂透镜体为主，局部也有黏砂互层或粉质黏土层。

本含气层天然气成分以甲烷（CH_4）为主，可占总含量的 85%以上，余为二氧化碳（CO_2）等。浅层天然气初始压力在 0.08~0.25MPa，其中以交互状的贝壳砂、贝壳砂层为主呈扁豆体出现的含气

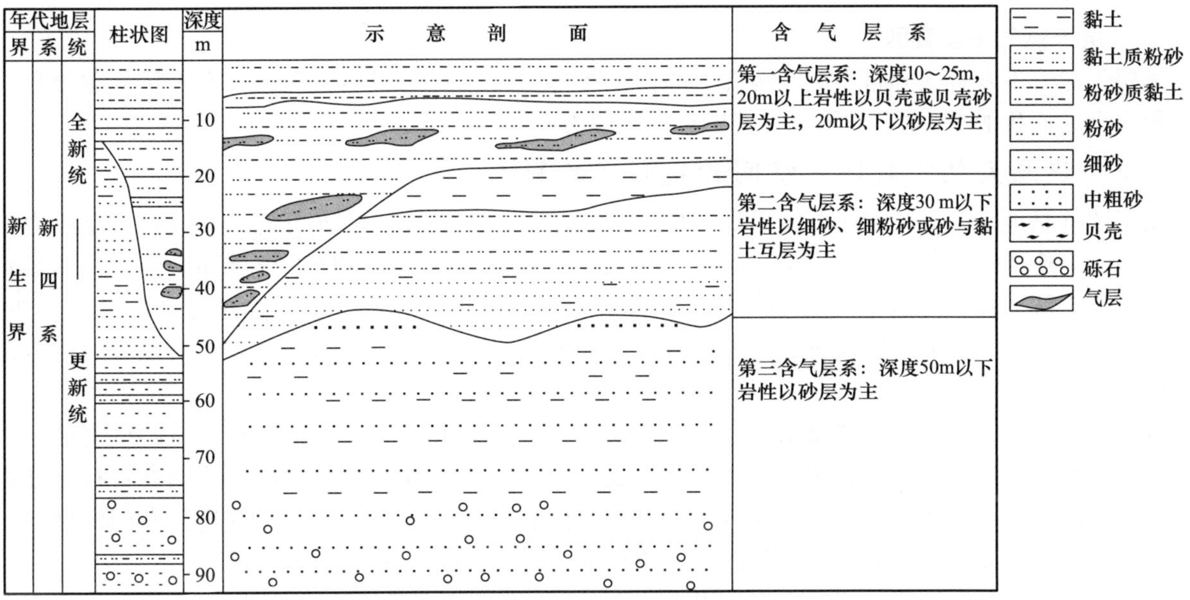

图 33-1　第四系含气层系示意剖面图

Fig. 33-1　Schematic section of the Quaternary strata bearing biogas

层初始压力在 0.08～0.12MPa；以砂透镜体为主出现的含气层初始压力一般在 0.10～0.20MPa，最大可达 0.25MPa，且气量较高，气流稳定；以黏砂互层为主出现的含气层初始压力一般在 0.15MPa；以粉质黏土层为主出现的含气层初始压力一般在 0.08～0.10MPa。

2. 第二含气层

第二含气层的分布范围较小，仅见于古河道地带。埋深一般在 30m 以下，海相层以充填的方式沉积嵌入陆相层内或与之相变。储气层主要为砂层或砂与黏土互层类型，一般呈透镜体出现。特点是干气层，压力较高，流量较大。

3. 第三含气层

第三含气层的分布范围较小，分布范围大致与古海岸线延伸方向一致。埋深较大（一般在 50m 以下，70m 左右），一般对地下空间的影响较小。

（三）浅层天然气发育程度分区

由于浅层天然气主要分布于东部的滨海平原区及河口砂岛区，所以本次只对这两区进行浅层天然气发育程度分区。浅层天然气发育程度分区以含气层的发育程度为主要原则进行划分，共划分为 5 个区：

1. 浅层天然气极发育区

浅层天然气极发育区分布于原南汇北部、宝山东部、长兴岛、中心城的龙华、漕河泾，3 个含气层均发育，气藏分布密集，含气面积在全区面积的 50%以上。

2. 浅层天然气发育区

浅层天然气发育区分布于浦东东部及崇明岛、长兴岛的部分地区，第一、第二含气层发育，其中埋深在 25m 左右的含气层最发育，本区气藏中 70%左右分布于该层。

3. 浅层天然气较发育区

浅层天然气较发育区分布于宝山北部、浦东合庆、蔡路及崇明北部地区，仅分布有第一含气层，10～20m 的气藏面积占全区面积的 15%左右，第二含气层有气显示，但不发育。

4. 浅层天然气较不发育区

浅层天然气较不发育区分布于闵行、黄浦江东侧、原南汇西部、奉贤东北部、宝山南部、嘉定东部及中心城。第一含气层中，10～20m 的气藏面积占全区面积的 10%左右，第二含气层有气显示，但不发育，呈零星分布。

5. 浅层天然气不发育区

浅层天然气不发育区分布于奉贤南部、原南汇东部，第一含气层不发育，仅有零星气藏分布，气藏面积占全区面积的5%左右，第二含气层在本区缺失。

（四）浅层天然气对地下空间开发利用的影响

浅层沼气是地下空间开发所可能遇到的地质灾害之一，由于浅层沼气可燃，并具有一定压力，工程若揭遇，会造成一定危险。特别是浅层沼气常赋存于浅层砂体中，气体逸出会加剧流砂，有经验教训表明，地下隧道作业时由于浅层沼气释放，将会造成下伏土层失稳，使已建隧道产生位移、断裂，造成无可挽回的重大损失。因而将其作为地质灾害种类加以分析评估是非常具有针对性，同时也是十分必要的。

浅层沼气对地下空间开发建设来说是一大地质灾害，对地下空间安全施工有严重的危害性，必须慎重对待，在含浅层沼气的土层中进行地下工程施工时，必须认真研究已有的地质勘察资料，分析在工程影响范围内是否存在浅层沼气，并对浅层气的成因、成分、分布及含气量等进行专门的勘察，然后提出处理方案和安全施工对策，如不注意，便有遭受浅层气危害的可能性。

浅层天然气对地下工程的影响主要有以下3个方面：①浅层天然气的存在会增加地下工程施工的难度；沼气的突然释放将威胁施工人员的安全，影响施工的质量，严重的甚至会造成停工、建（构）筑物的损毁、人员伤亡等事故；②已建成的建筑物下部若有浅层天然气存在，一旦天然气释放，会引起建筑物周围土体失稳，对建筑物造成毁灭性的破坏；③若天然气泄漏进地下建筑中，达到一定的浓度（一般超过1%人体会感到不适，超过5%有爆炸的危险）将威胁到建筑物中人员的生命安全。

第三节　中心城地下空间资源数量和质量

根据上海中心城地质条件及地下空间调查结果，结合上海城市2035年总体规划，区内地下空间资源可划分为浅层（0~15m）、中浅层（15~30m）、中层（30~50m）、深层适宜开发层（50~65m）、深层潜力开发层（65~100m）。

一、中心城地下空间资源数量

以上海外环线为边界，基于现有地质、地下空间数据，利用三维建模技术精细刻画地质结构及既有地下空间利用情况，并通过影响要素逐项排除法（地下空间资源的总蕴藏量中，排除地下隧道及其保护范围占用的地下空间、建筑物无桩基础及其保护范围占用的地下空间、建筑物桩基础及其保护范围占用的地下空间、城市道路和水体制约的空间等实际开发利用资源量），分别估算的浅层、中浅层、中层、深层适宜开发层及深层潜力开发层的地下空间资源数量（表33-2，表33-3）。

不同地下构筑物体积量算时，其保护区的设置原则为：无桩基础底面以下15m作为其保护距离；桩基础桩端以下取10m作为桩基础保护距离；城市道路取路面下5m作为受道路制约的空间；对于黄浦江其水面以下30m为制约空间，对于苏州河、川杨河、蕰藻浜均为12m，其他水系为8m；隧道上部至地表，隧道两侧取30m范围为其保护空间。

表33-2　平面各层次地下空间资源量统计

Table 33-2　Statistics of the horizontal underground space resources by depth interval

分层	总量面积/km^2	已开发利用面积/km^2	剩余可开发利用面积/km^2	剩余可开发利用面积占总量面积比例/%
0~15m	664	75.46	588.54	88.64
15~30m	664	26.79	637.21	95.97

续表

分层	总量面积/km^2	已开发利用面积/km^2	剩余可开发利用面积/km^2	剩余可开发利用面积占总量面积比例/%
30~50m	664	19.97	644.03	96.99
50~65m	664	5.76	658.24	99.13
65~100m	664	1.42	662.58	99.79

注：已开发利用面积不含道路占用面积。

表 33-3　100m 以浅地下空间体积计算表

Table 33-3　Calculations of the underground space volumes at levels shallower than 100m

分层	总量体积/m^3	已开发利用体积/m^3	影响及保护范围内体积/m^3	可充分开发利用量/m^3
0~15m	9.95×10^9	3.90×10^8	3.57×10^8	9.21×10^9
15~30m	9.95×10^9	2.95×10^8	3.95×10^8	9.26×10^9
30~50m	1.33×10^{10}	1.51×10^8	2.25×10^8	1.29×10^{10}
50~65m	9.95×10^9	3.06×10^7	4.33×10^7	9.88×10^9
65~100m	2.32×10^{10}	3.44×10^6	1.34×10^7	2.32×10^{10}
合计	6.64×10^{10}	8.70×10^8	1.03×10^9	6.44×10^{10}

二、中心城地下空间资源质量

在分析地下空间开发影响因素的基础上，建立质量评价指标体系，并采取层次分析法（AHP）和专家调查法获取各指标权重，构建评价模型，对中心城地下空间资源进行了分级评价。得到最终的分级评价结果并用三维属性建模技术将评价结果可视化。

（一）建立层次结构模型

选取城市地下空间资源质量为目标层（A），选择基岩条件、工程地质条件、水文地质条件、不良地质问题、地下空间开发现状等主要影响因素为准则层（B）层，将各因素中的若干影响因子作为指标层（X），建立层次结构模型如下（表 33-4）。

表 33-4　地下空间资源质量评价指标层次结构

Table 33-4　Hierarchy of the quality indexes for underground space resources

目标层（A）	准则层（B）	一级指标 X_j	二级指标 X_{jk}
地下空间资源质量 A	基岩条件 B1	基岩埋藏特征 x_1	基岩埋深 $x_{1,1}$
			基岩面起伏 $x_{1,2}$
		基底稳定性 x_2	活动断裂 $x_{2,1}$
			地震危险性 $x_{2,2}$
	工程地质条件 B2	土体埋藏特征 x_3	土层厚度 $x_{3,1}$
			土层面起伏 $x_{3,2}$
		土体物理性质 x_4	含水量 $x_{4,1}$
			压缩系数 $x_{4,2}$
			压缩模量 $x_{4,3}$
			黏聚力 $x_{4,4}$
			内摩擦角 $x_{4,5}$

目标层（A）	准则层（B）	一级指标 X_j	二级指标 X_{jk}
地下空间资源质量 A	水文地质条件 B3	含水层特征 x_5	沟通性 $x_{5,1}$
		水文地质参数 x_6	渗透系数 $x_{6,1}$
			单井涌水量 $x_{6,2}$
			承压水头 $x_{6,3}$
			地下水腐蚀性 $x_{6,4}$
	不良地质问题 B4	古河道 x_7	发育深度 $x_{7,1}$
		砂土液化 x_8	砂土液化程度 $x_{8,1}$
		地面沉降 x_9	地面沉降风险分区 $x_{9,1}$
			年沉降速率 $x_{9,2}$
		浅层气 x_{10}	浅层气分布 $x_{10,1}$
	地下空间开发现状 B5	地下空间开发利用情况 x_{11}	开发类型 $x_{11,1}$
			开发深度 $x_{11,2}$
			开发程度 $x_{11,3}$
	其他 B6	应急水源地 x_{12}	应急水源地分布 $x_{12,1}$
		浅层地温能 x_{13}	浅层地温能分布 $x_{13,1}$
		地下固体矿产 x_{14}	地下固体矿产分布 $x_{14,1}$
		地质遗迹 x_{15}	地质遗迹分布 $x_{15,1}$

（二）评价指标量化标准

根据各指标的作用性质及表达方式，分别按照以下原则进行指标量化处理：对于有国家标准的指标，按照国家标准进行等级划分；对于可度量的指标，将指标分为 4 个等级（表 33-5）；对于无法直接量化的定性指标，采用专家评分法来确定等级。

表 33-5 地下空间资源评估指标量化标准

Table 33-5 Quantification standard of the quality indexes for groundwater space resources

指标类型	指标 X		评估标准与定性描述			
			一级（好）	二级（较好）	三级（一般）	四级（差）
			[1.0-0.8]	(0.8-0.6]	(0.6-0.3]	(0.3-0.0]
基岩埋藏特征	基岩埋深/m		<200	200～300	>300	
	基岩面起伏		平缓	一般	较大	大
基岩稳定性	活动断裂		不发育	轻微	一般	严重
	地震危险性		无	轻微	一般	严重
土体埋藏特征	土层厚度/m	软土	<5	5～15	>15	
		黏性土	>15	5～15	<5	
		砂土	<5	5～30	>30	
	土层面起伏		平缓	一般	较大	大
土体物理性质	含水量/%		<20	20～30	30～40	>40
	压缩系数		<0.1	0.1～0.25	0.25～0.5	>0.5
	压缩模量/MPa		>7	5～7	3～5	<3
	黏聚力/kPa		>40	25～40	15～25	<15
	内摩擦角/（°）		<5	5～20	20～30	>30

续表

指标类型	指标 X	评估标准与定性描述			
		一级（好）	二级（较好）	三级（一般）	四级（差）
		[1.0–0.8]	(0.8–0.6]	(0.6–0.3]	(0.3–0.0]
含水层特征	沟通性	不沟通	Ⅰ+Ⅱ	Ⅱ+Ⅲ	Ⅰ+Ⅱ+Ⅲ
地下水特征	渗透系数	$<e^{-7}$	$e^{-7}\sim e^{-6}$	$e^{-6}\sim e^{-5}$	$>e^{-5}$
	单井涌水量/（$m^3\cdot d^{-1}$）	<500	500~1000	1000~3000	>3000
	承压水头高度/m	<20	20~30	30~40	>40
	地下水腐蚀性	微腐蚀	弱腐蚀	中腐蚀	强腐蚀
古河道	古河道发育深度/m	<5	5~15	15~25	>25
砂土液化	砂土液化等级	不液化	轻微	中等	严重
地面沉降	地面沉降危险分区	高	较高	中等	较低
	年沉降速率/（$mm\cdot a^{-1}$）	<10	10~30	30~50	>50
浅层气	浅层气分布	否	—	—	是
地下空间开发利用情况	开发类型	未开发	无桩基础	有桩基础	隧道
	开发深度/m	0~5	5~15	15~40	>40
	开发强度	很小	一般	较大	很大
应急水源地	应急水源地分布	否	—	—	是
浅层地温能	浅层地温能分布	否	—	—	是
地下固体矿产	地下固体矿产分布	否	—	—	是
地质遗迹	地质遗迹分布	否	—	—	是

（三）权重计算方法

采取层次分析法（AHP）和专家调查法来获取指标权重（表33-6~表33-10）。浅层与中浅层地下空间开发强度及规模相对较大，其对地下空间开发的影响程度也较大，故占有较大权重；而随着深度增加，工程地质条件、基岩条件、不良地质问题，尤其水文地质条件对地下空间开发的影响程度愈发明显，相对应的指标权重也较大。

表33-6　浅层（0~15m）地下空间资源质量评价各指标权重表

Table 33-6　Weights of the quality indexes for underground space resources at shallow levels (0–15 m)

准则层（B）	权重 ω_i	一级指标 X_j	内部权重 ω_j	二级指标 X_{jk}	内部权重 ω_{jk}	权重总序 ω_{Ajk}
基岩条件 B1	0.04	基岩埋藏特征 x_1	0.70	基岩埋深 $x_{1,1}$	0.42	0.0118
				基岩面起伏 $x_{1,2}$	0.58	0.0162
		基岩稳定性 x_2	0.30	活动断裂 $x_{2,1}$	0.50	0.0060
				地震危险性 $x_{2,2}$	0.50	0.0060
工程地质条件 B2	0.28	土体埋藏特征 x_3	0.40	土层厚度 $x_{3,1}$	0.66	0.0581
				土层面起伏 $x_{3,2}$	0.34	0.0299
		土体物理性质 x_4	0.60	含水量 $x_{4,1}$	0.20	0.0264
				压缩系数 $x_{4,2}$	0.20	0.0264
				压缩模量 $x_{4,3}$	0.20	0.0264
				黏聚力 $x_{4,4}$	0.20	0.0264
				内摩擦角 $x_{4,5}$	0.20	0.0264

续表

准则层（B）	权重 ω_i	一级指标 X_j	内部权重 ω_j	二级指标 X_{jk}	内部权重 ω_{jk}	权重总序 ω_{Ajk}
水文地质条件B3	0.08	含水层特征 x_5	0.35	沟通性 $x_{5,1}$	1.0	0.049
		水文地质参数 x_6	0.65	渗透系数 $x_{6,1}$	0.35	0.0319
				单井涌水量 $x_{6,2}$	0.33	0.0300
				承压水头 $x_{6,3}$	0.22	0.0201
				地下水腐蚀性 $x_{6,4}$	0.10	0.0091
不良地质问题B4	0.17	古河道 x_7	0.25	发育深度 $x_{7,1}$	1.0	0.0425
		砂土液化 x_8	0.20	砂土液化程度 $x_{8,1}$	1.0	0.0340
		地面沉降 x_9	0.40	地面沉降风险分区 $x_{9,1}$	0.55	0.0374
				年沉降速率 $x_{9,2}$	0.45	0.0306
		浅层气 x_{10}	0.15	浅层气分布 $x_{10,1}$	1.0	0.0255
地下空间开发现状B5	0.39	地下空间开发利用情况 x_{11}	1.0	开发类型 $x_{11,1}$	0.2	0.0780
				开发深度 $x_{11,2}$	0.4	0.1560
				开发程度 $x_{11,3}$	0.4	0.1560
其他B6	0.04	应急水源地 x_{12}	0.25	应急水源地分布 $x_{12,1}$	1.0	0.0100
		浅层地温能 x_{13}	0.25	浅层地温能分布 $x_{13,1}$	1.0	0.0100
		地下固体矿产 x_{14}	0.25	地下固体矿产分布 $x_{14,1}$	1.0	0.0100
		地质遗迹 x_{15}	0.25	地质遗迹分布 $x_{15,1}$	1.0	0.0100

表 33-7 中层（15~30m）地下空间资源质量评价各指标权重表

Table 33-7 Weights of the quality indexes for underground space resources at medium levels (15-30 m)

准则层（B）	权重 ω_i	一级指标 X_j	内部权重 ω_j	二级指标 X_{jk}	内部权重 ω_{jk}	权重总序 ω_{Ajk}
基岩条件B1	0.042	基岩埋藏特征 x_1	0.70	基岩埋深 $x_{1,1}$	0.40	0.0118
				基岩面起伏 $x_{1,2}$	0.60	0.0176
		基岩稳定性 x_2	0.30	活动断裂 $x_{2,1}$	0.50	0.0063
				地震危险性 $x_{2,2}$	0.50	0.0063
工程地质条件B2	0.208	土体埋藏特征 x_3	0.40	土层厚度 $x_{3,1}$	0.68	0.0566
				土层面起伏 $x_{3,2}$	0.32	0.0266
		土体物理性质 x_4	0.60	含水量 $x_{4,1}$	0.20	0.025
				压缩系数 $x_{4,2}$	0.20	0.0249
				压缩模量 $x_{4,3}$	0.20	0.025
				黏聚力 $x_{4,4}$	0.20	0.0249
				内摩擦角 $x_{4,5}$	0.20	0.025
水文地质条件B3	0.208	含水层特征 x_5	0.42	沟通性 $x_{5,1}$	1.0	0.0874
		水文地质参数 x_6	0.58	渗透系数 $x_{6,1}$	0.35	0.0422
				单井涌水量 $x_{6,2}$	0.22	0.0265
				承压水头 $x_{6,3}$	0.33	0.0398
				地下水腐蚀性 $x_{6,4}$	0.1	0.0121

续表

准则层（B）	权重 ω_i	一级指标 X_j	内部权重 ω_j	二级指标 X_{jk}	内部权重 ω_{jk}	权重总序 ω_{Ajk}
不良地质问题 B4	0.167	古河道 x_7	0.27	发育深度 $x_{7,1}$	1.0	0.0451
		砂土液化 x_8	0.17	砂土液化指数 $x_{8,1}$	1.0	0.0284
		地面沉降 x_9	0.40	地面沉降风险分区 $x_{9,1}$	0.55	0.0367
				年沉降速率 $x_{9,2}$	0.45	0.0301
		浅层气 x_{10}	0.16	浅层气分布 $x_{10,1}$	1.0	0.0267
地下空间开发现状 B5	0.333	地下空间开发利用情况 x_{11}	1.0	开发类型 $x_{11,1}$	0.20	0.0666
				开发深度 $x_{11,2}$	0.40	0.1332
				开发程度 $x_{11,3}$	0.40	0.1332
其他 B6	0.042	应急水源地 x_{12}	0.25	应急水源地分布 $x_{12,1}$	1.0	0.0105
		浅层地温能 x_{13}	0.25	浅层地温能分布 $x_{13,1}$	1.0	0.0105
		地下固体矿产 x_{14}	0.25	地下固体矿产分布 $x_{14,1}$	1.0	0.0105
		地质遗迹 x_{15}	0.25	地质遗迹分布 $x_{15,1}$	1.0	0.0105

表 33-8 （30~50m）地下空间资源质量评价各指标权重表

Table 33-8 Weights of the quality indexes for underground space resources at 30-50 m

准则层（B）	权重 ω_i	一级指标 X_j	内部权重 ω_j	二级指标 X_{jk}	内部权重 ω_{jk}	权重总序 ω_{Ajk}
基岩条件 B1	0.043	基岩埋藏特征 x_1	0.70	基岩埋深 $x_{1,1}$	0.40	0.0120
				基岩面起伏 $x_{1,2}$	0.60	0.0181
		基岩稳定性 x_2	0.30	活动断裂 $x_{2,1}$	0.50	0.0065
				地震危险性 $x_{2,2}$	0.50	0.0065
工程地质条件 B2	0.255	土体埋藏特征 x_3	0.40	土层厚度 $x_{3,1}$	0.68	0.0694
				土层面起伏 $x_{3,2}$	0.32	0.0326
		土体物理性质 x_4	0.60	含水量 $x_{4,1}$	0.20	0.0306
				压缩系数 $x_{4,2}$	0.20	0.0306
				压缩模量 $x_{4,3}$	0.20	0.0306
				黏聚力 $x_{4,4}$	0.20	0.0306
				内摩擦角 $x_{4,5}$	0.20	0.0306
水文地质条件 B3	0.298	含水层特征 x_5	0.52	沟通性 $x_{5,1}$	1.0	0.155
		水文地质参数 x_6	0.48	渗透系数 $x_{6,1}$	0.35	0.0501
				单井涌水量 $x_{6,2}$	0.22	0.0315
				承压水头 $x_{6,3}$	0.33	0.0472
				地下水腐蚀性 $x_{6,4}$	0.1	0.0143
不良地质问题 B4	0.128	古河道 x_7	0.30	发育深度 $x_{7,1}$	1.0	0.0384
		砂土液化 x_8	0.15	砂土液化指数 $x_{8,1}$	1.0	0.0192
		地面沉降 x_9	0.40	地面沉降防治分区 $x_{9,1}$	0.55	0.0282
				年沉降速率 $x_{9,2}$	0.45	0.0230
		浅层气 x_{10}	0.15	浅层气分布 $x_{10,1}$	1.0	0.0192

续表

准则层（B）	权重 ω_i	一级指标 X_j	内部权重 ω_j	二级指标 X_{jk}	内部权重 ω_{jk}	权重总序 ω_{Ajk}
地下空间开发现状 B5	0.255	地下空间开发利用情况 x_{11}	1.0	开发类型 $x_{11,1}$	0.20	0.051
				开发深度 $x_{11,2}$	0.40	0.102
				开发程度 $x_{11,3}$	0.40	0.102
其他 B6	0.021	应急水源地 x_{12}	0.25	应急水源地分布 $x_{12,1}$	1.0	0.0052
		浅层地温能 x_{13}	0.25	浅层地温能分布 $x_{13,1}$	1.0	0.0052
		地下固体矿产 x_{14}	0.25	地下固体矿产分布 $x_{14,1}$	1.0	0.0052
		地质遗迹 x_{15}	0.25	地质遗迹分布 $x_{15,1}$	1.0	0.0052

表 33-9 （50~65m）地下空间资源质量评价各指标权重表

Table 33-9 Weights of the quality indexes for underground space resources at 50-65 m

准则层（B）	权重 ω_i	一级指标 X_j	内部权重 ω_j	二级指标 X_{jk}	内部权重 ω_{jk}	权重总序 ω_{Ajk}
基岩条件 B1	0.085	基岩埋藏特征 x_1	0.70	基岩埋深 $x_{1,1}$	0.40	0.0238
				基岩面起伏 $x_{1,2}$	0.60	0.0357
		基岩稳定性 x_2	0.30	活动断裂 $x_{2,1}$	0.50	0.0127
				地震危险性 $x_{2,2}$	0.50	0.0127
工程地质条件 B2	0.255	土体埋藏特征 x_3	0.40	土层厚度 $x_{3,1}$	0.70	0.0714
				土层面起伏 $x_{3,2}$	0.30	0.0306
		土体物理性质 x_4	0.60	含水量 $x_{4,1}$	0.20	0.0308
				压缩系数 $x_{4,2}$	0.20	0.0308
				压缩模量 $x_{4,3}$	0.20	0.0308
				黏聚力 $x_{4,4}$	0.20	0.0308
				内摩擦角 $x_{4,5}$	0.20	0.0308
水文地质条件 B3	0.340	含水层特征 x_5	0.60	沟通性 $x_{5,1}$	1.0	0.204
		水文地质参数 x_6	0.40	渗透系数 $x_{6,1}$	0.33	0.0449
				单井涌水量 $x_{6,2}$	0.22	0.0299
				承压水头 $x_{6,3}$	0.35	0.0477
				地下水腐蚀性 $x_{6,4}$	0.1	0.0136
不良地质问题 B4	0.170	古河道 x_7	0.33	发育深度 $x_{7,1}$	1.0	0.0561
		砂土液化 x_8	0.12	砂土液化指数 $x_{8,1}$	1.0	0.0204
		地面沉降 x_9	0.45	地面沉降风险分区 $x_{9,1}$	0.55	0.0421
				年沉降速率 $x_{9,2}$	0.45	0.0344
		浅层气 x_{10}	0.10	浅层气分布 $x_{10,1}$	1.0	0.017
地下空间开发现状 B5	0.128	地下空间开发利用情况 x_{11}	1.0	开发类型 $x_{11,1}$	0.20	0.0256
				开发深度 $x_{11,2}$	0.40	0.0512
				开发程度 $x_{11,3}$	0.40	0.0512
其他 B6	0.021	应急水源地 x_{12}	0.30	应急水源地分布 $x_{12,1}$	1.0	0.0063
		浅层地温能 x_{13}	0.30	浅层地温能分布 $x_{13,1}$	1.0	0.0063
		地下固体矿产 x_{14}	0.30	地下固体矿产分布 $x_{14,1}$	1.0	0.0063
		地质遗迹 x_{15}	0.10	地质遗迹分布 $x_{15,1}$	1.0	0.0021

表 33-10 （65~100m）地下空间资源质量评价各指标权重表

Table 33-10 Weights of the quality indexes for underground space resources at 65-100 m

准则层（B）	权重 ω_i	一级指标 X_j	内部权重 ω_j	二级指标 X_{jk}	内部权重 ω_{jk}	权重总序 ω_{Ajk}
基岩条件 B1	0.115	基岩埋藏特征 x_1	0.70	基岩埋深 $x_{1,1}$	0.40	0.0313
				基岩面起伏 $x_{1,2}$	0.60	0.0469
		基岩稳定性 x_2	0.30	活动断裂 $x_{2,1}$	0.50	0.0184
				地震危险性 $x_{2,2}$	0.50	0.0184
工程地质条件 B2	0.232	土体埋藏特征 x_3	0.40	土层厚度 $x_{3,1}$	0.70	0.0650
				土层面起伏 $x_{3,2}$	0.30	0.0278
		土体物理性质 x_4	0.60	含水量 $x_{4,1}$	0.20	0.0280
				压缩系数 $x_{4,2}$	0.20	0.0278
				压缩模量 $x_{4,3}$	0.20	0.0278
				黏聚力 $x_{4,4}$	0.20	0.0278
				内摩擦角 $x_{4,5}$	0.20	0.0278
水文地质条件 B3	0.346	含水层特征 x_5	0.70	沟通性 $x_{5,1}$	1.0	0.2422
		水文地质参数 x_6	0.30	渗透系数 $x_{6,1}$	0.33	0.0343
				单井涌水量 $x_{6,2}$	0.22	0.0228
				承压水头 $x_{6,3}$	0.35	0.0363
				地下水腐蚀性 $x_{6,4}$	0.1	0.0104
不良地质问题 B4	0.192	古河道 x_7	0.15	发育深度 $x_{7,1}$	1.0	0.0288
		砂土液化 x_8	0.05	砂土液化指数 $x_{8,1}$	1.0	0.0096
		地面沉降 x_9	0.75	地面沉降防治分区 $x_{9,1}$	0.55	0.0792
				年沉降速率 $x_{9,2}$	0.45	0.0648
		浅层气 x_{10}	0.05	浅层气分布 $x_{10,1}$	1.0	0.0096
地下空间开发现状 B5	0.077	地下空间开发利用情况 x_{11}	1.0	开发类型 $x_{11,1}$	0.20	0.0154
				开发深度 $x_{11,2}$	0.40	0.0308
				开发程度 $x_{11,3}$	0.40	0.0308
其他 B6	0.038	应急水源地 x_{12}	0.40	应急水源地分布 $x_{12,1}$	1.0	0.0152
		浅层地温能 x_{13}	0.40	浅层地温能分布 $x_{13,1}$	1.0	0.0152
		地下固体矿产 x_{14}	0.10	地下固体矿产分布 $x_{14,1}$	1.0	0.0038
		地质遗迹 x_{15}	0.10	地质遗迹分布 $x_{15,1}$	1.0	0.0038

（四）质量评估计算模型及分级

采用多因素综合评价法：

$$S = \sum_{h=1}^{p} \left[\sum_{j=1}^{m} \left(\sum_{i=1}^{n} A_i B_i \right) C_j \right] D_h$$

式中：S 为评估目标总得分；A_i 为第 i 个单项指标的量化分值；B_i 为第 i 个单项指标的权重；C_j 为第 j 个子主题的权重；D_h 为第 h 个主题的权重。

将评估得分按 4 级进行等级划分，划分为好、较好、一般和差（表 33-11）。

（五）地下空间资源质量评估结果

根据上述公式进行模型属性运算，所述评估模型进行综合叠加评估计算，取得上海中心城 100m

以浅地下空间资源质量等级评估结果（图 33-2~图 33-6）。

表 33-11 地下空间资源质量评估等级设置

Table 33-11 Quality assessment rating of underground space resources

质量评估等级标度	评估分值区间	质量评价标准
Ⅰ级	(0.75, 1]	好
Ⅱ级	(0.55, 0.75]	较好
Ⅲ级	(0.35, 0.55]	一般
Ⅳ级	(0, 0.35]	差

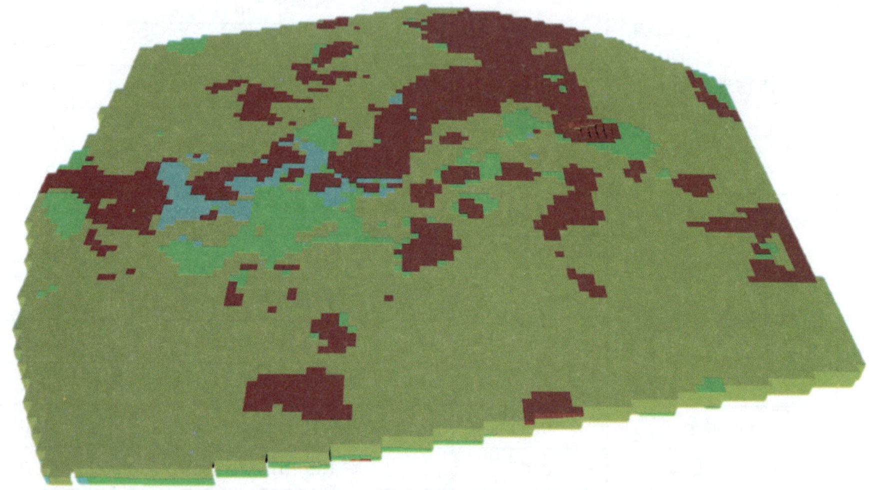

图 33-2 0~15m 地下空间资源质量评价结果

绿色—Ⅰ级；蓝色—Ⅱ级；黄色—Ⅲ级；红色—Ⅳ级

Fig. 33-2 Quality assessment result of underground space resources at 0-15m

green: grade Ⅰ; blue: grade Ⅱ; yellow: grade Ⅲ; red: grade Ⅳ

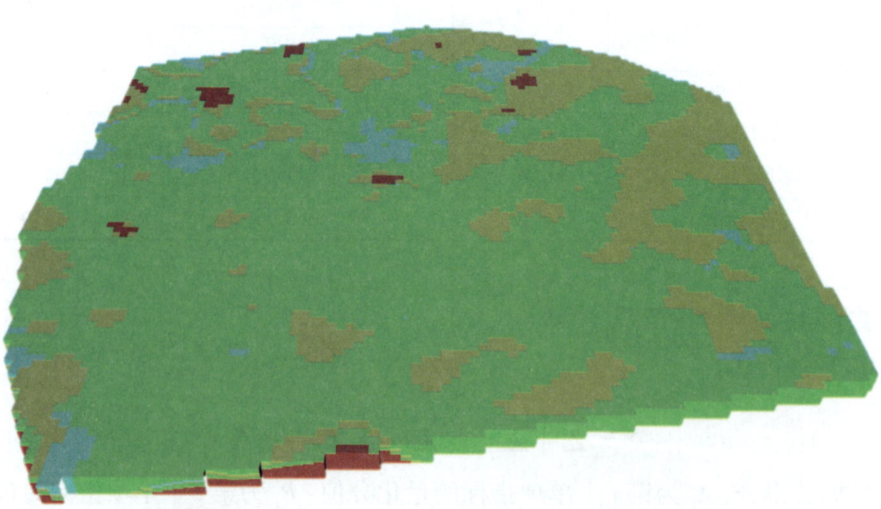

图 33-3 15~30m 地下空间资源质量评价结果模型

绿色—Ⅰ级；蓝色—Ⅱ级；黄色—Ⅲ级；红色—Ⅳ级

Fig. 33-3 Quality assessment result of underground space resources at 15-30 m

green: grade Ⅰ; blue: grade Ⅱ; yellow: grade Ⅲ; red: grade Ⅳ

图 33-4　30~50m 地下空间资源质量评价结果模型
绿色—Ⅰ级；蓝色—Ⅱ级；黄色—Ⅲ级；红色—Ⅳ级

Fig. 33-4　Quality assessment result of underground space resources at 30-50 m
green: grade Ⅰ; blue: grade Ⅱ; yellow: grade Ⅲ; red: grade Ⅳ

图 33-5　50~65m 地下空间资源质量评价结果模型
绿色—Ⅰ级；蓝色—Ⅱ级；黄色—Ⅲ级；红色—Ⅳ级

Fig. 33-5　Quality assessment result of underground space resources at 50-65 m
green: grade Ⅰ; blue: grade Ⅱ; yellow: grade Ⅲ; red: grade Ⅳ

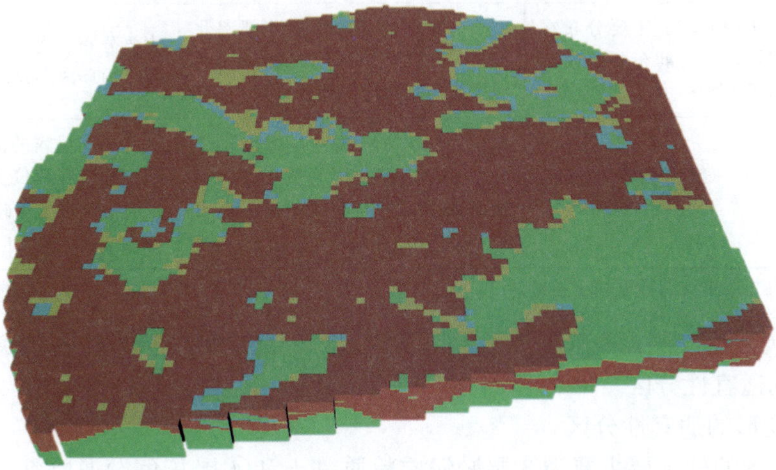

图 33-6　65~100m 地下空间资源质量评价结果模型
绿色—Ⅰ级；蓝色—Ⅱ级；黄色—Ⅲ级；红色—Ⅳ级

Fig. 33-6　Quality assessment result of underground space resources at 65-100 m
green: grade Ⅰ; blue: grade Ⅱ; yellow: grade Ⅲ; red: grade Ⅳ

第四节　中心城地下空间开发利用地质环境适宜性

中浅层地下空间开发利用适宜性主要根据地下水、软土和砂、粉性土等因素产生的基坑突涌问题、软土变形问题和流砂问题的严重程度，将中心城划分为问题轻微区（A）、中等区（B）和严重区（C）3类。深层地下空间优先开发层（40~65m）开发利用适宜性则以对深层地下空间开发影响较大的地下水和砂性土为主要影响因子，考虑地下空间开发现状、基岩埋深、地面沉降风险等因素进行适宜性综合分区评价，将中心城划分为较好、中等、较差、差4个等级。

一、中浅层地下空间开发利用适宜性分区

（一）基坑工程适宜性分区

对基坑影响较大的主要有砂、粉性土层（浅层砂性土（$②_3$）、中部砂性土（$⑤_2$）和下部砂层（⑦））、软土（③、④层）和地下水（主要是微承压水和第Ⅰ承压水），砂、粉性土层在基坑开挖中易产生流砂现象，致使边坡失稳和地面塌陷，地下水易引起基坑突涌，产生事故，软土则易产生变形，对周围建筑产生影响。针对以上3种问题对不同深度基坑分别进行适宜性分区。

1. 10m深基坑

对于中心城10m深基坑，开挖所揭露的土层一般为①层填土、$②_1$层褐黄—灰黄色黏性土、$②_3$层砂质粉土、③层淤泥质粉质黏土和④层淤泥质黏土。

（1）考虑砂、粉性土层产生的流砂问题的适宜性分区

基坑开挖深度范围内涉及的流砂层主要为$②_3$层，岩性以砂质粉土为主，局部为粉砂或黏质粉土，渗透系数较大，且中心城范围内变化不大，均易产生流砂现象。由于该层岩性、物理力学指标在中心城范围内变化不大，因此10m深基坑考虑流砂的适宜性分区则按照流砂层的埋藏分布特征进行分区：流砂问题轻微区无$②_3$层分布；中等区该层厚度一般小于等于6m；大于6m区则为流砂问题严重区（图33-7；表33-12）。

表33-12　上海中心城10m深基坑考虑流砂问题适宜性分区及评价

Table 33-12　Zoning and assessment of 10 m foundation pits in downtown Shanghai considering quicksand problems

分区	流砂问题轻微区（A区）	流砂问题中等区（B区）	流砂问题严重区（C区）
分区特征	无$②_3$层分布	$②_3$层厚度小于等于6m	$②_3$层厚度大于等于6m
分区范围	广泛分布于中心城，主要分布在浦西苏州河南部、浦东高桥南部及西南部部分地区	主要分布在浦西苏州河北部南北高架道路两侧，内环内黄浦江两侧及浦东高行局部亦有分布	主要分布在苏州河两侧及北部黄浦江两侧，内环内部分黄浦江两侧亦有分布
分区评价	基坑开挖时一般不会产生流砂现象	基坑开挖揭露的流砂层厚度小于6m，基坑底部土层为软土层，渗透系数低，为隔水层。应重视可能产生的流砂问题	基坑开挖揭露的流砂层厚度大于6m，大部分基坑底部位于透水层$②_3$层之上，施工时极易产生流砂问题，从而导致边坡失稳或地面塌陷，应特别重视流砂问题

注：③层淤泥质粉质黏土中多夹有砂质粉土透镜体$③_2$层，亦为流砂层，基坑开挖时应注意其可能产生的流砂问题。

由于10m深基坑开挖较浅，一般不涉及承压含水层，因此一般不会发生基坑突涌问题，因此不进行考虑地下水作用的适宜性分区。

（2）考虑软土变形的适宜性分区

10m深基坑所揭示的软土层主要为③层淤泥质粉质黏土和④层灰色淤泥质黏土，由于软土层具有含水量高、孔隙比大、压缩性高、强度低等不良工程地质特性，此外，该两层还具有流变和触变特性，在基坑开挖施工过程中，易引起边坡失稳。

10m深基坑考虑软土变形的适宜性分区主要依据软土层的埋藏分布特征进行划分：软土变形轻微

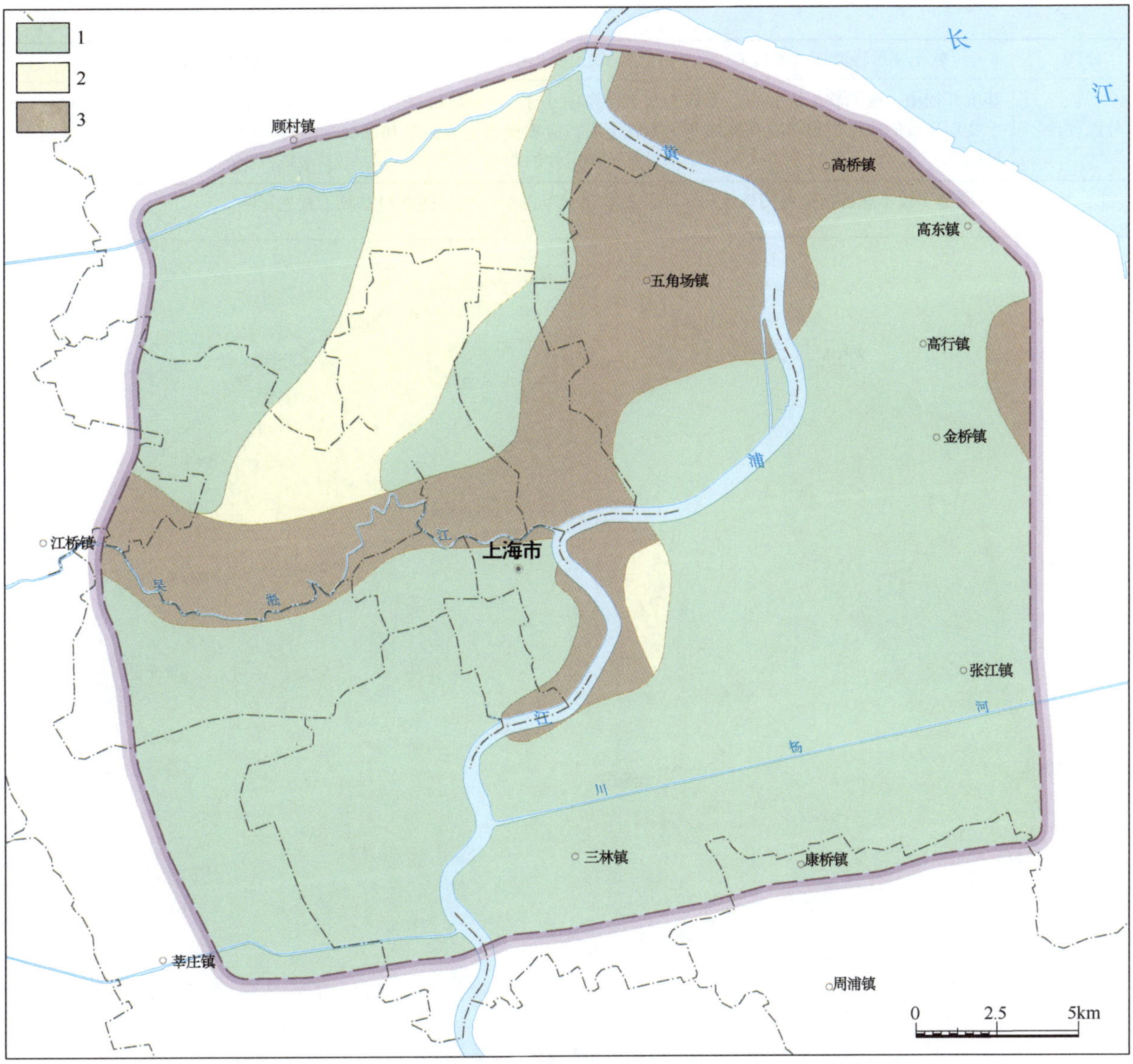

图 33-7 上海中心城 10m 深基坑考虑流砂问题适宜性分区示意图

Fig. 33-7 Suitability zoning of 10 m foundation pits in downtown Shanghai considering quicksand problems

1—流砂问题轻微区（基本为软土层分布区）；2—流砂问题中等区（$②_3$ 砂、粉性土厚度<5m 区）；3—流砂问题严重区（基本为$②_3$ 砂、粉性土分布区）

区基坑开挖深度范围内无软土层分布；中等区既有软土层分布亦有$②_3$ 层分布；开挖深度内都为软土层的为软土变形严重区（图 33-8；表 33-13）。

表 33-13 上海中心城 10m 深基坑考虑软土变形适宜性分区及评价

Table 33-13 Suitability zoning and assessment of 20m foundation pits in downtown Shanghai considering groundwater problems

分区	软土变形轻微区（A 区）	软土变形中等区（B 区）	软土变形严重区（C 区）
分区特征	基坑开挖深度范围内表土层下无软土层（③、④）分布，全部为$②_3$ 层	基坑开挖深度范围内表土层下既有软土层（③、④）分布，亦有$②_3$ 层分布	基坑开挖深度范围内表土层下都为软土层分布，所揭示的软土层厚度一般为 7m
分区范围	主要分布在苏州河两侧及北部黄浦江两侧，内环内部分黄浦江两侧亦有分布	主要分布在浦西苏州河北部南北高架道路两侧，内环内黄浦江两侧及浦东高行局部亦有分布	广泛分布于中心城，主要分布在浦西苏州河南部、浦东高桥南部及西南部分地区

续表

分区	软土变形轻微区（A区）	软土变形中等区（B区）	软土变形严重区（C区）
分区评价	基坑开挖中一般无软土层分布，无软土变形导致的边坡失稳问题和侧向变形问题	软土层厚度小于7m，基坑底部土层为软土层，应重视软土的流变和触变对基坑稳定性的影响	除表土层（$②_1$层）外均为软土层，厚度一般大于7m，应特别注意软土变形产生的边坡失稳问题

注：基坑底部或底部下较小深度范围内为软土层，应注意软土地基的竖向变形问题对基坑工程的影响。

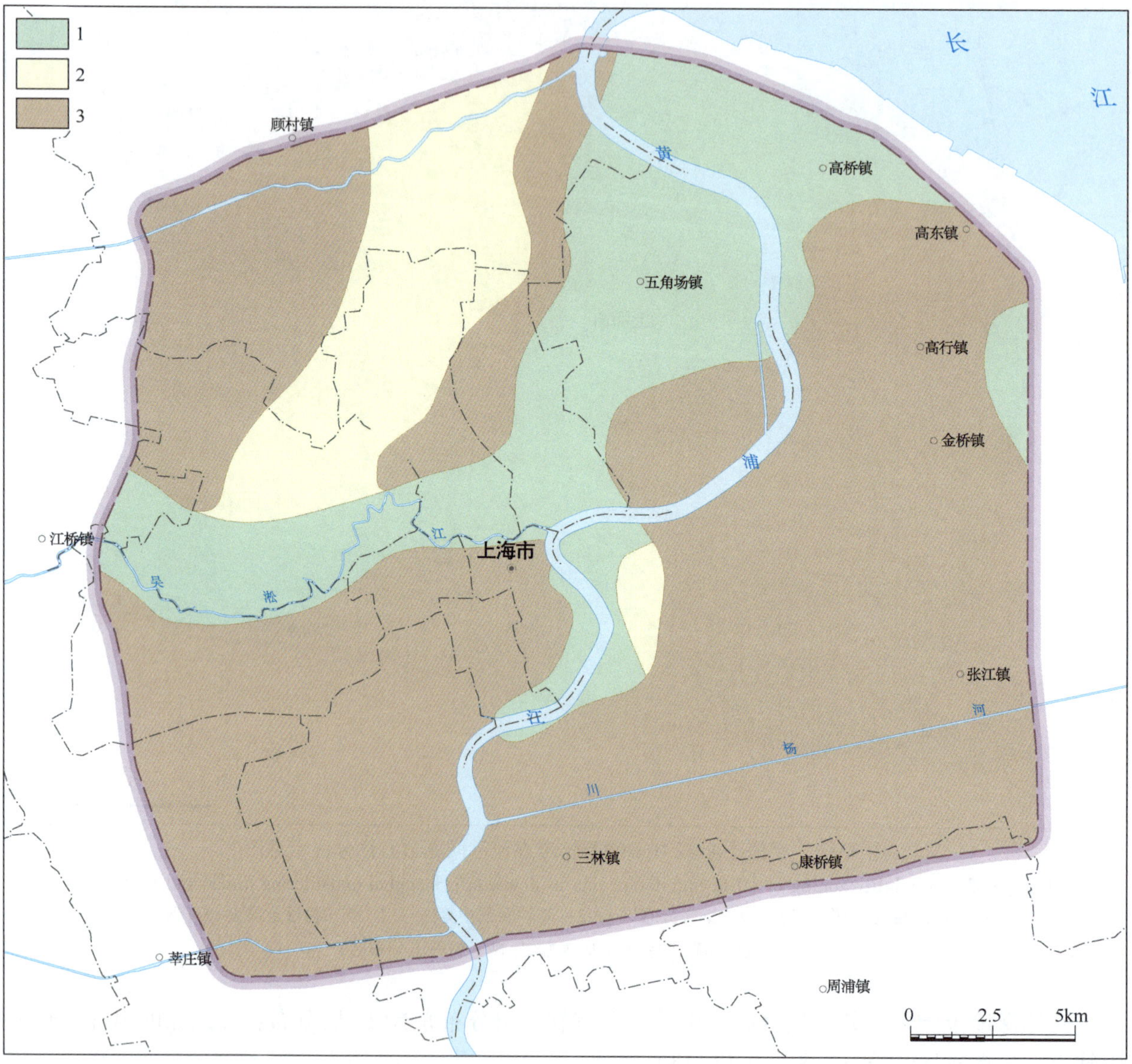

图33-8　上海中心城10m深基坑考虑软土变形问题适宜性分区示意图

Fig. 33-8　Suitability zoning map of 10m foundation pits in downtown Shanghai considering soft soil deformation problems

1—软土变形问题轻微区（基本为软土层分布区）；2—软土变形问题中等区（$②_3$砂、粉性土厚度<5m区）；

3—软土变形问题严重区（基本为$②_3$砂、粉性土分布区）

2. 20m深基坑

对于20m深基坑，开挖所揭露的土层除10m深基坑揭示的土层外，尚有$⑤_1$层黏性土。10m深基坑所考虑的问题在20m深基坑中同样要考虑，因此在该深度基坑的适宜性分区中主要对10～20m深度范围内土层对基坑的影响进行详细分析，并在此基础上进行适宜性分区。

（1）考虑砂、粉性土层产生的流砂问题的适宜性分区

20m深基坑所揭示的流砂层与10m深基坑基本相同，因此其适宜性分区与10m深基坑相同，但应

注意③层淤泥质粉质黏土中③$_{a}$ 层和④层淤泥质黏土中④$_{a}$ 层砂、粉性土透镜体的流砂问题。

（2）考虑地下水作用的适宜性分区

对于 20m 深基坑，由于开挖深度大，大都涉及承压水问题，包括微承压含水层（⑤$_{2}$ 层）和第 I 承压含水层（⑦层）。⑤$_{2}$ 层岩性一般为砂质粉土，部分地区为粉砂，渗透系数大，水位较高，对于 20m 深基坑，均易产生基坑突涌问题。⑦层岩性平面上变化不大，垂向上有一定差异，从上至下砂粒含量逐渐增加，粉粒含量逐渐减少，由于受到古河道切割，埋深、厚度变化较大。因此对于 20m 深基坑，西北部宝山局部地区由于⑦层埋藏浅，地下水位高，其他正常沉积区⑦层埋藏相对较深，但通过基坑突涌计算，亦易产生基坑突涌问题。而古河道切割区和部分缺失区则由于埋藏较深，一般不会产生基坑突涌问题。

根据以上分析，对 20m 深基坑考虑地下水作用的适宜性分区：⑦层埋藏深度大于 35m 区（主要为古河道切割区）和缺失区为地下水引起突涌问题轻微区；⑤$_{2}$ 层分布区和⑦层埋深小于 28m 区为地下水引起突涌问题严重区；其他地区为中等区（图 33-9；表 33-14）。

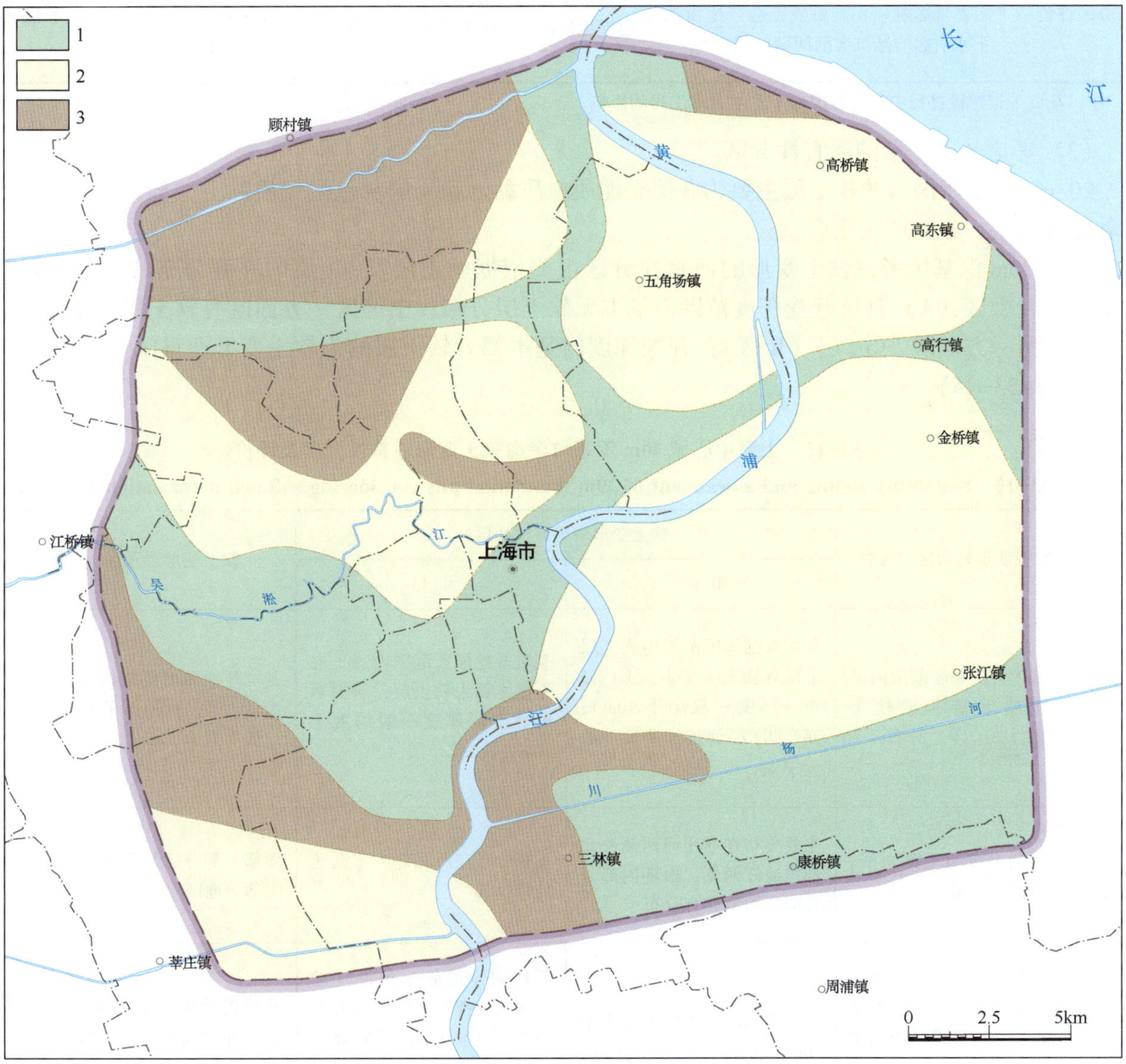

图 33-9 上海中心城 20m 深基坑考虑地下水问题适宜性分区示意图

Fig. 33-9 Suitability zoning map of 20m foundation pits in downtown Shanghai considering groundwater problems

1—20m 基坑突涌问题轻微区；2—20m 基坑突涌问题中等区；3—20m 基坑突涌问题严重区

表 33-14 上海中心城 20m 深基坑考虑地下水适宜性分区及评价

Table 33-14 Suitability zoning and assessment of 20 m foundation pits in downtown Shanghai considering groundwater problems

分区	地下水引起基坑突涌问题轻微区（A 区）	地下水引起基坑突涌问题中等区（B 区）	地下水引起基坑突涌问题严重区（C 区）
分区特征	第Ⅰ承压含水层（⑦层）埋藏深度大于 35m（主要为古河道切割区）或⑦层缺失	⑦层埋深大于 28m、小于 35m	$⑤_2$ 层分布或⑦层埋深小于 28m
分区范围	主要分布在南部地区的古河道内	主要分布在中北部和西南部	主要分布在西部宝山顾村、大场及西南部局部地区
分区评价	无浅部承压水或承压含水层埋藏深，对基坑影响较小，一般不会产生由地下水引起的基坑突涌问题	无微承压含水层分布，第Ⅰ承压含水层埋藏较浅，易产生基坑突涌事故	微承压含水层和第Ⅰ承压含水层均有分布，且埋藏浅，水位高，极易产生基坑突涌问题

注：基坑工程建设过程中均应考虑地下水浮力对工程的影响。

（3）考虑软土变形的适宜性分区

20m 深基坑所揭示的软土层主要为③层淤泥质粉质黏土和④层灰色淤泥质黏土上部，此外还有$⑤_1$层软黏性土层，亦易产生变形。

本次 20m 深基坑考虑软土变形的适宜性分区主要依据软土层及$⑤_1$ 层的埋藏分布特征进行划分：软土变形轻微区（A）基坑开挖深度范围内基本无软土层分布；中等区一方面既有软土层分布亦有$②_3$层分布，另一方面缺失$⑤_1$ 层（B 区）；开挖深度内基本都为软土层的为软土变形严重区（C 区）（图 33-10；表 33-15）。

表 33-15 上海中心城 20m 深基坑考虑软土变形适宜性分区及评价

Table 33-15 Suitability zoning and assessment of 30m foundation pits considering soft soil deformation problems

分区	软土变形轻微区（A 区）	软土变形中等区（B 区）		软土变形严重区（C 区）
		B_1 区	B_2 区	
分区特征	基坑开挖深度范围内表土层下基本无软土层（③、④）分布	基坑开挖深度范围内表土层下既有软土层（③、④）分布（厚度一般小于 10m），亦有$②_3$ 层分布（厚度一般大于 4m）	基坑开挖深度范围内表土层下基本都为软土层，所揭示的软土层厚度一般均大于 10m，缺失$⑤_1$ 层	基坑开挖深度范围内表土层下都为软土层，所揭示的软土层厚度一般均大于 10m
分区范围	仅局部分布于苏州河、黄浦江沿岸	主要分布在苏州河两侧及北部黄浦江两侧，内环内部分黄浦江两侧亦有分布	主要分布在西北部沪宁高速公路以北	南部大部分区域和北部闸北、普陀部分地区
分区评价	基坑开挖中一般无软土层分布，无软土变形导致的边坡失稳问题和侧向变形问题。但基坑底部为$⑤_1$ 层软黏性土层，应适当注意基坑底部的变形问题	开挖揭示的软土层厚度小于 10m，基坑底部土层为$⑤_1$层，应重视软土的流变和触变对基坑稳定性的影响	开挖揭示的软土层厚度较大，应重视软土的侧向变形问题，但缺失$⑤_1$ 层软黏性土层，且基坑底部为⑥层硬土层或压缩性低⑦层砂、粉性土层，因此基坑底部变形较小	开挖揭示的土层除表土层（$②_1$层）外均为软土层，厚度一般大于 10m，基坑底部为$⑤_1$ 层，应特别注意软土变形产生的边坡失稳问题

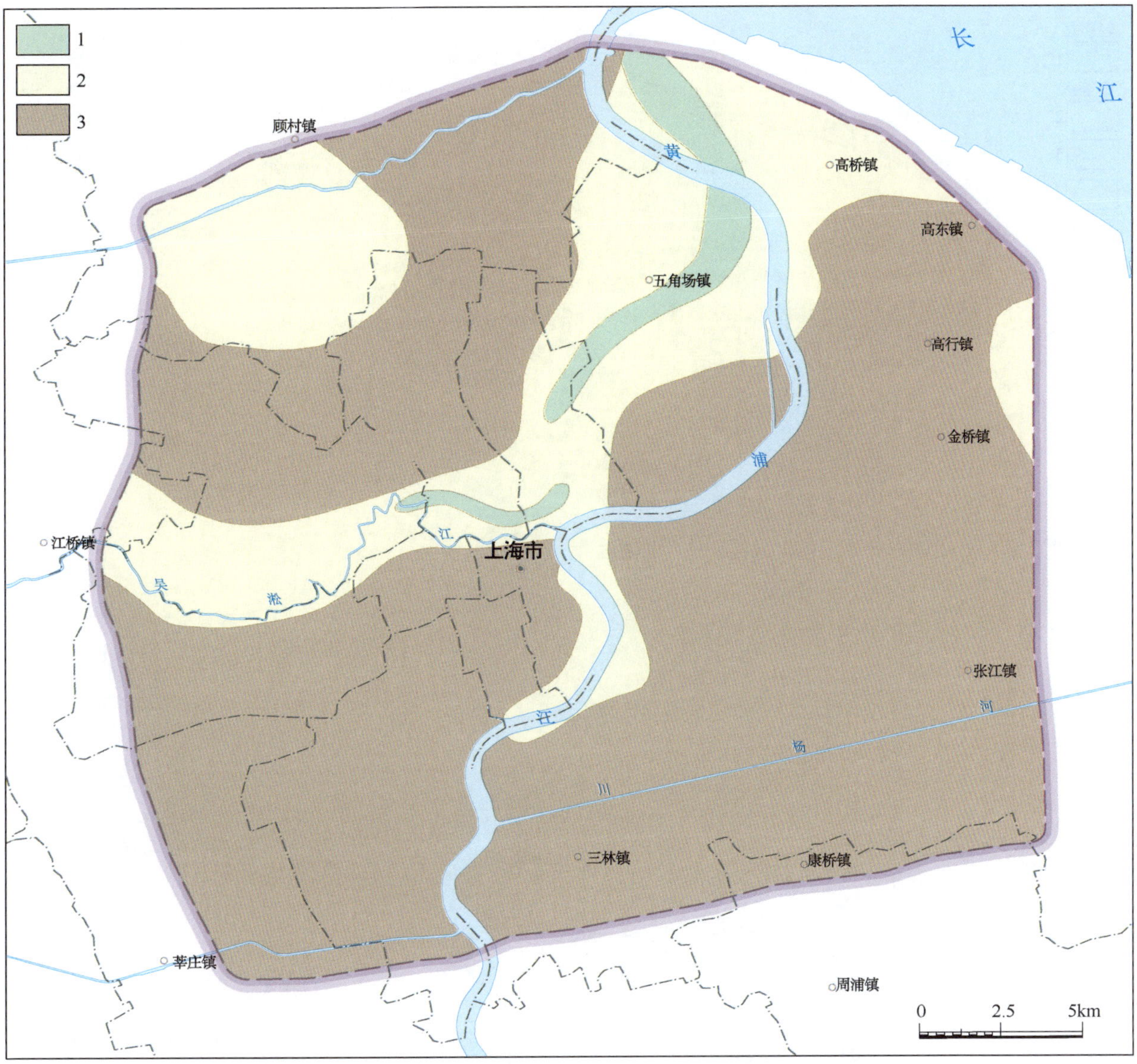

图 33-10　上海中心城 20m 深基坑软土变形问题适宜性分区图

Fig. 33-10　Suitability zoning map of 20m foundation pits in downtown Shanghai considering soft soil deformation problems

1—20m 深基坑软土变形轻微区（揭示土层基本以砂、粉性土为主，软黏性土厚度<2m）；

2—20m 深基坑软土变形中等区（揭示土层基本以砂、粉性土和软黏性土为主，软黏性土厚度<10m）；

3—20m 深基坑软土变形严重区（揭示土层基本以软黏性土为主，软黏性土厚度>10m）

3. 30m 深基坑

（1）考虑砂、粉性土层产生的流砂问题的适宜性分区

基坑所揭示的流砂层有 3 层，即浅部砂、粉性土层（$②_3$）、中部砂、粉性土层（$⑤_2$）和下部砂层（⑦），其中中部和下部砂层均为承压含水层。

考虑流砂问题的适宜性分区主要依据以上砂层分布情况进行：开挖深度范围内无砂层的为流砂问题轻微区（A）；有一层砂层分布的为中等区（B）；同时有两层砂层分布的为流砂问题严重区（C）。另外，根据分区内的砂层不同进行亚区划分（图 33-11；表 33-16）。

（2）考虑地下水作用的适宜性分区

由于开挖深度较大，含水砂层一般均已揭示，因此流砂问题较为严重，上面已有了详细叙述。在古河道切割区，下部承压含水层（⑦）埋藏变化大，埋深一般大于 30m，在此区域（图 33-12）应考虑基坑突涌问题，而其他地区则适当注意即可。

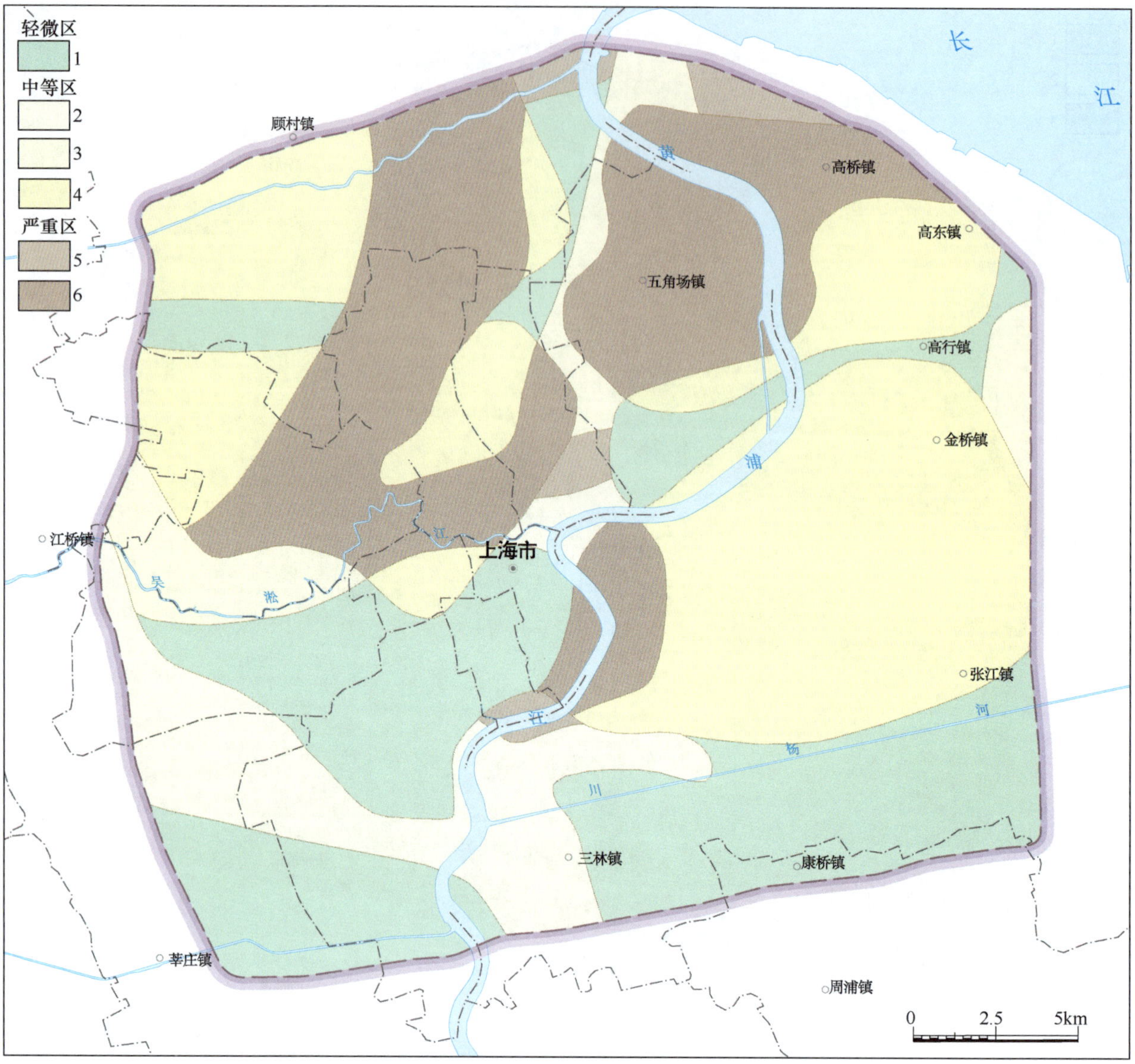

图 33-11　上海中心城 30m 深基坑流砂问题适宜性分区图

Fig. 33-11　Suitability zoning map of 30m foundation pits considering quicksand problems in downtown Shanghai

1—无②$_3$ 或⑤$_2$ 或⑦砂、粉性土层分布区或⑦层顶埋深>30m 区；2—仅有②$_3$ 砂、粉性土层分布区；3—仅有⑤$_2$ 砂、粉性土层分布区；4—仅有⑦砂、粉性土层分布区；5—②$_3$、⑤$_2$ 砂、粉性土层同时分布区；6—②$_3$、⑦砂、粉性土层同时分布区

表 33-16　30m 深基坑考虑流砂问题适宜性分区及评价

Table 33-16　Suitability zoning and assessment of 10-20 tunnels considering quicksand problems

分区	流砂问题轻微区（A 区）		流砂问题中等区（B 区）			流砂问题严重区（C 区）	
	A1 区	A2 区	B1 区	B2 区	B3 区	C1 区	C2 区
分区特征	基坑开挖深度范围内无砂层分布		开挖深度范围内有一层砂层分布			开挖深度范围内有两层砂层分布	
	基坑底部无⑦层分布	基坑底部有⑦层分布	有浅部砂层（②$_3$）分布	有中部砂层（⑤$_2$）分布	有下部砂层（⑦）分布	有浅部（②$_3$）和中部（⑤$_2$）砂层分布	有浅部（②$_3$）和下部（⑦）砂层分布
分区范围	仅在北部普陀和闸北零星分布	主要分布在南部徐汇和浦东川杨河以南	主要分布在浦西苏州河沿岸局部地区	主要分布在南部和浦东三林局部地区	主要分布在浦西北部的宝山和普陀，浦东大部地区	零星分布于市区中部	广泛分布于北部大部分地区和内环内黄浦江两岸部分地区

续表

分区	流砂问题轻微区（A 区）		流砂问题中等区（B 区）			流砂问题严重区（C 区）	
	A1 区	A2 区	B1 区	B2 区	B3 区	C1 区	C2 区
分区评价	基本无流砂问题	基本无流砂问题，但应注意底部基坑底部无⑦层的问题	应重视浅部砂层的流砂问题，中部和下部流砂问题轻微	流砂问题轻微，应重视中部砂层的流砂问题	应重视下部砂层的流砂问题，浅部和中部流砂问题轻微	应特别重视浅部和中部的流砂问题，下部流砂问题轻微	应特别重视浅部和下部的流砂问题，中部流砂问题轻微

注：该基坑还应注意③层淤泥质粉质黏土中$③_a$层和④层淤泥质黏土中$④_a$层砂、粉性土透镜体的流砂问题。

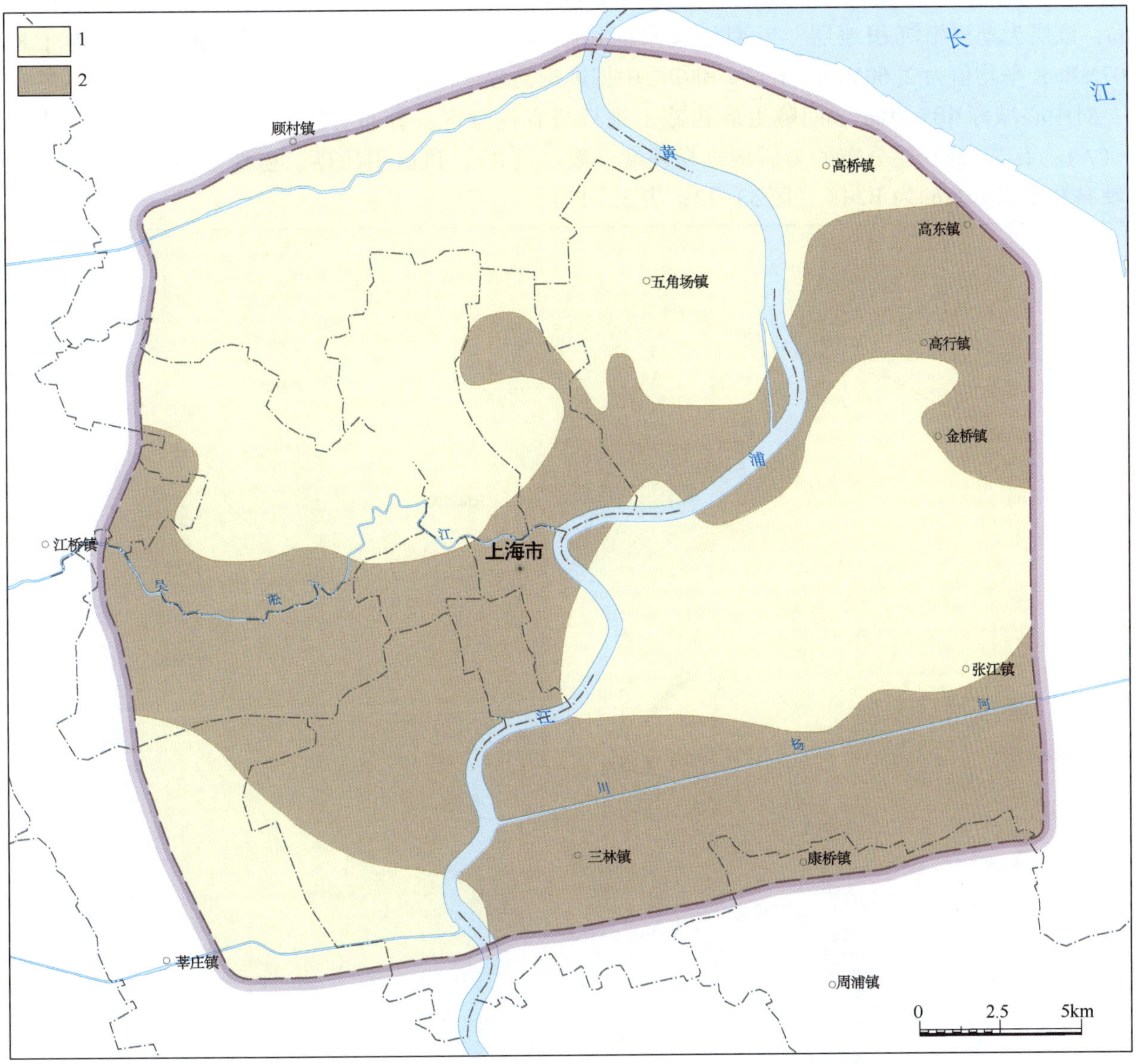

图 33-12　上海中心城 30m 深基坑突涌问题适宜性分区图

Fig. 33-12　Suitability zoning map of 30m foundation pits considering pit inrush in downtown Shanghai

1—适当注意地下水引起的基坑突涌问题；2—需特别重视地下水引起的基坑突涌问题

（3）考虑软土变形的适宜性分区

所揭示的软土层（③、④）和压缩性高的$⑤_1$层与 20m 深基坑相同，因此其适宜性分区基本同 20m 深基坑。

（二）隧道工程地质环境适宜性分区

对于隧道工程，施工中主要考虑的问题为盾构掘进面土层软硬的变化问题，一般情况下，在静力触探比贯入阻力（*ps*）小于1.5MPa的土层中掘进较为容易，在*ps*值为1.5~4MPa之间的土层中掘进难易程度一般，而在大于4MPa的土层中掘进则最为困难。

1. 深10~20m区间隧道盾构适宜性分区

揭露土层主要为软土层③、④和饱和黏性土层$⑤_1$层。其中，③层灰色淤泥质粉质黏土、④层灰色淤泥质黏土为软土，静力触探比贯入阻力*ps*值小于1MPa；$⑤_1$层为灰色黏性土，静力触探比贯入阻力*ps*值为0.85~1.32MPa，平均值为1.12MPa，在以上土层中进行掘进较为容易。浅部砂、粉性土层（$②_3$）厚度大，部分地区层底埋深超过10m，其岩性一般为砂质粉土，其静探比贯入阻力*ps*值为1.96~4.37MPa，平均值为2.95MPa。在西北部地区，由于第一硬土层⑥层埋藏较浅，一般在15~20m，该层为湖沼相沉积地层，后期经脱水固结，结构较密实，其静深比贯入阻力一般为1.83~3.07MPa，平均值为2.60MPa。在$②_3$和⑥层中掘进难易程度一般。

对中心城深10~20m区间隧道盾构施工进行适宜性分区：分布有③、④或$⑤_1$层的为盾构容易区（A）；有$②_3$或⑥层分布的为盾构难易程度一般区（B），该区中为砂、粉性土层分布的为B1区，为硬黏性土层分布的为B2区（图33-13；表33-17）。

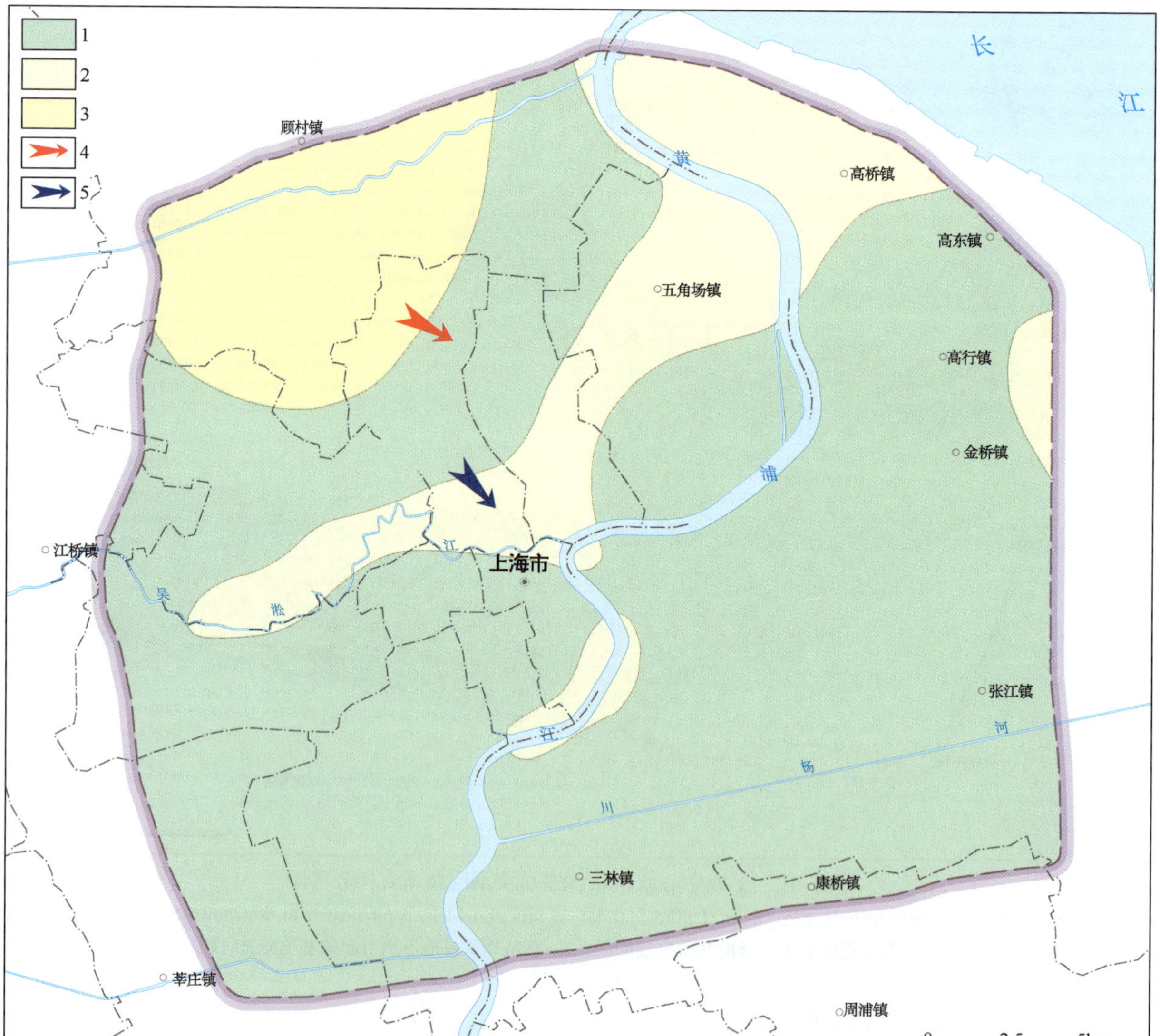

图33-13 上海中心城10~20m深隧道盾构地质环境适宜性分区图

Fig. 33-13 Geo-environmental suitability zoning map of 10-20m tunnel for shield construction in downtown Shanghai

1—盾构推进容易区（为软黏性土层分布区）；2—盾构推进一般区（有$②_3$砂、粉性土层分布）；3—盾构推进一般区（有⑥层硬黏性土层分布）；4—由硬黏性土层（⑥）向软性土层（③、④或$⑤_1$）过渡；5—由软黏性土层（③、④或$⑤_1$）向粉性土层（$②_3$）过渡

表 33-17　10~20m 深隧道工程盾构适宜性分区及评价

Table 33-17　Suitability zoning and assessment of 10-20 tunnels for shield construction

分区	盾构推进容易区（A 区）	盾构推进一般区（B 区）	
		B1 区	B2 区
分区特征	盾构区间范围内均为③、④或$⑤_1$层	盾构区间范围内均有$②_3$层分布	盾构区间范围内均有⑥层分布
分区范围	主要分布在南部及北部浦西部分地区	主要分布在吴淞江及北部黄浦江两岸	主要分布在西北部宝山顾村、大场附近
分区评价	盾构穿越土层基本为软黏性土层，比贯入阻力较小，且变化不大，因此适宜盾构推进	穿越土层有砂、粉性土层分布，且穿越厚度一般均大于 4m，其下为软黏性土层，水平方向上土层变化不大，但比贯入阻力较大，推进难易程度一般。垂向上应注意土性由上至下从砂、粉性土层向软黏性土层过渡	盾构穿越土层除软黏性土层外，尚有硬黏性土层，其厚度一般在 5m 左右，比贯入阻力较大，推进难易程度一般。垂向上应注意土性由上至下从软黏性土层向硬黏性土层过渡

注：应特别注意分区界线区间土层变化的影响。该区间范围内软土层较为发育，隧道盾构施工过程中和运营期间应注意软土地基变形对隧道工程的影响。

2. 深 20~30m 区间隧道盾构适宜性分区

正常沉积区土层主要为饱和黏性土层$⑤_1$、硬黏性土层⑥，西北部地区还有砂、粉性土层⑦。其中$⑤_1$和⑥层土特性上面已有论述，西北部地区的⑦层相对其他地区埋藏较浅，岩性一般为砂质粉土，部分地区为黏质粉土或粉砂，其比贯入阻力一般为 3.60~12.80MPa，平均值为 8.80MPa，在该层中掘进较为困难。古河道切割区深 20~30m 区间内土层注意为饱和黏性土层$⑤_1$层、中部砂层$⑤_2$层和一般黏性土层$⑤_3$层。其中$⑤_2$层岩性一般为砂质粉土，部分地区为粉砂，其比贯入阻力一般为 2.60~7.80MPa，平均值为 3.68MPa，在该层中掘进难易程度一般，$⑤_3$层为粉质黏土夹砂，土质不均，其比贯入阻力一般为 1.30~2.68MPa，平均值为 1.40MPa，在该层中掘进较为容易，但应注意其土质不均问题。

根据以上分析，对中心城 20~30m 深隧道盾构施工进行适宜性分区，由于该深度范围内土层垂向上有一定变化，正常沉积区 25m 以浅一般为软黏性土层，25~30m 为硬黏性土层，此外，西北部地区 20~30m 一般为砂、粉性土层（⑦）。因此该区间盾构适宜性分区按照 20~25m、25~30m 分别进行。

（1）20~25m 深隧道

可揭露有$⑤_1$层饱和黏性土层的为盾构容易区（A）；有$⑤_2$层分布的为盾构难易程度一般区（B）；有⑦层分布的为盾构困难区（C）（图 33-14；表 33-18）。

表 33-18　20~25m 深隧道考虑掘进面土层变化推进适宜性分区

Table 33-18　Suitability zoning of 20-25m tunnels considering soil strata changes on the driving face

分区	盾构推进容易区（A 区）	盾构推进一般区（B 区）	盾构推进困难区（C 区）
分区特征	该深度区间分布有$⑤_1$层饱和黏性土层	该深度区间分布有$⑤_2$层砂、粉性土层	该深度区间有⑦层砂性土层分布
分区范围	除西北部的大部分地区	零星分布	主要分布在西北部宝山顾村、大场镇附近地区
分区评价	盾构穿越土层为饱和黏性土层（$⑤_1$），比贯入阻力较小，且变化不大，因此适宜盾构推进	盾构穿越土层为砂、粉性土层（$⑤_2$），比贯入阻力较大，推进难易程度一般	盾构穿越土层为第Ⅰ承压含水层（⑦），且厚度大，比贯入阻力大于 4MPa，盾构推进较为困难

注：应特别注意分区界线区间土层变化的影响。该区间范围内软土层较为发育，隧道盾构施工过程中和运营期间应注意软土地基变形对隧道工程的影响。

（2）25~30m 深隧道

深 25~30m 区间内分布有$⑤_1$层或$⑤_3$层的为盾构容易区（A）；分布有⑥层或$⑤_2$层的为盾构难易

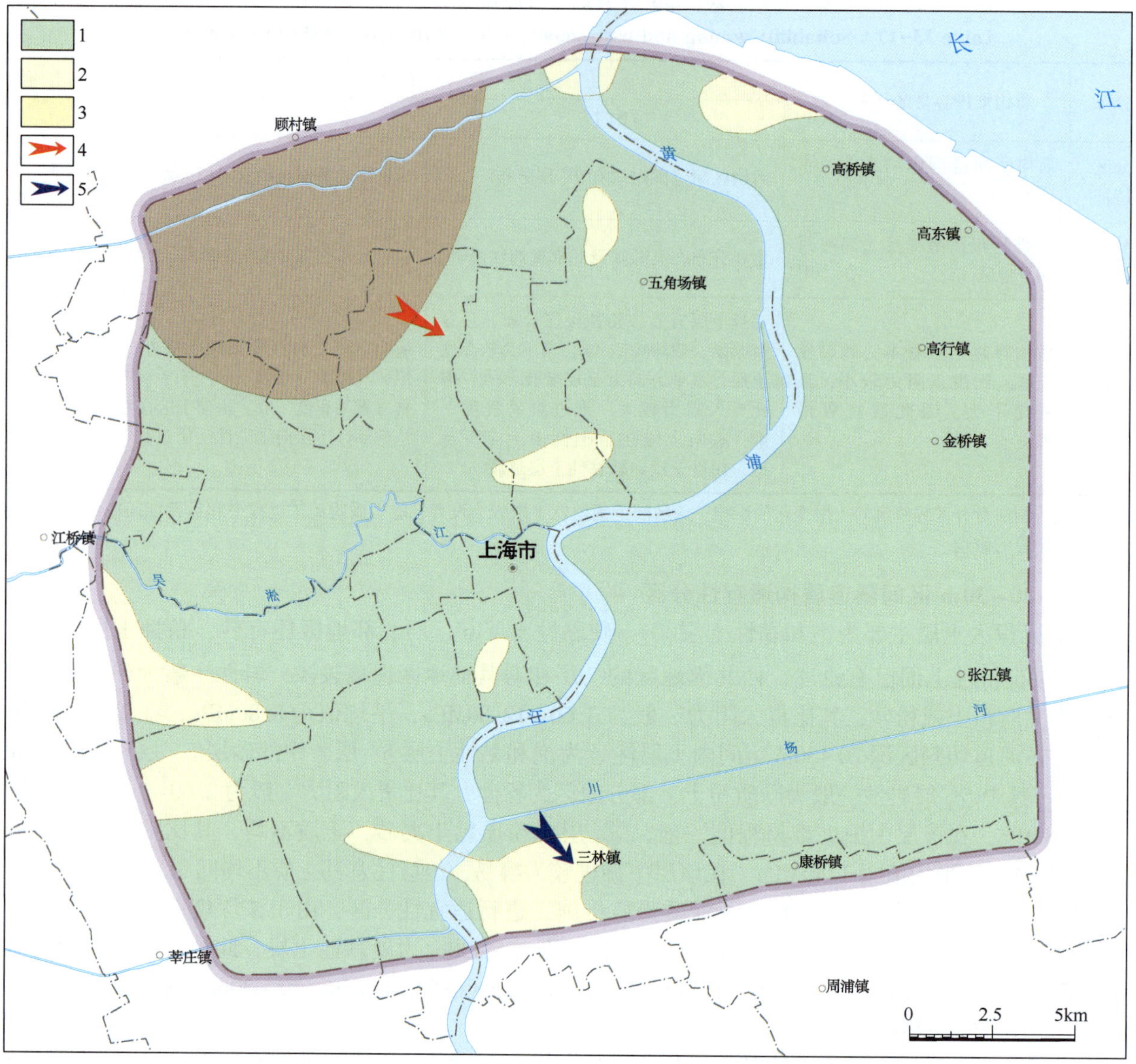

图 33-14　上海中心城 20~25m 深隧道盾构地质环境适宜性分区图

Fig. 33-14　Geo-environmental suitability zoning map of 20-25m tunnels for shield construction in downtown Shanghai

1—盾构推进容易区（为黏性土层（⑤）分布区）；2—盾构推进一般区（有$⑤_2$砂、粉性土层分布）；3—盾构推进困难区（有⑦砂、粉性土层分布）；4—由砂性土层（⑦）向黏性土层（$⑤_1$）过渡；5—由黏性土层（$⑤_1$）向砂、粉性土层（$⑤_2$）过渡

程度一般区（B 区），其中有⑥层分布的为 B1 区，有$⑤_2$层分布的为 B2 区；有⑦层分布的为盾构困难区（C 区）（表 33-19；图 33-15）。

表 33-19　25~30m 深隧道考虑掘进面土层变化推进适宜性分区

Table 33-19　Suitability zoning of 25-30m tunnels considering soil strata changes on the driving face

分区	盾构推进容易区（A 区）	盾构推进一般区（B 区）		盾构推进困难区（C 区）
		B1 区	B2 区	
分区特征	该深度区间分布有$⑤_1$层或$⑤_3$层一般黏性土层	分布有$⑤_2$层砂、粉性土层或⑥层硬黏性土层		该深度区间有⑦层砂性土层分布
		分布有⑥层硬黏性土层	分布有$⑤_2$层砂、粉性土层	

续表

分区	盾构推进容易区（A 区）	盾构推进一般区（B 区）		盾构推进困难区（C 区）
		B1 区	B2 区	
分区范围	呈条带状分布，东南部川杨河以南及徐汇和市区大部分地区	分布最为广泛，浦东川杨河以北，浦西吴淞江两侧及西南端	主要分布在西南部外环线附近	主要分布在西北部宝山顾村、大场附近
分区评价	盾构穿越土层为黏性土层（$⑤_1$ 和 $⑤_3$），比贯入阻力较小，且变化不大，因此适宜盾构推进	盾构穿越土层为硬黏性土层（⑥），比贯入阻力较大，推进难易程度一般	盾构穿越土层为砂、粉性土层（$⑤_2$），比贯入阻力较大，推进难易程度一般	盾构穿越土层为第Ⅰ承压含水层⑦层，且厚度大，比贯入阻力大于 4MPa，盾构推进较为困难

注：应特别注意分区界线区间土层变化的影响。该区间范围内软土层较为发育，隧道盾构施工过程中和运营期间应注意软土地基变形对隧道工程的影响。

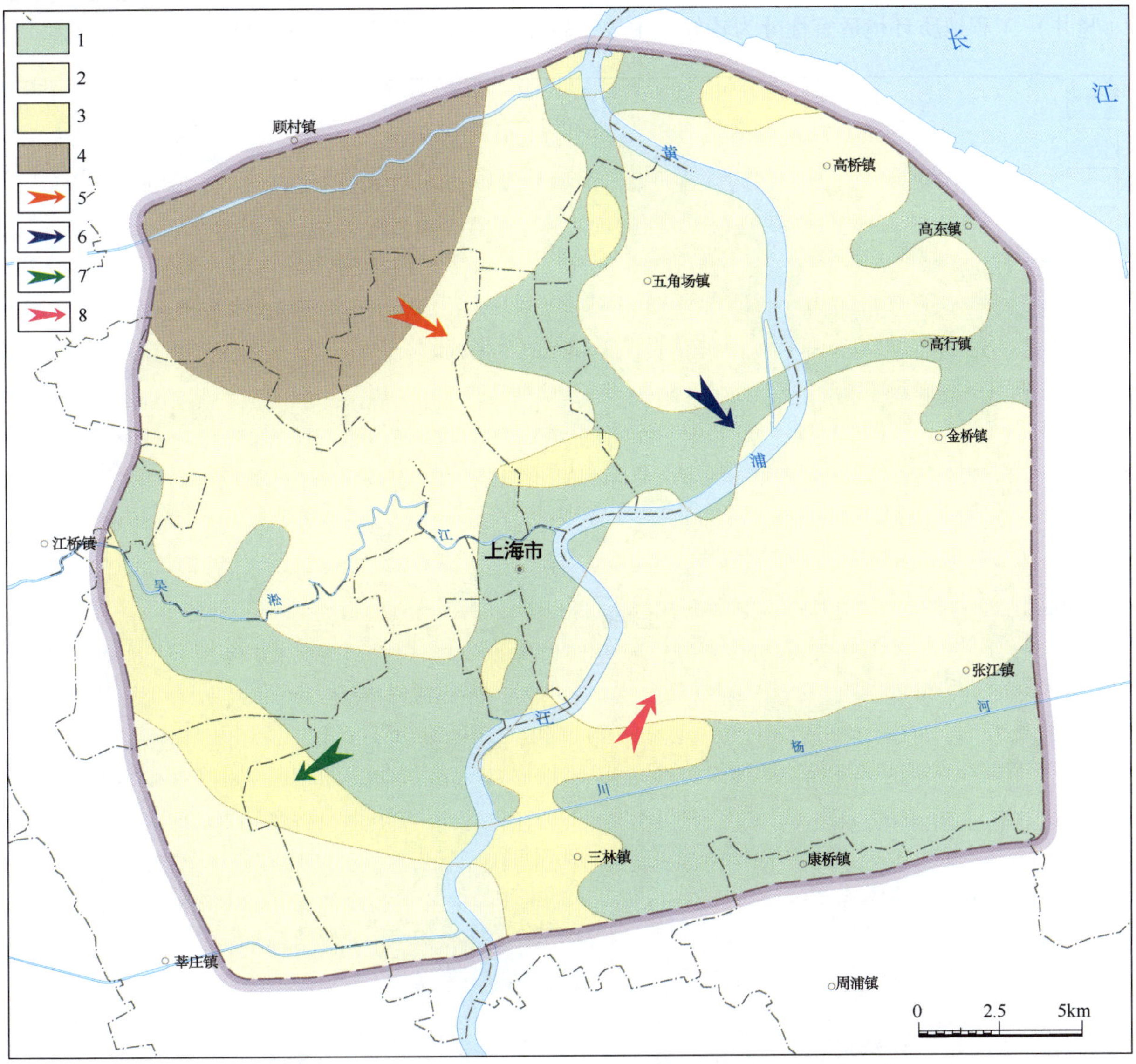

图 33-15　上海中心城 25～30m 深隧道盾构地质环境适宜性分区图

Fig. 33-15　Geo-environmental suitability zoning map of 25-30m tunnels for shield construction in downtown Shanghai

1—盾构推进容易区（为黏性土层（⑤）分布区）；2—盾构推进一般区（有硬黏性土层（⑥）分布）；3—盾构推进一般区（有$⑤_2$ 砂、粉性土层分布）；4—盾构推进困难区（有⑦砂、粉性土层分布）；5—由砂性土层（⑦）向硬黏性土层（⑥）过渡；6—由硬黏性土层（⑥）向黏性土土层（$⑤_1$）过渡；7—由黏性土土层（$⑤_1$）向砂、粉性土层（$⑤_2$）过渡；8—由砂、粉性土层（$⑤_2$）向硬黏性土层（⑥）过渡

二、深层地下空间开发利用适宜性分区与评价

（一）基坑（竖井）工程适宜性分区评价

上海市中心城地面沉降风险程度高—中等，对于40~65m深基坑（竖井）工程开挖施工揭遇的深部土层主要为⑦、⑧层，该深度区间内一般不会揭遇第⑨层。其中，⑦层为粉土、粉砂层，基坑开挖施工时易引发流砂和基坑突涌，⑧层为灰色黏性土层，软塑—可塑，高—中压缩性，基坑开挖施工时将引发一定的地基变形，但该土层土性尚可，基坑开挖时自稳性较好，总体对基坑边坡的稳定有利。但上海中心城第Ⅱ、第Ⅲ承压含水层均普遍分布，顶界埋深起伏较小，其承压水水头高，40~65m深基坑（竖井）工程施工易引发水土突涌，对基坑安全和周边地质环境影响较大，一般涉及对第Ⅱ、第Ⅲ承压含水层降排水问题。

综合考虑中心城地面沉降风险程度和地质结构特征，根据上述分区依据，将中心城优先开发层基坑（竖井）工程地质环境适宜性分为较好、中等、较差、差4个等级（图33-16；表33-20）。

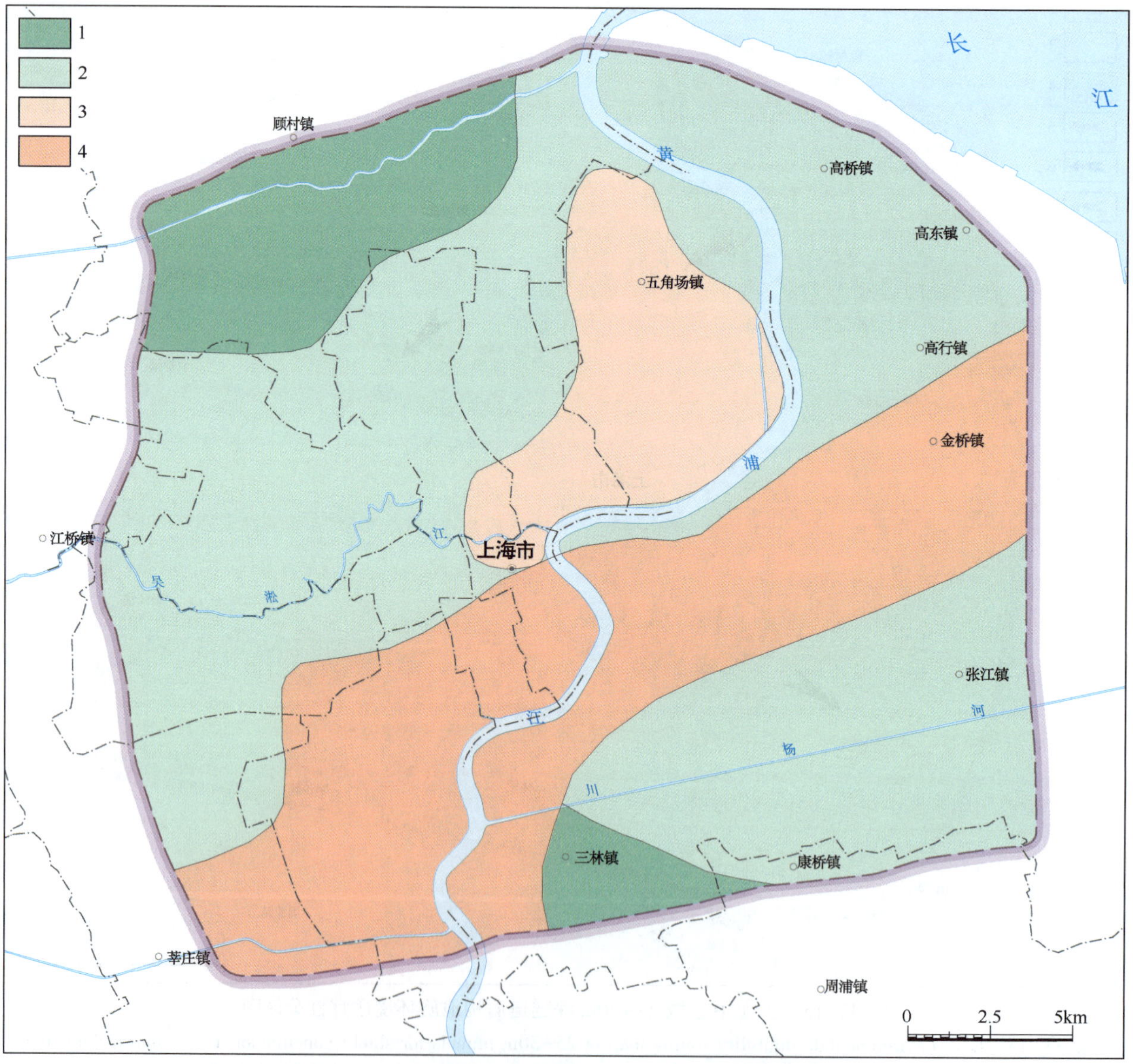

图33-16　上海中心城40~65m深基坑（竖井）工程适宜性分区图

Fig. 33-16　Suitability zoning map of 40-65m foundation pit (vertical shaft) projects in downtown Shanghai

1—适宜性较好区；2—适宜性中等区；3—适宜性较差区；4—适宜性差区

表 33-20 上海中心城 40~65m 深基坑（竖井）工程适宜性分区

Table 33-20 Suitability zoning of 40-65m tunnel projects in downtown Shanghai

分区	适宜性较好区	适宜性中等区	适宜性较差区	适宜性差区
分区特征	⑦层埋藏浅，厚度小，⑧层层顶埋深小，厚度大，基坑开挖深度区间内涉及的深部土层主要以⑧层或$⑤_3$层灰色黏性土为主	⑦层埋藏较浅，厚度一般，⑧层埋深较浅，厚度较大，基坑开挖深度区间内涉及⑦层、⑧层或$⑤_3$层	⑦层埋藏适中，厚度大，⑧层埋深深，厚度薄，基坑开挖深度区间内以⑦层为主，但隔水层⑧层有分布	⑧层灰色黏性土层缺失，⑦层厚度大，一般为15~30m，且与⑨层沟通，两层砂性土合计层厚大于60m
分区范围	北部宝山大场、庙行、张庙等区域	大柏树—上海北站—静安寺—七宝镇以北、以西大部，杨浦新江湾城，浦东陆家嘴—洋泾—金桥以北和以南	上海北站—大柏树—新江湾城一线至黄浦江西岸沿线大部	浦西闵行莘庄—梅陇—徐汇龙华以及浦东三林—杨思—严桥—洋泾—金桥沿线区域
分区评价	大深度基坑开挖时流砂风险小，应注意引发基坑突涌灾害的可能性。同时大量降排承压水将引发较大面积的不均匀地面沉降，地面沉降风险程度中等，地质环境条件相对有利	基坑开挖时易产生⑦层的流砂和基坑突涌灾害；应注意由⑨层、⑪层引发基坑突涌的可能性，工程建设引发的地面沉降范围较大。地面沉降风险程度中等—较高，地质环境条件一般	大深度基坑施工时易产生流砂和基坑突涌灾害，且⑨层、⑪层具有引发基坑突涌的风险，地面沉降风险高，地质环境条件一般	大深度基坑工程开挖易产生基坑突涌灾害；同时，由于基坑降排承压水时无法阻断含水层，降排水量大，将引发大面积的地面沉降和不均匀沉降，地面沉降风险高，地质环境条件相对较差

注：还应注意浅部粉、砂性土层（$②_3$）以及软土层（③、④）等对大深度竖井（基坑）工程产生的影响。

（二）深埋隧道工程适宜性分区评价

深埋隧道工程施工过程中深层地下水问题将直接影响隧道施工的难易程度和风险大小。在黏性土或黏性土与砂土互层的土层中掘进施工时，由于土层富水性差，渗透性弱，盾构掘进施工时地下水控制难度相对较小。而在砂土中掘进施工时，由于砂土富水性好，渗透性强，水头压力大，盾构掘进时产生流砂的可能性较大，且易引发盾构掘进面不稳及土层流失问题。

综合考虑中心城地面沉降风险程度和地质结构特征，将中心城 40~65m 深隧道工程地质环境适宜性分为较好、中等、较差、差 4 个等级（图 33-17；表 33-21）。

表 33-21 上海中心城 40~65m 深隧道工程适宜性分区

Table 33-21 Suitability zoning of 40-65m tunnel projects in downtown Shanghai

分区	适宜性较好区	适宜性中等区	适宜性较差区	适宜性差区
分区特征	⑦层厚度薄，一般小于10m，⑧层或$⑤_3$层埋藏浅，厚度大；深 40~65m 区间内以黏性土及黏性土夹砂为主	正常沉积区，⑧层层顶埋深为30~40m，厚度一般大于15m，局部受古河道切割，上部沉积较厚的$⑤_3$层	⑧层层顶埋深为40~50m，厚度一般大于10m，同时⑦层砂性土有分布	⑧层灰色黏性土层，⑦层和⑨层砂土层相互沟通，砂土层厚度大
分区范围	北部宝山大场、庙行、张庙等区域	中北部的白遗桥、江湾和高境、淞南东部、浦东高桥等区域，东南部康桥以东区域	中部的闵行虹桥、普陀长征、曹家渡、杨浦五角场、浦东新区高行等区域	浦西闵行莘庄—梅陇—徐汇龙华一线附近及浦东三林—杨思—严桥—洋泾—金桥一带
分区评价	盾构埋隧道工程施工难度较小，风险相对较低。盾构掘进施工及运营期间深层地下水控制难度相对较小。地面沉降风险程度中等，地质环境条件相对有利	隧道掘进施工过程中需注意隧道掘进面不稳等现象，防止在深层承压水的作用下引发水土突涌灾害。地面沉降风险程度较高，地质环境条件相对有利	在盾构掘进施工时易产生流砂及水土突涌问题，引发大面积的水土流失和地面沉降。地面沉降风险较高—高，地质环境条件一般	盾构掘进阻力大，工作面易产生流砂和水土突涌问题，深层地下水控制难度大，工程建设风险高。地面沉降风险较高—高，地质环境条件较差

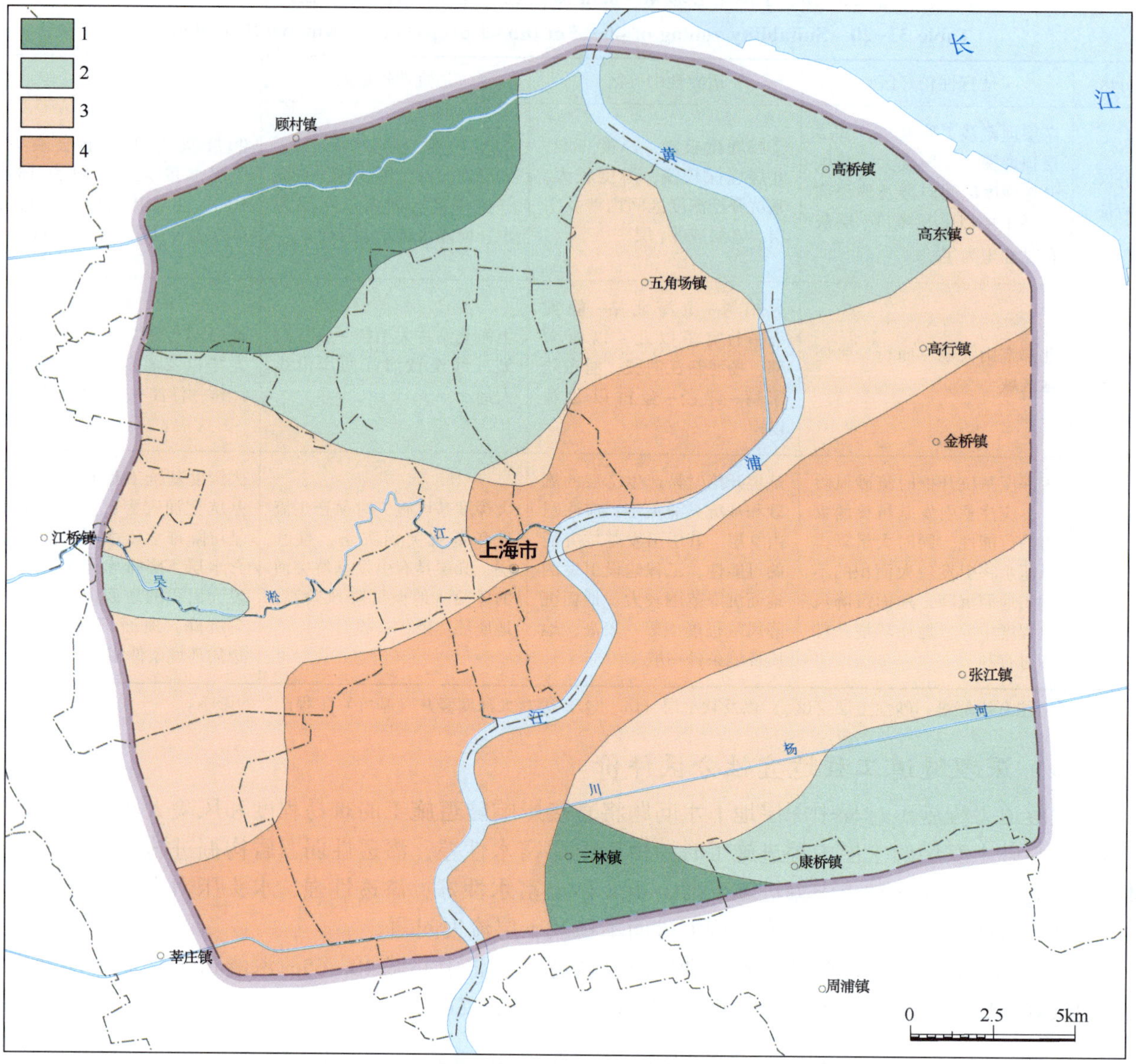

图 33-17 上海中心城 40～65m 深隧道工程适宜性分区图

Fig. 33-17 Suitability zoning map of 40-65m tunnel projects in downtown Shanghai

1—适宜性较好区；2—适宜性中等区；3—适宜性较差区；4—适宜性差区

第三十四章　滩涂资源

上海滩涂开发利用历史悠久，早在公元8世纪，就有古捍海塘，以后随着滩涂的淤涨，先后修筑了钦公塘、彭公塘、李公塘等，岸线不断外移。新中国成立以后，为了社会经济发展的需要，又围垦了大片滩涂，据统计，1953—2004年间，上海圈围滩涂976.22km²，年均19.1km²，其中1980—2004年间圈围435km²。新中国成立以后两个时间段圈围滩涂速度较快，第一是在20世纪60年代，围海造地252km²，年均圈围25.19km²；二是在21世纪初，圈围强度增加，2000—2004年间达到161.21km²，年均圈围40.3km²。圈围高程由高滩向中、低滩延伸，围垦方法采用先促淤，后吹沙填土，加速中低滩涂成陆。

2008年，国务院批准实施了《长江口综合整治开发规划》，规划提出将滩涂资源开发利用作为河势控制、航道整治辅助手段的理念。同时，《上海市土地利用总体规划（2006—2020）》明确："至2010年，结合长江口整治、港口建设、生态环境保护等，可在崇明东滩、长兴岛、横沙岛、南汇边滩、杭州湾北岸、铜沙、九段沙等促淤 80×10^4 亩，圈围 60×10^4 亩。"在此背景下，上海加快了滩涂资源的开发利用，自2000—2015年年底，上海共实施促淤工程面积 99.96×10^4 亩、圈围工程面积 55.75×10^4 亩（表34-1）。全区滩涂资源开发利用工作中始终坚持"开发利用与保护并重"，"先促后围，多促少围"等原则。在长江上游下泄泥沙的持续堆积作用下，虽经过了近15年高强度的滩涂开发利用，但区内滩涂资源总量仍然能够基本保持平衡。2000年-5m等深线（吴淞基面，下同）以上滩涂资源总

表34-1 "长江口综合规划"上海段滩涂圈围成陆完成情况

Table 34-1 Completeness of shoal enclosure projects in the Shanghai section under the "Comprehensive Planning of the Yangtze River Estuary"

地点	规划圈围面积/km²	实际圈围面积/km²		实施进度
		项目名称	规模	
中央沙	2.29	中央沙圈围工程	2.15	已基本完成
青草沙	9.51	青草沙水源地工程	7.47	已基本完成
堡镇沙、外团结沙	6.4			已纳入保护区，无圈围成陆可能
浦东边滩	0.81	浦东机场3#围区圈围工程	2.02	已完成
		浦东机场1#、3#围区圈围工程	0.48	已完成圈围，"十三五"完成吹填成陆
横沙东滩	12.5	横沙东滩三期圈围工程	2.6	已完成
		横沙东滩六期圈围工程	4.85	已完成
		横沙东滩七期圈围工程	2.02	已完成圈围，"十三五"完成吹填成陆
		横沙东滩八期圈围工程	6.4	未实施，"十三五"期间实施
南汇边滩	21.1	南汇东滩一、三期圈围工程	22.3	未实施，拟"十三五"实施圈围一期 8.6×10^4 亩
长江口北支边滩	21.1	崇明北沿圈围一期工程	4.51	已完成
		崇明北沿二期圈围成陆工程	1.82	已完成
		崇明北沿互花米草治理工程	3.65	已完成
		崇明北沿三期促淤圈围工程	1.13	已完成
总计	73.71		30.2 *	

注：①统计时间至2015年年底；②*为已圈围垦成陆的统计结果。

量为 350×10^4 亩，至 2015 年，-5m 等深线以上滩涂资源总量为 342.10×10^4 亩，仅减少了 2.56%；2000 年 0m 等深线以上滩涂资源总量为 100×10^4 亩，至 2014 年，0m 等深线以上滩涂资源总量为 117.16×10^4 亩，增加了 17.16%。

第一节　滩涂资源分布现状

根据近年来对上海河口和近海地质环境的调查发现，在长江口水域潜伏的沙体较为丰富，从长江口南支的白茆沙、扁担沙和新浏河沙，北支的崇头边滩和北沿黄瓜沙，南港的瑞丰沙，北港的中央沙、青草沙和六滧沙，南槽的江亚南沙，口门处存在一些较大型的沙体，如崇明东滩、北港北沙、横沙东滩、九段沙和南汇边滩，共计约 15 个较大型沙体（图 34-1）。上海潮滩有分布面积广、影响因素多的特点，其形态变化规律特别复杂。

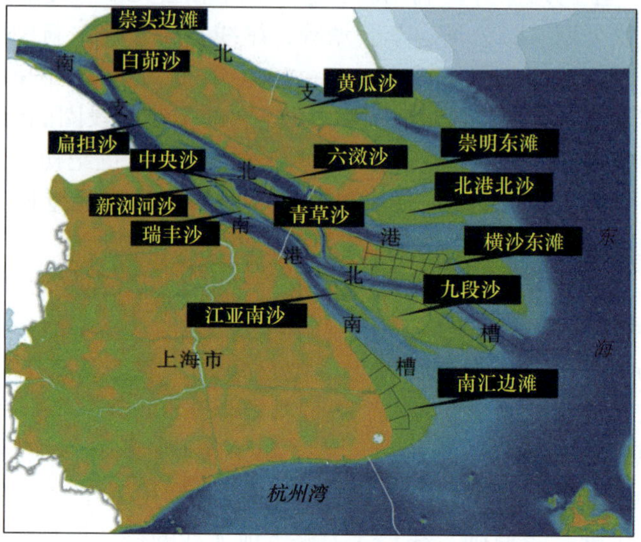

图 34-1　上海主要滩涂分布示意图

Fig. 34-1　Distribution of the main tidelands in Shanghai

一、崇明东滩

对崇明东滩 0m、-2m 和-5m 等深线进行对比，2011—2013 年间崇明东滩北侧 0m 和-2m 等深线略有侵蚀向南偏移，后 2013—2017 年逐渐向北偏移淤积，沙尾处 2011—2017 年间逐年向外延伸，其中 2013—2017 年淤积最为明显。在北港北汊道北侧口门处形成一些-2m 心滩；北港北沙南侧在 2013 年形成一个-2m 的阴沙心滩，在 2013—2017 年间逐渐淤积变大，有与北港北沙并靠的趋势；北港北沙变化相对较小；-5m 等深线变化相对较小，只在佘山岛附近略有侵蚀现象（图 34-2）。

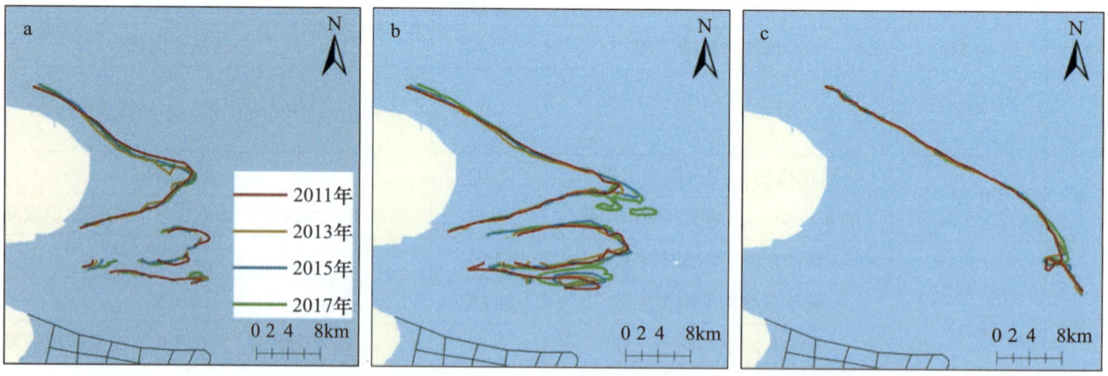

图 34-2　崇明东滩 0m（a）、-2m（b）和-5m（c）等深线变化情况

Fig. 34-2　Variations of the 0m (a), -2m (b) and -5m (c) isobaths in Chongming East Shoal

2011—2013 年间崇明东滩 0m、-2m 等深线的包络面积变小，但是在 2013—2017 年间包络面积逐渐增大，0m、-2m 等深线变化趋势基本一致；-5m 等深线变化很小，基本没有趋势性变化（图 34-3）。近年来，崇明东滩和北港北沙总体 0m 等深线包络面积为 115km^2；-2m 等深线包络面积约为 243km^2；-5m 等深线包络面积为 457km^2。

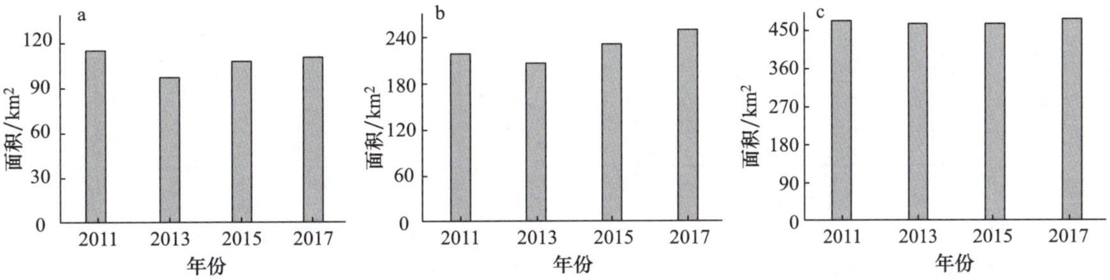

图 34-3　崇明东滩 0m（a）、-2m（b）和-5m（c）等深线包络面积变化情况

Fig. 34-3　Variations of the envelop sizes of the 0m (a), -2m (b) and -5m (c) isobaths in Chongming East Shoal

二、南汇边滩和九段沙

现今南汇边滩基本围垦成陆，在未围垦区内，2017 年南汇边滩 0m 等深线包络面积为 19.2km^2；-2m 等深线包络面积约为 64.8km^2；-5m 等深线包络面积为 276.5km^2。九段沙 0m 等深线包络面积为 63.2km^2；-2m 等深线包络面积约为 115.8km^2；-5m 等深线包络面积为 225.5km^2。

南汇边滩东侧由于围垦堤坝一侧小于 0m 等深线，所以只在南侧存在一些 0m 等深线，2011—2015 年将 0m 等深线向海方向淤涨，但是在 2015—2017 年间出现冲刷后退，而九段沙一侧上沙和中沙变化相对较小，下沙在 2011—2017 年间 0m 等深线发生冲刷后退（图 34-4a）。

南汇边滩东南角处在 2011—2017 年间-2m 等深线都向海推移，平均每两年推进 1km，2017 年新建设堤坝基本位于过去-2m 等深线处，但是 2017 年后新建设堤坝处的-2m 等深线消失，说明水深大于-2m，围垦堤坝的建设加大了河槽冲刷。九段沙一侧上沙和中沙-2m 等深线变化都较小，江亚南沙和九段沙之间河槽发生冲刷，江亚南沙在 2013—2015 年间发生冲刷，沙体顶部和沙尾都有侵蚀后退趋势。九段沙下沙沙尾发生明显冲刷（图 34-4b）。

南汇边滩北侧-5m 等深线在 2011—2017 年间发生明显侵蚀向陆侧推移，在 2011—2015 年间变化较大，2015—2017 年间变化相对较小。江亚南沙-5m 等深线包络沙体在沙尾处冲刷分裂出一个较小的沙体，并且沙尾向九段沙一侧逼近，九段沙在下沙也有明显的侵蚀后退。在 2013—2017 年间变化较为明显（图 34-4c）。

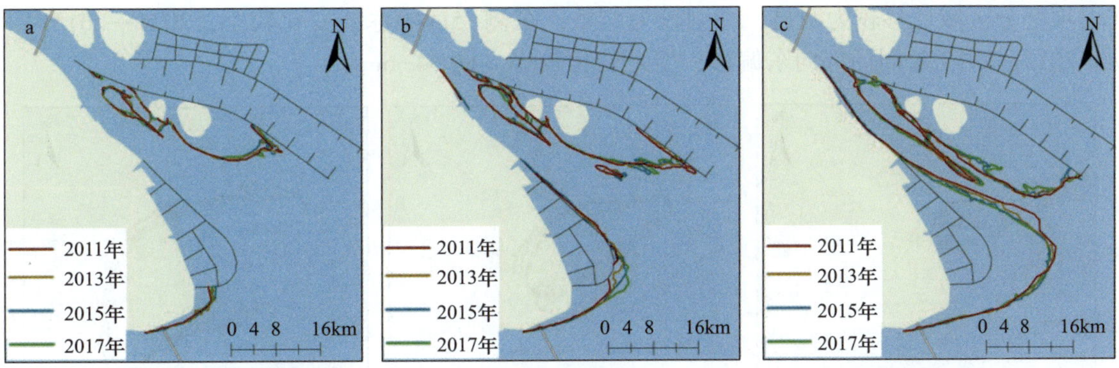

图 34-4　南汇边滩和九段沙 0m（a）、-2m（b）和-5m（c）等深线变化情况

Fig. 34-4　Variations of the 0m (a), -2m (b) and -5m (c) isobaths in Nanhui Shoal and Jiuduansha

计算南汇边滩和九段沙等深线包络面积，可知南汇东滩 0m 和-2m 等深线包络面积在 2011—2015 年间在扩大，处于淤涨态势，后 2015—2017 年间包络面积下降，处于冲刷态势。而-5m 等深线包络面积在 2011—2017 年间一直减小，处于冲刷态势。而九段沙除了在 2011—2013 年 0m 等深线包络面积在增加外，其余都处于逐年减小态势，说明九段沙整体为冲刷减小态势，尤其是 2015—2017 年间 0m 和-2m 等深线包络面积减小较为明显（图 34-5）。

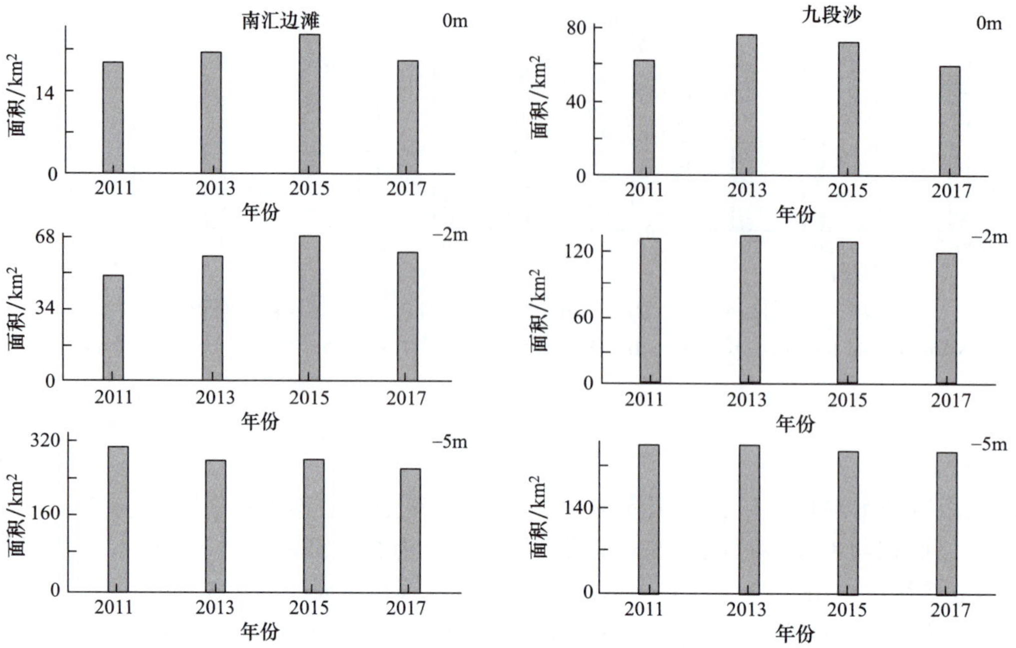

图 34-5　南汇边滩和九段沙 0m、-2m 和-5m 等深线包络面积变化情况

Fig. 34-5　Variations of the envelop sizes of the 0m, -2m and -5m isobaths in Nanhui Shoal and Jiuduansha

三、横沙浅滩

横沙浅滩靠北侧有两块 0m 等深线包络区，2011—2013 年西侧一块包络面积变大，东侧 0m 等深线包络面积减小；2013—2017 年两个 0m 等深线包络面积整体有变大趋势，并且沙体有向东侧发育变大的趋势。西侧 0m 等深线局部略有侵蚀，沙体整体向东迁移并变大（图 34-6a）。

横沙浅滩-2m 等深线包络沙体分布于调查区北侧，西侧和南侧有堤坝圈围的区域-2m 等深线变化较小，趋势不明显；2011—2017 年间北侧-2m 等深线有整体向南迁移趋势，东侧-2m 等深线向外延伸发展趋势明显，东南侧-2m 等深线侵蚀后退（图 34-6b）。

横沙浅滩在 2011—2017 年间北侧-5m 等深线向南移动，并趋势基本一致；南侧-5m 等深线向北移动，在堤坝部分区域迁移较为明显，冲刷较为严重；东侧-5m 等深线变化较小，2011—2015 年间向西侧移动，2015—2017 年间局部向东偏移，但是变化较小（图 34-6c）。

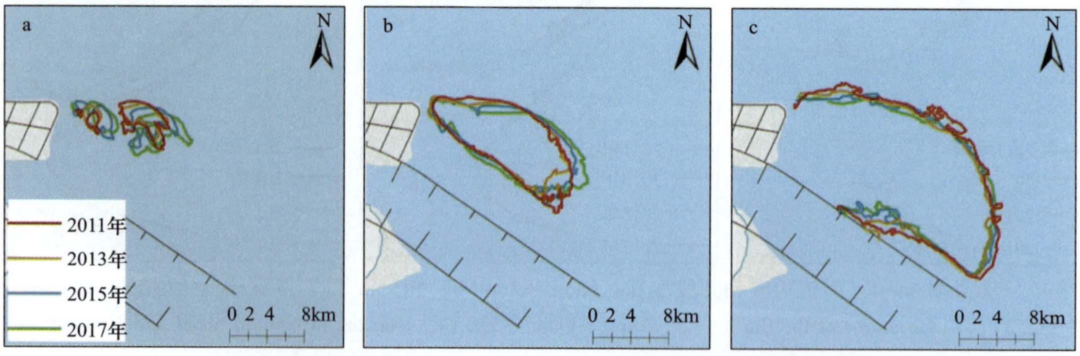

图 34-6　横沙浅滩 0m（a）、-2m（b）和-5m（c）等深线变化情况

Fig. 34-6　Variations ofthe 0m（a），-2m（b）and -5m（c）isobaths in Hengsha Shoal

计算横沙浅滩等深线包络面积，可知 2011—2013 年间 0m 和-2m 等深线包络面积都减小，后 2013—2017 年间 0m 和-2m 等深线包络面积逐年变大，但是 0m 等深线包络面积变化更为明显；-5m 等深线包络面积都在逐年减小，2011—2013 年减小幅度最为明显（图 34-7）。近年来横沙浅滩 0m 等

深线包络面积为 22km²；-2m 等深线包络面积约为 91km²；-5m 等深线包络面积为 255km²。

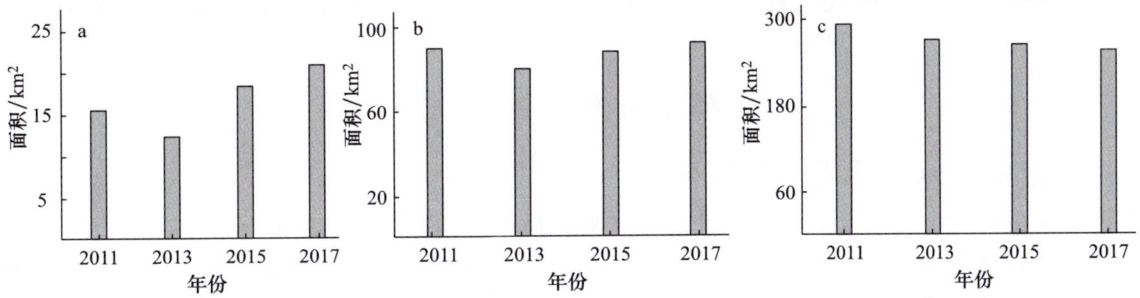

图 34-7　横沙浅滩 0m（a）、-2m（b）和-5m（c）等深线包络面积变化情况

Fig. 34-7　Variations of the envelop sizes of the the 0m（a），-2m（b）and -5m（c）isobaths in Hengsha Shoal

四、扁担沙

基于 2018 年地形资料统计，该区中高、低滩面积（-5m 等深线以上）为 121.18km²（表 34-2）。其中-2m 等深线以上的面积为 82.6km²，0m 等深线以上的面积为 49.3km²，2m 以上的面积为 9.4km²。

表 34-2　2018 年扁担沙面积统计

Table 34-2　Statistics of the area of Biandansha Sandbar in 2018

水位/m	面积/km²		
	上扁担沙	下扁担沙	扁担沙
2	4.93595	4.47275	9.40869
0	12.62174	36.65512	49.27686
-2	16.67547	65.87583	82.5513
-5	19.86662	101.31769	121.18431

对比扁担沙尾-2m 等深线变化，在 2011—2018 年间-2m 等深线变化复杂，基本以左右摆动为主，趋势性变化相对较小。中间阴沙分布较多，扁担沙尾南侧以侵蚀为主，北侧以淤积为主，基本是北淤南冲态势（图 34-8a）。-5m 等深线变化与-2m 规律相似，扁担沙尾发生局部侵蚀后退，基本态势以摆动为主（图 34-8b）。

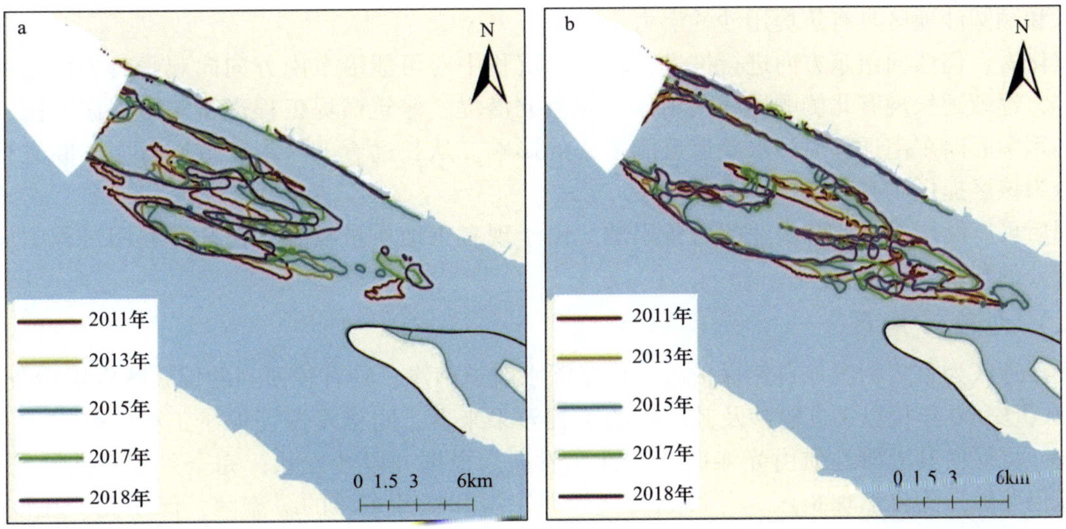

图 34-8　扁担沙-2m（a）和-5m（b）等深线分布情况

Fig. 34-8　Distribution of the -2m（a）and -5m（b）isobaths in Biandansha

第二节 滩涂资源开发利用和促淤圈围用砂潜力

一、滩涂资源开发利用潜力

（一）崇明东滩—北港北沙地区

在该区内具有两个自然保护区，面积较大，尤其是中华鲟国家自然保护区，其核心区几乎占据了崇明东滩的-5m等深线以上滩涂，包括缓冲区和试验区在内则还包括了北港北沙—崇明浅滩地区。从崇明东滩的演化趋势看，受进入北支的潮道和南侧北港北沙北潮道的限制，在非人为因素下其向东和南侧向的进积空间很小。向北的进积空间较大，与崇明北沿相连后成为重要的后备土地资源增长空间。可以通过实施高滩圈围、中滩生物促淤工程，加速滩涂淤涨，保护区面积短期内能够得到恢复。

北港北沙地区-5m等深线以上资源十分丰富，-2m等深线以上资源具有增多的趋势，且总体上仍然处于向南东的推进之中，是上海重要的一片后备土地资源区。从发育形态来看，与20世纪70年代的团结沙类型（90年代初圈围成陆）相似。团结沙促淤圈围过程中，促淤效果明显，并且并未因靠岸而严重地破坏东滩和临近水域的生态环境，相反促使了北港北沙和浅滩的发育。从这点上看，北港北沙地区完全可以复制团结沙的成陆过程。可以早期利用种青工程等加快淤积，北港北沙—崇明浅滩地区（Cr、S、Cd、Cu等元素含量较低）土壤几乎全部为一级质量区，相比较保护区核心区，土壤质量较高，单此因素表明加快淤积后可以用来替代保护区的核心区。如果该区能够将崇明东滩置换出来，将大大增大后者的综合利用潜力。或者在适当时候实施筑坝阻汊连沙靠岸工程使其成陆，促使其北缘和前缘加快淤积，向东和北部推进。六滧沙脊位于北港北沙西端，两者由北港北沙北汊道切割，为一条形近岸沙体，理论上可类似东风西沙一样促淤圈围成陆。

总体上从土壤环境质量角度来讲，该地区沉积物环境质量总体比较良好，可以用于自然保护区、集中式生活饮用水源地、茶园、牧场等保护区建设等，也适用于一般的农用地来进行使用。

（二）横沙浅滩地区

横沙浅滩地区低滩资源量极其丰富，高滩资源量相比崇明和南汇沿岸地区稍次。是目前大规模进行促淤圈围的地区。尽管低滩资源的促淤圈围成本较高，但长远看，该区是上海最具有潜力和面积最大的后备土地资源区。且从周围地理和地质条件看，促淤圈围的条件和开发潜力较高。近邻横沙岛，便于开发。南侧紧邻深水航道，可以利用航道疏浚挖沙进行吹填。从表层和沉积物来看，在临近的北港口地区和顾圆沙地区具有筑堤用沙的潜力。

从演化看，仍然向南东方向进行推进，较快的沉积中心可能沿东南方向向鸡骨礁方向迁移。可以通过种青、逐渐延长现有北侧顺堤和利用深水航道北潜堤、修筑横堤工程逐步实施促淤圈围工程。但要注意沉积中心向东南迁移对深水航道可能带来的影响。从长远看，深水航道运行中不断进行疏浚形成的泥沙为该区提供了有利条件。

圈围后可广泛用于工业和农业的后备用地。由于现有土地质量较好，在开发利用过程中应该注重保护和充分的利用。

（三）九段沙地区

九段沙地区为重要的湿地自然保护区，位于深水航道南缘，具有较便利的促淤圈围用沙条件。

从20世纪90年代以来九段沙及其周围的工程效果来看，间接发挥了促淤工程的作用，使其垂向沉积加快，也促使其潮滩性质由光滩向盐沼性质发展。根据九段沙现状，完全可以在上段筑堤圈围，所得土地仍然保持为自然保护区。并形成类似崇明东滩和横沙东滩的缓流区，促使泥沙在该区淤积。提高土地利用的社会效益。从沉积环境看，其前缘仍然主要处于斜坡沉积环境，长远来看仍然处于较高的沉积速率阶段。

（四）南汇边滩—杭州湾地区

南汇边滩地区经过大规模的促淤圈围，尽管高滩资源已经大幅度减少，但低滩资源十分丰富，仅次于横沙东滩地区。由于其历来是上海陆缘淤涨速度最快的地区，0m 等深线每年的外伸速率在 100m 左右，总体上不断地向东推进，通过种青等工程可大幅度在短期内增加高滩资源量。且该区从地理位置上看紧靠临港新城，交通条件比较便利，对后备土地资源的需求比较强烈。从实测资料和冲淤分析来看，垂向和侧向的沉积速率较快。由于本区来沙丰富和水体含沙量较高，根据河道整治指导线要求，可用人造导堤方式，在水下浅滩结合促淤筑坝分期逐步实现浅滩的淤涨成陆。

杭州湾地区，总体上处于不断的冲刷趋势中，综合来看，高、中低滩资源十分有限，开发利用的综合潜力很低。

沉积物质量较好，两区在综合利用上，可广泛用于工业和农业的后备土地用地。

（五）崇明北沿边滩地区

崇明北沿是本市可开发利用中高滩资源比较集中的区域，为目前重要的促淤圈围区域。并且由于北支处于不断的淤积和萎缩之中，临近崇明岛，该区综合利用的潜力较高。

崇明岛北沿潮道处于不断的萎缩之中，水体含沙量较高和落潮流南偏为该区的淤积提供了极好的水动力条件。局部促淤年淤积速率达 0.5～1m，效果极好。并且为边滩性质，促淤圈围成本较低，是重要的后备土地资源区。从长远来看，可能与崇明东滩相连。

北支主要受潮流影响，含盐量较高，在崇明岛北沿部分地区，尤其是下段，盐渍化程度较高，这些耕地后备资源区，随着围垦进度就要不断完善区内的水利设施，及时对区内土地进行脱盐改造。

二、促淤圈围用砂潜力

从区内长江口地区砂源地的分布来看，在扣除自然保护区、崇明北沿和南汇南滩光缆和管道登陆区以及《长江中下游干流河道采砂规划上海段实施方案》中的禁采区后，区内可用于未来重要促淤圈围区的袋状砂、堤心砂资源十分有限。总体上，在近岸地区有南支主槽—新浏河沙沙包、下扁担沙地区、北港潮流脊—六滧沙脊尾部、顾圆沙东北部、北港口—鸡骨礁地区、江亚南沙尾部、南汇边滩和杭州湾等 8 个地区还有一定的利用潜力，但从具体开采利用上还需要进行进一步论证。

南支主槽—新浏河沙沙包。沉积物较粗和较纯，呈现为以径流作用为主的分流河道沉积。新浏河沙沙包垂向上较厚并且粒度较粗，并且处于不断的下移之中；但在白茆沙北水道，受潮流作用影响，垂向上多表现为泥沙互层，尤其是上部。

下扁担沙地区。边缘砂体厚度均可达 12m。在靠近新桥通道的南缘和砂体中部，沉积物粒度较粗和较纯。但在靠近新桥水道的一侧，泥质层较多，构成泥沙互层或夹层。结合河道整治，可用于扁担沙、东风西沙和中央沙的促淤圈围工程。

北港潮流脊—六滧沙脊尾部。局部较粗，呈粉砂质砂。在六滧沙脊尾部砂层厚可达 11.2m，上部 10m 较纯，6～10m 偶含泥质团块、条带、薄层。沙脊向两侧，上部也见泥质夹层或互层，尤其是靠岸一侧。北港潮流脊一带厚在 6m 左右，5m 以下含泥质薄层和条带、团块较多。靠近南侧北港下段主流路的一侧，上部多呈现为泥沙互层。可以用北港北沙、六滧沙脊和横沙东滩地区促淤圈围的袋装砂和堤心砂。

顾圆沙东北部。为顾圆沙砂体的一部分，砂体较厚，可达 10m。尽管表层为中细砂，但垂向上看砂质不纯，总体上以粉砂、粉砂质砂为主。可用于崇明北沿和北港北沙地区的促淤圈围堤心砂。

北港口—鸡骨礁地区。北港口一带的砂总体上较粗和较纯。区内历史沉积速率较高，现阶段和未来数百年的沉积中心都位于鸡骨礁地区。并且水体泥沙含量较高，北港径流作用较强，泥沙源丰富，开挖后易于恢复。由于该区砂质优，分布广，储量大，即使将来促淤圈围，在筑堤工程中也可大量取砂，内砂外用。再寻找其他砂源进行吹填加高。资源量估算表明，北港口—鸡骨礁地区总的砂体体积约为 $3934\times10^6m^3$，可用于北港北沙和横沙东滩等邻区的袋装砂和堤心砂。除北港口地区外，南北港分流口以下的河道整治获取的砂泥资源只可用于吹填用砂。

江亚南沙尾部。江亚南沙南缘粒度较粗且尾部和南缘淤积较快，在江亚南沙未被划入九段沙保护区内的范围可以进行进一步的论证和调查，寻找袋装砂和堤心砂。

南汇边滩。沉积物较粗，但部分层位含泥量较高，砂体总厚度可达7.5m，较纯的砂性土可达数米。区内水域开阔，坡度平缓，水体近年来含沙量较高，经开挖取砂后，潮滩地形能较快得到恢复。该区曾在-2m等深线水深以内边滩取砂数千万立方米，经实测，经过围堤工程以及大量取砂开挖后，滩面仍处在不断的缓慢淤涨过程中，反映了在大规模围堤工程后长江来沙对堤外滩区继续得到较好的补给。可用于原南汇促淤圈围的袋装砂和堤心砂。

杭州湾地区。除下金山深槽一带存在浅部砂源外，总体上深部砂源潜力巨大，分布在20~25m以深层位，向西和南较浅。浦东南汇边滩地区埋深在25m左右。上部以粉砂质砂为主，细砂含量为9%~51.3%，极细砂含量为17.7%~48%。与现代长江三角洲前缘的砂体相比，该砂体略细，细砂含量较低，泥质夹层较少，较连续。

第三十五章　旅游地质

上海市内旅游地质资料丰富，包括多种地质遗迹资源和多处古文化遗址。主要有金山沿岸及其外侧海域的基岩残丘、古海岸线和不同历史时期的海堤遗迹、局部的古人类文化遗址，以及广布的现代滨海河口地貌和湿地资源、水域基岩岛屿的海蚀地貌等。此外，在金山张堰分布有上海唯一的隐伏型固体金属矿床。其中，地质遗迹资源极其具有特色，除观赏价值外，还具有重要的科学价值。

2000 年，上海市地质调查研究院开展了全市的地质遗迹调查，初步查明了上海地质遗迹的类型、景观、形成时代、分布现状和自然或人为因素对其的影响程度，开展了主要地质遗迹评价，初步提出地质遗迹保护设想，对遭破坏的地质遗迹提出抢救性保护办法。将上海地质遗迹按成因与形态原则划分为 2 类 6 种，一类是地质作用地质遗迹，另一类是人为-地质作用地质遗迹。

2003 年，为申报国家地质公园，上海市地质调查研究院和华东师范大学河口与海岸国家重点实验室联合开展了以“崇明地质——湿地公园”为主题的科学考察与研究，形成了《崇明东滩湿地国家地质公园综合考察报告》和《崇明东滩湿地国家地质公园总体规划报告》，系统地提出了崇明岛地质遗迹类型、内涵，并对在东滩湿地建设地质公园提出了布局规划。

2006 年，崇明区旅游局委托华东师范大学对崇明岛地质遗迹展开新一轮的调查，并制定了地质遗迹保护与地质旅游发展规划。此次调查在考察崇明东滩基础上，对崇明岛西端的西沙湿地地质遗迹景观区也进行了详细的调查。

2010—2013 年，上海市规划与国土资源管理局响应国土资源部地质公园管理的要求，编制崇明岛国家地质公园规划，规划将地质遗迹调查范围进一步扩大到崇明岛西北端的牛棚港湿地和北部的北湖。

2017 年，中国地质环境监测院委托上海市地质调查研究院参与华东地区开展重要地质遗迹调查，对上海陆域内地质遗迹进行了新一轮的核查。此次调查按照全国最新的地质遗迹调查规范，按照成因联系和组合对域内地质遗迹重新分类，确定重要地质遗迹 4 处，其中地貌景观类 3 处、地质灾害类 1 处。挖掘了上海各类地质遗迹的地学内涵和独特性，首次明确提出长江河口三角洲具有全球独特性，是上海域内唯一 1 处世界级地质遗迹。

第一节　地质遗迹类型及特征

按照地质遗迹调查规范，上海已有的地质遗迹可划分为两大类 4 类 4 亚类，即地貌景观大类和地质灾害大类两大类，水体地貌类、海岸地貌类、火山地貌类、地质灾害类 4 类，河流（景观带）、海积地貌、火山岩地貌和地面沉降 4 个亚类。根据地质遗迹典型现象组合，总结为 4 处地质遗迹，分别命名为上海长江河口三角洲、上海全新世古海岸、上海火山岩残丘和上海地面沉降遗迹（表 35-1）。

表 35-1　上海地质遗迹类型

Table 35-1　Geological relics types of Shanghai

大类	类	亚类	遗迹名称	遗迹特征
地貌景观	水体地貌	河流（景观带）	上海长江河口三角洲	三角洲地貌：河口汊道、河口沙岛、河口沙洲、河口半咸水湖、河口淤泥质潮滩； 三角洲地质：晚新生代松散沉积物厚覆盖
	海岸地貌	海积地貌	上海全新世古海岸	古贝壳沙堤、古沙堤等古海岸线
	火山地貌	火山岩地貌	上海火山岩残丘	火山岩小丘、火山岩岛、矿坑
地质灾害	地质灾害	地面沉降	上海地面沉降遗迹	外滩历史构筑物中存在的近百年来上海地面沉降遗迹

一、地质遗迹特征

（一）长江河口三角洲

由5个地质遗迹点组合而成，包括崇明东滩潮滩、崇明北湖、崇明西沙潮滩、崇明北支河口分汊景观点和九段沙河口沙洲。该遗迹点特征如下：

1）位于长江河口，地处中国海岸线中点，覆盖面积居世界第二，河口入海径流量居世界第五，年入海泥沙量居世界第四。

2）长江三角洲是典型的潮控型三角洲，平均潮差2.5m，在不规则半日潮作用下，形成河口独特的地貌、生态景观和水文泥沙现象。地表形成了“三级分汊、四口入海”的宏观地貌，可见沙岛、沙洲、汊道等地貌景观。其中，崇明岛是世界上最大的河口冲积沙岛。

3）在不规则半日潮作用下，河道边滩和河口沙岛、沙洲四周发育了独特而精美的河口淤泥质潮滩地貌，分为高、中、低潮滩和潮滩内蜿蜒曲折的潮沟体系。潮滩地貌的分带性产生了沉积物类型的分带性和生态环境的分带。高潮滩，沉积物以泥为主，是大型底栖动物（螃蟹）最佳繁殖区域，地表生物扰动最强烈，泥滩上形成了生物扰动地貌，也是泽林和芦苇的主要生长地带，也见碱蓬等盐生植被。中潮滩，沉积物以砂泥互层为主，也称砂泥混合滩，水动力相对变强，是海三棱藨草的主要生长地带；低潮滩以砂为主，也称光滩，只见低等藻类，在不同潮汐动力作用下，形成不同形态的波痕微地貌（沙波地貌）。潮沟同样塑造了潮滩地貌形态，在高潮滩溯源侵蚀，形成深而窄的潮沟和树枝状潮沟，在光滩潮沟变宽变浅流入河口，并在低潮时侵蚀光滩形成陡坎地貌。

（二）全新世古海岸

上海境内发育数列贝壳沙堤和沙堤。该遗迹仅在金山曹泾沙积村完整保留原生地貌形态，全新世古海岸遗迹特征如下：

1）上海古贝壳沙堤和沙堤发育于全新世6~7ka以来，一般厚0.5~1m，宽100~150m，在剖面上呈厚、宽比1∶100左右的薄层透镜体。贝壳沙堤主要由滨海相的贝壳和细砂组成。而沙堤主要由褐黄色、草黄色细砂、粉砂组成，有时含有较多的贝壳。这些贝壳沙堤和沙堤主要分布于崇明东部、“冈身带”和原南汇东部地区。

2）上海西部湖积平原边缘的“冈身带”发育3~4列贝壳沙堤，^{14}C测年结果表明其形成于6.5~3.0ka，最西端的吴塘冈—沙冈距今有6500年；中间的外冈—竹冈距今有5000年；东部的石冈—横泾冈距今有3000年。东部滨海平原浦东分布的两列古沙堤（潘泓沙和奉城沙堤）距今有600~700年，崇明裕安村分布的两列古沙堤距今有600~900年。

3）上海古海岸遗迹绝大多数已遭受破坏。目前，上海唯一较为完整的遗迹点在金山曹泾，遗迹特征如下：由海生贝壳类壳体和沙聚集而成的沙脊，称之为“冈身”地貌。根据南宋《云间志》记载，现存“冈身带”南北走向，长约40m，宽约20m，面积约800m^2，高出地面或伸入地表各1.5m，表土0.15m以下为泥沙夹层，0.3m以下见大量贝壳碎片。贝壳^{14}C测年结果为6024ka。

（三）火山岩残丘

该遗迹是指上海西部湖积平原和杭州湾及长江口散落分布的23个火山岩残丘。遗迹特征如下：

1）残丘岩性以安山岩为多，其次为凝灰岩、流纹岩、正长斑岩等，形成于距今约130Ma的早白垩世陆相火山喷发活动。

2）残丘规模均较小，23个残丘面积共计3.37km^2，占全市陆域面积的0.054%。残丘海拔也较低，最高仅逾100m。

3）近百年来，残丘多经历过矿石开采活动，遗留了较大的矿坑。

4）天马山残丘见有节理和断层等构造。

（四）地面沉降遗迹

指反映近百年来上海城市发展历程中地面沉降的标志性历史构筑物，包括外滩风景区内黄浦江大堤、历史建筑及苏州河历史桥梁。遗迹特征如下：

上海外滩防汛墙及沿线标志性历史构筑物记录了近百年来上海强烈的人为地下水利用开发活动导致的区域地面沉降灾害。外滩防汛墙全长 1.5km，南起延安东路，北至苏州河上的外白渡桥，东面即黄浦江，根据历史数据记载：1920—1993 年，防汛墙经过 4 次加高，高程从 3.5m 加高至 6.9m。1994 年以来进行常态化监测：1994—2004 年，累计沉降量平均为 273.92mm，平均沉降速率为 26.09mm/a；2004—2009 年，累计沉降量平均为 28.92mm，平均沉降速率为 6.08mm/a；2010—2013 年，地面呈回弹趋势，平均回弹 3.29mm，其中黄埔公园—南京东路段回弹 5.55mm。

外滩沿线的历史建筑多可见一楼窗户已接近地面，室内与室外形成明显倾斜，反映了百年来在当时桩基条件下建筑发生了整体的沉降。

外白渡桥、乍浦路桥等历史桥梁建成于 19 世末—20 世纪初，由于处于苏州河与黄浦江的交界处，因此成为连接黄浦与虹口的重要交通要道，曾具有通航重要功能。由于长期地面沉降，这些桥梁净空逐渐减少，目前观察桥墩净空不足 2m，无法用于通航。反映了沉降幅度应在 3m 以上。

二、地质遗迹分布规律

区内地质遗迹的空间分布主要受到地形地貌、断裂构造和人类历史活动的控制。依据地貌和地质构造特点，上海地质遗迹分布规律具有如下特征。

（一）上海长江河口三角洲

上海长江河口三角洲地貌遗迹景观集中分布于河口沙岛区，主要有 5 个景观点，分别是崇明东滩潮滩、崇明北湖、崇明西沙潮滩、崇明河口分汊和崇明九段沙河口沙洲。

1. 崇明东滩潮滩

崇明东滩位于崇明岛的最东端，主要由团结沙外滩、东旺沙外滩、北八滧外滩、潮间带滩涂湿地和河口水域（-6m 线以上）组成。崇明东滩分布着长江三角洲面积最大、保存最为完整的潮滩地貌。潮滩由潮上带、中潮滩、低潮滩和潮下带组成。潮上带指平均大潮高潮线以上的淤泥质沉积地带，潮间带指平均大潮高潮线和平均低潮线之间的滩涂，这两个区域的滩涂受周期性海洋潮汐影响。东滩潮下带非常宽，一直延伸至 20km 外的佘山岛，潮下带的水深约为 5m。东滩保护区地貌的一个重要特征是潮滩区中有大大小小众多发育良好的潮沟，潮沟是由于进潮和落潮时的潮水冲刷而成的，其主要作用是加速了潮水的涨落速度。潮沟在潮滩上的发育形成了众多的生态微环境，具有丰富的生物多样性。

2. 崇明北湖

崇明岛北部北支河道沙洲经过人工围垦并岸形成的特殊且唯一的近原生态半咸水湖泊景观。位于崇明岛北端，北湖及其滩涂东西长 17km，南北宽约 2km，湖泊面积为 18km^2。它是长江三角洲地区十分独特的半咸水湖，原先是由长江口冲积沙岛“黄瓜沙”与崇明岛主体部分合围而成的狭长形海湾，后经人工筑坝合围后形成了一个与河口具有潮汐通道相连的半咸水湖。目前北湖仍处于原始封闭的状态，几乎没有人为干扰，为理想的水鸟栖息环境。水面宽广，拥有滩涂、芦苇荡等绝佳的隐蔽栖息环境和各种鱼类、底栖动物等鸟儿丰富的食源。

3. 崇明西沙潮滩

崇明西沙潮滩位于崇明岛西南端淤泥质潮滩内，是全国唯一淡水的、具有大型潮汐现象的沼泽型国家级湿地。湿地内保存着多种由潮涨潮落等河口水文现象形成的地貌地质遗迹，有潮滩、潮沟、波痕、陡坎等，也保存着潮滩植被、鸟类、底栖动物等河口生态景观。

4. 崇明河口分汊

崇明河口分汊位于崇明岛西段北支河道边滩，是理想的观察长江河口一级分汊的理想地点，也发育着记录北支特殊动力条件下河口汊道演化的潮滩地貌。

5. 崇明九段沙河口沙洲

崇明九段沙河口沙洲是现代长江口第三代新生江心沙洲。九段沙南沿-6m 等深线与南汇边滩相连。

1950年代以前，九段沙是横沙东滩（铜沙浅滩）的组成部分。1954年的特大洪水使-5m等深线深槽贯穿整个串钩形成北槽，九段沙开始成为一个独立沙体。-5m等深线包络范围长约48.9km，南北宽约12.7m。2011年的监测结果表明，九段沙0m等深线以上面积为208.41km^2，0~-2m等深线包络面积为68.32km^2，-2~-5m等深线包络面积为137.08km^2。九段沙属于河口沙洲，海拔在2.5~3.5m。根据地势高低，其地貌形态可以划分为高潮滩、中潮滩、低潮滩和水下浅滩。微地貌形态可以分为规模不等的潮沟构成的近辐射状的退潮排水系统和沙洲中部的沼泽性浅洼地，而沙洲边缘的滩地则因涨潮流和落潮流的冲淤作用不同，微地貌形态也有差异。

（二）上海全新世古海岸

上海古海岸遗迹绝大多数已遭受破坏，目前唯一完整的遗迹点分布在西部湖积平原的金山曹泾沙积村，遗迹范围约为800m^2。据全新世岩相古地理环境，现长江南岸地区的古海岸线主要沿北西-南东向展布。

（三）火山岩残丘

上海火山岩残丘19个分布在陆域湖积平原，4个分布在杭州湾和长江口。大部分位于松江，少数分布在金山和青浦。

松江的15座山丘由东北向西南伸展，依次为北竿山、凤凰山、薛山、厍公山（苍公山）、小山、东佘山、西佘山、辰山、罗山、钟贾山、天马山、后山、机山、横山和小昆山。

金山陆域与杭州湾各有3个残丘和3个基岩小岛，由西北向东南依次为秦皇山、甸山、查山、小金山岛、大金山岛和浮山岛。

此外，青浦有一个淀山，长江口外有一岩石裸露的佘山岛。长江口还有一个鸡骨礁也是由基岩组成。

（四）地面沉降遗迹

上海仍存在地面沉降现象，但由地面沉降致灾的地面沉降遗迹主要分布在上海中心城外滩风景区内的黄浦江防汛墙、历史建筑及苏州河历史桥梁。

第二节　上海古文化遗址概述

经过考古调查和发掘，上海嘉定外冈、方泰，经闵行诸翟、俞塘和奉贤南桥，直抵杭州湾北岸的奉贤柘林一线以西，至今已发现25处古文化遗址。已发掘的有福泉山、崧泽、亭林、戚家墩及马桥等多处。按所处环境划分，有土墩遗址6处，山坡遗址1处，“冈身带”（高地）遗址2处，湖底遗址1处，平地遗址15处。

经发掘获得的地层资料和文化层由下而上的时代序列是：马家浜文化—崧泽文化—良渚文化—马桥类型—亭林类型—戚家墩类型—汉代文化。现将主要文化类型的绝对年代以及出土的代表器物介绍如下：

一、马家浜文化

以崧泽遗址下层为代表。出土器物有腰檐釜、敞口窄肩平底盆、鸡冠耳罐和黑衣灰陶罐残器、小镂孔豆把牛鼻形器耳、炉箅、弧刃梯形斧和玉块等。从灰坑中出土的木头，经^{14}C年代测定为距今5985±140年（曾用树木年轮校正，下同）。

二、崧泽文化

以崧泽遗址中层和汤庙村下层为代表。

崧泽遗址中层的出土器物主要有盆形粗把豆、花瓣形足杯、弧腹四横鼻圈足罐和弧腹凸棱平底罐等。同层位中的人骨经^{14}C年代测定为距今5180±140年。

汤庙村下层出土器物有扁平弧刃穿孔斧、折腹扁凿足鼎、侈口大陶罐、束腰把豆、直口弧肩罐、四系筒腹圈足罐、三耳穿孔盆、直口折腹盆、假腹盘、花瓣足筒腹杯、花瓣足凸棱筒腹杯、平底筒形杯、带钮盖等。同层3座墓的4件陶器经热释光测定平均年代为距今4950±245年。

三、良渚文化

以果园村和马桥遗址下层为代表。该层出土器物为平刃穿孔斧、有段锛、浅盘大镂孔高圈足豆、实足盉、高圈足筒形杯、扁三角形鼎足、黑衣灰陶细刻纹罐等。该层灰黑土层出土木头经^{14}C测年为距今4505±145年。

马桥遗址下层出土陶器，经热释光测定取得下列4个年龄数据：黑衣灰陶宽把杯残把4490年；锥刺纹橘黄陶罐口沿4510年；夹细沙黄褐陶鬶袋足4550年；黑衣灰陶实足盉口沿4100年。

四、马桥类型文化

以亭林文化遗址为其代表。该层出土器物有石镰、石耘田器、骨锥、扁侧足硬陶豆、圈凹底篮纹罐和方格纹罐、方格纹圜凹底盆、夹沙绳纹釜等。从灰坑中出土木头，经^{14}C年代测定为距今3730±150年。

上海古文化序列的建立以及大量出土器物的发掘和其年代的测定，为区内全新世中、晚期古地理面貌的恢复及详细的地层断代提供了重要的依据。

第三节　上海崇明岛国家地质公园概述

2005年9月19日，崇明岛国家地质公园通过国土资源部评审，正式获准成立。划分为东滩园区、西沙园区、牛棚港园区、北湖园区、东平园区共5个园区，属于中型地质公园。

公园地质遗迹景观主要包括，地下埋藏的漫长地质时期地壳升降和海面波动等内、外力共同作用下形成的第四系厚覆盖层和下伏基岩地层及隐伏断裂构造，地表长江河口复杂动力塑造的大河分流、淤泥质潮滩、湖泊水体等地貌和独特的潮滩植被、珍稀国际迁徙候鸟和以“活化石”中华鲟为代表的珍稀鱼类等生态景观，以及历史时期自然或人为作用形成的海岸线变迁遗迹。此外，公园内还分布着与河口沙岛演化密切关联的华东地区著名的水杉人工森林。主要特色有：

1）世界上最大的河口沙岛，是尖头状三角洲地貌体的一部分；

2）具有复杂河口动力体系作用下的大河分流地貌、淤泥质潮滩地貌、半咸水湖地貌及独特植被、鸟类、鱼类地质生态景观；

3）具有历史时期海岸线变迁遗迹；

4）具有典型潮控三角洲地层层序；

5）具有晚新生代以来地壳升降运动和海平面升降运动内、外力作用下形成的巨厚松散覆盖层，松散层以下深部地层蕴含着地热矿产资源；

6）历史悠久、生态优良，有“东海明珠”“瀛洲净土”的美誉；

7）有华东著名的人造森林——东平国家森林公园，其内广泛栽植了中国特产珍稀树种、具有植物“活化石”之称的水杉；有被国际组织列入国际湿地名录的东滩湿地，该湿地也是国家级鸟类自然保护区。

一、地质遗迹景观

公园所在地崇明岛是世界上最大的河口沙岛，属于长江三角洲地貌体的一个组成部分。现代的长江河口地区在地貌形态上呈现出“三级分叉、四口入海”的格局，地貌组成由陆向海分别为三角洲平原、三角洲前缘和前三角洲。在复杂河口动力体系塑造下，岛屿四周发育了大片淤泥质潮滩，潮滩内潮沟、波痕、陡坎地貌十分精致，植被、鸟类多样且独具特色，并且在自然淤涨和人工围垦共同作用下，岛内还形成了独特的内陆淡水湖和半咸水湖地貌。

按照地质遗迹类型划分表，结合公园地质遗迹成因和形态的特点，公园内的地质遗迹分为7大类9类11亚类23种：一是地质剖面大类，包括第四纪和基岩地层；二是地质构造大类，主要指区内隐伏断裂；三是矿物与矿产大类，主要指突出地方地质特色的地下水和地热；四是地貌景观大类，主要包括流水堆积地貌、海蚀海积地貌；五是水体景观大类，主要包括湖泊景观和风景河段；六是环境地质遗迹景观，主要包括岸线变迁与海岸工程遗迹和古文化遗址景观；七是其他类型遗迹景观，主要指长江口独特而复杂的水动力景观（表35-2，表35-3）。

表35-2　公园地质遗迹类型一览表

Table 35-2　Geological relics types in parks

大类	类	亚类	具体类型
一、地质剖面大类	1. 地层剖面	1. 地方标准剖面	1. 基岩地层
			2. 第四纪地层
二、地质构造大类	2. 构造形迹	2. 区域（大型）构造	3. 隐伏断裂
三、矿物与矿床大类	3. 典型矿物产地	3. 典型矿物产地	4. 地下水和地热
四、地貌景观大类	4. 流水地貌	4. 流水堆积地貌	5. 三角洲平原
			6. 拦门沙
			7. 三角洲前缘
			8. 前三角洲
	5. 海蚀海积地貌	5. 海蚀地貌景观	9. 侵蚀岸线
			10. 陡坎
			11. 潮沟
		6. 海积地貌景观	12. 古（贝壳）砂堤
			13. 潮滩
			14. 涨、落潮槽
五、水体景观大类	6. 湖沼景观	7. 湖泊景观	15. 半咸水湖
			16. 淡水湖
	7. 河流景观	8. 风景河段	17. 南北支和大河分流
六、环境地质遗迹景观	8. 地质灾害遗迹景观	9. 岸线变迁与海岸工程遗迹景观	18. 古海塘
			19. 避潮墩
		10. 古文化遗址景观	20. 新石器文化遗址
七、其他类型遗迹景观	9. 河口水动力景观	11. 河口水动力景观	21. 最大混浊带
			22. 盐水入侵
			23. 潮汐运动

表35-3　公园典型地质遗迹景观

Table 35-3　Typical geological relics landscapes in parks

地质遗迹景观	地质遗迹类型	特征	所在园区	级别
东滩湿地	海积地貌	多沼生草本及芦苇，潮间滩地发育树枝状潮沟	东滩园区	世界级
西沙湿地	海积地貌	具有自然潮汐现象和成片滩涂林地的自然湿地。周而复始的潮汐作用塑造了丰富而精美的潮滩地形地貌和生态系统，具有纤细的潮沟、精致的潮滩波痕、陡坎，广泛生长着高潮滩植被芦苇。潮汐的涨落还孕育了多样性的潮滩动物，具有迁徙到来的候鸟和随着潮汐而来的沙蟹	西沙园区	国家级

续表

地质遗迹景观	地质遗迹类型	特征	所在园区	级别
牛棚港湿地	海积地貌	位于河口分流区，地质遗迹体现了长江口分流的地质地貌变化过程，也是海陆相互作用敏感区，记录了盐水入侵、倒灌进入南支的过程。潮滩内发育了水下沙洲、分流水道、潮沟、波痕、侵蚀陡坎、古潮沟等地质地貌遗迹以及典型的芦苇、藨草等植被和鸟类等生态物种	牛棚港园区	省级
北湖	湖泊景观	是由长江口冲积半岛“黄瓜沙”与崇明岛主体部分合围而成的一个与部分潮汐相通的潟湖，目前仍处于原始封闭状态，是鸻鹬类和雁鸭类候鸟理想的栖息地	北湖园区	省级
明珠湖	湖泊景观	崇明岛最大的天然淡水湖泊，原是长江南支的一条支流，20世纪70年代人工筑堤围造而成，现建有进出口闸门，与长江保持水力联通	西沙园区	省级
裕安沙	海积地貌	分布在崇明新桥—牛棚一带，呈向东突出的弧形砂带，与现岸线平行，宽为100~200m，延长约为7.5km。砂中有夹层，却罕见海相贝壳碎片，表明沉积时水体环境不如滨海沙堤洁净稳定。露头较多，尚无定年资料	园区外围古沙堤保护点	省级
东滩历史海塘	岸线变迁与海岸工程遗迹景观	崇明东滩横亘数列历史人工海塘，主要包括98大堤、92大堤、68大堤等，代表了海岸线的变迁与人类活动的相互作用	园区外围古海塘保护点	省级
崇西大河分流头	风景河段	崇明岛最西端，是长江口南北支河道的分流点，游客在此处远眺能够同时观察到河口河流分流的奇特景观	园区外围地貌景观点	省级
基岩岩心	地方标准剖面	基岩地层，记录了第四纪以前的地质演化历史		
第四系钻孔		第四纪地层，记录第四纪以来的地质演化历史		

公园内的地质遗迹景观价值等级可划分出世界级1处，国家级1处，省级8处。

（一）淤泥质潮滩

东滩，位于崇明岛最东端，是公园内最典型的淤泥质潮滩，其内分布着精美的潮沟、波痕、陡坎等微地貌，具有优美、神秘、复杂多变的美感，记录了潮汐、波浪、风暴潮、植被、地形、泥沙等诸多河口动力地貌因素对地貌的塑造，是研究河口沉积动力体系的良好载体。东滩犹如物种基因库，生物多样性程度高。由于特殊的地理位置和快速淤涨的潮滩环境，东滩是国际候鸟和珍稀鱼类重要的栖息地和庇护所，每年有近百万只的水鸟在此过境、停歇、繁殖、越冬。东滩水域也是中华鲟、鳗鱼等许多珍稀鱼类周年性溯河和降河洄游的必经通道和承载其生活史的重要场所。2002年，东滩已被国际湿地秘书处正式接纳为国际重要湿地。

（二）半咸水湖

北湖，位于崇明岛北端，是长江三角洲地区十分独特的半咸水湖，原先是由长江口冲积沙岛“黄瓜沙”与崇明岛主体部分合围而成的狭长形海湾，后经人工筑坝合围后形成了一个与河口具有潮汐通道相连的半咸水湖。目前北湖仍处于原始封闭的状态，几乎没有人为干扰，是理想的水鸟栖息地。水面宽广，拥有滩涂、芦苇荡等绝佳的隐蔽栖息环境和各种鱼类、底栖动物等鸟儿丰富的食源，深受候鸟的喜爱，成为大批鸻鹬类以及雁鸭类候鸟春秋过境、冬季栖息最理想的“后花园”。

（三）第四系厚覆盖层

崇明岛及周边地区地表以下埋藏的巨厚第四纪松散沉积物是公园区别于许多大川名山、沙漠类地质公园的独特之处。尽管它的可视性较差，但具有全球代表性，即典型大河口地区沉积地质学过程。第四系厚覆盖层地学内涵在于记录了数百万年来新构造运动作用下地壳垂直升降和第四纪气候-海平面波动作用下的海陆交替变迁地质演化过程，同时也记录了长江作为一条全球性的大河何时到达现代河口贯通入海以及长江三角洲和河口沙岛的发育、演化过程，具有极高的科学研究价值。

二、生物及其他自然景观

公园内的海三棱藨草-藨草-芦苇潮滩植被群落、国际珍稀迁徙候鸟、中华鲟等珍稀鱼类以及水杉人工林构成了独特的生态景观。崇明东滩已被《湿地公约》列为国际重要湿地。

三、人文景观

崇明学宫、寿安寺、唐一岑墓、金鳌山等人文史迹，代表了崇明儒家文化历史和佛教历史，崇西水闸、长江边滩观光大堤等代表了河口水文化景观，瀛东村、前卫村、瑞华果园、江南三民文化村等代表了极富地方特色的生态农业文化。

第八篇
区域地质发展史

第三十六章　中元古代—青白口纪

根据建造、改造发育特点，区内地质发展史可划分出古元古代、中元古代—青白口纪、南华纪—中三叠世、晚三叠世—古近纪、新近纪和第四纪等5个演化阶段。

邻区资料显示，古元古代早期，随着上扬子太古宙陆核的孕育发展，华南地区地壳不断向周边增生（江西省地质矿产勘查开发局，2017）。区内金山群的Nd模式年龄和中新世玄武岩中的捕虏锆石U-Pb年龄都显示，大致在古元古代晚期，上海及邻区的原始陆壳雏形应该已基本形成。古元古代末，吕梁运动使扬子原始陆壳克拉通化。

中元古代—青白口纪为区内大规模地壳形成的时期。与华南中元古代、新元古代早期重要的地壳生长、改造期相对应（李献华，1993；郑永飞，2007）。

1900~1750Ma间，很可能与区域伸展构造有关的大规模玄武岩浆活动，引发了广泛的地壳物质重熔，形成花岗岩（李献华，1993），同时发生大规模的变质变形作用。形成的天井坪岩组、八都群原岩显示此时为岛弧、弧后盆地环境（付树超，1991；李献华，1999），可能与哥伦比亚超大陆有关的活动大陆边缘有关。中元古代，随着哥伦比亚超大陆的裂解，华南洋洋盆扩张（江西省地质矿产勘查开发局，2017），至晚期，形成弧后盆地环境的铁砂界岩组。该阶段，区内整体上表现为弧后盆地。由早期向晚期，海水似有加深趋势，形成一套具有复理石建造特征的浅海相沉积，伴随有间歇性的（海底）火山喷发活动，且岩浆作用似有加剧和“双峰式”特征。中元古代末期，区内已显示为造山带环境，发育有不成熟的岛弧，构造体制有由挤压环境向拉张环境过渡的特征，推测已转化为与罗迪尼亚超大陆有关的活动大陆边缘环境。

经过晋宁运动Ⅰ幕作用，区内中元古代建造褶皱变质，形成金山群。新元古代，区域上伴随着古华南洋板块向北持续的俯冲作用，扬子克拉通东南边缘形成江山-璜山-双溪坞岛弧。大约在850Ma，发生弧内裂解和弧后扩张，皖南伏川、皖赣交界等地形成局部洋壳堆积（丁炳华，2008）。830~820Ma，弧陆碰撞导致浙皖赣弧后盆地关闭（邓奇，2016；唐增才，2018）。该阶段，区内地壳似乎进入一个相对稳定阶段，火山活动基本停息，形成相对稳定的弧后浅海—次深海盆地环境，藻类植物开始大量出现。青白口纪晚期的晋宁运动Ⅱ幕造成基底的全面隆升，信钱地块与下扬子地块最后拼接，弧后盆地完全闭合（江西省地质矿产勘查开发局，2017）。区内中元古代—青白口纪早期地层，发生不同程度的褶皱、断裂等变形作用改造，形成岩性单一的“惠南板岩”。与此同时，金山群发生了局部退变质作用。约820Ma后，伴随着罗迪尼亚超大陆的裂解（邓奇，2016），开始形成碰撞后上叠裂谷盆地（张克信，2015），堆积了一套滨浅海相—陆相的火山-沉积建造，且火山作用日趋强烈。该阶段，区内堆积河上镇群他生陆源碎屑岩建造，上部层位含凝灰质或为熔结凝灰岩。

第三十七章　南华纪—中三叠世

青白口纪以后，区内进入以海相、海陆过渡相为主的盖层发展阶段，由中—新元古代的弧盆体系建造逐渐转化为扬子克拉通东南边缘的被动陆缘盆地、陆表海盆地建造（张克信，2015）。中、晚三叠世之交强烈的印支造山运动把区内带入陆相盆地建造发育与改造阶段。

一、南华纪—志留纪

区内南华纪建造未有揭露。区域上，南华纪早期开始海侵，先后形成一套滨岸相—浅海相的紫红色砂砾岩—紫红色和灰绿色砂岩、凝灰质砂岩、粉砂岩和沉凝灰岩韵律层—含黄铁矿灰黑色硅质岩—灰绿色凝灰质细砂岩和青灰色粉砂质泥岩、凝灰质泥岩、粉砂岩等夹较多沉凝灰岩建造（浙江省地质矿产局，1996）。南华纪中晚期，随着全球性大冰期的来临，进入冰期沉积，主要形成冰川海洋沉积，多为浅海环境，局部为滨海和湖盆（江西省地质矿产勘查开发局，2017）。

震旦纪开始，气候转暖，区内发生广泛海侵，接受了一套以镁、硅质碳酸盐岩为主的灯影组沉积。其中位于枫泾—川沙断裂以北的宝山大场一带，似属台地边缘浅滩相—开阔台地相沉积，其南的浦东坦直、三灶一带则为局限海台地相—蒸发岩台地相的滨海沉积。

在整个震旦纪时期，上海与西南侧的钱塘海盆无疑是相通的，两地应有类似的沉积建造。受上海中部基底隆起影响，上海古地势呈东南高、西北低的特征。随着晚震旦世晚期的海退，于隆起带西北侧的青浦、宝山和崇明等地形成滞水的海湾环境，沉积了薄层黑色硅质岩。隆起带南部的坦直—奉贤一带逐渐海水退出，接受剥蚀。

底寒武世—早寒武世早期，基本继承震旦纪晚期沉积环境，海水较震旦纪晚期加深。在隆起带北侧形成滞留深水平缓盆地，南侧推测暂未波及。此时，气候温暖，海湾中水流不畅，低级菌类植物迅速繁衍，有机物大量堆积，从而形成以碳质页岩为主，间夹薄层钙质页岩与硅质灰岩的超山组沉积。早寒武世晚期，沉积环境与早期相似，滞留还原环境已明显减弱，水动力作用增强，形成了大陈岭组的同生角砾状白云岩沉积。

中寒武世开始，发生明显海侵，形成开阔海台地环境，陆源碎屑物质增多，沉积了一套以泥灰岩及含粉砂质灰岩为主的磷酸盐岩建造。

至晚寒武世，海水变浅，气候转为炎热，形成台坪潟湖相厚层白云岩、同生角砾状白云岩及白云质灰岩建造。此时，由于海水咸度较高，生物不易繁衍、生存，层中化石稀少。

奥陶纪承袭了寒武纪的古地理和古构造条件，唯海域不断扩大，海水也较前加深。仑山组沉积时期，海侵初期形成开阔海台地相的碳酸盐岩建造。早奥陶世晚期—晚奥陶世早期，海水进一步加深，沉积了以台地斜坡建造为主的红花园组到黄花岗组，其由灰黑色、灰紫色含碳质碳酸盐岩、含泥、硅质条带的生物骨屑微晶灰岩、瘤状灰岩组成，部分层位出现角砾状结构和瘤状结构。尤其是牯牛潭沉积时期，层中富含生物化石碎片，含头足类、腹足类、三叶虫、海百合、牙形刺类古生物化石，其主要生物分子介于宁镇与浙西混生的生物群特征。表明水体仍然较浅，水动力能量较强。

至晚奥陶世中期，区域上海侵规模达到高峰，区内地势东南高、西北低，北部地区演化为陆棚环境，接受了一套巨厚的类复理石沉积，南部地区推测较薄，且层中富含笔石化石，其生物组合兼有华北型与东南型特征，属过渡型生物群。

区内的志留纪地层，目前仅见于崇明岛西北部，而且因揭露不全，难窥全貌。但从已揭露的岩性段判断，本区志留系和浙北、苏南类似，其下部有灰黑或灰绿色砂岩、粉砂岩及泥岩组成类复理石建造，其上部以紫红色含泥砾砂岩为主。由下而上，由潮下带演化为潮间带，构成一个明显的海退序列。

和浙北类似，加里东运动在本区主要表现为差异性隆起，并可能在北东向基底隆坳相间的构造格架基础上，继续发展并形成一些宽展型褶皱。

二、泥盆纪—中三叠世

本区迄未发现晚古生代—中三叠世地层，但据邻省资料分析，推测在上海北部，尤其是崇明岛西部地区完全可能有该阶段的沉积。

现仅根据邻省研究成果，将这一时期的地质发展历史推断、分析于下：

经过加里东运动，早、中泥盆世时期，上海可能和浙西北地区一起成为赣浙皖古陆的一部分，经受风化剥蚀。晚泥盆世—早石炭世，区域上自西向东发生海侵，形成滨海—浅海、河湖环境。其间，晚泥盆世末期曾发生海退。不同地区接受沉积的时间差别较大，部分地区至早石炭世接受沉积。中、晚石炭世，海侵范围急剧扩大，区域上已经演化为广阔的陆表浅海环境，并于浙西北沉积了稳定的镁质碳酸盐岩及生物碳酸盐岩建造，推测区内上海北部最晚在这个阶段应该开始接受沉积。

二叠纪，邻近诸省在总的海进序列中海水进退频繁。早—中二叠世以碳酸盐岩沉积为主，晚二叠世早期，植物繁茂，沉积了一套海陆交互相的碎屑岩相建造，并于浙北吴兴、长兴一带形成三角洲与潮坪相间的含煤碎屑岩。区内北部地区的构造，地貌环境与此相仿，估计也有可能发育含煤碎屑岩沉积；晚二叠世晚期，新的海侵开始并不断扩大，形成台地相碳酸盐岩建造为主的长兴组。这一时期的海侵作用可能一直沿至早三叠世。中三叠世开始海退，由浅海开始向陆相转变。

中三叠世末期的印支运动，结束了本区长期的海浸历史。经过这次运动，区域上加里东期形成的宽缓向、背斜，进一步发展成为紧密的复式线形褶皱。由于与此伴生的北东及北西向断裂构造十分发育，最后终于沿断裂形成了一系列紧密相间的断凸和断凹。

第三十八章　晚三叠世—新生代

一、晚三叠世—早白垩世

晚三叠世开始，区域上进入了强烈的陆内造山和改造阶段。该阶段为我国东部金属矿产的重要成矿期，在金山张堰形成铜矿。区内尚未发现晚三叠世—侏罗纪地层，总体的运动特征不明。仅根据邻区和区内早侏罗世末期、中侏罗世末期侵入岩形成构造环境推断，至少从早侏罗世晚期开始又进入强烈的陆内挤压造山阶段，形成吴淞同造山型花岗岩。中侏罗世末期，进入晚造山阶段，形成三林石英闪长岩岩珠。

早白垩世，区域上逐渐进入后造山—非造山阶段（毛建仁，2013），140Ma 左右全面进入伸展构造环境，发生了强烈的岩浆侵入和喷发活动（王芳，2016）。区内，该阶段的岩浆侵入和火山作用建造分布广泛，侵入活动与火山喷发旋回总体上显示出紧密的对应关系，在岩浆来源、演化趋势及空间分布上均有密切关系。相应形成了劳村组、黄尖组与寿昌组 3 个陆相火山喷发-沉积建造，市北花岗岩、市南花岗岩、堡镇花岗岩。岩浆作用经历了弱—强—弱几个演化阶段，130Ma 左右最为强烈，分布最广。早期阶段主要分布在庄行和金桥构造火山盆地；中期阶段火山作用则多见于天马山—青浦一带的火山隆起，侵入作用沿北东向、北西向和北北东向等岩石圈、基底断裂相互切割的地区分布，形成本区唯一的花岗岩岩基。晚期火山喷发作用微弱，并多分布于本区的北部。

二、晚白垩世—古近纪

从晚白垩世开始，区域应力场已转变为以引张作用为主（江西省地质矿产勘查开发局，2017）。本区在原有的构造体系上，沿北东向和北西向区域性断裂带沿线局部又发育了近东西向及北西向两组断裂，联合控制了晚白垩世—古近纪一系列断陷盆地的形成。在断陷盆地中先后沉积了上白垩统浦口组陆相紫色类磨拉石建造、赤山组红色碎屑岩建造、古近纪夏驾桥组棕色含钙细碎屑岩以及北桥组杂色碎屑岩建造。晚白垩世，地壳的断块差异趋于明显，岩浆作用逐渐减弱，主要表现为间歇性的火山作用和局部侵入作用，形成大量小的岩珠和次火山岩、脉岩。气候干热多雨，山麓及冲洪积环境发育，沉积物较粗，局部夹石膏层。岩浆活动微弱，末期见有基性岩浆喷溢。古近纪，气候以干旱为主，湖盆面积已明显缩小，沉积物较细。岩浆活动仍微弱，见有基性岩浆喷溢，似较晚白垩世增强。这些盆地呈箕状，沉积厚度较厚，其沉积中心均向主断裂一侧做单向迁移。盆地发育晚期，北桥断陷东南缘沉积可能超覆于控盆断裂之上。

三、新近纪—第四纪

区域上，始新世晚期开始发育的西太平洋沟弧盆体系逐渐形成，其中晚渐新世—早中新世是其边缘海盆的扩展高潮期（赵会民，2002）。区内在北西向与北东向断裂带的交汇处，玄武质岩浆间歇性喷溢，喷溢间断期形成湖湘堆积。气候由古近纪的干旱转为温暖渐趋寒冷，最终进入第四纪冰期。盆地范围普遍扩大，构成早期断陷盆地的坳陷沉积。区内，受基底北西向、近东西向及北东向断裂的控制，幔源型碱性橄榄玄武岩浆常沿断裂交汇处喷溢，分布于沿海地带。并在火山喷发间歇期，受温暖气候控制，形成未固结的湖泊堆积。

新近纪晚期之后，本区在原有的基底构造与古地貌的基础上，在北西向和北北西向断裂的改造下，最终成为隆坳相间的构造、地貌景观。在安（亭）—崇（明）凹陷及枫泾凹陷、浦东滨岸凹陷等低洼地区，沉积了较厚的崇明组冲洪积沉积物。西部隆起带则经受风化剥蚀。

进入第四纪，冰期和间冰期的交替引起了海平面周期性的升降。区内接受了厚达数百米的第四纪河湖相、河口三角洲相及海陆交互相沉积物，在此期间曾经过 5 次明显的寒冷期与温暖期交替，并伴有 7 次海侵作用。这一时期，高级植物及哺乳类动物迅速繁衍，特别是进入全新世以后，由于气候转暖，滨海平原上河湖密布，气候宜人，成为古人类生息、繁衍的良好场所。

进入第四纪，在构造沉降、气候的冰期和间冰期周期性交替以及海平面的周期性升降的耦合作用下，堆积了 200~400m 厚的松散沉积物，发育了江河湖海多种沉积体系。第四纪早—中期，在构造沉降的主导控制下，长江三角洲地区地形地貌发生了重大转变，沉积中心由苏北平原迁移至上海崇明岛一带的河口地区（Chen et al.，1995），区内环山麓冲洪积扇、辫状河逐渐转变为平原曲流河（Wang et al.，2005），沉积物源也由近距离的泥沙转变为远距离的长江泥沙（Yue et al.，2019），海面波动下的海水进退也逐渐影响区内沉积环境，气候波动下的化学风化强度影响着河道的摆动范围。第四纪晚期，气候和海平面波动主导控制了区内环境，海陆交替明显，海平面上升阶段，区内逐渐发生全面海泛，海平面下降阶段，区内发生海退，河流下切剥蚀而形成下切河谷。自晚更新世以来，区内发生了 3 次明显的大规模海侵，海侵期间区内呈滨浅海和河口环境。

末次盛冰期，海平面海拔约为-130m（Li et al.，2014），停顿在冲绳海槽陆坡，海水一度退出东海陆架，长江注入冲绳海槽。这个期间，上海发育了下切河谷体系，崇明岛一带河口地区发育了长江下切河谷，下切深度可达 60~80m，滨海平原发育长江支流小型河谷，下切深度可达 30~60m。末次盛冰期结束以后，气候转暖，海平面迅速上升，东海陆架迅速发生海侵，长江河口迅速后退。至 13~12ka，长江河口退至现代长江河口口外水下三角洲区域（Wang et al.，2018），发育了河口浅滩。这期间上海正值新仙女木气候凉干阶段（Yi et al.，2003），滨海平原小型下切河谷内呈湖沼环境（赵宝成等，2007）。进入全新世，随着海平面的不断快速上升，海泛作用使得长江河口不断快速向陆后退，下切河谷逐渐演变为巨大的河口湾。9~8ka，长江口逐渐退至镇江一带（Hori et al.，2001，2002；Li et al.，2002），上海滨海平原逐渐沦为滨浅海环境，上海西部湖积平原因地势较高而呈滨海湖沼环境（Wang et al.，2012）。

大约 8ka，海平面上升至-5m，长江三角洲开始发育（Li et al.，2002；Song et al.，2015）。7~6ka，海平面达到最高并保持稳定，随着入海泥沙的输入和水下可容空间的减少，三角洲开始快速向海推进，表现为长江河口和两翼海岸线不断向海迁移。Li et al.（2002）基于 600 个晚第四纪钻孔岩相分析，认为长江三角洲自发育以来不断向东南迁移，形成大致 5 期河口沙坝，呈雁状排列，分别是红桥沙（约 7ka）、黄桥沙（约 6ka）、金沙（约 4ka）、海门沙（约 3~2ka）、崇明—横沙（1.4ka 以来）。然而，最近 Wang et al.（2018）基于 344 个钻孔、658 个 ^{14}C 测年、28 个 OSL 测年，认为长江三角洲发育后并没有形成几期河口沙坝的模式，而是于 8~6ka 在金沙—海门一带形成潮流沙脊，4ka 在红桥—黄桥一带形成河口沙坝，2ka 河口沙坝迁移到海门一带及如东附近。海岸线迁移表现为 6~7ka 海面达到最高并稳定时在太湖平原东缘和里下河平原分别形成最古老的贝壳沙堤，随后随着河口迁移向今天岸线方向迁移。Hori et al.（2001）估算过去 5000 年三角洲主体平均进积速率达 50km/a。然而，实际上，由于水下地形、地貌及入海泥沙量的变化，长江河口及岸线向海迁移的速率不是均匀的。Wang et al.（2018）对比 8~6ka 和 6~2ka 两个阶段沉积通量发现，由于 6ka 以后东亚季风衰退，河口入海泥沙量明显减少，导致了三角洲进积减缓，这在三角洲南部平原贝壳沙堤分布可以得以印证，上海境内 6~3ka 的沙堤仅局限在大致 15~20km 狭窄范围内。长江河口海岸快速迁移发生自过去 2000 年以来，Hori et al.（2001）发现 2ka 前后三角洲主体的进积速率从 38km/a 突然增加至 80km/a，Wang et al.（2018）指出，2ka 以后沉积通量由大约 1×10^8t 增加至 1.62×10^8t。2ka 以来长江三角洲快速进积与强烈的人类活动有关，Wang et al.（2011）在长江口水下三角洲钻孔磁化率中识别出 2ka 以来长江流域强烈的人类活动导致土壤侵蚀加剧而贡献了大量入海泥沙。

近现代以来，随着人类活动的加剧，对区内地质环境演化产生了明显影响。在构造沉降和天然固结沉降基础上，人类过量抽取地下水资源加剧了地壳表层形变。地面沉降防治过程中的地下水回灌对原生水环境产生少量影响。流域水利工程实施，尤其是2003年三峡大坝合围以后，长江入海泥沙量锐减（Yang et al.，2011），且随着河口大量促淤圈围和河道治理工程的实施，长江三角洲的自然演化状态遭到明显改变。

结　　语

以第一代《上海市区域地质志》为基础，通过对区内近30多年来基础地质、城市地质调查和研究成果的收集、总结，结合邻区研究成果，综合运用板块地质学和新的研究方法对基础地质事实进行了重新认识，全面系统总结和研究了区内基岩地层、岩浆岩、断裂构造、新构造运动、岩石圈结构构造、第四纪地质等区域地质特征，并系统总结了城市地质调查和研究领域的成果，编制了系列地质图件，取得了以下几个方面的进展和认识。

一、主要进展和认识

（一）地层和沉积作用

1）将区内地层自下而上划分出33个岩石地层单位，其中，基岩地层有23个，松散层有10个。始新统—渐新统、上白垩统顶部、下—中侏罗统、上泥盆统—下三叠统和南华系—下震旦统作为暂未发现处理。本区综合区划属于华南地层大区、扬子地层区、江南地层分区、沪杭地层小区。自中元古代开始，可以划分为中元古代—青白口纪、南华纪—志留纪、白垩纪—古近纪、新近纪、第四纪等5个基本地层区划阶段。

2）将前震旦系划分为金山群、惠南板岩和河上镇群3个岩石地层单位，分别为古—中元古代、青白口纪早期和青白口纪晚期地层。金山群可与赣北的铁砂街岩组、田里岩片对比，惠南板岩与双桥山群可对比。

3）将寒武系超山组置于底—下寒武统，原超峰群划为超峰组，原奥陶系荆山群解体为仑山组、红花园组、牯牛潭组，志留系划分为河沥溪组和康山组。

4）劳村组、黄尖组和寿昌组为早白垩世建造，三者有可能存在同期异相关系。推测局部可能存在横山组，将原上白垩统松江组并入赤山组。重新定义了夏驾桥组，岩石组合以棕红色（局部砖红色）粉细砂岩、粉砂岩、泥质粉砂岩为主，局部含砾，夹薄层砂砾岩、粉砂质泥岩和泥砾，下部见夹薄层砂质灰岩。

5）白龙港玄武岩和崇明组可与江苏盐城组相当，其中，崇明组底界年龄推测大于6.3Ma。

6）第四系可划分为下更新统安亭组、中更新统嘉定组、上更新统川沙组和南汇组/滆湖组、全新统楼塘组、上海组和青浦组/如东组；其中，全新统3个地层单位可分别与江苏地区如东组的上、中、下3段相对应。

7）总结了33个岩石地层单位的沉积岩岩性组合特征和典型沉积岩岩相特征，综合沉积相标志等对不同时间段的岩相古地理特征进行了分析。

8）古—中元古代和青白口纪沉积主要形成于不同阶段活动陆缘体系的岛弧-弧后盆地环境。南华纪—中三叠世沉积形成于扬子克拉通南缘的被动陆缘盆地、陆表海环境。受中部隆起带影响，上海前中生代地势明显东南高、西北低，明显制约了海相盖层的分布特征。上海北部和杭嘉湖地区在整个古生代生物组合相似，具有华北型与东南型过渡区的特征。

（二）岩浆作用

1）将侵入作用划分为晋宁期、燕山期和喜马拉雅期等3期。其中，晋宁期和燕山期又分别细分为2个、4个阶段。将火山作用划分为晋宁早—中期、晋宁晚期、燕山晚期和喜马拉雅期等4期。其中，燕山晚期可细分为早、中、晚3个阶段，以中期活动最强烈。

2）晋宁期酸性岩为高钾钙碱性-钾玄岩系列，燕山期和喜马拉雅期侵入岩和火山岩多呈高钾钙碱性-钾玄岩系列，其中，基性侵入岩多呈碱性，中酸性侵入岩多呈钙碱性。

3）中元古代初期岩浆作用与破坏性活动板缘环境有关；中元古代晚期和新元古代早期与活动陆缘的岛弧或弧后盆地环境有关。早侏罗世末期仍显示为陆内挤压造山阶段，中侏罗世末期进入晚造山—后造山阶段。至早白垩世，区域性伸展构造活动逐渐强烈，但仍处于造山带环境，与火山弧有关；新生代岩浆活动主要形成于板内裂谷环境。

4）中元古代初期花岗斑岩和燕山期市北、堡镇岩体呈A型花岗岩特征，早侏罗世末期花岗岩具有I型向S型过渡的特征，其他岩体多为典型的I型花岗岩类，推测主要为地壳不同层次的部分熔融产物。

5）同一时期，岩浆侵入作用与火山作用的源区具有一定相似性，具有相似的地质环境和演化过程。中生代岩浆作用强度有由弱—强—弱的演化趋势，侵入岩和火山岩主要发育于130Ma左右，两者规模相当，波动趋势合拍。原始岩浆可能来自地壳深部，其中有部分上地幔成分掺入，在岩浆上升过程中，有壳源成分的不断加入。新生代岩浆作用主要表现为火山作用，岩浆起源于地幔源区。

6）早白垩世火山作用喷发类型兼有裂隙式及中心式两类，主要受北东向断裂的控制和北西向断裂的影响。根据火山活动及其在空间上的分布与组合特征，本区火山构造可划分为下列4级：火山岩区、火山喷发带、火山盆地和隆起、火山机构。

（三）变质岩及变质作用

1）将区域变质作用时间厘定为中、新元古代之交和青白口纪早、晚期之交。区域变质作用受到了后期多期构造热事件作用的叠加影响，尤其是在印支期和燕山期。

2）金山群原岩由古—中元古代岛弧-弧后盆地基性火山岩、部分中酸性火山岩、陆源碎屑岩及薄层碳酸盐岩建造组成；惠南板岩原岩为青白口纪新发育的弧后盆地细碎屑岩建造。

（四）地质构造

1）修订了区内构造单元划分方案。以湖苏断裂带为界划分为下扬子克拉通盆地和钦杭结合带2个三级构造单元；根据北西向断裂和中新生代建造差别划分出5个五级构造单元。

2）对断裂构造进行了修正，并将区内断裂构造划分为深断裂、基底断裂和盖层断裂3个级别。其中，北东—北东东向断裂最为发育，规模较大。

3）划分出新村-兴隆断陷盆地，认为该盆地受湖苏断裂带和后期新海农场-长征农场断裂、跃进农场-育新断裂控制。

4）构造可划分出基底形成（青白口纪晚期以前）、主要盖层形成（青白口纪晚期—中三叠世）和陆内改造（晚三叠世—新生代）等大阶段和晋宁期、扬子—加里东期、华力西期—印支期、燕山期和喜马拉雅期等构造阶段。总结了不同阶段的建造特征和构造变形机制。

5）系统总结了区内新构造运动标志和基本特征，新构造期的断裂活动性，区域性地震活动的震中分布、震源深度、震级与震源深度相关性特征，主要地震的震源机制解和区域应力场特征。

6）分古元古代—青白口纪、南华纪—中三叠世、晚三叠世以来3个阶段对区域地质发展史进行了总结。

（五）岩石圈结构构造

1）总结了区内岩石圈的速度场、重磁场、电性场、密度场和地温场等地球物理场特征。论述了莫霍面和地壳结构的区域变化特征。

2）初步分析了岩石圈结构构造特征的资源环境效应，总结了深部地质作用与成矿、石材资源、地热资源和地震活动的关系。

（六）第四纪地质与地貌

1）将陆域划分为4个三级地貌单元，分别是太湖堆积平原区、长江三角洲堆积平原区、杭州湾滨海平原区、剥蚀残丘和岛屿区。其中，太湖堆积平原区分为潟湖积平原、湖沼积低平原和沼积低平原

3 个四级地貌单元；长江三角洲堆积平原区分为河口沙岛、滨海平原及河流冲积平原 3 个四级地貌单元。近岸水域划分为长江三角洲水下三角洲 1 个三级地貌单元，细分为三角洲前缘区和前三角洲区 2 个四级地貌单元。

2）认为上新世以来经历了 18 个阶段的气候波动：上新世气候总体暖湿，下部暖热湿润，中间夹一次寒冷干燥期，命名为第一寒冷期，上部暖热湿润—温暖湿润，命名为第一温暖期；早更新世，气候开始明显变冷，但有波动，先后经历了 3 次冷期，中间有两次暖期交替，气候期先后命名为第二寒冷期、第二温暖期、第三寒冷期、第三温暖期、第四寒冷期；中更新世古气候先后经历了一次温暖期和一次寒冷期，分别命名第四温暖期和第五寒冷期；晚更新世 128～70ka 为一次暖期，命名第五温暖期，对应 MIS5 期；70～11ka 为寒冷期，命名第六寒冷期，气候总体温凉偏干，对应 MIS3 期；全新世为冰后期，气候温暖湿润，气候波动先后经历了温凉稍湿、温和偏干、暖热潮湿、温凉稍干、温暖湿润的变化，对应前北方期、北方期、大西洋期、亚北方期、亚大西洋期。

3）识别出早更新世中期如皋海侵和晚期周浦海侵，中更新世早期上海海侵和晚期嘉定海侵，晚更新世早期太湖海侵和中晚期滆湖海侵，全新世镇江海侵等 7 次海侵。其中，早更新世和晚更新世海侵较弱，海水主要沿古河道入侵，而晚更新世以来的 3 次海侵规模较大。

4）早更新世—更新世发育的东西向“枫奉古河道”和南北向“浏南古河道”构成区内重要的地下淡水资源存储区域。晚更新世以来发育了数条以东西向为主的古河道，对区内工程地质条件有着重要的影响。末次冰消期以来，晚更新世末下切古河道经历了河口湾相沉积充填，也是现代长江三角洲的主要发育区域。

（七）城市地质

1）系统总结了区内水文地质条件。区内地下水发育松散岩类孔隙水（第四系）、碎屑岩类孔隙水（新近系和白垩系）、碳酸盐岩类裂隙溶洞水（震旦系—古生界）和基岩裂隙水等 4 种类型，其中松散岩类孔隙水垂向上又可划分为潜水、微承压含水层、第Ⅰ承压含水层、第Ⅱ承压含水层、第Ⅲ承压含水层、第Ⅳ承压含水层和第Ⅳ承压含水层。松散岩类孔隙水不同含水层的埋藏条件、渗透性和富水性、矿化度、地下水质量和水化学类型差别较大。即使同一含水层，平面上也有明显差别。不同类型和不同含水层的补径排条件具有一定差异。因遭受不同程度的各种自然因素和人为作用影响，天然地下水流场也发生了不同程度变化。地下水位历史变化明显，且不同含水层特征不同。

2）结合上海地貌单元，综合含水层埋藏、富水性、水质、水位等分布总体特征，将水文地质条件可划分出河口—滨海平原水文地质亚区、剥蚀残丘—叉流河道水文地质亚区等两个亚区，并进一步可细分为 8 个块段。

3）以平均年度地面沉降量不大于 7mm，第Ⅱ、第Ⅲ、第Ⅳ、第Ⅴ承压含水层最大地下水位分别不低于-3m、-5m、-20m、-25m 为约束条件，全市松散层承压含水层可采资源总量为 2500m^3。其中，浦东地下水可开采量相对最大，中心城、宝山相对最小。

4）区内潜水和承压水具有不同程度的开发利用。潜水、第Ⅰ承压含水层地下水几十年来基本处于停采状态，而第Ⅱ、第Ⅲ、第Ⅳ、第Ⅴ承压含水层地下水已有近 160 年的开发利用历史，且不同含水层的开发历史具有明显阶段性。开发利用过程中，各种因素直接或间接地引发了诸多环境地质问题，最为显著的是原生水环境发生了变化，表现为区域地下水位下降、局部地区地下水遭受污染及之后的咸水界面迁移等环境地质问题。地下水开采利用过程中同时逐渐实施了人工回灌措施。人工回灌对水环境影响具有双重效应，在有效抬升区域地下水位或减缓地下水位下降速率的同时，局部原生地下水质量的稳定性也受到了一定程度影响。

5）提出了上海地下水可持续利用与保护规划方案，圈定了 4 个规划分区，包括地下水资源严格限制开采区、限制开采区、控制开采区和一般监测区。部署了中心城、宝山新城、嘉定安亭新城、浦东国际机场、惠南新城、临港新城、奉贤海滨农场、金山新城、松江新城、青浦新城、城桥新城、陈家镇新城、长兴岛、横沙岛等 14 个地下水应急水源地。

6）区内100m以浅可划分为10个工程地质层，即第①层填土、第$②_1$层褐黄—灰黄色粉质黏土，第$②_3$层粉性土层（浅部砂层），第③、④层灰色淤泥质黏性土层（软土层），第⑤层灰色黏性土层（局部夹有第$⑤_2$层砂、粉性土层），第⑥层暗绿—草黄色黏性土层（第一硬土层），第⑦层草黄—灰色砂、粉性土层，第⑧层黏性土夹粉性土层（局部夹第$⑧_{2-1}$层暗绿—草黄色黏性土硬土层），第⑨层灰色砂性土和第⑩层褐灰—蓝灰色黏性土层。

7）不同工程地质层的分布、埋深、厚度变化、岩性都有一定变化，且工程地质特性和产生的工程地质问题不同。主要有地基变形、砂土液化、水土突涌、边坡失稳、潜水腐蚀性等工程地质问题。第$②_1$层为俗称的“硬壳层”，可作为轻型建（构）筑物的天然地基持力层。第$②_3$层为浅部砂层，易产生震动液化和渗流液化现象。第③、④层为典型的软土层，是地基变形主要层次，也为区内第一压缩层。第⑤层为一般黏性土层，亦为第二压缩层，其中第$⑤_2$层为微承压含水层，基坑开挖时应注意由其引发的流砂、管涌等问题。第⑥层为第一硬土层，土性较好，可作为一般建（构）筑物的桩基持力层。第⑦层为下部砂、粉性土层，土性好，为中高层建筑和市政桥梁工程的良好桩基持力层，该层亦为第Ⅰ承压含水层，地下工程施工时以产生流砂和基坑突涌问题。桩基持力层为⑦层时，$⑧_1$层为桩基的主要压缩层，应注意由该层引起的桩基沉降和差异沉降问题。$⑧_{2-1}$层在西部湖沼平原区可与下部$⑧_{2-2}$层作为大型建筑物的桩基持力层。$⑧_{2-3}$层土性较好，中等压缩性，滨海平原区该层分布相对较为稳定，在第⑦层缺失时，可选择该层作为大型和超大型建筑物的桩基持力层。第⑨层土含水量低，压缩性低，强度高，土性好，可作为超大型建筑物的桩基持力层，但由于埋藏较深，如作为桩基持力层，经济成本较高。

8）陆域共划分为湖沼平原区、滨海平原区和河口砂岛区等3个工程地质结构区，前两者分别可细分为2个、4个工程地质结构亚区。

9）总结了区内天然地基和桩基条件。将天然地基工程建设适宜性划分为好区、一般区、较差区、最差区等4个区。将桩基条件划分为3个大区，A区为适宜性较好区，B区为适宜性较差区，C区为适宜性一般到差区。

10）总结了区内地面沉降的形成发展阶段、监测网络建设、地面沉降特征及影响因素、地面沉降规律分析及影响因素、工程建设引发地面沉降机理、地面沉降风险等成果。区内逐步形成了由精密水准监测网、地面沉降监测站、地下水动态监测网、重大基础设施地面沉降骨干监测网、GPS地面沉降监测网和InSAR技术地面沉降监测组成的地面沉降监测网络。地面沉降特征与趋势是随着地下水开采与人工回灌格局的调整、城市发展进程而不断发展的。不同阶段地面沉降空间分布、不同土层表现出不同的变形特征。将全区地面沉降危险性和易损性划分为高区、较高区、中等区、较低区和低区5个级别。

11）总结了元素基准值与背景值、元素地球化学分布特征、主要有机化合物分布特征、地球化学异常、环境质量、养分丰缺状况等土壤环境地球化学成果。全区深层土壤地球化学平均值、基准值与长江三角洲地区某些组分含量有所差异，但也具有较多的相似性。不同沉积地质背景的土壤地球化学基准值，表层土壤区域地球化学背景值具有差异性。空间上，土壤造岩元素含量、酸碱度、重金属元素含量分布具有明显差异。有机氯类农药在土壤中仍普遍存在，多环芳烃含量分布差异明显。土壤当中众多环境指标的高含量点主要集中在中心城（外环以内）及近郊一些工业区。全市农用地地区土壤质量整体良好。大多数可以安全利用，只存在个别风险点。全市土地作为一类用地，除个别地块，大多数可以安全利用；作为二类用地，As、Cd、Hg元素含量均处于筛选值范围内；Cu、Ni、Pb元素含量基本均处于筛选值范围内，仅个别点位含量大于管制值。对于大于筛选值的调查点，以农业利用为目标的土壤基本可划分为西部区、中部区、东部三岛区及海沿区等。

12）总结了重点岸带地区的冲淤演变特征和主要影响因素，滩涂资源分布、滩涂资源开发利用概况和影响因素、潜力等成果。区域上和不同区域的冲淤格局具有明显的阶段性，长江径流水量、来沙量，东海波浪、潮汐、潮流和输沙等因素都会影响长江口地形的发育和演变。重要滩涂地区中高滩资源比较有限，其中崇明北沿和横沙浅滩地区具有很大的促淤圈围空间。并且，区内可用于未来重要促淤圈围区的袋状砂、堤心砂资源十分有限。

13）总结了区内地下空间资源开发利用历史和现状，分析了地下空间开发利用影响因素，评价了地下空间资源量和开发地质环境适宜性。上海城市地下空间开发建设与利用先后经历了如下 3 个开发利用阶段，目前开发深度主要集中在中浅层（50m 以浅），以地下生产、生活服务设施、地下公共基础设施和轨道交通及附属设施开发利用为主。影响地下空间开发的因素主要有地下水、软土、砂和粉性土、浅层天然气。中心城可充分开发利用地下空间资源量为 $6.44\times10^{10}m^3$，中浅层地下空间开发利用适宜性可划分为问题轻微区（A）、中等区（B）和严重区（C）3 类。深层地下空间优先开发层（40～65m）开发利用适宜性可划分为较好、中等、较差、差 4 个等级。

14）总结了地质遗迹和古文化遗址等旅游地质资源条件。已有地质遗迹可划分为 2 大类 4 类 4 亚类，即地貌景观大类和地质灾害大类两大类，水体地貌类、海岸地貌类、火山地貌类、地质灾害类 4 类，河流（景观带）、海积地貌、火山岩地貌和地面沉降 4 个亚类。其空间分布主要受到地形地貌、断裂构造和人类历史活动的控制。自嘉定外冈、方泰，经闵行诸翟、俞塘和奉贤南桥，直抵杭州湾北岸的奉贤柘林一线以西，至今已发现 25 处古文化遗址。

二、存在问题和建议

（一）区域地质调查比较薄弱，有待加强

新中国成立以来，区内相继开展了石油和矿产资源、水工环地质和城市地质调查研究，但区域地质调查工作十分缺乏。相比较天津、北京先后完成 1∶25 万～1∶20 万、1∶5 万区域地质调查全覆盖的工作程度而言，区内仅完成了西部陆域 1∶25 万区域地质调查。该工作的缺乏必定会在一定程度上导致对区内岩石、地层、构造、沉积环境演化等地质条件认识的不足，进而制约对水热型和干热岩型地热资源条件、地震地质背景的认识和评价。

（二）基础地质问题认识不足，有待深化

由于深孔不足，区内已揭露的基岩地层剖面残缺不全，地层层序和空间分布、区域对比仍然存在很大不确定性。金山群 a、c 岩性段是上下叠置关系还是倒转背斜两翼的同一地层？暂未发现地层是否存在？前震旦系之间、寒武系与奥陶系、志留系与奥陶系、上白垩统与古近系等地层间是何种接触关系？上古生界是未发现还是未沉积，抑或是后期剥蚀缺失？

在不同的区域性构造图件上，区内断裂与邻区断裂连接时随意性较大。如，浙西北地区北东向的马金—乌镇断裂带、萧山—球川断裂带、常山—渚漓断裂带、江绍断裂带与上海张堰—南汇断裂带、枫泾—川沙断裂带如何相接？上海北北东向的大场—廊下断裂带、北西向的太仓—奉贤断裂带是否延入邻近浙江和江苏？

（三）新构造运动特征认识缺乏系统性

受基础地质条件的认识不足和工作手段限制，区内对新构造运动特征、区域地震地质条件的认识多年来未有进展。如，区内是否存在蠕滑性活动断裂？主要控震和发震构造特征及其地震危险性如何？新构造运动对松散层沉积厚度的控制作用如何？

参考文献

陈国光，奚小环，梁晓红，等 . 2008. 长江三角洲地区土壤地球化学基准值及其应用探讨 . 现代地质，22（6）：1041-1048.

陈沪生，周学清，李道琪，等 . 1993. 中国东部灵璧-奉贤（HQ-13）地学断面图及说明书 . 北京：地质出版社.

陈沪生，张永鸿，徐师文，等 . 1999. 下扬子及邻区岩石圈结构构造特征及油气资源评价 . 北京：地质出版社.

陈焕疆，景学立，张锡南，等 . 1990. 东海北部及邻区的基底和构造区划 . 中国海上油气（地质），4（6）：11-19.

陈吉余，蒋雪中，何青 . 2013. 长江河口发育的新阶段、上海城市发展的新空间 . 中国工程科学，15（6）：20-24.

邓奇，王剑，汪正江，等 . 2016. 江南造山带新元古代中期（830~750Ma）岩浆活动及对构造演化的制约 . 大地构造与成矿学，40（4）：753-771.

丁炳华，史仁灯，支霞臣，等 . 2008. 江南造山带存在新元古代（~850Ma）俯冲作用——来自皖南 SSZ 型蛇绿岩锆石 SHRIMP U-Pb 年龄证据 . 岩石矿物学杂志，27（5）：375-388.

都城秋穗（Akiho Miyashiro）. 1979. 变质作用与变质带 . 周云生译 . 北京：地质出版社

范代读，李从先，Yokoyama K，等 . 2004. 长江三角洲晚新生代地层独居石年龄谱与长江贯通时间研究 . 中国科学：D 辑，34（11）：1015-1022.

付树超，陈觉民，林文生 . 1991. 福建建宁西部上太古界天井坪组（Ar_2t）地质特征 . 福建地质，10（2）：103-113.

高林志，杨明桂，丁孝忠，等 . 2008. 华南双桥山群和河上镇群凝灰岩中的锆石 SHRIMP U-Pb 年龄 . 地质通报，27（10）：1744-1751.

高林志，刘燕学，丁孝忠，等 . 2013. 江山-绍兴断裂带铁砂街组变流纹岩 SHRIPM 锆石 U-Pb 测年及其意义 . 地质通报，32（7）：996-1005.

高林志，丁孝忠，刘燕学，等 . 2014. 江山-绍兴断裂带陈蔡岩群片麻岩 SHRIPM 锆石 U-Pb 年龄及其地质意义 . 地质通报，33（5）：641-648.

龚士良 . 2008. 上海地面沉降影响因素综合分析与地面沉降系统调控对策研究 . 上海：华东师范大学 .

龚士良，叶为民，陈洪胜，等 . 2008. 上海市深基坑工程地面沉降评估理论与方法 . 中国地质灾害与防治学报，19（4）：55-60.

顾斌，丁政 . 1987. 苏浙沪及邻区深部构造及其与地震活动性的关系 . 地震学刊，（3）：27-35.

顾澎涛，程之牧 . 1990. 也谈上海地区的断裂构造系统 . 上海地质，（4）：1-6.

顾知微，赵宗溥 . 1962. 中国中生代陆相地层 . 全国地层会议学术报告汇编 . 北京：科学出版社 .

顾志文 . 1997. 试论上海及邻区晚新生代沉积特征与断裂构造的关系 . 华东师范大学学报：自然科学版，（2）：68-75.

郭兴杰，严学新，黎兵，等 . 2016. 近年来长江口南槽河势演变过程及航道开发建议 . 海洋地质与第四纪地质，36（6）：63-70.

郭蓄民，许世远，王靖泰，等 . 1979. 长江河口地区全新统的分层与分区 . 同济大学学报，（2）：18-29.

郭玉贵，李延成，许东禹，等 . 1997. 黄东海大陆架及邻区大地构造演化史 . 地质学报，17（1）：1-12.

韩波，张训华，徐晓达，等 . 2011. 东海磁场特征及居里面分析 . 地球物理学报，26（2）：519-528.

何卫红，唐婷婷，乐明亮，等 . 2014. 华南南华纪—二叠纪沉积大地构造演化 . 海洋地质与第四纪地质，39（8）：929-953.

黄焕忠 . 1990. 上海地区地热问题的初步探讨 . 上海国土资源，（4）：16-20.

黄慧珍，唐保根，杨文达 . 1996. 长江三角洲沉积地质学 . 北京：地质出版社 .

黄鑫磊，占光辉，杨天亮 . 2015. 上海地区基坑减压降水地面沉降特征与防控措施初探 . 江西省地质学会 . 地学新进展——第十三届华东六省一市地学科技论坛文集 . 南昌：江西科学技术出版社 . 351-355.

黄兆熊，高德章 . 1989. 上海地区深部构造特征和地震活动趋势探讨 . 上海地质，（4）：48-57.

火恩杰，刘昌森，章振铨 . 2004. 上海市隐伏断裂及其活动性研究 . 北京：地震出版社.

鞠天吟 . 1979. 对苏杭地区早古生代地层的认识（苏杭扬子-江南型沉积过渡区的特征）. 地层学杂志，3（4）：294-303.

康清清，顾勤平，刘红桂，等．2013. 矩张量反演法在江苏及附近海域的应用．地震地磁观测与研究，34（3）：1-5.
李从先，汪品先，等．1998. 长江晚第四纪河口地层学研究．北京：科学出版社．
李毓尧，李捷，朱森．1935. 宁镇山脉地质．前国立中央研究院地质研究所集刊．第11号．
李四光，赵亚曾．1924. 峡东地质及长江之历史．中国地质学会志，3（3-4）：351-391.
李蔚秾．1965. 节头虫（*Arthricocephalus*）在浙西的发现．地质论评，23（6）：510-511.
李献华．1993. 华南地壳增长和构造演化的年代学格架与同位素体系制约．矿物岩石地球化学通报，13（3）：111-115.
李献华，王一先，赵振华．1998. 闽浙古元古代斜长角闪岩的离子探针锆石 U-Pb 年代学．地球化学，27（4）：327-334.
李献华，李寄嵎，刘颖，等．1999. 华夏古陆古元古代变质火山岩的地球化学特征及其构造意义．岩石学报，15（3）：364-371.
李祥辉，陈斯盾，罗家洪，等．2011. 浙江西部建德群锆石 LA-ICP-MS 之 U-Pb 同位素年龄及其地层学意义．地质论评，57（6）：825-835.
李祥根．2003. 中国新构造运动概论．北京：地震出版社.
李相传，孙柏年，肖良．2014. 浙江新近纪嵊县组地层特征及其化石研究进展．兰州大学学报：自然科学版，50（2）：145-153.
刘飞．2002. 上海市地面沉降变形特征研究．北京：中国地质大学（北京）.
刘金陵，叶萍宜．1977. 上海、浙江某些地区第四纪孢粉组合及其在地层和古气候上的意义．古生物学报，16（1）：1-10.
刘季辰，赵汝钧．1924. 江苏地质志．地质专报，甲种（第4号）.
刘若愚，朱荣生．1983. 对安甪盆地晚白垩世—早第三纪地层划分与对比的几点新认识．上海地质，（2）：8-21.
刘若愚，黄建国，吴惠兰，等．1991. 上海地区新构造隐伏活动断裂的标志——石灰华．上海地质，（4）：1-11.
卢衍豪，穆恩之，候祐堂，等．1955. 浙西古生代地层新见．地质知识，（2）：1-6.
卢衍豪，张日东，葛梅钰，等．1963. 浙江西部下古生代地层．全国地层委员会．全国地层会议学术报告汇编——浙西地层现场会议．北京：科学出版社．27-56.
陆志坚．1989. 从上海地区地下淡水古河道含水层的分布探讨嵊泗列岛的供水问题．上海地质，（1）：23-26.
骆祖江，张月萍，刘金宝．2007. 深基坑降水与地面沉降控制研究．沈阳建筑大学学报：自然科学版，23（1）：47-51.
吕炳全，方大卫．1991. 上海地块的地质属性及其地震活动特征．上海地质，（1）：8-15.
吕少伟，唐益群，2000. 上海地区浅层沼气地质灾害分析探讨．上海地质，（3）：25-28
马杏垣．1986. 中国及邻近海域岩石圈动力学图及说明书．北京：地质出版社.
毛建仁，叶海敏，厉子龙，等．2013. 钦杭结合带（东段）晚中生代挤压-伸展构造的岩浆记录与成矿记录．矿物学报，（增刊）：30-31.
茅志昌，虞志英．2014. 上海潮滩研究．上海：华东师范大学出版社．
闵秋宝，汪品先．1979. 论上海地区第四纪海侵．同济大学学报，（2）：109-128.
邱金波，李晓．2007. 上海市第四纪地层与沉积环境．上海：上海科学技术出版社．
钱勤．1996. 苏北盆地玄武岩地质年龄及地层时代．火山地质与矿产，（1）：86-93.
秦大正．1994. 上海覆盖区活断层．地震地质，16（2）：160-166.
戎嘉余，王怿，张小乐．2012. 追踪地质时期的浅海红层——以上扬子区志留系下红层为例．中国科学：地球科学，42（6）：862-878.
沈永盛．1984. 对上海及其邻区松散层厚度变化特征的初步分析．上海地质，（1）：43-47.
盛书中，万永革，李红光，等．2013. 应用综合震源机制解法推断南通市地壳应力场方向．地球物理学进展，28（3）：1297-1303.
史玉金，陈洪胜，杨天亮，等．2009. 上海市工程地质层厘定及工程地质条件分析，上海地质，89（1）：28-33.
宋先月，宋治平，张肃．2003. 上海及邻区的地壳形变速率与地震活动．华南地震，23（1）：13-19.
孙永福．1981. 上海地区全新世地层分层．上海地质，（3）：1-9
孙业君，刘红桂，江昊琳，等．2015. 江苏南部地区现今震源机制和应力场特征．地震研究，38（2）：203-210.
唐增才，汪发祥，周汉文，等．2020. 浙西开化地区新元古代（~800Ma）洋陆俯冲——来自活动陆缘弧火山岩序列组合的制约．地球科学，45（1）：180-193.
涂荫玖，唐良贵，陈成涛．1993. 安徽江淮地区北部构造应力场初步研究．安徽地质，3（1）：1-15.

万天丰 . 1993. 中国东部中、新生代板内变形构造应力场及其应用 . 北京：地质出版社.
万传彪，赵传本，乔秀云，等 . 2004. 中国早白垩世中、晚期被子植物花粉的特征及意义 . 大庆石油地质与开发，23（2）：21-24.
王椿镛，段永红，吴庆举，等 . 2016. 华北强烈地震深部构造环境的探测与研究 . 地震学报，38（4）：511-549.
王存智，邢光福，陈荣，等 . 2015. 浙江嵊州地区陈蔡岩群变质岩的锆石年代学研究：江绍结合带对华南印支期造山运动的响应 . 中国地质，42（6）：1700-1714.
王存智，姜杨，赵希林，等 . 2016. 陈蔡岩群下河图斜长角闪岩年代学：地球化学特征及其构造意义 . 岩石矿物学杂志，35（3）：425-442.
王芳 . 2016. 华南东北部中生代从挤压向伸展的构造转换 . 杭州：浙江大学
王寒梅 . 2013. 上海市地面沉降风险评价体系及风险管理研究 . 上海：上海大学 .
王寒梅，焦珣 . 2015. 海平面上升影响下的上海地面沉降防治策略 . 气候变化研究进展，11（4）：256-262.
王洪作，陈培荣，吴俊奇，等 . 2017. 浙江大桥坞铀矿床火山岩锆石形态学与年代学研究 . 地质学报，91（3）：571-587.
王加恩，王振，刘远栋 . 2016. 浙西建德群劳村组火山岩地质时代及构造环境研究 . 地层学杂志，40（3）：929-953.
王建秀，吴林高，朱雁飞，等 . 2009. 地铁车站深基坑降水诱发沉降机制及计算方法 . 岩石力学与工程学报，28（5）：1010-1019.
王开发，徐馨 . 1998. 第四纪孢粉学 . 贵阳：贵州人民出版社 .
汪品先，闵秋宝，卞云华，等 . 1979. 东海表层沉积中有孔虫、介形虫组合分布的初步研究 . 同济大学学报，（2）：90-108.
汪品先，章纪军，赵泉鸿 . 1988. 东海底质中的有孔虫和介形虫 . 北京：海洋出版社 .
汪庆华，董岩翔，郑文，等 . 2007. 浙江土壤地球化学基准值与环境背景值 . 地质通报，23（2）：81-88.
王小平，朱元清，于海英 . 2006. 用远震接收函数反演上海及其邻区地壳速度结构 . 地震研究，29（3）：239-244.
王小平，朱元清 . 2007. 上海及其邻近地区上地幔间断面接收函数的研究 . 地震地磁观测与研究，28（5）：150-154.
王小平，尹京苑，倪友忠，等 . 2010. 上海定点形变异常特征及机理分析 . 大地测量与地球动力学，30（s1）：50-52.
王孝磊，周金城，陈昕，等 . 2017. 江南造山带的形成与演化 . 矿物岩石地球化学通报，36（5）：714-735.
王永，迟振卿，李德贵，等 . 2004. 泥河湾盆地 0. 8Ma 以来的地磁场相对强度记录 . 科学通报，49（9）：879-882.
王卫平 . 1997. 上海地区区域稳定性评价 . 北京地质，（4）：9-16.
王维佳，蒋雪中，薛靖波，等 . 2014. 长江口横沙附近河势变化与可利用港航资源分析 . 长江流域资源与环境，23（1）：39-45.
王张华，赵宝成，陈静，等 . 2008. 长江三角洲地区晚第四纪年代地层框架及两次海侵问题的初步探讨 . 古地理学报，10（1）：99-110.
王争争，徐戈 . 2001. 常熟—太仓 M_s5. 1 地震精定位及震源机制解 . 防灾减灾工程学报，21（1）：1-3.
魏子新，翟刚毅，严学新，等 . 2010. 上海城市地质 . 北京：地质出版社 .
文子才，楼法生，黄志忠，等 . 2001. 赣东北地区浅变质岩中微体古植物及其形成时代讨论 . 现代地质 . 15（3）：281-289.
吴标云，李从先 . 1987. 长江三角洲第四纪地质 . 北京：海洋出版社 .
吴建中，王寒梅，杨天亮 . 2009. 浅层地下水人工回灌应用于上海市工程性地面沉降防治的试验研究 . 现代地质，23（6）：1194-1200.
谢建磊 . 2013. 上海南部齐贤桥辉石闪长岩体的锆石 U-Pb 定年及其地质意义 . 上海国土资源，34（4）：91-95.
谢建磊，张克信，马小林，等 . 2017. 长江三角洲上新世以来磁性地层及天文调谐年代标尺 . 地球科学：中国地质大学学报，42（10）：1760-1773.
许保桐 . 1981. 上海地区地层概述 . 上海地质，（1）：3-17.
许保桐，徐明德，等 . 2003. 上海志留纪、奥陶纪地层新发现化石及其地层意义 . 上海地质，（3）：7-10.
许保桐，蒋荣华 . 1995. 上海金山群变质岩 Rb-Sr 和 Sm-Nd 同位素年龄测定结果 . 上海地质，（3）：15-20.
徐磊，李三忠，刘鑫，等 . 2012. 华南钦杭结合带东段成矿特征与构造背景 . 海洋地质与第四纪地质，32（5）：57-66.
严学新，杨建刚，史玉金，等，2009. 上海市三维地质结构调查主要方法、成果及其运用，上海地质，（1）：22-27.
严学新，方正，曾正强，等 . 2004. 上海地下空间开发环境地质问题分析 . 上海地质，25（1）：1-5.
杨明桂，梅勇文 . 1997. 钦-杭古板块结合带与成矿带的主要特征 . 华南地质与矿产，（3）：52-59.

杨金玉，徐世浙，余海龙，等 . 2008. 视密度反演在东海及邻区重力异常解释中的应用 . 地球物理学报，51（6）：1909-1916.
杨天亮，严学新，王寒梅，等 . 2009. 基坑施工引发的工程性地面沉降研究 . 上海地质，30（2）：15-21.
杨天亮 . 2012. 深基坑减压降水引发的地面沉降效应分析 . 上海国土资源，33（3）：41-44.
杨天亮 . 2018. 深基坑减压降水地面沉降控制综合分区方法研究 . 上海国土资源，39（2）：64-69.
杨天亮，严学新，王寒梅，等 . 2010. 水位与沉降双控模式下浅层地下水压力回灌试验研究 . 上海地质，31（4）：12-17.
杨天亮，王寒梅，焦珣 . 2014. 上海地面沉降防治分区管控方法研究 . 上海国土资源，35（4）：105-109.
杨志坚 . 1989. 上海地区大地构造归属及其东延去向问题 . 上海地质，（4）：41-47.
姚保华 . 2005. 上海地区深部地震构造环境探测方法研究 . 上海：同济大学 .
姚保华，章振铨，王家林，等 . 2007. 上海地区地壳精细结构的综合地球物理探测研究 . 地球物理学报，50（2）：482-491.
叶为民，万敏，陈宝，等 . 2009. 深基坑承压含水层降水对地面沉降的影响 . 地下空间与工程学报，5（S2）：1799-1805.
尹延鸿，张训华，温珍河，等 . 2008. 中国东部海区及邻区区域构造图的编制方法及地质构造单元划分 . 海洋学报，30（6）：99-105.
于鹏，吴健生，王家林，等 . 2008. 利用长周期 MT 数据研究沪浙地区深部断裂结构 . 同济大学学报：自然科学版，36（4）：549-554.
于振江，张于平，王润华，等 . 2004. 长江三角洲（江南）地区新近纪地层划分及时代讨论 . 地层学杂志，28（3）：257-264.
袁宝印，朱日祥，田文来，等，1996. 泥河湾组的时代、地层划分和对比问题 . 中国科学：D 辑，26（1）：67-74.
袁峰，周涛发，范裕 . 2010. 安徽繁昌盆地中生代火山岩锆石 LA-ICP-MS U-Pb 年龄及其意义 . 岩石学报，26（9）：2805-2817.
张恒，李廷栋，高林志，等 . 2015. 赣东北中元古界铁砂街组石英角斑岩和流纹岩锆石 SHRIMP U-Pb 年龄、Hf 同位素及地球化学特征 . 地质论评，61（1）：65-78.
张宏良，吴林奎，潘厚耀，等 . 1999. 上海地质矿产志 . 上海：社会科学院出版社 .
张克信，潘桂堂，何卫红，等 . 2015. 中国构造-地层大区划分新方案 . 地球科学，40（5）：206-233.
张克信 . 2015. 中国沉积大地构造图（1：250000）及说明书 . 北京：地质出版社 .
张鸣韶，盛莘夫 . 1958. 川黔边境的奥陶纪地层 . 地质学报，38（3）：326-335.
张守信 . 2005. 程裕淇主编《中国地层典》记失 . 地质科学，40（1）：145-152.
张文堂，李积金，钱义元，等 . 1957. 湖北峡东寒武纪及奥陶纪地层 . 科学通报，（5）：145-146.
张信宝，刘彧，王世杰，等 . 2018. 黄河、长江的形成演化及贯通时间 . 山地学报，36（5）：3-10.
张永康 . 2008. 扬子板块东北缘中元古代的大地构造划分 . 资源调查与环境，29（3）：162-167.
赵宝成 . 2015. 长江三角洲晚更新世以来沉积地层及相对海平面变化环境响应 . 上海：华东师范大学 .
赵会民，吕炳全，孙洪斌，等 . 2002. 西太平洋边缘海盆的形成与演化 . 海洋地质与第四纪地质，22（1）：57-62.
章振铨，刘昌森，王锋 . 2004. 上海地区断裂活动性与地震关系初析 . 中国地震，20（2）：143-151.
郑家欣，周积元 . 1986. 苏南地区（包括上海市）新构造运动基本特征的探讨 . 同济大学学报，14（1）：123-131.
郑月军，张世民，崔效锋，等 . 2006. 地震震源机制解在华南及邻区潜源区长轴方向判定中的应用 . 中国地震，22（1）：24-33.
郑永飞，张少兵 . 2007. 华南前寒武纪大陆地壳的形成和演化 . 科学通报，52（1）：1-10.
周新华，蒋荣华，陈焕疆，等 . 1993. 上海金山群斜长角闪岩 Sm-Nd 和 Ar-Ar 同位素年龄测定 . 岩石学报，9（4）：418-421.
周春喜 . 1990. 舟山群岛 3. 2 级地震的震源机制及与海域地质构造的关系 . 浙江国土资源，（2）：52-56.
周涛发，范裕，袁峰，等 . 2008. 安徽庐枞（庐江-枞阳）盆地火山岩的年代学及其意义 . 中国科学：D 辑，38（11）：1342-1353.
朱光，徐嘉炜，刘国生，等 . 1999. 下扬子地区前陆变形构造格局及其动力学机制 . 中国区域地质，18（1）：73-79.
朱介寿，蔡学林，曹家敏，等 . 2002. 中国华南及东海地区岩石圈三维结构及演化 . 北京：地质出版社 .
朱履熹 . 1990. 上海及邻区新构造应力场初步研究 . 上海地质，（1）：45-55.
朱日祥，岳乐平，白立新 . 1995. 中国第四纪古地磁学研究进展 . 第四纪研究，15（2）：160-173.

朱子沾，徐关生 . 1990. 上海地区重磁场特征与断裂构造关系初探 . 上海地质，(2)：42-48.
竹淑贞，陈业裕，吕金荣，等 . 1983. 上海地区全新世地层与古地理 . 科学通报，(5)：296-299.
竹淑贞，陈业裕，孙永福，等 . 1980. 上海地区第四纪地层与古气候 . 科学通报，(5)：220-223.
竹淑贞，吕金荣，陈业裕，等 . 1984. 上海地区全新世沉积分区及其特征 . 东海海洋，2（4)：23-33.
江西省地质矿产勘查开发局 . 2017. 中国区域地质志 · 江西志 . 北京：地质出版社.
全国地层委员会 . 2017. 中国地层表（2014）说明书 . 北京：地质出版社 .
上海地质矿产局 . 1988. 上海市区域地质志 . 北京：地质出版社.
浙江省地质矿产局 . 1996. 浙江省岩石地层 . 武汉：中国地质大学出版社.
《中国地层典》编委会 . 2000. 中国地层典：第四系 . 北京：地质出版社 .
中国环境监测总站，国家环境保护局 . 1990. 中国土壤元素背景值 . 北京：中国环境科学出版社 .
Chen Z Y，Stanley DJ. 1995. Quaternary subsidence and river channel migration in the Yangtze delta plain，Eastern China. Journal of Coastal Research，11（3)：927-945.
Cox A. 1969. Geomagnetic reversals. Science，163：237.
Chen J，Wang Z H，Chen Z Y，et al. 2009. Diagnostic heavy minerals in Plio-Pleistocene sediments of the Yangtze Coast，China with special reference to the Yangtze River connection into the sea. Geomorphology，113（3)：129-136.
Gu J W，Chen J，Sun Q L，et al. 2014. China´s Yangtze delta：Geochemical fingerprints reflecting river connection to the sea. Geomorphology，227：166-173.
Hyndman D W. 1972. Petrology of igneous and Metamorphic Rocks. Newyork：McGraw-Hill Press.
Li C X，Wang P，Sun H P，et al. 2002. Late Quaternary incised-valley fill of the Yangtze delta（China)：its stratigraphic framework and evolution. Sedimentary Geology，152：133-158.
Li J X. 1996. Quaternary environmrnt in the eastern China. Beijing：Seismological press.
Lin J X，Dai L P，Yuan W S，et al. 1989. Quaternary marine transgressions in eastern China. Journal of Palaeogeography，1（2)：105-125.
Liu X B，Chen J，Maher B A，et al. 2018. Connection of the proto-Yangtze River to the East China Sea traced by sediment magnetic properties. Geomorphology，303：162-171.
Liu L，Xu X S，Xia Y. 2014. Cretaceous Pacific Plate movement beneath SE China：Evidence from episodic volcanism and related intrusions. Tectonophysics，614：170-184.
Li Z X，Li X H，Warto J A，et al. 2010. Magmatic and metamorphic events during the early Paleozoic Wuyi-Yunkai orogeny，southeastern south China：new age constrains and *P-T* conditions. GSA Bullletin，122（5-6)：772-793.
Mankinen E A，Dalrymple G B. 1979. Revised geomagnetic polarity time scale for the interval 0-5 m. y. B. P.. Journal of Geophysical Research Solid Earth，84（B2)：615-626.
Wang Z H，Jones B G，Chen T，et al. 2013. A raised OIS 3 sea level recorded in coastal sediments，southern Changjiang delta plain，China. Quaternary Research，79（3)：424-438.
Wang Z H，SaitoY，Zhan Q，et al. 2018. Three-dimensional evolution of the Yangtze River mouth，China during the Holocene：impacts of sea level，climate and human activity. Earth-Science Reviews 185：938-955.
Yang S Y，Li C Y，Cai J G. 2006. Geochemical compositions of core sediments in eastern China：Implication for Late Cenozoic palaeoenvironmental changes. Palaeogeography Palaeoclimatology Palaeoecology，229（4)：0-302.
Yue W，Jin BF，Zhao B C. 2018. Transparent heavy minerals and magnetite geochemical composition of the Yangtze River sediments：Implication for provenance evolution of the Yangtze Delta. Sedimentary Geology，364：42-52.
Yue W，Liu J T，Zhang D，et al. 2016. Magnetite with anomalously high Cr_2O_3 as a fingerprint to trace upper Yangtze sediments to the sea. Geomorphology，268：14-20.
YueW，YangS Y，ZhaoB C，et al. 2019. Changes in environment and provenance within the Changjiang（Yangtze River）Delta during Pliocene to Pleistocene transition. Marine Geology，416：105976.
ZhaoB C，WangZ H，ChenJ，ChenZ Y. 2008. Marine sediment records and relative sea level change during late Pleistocene in the Changjiang delta area and adjacent continental shelf. Quaternary International，186：164-172.
Zhou T F，FanY，YuanF，et al. 2011. Geochronology and signifinance of volcanic rocks in the Ning-Wu Basin of China. Science China（Earth Science)，54（2)：185-196.

中文摘要

上海市（简称沪，别称申）位于长江口与杭州湾之间，地势平坦，海拔一般在数米。仅在金山、松江境内有10余处小山丘，最高海拔约为100m。基岩面埋深起伏明显，整体上自西南向东北方向明显变大。松江、金山、青浦等地，埋深多在160~220m；浦东、宝山和嘉定等的沿江沿海地区埋深多在280~360m；河口砂岛区埋深总体大于400m，崇明岛局部可达近500m。

第一代地质志编纂以来，区内开展了大量地质调查和研究工作，在区域地质条件、城市地质等方面都取得了大量进展。在对近30年来各类资料进行搜集的基础上，以活动论、板块构造学说为指导，系统总结了区内地层及沉积作用、岩浆岩及岩浆作用、变质岩及变质作用、岩石圈结构构造、地质构造、第四纪地质与地貌等内容，并融合了城市地质调查和研究成果。编制了《上海市第四纪地质及地貌图》《上海市基岩地质图》《上海市地质构造图》《上海市布格重力异常图》《上海市航磁（ΔT）异常图》《上海市第Ⅳ承压含水层综合水文地质图》《上海市工程地质结构综合分区图》《上海市累计地面沉降等值线图（1921—2016年）》《上海市地质遗迹图》等9幅1∶250000系列地质图件。

上海地处扬子克拉通之下扬子陆块东南缘，紧邻华夏造山系。自古元古代以来，区内经历了地壳雏形形成、基底形成、盖层形成和陆内改造等4个大的构造演化阶段。迄今，未见晚古生代与三叠纪—侏罗纪地层。根据区域地层发育特点，本志暂将下—中泥盆统、中—上三叠统和上侏罗统做缺失处理，其他地层做未发现处理。受多次构造运动影响，地质构造复杂、岩浆活动频繁。

一、前第四系

上海主要发育有古—中元古界、青白口系、震旦系、寒武系、奥陶系、志留系、白垩系、古近系、新近系和第四系，暂未发现上古生界和三叠系。自下而上划分有33个岩石地层单位，其中基岩地层有23个。基岩地层的综合地层区划属于华南地层大区，扬子地层区，江南地层分区，沪杭地层小区。第四纪地层属于华南地层大区之中南—东南地层区。根据晚第四纪地层的发育特征，第四纪地层分区又可划分为长江三角洲和太湖平原两个沉积分区。

（一）古元古界—青白口系

古元古界—青白口系为地壳雏形和基底形成阶段建造，构成区内结晶或褶皱基底，可划分出金山群、惠南板岩和河上镇群。

金山群主要见于闵行华泾，浦东三林塘、天花庵、赵家宅、康家宅、沈庄以及金山朱泾、吕巷等地。为一套变质岩系，主要岩性为云母片岩、绿泥片岩、长石石英岩以及斜长角闪岩、斜长角闪片麻岩、角闪片岩、角闪斜长片麻岩、浅粒岩等。钻孔揭露厚度一般大于260m，未见顶、底。斜长角闪岩和糜棱花岗斑岩的锆石U-Pb年龄分别为1078±8.9Ma、1781±16Ma。

惠南板岩仅见于浦东惠南，未直接被松散层覆盖，上覆数十米的玄武岩。岩性单一，主要为灰色绢云母板岩，局部为板状泥岩。发育有*Leiominuscula minuta* Naum、*Lophominuscula* sp.、*Paleamorpha* sp.、*Dictyosphaera macroreticulata* Sin et Liu等化石。钻孔揭露厚度大于194.39m，未见底。

河上镇群仅见于金山钱圩，上部为以灰色、绿色为主的火山碎屑岩建造，岩性以熔结凝灰岩为主，下部为灰色、浅灰色的陆源碎屑岩建造，以含凝灰质砂岩、长石石英砂岩为主，少量含砂泥岩和泥质粉砂岩等，普遍见黄铁矿化。钻孔揭露厚度大于466.83m。锆石U-Pb年龄为774±12Ma和801±13Ma。

中元古代主要为具有类复理石建造特征的浅海相沉积，伴随有间歇性的（海底）火山喷发活动。

由早期向晚期，海水似有加深趋势，并且岩浆作用似有加剧特征。青白口纪早期，海水逐渐加深，属次深海沉积环境；晚期，海水逐渐退缩，逐渐转化为滨浅海—陆地环境，其间火山活动逐渐加剧。

（二）震旦系

震旦系主要分布于浦东坦直、三灶一带，在宝山罗店、大场、江湾等地钻孔中也有发现。在崇明港东一带也可能存在。整体上为一套以蚀变较强烈的灰岩、白云质灰岩、白云岩为主的地层，北部大场夹白云质泥岩、石英砂岩，顶部见硅质岩。北部似属台地边缘浅滩相—开阔台地相沉积，南部可能为局限海台地相—蒸发岩台地相沉积。钻孔揭露厚度在6~260m。未见底，

（三）寒武系

寒武系在下古生界中揭露最多，已于宝山大场和罗店、长宁北新泾—天山五村、奉贤齐贤桥—钱桥和青浦重固等地钻孔中揭遇。自下而上可分为：底—下寒武统超山组、大陈岭组，中寒武统杨柳岗组，上寒武统超峰组。未见顶。除超峰组与其上、下层位尚未见有接触关系外，其他各组间均为整合接触关系。

超山组以灰黑色碳质页岩为主，局部为泥岩，间夹有钙质页岩及含粉砂、细砂质页岩薄层。钻孔揭露厚度约为148m，具有南薄北厚之趋势。灰岩夹层中富含有三叶虫及腕足类等化石。与下伏地层灯影组呈整合接触。

大陈岭组主要由灰、灰黑色白云质灰岩、白云岩及灰岩组成，局部夹薄层硅质岩、碳质页岩和钙质砂岩。钻孔已揭露厚度在33~114m。顶、底地层均已揭露，

杨柳岗组主要以浅灰、深灰色泥灰岩、含泥质灰岩与含粉砂灰岩为主，局部夹钙质页岩或硅质灰岩薄层。泥质灰岩中可见有缝合线构造，含砂灰岩中粉砂常聚成团块状。钻孔揭露厚度在150m以上。顶部已剥蚀，出露不全，

超峰组主要由浅灰与青灰色白云岩、同生角砾状白云岩组成，局部夹灰白与灰黑色白云质灰岩、硅质岩。钻孔揭露厚度在248~762m，且尚未见顶、底。

区内寒武纪从早期—晚期，经历了由盆地相—开阔海台地相—局限台地潟湖相的全过程；海水深度发生浅—深—浅的变化。

（四）奥陶系

奥陶系分布局限，目前仅于市区北部、西南部和奉贤海湾西部等地钻孔中发现。地层层序较齐全，自下而上可分为：下奥陶统仑山组、红花园组，下—中奥陶统牯牛潭组，上奥陶统黄泥岗组和长坞组。钻孔厚度大于250m，未见顶、底，与上覆、下伏地层接触关系不明。

仑山组主要以灰色白云质灰岩与灰质白云岩为主，岩性单一，单层厚度较大。揭露钻孔最大厚度为198.42m。未见底。

红花园组以灰色、灰白色微晶灰岩为主，层间夹有蓝灰色含泥灰岩，呈互层状，本层最大特征是层内普遍含有泥质，分布极不规则，时而尖灭或消失。于顶部地层中尚见少量生物骨屑。已揭露钻孔最大厚度为176.99m。与下伏仑山组、上覆牯牛潭组呈整合接触。

牯牛潭组在奉贤南部地区主要为紫红色微晶碎屑生物灰岩夹微晶灰岩，顶部有角砾状结构。在市区主要为灰白色含生物碎屑灰岩。局部穿插有硅质条带或脉状体。含腹足类 *Umbonellina* cf. *inflata* Yu，*Maclurites* sp.，*Raphispira* sp.，*Liospira* sp.；含牙形石 *Periodon aculeatus* Hadding，*Protopanderodus rectus*（*Lindström*），*Walliserodusethingtom*（*Fahraeus*）。另外，还含有海百合茎、介形虫等化石碎片。钻孔揭露厚度多在十数米，个别至三四十米，最厚约为60m。与下伏红花园组呈整合接触。

黄泥岗组是一套厚度不大的瘤状灰岩层，岩性较为简单，以具有泥质或硅质瘤状体为其特征。钻孔揭露厚度达39.2m。与下伏牯牛潭组呈整合接触。

长坞组主要由黄绿色—灰黑色泥岩、泥质粉砂岩、粉砂质泥岩与粉—细砂岩等组成。沉积韵律清晰，具有类复理石沉积特征。层内富含以 *Climacograptus* 为主的笔石化石。已见钻孔厚度大于123.0m，

未见顶，与黄泥岗组呈整合接触。

奥陶系为海相沉积，由早期到晚期，水深增加，由继承寒武纪晚期的台地环境演化为台地边缘斜坡、陆棚浅海环境。

（五）志留系

志留系目前仅见于崇明岛西北部庙镇—草棚一带，是一套以砂岩、泥岩和页岩为主的碎屑岩建造。自下而上划分为河沥溪组和康山组。钻孔已揭露厚度约在656.7m。上未见顶，下未穿底。

河沥溪组以灰、微绿、灰绿色为主，局部呈灰黑和灰褐色，泥质砂岩、粉砂质泥岩互层出现，局部夹有页岩或石英砂岩薄层。层内普遍见有泥砾。钻孔揭露最大厚度仅500m左右，未见底。

康山组上部以紫红色、灰黑色岩屑石英杂砂岩、石英砂岩为主，下部以灰褐—灰色、灰黑色岩屑石英杂砂岩为主，夹有粉砂质泥岩、泥岩之薄层，普遍受到轻微变质影响。层内见有丰富藻类及孢子等微古植物。未见顶，厚度为大于40.0m。

志留系主要为滨海沉积，颜色由下部的灰绿、灰黑色往上变成紫红或暗红色，碎屑物粒度则由细变粗，显示出一个较完整的海退序列。

（六）白垩系

下白垩统分布范围最广，约占全市面积的70%以上。上白垩统分布范围局限，主要分布于西北部与江苏毗邻的安角盆地、中部的松江盆地、崇明岛西段的北缘地区。下白垩统为中、酸性火山岩及火山碎屑-沉积岩系，自下而上划分为劳村组、黄尖组和寿昌组。上白垩统为一套断陷盆地建造，自下而上划分为浦口组和赤山组。下白垩统和上白垩统揭露的钻孔最大厚度分别至1850.7m、789.56m，各组顶底不全。

劳村组分上、下两段。下段广布于上海东南部和奉贤、金山境内，主要由灰白、灰紫色相间的凝灰质砂岩、凝灰质砂砾岩、凝灰质砾岩、流纹质凝灰熔岩及砂岩组成。砾石成分复杂，分选差，呈次圆状到棱角状，局部见定向排列。层间有较多安山玢岩脉穿插，普遍硅化且较强烈。在金山松隐等处，见紫红与灰白色相间的凝灰熔岩。上段主要见于金山松隐、张堰。松隐主要为紫红色与灰白色相间的凝灰质砂岩，间夹薄层泥岩，层理清晰；张堰多为长石石英砂岩，基本上不含火山物质。已见钻孔厚度大于167.9m。

黄尖组主要分布上海中部和南部地区，岩性主要为英安岩、流纹英安岩、英安质含角砾熔结凝灰岩、流纹英安质熔结凝灰岩、熔结凝灰岩、凝灰熔岩等中酸性火山岩和火山碎屑岩。并往往和浅成、超浅成的次火山岩组成火山杂岩。在浦东金桥及松江天马山为安山玢岩、安山岩和英安玢岩。单孔所见最大厚度可达484.0m。与上覆地层寿昌组之间的接触关系不明。锆石U-Pb年龄为136.1±2.4Ma~129.0±0.8Ma。

寿昌组分为上段、中段、下段。顶、底出露不全，与上覆、下伏地层接触关系尚不清楚。下段为紫红色、青灰色、灰黑色粉细砂岩、砂岩和杂色砾岩组成的韵律层，夹灰黑色炭质条带和煤线，局部为泥灰岩，基本不含火山物质，层理发育。已见最大钻孔厚度为300.7m。见大量孢粉化石，以裸子植物花粉为主，裸子植物花粉中又以 *classopollis* 花粉含量最高，伴生的有 *Psophosphaera*、*Alisponites*、*Monocolpollenites*、*Araucariacites*、*Sphaeripollenites* 以及 *Podocarpidites* 花粉。$^{206}Pb/^{238}U$ 年龄比较分散，多集中在120Ma左右。中段主要由紫红色、青灰色、灰白色凝灰岩、凝灰质砂岩以及泥岩、粉砂岩、中粗砂岩组成，局部为含粉砂灰岩或晶屑玻屑白云岩。含较多植物碎片，层理清晰。已见最大钻孔厚度为256.1m。上段主要由球泡流纹英安岩、流纹英安岩、含角砾熔结凝灰岩、岩屑晶屑玻屑凝灰岩等中酸性熔岩与火山碎屑岩组成。已见钻孔最大厚度为326.9m。

浦口组在安角断陷盆地南缘分布广泛，以块状紫棕色、紫灰色、杂色砂砾岩、砾岩为主，夹有厚薄不一的红棕色粉砂岩、含砾粉—细砂岩和泥岩。砾石磨圆度差，分选性差，成分以火山岩为主。厚度变化极大，靠盆地边部显著变薄。与下伏下白垩统呈不整合接触。

赤山组主要分布在安角断陷盆地北部、松江和闵行北桥。嘉定安亭、南翔等地粒度变粗，以砂岩、砂砾岩、砂质泥岩等互层为主。闵行北桥主要由棕红色粉砂质泥岩与粉砂岩组成，夹紫红色细砾岩及含砾砂岩。松江以紫色粉砂质泥岩、泥质粉砂岩为主，夹中细砂岩及个别粗砂岩层。碎屑组分以石英为主，长石次之，并含云母、赤铁矿碎屑。分选性及滚圆度均较差。层中常可见薄层状或网脉状石膏。部分层位中可见钙质砾石、泥砾或炭质碎块。已见钻孔最大厚度为771.9m。玄武岩夹层K-Ar年龄为78.31±1.77Ma。

白垩系为陆相沉积，可分为冲积扇、河流、湖泊、沼泽等沉积类型。早白垩世，气候炎热，形成巨厚的陆相火山-陆源碎屑岩。在强烈火山喷发期，堆积以火山岩为主，火山喷发间歇期，在火山断陷洼地常常堆积河湖相火山碎屑岩、沉积岩；晚白垩世，在局部断陷盆地中充填巨厚的冲积扇和河湖相沉积物。

（七）古近系

古近系主要分布在安角断陷盆地及北桥断陷盆地。在崇明北缘盆地中推断也有分布。迄今仅发现古新统，自下而上划分为夏驾桥组及北桥组，为一套紫红色陆相沉积建造，已固结成岩。根据区域地层发育特点，将区内始新世、渐新世地层作为暂未发现处理。

夏驾桥组以棕红色（局部砖红色）粉细砂岩、粉砂岩、泥质粉砂岩为主，自下而上沉积变细，层内富含介形类及轮藻微体化石。介形虫以 *Cristocypridea-Proeucypris-Ruficypris* 为组合特征；轮藻以大量具顶孔的孔轮藻科宽轮藻属分子为主。钻孔最大厚度在130.0m左右，靠断陷盆地边缘和西部变薄。与下伏赤山组接触关系未见。

北桥组以棕红、暗（紫）红、灰褐色泥岩、粉砂质泥岩为主，局部含砾石，夹粉细砂岩和玄武岩为特征。钙质含量明显，局部含钙质团块、泥砾。介形虫种属比较单调，以 *Cypridea-Cyprois-Parailyocypris* 为组合特征；轮藻化石属种和数量十分丰富，以 *Gobichara deserta-Peckichara-Stephanocharajiangsuensis* 组合为特征；发现有孢粉化石，可划分出 *Ulmipollenites minor-Triporopollenites* 和 *Quercoidites-Ulmipollenites-Triporopollenites* 两个组合带。已见钻孔最大厚度为289.4m。与下伏夏驾桥组为整合渐变关系。

古新世早期基本继承了晚白垩世的沉积环境，在安角盆地和北桥盆地中，形成了一套湖相沉积，富含介形类及轮藻化石。

（八）新近系

由下部的玄武岩和上部的冲洪积和湖泊堆积两部分地层组成。下部地层称之“白龙港玄武岩”，属中新统，主要分布于东南部的长江河口滨岸地带；上部地层称之“崇明组”，属上新统，主要发育于上海晚新生代沉陷最深的崇明等河口地区，其次在宝山、浦东的滨岸地带及嘉定西北部等地凹陷区内也有分布。

白龙港玄武岩岩性主要为深灰色辉石玄武岩、橄榄玄武岩和灰绿色玻质玄武岩，夹有喷溢间歇期的松散堆积。最大钻孔厚度达260m，未见底。一般为23~69m。玄武岩全岩K-Ar法年龄为18.7~26.7Ma，与其下伏不同时代地层角度不整合接触。

崇明组在河口沙岛上，崇明岛东、西端地区发育较全，可见上、下两套地层：下部为砂砾石、含砾砂和灰绿色黏土频繁重现组成的辫状河流沉积，或以湖积的灰绿等色黏土为主夹洪积含泥砾质砂或砂砾石；上部以含砾砂与细砂互层为主，向上渐变成以黏土为主，形成具二元结构的曲流河道沉积；或以湖积黄绿、褐黄色黏土为主与洪积泥质砾石互层。厚度介于60~220m间。在宝山、浦东的滨岸地带和安亭凹陷内也有本组地层分布，其岩性、岩相序列与崇明所见相似：通常由下部的洪积灰黄色泥质砂砾石频夹黏土和上部的洪积—湖积灰绿色含砾黏土、黏土层组成。厚度大都为10~65m，唯在安亭凹陷内厚者可达120m。发现孢粉和介形虫化石，其中以孢粉为主。孢粉仅在少数钻孔中检出，以落叶栎（*Quercusdeciduous*）-常绿栎（*Quercus evergreen*）-莎草科（Cyperaceae）-凤尾蕨属（*Pteris*）组合为主。

中新世，区内西南部主要为广袤的剥蚀低山所占据，东部形成几个面积不等的火山熔岩盆地。在火山喷发间歇，火山盆地一度演化为湖盆。

上新世，上海中南部陆域的青浦、松江、金山、奉贤、闵行以及浦东与市区的接壤区仍为广袤的低山所占据，其间在闵行北桥和七宝—虹桥有山间湖盆和山上残积洼地嵌布。在现陆域滨岸的嘉定—宝山—市区东北部—浦东—南汇新城一带出现北西-南东向连续分布、波状起伏的沉降带，还有金山枫泾—朱泾与松江新浜、青浦蒸淀的山间沉降盆地，在这些沉降区内形成山前洪积平原和湖泊洼地。沉降盆地中心在崇明岛上。当时在崇明岛东端的向化、前哨农场及遁北一带、中部的城桥—富民农场以及西端的绿华—红星农场一带，在广阔的河间滩地之上出现数条河流。

二、岩浆岩及岩浆作用

上海岩浆作用时代跨度大，在晋宁期、燕山期和喜马拉雅期都有发育，以燕山晚期最为强烈。各期岩浆作用都以火山喷溢为主，同期侵入作用与喷溢作用的岩浆具有同源性。岩浆岩及与之相关的火山碎屑岩建造十分发育。岩石类型以中酸性岩、酸性岩为主，次为中性岩、中基性岩，基性岩较少。基性岩呈碱性，岩浆来源于地幔熔融。中酸性岩以钾玄岩-高钾钙碱性系列为主，多为I型和A型花岗岩类

（一）侵入岩及侵入作用

上海已发现各类侵入岩体约33个。酸性岩类规模较大，主要分布在上海的北部。中性岩类和基性岩类岩体较多且规模较小，多分布于上海的东部和南部。岩体产状以岩株、岩枝、岩脉为主，大于100km^{2}的岩基只有一个，即崇明花岗岩体。可划分出晋宁期、燕山期和喜马拉雅期等3期。

1. 晋宁期

晋宁期有两次侵入活动，分别在长城纪早期和蓟县纪晚期。主要见有辉绿岩、角闪辉绿岩、花岗岩和糜棱花岗斑岩。

辉绿岩、角闪辉绿岩见于浦东天花庵等地的金山群中，已发生变质。

花岗岩仅见于金山朱泾，为一小型岩珠。岩石分异程度较高，轻稀土元素富集，具有幔源花岗岩特征，可能与不成熟的岛弧或洋壳相关。锆石U-Pb年龄为1082.4±7.9Ma。

糜棱花岗斑岩主要见于浦东康家宅和奉贤肖塘等地，糜棱构造比较发育。处于破坏性活动板块边缘或同碰撞构造环境。锆石U-Pb年龄为1781±16Ma。

2. 燕山期

燕山期有4次侵入活动，中侏罗世1次，早白垩世2次，晚白垩世1次。

基性岩常见岩石有角岩化辉绿岩、角闪石辉绿岩及角岩化次闪石化辉绿玢岩等。分布于金山张堰、界山等地。岩体规模较小，多呈岩床、岩脉产出，多发生明显的角岩化。辉绿岩类蚀变明显，全岩K-Ar年龄为112.7Ma和106Ma。

中—酸性岩较多。其中，中性岩规模较小，主要有月浦闪长岩、坦直辉石闪长岩、齐贤桥辉石闪长岩、三林塘石英闪长岩和张堰石英闪长岩等。酸性岩规模较大，主要有崇明花岗岩、吴淞花岗岩、市区花岗岩等。燕山中期发育有石英闪长岩和花岗岩，岩浆源区具有一定的相似性。岩浆来源深度较深，且岩浆在结晶演化过程中发生了较低程度的地壳混染作用。吴淞花岗岩为同碰撞型花岗岩，明显不同于后造山期三林石英闪长岩的I型成因。吴淞花岗岩锆石U-Pb年龄为178.8±2.3Ma、三林石英闪长岩锆石U-Pb年龄为163.4±1.5Ma。燕山晚期侵入岩主要形成于后造山—非造山环境。坦直辉石闪长岩、沈庄辉石闪长岩都呈弱的铕负异常，具幔源型稀土元素曲线特征。锆石U-Pb年龄为106.9~104.2Ma。张堰石英闪长岩无铕异常，具有高Sr、低Y和Yb的特征，铕无亏损，显示岩浆来源较深，且未经历明显的斜长石分异作用。在其周围形成区内唯一矽卡岩型铜矿。崇明花岗岩和市区花岗岩属上地壳同熔型稀土元素曲线特征，为A型花岗岩成因。锆石U-Pb同位素年龄为132.9±1.1Ma。崇明花岗岩的钾长石K-Ar同位素年龄为107.1Ma。崇明花岗细晶岩，铕亏损强烈，是岩浆已进入晚期残余岩浆富碱阶段的产物。

此外，该阶段发育大量斑（玢）岩类。主要有角岩化次闪石化辉绿玢岩、煌斑岩、辉石闪长玢岩、石英闪长玢岩、闪长玢岩、安山玢岩和花岗斑岩、霏细斑岩、正长斑岩等。主要形成于燕山晚期，多呈岩脉产出。见有花岗斑岩、安山玄武玢岩等穿插石英闪长岩体；霏细斑岩、花岗斑岩、安山玢岩、煌斑岩、辉绿岩、安山玄武玢岩等穿插花岗岩体。花岗斑岩和霏细斑岩的全岩 K-Ar 同位素年龄在 81~89Ma。

3. 喜马拉雅期

目前所见都只是一些辉绿岩脉，分布广。并曾见其侵入于燕山晚期石英闪长岩体中。

（二）火山岩及火山作用

区内火山岩、火山碎屑岩分布极为广泛，但露头稀少。火山作用划分为晋宁早—中期、晋宁晚期、燕山晚期和喜马拉雅期等 4 期，暂未发现与中侏罗世和晚白垩世侵入作用相对应的喷溢作用记录。岩石类型复杂，基性、中性、中酸性、酸性岩类均有发现。其中以酸性、中酸性岩类分布面积较广。

1. 晋宁早—中期

火山岩主要发育在浦东周浦、金山张堰和钱圩等地金山群中。基性、中酸性及酸性火山岩发育，经过原岩恢复，其岩石类型主要为玄武质凝灰岩、流纹岩等，可能有流纹斑岩。经多次变质作用，成为斜长角闪岩、黑云母斜长角岩及长石石英角岩等区域变质岩系或热变质岩系。基性岩类主要为拉斑玄武岩系列，次为钙碱性玄武岩系列，铕轻微负异常。

2. 晋宁晚期

火山岩主要为青白口纪晚期建造，分布在金山钱圩、查山等上海西南部地区。中性、中基性火山岩、火山碎屑岩发育，主要有安山质熔结凝灰岩、角砾凝灰岩、凝灰岩等。属于高钾钙碱性系列。锆石 U-Pb 年龄为 774±12Ma、801±13Ma。

3. 燕山晚期

早白垩世火山作用发生于活动大陆边缘区，为后造山期挤压向伸展转换阶段的产物。可划分为 3 个亚旋回，第一亚旋回发生于初期，即劳村组下段形成时期，有酸性凝灰质和熔岩喷发；第二亚旋回发生于中期，相当黄尖组形成时期，此时火山活动强烈；第三亚旋回发生于末期，相当于寿昌组上段形成时期，此时火山活动渐趋减弱。呈现为弱—强—弱—强的喷发特征。晚白垩世火山作用形成于非造山阶段的伸展环境，玄武岩 K-Ar 年龄为 78. 31±1. 77Ma。

早白垩世火山作用形成一套高钾钙碱性—钾玄岩系列火山岩组合及相应的火山碎屑岩。分布最广，岩石类型最复杂。岩相发育较齐全，以溢流相和空落相为主，还可见沸溢相、火山碎屑流相、火山爆发崩塌相、火山喷发沉积相、火山颈和火山通道相、潜火山岩相等。火山岩主要有玄武安山岩、辉石安山岩、安山岩、角闪安山岩、石英安山岩、粗面岩、英安岩和流纹英安岩、流纹岩、霏细岩等，火山碎屑岩主要有向熔岩过渡类型（流纹凝灰熔岩、安山质凝灰熔岩、英安质凝灰熔岩、英安质角砾凝灰熔岩、安山质角砾凝灰熔岩等）、熔结凝灰岩类（晶屑熔结凝灰岩、玻屑熔结凝灰岩等）、凝灰岩类（流纹质凝灰岩、晶屑凝灰岩、含角砾凝灰岩等）、沉凝灰岩。此外，偶见火山角砾岩和集块岩。晚白垩世，仅在断陷盆地局部形成玄武岩夹层。

火山岩-火山碎屑岩的酸度整体上有高—低—高的波动。氧化系数在劳村组最高，黄尖组变化最大。不同层位火山岩是同源岩浆不同期次喷发的产物。原始岩浆可能来自地壳深部，其中有部分上地幔成分掺入，在岩浆上升过程中，有壳源成分的不断加入。早期旋回出现的玄武岩、玄武安山岩、辉石安山岩，其成分可能接近原始岩浆的成分。

4. 喜马拉雅期

新生代仅在局部地区有基性岩浆喷溢，形成了一套钙碱性-钾玄岩系列的玄武岩。岩相以溢流相为主。主要由基性熔岩组成，包括伊丁石化橄榄玄武岩、气孔状玄武岩、玄武岩、铁质玄武岩等。其中玄武岩分布最广，铁质玄武岩仅见于个别钻孔中。幔源成因，形成于大陆边缘和陆壳强烈拉张减薄的构造环境。

从活动时间上，可划分为两个阶段：古新世的碳酸盐化玄武岩主要分布在闵行北桥、马桥等地，夹于北桥组中；始新世末期—中新世的玄武岩成片分布。全岩 K-Ar 年龄在 38.57±2.05Ma~31.6±1.31Ma（崇明新海农场）、26.7Ma（南汇中学）、18.7Ma（白龙港）。

（三）火山构造

区内火山作用，尤其是早白垩世，喷发类型兼有裂隙式及中心式两类，主要受北东向断裂控制。此外，并受北西向断裂的影响。火山岩区和火山喷发带划分与浙西北地区相似。以枫径—川沙北东向断裂为界，又可划分为两个亚带，断裂带南部为金山—惠南火山喷发亚带，断裂带北部为青浦—宝山火山喷发亚带。主要的三级和四级火山构造有松隐火山盆地、查山锥状火山、庄行构造火山盆地、天马山构造火山隆起和重固—顾村构造火山盆地。

三、变质岩及变质作用

上海变质作用主要有晋宁中期、晋宁晚期和燕山期 3 期。晋宁中期和晚期以区域变质作用为主，形成金山群和惠南板岩；金山群和惠南板岩变质变形强度差别较大。在斜长角闪岩、糜棱岩化花岗斑岩中分别获得 1078±8.9Ma、1781±16Ma 的岩浆锆石 U-Pb 年龄。金山群后期受到了多期构造热事件作用的叠加影响，燕山期以接触变质作用为主，局部叠加后期的动力变质作用，形成构造角砾岩及糜棱岩化岩石。

（一）区域变质岩及变质作用

1. 晋宁中期

晋宁中期区域变质岩构成前震旦系金山群，断续分布于枫泾-川沙断裂带与张堰—南汇断裂带之间。根据变质岩自然共生组合，自下而上可以分为 a、b、c 3 个岩性段。a 岩性段见有浅粒岩、长石石英岩、石英岩、云母片岩、绿片岩组合及大理岩、绿片岩、石英岩组合等两种变质岩岩石自然共生组合。b 岩性段以角闪质类岩石夹浅粒岩、长石石英岩、石英岩及少量黑云母片岩、变粒岩组合为特征。c 岩性段以为云母片岩、云母石英片岩、绿片岩、大理岩、石英岩、浅粒岩组合为特征。

金山群 a、c 岩性段层状构造清晰，具有较明显的沉积岩特征。原岩由泥质岩、砂泥质岩、砂质岩及碳酸盐岩岩石组成。b 岩性段层状构造不明显，但条带构造、片状构造、片麻状构造较清晰，且条带和片理倾角与 a、c 岩性段中的片岩、大理岩的片理或层理倾角一致。兼具岩浆岩类及沉积岩类的构造特征。

变质作用类型为低压区域变质作用，形成低绿片岩相和高绿片岩相递增变质岩系。低绿片岩相与 a、c 岩性段的分布一致，相当于一般的绿片岩相（不包括绿帘石角闪岩相）。常见的矿物共生组合有白云母（绢云母）+绿泥石+石英、钠长石+阳起石+绿帘石+绿泥石+石英、白云母（绢云母）+斜长石+石英、斜长石+绿帘石+绿泥石+石英、方解石+斜长石+绿帘石+透闪石+石英。以钠长石+阳起石+绿帘石+绿泥石+石英矿物组合最为典型。阳起石、钠长石、绿帘石、绿泥石、白云母（绢云母）等为该变质相的特征变质矿物。推测形成于 350~450℃、2~4kbar 的温压条件下。高绿片岩相分布上与 b 岩性段一致。常见的矿物共生组合有普通角闪石+斜长石+绿帘石+石英、普通角闪石+斜长石+透辉石+绿帘石+石英、普通角闪石+斜长石+铁铝榴石+黑云母+石英、普通角闪石+斜长石+黑云母+绿帘石+石英、斜长石+钾长石+铁铝榴石+石英、斜长石+钾长石+黑云母+白云母+石英、斜长石+黑云母+绿帘石+石英。其中普通角闪石+斜长石+绿帘石+石英为特征矿物组合。推测形成于 450~570℃、3~6kbar 的温压条件下。

2. 晋宁晚期

晋宁晚期区域变质岩组成前震旦系惠南板岩，仅见于浦东惠南，且为新近系玄武岩覆盖。已揭示部分的变质岩由单一的低绿片岩相板岩组成。

由泥质、微细鳞片状绢云母、绿泥石、微粒状方解石及长英质粉砂组成。其中粉砂质点仍保留棱角—次棱角状沉积特征，无明显重结晶现象。绿泥石、绢云母则由泥质成分经变质结晶而成。镜下能见到较清晰的变余泥质结构及变余粉砂状结构。从岩石的矿物组分及残余结构可知，其原岩应为沉积成因的含钙质、粉砂质泥岩类岩石。

根据绿泥石+绢云母组合，初步认为惠南板岩所属变质相相当于低绿片岩相的下部。其变质作用类型相当于董申保提出的区域动力变质作用中的板岩千枚岩变质作用。惠南板岩变质程度远低于金山，所以形成的温压条件更低。

3. 大地构造环境

中元古代早期和中元古代末期，区内可能位于不同阶段超大陆活动大陆边缘，尤其是，在中元古代晚期，演化为洋内弧或具有洋壳的弧后盆地伸展环境。中元古代末期至晋宁运动Ⅰ幕发生之前，一直处于持续伸展的构造环境。青白口纪早期，为相对稳定的弧后浅海—次深海环境。由早期向晚期，海水似有加深趋势，并且岩浆作用似有加剧特征。

（二）接触变质岩及变质作用

该类变质岩主要形成于燕山晚期，分布上明显受燕山期岩体及围岩性质控制。目前已知规模最大的接触变质岩，分布在上海西南部金山境内，分布范围西起界山，东经时家、百家至戚家浜。下部层位原岩推测为元古宇—下古生界，上部层位原岩为劳村组，变质作用发生的上限应为早白垩世早期。

主要岩石类型为黑云斜长角岩、黑云堇青角岩、斜长角岩、透辉石榴矽卡岩、石榴石矽卡岩、绿泥金云石榴矽卡岩、绿帘石榴矽卡岩、大理岩、黑云斜长角岩、透辉斜长角岩、碳酸盐化石英石榴矽卡岩等。原岩为一套正常沉积岩系，由泥质岩、长英质碎屑岩及含白云质灰岩等组成，间有少量基性岩脉穿插。

典型矿物共生组合有斜长石+黑云母+角闪石+石英、斜长石+黑云母+堇青石+矽线石（少量）、红柱石+正长石+斜长石+黑云母+石英、方解石+白云石+石榴子石+透辉石+硅灰石（少数情况下可出现符山石）、石榴子石+透辉石+硅灰石+绿帘石等。界山—张堰变质岩所处变质相至少相当于角闪角岩相，其中的红柱石+正长石+斜长石+黑云母+石英组合已达辉石角岩相。

四、地质构造

上海位于江山—绍兴深断裂的北西侧，大地构造位置上属于扬子克拉通之下扬子陆块东南缘，紧邻华夏造山系。自古元古代以来，区内经历了地壳雏形形成阶段（古元古代）、基底形成阶段（中元古代—青白口纪早期）、盖层形成阶段（青白口纪晚期—中三叠世）和陆内改造阶段（晚三叠世以来）等4个大的构造演化阶段，整个地质构造演化经历了由活动—稳定—再活动的多旋回螺旋式发展过程。受多次构造运动影响，地质构造复杂。

（一）构造单元划分及基本特征

区内Ⅰ级构造单元为扬子克拉通，Ⅱ级构造单元为下扬子陆块。无论是从基底岩系与盖层的发育特征，或者是从陆缘活动阶段地质构造特征上看，上海均与下扬子陆块的特征一致。同时，作为二级构造单元的结合部位，具有鲜明的过渡性特征。

以湖苏断裂带、江绍断裂带为界划分为下扬子克拉通盆地和钦杭结合带两个Ⅲ级构造单元。两者在前南华纪的基底建造阶段和晚古生代陆表海演化阶段建造差别明显。

以北西向断裂和中生代火山、断陷盆地建造为依据，划分出吴江—南通断块区、杭嘉湖断块区2个四级构造单元，南通断隆区、金山—杭州湾断隆区、佘山—青浦火山隆起区、上海—长江口断隆区、长江口东断隆区和余杭—嘉兴断坳区6个Ⅴ级构造单元。不同构造单元的重力场和航磁场特征不同。

此外，在Ⅴ级构造单元中，由于燕山期以来断块活动差异和深部构造因素影响，形成一系列断凸和断凹、火山洼地和隆起等。根据其特征可以划分一系列更次级的构造单元。不同次级构造单元内，基岩面埋深、基岩地层组合差别较大。

（二）构造形变

多旋回、多期次的构造运动在本区地壳岩层中塑造了各种规模级别不等、形态复杂多样、有一定规律可循的构造形迹。构造形变以韧、脆性断裂为主。褶皱形态简单，下古生界倾角多在40°~65°，少量偏缓或偏陡；劳村组倾角多在20°~35°，少量偏陡；赤山组倾角多在10°左右，局部较陡，其倾角变化多受盆缘形态控制。

区内自西北向东南发育昆山—港西复背斜、青浦—宝山复向斜、朱泾—周浦复背斜和金山—大团复向斜。复式褶皱轴向总体呈北东向。受中生代火山构造盆地、断陷盆地的叠加和破坏，核部很不连续。局部又被巨厚的火山岩系及红层所覆盖。局部下古生界沉积盖层中发育小型的背斜和向斜构造，如罗店背斜、徐家汇—五角场背斜等。

断裂构造复杂，按其走向主要分为北东、北北东、北西—北北西、近东西向等4组，以北东向断裂最发育。其中北东、近东西向断裂一般属早期构造，北北东、北西—北北西向断裂为晚期构造，时代相对较新。根据切割深度，可分为深断裂、基底断裂和盖层断裂。深断裂包括岩石圈断裂和地壳断裂，指切穿岩石圈或地壳的断裂。深断裂包括湖苏断裂带、张堰—南汇断裂带、江绍断裂带；基底断裂包括枫泾—川沙断裂、太仓—奉贤断裂、罗店—周浦断裂和青浦—龙华断裂；其他都为盖层断裂。深断裂和基底断裂沿线，局部糜棱岩化比较发育。

燕山晚期断陷盆地主要有新村—兴隆断陷盆地、松江—北桥断陷盆地盆地、安角断陷盆地。盆地内上白垩统—古新统发育，在重力、航磁异常上明显呈低值区。

（三）构造阶段和建造、变形特征

在4大构造演化阶段的基础上，根据建造和后期改造特征，可进一步划分出晋宁期、扬子期—加里东期、华力西期—印支期、燕山期和喜马拉雅期等构造阶段。

晋宁期指中元古代—青白口纪这一构造阶段，是区内最早的大规模地壳建造发育和演化阶段。划分为晋宁早—中期和晋宁晚期。晋宁早—中期建造由金山群组成；晋宁晚期建造由惠南板岩和河上镇群组成。

扬子期—加里东期指南华纪—志留纪这一构造阶段，区内未进行细分。该阶段为区内沉积盖层的主要形成阶段，由上震旦统—志留系组成。

华力西期—印支期指泥盆纪—中三叠世末这一构造阶段，区内迄未发现该阶段建造。

燕山期指晚三叠世—白垩纪这一构造阶段。下—中侏罗统沉积建造目前未有发现，局部发育中侏罗世侵入岩建造，早白垩世以发育火山构造和强烈的火山活动为特征。晚白垩世以发育断陷盆地红层为特征，沉积物下部较粗，上部较细。

喜马拉雅期指古近纪的构造阶段，由于古近纪和新近纪建造差别较大，对应可分为早期和晚期两个阶段。古近纪继承晚白垩世沉积格局。中、上新统主要为河湖相杂色碎屑岩，间有大范围喷溢的玄武岩。第四系松散沉积物广覆全区，沉积环境复杂。

哥伦比亚超大陆和罗迪尼亚超大陆聚合、裂解是区内中、新元古代构造格局的形成演化构造背景。晋宁早—中期，湖苏断裂带、枫泾—川沙断裂带、江绍断裂带、太仓—奉贤断裂等北东和北西向断裂雏形形成。经过扬子期—加里东期的持续活动和改造，主要构造形迹于印支期—燕山期定型。喜马拉雅期构造运动主要表现为早期构造基础上的晚期构造及盆地叠加和活化，尤其是北东向、北西向、东西向断裂的活化。

（四）新构造运动特征

区内的新构造运动标志主要有地震活动、断裂活动、垂向变形和差异沉降、夷平面、节理和擦痕、地热活动等。

断裂活动有很强空间和时间差异性。北西—北北西向断裂自西向东活动性减弱或分段性不明显。北北东向断裂则反之。北东向断裂一般西段的活动性要较东段的稍强。北北西—北西向、北北东向、北东向断裂活动性总体由强至弱。一般来说，东西向、北东向断裂活动时间略早，北北东向其次，北西西向、北北西向较晚。最新活动时间一般为早更新世—中更新世，北北西向断裂的最新活动时间更为新些，一般为中更新世，有的可延续至晚更新世。晚更新世发育2条活动断裂，中更新世发育3条活动断裂。断裂的最大断距一般为5~8m、几十米，在断点剖面上显示，往上断距变小，有时代越新，位错幅度趋小的特点。太仓—奉贤断裂平均位移速率最大，在0.01~0.028mm/a之间。

地震活动在空间上具有很强的丛集性、重复性、分块性和条带性。时间上呈现出一定的周期性，表现出相对平静期和显著活动期的交互特征。震源深度多在5~20km，存在5~10km、15~20km两个优势层。弱震震级与震源深度无相关性，其他地震震源深度与震级总体呈一定线性关系。自西北向东南方向明显减少，总体呈北强南弱、东强西弱（或海域强、陆地弱）的特征。绝大部分地震受第四纪活动断裂，尤其是晚更新世活动断裂的控制。震源机制解释显示，两个节面通常呈北东向和北西向，少部分近东西向，节理面产状中等到近垂直，倾角集中在33°~87°。在多组断裂的交汇处、新生代断陷盆地的边界断裂，尤其是北北东（北东）向和北西向断裂、北东（北北东）向与东西向断裂交叉部位是主要发震部位。中强地震在条带相交区域的频度与强度相对较高。此外，地震与新生代盆地构造关系也很密切。

新构造期主压应力为北东东—近东西向，主压应力场为96°左右。现代构造应力场处于北东东向水平压应力与北北西向水平张应力下。区域主压应力方向大致与海岸轮廓线相垂直，主压应力轴方向在北东东—近东西向，优势平均方位角北东79°。上海最大剪应力方向约为北东55°和北西35°。

五、岩石圈结构构造

速度场、重力场、磁力场、电性场和地温场等地球物理场特征显示，区内岩石圈内部各向异性特征明显，具有明显的层块结构。

（一）地球物理场特征

速度场特征显示，岩石圈可划分出8个速度层，分别为松散层（P波波速为1.7~2.0km/s，下同）、古近系—新近系（2.2~3.5km/s）、中生界—志留系（3.8~4.25km/s）、震旦系—奥陶系碳酸盐岩地层（5.12~5.97km/s）、上地壳浅变质岩系（5.80~6.20km/s）、中地壳上部深变质岩系、中地壳下部深变质岩系、下地壳（6.63~7.05m/s）、岩石圈上地幔（8200m/s左右）。中、下地壳过渡区局部有壳内低速层。

重力场特征显示，除松江、青浦西侧和崇明堡镇一带外，基本上全部为正异常区。全区区内有两条重力高值带、一环状重力高值区和一低值区。分别为阳澄湖—崇明重力高值带、嘉兴—惠南高值带、昆山—真如—吕巷高值区和用直—松江重力低值区。局部重力高发育明显，主要有庙镇—港西重力高、吕巷重力高、嘉定—罗店重力高、宝山西重力高、重固重力高、北新径重力高、坦直—祝桥重力高；局部发育重力低，如庄行重力低、唐行重力低、松江重力低。

航磁场特征显示，异常明显呈带状分布，总体构成北东向和东西向醒目的异常带。自北向南，区内有3条醒目的北东东向航磁异常带，即宝山—崇明磁异常带、天马山—顾路磁异常带和张堰-惠南磁异常带。

电性场特征显示，壳幔总体上呈现出横向上被低阻带分割，纵深上低阻和高阻叠置的特征。在上海西部的中、下地壳交界地带或中地壳内部，发育低阻层。有新近系—第四系、白垩系—古近系、震旦系—下古生界、前震旦系和侵入岩类等5个典型电性层。

地温场特征显示，区内变温带深度在9.0~17.0m之间变化，平均值为13.3m；恒温带底界在17.0~27.0m之间变化，平均值为23.0m，以下为增温带。200m以浅地层温增温率在2.31~3.73℃/100m，平均值为3.04℃/100m，总体上呈西南部较高、东北部略低的变化趋势。地表大地热流值在40.67~68.37mW/m^2之间，区域上变化较大，平均值为52.62mW/m^2

（二）岩石圈结构构造

岩石圈属于下扬子强减薄型碎块状岩石圈，厚度变化较大，在青浦一带可达120~150km，松江—奉贤一带为50~70km，金山—浦东泥城一带多在100km左右。莫霍面埋深在30km左右。上地壳厚为12~14km，底界自西向东逐渐加深；中地壳厚度约为10km，埋深在22~24km，呈现出明显的地层物性或岩石结构的变化；下地壳厚为10~11km，与岩石圈地幔具有厚2~3.5km的壳幔过渡带。除上地壳底部、中地壳底部、莫霍面和岩石圈底等外，自上而下还发育印支侵蚀面、盖层底界、金山群顶面、金山群底面和下古生界软弱面（下寒武统底部、志留系）等界面。

六、第四纪地质与地貌

上海第四纪地层十分发育，厚度多在200~280m，局部下伏新近纪松散沉积物。属于华北地层大区的华北东部地层区，可进一步划分为长江三角洲沉积分区（三角洲主体河口沙岛沉积小区和三角洲侧翼滨海平原沉积小区）和太湖平原沉积分区（东部小区和西部小区）。

（一）多重地层划分

1. 磁性地层和年代地层

B/M界限埋深在130~150m，大致出现在中更新统底部砂层与下伏下更新统顶部杂色硬土层之间的假整合界面附近。G/M界限的深度为250~285m，更新统与上新统分界大都在G/M界限之下2~5m，推算的区内第四系底界年龄为2.61~2.65Ma，即大致为2.60Ma。G/G界限的深度为379m。此外，发育有Gotheborg、Blake、Jamaica、Big lost Emperor等负极性亚时或极性漂移事件。

更新统分为下更新统、中更新统和上更新统3个阶段、全新统划分出下全新统、中全新统和上全新统3个阶段。考虑同期异相沉积作用，共划分有9个岩石地层单位，自下而上为下更新统安亭组，中更新统嘉定组，上更新统南汇组和川沙组/滆湖组，全新统娄塘组、上海组和青浦组/如东组。川沙组下限年龄约为128ka B. P.；南汇组/滆湖组下段下限年龄约为75ka B. P.；南汇组/滆湖组中段下限年龄约为70ka B. P.；南汇组/滆湖组上段下限年龄约为20ka B. P.；娄塘组下限年龄为10~12ka B. P.；上海组下限年龄约为8ka；青浦组/如东组下限年龄为3~4ka。

2. 岩石地层

安亭组在上海广泛分布，为一套河流—泛滥湖泊相沉积，其岩性为以褐黄、褐色为主，有灰绿、黄绿、蓝灰色网纹、斑纹穿插的杂色网纹状、斑杂状黏土，大致由下而上呈坚硬—硬塑—可塑状变化，普遍以含有钙质结核、网脉和铁锰质结核为特征，与浅灰、灰黄等色含砾砂和砂砾石间互成层。一般形成粗、细相间3个沉积旋回，从而可将之相应划分为下、中、上3段。其埋深一般在150~290m之间，由西南的太湖沉积区向东北的河口沙岛区逐渐加大。厚度约占第四纪地层厚度的一半。与上覆和下伏地层呈平行不整合接触。

嘉定组在上海分布广泛，为一套三角洲河流—泛滥平原、湖泊相沉积，其岩性、岩相具有早、晚更新世地层之间过渡的特点。具特征的岩性为蓝灰、灰绿色黏土，褐黄杂蓝灰、灰绿色细网纹状、条纹状黏土、粉质黏土，也有灰色粉质黏土和粉土，多为可塑状，少部分为硬塑状；与灰、黄灰色含砾砂、粉细砂间互成层。东、西部地区地层发育有所差异。在“冈身带”以西的太湖沉积区，中更新统地层保存齐全，存在粗、细相间两个沉积旋回形成的上、下两段地层，其埋深通常在95~150m之间，层厚为50~60m。而在“冈身带”以东的滨海平原、河口沙岛等地区，由于上更新统地层底界的下切，中更新统上段地层因遭受侵蚀而断续出现，大都只保存下段地层，其埋深介于120~160m之间，厚度为25~35m。

川沙组在上海分布广泛，为一套河流—弱海陆过渡的三角洲河流相细砂和含砾砂。在东北部的崇明、长兴、横沙等河口沙岛上，沉积物最粗。在与浙江毗邻的青浦、金山、松江的接壤区，为河口湾堆积的灰、棕灰色粉质黏土与浅灰色粉砂、砂质粉土互层。其埋深由西南向东北逐渐加大：陆域西部埋深一般介于65~95m之间；陆域东部至崇明岛，其顶界埋深在75~90m，底界埋深在100~130m。

南汇组/滆湖组为末次冰期期间形成的陆相—海相/海、陆相交替—陆相沉积序列，地层相应分为下、中、上3段。下段，埋深大体上由西南向东北加大，底界埋深在52~90m，厚度0~17m不等。岩性、岩相与厚度变化较大，但典型岩性主要为泛滥平原—湖泊相的灰绿、灰黑、深灰色及局部的杂色黏土、粉质黏土和黏质粉土；局部夹砂质粉土或粉砂层。含钙质结核、腐木碎块，见有陆相软体动物化石碎片。中段，厚度占上更新统上部地层的大半。其埋深由西南向东北逐渐加大，底界埋深在 41~90m，顶界埋深在 10~60m，厚度 13~61m 不等。在现滨海平原区北部地区，主要为浅海—海湾相灰色粉质黏土频夹浅灰色细砂、砂质粉土纹层或成薄互层；现滨海平原区南部地区为滨海—浅海相灰、灰黄色粉细砂、砂质粉土互层夹粉质黏土。青浦及与之接壤的松江、金山的部分地区为滨海—潟湖相与河流边滩、泛滥平原—湖泊相交替的暗灰、褐黄和杂色黏土、粉质黏土，时夹黄色粉砂、砂质粉土或成互层。河口区，下部通常为灰、黄灰色中细砂、粉细砂夹粉土或成互层，底部或现砂砾层；上部多为青灰、灰色粉细砂夹褐灰色粉质黏土、粉土或成互层。以普遍含丰富的海相生物化石为特征。上段，下部通常为姜黄、褐黄色粉质黏土，局部夹砂质粉土或粉细砂；上部为暗绿、灰绿或黄绿色黏土和粉质黏土。可见鱼骨碎片、轮藻受精卵膜等淡水生物化石，通常不含有孔虫等海相生物化石，但在南汇嘴地区、市区和青浦练塘等上海南部区域的钻孔中陆续发现个别的海相生物化石。

娄塘组为一套海陆过渡相堆积，厚度具有西南部薄、东北部厚的变化规律，至现长江河口地区厚度明显加大。太湖沉积区厚度多为 3~6m，长江三角洲沉积区为 4~32m 不等，在河口沙岛分区普遍厚达 30m。地层剖面完整者，可分为上、下两段。长江三角洲沉积区多为河口相，在崇明岛西部为溺谷相堆积。岩性常为灰色砂质粉土、黏质粉土与灰白色粉砂互层，部分如浦东外高桥孔为灰色砂质粉土与粉质黏土互层。普遍含多少不一的广盐性有孔虫、少量浮游种和窄盐种，还有少量半咸水和海相介形虫，偶见陆相介形虫。太湖沉积区如金山朱泾、青浦赵巷等孔为湖沼堆积的灰黑、灰褐色粉质黏土，偶见半咸水介形虫，常含较多的植物碎片，局部形成泥炭或碳质黏土层。上段普遍为滨海堆积。大都为深灰、褐灰色（淤泥质）黏土、粉质黏土，具特征的标志是，可见大量的植物茎叶与黄色泥灰质结核，局部形成泥炭或碳质黏土层。发育少量广盐性有孔虫和很少的半咸水介形虫。与下伏南汇组为假整合接触。

上海组为一套滨海、浅海沉积。太湖沉积区，其顶、底界深度通常介于 2.5~15m，厚度为 7~12m；长江三角洲沉积区，包括“冈身带”，顶、底界深度通常介于 5~25m，厚度多为 10~18m。长江三角洲沉积区为滨岸浅海（下部）—河口滨海堆积（上部），地层保存完整的剖面可见上、下两段。下段为滨岸浅海堆积的深灰色淤泥质黏土和淤泥质粉质黏土；上段为滨海堆积的灰色砂质粉土、黏质粉土夹粉砂，其上部常为灰色砂质粉土与粉质黏土互层或为黏土、粉质黏土夹砂质粉土和粉砂。太湖沉积区为潟湖—滨海沼泽堆积，通常为较单一的深灰色淤泥质黏土、黏土和淤泥质粉质黏土。有孔虫、介形虫和孢粉丰富。“冈身带”的岩性、岩相特征具东、西两区之间的过渡性质。在梅陇以北地带主要为深灰色淤泥质黏土和淤泥质粉质黏土，中上部多夹粉砂纹层，常见贝壳碎片，偶尔形成贝壳砂层；下部常含植物碎片及黄色泥灰质结核。梅陇以南地带通常为灰色粉质黏土夹含贝壳碎片的粉砂层，反映为水动力较强的潮滩堆积。与下伏娄塘组为整合接触。

青浦组见于太湖沉积区，厚度一般为 2~4m，大致有从东向西，即从湖盆边缘向湖盆内部加厚的趋势。岩性和岩相有两种类型。主要类型为见于青浦西南部及与松江、金山的接壤区的夹有泥炭层的深灰、棕灰色淤泥质黏土、粉质黏土的湖沼沉积。次要类型为分布于上述地区以东至“冈身带”之间湖泊沉积的褐灰、褐黄色粉质黏土夹粉砂，局部可见泥炭或碳质黏土夹层。通常含有淡水软体动物、介形虫、轮藻等陆相生物化石，其下部偶见有孔虫。

如东组见于滨海平原区和河口沙岛区。岩性、岩相主要类型为河口湾—滨岸潮滩堆积的灰、灰黄色砂质粉土、黏质粉土夹粉砂。堆积厚度多为 5~8m，局部逾 10m。次要类型为三角洲支流河道堆积的青灰色粉砂和砂质粉土。堆积厚度为 2~10m 不等，具有西薄、东厚的变化规律。含有较丰富的广盐性有孔虫、少量浮游有孔虫和半咸水介形虫，也见个别的窄盐海相有孔虫、海相和陆相介形虫。

3. 生物地层

第四纪动物主要为淡水种、半咸水种、广盐滨岸种和盐度要求较高（多盐—真盐）的近岸浅海种和浮游类。海相动物群主要分布在 100m 以上地层中，100~180m 海相化石零星分布，大体上可以综合划分为 7 个带。第Ⅰ海相化石带位于安亭组中段，分布在 200m 以下，含有少量基本为广盐性的有孔虫及海相腹足类化石，仅在现今浦东沿江地带局部分布。第Ⅱ海相化石带，常出现在下更新统安亭组上段下部。安亭组上段上部地层内常见淡水软体动物化石，并产较丰富的陆相介形虫，在少数钻孔中还发现有海相化石。第Ⅲ海相化石带局部出现在中更新统嘉定组下段下部灰色细砂层内。第Ⅳ海相化石带主要在太湖堆积平原区中更新统嘉定组上段下部出现。第Ⅴ海相化石带分布在上更新统川沙组中，除崇明三岛和宝山、嘉定以外，在上海境内大都有所分布。第Ⅵ海相化石带位于上更新统南汇组中段，全区广泛分布。第Ⅶ海相化石带分布于全新统中，全区分布，自西而东化石组合特征揭示了海陆过渡相—海相—海陆过渡相序列，对应了娄塘组、上海组和如东组。

孢粉类型既有亚热带分子，也有温带成分，受局地沉积环境影响，常出现滨海及河流湖沼草本花粉、淡水藻类及海相沟鞭藻。上新世以来，大体可以划分出以下孢粉组合带。分别为落叶栎（*Quercus-deciduous*）-常绿栎（*Quercus evergreen*）-莎草科（Cyperaceae）-凤尾蕨属（*Pteris*）带、松属-云杉、冷杉属-落叶栎-蒿属（*Artimisia*）带、落叶栎-松属-禾本科带、松属-云杉、冷杉属-落叶栎-禾本科-凤尾蕨属带、落叶栎-松属-香蒲属-水龙骨科带、松属-落叶栎-禾本科-水龙骨科带、落叶栎-松属-禾本科-水龙骨科带、松属-落叶栎-桦属-禾本科-水龙骨科带、落叶栎-松属-枫香属-禾本科-水龙骨科带、落叶栎-松属-禾本科-凤尾蕨属带、松属-落叶栎-蒿属-水龙骨科带、落叶栎-松属-蒿属-铁线蕨属带、松属-落叶栎-香蒲属-凤尾蕨属带、落叶栎-松属-蒿属-凤尾蕨属带、蒿属-禾本科-松属-冷杉属-苔藓类孢粉带、松属-落叶栎-禾本科-凤尾蕨属带、落叶栎-松属-蒿属-凤尾蕨属带、落叶栎-常绿栎-松属-禾本科-凤尾蕨属带、松属-落叶栎-蒿属-凤尾蕨属带、落叶栎-松属-禾本科-凤尾蕨属带。

4. 矿物地层

重矿物共有30多种。含量较高的重矿物有角闪石、绿帘石、磁铁矿、钛铁矿、褐铁矿、赤铁矿、石榴子石、独居石、锆石等，含量较低的有金红石、白钛石、电气石等，较少见及的有榍石、锐钛矿、黄铁矿、磷灰石、重晶石、菱铁矿、软锰矿等，偶见辉石、蛇纹石、透闪石、十字石、硅线石、蓝晶石、刚玉、石棉，还有黄铜矿、方铅矿、闪锌矿、自然铅、自然铜、自然金、白钨矿、辰砂等。在单个砂样中出现的重矿物不过一二十种，中更新统以上地层中重矿物种类较少，其下较多。重矿物在分析砂样中所占质量一般约为1%，在下更新统上部及以上地层中含量较高，其下急剧降低。大致可以划分为角闪石-绿帘石-磁铁矿、钛铁矿-褐铁矿、赤铁矿-独居石带，绿帘石-角闪石-褐铁矿、赤铁矿-磁铁矿、钛铁矿-石榴子石、锆石带和褐铁矿、赤铁矿-绿帘石-锆石-磁铁矿、钛铁矿带等3个。

黏土矿物种类主要为伊利石、蒙脱石，其次为高岭石、绿泥石。上部，相当于全新统至中更新统，为伊利石（水云母）带。中部，相当于下更新统，为伊利石-蒙脱石带。下部，基本相当于上新统，略跨下更新统底部，为蒙脱石-伊利石带。

5. 气候地层

总体而言，上新世以温暖湿润为主，第四纪期间有5次明显的冷暖期交替。

上新统崇明组沉积物记录显示气候总体暖湿，下部暖热湿润，中间夹一次寒冷干燥期，命名为第一寒冷期，上部暖热湿润—温暖湿润，命名为第一温暖期。上新世气候记录在基岩面凹陷区钻孔上新世沉积物中多有揭示。

进入早更新世，气候开始明显变冷，但有波动，先后经历了3次冷期，中间有两次暖期交替，气候期先后命名为第二寒冷期、第二温暖期、第三寒冷期、第三温暖期、第四寒冷期。

中更新世古气候先后经历了一次温暖期和一次寒冷期，分别命名为第四温暖期和第五寒冷期，其中第五寒冷期经历了早期温凉偏干、中期暖热湿润、晚期寒冷干燥的变化。

12.8~7ka为一次暖期，命名为第五温暖期，对应MIS5期；7~1.1ka为寒冷期，命名为第六寒冷期，气候总体温凉偏干，对应MIS3期，在这个阶段大致5~3ka存在一次间暖期。

全新世为冰后期，气候温暖湿润，气候波动先后经历了温凉稍湿、温和偏干、暖热潮湿、温凉稍干、温暖湿润的变化，对应前北方期、北方期、大西洋期、亚北方期、亚大西洋期。

（二）沉积物成因类型及分布规律

沉积物主要成因类型包括流水沉积（洪积、冲积）、湖沼沉积、海陆交互沉积（河口湾、三角洲）、海洋沉积和人工成因5种类型。

洪积物堆积于2.58~1Ma期间基岩剥蚀区山麓低地内，出现在安亭组下段和中段。

冲积物是区内第四纪发育最普遍、占比最高的沉积物，几乎各阶段都有发育，出现在安亭组、嘉定组、川沙组、南汇组、如东组。以中更新世分布范围最广。

湖沼堆积物主要发育于早更新世早期和全新世晚期。早更新世早期，滨海平原北部、河口沙岛普遍发育滨湖、浅湖沉积，沉积物组成以杂色黏土为主。全新世晚期，湖沼堆积物普遍发育在太湖沉积区。

完整的河口湾堆积物主要形成于全新世早期（11~8ka）快速海侵过程中，娄塘组沉积物均属河口湾堆积物。此外，下更新统安亭组、中更新统嘉定组、上更新统川沙组常发现微弱海侵作用形成下的河口湾堆积物。

三角洲堆积物发育自全新世中期最大海泛以来，堆积在河口湾堆积物以上，分布在现今的河口沙岛、滨海平原。

滨浅海堆积物主要发育于上更新统中上部南汇组中段和全新统上海组。南汇组中段下部遍布全区，南汇组上段分布在滨海平原浦东南部。上海组主要分布在滨海平原地区。

人工填土分布在浅表，一般在1~2m，为人类工程活动所致，包括耕作层、杂填土、素填土、暗浜、冲填土。

（三）沉积物类型与时空分布规律

上海主要沉积物类型有黏土、粉砂质黏土、黏土质粉砂、粉砂、砂质粉砂、粉砂质砂、中细砂、含砾砂和砾质砂。黏土、粉砂质黏土主要分布于新近系与下更新统内。粉砂质黏土主要分布于全新统中下部和上更新统上部，其次分布于中、下更新统与新近系内。黏土质粉砂主要分布于全新统中下部和上更新统上部，其次分布于中、下更新统与新近系内。粉砂主要分布于全新统内。砂质粉砂主要分布于上更新统上部，其次分布于下更新统内。粉砂质砂主要分布于上更新统下部，其次分布于全新统与新近系内。砂主要分布于上更新统下部，其次分布于下更新统与新近系内。

在冲积扇、辫状河流沉积环境中，沉积物为分选较差的含砾泥质中粗砂或泥质砾砂夹黏土层，形成粗、细相间的互层特征。在曲流河、受海水影响的河流环境中，以粗粒的河道沉积与细粒的洪泛平原沉积构成河流二元结构为特征。海岸环境底部为灰黄色含砾中细砂，中部以粉砂、砂质粉土为主，至顶部为黏质粉土。三角洲环境中下段沉积为黏土质粉砂，上段为粉砂，呈现下细、上粗，分选性略有提高的特点。

（四）第四纪地貌

上海陆域进一步可划分为4个三级地貌单元，分别是太湖堆积平原区、长江三角洲堆积平原区、杭州湾滨海平原区、剥蚀残丘和岛屿区。上海近岸水域可划分为长江三角洲水下三角洲1个三级地貌单元，根据沉积特征，可进一步区分为三角洲前缘区和前三角洲区2个四级地貌单元。其中，太湖堆积平原区和长江三角洲堆积平原区是区内陆域最主要的地貌类型，分界线在江苏太仓—上海嘉定外冈—青浦徐泾—奉贤邬桥—金山漕泾一线，即以长江三角洲平原贝壳沙堤构成的“冈身带”西侧作为分界。

太湖堆积平原区分为潟湖积平原、湖沼积低平原和沼积低平原3个四级地貌单元，长江三角洲堆积平原区分为河口沙岛、滨海平原及河流冲积平原3个四级地貌单元，更次一级的微地貌单元可按地貌形成过程将河口沙岛划分为早、中、晚3期河口沙坝和潮滩，将滨海平原划分为早、中、晚3期滨海平原和潮坪。

除上述主要地貌类型外，太湖堆积平原与三角洲堆积平原区内的滨海平原交界处以及滨海平原内散布大致4~5列古贝壳砂堤，崇明岛东部散布1~2列古贝壳3砂堤。上海沿江沿海地带分布数列人工海塘。

（五）第四纪海侵海退事件

上海第四纪存在7次海侵事件，可与长江三角洲地区第四纪海侵对比。

如皋海侵——第Ⅶ海侵。发现于浦东滨岸地带，海侵层埋深在200m以下，位于安亭组中段，含有少量广盐性的有孔虫及海相腹足类化石，形成年代为2.00~1.85Ma。这是区内第四纪的最早一次海侵，海侵强度体现为河口—三角洲河流环境（以受海水影响的河流为主）。

周浦海侵——第Ⅵ海侵。由南汇嘴入侵，其北端沿浏（河）-南（汇）古河道抵达浦东海塘—横沔、周浦附近。海侵层位属下更新统安亭组上段，发现有窄盐性的有孔虫和钙质超微化石，形成年代主要为1.20~1.00Ma，海侵强度体现为河口—三角洲河流环境（以受海水影响的河流为主）。

上海海侵——第Ⅴ海侵层。在上海市区、浦东、嘉定和西南部的青浦、金山等地广泛分布。海侵层位属中更新统嘉定组下段的下亚段，通常赋存在灰色细砂层内。东部所见常为少量半咸水有孔虫或伴有陆相介形虫、钙质超微化石；西部一般见淡水介形虫，只出现个别的有孔虫且见冷水种。形成年代为0.78~0.65Ma。海侵程度为偶尔受强潮流影响的三角洲平原河湖、短暂海湾环境。

嘉定海侵——第Ⅳ海侵层。主要保存在太湖沉积区，其海侵层位属于中更新统嘉定组上段下部，而滨海平原东部滨岸地带和河口沙岛区则因多被侵蚀缺失而不明。见半咸水有孔虫或伴有陆相介形虫、钙质超微化石，淡水软体动物化石丽蚌等碎片。形成年代为0.26~0.20Ma。海侵程度为三角洲平原河流环境。

太湖海侵——第Ⅲ海侵层。除崇明三岛、宝山和嘉定以外，在上海境内大都有所分布。其海侵层位属于上更新统川沙组，形成年代为0.07~0.128Ma，海侵程度为间或受海水影响的三角洲河流环境，偶尔在局部可达三角洲前缘的河口环境。虽然在上海境内分布较广，但是仍同其前数个海侵层一样，属于海相性较弱的海侵层。

滆湖海侵——第Ⅱ海侵层。海侵强度为历次海侵之最。遍及全区，海侵层位属上更新统南汇组中段。按其海侵强度大体可分为两类地区：一是西部太湖沉积区及东北部长江三角洲河口沙岛沉积区分布的海相性弱的滨海、潟湖沉积，间有陆相河流沉积；二是东部长江三角洲滨海平原沉积区分布的海相性较强的浅海—海湾沉积。形成年代主要为20~55ka。

镇江海侵——第Ⅰ海侵层。在境内分布广泛，唯佘山、天马山附近的青浦西南部及与松江、金山接壤区未受到影响。形成年代始于11ka。在8~5ka期间进入海侵最盛期。区域性最大海侵时期形成最大海泛面约在6.5ka，为上海所见分划长江三角洲堆积平原与太湖堆积平原的首条贝壳沙堤——“竹冈”的构筑时期。在上海形成海陆过渡相—海相—海陆过渡相地层序列，构成一个完整的海侵旋回；相应由下而上组成娄塘组、上海组和如东组。以河口沙岛沉积区最为发育，向西经滨海平原沉积区至太湖沉积区渐薄以至尖灭。

不同阶段海侵范围有所差异。早更新世两期海侵微弱，如皋海侵踪迹仅在现今的浦东滨海局部地点发现。周浦海侵略有变强，海侵踪迹不仅在浦东个别地点发现，还可在青浦和金山等地发现。上海海侵范围明显较早更新世两次海侵明显变大，但仍局部分布，除了在浦东以及金山枫泾一带出现海侵踪迹，还在北部的嘉定出现。嘉定海侵的海相性程度又明显变高，海相地层发育也较厚，但仅在现今湖沼平原发现。晚更新世以来，全区普遍发生海侵。全新世海侵最大至湖沼平原和滨海平原的交界处，发育了古贝壳砂堤，在河口沙岛最强，其次在滨海平原，湖沼平原受到了海侵的微弱影响。

（六）岩相古地理演化

1. 早更新世

早更新世早期，区内普遍发生溯源侵蚀和河流下切，西南部低山丘陵的范围已收缩到青浦、松江、金山、闵行、奉贤境内，山势也已降低。陆域东、北部的嘉定、宝山、市区和浦东主要为一片湖盆。在湖滨地带内，即市区西南的大场—小闸—三林—坦直、三灶一带，分布着一串剥蚀低丘。浦东东南部与金山枫泾附近的低山区内，分别形成两个规模较大的山间冲洪积扇与扇上湖泊洼地。崇明区基本为河间滩地广布，主要堆积一套漫滩—泛滥湖泊沉积。只是在现崇明岛的周缘地带，主要如东南侧的前哨农场—奚家港和侯家镇—新河镇附近，还有北侧的红星农场和富民农场附近地段有曲流河道发育。长兴、横沙两岛地区普遍经历了曲流河道—边滩—泛滥平原的演化过程。浦东滨岸的龚路东侧—海塘—浦东机场—祝桥东渔场—老港及迤南一带，出现一条浦（东）—南（汇）古河道。祝桥以北属于河道分、合多变，河间滩地时常暴露的辫状河流沉积环境；以南已演化为比较稳定的河道环境。该阶段建造中所蕴含的地下水即为历来区内所称的狭义的第Ⅴ承压含水层。

早更新世中期，金山枫泾与闵行马桥、北桥以东的两条山谷经溯源侵蚀近于沟通，将原来连绵分布的低山分割为青浦—松江接壤区与金山—奉贤接壤区的两片低山，山势进一步下降。原来分布于宝山大场—三林—坦直、三灶的一列弧形低丘已大都湮没，只有坦直、三灶与市区西侧虹桥等数个低丘残留下来。在其东的浏（河）—南（汇）古河道为长江向东南的支流，开始形成一条摆动宽度有七八千米，明显受海水影响的三角洲河流沉积，堆积了浅灰、灰黄色中细砂和河床相含砾砂，成为陆域第$Ⅳ_3$承压含水层主要蕴藏地带。其西南方，在南、北两片低山、丘陵夹持的沟谷带内，出现一条源自松江和闵行北桥，流经下沙、大团的支流河道，至芦潮港汇入浏—南干流。在两片低山之间的垭口及东西两侧山谷内为泥质砂砾和含砾黏土充填形成的洪积扇、裙和扇上湖泊。西侧山谷内出现一条经金山枫泾向浙江流去的扇间河流。北部低山区北侧也可能出现一股支流，经嘉定黄渡、马陆，流向宝山刘行汇入浏—南干流河道，形成细砂和含砾砂夹粉土、粉质黏土堆积。在浏-南古河道与西南部低山之间的广阔地带为河流边滩和泛滥平原，广泛堆积了不同粒度的砂及黏性土与粉细砂的互层。在河口沙岛上，也出现一条由西北角红星农场附近流入，经庙镇、新河镇流向横沙岛的古河道，总体显示向下游加深、加厚的趋势，堆积了一套含砾中粗砂、砂砾石与中细砂的互层。崇明岛的大部和长兴岛，为古河道两侧的边滩或心滩及泛滥平原，堆积了一套细砂夹含砾砂及黏性土与粉细砂的互层。

早更新世晚期，陆域西南部的两片低山已成为散落于冲积平原上的丘陵。浏—南古河道向西摆动达嘉定—宝山月浦—市区真如—浦东周浦—芦潮港一线，摆动宽度达10~15km。堆积了一套砾质砂、含砾中粗砂与中细砂的互层。在浦东受到当时区内较早一次海侵的影响，形成了间或出现弱海陆过渡的溺谷河流沉积，成为上海的第Ⅵ海进层。陆域南部的枫—奉古河道形成，并至芦潮港附近与浏—南古河道汇合。下部堆积河床—沙坝相含砾中粗砂、含砾细中砂与细砂互层，上部堆积河漫滩相黏性土夹粉细砂层。在两条古河道的两侧分布中细砂、粉细砂夹粉土和粉质黏土的边滩沉积；在环绕低山丘陵的附近地区则为黏性土夹粉细砂组成的大片泛滥平原—湖泊洼地；在现崇明三岛上也为河流边滩和泛滥平原—湖泊洼地广布，大多堆积粉细砂夹粉土及黏性土与粉土、粉细砂的互层。

2. 中更新世

中更新世开始，区内进入温暖期。早期，广布于区内的河流受到海侵影响，普遍形成分布广泛的三角洲平原河流沉积。其后至中更新世晚期，区内古气候经历了一个较大的周期性更替旋回，广布的河道多受到海水的侵入形成三角洲平原河流堆积。市区东部、浦东和原南汇的大部以及河口三岛区由于后来晚更新世早期河流的强烈下切，致使中更新世晚期地层及相应的含水砂层被侵蚀而残缺不全。其余大部分地区较为完整地保存了这一期地层，主要为河道和边滩堆积的灰—黄灰色砂、含砾砂夹黏土，埋深在100~110m，厚度在10m左右。

3. 晚更新世

晚更新世早期，随着古气候回暖，区内进入强烈沉降阶段，开始广泛出现海陆过渡沉积环境。这一期间所形成的弱海陆过渡沉积三角洲河流相砂层构成了广布全区中深部位、厚度较大的川沙组地层。晚更新世晚期，在75~70ka B. P. 期间，上海处在类似现今暖温带的温凉气候下，演化为泛滥平原—泛滥湖泊环境，堆积了暗绿、灰绿、部分深灰、浅蓝灰、灰黄色的硬黏土，夹粉砂纹层，组成南汇组或滆湖组下段。在70~20ka B. P. ，气候转变为温暖湿润，海平面大幅度回升，海侵范围广泛波及长江三角洲和太湖地区。堆积了海相或海陆交替相地层南汇组中段或滆湖组中段，形成本区海相性最强的灰色黏土与粉砂交互地层。在23~20ka B. P. 期间，海平面开始下降导致海水逐渐向东海陆架退去，原上海境内的浅海和海湾演化为滨海—三角洲平原河流环境，堆积了南汇组中段上部的褐黄、锈黄色的粉砂、细砂。在20~11ka B. P. 期间，本区古气候变为凉冷干燥，海平面急剧下降。原局限于上海西北部发育的泛滥平原—湖泊已扩展到除现河口区外的上海整个陆域，广泛沉积了上更新统顶部的褐黄色粉砂质黏土和具特征的暗绿色黏土层，组成南汇组或滆湖组上段。在15~11ka B. P. 期间，随着海平面的持续下降，褐黄—暗绿色黏性土逐渐脱水成陆，经历了风化成壤的过程。

4. 全新世

全新世早期，在10. 3~9. 5ka B. P. 期间，海平面大幅上升，初期海水沿现河口区—浦东滨岸地带及原陆域内切割较深的支流河谷侵进，形成河口湾及溺谷堆积，主要为灰—灰白色黏土质粉砂、粉砂与细砂互层，构成了娄塘组下段，厚度在10~20m。在9. 5~8. 0ka B. P. 期间，海平面上升趋于缓慢，侵进的海水淹没了金山、松江、闵行、宝山及其以东地区，使现今上海的广大陆域沦为滨海沼泽和潮滩，堆积了富含芦苇根茎和黄色泥灰质结核的灰黑色黏土和泥炭夹层；在现崇明三岛和浦东则为河口湾和潮滩环境，堆积了深灰色粉砂质黏土夹灰白色粉砂。构成了娄塘组上段，厚度多为5~10m。

全新世中期，在8. 0~5. 0ka B. P. 期间，海平面再度显著上升，成为全新世海侵的最盛期，海侵范围直抵江苏镇江。全新世早期广布于现上海西北部的晚更新世暗绿色硬土台地已退缩至佘山、天马山附近的青浦及与松江、金山的接壤区。现陆域东部地带和河口地区沦为滨岸浅海环境，堆积了深灰色淤泥质黏土。在海侵最盛的中、后期，即在6. 5~5. 0ka B. P. 期间，海岸带长期稳定滞留于嘉定中部—闵行、奉贤的西部地带，先后构筑了紧邻的两道断续相连的贝壳沙堤，与相邻的滨海滩涂组成“冈身带”。其时，与“冈身带”毗邻的西部金山、松江及青浦、嘉定的部分地区演化为潟湖、沼泽，堆积了普遍含植物碎片和褐铁矿结核的灰色粉砂质黏土。在5. 0~3. 0ka B. P. 期间，古气候趋向温凉稍干，致使海平面略有下降，海岸线随之略为向东后撤，又陆续在嘉定城区至闵行颛桥一带构筑了两道贝壳沙堤。当时在“冈身带”以东的长江三角洲沉积区演化为河口滨海环境，主要堆积了灰色粉砂、黏土质粉砂；在“冈身带”之西的太湖沉积区则由前期的潟湖演化为后期的湖沼，堆积了深灰色粉砂质黏土夹粉砂。这一时期与上述海侵最盛期形成的以黏性土为特征的海相地层组成了上海组。其埋深和厚度均由西南太湖区向东北的长江三角洲沉积区增大，厚度多介于7~18m。

在3ka B. P. 之后的全新世晚期，古气候回复温暖湿润，此时海平面又略微上升并趋于稳定，但由于长江中上游携来泥沙显著增加，使长江三角洲前缘河口沙坝向海域的进积作用加快，导致海岸线迅速向东后撤。原“冈身带”及以东毗邻地区成为潮上泥坪，堆积了褐黄色粉砂质黏土夹粉砂层；其东的陆域则为潮间带沙泥混合坪和部分的潮上泥坪，主要堆积了褐黄色粉砂质黏土和粉砂的互层。在崇明等河口三岛上，为河口沙坝与汊流河道分布，在沙坝部位由下而上堆积灰色粉细砂和棕黄色粉砂质黏土与粉砂互层，汊流河道内则堆积了灰色粉细砂和粉砂。而在“冈身带”以西的青浦、松江、金山，此时已基本演化为淡水湖泊沼泽，超覆于原来的晚更新世台地之上，堆积出褐黄—深灰色黏土和淤泥质黏土以及浅部的泥炭层。另外从西侧的安亭、白鹤间流入吴淞江古河道，至市区后扩展为扇形，东流注入长江，成为三角洲平原支流河道，堆积了灰色粉砂和表层的褐黄色粉砂质黏土。这期间的沉积物组成上全新统如东组和青浦组，其厚度总体上也由西南太湖区向东北河口区增大，介于1~14m不等。

约在1. 2ka B. P. 以来的挽近历史时期，随着长江入海口逐渐向东南推移，长江输来的泥沙在河口不断进积，使上海的崇明、长兴、横沙等河口沙岛依次出现；近年，九段沙等水下沙洲也相继出水成陆。

七、城市地质

上海城市地质工作经过多阶段的发展，在水文地质、工程地质、环境地质、地质灾害和地质资源等城市地质条件方面获得了很多系统性的成果。

（一）水文地质条件

上海水文地质单元隶属于长江三角洲水文地质单元。除西部松江佘山等基岩残丘出露及浅埋区外，松散层中分布有潜水含水层和第Ⅰ、第Ⅱ、第Ⅲ、第Ⅳ、第Ⅴ承压含水层，潜水和承压水资源分布广泛，第Ⅱ、第Ⅲ、第Ⅳ、第Ⅴ承压含水层为地下水主采层。各含水层水文地质条件有其自身分布特点。埋藏条件、渗透性和富水性、矿化度、地下水质量和水化学类型差别较大。即使同一含水层，平面上也有明显差别。

各含水层水文地质条件明显受古地理环境约束。东部河口—滨海平原水文地质亚区（滨海平原、河口砂岛区）和西部剥蚀残丘—叉流河道水文地质亚区（湖沼平原地区）。

东部河口—滨海平原水文地质亚区，主要包括嘉定东部、宝山、中心城、浦东、崇明。含水层（组）齐全，砂层极其发育，厚度大，岩性以中粗砂含砾为主。地下水极丰富，各层（组）累计单井涌水量在7000～10000m^3/d。区域上广泛分布淡水资源，仅嘉定东部—浏河局部第Ⅱ承压含水层，以及浦东、崇明第Ⅱ、第Ⅲ承压含水层分布微咸水和咸水。

西部剥蚀残丘—汊流河道水文地质亚区，主要包括嘉定西部、青浦、松江、闵行、金山、奉贤。区域上不同叉流河道地带水文地质条件差异较大。松江佘山—青浦淀山湖，金山漕泾—山阳为基岩残丘零星分布及浅埋地区，含水层不发育或基本缺失。金山东部、奉贤，以及青浦南部、松江西部—闵行北部叉流河道地带，含水层不齐全，缺失第Ⅴ承压含水层，砂层不发育，或厚度薄，岩性以粉细砂和中细砂为主，各层累计单井涌水量在100～3000m^3/d，地下水在金山亭林以西、青浦城区—练塘一线以西各层主要为淡水，东侧主要分布微咸水和咸水，仅青浦县城—练塘一线以东第Ⅳ承压含水层为淡水。青浦北部—嘉定西部叉流河道地带以及金山北部—闵行南部叉流河道地带，含水层较齐全，砂层发育，厚度较大，以中细砂中粗砂含砾为主，各层累计单井涌水量在3000～8000m^3/d，由西往东递增。地下水在西部地区主要为淡水，往东过渡至微咸水和咸水。

潜水和承压水补径排条件有明显差异。潜水含水层主要的补给水源为大气降水，蒸发、蒸腾作用为主要排泄方式，农村地区的家庭洗涤用水、基坑施工过程中的潜水局部疏干，也是潜水人为排泄形式之一。第Ⅰ、第Ⅱ、第Ⅲ、第Ⅳ、第Ⅴ承压含水层地下水为封存水，与现代降水及地表水之间不存在直接补给关系。抽汲地下水为承压含水层地下水的主要排泄方式。工程降水为浅部承压含水层地下水的主要排泄方式。

（二）工程地质条件

上海工程地质条件总体变化不大，100m以浅土体共划分出11个工程地质层和40个亚层及次亚层。10个工程地质层分别为第①层填土、第$②_1$层褐黄—灰黄色粉质黏土，第$②_3$层粉性土层（浅部砂层），第③、第④层灰色淤泥质黏性土层（软土层），第⑤层灰色黏性土层（局部夹有第$⑤_2$层砂、粉性土层），第⑥层暗绿—草黄色黏性土层（第一硬土层），第⑦层草黄—灰色砂、粉性土层，第⑧层黏性土夹砂层（局部夹第$⑧_{2-1}$层暗绿—草黄色黏性土硬土层），第⑨层灰色砂性土，第⑩层褐灰—蓝灰色黏性土层和第⑪灰色粉细砂层。

不同工程地质层具有不同的工程地质特性，易产生不同的工程地质问题。其中，第$②_3$层为浅部砂层，埋藏浅，易产生震动液化和渗流液化现象；第③、第④层为典型的软土层，是天然地基变形主要层次，也为上海的第一压缩层；第⑤层为一般黏性土层，亦为第二压缩层，其中第$⑤_2$层为微承压含水层，基坑开挖时应注意由其引发的流砂、管涌等问题；第⑥层为第一硬土层，埋深从西部向东部逐渐加深，该层为更新统和全新统分界的标志层，在上海广受关注，其土性较好，对控制地基变形较为有利，可作为一般建（构）筑物的桩基持力层；第⑦层为下部砂、粉性土层，土性好，为中高层建筑和市政桥梁工程的良好桩基持力层，该层亦为第Ⅰ承压含水层，地下工程施工时易产生流砂和基坑突涌问题。

根据现代地貌类型、是否受古河道以及近期河道切割等特征，全区划分出3个工程地质结构区，8个工程地质结构亚区。其中，湖沼平原区（第Ⅰ工程地质结构区）以暗绿—褐黄色硬土层作为分区依据，可进一步划分为3个工程地质结构亚区（I_1、I_2、I_3）；滨海平原区（第Ⅱ工程地质结构区）以浅部砂层（$②_3$）及暗绿色硬土层（⑥）分布状况为依据，进一步划分为4个工程地质结构亚区（$Ⅱ_1$、$Ⅱ_2$、$Ⅱ_3$、$Ⅱ_4$）；河口砂岛区（第Ⅲ工程地质结构区）由于均受长江古河道切割，不进行亚区划分。

依照表土层（$②_1$）下卧层岩性及厚度将上海天然地基工程建设适宜性划分为天然地基条件好区（A区）、天然地基条件一般区（B区）、天然地基条件较差区（C区）、天然地基条件最差区（D区）4个区。"冈身带"、滨海平原区沿江、沿海和河口砂岛区为天然地基条件好区，滨海平原区和湖沼平原区东部地区为天然地基条件一般区，湖沼平原区西部为天然地基条件较差区，沿海地区如临港新城东部、崇明东滩地区等为天然地基条件最差区。

区内主要桩基持力层有中部砂层第$⑤_2$、第⑥层暗绿—褐黄色硬黏性土层、第⑦砂层、第$⑧_2$层粉质黏土夹粉土和第⑨层深部砂层。按照有两层硬土层、⑥层和无硬土层可划分为3个大区，其中，有第⑥、第$⑧_{2-1}$层同时存在的为A区，可选择的桩基持力层较多，且分布稳定。有第⑥层无第$⑧_{2-1}$层分布的为B区，桩基条件有一定变化。无硬土层分布的为C区，桩基埋深变化较大。

（三）地面沉降

上海地面沉降特征与趋势，总体上可分为地面沉降严重发展时期（1921—1965年）、地面沉降控制时期（1966年至今）等两个时期，并逐步形成了由精密水准监测网、地面沉降监测站、地下水动态监测网、重大基础设施地面沉降骨干监测网、GPS地面沉降监测网和InSAR技术地面沉降监测组成的地面沉降监测网络。

1. 全市地面沉降特征

1980—1995年，上海地面沉降比较明显，市区和虹桥—华漕是主要沉降漏斗区。市区平均地面沉降速率为5.6mm/a，最大沉降量为200mm。邻近江苏的嘉定娄塘和浙江的金山枫泾，形成局部沉降中心。崇明沉降发展缓慢，平均地面沉降速率只有3.7mm/a。

1996—2001年，地面沉降呈加速发展趋势，市区的平均地面沉降速率达到24.5mm/a。市区出现新的沉降漏斗，虹桥—华漕、娄塘、枫泾的沉降中心都明显扩展。崇明的前哨农场加速沉降，平均地面沉降速率达到10.1mm/a。

2002—2006年，地面沉降缓解，平均地面沉降速率为12.7mm/a。虹桥—华漕地面沉降中心已基本消失。娄塘沉降漏斗规模明显减缓，但枫泾仍保持较快的沉降速率（30mm/a）。崇明的地面沉降发展趋势基本保持不变，平均地面沉降速率为5.9mm/a，长江农场和绿华镇一带出现新的沉降漏斗。

2007—2011年，除局部地区出现沉降漏斗外，该阶段绝大部分地区累计沉降量较小，在0~25mm之间，西部、北部局部地区甚至出现微量回弹趋势。沉降漏斗主要分布在浦东、闵行、金山、崇明等局部地区。崇明东滩、崇明向化沉降最严重，漏斗中心累计沉降量超过200mm。

2012—2015年，绝大部分地区此阶段累计沉降量在0~30mm。崇明的东滩、南门、新海等地区沉降速率较上一阶段明显减小。浦东整体上地面沉降发育有所缓解，但在高桥、周浦—闵行杜行、大治河滨海等地出现沉降漏斗。闵行的大虹桥地区、徐汇漕河泾，金山的金山卫等地呈现小的沉降漏斗。西部青浦蒸淀—松江新浜、嘉定城区局部区域，出现小范围微量回弹迹象。

2002—2016年，岛域地区基本保持原有的沉降格局，崇明东滩地区受新近沉积土和吹填因素影响，地面沉降较为突出。陆域地区，全市仍发育中心城及宝山北部一带和嘉定黄渡—南翔一带两个较大范围的沉降漏斗，西南部金山吕巷—廊下一带发育新的沉降漏斗，省界嘉定竹桥—朱家桥—娄塘的沉降漏斗消失，闵行马桥的沉降漏斗基本消失，华亭—徐行—罗店—马陆—刘行—顾村一片出现回弹。浦东的沉降范围由原来的片状分布分解为若干个沉降漏斗，不均匀沉降现象突出。

2. 中心城地面沉降特征

1991—1995年，地面沉降速率相应明显增加，新出现了黄浦、闸北不夜城地区等数个新的沉降中心。形成了黄浦—闸北不夜城地区—杨浦控江路—五角场的地面沉降中心漏斗区。空间差异显著增大，杨浦、黄浦、虹口等地较明显。

1996—2000 年，随着本阶段地面沉降速率增大，市区沉降漏斗和西部沉降漏斗的形成，使中心城地面不均匀沉降特点更加突出。

2001—2005 年，平均地面沉降速率有所下降，中心城大范围地面沉降中心分解为孤立的沉降点。浦东形成了“六里—陆家嘴”和“东沟—高行—黄浦江”两个地面沉降中心。地面沉降不均匀的特点仍较为突出。

2006—2010 年，地面沉降总体上仍较发育，不均匀性沉降逐年加剧。地面沉降格局更加复杂，共分解为 6 个地面沉降影响区。中心城西部地区发育有闵行区的大虹桥区域和梅陇嘉定南翔以及中心城杨浦与虹口局部地区等沉降漏斗，漏斗中心极端累计沉降量在 300mm 以上。中心城东部地区、浦东成为的沉降主要发育地区，发育有张江、三林等较大范围的沉降漏斗，漏斗中心最大沉降量在 200mm 以上。

3. 地面沉降规律与成因

20 世纪 60—80 年代，地面沉降以第Ⅰ承压含水层及上部土层沉降为主；20 世纪 90 年代—21 世纪初，地面沉降主要以深部土层沉降为主。21 世纪以来，浅部土层成为地面沉降的主要发育层次，且在不同的阶段，不同土层表现出不同的变形特征。

21 世纪以来，第Ⅰ承压含水层水位整体抬升，但第Ⅰ承压含水层及其上覆软土层持续压缩变形。第Ⅰ承压含水层表现出弹性—弹塑性—塑性变形演化的不同特征，并出现了蠕变现象。第Ⅱ、第Ⅲ承压含水层地下水位呈现缓慢上升态势，含水层沉降速率明显减缓，并出现微量回弹态势。第Ⅱ、第Ⅲ承压含水层表现为压缩—回弹的弹性变形特征。第Ⅳ和第Ⅴ承压含水层变形速率显著减缓，部分地区第Ⅳ承压含水层出现回弹。第Ⅳ承压含水层呈现弹性—弹塑性—塑性变形或线弹性变形特征。

以地下空间开发为代表的深基坑工程是引发浅部土层沉降的一个重要因素。基坑工程地面沉降影响范围与基坑降水方式密切相关，且基坑 3H 以外的地面沉降主要受减压降水的影响。基坑工程减压降水以降水目的层上部软土层沉降变形为主。

高层建筑群诱发的地面沉降取决于建筑容积率和地质条件；浅部的淤泥质黏土层沉降最大；高层建筑群造成的区域性地面沉降呈不规则的凹凸起伏状；高层建筑群对地面沉降的影响与建筑物施工的先后顺序有关；密集高层建筑群之间存在明显的应力叠加效应。

4. 地面沉降风险

地面沉降危险性的综合评价结果显示，上海主要处于地面沉降危险性中等区，危险性高区主要分布于中心城及与江浙两省交界的嘉定、金山的局部地区，而地面沉降危险性低区主要分布在地质环境发育较好、地面沉降预测速率较低地区，以奉贤及青浦的局部地区为代表。此外，三岛区均属于地面沉降危险性中等区。

地面沉降易损性评价结果显示，中心城易损性最为严重，浦东东南部（原南汇）为较高易损性区，宝山和闵行归为较中等易损性区，松江、浦东东南部（原南汇）、嘉定归为较低易损性区，奉贤、崇明、金山、青浦归为低易损性区。

在上述地面沉降危险性及易损性评价基础上进行地面沉降风险评价结果显示，以中心城为中心形成地面沉降的 3 级风险体系，即中心城基本处于地面沉降风险高区，近郊区处于地面沉降风险较高至中等区，郊区处于地面沉降风险较低区。

（四）环境地球化学

1. 基准值和背景值

不同沉积地质背景中的土壤基准值具有不同特征。河口早期、晚期沉积基准值与上海全域比值的变化趋势和分布形式几乎一致。河道边滩沉积的图谱曲线在形态上和河口相图谱完全不重叠。滨海早、中期沉积基准值在分布模式上基本相似，而与晚期沉积则有明显差异。湖泊和湖泊沼泽沉积基准值基本展示出相类似的分布图谱，曲线走势大体一致。

土壤背景值表现出区域上显著的变异特征，大体划分出均匀分布类型、相对分异型、分异型、强分异型和极强分异型等 5 种类型。相当多的元素背景值接近于上海的基准值，且其中有铍、铈、氯、钴、镓、锗、碘等 26 种元素含量偏低于基准值。对环境敏感、化学活动性大的元素普遍富集，包括银、铋、铜、钼、铅、硫、磷、锑、金、溴、镉、汞、氮、硒、锡、碳、有机碳等，其富集系数均在 1.1 以上。其中金、镉、碳表层富集系数为深层的 1.5 倍以上，汞、溴、氮、硒、锡、有机质富集系数达 1.8~2.45 倍。不同沉积区具有不同类型的土壤，土壤类型不同，元素分布型式具有差别。

2. 土壤元素空间分布

表层和深层土壤当中造岩元素总的平均含量分别为 85.13%、86.35%，两者平均含量基本一致，不同造岩元素的高值区、低值区具有不同的分布特征。

表层土壤酸碱度为 4.56~8.96，平均含量为 7.61，以碱性和中性为主，在西部的个别地方存在弱酸性土壤。深层土壤酸碱度为 6.55~8.88，平均含量为 8.16，以碱性为主。酸碱度低的地方主要集中分布在西部及中部，而在东部及东南部地区则较高。

表层土壤有机碳含量范围为 0.19%~5.78%，平均含量为 1.24%。深层土壤有机碳含量范围为 0.24%~1.57%，平均含量为 0.53%。西部地区普遍较高，东部地区较低。表层土壤在中心城也明显为高含量区。

表层土壤中的 Cd、Hg、Pb、Zn 等重金属元素，均以上海城区为中心，形成似环状的异常—高背景分布区。在分布的高背景区内亦出现宝山、安亭、嘉定、松江、吴泾等城镇中心上的异常区。崇明、长兴、横沙三岛及杭州湾滨海地带和上海城区基本可形成相反的对比，在这些地区，4 种元素全部处于低量的分布态势。Cd、Pb、Zn 在上海陆域城区、交通要道之外地区，仍保持着偏低背景分布的态势。

深层土壤中 Cd、Hg、Zn、Pb 分布相对较均匀，背景的起伏变化并不大。Cd 在上海东北部，即上海市区、宝山吴淞及崇明岛北西段，表现为中高背景的含量分布。Hg 在除崇明、长兴、横沙三岛及杭州湾沿江地带外，基本上为一个中高背景含量分布区。Pb、Zn 的含量分布表现为以市区为中心，其含量向周边地区呈梯度扩散的分布模式。

表层土壤氮含量范围为 209~3623mg/kg，平均含量为 1351mg/kg。上海深层土壤氮含量范围为 268~1382mg/kg，平均含量为 577mg/kg。高含量地区主要集中在西部湖沼相沉积地区。在三岛等沿江沿海地区为氮相对低含量地区。

表层土壤磷含量范围为 361~2718mg/kg，平均含量为 964mg/kg。深层土壤磷含量范围为 264~1064mg/kg，平均含量为 615mg/kg。深层土壤中高含量地区主要分布在宝山吴淞及宝山工业区。在西部湖沼相沉积区含量普遍较低，其他地区含量中等。表层土壤中的高含量地区主要集中在中心城及其外围附近，而崇明东滩和南汇边滩及淀山湖含量普遍较低。

3. 主要有机化合物分布特征

土壤中的主要有机化合物包括六六六（BHC）、滴滴涕（DDT）和多环芳烃类。六六六残留量较 30 年前显著下降，均在风险筛选值范围内，但检出率较高。浦东东南部（原南汇）和嘉定的个别地区土壤中六六六残留量相对较高，崇明土壤中的六六六残留量最低。滴滴涕残留量最低小于检出限，最高 736.7μg/kg，较 30 年前也有了显著降低；但在青浦、嘉定、宝山和浦东的局部地区超标。多环芳烃含量分布差异明显，崇明、奉贤相对较低，其他区均比较高。

4. 土壤地球化学异常和环境质量

外环以内以及在外环线附近的城市化较高的近郊地区、中环与外环间和中环及内环间等地平均含量较高的元素不同。众多环境指标的高含量点主要集中在中心城（外环以内）及近郊一些工业区。在外环以内及松江，初步形成面状区域，而在上海其他地方有局部的点状分布。汞、铬、铅、钴、钼等主要集中在宝钢及吴淞工业区附近。镉、镍、砷等主要集中在桃浦工业区附近。江南造船基地等附近地区镉、铬、铜、锌及铅等含量较高。金山石化、吴泾及外高桥等工业区附近土壤当中有机污染物含量相对较高。

农用地土壤质量整体较好，大多数可以安全利用。除了有 2 个调查点污染指标超过了土壤污染风险管控值之外，还有部分点位（以 Cu、Ni、Zn）超过了土壤污染风险筛选值，安全利用受到了影响，需要加强土壤环境监测和农产品质量协同监测，限制其利用方式。

建设用地作为一类用地，除了个别地块，大多数可以安全利用；作为二类用地，As、Cd、Hg 含量均处于筛选值范围内；Cu、Ni、Pb 含量基本均处于筛选值范围内，仅个别点位元素含量大于管制值。对于大于筛选值的调查点，在开发利用的过程中需要进一步详细调查。

5. 土壤养分质量丰缺

土壤中不同养分指标区域上具有不同的发育水平。有机质仅处于中等含量水平。其中西部地区、中心城的含量较低，其他地区属于中等含量水平。氮元素可划分为缺乏、相对缺乏、适度、富足 5 种分级区。其中，崇明、长兴、横沙三岛和长江以南沿江、沿海地带隶属于缺乏—相对缺乏区，而以崇明岛的滨海盐土最差；而在长江以南，从城区及周缘向西部湖区，属于适度—富足过渡区。全磷高，有效态磷高，在川沙北部和长兴、横沙、崇明三岛为相对缺乏区，西部湖区、城区周缘为富足区。钾元素在北部及西南部地区较富足，在原南汇大部、奉贤和金山沿海地区的土壤中交换性钾相对缺乏。崇明三岛及除沿金山—奉贤—原南汇一线以东的滨海地区外，上海大部分地区土壤不存在 K 的缺乏。

以农业利用为目的的土壤基本可划分为西部湖积区、中部区和东部三岛区及海沿区等 3 类区。西部湖积区土壤低钙、镁，高硅，富有机质，土壤养分多表现为适度或者富足，但多因土壤渍水严重使土壤利用效益降低。中部区土壤以偏高的钙、镁、钾、铁、铝和微量元素碘，土壤养分多为适度，土质黏重适度，透气爽水，为优质土壤区。东部三岛区及海沿区土壤高钙，土壤养分显示为磷、氮、锌、铜、钼、硼相对缺乏，加之滨海盐渍，只能属于农业上的一般性土壤。

（五）岸带冲淤

1. 重点岸段冲淤特征

岸带冲淤是上海重要的地质灾害之一。不同岸段、不同阶段呈不同的冲刷和淤积特征。

崇明东滩不断淤涨，早在 1842 年，即呈向东延伸的“鸭舌状”边滩，之后，在其东南侧先后出现团结沙和北港北沙，但是近年来淤积速率逐渐趋缓。2011—2013 年，崇明东滩出现小幅度冲刷，尤其是在崇明东滩北侧区域，冲刷较为明显。

横沙浅滩是早期铜沙浅滩的一部分，20 世纪 80 年代后，在滩中间形成一条连通北槽和北港的串沟。西侧的横沙东滩在 1998 年后逐渐围垦成陆。2011—2013 年，横沙东滩围垦区前沿整体为冲刷态势。2013—2015 年，两侧出现冲刷，中间靠近东侧发生较为明显的淤积。2015—2017 年，淤积面积继续增加。

南汇边滩为一向海突出的梨头状浅滩，位于涨落潮流交汇的缓流区，极易泥沙堆积，长期以来南汇边滩为长江河口淤涨速率最快的岸滩之一。自公元 8 世纪—20 世纪 60 年代，海岸线年平均外涨 22. 3m。713—1172 年年均外延 34. 85m，1172—1584 年间外涨速度仅为 3. 4m/a，1584—1884 年间外涨速度增加至 24m/a，1884—1960 年涨速扩大至 41. 3m/a。1997—2009 年，除区域北部和南部小范围为冲刷外，大部分区域为冲淤平衡或淤积状态。由大治河以北沿岸斜向东南方向，淤积带宽由沿岸约 3km 逐步扩大至 6km 左右。

九段沙是长江口第三代新生沙洲，由上沙、中沙和下沙组成，上沙与江南亚沙毗邻。自 1958 年成为独立沙体以来，滩地面积一直处于增长状态。特别是 2000 年以来，滩地面积更是加速增长。1997—2002 年，南侧的江亚南沙显著淤积，而九段沙下沙出现轻微冲刷。2002—2008 年，江亚南沙继续显著淤积，而九段沙中、下沙大范围淤积。

青草沙水库区域指长兴岛西北侧南北港分流附近。近 10 年来，受河势整治工程、青草沙水库工程等影响，青草沙附近河势调整明显。不同区域的年际冲淤变化特征十分明显。

金山深槽始于大、小金山，向西延伸。深槽的底部由一系列深潭串联而成，一般深 30~40m，以大、小金山附近为最深。2012 年以前总体保持稳定，2013 年之后基本以冲刷为主，局部有淤积。

2. 岸带冲淤规律

近 10 年来，-10m 等深线在北支和北港口外变化很小，基本趋势不变；在南汇边滩和南槽外缘变化较为明显。口内、口外和杭州湾地区表现出不同的冲淤特征。长江口、杭州湾北岸基本还以冲刷为主，北支下段、崇明东滩和横沙浅滩以及南汇南滩外侧基本以淤积为主。南北支、南北港和南北槽分流口处，冲淤特征整体表现为滩淤槽冲的特征，但有所差异。

（六）地下水资源

1. 地下水资源及开发利用

潜水可采资源以调节储存量表征。1981 年、1951—1999 年、1980—2000 年 3 个阶段评价年计算评价结果显示，天然资源总量分别为 $28.266\times10^8m^3/a$、$12.938\times10^8m^3/a$、$12.060\times10^8m^3/a$，调节储存量分别为 $8.69\times10^8m^3/a$、$2.03\times10^8m^3/a$、$2.03\times10^8m^3/a$。

承压水可采资源以可开采量表征。针对上海不同阶段城市对地下水资源的需求以及地质环境保护目标，分别进行了多轮承压含水层地下水主采层（第Ⅱ、第Ⅲ、第Ⅳ、第Ⅴ承压含水层）可开采量计算。1987 年、1990 年、1995 年、2000 年、2005 年，曾先后采用比拟法、相关法、二维有限元法、准三维有限元法、真三维有限元法等进行计算评价，地下水可开采量计算结果分别为 $4.39\times10^8m^3/a$、$2.23\times10^8m^3/a$、$1.45\times10^8m^3/a$、$1.24\times10^8m^3/a$、$0.25\times10^8m^3/a$。

上海潜水和浅部第Ⅰ承压含水层地下水，长期来不予集中开发利用。深部第Ⅱ、第Ⅲ、第Ⅳ、第Ⅴ承压含水层地下水资源作为城市供水补充水源。自 20 世纪 50 年代开始实施开发利用，1949—1964 年和 1978—1996 年为历史上地下水开采强度最大的两个时期，年度开采量大于 $1\times10^8m^3$，最大在 $2\times10^8m^3$。2013 年以来年度开采量已低于 $800\times10^4m^3$，2018 年已降至 $120\times10^4m^3$ 左右，并有进一步减少趋势。

2. 地下水环境

地下水的长期开发利用，使得地下水环境也发生了规律性动态变化。

自然状态下的第Ⅱ、第Ⅲ、第Ⅳ、第Ⅴ承压含水层地下水位处于正标高状态。多年地下水位监测数据显示，第Ⅱ、第Ⅲ、第Ⅳ、第Ⅴ承压含水层地下水在不同强度的开采过程中，地下水位呈现不同幅度的下降、上升交替现象。开采强度大的阶段，地下水位处于相对低的状态，开采强度小的阶段，地下水位处于相对高的状态。历史上，第Ⅱ、第Ⅲ、第Ⅳ、第Ⅴ承压含水层地下水位最大降至-34m、-37m、-68m、-62m 左右。2018 年，第Ⅱ、第Ⅲ、第Ⅳ、第Ⅴ承压含水层地下水最低水位分别在-10m、-10m、-16m、-16m 左右。

原生状态下的第Ⅱ、第Ⅲ、第Ⅳ、第Ⅴ承压含水层地下水有Ⅰ类、Ⅱ类、Ⅲ类、Ⅳ类、Ⅴ类。多年地下水质量监测数据显示，在几十年来地下水开采过程中，原生地下水质量保持稳定。

3. 地下水回灌

上海通过地下水采灌井、专门地下水回灌井、应急回灌井等管井（回灌井）开展地下水人工回灌工作。由早期的日常采灌结合为主，逐渐转变成日常（应急）采灌、专门回灌等多种形式并存的特点。第Ⅱ、第Ⅲ、第Ⅳ、第Ⅴ承压含水层为地下水人工回灌主要层次，且在不同阶段的回灌区也发生一定的变化。随着阶段地面沉降防治目标的不同，回灌强度呈现明显的阶段性，1990—1999 年是历史上地下水人工回灌强度相对大的时期。

地下水人工回灌对水环境影响具有双重效应，在有效抬升区域地下水位或减缓地下水位下降速率的同时，局部原生地下水质量的稳定性也受到了一定程度影响，且不同背景（淡水、或咸水）的原生地下水影响特征或变化趋势有一定差异。

4. 地下水开发利用与保护规划

根据地面沉降约束条件及地下水可开采量，将地下水圈定了 4 个规划分区，包括地下水资源严格限制开采区、限制开采区、控制开采区和一般监测区。严格限制开采区为中心城。限制开采区包括近郊的宝山、浦东、闵行、嘉定，及远郊的青浦、金山等地。控制开采区包括远郊的崇明、奉贤、松江等地。

确定了城市中心地区、岛域地区和城市规划新城区等3大类重点部署区。基于2020年城市总体规划，根据城市规划所确定的城市重点规划建设地区以及人口、产业导向，总体部署中心城、宝山新城、嘉定安亭新城、浦东国际机场、惠南新城、临港新城、奉贤海滨农场、金山新城、松江新城、青浦新城、城桥新城、陈家镇新城、长兴岛、横沙岛等14个地下水应急水源地。

（七）地下空间资源

上海城市地下空间开发建设与利用始于1862年的市政管线建设，先后经历了以箱形基础为特征的地下空间开发（1949年以前）、平站结合的人防地下空间开发（1949—1990年）、以地下轨道交通设施为典型代表的地下空间开发（1990年以后）3个阶段。

截至2016年，全市已建成地下工程共有36809个，总建筑面积为$8186\times10^4m^2$，比2011年前增加建筑面积$2487\times10^4m^2$，地下空间开发量以每年约$1000\times10^4m^2$的速度增长。其中，地下生产、生活服务设施占地下工程总量的95%；地下公共基础设施占3%；轨道交通及附属设施占2%；上海地下空间开发规模已位居国际大城市前列。自21世纪初以来，上海地下空间开发数量和规模逐年增加并呈现爆炸性的增长态势，地铁、越江隧道、地下道路、城市地下综合体、大型深埋地下市政设施等地下工程不断增多。

上海地下空间开发利用总体上具有：区域分布上与土地价值呈现较强的相关性、开发深度上呈现较明显的分层利用、开发利用趋势上从浅层和中层向深层发展的3大典型特征。

地下空间开发利用过程中，地下水、软土、砂土、浅层天然气是影响地下空间开发利用的4大主要地质因素。

基于现有数据水平利用三维建模技术精细刻画地质结构及既有地下空间利用情况，以上海外环线为边界，利用几何方法和逻辑运算对浅层、中浅层、中层、深层适宜开发层及深层潜力开发层的地下空间资源数量进行估算，估算其地下空间总量、考虑既有地下空间条件下地下空间的余量，估算可开发利用地下空间的总量及可有效开发利用地下空间的数量。

针对基坑和隧道两种地下空间施工工艺，结合对地下空间影响较大的地下水、软土和砂、粉性土等地质因素以及其产生的基坑突涌问题、软土变形问题和流砂问题的严重程度，将上海中心城中浅层地下空间开发适宜划分为问题轻微区（A）、中等区（B）和严重区（C）3类；以工程建设引发水土突涌、流砂和地面沉降等环境地质问题的严重程度为主要评价依据，结合对深层地下空间开发影响较大的地下水和砂性土为主要影响因子，考虑地下空间开发现状、基岩埋深、地面沉降风险等因素将上海深层地下空间开发适宜性划分为好、中等、较差和差4个等级。

区内地下空间资源可划分为浅层（0~15m）、中浅层（15~30m）、中层（30~50m）、深层适宜开发层（50~65m）、深层潜力开发层（65~100m）。中心城地下空间总量体积达$6.64\times10^{10}m^3$，30m以浅的地下空间资源质量较高。

（八）滩涂资源

在长江口地区分布有白茆沙、扁担沙、新浏河沙，北支的崇头边滩和北沿黄瓜沙，南港的瑞丰沙，北港的中央沙、青草沙和六滧沙，南槽的江亚南沙，口门的崇明东滩、北港北沙、横沙浅滩、九段沙和南汇边滩等15个较大型沙体区，构成区内滩涂资源的主体。

截至2017年，崇明东滩和北港北沙-5m等深线包络面积约为$457km^2$。南汇边滩-5m等深线包络面积为$276.5km^2$。九段沙-5m等深线包络面积为$225.5km^2$。横沙浅滩-5m等深线包络面积为$255km^2$。

崇明东滩向北的进积空间较大，与崇明北沿相连后成为重要的滩涂增长空间。北港北沙与早期完成促淤圈围工程的团结沙类似，总体上仍然处于向东南的推进之中。横沙浅滩仍然向东南方向推进，较快的沉积中心可能沿东南方向向鸡骨礁方向迁移；但要注意沉积中心向南东迁移对深水航道可能带来的影响。九段沙为重要的湿地自然保护区，整体处于比较高的淤积速率阶段。南汇边滩经过大规模的促淤圈围，尽管高滩资源已经大幅度减少，但低滩资源十分丰富，且总体上仍不断地向东推进，仅次于横沙东滩。杭州湾总体上处于不断的冲刷趋势中，综合来看，高、中低滩资源十分有限，开发利用的综合潜力很低。崇明北沿为中高滩资源比较集中的区域，且由于北支处于不断的淤积和萎缩之中，临近崇明岛，该区综合利用的潜力较高。

长江口可用于促淤圈围的袋状砂、堤心砂资源十分有限。在南支主槽—新浏河沙沙包、下扁担沙、北港潮流脊—六滧沙脊尾部、顾圆沙东北、北港口—鸡骨礁、江亚南沙尾部、南汇边滩和杭州湾等8个地区还有一定的利用潜力。

（九）旅游地质

旅游地质资源丰富，包括多种地质遗迹资源和多处古文化遗址。主要有金山沿岸及其外侧海域的基岩残丘、古海岸线和不同历史时期的海堤遗迹、局部的古人类文化遗址，以及广布的现代滨海河口地貌和湿地资源、水域基岩岛屿的海蚀地貌等。此外，在金山张堰分布有上海唯一的隐伏型固体金属矿床。

地质遗迹资源极其具有特色，除观赏价值外，同时具有重要的科学价值。可划分为2大类4类4亚类，即地貌景观大类和地质灾害大类两大类，水体地貌类、海岸地貌类、火山地貌类、地质灾害类4类，河流（景观带）、海积地貌、火山岩地貌和地面沉降4个亚类。依托崇明岛设立的崇明岛国家地质公园特色鲜明，公园内的地质遗迹景观价值等级可划分出世界级1处，国家级1处，省级8处。

八、区域地质演化史

综观，区内可划分出古元古代、中元古代—青白口纪、南华纪—中三叠世、晚三叠世—古近纪、新近纪和第四纪等5个地质演化阶段。大致在古元古代晚期，区内原始陆壳雏形已基本形成。古元古代末，吕梁运动使扬子原始陆壳克拉通化。

（一）中元古代—青白口纪

中元古代期间，区内整体上表现为弧后盆地。由早期向晚期，海水似有加深趋势，形成一套具有复理石建造特征的浅海相沉积，伴随有间歇性的（海底）火山喷发活动，且岩浆作用似有加剧和“双峰式”特征。中元古代末期，区内呈活动大陆边缘的造山带环境，发育有不成熟的岛弧，构造体制有由挤压环境向拉张环境过渡的特征。

晋宁运动Ⅰ幕使中元古代建造褶皱变质，形成金山群。青白口纪早期，区内处于相对稳定的弧后浅海—次深海盆地环境，藻类植物开始大量出现。地壳似乎进入一个相对稳定阶段，火山活动基本停息。晋宁运动Ⅱ幕造成基底的全面隆升，青白口纪及中元古代地层发生不同程度的褶皱、断裂等变形作用改造，一些地区发生明显的低绿片岩相变质作用，形成“惠南板岩”。与此同时，金山群发生了局部退变质作用。之后，碰撞后上叠裂谷盆地开始发育，区内堆积河上镇群陆源碎屑岩建造，且火山作用逐渐强烈。

（二）南华纪—中三叠世

青白口纪早期以后，区内进入海相、海陆过渡相为主的盖层发展阶段，由中—新元古代的弧盆体系建造逐渐转化为扬子克拉通东南边缘的被动陆缘盆地、陆地表海盆地建造。

区内南华纪建造未有揭露。震旦纪开始，气候转暖，发生广泛海侵，区内形成一套以镁、硅质碳酸盐岩为主的灯影组沉积。宝山大场一带，似属台地边缘浅滩相—开阔台地相；其南的浦东坦直、三灶一带则为局限海台地相—蒸发岩台地相的滨海沉积。整个震旦纪，受中部基底隆起影响，区内古地势呈南东高、西北低的特征。随着晚震旦世晚期的海退，于隆起带西北侧青浦、宝山和崇明等地形成滞水的海湾环境，沉积了薄层黑色硅质岩。隆起带南部的浦东坦直—奉贤一带逐渐海水退出，接受剥蚀。

底寒武世到早寒武世早期，基本继承震旦纪晚期沉积环境，海水较震旦纪晚期加深。在隆起带北侧形成滞留深水平缓盆地，南侧推测暂未波及。此时，气候温暖，海湾中水流不畅，低级菌类植物迅速繁衍，有机物大量堆积，从而形成以碳质页岩为主，间夹薄层钙质页岩与硅质灰岩的超山组沉积。早寒武世晚期，沉积环境与早期相似，滞留还原环境已明显减弱，水动力作用增强，形成了大陈岭组的同生角砾状白云岩沉积。

中寒武世开始，发生明显海侵，形成开阔海台地环境，陆源碎屑物增多，沉积了一套以泥灰岩及含粉砂质灰岩为主的磷酸盐岩建造。

至晚寒武世，海水变浅，气候转为炎热，形成台坪潟湖相厚层白云岩、同生角砾状白云岩及白云质灰岩建造。此时，由于海水咸度较高，生物不易繁衍、生存，层中化石稀少。

奥陶纪承袭了寒武纪的古地理和古构造条件，唯海域不断扩大，海水也较前加深。仑山组沉积时期，海侵初期形成开阔海台地相的碳酸盐岩建造。早奥陶世晚期—晚奥陶世早期，海水进一步加深，沉积了以台地斜坡建造为主的红花园组—黄花岗组，由灰黑色、灰紫色含碳质碳酸盐岩和含泥、硅质条带的生物骨屑微晶灰岩、瘤状灰岩组成，部分层位出现角砾状结构和瘤状结构。尤其是牯牛潭组沉积时期，层中富含生物化石碎片，含头足类、腹足类、三叶虫、海百合、牙形刺类，其主要生物分子介于宁镇与浙西混生的生物群特征。表明水体仍然较浅，水动力能量较强。

至晚奥陶世中期，海侵规模达到高峰，区内地势东南高、西北低，北部地区演化为陆棚环境，接受了一套巨厚的类复理石沉积；南部地区推测较薄，层中富含笔石化石，其生物组合兼有华北型与东南型特征，属过渡型生物群。

志留系下部有灰黑或灰绿色砂岩、粉砂岩及泥岩组成类复理石建造，其上部以紫红色含泥砾砂岩为主。由下而上，由潮下带演化为潮间带，构成一个明显的海退序列。

迄今，区内未发现晚古生代—中三叠世地层。推测在上海北部，尤其是崇明岛西部地区完全可能有该阶段的沉积。

（三）晚三叠世—古近纪

中、晚三叠世之交强烈的印支造山运动把区内带入陆相盆地建造发育与改造阶段。

迄今，区内尚未发现晚三叠世—侏罗纪地层。根据岩浆岩特征，推测至少从早侏罗世晚期开始区内又进入强烈的陆内挤压造山阶段，形成吴淞同造山型花岗岩。中侏罗世末期，进入晚造山阶段，形成三林石英闪长岩岩珠。

早白垩世，区内逐渐进入后造山—非造山阶段。岩浆侵入和火山喷发建造分布广泛，经历了弱—强—弱几个演化阶段，在130Ma左右最为强烈，分布最广。形成了劳村组、黄尖组与寿昌组3个陆相火山喷发-沉积建造，市北花岗岩、市南花岗岩、堡镇花岗岩。

从晚白垩世开始，区域应力场已转变为以引张作用为主。地壳断块差异趋于明显，受北东向、近东西向及北西向断裂控制，形成一系列断陷盆地。发育浦口组陆相紫色类磨拉石建造、赤山组红色碎屑岩建造、夏驾桥组棕色含钙细碎屑岩建造以及北桥组杂色碎屑岩建造。岩浆作用逐渐减弱，主要表现为间歇性的火山喷发和局部岩浆侵入作用，形成大量小的岩珠和次火山岩、脉岩。晚白垩世，气候干热多雨，山麓及冲洪积环境发育，沉积物较粗，局部夹石膏层。古近纪，气候以干旱为主，湖盆面积已明显缩小，沉积物较细。

（四）新近纪—第四纪

气候由古近纪的干旱转为温暖渐趋寒冷，最终进入第四纪冰期。盆地范围普遍扩大，构成早期断陷盆地的坳陷沉积。区内，受基底北西向、近东西向及北东向断裂的控制，幔源型碱性橄榄玄武岩浆常沿断裂交汇处喷溢，分布于沿海地带。并在火山喷发间歇期，受温暖气候控制，形成未固结的湖泊堆积。

新近纪晚期之后，在原有的基底构造与古地貌的基础上，在北西向和北北西向断裂的改造下，最终成为隆坳相间的构造、地貌景观。在安（亭）—崇（明）凹陷及枫泾凹陷、浦东滨岸凹陷等低洼地区，沉积了较厚的崇明组冲洪积沉积。西部隆起带则经受风化剥蚀。

进入第四纪，冰期和间冰期的交替引起了海平面周期性的升降。区内接受了厚达数百米的第四纪河湖相、河口三角洲相及海陆交互相沉积，在此期间曾经过5次明显的寒冷期与温暖期交替，并伴有7次海侵作用。这一时期，高级植物及哺乳类动物迅速繁衍，特别是进入全新世以后，由于气候转暖，海平面上升，长江河口不断快速向陆后退，下切河谷逐渐演变为巨大的河口湾。海平面达到最高并保持稳定之后，随着入海泥沙的输入和水下可容空间的减少，三角洲开始快速向海推进，表现为长江河口和两翼海岸线不断向海迁移。此后，区内河湖密布，气候宜人，成为古人类生息、繁衍的良好场所。

近现代以来，人类活动对区内地质环境演化形成了明显影响，长江三角洲的自然演化状态遭到明显改变。

Abstract

Shanghai (referred to as "Hu" or "Shen" for short), is a city situated between the Yangtze River Estuary and the Hangzhou Bay. Here the terrain is flat. The altitude is generally on the order of a few meters except some ten-plus small hills in the Jinshan and Songjiang Districts. The maximum altitude is around 100 m. The buried depth of the bedrock surface is obvious undulated, showing a distinct SW-NS increase. In Songjiang, Jinshan, and Qingpu Districts, the buried depth is mostly in a range of 160-220 m; in the coastal districts of Pudong, Baoshan, and Jiading, the buried depth is mostly 280-360 m; in the estuarine sand island areas, the buried depth is generally greater than 400 m; at local points of the Chongming Island, it can be as great as almost 500 m.

Since the compilation of the first generation of the geological annals, extensive geological survey and research has been carried out in the area, having made substantial progress in regional geological conditions and urban geology. After collecting the various data acquired over the past three decades, a systematic summary has been conducted on the stratigraphic and sedimentary processes, magmatic rock and magmatism, metamorphic rock and metamorphism, lithospheric composition, geological structure, Quaternary geology and geomorphology of the area under the inspiration of activity theory and plate tectonics. Urban geological survey and research results have also been integrated. A series of nine 1 : 250000-scale geological maps has been compiled. These include the *Quaternary Geological and Geomorphologic Map of Shanghai Municipality*, *Bedrock Geological Map of Shanghai Municipality*, *Geological Structure Map of Shanghai Municipality*, *Bouguer Gravity Anomaly Map of Shanghai Municipality*, *Aeromagnetic (ΔT) Anomaly Map of Shanghai Municipality*, *Integrated Hydrogeological Map of the Ⅳ Confined Aquifer of Shanghai Municipality*, *Integrated Engineering Geological Structure Zoning Map of Shanghai Municipality*, *Cumulative Land Subsidence Contours Map of Shanghai Municipality* (1921-2016), and *Geological Relics Map of Shanghai Municipality.*

Shanghai lies on the southeastern margin of the Yangtze Craton, in close adjacency to the Cathysian Orogene. Since the Paleoproterozoic, the area has undergone four major tectonic evolution stages: formation of the crustal rudiment, formation of the basement, formation of the caprock, and intracontinental transformation. So far, no Late Paleozoic and Triassic-Jurassic formations have been detected in Shanghai. According to the stratigraphic development of the area, we assume that the Lower-Middle Devonian, Middle-Upper Triassic and Upper Jurassic formations are all absent and the other formations are undiscovered. Affected by multiple tectonic movements, the area characterizes complicated geological structure and frequent magmatism.

1. Prequaternary

Shanghai is host to the Paleoproterozoic-Mesoproterozoic, Qingbaikouan, Sinian, Cambrian, Ordovician, Silurian, Cretaceous, Paleocene, Neogene, and Quaternary formations. No Paleozoic or Triassic formations have been detected yet. From the bottom up, there are 33 rock stratigraphic units, including 23 bedrock formations. By comprehensive zoning, these bedrock formations are part of the Huhang stratigraphic minor region, Jiangnan stratigraphic subregion, Yangtze stratigraphic region, South China stratigraphic province. The Quaternary formations belong to the Central South-Southeast stratigraphic region of the South China stratigraphic province. According to the development characteristics of the late Quaternary formations, the Quaternary stratigraphic subregion can be further divided into two subregions, namely, the Yangtze Delta and Taihu Plain sedimentary subregions.

1.1 Paleoproterozoic–Qingbaikouan

The Paleoproterozoic–Qingbaikouan system consists of formations from the crustal rudiment and basement formation period, which make up the crystallized or folded basement of the area. It covers the Jinshan Group, Huinan slate, and Heshangzhen Group.

The Jinshan Group is primarily found in Minhang District's Huajing; Pudong District's Sanlintang, Tianhua'an, Zhaojiazhai, Kangjiazhai, Shenzhuang; and Jinshan District's Zhujing and Luxiang. It is a metamorphosed metarock system comprised of mica schist, chlorite schist, feldspar quartzite, as well as amphibolite, amphibolite gneiss, amphibolite plagiogneiss, and leptite. The drilling revealed thickness is generally greater than 260 m. The top and bottom of this series have not been encountered. The zircon U–Pb ages of the amphibolite and mylonite granite porphyry are 1078 ± 8.9 and 1781±16 Ma, respectively.

The Huinan slate is found only in Huinan Town. Instead of being directly overburdened by unconsolidated strata, this rock series is overlain by tens of meters of basalt. The lithology is monotonously gray sericite slate, with tabular mudstone observed at local points. The rocks contain fossilized *Leiominuscula minuta* Naum, *Lophominuscula* sp., *Paleamorpha* sp., and *Dictyosphaera macroreticulata* Sin et Liu. The drilling revealed thickness is greater than 194.39 m. The bottom of this series has not been encountered.

The Heshangzhen Group is found only in Jinshan District's Qianwei area. The upper part of this group consists of gray or green volcanoclastic formations that mainly contain welded tuff. The lower part consists of gray or light gray allogenic terrigenous clastic formations mainly containing tuffaceous sandstone and feldspar quartzite, with lesser amounts of sand–bearing mudstone and muddy siltstone. The rocks commonly show signs of pyritization. The drilling revealed thickness is greater than 466.83 m. The zircon U–Pb ages are 774±12 and 801±13 Ma.

The Mesoproterozoic consists mainly of neritic sediments bearing the characteristics of flyschoid formations, associated with intermittent (seafloor) eruption activity. From the early to the late stage, the seawater seems to have been deepening and the magmatism seems to have been intensifying. In the early Qingbaikouan, the seawater gradually deepened and rendered a subabyssal sedimentary environment; in the late Qingbaikouan, the seawater gradually regressed, rendering a littoral neritic–terrestrial environment. The volcanic activity gradually intensified over this period.

1.2 Sinian

The Sinian is typically distributed in Pudong New District's Tanzhi and Sanzao. It is also found in boreholes drilled in Baoshan District's Luodian, Dachang and Jiangwan. Besides, and some may also be present in Chongming District's Gangdong Town. Overall, it is a formation of fairly strongly altered limestone, dolomitic limestone, and dolomite. In the northern part's Dachang area, this series is interrupted by dolomitic mudstone and quartzite sandstone, with siliceous rock observed at the top. The northern part seems to be comprised of platform margin shoal facies–open platform facies sediments. The southern part is probably comprised of restricted sea platform–evaporite platform facies sediments. The drilling revealed thickness is approximately 6–260 m. The bottom of this series has not been encountered.

1.3 Cambrian

The Cambrian system is revealed most extensively in the Lower Paleozoic. This system has been encountered in Baoshan District's Dachang and Luodian, Changning District's Beixinjing to Tianshanwucun, Fengxian District's Qixianqiao–Qianqiao, and Qingpu District's Chonggu Town. From the bottom up, the strata include the Eocambrian-Lower Cambrian Chaoshan Formation, Dachenling Formation, Middle Cambrian Yanliugang Formation, and Upper Cambrian Chaofeng Formation. The top of this series has not been encountered. Except the Chaofeng Formation which has not been observed to be in contact with either the underlying or overlying formation, all the other formations lie in conformity with one another.

The Chaoshan Formation consists mainly of grayish black carbonaceous shale or, locally, mudstone, interrupted by carbonaceous shale and thin layers of silt-bearing, fine sandy shale. The drilling revealed thickness is approximately 148 m; the thickness appears to be thin in the south but thick in the north. The limestone interlayers are rich in fossilized *Trilobita* and *Brachiopoda*. The formation lies in conformity with the underlying Dengying Formation.

The Dachenling Formation consists mainly of gray or grayish black dolomitic limestone, dolomite and limestone, locally interrupted by thin layers of siliceous rock, carbonaceous shale, and calcareous sandstone. The drilling revealed thickness spans from 33 to 114 m. Both the top and bottom strata have been revealed.

The Yangliugang Formation consists mainly of light gray or dark gray marlite, muddy limestone, and silt-bearing limestone, locally interrupted by thin layers of carbonaceous shale or siliceous limestone. Stylolitic structures can be observed in the muddy limestone; the silt in the sand-bearing limestone is frequently aggregated into lumps. The drilling revealed thickness is greater than 150 m. The top of this formation has been denuded and is not fully outcropped.

The Chaofeng Formation consists mainly of light gray and steel gray dolomite and contemporaneous brecciated dolomite, locally interrupted by grayish white and grayish black dolomitic limestone and siliceous rock. The drilling revealed thickness spans from 248 to 762 m. The top and bottom of this formation have not been encountered yet.

From the early to the late stage, the Cambrian system in the area underwent the entire environmental evolution from basin facies to open sea platform facies torestricted platform lagoon facies. The seawater depth displays a shallow-deep-shallow rhythm.

1.4 Ordovician

The Ordovician system is distributed in limited areas. So far, this system has been discovered in boreholes drilled in the north and southwest of the downtown and the west of the Fengxian Haiwan Town. The stratigraphic sequence is quite complete. From the bottom up, it includes the Lower Ordovician Lunshan and Honghuayuan Formations, Lower to Middle Ordovician Guniutan Formation, and Upper Ordovician Huangnigang and Changwu Formations. The drilling revealed thickness is greater than 250 m. The top and bottom of this series have not been encountered. Its contact with the overlying and underlying formations is unclear yet.

The Lunshan Formation consists mainly of gray dolomitic limestone and limy dolomite. The lithology is monotonous; the per-layer thickness is quite great. The drilling revealed maximum thickness is a 198.42 m. The bottom of this series has not been encountered.

The Honghuayuan Formation consists mainly of gray or grayish white micrite, alternated by bluish gray marlite within the strata. The most prominent feature of this formation is the commonpresence of mud in the strata, which is distributed very irregularly, pinching out at one point and disappearing at another. A small amount of bio-skeletal fragments is observed in the top stratum. The drilling revealed maximum thickness is 176.99 m. The formation lies in conformity with both the underlying Lunshan Formation and the overlying Guniutan Formation.

In the southern part of Fengxian District, the Guniutan Formation consists mainly of purplish red microcrystalline detrital biolimestone interrupted by micrite, with brecciated textures at the top. In the downtown, it consists mainly of grayish white bioclast-bearing limestone, locally interluded by siliceous strips or veins. The rocks contain gastropods like *Umbonellina* cf. *inflata* Yu, *Maclurites* sp., *Raphispira* sp., *Liospira* sp.; conodonts like *Periodon aculeatus* Hadding, *Protopanderodus rectus* (*Lindström*), *Walliserodusethingtom* (*Fahraeus*), as well as fossilized *Metacrinus* and *Ostracoda* debris. The drilling revealed thickness is mostly tens of meters. Some can be as thick as 30–40 m. The maximum thickness is about 60 m. The formation lies in conformity with the underlying Honghuayuan Formation.

The Huangnigang Formation is a nodular limestone series with limited thickness. The lithology is fairly simple, characterizing muddy or siliceous nodules. The drilling revealed thickness is 39.2 m. The formation lies in conformity with the underlying Guniutan Formation.

The Changwu Formation consists mainly of yellowish green-grayish black mudstone, muddy siltstone, silty mudstone, and silty-fine sandstone. It features clear sedimentary rhythm and flyschoid sediments. The rocks contain rich fossilized graptolite, typically *Climacograptus*. The drilling revealed thickness is greater than 123.0 m. The top of this series has not been encountered. The formation lies in conformity with the Huangnigang Formation.

The Ordovician is a marine sedimentary series. From the early to the late stage, as the seawater depth increased, the area evolved from a platform environment inherited from the Late Cambrian to a platform margin slope or continental shelf neritic environment.

1.5 Silurian

The Silurian is found only in Miaozhen to Caopeng of the northwestern Chongming Island. It is a clastic formation comprised of sandstone, mudstone, and shale. From the bottom up, this system includes the Helixi and Kangshan Formations. The drilling revealed thickness is approximately 656.7 m. The top and bottom of this series have not been encountered.

The Helixi Formation consists mainly of gray, slightly green or grayish green, and locally grayish black and grayish brown, alternations of muddy sandstone and silty mudstone, locally interrupted by thin layers of shale or quartz sandstone. Mud conglomerates are commonly present in the strata. The drilling revealed maximum thickness is only around 500 m. The bottom of this series has not been encountered.

The Kangshan Formation consists of purplish red or grayish black lithic quartz wacke and quartz sandstone in the upper part, and grayish brown-gray or grayish black lithic quartz wacke in the lower part, interrupted by thin layers of silty mudstone or mudstone. The formation is commonly affected by slight metamorphism. Abundant microfossil plants like algae and spores are observed inside the strata. The top of this series has not been encountered. The thickness is greater than 40.0 m.

The Silurian consists mainly of littoral sediments. The color of the sediments turn from grayish green or grayish black in the lower part toward purplish red or dark red in the upper part; the grain size of the clasts also change from fine to coarse, showing a relatively complete regression sequence.

1.6 Cretaceous

The Lower Cretaceous is the most widely distributed, covering approximately 70% or more of the total area of the city. The Upper Cretaceous is distributed in limited areas, typically in the northwestern part's Anlu Basin adjacent to Jiangsu Province, the central part's Songjiang Basin, and the northern margin of the western section of the Chongming Island. The Lower Cretaceous consists of intermediate or acidic volcanic and volcanoclastic-sedimentary rock system, which, from the bottom up, includes the Laocun, Huangjian, and Shouchang Formations. The Upper Cretaceous consists of a series of fault basin formations, including the Pukou and Chishan Formations. The drilling revealed maximum thicknesses of the Lower and Upper Cretaceous formations are 1850.7 and 789.56 m, respectively. The top and bottom are incomplete for both formations.

The Laocun Formation includes the upper and the lower members. The lower member is widespread in southern downtown, Fengxian, and Jinshan. It is mainly comprised of alternately grayish white and grayish purple tuffaceous sandstone, tuffaceous gritstone, tuffaceous conglomerate, rhyolitic tuff lava, and sandstone. The gravels feature a complicated composition, poor sortability, subrounded to angular shape, and local directional alignment. The strata are frequently interluded by andesite porphyrite veins. The rocks are commonly strongly silicified. In Songyin, alternately purplish red and grayish white tuff lava is observed. The upper member is typically found in Songyin and Zhangyan. In Songyin, the Upper Laocun Formation is mainly comprised of alternately purplish red and grayish white tuffaceous sandstone, occasionally interrupted by thin layers of mudstone, showing clear stratification. In Zhangyan, it is mostly comprised of feldspar quartz sandstone with basically no volcanic content. The drilling revealed thickness is greater than 167. 9 m.

The Huangjian Formation is mainly distributed in the central and southern part of Shanghai. This formation consists mainly of intermediate-acidic volcanic rocks and volcanoclastic rocks like dacite, rhyodacite, dacitic breccia-bearing welded tuff, rhyodacitic welded tuff, welded tuff, and tuff lava, and volcanoclastic rocks, which have usually assembled with hypabyssal or ultrahypabyssal volcanic rocks into volcanic complexes. In Jinqiao and Tianmashan, the formation consists of andesitic porphyrite, andesite, and dacitic porphyrite. The maximum thickness per borehole is 484. 0 m. Its contact relationship with the overlying Shouchang Formation is unclear. The zircon U-Pb age is in the range of 136. 1±2. 4-129. 0±0. 8 Ma.

The Shouchang Formation is divided into the upper, middle, and lower members. Its contact relationship with the overlying and underlying formations is unclear yet. The lower member consists of cyclothems of purplish red, steel gray or grayish black silty-fine sandstone, sandstone, and particolored conglomerate, interrupted by grayish black carbonaceous strips, with basically no volcanic contents and well developed stratification. The maximum drilling revealed thickness is 300. 7 m. A large amount of fossilized *Sporopollen*, predominantly gymnosperm pollen, is present. The gymnosperm pollen is the richest in *Classopollis* pollen, associated with *Psophosphaera*, *Alisponites*, *Monocolpollenites*, *Araucariacites*, *Sphaeripollenites*, and *Podocarpidites* pollen. The $^{206}Pb/^{238}U$ age is quite dispersed, but is mostly on the order of 120 Ma. The middle member consists mainly of purplish red, steel gray, or grayish white tuff, tuffaceous sandstone, as well as mudstone, siltstone, and medium-coarse sandstone, with silt-bearing limestone or crystalline or vitric dolomite. It contains quite a large amount of plant debris and is clearly stratified. The maximum drilling revealed thickness is 256. 1 m. The upper member consists mainly of intermediate-acidic lavas and volcanoclastic rocks like bubble rhyolitic dacite, rhyolitic dacite, breccia-bearing welded tuff, and lithic or crystalline or vitric tuff. The drilling revealed maximum thickness is 326. 9 m.

The Pukou Formation is widespread on the southern margin of the Anlu fault basin. This formation consists mainly of massive purplish brown, purplish gray, or particolored glutenite-conglomerate, interrupted by reddish brown siltstone, pebbly silty-fine sandstone and mudstone of varying thicknesses. The gravels feature poor psephicity and poor sortability, containing predominantly volcanic components. The thickness varies greatly, reducing significantly toward the edge of the basin. The formation lies in unconformity with the underlying Lower Cretaceous.

The Chishan Formation is typically distributed in the northern part of the Anlu fault basin, Songjiang, and Beiqiao. In Anting and Nanxiang, the formation coarsens, consisting mainly of sandstone, glutenite, and sandy mudstone. In Beiqiao, it consists of brownish red silty mudstone and siltstone, interrupted by purplish red fine conglomerate and pebbly sandstone. In Songjiang, it consists of purple silty mudstone and muddy siltstone, interrupted by medium-fine sandstone and individual gritstone layers. The detrital contents are primarily quartz and secondarily feldspar, with some mica and hematite debris. The sortability and psephicity are both relatively poor. Thin-layered or stockwork gypsum is frequently observed in the strata. In some of the formations, calcareous gravel, boulder clay or carbonaceous lumps can be seen. The drilling revealed maximum thickness is 771.9 m. The K-Ar age yielded for the basalt interlayers is 78.31±1.77 Ma.

The Cretaceous system consists of continental sediments of alluvial fan, fluvial, lacustrine, and paludal facies. In the Early Cretaceous, the hot climate gave rise to very thick continental volcanic to terrestrial clastic rocks. During intense volcanic eruption, mainly volcanic rocks were accumulated. During the hiatus of volcanic eruption, fluvial-lacustrine volcanoclastic and sedimentary rocks were often accumulated in the volcanic rift depressions. In the Late Cretaceous, very thick alluvial fan and fluvial-lacustrine volcanoclastic sediments infilled local fault basins.

1.7 Paleogene

The Paleogene system is mainly distributed in the Anlu fault basin and the Beiqiao fault basin. Presumably, it is also present in the basin in the northern margin of Chongming. So far, only the Paleocene has been discovered. From the bottom up, this system includes the Xiajiaqiao and Beiqiao Formations, consisting of a purplish red terrestrial sedimentary formation that has been consolidated into rocks. According to the regional stratigraphic development, the Eocene and Oligocene formations in the area are deemed undiscovered.

The Xiajiaqiao Formation consists mainly of brownish red (locally brick-red) silty-fine sandstone, siltstone, and muddy siltstone. The deposition thickness reduces upward. The strata contain rich *Ostracoda* and *Stonewort* microfossils. The *Ostracoda* features a *Cristocypridea-Proeucypris-Ruficypris* assemblage; the *Charophytes* features large amounts of *Characeae Chlamydomonas* components with apical pores. The drilling revealed maximum thickness is around 130.0 m. The thickness reduces toward the margin and the west of the fault basin. The contact relationship with the underlying Chishan Formation is not detected yet.

The Beiqiao Formation consists mainly of brownish red, dark (purplish red) or grayish brown mudstone and silty mudstone, with gravel and interruptions of silty-fine sandstone and basalt at local points. The calcareous content is significant. Calcareous lumps and boulder clay are detected at local points. The *Ostracoda* species is quite monotonous, featuring a *Cypridea-Cyprois-Parailyocypris* assemblage. The *Stonewort* fossils have a rich variety of species and a great quantity, featuring a *Gobichara deserta-Peckichara-Stephanocharajiangsuensis* assemblage. *Sporopollen* fossils are detected either in a *Ulmipollenites minor-Triporopollenites* assemblage zone or a *Quercoidites-Ulmipollenites-Triporopollenites* assemblage zone. The drilling revealed maximum thickness is 289.4 m. The formation lies in conformable transition with the underlying Xiajiaqiao Formation.

The Early Paleocene virtually inherited the sedimentary environment of the Late Cretaceous. In the Anlu and Beiqiao Basins, a lacustrine sedimentary formation emerged, containing rich *Ostracoda* and stonewort fossils.

1.8 Neogene

This system is made up of two parts: the basalt in the lower part and the alluvial and lacustrine accumulations in the upper part. The lower part is called the "Bailonggang basalt", which is a Miocene formation typically found in the shoreland zone of southeastern Shanghai's Yangtze River Estuary. The upper part is called the "Chongming Formation", which is a Pliocene formation typically found in Chongming and other estuary areas inside Shanghai where the Late Cenozoic is the most deeply settled. This formation is also detected in the shoreland zones of Baoshan and Pudong Districts, as well as in the depression regions in the northwestern part of Jiading District.

The Beilonggang basalt consists mainly of dark gray pyroxene basalt, olivine basalt, and grayish green vitric basalt, interrupted by unconsolidated accumulations from effusion hiatus. The drilling revealed maximum thickness is 260 m. The bottom of this series has not been encountered. The thickness is generally 23-69 m. The total rock K-Ar age of the basalt is 18.7-26.7 Ma. The formation lies in angular unconformity with the underlying formations of different ages.

The Chongming Formation is quite completely developed on the estuarine sand islands and at the eastern and western ends of the Chongming Island. It contains the upper and lower formations. The lower part consists of braided fluvial sediments resulted from frequent recurrence of sand gravel, pebbly sand, and grayish green clay, or of lacustrine grayish green clay interrupted by mud-bearing gravelly sand or sand gravel. The upper part consists mainly of pebbly sand and fine sand alternations which evolve upward into dual-structured clay-dominated meandering channel sediments, or of lacustrine yellowish green or brownish yellow clay alternated with muddy gravel. The thickness ranges from 60 to 220 m. This formation is also found inside the shoreland zones of Baoshan and Pudong, and in the Anting Sag. There, its lithological and lithofacies sequences are similar to what is observed in Chongming: It is generally a combination of proluvial grayish yellow muddy sand gravel frequently interrupted by clay in the lower part, and proluvial-lacustrine grayish green pebbly clay and clay layers in the upper part. The thickness is mostly 10-65 m except in the Anting Sag, where the thickness can be 120 m. Fossilized *Sporopollen* and *Ostracoda*, predominantly *Sporopollen*, are present. *Sporopollen* has been detected only in a few boreholes. It features a *Quercusdeciduous* - *Quercus evergreen* - Cyperaceae - *Pteris* assemablage.

In the Miocene, the southwestern part of the area was mainly occupied by vast stretches of denudation low hills; in the east, a few volcanic lava basins of varying areas took shape. During the hiatus of volcanic eruption, the volcanic basins once evolved into lake basins.

In the Pliocene, the land area in central southern Shanghai land area's Qingpu, Songjiang, Jinshan, Fengxian, and Minhang Districts, and the bordering area between Pudong and the downtown were still occupied by vast stretches of low hills. In Minhang District's Beiqiao and Qibao, these low hills were intercalated by intermontane lake basins and hilltop eluvial depressions. NS-trending, continuously distributed, wavy subsiding zones occurred along the current land area's Jiading-Baoshan-northeastern downtown-Pudong-Nanhui New City line; intermontane subsiding basins also emerged in Jinshan District's Fengjing, Songjiang District's Xinbang, and Qingpu District's Zhengdian, giving rise to piedmont alluvial plains and lacustrine depressions in these subsiding regions. The subsiding basin center fell on the Chongming Island. In Xianghua, Qianshao and Yibei areas at the eastern end, Chengqiao Town-Fuming Farm in the central part, and Luhua-Hongxing Farm at the western end, of the Chongming Island, a number of rivers appeared on the capacious interchannel beachland.

2. Magmatic Rock and Magmatism

The magmatism in Shanghai spans a wide spectrum of time, with presence in the Jinningian, Yanshanian, and Himalayan periods, although it was the strongest in the Late Yanshanian. All stages of magmatism there have been manifested mainly by volcanic effusion; contemporaneous intrusion and volcanism share the same origin. Magmatic rocks and related volcanoclastic formations are well developed. The rocks are primarily intermediate-acidic rocks and acidic rocks, and secondarily intermediate and intermediate-basic rocks; basic rocks take a smaller proportion. The basic rocks are alkaline, with magma originating from mantle melting. The intermediate-acidic rocks are predominantly shoshonite-high-K calc-alkaline series, principally type I and type A granites.

2.1 Intrusive rock and intrusion

About 33 intrusive rock bodies have been detected in the area. The acidic rocks exist on a larger scale, typically found in the northern part of the area. The intermediate and basic rocks exist in a greater variety but on a smaller scale, typically found in the eastern and southern parts of the area. The rock bodies occur mainly in forms of stocks, apophyses, or dykes. Only one batholith, namely, the Chongming granite body, is found to measure larger than 100 km^2. Three intrusion stages, namely, Jinningian, Yanshanian, and Himalayan, can be identified.

2.1.1 Jinningian

The Jinning period witnessed two intrusion events, one in the Early Changchengian, the other in the Late Jixianian. The lithology is mainly represented by diabase, hornblende diabase, granite, and molonite granite porphyry.

The diabase and hornblende diabase are detected in the Jinshan Group in Tianhua'an. The rocks have been metamorphosed.

The granite is seen only in Zhujing, consisting of a small-sized stock. The rocks are relatively highly differentiated with rich LREEs and characterize mantle-derived granite. This should be attributable to an immature island arc or oceanic crust. The zirconU-Pb age is 1082.4±7.9 Ma.

The molonite granite porphyry is typically found in Pudong District's Kangjiazhai and Fengxian District's Xiaotang, with relatively well developed mylonitic structure. The intrusion appears to have occurred in a destructive active plate margin or syn-collision tectonic environment. The zircon U-Pb age is 1781±16 Ma.

2.1.2 Yanshanian

The Yanshanian period witnessed four intrusion events, including one in the Middle Jurassic, two in the Early Cretaceous, and one in the Late Cretaceous.

Basic rocks commonly seen include hornfelsed diabase, hornblende diabase, and hornfelsed sub-amphibolized diabase porphyrite. These rocks are present in Jinshan District's Zhangyan and Jieshan. The rock bodies are small in size, mostly occurring in forms of sills or dykes. Most of them are obviously hornfelsed. The diabase is obviously altered, with total rock K-Ar age of 112.7 and 106 Ma.

Intermediate-acidic rocks make up a large proportion. The intermediate rocks exist on a smaller scale, consisting of the Yuepu diorite, Tanzhi pyroxene diorite, Qixianqiao pyroxene diorite, Sanlintang quartz diorite, and Zhangyan quartz diorite. The acidic rocks exist on a larger scale, consisting of the Chongming granite, Wusong granite, and downtown granite. In the Middle Yanshanian, quartz diorite and dioritedeveloped; the magma provenances show a degree of similarity. The magma origin was relatively deep and the magma received a low degree of crustal hybridization during crystallization. The Wusong granite, as a syn-collisional granite, is distinctly different from the type I genesis of the post-orogenic Sanlin quartz diorite. The zircon U-Pb age of the Wusong granite is 178.8±2.3 Ma; that of the Sanlin quartz granite is 163.4±1.5 Ma. The Late Yanshanian intrusive rocks appear to have emerged in a post-orogenic-non-orogenic environment. The Tanzhi diabase diorite and Shenzhuang diabase diorite both show weak Eu negative anomaly, with a curve profile of mantle-derived REEs. The zircon age is 106.9-104.2 Ma. The Zhangyan quartz diorite does not show Eu anomaly. It features high Sr, low Y and low Yb, with no Eu depletion. This suggests a relatively deep magma origin and no obvious plagioclase differentiation. Around these rocks, the only skarn-type copper deposit in the area has been formed. The Chongming granite and downtown granite show a curve profile of upper crust eutectic REEs, suggesting type A granite genesis. The zircon U-Pb isotopic age is 132.9±1.1 Ma. The K-feldspar K-Ar isotopic age of the Chongming granite is 107.1 Ma. The Chongming granite aplite is strongly Eu depleted. It is the product of the magma's entry into a late-stage residual magma alkali enrichment period.

Meanwhile, this period witnessed the extensive development of porphyry and porphyrite, which mainly include hornfelsed sub-amphibolized diabase porphyrite, lamprophyre, diabase diorite porphyrite, quartz diorite porphyrite, diorite porphyrite, andesite porphyrite, and granite porphyry, felsite porphyry, and orthophyre. These rocks mainly emerged in the Late Yanshanian, mostly in forms of dykes. Granite porphyry and andesite basalt porphyrite are observed to interlude the quartz diorite bodies; felsite porphyry, granite porphyrite, andesite porphyrite, lamprophyre, diabase, and andesite basalt porphyrite are seen to interlude the granite bodies. The total rock K-Ar isotopic age of the granite porphyry and felsite porphyry is approximately 81-89 Ma.

2.1.3 Himalayan

What have been seen so far are unexceptionally some wide extensions of pyroxene dykes. Some are observed to have intruded into the Late Yanshanian quartz diorite bodies.

2.2 Volcanic rock and volcanism

Volcanic and volcanoclastic rocks are very widely present in the area, but they are rarely outcropped. By volcanism, these rocks can be dated to four periods: Early to Late Jinningian, Late Jinningian, Late Yanshanian, and Himalayan. No record of volcanism corresponding to the Middle Jurassic and Late Cretaceous intrusions has been discovered yet. The rocks exist in a wide variety, encompassing basic, intermediate, intermediate-acidic, and acidic rocks. The acidic and intermediate-acidic rocks are more widely present.

2.2.1 Early-Middle Jinningian

This is mainly found in the Jinshan Group in Zhoupu, Zhangyan, and Qianwei. There, basic, intermediate, and acidic rocks are developed. Protolith restoration has revealed that the rocks are mainly basaltic tuff, rhyolite, and possibly rhyolite porphyry. After multiple metamorphic processes, they have become regional metamorphic series or thermal metamorphic series like plagioclase amphibolite, biotite plagioclase hornfels, and feldspar quartz hornfels. The basic rocks are primarily tholeite series and secondarily calc-alkaline basalt series with slight Eu negative anomaly.

2.2.2 Late Jinningian

This is mainly a Late Qingbaikouan formation present in southwestern Shanghai's Qianwei and Chashan. Intermediate and intermediate-acidic volcanic and volcanoclastic rocks are well developed, consisting of andesitic welded tuff, breccia tuff, and tuff. It is a high-K calc-alkaline series. The zircon U-Pb ages are 774 ± 12 and 801 ± 13 Ma.

2.2.3 Late Yanshanian

Early Cretaceous volcanism took place in the active continental margin. It was the product of the transition from post-orogenic extrusion to extension. Three subcycles can be identified. The first subcycle occurred in the beginning of this period, namely, the period of the Lower Laocun Formation when acidic tuffaceous and lava erupted. The second subcycle occurred in the middle stage, which was equivalent to the period of the Huangjian Formation when volcanic activity was intense. The third subcycle occurred at the end of this period, equivalent to the period of the Upper Shouchang Formation. The volcanic activity weakened down at that time. The effusion displays a weak-strong-weak-strong rhythm. The Late Cretaceous volcanism was formed in a non-orogenic extensional environment. The basalt K-Ar age is 78.31 ± 1.77 Ma.

The Early Cretaceous volcanism gave rise to a high-K calc-alkaline-shoshonite series volcanic association andrelated volcanoclastic rocks. These rocks are the most widely spread with the most complicated rock types. The lithofacies are quite complete, mainly including effusive and fallout facies, as well as boil-over, volcanoclastic flow, volcanic eruption collapse, volcanic eruption sedimentary, volcanic neck and channel, and subvolcanic facies. The volcanic rocks mainly include basalt andesite, pyroxene andesite, andesite, hornblende andesite, quartz andesite, trachyte, dacite, and rhyolite dacite, rhyolite, and felsite. The volcanoclastic rocks mainly include transitions toward lava (rhyolite tuff lava, andesitic tuff lava, dacitic tuff lava, dacitic breccia tuff lave, and andesitic breccia tuff lava), welded tuffs (crystalline welded tuff, vitric welded tuff), tuffs (rhyolite tuff, vitric tuff, breccia-bearing tuff), and tuffite. Volcanic breccia and agglomerate are occasionally observed, too. In the Late Cretaceous, only basalt interlayers were formed at local points of fault basins.

The acidity of the volcanic-volcanoclastic rocks shows a general high-low-high fluctuation. The oxidation coefficient is the highest in the Laocun Formation and the most variable in the Huangjian Formation. The volcanic rocks in different formations are resulted from different eruption events of comagma. The original magma should have come from deep in the crust, with a portion of upper mantle contents. During the elevation of the magma, it was continuously joined by crust-derived contents. The basalt, basalt andesite, and pyroxene andesite were formed in the early cycle should be composed of contents approximating the primitive magma.

2.2.4 Himalayan

The Cenozoic recorded only basic magma effusions at local points, giving rise to a series of calc-alkaline-shoshonite basalt. The lithofacies is mainly effusive. This rock series is composed of basic lava, covering iddingitized olivine basalt, vesicular basalt, basalt, and ferruginous basalt. The basalt is the most widely present among all rocks; the ferruginous basalt is observed only in a couple of the boreholes. The rocks are mantle derived and originated in a tectonic context where the continental margin and continental crust were intensely tensioned and thinned.

In terms of the activity, two periods can be identified. The Paleocene carbonatized basalt is typically found in Beiqiao and Maqiao and held in the Beiqiao Formation. The End of Eocene to Miocene basalt exists in patchwork distribution. The total K-Ar ages include 38.57±2.05 to 31.6±1.31 Ma (Chongming's Xinhai Farm), 26.7 Ma (Nanhui Middle School), and 18.7 Ma (Bailong Port).

2.3 Volcanic structure

The volcanic eruption types of the area, especially during the Early Cretaceous, included both fissure type and central type. The volcanism was mainly controlled by NE faults. Besides, NW faults also made some contribution. The division of volcanic rock regions and volcanic eruption zones is similar to that for northwestern Zhejiang Province. Using the Fengjing-Chuansha NE fault as the dividing line, two subzones can be further identified. South of the fault belt is the Jinshan-Huinan volcanic eruption subzone; north of the fault belt is the Qingpu-Baoshan volcanic eruption subzone. Main third and fourth-order volcanic structures include the Songyin volcanic basin, Chashan cone volcano, Zhuanghang tectono-volcanic basin, Tianmashan tectono-volcanic uplift and Zhonggu-Gucun tectono-volcanic basin.

3. Metamorphic Rock and Metamorphism

The metamorphism in Shanghai mainly occurred in the Middle Jinningian, Late Jinningian, and Yanshanian period. The volcanism in the Middle and Late Jinningian mainly consisted of regional metamorphism, resulting in the Jinshan Group and Huinan slate. The Jinshan Group and Huinan slate are highly different in terms of degree of metamorphism and deformation. In the plagioclase amphibolite and mylonitized granite porphyry, the magma zirconU-Pb ages of 1078±8.9 and 1781±16 Ma have been yielded, respectively. The Jinshan Group has been subject to the superposition of multiple tectonic thermal events later on. In the Yanshanian, the metamorphism mainly included contact metamorphism, superposed by subsequent dynamic metamorphism at local points. This gave rise to tectonic breccia and mylonitized rocks.

3.1 Regional metamorphic rock and metamorphism

3.1.1 Middle Jinningian

The Middle Jinningian regional metamorphic rocks constitute the Presinian Jinshan Group, which is intermittently distributed between the Fengjing-Chuansha fault belt and the Zhangyan-Nanhui fault belt. According to the natural paragenetic combination of metamorphic rocks, from the bottom up, three lithologic members, namely members a, b, and C, can be identified. Lithological member a contains two natural paragenetic combinations: a combination of leptite, feldspar quartzite, mica schist, and greenschist, and a combination of marble, greenschist, and quartzite. Lithological member b features a combination of hornblende rocks interrupted by leptite, feldspar quartzite, quartzite, and lesser amounts of biotite schist and granulite. Lithological member c features a combination of mica schist, mica quartz schist, greenschist, marble, quartzite, and leptite.

Lithological members a and c of the Jinshan Group have clear stratified structure and distinctcharacteristics of sedimentary rocks. The primary rock is comprised of mud, sandy mud, sand, and carbonate rocks. Lithological member b does not show obvious stratified structure, but has fairly clear banded structure, schistose structure, and gneissic structure; the dip angle of the bands and schistosity is also consistent with the schistosity or stratification of the schist and marble in lithological members a and c. The rocks bear the tectonic characters of both magmatic rocks and sedimentary rocks.

The metamorphic type is low-pressure regional migmatization, which gave rise to an incremental metamorphic system of low greenschist facies and high greenschist facies. The low greenschist facies occurs in the same way as lithological members a and c; it is equivalent to a general greenschist facies (other than the epidote amphibolite facies). Commonly seen mineral paragenetic assemblages combinations include the muscovite (sericite+ chlorite+quartz, albite+actinolite+epidote+chlorite+quartz, muscovite (sericite) +plagioclase+quartz, plagioclase+epidote+chlorite+quartz, and calcite+plagioclase+epidote ± tremolite +quartz combinations. Among them, the albite+actinolite+epidote+chlorite+quartz mineral combination is the most typical. Actinolite, albite, epidote, chlorite, and muscovite (sericite) are the characteristic metamorphic minerals of this metamorphic facies. The formation probably occurred under the temperature of 350-450 ℃ and the pressure of 2-4 kPa. The distribution of the high greenschist facies is consistent with that of lithological member b. Commonly seen mineral paragenetic combinations include the common hornblende+plagioclase+epidote+quartz, common hornblende+plagioclase+diopside+epidote+quartz, common hornblende+plagioclase+almandine+biotite+quartz, common hornblende+plagioclase+biotite +epidote+quartz, plagioclase+K-feldspar+almandine+quartz, plagioclase+K-feldspar+biotite+muscovite+quartz, and plagioclase+biotite+epidote+quartz combinations. Among them, the common hornblende+plagioclase+epidote+quartz combination is the characteristic mineral combination. The formation probably occurred under the temperature of 450-570 ℃ and the pressure of 3 to 6 kPa.

3.1.2 Late Jinningian

The Late Jinningian regional metamorphic rocks constitute the Presinian Huinan slate. This rock series is observed only in Huinan Town and is overburdened by the Paleogene basalt. The metamorphic rocks revealed so far consists solely of low greenschist facies slate.

The Huinan slate is comprised of mud, fine scaly sericite, chlorite, microgranular calcite, and felsic silt. The silt particles still maintain the characteristics of angular to subangular deposition with no detectable recrystallization. The chlorite and serite are resulted from metamorphic crystallization of mud contents. Under the microscope, fairly clear blastopelitic texture and blastoaleuritic texture can be observed. From the mineral composition and relict texture of the rocks, the primary rocks should be sedimentogenic rocks containing calcium and silty mudstone.

According to the chlorite+ sericite combination, we assume that the metamorphic facies of the Huinan slate is equivalent to the lower part of the low greenschist facies. Its metamorphic type is equivalent to the slate phyllite metamorphism in the regional dynamic metamorphism proposed by Dong Shenbao. The metamorphic degree of the Huinan slate is far lower than that of Jinshan Group, hence the temperature and pressure supporting its formation should have been even lower.

3.1.3 Geotectonic environment

In the Early and the last stages of the Mesoproterozoic, the area was possibly located in a supercontinental or active continental margin of different stages. In the Late Mesoproterozoic, in particular, the area evolved into an intraoceanic arc or a backarc basin extensional environment with oceanic crust. From the last stage of the Mesoproterozoic to episode I of the Jinningian movement, the area had remained a continuously extending tectonic environment. In the early stage of the Qingbaikouan, it was a relatively stable backarc neritic to bathyal environment. From the early to the late stage, the seawater seems to have deepened; the magmatism appears to have intensified, too.

3.2 Contact metamorphic rock and metamorphism

This was mainly formed in the Late Yanshanian. Its distribution is obviously controlled by the properties of the Yanshanian rock bodies and the surrounding rocks. As the largest-sized contact metamorphic rock series known so far, it is distributed in northwestern Shanghai's Jinshan District, starting from Jieshan in the west and terminating at Qijiabang by way of Shijia and Baijiacun. The primary rock of the underlying formation is probably the Proterozoic to Lower Paleozoic series; that of the overlying formation is probably the Laocun Formation. The metamorphism should have taken place in the early stage of the Early Cretaceous at the earliest.

The main rock types include biotite plagioclase hornfels, biotite cordierite hornfels, plagioclase hornfels, diopside garnet skarn, garnet skarn, chlorite phlogopite garnet skarn, epidote garnet skarn, marble, biotite plagioclase hornfels, diopside plagioclase hornfels, and carbonatized quartz garnet skarn. The primary rock is a normal sedimentary rock series comprised of pelite, felsic clasolite, and dolomite-bearing limestone, occasionally interluded by a small fraction of basic rock dykes.

Typical mineral paragenetic combinations include the plagioclase+biotite+hornblende+quartz, plagioclase+biotite+codierite+fibrolite (a small amount) +andalusite+orthoclase+plagioclase+biotite+quartz, calcite+dolomite+garnet+diopside+wollastonite (on very few occasions, possibly idocrase), and garnet+diopside+wollastonite+epidote combinations. The metamorphism in the place of the Jieshan-Zhangyan metamorphic rocks is at least equivalent to the hornblende hornfels facies. The andalusite+orthoclase+plagioclase+biotite+quartz combination is already up to the pyroxene hornfels facies.

4. Geological Structure

Shanghai lies to the northwest of the Jiangshan-Shaoxing deep fault. Tectonically, it is in the southeastern margin of thelower Yangtze block of the Yangtze craton, in close adjacency to the Cathysian orogene. Ever since the Paleoproterozoic, the area has undergone four major tectonic evolution stages, namely, formation of crustal prototype (Paleoproterozoic), formation of the basement formation of the basement (Mesoproterozoic to Early Qingbaikouan), formation of the caprock (Early Qingbaikouan to Middle Triassic), and intracontinental transformation (since Late Triassic). The entire geological structural evolution constitutes an active-stable-active again multicycle development. Affected by the multiple tectonic movements, the geological structure is complex.

4.1 Division of tectonic units and their basic characteristics

The first-order tectonic unit of the area is the Yangtze craton; the second-order tectonic unit is the lower Yangtze block. Shanghai is consistent with the Yangtze Block both in terms of the development characteristics of the basement rock series and caprock, and the geological structural characteristics during the continental margin activity periods. Also, as the combining site between the two orders of tectonic units, the area bears distinct transitional characteristics.

The Husu Fault and Jiangshao Fault divide the area into two third-order tectonic units: the Yangtze craton and the Qinhang belt. These two units contain notably different fromations from the Prenanhuan basement period and the Late Paleozoic epicontinental sea evolution period.

Based on the NW faults and the Mesozoic volcanic and fault basin formations, two fourth-order tectonic units, namely, the Wujiang-Nantong faulted block region and the Hangzhou-Jiaxing-Huzhou faulted block region, and six fifth-order tectonic units, namely, the Nantong faulted uplift region, Jinshan-Hangzhou Bay faulted uplift region, Sheshan-Qingpu volcanic uplift region, Shanghai-Yangtze Estuary faulted uplift region, East Yangtze Estuary faulted uplift region, and Yuhang-Jiaxing faulted depression region, are identified. The gravity field and airborne magnetic characters vary from one tectonic unit to another.

Also, within the fifth-order tectonic units, the activity divergence of the faulted blocks as well as deep tectonic factors has also accounted for a series of faulted bulges and sags, and volcanic depressions and uplifts. According to their characteristics, a series of suborder tectonic units can be further identified. Among the different orders of tectonic units, the buried depth of the bedrock surface and the stratigraphic assemblage of the bedrock vary substantially.

4.2 Tectonic deformation

Multicycle, multistage tectonic movements have created tectonic features of varying sizes, varying orders, varying morphologies in the crustal rock strata, with a degree of traceability. The tectonic deformation is principally ductile or brittle faulting. The folding morphology is simple. The dip angle of the Lower Paleozoic is mostly on the order of 40-65° , with a few exceptions. In the Laocun Formation, the dip angle is mostly 20-35 °; in the Chishan Formation, it is mostly around 10 °, except at local points where the dip angle is greater. The dip angle variation is mostly controlled by the morphology of the basin margin.

From northwest to southeast, the area contains the Kunshan-Gangxi anticlinorium, Qingpu-Baoshan synclinorium, Zhujing-Zhoupu anticlinorium, and Jinshan-Datuan synclinorium. The axial direction of the complex folds is generally NE-oriented. Under the superposition and disruption of the Mesozoic volcano-tectonic basins and fault basins, the core is very discontinuous. Locally, it is also overburdened by a very thick volcanic series and red beds. Locally, small-sized anticline and syncline structures are present in the Lower Paleozoic sedimentary caprock, such as the Luodian anticline and the Xujiahui-Wujiaochang anticline.

The fault structure is complicated. By strike, four groups of faults, namely, NE, NNE, NW-NNW, and near EW faults, are identified, with the NE faults being the best developed. The NE and nearly EW faults are generally early structures, whereas the NNE and NW-NNW ones are younger ones. By cutting depth, they can be divided into deep faults, basement faults, and caprock faults. The deep faults cover lithospheric faults and crustal faults, meaning the faults cutting through the lithosphere or crust. The deep faults include the Husu fault belt, Zhangyan-Nanhui fault belt, and Jiangshao fault belt; the basement faults include the Fengjing-Chuansha fault, Taicang-Fengxian fault, Luodian-Zhoupu fault, and Qingpu-Longhua fault. The remaining faults are all caprock faults. Along the deep and basement faults, local mylonitization is quite widely present.

The Late Yanshanian fault basins mainly include the Xincun-Xinglong fault basin, Songjiang-Beiqiao fault basin, and Anlu fault basin. These basins contain well-developed Upper Cretaceous to Paleocene Series. They are distinctly in low-value regions in terms of gravity and aeromagnetic anomaly.

4.3 Tectonic stages, construction and deformation

On the basis of the four tectonic evolution stages, according to the characteristics of construction and later reconstruction, the tectonic stages can be further divided into Jinningian, Yangtzeian-Caledonian, Warrissian-Indosinian, Yanshanian and Himalayan.

Jinningian refers to the tectonic stage from Mesoproterozoic to Qingbaikouan, which is the earliest stage of large-scale crustal formation development and evolution in the area. It is divided into early-middle and late Jinningian. Early-middle Jinningianconstructionconsists of Jinshan Group; Late Jinningianconstruction is composed of Huinan Slate and Heshangzhen Group.

Yangtzeian-Caledonianrefers to the tectonic stage from Nanhuan to Silurian. There is no subdivision for this period. This stage is the main forming stage of sedimentary cover in the area, which is composed of Upper Sinian and Silurian.

Warrissian-Indosinianrefers to the tectonic stage from Devonian to Middle Triassic. No construction at this stage has been found in this area so far.

Yanshanian refers to the tectonic stage from Late Triassic to Cretaceous. Sedimentary formation of the Lower Middle Jurassic has not been foundso far, while intrusive rock formation of the Middle Jurassic is partially developed. The Early Cretaceous is characterized by volcanic structure and strong volcanic activity. The Late Cretaceous is characterized by the development of red bed deposits in faulted basins. The lower part of the deposits is coarser and the upper part is finer.

Himalayan refers to the tectonic stage since the Paleogene, which can be divided into early and late stages due to the great difference between Paleogene and Neogene. The Paleogene inherited thelate Cretaceous sedimentary pattern. Miocene-Pliocene consists of variegated clastic rocks of fluvial-lacustrine facies, with a large range of basalt spouts. Quaternary loose sediments cover the whole area and the sedimentary environment is complex.

The convergence and breakup of theColumbia supercontinent and Rodinia supercontinent is the tectonic background of the Mesoproterozoic and Neoproterozoic tectonic framework in the area. In the early to middle Jinningia, the rudiments of Husu fault, Fengjing-Chuansha fault, Jiangshao fault, Taicang-Fengxian fault and other NE- and NW- trending faults were formed. After the continuous activity and reformation from Yangtzeian to Caledonian, the main structural features were formed from Indosinian to Yanshanian. The Himalayan tectonic movement is mainly characterized by the superposition and activation of the late structures and basins on the basis of the early structures, especially the activation of the NE-, NW- and EW- trending faults.

4.4 Characteristics of neotectonic movement

The neotectonic movement markers in the area mainly include seismic activity, fault activity, vertical deformation and heterogeneous subsidence, planation surfaces, joints and striations, and geothermal activity.

The fault activity displays very strong spatial and temporal divergence. From west to east, the NW-NNW faults becomes less active or the segmentation of their activity is not obvious. The NNE faults are just the opposite. The NE faults are generally more active in the western section than in the eastern section. The activity of the NNE-NNW, NNE, and NE faults changes from strong to weak at large. Generally speaking, the EW and NE faults became active a little earlier, followed by the NNE faults; the WNW and NNW faults were active later. The latest activity of these faults generally fell in the Early-Middle Pleistocene; the latest activity of the NNW faults fell later, generally in the Middle Pleistocene, although it could continue to the Late Pleistocene for some of the faults. Two Late Pleistocene active faults and three Middle Pleistocene faults are developed. The maximum fault throw is generally 5-8 m or tens of meters. As shown in the breakpoint section, the fault throw reduces upward. It appears that the later the time, the smaller the dislocation amplitude. The Taicang-Fengxian fault boasts the greatest average displacement rate at 0.01-0.028 mm/a.

Spatially, the seismic activity is strongly clustered, repetitive, fragmented and banded. Temporally, it displays a degree of cyclicity, as evidenced by the alternation between a relative calm period and a significantly active period. The focal depth is mostly in the range of 5 to 20 km, with two dominant layers: the 5-10 km and the 15-20 km layers. The magnitude of weak earthquakes is irrelevant to the focal depth, while that of the other earthquakes is linearly relevant to the focal depth at large. From northwest to southeast, the seismic activity reduces significantly. Overall, the activity is strong in the north and east but weak in the south and west (or stronger in the sea than on the land). The overwhelming earthquakes are controlled by the Quaternary active faults, especially the Late Pleistocene faults. The focal mechanism solution indicates that the two nodal planes are usually NE or NW oriented; a small part are nearly EW oriented. The attitude of the nodal planes is moderate to nearly vertical; the dip angle is centralized at 33 to 87 °. The junction between a number of faults, the boundary faults in the Cenozoic fault basins, especially the intersection between the NNE (NE) and NW faults, and between the NE (NNE) and EW faults, mark the main earthquake-generating points. The moderately strong earthquakes have higher frequency and intensity at band intersections. Besides, the earthquakes are also closely relevant to the Cenozoic basin structures.

The Neotectonic principal compressive stress is the NEE to nearly EW principal compressive stress; the direction of the stress field is around 96°. The recent tectonic stress field falls under the NEE horizontal compressive stress and NNW horizontal tensile stress. The direction of the regional principal compressive stress is roughly perpendicular to the coastal contour line. The direction of the principal compressive stress axis is NEE to nearly EW; the dominant average azimuth angle is NE79°. In Shanghai, the direction of the maximum shear stress is approximately NE55° and NW35°.

5. Lithospheric Structure

The characteristics of geophysical fields, including velocity field, gravity field, magnetic field, resistivity field, show obvious anomalies in the lithosphere of the area with distinct layer and block textures.

5.1 Geophysical field characteristics

In terms of velocity field, the lithosphere contains eight velocity layers, namely, the unconsolidated formation (1700-2000 m/s), Paleogene to Neogene (2200-3500 m/s), Mesozoic to Silurian (3800-4250 m/s), Sinian to Ordovician carbonate strata (5.12-5.97 km/s), upper crust epimetamorphic rock series (P-wave velocity 5.80-6.20 km/s), upper middle crust hypometamorphic rock system, lower middle crust hypometamorphic rock series, lower crust (P-wave velocity 6.63-7.05 m/s), and upper lithospheric mantle (P-wave approximately 8200 m/s). Local intracrustal low velocity layers exist in the transition between the middle and lower crust.

In terms of gravity field, except the westerly part of Songjiang and Qingpu Districts and Baozhen Town, all other parts of Shanghai are positively anomalous. The area contains two gravity high-value regions, one annular gravity high-anomaly region, and one gravity negative-anomaly region. They are: the Yangchenghu-Chongming gravity high-value zone, Jiaxing-Huinan high-value zone, Kunshan-Zhenru-Luxiang high-value zone, and and Tanzhi-Songjiang gravity low-value region. Local gravity high anomalies are obviously present. These include the Miaozhen-Gangxi, Luxiang, Jiading-Luodian, western Baoshan, Zhonggu, Beixinjing, and Tanzhi-Zhuqiao. Gravity low anomalies are present, such as the Zhuanghang gravity low anomaly, Tanhang gravity low anomaly, and Songjiang gravity low anomaly.

In terms of airomagnetic field, the anomalies occur in an obviously zonal form, making up striking NE and EW anomaly zones at large. From north to south, there are three striking NEE airborne magnetic anomaly zones: the Baoshan-Chongming airomagnetic anomaly zone, Tianmashan-Guluairomagnetic anomaly zone, and Zhangyan-Huinan airomagnetic anomaly zone.

In terms of electrical field, at large, the crust-mantle is incised by low-resistivity zones in the horizontal direction and displays an overlap of low resistivity with high resistivity in the longitudinal direction. In western

Shanghai, low-resistivity layers are present at the junction between the middle and lower crust or inside the middle crust. There are four typical resistivity layers, namely, the Neogene-Quaternary, Cretaceous-Paleogene, Sinian-Lower Paleozoic, and Presinian and intrusive rocks.

In terms of geothermal field, the depth of the variable temperature zone of Shanghai ranges between 9. 0 and 17. 0 m, with an average of 13. 3 m; the bottom boundary of the constant temperature zone ranges between 17. 0 and 27. 0 m, with an average of 23. 0 m. Below this level is the increasing temperature zone. In the strata shallower than 200 m, the warming rate is 2. 31-3. 73 ℃/100 m, with an average of 3. 04 ℃/100 m. Overall, the geothermal temperature is high in the southwest but low in the northeast. The surface terrestrial heat flow varies significantly from 40. 67 to 68. 37 mW/m^2, with an average of 52. 62 mW/m^2.

5. 2 Lithospheric texture and structure

The lithosphere is defined as the Lower Yangtze strongly thinned fragmentary lithosphere. The thickness of the lithosphere varies significantly. It is 120-150 km in Qingpu, around 50-70 km in Songjiang-Fengxian, and mostly on the order of 100 km in Jinshan-Nicheng. The buried depth of the moho is around 30 km. The upper crust is 12-14 km thickness; the bottom boundary gradually deepened from west to east. The middle crust is about 10 km thick; the buried depth is approximately 22-24 km, displaying obvious stratigraphic physical or rock texture variation. The lower crust is 10-11 km thick, with a 2-3. 5 km crust-mantle transition with the lithospheric mantle. Besides the bottom of the upper crust, the bottom of the middle crust, the moho, there are other boundaries, too. These include, From the bottom up, the Indosinian erosion surface, bottom boundary of the caprock, top surface of the Jinshan Group, bottom surface of the Jinshan Group, and Lower Paleozoic weak surface (bottom of Lower Cambrian, Silurian).

6. Quaternary Geology and Geomorphology

Quaternary formations are well developed in Shanghai. These formations are mostly 200-280 m thick and are locally underlain by Neogene loose sediments. The area belongs to the eastern North China stratigraphic region of the North China stratigraphic province. It can be further divided into a Yangtze Delta sedimentary subregion (the delta main estuarine sand island sedimentary minor region and delta flank littoral plain sedimentary minor region) and Taihu Plain sedimentary subregion (eastern minor region and western minor region).

6. 1 Multiple stratigraphic division

6. 1. 1 Magnetostratigraphy and chronostratigraphy

The buried depth of the B/M boundary is approximately 130-150 m. It occurs somewhere near the disconformity boundary between the current sand bed at the bottom of the Middle Pleistocene and the underlying parti-colored hard soil at the top of the underlying Lower Pleistocene. The G/M boundary is 250-280 m deep. The interface between the Pleistocene and the Pliocene is mostly 2-5 m below the G/M boundary. The bottom boundary of the Quaternary in the area is estimated at 2. 61-2. 65 Ma. Namely, its age is approximately 2. 60 Ma. The G/G boundary is 379 m deep. Besides, Gotheborg, Blake, Jamaica, and Big lost Emperor negative polarity subtime or polarity drift events are also developed.

The Pleistocene includes three periods, namely, the Lower Pleistocene, Middle Pleistocene, and Upper Pleistocene. The Holocene also includes three periods: the Lower Holocene, Middle Holocene, and Upper Holocene. Considering contemporaneous heterotopic facies deposition, nine rock stratigraphic units are identified. These include, From the bottom up, the Lower Pleistocene Anting Formation, Middle Pleistocene Jiading Formation, Upper Pleistocene Nanhui and Chuansha/Gehu Formations, and Holocene Loutang, Shanghai and Qingpu/Rudong Formations. The minimum age is about 128 ka B. P. for the Chuansha Formation; 75 ka B. P.

for the lower Nahui/Gehu Formation, 70 ka B. P. for the middle Nanhui/Gehu Formation, 20 ka B. P. for the upper Nanhu/Gehu Formation; 10–12 ka B. P. for the Loutang Formation; 8 ka for the Shanghai Formation; and 3–4 ka for the Qingpu/Rudong Formation.

6. 1. 2 Lithostratigraphy

The Anting Formation is widespread throughout the area. It is a fluvial–flood lacustrine sedimentary system consisting mainly of particolored reticulated or taxitic clay that is mainly brownish yellow or brown in color, interluded by grayish yellow, yellowish green, or bluish gray. From the bottom up, the plasticity roughly changes from hard to hard plastic to plastic. It commonly contains calcareous nodules, stockworks and ferromanganese nodules and is alternated with light gray or grayish yellow pebbly sand and sand gravel. The clay generally consists of three sedimentary cycles with alternate coarse and fine sizes, corresponding to the lower, middle, and three "members" . The buried depth is generally in the range of 150–290 m, gradually increasing from the southwestern part's Taihu sedimentary region to the northeastern part's estuarine sand island region. Its thickness accounts for roughly a half of the Quaternary formations. The formation lies in parallel unconformity with the overlying and underlying formations.

The Jiading Formation is widespread throughout the area. It is a delta fluvial–floodplain and lacustrine sedimentary system, with lithology and lithofacies recording the transition between the Early and Late Pleistocene formations. The lithology is characterized by bluish gray or grayish green clay, brownish yellow cum bluish gray or grayish green fine reticulated or striated clay or silty clay, as well as gray silty clay and silt. The plasticity is mostly plastic except a small fraction that is hard plastic. The formation is alternated with gray or yellowish gray pebbly sand or silty fine sand. The stratigraphic development varies slightly between the east and the west. In the Taihu sedimentary region west of the shore body, the Middle Pleistocene formations are well preserved. It contains an upper and a lower formation comprised of two sedimentary cycles with alternate coarse and fine sizes. The buried depth is usually 95–150 m; the layer thickness is 50–60 m. In the littoral plain and estuarine sand island regions east of the shore body, due to the undercutting by the bottom boundary of the Upper Pleistocene formations, the upper formation of the Middle Pleistocene have been exposed as a result of erosion. Hence only the lower member has been preserved in most places. The buried depth is about 120–160 m; the thickness is 25–35 m.

The Chuansha Formation is widespread throughout Shanghai. It is a fluvial–weak marine–continental delta fluvial fine sand and pebbly sand system. On the estuarine sand islands in the northeastern part's Chongming, Changxing, and Hengsha, the sediments are the coarsest. In the junction among Qingpu, Jinshan, and Songjiang that borders Zhejiang, this formation consists of firth accumulations of gray or brownish gray silty clay, alternated with light gray silt or sandy silt. From southeast to northeast, the buried depth gradually increases. In the western part of the land area, the buried depth is generally between 65 and 95 m; from the eastern part of the land area to the Chongming Island, the buried depth is 75–90 m for the top boundary and 100–130 m for the bottom boundary.

The Nahui/Gehu Formation is a continental–marine/alternate marine and continental–continental sedimentary sequence formed during the last glacial period. Accordingly, it contains a lower, a middle, and an upper member. In the lower member, the buried depth roughly increases from southwest to northeast; the buried depth of the bottom boundary is 52–90 m; the thickness is 0–17 m. The lithology, lithofacies, and thickness are significantly variable, but the typical lithology is mainly represented by the floodplain–lacustrine grayish green, grayish black or dark gray and locally gray clay, silty clay and clayey silt, interrupted by sandy silt or silt layers. The formation contains calcareous nodules and rotten wood pieces, as well as fossilized continental mollusk debris at local points. In the middle member, the thickness accounts for the greater part of the upper formation of the Upper Pleistocene. The buried depth gradually increases from southwest to northeast, ranging from 41–90 m for the bottom boundary to around 10–60 m for the top boundary. The thickness varies from 13 to 61 m. In the northern part of the current littoral plain region, this formation consists mainly of neritic–estuarine gray silty clay frequently interrupted by or thinly alternated with light gray fine sand or sandy silt lamina. In the southern part of the current littoral plain region, it consists of a littoral–neritic gray or grayish yellow silty fine sand and sandy silt alternations interrupted by silty clay. In Qingpu District and the bordering part of Songjiang and Jinshan Districts, the formation is represented by alternate littoral–lagoonal and riverside beach or flood-plain–lacustrine dark gray, brownish yellow and particolored clay or silty clay, sometimes interrupted by or alternated with yellow silt or sandy silt. In the estuaries, the lower part of this formation is usually comprised of gray or yellowish gray medium fine sand, silty fine sand interrupted by or alternated with brownish gray silty clay or silt, probably with sand gravels at the bottom. The upper part is mostly comprised of steel gray or gray silty fine sand interrupted by or alternated with brownish gray silty clay or silt. This part is commonly characterized by rich contents of fossilized marine organisms. In the upper member, the lower part is normally comprised of ginger yellow or brownish yellow silty clay locally interrupted by sandy silt or silty fine sand; the upper part is comprised of dark green, grayish green or yellowish green clay and silty clay. Fossilized freshwater organisms such as fish bone fragments and fertilized stonewort egg membranes can be observed. Normally no fossilized marine organisms such as foraminifers are contained, but a couple of fossilized marine organisms have been successively discovered in boreholes in southern Shanghai's Nanhuizui, downtown, and Qingpu District's Liantang.

The Loutang Formation is a marine–continental transitional accumulation series that is thin in the southwest and thick in the northeast. Its thickness becomes distinctly greater toward the current Yangtze estuary region. In the Taihu sedimentary region, it is mostly 3–6 m thick; in the Yangtze Delta sedimentary region, its thickness varies from 4 to 32 m; in the estuarine sand island subregion, its thickness is commonly 30 m. Where the stratigraphic section is complete, this formation can be divided into an upper and a lower member. In the Yangtze Delta sedimentary region, it is mostly estuarine facies; in the western Chongming Island, it consists of liman accumulations. The lithology is often represented by gray sandy silt or clayey silt alternated with grayish white silt; in some places, such as the boreholes in Waigaoqiao, the lithology is grayish sandy silt alternated with silty clay. The formation commonly contains varying amounts of euryhaline*Foraminfera*, small amounts of planktonic and stenohaline species, as well as small amounts of semi–saline and marine *Ostracoda*; continental *Ostracoda* is occasionally observed. In the Taihu sedimentary region, such as in the boreholes in Jinshan District's Zhujing and Qingpu District's Zhaoxiang, the lithology is represented by limnetic accumulations of grayish black or grayish brown silty clay, with occasional presence of semi–saline Foraminifera. The formation often contains quite a large amount of plant debris, with peat or carbonaceous clay layers at local points. The upper member is commonly occupied by littoral accumulations, mostly consisting of dark gray or brownish gray (mucky) clay or silty clay. The characteristic indicator is the presence of large amounts of plant stems and leaves and yellow argillo–calcareous nodules, with peat or carbonaceous clay layers at local points. A small amount of euryhaline *Foraminfera*, and a very small amount of semi–saline *Foraminfera* are present. The formation lies in disconformity with the underlying Nanhui Formation.

The Shanghai Formation is a littoral–neritic sedimentary formation. In the Taihu sedimentary region, its top and bottom boundaries are normally 2.5–15 m deep; its thickness is 7–12 m. In the Yangtze Delta sedimentary region, including the "shorebody belt", its top and bottom boundaries are normally 5–25 m deep; its thickness is mostly 10 – 18 m. In the Yangtze Delta sedimentary region, this formation is a shoreland neritic (lower part) –estuarine littoral (upper part) accumulation series. Where the stratigraphic section is complete, it contains an upper and a lower member. The lower member consists of shoreland neritic accumulations of dark gray mucky clay and mucky silty clay. The upper member consists of littoral accumulations of gray sandy silt or clayey silt interrupted by silt, with its upper part often occupied by gray sandy silt alternated with silty clay or by clay or silty clay interrupted by sandy silt and silt. In the Taihu sedimentary region, this formation is a lagoonal–coastal marsh accumulation series, normally consisting of relatively monotonous dark gray mucky clay, clay, and mucky silty clay. The formation contains rich *Foraminifera*, *Ostracoda*, and *Sporopollen*. In the "shore body belt", the lithology and lithofacies of this formation shows the characteristics of a transition between the eastern and western areas. North of Meilong, the formation is mainly represented by dark gray mucky clay and mucky silty clay. The middle and upper parts are mostly interrupted by silt lamina, with shell debris frequently observed and shell sand layers occasionally detected; the lower part often contains plant debris and yellow argillo–calcareous nodules. South of Meilong, the formation is mainly represented by gray silty clay interrupted by silt layers containing shell debris. They appear to be highly hydrodynamic tidal flat accumulations. The formation lies in conformity with the underlying Loutang Formation.

The Qingpu Formation is found in the Taihu sedimentary region. The thickness of this formation is generally 2–4 m, showing an approximate east to west increase. That is, it increases from the margin toward the inside of the lake basin. Two different types of lithology and lithofacies have been detected. The primary type is represented by the limnetic accumulations of peat–interrupted dark gray or brownish gray mucky clay or silty clay present in the southwestern part of Qingpu District and in the area where Songjiang and Jinshan Districts are bordered. The secondary type is represented by the lacustrine accumulations of brownish gray or brownish yellow silty clay interrupted by silt, as well as peat or carbonaceous clay interlayers at local points, from east of the above–mentioned areas to the "shore body belt". The formation normally contains fossilized continental organisms such as freshwater mollusk, *Ostracoda*, and stonewort. *Foraminifera* is occasionally observed in its lower part.

The Rudong Formation is found in the littoral plain region and estuarine sand island region. The primary type of lithology and lithofacies is represented by the firth–shoreland tidal flat accumulations of gray or grayish yellow silt or clayey silt interrupted by silt. The accumulation thickness is 5–8 m and can be greater than 10 m at local points. The secondary type is represented by the delta tributary channel accumulations of steel gray silt and sandy silt. The accumulation thickness spans from 2 to10 m and is thin in the west but weak in the east. The formation contains relatively rich euryhaline *Foraminfera* and small amounts of planktonic *Foraminfera* and semi–saline *Ostracoda*. Individual stenohaline *Foraminifera* and marine and continental *Ostracoda* fossils have also been observed.

6.1.3 Biostratigraphy

The Quaternary animals mainly include freshwater species, semi–saline species, euryhaline shoreland species, and offshore neritic species and planktons requiring a high level of salinity. The marine fauna is mainly distributed in strata shallower than 100 m. In the 100–180 m interval, marine fossils are sporadically distributed. Roughly seven zones can be identified. Marine fossil zone Ⅰ lies in the middle member of the Anting Formation, at a level below 200 m. This zone contains a small amount of basically euryhaline Foraminifera and marine Gastropoda fossils. Its distribution is limited only to local points in the current along–the–river area of Pudong District. Marine fossil zone Ⅱ is often found in the lower part of the upper member of the Lower Pleistocene Anting Formation. In the upper part of the upper member of the Anting Formation, freshwater mollusk fossils are frequently observed, together along quite an abundance of continental Ostracoda. Marine fossils have also been detected in a few boreholes. Marine fossil zone Ⅲ is locally found in the gray fine sand layers in the lower part of the lower member of the current Middle Pleistocene Jiading Formation. Marine fossil zone Ⅳ is mainly present in the lower part of the upper member of the Middle Pleistocene Jiading Formation in the Taihu accumulation plain region. Marine fossil zone V is distributed in the Upper Pleistocene Chuansha Formation. Except the three Chongming islands, Baoshan, and Jiading districts, it is present in almost all other parts of Shanghai. Marine fossil zone Ⅵ lies in the middle member of the Upper Pleistocene Nanhui Formation and it is widely present throughout the area. Marine fossil zone Ⅶ is distributed in the Holocene formations and it is present throughout the area. From the bottom up, the fossil assemblage reveals a marine–continental–marine–marine–continental succession, corresponding to the Loutang, Shanghai, and Rudong Formations.

The *Sporopollen* type includes both subtropical and temperate components. Affected by the local sedimentary environment, littoral and fluvial/lacustrine herbaceous pollen, freshwater algae, and marine *Dinoflagellate* are often present. From the Pliocene upward, the following *Sporopollen* assemblage zones can be roughly identified: the *Quercus deciduous–Quercus evergreen–Cyperaceae–Pteris* zone, *Pinus–Picea* zone, *Abies–Quercus deciduous–Artimisia* zone, *Quercus deciduous–Pinus–Quercus deciduous–Pinus–Gramineae* zone, *Pinus–Picea*, *Abies–Quercus deciduous–Gramineae–Pteris* zone, *Quercus deciduous–Pinus–Typha–Polypodiaceae* zone, *Pinus–Quercus deciduous–Gramineae–Polypodiaceae* zone, *Quercus deciduous–Pinus–Gramineae–Polypodiaceae* zone, *Pinus–Quercus deciduous–Betula–Gramineae–Polypodiaceae* zone, *Quercus deciduous–Pinus–Liquidambar–Gramineae–Polypodiaceae* zone, *Quercus deciduous–Pinus–Gramineae–Pteris*zone, *Pinus–Quercus deciduous–Artimisia–Polypodiaceae* zone, *Quercus deciduous–Pinus–Artimisia–Adiantum* zone, *Pinus–Quercus deciduous–Typha–Pteris* zone, *Quercus deciduous–Pinus–Artimisia–Pteris* zone, *Artimisia–Gramineae–Pinus–Abies–Bryophyte–Sporopollen* zone, *Pinus–Quercus deciduous–Gramineae–Pteris* zone, *Quercus deciduous–Pinus–Artimisia–Pteris* zone, *Quercus deciduous–Quercus evergreen–Pinus–Gramineae–Pteris* zone, *Pinus–Quercus deciduous–Artimisia–Pteris* zone, and *Quercus deciduous–Pinus–Gramineae–Pteris* zone.

6.1.4 Minerostratigraphy

There are more than 30 kinds of minerals. Higher–content heavy minerals include hornblende, epidote, magnetite, ilmenite, limonite, hematite, garnet, monazite, and zircon. Lower–content ones include rutile, leucoxene, and tourmaline. Rarely seen ones include sphene, anatase, pyrite, apatite, barite, siderite, and pyrolusite. Occasionally present ones include pyroxene, serpentine, tremolite, grenatite, sillimanite, kyanite, corundum, asbestos, as well as chalcopyrite, galena, sphalerite, native lead, native copper, native gold, scheelite, and cinnabarite. Within one single sand sample, not more than 10 to 20 kinds of heavy minerals are contained. Fewer kinds of heavy minerals are contained in formations above the Middle Pleistocene than in formations below. Among the sand samples analyzed, heavy minerals generally account for approximately 1% by weight. Their proportion is higher in and above the upper part of the Lower Pleistocene but reduces sharply in formations below. Three mineral zones can be roughly identified: the hornblende–epidot–magnetite/ilmenite–limonite/hematite–monazite zone, epidote–hornblende–limonite/hematite–magnetite/ilmenite–garnet/zircon zone, and limonite/hematite–epidote–zircon–magnetite/ilmenite zone.

Clay minerals primarily include illite, smectite, and secondarily kaoline and chlorite. The upper part is equivalent to the Holocene to Middle Pleistocene; it is an illite (hydromica) zone. The middle part is equivalent to the Lower Pleistocene; it is an illite–smectite zone. The lower part is basically equivalent to the Pliocene but slightly spans the bottom part of the Lower Pleistocene; it is a smectite–illite zone.

6.1.5 Climatostratigraphy

Overall, the Pliocene climate was warm–humid. The Quaternary climate underwent five distinct cold–warm alternations.

The Pliocene Chongming Formation sediments record an overall warm–humid climate. The lower part was hot and humid, interrupted by a cold and dry period designated as the first cold period. The upper part was hot–

humid to warm-humid, designated as the first warm period. The Pliocene climate is recorded in the bedrock depressions and is frequently revealed in the Pliocene sediments in the boreholes.

Upon entry into the Early Pleistocene, the climate became obviously cold but there were fluctuations. There were three cold periods, alternated by two warm periods in between. From the early to the late stage, the climate periods are designated as the second cold period, second warm period, third cold period, third warm period, and fourth cold period.

The Middle Pleistocene paleoclimate underwent a warm period and a cold period, designated as the fourth warm period and fifth cold period. The fifth cold period had a history of warn, cool and fairly hot in the early stage, hot and humid in the middle stage, and cold and dry in the late stage.

The period 12.8-7 ka marked a warm period designated as the fifth warm period, which corresponds to MIS5. The period 7-1.1 ka marked a cold period designated as the sixth cold period. During this time, the climate was generally cool and fairly dry. It corresponds to MIS3. During this time, there was an interwarm period approximately from 5 to 3 ka.

The Holocene was a post-glacial period when the climate was warm-humid. The climate fluctuated from cool and slightly humid to warm and fairly dry to hot and humid to cool and slightly dry to warm and humid. These fluctuations correspond to the Preboreal, Boreal, Atlantic, Subboreal, and Subatlantic periods.

6.2 Genetic type and distribution pattern of sediments

The main genetic type combinations of sediments include fluvial sediments (proluvial, alluvial), limnetic sediments, martine-continental sediments (firth, deltaic), marine sediments, and artificial causes.

The proluvial sediments were deposited in the piedmont lowlands of bedrock denudation regions approximately 2.58-1 Ma. They appear in the lower and middle members of the Anting Formation.

Alluvial sediments are the most widespread sediments in the Quaternary of the area with the highest proportion, with presence in almost all periods. They appear in the Anting, Jiading, Chuansha, Nanhui, and Rudong Formations. The Middle Pleistocene accumulations are the most widely distributed.

Limnetic accumulations mainly occurred in the early stage of the Early Pleistocene and the late stage of the Holocene. In the early stage of the Early Pleistocene, the northern part of the costal plains and the estuarine sand islands were commonly host to littoral and shallow lacustrine sediments, which are mainly comprised of particolored clay. In the late stage of the Holocene, limnetic accumulations commonly occurred in the Taihu sedimentary region.

Complete firth accumulations were mainly formed during the rapid transgression in the early stage of the Holocene (c. 11-8 ka). The Loutang Formation sediments are all estuarine accumulations. Besides, in the Early Pleistocene Anting Formation, Middle Pleistocene Jiading Formation, the Late Pleistocene Chuansha Formation, estuarine accumulations resulted from weak transgress have frequently been detected.

Deltaic accumulations developed following the greatest sea flooding in the middle stage of the Holocene, accumulating upon the firth accumulations. They are distributed in the current estuarine sand islands and littoral plains.

Littoral-neritic accumulations mainly developed in the middle member of the middle and late Late Pleistocene Nanhui Formation and Holocene Shanghai Formation. The lower part of the middle member of the Nanhui Formation is distributed throughout the area. The upper member of this formation is found in the southeastern part of the littoral plain's Pudong District. The Shanghai Formation is mainly distributed in the littoral plain region.

Artificial fills are distributed at the superficial level, standing 1-2 m thick. As a product of human engineering activities, they include cultivated plants, miscellaneous fills, plain fills, underground silts, and dredger fills.

6. 3 Type and spatio-temporal distribution of sediments

The main types of sediments include clay, silty clay, clayey silt, silt, sandy silt, silty sand, medium fine sand, pebbly sand, and gravelly sand. The clay and silty clay is typically found in the Neogene and Lower Pleistocene. The silty clay is typically found in the middle and lower parts of the Holocene and the upper part of the Upper Pleistocene; some is also seen in the Middle and Lower Pleistocene and the Neogene. The clayey silt is typically found in the lower part of the Holocene and the upper part of the Upper Pleistocene; some is also seen in the Middle and Lower Pleistocene and the Neogene. The silt is mainly distributed in the Holocene. The sandy silt is mainly distributed in the upper part of the Upper Pleistocene; some is also seen in the Lower Pleistocene. The silty sand is mainly distributed in the lower part of the Upper Pleistocene, but some is also seen in the Holocene and Neogene. The sand is mainly distributed in the lower part of the Upper Pleistocene; some is also seen in the Lower Pleistocene and the Neogene.

In an alluvial fan or braided river sedimentary environment, the sediments are poorly sorted, pebbly muddy medium-coarse sand or muddy gravel interrupted by clay beds, producing alternations of coarse and fine layers. In a meandering river or seawater-affected fluvial environment, the sediments are characterized by fluvial binary structure composed of fine-grained channel sediments and fine-grained floodplain sediments. In a coastal environment, the bottom part consists of grayish yellow pebbly medium-fine sand; the middle part consists mainly of silt or sandy silt; the top part consists of clayey silt. In a deltaic environment, the middle and lower members consist of clayey silt; the upper member consists of silt, producing a succession that is fine at the bottom and coarse at the top with slightly increasing sortability.

6. 4 Quaternary geomorphology

The land area of Shanghai can be further divided into four third-order geomorphologic units. They are the Taihu accumulation plain region, Yangtze Delta accumulation plain region, Hangzhou Bay littoral plain region, and the denudation monadnock and island region. The offshore waters of the city can be identified as one third-order geomorphologic unit, namely, the Yangtze Delta subaqueous delta. By sedimentology, this unit can be further divided into two fourth-order geomorphologic units: the delta front region and the prodelta region. The Taihu accumulation plain region and the Yangtze Delta accumulation region are the most principal geomorphologic types in the land area of Shanghai. These two types are separated by the Jiangsu's Taicang-Shanghai's Waigang-Xujing-Wuqiao-Caojing line. That is, the western side of the shore body belt composed of the Yangtze Delta plain shell sandbanks serve as the dividing line between them.

The Taihu accumulation plain region is subdivided into three fourth-order geomorphologic units: the lagoonal plains, limnetic low plains, and palustrine plains. Microgeomorphologic units can be identified according to the forming process of the geomorphology. In this case, the estuarine sand islands can be subdivided into the early, middle, and late stage estuarine sandbars. Accordingly, the littoral plains can be subdivided into the early, middle, and late stage littoral plains and tidal flats.

Besides the main geomorphologic types above, approximately 4-5 rows of paleoshell sandbanks are scattered at the angular boundaries of the littoral plains inside the Taihu accumulation plain and deltaic accumulation plains and in the littoral plains; 1-2 rows of paleoshell sandbars are scattered in the eat of the Chongming Island. A few rows of artificial seawalls are distributed in the riverside and seaside areas of Shanghai.

6. 5 Quaternary transgression and regression events

During the Quaternary, there have been seven transgression events in Shanghai, which can be correlated to the Quaternary transgressions in the Yangtze Delta.

Rugao Transgression—Transgression layer Ⅶ: This is discovered in the shoreland area of Pudong District. The buried depth of the transgression layer is less than 200 m. It lies in the middle member of the Anting Formation, containing small amounts of fossilized euryhaline *Foraminifera* and marine *Gastropoda*. The forming age was approximately 2.00–1.85 Ma. This was the earliest Quaternary transgression in the area. The transgression intensity suggests an estuarine-deltaic fluvial environment (predominantly seawater-affected rivers).

Zhoupu Transgression—Transgression layer Ⅵ: The transgression started from Nanhuizui. Its northern end stretched along the Liuhe-Nanhui paleochannel to near Pudong District's Haitang-Hengmian and Zhoupu. The transgression layer lies in the upper member of the Lower Pleistocene Anting Formation. Stenohaline *Foraminifera* and calcareous ultramicrofossils have been discovered. The forming age is typically around 1.20–1.00 Ma. The transgression intensity suggests an estuarine-deltaic fluvial environment (predominantly seawater-affected rivers).

Shanghai transgression—Transgression layer Ⅴ: This has been widely discovered in the downtown, Pudong, Jiading, and the southwestern part's Qingpu and Jinshan Districts of Shanghai. The transgression layer lies in the lower submember of the lower member of the Middle Pleistocene Jiading Formation, normally hosted in the gray fine sand beds. In the east, contains a small amount of semi-saline *Foraminifera* is often seen, sometimes associated with continental *Ostracoda*, calcareous ultramicrofossils; in the west, freshwater *Ostracoda* is generally detected, with only individual *Foraminifera* fossils along with cold water species. The forming age was approximately 0.78–0.65 Ma. The transgression intensity suggests a deltaic plain fluvial/lacustrine and short-term gulf environment occasionally affected by strong tide currents.

Jiading Transgression—Transgression layer Ⅳ: This is mainly preserved in the Taihu sedimentary region. The transgression layer lies in the lower part of the upper member of the Middle Pleistocene Jiading Formation; in the shoreland area in the east of the littoral plains or in the estuarine sand islands, the whereabouts of this layer is unclear as it is mostly eroded. semi-saline *Foraminifera* has been detected, sometimes associated with continental *Ostracoda*, calcareous ultramicrofossils, and fossilized debris of freshwater mollusks like *Lamprotula*. The forming age was approximately 0.26–0.20 Ma. The transgression intensity suggests a deltaic plain fluvial environment.

Taihu Transgression—Transgression layer Ⅲ: Except the three Chongming islands, Baoshan and Jiading, this is found almost in all other parts of Shanghai. The transgression layer lies in the Upper Pleistocene Chuansha Formation. The forming age was approximately 0.07–0.128 Ma. The transgression intensity suggests a deltaic fluvial environment intermittently affected by seawater or, occasionally at local points, a delta front estuarine environment. Despite its wide presence in Shanghai, like the transgression layers described above, it is still a transgression layer with relatively weak marine property.

Gehu Transgression—Transgress layer Ⅱ: As the strongest transgression among all others, it is ubiquitous throughout the area. The transgression layer lies in the middle member of the Upper Pleistocene Nahui Formation. The transgression intensity suggests two types of sedimentary environment. One is a littoral, lagoonal, and sometimes continental, fluvial environment with relatively weak marine property in the western part's Taihu sedimentary region and the eastern part's Yangtze Delta estuarine sand island sedimentary subregion; the other is the neritic-esturine environment with relatively strong marine property in the eastern part's Yangtze Delta littoral plain sedimentary region. The forming age was principally 20–55 ka.

Zhenjiang Transgression—Transgression layer Ⅰ: This is widespread throughout the area except the southwestern part of Qingpu District near the Sheshan and Tianmashan Mountains, and the junction between Songjiang and Jinshan Districts. The forming age began in 11 ka. The period 8–5 ka marked the heyday of this event. The maximum flooding surface was formed during the regional maximum transgression dates back to approximately 6. 5 ka, which was the period when the first shell sandbank seen in Shanghai that divides between the Yangtze Delta accumulation plains and the Taihu accumulation plains – the Zhugang shorebank. Within Shanghai, the transgression gave rise to a marine – continental – marine – marine – continental stratigraphic sequence, making up a complete transgressive cycle. The resulted formations are, From the bottom up, the Loutang, Shanghai, and Rudong Formations. The transgression layer is the best developed in the estuarine sand island sedimentary subregion; westward, from the littoral plain sedimentary subregion to the Taihu sedimentary region, it gradually thins up and finally pinches out.

The coverage of the transgressions varies slightly from one period to another. The two Early Pleistocene transgressions were weak; traces of the Rugao Transgression have been discovered only at local points of the current coastal area of Pudong District. The Zhoupu Transgression was marginally stronger; its traces have been discovered not only at a couple of points in Pudong, but also in Qingpu and Jinshan. The coverage of the Shanghai Transgression was obviously larger than the two Early Pleistocene transgressions, but is still limited to local points; besides Pudong and Jinshan's Fengjing, its traces have also been spotted in the northern part's Jiading. The marine property of the Jiading Transgression was again obviously higher; the marine strata thickness was quite thick, too, but it has been discovered only in the current limnetic plains. Transgression has been universal since the Late Pleistocene. In the Holocene, the transgression reached the interface between limnetic plains and littoral plains at the largest, giving rise to paleoshell sandbanks. The transgression intensity was the strongest in the estuarine sand islands, followed by the littoral plains; the limnetic plains was only weakly affected.

6. 6 Lithofacies paleogeographic evolution

6. 6. 1 Early Pleistocene

In the early stage of the Early Pleistocene, retrogressive erosion and river undercutting commonly occurred in the area. The scope of the low mountains and hills in the southwest had been contracted to within Qingpu, Songjiang, Jinshan, Minhang, and Fengxian Districts; the mountains had been lowered, too. The eastern and northern land area's Jiading, Baoshan, downtown, and Pudong were principally a stretch of lake basins. In the lakeside areas, namely, the southwestern downtown's Dachang–Xiaozhazhen–Sanlin–Tanzhi and Sanzao, there was a string of denudation low hills. In the low mountain areas in the southeast of Pudong and near Jinshan's Fengjing emerged two relatively large intermontane alluvial fans and superfan lacustrine depressions. In Chongming District, the area was basically widely covered by interfluvbeachland, mainly consisting of an accumulation series of floodplain–flood lacustrine sediments. Meandering channels were existent only on the periphery of the current Chongming Island, primarily near Qianshao Farm – Xijia Port and Houjia Town – Xinhe Town south of Rudong, and near Hongxin and Fumin Farms north of Rudong. The Changxing and Hengsha Islands commonly underwent a meandering channel – nearshore – floodplain evolution. Along the eastern Gonglu Town – Haitang–Pudong Airport–Zhuqiao East Fishing Ground–Laogang and Yinan in the shoreland of Pudong, a "Pudong–Nanhui paleochannel" appeared. North of Zhuqiao, it was a braided river sedimentary environment characterized by repeated branching and recombination of channels and frequent exposure of interfluvbeachland. South of Zhuqiao, the area had evolved into a relatively stable channel environment. The groundwater contained in the formations from this period is the narrower–sense aquifer V of the area.

In the middle stage of the Early Pleistocene, the two valleys in Jinshan's Fengjing and to the east of Minhang's Maqiao and Beiqiao were almost interconnected by retrogressive erosion. As a result, the originally contiguous low mountains were separated into two tracts of low mountains, one in the Qing–Songjiang bordering area, the other in the Jinshan – Fengxian bordering area. The mountains were further lowered. The row of arc – shaped low hills originally existing in Dachang–Sanlin–Tanzhi and Sanzao had been mostly inundated, leaving only a few low hills in Tanzhi, Sanzao, and the western downtown' s Hongqiao. The Liuhe–Nanhui paleochannel to its east, as a southeastern tributary of the Yangtze River, began to evolve into a deltaic fluvial sedimentary band obviously affected by seawater, with swing width of 7 or 8 km. This resulted in the accumulation of light gray or grayish yellow medium–fine sand and riverbed facies pebbly sand, making up the main host formation to the lower aquifer Ⅳ in the land area. To its southwest, inside the ravine zone held between the southern and northern tracts of low mountains and hills, there appeared a tributary channel that originated from Songjiang and Minyhang/Beiqiao, flowed past Xiasha and Datuan and finally poured in the Liuhe–Nanhui mainstream at the Luchao Port. The bealock between the two tracts of low mountains and the valleys to the east and south were covered by proluvial fans/skirts and superfan lakes composed of muddy sand gravel and pebbly clay infills. In the valley to its west appeared an interfan river that flowed to Zhejiang by way of Fengjing. To the north of the northern part' s low mountain area, a tributary may also have appeared, which flew to Baoshan' s Liuhang by way of Jiading' s Huangdu and Malu and finally poured into Zhejiang. This gave rise to accumulation of fine sand and pebbly sand interrupted by silt and silty clay. The vast area between the Liuhe–Nanhui paleochannel and the low mountains in the southwest was comprised of fluvial nearshores and floodplains, resulting in the accumulation of varying sized sand and clayey soil alternated with silty fine sand. On the estuarine sand islands also appeared a paleochannel that entered from near the northwestern corner' s Hongxing Farm and flew to the Hengsha Island by way of Miaozhen and Xinhe Town. This channel seems to have deepened and thickened toward the lower reaches, resulting in the accumulation of a series of pebbly medium–coarse sand and sand gravel alternated with medium–fine sand. In the greater part of the Chongming Island and in the Changxing Island, the area was comprised of nearshores or central bars and floodplains, where a series of fine sand interrupted by pebbly sand and clayey soil alternated with silty fine sand was deposited.

In the late stage of the Early Pleistocene, the two tracts of low mountains in the southwestern land area had become hills scattered on the alluvial plains. The Liuhe–Nanhui paleochannel swung westward to the Jiading–Baoshan' s Yuepu–downtown' s Zhenru–Zhoupu–Luchao Port line, with swing width of 10–15 km. This resulted in the accumulation of a series of gravelly sand and pebbly medium–coarse sand alternated with medium–fine sand. In Pudong District, affected by an earlier transgression in the area appeared a series of liman fluvial sediments, sometimes with weak marine–continental transition, which became transgression layer Ⅵ of Shanghai. In the southern land area, the Fengjing–Fengcheng paleochannel emerged and joined with the Liuhe–Nanhui paleochannel somewhere nearby Luchao Port. The lower part was occupied by riverbed–sandbar pebbly medium–coarse sand and pebbly fine–medium and alternated with fine sand; the upper part was occupied by floodplain clayey soil interrupted by silty fine sand beds. The sides of the two paleochannels were lined up with nearshore sediments of medium–fine sand, silty fine sand and silty clay; near areas surrounding low mountains and hills, the area was covered by large tracts of floodplain–lacustrine depressions composed of clayey soil interrupted by silty fine sand; the current three Chongming Islands was also widely covered by river nearshores and floodplain–lacustrine depressions, which were mostly occupied by silty fine sand interrupted by silt, and silt alternated with silty fine sand.

6.6.2 Middle Pleistocene

The beginning of the Middle Pleistocene marked the entry into a warm period for Shanghai. In the early stage, affected by transgression, transgression resulted in universal, widely distributed deltaic plain fluvial sediments in the rivers widespread throughout the area. Toward the late stage of the Middle Pleistocene, as a result of a considerable periodic replacement cycle of the paleoclimate in the area, the widespread river channels were mostly transgressed by seawater, giving rise to deltaic plain fluvial accumulations. In the eastern downtown, the greater part of Pudong and Nanhui, and the three estuarine islands, the strong river undercutting in the subsequent early stage of the Late Pleistocene caused the formations of the late stage of the Middle Pleistocene and the associated water-bearing sand beds to become incomplete as a result of erosion. In most of the other areas, this formation has remained relatively intact. It consists mainly of channel and nearshore accumulations of gray to yellowish gray sand or pebbly sand interrupted by clay. The buried depth is approximately 100-110 m; the thickness is on the order of 10 m.

6.6.3 Late Pleistocene

In the early stage of the Late Pleistocene, as the paleoclimate became warm again, the area entered a period of intense subsidence, a marine-continental sedimentary environment began to emerge extensively. The weakly marine-continental sedimentary deltaic fluvial sand beds emerged during this period made up the relatively thick Chuansha Formation that is widespread in the middle to deep level of the area. In the late stage of the Late Pleistocene, during 75-70 ka B. P., when Shanghai was under a warm-cool climate similar to that in the current warm temperate zone, the area evolved into a floodplain-flood lacustrine environment. Dark green, grayish green, partly dark gray, light bluish gray or grayish yellow hard clay interrupted by silty lamina was deposited, making up the lower member of the Nanhui or Gehu Formation. During 70-20 ka B. P., the climate changed to warm and humid, the sea level went back up steeply; the transgression widely covered the Yangtze Delta and Taihu. The middle member of the marine or alternate marine-continental Nanhu or Gehu Formation was deposited, making up a formation of gray clay alternated with silt boasting the greatest marine property in the area. During 23-20 ka B. P., falling sea level caused the seawater to gradually regress toward the East China Sea shelf. The original shallow seas and bays in Shanghai evolved into a littoral-deltaic plain fluvial environment. This resulted in the accumulation of the brownish or rust yellow silt and fine sand in the upper part of the middle Nanhui Formation. During 20-11 ka B. P., the paleoclimate of the area changed to cold and dry; the sea level dropped sharply. The floodplains-lakes originally limited to northwestern Shanghai had expanded to the entire land area of Shanghai except only for the current estuarine area. This resulted in the accumulation of the brownish yellow silty clay and characteristic dark green clay beds at the top of the Upper Pleistocene, making up the upper member of the Nanhui or Gehu Formation. During 15-11 ka B. P., as the sea level continued to fall, the brownish yellow-dark yellow clayey soil was gradually dehydrated into land, undergoing a weathering-pedogenic process.

6.6.4 Holocene

In the early stage of the Holocene, during 10.3-9.5 ka B. P., the sea level went up steeply. In the beginning, the seawater transgressed along the current estuarine area-Pudong's shoreland area and the deeply cut tributary valleys in the original land area, resulting in firth and liman accumulations. The sediments are mainly gray to grayish white clayey silt or silt alternated with fine sand, making up the lower member of the Loutang Formation which is 10-20 m thick. During 9.5-8.0 ka B. P., the sea level rose at a lower rate. The transgressed seawater submerged Jinshan, Songjiang, Minhang, Baoshan and the area to its east, turning the vast land area of today's Shanghai into littoral swamps and tidal flats. This resulted in the accumulation of grayish black clay containing rich reed rootstock and yellow argilla-calcareous nodules, and white silt. In the current three Chongming islands and Pudong District, the area was an firth and tidal flat environment. There, dark gray silty clay interrupted by grayish white silt was accumulated, making up the upper member of the Loutang Formation that is mostly 5-10 m thick.

In the middle stage of the Holocene, during 8.0-5.0 ka B. P., the sea level rose significantly again, marking the heyday of the Holocene transgression, which extended straight through to Jiangsu's Zhenjiang. The Late Pleistocene dark green hard soil platforms widespread in the current northwestern Shanghai in the early stage of the Holocene had retreated to Qingpu near the Sheshan and Tianma Mountains and in the Songjiang-

Jinshan bordering area. The current eastern land area and estuarine area turned into a shoreland neritic environment, where dark gray mucky clay was accumulated. In the middle and late stages when transgression peaked, namely, during 6.5–5.0 ka B.P., the coastal zone stayed constantly and stably in the central Jiading–the western part of Minhang and Fengxian, giving rise to two neighboring, discontinuously connected shell sandbanks, which made up a shorebody belt together with the neighboring littoral tideland. By then, the western part's Jinshan and Songjiang adjacent to the shorebody belt, and part of Qingpu and Jiading had evolved into lagoons and swamps, where gray silty clay commonly containing plant debris and hematite nodules had been accumulated. During 5.0–3.0 ka B.P., the paleoclimate changed toward warm–cool and slightly dry, causing the sea level to fall marginally and the coastal line to retreat eastward and backward accordingly, giving rise to another two shell sandbanks from downtown Jiading to Minhang's Zhuanqiao. The Yangtze Delta sedimentary region then located east of the shorebody belt evolved into an estuarine littoral environment, where mainly gray silt and clayey silt were accumulated; the Taihu sedimentary region located west of the shorebody belt changed from a lagoon environment in the early years to a limnetic environment in the later years, where dark gray silty clay interrupted by silt was accumulated. This, together with the clayey soil–dominated marine formation emerged in the heyday of transgression described above, makes up the Shanghai Formation. Its buried depth and thickness both increase from the southwestern part's Taihu District to the northeastern part's Yangtze Delta sedimentary region. The thickness is mostly in the range of 7–18 m.

In the late stage of the Holocene after 3 ka B.P., the paleoclimate changed back to warm and humid. The sea level went up marginally again and gradually stabilized. However, the remarkably increased amount of sediments from the middle and upper reaches of the Yangtze River caused the progradation of the estuarine sandbars at the front of the Yangtze Delta to speed up, forcing the coastal line to retreat backward rapidly. The original shorebody belt and the neighboring area to its east became a tidal mud flat, where brownish silty clay interrupted by silt beds was accumulated; the land area to its east became an intertidal zone sand–mud flat and partial supertidal mud flats, where mainly brownish silty clay alternated with silt was accumulated. The estuarine sand islands and other places of Chongming were occupied by estuarine sandbars and branching channels. At the sandbars, gray silty fine sand and brownish yellow silty clay alternated with silt were accumulated; in the branching channels, gray silty fine sand and silt were accumulated. To the west of the shorebody belt, Qingpu, Songjiang and Jinshan had basically evolved into a freshwater lake swamp that overlapped the original Late Pleistocene platform. There, brownish yellow–dark gray clay and mucky clay together with peat layers at a shallower level were accumulated. The Wusong River paleochannel from between the western side's Anting and Baihe expanded into a fan shape after it reached the downtown area and flew east into the Yangtze River, becoming a deltaic plain tributary channel. This resulted in the accumulation of gray silt and the superficial brownish yellow silty clay. The sediments from this period makes up the Upper Holocene Rudong and Qingpu Formations. Its thickness also increases from the southeastern part's Taihu District toward the northeastern part's estuarine area, spanning from 1 to 14 m.

In the recent historic period since approximately 1.2 ka B.P., as the Yangtze River Estuary gradually progrades southeastward, the sediments from the Yangtze River has continued to prograde at the estuaries, resulting in the emergence of estuarine sand islands in Shanghai's Chongming, Changxing, and Hengsha. Recently, subaqueous sandbars like the Jiuduansha Sandbar have also come out of water and become land.

7. Urban Geology

The urban geological work of Shanghai has harvested many systematic results with respect to urban geological conditions such as hydrological geology, engineering geology, environmental geology, geohazards, and geological resources after multistage development.

7.1 Hydrological geological conditions

The hydrological geological unit of Shanghai belongs to the Yangtze Delta hydrological geological unit. Except in the Sheshan Mountain and other places in the western part's Songjiang District where the bedrock monadnocks are exposed or shallowly buried, the unconsolidated formations are home to the phreatic aquifer and confined aquifers Ⅰ, Ⅱ, Ⅲ, Ⅳ, and Ⅴ, with widespread phreatic and confined water resources. Confined aquifers Ⅱ, Ⅲ, Ⅳ, and Ⅴ are the main groundwater mining layers. Each of the aquifers have their particular hydrological geological conditions, varying considerably in terms of burial conditions, permeability, water abundance, mineralization degree, groundwater quality, and hydrochemical type. Even within the same aquifer, the hydrological geology varies distinctly on the horizontal dimension.

The hydrological geology of the aquifers is obviously constrained by the paleogeographic environment. Overall, Shanghai can be divided into two subregions: the eastern estuarine-littoral plain hydrological geological subregion (the littoral plain and estuarine sand island area) and the western denudation monadnock-branching channel hydrological geological subregion (the limnetic plain area).

The eastern estuarine-littoral plain hydrological geological subregion mainly covers eastern Jiading District, Baoshan District, central city, Pudong New District, and Chongming District. There, all aquifers are existent; sand beds are extremely well develop with great thickness. The lithology is mainly represented by medium-coarse sand with pebbles. Groundwater is extremely abundant, with single-well cumulative water inflow of approximately 7000-10000 m^3/d for the individual aquifers. Freshwater resources are widely distributed throughout the area. semi-saline and saline water exists only in confined aquifer Ⅱ at local points in the eastern part of Jiading District-local points of Liuhe, and in confined aquifers Ⅱ and Ⅲ of Pudong New District and Chongming District.

The western denudation monadnock-branching channel hydrological geological subregion mainly covers western Jiading, Qingpu, Songjiang, Minhang, Jinshan, and Fengxian Districts. Within the area, different branching channels differ considerably in terms of hydrological geological conditions. Songjiang District's Sheshan-Qingpu District's Dianshanhu, and Jinshan District's Caojing-Shanyang are locations where bedrock monadnocks are sporadically distributed and shallowly buried. Aquifers are either not developed or basically absent. In eastern Jinshan District and Fengxian District, and in the branching channel areas of southern Qingpu and western Songjiang-northern Minhang, the aquifers are incomplete with no confined aquifer V. There, sand beds are not developed or exist in limited thickness. The lithology is mainly represented by silty fine sand and medium-fine sand. The single-well cumulative water inflow is approximately 3000-8000 m^3/d for the individual aquifers. The groundwater is mainly freshwater west of Tinglin Town and west of the Qingpu county town-Liantang Town line; to the east, the groundwater is typically brackish and saline water, except east of the Qingpu county town-Liantang Town line, where confined aquifer Ⅳ is freshwater. In the branching channel area in northern Qingpu-western Jiading, and the branching channel area in northern Jinshan-southern Minhang, the aquifers are quite complete; sand beds are develop and exist in relatively great thickness, consisting mainly of medium-fine sand and medium-coarse sand with pebbles. The single-well cumulative water inflow is approximately 3000-8000 m^3/d for the individual aquifers, increasing from west to east. The groundwater is typically freshwater in the west and gradually changes to brackish and saline water toward the east.

The phreatic water and confined water vary significantly in terms of recharge, runoff and discharge conditions. The phreatic aquifers are mainly recharged by atmospheric precipitation and discharged mainly through evaporation and transpiration. Household washing water in rural areas, and local drainage of phreatic water during foundation pit construction also constitute a form of artificially discharging phreatic water. The groundwater of confined aquifers Ⅰ, Ⅱ, Ⅲ, Ⅳ, and Ⅴ is pent-up water that has no direct recharge relationship with modern precipitation and surface water. Groundwater withdrawal is the main form of discharging groundwater of confined aquifers. Engineering dewatering is the main form of discharging groundwater of shallow confined aquifers.

7.2 Engineering geological conditions

The engineering geological conditions in Shanghai does not vary significantly at large. The soil layers at levels shallower than 100 m can be divided into 10 engineering geological layers and 40 sublayers and sub-sublayers. The 10 engineering geological layers include layer ①, fill; layer $②_1$, brownish yellow-grayish yellow silty clay $②_3$, silty soil (shallow sand layer); layers ③ and ④, gray mucky clayey soil (soft soil layer); layer ⑤, gray clayey soil (locally interrupted by the sand/silty soil of layer $⑤_2$); layer ⑥, dark green-straw yellow clayey soil (first hard soil layer); layer ⑦, straw yellow-gray sand/silty soil; layer ⑧, clayey soil interrupted by sand (locally interrupted by the dark green-straw yellow clayey soil hard soil of layer $⑧_{2-1}$); layer ⑨, gray sandy soil; and layer ⑩, brownish gray-bluish gray clayey soil.

Different engineering geological layers have different engineering geological conditions, giving rise to different engineering geological problems. Layer $②_3$ is a shallow sand layer with limited buried depth. Hence it is prone to vibration liquefaction and seepage liquefaction. Layers ③ and ④ are typical soft soil layers. They are the principal layers for natural foundation deformation as well as the first compression layer of Shanghai. Layer ⑤ is a general clayey soil layer as well as the second compression layer. When excavation foundation pit, attention should be paid to potential quicksand and pipe gush; layer $⑤_2$ is a microconfined aquifer. Hence during foundation pit excavation, attention should be paid to potential quicksand and pipe gush. Layer ⑥ is the first hard soil layer; its buried depth gradually increases from west to east. As is the marker dividing between the Pleistocene and Holocene, this layer is a focus of concern in Shanghai. Its good soil property is quite conducive to controlling foundation deformation, hence can serve as the pile bearing stratum for general buildings or structures. Layer ⑦ is the underlying sand and silty soil layer with good soil property, constituting a good pile bearing stratum for medium or tall buildings and municipal bridge projects. This is also the first confined aquifer. During construction of underground projects, attention should be paid to potential quicksand and sudden inrush in the foundation pit.

According to the modern geomorphologic type and the exposure to the cutting by paleo- and recent channels, three engineering geological structural regions and eight engineering geological structural subregions are identified. Using the darn green-brownish yellow hard soil layer as the basis, the limnetic plain region (Engineering geological structural region Ⅰ) can be further divided into three engineering geological structural subregions ($Ⅰ_1$, $Ⅰ_2$, $Ⅰ_3$). Using the shallow sand ($②_3$) and the dark green hard soil (⑥) as the basis, the littoral plain region (Engineering geological structural region II) can be further divided into four engineering geological structural subregions ($Ⅱ_1$, $Ⅱ_2$, $Ⅱ_3$, $Ⅱ_4$). The estuarine sand island region is not subdivided as all the sand islands have been cut by the Yangtze River paleochannel.

Based on the lithology and thickness of the formation underlying the surface soil layer ($②_1$), four different kinds of natural foundation engineering construction suitability regions can be categorized: good natural foundation regions (A), average natural foundation regions (B), poor natural foundation regions (C), and worst natural foundation regions (D). The shorebody, the riverside, seaside and estuarine sand island areas of littoral plain regions are good natural foundation regions; the eastern parts of the littoral and limnetic plain regions are average natural foundation regions; the western parts of limnetic plain regions are poor natural foundation regions; the seaside regions, such as the east of Lingang New City and Chongming East Shoal, are the worst natural foundation regions.

The main pile foundation bearing layers include the ⑤$_2$middle sand layer, ⑥ dark green-brownish yellow hard cohesive soil layer, ⑦ sand layer, ⑧$_2$silty clay with silt layer and ⑨ deep sand layer. According to the three conditions of two hard soil layers, ⑥ layer and no hard soil layers, three different kinds of pile foundation condition regions can be categorized: has region A both ⑥ and ⑧$_{2-1}$ layer. Many pile bearing strata are available for choice and their distribution is stable, too; has region B layer ⑥ but no layer ⑧$_{2-1}$, pile foundation conditions has certain changes. Regions C contains no hard soil layer, the pile bearing strata surface fluctuates significantly.

7.3 Land subsidence

Based on the characteristics and tendency of land subsidence, contemporarily, Shanghai has undergone two major land subsidence periods. One is the severe land subsidence development period (1921-1965); the other is the land subsidence control period (1966-present). A land subsidence monitoring network system composed of the precision water level monitoring networks, land subsidence monitoring stations, groundwater dynamic monitoring networks, backbone major infrastructure land subsidence monitoring networks, GPS land subsidence monitoring networks, and InSARland subsidence monitoring networks.

7.3.1 Citywide land subsidence

From 1980 to 1995, land subsidence was quite obvious. The downtown and Hongqiao-Huacao were the main land subsidence funnels. In the downtown, the average land subsidence rate was 5.6 mm/a; the maximum subsidence was 200 mm. The neighboring Loutang and Zhejiang's Fengjing became local subsidence centers. In the former Chongming County, the land subsidence proceeded slowly; the average land subsidence rate was only 3.7 mm/a.

From 1996 to 2001, land subsidence began to accelerate. In the downtown, the average land subsidence rate reached 24.5 mm/a. New subsidence funnels appeared in the downtown. The original subsidence centers in Hongqiao-Huacao, Loutang, and Fengjing all expanded appreciably. Qianshao Farm of the former Chongming County subsided more rapidly at an average land subsidence rate of 10.1 mm/a.

From 2002 to 2006, the land subsidence slowed down; the average land subsidence rate reduced to 12.7 mm/a. the Hongqiao-Huacaoland subsidence center almost disappeared. The subsidence funnel in Loutang shrank significantly, but that in Fengjing was still subsiding at a high rate (30 mm/a). In Chongming District, the land subsidence development remained basically unchanged; the average land subsidence rate was 5.9 mm/a. In Changjiang Farm and Luhua Town, a new subsidence funnel appeared.

From 2007 to 2011, except for localized subsidence funnels, the cumulative subsidence in most areas during this phase was small, ranging from 0-25 mm. Subsidence funnels were mainly distributed in localized areas in Pudong, Minhang, Jinshan and Chongming. The subsidence is most severe in Dongtan and Xianghua area of Chongming District, with cumulative subsidence in the center of funnel exceeding 200 mm. Minor resilience appeared in the small area in the western and northern Shanghai.

From 2012 to 2015, the cumulative subsidence in the overwhelming majority of Shanghai was in the range of 0-30 mm. In Chongming East Shoal, Nanmen and Xinhai, the subsidence rate was appreciably smaller than in the previous period. In Pudong New District, the land subsidence slowed down a little at large, but minor subsidence funnels still appeared in Gaoqiao, Zhoupu-Duhang, and the coastal area of Dazhihe. In Minhang District's Dahongqiao, Caohejing, and Jinshan District's Jinshanwei, minor subsidence funnels appeared, too. At local points of Qingpu District's Zhengdian-Songjiang District's Xinbang and Jiading District in the west, there was small-area minor resilience.

From 2002 to 2016, the island area basically maintained its original subsidence pattern. In Chongming East Shoal, affected by recently deposited soil and hydraulic reclamation, the subsidence was quite prominent. In the land area, there were still two large-area land subsidence funnels, one in the central city and northern Baoshan, the other in Huangdu-Nanxiang. A new subsidence funnel was developed in the southwestern part's Luxiang Town-Langxia Town. The subsidence funnel along the provincial border's Zhuqiao-Zhujiaqiao-Loutang disappeared. The original land subsidence funnel in Minhang District's Maqiao was basically gone. There was resilience in Huating-Xuhang-Luodian-Malu-Liuhang-Gucun. The subsidence range of in Pudong New District decomposed from the original patch distribution into a number of subsidence funnels; heterogeneous subsidence was prominent.

7.3.2 Land subsidence in the central city

From 1991 to 1995, the land subsidence rate increased obviously; a number of new subsidence centers appeared in Huapu District and Zhabei District's Buyecheng. A Huangpu District-Zhanbei District's Buyecheng-Yangpu District's Kongjiang Road-Wujiaochangland subsidence center funnel took shape. The spatial divergence increased significantly, especially in Yangpu, Huangpu, Hongkong, and Luwan Districts.

From 1996 to 2000, increased land subsidence rate and the formation of the downtown land subsidence funnel and western downtown funnel made the land subsidence in the central city even more heterogeneous.

From 2001 to 2005, the average land subsidence rate dropped a little. The large-area land subsidence center in the central city broke down into isolated subsidence spots. In Pudong District, two land subsidence centers, the Liuli-Lujiazui and the Donggou-Gaohang-Huangpujiangland subsidence centers, took shape. The heterogeneity of land subsidence was still quite prominent.

From 2006 to 2010, land subsidence was still quite well developed; its heterogeneity increased year by year. The land subsidence pattern became more complicated. The area was decomposed into six land subsidence exposed regions. In the western part of the central city, land subsidence funnels existed in Dahongqioa, Jiading's Nanxiang, Minhang's Meilong, and at local points of the central city's Yangpu and Hongkou; the extreme cumulative subsidence at the funnel centers was greater than 300 mm. In the eastern part of the central city, Pudong became the main host area to land subsidence in the central city. There, relatively large-area land subsidence funnels were developed; the maximum subsidence at the funnel center was more than 200 mm.

7.3.3 Pattern and cause of land subsidence

From the 1960s to the 1980s, the land subsidence mainly covered confined aquifer Ⅰ and overlying soil strata. From the 1990s to the beginning of the 21st century, the land subsidence mainly covered the deep soil strata. Since the beginning of the 21st century, the shallow soil strata have become the main host strata for land subsidence. Among the different periods, different soil strata displayed different forms of deformation.

Since the beginning of the 21st century, the water table of confined aquifer Ⅰ elevated in its entirety, but confined aquifer Ⅰ and the overlying soil strata were subject to continuous compressive deformation. Confined aquifer Ⅰ showed an elastic-elastoplastic-plastic deformation profile, along with creepage, too. The water table of confined aquifers Ⅱ and Ⅲ began to rise slowly; the subsidence rate of the aquifers slowed down appreciably and there was minor resilience, too. Confined aquifers Ⅱ and Ⅲ showed an elastic deformation pattern of compression-resilience. The deformation rate of confined aquifers Ⅳ and Ⅴ was remarkably lower; in some areas, there was a resilience in confined aquifer Ⅳ. Confined Ⅳ showed an elastic-elastoplastic-plastic or linear-elastic deformation profile.

Deep foundation pit engineering represented by underground space development constitutes an important contributor to shallow soil strata subsidence. The coverage of land subsidence induced by foundation pit engineering is closely relevant to the dewatering form of the foundation pit; the land subsidence beyond 3H of the foundation pit is mainly attributable to depressurization dewatering, which typically gives rise to subsidence deformation of the soft soil strata overlying the target strata of dewatering.

The land subsidence induced by high-rise building complexes is dependent on the building plot ratio and geological conditions. The shallow mucky clay strata boast the greatest subsidence. Regional land subsidence caused by high-rise building complexes is irregularly uneven in shape. The impact of high-rise building complexes on land subsidence is related to the sequence of building construction. Between densely distributed high-rise building complexes, there is obvious stress superposition.

7.3.4 Land subsidence risks

Comprehensive land subsidence hazard assessment shows that Shanghai is mainly a moderate-hazard region. High-hazard regions mainly include the central city and local points of the Jiangsu-Zhejiang bordering area's Jiading and Jinshan Districts; low-hazard regions are mainly places where the geological environment is good and the predicted land subsidence rate is low, such as Fengxian and local places of Qingpu. The three island areas are all moderate-hazard regions.

Land subsidence vulnerability assessment shows that the central city is the highest-vulnerability region; the former Pudong New District is a relatively high-vulnerability region; Baoshan and Minhang Districts are moderate-vulnerability regions; Songjiang, the former Nanhui District, and Jiading Districts are relatively low-vulnerability regions; Fengxian District, Chongming County, Jinshan District, and Qingpu District are low-vulnerability regions.

Land subsidence rick assessment based on the land subsidence hazard and vulnerability assessment results above reveals a three-level land subsidence risk system centering round the central city. The central city is basically a high-risk region; the near suburbs are a relatively high to moderate-risk region; the outer suburbs are a lower-risk region.

7.4 Environmental geochemistry

7.4.1 Reference and background values

Soil from different sedimentary geological backgrounds has different reference values. The reference values of the early and late-stage estuarine sediments show almost the same variation and distribution as the regional ratio of Shanghai. The spectral curve of channel nearshore sediments is totally not coincident with that of the estuarine sediments. The reference values of the early and middle-stage littoral sediments show basically the same distribution pattern, but are significantly different from that of the late-stage sediments. The reference values of the lacustrine and limnetic sediments exhibit roughly the same distribution spectra with roughly the same curve profile.

The soil background value indicates significant variability. Five different types can be roughly identified. They are: homogeneous distribution type, relatively heterogeneous distribution type, heterogeneous type, strongly heterogeneous type, and very strongly heterogeneous type. Quite many element background values are close to the reference values of Shanghai; 26 of the elements, including Be, Ce, Cl, Co, Ge, and I, are lower than the reference values. Environmentally sensitive, high chemical-activity elements, including Ag, Bi, Cu, Mo, Pb, S, P, Sb, Au, Br, Cd, Hg, N, Se, Sn, C, and organic carbon, are commonly concentrated; their concentration values area all 1.1 or higher. The surface concentration of Ag, Cd, and C are all more than 1.5 times as high as their deep concentration values; the surface concentrations of Hg, Br, N, Se, Sn, C, and organic carbon are all more than 1.8-2.45 times as high. Different sedimentary regions have different types of soil; as the soil type varies, so does the distribution pattern of the elements.

7.4.2 Spatial distribution of soil elements

The total average content of rock-forming elements in the surface and deep soils are 85.13% and 86.35%, respectively; the total average contents are roughly the same. As the rock-forming element varies, so does the distribution of the high and low-value regions.

The pH of the surface soil is 4.56–8.96 with an average of 7.61. The soil is predominantly alkaline and neutral; weakly acidic soil exists at local points in the west. The pH of the deep soil is 6.55–8.88 with an average of 8.16; the soil is predominantly alkaline. Low pH levels are mainly found in the west and center. In the east and southeast, the pH is quite high.

The TOC of the surface soil is 0.19%–5.78% with an average of 1.24%. The TOC of the deep soil is 0.24%–1.57% with an average of 0.53%. The TOC level is commonly high in the west but quite low in the east. In the central city, the surface soil also has a high TOC level.

Heavy metal elements, namely, Cd, Hg, Pb, and Zn, in the surface soil all produce a quasi–annular anomaly–high background region centering round the urban area of Shanghai. In the high background regions, anomaly regions also appear in the urban centers of Baoshan, Anting, Jiading, Songjiang, and Wujing. Chongming, Changxing, Hengsha Island, and the littoral area of the Hangzhou Bay can be negatively correlated to the urban area of Shanghai. In these areas, all four elements are present in low contents. Cd, Pb, and Zn still maintain a slightly low–background distribution in areas other than the onshore urban area and important transport lines of Shanghai.

In the deep soil, Cd, Hg, Zn, and Pb are more homogeneously distributed, with limited background fluctuation. The Cd content shows a high background distribution in northeastern Shanghai, namely, in the downtown, Baoshan's Wusong, and northwestern Chongming Island, Cd; the Hg content shows basically a medium–high background distribution in all parts of Shanghai except Chongming, Changxing, Hengsha Island and the riverside area of the Hangzhou Bay; the Pb and Zn concentrations seem to radiate from the downtown outward in a trapezoidal form.

In the surface soil, the N content spans from 209 to 3623 mg/kg with an average of 1351 mg/kg. In the deep soil of Shanghai, the N content spans from 268 to 1382 mg/kg with an average of 577 mg/kg. High N levels are mainly found in the western part's lacustrine sedimentary regions. The three islands and other riverside/seaside areas are low N content regions.

In the surface soil, the P content spans from 361 to 2718 mg/kg with an average of 964 mg/kg. In the deep soil, the P content spans from 264 to 1064 mg/kg with an average of 615 mg/kg. Deep–soil high P levels are mainly found in Wusong and Baoshan industrial zone. In the western part's lacustrine sedimentary regions, the P content is commonly low. In the other parts of Shanghai, the P content is moderate. Surface–soil high P levels are found mainly in and around the central city. In Chongming East Shoal, Nanhui Nearshore, and the Dianshan Lake, the P content is commonly low.

7.4.3 Distribution of main organic compounds

Main organic compounds in the soil include BHC, DDT, and PAH. The residual BHC level has reduced noticeably from 30 years ago. Throughout the area, the residual BHC level is within the risk screening limit, but the detection rate is quite high. At local points of Pudong (formerly Nanhui) and Jiading, the residual BHC level in the soil is relatively high; in Chongming, the value is the lowest. The residual DDT level ranges from lower than the detection limit up to 736.7 μg/kg, which is significantly lower than 30 years ago. However, at local points of Qingpu, Jiading, Baoshan, and Pudong Districts, the value is higher than the specified limit. The distribution of the PAH content varies noticeably. It is quite high in all other parts of Shanghai except Chongming and Fengxian Districts.

7.4.4 Soil geochemical anomalies and environmental quality

Within the outer ring and the highly urbanized near suburbs near the outer ring line, between the middle and outer rings, and between the middle and inner rings, the high average–content elements are different. Many high levels of environmental indexes fall in the central city (within the outer ring) and some industrial zones in the near suburbs. Within the outer ring and in Songjiang, areal regions have preliminarily taken form; in other parts of Shanghai, the element contents occur in a point form. Hg, Cr, Pb, Co, and Mo are mainly concentrated near Baosteel and near Wusong industrial zone. Cd, Ni, and As are mainly concentrated near Taopu industrial zone. Near Jiangnan shipbuilding base, high-content elements mainly include Cd, Cr, Cu, Zn, and Pb. Near Jinshan District's petrochemical, Wujing, and Waigaoqiao industrial zones, the contents of organic pollutants in the soil are relatively high.

The soil quality of agricultural land is quite good at large. Most of the soil can be utilized safely. Apart from two survey spots that were higher than the soil contamination risk control limits, some other spots were also higher than the soil risk screening limits (in terms of Cu, Ni, and Zn). The safety utilization is affected. Hence it would be necessary to strengthen collaborated monitoring of soil environment and farm produce to restrict the form of soil utilization.

The construction land, when used as class 1 land, can be utilized safely for the majority of the area except a couple of the land plots; when used as class 2 land, the As, Cd, and Hg levels are within the screening limits; the Cu, Ni, and Pb levels are basically within the screening limits except a couple of spots that are higher than the control limits. For survey spots higher than the screening limits, further detailed survey should be made during exploitation.

7.4.5 Quality and abundance of soil nutrition

Different nutrition indexes in the soil differ from one another throughout the area. The organic content is only at the moderate level. In the western part and the central city, the organic content is quite low; in the other parts of the city, it is at the moderate level. In terms of the N content, five levels can be identified: deficient, relatively deficient, moderate, and abundant. In Chongming, Changxing, Hengsha, and the riverside/seaside area south of the Yangtze River, the P content is deficient to relatively deficient, with the saline soil in the littoral area of Chongming Island being the worst. South of the Yangtze River, from the downtown and its periphery to the western part's lake region, the P content is a moderate-abundant transition with high total P and high available P. In northern Chuansha, Changxing, Hengsha, and Chongming, the P content is relatively deficient. In the western part's lake region and the periphery of the urban area, the P content is abundant. K is quite abundant in the north and southwest. In the greater part of Nanhui, and the seaside area of Fengxian and Jinshan, the exchangeable K in the soil is relatively deficient. Except the littoral region east of the Jinshan-Fengxian-Nanhui line, K deficiency does not exist in the soil in all other parts of Shanghai.

The soil targeted for agricultural utilization can basically be divided into three regions: the western lacustrine region, central region, and eastern three islands and seaside region. In the western lacustrine region, the soil is low in Ca and Mg, high in Si, and rich in organic matter. The soil nutrition is mostly moderate or abundant. Yet the soil utilization is mostly not so efficient due to heavy waterlogging. The central region is characterized by overabundant Ca, Mg, K, Fe, Al, and trace element I. The soil nutrition is mostly moderate; the soil thickness is moderate and permeable, making a high-quality soil region. In the eastern three islands and seaside region, the soil is relatively deficient in P, N, Zn, Cu, Mo, and B, coupled with the exposure to salinization in the littoral area. Agriculturally, it can only be average soil.

7.5 Coastal erosion and deposition

7.5.1 Erosion and deposition of key coastal sections

Coastal erosion and deposition represent one of the key geohazards in the area. Different erosion and deposition occur in different coastal zones during different periods of time.

Chongming East Shoal has been silting continuously. As early back as in 1842, the area was already an eastward-extending duck's tongue-shaped nearshore. After that, the Tuanjiesha and Beigangbeisha Sandbars appeared to its southeast. However, the siltation rate has been slowing down in recent years. In 2011–2013, minor erosion occurred in Chongming East Shoal, particularly in the northerly part of this area.

The Hengsha shoal area is part of the Tongsha shoal in the earlier years. After the 1980s, a bifurcation ditch bridging between the North Passage and the North Channel was formed in the middle of the shoal. The Hengsha East Shoal to the west gradually reclaimed into a land area after 1998. In 2011–2013, the front of the Hengsha East Shoal reclamation region was entirely in an eroded state. In 2013–2015, erosion occurred at both sides, while observable deposition occurred in the center to eastern part. In 2015–2017, the deposition area continued to increase.

The Nanhui Nearshore features a seaward–protruding plowshare–like shoal. It lies in a slackwater area at the junction between ebb and flow currents, which is very prone to sediment accumulation. For many years, the Nanhui Nearshore has been one of the coastal shoals with the greatest siltation rate. From the 8th century AD to the 1960s, the coastal line had risen by an average of 22. 3 m per year. From 713 to 1172, it extended outward by an average of 34. 85 m per year. From 1172 to 1584, the extension rate was only 3. 4 m/a. From 1584 to 1884, the extension rate increased to 24 m/a. From 1884 to 1960, the extension rate increased to 41. 3 m/a. From 1997 to 2009, except small locations in the north and southern parts of the area where erosion occurred, most of the area was in an erosion–deposition balanced or deposited state. From the coastal area north of the Dazhihe River to the southeast, the width of the deposition belt gradually increased from 3 km in the coastal area to around 6 km.

The Jiuduansha Sandbar is a third–generation new sandbar in the Yangtze River estuary. This sandbar is composed of the upper sand, middle sand and lower sand. The upper sand borders the Jiangyanansha Sandbar. Since it became an independent sand body in 1958, especially since 2000, the beachland area has remained on the rise. From 1997 to 2002, the Jiangyanansha Sandbar to the south deposited significantly, whereas the lower sand of the Jiuduansha Sandbar was slightly eroded. From 2002 to 2008, the Jiangyanansha Sandbar continued to deposit significantly, whereas the middle and lower sand of the Jiuduansha Sandbar was extensively deposited.

The Qingcaosha reservoir area is the place near the split between the South and North Channel northwest of the Changxing Island. Over the past ten years, affected by the river regime rectification project and the construction of the Qingcaosha reservoir, the river regime near the Qingcaosha Sandbar has been prominently modified. The interannual erosion and deposition pattern varies very distinctly from one area to another.

The Jinshan Deep Trough begins from the Dajinshan and Xiaojinshan island and extends westward. The bottom of the deep trough is comprised of a series of deep pools connected together. The depth is generally −30–−40 m and is the greatest near the Dajinshan and Xiaojinshan island 2012, the trough had remained stable at large. After 2013, erosion has been playing the leading role, although deposition has also occurred at local points.

7. 5. 2 Coastal erosion and deposition pattern

Over the past decade, the −10 m isobath has varied very little with roughly no change outside the North Branch and North Channel, but has changed quite significantly in the Nanhui Nearshore and the outskirts of the South Passage. The erosion and deposition pattern differs among the area inside and outside the estuary and in the Hangzhou Bay. In the Yangtze River Estuary and the northern bank of the Hangzhou Bay, the bank has mainly suffered erosion. In the lower section of the North Branch, Chongming East Shoal and Hengsha Shoal, and on the outer edge of Nanhui South Shoal, the bank has basically been deposited. At the split between the Southern and North Branches, the South and North Channel, and the South and North Passages, the shoals have been deposited and the passages have been eroded, but these processes still differ.

7.6 Groundwater resources

7.6.1 Groundwater resources and exploitation

The exploitable phreatic water resources are represented by the regulated storage capacity. Calculation of the assessment years of 1981, 1995–1999, and 1980–2000 shows that the total amounts of natural resources over these periods are 28.266×10^8, 12.938×10^8, and 12.060×10^8 m^3/a; the corresponding regulated storage capacities are 8.69×10^8, 2.03×10^8, and 2.03×10^8 m^3/a, respectively.

The exploitable confined water resources are represented by the exploitable amount. In view of the urban demand for groundwater resources and geological environmental protection target of Shanghai during different periods of time, multiple rounds of calculation have been conducted on the exploitable amounts of confined water resources in the main groundwater exploitation confined aquifers (confined aquifers Ⅱ, Ⅲ, Ⅳ, and Ⅴ). In 1987, 1990, 1995, 2000, and 2005, the exploitable confined water resources were calculated using analog method, correlation method, 2D FEM, quasi-3D FEM, and true 3D FEM. The resulted exploitable amounts of groundwater resources are 4.39×10^8, 2.23×10^8, 1.45×10^8, 1.24×10^8, and 0.25×10^8 m^3/a, respectively.

In Shanghai, the phreatic water and groundwater in the shallow level's confined aquifer Ⅰ have long been excluded from centralized exploitation. The groundwater in the deep level's confined aquifer Ⅱ, Ⅲ, Ⅳ, and Ⅴ, as the supplemental source for urban water supply, has been exploited since the 1950s. The periods 1949–1964 and 1978–1996 marked the most intense groundwater exploitation in history, with annual exploitation greater than 100 million cubic meters and the maximum of 200 million cubic meters. After 2013, the annual exploitation was already lower 8 million cubic meters. In 2018, the exploitation dropped to around 1.2 million cubic meters. Up to the present day, the annual exploitation has been on the fall.

7.6.2 Groundwater environment

Constant exploitation of groundwater has led to rhythmic dynamic changes in the groundwater environment.

In the natural state, the groundwater tables of confined aquifers Ⅱ, Ⅲ, Ⅳ, and Ⅴ are at positive elevations. Multiyear groundwater table monitoring data indicate that the groundwater tables of confined aquifers Ⅱ, Ⅲ, Ⅳ, and Ⅴ have been falling and rising alternately depending on the intensity of exploitation. During high-intensity exploitation, the groundwater table was lower; during low-intensity exploitation, it became higher. In the past, the groundwater tables of confined aquifers Ⅱ, Ⅲ, Ⅳ, and Ⅴ once dropped to a maximum of around −34, −37, −68, and −62 m, respectively. In 2018, the minimum groundwater tables of these aquifers stood at around −10, −10, −16, and −16 m, respectively.

In the primary state, the groundwater quality of confined aquifers Ⅱ, Ⅲ, Ⅳ, and Ⅴ can be classified as type Ⅰ, type Ⅱ, type Ⅲ, type Ⅳ, and type Ⅴ. Multiyear groundwater quality monitoring data indicate that, over the decades of groundwater exploitation, the primary groundwater quality has remained stable.

7.6.3 Groundwater recharge

In Shanghai, groundwater is artificially recharged through groundwater mining and recharging wells, separate groundwater recharging wells, and emergency recharging wells. The form of groundwater recharging has evolved from a combination of normal mining and recharging in the early years to a combination of normal (emergency) mining and recharging and separate recharging today. Confined aquifers Ⅱ, Ⅲ, Ⅳ, and Ⅴ are the main aquifers for artificial groundwater recharge; the exact recharging points may vary among different periods of time. As the land subsidence control target differs from one period to another, so does the intensity of recharge. The period 1990–1999 marked a high intensity of artificial groundwater recharge in history.

Artificial groundwater recharge affects the water environment in both ways. While effectively raising the regional groundwater table or decelerating the groundwater table reduction, it also compromises the stability of the primary groundwater quality at local points, although this impact can vary across different backgrounds (freshwater or saline water).

7.6.4 Groundwater exploitation and protection planning

Based on the land subsidence constraints and the exploitable amount of groundwater, four groundwater planning regions, including the strictly restricted exploitation region, restricted exploitation region, controlled exploitation region, and general monitoring region, are identified. The strictly restricted exploitation region is the central city; the restricted exploitation region includes the near suburbs' Baoshan District, Pudong New District, Minhang District, Jiading District, and the outer suburbs' Qingpu and Jinshan Districts. The controlled exploitation region includes the outer suburbs' Chongming, Fengxian, and Songjiang Districts.

Three key deployment regions, including the central city, island area, and urban planning New District, are defined. Based on the general urban plan of 2020, according to the key urban construction areas determined by the urban plan and their populations and industry orientations, 14 groundwater emergency water sources, namely, the central city, Baoshan New City, Jiading - Anting New City, Pudong International Airport, Huinan New City, Lingang New City, Fengxian Seaside Farm, Jinshan New City, Songjiang New City, Qingpu New City, Chengqiao New City, Chenjiazhen New City, Changxing Island, and Hengsha Island, are deployed.

7.7 Underground space resources

Starting with the municipal pipeline construction in 1862, the urban underground space development in Shanghai has undergone three stages: underground space development characterized by box foundation (before 1949), civil air defence underground space development featuring a combination of peacetime and wartime usages (1949-1990), and underground space development represented by underground transit facilities (1990 onward).

As of 2016, a total of 36809 underground projects had been completed, with total floor area of 81.86 million square meters, an increase of 24.87 million square meters from before 2011; the underground space development area had increased at a rate of approximately 10 million square meters per year. Underground production and living service facilities accounted for 95% of the total area of underground projects; underground public infrastructure accounted for 3%; rail transit and associated facilities accounted for 2%. The underground space development of Shanghai is already among the world leading municipalities. Since the beginning of this century, the underground space development amount and scale of Shanghai have been increased year by year, showing a trend of explosive growth. Underground projects like subways, cross-river tunnels, underground roads, urban underground complexes, and large deeply buried underground municipal facilities have been increasing continuously.

Overall, the underground space exploitation in Shanghai has three typical features: its regional distribution is strongly correlated to the land value; its development depth is obviously stratified; its development gradually advances from the shallow, medium level to the deep level.

Groundwater, soft soil, sand soil, and shallow natural gas are the four major geological constraints for underground space exploitation.

Based on available data, geological structure and existing underground space utilization were finely depicted using 3D modeling. Within the outer ring line of Shanghai, the amounts of underground space resources at the shallow, medium-shallow, medium, and deep exploitable levels, and at the deep potentially exploitable levels were estimated using geometric method and logic operation. The estimation covered the total area of underground space, as well as the total areas of the underground space available for exploitation and available for effective exploitation, taking into account the allowance available under the existing underground space conditions.

Based on the two underground space construction processes of pitting and tunneling, taking into account geological factors greatly affecting the underground space such as groundwater, soft soil, sand, and silty soil, as well as the severity of the resultant pit inrush, soft soil deformation and quicksand, the shallow underground space development suitability in the central city of Shanghai can be categorized into three different regions: slightly problemic regions (A), moderately problemic regions (B), and severely problemic regions (C). Based on the severity of potential water and soil gush, quicksand, and land subsidence induced by engineering construction, using the geological constraints that largely affect deep underground space development such as groundwater and sandy soil, taking into account the current underground development situation, the buried depth of the bedrock, and potential land subsidence, the deep underground space development suitability of Shanghai is categorized into four grades, namely, good, average, relatively poor, and poor.

Within the area, the underground space resources can be divided into the shallow level (0-15 m), medium-shallow level (15-30 m), medium level (30-50 m), deep exploitable level (50-65 m), and deep potentially exploitable level (65-100 m). The total volume of the underground space in the central city is 6.64×10^{10} m^3. The quality of the underground space resources is relatively high at levels shallower than 30 m.

7.8 Tideland resources

In the Yangtze River Estuary, 15 large-sized sand body regions make up the principal part of the tideland resources in Shanghai. These regions are: the Baimaosha Sandbar; Biandansha Sandbar; Xinliuhesha Sandbar; the North Branch's Chongtou Nearshore and its northern extension, the Huangguasha Sandbar; the North Channel's Zhongyangsha, Qingcaosha and Liuxiaosha Sandbars; the South Passage's Jiangyanansha Sandbar; the Channel Entrance's Chongming East Shoal, Beigangbeisha Sandbar, Hengsha Shoal, Jiuduansha Sandbar, and Nanhui Nearshore.

As of 2017, the envelop size of the −5 m isobath was approximately 457 km^2 for the Chongming East Shoal and Beigangbeisha Sanba; 276.5 km^2 for the Nanhui Nearshore; 225.5 km^2 for the Jiuduansha Sandbar; and 255 km^2 for the Hengsha Shoal.

The Chongming East Shoal has considerable potential for northward progradation. It became an important tideland increase space after connected to the Chongming North Border. Similar to the Tuanjiesha Sandbar where deposition-promoting enclosure was completed earlier, the Beigangbeisha Sandbar is still in the process of southeastward progradation at large. The Hengsha Shoal is still prograding southeastward; the faster depocenter may have been migrating in the southeast direction toward the Jigujiao Reef. However, attention should be paid to the potential impact of this southward migration on the deepwater channels. The Jiuduansha Sandbar, as an important wetland natural reserve, is generally in a stage of a relatively high deposition rate. The Nanhui Nearshore has received large-scale deposition-promoting enclosure. There, although the high-beach resources have been largely reduced, the low-beach resources are still very abundant and the area is still prograding eastward at large. Its tideland resources are second only to the Hengsha Shoal. The Hangzhou Bay is generally in the process of continuous erosion. Overall, the high and medium-low-beach resources are very limited; the comprehensive exploitation potential is very low. The Chongming North Beach is an area with relatively concentrated medium-high-beach resources. Also, as the North Branch is in the process of continuous deposition and shrinkage and is close to the Chongming Island, this area has relatively high potential for comprehensive utilization.

In the Yangtze River Estuary, bagged sand and dike core sand resources available for deposition-promoting enclosure are very limited. There are some utilization potentials in the eight areas of the South Branch Main Passage-Xinliuhe Sand, Lower Biandansha Sandbar, North Port Tidal Ridge-tail of the Liuxiaosha Sand Ridge, northeast of the Guyuansha Sandbar, North Channel Entrance-Jigujiao Reef, tail of the Jiangyanansha Sandbar, Nanhui Nearshore, and Hangzhou Bay.

7.9 Tourist geology

The tourist geological resources are abundant, covering a good variety of geological relic resources and a great number of paleocultural sites. These mainly include the bedrock monadnocks, paleocoastal lines, the seawall relics from different historical periods and local ancient human cultural sites in the coastwise area of Jinshan and in the waters off the shore, as well as the widespread modern littoral and estuarine landforms, wetland resources, and the marine erosional landforms of bedrock islands in the waters. Besides, in Jinshan's Zhangyan Town lies the only hidden solid metal deposit in Shanghai.

The geological relic resources are extremely distinctive. Besides being visually significant, they are also scientifically significant. These resources can be categorized into two major types, namely, geomorphologic landscapes and geohazards; four types, namely, water landforms, coastal landforms, volcanic landforms, and geohazards; and four subtypes, namely, river (landscape) zones, marine accumulation landforms, volcanic rock landforms, and land subsidences. The Chongming Island National Geopark based on the Chongming Island has distinctive features. According to the landscape value of the geological relics in this park, there are one world-class, one national-class, and eight provincial-class geological relic spots.

8. Regional Geological Evolution

Overall, the area has undergone five geological evolution periods, namely, the Paleoproterozoic, Mesoproterozoic-Qingbaikouan, Nanhua-Middle Triassic, Late Triassic-Paleogene, Neogene, and Quaternary. Approximately by the late stage of the Paleoproterozoic, the rudiment of the primitive continental crust of the area had basically taken form. At the end of the Paleoproterozoic, the Luliang Movement caused the Yangtze primitive continental crust to be cratonized.

8.1 Mesoproterozoic-Qingbaikouan

During the Mesoproterozoic, the area was a back arc basin at large. From the early to the late stage, the seawater seems to have been deepening, giving rise to a series of neritic sediments featuring flysch formations. The deposition was accompanied by intermittent (seafloor) volcanic eruptions, and magmatism seems to have been intensifying and displaying a bimodal pattern. In the last stage of the Mesoproterozoic, the area was an active continental margin orogenic environment. Immature island arcs were developed. The structural regime seems to have been transitioning from an extrusional environment toward an extensional environment.

Episode Ⅰ of the Jinning Movement caused the Mesoproterozoic formations to be folded, producing the Jinshan Group. In the early stage of the Qingbaikouan, the area was a relatively stable back arc neritic-hypabyssal basin environment. Algae plants began to emerge in large quantities. The crust seems to have entered a relatively stable stage; the volcanic activity had basically drawn to a halt. Episode Ⅱ of the Jinning Movement led to an overall uplifting of the basement. The Qingbaikouan and Mesoproterozoic formations were folded or faulted to varying degrees. Some areas, noticeable low greenschist facies metamorphism took place, resulting in the formation of the Huinan Slate. Meanwhile, the Jinshan Group also suffered local degradation or metamorphism. After that, post-collisional superposed rift basins began to occur. The area received the accumulation of the Heshanzhen Group allogenic terrigenous clastic formations. The volcanic activity also grew intense.

8.2 Nanhua-Middle Triassic

After theEarly Qingbaikouan, the area entered a caprock development period featuring marine or marine-continental faces. The strata changed from the Meso to Neoproterozoic arc basin system formations into passive continental margin basin or epicontinental basin formations in the southeastern margin of the Yangtze craton.

The Nanhua formations of the area have not been disclosed. Starting from the Sinian, the climate became warm. Widespread transgression took place, giving rise to the Dengying Formation sediments consisting mainly of magnesian or siliceous carbonate rocks. Baoshan and Dachang seem to have been a platform margin shoal facies-open platform environment. To its south, Tanzhi and Sanzao were comprised of restricted sea platform facies-evaporite platform facies littoral sediments. Throughout the Sinian, affected by the basement uplifting in the center, the paleoterrain of the area was higher in the southeast than in the northwest. With the regression in the late stage of the Late Sinian, a backwater gulf environment was formed in Qingpu, Baoshan and Chongming northwest of the uplift zone. Thin layers of black siliceous rocks deposited there. In the southern part of the uplift zone's Tanzhi-Fengxian, the seawater gradually retreated and the area began to receive denudation.

From the Eocambrian to the early stage of the Early Cambrian, the area basically succeeded with the sedimentary environment of the late stage of the Sinian. The seawater was deeper than in the late stage of the Sinian. To the north of the uplift zone, a stagnant deepwater flat basin was formed. To its south, the environment should not have changed. At that time, the climate was warm. The water in the gulf could not flow smoothly. Lower fungi plants multiplied rapidly and organic matter accumulated in large quantities. This gave rise to the Chaoshan Formation sediments of carbonaceous shale sometimes interrupted by thin layers of calcareous shale and siliceous shale. In the late stage of the Early Cambrian, the sedimentary environment was similar to that in the early stage. The stagnant reducing environment had weakened noticeably and the hydrodynamism intensified distinctly. This resulted in the deposition of the Dachenling Formation contemporaneous brecciform dolomite.

At the beginning of the Middle Cambrian, obvious transgression took place, producing an open sea platform environment. Terrestrial detrital materials increased. A series of phosphorite formations of marlite and silty limestone deposited.

In the Late Cambrian, the seawater shallowed and the climate turned hot, giving rise to a series of terrace lagoonal thin layers of dolomite, contemporaneous dolomite and dolomitic limestone formations. At that time, as the seawater salinity was quite high, it was hard for organisms to multiply and survive. Hence fossils are rarely seen in the strata.

The Ordovician inherited the paleogeographic and paleotectonic conditions of the Cambrian, except that the sea area continued to expand and the seawater was also deeper than before. During the deposition of the Lunshan Formation, open sea platform carbonate formations emerged at the beginning of the transgression. From the late stage of the Early Ordovician to the early stage of the Late Ordovician, the seawater continued to deepen. A series of platform slope formations comprised of grayish black or grayish purple bioclastic microcrystalline limestone and nodular limestone containing carbonaceous carbonate rocks, mud, and siliceous bands deposited from the Honghuayuan to the Huanghuagang Formations. During the deposition of the Guniutan Formation, in particular, the strata contained rich biofossil debris, including *Cephalopoda*, *Gastropoda*, *Trilobita*, *Crinodea*, and *Conodont*. The main biological components appear to be a hybrid biota between Ningzhen and western Zhejiang. This suggests that the water body was still quite shallow and the hydrodynamic energy was quite strong.

In the middle stage of the Late Ordovician, the transgression range reached its maximum. Within the area, the terrain was high in the southeast but low in the northwest. The northern part of the area evolved into a continental shelf environment and received a series of very thick flysch sediments. The southern part should have been thinner and the strata contain rich *Graptolite* fossils. The biological assemblage bears the characteristics of both the North China type and the Southeast Type, suggesting a transitional biota.

The Silurian formations are underlain by sub-flysch formations comprised of grayish black or grayish green sandstone, siltstone, and mudstone, and overlain mainly by purplish red sandstone containing boulder clay. From the bottom up, the environment evolved from a subtidal zone to an intertidal zone, making up an obvious regression sequence.

So far, no Late Paleozoic–Middle Triassic formations have been detected. However, it is totally possible that sediments from this period exist in northern Shanghai, especially the western part of the Chongming Island.

8.3 Late Triassic–Paleogene

The intense Indosinian orogeny at the transition between the Middle and Late Triassic brought the area into a period of development and transformation for continental basin formations.

So far, no Late Triassic–Jurassic formations have been detected in the area. According to the characteristics of the magmatic rocks, the area should have entered a period of intense intracontinental extrusion at least in the late stage of the Early Jurassic, which gave rise to the Wusong synorogenic type granite. The final stage of the Middle Jurassic marked the entry into a late orogenic period, resulting in the formation of the Sanlin quartz diorite stocks.

In the Early Cretaceous, the area gradually entered a post–orogenic–non–orogenic period, resulting in widespread magma intrusion and volcanism formations. These activities experienced a weak–strong–weak evolution, peaking in around 130 Ma with the widest distribution. This resulted in the emergence of the three continental eruptive–sedimentary formations of the Laocun, Huangjian, and Shouchang Formations, as well as the northern Shanghai granite, southern Shanghai granite, and Baozhen granite.

Since the Late Cretaceous, the regional stress changed to be primarily tensional type. The crustal fault blocks became obviously differential. Controlled by the NE, nearly EW, and NW faults, a series of fault basins took shape. The Pukou continental purple quasi–molasse formation, Chishan red clastic formation, Paleogene Xiajiaqiao brown calcium–bearing fine clastic formation and Beiqiao particolored clastic formation were developed. The magmatism gradually weakened. The area was mainly in the process of intermittent volcanism and local intrusion, which resulted in the formation of a great amount of small rock stocks and subvolcanic and dike rocks. In the Late Cretaceous, the climate was dry, hot and rainy. Piedmont and alluvial–proluvial environment prevailed. The sediments are quite coarse in size and are locally interrupted by gypsum layers. In the Paleogene, the climate was typically arid. The area of the lake basin reduced significantly. The sediments are quite fine.

8.4 Neogene–Quaternary

The climate changed from the arid in the Paleogene to warm and cooling, eventually entering the Quaternary glacial period. The basins commonly enlarged, resulting in the depression sediments in the early fault basins. Within the area, controlled by the basement NW, nearly EW, and NE faults, mantle–derived alkaline olivine basaltic magma often effused along the junctions between faults, occurring in the coastal area. In the hiatus of volcanic eruption, controlled by the warm climate, unconsolidated lacustrine sediments took shape.

After the late stage of the Neogene, the original basement structures and paleogeomorphy was reworked by the NW and NNW faults into structures and geomorphological landscapes of alternate uplifts and depressions. In the low–laying areas such as the Anting–Chongming sag, Fengjing sag, and Pudong's Binbin sag, the relatively thick Chongming Formation alluvial–proluvial sediments were deposited. The uplift zone in the west was weathered and denuded.

At the beginning of the Quaternary, the glacial–interglacial alternation led to periodical fluctuation of the sea level. The area received the deposition of hundreds of meters thick Quaternary fluvial–lacustrine, estuarine deltaic, and alternate marine–continental sediments. During this period, there were five distinct cold–warm alternations, accompanied by seven transgressions. Higher plants and mammals multiplied rapidly. After the beginning of the Holocene, in particular, as the climate warmed and the sea level rose, the Yangtze River Estuary continued to retreat toward the land. The undercut valley gradually evolved into a huge firth. After the sea level reached its maximum and remained stable, with the input of sediments into the sea and the reduction of accommodation space underwater, the delta began to advance quickly seaward, as manifested by the continuous seaward migration of the Yangtze River Estuary and the coastal lines in the flanks. After that, the area was densely covered by rivers and lakes and enjoyed pleasant climate, becoming a good place for the ancient humans to live and multiply.

Since the beginning of the modern age, human activities have exerted obvious impact on the geological environmental evolution of the area. The natural evolution of the Yangtze Delta has been remarkably changed.

图版及说明

图版 Ⅰ

1—*Redlichia shangaiensis* Ju et Xu（sp. nov.）

1a，1b—头盖，外模×2。大场岩溶 8#孔（深度 475.5m），底—下寒武统中、上部。

2，3，4，5—*Shanghaia* Ju et Xu（gen. et sp. nov.）

2a，2b，2c—头盖，侧视及外模×12；3—幼虫×12；4—幼虫×12；5—尾部×12。大场岩溶 8#孔（深度 478.9m），底—下寒武统超山组中、上部。

6—*Shanghaia* sp.

头盖，外模×8。大场岩溶 8#孔（深度 478.9m），底—下寒武统中、上部。

7，8，9—*Shanghaia spisusualis* Ju et Xu（gen. et sp. nov.）

头盖，侧视及外模×8。大场岩溶 8#孔（深度 478.9m），底—下寒武统中、上部。

图版 Ⅱ

1，2—*Shanghaia spislatilimbatus* Ju et Xu（gen. et sp. nov.）

头盖，外模×8。大场岩溶 8#孔（深度 478.9m），底—下寒武统中、上部。

3，4—*Shanghaia spisdecipiens* Ju et Xu（gen. et sp. nov.）

头盖，侧视×12。大场岩溶 8#孔（深度 478.9m），底—下寒武统中、上部。

5—*Homotreta*（?）sp.

5a，5b—×12。大场岩溶 8#孔（深度 469.14~478.9m），底—下寒武统中、上部。

6—*Homotreta*（?）sp.

6a，6b—×3。大场岩溶 8#孔（深度 469.14~478.9m），底—下寒武统中、上部。

7，8—不明管状物?（痕迹化石）

7a，7b—×3；8—×9。大场岩溶 8#孔（深度 472.22~478.9m）。

9—*Laminarites* sp.

×600。大场岩溶 8#孔（深度 469.9m）。

图版 Ⅲ

1—*Laminarites* sp.

1a，1b—×600。大场岩溶 8#孔（深度 469.9m）。

2—*Orthograptus gracilis*（Roemer）

×10。上海市酒精一厂孔（深度 449.7m），上奥陶统长坞组。

3—*Climacograptus typicalis* Hall

×3。上海市酒精一厂孔（深度 342.8m），上奥陶统长坞组。

4—*Climacograptus latus* Eulles et Wood

×10。上海市酒精一厂孔（深度 348.3~359.2m），上奥陶统长坞组。

5—*Climacograptus* sp.

×10。上海市酒精一厂孔（深度 332. 30m），上奥陶统长坞组。

图版Ⅳ

1—*Leiominuscula minuta* Naum

×800。南汇中学孔（深度 356. 42~356. 52m，384. 15~384. 20m），惠南板岩。

2—*Margominuscula* aff. *tennela* Naum

×800。南汇中学孔（深度 384. 15~384. 20m），惠南板岩。

3—*Margominuscula* sp.

×800。南汇中学孔（深度 356. 42~356. 52m），惠南板岩。

4—*Lophominuscula* sp.

×800。南汇中学孔（深度 356. 42~356. 52m，384. 15~384. 20m），惠南板岩。

5—*Leiopsophosphaera solida* Liu et Sin

×800。南汇中学孔（深度 356. 42~356. 52m），惠南板岩。

6—*Asperatopsophosphaera umishanensis* Sin et Liu

×800。南汇中学孔（深度 356. 42~356. 52m，384. 15~384. 20m），惠南板岩。

7—*Asperatopsophosphaera umishanensis* var. *minor* Sin

×800。南汇中学孔（深度 384. 15~384. 20m），惠南板岩。

8—*Dictyosphaera simca* Sin et Liu

×800。南汇中学孔（深度 384. 15~384. 20m），惠南板岩。

9—*Dictyosphaera* sp.

×800。南汇中学孔（深度 384. 15~384. 20m），惠南板岩。

10—*Paleamorpha* sp.

×800。南汇中学孔（深度 356. 42~356. 52m，384. 15~384. 20m），惠南板岩。

11—*Asperatopsophosphaera* sp.

×800。南汇中学孔（深度 384. 15~384. 20m），惠南板岩。

图版Ⅴ

1，6，8—节球孢未定种（*Nodospora* sp.）

×665。新庙镇孔（深度 407. 05m、410. 17m），康山组。

2，12—光面球藻未定种（*Leiosphaeridia* sp.）

×665。新庙镇孔（深度 407. 05m、410. 17m），康山组。

3—连球藻未定种（*Synsphaeridium*? sp.）

×665。新庙镇孔（深度 407. 05m），康山组。

4—四胞体未定种（*Tetrahedraletes* sp.）

×665。新庙镇孔（深度 407. 05m），康山组。

5，9，10，11—卡博特指棱藻（*Dactylofusacabottii*）

×665。新庙镇孔（深度 407. 05m、410. 17m），康山组。

7—网面球藻未定种（*Dictyotidium* sp.）

×665。新庙镇孔（深度 402. 63m），康山组。

图版Ⅵ

奉新星火交管站 B52 孔（岩 2）（深度 171.58m），中—上奥陶统：

1，2—埃氏瓦利塞刺（*Walliserodus ethingtoni*）
　侧视，×100。

3，14—直原潘德尔刺（*Protopanderodus rectus*）
　侧视，×60。

4，6，7，9—刺状围刺（*Periodon aculeatus*）
　分子之侧视，×60，×60，×80，×60。

5，10—刺状围刺（*Periodon aculeatus*）
　分子之侧视，×66，×80。

8，11，12，13—刺状围刺（*Periodon aculeatus*）
　分子之侧视，×66，×100，×100，×120。

图版Ⅶ

奉新 B52 孔（深度 171.58m），中奥陶统：

1，2—马氏螺未定种（*Maclurites* sp.）
　1—底视，×100；

2—口视，×100。

3—缺凹螺未定种（*Sinuites* sp.）
　顶视，×60。

4—膨胀似昌螺比较种（*Umbonellina* cf. *inflata*）
　口视，×80。

5—脊旋螺未定种（*Raphispira* sp.）
　口视，×87。

6—滑旋螺未定种（*Liospira* sp.）
　口视，×57。

图版Ⅷ

书院镇 SG7 孔（深度 316~318.44m），北桥组：

1，2，3—稀刺铁杉粉（*Tsugaepollenites spinulosus*）

4，5—云杉粉未定种（*Piceapollis* sp.）

6，7，8—罗汉松粉未定种（*Podocarpidites* spp.）

9，10，12，13—雪松粉未定种（*Cedripites* spp.）

11—小型油杉粉（*Keteleeriaepollenites minor*）

14—铁坚杉型油杉粉（*Keteleeriaepollenites davidianaeformis*）

15—盘星藻未定种（*Pediastrum* sp.）

16—微点木兰粉（*Magnolipollis micropunctatus*）

17，18，19—单/双束松粉未定种（*Abietineaepollenites*/*Pinuspollenites*）

图版 I

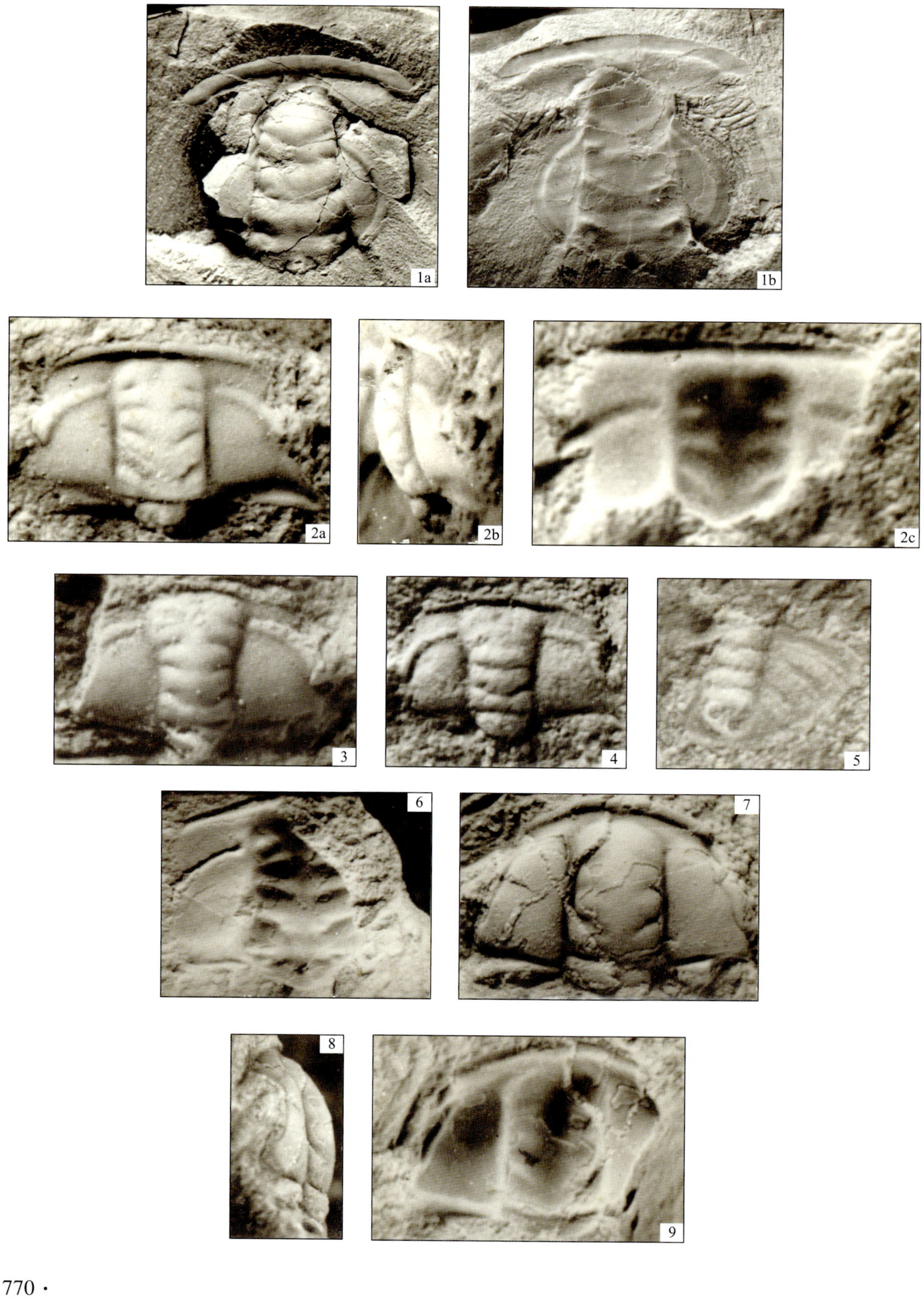

图版 II

图版Ⅲ

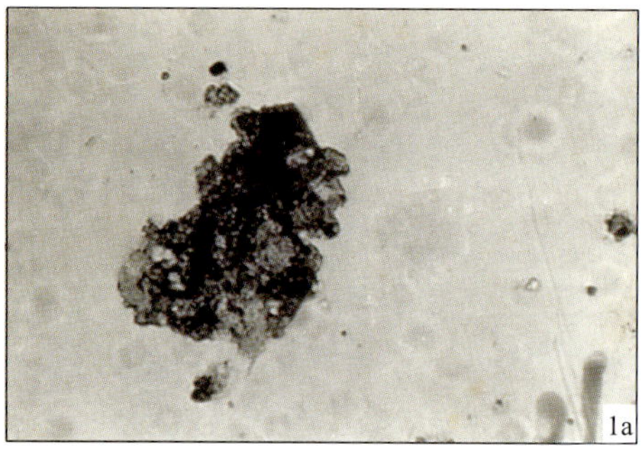

1a

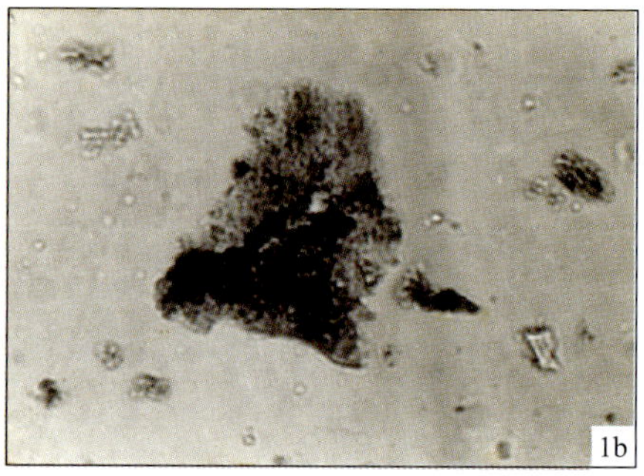

1b

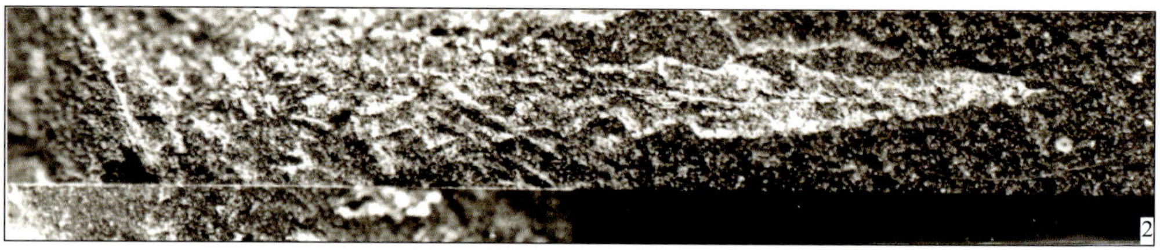

2

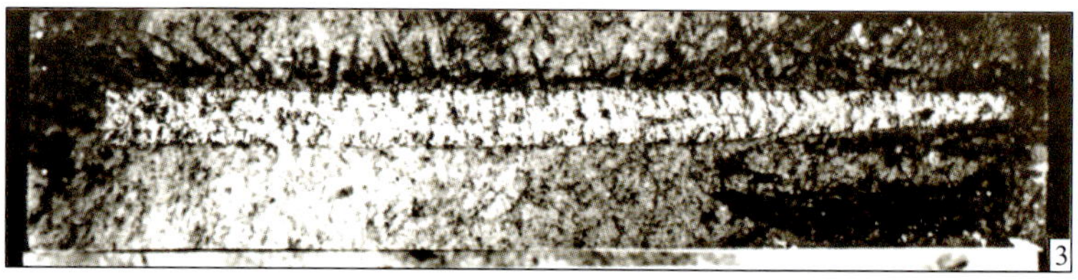

3

4

5

图版Ⅳ

1

1

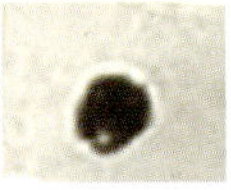
1

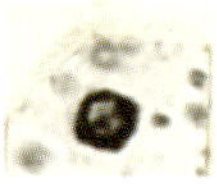
2

3

3

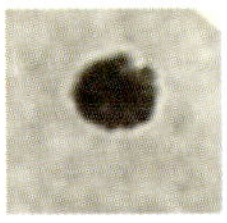
4

5

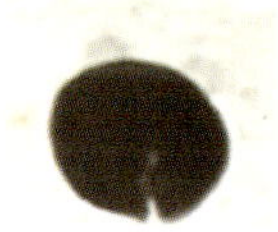
5

5

6

7

7

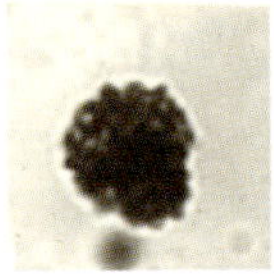
7

7

8

9

11

11

11

10

10

图版Ⅴ

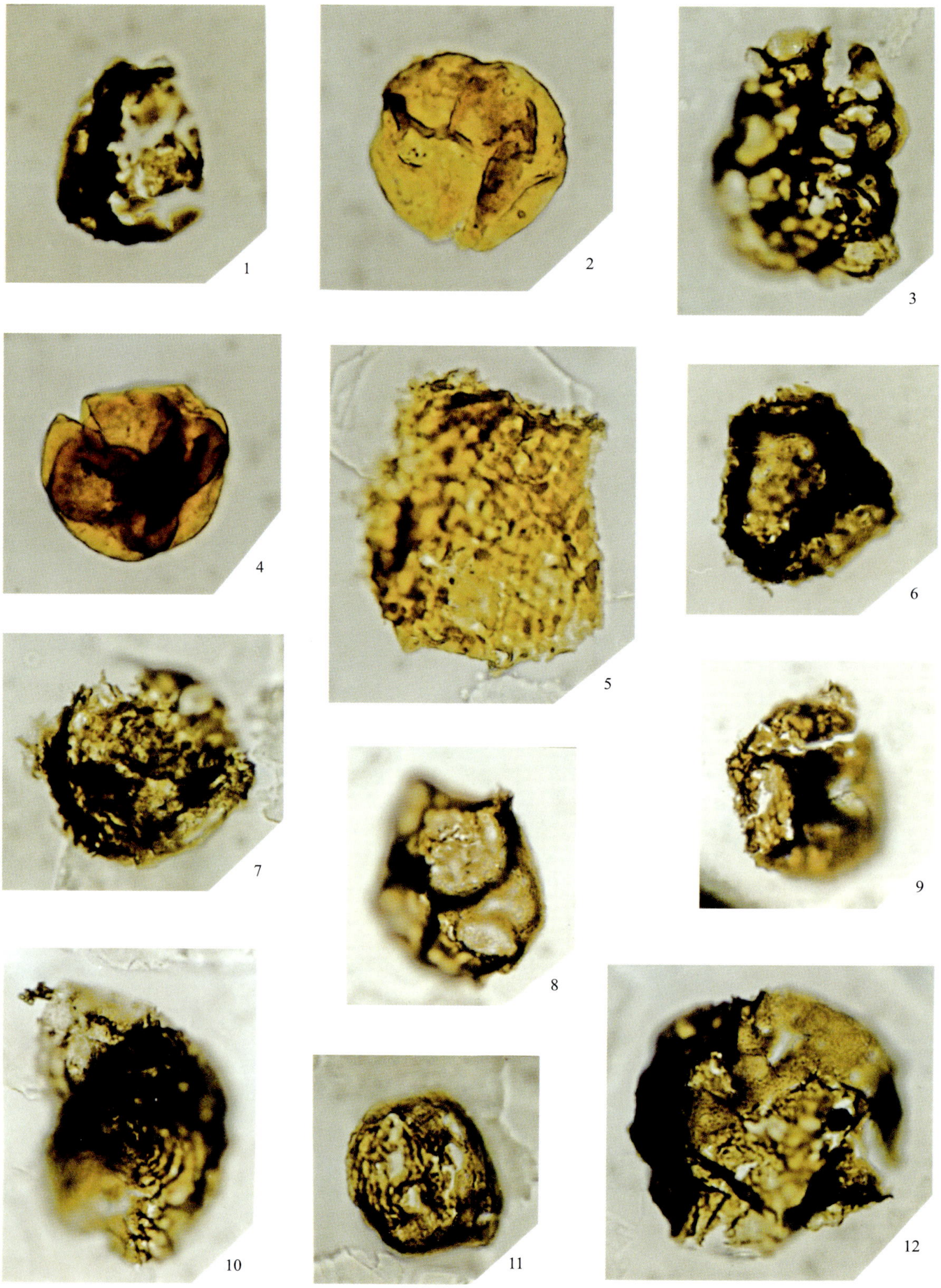

图版Ⅵ

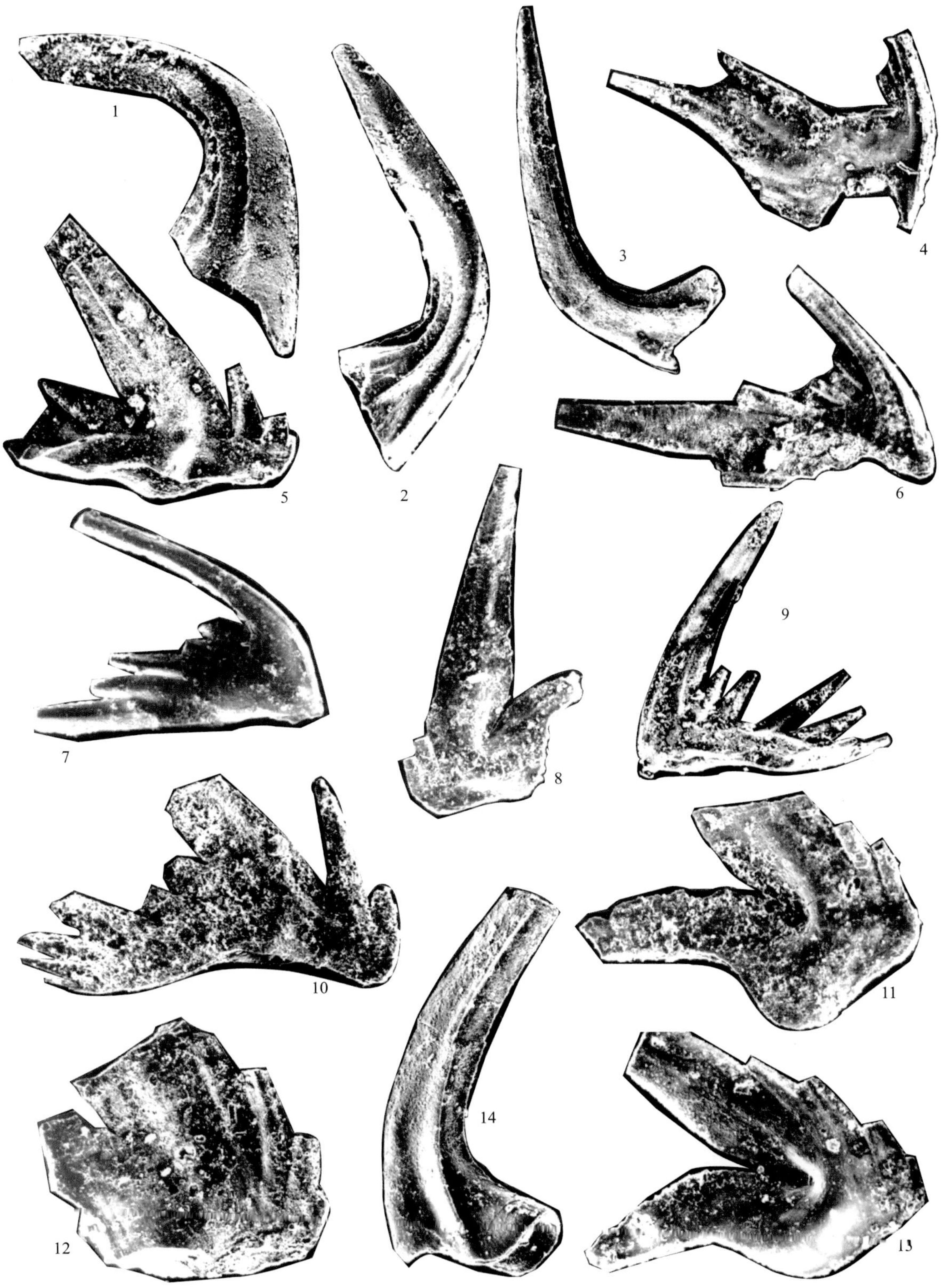

图版Ⅶ

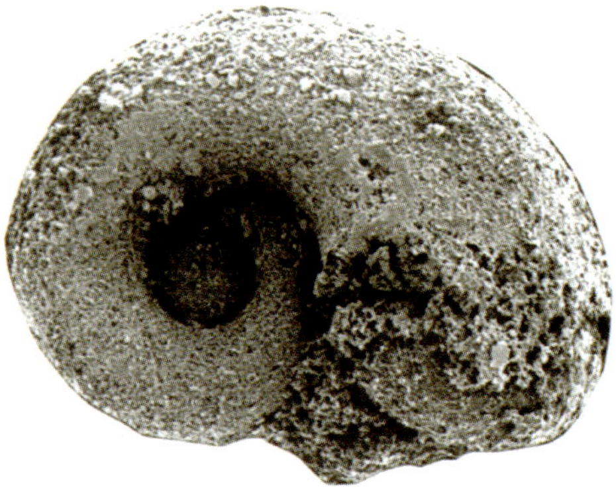

1

2

3

4

5

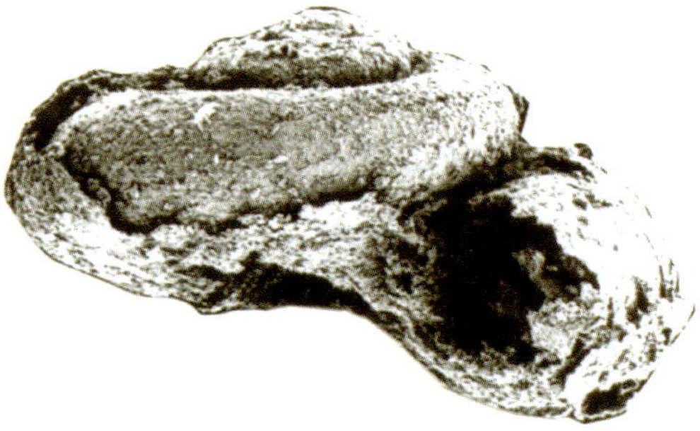

6

图版Ⅷ

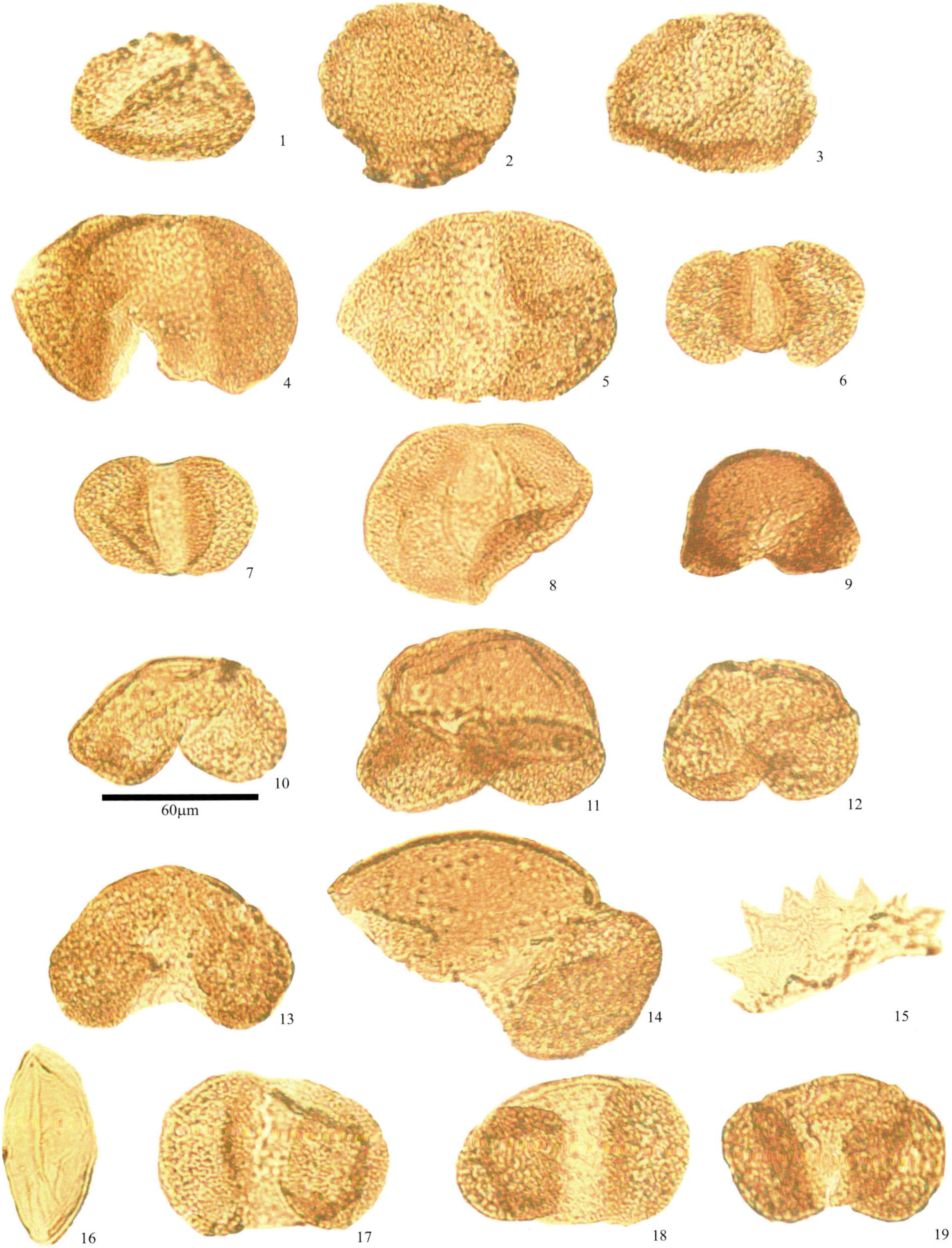

图版IX

角岩化辉绿岩（角岩化辉绿结构）
正交偏光×20　金山张堰

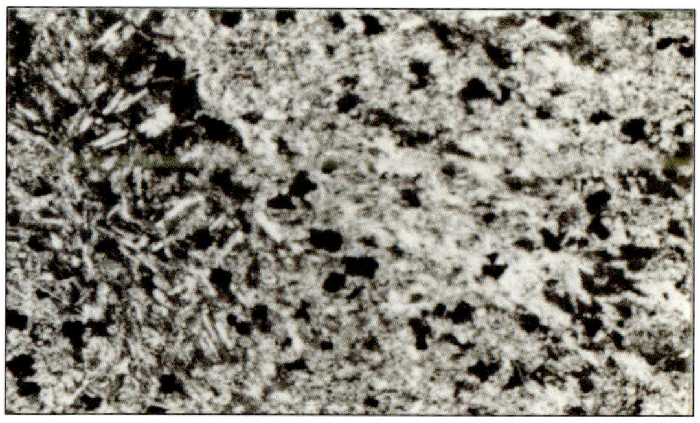

次闪石辉绿玢岩（角岩化含长嵌晶结构）
正交偏光×20　金山张堰

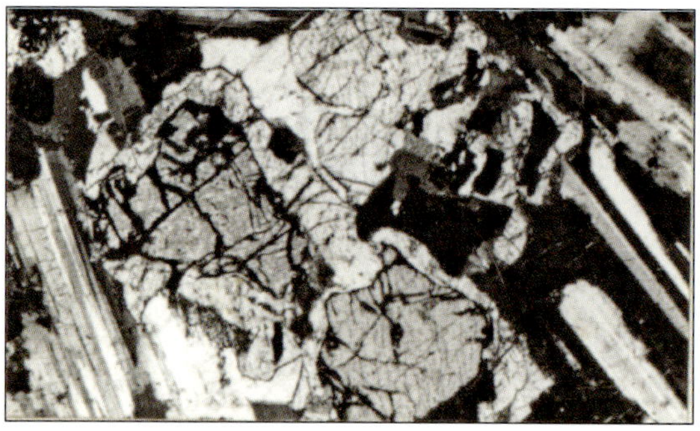

辉石闪长岩（辉石具角闪石反应边）
正交偏光×20　金山沈庄

辉长岩
正交偏光×20　浦东坦直

石英闪长岩（斜长石具环带构造）
正交偏光×20　金山张堰

花岗岩（不等粒结构）
正交偏光×20　上海北部

花岗斑岩（文象结构）
正交偏光×20　金山张堰

花岗细晶岩（细晶结构）
正交偏光×20　上海北部

图版X

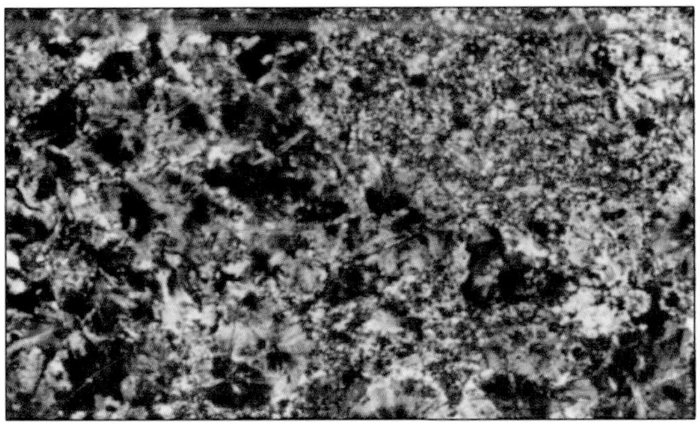

霏细斑岩（球粒结构）
正交偏光×20　金山张堰

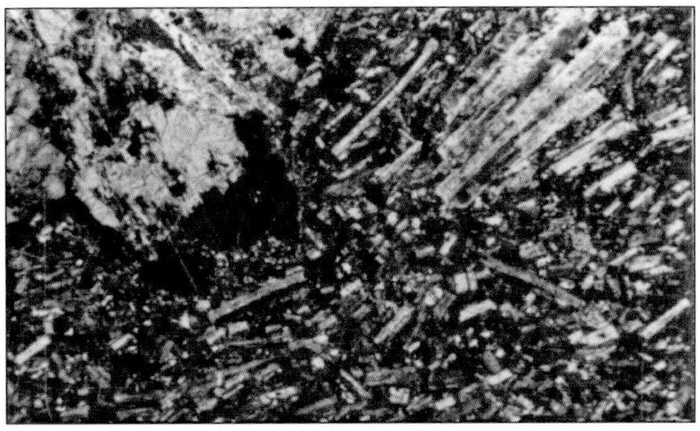

安山岩（交织结构）
正交偏光×20　查山

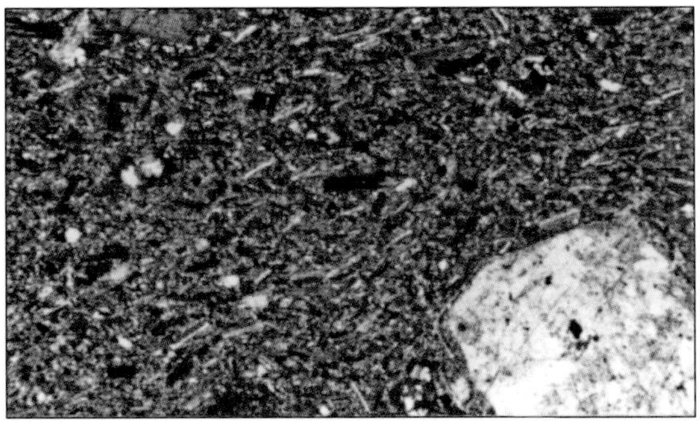

安山岩（交织-半粗面结构）
正交偏光×20　查山

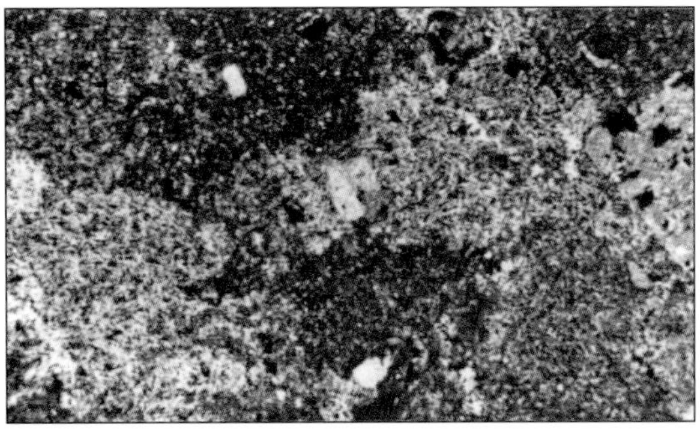

英安岩
正交偏光×20　天马山

流纹英安岩（斑状结构和流纹构造）
正交偏光×20　佘山

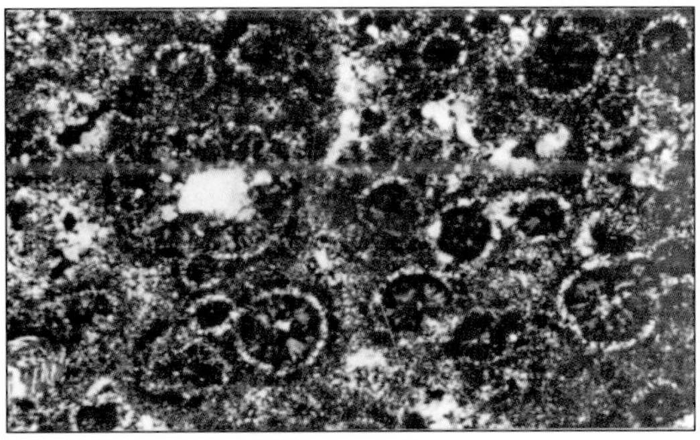

球粒流纹岩（球粒结构）
正交偏光×20　金山张堰

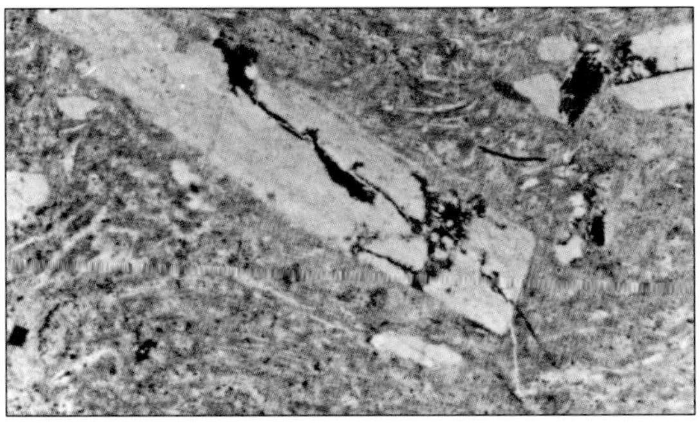

流纹质玻屑熔结凝灰岩（玻屑塑变结构）
正交偏光×20　辰山

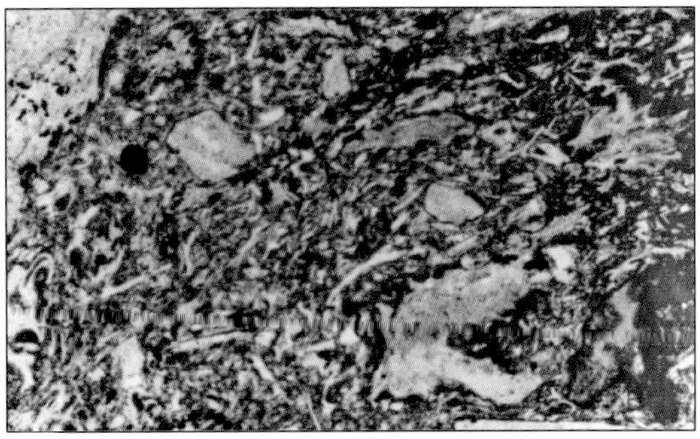

凝灰岩（玻屑凝灰结构）
正交偏光×20　天马山

图版XI

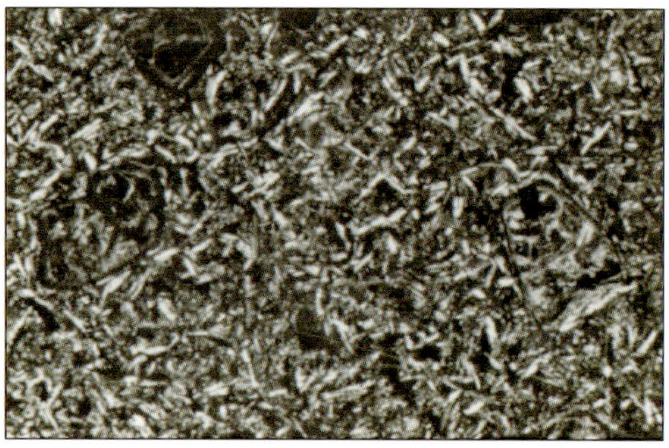

橄榄玄武岩（气孔状）
正交偏光×20　惠南

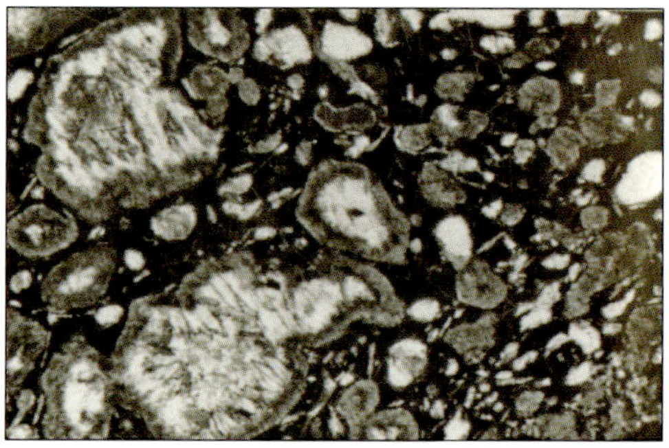

玄武岩（玻晶交织结构）
正交偏光×20　北蔡

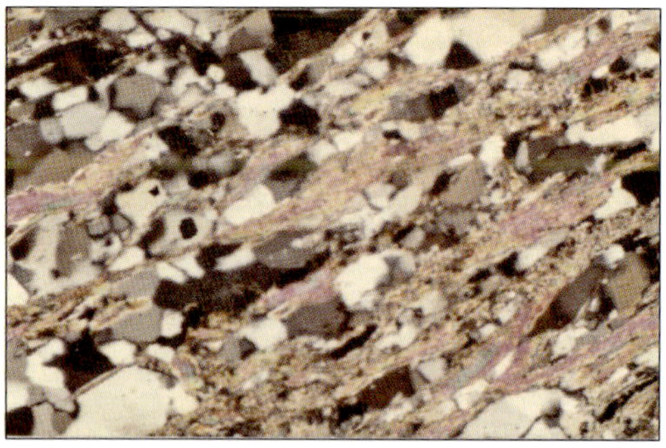

白云母石英片岩
正交偏光×50

黑云母片岩
正交偏光×20

绿帘阳起片岩
正交偏光×20

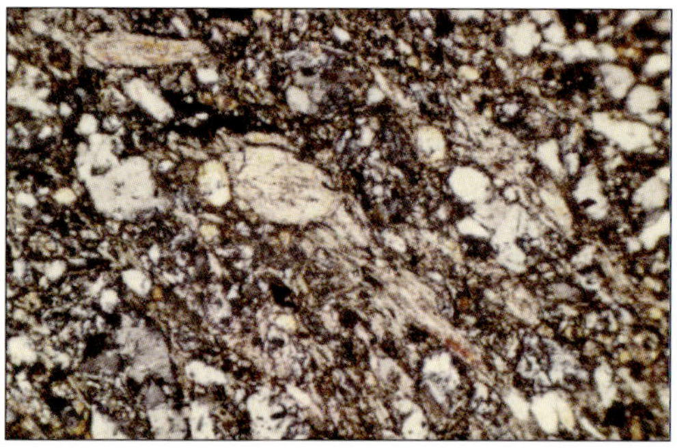

绿帘阳起片岩
正交偏光×50

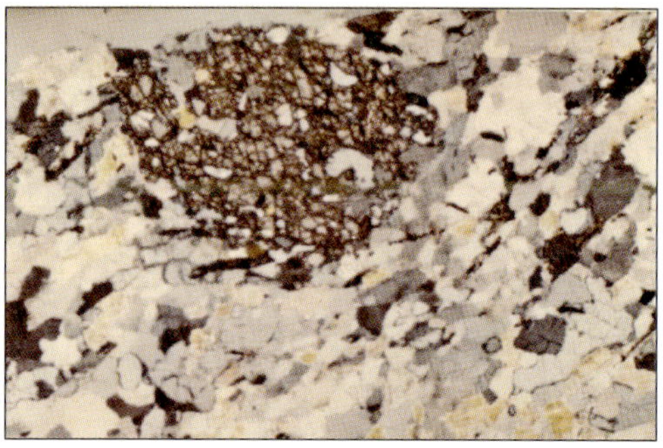

含榴浅粒岩（铁铝榴石变斑）
正交偏光×20

含榴浅粒岩
正交偏光×50

图版XII

石英岩
正交偏光×50

斜长角闪岩（粒状柱状变晶结构）
正交偏光×20

斜长角闪岩（粒状柱状变晶结构、片状构造）
正交偏光×20

条带状斜长角闪岩
正交偏光×20

角闪斜长片麻岩
正交偏光×20

斜长角闪片麻岩
正交偏光×20

含榴斜长角闪片麻岩
正交偏光×20

含石英大理岩
正交偏光×50

图版 XIII

糜棱岩化花岗斑岩
正交偏光×20

糜棱岩化花岗斑岩
正交偏光×50

钙质绢云母板岩
正交偏光×50

黑云斜长角岩
正交偏光×50

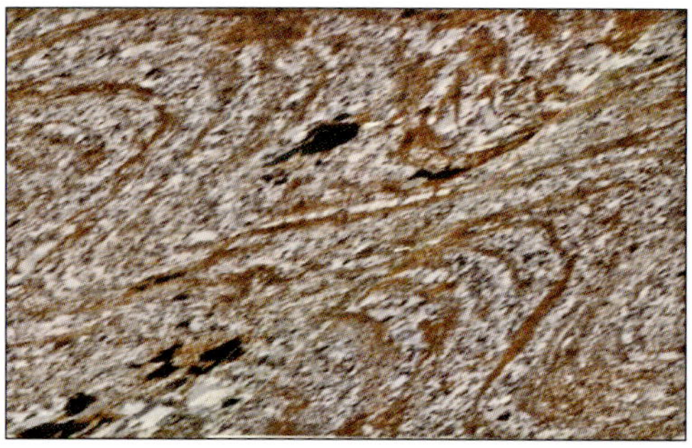

黑云斜长角岩褶皱
正交偏光×20

黑云斜长角岩
正交偏光×50

透辉斜长角岩
正交偏光×50

黑云堇青角岩
正交偏光×20

图版XIV

崇明东滩高潮滩和芦苇植

崇明东滩高潮滩盐生植被

崇明东滩中潮滩和海三棱藨草植被

崇明东滩高潮滩潮沟

图版XV

崇明东滩光滩波痕地貌

曹泾古贝壳沙堤地表出露的贝壳碎片

佘山及山前平原地貌景观

辰山矿坑岩石露头

图版 XVI

20 世纪初上海外滩大堤

现代上海外滩大堤

上海外滩历史建筑地面沉降现象（一楼窗口接近地表）